Sources and S
in the History of M
and Physical Sc

For further volumes:
http://www.springer.com/series/4142

Mathieu Ossendrijver

Babylonian Mathematical Astronomy: Procedure Texts

Springer

Mathieu Ossendrijver
Institute for Ancient Near Eastern Studies (IANES)
University of Tübingen
Tübingen
Germany

ISBN 978-1-4614-3781-9 ISBN 978-1-4614-3782-6 (eBook)
DOI 10.1007/978-1-4614-3782-6
Springer New York Heidelberg Dordrecht London

Library of Congress Control Number: 2012935875

Printed on acid-free paper

Springer is part of Springer Science+Business Media (www.springer.com)

To the memory of John Britton

Preface

This study of the procedure texts of Babylonian mathematical astronomy is aimed at Assyriologists, historians of science, astronomers and others with an interest in Babylonian astronomy. I have organised it in a manner prompted by methodological as well as didactical considerations, in the hope that this will increase its accessibility to readers with different backgrounds. Chapter 1 includes a brief introduction to Babylonian mathematical astronomy, a history of research on this topic and sections devoted to documentary, historical, contextual and methodological aspects. Chapter 2 offers a semantic analysis of the procedures and a systematic investigation of the underlying mathematical concepts. Chapters 3 and 4 are devoted to the algorithms for the planets and the Moon, respectively. While some of the algorithms are undeniably complex, many aspects of the procedure texts can be explored without mathematical or astronomical training. Chapters 2–4 are structured in a bottom-to-top fashion, setting out from elementary concepts (numbers and arithmetic operations) to more complex ones (algorithms and computational systems). Most readers will find little or no difficulty in coming to terms with Chapter 1 and Chapter 2 up to §2.6.7. For those with some elementary mathematical training the rest of Chapter 2 and Chapter 3 should also pose few difficulties. More audacious readers may venture into Chapter 4, which deals with the algorithms for the Moon, by far the most complicated part of Babylonian mathematical astronomy. Critical editions of all known cuneiform tablets and fragments of the corpus can be found in Chapter 5. Many other interesting aspects of Babylonian mathematical astronomy, including its social and institutional context, purpose and applications, empirical foundation, evolution and relationship with astronomical diaries and other observational texts from Babylonia, could not be addressed to any depth in this book. Some of these issues have been or will be addressed in separate publications. An upcoming second volume will deal with the tabular texts of mathematical astronomy. Apart from occasional remarks I also do not address the issue of how accurately the algorithms reproduce the phenomena they aim to predict — for this topic cf. Aaboe (1958) and Swerdlow (1998).

Mathieu Ossendrijver
Berlin
February 14, 2012

Acknowledgements

This book was written in the framework of a research project supported by the Deutsche Forschungsgemeinschaft (DFG) and based at the Institute for Ancient Near Eastern Studies (IANES) of the Eberhard-Karls-Universität Tübingen. It was accepted as a dissertation by the former Fakultät für Kulturwissenschaften of the Eberhard-Karls-Universität Tübingen (2010). My supervisor Konrad Volk (IANES) is gratefully acknowledged for enabling the project and supporting it in all possible ways. The project was prepared while I was a research associate in astrophysics at the Kiepenheuer Institute for Solar Physics (Freiburg i.Br.) and pursuing Assyriological studies at the Orientalisches Seminar of the Albert-Ludwigs-Universität (Freiburg i.Br.). Part of the work was carried out at the Nederlands Instituut voor het Nabije Oosten (NINO) in Leiden during a visiting scholarship (December 2005 – May 2006) provided by the Nederlandse organisatie voor Wetenschappelijk Onderzoek (NWO), for which I thank Wilfred van Soldt (University of Leiden) and Teije de Jong (University of Amsterdam). Eva Cancik-Kirschbaum is acknowledged for her hospitality during numerous stays at the Institut für Altorientalistik of the Freie Universität (Berlin). The manuscript was prepared for publication during a visiting research scholarship at the Institute for the Study of the Ancient World (ISAW), New York University (August 2010 – September 2011). I thank the Trustees of the British Museum, the Vorderasiatisches Museum (Berlin), the Birmingham City Museum, the Oriental Institute (Chicago), the Rare Books Collection of the Columbia University Library (New York), the Metropolitan Museum of Art (New York), the Yale Babylonian Collection (New Haven) and the University of Pennsylvania Museum of Archaeology and Anthropology (Philadelphia) for providing access to their cuneiform collections, and for permission to publish the texts. Nearly all tablets were photographed with a digital camera purchased with financial means provided by the late John Britton. I wish to acknowledge support and input from numerous other individuals, foremost my second supervisor Hermann Hunger, Christopher Walker, Lis Brack-Bernsen, Geert de Breucker, Gregory Chambon, Karine Chemla, Gerd Graßhof, Jens Høyrup, Peter Huber, Alexander Jones, Stefan Maul, Joachim Marzahn, Joachim Oelsner, Christine Proust, Francesca Rochberg, Hein Stadhouders, John Steele, Horst Steible, Marten Stol, Jonathan Taylor and Cornelia Wunsch.

Contents

Abbreviations and symbols

Bibliographical abbreviations

ACT	*Astronomical Cuneiform Texts* (Neugebauer 1955)
ADOG	*Abhandlungen der Deutschen Orient-Gesellschaft*
ADRT	*Astronomical Diaries and Related Texts from Babylonia* (I–III: Sachs & Hunger 1988–1996; V: Hunger & Sachs 2001; VI: Hunger 2006)
AfO	*Archiv für Orientforschung*
AHES	*Archive for History of Exact Sciences*
AHw	*Akkadisches Handwörterbuch*
AIHS	*Archive Internationale d'Histoire des Sciences*
AnOr	*Analecta Orientalia*
AOAT	*Alter Orient und Altes Testament*
AoF	*Altorientalische Forschungen*
AUWE	*Ausgrabungen in Uruk-Warka. Endberichte*
BAK	*Babylonische und assyrische Kolophone* (Hunger 1968)
BaM	*Baghdader Mitteilungen*
BaM Beih	*Baghdader Mitteilungen Beihefte*
BBVO	*Berliner Beiträge zum Vorderen Orient*
BiOr	*Bibliotheca Orientalis*
CDLJ	*Cuneiform Digital Library Journal*
CM	*Cuneiform Monographs*
CT	*Cuneiform Texts from Babylonian Tablets in the British Museum*
CTMMA	*Cuneiform Texts from the Metropolitan Museum of Art*
DIO	*The International Journal of Scientific History*
FAOS	*Freiburger Altorientalische Studien*
Fs Hunger	*Festschrift Hunger* (*WZKM* 97, 2007)
Fs Oelsner	*Festschrift Oelsner* (Marzahn & Neumann 2000)
Fs Slotsky	*Festschrift Slotsky* (Ross 2008)
GAG	*Grundriß der akkadischen Grammatik* (von Soden 1969)
HAMA	*History of Ancient Mathematical Astronomy* (Neugebauer 1975)
JAOS	*Journal of the American Oriental Society*
JCS	*Journal of Cuneiform Studies*

JHA	*Journal for the History of Astronomy*
JNES	*Journal of Near Eastern Studies*
KDVSMM	*Kongelige Danske Videnskabernes Selskab Matematisk-fysiske Meddelelser*
LBAT	*Late Babylonian Astronomical Texts* (Pinches & Sachs 1955)
MCT	*Mathematical Cuneiform Texts* (Neugebauer & Sachs 1986)
MKT	*Mathematische Keilschrifttexte* (Neugebauer 1935–1937)
NABU	*Nouvelles Assyriologiques Brèves et Utilitaires*
OECT	*Oxford Editions of Cuneiform Texts*
PAPS	*Proceedings of the American Philosophical Society*
QS	*Quellen und Studien zur Geschichte der Mathematik, Astronomie und Physik*
RA	*Revue d'Assyriologie et d'Archéologie Orientale*
RlA	*Reallexikon der Assyriologie und Vorderasiatischen Archäologie*
SpTU	*Spätbabylonische Texte aus Uruk*
SSB	*Sternkunde und Sterndienst in Babel* (I: Kugler 1907; II: Kugler 1909, 1912–1924)
TAPS	*Transactions of the American Philosophical Society*
TEBR	*Textes babyloniens d'époque récente* (Durand 1981)
TMS	*Textes Mathématiques de Suse* (Bruins & Rutten 1961)
TU	*Tablettes d'Uruk* (Thureau-Dangin 1922)
UOS	*Under One Sky. Astronomy and Mathematics in the Ancient Near East* (Steele & Imhausen 2002)
UVB	*Uruk, vorläufige Berichte*
WZKM	*Wiener Zeitschrift für die Kunde des Morgenlandes*
ZA	*Zeitschrift für Assyriologie*

Assyriological abbreviations

A	tablet siglum of the Oriental Institute museum (Chicago)
Akk.	Akkadian
AO	tablet siglum of the Louvre (Paris)
AT	auxiliary table
acc.	accusative
adv.	adverb(ial)
BCM	tablet siglum of the Birmingham City Museum
BE	tablet siglum of the Koldewey excavations at Babylon
BM	tablet siglum of the British Museum (London)
CBS	tablet siglum of the University of Pennsylvania Museum of Archaeology and Anthropology (Philadelphia)
c.	common (gender)
col.	colophon(s)
conj.	conjunction
DN	name of a deity
DT	1. daily motion table
	2. tablet siglum of a collection in the British Museum (London)
E	edge

EAE	*Enūma Anu Enlil*
ED	Early Dynastic (2900–2350 BC)
f.	feminine
gen.	genitive
inf.	infinitive
inv.	invocation
LB	Late Babylonian (539 BC–100 AD)
LE	lower edge
lit.	literally
loc.	locative
MB	Middle Babylonian (1500–1100 BC)
MLC	tablet siglum of the Morgan Library Collection (Yale)
MMA	tablet siglum of the Metropolitan Museum of Art (New York)
MN	month name
Ms.	manuscript
m.	masculine
NA	Neo Assyrian (900–612 BC)
NB	Neo Babylonian (750–539 BC)
NMAT	non-mathematical astronomical text(s)
nom.	nominative
O	obverse
OB	Old Babylonian (1800–1600 BC)
P	procedure
PN	personal name
PT	procedure text
p.	person
pl.	plural
poss. suff.	possessive suffix
pres.	present tense
pret.	preterite tense
ptc.	participle
QN	name of a quantity
R	reverse
Rm	tablet siglum of the Rassam collections in the British Museum (London)
SN	name of a synodic phenomenon
Sp	tablet siglum of the Spartali collections in the British Museum (London)
ST	synodic table
Sum.	Sumerian
sg.	singular
stat.	stative
T	table
TT	template table

U	tablet siglum of the Uruk collection in the Eski Şark Eserleri Müzesi (Istanbul)
UE	upper edge
VAT	tablet siglum of the Vorderasiatisches Museum (Berlin)
verb. adj.	verbal adjective
W	tablet siglum of the German excavations at Uruk

Astronomical abbreviations and symbols

Unless stated otherwise the conventions are the same as in *ACT* and *HAMA*.

A	system A	
A	synodic arc of the Moon and the Sun (Moon system B)	
AR	Acronychal Rising (outer planets)	[*ACT*, *HAMA*: Θ]
B	system B	
B	zodiacal position (lunar and planetary systems)	
b_j	'trailing' boundary of zone j of step function (type-A systems)	
^{b}	*bēru*, 'mile' (unit of arc and time)	
C	system C (Venus)	
C	duration of daylight (Moon systems A, B, K)	
C'	correction to the time of the lunation (Moon system A)	
CO	Conjunction	
c_j	transition coefficient for step function in zone j (type-A systems)	
c'_j	transition coefficient for generalised step function in zone j (type-A systems)	
c, c_k, c'_k	interpolation coefficient (for interval k)	
c_Ψ	maximum value of eclipse magnitude (parameter of Ψ, Ψ' and Ψ'' in Moon systems A, B)	
c_τ	constant difference between synodic time τ and synodic arc σ (planetary systems)	[*ACT*, *HAMA*: c]
c'_τ	constant difference between synodic time τ and total synodic arc Σ (planetary systems)	
$c'_{\delta\tau}$	constant difference between temporal push $\delta\tau$ and angular push $\delta\Sigma$ (Mars system A)	
^{c}	cubit (unit of arc)	
D	duration of the night (Moon systems A, B)	[*ACT*, *HAMA*: D']
D'	half the duration of the night (Moon system B)	[*ACT*, *HAMA*: D]
d	difference of a function for 1 synodic cycle [$= df(1,0)$]	
$dB(s)$	or $dB(s,t)$: net displacement for s synodic cycles, where t is nearest whole number of revolutions	
$dB_j(s)$	or $dB_j(s,t)$: same, for zone j (type-A systems)	
$df(s)$	or $df(s,t)$: net difference of f for s synodic cycles, where t is nearest whole number of oscillations of f	
$df_j(s)$	or $df_j(s,t)$: same, for zone j (type-A systems)	
d	(whole) day number [1–30] (Moon systems A, B, K)	
^{d}	day	
E	distance to the ecliptic (Moon systems A, B, K; Jupiter; Saturn)	
EF	Evening First (inner planets)	[*ACT*, *HAMA*: Ξ]
EL	Evening Last (inner planets)	[*ACT*, *HAMA*: Ω]
ES	Evening Station (inner planets)	[*ACT*, *HAMA*: Ψ]
e	yearly epact (length of the year in excess of 12^{m})	

F	the Moon's daily displacement along the zodiac measured in degrees per day (Moon systems A, B)	
F'	idem, measured in degrees per time degree	
FA	First Appearance (outer planets)	[*ACT*, *HAMA*: Γ]
FM, fm	Full Moon (opposition)	[*ACT*, *HAMA*: index 2]
f	unspecified function	
^{f}	finger (unit of arc)	
G	duration of synodic month -29^{d} (Moon systems A, B, K)	
GI_6	Lunar Six interval near Full Moon: sunset to first moonrise after sunset (Moon systems A, B, K)	
GY	Goal Year	
H	monthly difference of J (Moon system B)	
h	astronomical altitude (height above horizon)	
IC	Inferior Conjunction	
i	index for the event number of a synodic phenomenon, lunation number, day number or tithi number	
J	zodiacal correction to G (Moon systems A, B)	
j	index for the zone of a step function	
K	system K (Moon)	
K	time interval between successive lunations -29^{d} (Moon systems A, B)	
KUR	Lunar Six interval near New Moon: last visible moonrise before sunrise to sunrise (Moon systems A, B, K)	
k	index for the control values of an interpolation scheme	
L	time of lunation with respect to preceding midnight (Moon system B)	
LA	Last Appearance (outer planet)	[*ACT*, *HAMA*: Ω]
M	1. maximum of a function	
	2. time between lunation and sunset or sunrise (Moon systems A, B, K)	
ME	Lunar Six interval near Full Moon: last moonrise before sunset to sunset (Moon systems A, B, K)	
MF	Morning First (inner planets)	[*ACT*, *HAMA*: Γ]
ML	Morning Last (inner planets)	[*ACT*, *HAMA*: Σ]
MS	Morning Station (inner planets)	[*ACT*, *HAMA*: Φ]
m	1. minimum of a function	
	2. number of additional full revolutions of the Sun for every synodic cycle	
^{m}	synodic month (unit of the event frame)	
$^{\overline{\text{m}}}$	mean synodic month (unit of time)	
N	1. time interval between lunation and Lunar Six event (Moon systems A, B)	
	2. number of elementary steps contained in total synodic arc Σ (type-A systems)	
NA	Lunar Six interval near Full Moon: sunrise to first moonset after sunrise (Moon systems A, B, K)	
NA_1	Lunar Six interval near New Moon: sunset to first visible moonset after sunset (Moon systems A, B, K)	
NM, nm	New Moon (conjunction)	[*ACT*, *HAMA*: index 1]
n	whole number of additional revolutions of the planet per synodic cycle	
^{nin}	*nindanu*, 'rod' (unit of arc and time)	
O	corrected elongation of Moon from Sun or anti-Sun (Moon system B)	
OP	Opposition (Moon, outer planets)	
P	period measured in fractional synodic events after which a function returns to the same value	

p	1. period, measured in synodic events, of more rapidly varying function underlying tabulated function	
	2. index for different pushes within 1 synodic cycle (planetary systems)	
Q	1. contribution to Lunar Six from distance along the ecliptic (Moon systems A, B)	
	2. unspecified quantity	
QN	name of an unspecified quantity	
q	ratio between rising or setting time and distance along the zodiac (Moon systems A, B)	
R	1. contribution to Lunar Six from the Moon's distance to the ecliptic (Moon systems A, B)	
	2. refraction angle	
RA	right ascension ($= \alpha$)	
r	ratio between rising or setting time and distance to the ecliptic (Moon systems A, B)	
r_j	transition coefficient for step function in zone j	
S_1, S_2...	schemes for the subdivision of the synodic cycle	
S1	First Station (outer planet)	[*ACT*, *HAMA*: Φ]
S2	Second Station (outer planet)	[*ACT*, *HAMA*: Ψ]
SC	Superior Conjunction (inner planet)	
SIGN	name of unspecified zodiacal sign	
SN	name of an unspecified synodic phenomenon	
s	whole number of synodic events	
$ŠU_2$	Lunar Six interval near Full Moon: last moonset before sunrise to sunrise (Moon systems A, B, K)	
T	1. column containing year number and month name (Moon systems A, B)	
	2. time (planetary systems)	
T_{rise}	time interval between rising of planet at FA and sunrise (Jupiter systems A, A''')	
T_{set}	time interval between sunset and setting of planet at LA (Jupiter systems A, A''')	
t	nearest whole number of oscillations of a function for a close return to the same value	
UT	Universal Time	
v	'daily' displacement along the zodiac	
W	duration of 6 synodic months modulo a whole number of days (Moon system A)	
w	generalised step function (Moon system A)	
w_j	preliminary value of generalised step function in zone j (Moon system A)	
X	unidentified computational system	
X_m	lunar variation of the Moon's net displacement for 1^m? (Moon system A)	
X_s	lunar variation of the Moon's net displacement for 223^m? (Moon system A)	
Y	1. whole number of years after which a function returns to exactly the same value	
	2. zodiacal correction to Λ (Moon system A)	
y	whole number of years after which a function returns to nearly the same value ('close return')	
Z	1. whole number of oscillations after which a function returns to exactly the same value	
	2. zodiacal correction to W (Moon system A)	
zod.	zodiacal	
α	right ascension	
α_j	length of zone j of step function	
β	distance to the ecliptic [degrees] (Moon systems A, B)	
β_0	distance to the ecliptic at lunation [degrees] (Moon systems A, B)	

Δ	1. amplitude of a function ($= M - m$)	
	2. disk correction (Moon system A)	
ΔB	zodiacal displacement (other than synodic arc σ, push $\delta\Sigma$, or net displacement dB)	
δ	1. declination	
	2. elementary step (of zigzag function)	
δ_j	elementary step of step function	[*HAMA*: I_j]
δN_p	number of elementary steps contained in angular push p (type-A systems)	[*HAMA*: s_p]
$\delta\Sigma$	angular push (subdivision of Σ)	[*ACT*: Δ; *HAMA*: $\delta\lambda$]
$\delta\Sigma_p$	angular push p	
$\delta\Sigma_{pj}$	angular push p in zone j (type-A systems)	
$\delta\tau$	temporal push (subdivision of τ)	
$\delta\tau_p$	temporal push p	
η	elongation	
ε	1. obliquity of the ecliptic	
	2. correction to the lunar elongation (step 3 of Lunar Six module of Moon system B)	
θ	sidereal time	
κ	half width of nodal region (Moon system A)	
Λ	duration of 12 synodic months -354^d (Moon system A)	
λ	ecliptical longitude	
μ	mean value of a function	
ν_j	number of elementary steps contained in zone j (step function)	[*HAMA*: ν_j, π_j]
Π	number period (smallest whole number of events after which a function returns to exactly the same value)	
Φ	duration of 223 synodic months -6585^d (Moon system A)	
ϕ	1. geographical latitude	
	2. basic interval between adjacent Φ_k in the interpolation schemes for G, W and Λ (Moon system A)	
Ψ, Ψ', Ψ''	eclipse magnitude and its variants (Moon systems A, B)	
Σ	total synodic arc	[*ACT*: $\Delta\Lambda$]
σ	synodic arc	[*ACT*: $\Delta\lambda$]
σ_j	preliminary value of step function for synodic arc in zone j (type-A systems)	[*ACT, HAMA*: w_j]
τ	1. synodic time	[*ACT, HAMA*: $\delta\tau$]
	2. hour angle	
τ_j	synodic time in zone j (planetary systems of type A, C)	
$^{\tau}$	(real) tithi (unit of the event frame)	
$\bar{\tau}$	mean tithi (unit of time)	[*ACT, HAMA*: $^{\tau}$]

Names of the planets, zodiacal signs, months, regnal years and units

names of the planets			
transliteration	Akkadian (literal translation)	translation	symbol
[^{d}]*sin*(30)	*Sîn*	Moon	☽
šamaš$_2$(20), ^{d}**utu**	*Šamaš*	Sun	☉
[^{d}]**gu$_4$.ud**	*Šiḫṭu* ('Attack; Jump')	Mercury	☿
[^{d}]*dil-bat*	*Dilbat*	Venus	♀
AN	*Ṣalbatānu*	Mars	♂
mul$_2$.babbar	*Peṣû* ('The white one')	Jupiter	♃
genna(TUR.DIŠ)	*Kajjamānu* ('The steady one')	Saturn	♄

Occasionally used alternative names: *Sagmegar*(**sag.me.gar**) = Jupiter, *Kajjamānu*(**sag.uš**) = Saturn.

signs of the zodiac						
transliteration	older form	Akkadian	literal translation	sign	abbr.	symbol
ḫun, lu	$^{mul.lu_2}$**ḫun.ga$_2$**	*agru*	hireling	Aries	Ari	♈
mul$_2$[**.mul$_2$**]	**mul.mul**	*zappu*	stars (Sum.); bristle (Akk.)	Taurus	Tau	♉
maš[**.maš**]	^{mul}**maš.tab.ba**	*māšu, tū'amū*	twins	Gemini	Gem	♊
alla	^{mul}**al.lul**	*alluttu*	crab	Cancer	Cnc	♋
a	^{mul}**ur.a**	*nēšu*	lion	Leo	Leo	♌
absin, absin$_0$(KI)	^{mul}**ab.sin$_2$**	*šer'u*	furrow	Virgo	Vir	♍
rin$_2$	^{mul}**giš.rin$_2$**	*zibānītu*	scales	Libra	Lib	♎
gir$_2$[**.tab**]	^{mul}**gir$_2$.tab**	*zuqaqīpu*	scorpion	Scorpius	Sco	♏
pa	^{mul}**pa.bil.sag**	*Pabilsag*	Pabilsag	Sagittarius	Sgr	♐
maš$_2$	^{mul}**suḫur.maš$_2$**	*suḫurmāšu*	goat-fish	Capricorn	Cap	♑
gu	^{mul}**gu.la**	*Gula*	Gula	Aquarius	Aqr	♒
zib[**.me**]	^{mul}**kun.meš**	*zibbātu*	tails	Pisces	Psc	♓

Also listed are the official abbreviations which are used in all translations.

month names			
transliteration	OB Nippur	Akkadian (+literal translation)	translation
bar	^{iti}**bara$_2$.zag.gar**	*Nisannu*	I
gu$_4$	^{iti}**gu$_4$.si.su$_3$**	*Ajjaru*	II
sig	^{iti}**sig$_4$.^{giš}u$_5$.šub.ba.ga$_2$.gar**	*Simānu*	III
šu	^{iti}**šu.numun**	*Dûzu*	IV
ne	^{iti}**ne**.IZI.**gar**	*Abu*	V
kin	^{iti}**kin.^{d}inana**	*Ulūlu*	VI
kin.2[**.kam**]		*Ulūlu arkû* ('second Ulūlu')	VI$_2$
du$_6$	^{iti}**du$_6$.ku$_3$**	*Tašrītu*	VII
apin	^{iti}**apin.du$_8$.a**	*Araḫsamna*	VIII
gan	^{iti}**gan.gan.e$_3$**	*Kislīmu*	IX
ab	^{iti}**ab.e$_3$**	*Ṭebētu*	X
ud$_2$(ZIZ$_2$)	^{iti}**ud$_2$.duru$_5$**	*Šabāṭu*	XI
še	^{iti}**še.kin.ku$_5$**	*Addaru*	XII
diri[**.še**]	^{iti}**diri.še.kin.ku$_5$**	*Addaru arkû* ('second Addaru')	XII$_2$

All month names are translated with Roman numbers. The cuneiform sign for month XI is transliterated as ZIZ$_2$ even though the month names of OB Nippur (Cohen 1993) imply that the proper reading is **ud$_2$**.

regnal years		
ruler	regnal years	Julian yr (BC)
Xerxes	1–21	485/4
Artaxerxes I	1–41	464/3
Darius II	1–19	423/2
Artaxerxes II	1–46	404/3
Artaxerxes III	1–21	358/7
Arses	1–2	337/6
Darius III	1–5	335/4
Alexander the Great	7–14	330/329
Philip Arrhidaeus	2–8	322/1
Alexander IV	2–5	315/4
Seleucid Era (SE)	1–	311/0
Arsacid Era (AE)	1–	247/6

Late Achaemenid and early Seleucid rulers. The third column contains the Julian equivalent of the first regnal year.

units of measurement						
transliteration	Akkadian	translation	symbol	internal equivalence	modern equivalence angular	temporal
danna	*bēru*	'mile'	b	$1^b = 30^\circ$	30°	2 hrs.
UŠ	?	(time) degree	°	$1^\circ = 60^{nin}$	1°	4 min.
ninda(GAR)	*nindanu* (?)	'rod' (?)	nin		0;1° = 1' (arcmin)	4 sec.
kuš₃	*ammatu*	cubit	c	$1^c = 24^f = 2^\circ$	2°	
[**šu**].**si**, *u*	*ubānu*	finger	f	$1^f = 6^{še}$	0;5°	
še	*uṭṭatu*	barleycorn	še		0;0,50°	

Top: the 'degree system', used for angles and time intervals. Bottom: the 'cubit system', used for angular distance to the ecliptic and eclipse magnitude.

Factor diagram for the 'degree system':

$$1 \text{ 'mile' } (^{b}) \xleftarrow{30} 1 \text{ degree } (^{\circ}) \xleftarrow{60} 1 \text{ 'rod' } (^{nin})$$

Factor diagram for the 'cubit system':

$$1 \text{ cubit } (^{c}) \xleftarrow{24} 1 \text{ finger } (^{f}) \xleftarrow{6} 1 \text{ barleycorn } (^{\check{s}e})$$

Chapter 1
Procedure texts

1.1 The corpus of mathematical astronomy

The corpus of Babylonian mathematical astronomy comprises about 440 cuneiform tablets and fragments from Babylon and Uruk from the period 450–50 BC. A distinctive feature of mathematical astronomy, which represents a small fraction (perhaps 5%) of the Babylonian astronomical corpus, is that astronomical quantities are computed with mathematical algorithms. The corpus can be divided into procedure texts (Akk. *epūšu* = 'procedure'), currently numbering about 110, and tabular texts (*tērsītu* = 'computed table'). This volume is concerned with the procedure texts, which contain instructions for computing astronomical quantities. Tabular texts contain numerical tables arranged in rows and columns with computed astronomical quantities for the Moon or a planet.[1] There are four kinds of tabular texts. Their modern designations in this study, and the approximate number of tablets for each type, are synodic tables (230), template tables (50), daily motion tables (30) and auxiliary tables (20). The term *tērsītu*, attested in the colophons of several synodic tables for the Moon, a daily motion table for Jupiter and in administrative documents concerning astronomers, can be assumed to denote any kind of tabular text.[2] Synodic tables and daily motion tables are the end products of Babylonian mathematical astronomy. In a synodic table consecutive lines (rows) correspond to successive events of a synodic phenomenon,[3] in a daily motion table to successive days or tithis.[4] Each column contains numerical values of an astronomical quantity, e.g. zodiacal positions or times, and sometimes also other information pertaining to that quantity. Template tables differ from synodic tables in that they contain only a selection of columns corresponding to some intermediate stage in the production of a synodic table.[5] Auxiliary tables provide coefficients that are needed for computing certain columns of the synodic, template or daily motion tables. Most tablets with mathematical astronomy contain either procedures or tables, but some contain both. Behind that separation there are obviously practical considerations, because procedures are meant to be applied again and again with different initial data, whenever the astronomer wished to compute a synodic table, template table or daily motion table.

1.1.1 History

Babylonian mathematical astronomy was developed between 450 BC and 350 BC, which makes this the earliest known form of mathematical astronomy of the ancient world. Relatively few secure facts are known about its initial stages. Both the empirical origin of the astronomical parameters and the evolution of the algorithms still await a satisfying and comprehensive reconstruction.[6] Since these issues fall outside the scope of the present investigation, the aim of this section is only to briefly summarise the relevant textual evidence. Nearly all datable tablets of the corpus were written between 320 BC and 50 BC, which cover the Seleucid period and part of the Arsacid period. Only a few tablets, all of which are tabular texts, are securely dated to the preceding Achaemenid era. The oldest of these were written between 450 BC and

---

1 For new editions of the tabular texts cf. Volume II; for previous editions cf. *ACT*.

2 Synodic tables: *ACT* 18, 122, 123a (cf. also p. 115); daily motion table: *ACT* 80; for references to the administrative documents cf. p. 8.

3 I follow van der Waerden (1965) in avoiding the term 'ephemerides' for the synodic tables, since it wrongly suggests that the rows pertain to successive days.

4 For the synodic phenomena cf. §3.1.2 (planets) and §4.1.2 (Moon); for the tithi cf. §2.5.2.3.

5 In *ACT* all template tables are classified as auxiliary tables. Some template tables are also known as 'dateless synodic tables', because they lack a column for the time.

6 For the planetary systems cf. Swerdlow (1998); for the lunar systems: Brack-Bernsen (1990, 1997); Brack-Bernsen & Schmidt (1994); Britton (1990, 1999, 2007a, 2009, 2010).

the beginning of the 4th c. BC, when Babylonia was successively ruled by Artaxerxes I, Darius II and Artaxerxes II.[7] No later than 400 BC Babylonian astronomers had introduced the zodiac (cf. §2.5.3), which became the coordinate system of mathematical astronomy, and to some extent also of the astronomical diaries and related observational texts.[8] Along with the introduction of the zodiac, new forms of astrology were invented in which the zodiac plays a central role. This includes Babylonian 'horoscopes', whereby the fate of a newborn is derived from the zodiacal positions of the Moon and the planets computed for the day of birth.[9]

The oldest tablets with mathematical astronomy partly reflect a different stage of development compared to the texts from the Seleucid era. The algorithms underlying the earliest known tabular texts for the Moon[10] clearly predate the fully developed lunar systems A and B. Also lunar system K, preserved in one procedure text (No. **52**) probably written near 320 BC, is likely a remnant from an earlier stage of mathematical astronomy. The earliest synodic table[11] exhibiting all essential features of the type-A systems as we know them from the Seleucid era contains Mercury data covering year 41 of Artaxerxes I until year 2 of Artaxerxes II. Since the same tablet also contains a lunar table with data up to year 24 of Artaxerxes II, the tablet was probably written close to the latter year (381/0 BC). It therefore seems that at least some of the planetary systems of type A had reached their final form no later than 380 BC. The lunar algorithms probably continued to develop for some time after that, which is not surprising given their far greater complexity.[12] The earliest lunar tables computed fully in accordance with the fully developed lunar system A date to years 6–7 of the reign of Philip Arrhidaeus (318–316 BC). Near the beginning of the Seleucid era lunar systems A and B, the latter perhaps a few decades later than the former, had thus reached their final form. After that, there is very little evidence of further change.

1.1.2 Purpose and applications

It may come as a surprise that the question of what is the purpose – or are the purposes – of Babylonian mathematical astronomy has not yet been answered in a satisfying manner. Since this will be addressed in a separate investigation, I will here only briefly explore some of the issues that are involved. The texts themselves do not mention the purpose of the tables. Various suggestions have been made in the literature, e.g. astrology, the calendar or the production of horoscopes, but a concrete and comprehensive scenario for the practical application of the tables has not been formulated. All of the mentioned purposes are of a general nature, and if one closely examines their merit they turn out to be deficient in one way or another. In order to proceed it is useful to introduce different levels of application pertaining to the immediate astronomical purposes and to the practical purposes of the tabular texts. The former can be established by identifying those quantities in the tabular texts that are 'final' in the sense that they are not auxiliary to the computation of other quantities. For the planetary tables these are synodic phenomena and daily positions (cf. §3.1); for the lunar tables 'Lunar Six' intervals, lunar and solar eclipses and daily positions (cf. §4.1). The next level concerns the practical applications of these quantities. They can be established by exploring the relations between mathematical astronomy and the rest of the astronomical corpus, which is numerically dominated by astronomical diaries and related observational texts.[13] The observational texts are the result of a systematic program of observation believed to have been carried out more or less continuously from the 8th c. BC until the end of cuneiform writing in the first c. AD. Since the observational texts and mathematical astronomy are concerned with essentially the same group of astronomical phenomena, the former should provide clues to our question. The immediate purposes of the observational texts are relatively well understood due to the rediscovery of the Goal-Year method, a predictive method which utilises data from the astronomical diaries, and recent significant advances in its reconstruction made by L. Brack-Bernsen.[14] This method, which predates mathematical astronomy by at least 150 years, operates by projecting past astronomical phenomena – precisely those that are recorded in the observational texts – into the future by exploiting their periodicity. In order to identify the applications of mathematical astronomy one should therefore first explain the purpose of the astronomical predictions made with the Goal-Year method. One possibly important clue is that the observational texts contain not only astronomical data but also market prices, river levels, weather phenomena and events that occurred in the city. This systematic juxtaposition of astronomical and terrestrial data, already apparent in the earliest known diaries from the 7th c. BC, suggests a connection between astronomical prediction and the prediction of terrestrial events. Explicit proof for this hypothesis is found in several LB

7 Aaboe & Sachs (1969), Texts A–D; Aaboe et al. (1991), Text E. According to Britton (2010), all of these tablets were written after 400 BC.

8 Britton (2010) has presented strong arguments for the zodiac having been introduced within a narrow window centred at 400 BC, some 50 years later than the conventional date for this event.

9 Rochberg (1998, 2004).

10 Aaboe & Sachs (1969), Texts A–D; presumably also the undated BM 45861 (Steele 2002c).

11 BM 36651+ (Aaboe et al. 1991, Text E).

12 For the chronology of the lunar systems cf. also §4.2.

13 For critical editions of these texts cf. *ADRT* I–III, V–VI.

14 Brack-Bernsen (1997); Brack-Bernsen & Hunger (2002). For editions of the Goal-Year texts cf. Hunger (2006).

astrological texts in which various planetary phenomena are correlated with terrestrial phenomena.[15] Another application of mathematical astronomy is connected with Babylonian 'horoscopes', for which the daily motion tables may have provided data.[16] Synodic tables for the Moon may also have been relevant for the lunar calendar.[17]

1.1.3 Discovery and historiography

Before the cuneiform script was deciphered in the middle of the 19th century the only available sources of information concerning Babylonian astronomy were the Bible and the works of several Greek and Roman scholars. Numerous passages make mention of the Chaldaean astrologers and their preoccupation with observing, interpreting and predicting celestial phenomena, reflections of a widespread perception that the astral sciences — astronomy and astrology — were highly developed in Babylonia. However, detailed information about the Babylonian methods, in particular mathematical astronomy, is completely lacking in these sources.[18]

Phase one: the first translations (1881–1935)

In the last decades of the 19th century, shortly after the decipherment of the cuneiform script, numerous astronomical and astrological tablets were excavated in Nineveh and Babylon. Large numbers of them were acquired by the British Museum through various explorers and antiquities dealers. Among the approximately 20,000 tablets and fragments forming the 'Babylon collection' in the British Museum (cf. below) Joseph Epping (1835–1894) and Johann N. Strassmaier (1846–1920), German Jesuit priests with a keen interest in ancient astronomy, rediscovered Babylonian mathematical astronomy near 1880. This marked the beginning of a period of intense research by Epping and Strassmaier, resulting in the first modern publication (Epping & Strassmaier 1881) and two further papers (Epping 1889, 1890) concerning Babylonian mathematical astronomy. While Strassmaier produced copies of the tablets, the astronomical interpretation was mainly the work of Epping, who was an astronomer. In the 1890s Franz Xaver Kugler (1862–1929), also a Jesuit priest from Germany, entered the field. His efforts culminated in two comprehensive works: *Babylonische Mondrechnung* (1900) and *Sternkunde und Sterndienst in Babel* (1907–1924).[19] The achievements of Epping and Kugler cannot be overestimated, especially considering the fact that the tabular texts contain mainly numbers lacking any indication about their astronomical significance. The work of Kugler was continued by Johann Schaumberger, a German priest of the Redemptorist Order, whose main contribution in the field of Babylonian mathematical astronomy is his reconstruction of the algorithms for the 'Lunar Six intervals' (Schaumberger 1935).[20] Other scholars who were active in this phase were the Dutch astronomer Anton Pannekoek (1873–1960) and the German orientalist Paul Schnabel (1887–1947).[21]

Even though the bulk of the corpus of mathematical astronomy had been available in museums since the end of the 19th century, only very few tablets were published in this phase (for the procedure texts cf. Table 1.1). The earliest edition of a procedure text, the fragment BM 32651 (No. **53**), appeared in Kugler's *Babylonische Mondrechnung* (1900). This tablet happens to be the largest and most important procedure text for lunar system A even today.[22] The numerous occasions where Kugler quotes this text reveal that he fully understood its importance for reconstructing lunar system A ('System II' in his terminology). Kugler classified it as a 'Lehrtafel' (instructional tablet) or 'Lehrtext' (instructional text), terms that were subsequently taken up by Schnabel, Schaumberger and Neugebauer. In his later English publications Neugebauer introduced the term 'procedure text' which has become the standard designation for these texts.

Phase two: the standard editions (1935–1990)

A second phase began near 1935, when Otto Neugebauer (1899–1990) and Bartel L. van der Waerden (1903–1996) made their entry into the field of Babylonian mathematical astronomy. Over a period of about 20 years Neugebauer carried out an ambitious program of translation and analysis culminating in *ACT* (*Astronomical Cuneiform Texts*, 1955), the first

[15] E.g. market prices (Hunger 1976a, Text 94) and weather phenomena (Hunger 1976b).

[16] For Babylonian 'horoscopes' and their sources cf. Rochberg (1998, 2004). For a tablet with a compilation of computed data that may have been a source for horoscopes cf. Steele (2000c).

[17] However, Steele (2007b) has argued that, during the Seleucid era, the first day of the month was determined in advance and not by observing the first crescent, and that this was achieved with the Goal-Year method and not with mathematical astronomy (cf. also §4.1.4).

[18] References to Babylonian astronomy and astrology in classical sources are discussed in Rochberg (1988), pp. 2–5; for the reception of Babylonian astronomy by historians of science cf. Rochberg (2004), Ch. 1.

[19] During the 'Babel Bible' controversy that took place in Germany at the beginning of the 20th century. Kugler used his knowledge of Babylonian astronomy to argue emphatically against the 'astro-mythological' school of orientalists (Kugler 1910).

[20] Cf. the recension in Neugebauer (1936). For the Lunar Six intervals cf. §4.1.4.

[21] For a comprehensive survey of the literature on Babylonian astronomy up to 1998 cf. Hunger & Pingree (1999).

[22] Kugler only knew the main fragment 76–11–17(=S+),2418 (BM 32651); cf. *BMR*, Pls. v–vi, and the references on pp. xi–xv ('System II'). P16' and P17' were not discussed by Kugler.

publication	procedure text	content
Kugler (1900) = *BMR*	No. **53** (copy; partial edition)	Moon system A
Kugler (1907) = *SSB* I	No. **32** (copy; edition);	Jupiter system A'
	No. **18** (partial edition)	Jupiter systems A',A
	BM 45728 (copy; edition)	NMAT (Goal-Year astronomy)
Thureau-Dangin (1922) = *TU*	No. **42** (copy);	Mercury system A_1; Saturn systems A,B
	AO 6455 = *TU* 11 (copy)	NMAT (Goal-Year astronomy); omens
Kugler (1924) = *SSB* II	No. **42** (partial edition)	Mercury system A_1; Saturn systems A,B
Schnabel (1924)	No. **42** (partial edition)	Mercury system A_1; Saturn systems A,B
Neugebauer (1953)	No. **53** (partial edition of P17')	Moon system A

Table 1.1: Publications of astronomical procedure texts that appeared before *ACT* (1955).

and standard edition of the corpus of mathematical astronomy. In a programmatic paper written near the beginning of his project, Neugebauer (1938a) announced its goals and methods. A notable feature is his focus on the mathematical analysis, dating and reconstruction of tabular texts, for which he developed a range of sophisticated analytical tools. This probably explains why Neugebauer, unlike Kugler, initially displayed little interest in the procedure texts. It also informed his method of translation, as will be argued in §1.6.1. Apart from *ACT*, two other major works on Babylonian mathematical astronomy were produced in this phase. Van der Waerden's *Erwachende Wissenschaft II. Die Anfänge der Astronomie* (van der Waerden 1965)[23] includes an overview of Babylonian mathematical astronomy with selected translations. *History of Ancient Mathematical Astronomy* (Neugebauer 1975) offers a detailed account of mathematical astronomy in Babylonia and other ancient cultures. This opus major does not contain primary sources, but instead a thorough analysis of the reconstructed algorithms using a fully modern formalism. Some of the most intricate algorithms of lunar system A (Φ, G, W, Λ, Y, Z) were reconstructed and interpreted by A. Aaboe (1922–2007). P. Huber made important contributions to the reconstruction and interpretation of the algorithms for the planets.[24]

Phase three: new approaches and broadening views (1990–)

A third phase began near 1990 with investigations by F. Rochberg, which involved a critical evaluation of the historiography of Mesopotamian astronomy and astrology, using insights from the philosophy and history of science.[25] As a result, common perceptions of a dichotomy between scientific practices in ancient oriental cultures and classical Greece, as can be found in general works on the history of science, turned out to be untenable. Second, the internalist approach typical of the previous phase, with its strong focus on the reconstruction of astronomical algorithms, began to make way for a more holistic one that aims to explain Babylonian astronomy in its institutional, political, religious and social contexts. In this approach the intricate link between astronomy and astrology is fully acknowledged, and astronomy is viewed as a discipline sharing essential features with other forms of Babylonian scholarship. In a similar spirit D. Brown (2000) proposed a detailed scenario explaining how celestial divination developed in ways that led to the emergence of a paradigm of predictive astronomy in Mesopotamia. These developments have more or less coincided with the gradual publication of the Babylonian astronomical diaries and related texts by H. Hunger, initially in collaboration with A. Sachs, in the series *ADRT* (1988–). The enhanced accessibility of this vast corpus, which had remained almost completely untranslated for over a century, has led to a much better understanding of the goals and methods of Babylonian observational astronomy and the associated predictive Goal-Year method, in particular due to the work of L. Brack-Bernsen[26] and J. Steele.[27] At the same time L. Brack-Bernsen, J. Britton and N. Swerdlow have explored the empirical foundations of Babylonian mathematical astronomy and formulated scenarios by which some of the algorithms may have been derived from observations.[28] Without these broadening insights into different aspects of the astral sciences in Babylonia there would be little hope of achieving a comprehensive understanding of Babylonian mathematical astronomy.

Babylonian mathematics

The historiography of Babylonian mathematical astronomy partly mirrors that of Babylonian mathematics, which was vividly sketched by J. Høyrup (1996, 2010). Since this brings to the surface a number of methodological issues that are of fundamental importance for the present study, it is worthwhile to compare the development of both disciplines in some

[23] A modified English translation appeared in 1974 as *Science Awakening II. The Birth of Astronomy*.

[24] Huber (1957); Aaboe (1958); Aaboe & Huber (1977).

[25] Rochberg (1991, 1999, 2004).

[26] Brack-Bernsen (2002); Brack-Bernsen & Hunger (2002, 2008, 2010).

[27] Steele (2000a, 2000b, 2002c, 2007b).

[28] Brack-Bernsen (1990, 1997); Britton (1999, 2007a, 2009, 2010); Swerdlow (1998).

detail.[29] Høyrup distinguishes three stages which he interprets as the phases of a Kuhnian paradigm (Kuhn 1970). In the 'heroic', pioneering phase (1930–1940) O. Neugebauer (1935–1937) and F. Thureau-Dangin (1938) published their first translations of cuneiform mathematical texts, mainly from the Old Babylonian period (1800-1600 BC). Both propogate the view that the geometric formulation characteristic of many Old Babylonian mathematical problem texts is a superficial feature hiding the supposedly algebraic nature of these texts. This view, now known to be anachronistic, was solidified in the 'orthodox' phase (1940–1975) characterised by 'normal science' in the Kuhnian sense. As argued by Høyrup, the anachronisms become more pronounced in this phase. Babylonian mathematics is now construed as the manifestation of a universal mathematics and described in modern categories. While these views are implied and partly made explicit in the works of the pioneers themselves, they are most clearly pronounced in the secondary literature. The third phase (1971–) began when J. Høyrup and J. Friberg carried out a critical evaluation of the methodological foundations of the previous interpretations. The outcome of these reflections has been the formulation and execution of a program of translation based on innovative principles, which will be addressed in more detail in §1.6.1.

A comparison of the two cases reveals obvious similarities, but also notable differences. First of all, by the time that research on Babylonian mathematics began to take shape, the pioneering phase of research on mathematical astronomy was already over. Neugebauer was a central figure in both disciplines, but in the field of Babylonian mathematical astronomy he was a successor, after having been a pioneer in the field of OB mathematics. Second, although the pioneering phase of mathematical astronomy ended in the 1930s, the standard edition of the corpus (*ACT*) appeared only in the second phase — 75 years after mathematical astronomy was rediscovered by Epping and Strassmaier. Third, mathematical astronomy has not yet benefited from the new approaches that were developed for translating and interpreting OB mathematics. An important aim of the present study is therefore to evaluate the methodological foundations of the translations in *ACT* and elsewhere in the light of the approach of Høyrup and Friberg, and formulate a more refined method for translating astronomical procedure texts (§1.6.1).

1.2 Selection of the texts

The corpus of procedure texts as defined in the present study comprises 108 tablets and fragments, which include all published procedure texts as well as newly discovered unpublished tablets and fragments from the British Museum. All of the selected tablets and fragments contain at least one procedure with mathematical astronomy or some early stage of it, or astronomical computations that (may) have a close affinity with mathematical astronomy. These criteria are more inclusive than the ones underlying *ACT*; in particular I consider some of the so-called 'atypical procedure texts' (Neugebauer & Sachs 1967, 1969) as clearly belonging to mathematical astronomy. Several unpublished tablets and fragments in the British Museum were pointed out to me by C.B.F. Walker; others were found in a systematic search of the Babylon collection made possible by C.B.F. Walker's catalogue of astronomical fragments, which was graciously made available to me. It is usually straightforward to determine whether the selection criterion is satisfied. Problems of demarcation arise mainly in the case of small fragments, for which the main alternative identifications are (i) tabular texts of mathematical astronomy, (ii) procedure texts with non-mathematical astronomy (NMAT), (iii) mathematical problem texts and mathematical tabular texts. Most of the dubious cases concern the second group. NMAT procedure texts (§1.7.2) share essential features with the procedure texts of mathematical astronomy, since both contain arithmetical operations involving astronomical quantities. Hence the final identification of these fragments depends on the astronomical interpretation, which may be difficult to establish if the fragment is small. A more fundamental difficulty results from the assumed separation between mathematical astronomy and non-mathematical astronomy. For some procedure texts this distinction is open to debate, and one may argue that it necessarily breaks down at some point in the formative stage of mathematical astronomy. I have therefore included (§5.3.4) procedure texts that probably belong to a formative stage of mathematical astronomy,[30] and excluded only those that clearly belong to NMAT (e.g. procedures for the Goal-Year method or concerning solstices and equinoxes). About 30 small fragments of (candidate) procedure texts were eventually not included because they were too small or strongly damaged to warrant a translation.

[29] This point was raised with more humour by D. Brown (2001).

[30] Newly discovered texts in this category are No. **100** and No. **101**.

1.3 Archaeological and archival aspects

All known texts with mathematical astronomy originate from Babylon and Uruk (Fig. 1.1). The only other site in Babylonia that has produced astronomical tablets from the NB or LB period is Nippur,[31] but mathematical astronomy has not (yet) been discovered there. This confirms an account by Pliny the Elder (23–79 AD), according to which Babylon, Orchenus (Uruk) and Hipparenum (Nippur) were famous centres of the astral sciences.[32] The distribution of the tablets with mathematical astronomy over Babylon and Uruk is strongly skewed: about 80% of the tabular texts and 90% of the procedure texts originate from Babylon, the rest from Uruk. Also their temporal distribution is uneven. While the datable fragments with mathematical astronomy cover the period from 450 BC (reign of Artaxerxes I) until 49 BC (Parthian era), the bulk are from SE 120–210 (192–102 BC). However, most procedure texts cannot be dated to any precision, since only few preserve a colophon mentioning the scribe and a date. Sometimes an approximate date can be inferred from the biography of the scribe or, if the procedure accompanies a synodic table, the initial date of the table.[33] Taking into account all the available information, the datable procedure texts cover the period between year 5 of Philip Arrhidaeus and SE 202 (319–110 BC), a narrower range than that of the tabular texts. In particular, procedure texts demonstrably written in the Achaemenid era are lacking.

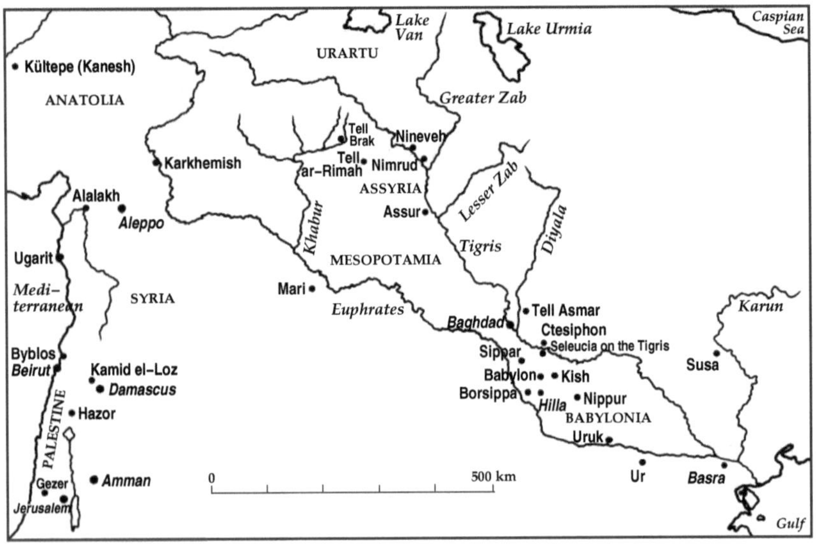

Fig. 1.1: Mesopotamia

1.3.1 Babylon

Table 1.2 lists all collections with procedure texts from Babylon. R. Koldewey's excavations in Babylon between 1899 and 1917 produced only about 10 tablets (siglum BE or Bab) with a known or possible astronomical content, including a single procedure text (No. **81**).[34] These are the only astronomical tablets from Babylon for which the archaeological context is reliably known. All originate from a single private house[35] in the area known as Amran, about 60 m to the west of the Išḫara temple (library N19 in Pedersén 2005).[36] This location is not far south of the Esagila, the temple of Bēl (Marduk) and his spouse Bēltu and main cult centre of Babylon. Although the other astronomical tablets do not seem to originate from within the Esagila,[37] the astronomical activities in Babylon are believed to have been closely connected with it.[38] Textual evidence from administrative documents suggests that the astronomers were priests employed by the Esagila, at least from the Achaemenid era onwards (Beaulieu 2006).

31 Three observational reports: *ADRT* V 57, with Lunar Six data, solstices and equinoxes for year 40 of Artaxerxes I (425/4 BC); *ADRT* V 63, with planetary data for years 41–42 of Artaxerxes II (364/2 BC); Huber & Steele (2007) Text B, with Lunar Six data, probably written before the Seleucid era; one horoscope: Rochberg (1998) Text 1, from year 13 of Darius II (410 BC).

32 *Natural History*, Book 6, Chapter 30. For the identification of Hipparenum with Nippur cf. Oelsner (1982) and (1986), p. 108.

33 The latter option must be used with care, because some synodic tables, e.g. A 3405 (Steele 2000c), were demonstrably written much later than the initial date of the predictions.

34 Pedersén (2005), library N19. Apart from the procedure text, first published as *ACT* 207e, none of these tablets has been published. Their current location is unknown except for that of BE 15617a = VAT 17343, which is kept in Berlin.

35 The interpretation as a private house has been challenged by Clancier (2009, p. 150), who suggests that the structure is more monumental.

36 A scientific report of the excavations in this part of Amran (square 35x = 35af) was never published.

37 Only a very small portion of the Esagila was excavated, partly by means of tunnels. Among the eight tablets found within the Esagila, all still unpublished, none are believed to have an astronomical content (cf. Pedersén 2005, library N20).

38 Cf. the discussion in Clancier (2009), pp. 200–214.

collection	# astro	# PT	Nos.	NMAT procedures	astro joins
BE, Bab	10	1	**81**		
DT	10	1	**23**		
76–11–17	260	5 (6)	**10, 48, 49, 53, 62**	BM 32236+32336	(Sp2, 82–9–18, 1958–4–12)
Rm	134	1	**54**		(81–6–25, Sp)
78–5–31	39	1	**75**		
78–10–15	12	1	**43**		(81–7–6)
Sp	223	7	**11, 18, 46, 57, 58, 65, 70**		Sp2, 81–6–25, 81–7–1, 81–7–6, (Rm)
Sp2	542	22 (26)	**13, 16, 17, 18, 24, 27, 28, 34, 35, 36, 38, 45, 46, 55, 61.AEF, 63, 65, 68, 71, 98**	BM 34560, BM 34639, BM 34790, BM 43418	Sp, Sp3, 81–7–1, 81–7–6, low VAT, (76–11–17, 81–2–1, 83–6–30, 84–2–11, 89–4–26)
Sp3	395	7	**7, 16, 19, 36, 72, 79, 99**		Sp2, 81–2–1, 81–4–28, 81–7–6, 82–7–4
Rm4	120	8 (10)	**9, 32, 44, 50, 56, 61.D, 64, 88**	BM 33564, BM 33739	81–11–3
80–6–17	448	20 (29)	**2, 5, 14, 21, 25, 29, 47, 51, 52, 74, 78, 80, 82, 83, 84, 85, 92, 100, 101, 102**	BM 36747+37018, BM 36760, BM 36766, BM 36782, BM 36854, BM 36855, BM 37056+37074, BM 37110, BM 37332	81–2–1, (81–11–3)
81–2–1	34	3	**7, 52, 79**		Sp3, 80–6–17, 81–7–6, (Sp2)
81–4–28	174	4 (5)	**22, 26, 30, 72**	BM 41004	Sp3, 82–7–4
81–6–25	302	3	**57, 73, 86**		Sp, (77–2–22, 1958–4–12, Rm)
81–7–1	442	3 (4)	**8, 18, 91**	BM 42282+42294	Sp, Sp2, 81–7–6
81–7–6	393	9 (11)	**1, 7, 18, 36, 39, 46, 59, 69, 94**	BM 45728, BM 45990	Sp, Sp2, Sp3, 81–2–1, 81–7–1, (78–10–15, 82–7–4, 84–2–11)
81–11–3	229	2 (3)	**61.BD**	BM 47723	Rm4, (80–6–17)
82–5–22	22	1	**37**		
82–7–4	64	6 (7)	**15, 20, 60, 72, 95, 96**	BM 57980	Sp3, 81–4–28, (81–7–6)
83–1–18	22	(2)		BM 76488, BM 76703	
83–1–21	79	1	**40**		83–9–28
83–6–30	15	1	**87**		83–9–28, (Sp2)
83–9–28	13	2	**40, 87**		83–1–21, 83–6–30
84–2–11	26	1	**76**		(Sp2, 81–7–6, 89–4–26)
BCM	5	1	**97**		(1958–4–12)
low VAT	5	2	**35, 66**		Sp2, (MMA)
CBS	?	1	**77**		
MMA	10	1	**61.G**		(low VAT)

Table 1.2: Babylon: tablet collections with (fragments of) astronomical procedure texts. Collections of the British Museum (DT until 84–2–11) are arranged chronologically according to their accession date. Estimates in the second column derive from C.B.F. Walker's catalogue of astronomical fragments (unpublished). The third column contains the number of procedure texts with mathematical astronomy (in brackets the total number including NMAT procedures). The final column lists collections containing fragments that join a procedure text in the given collection (in brackets if the join concerns other astronomical texts). Published editions of the NMAT procedure texts, if existent, are listed in Appendix A.

Astronomical tablets in the 'Babylon collection' of the British Museum

Except for the small number of tablets discovered by Koldewey, all known astronomical tablets and fragments from Babylon were excavated unscientifically by local inhabitants and Hormuzd Rassam between 1876 and the end of the 19th century. The bulk of these tablets were acquired by the British Museum, whose 'Babylon collection' comprises approximately 20,000 tablets and fragments dating from the 7th c. BC to the first c. AD.[39] As much as 15–20% of the tablets in the Babylon collection concern astronomy and astrology. The overwhelming majority of these (perhaps 90%) are astronomical diaries and related observational texts, perhaps 5% belongs to mathematical astronomy, the rest containing zodiacal astrology and astrological omens. Almost nothing is known about the archaeological context of the Babylon collection. Much of it probably originates from the areas Amran and Jumjuma, where Rassam is believed to have carried out his excavations for the British Museum between 1879 and 1881. As suggested by Pedersén (2005), p. 276, the mentioned library N19, also located in Amran, may represent a small remnant of the scholarly libraries that make up the Babylon collection. Some hints about the provenience of the tablets might be inferred from the 'accession numbers' assigned upon their arrival in the British Museum (e.g. 76–11–17,2 = 17 November 1876, tablet 2). Apart from the collections that can be traced back to Rassam, other collections were acquired from various antiquities dealers

[39] Roughly BM 30000–50000. For a history of the Babylon collection cf. Leichty (1986), pp. xiii–xxxvi; Leichty et al. (1988), pp. xi–xxv.

in Baghdad and London.[40] The large number of joins among fragments from all of these collections suggests that they originate from the same area, perhaps from one or more astronomical libraries in private houses similar to library N19. Some systematic features become apparent if one considers the content of each collection. For instance, 80–6–17 includes some of the oldest known tabular texts dating from the reign of Artaxerxes I, but also very late tablets.[41] Lunar system B, completely lacking in 80–6–17, is represented in 82–7–4 by two procedure texts (Nos. **95** and **96**), and in 81–7–6 by numerous tabular texts. It is difficult to assess whether these features are significant enough to conclude that 82–7–4 and 81–7–6 belonged to a different library or dossier than 80–6–17.

Tablets in the Vorderasiatisches Museum (Berlin) and other museums

The same general provenience from the areas Amran and Jumjuma can be assumed for the astronomical tablets from Babylon in the Berlin museum with 'low VAT numbers' (below 600 or in the range 1700–2200), those in the Metropolitan Museum (siglum MMA), the University Museum in Philadelphia (CBS) and the Birmingham Museums and Art Gallery (BCM). The 'low VAT' astronomical tablets from Berlin were acquired by the then Königliche Museen (now the Vorderasiatisches Museum) in 1886 after having been excavated in Babylon. That they originate from the same locations as the astronomical tablets in the British Museum is proven by several joins with fragments from the Sp2 collection. Other 'low VAT' fragments have been joined with fragments in the Metropolitan Museum, including the latest known synodic table for the Moon, *ACT* 18, which was written in SE 263 (49 BC)[42] and, outside the astronomical corpus, several bilingual hymns.[43] For the astronomical tablets in the Birmingham Museums and Art Gallery the same is suggested by an astronomical diary joining a fragment from the 1958–4–12 collection of the British Museum (Horowitz 2000).

The astronomers in Babylon

The collection 81–7–6 includes several fragments written by astronomers from the Egibi clan (No. **7** and *ACT* 122). Other tablets in that collection and the collections 80–6–17 and Rm4 were written by members of the Mušēzib clan (Nos. **5**, **9**, **21**, **44**, **47**, **52**), whose involvement with astronomy can be traced over at least seven generations.[44] Astronomers from that clan are also mentioned in several Parthian era tablets from the Babylon collection and the Bodleian Library (Oxford) with protocols of the decisions of the Esagila temple council.[45] These important documents inform us about the employment of the astronomers at the Esagila, in which connection they mention that their duties include producing 'computed tables' (*tērsītu*), i.e. the tabular texts of mathematical astronomy, 'keeping the watch' (*naṣāru ša naṣāri*), a reference to astronomical diaries and related reports, and writing 'measurements' (*mešḫu*), which refers to Almanacs and Normal Star Almanacs (also observational texts).[46] The Babylonian term for astronomer was 'scribe of *Enūma Anu Enlil*' (*ṭupšar Enūma Anu Enlil*), after the incipit of the astrological omen series,[47] but not all astronomers used that title. The joins that were established between the 'low-VAT' tablets in Berlin and fragments in the Metropolitan museum have in common that they were written by members of the Nanna-utu clan, who were lamentation priests (*kalû*) as well as astronomers. Many of the mentioned bilingual hymns were written in SE 177–180 by the lamentation priest and astronomer Marduk-zēra-ibni/Ea-balāssu-iqbi, who also wrote the Jupiter text No. **35**.

1.3.2 Uruk

Nearly all known astronomical tablets from Uruk belonged to a scholarly library in the Rēš temple (*bīt Rēš*), the sanctuary of the sky god Anu and his spouse Antu. In the Seleucid era this main cult centre of Uruk became one of the largest temples ever built in Mesopotamia — a testimony to the economic and cultural vitality of Hellenistic Babylonia. Only few astronomical tablets were found at other sites in Uruk, namely in living quarters in the 'Parthian' excavation area U/V 18.

---

40 Rassam: Rm4, 80–6–17, 81–4–28, 81–7–1, 81–11–3, 82–5–22, 83–1–18, 83–1–21. The collections 76–11–17 (called S+ in the older literature) and Rm were acquired from the dealer Maroni, the latter by Rassam in 1877; 78–5–31 from Cutter; 78–10–15, 81–7–6, 83–6–30 and 83–9–28 from Shemtob, and Sp (1878), Sp2 (1879), Sp3 (1880), 81–7–1, 81–6–25, 82–7–4 and 84–2–11 from Spartali.

41 Early: Aaboe & Sachs (1969) Texts A–D; Aaboe et al. (1991) Text E; late: *ACT* 18a (SE 266).

42 VAT 209 + MMA 86.11.405.

43 For editions of these hymns cf. *CTMMA* II.

44 Oelsner (2000). To these one may tentatively add an eighth generation in the person of Iddin-Bēl/Bēl-aḫḫē-uṣur, the scribe of *ACT* 123, a synodic table for lunar system B for SE 235 (77/76 BC), and probably a member of the Mušēzib clan.

45 For this group of 13 tablets cf. also Jursa (2005), p. 75. For translations of some of the tablets cf. McEwan (1981), pp. 18–21; van der Spek (1985); Rochberg (2000); Clancier (2009), pp. 175–6.

46 Editions of Almanacs and Normal Star Almanacs are to appear in a future volume of *ADRT*.

47 See for instance the colophon of No.**7**.

The library of the Rēš temple

In 1912 the German archaeologist J. Jordan initiated scientific excavations in Uruk which continued, with interruptions, until the 1990s. Already during the first campaign cuneiform tablets from the Seleucid era, including astronomical ones, were found near the south-eastern gate of the Rēš temple.[48] These tablets were divided between Berlin, where they were assigned sigla in the range VAT 7000-9200 ('high VAT numbers'), and Istanbul (sigla U 1–600).[49] Most of the astronomical fragments ended up in Istanbul. They include only three procedure texts (Table 1.3); among the tablets in Berlin there are none. However, numerous other Seleucid tablets were excavated in Uruk before or near 1912 by local inhabitants, who sold them to various museums in Europe and the United States. While the bulk of these tablets are legal documents, they also include several astronomical and other scholarly tablets. Joins with fragments from Berlin and Istanbul (Table 1.3) prove that they originate from the same library in the Rēš temple.[50]

siglum	# tablets	# astro	# math. astron.	# PT	Nos.	NMAT; astrological procedures	astro joins
A	?	?	23	3	**33, 41, 93**		(U, AO, high VAT)
MLC	?	?	2	1	**67**		—
AO (*TU*)	58	16	7	1 (4)	**42**	AO 6449, AO 6455, AO 6488	(A, U)
U	600	> 100	73	3	**6, 12, 31**		(A, AO, high VAT)

Table 1.3: Uruk: tablet collections with (fragments of) astronomical procedure texts. Sigla: A = Oriental Institute (Chicago); MLC = Morgan Library Collection (kept at Yale); AO = Louvre (Paris); U = Eski Şark Eserleri Müzesi (Istanbul). The final column lists the sigla of the joined fragments; none of these concern procedure texts (all are tabular texts).

The archaeological context of these tablets was further clarified in 1959/60 when, during excavations led by H.J. Lenzen, 158 mostly scholarly tablets and fragments, obviously the remains of a temple library ('Uruk 4' in Pedersén 1998), were uncovered in a small brick-paved room ('79b') adjacent to the south-eastern gate.[51] Some of the tablets were found in their original location, in niches in the walls,[52] but most had been scattered. The tablets include numerous rituals and lamentations, and smaller numbers of omen texts and astronomical and astrological texts. Mathematical astronomy is represented only by a single tabular text.[53] The presence of disturbances resulting from previous illegal digs[54] supports the conclusion that many of the scholarly tablets from Uruk for which the archaeological context is unclear originate from this location. The dates of the tablets cover the period SE 15–150, with few exceptions.[55] The excavations revealed that room 79b was created during a reconstruction of the south-eastern gate probably undertaken in 201 BC by the governor Anu-uballiṭ-Kephalon. On that occasion the south-eastern gate may have obtained its Sumerian name **ka$_2$.sag**, 'Main Gate'.[56] Before that time room 79b did not exist, since the gate alley was located there.

Several suggestions have been made regarding the function of this library. According to Weidner (1925) and Lenzen (1962) it was the scribal school of the Rēš temple. Van Dijk (1962) proposed that it was the private library of the family of the astronomer Anu-bēlšunu from the Sîn-lēqi-unninni clan, because he, his son Anu-aba-utēr and other relatives are mentioned in many of the colophons. However, the colophons also mention individuals from three other clans, Ekur-zākir, Aḫûtu and Ḫunzû, so that this interpretation can be ruled out. A unique feature of the colophons is that nearly all mention two individuals: in first position the 'owner', indicated by the phrase 'tablet of PN$_1$' (*ṭuppi* PN$_1$), followed by the scribe, indicated by the phrase 'hand of PN$_2$' (*qāt* PN$_2$). As I have argued elsewhere (Ossendrijver 2011a,b), the functions of 'scribe' and 'owner' correspond to strictly successive, non-overlapping phases in the career of a scholar. Virtually all scholarly tablets of the Rēš library are therefore the product of a collaboration between a young, but obviously advanced scribe, and an experienced scholar (the 'owner' of the tablet). Often the 'owner' is the father of the 'scribe', suggesting that the relation between 'owner' and 'scribe' is that between teacher and student. Moreover, since the tablets were kept in

48 A comprehensive list of the cuneiform tablets from Hellenistic Uruk is provided by Oelsner (1986). For a list of the scientifically excavated tablets and fragments cf. Lindström (2003).

49 A description of the Uruk collection in Istanbul is provided by Kraus (1948) and Oelsner (1986), p. 144. U 91–284 are scholarly tablets; U 93–180 are astronomical; the content of numerous other tablets remains unknown.

50 Van Dijk (1962), p. 43; Pedersén (1998), p. 210, n. 99.

51 W 20030, 1–158 (copies: *Bam Beih* 2). For a list of tablets found in room 79b cf. Lindström (2003), pp. 211–216. For a description of the findspot cf. Lindström (2003), pp. 71–72, and Kose (1998), p. 193 (area Le 16-3); for photographs of the excavated room cf. Kose (1998), Pls. 11d–f. The tablets are currently kept in Heidelberg and Baghdad.

52 W 20030 1, 4, 6, 7, 18, 108, 109 (Kose 1998, pp. 193, 468, 469: excavation numbes 652–654, 660–663). Cf. also van Dijk *UVB* 18 (1962), pp. 43–61, and *AfO* 20 (1963), pp. 217–218.

53 *Bam Beih* 2 86 = W 20030, 111 (Steele 2005b), an auxiliary table for Mars.

54 Lindström (2003), p. 66.

55 *Bam Beih* 2 84, an astronomical diary for year 1 of Artaxerxes I (464/3 BC) and *Bam Beih* 2 17, a lamentation from year 2 of an unidentified king, probably Philip Arrhideus (322 BC).

56 Kose (1998), p. 189. The nocturnal festival described in *TU* 41 (AO 6460) implies that the guardian god of this gate was Usmû, the vizier of Ea (Falkenstein 1941, p. 11). For an edition of this text cf. Linssen (2004), pp. 245–251.

the library, the 'owners' do not appear to have been in physical possession of the tablets. Perhaps the phrase 'tablet of PN_1' only reflects that the tablet was written under the supervision of PN_1. These considerations confirm the original suggestion by Weidner and Lenzen that the library of the Rēš is the product of scribal education. By following the collaborations between 'owner' and 'scribe' it is possible to partly reconstruct the network of the astronomers in Uruk (Ossendrijver 2011a,b).

It is striking that very few observational texts (astronomical diaries and related texts) were found in Uruk, while they constitute the overwhelming majority of the astronomical tablets from Babylon.[57] Since it can be assumed that the astronomers from Uruk performed astronomical observations like their colleagues in Babylon, there must have been a sizeable corpus of diaries and related texts. Perhaps these were kept in another library, or in a private house.

It appears that very few tablets from the Rēš library, if any, were written after 161 BC (SE 150). This possible end of the library has been correlated with archaeological evidence of a destructive fire in the temple. Shortly thereafter, in 141 BC (SE 170), Babylonia was conquered by the Parthians. Until recently it was thought that the temple ceased to function by that time, but a legal document from SE 203 (108 BC) mentioning the Rēš (Kessler 1984) implies that it continued to exist, in some form, until about 100 BC.[58]

Other libraries with astronomical texts

A few astronomical texts, including tabular texts with mathematical astronomy but no procedure texts, were discovered outside the Rēš, in the living quarters in the 'Parthian' area U/V 18. Two early Seleucid synodic tables[59] were excavated in a house in area Ue 18–1. Several NMAT-type and astrological procedures were also found there.[60] Its location is close to the house of the exorcist Iqīša from the Ekur-zākir clan, who lived during the reign of Philip Arrhidaeus and left a large scholarly library containing several astrological texts (but no astronomical ones).[61]

1.4 Physical characteristics

Given the fragmentary state of most procedure texts, only few of them preserve the original width, height and thickness of the tablet. For some fragments the original measures of the tablet could be estimated from textual restorations. The following account of the physical characteristics of the procedure texts is therefore necessarily approximative and preliminary. Some procedures represent only a small portion of a tablet, the rest being inscribed with tables. They are treated separately, since their physical characteristics are similar to those of other tabular texts.

Tablets with mainly procedures

Most procedure texts contain only procedures. Their sizes cluster into a relatively small number of types, denoted A–H (Table 1.4, Fig 1.2).[62] On some tablets procedures appear together with auxiliary tables; as far as their sizes are known they overlap with those of the tablets containing only procedures.[63] For some adjacent types (C and D; D and F) the size ranges are not separated by a significant gap, so that they might merge into a single type if the sample of tablets were to increase.[64] Tablets wider than about 12 cm (types A–C) are usually divided into columns. As a rule of thumb, the width of a column is rarely outside the range 4–12 cm, but Nos. **2**, **43** and **50** are planetary procedure texts with unusually narrow columns of about 2 cm.[65] The most popular type is F; many of these tablets are lunar procedures. Types E and G are predominantly planetary procedures; for these types the width may exceed the height. The oval type H is represented by

[57] Diaries: *ADRT* I -463 = W 20030,142 (excavated in the Rēš, room 79b), *ADRT* III -99C = BM 140677 (perhaps from Uruk but cf. below); planetary observations: *LBAT* 1377 = A 3456 (Hunger 1988), *ADRT* V 42, 82; Normal Star Almanacs (to appear in a future volume of *ADRT*): *LBAT* 1004*, 1025*, 1030*a, 1031*, 1032*; Almanacs (to appear in a future volume of *ADRT*): *LBAT* 1124*.

[58] Cf. also Kose (1998), pp. 49–51. If the astronomical diary BM 140677 = *ADRT* III, -99.C, with observations for SE 212–214 (99–97 BC), originates from Uruk, as suggested by H. Hunger but doubted by C.B.F. Walker (private communication), astronomical activities in Uruk also continued into the first c. BC.

[59] W 22342 = *SpTU* I 98 (lunar system A), written SE 40; W 22755/3 = *SpTU* III 170 (Saturn system B or B'), perhaps for the years SE 1–21.

[60] NMAT: W 22554/6 (*SpTU* V 269); zodiacal astrology: W 23293/14 (*SpTU* IV 161).

[61] E.g. W 22246a (*SpTU* I 94), an astrological procedure for predicting market prices.

[62] The following tablets form the basis of this typology: **A**: Nos. **18**, **46**, **53**, **79**?, **97**?; **B**: Nos. **48**, **49**; **C**: Nos. **60**?, **72**?, **95**?, **96**?; **D**: Nos. **13**, **52**, **74**, **82**; NMAT: *TU* 11; **E**: Nos. **32**, **42**, **83**; **F**: Nos. **41**, **55**?, **56**, **61.ABDE?F?G**, **65**, **68**?, **69**?, **70**?, **76**?, **92**?; NMAT: BM 41004; **G**: Nos. **5**, **9**, **37**, **47**?, **57**?, **102**; **H**: No. **67**. Each measure represents the approximate range of attested values for that type (the range of thicknesses is based on the maximal thickness of each tablet).

[63] No. **17**: **C**?; No. **79**: **A**?

[64] Type B is represented only by two badly preserved tablets of unknown height whose missing parts probably contained a synodic table, so that they might also belong to type i of Table 1.5.

[65] None of them are listed in Table 1.2 because their original size could not be established.

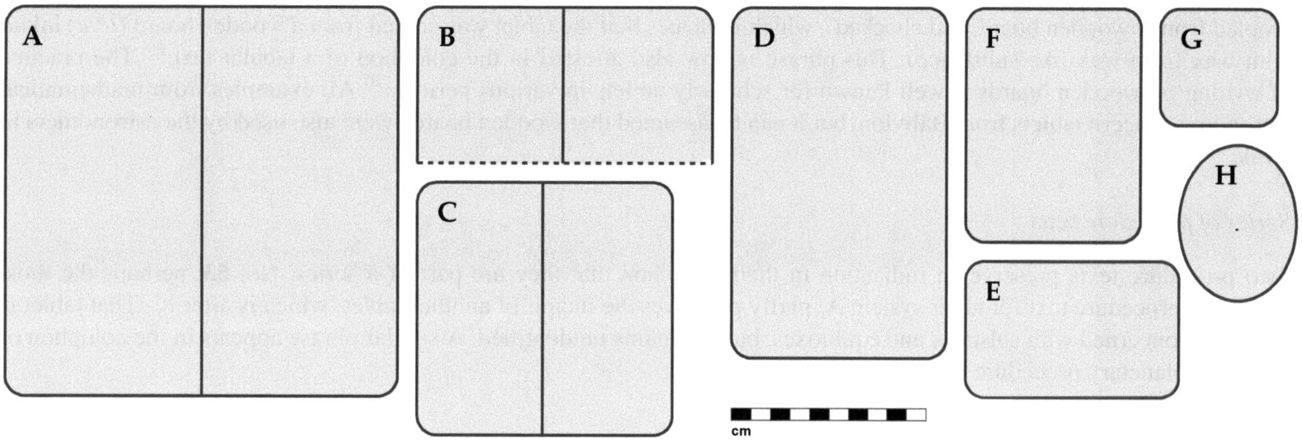

Fig. 1.2: Schematic drawings of types A–H.

size	# tablets	width	height	thickness	# columns	# procedures	remarks
A	5	18–23	20	2–3	2–3	18–32	
B	2	15–16	>8	2–3	2	>7	
C	4	12–14	12–14	2.5–4	1–2	2–8	
D	5	9–13	15–20	2–3	1	1–15	
E	3	10–11	6–8	2–3	1	1–8	
F	17	7–10	8–14	2–3	1	1–8	
G	6	5–8.5	4–7	1–2	1	1–6	
H	1	8	9	2	1	1	oval shape

Table 1.4: Sizes of tablets with mainly procedures.

a single tablet from Uruk. All tablets of types A–C, most tablets of types D–G, and half of the tablets of type H contain a collection of procedures. Given the small number of procedure texts from Uruk whose measures are known (three in Table 1.2) very little can be said about local variations. All in all the procedure texts thus form a rather heterogeneous group in terms of their physical characteristics.

Tablets with procedures and tabular content

On 21 tablets procedures occur together with synodic tables or template tables. For 11 of them the original size could be measured or estimated. Since the layout of these tablets is determined by the tables which occupy most of the surface, their physical characteristics (Table 1.5) are similar to those of other tabular texts, and different from tablets inscribed with only procedures.[66] All of these tablets concern planets (cf. §3.2); the combination of procedures and synodic tables or template tables is not attested for the Moon.

type	# tablets	width	height	thickness	# columns	# procedures
i	1	13–15	18–19	2–3	1	1
ii	4	17–21	9–12	2.5–3	1, 7	1, 14
iii	2	20–25	14–15	3	1	1, 2
iv	4	25–30	9–11	3	1, 4	1, 4

Table 1.5: Tablets containing procedures and tables. The number of columns refers to the procedures only.

References to tablets and wooden boards

In the colophons of the procedure texts three different Akkadian designations for tablets are found: *ṭuppu*(**im**, **dub**, **im.dub**) = 'tablet'; **im.gid$_2$.da** = *giṭṭû* or *imgiddû*, and *lē'u*(^{giš}**da**), 'wooden board'. 'Tablet' (*ṭuppu*) is a general term that cannot be correlated with any physical characteristics. It appears in the phrase *ṭuppi* [*ša$_2$*] PN, 'tablet of PN', by which the 'owner' of a tablet is identified in colophons. The term *giṭṭû* or *imgiddû*, lit. 'long tablet', designates one-columnar tablets with a literary or scholarly content.[67] It is attested in the colophons of Nos. **9** and **44**, both small tablets of type G or H. Several procedure texts[68] include a colophon with the phrase *ultu*(**ta**) *muḫ-ḫi lē'i*(^{giš}**da**) *šaṭir*(**sar**)-*ma bari*(**igi.tab**),

[66] The tablet sizes are preliminarily labeled i to iv. The following tablets form the basis of this typology: i: No. **1**; ii: Nos. **7**, **8**?, **17**?, **28**?; iii: Nos. **12**, **35**?; iv: Nos. **10**, **20**?, **24**?, **36**? A detailed study of the typology of the tabular texts is outstanding (Volume II).

[67] In some literary corpora it designates an excerpt; on school tablets possibly an exercise tablet. Cf. Hunger *BAK*, pp. 7, 162–3; *CAD*: 'one-columnar (literary) tablet'; *AHW*: excerpt tablet. Alternative reading: *liginnu* = 'excerpt tablet' or 'exercise tablet'.

[68] Nos. **44**, **53**. NMAT: BM 41004, *LBAT* 1571a, BM 45990, BM 76488. Editions of the NMAT, if existent, are listed in Appendix A (p. 523).

'copied from a wooden board, and checked', which indicates that the tablet was copied from a wooden board (*lē'u*) inlaid with wax (or a wax-like substance). This phrase is now also attested in the colophon of a tabular text.[69] The practice of writing on wooden boards is well known for scholarly tablets in various periods.[70] All examples from mathematical astronomy concern tablets from Babylon, but it can be assumed that wooden boards were also used by the astronomers in Uruk.

Series of procedure texts

Two procedure texts preserve an indication in their colophon that they are part of a series. No. **53**, perhaps the most important procedure text for lunar system A, partly preserves the incipit of another tablet 'which is after it'. That tablet is probably concerned with solstices and equinoxes, but it remains unidentified. A similar phrase appears in the colophon of No. **45**, a planetary procedure text, but the incipit is broken.

Composition of the tablets

The number of procedures per tablet varies considerably, between 1 and more than 30. Only few systematic features of composition can be identified. For the planetary texts these are discussed in §3.2, for lunar system A in §4.4.1, and for lunar system B in §4.5.1. Some procedure texts with mathematical astronomy include procedures that belong to non-mathematical astronomy (NMAT); for instance No. **53** (lunar system A) includes a procedure for solstices and equinoxes. Finally, several tablets with procedures for Jupiter include 'mathematical' procedures involving trapezoids (cf. §3.7.10.3), but it cannot be ruled out that they have some hitherto unrecognised connection with Jupiter.

1.5 Transmission and evolution

Our ability to trace the transmission and evolution of the procedure texts is limited by the narrow temporal distribution of the tablets. All datable procedure texts were written between year 5 of Philip Arrhidaeus (319/8 BC) and the year SE 202 (110/109 BC) in the Parthian era. The tabular texts cover a wider span of time, but most of them date from SE 120–210. Most undated procedure texts were probably written in the same period.[71] However, since the underlying algorithms appear to have reached their final form at the end of the Achaemenid era, many of the extant procedure texts may be copies of originals composed before the Seleucid era. Explicit evidence of copying activities are the mentioned references to wooden boards (§1.4), and several tablets containing the remark 'break' (Akk. *ḫīpu*, written *ḫi-pi*) or 'new break' (*ḫi-pi₂ eš-šu*),[72] which signifies that the tablet was copied from a damaged original. Copying activity is also implied by the existence of duplicates.[73] Moreover, many procedures are copies in the sense that their formulation is based on a relatively small number of 'template procedures' (§2.6.4.2). Procedures copied from the same template differ only with regard to the numerical parameters. The composition of a tablet can also provide clues about copying activities. No. **52** contains procedures for lunar system K, which is believed to reflect a pre-Seleucid stage of development, but other procedures on that tablet belong to lunar system A. It can therefore be assumed that the system-K procedures were copied from an older tablet, perhaps because they had an antiquarian value.

It is clear from these considerations that many of the extant procedure texts are copies transmitted from one generation of astronomers to the next, but the lack of securely dated tablets from different periods makes it difficult to reconstruct the transmission and evolution of their content in any detail. It would be of interest to explore how the algorithms and their formulation changed in the course of time, and to determine when they reached their final stage. One of the few identifiable trends is that the example-based formulation occurs mainly in texts from the early Seleucid era, while the later texts favor an abstract formulation; for this issue cf. §2.6.3.1.

1.6 Approaching procedure texts

The motivation for producing this new edition and investigation of the procedure texts partly derives from methodological questions concerning translation and other aspects suggested by recent developments in the historiography of science.

---

[69] BM 37021 (Steele 2006a, Text B), a synodic table from Babylon with data for SE 45–48, computed with lunar system A.

[70] Hunger *BAK*, pp. 7–8.

[71] A date of composition in the Parthian era has been suggested for No. **95** on account of a parameter mentioned in one of its procedures; cf. the commentary.

[72] No. **5** R8; No. **92** O7; NMAT: BM 41004 (Brack-Bernsen & Hunger 2005/6) R3; *TU* 11 O2,3, R24,26,28.

[73] For instance among the procedures for computing G from Φ (lunar system A).

Although it has always been obvious that Babylonian mathematical astronomy can be investigated only with an interdisciplinary approach incorporating up-to-date methods from all involved disciplines, these recent developments have barely been applied to Babylonian mathematical astronomy.

1.6.1 Translation issues

One aim of the present study has been to evaluate the methodological foundations of the translations in *ACT* in the light of the innovative approach to OB mathematics of J. Høyrup and G. Friberg[74] and, on that basis, formulate a refined method tailored to the procedure texts.

Translating Babylonian mathematics and mathematical astronomy

The program of translation initiated by Høyrup and Friberg sets out from the principle that the original texts are decisive for the interpretative effort.[75] This is not as obvious as it seems, since important aspects of the original text may be lost in a translation, which those who interpret it may neither be aware of, nor have the ability to verify. The solution proposed by Høyrup and Friberg is that terminology should not be transformed by translation. Particularly harmful transformations are those whereby different terms that *appear* to have the same functional meaning are replaced by a single modern term. In order to avoid these and other anachronistic tendencies they propose the 'conformal translation', which aims to maintain a high degree of conformity between original text and translation. Hence Akkadian word order is usually maintained, even at the cost of English grammar, and each Akkadian word is always translated by the same English word; if no suitable English word exists a neologism is introduced. With that approach, in combination with a careful semantic analysis, Høyrup rediscovered the full extent of the original, geometric formulation in OB mathematics, which had been obscured in the previous translations by numerous violations of the principle of conformity. For instance, different words for what appeared to be equivalent additions were routinely translated by the same English verb 'to add'.

Turning to the procedure texts, the obvious question is whether similar distortions may be identified in the previous translations. This is indeed the case, as will become apparent in the following chapters. However, they are usually of a different kind than in OB mathematics. In particular, the geometrical terminology of the OB mathematical corpus is largely absent from the astronomical procedure texts, and appears to have become extinct in the LB period (§2.2). Since there is no hidden geometric formulation to be unveiled, one might therefore suspect that a new translation of the procedure texts will not result in significantly different interpretations. However, it turns out that the anachronistic tendencies in the translations are to some extent stronger than in OB mathematics. This is partly because, for some reason, Neugebauer was less hesitant with the astronomical texts than with the mathematical texts to introduce modern concepts into the translations and the commentaries, including 'negative numbers' and 'velocity'.[76] The translations in *ACT* may thus be characterised as pragmatic, but sometimes inadequate in a semantic sense. It is a testimony to the genius of Neugebauer that his astronomical interpretations are barely affected by these anachronistic tendencies, with a few notable exceptions.[77]

Also the conformal translations exhibit certain problematic features. The demand of strict conformity with the original text results in translations that are sometimes difficult to understand, because Akkadian word order is maintained even if it violates English grammar, and many technical terms are translated with neologisms. Another possibly problematic aspect is the use of etymology for constructing English neologisms that serve as translations of certain technical terms but also of common words. The evidence from practical linguistics suggests that the meaning of a word cannot be established through etymology, but only through the analysis of context and usage. That interpretative process does not necessarily become more objective if well-established meanings are replaced by neologisms. The conformal translations succeed in a strictly semantic sense, but they sometimes fail to make a text accessible to readers who do not know the source language.

A perspective from translation theory

In order to come up with a practical solution to the problem of translation it is helpful to explore the insights of modern translation theory,[78] and seek an answer to the following questions. First, what are the criteria by which the adequacy of a translation can be judged? Second, what does translation theory tell us about the 'orthodox' translations and the conformal translations? According to translation theory, translation involves a sender, an original text, a translator, a goal text and a recipient. Instead of a notion of equivalence between original and goal text, translation theory recognises that

[74] Exemplified in Høyrup (2002) and Friberg (2007).

[75] For a recent discussion of methodological issues concerning the translation of OB mathematics cf. Høyrup (2010).

[76] In certain respects the translations are also more modernising than those of Kugler, who, for instance, did not speak of 'negative/positive latitude', but used the less anachronistic 'southern/northern latitude'; cf. §2.5.3.2.

[77] E.g. in step 6 of the Lunar Six module for lunar system A (No. **61**).

[78] Cf. for instance Bassnett (2002).

all partners in the translation process may affect the notion of equivalence. After selecting those aspects in which an optimal equivalence is desired, at best a partial equivalence can be achieved, necessarily at the cost of other types of equivalence. Two relevant forms are semantic equivalence, which is text-oriented, and pragmatic or functional equivalence, which is recipient-oriented. To start with the former, a complete semantic equivalence cannot be attained between natural languages, with the exception of neologisms that are introduced into a translation for that purpose, at the cost of the immediate comprehensibility of the translation. Pragmatic or functional equivalence aims to produce a translation that has the same function for its recipient as the original text had for its respective audience. This is important for translations of modern literature, handbooks, etc., but less relevant for scholarly editions of ancient texts. There is very little hope of achieving functional equivalence when translating Akkadian texts, since there are no modern equivalents for many of the concepts in these texts. More importantly, the aim of such editions is precisely to reveal the ancient concepts and not to replace them by a functional equivalent.

Implications for the present study

This digression into translation theory reveals that, to some extent, the 'orthodox' translations and the conformal translations correspond to opposite, extreme positions in the spectrum of translation methods. Whereas the early, modernising translations of Babylonian mathematical and astronomical texts can be characterised as functional and pragmatic, the conformal translation aims for strict semantic equivalence. While I agree that a scholarly edition should aim to optimise semantic equivalence and not pragmatic equivalence, the approach in this volume is a compromise between both approaches, in that I also aim for optimal semantic equivalence, but not to the full extent of the conformal method. In particular, I follow the conformal method in trying to preserve any terminological differentiation in the translation, and use the same English word for the same meaning of an Akkadian word as much as possible.[79] However, I do not maintain Akkadian word order, and make only limited use of etymology for coining translations of technical terms. Practical implications are discussed in the introduction to the critical editions, §5.1.2 (p. 205). Other, more pragmatically oriented representations of procedure texts in the present study include flow charts, formulas and graphical representations. Given the complexity of Babylonian mathematical astronomy, with its numerous interrelated functions, these modern tools are practically indispensable for making the astronomical algorithms accessible to a modern audience and to modern forms of analysis. They must obviously be interpreted with care in order to avoid anachronistic interpretations; cf. the discussion in §2.6.4 (p. 37).

1.6.2 Procedure texts and 'scientific discourse'

In general works on the history of science, Babylonian astronomy is sometimes classified as 'instrumentalist', because its purpose is, supposedly, restricted to the prediction of phenomena, and the cuneiform texts contain no hint of any intention to explain the phenomena that are predicted, which is considered to be an essential feature of (modern) science.[80] Babylonian astronomy, as opposed to those forms of Greek mathematical astronomy developed by Hipparchos and Ptolemy, is said to lack an underlying 'theory' involving concepts and causal relations that are capable of explaining the phenomena. By contrast, Greek mathematical astronomy is seen as achieving its predictions on the basis of models, derived from physical concepts with the axiomatic-deductive method also known from the Elements of Euclid. Recent investigations have resulted in a more diverse picture of Greek astronomy, in particular the discovery of Babylonian methods in the Greek astrological papyri from Oxyrhynchus (Jones 1999). Moreover, even though deductive logic and explanation — ingredients commonly associated with the concept of theory in modern science — are not explicitly formulated in the Babylonian texts, it is clear that they do contain a considerable body of implicit theoretical knowledge.[81] The procedure texts, as verbal representations of algorithms, exhibit numerous innovations in terms of mathematical representation, which are explored in the next chapter. Many of these innovations can be interpreted as steps in a direction of increased abstraction, and seem to have been invented specifically in order to cope with the complexity of mathematical astronomy. Another aspect of theoretical knowledge in mathematical astronomy that is overlooked in general works on the history of science concerns the underlying mathematical structures hidden within the computational systems. This is especially true for lunar system A, whose mathematical structure exhibits a rigorous internal coherence combined with a surprising complexity that is not immediately apparent from the texts, and comes to the surface only by a thorough analysis of the algorithms, as has been pointed out repeatedly by J. Britton.[82]

79 Obviously not necessarily the same English word for every instance of the same Akkadian word.

80 For an account of the reception of Babylonian astronomy by historians of science including numerous illuminating quotations cf. Rochberg (2004).

81 As remarked for the mathematical texts by Friberg in Waschkies (1989), p. 11.

82 This concerns in particular the interrelated functions Φ, G, Λ and W, and the way in which they are constructed (Britton 1999, 2007a, 2009).

1.7 Didactical, rhetorical and comparative aspects

The existence of procedure texts can be viewed as an obvious consequence of the complexity of mathematical astronomy. We may imagine that the evolution of mathematical astronomy eventually led to a situation where the algorithms began to exceed what can reasonably be memorised, so that it became necessary to write them down in the form of verbal instructions. Even if one assumes that the algorithms of mathematical astronomy can be memorised, it would be advantageous for a Babylonian astronomer to have them available in written form.

The number of procedures per tablet varies widely – from 1 to about 30. Some tablets contain a complete collection of procedures providing algorithms for all columns of a synodic table for the Moon or a planet. However, closer inspection reveals that important computational as well as non-computational aspects are often not explained. Usually 'meta-instructions' about the proper way to apply the procedures are lacking. In the case of lunar system A (§4.4) with its numerous functions, the procedures tend to be arranged in the same order as the columns of a synodic table (§4.4.1), but explicit information about the proper sequence of the procedures is not provided. Neither is the user informed about the initial values to be used in the computations and other practical issues. Procedure texts are therefore not self-explanatory didactic handbooks for producing astronomical tables. This lack of comprehensiveness is aggravated by the fact that, as representations of algorithms, procedures are often, strictly speaking, deficient in one way or the other (§2.6.3.2).

1.7.1 Rhetorical aspects

Most procedure texts consist of complete and grammatically correct sentences conforming fully with the syntactical rules of Late Babylonian. However, the range of grammatical forms and syntactical structures is rather limited compared to literary texts. Nearly all procedures are formulated in the second person singular of the present tense (e.g. 'you multiply ..., you add to it ...'). Only planetary procedures for the subdivision of the synodic cycle (§3.3.3) regularly employ the third person singular of the present tense when describing the motion of a planet (e.g. 'for 3 months it moves 0;4,10 per day, then it becomes stationary...'). The only other significant deviation from the basic pattern is found in a few texts in which the instructions are embedded in a dialogue, as in the following example:[83]

mi-nu-u$_2$ **a.ra**$_2$ 2.13 / **igi.meš** *lu*-**du**-*ma lu*-⌜*u*$_2$⌝ 15.6 6.48.43.18.30 GAM 2.13 **igi.meš du**-*ma* / 15.6	What should I multiply by 2,13 / appearances so that it is 15,6,0? You multiply 6,48;43,18,30 by 2,13 appearances, it is / 15,6,0.	No. **13** R33'–35'

The only known example involving subtraction is the following:

[*mi-nu-u*$_2$] ⌜**ta**⌝ **ša**$_3$ *lu-us-suḫ* / *zib* 18 **zi** ...	What should I 'tear out' from it? Psc: you 'tear out' 18 ...	No. **52** Ri25–26

These phrases represent fictitious dialogues between master and student, a rhetorical device that is rare in the astronomical procedure texts but well known from mathematical problem texts from both the OB and the LB period. In the mathematical texts the phrase *aššu*(**mu**) *lā*(**nu**) *tīdû*(**zu**-*u*$_2$), 'since you do not know it', is usually placed in between the question and the procedure. Although this phrase is rarely used in the astronomical procedures,[84] it does clarify their rhetorical structure. The user utters a question like a student facing a master, but since he/she 'does not know' the answer, the master provides it in the form of a procedure.[85] It seems reasonable to assume that this rhetorical model underlies all astronomical procedures, even though it is usually not made fully explicit. The extent to which this reflects an actual didactic practice is not clear, neither for mathematical astronomy nor for the mathematical texts.

The preterite does occasionally occur in the procedure texts, but only in subordinate clauses (e.g. 'the elongation which you had put down ...' in step 8 of No. **61** P1–P4). This limited usage of tense and persons, which is typical for all instructional texts from the LB period,[86] reflects their predominantly practical purpose, which in the case of mathematical astronomy is to provide instructions for computing tables. Even those procedure texts of mathematical astronomy whose purpose is not directly connected with the production of tabular texts[87] are indistinguishable from the more practically oriented texts in terms of syntax and choice of grammatical forms. However, it would be wrong and anachronistic to view

[83] Other examples: No. **6** R2',4'; No. **46** Oi1'; No. **96** R11.

[84] It may be partly preserved in No. **58** O6.

[85] Due to the logographic writing it is usually not obvious that the phrases following the question are formulated in the second person, but this is certain in the LB mathematical problem text BM 34568 Oii3–5, 9–11 (*MCT* Text Y).

[86] The imperative, common in OB mathematical problem texts, is not attested in the procedure texts.

[87] E.g. Nos. **82**, **102**.

this as a decisive limitation that might explain why Babylonian mathematical astronomy lacks 'theoretically' oriented texts informing us about its underlying principles, as they exist in Greek astronomy.

1.7.2 Other Mesopotamian instructional texts from the first millennium BC

Many features of the procedure texts with mathematical astronomy are shared by other Babylonian and Assyrian instructional texts from the first millennium BC. A list of such texts, without any claim to completeness, can be found in Appendix A. Arguably the first text that should be mentioned in this connection is the astronomical compendium Mul.Apin ('Plough Star'),[88] which is preserved on numerous tablets and fragments from Assyria and Babylonia dating between the 7th c. BC and the Seleucid era. Although Mul.Apin is dominated by lists and omen-like phrases, it includes several instructions in which the reader is addressed in the second person. Apart from instructions related to astronomical observation[89] three procedures deal with computation.[90] Their formulation is similar to that of other astronomical procedure texts.

The closest affinity with the procedure texts of mathematical astronomy is found among the procedure texts of 'non-mathematical' astronomy (NMAT). By and large they employ the same grammatical forms, syntactical structures and terminology, both arithmetical and astronomical. Most of the NMAT procedure texts also date from the same period and originate from the same locations in Babylon and Uruk where mathematical astronomy was practised. The main representative of this group is *TU* 11 (Brack-Bernsen & Hunger 2002), very likely from the library of the Rēš temple in Uruk. This large tablet contains Goal-Year procedures for predicting month lengths, but also astrological procedures and omens. W 20030/16 is another Goal-Year procedure text from the same library that partly duplicates *TU* 11. Some NMAT procedures contain instructions concerning the observation of planetary phenomena, *ziqpu* stars or other stars, as can be found in Mul.Apin. Another group of NMAT procedures, thus far known only from Babylon, deal with the construction of what appears to be a gnomon-like instrument. Unfortunately they are difficult to understand.[91] From a formal viewpoint they are similar to other astronomical procedure texts. For instance, they begin with a statement of purpose *uskara*(**u$_4$.sakar**) *ana* **du$_3$**-*ka*, 'For you to construct a 'crescent'', apparently the name of the instrument, and all instructions employ the second person of the present tense, features well known from other procedure texts (cf. §2.6.2). Several astrological tablets from the same period, listed in Appendix A, contain instructions pertaining to the prediction of weather and market prices. All known examples originate from Seleucid Uruk. The Goal-Year procedure text *TU* 11 mentioned earlier includes a similar astrological procedure. A common feature of these procedures is that the actual method of prediction, which involves the Goal-Year periods of the planets, is alluded to but not spelled out, so that they are not immediately comprehensible to us without a significant interpretative effort.

The small corpus of LB mathematical problem texts also shares many features with the astronomical procedure texts, in particular with regard to the arithmetical terminology (§2.2). However, there are differences as well, especially in W 23291 and W 23291-x, two early Seleucid problem texts from Uruk. While the arithmetical terminology is virtually the same as in the astronomical procedure texts, the syntactical structures are more varied, although they share the preference for the second person of the present tense. For instance, the mathematical problems regularly include the mentioned phrase 'since you do not know it'. AO 6484, a tablet with geometrical problems, contains procedures with highly abbreviated multiplications of the kind *mi* GAM *mi lu*-**du** *lu* 16, 'What should I multiply times what so that it is 16?', where *mi* is a phonetic abbreviation of *minû* that is not yet attested in the astronomical procedures.

In this connection mention must also be made of the sizeable corpus of ritual instructions (*nēpešu* = 'ritual') for the cult in the very temples of Babylon (the Esagila) and Uruk (the Rēš) where the astronomers were employed (Linssen 2004). Although some of these elaborate texts include instructions in the second person of the present tense, others are formulated in the third person, with the priest ('he') appearing as the subject of the ritual activities. The ones from Uruk very likely originate from the same library within the Rēš as the astronomical tablets, and some were written by scholars who were active in mathematical astronomy.[92]

88 Standard edition: Hunger & Pingree (1989). The present discussion has benefited from a recent analysis of Mul.Apin by Watson & Horowitz (2011).

89 Mul.Apin a.Iiv7–30: on the usage of *ziqpu* stars; g.IIi22–43: first intercalation rule, hemerologies; h.IIi68–71: observation of winds; j.IIii7–17: second intercalation rule.

90 Mul.Apin j.IIii7–17: second intercalation rule; k.IIii41–42: computing shadow lengths from the length of daylight; l.IIiii13–15: computing lunar visibility intervals from the length of daylight.

91 An edition is being prepared by J. Steele and C.B.F. Walker.

92 Cf. for instance the scribal biography of Šamaš-ēṭir in Ossendrijver (2011a).

Chapter 2
Mathematical concepts – from numbers to computational systems

Mathematical astronomy can be defined as a technique for predicting astronomical phenomena using mathematical algorithms. Procedure texts contain verbal representations of such algorithms. This chapter offers a systematic analysis of the concepts and algorithms of Babylonian mathematical astronomy and their verbal representations in the procedure texts, using the tabular texts as a complementary source of information.[93] Setting out from the most elementary concepts to more complex ones, the analysis results in a hierarchy of concepts comprising three levels (Fig. 2.1). In order of increasing complexity they are defined by (i) numbers (§2.1), arithmetical and other elementary operations (§2.2–§2.4) and coordinate systems (§2.5); (ii) algorithms and functions (§2.6) and (iii) computational systems (§2.7). §2.6 is concerned only with generic algorithms; specific implementations of these algorithms, and other algorithms that are unique to the planets or the Moon, are discussed in Ch. 3 and 4, respectively.

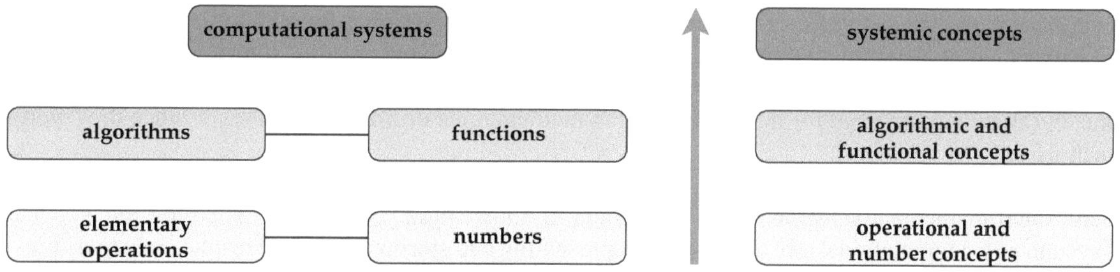

Fig. 2.1: Hierarchy of the mathematical concepts of Babylonian mathematical astronomy.

2.1 The sexagesimal place-value system

The first level of concepts to be discussed concerns numbers and arithmetic operations: addition, subtraction, multiplication and division. Babylonian mathematical astronomy is based on the sexagesimal place-value system, i.e. numbers are represented as sequences of digits using base number 60 ('sexagesimal'), such that the value of each digit depends on its position within the sequence ('place-value' or 'positional' system). The resulting number system operates in a manner entirely analogous to our own decimal system, as may be illustrated by an example:

$$\begin{aligned} 10{,}0{;}6 &= 10 \cdot 60^1 + 0 \cdot 60^0 + 6 \cdot 60^{-1} \\ &= 6 \cdot 10^2 + 0 \cdot 10^1 + 0 \cdot 10^0 + 1 \cdot 10^{-1} = 600.1 \end{aligned}$$

In the modern notation for sexagesimal numbers, individual digits 0–59 are separated by commas, except for those pertaining to $60^0 = 1$ and $60^{-1} = 1/60$, which are separated by a semicolon (;). This is called the absolute notation, because each digit is multiplied by a unique power of 60. In cuneiform, sexagesimal numbers are written using a relative notation, because there is no sign equivalent to the semicolon that would indicate which digit pertains to 1. Hence the power of 60 corresponding to a given digit is defined only in relation to the other digits. In transliterations this feature is conserved by using the period (.) as a neutral separator between all digits. The cuneiform signs representing 10,0;6 are therefore transliterated as 10.0.6.[94] Also 10;0,6 or 0;10,0,6 are possible absolute values of that cuneiform number, whose correct

[93] New editions and analyses of the tabular texts will be presented in Volume II. Previous editions of most tabular texts can be found in *ACT*.
[94] For these and other practical aspects of transliteration and translation cf. §5.1.

interpretation can be determined only from context.[95] In order to do basic calculations in the sexagesimal positional system it was necessary to master the elementary algorithms for adding, subtracting, multiplying and dividing multidigit sexagesimal numbers.[96] All numbers in Babylonian mathematics and mathematical astronomy are finite sexagesimal numbers, which means that they can be represented as a finite sequence of sexagesimal digits. Apart from whole numbers this includes numbers containing fractions smaller than 1, provided that they do not have any other dividers than 2, 3 and 5, the dividers of 60. Therefore, all finite sexagesimal numbers can be factorised as

$$\frac{n}{2^p \cdot 3^q \cdot 5^r}, \tag{2.1}$$

where n, p, q, r are (non-negative) whole numbers. Examples of numbers that cannot be represented in this way are 1/7, 1/11, 1/13, 1/17, 1/19, etc.

The origin of the sexagesimal place-value system has not been fully explained. Already long before its introduction the number 60 featured prominently in various Mesopotamian number systems, but these were not of the positional type. Apart from mathematical astronomy the sexagesimal place-value system is mainly known from the OB mathematical corpus, but the earliest textual evidence for it dates from the Ur-III period (2100–2000 BC).[97] The sexagesimal notation of mathematical astronomy incorporates a few minor innovations with respect to that of OB mathematics: the sign GAM now represents empty digits (cf. below), and the digit 9 is represented by a simplified sign. Outside the context of scholarship and the scribal school the sexagesimal system was rarely used in writing. In administrative and other non-scholarly documents, numbers continued to be written in traditional non-positional systems until the very end of cuneiform writing. However, the practical computations in these texts are believed to be based on sexagesimal calculus as an intermediate step. This may involve an abacus-type device on which the elementary operations for addition, subtraction and multiplication could be executed by manipulating physical objects representing the sexagesimal digits. For non-scholarly purposes the scribes would always write down the final results in the traditional number systems, which were readily understood by everyone, but there was no need for that when writing mathematical or astronomical texts, since they were read only by fellow scholars.

There is little doubt that the invention of the sexagesimal system was a major factor contributing to the development of mathematics and mathematical astronomy in Babylonia. It allowed the scribes to fully exploit the advantages of the positional system as a computational tool and also for representing and storing numerical results and algorithms in written form. Several practical advantages of the sexagesimal positional system over non-positional number systems as well as positional systems with other base numbers may be mentioned. First, in a positional system a single set of signs for the digits between 0 and the base number minus 1 (0...59) suffices for representing arbitrary numbers, as opposed to non-positional systems, in which every base number is represented by a separate sign (e.g. Sumerian numerals). Second, computations proceed more efficiently because all digits are manipulated in the same manner. Third, the sexagesimal system has a specific advantage, because 60 has many dividers (2, 3, 4, 5, 6, 10, 12, 15, 20 and 30). Hence more fractions can be represented as a finite sequence of digits (e.g. 1/3 = 0;20) than in other systems including the decimal one (where 1/3 = 0.33333...). While it is clear that these factors enhanced the ability of the Babylonian scholars to solve complex problems, their historical relevance is difficult to assess.

The zero and its manifestations

The sexagesimal system of mathematical astronomy includes a special sign (GAM) for the digit 0, which is attested for the first time during the Achaemenid era.[98] GAM is used for marking the absence of an initial or intermediate digit, but not of a final digit. This usage of GAM can be traced back to its role as a separator in literary texts ('Glossenkeil'). Also in mathematical astronomy GAM is sometimes used in order to mark a separation between signs, words or numbers (transliterated as :). In Uruk GAM is commonly used within sexagesimal numbers for separating multiples of 10 and digits 1–9 in order to prevent them from being interpreted as a single digit.[99]

A unified notation for the vanishing of a digit and the vanishing of a number was not developed. The number zero is not represented as GAM, but usually as 'it does not exist' (**nu tuk**),[100] sometimes as 'nothing' (*jānu*).[101] These representations

95 In Babylonian mathematical astronomy it is meaningful to speak of the 'correct interpretation' of numbers, because they nearly always represent quantities measured in certain empirical units (degrees of arc, time degrees, barleycorns, mean tithis, etc.). This is not the case in some mathematical texts containing numerical computations devoid of context, e.g. multiplication tables.

96 For these algorithms cf. for instance Friberg (2007), pp. 6–8.

97 E.g. HS 201, a table of reciprocals from Nippur (Oelsner 2001).

98 In OB mathematics the vanishing of a digit was sometimes marked by an empty space.

99 Cf. Neugebauer (1941). In Babylon the scribes achieved this by leaving a bit of extra space. In the OB mathematical text MS 2731 (Friberg 2007, p. 40) there is an instance where GAM separates two digits that are both less than 10, perhaps in order to mark the absence of a multiple of 10?

100 Akkadian reading probably *ul ibašši*, as proven by the occasional **nu tuk**-*ši* [*ACT* 135, Ri24'].

101 No. **52** Ri9: *bi-rit* **tab** *ana* **tab** *ia-a-nu*, 'the distance from addition to addition is zero'.

of the number zero are used only in certain contexts. They can stand by themselves like other numbers, or appear as predicates of quantities (‘QN does not exist/is zero’), but are rarely attested as operands in arithmetical operations.[102] In the tabular texts the empty space also serves as a representation of the number zero, e.g. in column *J* of lunar system A (§4.4.14).

2.2 Arithmetical operations

Astronomical procedures are essentially sequences of arithmetical operations linked together in a verbal structure. The purpose of this section is to present a lexicographic and semantic analysis of the arithmetical operations. There are several reasons why such an analysis is necessary. First, some of the logograms representing arithmetical operations are attested only in LB astronomical and mathematical texts and their Akkadian reading is not well established. Second, there are different terms for apparently equivalent operations. As Høyrup (2002) has demonstrated for OB mathematical problem texts, this might reflect certain semantic differentiations that have hitherto remained unnoticed. A third aim is to compare the arithmetical terminology of mathematical astronomy with that of LB mathematical problem texts and OB mathematics. Apart from arithmetical operations the procedures contain other elementary operations which are discussed in §2.3.

2.2.1 Identity of quantities and symmetry of operations

A notable feature of the procedure texts is the existence of apparently synonymous words or phrases for the same arithmetical operation. It is known from OB mathematical problem texts that some of these apparently synonymous terms reflect a differentiation with regard to the *identity* of the involved quantities and/or the *symmetry* of the operation.[103] Both concepts manifest themselves most clearly in the additive operations. For instance, addition by means of ‘appending’ (in OB mathematics: *waṣābu*; in mathematical astronomy: *ṭepû*) implies that the sum inherits the identity of the summand to which something is ‘appended’, i.e. this type of addition conserves the identity of a quantity. By contrast, addition by means of ‘accumulation’ (*kamāru*) implies that the identity of the sum is different from that of any summand, so that there is a loss of identity. Identity-conserving addition therefore occurs in situations where a difference or increment is added to something — for instance a displacement to a position, resulting in a new position. In mathematical astronomy, most computations involve time intervals, angular distances or other empirical quantities having a well-defined identity, so that one can speak meaningfully about identity conservation.[104] The concept of identity conservation is potentially also relevant for subtractions and perhaps even multiplication and division.[105] However, in practice these other arithmetical operations do not exhibit a similarly clear differentiation with respect to identity conservation.

An arithmetical operation is called symmetric if the operands can be exchanged without a change of meaning. This feature is also mainly relevant in connection with addition. In the case of subtraction or division the question about symmetry or asymmetry is mute, because these operations are intrinsically asymmetric: $x - y$ is not equivalent with $y - x$, and a similar argument applies to division. With addition however, the sum $Q_1 + Q_2$ is numerically the same as $Q_2 + Q_1$, but there is a conceptual difference between ‘appending’ Q_1 to Q_2 and ‘appending’ Q_2 to Q_1, because in the former case the sum inherits the identity of Q_2, in the latter case the identity of Q_1.

2.2.2 Addition

Three different verbs are used for addition: *ṭepû*(**tab**), ‘to append’, *kamāru* (GAR.GAR), ‘to accumulate’ and *(w)aṣābu* (**daḫ**), ‘to append’.[106] In the translations the distinctness of these verbs is preserved by translating *ṭepû* as ‘to add’, *(w)aṣābu* as “to append” (in quotation marks) and *kamāru* as ‘to accumulate’.

---

[102] One example, ‘you accumulate nothing and 10^{nin}’ (**nu tuk** *u₃* 10 **ninda** GAR.GAR), occurs in No. **25** O7’. Some interpolation algorithms contain an interval where the interpolation coefficient is zero (e.g. *ΦG* scheme 1 in §4.4.11). In these intervals the interpolation procedure is formulated differently from the others, such that multiplication by zero is avoided.

[103] Høyrup (2002), pp. 18–20.

[104] This is not the case in purely numerical operations as they occur for instance in multiplication tables.

[105] For instance, if one multiplies the daily displacement of a planet by a time interval expressed in days, one obtains a total displacement, so that the identity of one factor is passed on to the product.

[106] The verb *ruddû*(**uš**), ‘to extend’, is used for addition in OB mathematics and NB administrative documents, but it is not attested in the astronomical texts, except for its cognate noun *tardītu*, ‘addition’ (lit. ‘extension’), which is mentioned a few times (No. **93** O7’; No. **100** O6’).

2.2.2.1 'To add'

Nearly all instances of addition in the procedure texts involve the verb *ṭepû*(**tab**), literally 'to append, attach' (translated as 'to add').[107] Also the cognate noun *ṭīpu*(**tab**), 'addition' (< 'attachment'), occurs in the procedure texts. The earliest evidence for *ṭepû* as a verb for addition is found in LB astronomical and mathematical texts. The verb itself is attested in the OB period, but not with an arithmetical meaning, and not in the mathematical corpus.[108] However, there was a semantic equivalent of *ṭepû* in OB mathematics, *waṣābu*(**daḫ**), with the same basic meaning 'to append'. Additions with *ṭepû* are based on the following formulations:

Q_2 *itti*(**ki**) Q_1 *ṭepû*(**tab**)	to add Q_2 with Q_1	(TAB.1)
Q_2 *ana* [*muḫḫi*(**ugu**)] Q_1 **tab**	to add Q_2 to Q_1	(TAB.2)
Q_2 *ana muḫḫi* **tab**	to add Q_2 to it	(TAB.3)
Q_1 *u* Q_2 {*itti*|*ana muḫḫi*} *aḫāmiš* **tab**	to add Q_1 with|and Q_2 together	(TAB.4)

There appears to be no difference in meaning between 'adding with' (*itti*) or 'adding to' (*ana* [*muḫḫi*]). TAB.3 is used if something is added to the result of an immediately preceding computation. The applications of TAB.1–3 imply two basic properties of *ṭepû*. First, addition with *ṭepû* is asymmetric, i.e. the summands are not interchangeable. This is already suggested by the literal meaning, since Q_2 being 'appended' to Q_1 is not the same as Q_1 being 'appended' to Q_2. Second, the sum $Q_1 + Q_2$ inherits the identity of Q_1. The same properties apply to *waṣābu*, the OB equivalent of *ṭepû*. The following numerical example illustrates this:

[10 **ḫun** 3 *ša*$_2$] ⌜*al* 10 **ḫun diri**⌝ **a.ra**$_2$ 40 **du ki** 3 *teṭeppi*(**tab**)	[10 Ari: 3,0. (The amount) by which it] ⌜exceeds 10 Ari⌝ you multiply by 0;40, add with 3,0.	No. **53** Oia2'

The conservation of identity is consistent with the interpretation of the algorithm (§4.4.5).[109] The following quotation contains a fully algebraic example:

nisḫa(**zi**) *ša*$_2$ *sin issuḫu*(**zi**) **ki ki** *sin* **tab**	The distance by which the Moon moved you add with the position of the Moon.	No. **61**.**A** O19

In this case the asymmetric, identity-conserving nature of the addition is immediately obvious, because a displacement is added to the position of the Moon, which results in an updated position of the Moon (§4.4.19.4). One should not expect every single instance of TAB.1–3 to respect identity conservation and asymmetry, but the exceptions are in fact very few. In the following example of TAB.1 it appears that the identity of Q_1 is not passed on to the sum $Q_1 + Q_2$:

[*ki*] *šaqû*(**nim**) ⌜**ki** BE **tab** *ki*⌝ *šaplu*(**sig**) **ta** BE *tanassaḫ*(**zi**-*aḫ*)	If (the Moon) is 'high' you add it (R) with the elongation (Q), if it is 'low' you 'tear' it out of the elongation.	No. **53** Rii4'

The astronomical interpretation of this algorithm (§4.4.19.11) implies that the outcome is not an 'elongation' (Q), so that the identity is not conserved.[110]

In TAB.4 *ṭepû* is qualified by the adverb *aḫāmiš*, 'together'. The procedure texts imply that these additions are also asymmetric, but the sum $Q_1 + Q_2$ does not inherit the identity of Q_1 (nor that of Q_2). This is illustrated by the following example from a procedure for lunar system A (step 1 of the Lunar Six module; cf. §4.4.19.1):

gi$_6$ **du** *u* LA$_2$ *ša*$_2$ ⌜**me**⌝ **ki** *a-ḫa-miš*$_2$ **tab**	(The time by which) the night has progressed (M) and the length of ⌜daylight⌝ (C) you add together.	No. **61**.**D** O3

The use of *ṭepû* instead of *kamāru*, 'to accumulate', is prompted by the asymmetry of the summands, M and C. Moreover, the identity of the sum differs from that of either summand, which explains the presence of *aḫāmiš*, 'together'. This interpretation is confirmed by other examples.[111] The textual evidence can thus be summarised as follows: *ṭepû* in its bare form (TAB.1–3) usually expresses asymmetric, identity-conserving addition. If $Q_1 + Q_2$ does not have the same identity as Q_1 then the addition is not a pure case of 'appending', which is made explicit by including *aḫāmiš*, 'together' (TAB.4).

[107] Besides 'to add' *ṭepû* can also mean 'to be increasing' or 'to be additive'; cf. the Glossary.

[108] *CAD* Ṭ sub *ṭepû*. Moreover, in OB mathematics **tab** represents *eṣēpu*, 'to double; to repeat'; cf. Proust (2009b), pp. 183–4.

[109] This is an interpolation rule for the duration of daylight (C) for lunar system A.

[110] However, one might justify the use of *ṭepû* in its bare form by the fact that all involved quantities are time intervals. It may also be significant that the addition and the subtraction are embedded in a conditional phrase, since this may necessitate a more concise formulation.

[111] No. **16** Xi'10', No. **18** Ri15', No. **38** Y5', No. **53** Oi31', No. **61**.**A** O3, No. **61**.**D** O3 and the mathematical problem text BM 34568 Ri8.

The concept of identity conservation also sheds light on several badly understood phrases involving TAB.2 and the 'value', *meḫertu*(**gaba.ri**), of a quantity:[112]

[10.5]3.52.42 **u₄.meš** *ana* *meḫret*(**gaba.ri**) *šatti*(**mu.an.na**) *ša₂* ᵈ**utu tab**	[10;5]3,52,42 days you add to the value of the year of the Sun.	No. **95** Oii13'

A number of days is added to the 'year of the Sun', the result being another time interval, which in subsequent lines is called the 'year of the Moon'. Hence $Q_1 + Q_2$ has a different identity than Q_1, but nevertheless *ṭepû* is not accompanied by *aḫāmiš*, 'together'. I propose that this can be explained if one assumes that the word 'value' is inserted before 'year' in order to avoid the wrong impression that the outcome is a 'year of the Sun'. This explanation assumes that 'value of QN' effectively masks the identity of QN, as if the addition is applied not to the quantity itself but to its numerical value.

2.2.2.2 'To append'

A small number of procedure texts and LB mathematical problem texts employ *(w)aṣābu*(**daḫ**), 'to append'.[113] Various features of these texts suggest that they (or their originals) predate the Seleucid period, or are at least older than the bulk of the corpus of mathematical astronomy. In OB mathematics *(w)aṣābu* was the most common verb for addition.[114] In the LB period its role has been taken over by *ṭepû*(**tab**), which has the same basic meaning.[115] The additions with *(w)aṣābu* are formulated as follows:

Q_2 *ana* Q_1 *(w)aṣābu*(**daḫ**)	to 'append' Q_2 to Q_1
Q_2 [*ana muḫḫi*] **daḫ**	to 'append' Q_2 [to it]

The second variant is used if something is 'appended' is the outcome of an immediately preceding computation. The following example occurs in an early Seleucid procedure text for the Moon:

40 **ninda** *a-na* 6.4.30 **daḫ**-*ma* 6 ⸢**uš** 5.10⸣	40^{nin} you 'append' to 6,4;30, it is 6 · ⸢60 (and) 5;10⸣.	No. **102** O3'

The astronomical interpretation (cf. the textual commentary) confirms the asymmetric, identity-conserving nature of *(w)aṣābu*.

2.2.2.3 'To accumulate'

'To accumulate', *kamāru*, is well known from OB mathematics, where it is used for symmetric addition of two or more quantities.[116] Basically the same applies in the astronomical procedure texts:

Q_1 *u* Q_2 [{*ana muḫḫi*|*itti*} *aḫāmiš*] *kamāru* (GAR.GAR)	to accumulate Q_1 and Q_2 [together]

Addition with *kamāru* is symmetric in the sense that the summands are always interchangeable. It is therefore applied only if the summands are in one way or another similar quantities. The 'accumulated' sum does not inherit the identity of any summand, which may explain why *kamāru* is often accompanied by *aḫāmiš*, 'together'. Most examples of *kamāru* in the procedure texts concern the averaging of the coefficients q and r in the Lunar Six module of lunar system A:

ana **ugu** *a-ḫa-miš₂ ša₂ sin u šam*[*aš₂ takam-mar*(GAR].GAR)-*ma* $\frac{1}{2}$-*šu₂ tanašši*(GIŠ)	You accumulate (the coefficients) for the Moon and the Sun together and you compute half of it.	No. **53** Rii8'

The summands are similarly defined coefficients for the Moon and the Sun ($q_☽$ and $q_☉$; cf. §4.4.19.5). There is loss of identity, because the outcome is a mean coefficient which is neither a coefficient for the Moon nor for the Sun. Another

[112] Other examples: No. **18** Ri7', No. **95** Oii14'.

[113] Mathematical astronomy: No. **102** O3',16',R4; Goal-Year astronomy: *TU* 11 O11,23,37 (Brack-Bernsen & Hunger 1999); other NMAT: BM 36414+ O4',5',11'; BM 43418 O10'; BM 77250 Xi'3' (all three unpublished); BM 45821+ R15,22 (Al-Rawi & George 1991/2); mathematical texts: W 23291 Oi23', 33' (Friberg 1997).

[114] Høyrup (2002), p. 19.

[115] I reserve the translation 'to append' (in quotation marks) for *(w)aṣābu* in order to maintain a distinction between these verbs.

[116] Besides GAR.GAR, UL.GAR and GAR are also attested as logograms for *kamāru* in OB mathematical problem texts; cf. the index in Høyrup (2002).

application of *kamāru* is encountered in the following Goal-Year procedure, in which pairs of Lunar Six intervals are added together:[117]

šu₂ *u* **na** *ša₂* **iti.du₆** *tanašši*(GIŠ)-*ma ana* **ugu** *a-ḫa-miš₂ takammarma*(GAR.GAR)-*ma*	You compute ŠU₂ and NA for month VII, and you 'accumulate' them together.

Also here *kamāru* is appropriate because the Lunar Six intervals ŠU₂ and NA have a similar definition (symmetry), and their sum is not a Lunar Six interval (loss of identity).[118] For a more complicated example in which *kamāru* is used for computing the sum of a series cf. §4.4.11.

2.2.3 Subtraction

Compared to the relatively clear semantic distinctions between the different terms for addition, subtraction offers a less clear picture. There is at best a weak tendency for semantic differentiations between some of the verbs. Their investigation is complicated by lexical difficulties. Of the four logograms that represent subtraction, **la₂**, **nim**, **e₁₁** and **zi**, only the latter two have a well-established Akkadian reading: **e₁₁** = *šūlû*, 'to deduct', and **zi** = *nasāḫu*, 'to tear out'. Moreover, the most common logogram in subtractions, **la₂**, has two different Akkadian readings: *naḫāsu*, 'to deduct', and *muṭṭû*, 'to diminish', apart from several other readings that do not represent subtractions.

2.2.3.1 'To tear out'

The least common verb for subtraction in the procedure texts is *nasāḫu*(**zi**), 'to 'tear out'' (thus translated, including quotation marks):

Q_2 *ina*|*ultu*(**ta**) Q_1 *nasāḫu*(**zi**)	to 'tear out' Q_2 from Q_1
Q_2 *ina*|*ultu*(**ta**) *libbi*(**ša₃**) **zi**	to 'tear out' Q_2 from it.

The preposition 'from' is usually written logographically (**ta**), sometimes phonetically (*ina*), which may be the usual reading of **ta**, although *ultu* (likewise meaning 'from') cannot be excluded. This kind of subtraction was very common in OB mathematics, where it served as the subtractive, identity-conserving counterpart of *(w)aṣābu*(**daḫ**), 'to append' (Høyrup 2002, p. 20). One factor that may have contributed to the decline of *nasāḫu* for subtraction is the fact that this verb is commonly used in the astronomical texts for describing motion along the zodiac ('to displace itself'). The following example of *nasāḫu* occurs in a (probably) early Seleucid lunar procedure text:

[*ki*] *šaqû*(**nim**) ⌜**ki** BE **tab** *ki*⌝ *šaplu*(**sig**) **ta** BE *tanassaḫ*(**zi**-*aḫ*)	If it (= the Moon) is 'high' you add it with the elongation, if it is 'low' you 'tear' it out from the elongation.	No. **53** Rii4'

The semantic restriction to identity-conserving subtraction known from OB mathematics is mostly confirmed by the procedure texts, but the quoted example may constitute a violation, since the outcome of the computation is not an 'elongation' (for the interpretation of this algorithm cf. the commentary and §4.4.19.11). As far as the instances of 'tearing out' can be dated, they tend to appear only in procedure texts from the early Seleucid era. Nevertheless, the subtractive use of *nasāḫu* is not merely a fossilised remainder of OB practices. The cognate noun *nisḫu*(**zi**), 'subtraction', becomes a subtractive number of undetermined magnitude, a concept unknown from OB mathematics (cf. §2.4).

2.2.3.2 'To subtract'

One of the more common verbs for subtraction in mathematical astronomy is *naḫāsu*(**la₂**), 'to subtract', which occurs in the following phrases:

Q_2 *ina*(**ta**) Q_1 *naḫāsu*(**la₂**)	to subtract Q_2 from Q_1
Q_2 *ina*(**ta**) *libbi*(**ša₃**)[-*šu₂*] **la₂**	to subtract Q_2 from it.

The subtractive sense of **la₂** in these phrases has long been known, since it is implied by the procedures in which they are embedded. However, it was not recognised that **la₂** is a form of *naḫāsu*. There are two reasons for this: first, there are no

[117] BM 42282+42294 O7 (Brack-Bernsen & Hunger 2008).

[118] In fact ŠU₂ + NA is the increment for another Lunar Six interval; cf. Brack-Bernsen & Hunger (2008).

phonetically written examples of *naḫāsu*, ‘to subtract’, in the procedure texts. Second, the transitive meaning ‘to subtract’ is a semantic innovation of the NB period deriving from the more common and original intransitive meaning ‘to recede’, also written **la**$_2$, in which sense this verb is frequently used in the planetary procedure texts for describing retrograde motion (§3.3.3.2). Phonetically written attestations of transitive *naḫāsu* were not readily available prior to the appearance in 1980 of volume N I of the *CAD*.[119] The following quotation from a procedure for *J* (lunar system A) proves beyond doubt that **la**$_2$ = *naḫāsu*, ‘to subtract’:[120]

ta 25.7.30 **ḫun en** 1[3 **absin** xxxxxxx] /**ta** *si-man*.**meš** *tanaḫḫis*(**la**$_2$-*is*)	From 25;7,30 Ari until 1[3 Vir] /you subtract [xxxxxxx] (|*J*|) from the durations (*G*).	No. **53** Oii24’–25’

In this example $Q_1 - Q_2$ preserves the identity of Q_1, since *J* is a correction to *G*, the duration of the month (§4.4.14). Also in the following example the identity of Q_1 is conserved:

10 **rin**$_2$ 3 *ša*$_2$ *al* 10 **rin**$_2$ **diri a.ra**$_2$ 40 **du ta** 3 **la**$_2$	10 Lib: 3,0. (The amount) by which it exceeds 10 Lib you multiply by 0;40 und subtract from 3,0.	No. **53** Oia8’

However, in other examples the identity of Q_1 is not conserved,[121] so that this is at best a rule of thumb.

2.2.3.3 ‘To diminish’

The same logogram **la**$_2$ also appears in subtractions in conjunction with the preposition *ana*, ‘to’, which precludes an interpretation as *naḫāsu*, ‘to subtract’. As I will argue, these instances of **la**$_2$ represent *muṭṭû*, ‘to diminish’. They are formulated as follows:

Q_1 *ana* Q_2 *muṭṭû*(**la**$_2$)	to diminish Q_2 by Q_1

The evidence for interpreting **la**$_2$ as *muṭṭû* is as follows. First, this phrase appears as the subtractive counterpart of *ṭepû*(**tab**) ‘to add’:

ki **nim** *ana* BE *tumaṭṭa*(**la**$_2$) *ki* [**sig**] *ana* BE **tab**	If it is ‘high’ you diminish the elongation by it, if [it is ‘low’] you add it to the elongation.	No. **61.A** O11

This pair of logograms can be identified with the following pair of verb forms in a Goal-Year procedure first edited by Kugler:[122]

4 *u*$_4$-*mu*.**meš** *tu-maṭ-ṭa* **igi**-*mar*	You diminish (the date of appearance 8^{yr} earlier) by 4 days, (then) you see it (= Venus).	BM 45728 O6
⌜1⌝2 **u**$_4$ *it-ti* **igi.du**$_8$.**a** *ta-ṭep-pi-ma* **igi.la**$_2$	You add 12 days with the appearance (47^{yr} earlier), then it (= Mars) appears (again).	BM 45728 O12

Here *tumaṭṭa*, ‘you diminish’, is the subtractive counterpart of *ṭeṭeppi*, ‘you append’. Since the latter is written **tab**, this strongly suggests that *tumaṭṭa* is the Akkadian reading of **la**$_2$ in Q_1 *ana* Q_2 **la**$_2$. Kugler had already recognised that *tumaṭṭa* is a subtractive operation, but this is not mentioned in the dictionaries. In OB mathematics *muṭṭû* is not attested;[123] all instances date from the LB period. Probably *muṭṭu* also accounts for many other passages in the procedure texts where **la**$_2$ appears as the subtractive counterpart of *ṭepû*, but a preposition is lacking because the quantity from which something is subtracted is not mentioned, as in the common phrase *x* **tab** *u* **la**$_2$, ‘you add and subtract *x*’:[124]

epūš(**du**$_3$-*uš*) *ša*$_2$ **zi** *sin* **ab**$_2$ *ana* **ab**$_2$ 42 *ṭeṭeppi*(**tab**) *u tumaṭṭa*(**la**$_2$)	Procedure for the displacement of the Moon: month by month you add and subtract (‘diminish by’) 0;42.	No. **53** Oi14’

[119] The *AHw* does not mention a transitive meaning of *naḫāsu*. The passage No. **53** Oii24’–25’ is quoted in the *AHw*, but incorrectly translated with an intransitive form. *CAD* N I sub *naḫāsu* A4 (p. 130) does provide phonetically written examples of transitive *naḫāsu* from NB and LB administrative documents, but not the instances from the LB astronomical and mathematical corpus written with **la**$_2$.

[120] Apart from the phonetic complement, also the preposition *ina*(**ta**), ‘from’, rules out that **la**$_2$ = *muṭṭû*, ‘to diminish’, since that verb requires *ana*, ‘to’.

[121] E.g. No. **61.A** R11,13.

[122] Kugler *SSB* I (1907), p. 45; Britton (2003).

[123] However, the G stem *maṭû*, ‘to be lacking’, is common in OB mathematics.

[124] In order to avoid the cumbersome ‘you add and diminish by *x*’ I maintain the conventional translation ‘you add and subtract’ for this phrase.

2.2.3.4 'To deduct'

A third verb for subtraction in the procedure texts is *šūlû*(**e**$_{11}$), 'to deduct', which occurs in the following phrase:

Q_1 *ina*(**ta**) Q_2 *šūlû*(**e**$_{11}$)	to deduct Q_1 from Q_2

The earliest attestations of *šūlû* for subtraction occur in administrative documents from the MB era.[125] The basic meaning of the verb is 'to remove', which derives from the literal meaning 'to raise'. In the OB mathematical corpus *šūlû* is not used for subtraction, and the logogram **e**$_{11}$ is not attested.[126] The following example is taken from a procedure text for lunar system A:

[LA$_2$ *ša*$_2$ **gi**$_6$ **ta**] / ⌜**gi**$_6$⌝ **du** *tušelli*(**e**$_{11}$)	You deduct [the length of night] from /(the time by which) the night has progressed.	No. **61.D** O1–2

Although the interpretation of the algorithm (§4.4.19.1) implies that $Q_2 - Q_1$ does not inherit the identity of Q_2 in this example, there are as many cases where it does. Hence *šūlû* does not appear to reflect a semantic differentiation with respect to other words for subtraction.

A much more commonly used logogram for subtraction in procedure texts and mathematical texts from the LB period is **nim**, which occurs in basically the same contexts as **e**$_{11}$:

Q_1 *ina*(**ta**) Q_2 **nim**	to deduct Q_1 from Q_2

Very likely **nim** represents the same verb *šūlû* but a definite proof is lacking.[127] The following example is taken from a procedure text for lunar system A:

zi *ša*$_2$ *šamaš*$_2$ **zi ta ki** *šamaš*$_2$ **nim**	The displacement by which the Sun moved you deduct from the position of the Sun.	No. **61.A** O5

In this case the subtraction is obviously identity-conserving, but a wider investigation indicates that, as with **e**$_{11}$, no semantic differentiation with respect to the other terms for subtraction can be established. The usage of *šūlû*(**nim**, **e**$_{11}$) is not fully equivalent with that of *naḫāsu*(**la**$_2$) or *nasāḫu*(**zi**), since **nim** and **e**$_{11}$ are not attested as markers of subtractivity in subtractive numbers, nor are they used in the sense of a subtractive number of undetermined magnitude (for these issues cf. §2.4).[128]

2.2.4 Multiplication

Multiplication is the third most common arithmetic operation in the procedure texts. With very few exceptions multiplication is realised by *alāku*(**du**), 'to go'. Besides that, there are a few examples of multiplication with *našû*, 'to raise'.

2.2.4.1 'To go Q_1 times Q_2'

Multiplication with *alāku* is formulated as follows:

Q_1 **a.ra**$_2$|GAM|GAM$_0$ Q_2 *alāku*(**du**)	to multiply ('go') Q_1 by ('times') Q_2

Although this phrase goes back to OB mathematics, very few OB problem texts contain the complete phrase *x* **a.ra**$_2$ *y* *alāku*(**du**).[129] Virtually all instances of **a.ra**$_2$, 'times', in OB mathematics occur in multiplication tables in the phrase *x* **a.ra**$_2$ *y z*, '*x* times *y* is *z*'. It thus appears that **a.ra**$_2$ 'times' was almost exclusively used for purely numerical (context-free)

[125] *CAD* E, p. 133, *elû* 11a (Š).

[126] In OB mathematics **nim**, perhaps to be read *šūlû*, is used for multiplication.

[127] The Akkadian equivalents of **nim** mentioned in the dictionaries include *elû* and *šaqû*, both meaning 'to be/become high'. In principle the Š-stems of both verbs (lit. 'to raise') are legitimate candidates for the Akkadian reading of **nim** in subtractions, but since the usage of *šūlû* for subtraction is proven while there is no such evidence for *šušqû*, I assume that **nim** = *šūlû*.

[128] In fact a cognate noun of *šūlû* that might serve that purpose does not appear to exist.

[129] E.g. *UET* V 864: O8, R11–13 (for a recent translation of this Sumerian text from Ur cf. Høyrup (2002), p. 251). If one takes *ana* to be the phonetic reading of **a.ra**$_2$ then *TMS* IX R21 and *TMS* XII O1,2,5 are further examples.

multiplications, and that any trace of an originally geometrical connotation of x **a.ra₂** y, if it ever existed, was lost at an early stage.[130] By contrast, usage of Q_1 **a.ra₂**|GAM|GAM$_0$ Q_2 *alāku* in astronomical procedure texts and LB mathematical problem texts covers all possible types of multiplication. This includes the geometric multiplicative operations for which OB mathematics employed a special terminology.[131] Another innovation of the LB period is that Q_1 **a.ra₂**|GAM|GAM$_0$ Q_2 *alāku* is now applied to abstract quantities of undetermined magnitude, as in the following example from a procedure text for lunar system A:

BE GAM$_0$ *ṣilipti*(**bar.nun**) *tallak*(**du**)	You multiply the elongation by the *ṣiliptu*-coefficient.	No. **61.A** O11

2.2.4.2 'To raise'

In rare cases multiplication is formulated in terms of *našû*, 'to raise':[132]

Q_1 *ana* Q_2 *našû*(**il₂**)	to 'raise' Q_1 to Q_2
Q_1 **a.ra₂** Q_2 *našû*(**il₂**)	to 'raise' Q_1 times Q_2

This type of multiplication is well known from OB mathematics. In the LB procedure texts and the mathematical problem texts *našû* is predominantly used in another sense, namely 'to compute', written GIŠ, less often **il₂**, usually with a fraction or the name of a quantity as the object. That usage of *našû* is unknown from OB mathematics and appears to be an innovation of the LB period.

2.2.5 Division and reciprocals

Explicit non-trivial divisions are rare in the procedure texts. The few known examples are formulated as follows:[133]

x ana y aḫḫê(**šeš.meš**) *zâzu*(**bar**, SE$_3$)	To divide x into y parts

This phrase is unknown from OB mathematics and appears to be an innovation of the LB period. The use of SE$_3$ instead of **bar** is attested only in No. **52**; it is assumed here that the Akkadian reading is likewise *zâzu*.[134] None of the procedures in which the phrase occurs have a practical purpose connected with the computation of synodic tables. All are unusual in one way or another, and involve concrete numbers; i.e. abstract quantities are not attested. For trivial divisions by small whole numbers there is a separate phrase involving the verb *našû*, 'to compute':

n-šu₂ tanašši(GIŠ)	You compute $1/n$ of it.

Here n is an ordinal number ('third', 'fourth', etc.), but for $n = 2$ the word *mišlu*(**bar**), 'half', is used. This type of division, rather common in the procedure texts, is not attested in OB mathematics. Even though non-trivial divisions are rare, many multiplications in the procedure texts involve factors that are actually reciprocals ($1/y$), so that they can be viewed as reformulated divisions (x/y).[135] The reciprocal operation itself has exactly the same form as in OB mathematics:

---

[130] Sumerian **a.ra₂** has several distinct meanings, all deriving from the basic meaning 'to go' of the cognate verb DU. The *PSD* (A sub **a.ra₂**) lists A: 1. 'occasion'; 2. 'times' (in multiplication); B: *alaktu* = 'way'. There is no textual evidence for a meaning 'step' proposed by Høyrup (2002), p. 22 and Friberg (2007), p. 72 (the common word for 'step', **ki.us** = *kibsu*, is not used in multiplications). The Akkadian reading of **a.ra₂** has thus far not been established beyond doubt, but it is probably *ana*. First, two OB mathematical problem texts from Susa suggest this: *TMS* IX R21: 30 *a-na* 17 *a-li-ik-ma* 8.30 *ta-mar*, 'multiply 30 times 17, you see 8.30', and *TMS* XII O1: *ma-na-at* **uš** *a-na* **sag** *i-la-ku*, 'the number that the length goes 'to' (= times) the width...' (= length times width). Second, in the LB problem text BM 78822 (Jursa 1993/4) one multiplication is formulated as *x a-na y alāku*. (Conversely, in No. **18** P20 **a.ra₂** replaces *ana*, 'for'; compare the duplicate No. **46** P10', but these are not multiplications.) In cases where **a.ra₂** represents 'n-fold', 'for the nth time', it is sometimes replaced by *adi* (*CAD* A sub *adi* 4), but *adi* is not attested in multiplications $x \cdot y$. One might also consider the possibility that **a.ra₂**, GAM and GAM$_0$ were pronounced as /ar(a)/, since GAM has a rare phonetic value *ar₅* (e.g. in the colophon of *ACT* 18).

[131] E.g. *šutakūlu*(**gu₇.gu₇**) = 'to make hold one another' for the construction of a rectangle from two unequal sides ('rectangularisation') and *šutamḫuru*(NIGIN) = 'to make encounter' for the construction of a square from two equal sides ('squaring'); cf. Høyrup (2002), pp. 23–25.

[132] With *ana*: No. **102** O7',9',15',R3; with **a.ra₂**: BM 45821+ R9.

[133] With SE$_3$: repeatedly in No. **52** but nowhere else; with **bar**: in No. **82** R11 and No. **102** R5.

[134] This reading of SE$_3$ is not attested elsewhere in the cuneiform literature as far as I know.

[135] For instance, many interpolation algorithms involve coefficients which are computed from a division (§2.6.7).

igi *x* **gal$_2$.bi** *y*	The reciprocal of *x* is *y*.

It rarely appears in the procedure texts: the only known examples occur in No. **102**, an unusual early Seleucid procedure text, but it is not uncommon in the LB mathematical problem texts.[136] The usage of this operation is the same as in OB mathematics: it is not applied to abstract quantities, only to concrete numbers for which the reciprocal exists as a finite sexagesimal number.

2.2.6 The copula u, 'and', as a placeholder for arithmetical operations

In some texts the copula *u*, 'and', is used as a placeholder for various arithmetical operations — not only additions as one might expect. In the translations these instances of 'and' are placed in quotation marks. The following example from a Goal-Year procedure concerns addition:

šu$_2$ *u* **na** *ša$_2$* **du$_6$** *tanašši*(GIŠ)	You compute ŠU$_2$ 'and' NA for month VII.	*TU* 11 O36

The algorithmic interpretation of this procedure (Brack-Bernsen & Hunger 2002) implies that ŠU$_2$ + NA is meant. Examples where *u* replaces other operations are found in No. **61**, a procedure text for the Lunar Six intervals (§4.4.19):

BE *u* *ṣiliptu*(**bar.nun**)	The elongation 'and' the *ṣiliptu*-coefficient.
BE *u* **bar.nun** *u* 2 ḪAB-*rat*	The elongation 'and' the *ṣiliptu*-coefficient 'and' the 2 for the disk

Both phrases appear repeatedly in the text. The algorithmic interpretation leaves no doubt which operation is hidden behind the copula *u*. In both phrases the first instance of 'and' replaces multiplication. In the second phrase there is a second instance of 'and', which represents either addition (cf. steps 10–12 in the procedure for ME), or subtraction (cf. steps 10–12 in the procedure for GI$_6$). Hence *u* is attested as a placeholder for the three main elementary arithmetical operations.

2.2.7 Diachronic overview of arithmetical terms and a comparison with mathematical texts

As demonstrated in the previous sections, the arithmetical terminology of the astronomical procedure texts represents a significant break with OB mathematics. A diachronic overview is shown in Table 2.1. Only a small subset of the terminology of OB mathematics is still regularly used in mathematical astronomy, namely *kamāru* ('to accumulate'), *x* **a.ra$_2$** *y alāku* (to 'go' *x* times *y*) and *nasāḫu* ('to tear out'). A few other OB terms have not completely disappeared but become very rare in mathematical astronomy. The most common term for addition, *ṭepû* = 'to append', is an LB innovation, but its semantic function in relation to *kamāru*, 'to accumulate', is basically the same as that of its OB counterpart *(w)aṣābu*. Various new terms for subtraction (*naḫāsu* and *šūlû*) seem to be attested for the first time in MB or NB administrative texts, suggesting an origin in accounting practices. The sole surviving term for multiplication (*alāku*) is not really new, but it has a much wider application than in OB mathematics.

This raises the question of to what extent the terminological innovations are a general phenomenon affecting both the astronomical and the mathematical texts of the LB period. One superficial distinction between astronomical procedures and mathematical problem texts is that the former contain only additions, subtractions and multiplications, with very few exceptions. Explicit divisions are virtually absent because they have been converted into multiplications by reciprocals. A comparison is obviously possible only for those operations that are common in both corpora. Indeed most tablets of the small corpus of LB mathematical problem texts[137] employ exactly the same arithmetical terms for addition, subtraction and multiplication as the astronomical procedure texts. This may be illustrated by the following problem from BM 34568, which contains a numerical example of what we know as the theorem of Pythagoras:

Oi6[4 u]š *u$_3$* 5 *ṣiliptu*(**bar.nun**) **en sag** *aššu*(**mu**) *lā*(**nu**) *tīdû*(**zu**-*u$_2$*) 4 GAM 4 7[1]6 5 GAM 5 25 16 **ta** 25 **nim**-*ma re-ḫi* 9 8*mi-nu-u$_2$* GAM *mi-ni-i lu*-**du**-*ma lu* 9 3 GAM 3 9 8a3 **sag**	Oi6The length is 4 and the diagonal is 5, what is the width? Since you do not know it: 4 times 4 is 716. 5 times 5 is 25. You deduct 16 from 25, 9 remains. 8How much times how much should I multiply so that it is 9? 3 times 3 is 9: 8athe width is 3.

136 AO 6484 (*MKT* I, pp. 96–107); AO 7848 (*MCT* Text Y); BM 34568 (*MKT* III); W 23291–x (Friberg, Hunger & al-Rawi 1990).

137 For a list of tablets of this corpus cf. Friberg (1997), pp. 356–357.

addition		OB	LB
to ‘append’	*(w)aṣābu*(**daḫ**)		
to add	*ṭepû*(**tab**)		
to accumulate	*kamāru*(GAR.GAR; OB: GAR, UL.GAR)		
subtraction			
to ‘tear out’	*nasāḫu*(**zi**)		
to subtract	*naḫāsu*(**la**$_2$)		
to deduct	*šūlû*(**e**$_{11}$, **nim**)		
to diminish	*muṭṭû*(**la**$_2$)		
multiplication			
to ‘go’ Q_1 times Q_2	Q_1 **a.ra**$_2$|GAM|GAM$_0$ Q_2 *alāku*(**du**)		
to ‘lift’	*našû*(**il**$_2$)		
division and reciprocals			
to divide into parts	*ana aḫḫē zâzu*(**bar**, SE$_3$)		
reciprocal of *x*	**igi** *x* **gal**$_2$**.bi**		

Table 2.1: Diachronic overview of the arithmetical terminology.

As mentioned earlier, OB mathematical problem texts of this geometrical type usually employ a special terminology including *nasāḫu*, ‘to tear out’, for subtraction, *šutakūlu*(**gu**$_7$**.gu**$_7$), ‘to make hold one another’, for the construction of a rectangle and *šutamḫuru*(NIGIN), ‘to make encounter’, for the construction of a square.[138] In the LB period virtually nothing remains of this, apart from the occasional use of *nasāḫu*, and even purely geometric problems are now formulated with the new ‘general’ terminology.[139] This may be interpreted as a manifestation of a new level of abstraction, in the sense that a uniform terminology is introduced for all arithmetical operations, irrespective of the nature of the quantities to which they are applied (geometric objects, weights, astronomical quantities, etc.). One remaining fundamental diffence between the astronomical procedures and the (extant) LB mathematical problem texts is that the latter are always formulated in terms of numerical examples, while most astronomical procedures employ an abstract formulation involving named quantities of undetermined magnitude. Hence the turn from an example-based formulation to a fully abstract formulation which can be observed in the procedure texts is not (yet) apparent in the mathematical corpus, but otherwise the similarities with regard to the terminology and the representation of algorithms between procedure texts and mathematical problem texts are striking. This suggests that the influence between Babylonian mathematics and astronomy went in both directions. Mathematical astronomy could not have been invented in the absence of the computational techniques and terminology of OB mathematics. On the other hand one may suspect that the complexity of mathematical astronomy is behind some of the developments in mathematical representation, which also found their way into the mathematical texts of the LB period.[140]

2.3 Other elementary operations

Several other, non-arithmetical elementary operations are found in the procedure texts, e.g. conditions, comparisons, coordination and storage of information. They are essential for constructing complex procedures.

2.3.1 Introducing initial data

Some procedures include a statement by which initial data are introduced into the algorithm. One formulation involves the verb ‘putting down’ (*šakānu*):

qaqqara(**ki**) *tašakkan*(**gar**)	You put down the position.	No. **14** Oi3’ (Mars)
qaqqar(**ki**) *sin u* **ki** *šamaš*$_2$ **gar**-*an*	You put down the position of the Moon and the Sun.	No. **61** P1–P4 (step 4)
šūqa(**nim**) *u šupla*(**sig**) *ša*$_2$ *sin* **gar**-*an*	You put down the Moon’s ‘height and depth’.	No. **61** P1–P4 (step 6)

[138] Høyrup (2002), pp. 23–25.

[139] Cf. for instance the late Achaemenid problem texts W 23291 (Friberg, Hunger & al-Rawi 1990) and W 23291-x (Friberg 1997).

[140] A similar conclusion is drawn by Høyrup (1996), who states in a footnote that ‘the rather few mathematical texts which we know from the late period [...] were written by and for members of the astronomical environment, a context that seems to have influenced the mathematical mode of thought’.

In all examples quoted here the quantity that is 'put down' is input for the procedure.[141] Occasionally the same thing is formulated in terms of 'holding' (*kullu*) or 'taking' (*ṣabātu*) 'in the hands':

u$_4$.meš *u* **ki.meš** *ana igi-gub-u$_2$* [x] / *ina qātē*(**šu**.2)-*ka tu-kal*	You hold the times and the positions for the *igigubbû*-coefficients [...] / in your hands.	No. **46** Rii1–2
mu-*ka ina* **šu**.2-*ka tu-kal*	You hold your year in your hands.	BM 42282+42294 O6
si-man ša$_2$ **gub** *u* **gur il$_2$.a** *ina* **šu**.2 *taṣabbat*(**dib**)	You take the duration of the full month and the hollow month in (your) hands.	No. **53** Rii17

In BM 42282+42294, a Goal-Year procedure text (Brack-Bernsen & Hunger 2008), the object of the 'holding' is the Goal Year for which the astronomer wishes to make a prediction.

2.3.2 Conditions

While some algorithms are simple, linear chains of operations, many have a more complex structure. Often there is a branching point where the algorithm chooses between two or more options, depending on a condition. Two basic types of conditions can be distinguished: those involving a comparison between a quantity and a threshold value, and those involving an evaluation of a quantity's change (increasing/decreasing or ascending/descending) or relative position (above/below the ecliptic).

2.3.2.1 Conditions involving a threshold value

The course of an algorithm may depend on whether a control quantity, say Q, is less than or exceeds a threshold value Q_0, i.e. $Q < Q_0$ or $Q > Q_0$. Sometimes the comparison is formulated explicitly as a conditional clause involving the verbs *atāru*(**diri**), 'to exceed', and *maṭû*(**la$_2$**), 'to lack', rarely *iṣû*, 'to be small(er)':

ki-i Q al-la Q_0 atru(**diri**) ...	If Q exceeds Q_0 ...
ki-i Q al-la Q_0 maṭû(**la$_2$**)|*i-ṣi* ...	If Q is less|smaller than Q_0 ...

Often the comparison is combined with an immediately following algorithm involving the difference between Q and Q_0. Since Babylonian calculus does not know negative numbers, this is computed as $Q - Q_0$ if $Q > Q_0$, as $Q_0 - Q$ if $Q < Q_0$. Most examples concern the reflection rule of the zigzag function (§2.6.8) or the transition rule of the step function (§2.6.10). They exhibit two different but equivalent formulations. The most complete version is as follows:

ša$_2$ al-la Q_0 **diri**|**gal** *Q_0* **ta ša$_3$**-*šu$_2$* **la$_2$**|**e$_{11}$**|**zi** ...	That which {exceeds|is larger than} Q_0: you {subtract|deduct|'tear out'} Q_0 from it, ...
ša$_2$ al-la Q_0 {**la$_2$**-*u$_2$*|**tur**} **ta** *Q_0* **la$_2$**|**e$_{11}$**|**zi** ...	That which is less|smaller than Q_0: you {subtract it|deduct it|'tear it out'} from Q_0, ...

Here the subtractions $Q - Q_0$ and $Q_0 - Q$ are carried out explicitly. Most procedures employ *atāru*(**diri**), 'to exceed', and *maṭû*(**la$_2$**), 'to be less, deficient', but some use *rabû*(**gal**), 'to be larger', and *ṣeḫēru*(**tur**), 'to be smaller'. In the shorter version, which begins with exactly the same phrase, the subtraction is not spelled out, but the algorithmic interpretation usually implies it, so that one has to translate as follows:

ša$_2$ al-la Q_0 **diri**|**gal** ...	(The amount by) which it {exceeds|is larger than} Q_0 ...
ša$_2$ al-la Q_0 {**la$_2$**-*u$_2$*|**tur**} ...	(The amount by) which it is less|smaller than Q_0 ...

This usage of *(w)atāru*(**diri**) and *maṭû*(**la$_2$**) is well known from OB mathematics.[142]

2.3.2.2 Conditions involving the change of a quantity or the relative position

The second type of condition involves clauses of the kind

[141] The alternative interpretation, that these statements anticipate the outcome of the procedure, is problematic because a quantity cannot be 'put down' unless it is available.

[142] In OB mathematics *alla*, 'beyond', is not used. Instead, *(w)atāru* is followed by *eli*(**ugu**), 'above', and *maṭû* by *ana*, '(compared) to'.

ki-i ṭepû(**tab**) ... *ki-i maṭû*(**la$_2$**) ...	If it is increasing ... if it is decreasing ...
ki-i šaqû(**nim, la$_2$**) ... *ki-i šaplu*(**sig, bur$_3$**) ...	If it is 'high' ... if it is 'low' ...
ki-i išaqqû(**nim, la$_2$**) ... *ki-i išappilu*(**sig, bur$_3$**) ...	If it is ascending ... if it is descending ...

That **tab** and **la$_2$** in the first phrase can be interpreted as statives (ideally in the subjunctive) is implied by rare instances where **la$_2$** is rendered phonetically as *ma-ṭi* (e.g. No. **53** Ri8), but a present tense is also possible in principle. The quantity which is either 'high' or 'low' in the second phrase, and which is ascending or descending in the third phrase, is in both cases the distance of the Moon or the planet to the ecliptic (§2.5.3.2). Sometimes the conditional clause introduced by *kī*, 'if' (or 'when') is placed in the middle of the sentence, as in the following example:[143]

ki nim *u* **sig** *ša$_2$* ^{d}*sin ki-i* **tab la$_2$** *ki-i* **la$_2$ tab**	If it is increasing you diminish the Moon's 'height and depth' (by it), if it is decreasing you add it with it.	No. **61.G** R11'–12'

For the interpretation of this procedure, which is part of the Lunar Six module of lunar system A, cf. §4.4.19.11.

2.3.3 Coordination

Procedures are essentially sequences of arithmetical operations, conditions and comparisons to be executed in a certain order. In the case of a linear chain of operations the order is implied by the sequence itself. The most common, trivial form of coordination in such sequences is achieved asyntactically, e.g. 'you add *x* to *y*, you multiply it by *z*,...', but often the operations are coordinated by the particle *-ma*, 'and then'. A related function of *-ma* in the procedure texts is to introduce the result of a computation (...'and it is'), in which case it plays a similar role as our equal sign (=). In linear sequences of operations the outcome of one operation is usually passed on to the immediately following one. However, *-ma* is also used for coordinating operations that are performed on different quantities, so that there is no flow of data between them.[144] An explicit form of coordination is achieved by the following phrases:

ša$_2$ reḫi(*re-ḫi*, **tag$_4$**)	what remains
ša$_2$ illi(**e$_{11}$**)[*-ka*]	what comes out [for you]

The first is attested only after subtractions, the second one predominantly after multiplications, rarely after subtractions. If an operation or subalgorithm within a procedure depends on previously computed quantities, this is sometimes made explicit by qualifying the name of that quantity by the phrase 'which you had put down'. Accordingly, the subalgorithm in which that quantity was computed terminates with a statement of the kind '... and you put it down', by which the result is, so to speak, stored for later usage.

2.4 Additive and subtractive numbers

In the previous sections numbers and arithmetical operations were treated as separate, complementary entities. A notable feature of mathematical astronomy is the existence of additive and subtractive numbers, a construction in which a number and an additive or subtractive operation are intricately linked. The formulation of these numbers in the procedure texts and the tabular texts exhibits several innovations compared to OB mathematics. This is not surprising, because it reflects the central role of the difference concept in Babylonian mathematical astronomy. In the modern sense of the word most algorithms of Babylonian mathematical astronomy are 'difference schemes'. Ultimately this is because Babylonian astronomy is mainly concerned with synodic phenomena, a restriction that almost automatically leads to a mathematical formalism whereby the angular and temporal coordinates of the planet, the Moon or the Sun are updated from one to the next phenomenon by applying additive or subtractive differences.

While additive numbers are usually taken for granted, the notion of subtractive numbers has attracted interest in general works on the history of mathematics on account of their possible relation to the modern concept of negative numbers. Neugebauer did not hesitate to use the term 'negative' when translating or discussing subtractive numbers in the context of Babylonian astronomy; the same applies to the Moon's distance below the ecliptic, which he refers to as 'negative

[143] According to Dietrich (1969) and Hackl (2007), this usage of the subjunction *kī* in LB Akkadian is typical for temporal clauses but not for conditional clauses, which may suggest that 'when' is a more appropriate translation.

[144] Examples of this can be found in the Lunar Six module of lunar system A (§4.4.19).

latitude'.[145] However, this usage is incorrect and misleading, because 'negative' numbers in the modern sense do not exist in Babylonian mathematical astronomy, nor in the mathematical texts. This is already apparent from the fact that subtractions $x-y$ are avoided when $y>x$, as will be confirmed again and again in the procedure texts. Therefore subtractive numbers (and the same holds for distance below the ecliptic) are, in the modern sense, not negative numbers but (positive) magnitudes accompanied by a subtractive operation. This becomes particularly clear in procedures where something is added to or subtracted from a subtractive number (cf. below).

Høyrup (1993) makes the fundamental point that the only meaningful way to address questions about the existence of negative numbers and other modern concepts in ancient mathematics is to phrase them in terms of functional equivalence.[146] With that in mind I will explore various manifestations of additive and subtractive numbers in mathematical astronomy. As will become apparent, certain constructions involving subtractive numbers and other pairs of numbers can be viewed as a step in the direction of a functional equivalence to our positive and negative numbers.

History of subtractive numbers

The history of subtractive numbers in cuneiform can be traced back to the ED III period (2600–2350 BC), when Sumerian **la₂**, 'to be lacking', appears in so-called subtractively formed numbers. This is a notation whereby numbers ending with a digit 7, 8 or 9 are written as multiples of 10 and a subtractive correction, e.g. 28 = 30 **la₂** 2 = '30 lacking 2'.[147] After the second millennium this practice becomes rare and seems to have been eventually abandoned. In subtractively formed numbers the subtraction is always applied to a concrete number. The first evidence of subtractive numbers being applied to quantities of undetermined magnitude is found in OB mathematical problem texts, e.g.[148]

igi.5.gal₂ a.ša₃ GAR.GAR **uš sag a.na uš u.gu₃ sag diri** 1.40 **ba.la₂**	1/5 of the area of the 'accumulation' of length and width is 1,40 less than whatever the length exceeds the width.

In *MKT* (Nr. 45, p. 455) this problem is represented in the form of the modern equation $(x+y)^2/5-1{,}0\cdot(x-y)=-1{,}40$. Some later investigators concluded from such representations that there are negative numbers in OB mathematics.[149] As pointed out by Høyrup (1993), the examples from OB mathematics only prove the existence of subtractive numbers, and they are not equivalent to the modern concept of negative numbers. The use of modern equations for representing Babylonian algorithms always carries a risk of anachronistic interpretations since they are ill-suited for this, as will be argued in §2.6.4.

Additive and subtractive numbers as numerical differences

The most common manifestation of additive and subtractive numbers in mathematical astronomy is that of concrete numbers followed by a logogram representing addition or subtraction. All additive and subtractive numbers in the tabular texts are of this type. They are especially common in the lunar tables, but they also occur in planetary tables and in procedure texts. In additive numbers **tab** follows the number, in subtractive numbers **la₂**, rarely **zi**. Other logograms for addition or subtraction are not attested in additive and subtractive numbers. The essential feature of these numbers is that they appear isolated from the quantity to which they are applied, i.e. $\pm x$ is abstracted from the application in an actual operation $y\pm x$. While the purpose of additive and subtractive numbers is clear, namely to indicate how the preceding number is applied to other numbers, their correct Akkadian reading is not known, apart from the fact that **tab** is a form of *ṭepû*, 'to add', **la₂** a form of *maṭû*, 'to be lacking', and **zi** a form of *nasāḫu*, 'to tear out'.[150] Without this innovation it is difficult to imagine how a computational system as complex as lunar system A or B with its numerous additive and subtractive functions could have been formulated.

[145] Cf. for instance the entries 'lal' and 'Negative'(= **la₂**) in *ACT*'s index (pp. 481, 501).

[146] In Høyrup's words: 'Asking whether the Babylonians had discovered, or not discovered, negative numbers, is as meaningful as asking whether they had discovered the Eiffel tower or not; the meaningful questions concerning the Eiffel tower would be whether they made constructions that in one way or the other expressed similar aspirations as those of the illustrious engineer, and whether they had created the techniques of which he made use'.

[147] The earliest known examples of subtractively formed numbers are found in A 681, an ED III text from Adab listing areas of rectangles. For an edition cf. Edzard (1969), more recently Friberg (2007). Ist T 7375 (from Girsu; Ur-III period) is a table of reciprocals in sexagesimal place-value notation in which e.g. 59 is written as 1 **la₂** 1 = '1,0 lacking 1' (Friberg 2007).

[148] YBC 4668 Oiii30–33 (Nr. 45), *MKT*, pp. 427–466.

[149] E.g. Goetsch (1968), who claims (p. 83) that '(the Babylonians) did not succeed in unifying the [...] newly formed set of negative rational numbers with the original set in such a way, that in the combined set the arithmetic operations are equally applicable to both' (translated from the original German).

[150] Occasional phonetic writings suggest that **tab** and **la₂** in additive and subtractive numbers may be statives (*ṭepi* = 'additive', *maṭi* = 'subtractive', 'lacking'). Second, they may be cognate nouns (*ṭīpu* = 'addition'; *mīṭu*, *nisḫu* = 'subtraction'). Third, in tabular texts they might be present tenses (**tab** = *teṭeppe* = 'you add', **la₂** = *tumaṭṭa* = 'you diminish' or *tanaḫḫis* = 'you subtract', **zi** = *tanassaḫ*, 'you tear out'). A fourth possibility, which may apply in tabular texts, is that they were not read as Akkadian words, but functioned more or less as symbols.

Additive and subtractive numbers of undetermined magnitude

Another innovation of the LB period concerns the introduction of a notation for additive and subtractive numbers of undetermined magnitude. This involves the cognate nouns of the mentioned verbs, i.e. *ṭīpu*(**tab**) = 'addition' and *mīṭu*(**la**$_2$) or *nisḫu*(**zi**) = 'subtraction'. The composite term *ṭīpu*(**tab**) *u mīṭu*(**la**$_2$), 'addition and subtraction', is used in the sense of a difference of undetermined magnitude and undetermined additive or subtractive nature.[151] As far as known an equivalent notation did not exist in OB mathematics, where additive and subtractive numbers are always concrete numbers.

Arithmetical operations performed on additive and subtractive numbers

A third innovation concerns the explicit formulation of arithmetical operations involving additive or subtractive numbers.[152] In the following example from a procedure text for lunar system K an arithmetical operation is performed on a subtractive number:

32 **ta** 22 *nisḫi*(**zi**) *ša*$_2$ **ḫun** *tanassaḫ*(**zi**)-*ma* 21.28	You 'tear out' 0;32 from 22, the subtraction ('tearing out') for Ari, it is 21;28.	No. **52** Oi35

Something is subtracted from the subtractive number, resulting in a subtractive number with a smaller magnitude. The same text contains an analogous example where something is added to a subtractive number, resulting in a subtractive number with a larger magnitude (Ri15–16). This makes it abundantly clear why it is anachronistic and misleading to refer to subtractive numbers as negative numbers. Arithmetical operations on subtractive numbers marked with **la**$_2$ are found in several procedures for lunar system A, e.g. the following multiplication, which occurs in a procedure for computing J from B (for this algorithm cf. §4.4.14):

2.1.44 **la**$_2$ **a.ra**$_2^!$ 28.7.30 **ki** *ša*$_2$ *šamaš*$_2$ **du** 57.3.45 **la**$_2$	You multiply 2;1,44, subtractive, by 28;7,30, the position of the Sun, (it is) 57;3,45, subtractive.	No. **53** Oii26'

Here J is computed for an exemplary zodiacal position of the Moon (B = 25;7,30° Ari) in a section of the zodiac where J is subtractive and linearly increasing with B. The subtractive number 2;1,44 is an interpolation coefficient which is multiplied by 28;7,30°, a distance measured along the ecliptic.[153] More complex examples are found in procedures with interpolation rules, also belonging to lunar system A, for computing Λ, which assumes additive as well as subtractive values, from Φ (cf. §4.4.13). No. **79** contains two representations of the interpolation scheme: one in tabular form (T1') and one in verbal form (P3'). The tabular entry for interval 2 (Oii12') is as follows:

[2.0.20.2]2.13.20 **la**$_2$ 1⸢9.5⸣5.33.20 **la**$_2$ **a.ra**$_2$ 3 **du**	[2,0;20,2]2,13,20, decreasing: 1⸢9;5⸣5,33,20, subtractive, you multiply by 3.

The first number is the control value Φ_k ($k = 2$), the second one is the corresponding Λ_k, which is subtractive, and the third number is the interpolation coefficient c_k. The interpolation procedure for this interval is written out fully in P3' (Rii3–4), which is damaged but can be restored as follows:

^{3}⸢*ana*⸣ [*tar* 2.0.20.22].13.20 **la**$_2$ 19.55.33.20 **gar** ^{4}[*ša*$_2$ *al* 2.0.20.22.13.20] ⸢**la**$_2$⸣ **la**$_2$ *adi*(**en**) 2.0.2.3⸢5⸣.[33.20 **la**$_2$ **a.ra**$_2$ 3 **du**] ⸢**ta**⸣ 19.55.33.20 **nim gar**	^{3}⸢Op⸣[posite 2,0;20,22],13,20, decreasing, you put 19;55,33,20. ^{4}[(The amount) by which] it is less than [2,0;20,22,13,20], ⸢decreasing⸣, until 2,0;2,35,[33,20, decreasing, you multiply by 3], deduct ⸢from⸣ 19;55,33,20, put down.

This procedure is an example of the interpolation template IP.B.2 (§2.6.7). Whereas 19;55,33,20 is marked as subtractive in the table, this is not done in the procedure, perhaps because the scribe did not want to overburden the formulation with yet another instance of **la**$_2$.[154]

The examples demonstrate that additive and subtractive numbers in mathematical astronomy are subjected to addition, subtraction and multiplication — all three elementary arithmetical operations. This phenomenon is unknown from OB mathematics. One limitation that may be noted is that in all cases the other number is a bare number not marked as additive or subtractive. In particular there is no explicitly formulated 'sign rule' of the kind 'subtractive times subtractive is additive', or 'subtractive times additive is subtractive'.

[151] A similar construction involving *nisḫu*(**zi**) or any other term for subtraction (**nim**, **e**$_{11}$) is not attested.

[152] I.e. operations other than the addition implied by an additive number, or the subtraction implied by a subtractive number.

[153] The distance between B and the control point 27° Psc; cf. the commentary and §4.4.14.

[154] Unfortunately the interpolation rule for intervals 9–10 or 53–54 where Λ changes from subtractive to additive or vice versa are not preserved. It may be assumed that the subtractive or additive nature of Λ does not remain unmentioned there.

2.5 Coordinate systems

Coordinates are elementary concepts of a special kind. On the one hand they belong to the fundamental concepts since all functions and parameters are defined in relation to certain coordinate systems. On the other hand they also belong to the second, algorithmic level, because the angular and temporal coordinates of the Moon, the Sun and the planets are computed with algorithms. All coordinate systems are discussed here, together with what I call the 'event frame'.

2.5.1 The event frame

The 'event frame' defined by the successive occurrences of a synodic phenomenon (Moon: New Moon, Full Moon; planets: first appearances, stations, acronychal risings, last appearances)[155] can be viewed as the fundamental reference frame of Babylonian mathematical astronomy. All angular and temporal coordinates are defined on that frame. The only 'coordinate' of the event frame is the event number, i, which corresponds to the lines (rows) of a synodic table. The distinction between the event frame and the coordinate systems for positions and times is of fundamental importance for a proper understanding of Babylonian mathematical astronomy. For instance, the time between successive occurrences of a synodic phenomenon varies from one to the next occurrence, an empirical fact respected in nearly all computational systems. Therefore, even though one may be tempted to view the event frame as being primarily associated with time, it is not a fixed time frame. The event frame of the Moon, which is defined by New Moons and Full Moons, is of special significance because of its connection with the Babylonian calender. Moreover, the Babylonian astronomers constructed from this an artificial coordinate system for the time of the planetary phenomena, by defining the mean synodic month as the basic unit of time, and dividing this into 30 artificial 'days', which will be referred to as 'mean tithis' (cf. below).

2.5.2 Temporal coordinates

2.5.2.1 The calendar

Babylonian dates are expressed by year, month and day. The month names in the astronomical texts derive from the calendar of OB Nippur (p. xxv). In this study they are translated with Roman numerals (I–XII plus VI_2 and XII_2 for the intercalary months). Modern equivalents of Babylonian dates are expressed in the Julian calender.[156] Texts written before the Seleucid era are dated by regnal years of the ruling king (cf. p. xxvi for a list of rulers). In synodic tables from that period with predicted data, year numbers are obviously hypothetical future regnal years of the current ruler. If the year numbers include a historically attested change of rule the data must have been computed after the change of rule. When mathematical astronomy was invented, the 19^{yr} ('Metonic') intercalation cycle was already in place, so that all intercalations follow a fixed and well-known pattern involving the insertion of one extra month (either a VI_2 or a XII_2) in 7 out of 19 years, resulting in a cycle of 235 months. During the period of concern (450 BC to 50 BC) this had the effect of confining the Babylonian New Year, day 1 of the month Nisan, to the months March – April of the Julian calendar. The Babylonian day begins at sunset, a commonly used reference point for timing astronomical phenomena.

2.5.2.2 Time degrees

The fundamental unit of time in mathematical astronomy is the time degree (UŠ), which is part of the 'degree system':[157]

$$1\ \bar{u}mu(\mathbf{me, u_4}) \xleftarrow{12} 1\ b\bar{e}ru(\mathbf{danna}) \xleftarrow{30} 1\ \text{UŠ} \xleftarrow{60} 1\ nindanu(\mathbf{ninda}) \tag{2.2}$$

or, in translation:

$$1\ \text{day}\ (^{d}) \xleftarrow{12} 1\ \text{`mile'}\ (^{b}) \xleftarrow{30} 1\ \text{degree}\ (^{\circ}) \xleftarrow{60} 1\ \text{`rod'}\ (^{nin}) \tag{2.3}$$

The unit UŠ is usually omitted and must be inferred from the context.[158] The earliest evidence of the 'degree system' is found in Tablet 14 of the astrological omen series *Enūma Anu Enlil*, which is dated to appr. 1200 BC.[159] In that text,

[155] For the synodic phenomena cf. §3.1.2 (planets) and §4.1.2 (Moon).

[156] In modern publications on ancient astronomy the notation with BC is often replaced by negative year numbers, such that 0 corresponds to 1 BC, -1 to 2 BC, etc; i.e. $-n$ corresponds to $n+1$ BC.

[157] The Akkadian reading of UŠ has not been established.

[158] In publications by Neugebauer, Aaboe and others up to the 1970s, time intervals were expressed in 'large hours' (^{H}), a hypothetical unit consisting of 60°, so that $6^{H} = 1$ day. Since this is not a Babylonian unit, its usage has been largely abandoned in favour of the time degree.

[159] For an edition of this tablet cf. Al-Rawi & George (1991/2).

and others that predate the 5th c. BC, the degree system is used for time intervals, and not yet for angular distance. The time degree has a constant duration corresponding to 1/360th of 1^d (defined from midnight to midnight), so that its modern equivalent is 4 minutes. The traditional division of the night into three watches (**en.nun** = *maṣṣartu*), well known from astronomical diaries and astrological texts,[160] is not used in mathematical astronomy. However, the word 'watch' is occasionally attested with a meaning 'duration' (e.g. of the *entire* night in No. **52**). As mentioned, time intervals are usually defined in relation to sunset, sunrise or midnight.

2.5.2.3 Mean tithis

Kugler (1907)[161] led the way to Pannekoek's discovery (1916) that the planetary texts employ an artificial system for measuring time, in which the basic units are the *mean* synodic month and 1/30 of that. The latter unit was called 'Mondtage' ('month days') by van der Waerden (1941), but nowadays they are commonly known as tithis, a Sanskrit term from Indian astronomy. I will systematically call this unit the *mean* tithi ($^{\bar{\tau}}$) in order to distinguish it clearly from the (real) tithi ($^{\tau}$) which is used in the daily motion tables of lunar system A (cf. §4.4.2.2). The only named units of the mean-tithi system are the mean synodic month (*arḫu* = 'month') and the mean tithi (*ūmu* = 'day'). A terminological distinction between real days and (mean) tithis is not apparent in the texts. In the translations 'day' is put in quotation marks if it denotes a (mean) tithi. Smaller units such as 1/60 of the mean tithi are not explicitly named. Time intervals measured in mean tithis are usually defined in relation to the beginning of the year (1 Nisan), or the beginning of the month. The relation between days and mean tithis is not formulated explicitly in the texts. An analysis of Babylonian values of the mean synodic time ($\bar{\tau}$) for various planets,[162] which are expressed in mean tithis, and modern empirical values expressed in days, implies that the ratio of the mean tithi to the day is always close to

$$\frac{1^{\bar{\tau}}}{1^{d}} = \frac{1^{\overline{m}}}{30^{d}} \approx \frac{29;31,50,8,20^{d}}{30^{d}} \approx 0;59,3,40,16,40. \tag{2.4}$$

This can be used for converting between degrees and mean tithis, the reciprocal being $1^d/1^{\bar{\tau}} \approx 1;0,57,13,26,40$. The underlying value of the mean synodic month, $\overline{m}_{\text{syn}} = 29;31,50,8,20^d$, is also embedded in lunar system B (function G; cf. §4.5.12) and, to good approximation, lunar system A (cf. No. **82** P2.b). For the advantages of the mean-tithi system cf. §3.3.2.

2.5.3 *Angular coordinates*

The positions of the Moon, the Sun and the planets are expressed in a two-dimensional coordinate system based on the zodiac, a circular band whose centre is defined by the orbit of the Sun. The Sun's path was divided into 12 sections of 30° that were named after nearby constellations (p. xxv). The earliest textual evidence for the zodiac dates to about 450 – 400 BC.[163] Cuneiform texts do not mention the width of the zodiac, but in practice it does not extend much beyond the path of the Moon, which reaches to about 5° above or below the ecliptic. The division into 12 sections of 30° was likely conceived in analogy to the existing division of the day into 12 'miles' (*bēru*) of 30 time degrees, and the ancient schematic division of the year into 12 months of 30 days (Brack-Bernsen 2007). As in the degree system for measuring time, 1° of angle (UŠ) is divided into 60^{nin} ('rods'):

$$1^{\circ}\ (\text{UŠ}) \xleftarrow{60} 1^{\text{nin}}\ (\textbf{ninda} = \textit{nindanu}) \tag{2.5}$$

2.5.3.1 Zodiacal position (*B*)

terminology	*qaqqaru*(**ki**) = 'position'; *lumāšu* = 'zodiacal sign'
unit	zodiacal sign and degrees within it [0–30]
astronomical meaning	zodiacal position

Zodiacal positions of the Moon, the Sun and the planets (conventional modern symbol: *B*) are expressed as a zodiacal sign and a number of degrees measured from the beginning of that sign (Fig. 2.2).[164] Babylonian astronomers usually

[160] E.g. in lunar eclipse omens (*EAE* Tablets 15–22; cf. Rochberg 1988).

[161] Kugler *SSB* I (1907), pp. 167–169

[162] Neugebauer (*HAMA*, p. 1070). For the synodic time, τ, cf. §3.3.2.1.

[163] Cf. Rochberg (1998), p. 30, and Britton (2010) who argues for a date near 400 BC.

[164] In formulaic representations the zodiacal sign and the position within it are usually combined in a single notation, thus allowing B to assume values between 0° and 360°.

referred to the boundary between adjacent signs as 30° of the preceding sign, thus avoiding the zero for denoting zodiacal positions.[165] Babylonian zodiacal positions are sidereal, i.e. fixed with respect to the stars, as opposed to modern ecliptical coordinates which are tropical, i.e. defined with respect to the moving vernal equinox.[166]

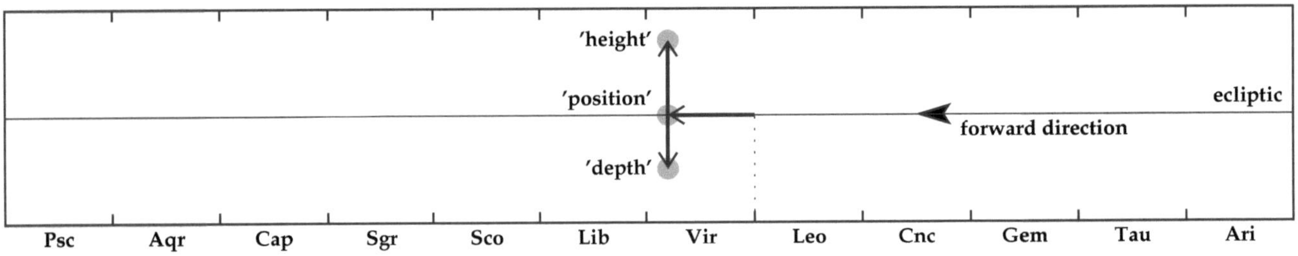

Fig. 2.2: Reference frame for the angular coordinates of the Moon, the Sun and the planets as viewed against the sky.

2.5.3.2 Distance to the ecliptic (E)

terminology	*šūqu*(**nim**, **la₂**) = 'height' ($E_{\uparrow}$) *šuplu*(**sig**, **bur₃**) = 'depth' ($E_{\downarrow}$)
units	barleycorn (^{še}), cubits (^{c}), fingers (^{f}) or degrees (°)
astronomical meaning	distance above or below the ecliptic

Lunar systems A, B and (probably) K, and some planetary procedure texts, include algorithms for the Moon or the planet's distance to the ecliptic (conventional modern symbol: E). The Akkadian terms for this quantity are 'height' = *šūqu*(**nim**, **la₂**) and 'depth' = *šuplu*(**sig**, **bur₃**), cognate nouns of *šaqû*, 'to be high' and *šapālu*, 'to be low' (Fig. 2.2).[167] In lunar system A the algorithms for E imply that its reference points form a circle in the middle of the zodiac similar to the modern ecliptic. Their precise location is not explicitly defined in texts, but the algorithms for eclipse magnitude and the Lunar Six intervals imply that they coincide with the annual course of the Sun's centre. The reference circles for the planets must also be near the middle of the zodiac. For lack of an alternative term, they will be called 'ecliptics', even though there are some conceptual differences with the modern ecliptic.[168] For instance, the Babylonian ecliptic is, in modern terms, sidereal, since the boundaries of the zodiacal signs are fixed with respect to the stars. The value of E is usually expressed in barleycorns (^{še}), sometimes in cubits (^{c}) and fingers (^{f}), all of which are units of the cubit system:

$$1\ \textit{ammatu}(\textbf{kuš}_3) \xleftarrow{24} 1\ \textit{ubānu}([\textbf{šu.}]\textbf{si}, \textit{u}) \xleftarrow{6} 1\ \textit{uṭṭatu}(\textbf{še}) \tag{2.6}$$

In translation:

$$1\ \text{cubit}\ (^{c}) \xleftarrow{24} 1\ \text{finger}\ (^{f}) \xleftarrow{6} 1\ \text{barleycorn}\ (^{še}) \tag{2.7}$$

In lunar system A the numbers in column E are usually expressed in barleycorns (^{še}). If E exceeds $6^{še}$ it is not converted into fingers plus a remainder in barleycorns. In the Lunar Six modules E is expressed in degrees; in the planetary procedures in fingers and/or cubits, such that $1^{c} = 24^{f}$, the same ratio known from astronomical diaries, other observational texts and NMAT procedures.[169] It will be clear from this discussion that the cubit is rarely used in mathematical astronomy. The procedures for E imply that $1^{c} = 2°$;[170] there are also traces of evidence for a 'large cubit' (*ammatu rabû*) equivalent to 2;30° and presumably containing 30^{f}.[171]

---

165 However, fractional positions within the first degree of a sign are expressed as 0;...°.

166 Due to precession (and nutation) tropical coordinates change slowly with time, a phenomenon unknown to the Babylonian astronomers (Neugebauer 1950). By analysing star positions and planetary data in the observational texts, Britton (2010) established that the difference between Babylonian (sidereal) and modern (tropical) longitudes is described by the formula $\Delta\lambda = 3.20° - 1.3828 \cdot Y$, where Y measures Julian centuries from 1 BC, a slight modification of an earlier result by Huber (1958).

167 For **nim** and **sig** in the Sumerian tradition cf. Balke (2002).

168 As pointed out by Steele (2003; 2007a) there is no compelling evidence that the Moon's ecliptic coincides exactly with any of the planetary ecliptics. However, evidence that they are significantly different is also lacking.

169 E.g. BM 41004 O13 (Brack-Bernsen & Hunger 2005/6). The only other astronomical quantity measured in fingers is eclipse magnitude (Ψ etc. in lunar systems A and B). Again, if the eclipse magnitude exceeds 24^{f} it is not converted into cubits plus a remainder in fingers.

170 In the observational texts $1^{c} \approx 2.2°$ (Fatoohi & Stephenson 1997/98; Jones 2004).

171 The existence of the 'large cubit' is suggested by No. **9** O7 (cf. the commentary). While the 2° cubit of mathematical astronomy goes back to the NB cubit, a unit of length likewise divided into 24^{f}, the 'large cubit' may go back to the OB cubit, a unit of length containing 30^{f} (*RlA*, p. 462); less likely the Kassite 'large cubit' amounting to 1 1/2 OB cubits, which was divided into 24^{f} (*RlA*, p. 469).

Kugler (1900) did not know the Akkadian reading of the logograms for 'height' and 'depth', but he did establish the essentially correct interpretation as 'northern latitude' and 'southern latitude',[172] a terminology borrowed from the modern astronomical concept of (ecliptical) latitude. Since Neugebauer, the conventional translation has been 'positive latitude' and 'negative latitude' (e.g. *ACT*, *HAMA* and elsewhere). However, compared to Kugler's interpretation, that translation is anachronistic and potentially misleading, since there are no negative numbers in Babylonian calculus (§2.4), and 'height' and 'depth' are positive quantities. It is difficult to avoid confusion and wrong interpretations if one translates 'negative latitude' instead of 'depth' in procedures where something is added to or subtracted from this. If something is added to the Moon's 'depth' then the Moon's distance to the ecliptic increases, but if one translates 'negative latitude' instead of 'depth' then the addition corresponds to a decrease of this distance. In order to obtain a consistent translation one would then have to go as far as translating the Babylonian additions as subtractions in all cases where something is added to 'negative latitude', and vice versa for 'positive latitude'. Such problems are avoided if one adopts the semantically more accurate translations 'height' and 'depth'.

2.6 Procedures and algorithms

The elementary operations and other basic concepts discussed in the previous section are the building blocks of procedures, which brings us to the next level of concepts.

2.6.1 Composite procedures and subprocedures

Individual procedures are identified by means of visual markers on the tablets, such as horizontal dividing lines, spacing and sometimes columns. However, if one uses only criteria based on visual markers this often results in composite procedures that are not strictly coherent or devoted to a single algorithm. I therefore introduce an extra layer of structure such that procedures, as defined by visual markers, may consist of subprocedures, each representing a coherent set of instructions devoted to the computation of a single quantity, or several related quantities, or a subalgorithm of a larger algorithm. In order to maintain consistency with the established numbering of procedures in previous editions, the subprocedures are labeled a,b,c,..., e.g. P2.a is subprocedure a of the (composite) procedure P2. Sometimes the correctness of a division into subprocedures is confirmed by duplicates containing the same sequence of procedures but separated by horizontal lines.

2.6.2 Initial and final statements of a procedure

Many procedures begin with a statement indicating the purpose or subject of the procedure. A common formulation of these statements also known from NMAT procedures involves the verb *epēšu*, 'to construct', or its cognate noun *epūšu*, 'procedure':

epūšu(**du$_3$**-*šu$_2$*) *ša$_2$* 2.13.20	Procedure for 2,13;20 (= Φ).	No. **53** Oi1
du$_3$-*uš ša$_2$* **zi** *sin*	Procedure for the displacement of the Moon.	No. **53** Oi14'
e-pe-šu$_2$ ša$_2$ **igi.meš** *an-ne$_2$-e-tu$_4$*	Procedure for these appearances (of Mercury).	No. **1** R1
epūšū(**du$_3$.meš**) *an-nu-tu$_2$ ša$_2$ kun-nu u tur-ru u ūm bubbuli*(**u$_4$.na$_2$.am$_3$**)	These procedures are for hollow and full months and the 'day of disappearance'.	No. **53** Rii39–40

All but the last example appear at the beginning of the procedure. The third example makes reference to a synodic table on the obverse of the same tablet. The last example is placed at the end of a procedure (No. **53** P18'). The mentioned purposes are in fact those of the tablet as a whole.[173] In other procedures only the name of the planet is mentioned:

ša$_2$ ^{d}**gu$_4$.ud igi** *ša$_2$* **kur** *a-na* **igi** *ša$_2$* **kur**	For Mercury, eastern appearance (MF) to eastern appearance.	No. **42** O1
ša$_2$ **mul$_2$.babbar**	For Jupiter.	No. **18** Rii22'
ša$_2$ **genna**	For Saturn.	No. **42** O18

Also known from NMAT procedure texts and mathematical problem texts are statements of purpose involving an infinitive construction:

[172] 'Nördliche Breite' and 'südliche Breite' (Kugler 1900, pp. 37, 136).

[173] Similar phrases describing the purpose of an entire tablet are well known from the observational texts, where they have a highly standardised form. Diaries: 'Regular watch from month MN$_1$ until the end of MN$_2$ of year *x*, king PN'; Goal-Year texts (at the end of the tablet): 'First days, appearances, passings and eclipses which were established for year *x*, king PN' (e.g. *ADRT* VI 69).

šu$_2$ *ana* **du**$_3$-*ku*	For you to 'construct' ŠU$_2$.	No. **61**.**D** O1
si-man qa-tu-u$_2$ *ana* **du**$_3$-*ka*	For you to 'construct' the final duration.	No. **53** Oii27'
[u$_4$.n]**a**$_2$.**am**$_3$ *ana* **du**$_3$-*ka*	For you to 'construct' the 'day of disappearance'.	No. **53** Rii17
nim *u* **sig** *ana* **du**$_3$-*ka*	For you to 'construct' the 'height and depth' (E).	No. **56** R17
nisḫa(**zi**) *rabû*(**gal**-*u*$_2$) *ana* **igi**-*ka*	For you to see the largest displacement (F).	No. **53** Oii16'
[z]**i** *ša*$_2$ **an** *ša*$_2$ *kal* **mu**.⌜**an.na**⌝ *ana* **igi**-*ka*	For you to see the displacement of Mars for the whole year.	No. **13** O24

There does not appear to be a difference between 'constructing' (*epēšu*) and 'seeing' (*amāru*) a quantity.[174] The following examples involving the verb 'to produce' (*šūṣû* = 'let come forth') concern procedures for lunar system A:

nim *u* **sig ta** ḪAB-*rat*.**meš** *ana šu-ṣu-u*$_2$	In order to produce the 'height and depth' (E) from the 'disks' (Ψ).	No. **65** O15'
nim *u* [**sig**] **ta** ḪAB-*rat ana* **e**$_3$-*u*$_2$	In order to produce the 'height and depth' from the 'disk'.	No. **53** Oii15'
[**zi** *sin* **ta** 2.13].20 *ana šu-ṣu-u*$_2$	In order to produce [the Moon's displacement (F) from 2,13];20 (Φ).	No. **65** R1

The second example is placed at the very end of the procedure (No. **53** P11'). In another procedure for G the transformation is represented by the bare preposition 'from':

si-man ina(**ta**) *nisḫi*(**zi**) *našû*(GIŠ-*u*$_2$)	The duration (G) from the computed displacement (F).	No. **96** O9'

Finally, some procedures begin with the word 'Secondly' (*šanîš*), indicating that they are alternative to another procedure, usually the preceding one.

2.6.3 Procedures as verbal representations of algorithms

In this study the term procedure is used in the sense of a verbal representation of an algorithm. Maintaining a distinction between procedures and algorithms is important for several reasons. First, the reconstruction of algorithms is an interpretative process requiring lexical, semantic, mathematical and astronomical analysis of procedure texts (and tabular texts). Second, the representation of algorithms in procedure texts is often incomplete or deficient in one way or another (cf. below). In this view procedures are actual verbal instructions, while algorithms are the underlying complete sequences of operations reconstructed from procedures, sometimes in conjunction with analysis of tabular texts. The reconstructed algorithm will be referred to as the 'algorithmic interpretation' of the procedure.

2.6.3.1 Example-based and abstract formulation

The procedure texts employ two different methods for representing algorithms: by offering numerical examples ('example-based formulation'), or general rules involving named quantities of undetermined magnitude ('abstract formulation'). In the example-based formulation an algorithm is demonstrated for specific numerical values of the involved functions. In the abstract formulation only the defining parameters of the algorithms are given numerical values, but the quantity that is computed does not assume a numerical value. With the example-based formulation, the user must first extract a general rule before being able to apply the algorithm with different initial values than the ones that happened to be used in the procedure. With the abstract formulation this interpretative step is no longer necessary, because the procedure is valid for all possible initial values of the input functions. This obviously reduces the chance that the user of the procedure makes an error. The abstract formulation necessarily involves operations on abstract quantities of undetermined magnitude, which constitutes a point of similarity with the modern concept of a variable.

The available evidence suggests that the example-based formulation occurs mainly in early procedure texts, and that it became less popular after the beginning of the Seleucid era. Perhaps No. **53**, a large procedure text for lunar system A, reflects the transition from the example-based to the abstract formulation, since it contains a mixture of both formulations. Interestingly, nearly all algorithmic interpretations derived from the procedures in No. **53** fully conform with lunar system A in its final stage. This suggests that the formulation of the procedures underwent further changes even after the algorithms themselves had reached their final form.

[174] Another verb with a similar meaning as *epēšu* is *našû* (GIŠ, **il**$_2$), 'to compute'. It is used in contexts where the focus is less on the procedural aspect of an algorithm and more on its outcome. It is not attested in statements of purpose.

2.6.3.2 Deficient procedures

That procedures are best interpreted as representations of algorithms is particularly clear from various kinds of deficiencies. If one closely examines a procedure it is usually the case that at least some step or subalgorithm is strictly speaking not completely represented, although usually clearly implied. Occasionally these omissions are so grave that it is difficult to understand how the procedure could have been used at all. For instance, No. **53** P17' is incomplete and formulated in an opaque manner, so that it is at best a 'mnemonic device' reminding the user about certain steps, without actually providing a coherent set of instructions. Moreover, in this procedure the sequence of the instructions does not agree with the order of the subalgorithms that they represent. Another reason why procedure texts do not provide a complete representation of algorithms derives from their limited purpose. Most procedure texts are mainly concerned with the practical aspects of producing astronomical tables. There may be other, more fundamental algorithms that, according to a closer analysis of the practical algorithms, underlie the computational systems. These hidden levels of mathematical structure are hardly represented in the procedure texts.

2.6.4 Representing procedures, algorithms and functions

In this section I discuss how procedures, algorithms and functions are represented in the present study. Each representation has certain limitations which must be kept in mind in order to interpret them in a proper way.

2.6.4.1 Columns, functions and parameters

The columns of the tabular texts are conventionally denoted by modern symbols, e.g. 'column Φ'.[175] I use the word function for any quantity that is determined by an algorithm, and reserve the word column for an actual column in a tabular text. This convention differs slightly from that of Neugebauer and others, who use the term column as (nearly) synonymous with function.[176] The distinction between function and column is important for two reasons. First, not all functions attested in procedure texts are represented in the tabular texts, e.g. the whole of system K. Second, some columns contain not only numerical values of a function but also other numerical or non-numerical information associated with that function.[177] Functions are also distinct from parameters, which do not have an argument, and do not vary within a computational system. The conventional modern terminology for the parameters of Babylonian mathematical astronomy is largely maintained here; e.g. m is the minimum of a function, M the maximum, d the difference for 1 synodic cycle, P the period, etc. The meaning of these and other parameters is explained in the following chapters; for a list of (modern) symbols cf. p. xxi.

2.6.4.2 Template procedures

Template procedures are essentially quotations from actual procedures in which the numerical values of functions and parameters are replaced by the corresponding modern symbols. Many actual procedures can be viewed as specific implementations of an underlying template. Templates are therefore useful in order to identify generic elements of the procedures.[178] They are derived in accordance with the following rules:

(i) Numbers are replaced by modern symbols of the corresponding quantities. These symbols always correspond to numerical values in the actual procedures (i.e. not abstract quantities of undetermined magnitude).
(ii) The Babylonian name of a quantity is either translated into English, or replaced by QN (= 'name of a quantity').
(iii) Optional elements, i.e. words or phrases that are not always present, are enclosed in square brackets, [...]
(iv) Alternative signs, words or phrases are separated by vertical bars (|) and ordered according to decreasing frequency. If necessary the alternatives are enclosed in accolades, {...|...|...}.[179]

An example may illustrate how templates are extracted from the procedure texts. No. **37** P1 is a procedure for updating the time of a synodic phenomenon (§3.3.2.1) by means of a zigzag function (§2.6.8). The algorithm includes the following 'reflection rule' near the maximum of the zigzag function:

[175] The currently used notation for the columns was established by Neugebauer (*ACT*, *HAMA*), who modified Kugler's notation in order to remove certain inconsistencies.
[176] In his early, German papers, Neugebauer did use the term function ('Funktion') much in the same sense as is done here.
[177] For instance, column M of lunar system A contains the value of M and the date of the lunation.
[178] Kugler had already reconstructed template procedures, e.g. in connection with the ΦG interpolation scheme of lunar system A (Kugler 1900, p. 171).
[179] Purely orthographic variations involving phonetic complements etc. are usually not taken into account.

ša$_2$ *al-la* 50.07.15 *atrat*(**diri**) *ina*(**ta**) 1.40.14.30 *tušelli*(**e**$_{11}$)-*ma* **gar**	That which exceeds 50;7,15 you deduct from 1,40;14,30 and put down.

Here 50.07.15 represents the maximum $M = 50;7,15^{\bar{\tau}}$. If we replace this by M and incorporate other implementations of this rule we obtain the following template:

ša$_2$ *al-la* M {**diri**|**gal**-*u*$_2$} **ta** $2M$ **e**$_{11}$|**nim**|**la**$_2$[-*ma* **gar**]	That which {exceeds|is larger than} M you deduct|deduct|subtract from $2M$ [and put down.]

Some templates are generalised one step further, in the sense that

(v) the numerical value or name of a function or parameter is replaced by the symbol Q (in the case of multiple quantities Q_1, Q_2, etc.).

This is mainly used for arithmetical operations, which are applied to both numbers as well as abstract quantities of undetermined magnitude (§2.2).

2.6.4.3 Formulaic and graphical representations

Formulaic representations

Formulaic representations have a long history in the literature on Babylonian mathematical astronomy that goes back to the very first publication by Epping and Strassmaier (1881). As in other branches of the history of mathematical astronomy it is common practice to use modern symbols for representing functions and parameters of Babylonian astronomy. This might be criticised as anachronistic, because there is no Babylonian equivalent of symbolic manipulation, neither in mathematical astronomy, nor in the mathematical corpus. In the procedure texts all manipulations are represented in verbal form. Nevertheless, formulaic representations, if interpreted with care, are useful and virtually indispensable tools in the context of Babylonian mathematical astronomy, if only because many of its quantities (for instance those in lunar system A) have a complex astronomical interpretation that is not readily captured in words. Moreover, the Babylonian names for these quantities, if known, often reveal little about their precise astronomical meaning. The modern symbols provide us with convenient labels for referring to these quantities (e.g. to speak of 'function Φ' etc.), representing their algorithms in the form of equations, and analysing their properties.

As representations of algorithms modern equations must be interpreted with care. Equations are capable of representing the outcome of algorithms, but they provide only limited information about the way in which this outcome is achieved, i.e. by which precise sequence of steps. Modern equations are thus in fact unsuitable for representing algorithms, because any modern equation may represent several different, mathematically equivalent algorithms. Another fundamental objection to the use of formulaic representations stems from the fact that modern equations imply rules, for manipulating identities, that do not always have a counterpart in Babylonian mathematical astronomy. In particular there is often no Babylonian equivalent to the operation whereby a quantity is transferred to the other side of an equation, because this involves transforming positive into negative and vice versa. Since there is no concept of negative numbers in Babylonia, this operation is possible only if the term as a whole remains positive. Modern equations are therefore not well suited for representing relations between quantities as they are formulated in the procedure texts, and the same applies to OB mathematical problem texts.

Graphical representations

For the convenience of the modern reader many algorithms are accompanied by graphical representations in which the computed function is plotted against its argument, usually the zodiacal position, event number, day number or tithi number. Also graphical representations have a long history in the field of Babylonian mathematical astronomy, and likewise they have profoundly shaped the modern discourse about its concepts. Two obvious examples are the terms zigzag function and step function, which originate from modern graphical representations of the Babylonian algorithms. However, there is no evidence that the Babylonian astronomers conceptualised these algorithms in graphical terms.

As is true for equations, mathematical graphs are mainly suitable for representing the outcome of an algorithm, but not the algorithm itself. Only for some simple algorithms such as linear interpolation (Fig. 2.4, p. 41) or the zigzag function (Fig. 2.6, p. 45), do the figures in fact represent all steps of the algorithm. Similarly, all representations of step functions (Fig. 2.10, p. 49) include the preliminary values of the step function as a dotted curve. In this connection note also that in most graphical representations the argument of a function (the coordinate) increases from left to right in agreement with the modern convention, whereas the normal apparent motion of a planet, the Moon or the Sun if viewed against the background sky proceeds from right to left for an observer in Babylonia or elsewhere in the northern hemisphere.

2.6.4.4 Flow charts

My use of flow charts is inspired by recent investigations of algorithms in Old Babylonian and ancient Egyptian mathematics.[180] Up to now flow charts have hardly been used in the context of Babylonian mathematical astronomy.[181] The main advantage of flow charts in comparison with other representations is that they can preserve all essential features of an algorithm including the temporal and logical sequence of the involved steps (subalgorithms) and the flow of information (input quantities, initial values, output quantities, parameters). Much of that information is lost when algorithms are represented as equations. Flow charts are particularly useful for representing highly complex algorithms as they occur in Babylonian mathematical astronomy. The form of the flow chart and its constitutive graphemes (Fig. 2.3) have been slightly adapted in order to cope with that complexity.[182] In order to interpret a flow chart it suffices to keep in mind the following rules, which are basically self-evident:

- a rectangular box represents an operation or subalgorithm ('computational step')
- a diamond-shaped box represents a choice between two or more options, each of which involves a condition
- a line connected to a box or diamond represents a quantity (function of parameter)
- a solid arrow indicates the flow of information (principally from left to right)
- occasionally a disconnected open arrow indicates the sequence of two operations without a transfer of information.

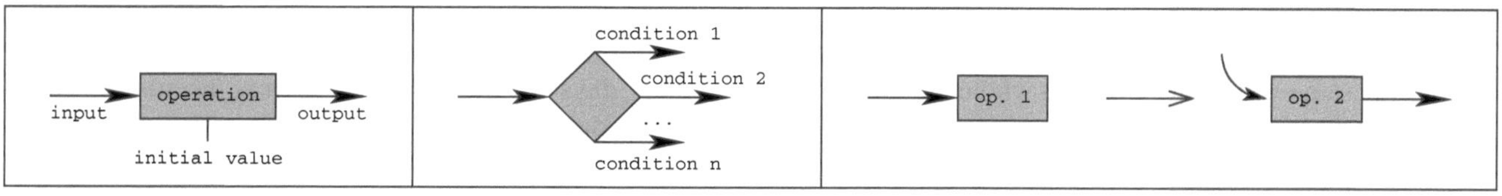

Fig. 2.3: Elementary graphemes of the flow charts. **Left**: operations and subalgorithms. The optional vertical line below the box represents an initial condition. **Middle**: conditions. **Right**: sequence of two operations without transfer of information.

2.6.5 Purposes of the algorithms

Having clarified the relation between procedures and algorithms, we are in a position to classify the purposes of the algorithms.

2.6.5.1 Computing or updating a function

Most of the algorithms are aimed at computing (in Akkadian: to 'construct' = *epēšu*) the columns of the tabular texts (or intermediate quantities not represented by a column). These algorithms are of two types:

(i) updating a function by applying a difference to the value at the previous synodic event, day or (mean) tithi, i.e. $f_i = f_{i-1} \pm d$. The difference d can be constant (e.g. in the case of a zigzag function) or variable (e.g. if d is modeled as a step function). Sometimes the updating involves several differences which are combined, i.e. $f_i = f_{i-1} \pm d_1 \pm d_2$....

(ii) nested functions, i.e. $g(f)$. This concerns algorithms by which one function (g) is computed from another function (f). The updating of g then proceeds indirectly, namely by updating f_{i-1} to f_i and computing $g(f)$. Very often f is the zodiacal position; a more complex example is the interpolation algorithm for computing G from Φ in lunar system A (§4.4.11).

[180] Robson (1997); Ritter (1998, 2010); Imhausen (2002, 2003).

[181] E.g. Aaboe (2001), p. 60, and Britton (2007a), who use flow charts for representing the sequence of algorithms for computing a synodic table of lunar system A.

[182] I use a more compact notation than Imhausen (2003), such that input and output quantities are not represented by separate graphemes. Furthermore, I order the operations from left to right, not from top to bottom.

2.6.5.2 Verification

There are also procedures that can be used for verifying whether the column of a tabular text has been computed correctly. Usually this verification is achieved by deriving from the basic algorithm for computing or updating the function an algorithm for the net difference of that function for intervals longer than 1 synodic cycle. By applying the net difference, say for n cycles (corresponding to n lines in the table), to a function value in the table, say in line 1, this should ideally yield the value in line $n+1$. If not, then an error was committed somewhere in between. This kind of verification is possible only if a simple expression for the net difference can be derived, which is true for zigzag functions and step functions. However, some of the lunar functions are sufficiently complicated that there is no simple expression for the net differences (e.g. M in lunar system A). It is therefore not surprising that computational errors are especially common in these columns.

2.6.5.3 Theoretically oriented procedures

A small number of tablets contain procedures involving astronomical quantities that are known from the tabular texts or other procedures, but the algorithms do not have any obvious connection with the production or verification of tables. Some of these procedures contain rather sophisticated computations that reveal a deep insight into the astronomical significance of, or the relations between various functions or parameters (e.g. No. **82** and No. **102**). The 'theoretical' orientation and apparent lack of practical use of these procedures can be viewed as explicit evidence against the 'instrumentalist' interpretation of Babylonian mathematical astronomy that one sometimes finds in general works on the history of science.

2.6.6 The basic period relation of a function

All functions of Babylonian mathematical astronomy are periodic in the event frame, which means that they return to *exactly* the same value after a certain whole number of synodic events, days or tithis. This is known as the number period Π. A second basic parameter characterising the periodic behaviour of a Babylonian function is the corresponding number of oscillations Z, i.e. the number of times that the function assumes its minimum or maximum value in the course of Π events. They constitute the basic period relation of any function:

$$\Pi \text{ events of the synodic phenomenon } \sim Z \text{ oscillations of the function,} \tag{2.8}$$

where the symbol $\sim$ denotes exact correspondence. Several procedure texts mention period relations of this kind, using the term *palû*(**bala**), 'turn', to describe Z.[183] The ratio of Π and Z defines the 'period',

$$P = \frac{\Pi}{Z}. \tag{2.9}$$

Since Π events correspond to Z oscillations of the function, P is the (fractional) number of events after which the function completes exactly 1 oscillation. Unlike Π and Z, the period P is rarely a whole number; often it is not even a finite sexagesimal number.[184] For a proper understanding of P it is also important to recall that the event frame is not an absolute time frame. The actual time corresponding to P synodic events (or any other number of events) is determined by the algorithm for the synodic time or its lunar equivalent, the synodic month.[185]

2.6.7 Interpolation

Interpolation is one of the most common algorithms of Babylonian mathematical astronomy. It is used for computing the value of a goal function, say g, from a given value of the source function, say f. This is achieved by interpolating linearly between control values of f and g. Some algorithms for computing daily positions of the planets may be interpreted as quadratic interpolation schemes, since the planet's zodiacal position (B) assumes a quadratic dependence on the day or tithi number (i). However, this quadratic dependence is merely a consequence of the linear dependence of the daily displacement v on i. As far as known an explicit concept of quadratic interpolation is not formulated anywhere.

[183] E.g. No. **13** P11, No. **18** P22, No. **41** P4, No. **45** P1'. This usage of *palû* is not mentioned in the dictionaries.

[184] In practice this is not a problem, because P is not a fundamental parameter like Π and Z. Nevertheless some values of P are mentioned in the procedure texts.

[185] Nevertheless, one can occasionally find incorrect statements in the literature suggesting that P is expressed in multiples of the mean synodic time $\overline{\tau}$; e.g. Schmidt (1969).

The zodiacal position (B) as source function

The most common class of interpolation procedures includes those involving the zodiacal position of the planet or the Moon (B) as the source function. They are based on the following templates:

[*ana tar-ṣi*] B_k g_k [**gar**-*an*] [*mim-ma*] *ša*₂ *al-la* B_k **diri a.ra**₂|GAM|GAM₀ c_k **du**-*ma* **ki** g_k **tab**[-*ma* **gar**-*an*]	[Opposite] B_k [you put down] g_k. [Whatever] (the amount) by which it exceeds B_k you multiply by c_k, add to g_k [and put down].	(IP.A.1)
[*ana tar-ṣi*] B_k g_k [**gar**-*an*] [*mim-ma*] *ša*₂ *al-la* B_k **la**₂ **a.ra**₂|GAM|GAM₀ c_k **du**-*ma* **ta** g_k **la**₂|**nim**[-*ma* **gar**-*an*]	[Opposite] B_k [you put down] g_k. [Whatever] (the amount) by which it is less than B_k you multiply by c_k, subtract|deduct from g_k [and put down].	(IP.A.2)

A formulaic representation is easily derived. If k labels the control values (a convention used throughout this study), and B is situated between B_k and B_{k+1}, then

$$g = \begin{cases} g_k + c_k \cdot (B - B_k) & \text{if } g \text{ is increasing} \\ g_k - c_k \cdot (B - B_k) & \text{if } g \text{ is decreasing.} \end{cases} \tag{2.10}$$

For a graphical representation cf. Fig. 2.4. In templates IP.A.1,2 the subtractions $B - B_k$ are not performed explicitly but implied (§2.3.2). They have numerous applications, e.g. for computing a zigzag function (§2.6.9) or other functions of the zodiacal position such as the length of daylight (C in lunar systems A and B). For the zigzag function it would suffice to use the extrema m and M as the only control points, but the procedure texts usually employ 12 control points — one for each zodiacal sign. In the lunar systems, they are anchored at the same number of degrees within the zodiacal signs as the equinoxes in Ari and Lib, i.e. 8°, 10° or 12° for systems B, A and K, respectively.

A more elaborate, probably older formulation of the same algorithm is employed in No. **52** (lunar system K). Here several functions are computed from control values (anchored at 12° of the zodiacal signs), but a general procedure is not offered, all computations being presented as numerical examples. For the increasing branch of g the algorithm can be represented as

$$g = g_k + 0;2 \cdot (g_{k+1} - g_k) \cdot (B - B_k). \tag{2.11}$$

Hence the interpolation proceeds without any explicit reference to coefficients c_k, since these are computed separately for every instance of interpolation as $(g_{k+1} - g_k)/30$.

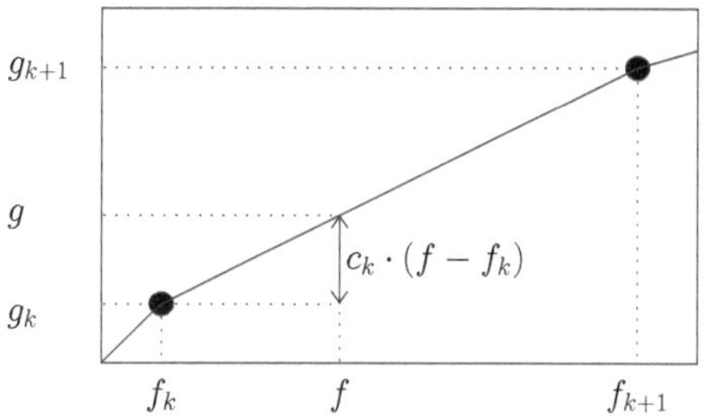

Fig. 2.4: Graphical representation of the linear interpolation algorithm for the case when f and g are both increasing.

Source functions with increasing and decreasing values

If f is a function with increasing as well as decreasing values (e.g. a zigzag function) there are four different interpolation rules:

$$g = \begin{cases} g_k + c_k \cdot (f - f_k) & \text{if } f \text{ increasing, } g \text{ increasing} \\ g_k - c_k \cdot (f - f_k) & \text{if } f \text{ increasing, } g \text{ decreasing} \\ g_k + c_k \cdot (f_k - f) & \text{if } f \text{ decreasing, } g \text{ increasing} \\ g_k - c_k \cdot (f_k - f) & \text{if } f \text{ decreasing, } g \text{ decreasing.} \end{cases} \tag{2.12}$$

A novel feature compared to the previous case is that g depends not only on the magnitude of f, but also on its direction (increasing or decreasing). The main applications are the algorithms for computing G, Λ and W from Φ in lunar system A (§4.4.11). They tend to be formulated in a highly compact form. The most common template is as follows. If f_k ($= \Phi_k$) and g_k ($= G_k$, W_k or Λ_k) are both increasing in magnitude, then

ana tar-ṣi f_k **tab**-*u*₂ g_k **gar**-*an* [*mim-ma*] *ša*₂ *al-la* f_k **tab**-*u*₂ **diri** [**en** f_{k+1} **tab**-*u*₂] **a.ra**₂ c_k **du**-*ma* **ki** g_k **tab**-*ma* **gar**-*an*	Opposite f_k, increasing, you put down g_k. [Whatever] (the amount) by which it exceeds f_k, increasing, [until f_{k+1}, increasing,] you multiply by c_k, and add with g_k, and put down.	(IP.B.1)

The corresponding template for the case when f_k and g_k are both decreasing is

ana tar-ṣi f_k **la**$_2$-*u*$_2$ g_k **gar**-*an* [*mim-ma*] *ša*$_2$ *al-la* f_k **la**$_2$-*u*$_2$ **la**$_2$-*u*$_2$ [**en** f_{k+1} **la**$_2$-*u*$_2$] **a.ra**$_2$ c_k **du**-*ma* **ta** g_k **nim**|**e**$_{11}$-*ma* **gar**-*an*

Opposite f_k, decreasing, you put down g_k. [Whatever] (the amount) by which it is less than f_k, decreasing, [until f_{k+1}, decreasing,] you multiply by c_k, and deduct from g_k, and put down. (IP.B.2)

Corresponding templates for the case when f_k is increasing and g_k is decreasing, or vice versa, are constructed analogously. Rarely the element 'until f_{k+1}, increasing/decreasing' is omitted (No. **72** P1). In one variant of these templates, attested only in No. **72** P1 and No. **97** P6', the phrase *ša*$_2$ *al-la* ... **diri**|**la**$_2$ is followed by *tanašši*(GIŠ), resulting in 'you compute (the amount) by which it exceeds|is less than f_k'. The following alternative but equivalent template uses a more elaborate formulation in which the subtractions $f - f_k$ or $f_k - f$ are explicitly mentioned (for applications cf. Table 4.21, p. 146):

ana tar-ṣi f_k **tab**-*u*$_2$ g_k **gar**-*an* [*mim-ma*] *ša*$_2$ *al-la* f_k **tab**-*u*$_2$ **diri en** f_{k+1} **tab**-*u*$_2$ f_k **ta ša**$_3$ **nim a.ra**$_2$ c_k **du**-*ma* **ki** g_k **tab**-*ma* **gar**-*an*

Opposite f_k, increasing, you put down g_k. What[ever] exceeds f_k, increasing, until f_{k+1}, increasing: you deduct f_k from it, you multiply it by c_k, and you add it with g_k, and you put it down. (IP.C.1)

ana tar-ṣi f_k **la**$_2$-*u*$_2$ g_k **gar**-*an* [*mim-ma*] *ša*$_2$ *al-la* f_k **la**$_2$-*u*$_2$ **la**$_2$-*u*$_2$|*i-maṭ-ṭu-u*$_2$ **en** f_{k+1} **la**$_2$-*u*$_2$ **ta** f_k **e**$_{11}$|**nim**-*ma* *ša*$_2$ *re-ḫi*|**tag**$_4$ **a.ra**$_2$|GAM c_k **du**-*ma* **ta** g_k **nim**|**e**$_{11}$-*ma* **gar**-*an*

Opposite f_k, decreasing, you put down g_k. What[ever] is less than f_k, decreasing, until f_{k+1}, decreasing, you deduct from f_k, what remains you multiply by c_k, and you deduct it from g_k, and put it down. (IP.C.2)

If f and g are zigzag-type functions (§2.6.8) that share the same period and are also exactly in phase (e.g. Φ and $F^{(1)}$ in lunar system A) then a simplified version of this template is used, with the extrema (m, M) as the only control points (No. **65** P5):

ana tar-ṣa m_f m_g **gar**-*an* *ša*$_2$ *al* m_f **diri** m_f *ina* **ša**$_3$ M_f **la**$_2$ *ša*$_2$ **tag**$_4$ GAM c **du ki** m_g **tab**-*ma* **gar**-*an*

Opposite m_f you put down m_g. That which exceeds m_f: you subtract m_f from it, what remains you multiply by c, you add it with m_g, and you put it down. (IP.D.1)

ana tar-ṣa M_f M_g **gar**-*an* *ša*$_2$ *al* M_f **la**$_2$ **ta ša**$_3$ M_f **nim** *ša*$_2$ **tag**$_4$ GAM c **du** *ša*$_2$ **tag**$_4$ **ta** M_g **nim**-*ma* **gar**-*an*

Opposite M_f you put down M_g. That which is less than M_f you deduct from M_f, what remains you multiply by c, what remains you deduct from M_g, and you put it down. (IP.D.2)

Finally, the following template is thus far attested only in No. **77** for the case where g is increasing:

ana tar-ṣi f_k g_k **gar**-*an* **ta** f_k **tab**|**la**$_2$ **en** f_{k+1} **tab**|**la**$_2$ **a.ra**$_2$ c_k **du**-*ma* **ki** g_k **tab**

Opposite f_k you put down g_k. From f_k, increasing|decreasing, until f_{k+1}, increasing|decreasing: you multiply by c_k, and add it with g_k. (IP.E)

Goal functions with additive and subtractive values

The next level of complexity of the linear interpolation algorithms involves goal functions g that assume subtractive or additive values depending on the value and direction of f. One example is the algorithm for Λ (lunar system A), which is computed from Φ (§4.4.13). The procedures employ basically the same templates.

2.6.8 *Zigzag functions of the event number*

Babylonian mathematical astronomy makes abundant use of zigzag functions. There are two kinds of zigzag functions, each computed with a different algorithm. Zigzag functions of the event number are updated with a constant difference from one to the next event of a synodic phenomenon. Zigzag functions of the zodiacal position (§2.6.9) are computed from the zodiacal position of the planet, the Moon or the Sun by means of linear interpolation. Both share the basic feature that the function increases and decreases linearly between a minimum and maximum. A Babylonian term for zigzag function is not attested. In the procedure texts the algorithm is formulated purely arithmetically, without any indication of an underlying graphical concept as suggested by the modern name.

2.6.8.1 Templates and algorithm

Zigzag functions of the event number appear in a wide range of computational systems, but they are a defining feature of type-B systems, where this algorithm is used for the synodic arc (for the planets cf. §3.3.1.1; for the Moon §4.1.2). Some of the procedures provide only the defining parameters of the zigzag function (Table 2.2). They employ the following templates:[186]

M šūqu(**nim**|**la₂**) *m šuplu*(**sig**) *d tašpiltu*(*taš*)	*M* is the 'height', *m* is the 'depth', *d* is the difference.	(ZZ.A.1)
M **la₂** : *m* **sig** : SN *ana* SN *d taš*	*M* is the 'height', *m* is the 'depth', (from) SN to SN *d* is the difference.	(ZZ.A.2)
B_M *M* **la₂** B_m *m* **sig** *d taš* SN *ana* SN	B_M, *M*: the 'height'; B_m, *m*: the 'depth'; *d*: the difference (from) SN to SN.	(ZZ.A.3)
ta *M* **en** *m* **iti** *ana* **iti** *d* **tab** *u* **la₂**	From *M* until *m* month by month you add and subtract *d*	(ZZ.A.4)
M rabītu(**gal**) *m ṣeḫertu*(**tur**) **ab₂** *ana* **ab₂** *d* **tab** *u* **la₂**	*M* is the largest one, *m* is the smallest one, month by month you add and subtract *d*	(ZZ.A.5)

Templates ZZ.A.1–3 are little more than lists of parameters. Some other procedures provide only the numerical values of m, B_m, M, B_M and d without any description.[187] The extrema are referred to either as 'height' and 'depth' or 'large one' and 'small one'. According to ZZ.A.3 the extrema occur at certain zodiacal positions, here represented as B_M and B_m. In reality the extrema of a zigzag function of the event number cluster near a certain position, but they are not fixed; this is how B_m and B_M must here be understood. ZZ.A.4–5, which are attested in lunar procedures, include a rudimentary instruction for updating the zigzag function.

symbol	meaning	Akkadian term	translation
m	minimum	*ṣeḫertu*(**tur**)	the smallest one
		šuplu(**sig**)	'depth'
M	maximum	*rabītu*(**gal**)	the largest one
		šūqu(**nim**, **la₂**)	'height'
d	difference	*tašpiltu*(*taš*) [SN *ana* SN]	difference [(from) SN to SN]
		ṭīpu(**tab**) *u mīṭu*(**la₂**)	addition and subtraction

Table 2.2: Defining parameters of a zigzag function.

The templates ZZ.A.1–5 do not explain when to add d or when to subtract d, and what to do near the extrema. More complete instructions for updating the zigzag function from event to event (or day to day) are provided by other templates, for instance:[188]

d **tab** *u* **la₂**	You add and subtract *d*.	(ZZ.B.1)
[**en** *M* **gal**] *ša₂ al-la M* **diri**|**gal**-*u₂* **ta** 2*M* **la₂**|**e₁₁**|**nim**[-*ma* **gar**]	[Until *M*, the large one;] that which exceeds|is larger than *M* you subtract|deduct from 2*M* [and put down].	
[**en** *m* **tur**] *ša₂ al-la m* **tur**-*er* **ta** 2*m* **la₂**|**e₁₁**|**nim**[-*ma* **gar**]	[Until *m*, the small one;] that which is smaller than *m* you subtract|deduct from 2*m* [and put down].	

Before discussing the variants we shall analyse the underlying algorithm. The function value at event i, say f_i, is obtained in two steps. All templates begin with the phrase **tab** *u* **la₂**, which usually represents *ṭeṭeppe u tumaṭṭa*, 'you add and subtract'.[189] Hence a preliminary update is achieved by adding d to, or subtracting it from f_{i-1}, depending on whether f_{i-1} is on the increasing or the decreasing branch:

$$f_i = f_{i-1} \pm d. \tag{2.13}$$

If the result lies outside the range $[m, M]$ a second step is necessary, which involves a reflection rule:

$$f_i = 2M - d - f_{i-1} \tag{2.14}$$

if f_{i-1} is on the increasing branch ('that which exceeds M' is $f_{i-1} + d$), and

$$f_i = 2m + d - f_{i-1} \tag{2.15}$$

if f_{i-1} is on the decreasing branch ('that which is less than m' is $f_{i-1} - d$). The corresponding flow chart is shown in Fig. 2.5; a graphical representation of a zigzag function is shown in Fig. 2.6. Mathematically speaking, f_i is completely determined by m, M, d and one initial value. Another version is attested in the lunar procedures (where i labels successive lunations):[190]

[186] E.g. ZZ.A.1: No. **41** P5,P6 (Saturn systems B,B"); No. **42** P7,P8; ZZ.A.2: No. **33** P1 (Jupiter system B'); ZZ.A.4: No. **83** P5' (Moon system A); ZZ.A.5: No. **52** P7.a (unknown lunar system).

[187] E.g. No. **97** P5' (lunar system B).

[188] No. **55** P1.a, No. **60** P2' (Moon system A); No. **24** P1, No. **36** P1, No. **37** P1 (Jupiter system B).

[189] An exception is No. **58** P2, which has *d* **tab** *u* **la₂**-*ka*; this implies a translation 'your addition and subtraction is *d*', i.e. **tab** = *ṭīpu*, **la₂** = *mīṭu*.

[190] E.g. No. **53** P1.a.

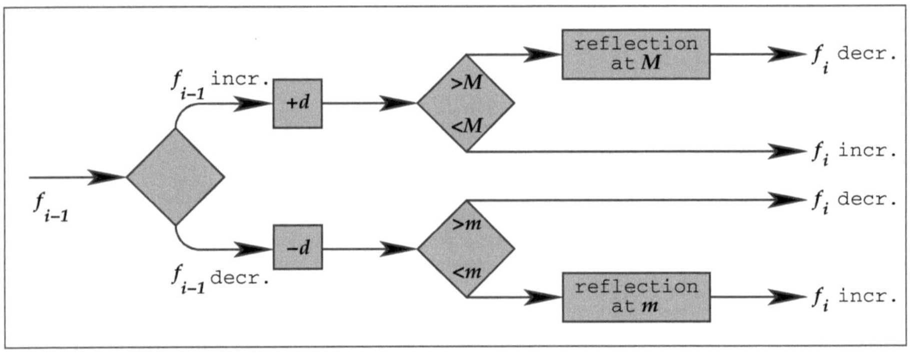

Fig. 2.5: Flow chart of the algorithm for updating a zigzag function of the event number.

d **ab₂** *ana* **ab₂ tab** *u* **la₂** *ša₂ al-la M* **diri** *M ina lib₃-bi-šu₂* **e₁₁** *ša₂* **tag₄** *ina M* **e₁₁** *ša₂ al-la m* **la₂**-*u₂ ina m* **e₁₁** *ša₂* **tag₄ ki** *m* **tab**-*ma* **gar**	Month by month you add and subtract *d*. That which exceeds *M*: you deduct *M* from it, what remains you deduct from *M*. That which is less than *m*: you deduct from *m*, what remains you add with *m* and put down.	(ZZ.B.2)

In ZZ.B.1 and ZZ.B.2 the difference between f_{i-1} and the extremum is calculated explicitly. In the following templates it is only implied:

d **ab₂** *ana* **ab₂ tab** *u* **la₂** *ša₂ al-la M* **diri ta** *M* **la₂** *ša₂ al-la m* **la₂**-*u₂* **ki** *m* **tab**-*ma* **gar**	Month by month you add and subtract *d*. (The amount) by which it exceeds *M* you subtract from *M*. (The amount) by which is less than *m* you add with *m* and put down.	(ZZ.B.3)

This also applies to the following template, which is otherwise similar to ZZ.B.1:[191]

ab₂ *ana* **ab₂** *d* **tab** *u* **la₂** *lib₃-bu-u₂ ša₂ M* **kur**-*ad₂ ša₂ al M* **diri ta** *M* **la₂** *lib₃-bu-u₂ ša₂ m* **kur**-*ad₂ ša₂ al m* **la₂**-*u₂* **ki** *m* **tab**	Month by month you add and subtract *d*. By means of which you reach *M*; (the amount by) which it exceeds *M* you subtract from *M*. By means of which you reach *m*; (the amount by) which it is less than *m* you add with *m*.	(ZZ.B.4)

Eqs. (2.14–2.15) are retrieved by noting that '(the amount by) which it (the preliminary new value $f_{i-1} \pm d$) exceeds *M*' is $M - (f_{i-1} + d)$, and '(the amount by) which it is less than *m*' is $m - (f_{i-1} - d)$. Note that even templates ZZ.B.1–4 do not explicitly mention that the additive or subtractive sense of *d* is reversed after each reflection.

2.6.8.2 Period relations

Function values computed with the zigzag algorithm satisfy the usual period relation (Eq. 2.9), which assumes the form

$$P = \frac{\Pi}{Z} = \frac{2\Delta}{d}, \tag{2.16}$$

where

$$\Delta = M - m \tag{2.17}$$

is the amplitude of the zigzag function. Since 2Δ is the total change of f over one full oscillation (Fig. 2.6) and d the difference from event to event, their ratio P is the fractional number of synodic events after which f returns to the same value. This relation determines both Π and Z (Eq. 2.9), because they are defined as smallest possible whole numbers. Recall that Π is the whole number of synodic events after which f returns to exactly the same value (§2.6.6), as illustrated in Fig. 2.6, i.e.

$$f_{i+\Pi} = f_i, \tag{2.18}$$

and Z is the corresponding whole number of oscillations (Akk.: **bala** = *palû*, 'turn') of the zigzag function.

[191] No. **53** P7' (lunar system A). For the assumed instrumental sense of *libbû ša*, which is not mentioned in the dictionaries, cf. the Glossary.

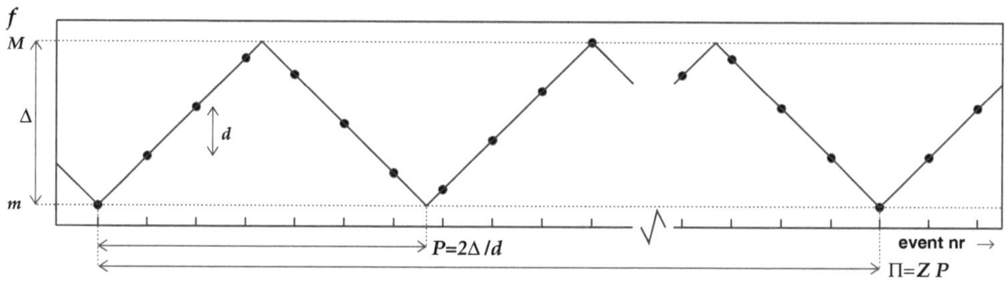

Fig. 2.6: Graphical representation of the zigzag function of the event number.

Underlying zigzag functions

For some zigzag functions it is meaningful to consider the values in the tabular texts as being sampled from a more rapidly oscillating underlying zigzag function.[192] In principle there are infinitely many such underlying zigzag functions with progressively smaller periods, but in practice it is sufficient to consider the next in sequence. This is defined by having one additional oscillation for each event, so that after Π events there have been $Z+\Pi$ oscillations as compared to Z for the tabulated function, resulting in a period

$$p = \frac{\Pi}{Z+\Pi} = \frac{P}{P+1}, \tag{2.19}$$

measured in fractional synodic events (Fig. 2.7). Conversely, the period of the tabulated function can be expressed in terms of the underlying function,

$$P = \frac{p}{1-p}. \tag{2.20}$$

This concept is relevant mainly for functions representing the lunar variation (cf. §4.1.2) such as Φ and F of lunar system

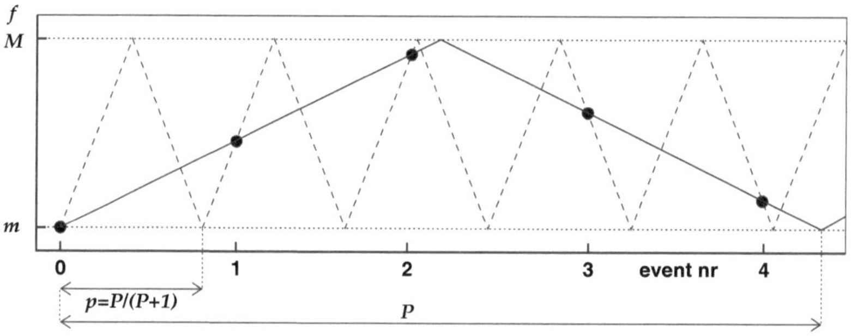

Fig. 2.7: Zigzag function of the event number (solid line), and the underlying more rapidly varying zigzag function (dashed line).

A. Whereas the period of the lunar variation is about 27.56^d (the anomalistic month), these functions are evaluated in the synodic tables at successive lunations separated by the synodic month, whose mean value is 29.53^d. Hence the tabulation interval slightly exceeds the intrinsic period of the functions, the ratio being $p \approx 0.9333$. As a result, the tabulated functions appear to vary with a much longer period $P \approx 13.99$ expressed in fractional synodic events.

2.6.8.3 Elementary steps

The period relation implies that the zigzag function assumes exactly Π different values, which are equidistantly distributed over the interval 2Δ. Hence there is a smallest possible difference separating two arbitrary values of a zigzag function. This difference, δ, referred to as the 'elementary step' of the zigzag function, equals

$$\delta = \frac{2\Delta}{\Pi} = \frac{d}{Z}. \tag{2.21}$$

The elementary step can be used for assessing whether or not two given numbers are connectible by the same zigzag function. Furthermore, the net difference of a zigzag function for intervals longer than 1 synodic cycle always corresponds to a whole number of elementary steps (cf. below). However, it is not clear to what extent, if at all, δ is a Babylonian concept.

[192] Neugebauer (1938c), pp. 215–229; *ACT*, p. 31. In modern data analysis this is known as the aliasing phenomenon, which occurs when a signal is sampled at an interval longer than the period of the signal.

2.6.8.4 Net differences for intervals longer than 1 synodic cycle

Zigzag functions of the event number are primarily used for computing quantities from one to the next occurrence of a synodic phenomenon. However, it is also possible to update a zigzag function over s synodic cycles in a single operation by adding the appropriate net difference, $df(s) = f_{i+s} - f_i$. Of particular significance are net differences for close returns to the same value, which reflect approximate period relations for the phenomenon under consideration. They are a recurrent subject of the procedure texts because they can be used for verifying whether a column with successive values of f was computed correctly. These procedures usually list numerical values of $df(s)$ for selected s, without any explanation of their origin. Values of $df(s)$ can be computed as follows. If we momentarily ignore the reflections that occur at the extrema of f, then s synodic cycles cause f to change by the amount sd. From this we obtain the net difference $df(s)$ by subtracting

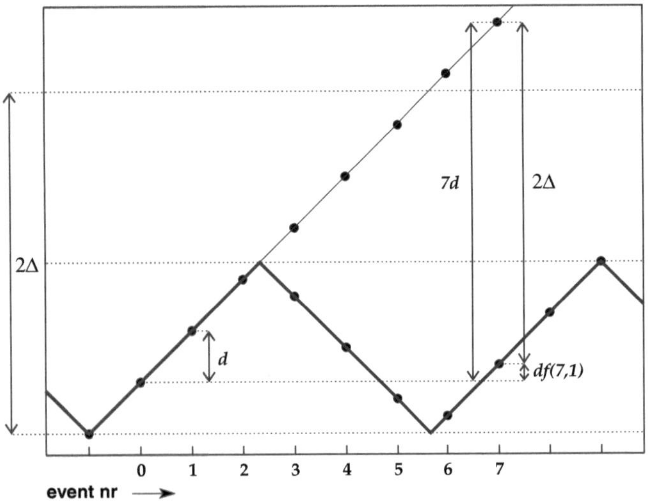

Fig. 2.8: Graphical representation of the computation of the net difference, $df(s,t)$, for a zigzag function of the event number. The example concerns $df(7,1) = 7d - 2\Delta$.

2Δ as many times, say t, as needed for obtaining a value of $sd - 2t\Delta$ within the range $[-\Delta, \Delta]$, i.e.

$$df(s,t) = sd - 2t\Delta. \tag{2.22}$$

Fig. 2.8 shows hows this works for $df(7,1)$. Note that for close returns the value of t is always uniquely defined by s. This allows us to omit the argument t from the formulaic representation, leading to $df(s)$. Further insight into the properties of $df(s)$ is obtained by expressing it in terms of elementary steps, which is achieved by inserting Eq. (2.21):

$$df(s) = (sZ - t\Pi) \cdot \delta. \tag{2.23}$$

Hence $df(s)$ always contains a whole number, $sZ - t\Pi$, of elementary steps.

2.6.8.5 Checking rule for function values on opposite branches

Some procedure texts preserve rules connecting one value of a zigzag function, say f_0, located on one branch (increasing or decreasing), to the value f_s separated from it by s events and located on an opposite branch. The latter implies that $f_0 + f_s$ is constant, i.e. independent of f_0. This can be readily seen from Fig. 2.6: if one updates $f_0 + f_s$ to $f_1 + f_{s+1}$, and assumes that both instances of f remain on the same branch, then one instance of f increases by d while the other decreases by d. The constant can be computed by updating f_0 with the net difference $df(s)$. Assuming that an odd number of extrema are passed between f_0 and f_s, so that f_s is on an opposite branch as required, and applying the appropriate reflection rule (Eqs. 2.14–2.15) to $f_0 + df(s)$, it follows that

$$f_0 + f_s = \begin{cases} 2M - df(s) & (f_0 \text{ increasing}) \\ 2m + df(s) & (f_0 \text{ decreasing}). \end{cases} \tag{2.24}$$

Hence $f_0 + f_s$ is a constant whose value depends only on whether f_0 is increasing or decreasing.

2.6.8.6 Empirical aspects and the construction of a zigzag function

Zigzag functions are obviously a coarse tool in terms of their ability to follow the temporal evolution of periodically varying empirical quantities, which tend to behave in a smooth sinusoidal fashion. It may therefore seem odd that m and

M are often defined with many sexagesimal digits. As pointed out by Neugebauer and others, this does not reflect accuracy with regard to empirical data. While the mean value of the zigzag function, $\mu = (m+M)/2$, is usually close to the mean value of the empirical quantity, the extrema tend to deviate considerably from their empirical values. It is therefore clear that the purpose of a zigzag function is not to exactly, or even closely, reproduce the extrema, or any other specific value of the empirical quantity apart from the mean value. The more important property of zigzag functions is that they satisfy an accurate period relation (P), and enable one to compute a function for arbitrary future (or past) occurrences of a synodic phenomenon without introducing round-off errors. As a result of these two features the deviations between the zigzag function and the empirical quantity vary according to a fixed pattern, and they do not accumulate on the time scale on which P is accurate. The accuracy of P depends on that of Δ/d (Eq. 2.16), and not on the number of digits to which m and M are individually defined. In practice at most the first few digits of m, M and d are empirically based, whereas the higher digits are chosen in such a way that f satisfies the given period relation. The procedure texts offer little insight about the derivation of algorithms from empirical data. However, No. **13** P11'.b presents a derivation of the defining parameters of the zigzag function for the synodic arc for Mars system B (cf. §3.6.3).

2.6.9 Zigzag functions of the zodiacal position

Zigzag functions of the zodiacal position are applied in many computational systems, both lunar and planetary (e.g. Fig. 3.15, p. 87, and Fig. 4.33, p. 168). Formally this algorithm is fully defined by the minimum m, the maximum M and the zodiacal position of m or M. The value of the zigzag function, say f, is computed through extrapolation with respect to control points B_k and f_k (§2.6.7). A common template is as follows:[193]

ša₂ al-la B_k *lu-maš* **diri a.ra₂**|GAM₀ c **du**	(The amount) by which (the position) exceeds B_k (°) of the zodiacal sign you multiply by c.	(ZZ.C.1)
ki f_k *ki-i* **tab tab** *ki-i* **la₂ la₂**	You add it with f_k if it is increasing, subtract if it is decreasing.	

A formulaic representation is

$$f = f_k \pm (B - B_k) \cdot c, \tag{2.25}$$

where c, the interpolation coefficient, equals

$$c = \frac{M-m}{3{,}0}. \tag{2.26}$$

In ZZ.C.1 the subtraction $B - B_k$ is not mentioned explicitly (§2.3.2). Usually ZZ.C.1 involves 12 control points, one for each zodiacal sign.[194] In the following equivalent templates the zigzag function is defined with only two control points, the extrema of the zigzag function:[195]

ina B_M M [*ina* B_m m **ta** B_M **en** B_m] *ana* 1 UŠ c **la₂** **ta** B_m **en** B_M *ana* 1 UŠ c **tab**	In B_M M, in B_m m. From B_M to B_m for 1° you subtract c, from B_m to B_M for 1° you add c.	(ZZ.C.2)
mim-ma ša₂ al-la B_M **diri a.ra₂** c **du ta** M QN **e₁₁** **ta** B_M **en** B_m **ta** QN **la₂ ta** B_m **en** B_M **ki** QN **tab**	(The amount) by which it (B) exceeds B_M you multiply by c, deduct from M, QN. From B_M until B_m you subtract from QN, from B_m until B_M you add with QN.	(ZZ.C.3)

2.6.10 Step functions for the synodic arc

The step function is a fundamental algorithm unique to type-A systems. It is mainly used for modeling the synodic arc σ, which is the net displacement of the planet, the Moon or the Sun along the zodiac between successive events of a synodic phenomenon, i.e.

$$B_i = B_{i-1} + \sigma(B_{i-1}). \tag{2.27}$$

In lunar system A, step functions are also used for modeling several other differences, including the monthly difference of the Moon's distance to the ecliptic (E). For these 'generalised step functions' cf. §4.4.6.

[193] Numerous examples can be found in No. **61** P1–P4 (steps 5 and 7).

[194] In the lunar systems these are always located at the same position within the zodiacal signs as the equinoxes in Ari and Lib, i.e. at 8° (system B), 10° (system A) or 12° (system K).

[195] ZZ.C.2: No. **96** P7'; ZZ.C.3: No. **53** P17' (step 5).

2.6.10.1 Updating the zodiacal position with a step function for the synodic arc

In the procedure texts, the computation of σ is not formulated as a separate algorithm, but as an integral part of the algorithm for updating the zodiacal position (B). The most common template is as follows:[196]

ta b_j **en** b_{j+1} σ_j **tab** *ša₂ al-la* b_{j+1} **diri** **a.ra₂**|GAM r_j **du** [**ki** b_{j+1} **tab**-*ma* **gar**-*an*]	From b_j to b_{j+1} you add σ_j. (The amount by) which it exceeds b_{j+1} you multiply by r_j, [add to b_{j+1} and put down.]	(STEP.A.1)

The modern name step function reflects that σ assumes different constant values, σ_j, in different parts of the zodiac ('zones'), resulting in a steplike appearance. Attested values of the number of zones, n, range from 1 (Venus system A_0) to 6 (Mars system A). The step function is fully defined by the boundaries of the zones, b_j (where b_j is the left or 'trailing' boundary of zone j), the preliminary values of the synodic arc, σ_j,[197] and the transition coefficients, r_j. The unnamed quantity to which σ_j is added is the previous zodiacal position B_{i-1}. If the provisional new position $B_{i-1}+\sigma_j$ is in the next zone ($j+1$), which occurs if B_{i-1} is within a distance σ_j of b_{j+1}, the 'transition rule' is invoked. This means that the distance by which b_{j+1} is transgressed is multiplied by r_j and added to the distance traveled up to b_{j+1} (cf. the flow chart in Fig. 2.9). The transition rule causes the final value of the synodic arc, σ, to differ from the preliminary value σ_j in some portions of the zodiac (Fig. 2.10 and Table 2.3). This effect is particularly strong if σ_j is not small compared to α_j (cf. below).

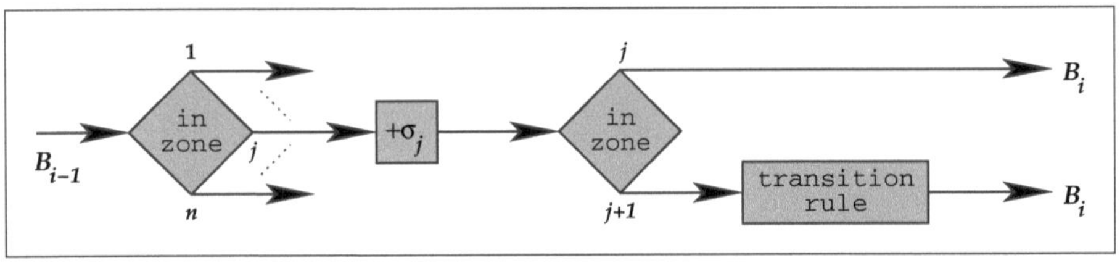

Fig. 2.9: Flow chart of the algorithm for updating B with a step function for the synodic arc.

j	1...n	index of zone
b_j	[°]	zodiacal position of 'trailing' boundary of zone j
α_j	[°]	length of zone j
σ_j	[°]	preliminary synodic arc in zone j

Table 2.3: Parameters of a step function for the synodic arc.

Formulaic representation

We now have sufficient information to derive a formulaic representation of σ (Eq. 2.27). The procedure begins with a preliminary update $B_{i-1}+\sigma_j$. If the distance from the old position to b_{j+1}, say

$$x = b_{j+1} - B_{i-1}, \tag{2.28}$$

is larger than σ_j then the new position is still in zone j, in which case $\sigma = \sigma_j$. If however $x < \sigma_j$ then the preliminary new position is in the next zone, in which case a transition rule is applied, i.e.

$$\sigma = \begin{cases} \sigma_j & \text{if } x \geq \sigma_j, \\ x + y \cdot r_j = \sigma_j + y \cdot c_j & \text{if } x < \sigma_j, \end{cases} \tag{2.29}$$

where

$$r_j = \frac{\sigma_{j+1}}{\sigma_j}, \tag{2.30}$$

$$c_j = \frac{\sigma_{j+1} - \sigma_j}{\sigma_j} = r_j - 1 \tag{2.31}$$

are transition coefficients, and y is the distance by which the *preliminary new* position exceeds b_{j+1}, i.e.

$$y = B_{i-1} + \sigma_j - b_{j+1} = \sigma_j - x. \tag{2.32}$$

[196] E.g. No. **28** P1, No. **35** P1.a, No. **53** P5', No. **55** P2, No. **62** P3'. In No. **1** P1.a,c and No. **32** P1 'you add to b_{j+1} and put down' is omitted. Some badly preserved candidates: No. **2** P2', No. **4** P1', No. **17** P1, No. **20** P1'.b.

[197] This notation differs from that of Neugebauer, who uses w_j instead of σ_j, and $\Delta\lambda$ instead of σ. I reserve w_j for the preliminary values of a 'generalised' step function, i.e. one that does not represent a synodic arc.

The representation based on r_j underlies template STEP.A.1; the one based on c_j is attested in several other templates to be discussed below. The numerical values of r_j and c_j provided in the procedure texts or derived from the tabular texts always satisfy Eqs. (2.30–2.31). Mathematically speaking σ is therefore completely defined by σ_j and b_j. The reason why r_j and c_j are defined in this way becomes clear through a further analysis of the step function algorithm (§2.6.10.3). One effect of the transition rule is that σ deviates from the preliminary values σ_j in a region of width σ_j at the end of zone j, where σ is linearly interpolated between σ_j and σ_{j+1} (Fig. 2.10).

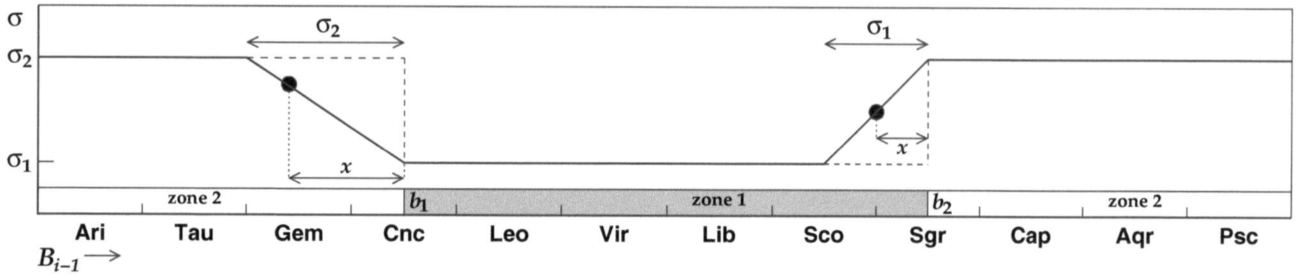

Fig. 2.10: Graphical representation of the step function for σ against the 'old' position B_{i-1}. In this example there are two zones. In the transition regions next to the zonal boundaries σ differs from the preliminary values σ_j (dashed line). Also shown is the distance x between B_{i-1} and the next zonal boundary for two exemplary positions in the transition regions.

Alternative templates

We now consider some alternative but equivalent templates. Sometimes the excess y is computed explicitly (No. **56** P2):

ta b_j **en** b_{j+1} σ_j **tab** *ša*$_2$ *al-la* b_{j+1} **diri** b_{j+1} **ta** **ša**$_3$ **nim a.ra**$_2$ r_j **du** [**ki** b_{j+1} **tab**-*ma* **gar**-*an*]

From b_j to b_{j+1} you add σ_j. That which exceeds b_{j+1}: you deduct b_{j+1} from it, you multiply by r_j, [add it to b_{j+1} and put down.] (STEP.A.2)

In some computational systems (e.g. Mars system A) each zone comprises a whole number of zodiacal signs. The phrase 'From b_j to b_{j+1}' is then usually modified as follows:[198]:

[*ina*] SIGN_1 SIGN_2 σ_j [**tab**] *ša*$_2$ *al* 30 SIGN_2 **diri** **a.ra**$_2$ r_j **du**

[In] SIGN_1, SIGN_2 [you add] σ_j. (The amount) by which it exceeds 30 SIGN_2 you multiply by r_j. (STEP.A.3)

A more significant group of variants employs the alternative transition coefficients c_j, e.g. (No. **42** P1.a,P2.a):

ta b_j **en** b_{j+1} σ_j **tab** b_{j+1} **dib**-*iq* GAM $|c_j|$ **du**-*ma* **tab**|**la**$_2$

From b_j to b_{j+1} you add σ_j. It passes b_{j+1}, you multiply it (= y) by $|c_j|$ and add|subtract it. (STEP.B)

Here $y \cdot |c_j|$ is added or subtracted in accordance with Eq. (2.29). The use of c_j instead of r_j necessitates the consideration of 2 cases, because $y \cdot c_j$ can be additive or subtractive, whereas $y \cdot r_j$ is always additive. Another formulation (No. **31** P2'.b) involving c_j is as follows:

tur-*tu*$_2$ σ_j **tab** b_{j+1} **dib** $|c_j|$ **tab** **murub**$_4$-*tu*$_2$ *ip-pal-ka*

The small one: you add σ_j. It passes b_{j+1}, you add $|c_j|$, you get the middle one. (STEP.C.1)

gal-*tu*$_2$ σ_j **tab** b_{j+1} **dib** $|c_j|$ **la**$_2$ **murub**$_4$-*tu*$_2$ *ip-pal-ka*

The large one: you add σ_j. It passes b_{j+1}, you subtract $|c_j|$, you get the middle one. (STEP.C.2)

The phrase 'It passes b_{j+1}, you add|subtract $|c_j|$...', makes sense if one interprets 'you add|subtract $|c_j|$' as the same instruction to add or subtract $y \cdot |c_j|$ as in STEP.B, i.e. '(for every degree by which) it passes b_{j+1} you add|subtract $|c_j|$'. The terms 'small/middle/large one' make sense if interpreted as references to the zones (j) rather than to the synodic arcs (σ_j) themselves.[199] In the following variant of STEP.C σ_j is omitted, leaving only the transition rules:

[198] No. **15** P1, No. **43** P6'.

[199] Other instances where that meaning is obvious are discussed following STEP.F.

zi *gab-ba ša*$_2$ **tur**-*ti* **a.ra**$_2$ c_j **du**-*ma* **ki**-*šu*$_2$ **tab**-*ma* **murub**$_4$-*ti ip-pal-ka*	All displacements for the small one you multiply by c_j, and you add it with it, and you get the middle one.	(STEP.D.1)
zi *gab-ba ša*$_2$ **gal**-*ti* **a.ra**$_2$ c_j **du**-*ma* **ta** **ša**$_3$ **la**$_2$-*ma* **murub**$_4$-*ti ip-pal-ka*	All displacements for the large one you multiply by c_j, and you subtract from it, and you get the middle one.	(STEP.D.2)

In other procedures the transition rules are omitted,[200] and one is left with

ta b_j **en** b_{j+1} σ_j **tab**	From b_j to b_{j+1} you add σ_j.	(STEP.E)

There are indeed a few computational systems in which B is updated without transition rules (the 'pseudo' step functions discussed below), but the synodic tables imply that in all other type-A systems this involves transition rules. Finally, some procedures provide only the zonal boundaries b_j:[201]

ta b_j **en** b_{j+1} **tur**-*tu*$_2$|**murub**$_4$-*tu*$_2$|**gal**-*tu*$_2$	From b_j to b_{j+1}: the small|middle|large one.	(STEP.F)

The 'small|middle|large one' are references to the σ_j, but sometimes (e.g. in STEP.C) they are actually used as labels for the zones (j).[202] A general word for 'zodiacal zone' is not attested in the procedures for updating B, but a few previously unexplained procedures for Jupiter (§3.7.2.5) employ the word 'ward; neighbourhood' (*bābtu*), such that 'the 'ward' of σ_j' (*bābtu ša*$_2$ σ_j) denotes zone j.

Multiple transitions of the zonal boundaries during 1 synodic cycle

Until now it has been tacitly assumed that $\sigma_j < \alpha_j$, so that at most one zonal boundary is crossed between events $i-1$ and i. For Mercury and Mars this condition is not always satisfied, and two zonal boundaries may be crossed at once. The formulation of the algorithm for σ in the procedure texts is such that no special treatment is necessary in order to cope with that situation. All that is required is repeated application of the transition rules until the final zone is reached. The effects of these multiple transitions on σ are explored in Appendix C.

Constructing a step function from empirical data

The construction of a step function is obviously constrained by empirical data (the period P and the synodic arc's mean value and variation along the zodiac). However, the freedom of choice with regard to the values of σ_j, the number of zones and the location of the boundaries (b_j) is also constrained by the condition that all r_j should be conveniently simple finite sexagesimal numbers. This greatly limits the number of possible implementations of a step function.[203]

'Pseudo' step functions

In some planetary systems (Venus system A$_1$/A$_2$) the algorithm for updating B is similar to the step function, but it lacks transition rules (as proven by the synodic tables). For Venus system A$_1$/A$_2$ that algorithm is also used for updating T. Moreover, in these systems the zones always comprise a whole number of zodiacal signs. The corresponding template for updating B or T is as follows:[204]

SIGN$_1$ SIGN$_2$ [SIGN$_3$] x **tab**	SIGN$_1$ SIGN$_2$ [SIGN$_3$]: you add x.	(PSTEP)

Here x is either the synodic arc σ_j or the synodic time τ_j (cf. §3.3.2.1).

[200] No. **18** P7,P8; No. **19** P1; No. **21** P2',P3'.a; No. **44** P1.a. This appears to be common when the instruction for updating B occurs within a procedure for the subdivision of the synodic cycle (No. **18** P2.a, No. **23** P2'.a and No. **25** P2'a) or the synodic time (No. **18** P14–P16).

[201] E.g. No. **18** P9,P10.a,P17,P25; No. **31** P2'; No. **32** P2; No. **41** P1; No. **42** P3; No. **46** P3.

[202] This is also obvious in the phrase *ina* **tur**-*tu*$_2$ **ki** *šamaš*$_2$ *ša*$_2$ **me** 5 **zi**-*šu*$_2$, 'In the 'small one' the displacement of the Sun's position is 5 per day' (No. **42** O20).

[203] Different strategies by which one may construct step functions for the synodic arc from empirical data have been proposed by Aaboe (1968) and Swerdlow (1998).

[204] E.g. No. **7** P1–P12.

2.6.10.2 Period relation and mean synodic arc

The algorithm for updating B with a step function for σ results in zodiacal positions B satisfying the period relation

$$P = \frac{\Pi}{Z} = \sum_j \frac{\alpha_j}{\sigma_j}. \tag{2.33}$$

Recall that Π is the smallest number of synodic events corresponding to a whole number Z of revolutions around the zodiac, and P is the fractional number of synodic events after which the synodic phenomenon performs exactly one revolution. For a given step function, defined by values of α_j and σ_j for every zone, this relation simultaneously determines Π and Z — since α_j and σ_j are finite-digit sexagesimal numbers, there is always such a pair. The actual proof of Eq. (2.33) is postponed to §2.6.10.3. It can be understood intuitively by noting that α_j/σ_j may be viewed as the 'time', expressed in fractional synodic events, that it takes a synodic phenomenon, moving a distance σ_j per synodic cycle, to cross zone j. Hence the sum of these ratios is the 'time' needed to complete one revolution around the zodiac, expressed in fractional synodic events. P need not be a finite sexagesimal number and is rarely mentioned in procedure texts. From the period relation one can derive the mean synodic arc as

$$\overline{\sigma} = \frac{6{,}0}{P} = \frac{6{,}0 \cdot Z}{\Pi}. \tag{2.34}$$

For this computation to be possible in the Babylonian sense it is usually necessary to first round off P or approximate $1/\Pi$.

2.6.10.3 Elementary steps

Our understanding of the step function for the synodic arc was greatly enhanced after van der Waerden (1957) introduced the concept of the (elementary) step, and Aaboe (1964a) explored in more detail its consequences.[205] As discovered by van der Waerden, the existence of a number period, Π, after which there is a first exact return of the synodic phenomenon, implies that there are exactly Π different positions on the zodiac that can be assumed by the planet. Therefore the zodiac can be construed as being divided into Π elementary intervals of variable length

$$\delta_j = \frac{\sigma_j}{Z}, \tag{2.35}$$

so that σ_j always contains exactly Z elementary steps. Since this applies to all zones, σ also contains Z elementary steps if a zonal boundary is crossed between B_{i-1} and B_i. By introducing elementary steps the algorithm for updating B with the step function for σ is thus summarised in a single rule, namely that the planet always proceeds by Z steps from one to the next occurrence of a synodic phenomenon. In particular, the transition coefficients

$$r_j = \frac{\sigma_{j+1}}{\sigma_j} = \frac{\delta_{j+1}}{\delta_j} \tag{2.36}$$

turn out to be a simple consequence of that single rule, because they are the ratios of the elementary steps in adjacent zones. If the elementary step is the fundamental concept underlying the step function, one may also expect α_j to contain a whole number of them. Indeed,

$$\nu_j = \frac{\alpha_j}{\delta_j} \tag{2.37}$$

is usually a whole number, with few exceptions (e.g. Jupiter system A"). By definition, the total number of steps contained in the zodiac is

$$\sum_j \nu_j = \Pi. \tag{2.38}$$

Analogous to δ for the zigzag functions, δ_j represents the smallest possible difference between two zodiacal positions in zone j. Consequently, two positions are connectible by the same step function σ if and only if they are separated by a whole number of elementary steps. As pointed out by van der Waerden (1965), p. 187,[206] the clearest evidence that the elementary step is not merely an implicit concept reconstructed from the step function algorithm but an actual Babylonian concept is found in No. **13**, a procedure text for Mars system A. Apart from that extraordinary text the elementary steps remain an implicit and largely elusive concept, and the full extent to which the Babylonian astronomers made use of them remains to be established.

[205] I use the term 'elementary step' instead of 'step' in order to distinguish it more clearly from 'step function'. This ambiguity does not exist in the original German terminology (Schritt = elementary step; Stufenfunktion = step function).

[206] English edition: Van der Waerden (1974), p. 265.

2.6.10.4 Net displacements for intervals longer than 1 synodic cycle

By repeatedly updating B with σ we obtain positions for successive events of a synodic phenomenon. Babylonian astronomers also considered net displacements for intervals longer than 1 synodic cycle. The net displacement for s events will be denoted as $dB_j(s,t)$, where t is the closest whole number of revolutions of the synodic phenomenon, and j is the zone. By definition $\sigma_j = dB_j(1,0)$. Of particular interest are values of s for which the planet or the Moon returns to nearly the same position, as illustrated in Fig. 2.11. Numerous procedures, both planetary and lunar, mention net displacements for such close returns, because they can be used for verifying the correctness of column B. Note that for close returns the argument t in $dB_j(s,t)$ can be omitted, because t is uniquely defined by s, resulting in $dB_j(s)$. The main templates are as follows:[207]

[**ta** b_j **en** b_{j+1}] [*ina*|*ana*|**a.ra₂** *y* **mu.meš**] dB_j [**ki**-*šu₂*] **tab**	[From b_j until b_{j+1}] [in|for *y* years] you add dB_j [with it].	(DIFF.A.1)
[**ta** b_j **en** b_{j+1}] [*ina*|*ana*|**a.ra₂** *y* **mu.meš**] dB_j **la₂**	[From b_j until b_{j+1}] [in|for *y* years] you diminish it by dB_j.	(DIFF.A.2)
ta b_j **en** b_{j+1} *ana s* **ab₂** dB_j **mu.du**|**tab**|**la₂**	From b_j until b_{j+1} for *s* months {it proceeds|you add|you subtract} dB_j.	(DIFF.A.3)
ša₂ {*s* **ab₂** | *y* **mu**} **ta** b_j **en** b_{j+1} dB_j **tab**|**la₂**	For {*s* months|*y* years} from b_j until b_{j+1} you add|subtract dB_j.	(DIFF.A.4)
ina y **mu.meš** dB_j [**ki**-*šu₂*] *ana* **nim** *illak*(**du**)	In *y* years [its position] proceeds dB_j to the east.	(DIFF.B.1)
ina y **mu.meš** dB_j [**ki**-*šu₂ ana* **šu₂**] *inaḫḫis*(**la₂**)	In *y* years [its position] recedes dB_j [to the west].	(DIFF.B.2)

The use of **a.ra₂** in DIFF.A.1–2, which usually means 'times' but must represent *ana* = 'to' in this context, is quite unexpected.[208] Templates DIFF.A.3–4 are unique to lunar system A. The following template is attested in No. **5** P4,P5:

ana na-bal-kat₂-tu₄ ša₂ y **mu.an.na.meš** SN **ta** b_j **en** b_{j+1} dB_j **tab**|**la₂**	For the transgression of *y* years of SN, from b_j until b_{j+1} you add|subtract dB_j	(DIFF.C)

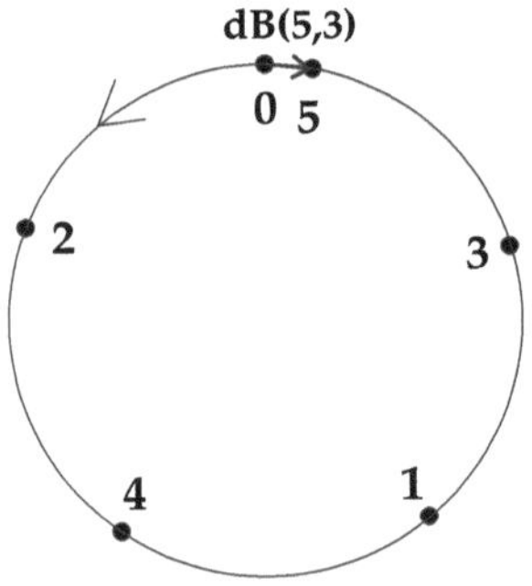

Fig. 2.11: Net displacement of a planet for a close return to the same zodiacal position. The normal, forward motion along the zodiac corresponds to the anticlockwise direction (large arrow). Dots denote successive events of the same synodic phenomenon. After $s = 5$ synodic cycles the planet completes $t = 3$ revolutions except for a small net displacement $dB(5,3)$, which is subtractive in this example.

Nearest whole number of years y corresponding to a close return

In the procedure texts $dB(s)$ is usually identified by the nearest whole number of years (y) corresponding to s synodic cycles, and not by s itself. The relation between y and s differs from planet to planet and can be established from elementary considerations. Since each occurrence of the same synodic phenomenon corresponds to approximately the same configuration of the planet with the Sun, a close return of the planet implies a close return of the Sun. The value of y is therefore obtained by counting the total number of revolutions performed by the Sun. For every revolution of the synodic phenomenon, of which there are t, the Sun also performs 1 revolution. On top of that the Sun performs m additional revolutions for each synodic cycle, where $m = 0$, 1 or 2 (Table 3.4). Hence

$$y = t + ms. \qquad (2.39)$$

By the same token, the number of years corresponding to an *exact* return, which occurs after Z revolutions of the phenomenon, equivalent to Π synodic cycles, is $Y = Z + m\Pi$ (cf. also §3.3.1.4).

Methods for computing net displacements

How the Babylonian astronomers computed the net displacements is not revealed by the procedure texts. An easy, heuristic method would be to identify in a synodic table suitable positions B_i and B_{i+s}, separated by s lines (synodic events) and

[207] Examples: DIFF.A.1,2: No. **18** P1.b',P20; No. **23** P1'.b'; No. **25** P1'.b; No. **26** P1'.c; No. **44** P3'; No. **46** P10'. DIFF.A.3: No. **55** P2.b; DIFF.A.4: No. **65** P7.b; DIFF.B.1: No. **18** P28. DIFF.B.2: No. **44** P3'.

[208] Attested only in No. **18** P20.

both located in the same zone j, and compute the net displacement for each zone j directly as $dB_j(s) = B_{i+s} - B_i$. A somewhat more efficient approach would be to compute B for s consecutive events setting out from an initial position B_0 in zone j sufficiently far away from the zonal boundaries, so that B_s is also in zone j, compute $dB_j(s) = B_s - B_0$, and repeat this for every zone.[209] Finally, as will be shown in the next paragraph, net displacements may also be computed directly from the elementary steps δ_j.

Net displacements in terms of elementary steps

An alternative method for computing net displacements exploits that they always contain a whole number of elementary steps, analogous to the synodic arc. In order to prove this suppose that B_i and B_{i+s} are both in zone 1 and that the closest whole number of revolutions, t, is known.[210] The total distance covered by the planet is $s \cdot \sigma_j$, which contains sZ elementary steps (§2.6.10.3). Exactly $sZ - t(\Pi - v_1)$ of these steps are in zone 1, which corresponds to a distance $[sZ - t(\Pi - v_1)] \cdot \delta_1$. Since the distance traveled by the planet after passing t times through zones 2,...,n equals $t \cdot (6{,}0 - \alpha_1)$, the total distance is $t \cdot (6{,}0 - \alpha_1) + [sZ - t(\Pi - v_1)] \cdot \delta_1$. Inserting $\alpha_1 = v_1 \delta_1$ this reduces to $t \cdot 6{,}0 + (sZ - t\Pi) \cdot \delta_1$. Hence the planet's net displacement, defined as the excess over t full revolutions, is $dB_1(s) = (sZ - t\Pi) \cdot \delta_1$. Generalised to an arbitrary zone j this yields

$$dB_j(s,t) = (sZ - t\Pi) \cdot \delta_j. \tag{2.40}$$

It follows that $dB_j(s,t)$ contains $sZ - t\Pi$ elementary steps, regardless of the zone.[211] As expected $dB_j(1,0) = \sigma_j$ and $dB_j(\Pi, Z) = 0$. Although it was assumed that $dB_j(s,t)$ lies entirely within one zone, the number of elementary steps remains the same if the initial and final positions are in different zones. Again the formalism of elementary steps leads to a concise and convenient description of the planet's motion.

Furthermore, since σ_j and $dB_j(s,t)$ each contain a constant number of elementary steps, the ratio between σ_j and $dB_j(s,t)$ is also constant. This can be seen by combining Eqs. (2.40) and (2.35):

$$\frac{dB_j(s,t)}{\sigma_j} = \frac{sZ - t\Pi}{Z} = s - tP. \tag{2.41}$$

This property can be used for verifying whether given values of $dB_j(s,t)$ are consistent with given values of σ_j, or for reconstructing σ_j from given values $dB_j(s,t)$.[212]

Updating B with a step function for the net displacement

While the usual purpose of the net displacements appears to be verification (§2.6.5.2), they can in principle also be used for updating the zodiacal position of the planet in the same manner as is done with σ. To that end the net displacements $dB_j(s)$ computed from the synodic arc are considered as the preliminary values, say σ'_j, of a step function σ'.[213] Precisely that strategy is followed in system A$_3$ of Mercury, where $\sigma'_j = dB_j(3)$ (§3.4.4). In analogy to σ, the period of σ' is characterised by parameters P', Π', Z'. Their relation to P, Π and Z can be derived by reformulating Eq. (2.41) as

$$\sigma'_j = (s - tP)\,\sigma_j. \tag{2.42}$$

The period of the step function for σ' is therefore

$$P' = \sum_j \frac{\alpha_j}{\sigma'_j} = \frac{1}{s - tP} \sum_j \frac{\alpha_j}{\sigma_j} = \frac{P}{s - tP} = \frac{\Pi}{sZ - t\Pi} = \frac{\Pi'}{Z'}. \tag{2.43}$$

In this derivation it is assumed that the same values of α_j apply to σ and σ'. The original period P is reduced by a factor $s - tP$, as expected. The correct values of Π' and Z' are the smallest whole numbers Π' and Z' for which $\Pi'/Z' = P'$. In the penultimate expression P' is represented as a ratio of two whole numbers Π and $sZ - t\Pi$, but it would be premature to identify Π' with Π, and Z' with $sZ - t\Pi$, because they may still share a common factor. Conversely one may use the inverse of this expression for deriving the period P if one happens to know the step function for σ' but not that for σ:

$$P = \frac{sP'}{1 + tP'}. \tag{2.44}$$

[209] Or do this for only one zone, say j, and compute the values for the other zones as $dB_{j+1}(s) = r_j \cdot dB_j(s)$.

[210] It can be found from the requirement $s \cdot \overline{\sigma} \approx t \cdot 6{,}0°$ or, equivalently, $s/t \approx 6{,}0/\overline{\sigma} = P = \Pi/Z$.

[211] Cf. also *HAMA*, p. 433.

[212] Although $dB_j(s,t)$ and σ_j are finite sexagesimal numbers, this need not be true for their ratio $s - tP$.

[213] Cf. also *HAMA*, p. 468.

2.7 Computational systems

In this study a computational system is pragmatically defined as a distinct collection of algorithms for computing (and verifying) the columns of a synodic table or a daily motion table, or auxiliary functions for such tables. I reserve the term 'scheme' for algorithms concerned with interpolation or the subdivision of the synodic cycle of a planet. Most modern names of the computational systems were coined by Neugebauer, who modified and systematised the older notation handed down from Kugler, introducing new names for previously unknown systems. The resulting nomenclature is unsatisfying but widely used, so that any attempt to improve it would only add to the confusion. One problematic feature is that it does not allow for a distinction between individual computational systems and families of computational systems. In order to alleviate this problem, families of systems will be referred to as type A, type B, etc. Within these families, individual systems are named lunar system A, Mercury system A_2, Jupiter system A', etc. The main distinction between type-A and type-B systems concerns the algorithm for the synodic arc (σ), which is a step function of the zodiacal position (A) or a zigzag function of the event number (B); cf. §3.3.1.1. In the planetary systems (Ch. 3), most of which offer only algorithms for computing zodiacal positions and times, this is the only significant difference between type-A and B systems. Lunar systems A and B (Ch. 4) are far more complex than their planetary counterparts, many lunar quantities having no parallel in the planetary systems. The differences between lunar systems A and B are also more profound and numerous than for the planetary systems of type A and B. Note that the coexistence of different computational systems for the same planet or the Moon is an interesting aspect of Babylonian mathematical astronomy whose significance is not really understood.

The procedure texts usually respect some sort of separation between different computational systems. Often all procedures on a single tablet belong to one and the same system; otherwise those belonging to one system are usually grouped together. Procedures belonging to different computational systems are sometimes introduced by the word 'secondly' (*šanîš*) or 'thirdly' (*šalšiš*). The algorithms constituting a computational system are not always coherent in the mathematical sense. For instance, many planetary systems incorporate two complementary methods for computing zodiacal positions that need not be mutually consistent: one based on the synodic arc and one based on pushes (§3.3.1).

Chapter 3
Planets

About half the corpus of mathematical astronomy — 52% of the procedure texts, 41% of the tabular texts — is concerned with the planets Mercury, Venus, Mars, Jupiter and Saturn. §3.1 introduces the planetary phenomena and the terminology of the planetary texts. In §3.2 the composition of the procedure texts for the planets is briefly explored. §3.3 discusses generic elements of the computational systems, i.e. those concepts, parameters, period relations and algorithms that are common to all planets. §3.4–§3.8 are concerned with the algorithms for the individual planets. Critical editions of the planetary procedure texts are found in §5.2.[214]

3.1 Planetary phenomena

3.1.1 Apparent motion

The planets move with respect to the stars through a narrow band centred at the ecliptic. Periods of normal, forward ('prograde') motion, when the planet passes through the zodiacal signs in the sequence Ari, Tau, Gem, etc., are interrupted by intervals of retrograde motion, during which it moves in the opposite direction (Fig. 3.1). According to a convention

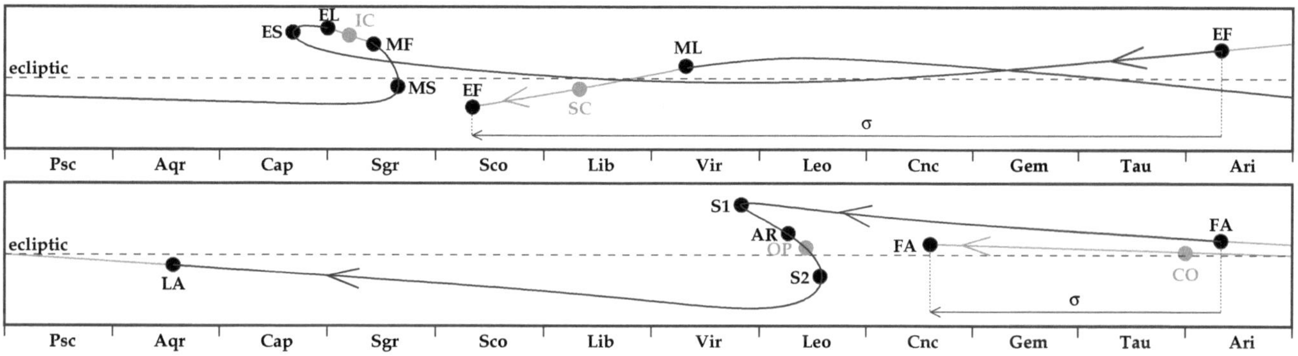

Fig. 3.1: View of the sky showing schematically the motion of an inner planet between two occurrences of the synodic phenomenon EF (top), and that of an outer planet between two occurrences of FA (bottom). The normal motion along the zodiac proceeds from right to left. Between the stations the planet moves in the retrograde direction, from left to right. Grey segments are intervals of invisibility. The synodic arc (σ) is the net displacement along the zodiac during one synodic cycle. In these examples, the planet performs one full revolution in addition to σ, so that the *total* synodic arc is $\Sigma = \sigma + 6{,}0°$.

shared by Babylonian and modern astronomers alike, the forward motion corresponds to an increase of the zodiacal position. To an observer in Mesopotamia or elsewhere in the northern hemisphere, the normal motion proceeds from right (west) to left (east) with respect to the stars. Since this is roughly opposite to the daily rotation of the sky, which proceeds from east (sunrise) to west (sunset), the normal motion is also referred to as 'eastward', by both modern and Babylonian astronomers, the retrograde motion as 'westward'. The planets also perform a motion perpendicular to the ecliptic ('latitudinal motion'). A few procedure texts are concerned with that topic, but planetary tables with a column for the distance to the ecliptic have not yet been discovered.

[214] Most of the algorithms are attested in procedure texts, but for some the tabular texts are the only available source of information. Since new editions of the tabular texts will appear in Volume II, some of the results are preliminary.

3.1.2 Synodic phenomena

The planetary texts are mainly concerned with synodic phenomena. Their terminology, ancient and modern, is compiled in Tables 3.1–3.2.[215] Synodic phenomena are conspicuous events in terms of the planet's visibility (first and last appearances), apparent motion (stations), or elongation from the Sun (conjunctions and oppositions). They occur in a fixed sequence (Figs. 3.1–3.2) which is different for inner and outer planets. All synodic phenomena — not only conjunctions and oppositions — satisfy a 'solar-distance principle' in the sense that they occur at a typical elongation (angular distance) from the Sun (§3.3.1.3). With each synodic cycle the position of the synodic phenomenon shifts along the zodiac by a distance known as the synodic arc, for which I employ the symbol σ. It is also useful to introduce the planet's *total* synodic arc, Σ, which may exceed σ by a whole number of revolutions around the zodiac. The time that it takes for completing one synodic cycle is known as the synodic time or the synodic period, for which I employ the symbol τ.[216] For a given planet the empirical values of Σ and τ vary from one cycle to the next. Modern formulae relating the mean values $\overline{\Sigma}$ and $\overline{\tau}$ to the sidereal period of the planet and that of the Earth (1^{yr}) are provided in Appendix B.

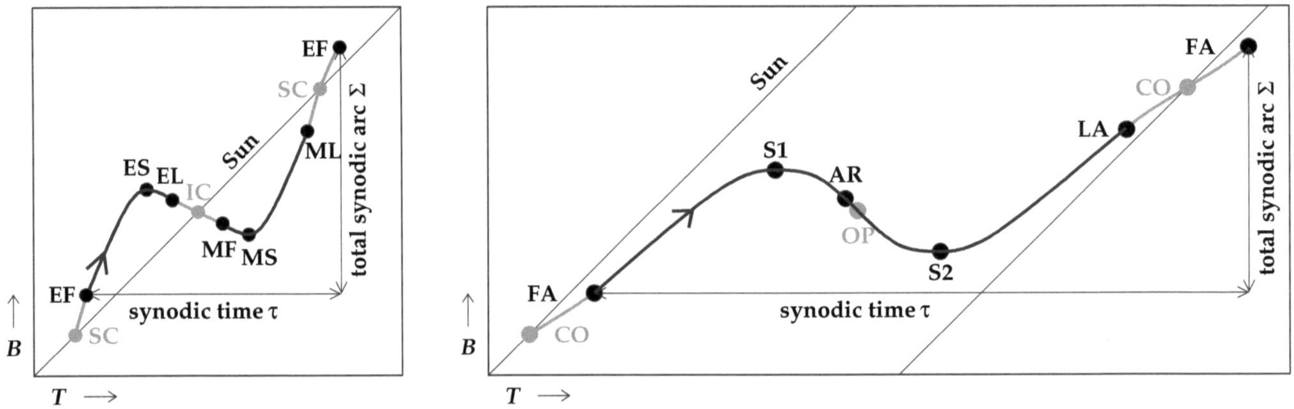

Fig. 3.2: Schematic graphs of the zodiacal position of the planets and the Sun against time for 1 synodic cycle. **Left**: inner planets (Mercury, Venus). **Right**: outer planets (Mars, Jupiter, Saturn). Grey segments are intervals of invisibility. The vertical arrow represents the total synodic arc Σ for EF and FA, respectively; the horizontal arrow the corresponding synodic time (τ). For each synodic phenomenon there is a characteristic elongation from the Sun (distance along the zodiac between the planet and the Sun).

Some features of the synodic cycle are common to all planets. At the last appearances (EL, ML, LA), called 'setting' in Akkadian (**šu**$_2$ = *rabû*), the planet is visible for the last time.[217] During the following period of invisibility the planet is said to be 'with the Sun' (*itti Šamaš*) or 'stay away in heaven' (**zal** *ina* **an**-*e* = *uḫḫuru ina šamê*). At the conjunction (IC, SC, CO) the planet has the same zodiacal position as the Sun (zero elongation), but this phenomenon is not observable and was ignored. At the first appearances (MF, EF, FA), called 'appearance' (**igi** = *nanmurtu*), the planet is far enough from the Sun to become visible again. All planets experience intervals of retrograde motion, called 'turning back' (**gur** = *târu*) or 'receding' (**la**$_2$ = *naḫāsu*). They are delimited by stations (**uš** = *nēmettu*), at which the planet seems to come to a standstill (**uš** = *nenmudu*, 'to become stationary').[218] Since the planet's motion is very slow near a station, its precise moment is difficult to establish through observation.

First and last appearances form three groups in terms of the order in which the planet and the Sun rise or set: at EF, EL and LA the planet sets just after sunset, and at MF, ML and FA the planet rises just before sunrise (Fig. 3.3). LA and EL are thus equivalent to the heliacal setting of a star, FA and MF to the heliacal rising. A newly discovered class of procedures for Jupiter (cf. §3.7.2.5 and §3.7.6.2) concerns the time interval between the setting of the planet and that of the Sun at LA, and that between the rising of the planet and that of the Sun at FA. The acronychal rising (AR) is the last visible rising of an outer planet after sunset. At AR the Sun sets near the opposite point on the horizon from where

215 In the modern literature synodic phenomena are also referred to as 'phases', in the context of ancient astronomy as 'Greek letter phenomena' (the third column in Tables 3.1–3.2), after a symbolic notation introduced by Neugebauer (1954).

216 In *ACT* and *HAMA* the synodic arc is $\Delta\lambda$, the synodic time $\Delta\tau$.

217 Huber (1982) has argued that, at least for some planets, the 'setting' (last appearance) actually refers to the day of first invisibility, one day after the last appearance. However, as pointed out by Mebert (2010), some observational reports (e.g. A 3456, a tablet with observations of Mercury, presumably excerpted from diaries; cf. Hunger 1988) mention the planet's position or the time during which it is visible above the horizon on the day of EL or ML, which rules out an interpretation as (day of) first invisibility. This is also implied by a newly discovered group of procedures concerning Jupiter's setting time at LA (cf. §3.7.2.5). At least for some planets and in some contexts the 'setting' is therefore certainly the last appearance and not the day of first invisibility, but for other planets and other contexts the alternative interpretation cannot be ruled out. Since this issue is of little concern for the present study I maintain the conventional terms (Tables 3.1 and 3.2).

218 The corresponding verbal phrase is usually also written logographically (**uš**), but occasionally phonetically *nen-mud* = *nenmud* (stat. 3 m. sg. N), 'it is stationary'. A 3 c. sg. pres. N, *innemmed*, 'it becomes stationary', also seems a possible reading of these instances of **uš**.

the planet rises. It is therefore most conveniently visualised by displaying the planet together with the imaginary point diametrically opposite to the Sun, the so-called 'anti-Sun' (Fig. 3.3).

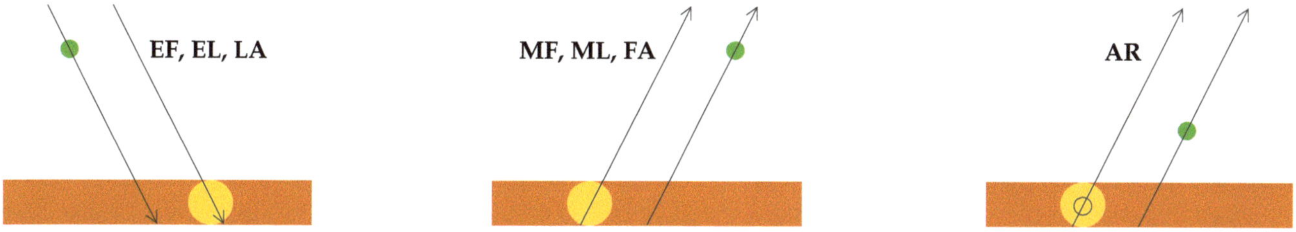

Fig. 3.3: Schematic views of the horizon near the moment of sunset or sunrise showing the planet (small disk) and the Sun (large disk), for AR the point diametrically opposite to the Sun (the 'anti-Sun', large disk with circle). The arrows indicate the direction of rising or setting parallel to the celestial equator. At EF, EL (inner planets) and LA (outer planet) the planet sets in the west just after sunset; at MF, ML (inner planets) and FA (outer planet) the planet rises in the east just before sunrise; at AR the outer planet rises in the east just after sunset.

synodic phenomenon			noun			verbal phrase	
			transliteration	Akkadian	translation	transliteration + Akk.	translation
Superior Conjunction	SC						
Evening First	EF	Ξ	**igi** *ša*$_2$ **šu**$_2$	*nanmurtu ša rabê*	'western appearance'	*ina* **šu**$_2$ *innammer*(**igi**)	'it appears in the west'
			kur *ša*$_2$ **šu**$_2$	*napāḫu ša rabê*	'western rising'	*ina* **šu**$_2$ *inappaḫ*(**kur**)	'it rises in the west'
			igi	*nanmurtu*	'appearance'		
Evening Station	ES	Ψ	**uš** *ša*$_2$ **šu**$_2$	*nēmettu ša rabê*	'western station'	*ina* **šu**$_2$ *nenmud*(**uš**)	'it is stationary in the west'
			uš igi-*tu*$_2$	*nēmettu maḫrītu*	'first station'		
Evening Last	EL	Ω	**šu**$_2$ *ša*$_2$ **šu**$_2$	*rabû ša rabê*	'western setting'	*ina* **šu**$_2$ *irabbi*(**šu**$_2$)	'it sets in the west'
			šu$_2$	*rabû*	'setting'		
Inferior Conjunction	IC						
Morning First	MF	Γ	**igi** *ša*$_2$ **kur**	*nanmurtu ša šadî*	'eastern appearance'	*ina* **kur igi**	'it appears in the east'
			kur *ša*$_2$ **nim**	*napāḫu ša elî*	'eastern rising'	*ina* **nim kur**	'it rises in the east'
			kur	*napāḫu*?	'rising'?		
Morning Station	MS	Φ	**uš** *ša*$_2$ **kur**	*nēmettu ša šadî*	'eastern station'	*ina* **kur uš**	'it is stationary in the east'
			uš *ar*$_2$-*tu*$_2$	*nēmettu arkītu*	'second station'		
Morning Last	ML	Σ	**šu**$_2$ *ša*$_2$ **kur**	*rabû ša šadî*	'eastern setting'	*ina* **kur šu**$_2$	'it sets in the east'
			šu$_2$ *ša*$_2$ **nim**	*rabû ša elî*	'eastern setting'	*ina* **nim šu**$_2$	'it sets in the east'
			DU	?	?		

Table 3.1: Terminology of the synodic phenomena for the inner planets (Mercury and Venus). In the procedure texts both nouns and verbal constructions are used when referring to a synodic phenomenon. The Greek letter symbols were introduced by Neugebauer (*ACT*, *HAMA*).

Inner planets

The Akkadian terms for the synodic phenomena imply an awareness that the synodic cycle is different for inner and outer planets. However, no cuneiform source informs us how this might have been interpreted by the Babylonian astronomers. Since the inner planets remain close to the Sun, they are never in opposition, and their synodic phenomena occur just before sunrise near the eastern horizon, or just after sunset near the western horizon (Fig. 3.3). Furthermore, they experience two conjunctions during each cycle. These features are reflected in the Akkadian terminology.[219] As a morning star (MF–ML) the planet's rising can be observed daily before sunrise at the eastern horizon; as an evening star (EF–EL) the planet's setting can be observed daily after sunset at the western horizon.[220] The interval between EL and MF is shorter than that between ML and EF, because the planet is brighter near IC than near SC. In the case of Mercury some or all synodic phenomena may be unobservable.

Outer planets

Outer planets can assume any angular distance from the Sun, and they experience only one phase of invisibility per synodic cycle. After FA the planet's daily rising occurs earlier every day. Accordingly, the fraction of night during which the planet is visible above the horizon increases from day to day. The stations are referred to as first (literally: front) station

[219] In the Akkadian terms for the synodic phenomena for the inner planets the word 'setting' (*rabû*) appears with two distinct meanings: 1) the western horizon, where the Sun and the planet set; 2) disappearance, in the sense that the rising (ML) or setting (EL) can no longer be observed. The use of DU (Akk. reading unclear) for ML is attested only in No. **10**.

[220] The planet's rising as an evening star and its setting as a morning star occur during day time, so they are not observable.

(*nēmettu maḫrītu*) and second (literally: rear) station (*nēmettu arkītu*), terms that reflect the position of the stations with respect to the forward motion along the zodiac.[221] During opposition (OP) the planet stands diametrically opposite to the Sun, so that it is visible all night. That night the planet's rising and setting cannot be observed, because they coincide with sunset and sunrise, respectively. The exact moment of opposition itself is difficult to observe and was ignored by the Babylonian astronomers, but they did consider the associated phenomenon of acronychal rising (AR), which occurs about 1 day before OP, when the planet can be seen rising for the last time just after sunset.[222] After OP the planet has already risen when the Sun sets. From now on the planet's setting can be observed daily.[223] The time of setting shifts away from sunrise to an ever earlier time of night, so that the fraction of night during which the planet is visible decreases again, until it can no longer be seen (LA).

<table>
<tr><th colspan="3">synodic phenomenon</th><th colspan="3">noun</th><th colspan="3">verbal phrase</th></tr>
<tr><th></th><th></th><th></th><th>transliteration</th><th>Akkadian</th><th>translation</th><th>translit.</th><th>Akkadian</th><th>translation</th></tr>
<tr><td>Conjunction</td><td>CO</td><td></td><td></td><td></td><td></td><td></td><td></td><td></td></tr>
<tr><td>First Appearance</td><td>FA</td><td>Γ</td><td>igi</td><td>nanmurtu</td><td>'appearance'</td><td>igi</td><td>innammer</td><td>'it appears'</td></tr>
<tr><td>First Station</td><td>S1</td><td>Φ</td><td>uš igi-tu$_2$</td><td>nēmettu maḫrītu</td><td>'first station'</td><td>uš</td><td>nenmud</td><td>'it is stationary'</td></tr>
<tr><td>Acronychal Rising</td><td>AR</td><td>Θ</td><td>E|A me</td><td>elê|aṣê$^?$ ūmi</td><td>'daylight rising'</td><td>ana me E|A</td><td>ana ūmi illi|uṣṣi$^?$</td><td>'it rises to daylight'</td></tr>
<tr><td>Opposition</td><td>OP</td><td></td><td></td><td></td><td></td><td></td><td></td><td></td></tr>
<tr><td>Second Station</td><td>S2</td><td>Ψ</td><td>uš egir-tu$_2$|ar$_2$-tu$_2$</td><td>nēmettu arkītu</td><td>'second station'</td><td>uš</td><td>nenmud</td><td>'it is stationary'</td></tr>
<tr><td>Last Appearance</td><td>LA</td><td>Ω</td><td>šu$_2$</td><td>rabû</td><td>'setting'</td><td>šu$_2$</td><td>irabbi</td><td>'it sets'</td></tr>
</table>

Table 3.2: Terminology of the synodic phenomena for the outer planets (Mars, Jupiter and Saturn).

3.2 Composition of the tablets

The tablets with planetary procedures exhibit so much variation in terms of composition that only few systematic features can be identified. On some tablets for type-A systems the following sequence of procedures is rather common: (1) updating B for synodic phenomenon 1, (2) updating T for synodic phenomenon 1, (3) idem for synodic phenomenon 2, etc., sometimes followed by procedures for the subdivision of the synodic cycle. For type-B systems the reverse sequence is more common, i.e. (1) updating T, (2) updating B. The highest degree of regularity can be observed among procedures that accompany a synodic table (Table 3.3). Nearly all of them are placed after the table at the end of the reverse of the tablet, and their purpose is always closely connected with the production of the table. Among the procedure texts for Jupiter there are six tablets (cf. Table 3.35 on p. 90) that preserve various segments of a fixed sequence of procedures devoted to a wide range of topics. While most of them are clearly associated with system A, this is not obvious for some (e.g. a procedure for Jupiter's distance to the ecliptic).

3.3 Algorithms for the planets

The computational systems for the planets (§3.4–3.8) are constructed from a rather small core of generic algorithms which are explored in the present section. With only two main functions — the zodiacal position (B) and the time (T) — the planetary systems have a much simpler structure than the lunar systems. On the other hand, the synodic phenomena of the planets are more numerous, and their distribution (angular and temporal) over the synodic cycle is an important subject of the planetary texts. Beyond that a few procedure texts are concerned with distance to the ecliptic (E), but this function is not yet attested in the tabular texts. Most of the planetary systems are strictly type A or B. Exceptions are various systems for Venus, and some schemes for the subdivision of the synodic cycle that are not strictly associated with a particular computational system.

---

221 In No. **39** O9',11' the verbal expression for S1 is **uš igi** *nen-mud* = *nēmetta maḫrīta nenmud*, 'it is stationary at the first station', in which the 3 m. sg. stat. N of *emēdu* is preceded by the paronomastically used cognate noun in the accusative.

222 In some astronomical diaries AR is written *ana* **me e**$_{11}$ (e.g. *ADRT* I, -567). I therefore follow H. Hunger's suggestion (*ADRT* VI, p. X) that **me** = *ūmu*, 'day(light)', **e** is an abbreviation for *elû*(**e**$_{11}$), 'to ascend' (either logographic or phonetic); A may stand for *aṣû*, 'to appear'. In the procedure texts E|A **me** functions as a noun, 'daylight rising', where E and A are probably infinitives in the construct state, while *ana* **me** E|A is the corresponding verbal phrase, 'it rises to daylight', in which E = *illi* (3 c. sg. pres. G of *elû*) and A is presumably *uṣṣi* (3 c. sg. pres. G of *aṣû*). There are very few exceptions to these conventions; for instance E.A (= *elû*) appears as a noun for AR in No. **37** O2, apparently an abbreviation of E.A **me**.

223 The first visible setting after OP, known by the modern term cosmical setting, occurs just before sunrise about 1 day after OP, but this event was ignored by the Babylonian astronomers.

	size	system		content of the table	procedures
No. **1**	i	Mercury A_1	ST	T, B: EF, EL, MF, ML	updating B, T: MF, EF, ML, EL
No. **7**	ii	Venus A_1,A_2	ST	T, B: EF, ES, EL, MF, MS, ML	updating B, T: EF, ES, EL, MF, MS, ML; leap years
No. **8**	ii?	Venus A_2	ST	T, B: EF, ES, EL, MF, [MS, ML]	updating B, T: EF, ES or EL? [...]
No. **10**	iv	Venus A_3	ST	B: EL, MF, ML, EF	updating B with $dB_j(5)$: EF, ML,MF, EL
No. **12**	iii	Mars A	AT	$\delta\Sigma$ (S1 to AR), scheme S	$\delta\Sigma$ (S1 to AR), scheme S or R
No. **15**	?	Mars A	ST	T, B: [...], AR, S2	updating B, [...], subdivision synodic cycle
		Venus A_1/A_2	ST	T, B: [EF, ES, EL, MF]?, MS, ML	[?]
No. **17**	C?	Jupiter A	ST	T, B: [FA, S1], AR, S2, LA	updating B, T
No. **20**	iv?	Jupiter A'	ST	T, B: FA, S1, [AR, S2, LA?]	[...] updating B, [...]
No. **24**	iv?	Jupiter B	ST	τ, T, σ, B: [FA, S1], AR, S2, LA	updating T, B
No. **28**	ii?	Jupiter A	ST	T, B: [FA], S1, AR, S2, [LA]	updating B, T
No. **33**	?	Jupiter B'	ST	τ, T, σ, B: S2	updating T, B; net displacements for 12^{yr}
No. **34**	?	Jupiter A'	ST	T,[B?]: FA, S1, AR, S2, LA	subdivision synodic cycle
No. **35**	iii?	Jupiter A'	ST	T, B: FA, S1, AR, S2, [LA]	updating B, T
No. **36**	iv?	Jupiter B	ST	τ, T, σ, B: [FA, S1], AR, S2, LA	updating T, B
No. **45**	?	Saturn?	ST	unclear	period relations

Table 3.3: Planetary texts with procedures and synodic tables (ST) or auxiliary tables (AT).

3.3.1 Algorithms for the zodiacal position (B)

The zodiacal position of a planet (in Akkadian: *qaqqaru*(**ki**) = 'position'; modern symbol: B) is expressed by the zodiacal sign and the number of degrees [0–30] within it (§2.5.3). Consecutive values of B in a synodic table pertain to successive occurrences of the same synodic phenomenon. They are computed either from the synodic arc (§3.3.1.1), or from pushes (§3.3.3). Consecutive values of B in a daily motion table pertain to successive days or (mean) tithis. They are computed from the daily displacement v (§3.3.3).

3.3.1.1 Updating B with the synodic arc (σ)

The basic method for updating the zodiacal position of a synodic phenomenon involves the synodic arc σ,

$$B_i = B_{i-1} + \sigma, \tag{3.1}$$

where i numbers successive occurrences of the same synodic phenomenon. Synodic phenomena whose positions are computed in this manner will be referred to as *primary phenomena*. In some planetary systems all synodic phenomena fall in that category, but more often the positions of some phenomena, referred to as *secondary*, are computed with the method based on pushes (§3.3.3). Two Akkadian terms for the synodic arc are attested: *birīt* SN *ana* SN = 'distance of SN to SN', and occasionally *nisḫu*(**zi**) = 'displacement'.

Synodic arc in systems of type A, B, pseudo-A and C

The algorithms for updating B with the synodic arc form four groups. In type-A systems σ is modeled as a step function of the zodiacal position. For this algorithm and its formulation in the procedure texts cf. §2.6.10. In type-B systems σ is a zigzag function of the event number; cf. §2.6.8. In only a few planetary systems the algorithm for σ differs from these two. In Venus systems A_1/A_2 (§3.5.4) a unique approach is followed whereby σ and τ alternate cyclically between five different constant values. In the Venus systems A_0, and probably C_3 and C_4, σ is constant. For some other systems, e.g. Jupiter system A_6, the algorithm for σ could not be reconstructed.

3.3.1.2 Total synodic arc (Σ)

The planet's total displacement after 1 synodic cycle may differ from σ by a whole number of revolutions, i.e.

$$\Sigma = \sigma + n \cdot 6{,}0°, \tag{3.2}$$

where n depends on the planet. The distinction between synodic arc and total synodic arc is important only for Venus, Mars and the Moon; for the other planets (Mercury, Jupiter and Saturn) $\Sigma = \sigma$ (cf. Table 3.4). The total synodic arc is important for two reasons. First, the synodic time τ is computed from it (§3.3.2.1). Second, it plays a role in the subdivision of the synodic cycle, because the angular pushes, $\delta\Sigma$, of 1 synodic cycle ideally add up to Σ (§3.3.3).

Number of elementary steps contained in Σ (type-A systems)

Recall that in type-A systems σ always contains Z elementary steps δ_j (§2.6.10.3). Since the zodiac contains Π elementary steps (Eq. 2.38), the number of elementary steps contained in Σ equals

$$N = Z + n\Pi, \tag{3.3}$$

which is also the whole number of revolutions of the planet corresponding to $\Pi \cdot \Sigma$ (Eq. 3.7). This result is important mainly for system A of Mars.

3.3.1.3 Solar-distance principle

Synodic phenomena correspond to distinct configurations of the planet and the Sun (§3.1). Consequently the planet's elongation from the Sun assumes nearly or exactly the same value after one full synodic cycle. This so-called solar-distance principle, a term introduced by van der Waerden (1941), manifests itself in various ways. For conjunctions and oppositions there is a strict solar-distance principle, meaning that the Sun travels exactly the same *net* distance as the planet between two occurrences of the phenomenon. The corresponding *total* distance traveled by the Sun may differ from σ by a whole number of revolutions, say m, so that

$$\Sigma_{\odot} = \sigma + m \cdot 6{,}0°. \tag{3.4}$$

The value of m depends on the planet (Table 3.4; for their derivation cf. Appendix B) and is connected to n (§3.3.1.2) by a simple relation. For the inner planets $m = n$, because their motion is linked to that of the Sun. For the outer planets $m = n + 1$, because the Sun performs one additional revolution for every revolution of a synodic phenomenon. For the Moon $m = n - 1 = 0$, since the Moon performs one additional revolution for every lunation. The strict solar-distance principle applies only at conjunctions and oppositions, when the planet's elongation from the Sun is exactly 0° or 180°, respectively. At first and last appearances and stations the planet's elongation from the Sun is not always exactly the same constant, but varies about a mean value depending on the planet and the synodic phenomenon. The positions of these planetary phenomena therefore do not satisfy a strict solar-distance principle, neither empirically nor according to the Babylonian algorithms, as will become apparent in §3.4–3.8. However, van der Waerden (1941) discovered that Eq. (3.4) does underlie the algorithm for the time of all synodic phenomena (§3.3.2.1).

	Moon	Mercury	Venus	Mars	Jupiter	Saturn
m	0	0	1	2	1	1
n	1	0	1	1	0	0

Table 3.4: The number of whole revolutions of the Sun (m) and the Moon or planet (n) per synodic cycle.

3.3.1.4 Period relations and mean synodic arc

Each algorithm for updating B implies a period relation defined by a number period Π and the associated whole number of revolutions of the synodic phenomenon, Z (§2.6.6). For the step function (system A) they are obtained through Eq. (2.33), for the zigzag function (system B) through Eq. (2.16). The status of this period relation is slightly different for both systems. In type-A systems Π corresponds to an *exact* return of both σ and B. In type-B systems σ also returns to exactly the same value after Π synodic events, but in order for B to do the same the mean value $\mu = (m+M)/2$ of the zigzag function for σ must be exactly the same as

$$\overline{\sigma} = \frac{6{,}0}{P} = \frac{6{,}0 \cdot Z}{\Pi}, \tag{3.5}$$

where $P = \Pi/Z$ is the period of the zigzag function for σ. Since μ is a finite sexagesimal number while P and $1/P$ need not be, μ is usually not exactly the same as $\overline{\sigma}$ (in fact this is true only for Saturn system B), but they are always very close, differing at most in the last sexagesimal digit. The small numerical difference between them is one reason why in type-B systems a planet does not necessarily return to *exactly* the same position after Π synodic events.[224]

Fundamental period relation of a planet

Further period relations for B are obtained by considering the planet's total synodic arc Σ. For each synodic cycle the planet covers n extra revolutions in addition to the synodic arc σ (Eq. 3.2). Hence Π synodic cycles result in $n\Pi$ additional

[224] Another reason is that if σ is a zigzag function then the planet's net displacement after Π events, $dB(\Pi)$, is dependent on the initial condition, an effect that does not occur if σ is a step function. For this issue cf. *HAMA*, pp. 382–386 and Neugebauer (1948b).

revolutions with respect to the Z revolutions of the synodic phenomenon, so that the planet performs $Z+n\Pi$ revolutions in total.[225] Secondly, since for each of the Π synodic cycles the Sun performs m additional revolutions (Table 3.4), the corresponding number of years is

$$Y = Z + m\Pi. \tag{3.6}$$

Together, these two relations constitute what Neugebauer (*ACT*, p. 282) calls the 'fundamental period relation' of a planet,

$$\Pi \text{ events } \sim Z + n\Pi \text{ revolutions of the planet } \sim Z + m\Pi \text{ years,} \tag{3.7}$$

where $\sim$ expresses exact correspondence (only nearly exact in type-B systems).

3.3.1.5 Net displacements for various intervals

Apart from the exact return after Π synodic cycles, most period relations imply the existence of intervals comprising $s < \Pi$ whole synodic cycles, corresponding to $y < Y$ whole years, after which the planet returns to nearly the same position. Numerous procedures are concerned with the planet's net displacement along the zodiac, say $dB(s)$, for such intervals. Often these procedures list values of $dB(s)$ for a selection of intervals, whose length is usually expressed as the closest whole number of years y. For a general discussion of this topic and the relation between y and s cf. §2.6.10.4 (type-A systems) and §2.6.8.4 (type-B systems).

3.3.2 *Algorithms for the time (T)*

The times of the planetary phenomena are expressed as a month and a number of 'days' (*ūmu*) [0–30] within it. Through the cumulative efforts of Kugler (1907), Pannekoek (1916) and van der Waerden (1941) it was established that these 'months' and 'days' are not real months and days but artificial units corresponding to the mean synodic month and the mean tithi defined as 1/30 of the mean synodic month (§2.5.2.3). Babylonian astronomers did not make an explicit terminological distinction between the two, for which reason I translate 'day' (*ūmu*) in quotation marks if the context implies that a mean tithi is meant. In order to identify the computed mean times T with real dates of the Babylonian calender, the astronomers interpreted the mean synodic month as an actual month, and the number of mean tithis within it as the day number in that month. By doing so an error of at most 1 day is committed, since a real month has either 29 or 30^d. The resulting inaccuracy is outweighed by a significant advantage, because the use of the mean time frame eliminates the need for computing in advance the actual month lengths, which would require an enormous effort.[226]

Depending on the type of table, T pertains to successive synodic phenomena (synodic tables and template tables), mean tithis (daily motion tables) or, occasionally, real days of the civil calendar (some daily motion tables). The algorithms for T mirror the ones for B. The main method is based on the synodic time (τ), which is the time interval between successive occurrences of the same synodic phenomenon (§3.3.2.1). A second method is based on temporal pushes ($\delta\tau$), which are time intervals between different synodic phenomena within the same synodic cycle (§3.3.3). Finally, in the daily motion tables T is trivially updated from day to day, or from mean tithi to mean tithi.

3.3.2.1 Updating T with the synodic time (τ)

The time of the primary phenomena is updated from one to the next event by adding the synodic time τ:

$$T_i = T_{i-1} + \tau. \tag{3.8}$$

In most computational systems[227] τ is related to σ by the following expression:

$$\tau = \sigma + c_\tau. \tag{3.9}$$

Since τ is expressed in mean tithis, this involves a reinterpretation of σ, which is measured in degrees, as a number of mean tithis.

---

[225] The same result is found by constructing the total displacement for Π cycles as $\sum_{i=1}^{\Pi} \Sigma_i = \sum_{i=1}^{\Pi} (\sigma_i + n \cdot 6{,}0) = (Z + n\Pi) \cdot 6{,}0°$.

[226] For this reason synodic tables containing a column T with times expressed in actual months and days of the Babylonian calender were likely computed after the final date of the table, when the lengths of the preceding months were known.

[227] Exceptions are systems A_1/A_2 for Venus and system B' for Jupiter.

Type-A systems

The status of Eq. (3.9) is somewhat different in type-A and type-B systems. In most type-A systems it is the actual algorithm by which τ is computed. The corresponding templates in the procedure texts are as follows:[228]

[**u₄.meš**] *bi-rit* SN *ana* SN GIŠ.GIŠ.ḪI|*tanaššî*(GIŠ)-*ma* c_τ *itti*(**ki**)-*šu*₂ *ṭeṭeppi*(**tab**)-*ma* *ūmē*(**u₄.meš, me.meš**) *taqabbi*(**e**-*bi*)	[The 'days':] the distance from SN to SN (σ) you compute, and c_τ you add with it, and you predict the 'days' (T_i).	(T.1)
[**u₄.meš**] *bi-rit* SN *ana* SN GIŠ c_τ **ki**-*šu*₂ **tab**-*ma* **ki** *u*₄-*mu* SN **tab**	[The 'days':] the distance from SN to SN (σ) you compute, you add c_τ with it, you add it with the 'day' of SN (T_{i-1}).	(T.2)

T.1 corresponds to Eq. (3.9) but does not actually mention how to update T, unlike T.2, which combines Eqs. (3.8–3.9) in a single procedure.

Type-B systems

In type-B systems τ and σ are both modeled as zigzag functions of the event number. Hence Eq. (3.9) is exactly satisfied if and only if Δ, d, and hence the period P, of both zigzag functions are identical, which is the case for Jupiter system B and Saturn system B. In No. **18** P13 and No. **46** P2' (Jupiter system B) τ is indeed derived from σ in accordance with Eq. (3.9). However, the existence of procedures for updating T with a zigzag function for τ suggests that, at the practical level pertaining to the computation of synodic tables, τ is usually computed from its own zigzag function, and not from σ by invoking Eq. (3.9). That is certain for Saturn system B', because Δ is slightly different for σ and for τ in that system, so that Eq. (3.9) is no longer satisfied. Consequently B and T, each updated with its own zigzag function, do not remain exactly in phase.

Derivation of the algorithm for τ (Eq. 3.9)

Van der Waerden (1941, 1957) demonstrated that Eq. (3.9) can be derived from the following two assumptions:

(i) a strict solar-distance principle, and
(ii) only the mean motion of the Sun is taken into account.

Assumption (i) implies that the distance $\Sigma_\odot$ covered by the Sun during 1 synodic cycle satisfies Eq. (3.4).[229] Assumption (ii) implies that the variation of the Sun's daily displacement along the zodiac is ignored.[230] With these assumptions the computation of τ reduces to the relatively simple problem of establishing the time in which the mean Sun covers the distance $\Sigma_\odot$ as defined by Eq. (3.4). For one revolution around the zodiac the Sun needs 1^{yr}, which can be expressed as $6{,}0^{\bar{\tau}} + e$, where e is the excess of 1^{yr} over $6{,}0^{\bar{\tau}} = 12$ (mean) synodic months, also referred to as the yearly epact,[231]

$$1^{\text{yr}} = 6{,}0^{\bar{\tau}} + e. \tag{3.10}$$

Attested values of e, expressed in mean tithis and, for later purposes, also multiplied by 0;0,10, are compiled in Table 3.5. Most instances of column T in the synodic tables are based on variant 1, which is also explicitly mentioned in several

variant	e	0;0,10·e
1	11;4$^{\bar{\tau}}$	0;1,50,40°
2	11;3,20$^{\bar{\tau}}$	0;1,50,33,20°
3	11$^{\bar{\tau}}$	0;1,50°
4	11;3,55$^{\bar{\tau}}$	0;1,50,39,10°

Table 3.5: Attested values of the yearly epact e.

procedure texts.[232] The other variants are attested only in procedure texts.[233] Eq. (3.10) implies that it takes the Sun on

[228] T.1: No. **13** P3' (Mars system A); No. **42** P1.b, P2.b (Mercury systems A_1); perhaps also No. **17** P1.b (Jupiter system A) and No. **28** P1.b. T.2: No. **1** P1.b, d (Mercury system A_1). A similar instruction is partly preserved in No. **35** P1.b (Jupiter system A').

[229] This was already noticed by Kugler (*SSB* I, p. 193) in connection with system A_1 of Mercury.

[230] The mean-motion approximation is not bad for the Sun, since its velocity varies only slightly as a function of the zodiacal position ('zodiacal anomaly').

[231] As mentioned in *ACT* and *HAMA*, the Sun's mean displacement per mean tithi is therefore $\bar{v}_\odot = 6{,}0/(6{,}0+e)$, but this quantity is not needed in the derivations, and in general it cannot be represented as a finite sexagesimal number.

[232] No. **13** R7',8',12',16',18',21'; No. **18** Ri6',7'; No. **46** Oi7',8'.

[233] Variant 2: No. **18** Ri10',11',15',19'; No. **29** R3'. Variant 3 is mentioned only in the form 0;0,10·e = 0;1,50 in No. **18** Ri14'. The context implies that it is a rounded value of variant 2. Variant 4 may be mentioned in No. **97** P2', which belongs to lunar system B.

average $(6{,}0+e)/6{,}0 = 1+0;0{,}10\cdot e$ mean tithis to travel 1° along the zodiac, from which the time needed for traveling the total distance $\Sigma_{\odot}$ (Eq. 3.4) is found as

$$\tau = \Sigma_{\odot}\cdot(1+0;0{,}10\cdot e) = m\cdot 6{,}0\cdot(1+0;0{,}10\cdot e)+\sigma+0;0{,}10\cdot e\,\sigma. \tag{3.11}$$

Here m is the number of additional revolutions of the Sun around the zodiac per synodic cycle (Table 3.4), $\sigma+0;0{,}10\cdot e\,\sigma$ is the time in which the mean Sun covers the distance σ, and $m\cdot 6{,}0\cdot(1+0;0{,}10e)$ is the time needed for the additional m full revolutions. As mentioned, this result is an approximation in that the variation of the solar velocity along the zodiac is neglected. The Babylonian astronomers made a further approximation by exploiting that $0;0{,}10\cdot e$ is small compared to 1. This means that only a small error is committed if one replaces the actual synodic arc σ in the final term by its mean value $\overline{\sigma}$, resulting in

$$\tau = \sigma+m\cdot(6{,}0+e)+0;0{,}10\cdot e\overline{\sigma}. \tag{3.12}$$

We therefore arrive at the following expression for c_τ:

$$c_\tau = m\cdot(6{,}0+e)+0;0{,}10\cdot e\overline{\sigma}. \tag{3.13}$$

Most of the attested values of c_τ can be explained by this expression if one inserts for e the usual value $11;4^{\tau}$ and for $\overline{\sigma}$ the value implied by the algorithm for σ, or some approximation of it. Note that for Venus, Mars, Jupiter and Saturn, c_τ is dominated by the term $m\cdot 6{,}0$, which amounts to $12m$ mean synodic months, or m (non-intercalary) years. For these planets the procedure texts usually do not give c_τ itself, but the excess of c_τ over this whole number of years, i.e. $c_\tau - m\cdot 6{,}0 = e\cdot(m+0;0{,}10\cdot\overline{\sigma})$. The time of the synodic phenomenon, T_i, is then obtained by adding $c_\tau - m\cdot 6{,}0$ to the number of mean tithis of T_{i-1}, and updating the month and year number by adding $12m$ mean synodic months.

Computing the synodic time from the total synodic arc

In some procedure texts the synodic time τ is not computed from σ (Eq. 3.9), but from the total synodic arc Σ (Eq. 3.2), i.e.

$$\tau = \Sigma + c'_\tau, \tag{3.14}$$

which involves a modified constant

$$c'_\tau = (m-n)\cdot(6{,}0+e)+0;0{,}10\cdot e\overline{\Sigma}. \tag{3.15}$$

The contribution $\Sigma+0;0{,}10\cdot e\overline{\Sigma}$ can be interpreted as the approximate time in which the mean Sun covers the distance Σ, and $(m-n)\cdot(6{,}0+e)$ as the time in which it covers the $m-n$ full revolutions in excess of Σ. The latter amounts to a distance $(m-n)\cdot 6{,}0°$, which is the total change of the planet's elongation from the Sun over 1 synodic cycle. Unlike Eq. (3.12), this representation of τ can be easily adapted to the computation of the time $\delta\tau$ in which the planet moves across subintervals of Σ known as angular pushes ($\delta\Sigma$), as will be discussed in the next section. The corresponding value of $c'_{\delta\tau}$ can be obtained by replacing $\overline{\Sigma}$ in Eq. (3.15) by $\overline{\delta\Sigma}$, and $(m-n)\cdot(6{,}0+e)$ by a certain fraction of it. This adaptation of the algorithm for τ is attested in the remarkable procedure text No. **13**, which belongs to system A of Mars (§3.6.2.3). As argued by van der Waerden (1957), this also suggests that the Babylonian astronomers interpreted $(m-n)\cdot(6{,}0+e)$ as the time that it takes for the planet's elongation to change by the amount $(m-n)\cdot 6{,}0°$.

3.3.3 Subdivision of the synodic cycle

Numerous procedure texts are concerned with the question of how the zodiacal positions and times of different synodic phenomena are distributed within the synodic cycle, and how the planet moves in between the synodic phenomena. In order to address these issues the Babylonian astronomers formulated algorithms involving pushes and daily displacements, which are explored in the present section.

3.3.3.1 Pushes and daily displacements

Push is the modern term for the angular or temporal interval between different synodic phenomena within the synodic cycle.[234] If we denote the involved phenomena as SN_1 and SN_2, then the pushes between them can be used for computing the coordinates of SN_2:

$$\begin{cases} B_i(SN_2) = B_i(SN_1)+\delta\Sigma\,(SN_1 \text{ to } SN_2),\\ T_i(SN_2) = T_i(SN_1)+\delta\tau\,(SN_1 \text{ to } SN_2). \end{cases} \tag{3.16}$$

[234] Derived from the originally German term Stoß, which was introduced by Kugler.

For obvious reasons SN_1 is also referred to as the parent phenomenon, and SN_2 as the satellite phenomenon. The symbols $\delta\Sigma$ and $\delta\tau$ which I employ throughout the present study reflect that the angular pushes are subintervals of the total synodic arc Σ, and the temporal pushes of the (total) synodic time τ. Since SN_1 always precedes SN_2, $\delta\tau$ is always additive, but $\delta\Sigma$ can be additive or subtractive, depending on whether the motion from SN_1 to SN_2 is prograde or retrograde. The Akkadian terms for the pushes are constructed from the general word for distance (*birītu*) or time (*ūmū* = 'days') and the names of the involved phenomena (Table 3.6). Pushes may also be divided into smaller intervals, for which I employ the same symbols $\delta\Sigma$ and $\delta\tau$. Usually the subintervals of a temporal push are multiples of $10^{\overline{\tau}}$ or $1^{\overline{m}}$. In the procedure texts $\delta\tau$ is usually expressed as x 'months' (= mean synodic months) and/or y 'days' (= mean tithis), so that the value in mean tithis equals

$$\delta\tau = 30 \cdot x + y. \tag{3.17}$$

Although pushes may be construed between arbitrary pairs of synodic phenomena, the procedure texts are usually concerned with pushes for which SN_1 is a primary phenomenon (i.e. its positions are updated with the synodic arc), but SN_2 is not. In that case the procedures for the subdivision of the synodic cycle are complementary to the procedures for updating the coordinates of the primary phenomena. Together they constitute a complete set of algorithms for the coordinates of the synodic phenomena. However, quite a few procedures for Mars and Jupiter (e.g. system A) are concerned with pushes between primary phenomena (cf. §3.3.3.3).

$\delta\Sigma$	astronomical meaning	angular push = displacement along the zodiac
	terminology	*bi-rit ša₂* SN_1 *ana* SN_2 = 'distance for SN_1 to SN_2'
	unit	degrees [°]
$\delta\tau$	astronomical meaning	temporal push = corresponding time interval
	terminology	**u₄.meš|me.meš** *ša₂* SN_1 *ana* SN_2 = "days' for SN_1 to SN_2'
	unit	mean synodic months [$\overline{m}$] and mean tithis [$\overline{\tau}$]
v	astronomical meaning	'daily' displacement along the zodiac
	terminology	*nisḫu*(**zi**) *ša₂ ūmi*(**me**) = 'displacement per 'day''
	unit	degrees per mean tithi [°/$\overline{\tau}$], sometimes per day [°/d]

Table 3.6: Pushes and daily displacements: terminology and units.

Often the procedure texts also provide or imply values for the planet's 'daily' displacement along the zodiac, v, for each push or subinterval, by which zodiacal positions can be computed for successive 'days' (usually to be interpreted as mean tithis). The abundance of procedures for the subdivision of the synodic cycle and the considerable number of extant daily motion tables indicate the importance of this topic for the Babylonian astronomers. Usually v is assumed to be constant within the push or subinterval, as illustrated in Fig. 3.4, but there are also more sophisticated models in which v changes linearly or even quadratically from 'day' to 'day'.

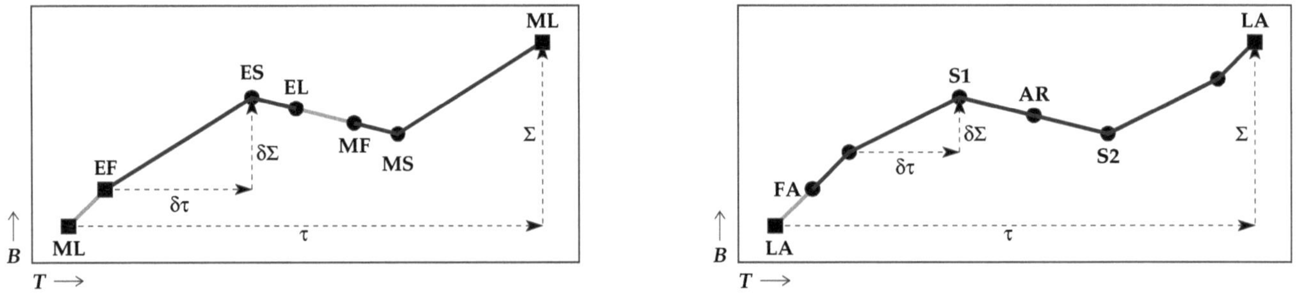

Fig. 3.4: Schematic representation of the subdivision of the synodic cycle for an inner (left) and outer planet (right) involving constant values of the 'daily' displacement v. Squares are primary phenomena, circles secondary phenomena.

Schemes for the subdivision of the synodic cycle in the tabular texts

Some schemes for the subdivision of the synodic cycle are attested only in the tabular texts and not in procedure texts. Most of these are not discussed in the present volume; they will be dealt with in Volume II. The most elaborate schemes for the subdivision of the synodic cycle are contained in the daily motion tables, some of which feature intervals in which v changes linearly with time. Those schemes for the subdivision that are implied by the synodic tables are always fully characterised by $\delta\Sigma$ and $\delta\tau$. The corresponding values of v that may be computed from $\delta\Sigma$ and $\delta\tau$ are rarely finite or simple sexagesimal numbers, because $\delta\tau$ usually deviates from the schematic values mentioned in the procedure texts. This is of no concern, since these hypothetical values of v were not used for computing daily positions.

Typology of the procedures for the subdivision of the synodic cycle

The procedures for the subdivision of the synodic cycle exhibit a large variety, but they can be roughly divided into the following two groups:

(i) Subdivisions of the synodic cycle for successive pushes or subintervals, usually for all constituent pushes of 1 cycle. The procedures may involve several different templates by which values of v, $\delta\tau$ and/or $\delta\Sigma$ are provided for each push or subinterval (cf. below).
(ii) Lists of control values of v, $\delta\tau$ or $\delta\Sigma$ for a single push. Usually they contain 12 values, one for each zodiacal sign, which form a zigzag function of the zodiacal position of the parent phenomenon, but more irregular dependencies on the zodiacal position are also attested.

Along with procedure texts there are also auxiliary tables in which pushes are tabulated as a function of the zodiacal position of the parent phenomenon, usually one value for every zodiacal sign, sometimes at intervals of 1° (Mars system A). In the simplest schemes of type (i), $\delta\Sigma$, $\delta\tau$ and v are independent of the planet's zodiacal position, and they do not vary from cycle to cycle. Such schemes are in fact rare, but several exist for Venus (cf. §3.5.5, §3.5.6). It is much more common that $\delta\Sigma$ and v exhibit some type of variation. In type-A systems $\delta\Sigma$ is often modeled as a step function of the zodiacal position, sometimes in a manner (including transition rules) fully consistent with the step function for σ (§3.3.3.3).

Given the large variety of schemes for the subdivision of the synodic cycle it is not always clear to which computational system a given scheme belongs, especially if there is no application of it in the tabular texts. The schemes within a computational system are labeled S_1, S_2, etc.; if necessary the computational system is attached to this label, e.g. A.S_1 for scheme S_1 of system A (e.g. for Jupiter), or X.S_1 for a scheme S_1 that is not associated with a known computational system. Procedures for the subdivision of the synodic cycle do not appear to exist for type-B systems, probably because in these systems there are only primary phenomena whose zodiacal positions are computed from zigzag functions for σ. Hence there is little need for such procedures, except perhaps for determining the subdivision of the initial cycle for a synodic table.

3.3.3.2 Subdivisions involving constant values of the daily displacement

Having laid out the basic properties of pushes and daily displacements we are in a position to explore the various ways in which the subdivision of the synodic cycle is formulated in the procedure texts. We shall first consider the simplest case where v is constant in each push or subinterval. The most complete formulation of this algorithm involves $\delta\tau$, $\delta\Sigma$ as well as v. The procedure texts exhibit the following templates:[235]

ina $\delta\tau$ **me** *ša*$_2$ **me** v **zi** $\delta\tau$ GAM v *tallak*(**du**)-*ma* $\delta\Sigma$	In $\delta\tau$ 'days' it moves v per 'day'. You multiply $\delta\tau$ by v, it is $\delta\Sigma$.	(PUSH.A.1)
ultu(**ta**) $\delta\tau_0$ **me** *ušallimu*(**gi**) **en** $\delta\tau$ **me** v v **a.ra**$_2$ $\delta\tau$ **me du**-*ma* $\delta\Sigma$	After completing $\delta\tau_0$ 'days', until $\delta\tau$ 'days' v. You multiply v by $\delta\tau$ 'days', it is $\delta\Sigma$.	(PUSH.A.2)

The rare template PUSH.A.2, attested only in two procedures for Jupiter (No. **21** P1, No. **22** P1) concerned with scheme X.S_1 (§3.7.10.1), was not fully understood until the present study. It begins with a reference to the previous interval ($\delta\tau_0$), which the planet is said to have 'completed' (*ušallim*). After that the angular push $\delta\Sigma$ is computed as the product of $\delta\tau$ and v. The algorithm for updating the coordinates implied by these templates can be represented as follows:

$$\begin{cases} B_i = B_{i-1} \pm v &= B(\mathrm{SN}_1) \pm i \cdot v \\ T_i = T_{i-1} + 1^{\overline{\tau}} &= T(\mathrm{SN}_1) + i, \end{cases} \tag{3.18}$$

where $i = 0, 1, ..., \delta\tau$ labels mean tithis, sometimes days, between SN_1 and SN_2.[236] For normal, forward motion v is added; for retrograde motion it is subtracted. The angular push $\delta\Sigma$ satisfies the relation

$$\delta\Sigma = v \cdot \delta\tau. \tag{3.19}$$

This relation is satisfied in most, but not all schemes that are mentioned in the procedure texts. If it is violated then the planet, moving from SN_1 to SN_2 at v degrees per mean tithi, cannot reach the exact position of SN_2 defined by Eq. (3.16).[237]

[235] Examples of PUSH.A.1: No. **19** P5' (Jupiter scheme A.S_2?); No. **27** P2',P3' (Jupiter scheme A.S_1); PUSH.A.2: No. **21** P1; No. **22** P1 (Jupiter scheme X.S_1). In these examples all preserved temporal pushes are expressed in 'days'.

[236] I assume for the moment that $\delta\tau$ is a whole number. If $\delta\tau$ is a subinterval of a push then SN_1 and/or SN_2 are intermediate points between synodic phenomena.

[237] If the planet's position at SN_2 is computed from Eq. (3.16) then no errors will accumulate as a result of this anomaly, but the displacement during the last mean tithi before SN_2 will differ from v.

Templates for pushes involving $\delta\tau$ and v

We may now consider the other templates concerning pushes. Many procedures provide only $\delta\tau$ and v. The following templates are mainly used for prograde pushes:[238]

<table>
<tr><td>[ultu(ta)|arki(egir, ar$_2$) SN_1] x iti y me ša$_2$ me v inassaḫ(zi)[-ma SN_2]</td><td>[From|After SN_1] x 'months' y 'days' it moves v per 'day' [(then) SN_2].</td><td>(PUSH.B.1)</td></tr>
<tr><td>x iti y me ina igi-at$_2$ SN_2 ša$_2$ me v zi[-ma SN_2]</td><td>x 'months' y 'days' before SN_2 it moves v per 'day' [(then) SN_2].</td><td>(PUSH.B.2)</td></tr>
<tr><td>[ta|egir|ar$_2$ SN_1] x iti y me ša$_2$ me v nisiḫ(zi)-šu$_2$</td><td>[From|After SN_1] x 'months' y 'days' its displacement is v per 'day'.</td><td>(PUSH.B.3)</td></tr>
</table>

PUSH.B.1 also appears as an integral part of PUSH.A.1. The following templates are used exclusively for retrograde pushes:[239]

<table>
<tr><td>[ta|egir|ar$_2$ SN_1] x iti y me ša$_2$ me v zi-ma itâr(gur)-ma SN_2</td><td>[From|After SN_1] x 'months' y 'days' it moves backwards v per 'day', then SN_2.</td><td>(PUSH.B.4)</td></tr>
<tr><td>[ta|egir|ar$_2$ SN_1] x iti y me ša$_2$ me v gur-ma SN_2</td><td>[From|After SN_1] x 'months' y 'days' it turns back v per 'day', then SN_2.</td><td>(PUSH.B.5)</td></tr>
<tr><td>[ta|egir|ar$_2$ SN_1] x iti y me ša$_2$ me v ana ar$_2$-ki-šu$_2$ inaḫḫis(la$_2$)[-ma SN_2]</td><td>[From|After SN_1] x 'months' y 'days' it recedes backwards v per 'day', [then SN_2].</td><td>(PUSH.B.6)</td></tr>
</table>

The corresponding value of $\delta\Sigma$ can be trivially obtained from Eq. (3.19).

Templates for pushes involving $\delta\tau$ and $\delta\Sigma$

The following templates provide only $\delta\tau$ and $\delta\Sigma$:[240]

<table>
<tr><td>[ta SN_1] ina $\delta\tau$ me $\delta\Sigma$ qaqqaru(ki) illak(du)[-ma SN_2]</td><td>[From SN_1] in $\delta\tau$ 'days' the position proceeds $\delta\Sigma$, [then SN_2].</td><td>(PUSH.C.1)</td></tr>
<tr><td>[ta|ar$_2$ SN_1] ina $\delta\tau$ me $\delta\Sigma$ ki [ana ar$_2$-šu$_2$] la$_2$ [SN_2]</td><td>[From SN_1] in $\delta\tau$ 'days' the position recedes [backwards] $\delta\Sigma$, [(then) SN_2].</td><td>(PUSH.C.2)</td></tr>
</table>

Grammatical considerations suggest that *qaqqaru*(**ki**), 'position', is the subject, and this has been assumed in all translations.[241] Even though v is not provided one may use Eq. (3.19) to derive v from $\delta\Sigma$ and $\delta\tau$. If v turns out to be a reasonably short sexagesimal number, this is probably a deliberate feature, which suggests that the computation of daily positions is a purpose of that scheme.

Templates for pushes involving only $\delta\Sigma$

The following templates provide only $\delta\Sigma$:[242]

<table>
<tr><td>ta SN_1 ana SN_2 $\delta\Sigma$ ki du[-ma SN_2]</td><td>From SN_1 to SN_2 the position proceeds $\delta\Sigma$, [then SN_2].</td><td>(PUSH.D.1)</td></tr>
<tr><td>ta SN_1 ana SN_2 $\delta\Sigma$ ki gur[-ma SN_2]</td><td>From SN_1 to SN_2 the position turns back $\delta\Sigma$, [then SN_2].</td><td>(PUSH.D.2)</td></tr>
<tr><td>ta SN_1 ana SN_2 $\delta\Sigma$ ki ana ar$_2$-šu$_2$ inaḫḫis(la$_2$-is)-ma SN_2</td><td>From SN_1 to SN_2 the position recedes backwards $\delta\Sigma$, then SN_2.</td><td>(PUSH.D.3)</td></tr>
</table>

Since the temporal pushes $\delta\tau$ are essential for computing the time of SN_2 these templates are useful only in conjunction with procedures for $\delta\tau$ and/or v. This is indeed the case in the composite procedure No. **18** P9a.

Templates for pushes involving only $\delta\tau$

Some procedures provide only $\delta\tau$, using the following template:[243]

---

[238] Examples of PUSH.B.1: No. **18** P9a; No. **32** P3–P6; No. **41** P2,P3 (Saturn system A); No. **46** P4' (Jupiter system A'); PUSH.B.2: No. **13** P10'.b (Mars system A); No. **18** P9a; No. **32** P4; No. **41** P2,P3; PUSH.B.3: No. **41** P2; No. **42** P4 (Saturn system A). PUSH.B.3 is occasionally used for retrograde pushes: No. **21** O4; No. **22** O4; No. **23** O13'.

[239] Examples of PUSH.B.4: No. **18** P9a,P18; No. **32** P4–P6; No. **46** P4'; PUSH.B.5: No. **41** P2; PUSH.B.6: No. **13** P10'.b.

[240] Examples: PUSH.C.1: No. **9** P1 (Venus system C_3), No. **18** P11 (Jupiter scheme $A.S_5$), No. **19** P2 (Jupiter scheme $A.S_1$?), P5' (Jupiter, variant of scheme $A.S_1$), No. **39** P3'; PUSH.C.2: No. **18** P11.a,b (Jupiter scheme $A.S_5$), No. **27** P2' (Jupiter scheme $A.S_1$).

[241] In *ACT*, Neugebauer ignored the word *qaqqaru* when translating these phrases, assuming that the planet is the implicit subject.

[242] Examples: PUSH.D.1: No. **18** P2.b, P9a.b, P9a.d, P23.a, P24, P31; PUSH.D.2: No. **18** P24, P31; D3: No. **18** P2.b, **23** P3'.

[243] No. **18** P23.b, P30 (Jupiter system A?).

ta SN_1 *ana* SN_2 [*x* **iti**] *y* **u₄.meš|me tab**	From SN_1 to SN_2 you add [*x* 'months'] *y* 'days'.	(PUSH.E)

This template is always used in conjunction with other procedures that provide the corresponding angular pushes.[244]

3.3.3.3 Angular pushes between primary phenomena in type-A systems

Some procedures for Mars system A and Jupiter system A are concerned with angular pushes between primary phenomena. Since the positions of these phenomena are updated with σ, the angular pushes between them are fully determined for all cycles once the initial positions are known. These procedures were therefore probably not used for computing column *B*, but for establishing the subdivision of the initial cycle, or verifying previously computed positions. It is not a priori obvious that the angular pushes provided in these procedures are fully consistent with the positions obtained with the step function for σ, but this can be shown to be the case if each angular push can be defined in terms of elementary steps. A necessary condition is that one and the same step function for σ is used for all primary phenomena. Full consistency is then achieved by dividing the total synodic arc Σ, which contains $N = Z + n\Pi$ elementary steps of size δ_j (§3.3.1.2), over the angular pushes $m = 1,...,M$ according to

$$\sum_{m=1}^{M} \delta N_m = N. \tag{3.20}$$

This implies $\Sigma = \delta\Sigma_1 + \cdots + \delta\Sigma_M$. Furthermore, completely analogous to σ we can define preliminary values of the angular pushes as

$$\delta\Sigma_{mj} = N_m \delta_j, \tag{3.21}$$

i.e. $\delta\Sigma_{mj}$ is the preliminary value of $\delta\Sigma_m$ in zone j, from which the final value $\delta\Sigma_m$ is obtained by applying the usual transition rules (§2.6.10). For a given push m, the ratio of $\delta\Sigma_{mj}$ in adjacent zones becomes

$$\frac{\delta\Sigma_{m,j+1}}{\delta\Sigma_{mj}} = \frac{\sigma_{j+1}}{\sigma_j}. \tag{3.22}$$

If the $\delta\Sigma_{mj}$ are known and they satisfy this equation, that is a strong indication that they are modeled as fully fledged step functions, even if the procedure text may not mention transition rules. In practice full consistency between the angular pushes and the synodic arc is rarely achieved, one example being Mars system A. In most computational systems the necessary conditions are not satisfied, for instance because different step functions for σ are used for different primary phenomena.

3.3.3.4 Interpolation schemes for v

Some sophisticated schemes for the subdivision of the synodic cycle employ linear interpolation for computing the daily displacement v at successive days or mean tithis between SN_1 and SN_2. The earliest known scheme of this kind was discovered by Kugler in a daily motion table for Jupiter.[245] The same approach is attested in a daily motion table for Mercury (*ACT* 310), and in the procedure text No. **9**, first published by Britton & Walker (1991), which belongs to system C_3 for Venus. In the present study, another example has been discovered in No. **38** P3', a procedure for Jupiter. It therefore appears that this approach is more common than previously assumed. The algorithms can be represented as

$$v_i = v(\text{SN}_1) \pm i \cdot d, \tag{3.23}$$

where $i = 1,...,\delta\tau$ labels the days or mean tithis between SN_1 and SN_2, and $v(\text{SN}_1)$ is the value at SN_1. The 'daily' difference d is given by

$$d = \frac{v(\text{SN}_2) - v(\text{SN}_1)}{\delta\tau}. \tag{3.24}$$

For this algorithm to be of practical use the values of $v(\text{SN}_1)$, $v(\text{SN}_2)$ and $\delta\tau$ must be chosen in such a way that d is a convenient sexagesimal number. The usual relation $\delta\Sigma = v \cdot \delta\tau$ (Eq. 3.19) remains valid, if one replaces v by the mean value $\bar{v} = [v(\text{SN}_1) + v(\text{SN}_2)]/2$. It is obviously more difficult to implement the interpolation algorithm if $\delta\Sigma$ is not constant, and there do not appear to be any examples of that.

---

[244] Cf. No. **18** P24,P31.

[245] *ACT* 654. Cf. Kugler *SSB* 2, pp. 525–530; *HAMA*, pp. 413–418. For a comprehensive study of the daily motion tables for Jupiter cf. Huber (1957).

3.3.4 Distance to the ecliptic

While distance to the ecliptic plays a central role in the lunar systems, it is rarely addressed in the planetary texts, and was largely unknown to Neugebauer at the time when *ACT* and *HAMA* appeared. Since then, several procedures for Jupiter and Saturn concerned with distance to the ecliptic were discovered by Steele (2005a), to which the present study adds a further procedure. For this topic cf. §3.7.10.2 for Jupiter and No. **47** P2' for Saturn.

3.4 Mercury

Mercury is represented by three computational systems, A_1, A_2 and A_3, and several procedures whose computational system has not been identified. The Mercury texts for systems A_2 and A_3 are among the oldest of the corpus of mathematical astronomy. In systems A_1 and A_2 there are two primary phenomena, each serving as parent for another phenomenon. In system A_1 the primary phenomena are EF and MF and their satellites are EL and ML, respectively; in system A_2 the primary phenomena are EL and ML and their satellites are MF and EF. System A_3 is based on a unique strategy whereby the zodiacal positions of two successive events of a synodic phenomenon are computed from the synodic arc, while the position of every third event is obtained from the net displacement for 3 cycles. Stations are attested only in system A_3. To the previously known procedure texts this study adds four, perhaps five newly discovered fragments: No. **2** (A_1, A_2), No. **3** (A_1?), No. **4** (A_2), No. **43** (unknown system) and perhaps No. **49** (Mercury?, unknown system).

3.4.1 Synodic cycle

Mercury's synodic cycle exhibits unique features resulting from its proximity to the Sun and its rapid motion (the latter may explain Mercury's Akkadian name *Šiḫṭu*, 'Attack, Jump'). It performs one revolution around the Sun in 87.969^d, which results in a mean synodic time $\overline{\tau} \approx 115.877^d \approx 1,55;52,37^d \approx 1,57;43,8^{\overline{\tau}}$, and a mean total synodic arc $\overline{\Sigma} \approx 7,54;12,43°$.[246] The time between MF and MS and between ES and EL can be very short, which may explain why the stations were usually ignored (but not in system A_3). Moreover, some synodic phenomena are not observable at all, because Mercury does not always reach a sufficient altitude if its orbit is flat with respect to the horizon, even at maximum elongation from the Sun. For the morning phenomena this happens in April, for the evening phenomena in September. The effect on the synodic cycle is illustrated in Fig. 3.5. With increasing flatness the dates of EL, MF, ML and EF shift away from the conjunction, until eventually Mercury remains invisible for the entire synodic cycle (not shown). Since these omissions remain unnoticed in the algorithms for T and B, they are covered by separate rules, which are preserved in the procedure texts for system A_1.

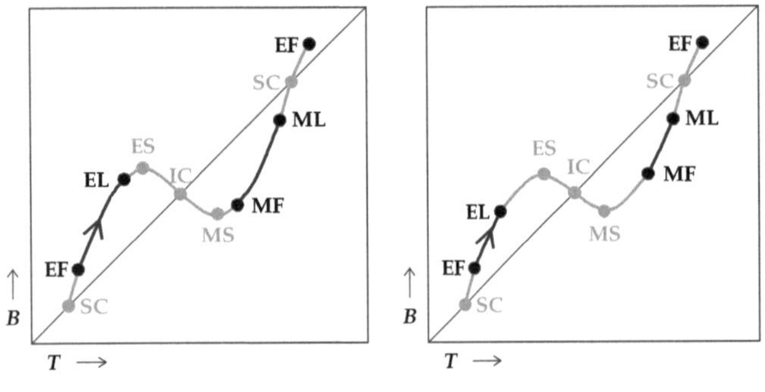

Fig. 3.5: Mercury: omitted phenomena. **Left**: stations no longer observable, EL to MF still retrograde. **Right**: EL to MF is prograde.

Composition of the tablets

All tablets with Mercury procedures are collections of procedures. No. **42**, No. **43**, perhaps also No. **2**, contain procedures for (at least) two planets, Mercury being the first one. In No. **42** Mercury is followed by Saturn, in No. **43** by Mars, in No. **2** perhaps by Venus. The other tablets deal exclusively with Mercury: No. **1** accompanies a synodic table; No. **2** is a collection of procedures. Finally, No. **6** contains at least one procedure for Mercury, but the other procedures are badly damaged and therefore difficult to interpret.

[246] For a derivation of these parameters cf. Appendix B (p. 525).

3.4.2 System A_1

System A_1, discovered by Kugler,[247] is represented by three procedure texts, nine synodic tables[248] and five auxiliary tables concerned with pushes.[249] Two procedure texts were edited previously by Neugebauer; the newly discovered fragment No. **2** is published for the first time. Two of the tablets contain a collection of procedures including one or more belonging to another system of Mercury or another planet: in the case of No. **2** a procedure for system A_2 and one for an inner planet, in the case of No. **42** procedures for Saturn. The procedures of No. **1** accompany a synodic table. None of the tablets preserves a date in the colophon. The initial dates of the predictions in the synodic tables cover the years SE 118–224. If this provides a rough indication of the time when system A_1 was in use, then system A_1 is more recent than A_2 and A_3.

MF: updating B with synodic arc	No. **1** P1.a, No. **42** P1.a
MF: updating T with synodic time	No. **1** P1.b, No. **42** P1.b
EF: updating B with synodic arc	No. **1** P1.c, No. **42** P2.a
EF: updating T with synodic time	No. **1** P1.d, No. **42** P2.b
rules for omitted phenomena	No. **42** P1.c, P2.c
list of angular pushes, MF to ML	No. **1** P1.e
list of temporal pushes, MF to ML	No. **1** P1.g
list of angular pushes, EF to EL	No. **1** P1.f
list of temporal pushes, EF to EL	No. **1** P1.h, No. **2** P1'?

Table 3.7a: Mercury system A_1: procedures.

	j	b_j	α_j	σ_j	r_j	δ_j	v_j
MF	1	1° Leo	2,45°	1,46°	1;20	0;7,30°	22,0
	2	16° Cap	2,14°	2,21;20°	0;40	0;10°	13,24
	3	30° Tau	1,1°	1,34;13,20°	1;7,30	0;6,40°	9,9
EF	1	6° Cnc	1,50°	2,40°	0;40	0;20°	5,30
	2	26° Lib	2,14°	1,46;40°	0;54	0;13,20°	10,3
	3	10° Psc	1,56°	1,36°	1;40	0;12°	9,40

	Π	Z	P	$\overline{\sigma}$
MF	44,33	14,8	3;9,7,38,...	1,54;12,31,...°
EF	25,13	8,0	3;9,7,30	1,54;12,36,...°

Table 3.7b: Mercury system A_1: parameters of the step functions for σ for MF and EF.

3.4.2.1 Updating B with the synodic arc

The synodic tables and the procedure texts imply that EF and MF are primary phenomena whose zodiacal positions are updated with step functions for σ based on three zones (Table 3.7b). Different step functions are used for EF and MF.[250] The listed procedures with rules for updating B are based on different templates: No. **1** P1.a,c follow STEP.A.1, and No. **42** P1.a,P2.a follow STEP.B (§2.6.10). The former is based on transition coefficients r_j, the latter on c_j. Due to Mercury's rapid motion, $\sigma_j > \alpha_j$ in 2 of the 3 zones, which causes σ to deviate considerably from the preliminary values σ_j, as illustrated in Fig. 3.6.

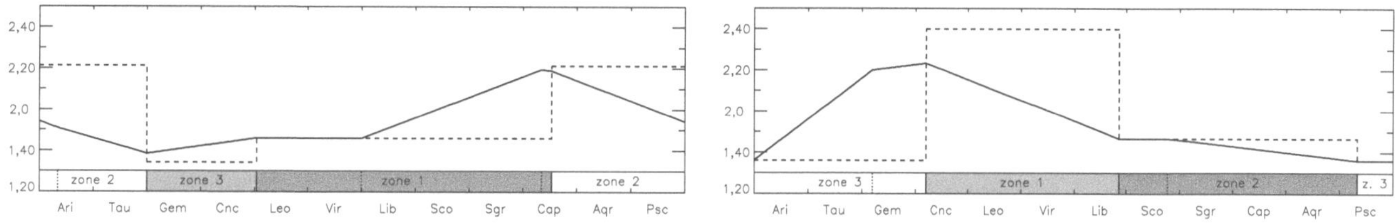

Fig. 3.6: Mercury system A_1: synodic arc σ for MF (left) and EF (right) as a function of B_{i-1}. The dashed lines represent the preliminary values σ_j.

3.4.2.2 Updating T with the synodic time

From the synodic tables it is inferred that the times of EF and MF are updated with the synodic time τ in accordance with Eq. (3.9). Corresponding instructions are contained in No. **42** P1.b, P2.b (template T.1; §3.3.2.1) and No. **1** P1.b,d (template T.2). Each of these tablets employs slightly different values of c_τ, namely

$$c_\tau = \begin{cases} 3;30,39,4,20^{\overline{\tau}} & \text{(No. \textbf{42})} \\ 3;30,39^{\overline{\tau}} & \text{(No. \textbf{1})}. \end{cases} \tag{3.25}$$

[247] Kugler *SSB* I, pp. 178–186, 197–200. For an account of the efforts to reconstruct system A_1 cf. Neugebauer (1951).

[248] *ACT* 300–305, 303a,b, BM 48147 (Aaboe 1977).

[249] *ACT* 800a–e.

[250] The step function for the synodic arc of MF was first reconstructed by Kugler (*SSB* I, p. 192).

Both are essentially reproduced by Eq. (3.13) with $m = 0$, if one adopts for the yearly epact e the usual value 11;4$^{\bar{\tau}}$, and for $\overline{\sigma}$ either value from Table 3.7b rounded to 1,54;12,30°. Most synodic tables appear to employ the second, rounded value of $c_{\bar{\tau}}$. This can also be verified with the help of Eq. (3.13) by deriving the mean synodic arc $\overline{\sigma}$ underlying $c_{\bar{\tau}}$, assuming e = 11;4$^{\bar{\tau}}$, which results in $\overline{\sigma}$ = 1,54;12,30,...° (No. **42**) and 1,54;12,28,...° (No. **1**).

3.4.2.3 Subdivision of the synodic cycle

The coordinates of EL and ML are computed by applying pushes to the parent phenomena EF and MF, respectively (Fig. 3.7). The stations are ignored, i.e. the retrograde arcs from ES to EL and from MF to MS are subsumed in the prograde pushes from EF to EL and from MF to ML. Apart from the procedures listed above, five auxiliary tables (*ACT* 800a–e) provide control values of the pushes as a function of the zodiacal position of the parent phenomenon at intervals of 1° (!). From the procedures and the auxiliary tables it is inferred that the pushes are computed by means of linear interpolation with respect to 12 control values pertaining to 15° of the zodiacal signs (Table 3.8).

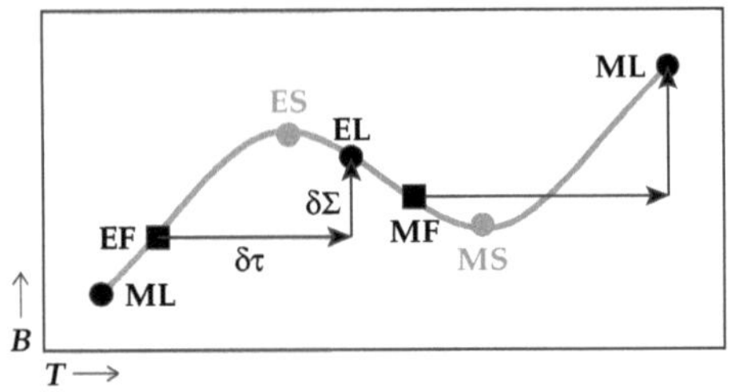

Fig. 3.7: Mercury system A_1: subdivision of the synodic cycle.

Control values for the pushes from MF to ML are contained in No. **1** P1.e,g, and in the auxiliary tables *ACT* 800a,b and perhaps 800e; for the pushes from EF to EL in No. **1** P1.f,h, No. **2** P1', and in the auxiliary tables *ACT* 800c,d, perhaps also 800e. No. **1** P1.e,f originally contained one control value of $\delta\Sigma$ for each zodiacal sign, but only some are preserved. No. **1** P1.g,h preserve $\delta\tau$ for Ari. Other control values (in italics) were reconstructed by Neugebauer from the synodic tables; some are only approximately known.[251] The newly discovered No. **2** P1' contains values of $\delta\tau$ (EF to EL) for Cap, Aqr and Psc that confirm Neugebauer's reconstruction. Perhaps No. **2** P4'–P6' also preserves elements of a scheme for the subdivision of the synodic cycle connected with system A_1 (cf. the commentary).

	MF to ML				EF to EL			
B	$\delta\Sigma$	*c*	$\delta\tau$	*c*	$\delta\Sigma$	*c*	$\delta\tau$	*c*
15° Ari	*12°*	*0;4*	14$^{\bar{\tau}}$	*0;4*	*36°*	*0;12*	36$^{\bar{\tau}}$	*0;12*
15° Tau	*14°*	*0;8*	*16$^{\bar{\tau}}$*	*0;6*	*42°*	*0;6*	*42$^{\bar{\tau}}$*	*0;12*
15° Gem	18°	0;8	19$^{\bar{\tau}}$	0;10	*45°*	*−0;6*	*48$^{\bar{\tau}}$*	*−0;8*
15° Cnc	22°	0;8	24$^{\bar{\tau}}$	0;6	42°	−0;12	*44$^{\bar{\tau}}$*	*−0;12*
15° Leo	26°	0;8	27$^{\bar{\tau}}$	0;6	36°	−0;28	38$^{\bar{\tau}}$	−0;36, *−0;16*
15° Vir	30°	0;8	30$^{\bar{\tau}}$	*0;12*	22°	−0;16	20$^{\bar{\tau}}$, *30$^{\bar{\tau}}$*	*−0;10, −0;30*
15° Lib	34°	0;20	*36$^{\bar{\tau}}$*	*0;20*	14°	*0*	*15$^{\bar{\tau}}$*	*0*
15° Sco	44°	0	*46$^{\bar{\tau}}$*	*0*	*14°*	*0;4*	*15$^{\bar{\tau}}$*	*0;2*
15° Sgr	44°	−0;4	46$^{\bar{\tau}}$	−0;4	*16°*	*0;8*	*16$^{\bar{\tau}}$*	*0;12*
15° Cap	42°	−0;24, −0;16	44$^{\bar{\tau}}$	−0;20	20°	0;4	22$^{\bar{\tau}}$	0;4
15° Aqr	30°, 34°	−0;12, −0;20	34$^{\bar{\tau}}$	−0;20	22°	0	24$^{\bar{\tau}}$	0
15° Psc	24°	*−0;24*	24$^{\bar{\tau}}$	−0;20	22°	*0;28*	24$^{\bar{\tau}}$	0;24

Table 3.8: Mercury system A_1: control values of the pushes $\delta\Sigma$ and $\delta\tau$ from MF to ML and from EF to EL, and interpolation coefficients *c*. Italic numbers are reconstructions derived by Neugebauer from the synodic tables; they are not attested yet in procedure texts or auxiliary tables. Further comments: MF to ML: $\delta\Sigma$ (15° Aqr) = 30° according to No. **1**, but 34° in the auxiliary table *ACT* 800a. EF to EL: $\delta\tau$ (15° Vir) = 20$^{\bar{\tau}}$ in the auxiliary table *ACT* 800d, but the synodic tables imply 30$^{\bar{\tau}}$.

3.4.2.4 Omitted phenomena

No. **42** P1.c and P2.c are the only extant procedures with rules for Mercury's omitted phenomena. Although they are embedded in procedures for system A_1, there is little that suggests an intimate connection with system A_1. The omissions are formulated in terms of 'passing by' (*etēqu*), as is done in the astronomical diaries. EF and EL are said to be omitted when Mercury is between 10° Ari and 20° Tau, and the same for MF and ML when Mercury is between 30° Vir and 5° Sco. As explained above, one expects the pushes from MF to ML and from EF to EL to decrease when Mercury

[251] Neugebauer discovered that the values of $\delta\Sigma$ and $\delta\tau$ derived from the synodic tables for Mercury system A_1 are not always fully consistent with exact linear interpolation. Sometimes the zodiacal position of the parent phenomenon appears to have been rounded to the nearest degree, or the nearest control value of $\delta\Sigma$ or $\delta\tau$ was used without any interpolation (cf. also Swerdlow 1998, p. 106).

approaches the region where the omissions occur. This is approximately satisfied (Table 3.8), because the minima of $\delta\Sigma$ and $\delta\tau$ for the push from MF to ML occur near Ari, and of the push from EF to EL near Lib–Sco.

3.4.2.5 Net displacements for various intervals

Procedures concerning net displacements for intervals longer than 1 synodic cycle are not attested for system A_1.[252] Values of $dB_j(s)$ for several intervals, where s is the number of synodic cycles and t the approximate number of years (Eq. 2.39), are listed in Table 3.9.

		EF				MF			
s	t	dB_1	dB_2	dB_3	dB_j/δ_j	dB_1	dB_2	dB_3	dB_j/δ_j
3	1^{yr}	−24;20°	−16;13,20°	−14;36°	−1,13	−16;7,30°	−21;30°	−14;20°	−2,9
63	20^{yr}	−6;40°	−4;26,40°	−4°	−20	−4;30°	−6°	−4°	−36
145	46^{yr}	0;40°	0;26;40°	0;24°	2	0;15°	0;20°	0;13,20°	2

Table 3.9: Mercury system A_1: net displacements for various intervals.

3.4.3 System A_2

System A_2, discovered by Neugebauer (1954), is represented by two synodic tables (*ACT* 300a,b) and three newly discovered fragments of procedure texts.[253] The procedure texts are of little use for dating system A_2, because their colophons are not preserved. The initial dates of the synodic tables cover the years SE 4–10, which gives a rough indication of the date when they were written. Hence system A_2 of Mercury is among the oldest known planetary systems. Procedures for net displacements are not yet attested.

ML: list of synodic arcs	No. **3** P1'?
EL: list of synodic arcs	No. **2** P2'; No. **4** P1'
temporal pushes EL to MF	No. **2** P3'
subdivision synodic cycle	No. **2** P4',P5'?,P6'?; No. **4** P2'?

	Π	Z	P	$\overline{\sigma}$
ML	20,23	6,28	3;9,7,25,...	1,54;12,39,...°
EL	11,24	3,37	3;9,7,28,...	1,54;12,37,...°

Table 3.10a: Procedures for Mercury system A_2, period relations and mean synodic arcs.

	j	b_j	α_j	σ_j	r_j	δ_j	v_j
ML	1	30° Gem	1,30°	1,47;46,40°	1;12	0;16,40°	5,24
	2	30° Vir	1,36°	2,9;20°	0;45	0;20°	4,48
	3	6° Cap	1,29°	1,37°	1;20	0;15°	5,56
	4	5° Ari	1,25°	2,9;20°	0;50	0;20°	4,15
EL	1	30° Gem	3,0°	1,48;30°	1;6,40	0;30°	6,0
	2	30° Sgr	1,0°	2,0;33,20°	0;54	0;33,20°	1,48
	3	30° Aqr	1,0°	1,48;30°	1;15	0;30°	2,0
	4	30° Ari	1,0°	2,15;37,30°	0;48	0;37,30°	1,36

Table 3.10b: Mercury system A_2: parameters of the step functions for σ for ML and EL.

3.4.3.1 Updating *B* with the synodic arc

The synodic tables imply that system A_2 follows the reverse scenario compared to system A_1: the zodiacal positions of ML and EL are updated with the synodic arc (Eq. 3.1) while MF and EF are treated as their satellites. As in system A_1 different step functions for σ are used for the two primary phenomena, and the stations appear to be ignored. Two newly discovered procedures preserve values of σ_j for EL. The algorithm for updating *B* is partly preserved in No. **4** P1', but the template (§2.6.10) cannot be identified. The value of σ_1 is preserved in No. **2** P2; those of σ_3 and σ_4 in No. **4** P1' (they are not fragments of the same tablet). They confirm Neugebauer's reconstruction of σ in *ACT*. A third newly discovered small fragment, No. **3** P1', preserves two zonal boundaries that are consistent with the step function for σ for ML. For a graphical representation cf. Fig. 3.8. Since $\sigma_j > \alpha_j$ in 3 of the 4 zones for ML, and in all zones for EL, σ deviates considerably from the preliminary values σ_j.

[252] 0;13,20° = $dB_3(145)$ might be mentioned in No. **6** P2', a procedure for Mercury whose purpose is not fully clear.

[253] Note that in *ACT*, p. 296, and *HAMA*, pp. 402, 466, *P* is erroneously given as 3;9,7,32,...

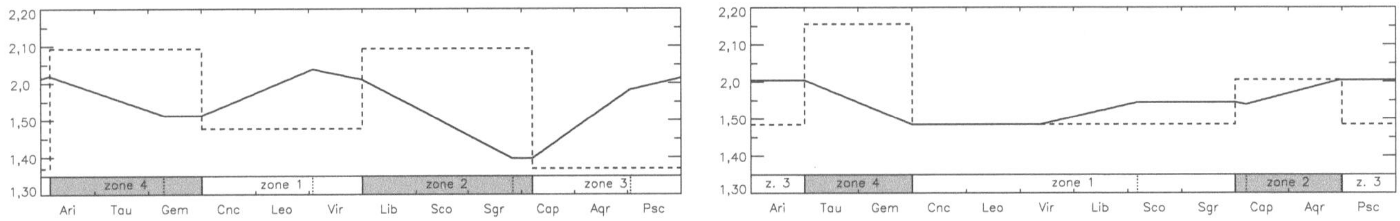

Fig. 3.8: Mercury system A_2: graphical representation of σ as a function of B_{i-1} for ML (left) and EL (right). The dashed lines represent the preliminary values σ_j.

3.4.3.2 Updating T with the synodic time

Procedures for updating T with the synodic time are not preserved for system A_2. Neugebauer points out that the values of T_i in the synodic tables cannot be exactly reproduced by the usual algorithm based on the synodic time (Eq. 3.9). The algorithm that does reproduce column T remains to be reconstructed, which may be achieved in a future investigation.

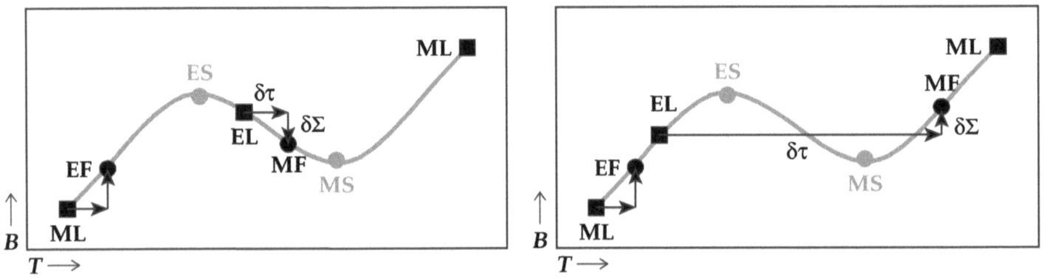

Fig. 3.9: Mercury system A_2: subdivision of the synodic cycle. As in system A_1 the stations are ignored. Right: the push from EL to MF can be forward (prograde) under special circumstances (§3.4.1).

3.4.3.3 Subdivision of the synodic cycle

The coordinates of MF and EF are obtained by applying pushes to those of EL and ML, respectively (Fig. 3.9). Each push is computed from the zodiacal position of its parent phenomenon by means of linear extrapolation. In the absence of procedure texts and auxiliary tables as they exist for system A_1, Neugebauer (*ACT*, *HAMA*) and Swerdlow (1998) reconstructed approximate pushes from the synodic tables (Table 3.11). As in system A_1 they do not form a zigzag function or some other simple pattern. The angular push from EL to MF is mostly subtractive (Fig. 3.9, left), but additive (Fig. 3.9, right) if EL occurs between 30° Psc and 5° Gem, as explained in §3.4.1. As expected, the largest values of $\delta\tau$ occur when $\delta\Sigma$ (EL to MF) is additive. The newly discovered No. **2** P3' preserves four numbers, each associated with a zodiacal sign, which are probably values of $\delta\tau$ (EL to MF). Three of them differ by only $1^{\bar{\tau}}$ from the values reconstructed by Neugebauer.[254] Perhaps also No. **2** P4'–P6' partly preserve schemes for the subdivision of the synodic cycle for system A_2 (cf. the commentary).

	ML to EF					EL to MF				No. **2** P3'
B_{ML}	$\delta\Sigma$	c	$\delta\tau$	c	B_{EL}	$\delta\Sigma$	c	$\delta\tau$	c	$\delta\tau$
15° Ari	*58;30°*	*−0;8*	$30^{\bar{\tau}}$	*−0;7,30*	30° Psc	*0°*	*0;24*	$28;30^{\bar{\tau}}$	*0;38*	
15° Tau	*54;30°*	*−0;4*	$26;15^{\bar{\tau}}$	*−0;8*	15° Ari	*6°*	*−0;4*	$38^{\bar{\tau}}$	*−0;6*	
15° Gem	*52;30°*	*0;30*	$22;15^{\bar{\tau}}$	*0;27*	15° Tau	*4°*	*−0;12*	$35^{\bar{\tau}}$	*−0;13*	
15° Cnc	*1,7;30°*	*0;19*	$35;45^{\bar{\tau}}$	*0;27*	15° Gem	*−2°*	*−0;12*	$28;30^{\bar{\tau}}$	*−0;13*	
15° Leo	*1,17°*	*0;6*	$49;15^{\bar{\tau}}$	*0;6*	15° Cnc	*−8°*	*0*	$22^{\bar{\tau}}$	*0*	
15° Vir	*1,20°*	*0;10,30*	$52;15^{\bar{\tau}}$	*−0;15*	15° Lib	*−8°*	*−0;4*	$22^{\bar{\tau}}$	*−0;16*	
15° Lib	*1,14;45°*	*−0;15*	$44;45^{\bar{\tau}}$	*−0;15*	15° Sco	*−12°*	*0*	$14^{\bar{\tau}}$	*−0;3,12*	$14^{\bar{\tau}}$
15° Sco	*1,7;15°*	*−0;15*	$37;15^{\bar{\tau}}$	*−0;15*	15° Sgr	*−12°*	*0*	$12;24^{\bar{\tau}}$	*−0;3,12*	
15° Sgr	*59;45°*	*−0;2,30*	$30^{\bar{\tau}}$	*0*	15° Cap	*−12°*	*0*	$10;48^{\bar{\tau}}$	*−0;3,12*	$9^{\bar{\tau}}$
15° Cap	*58;30°*	*0*	$30^{\bar{\tau}}$	*0*	30° Cap	*−12°*	*0*	$10^{\bar{\tau}}$	*0;12*	
					15° Aqr	*−12°*	*0*	$13^{\bar{\tau}}$	*0;12*	$14^{\bar{\tau}}$
					15° Psc	*−12°*	*0;48*	$19^{\bar{\tau}}$	*0;38*	$20^{\bar{\tau}}$

Table 3.11: Mercury system A_2: control values and interpolation coefficients (c) for $\delta\Sigma$ and $\delta\tau$ from ML to EF and from EL to MF. In italics: approximate values reconstructed from synodic tables.

[254] A future investigation may establish whether or not they are consistent with the synodic tables.

3.4.3.4 Net displacements for various intervals

Procedures concerning net displacements have not been discovered yet for system A_2. Values of $dB_j(s)$ for several intervals of s synodic cycles corresponding to t whole years (Eq. 2.39) are listed in Table 3.12.

		EL				ML			
s	t	$dB_1 = dB_3$	dB_2	dB_4	dB_j/δ_j	dB_1	$dB_2 = dB_4$	dB_3	dB_j/δ_j
3	1^{yr}	−16;30°	−18;20°	−20;37,30°	−33	−16;23,20°	−19;40°	−14;45°	−59
63	20^{yr}	−4;30°	−5°	−5;37,30°	−9	−4;26,40°	−5;20°	−4°	−16
145	46^{yr}	0;30°	0;33,20°	0;37,30°	1	0;33,20°	0;40°	0;30°	2

Table 3.12: Mercury system A_2: net displacements for various intervals.

3.4.4 System A_3

System A_3, discovered by Neugebauer (*ACT*), is represented by one procedure text (No. **5**) and one synodic table.[255] No. **5** does not preserve a date, but its scribe is known to have lived near the beginning of the Seleucid era. The data in the synodic table pertain to Artaxerxes I and II. This makes system A_3 the earliest known planetary system of mathematical astronomy.

ES, MS: updating B with the synodic arc	No. **5** P3
EL, MF: net displacements for 3 cycles	No. **5** P1
EF: net displacements for 3 cycles	No. **5** P2
ES, MS: net displacements for 3 cycles	No. **5** P3
EL, MF: net displacements for 63 cycles	No. **5** P4
EF: net displacements for 63 cycles	No. **5** P5

	Π	Z	P	$\overline{\sigma}$
EL, ES, MS, MF?	18,39	5,55	3;9,7,36,...	1,54;12,32,...°

Table 3.13a: Mercury system A_3: procedures; period relations.

	j	b_j	α_j	σ_j	r_j	δ_j	v_j
EL, ES, MS, MF?	1	30° Ari	1,20°	2,11;28,53,20°	0;48	0;22,13,20°	3,36
	2	20° Cnc	40°	1,45;11,6,40°	1;3,16,52,30	0;17,46,40°	2,15
	3	30° Leo	4,0°	1,50;56,15°	1;11,6,40	0;18,45°	12,48
EF	1	30° Gem	1,0°	*2,11–2,16°*	*1;24*	?	?
	2	30° Leo	30°	*3,4–3,9°*	*0;49,20*	?	?
	3	30° Vir	30°	*2,31–2,36°*	*0;46,50*	?	?
	4	30° Lib	30°	*1,58–2,2°*	*0;46,45*	?	?
	5	30° Sco	1,0°	*1,32–1,35°*	*1;8,40*	?	?
	6	30° Cap	2,30°	*1,45–1,49°*	*1;14,50*	?	?

Table 3.13b: Mercury system A_3: parameters of the step functions for σ. Italic numbers are approximate reconstructions.

3.4.4.1 Updating B with the synodic arc

A complete set of rules for updating the positions of the synodic phenomena is not preserved in No. **5**. The mentioned synodic table contains only zodiacal positions of EL, which are computed in sequences of 3 cycles. For 2 successive cycles B is updated as usual with a step function for the synodic arc, which is based on 3 zones (Table 3.13b and Fig. 3.10, left).[256] The same values of σ_j are mentioned in No. **5** P3, which concerns ES and MS, with no mention of EL. Perhaps they also apply to MF, because the net displacements of MF for 3 and 63 cycles are likewise defined for 3 zones in No. **5** P1,P4 (Table 3.14). The formulation of No. **5** P3 does not comply with any of the known templates (§2.6.10.1).

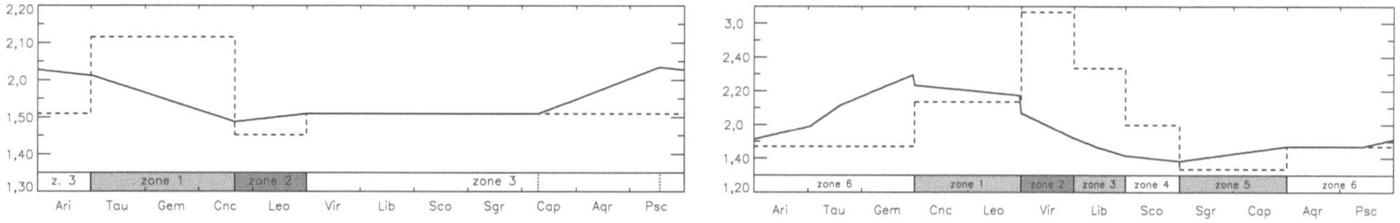

Fig. 3.10: Mercury system A_3: graphical representation of σ as a function of B_{i-1}. **Left**: EL (and probably ES, MS and MF). **Right**: EF (approximate reconstruction). Dashed lines: preliminary values σ_j.

The synodic arcs for ML and EF are also not mentioned in No. **5**, but for EF they can be estimated, following Swerdlow (1998), from the net displacements for 3 cycles (corresponding to 1 revolution) and 63 cycles (20 revolutions) mentioned

[255] Aaboe et al. (1991), Text E; cf. also *HAMA*, pp. 469–471.

[256] In Aaboe et al. (1991) this step function is called system $A_{3.II}$.

in P2 and P5 (Table 3.15), if one assumes that the step function for σ is based on the same 6 zones. The derivation sets out from Eq. (2.41), which can be reformulated as

$$\sigma_j = \frac{dB_j(s)}{s - tP}. \tag{3.26}$$

If we approximate P by the value 18,39/5,55 (Table 3.13a), then $1/(s - tP)$ is a convenient sexagesimal number for both sets of net displacements, namely −6;34,26,40 ($s = 3, t = 1$) and −23;40 ($s = 63, t = 20$). Table 3.13b shows the resulting pairs of σ_j rounded to whole degrees.[257] In each zone the two values are similar but not identical, which implies that some of the $dB_j(3)$ and $dB_j(63)$ mentioned in P2 and P5 are rounded or truncated. Hence we cannot derive the exact values of σ_j, but the similarity of both reconstructions suggests that they are not much different.[258] A graphical representation based on the mean values is shown in Fig. 3.10, right.

3.4.4.2 Updating B with the net displacement for 3 cycles

A unique feature of system A_3 is that the position of every third event of EL is obtained by applying a step function for the net displacement for 3 cycles, say σ', to the position preceding it by 3 cycles. Both the synodic table and No. **5** P1,P3 imply the same step function (Table 3.14).[259] It may be assumed that this function was also applied to ES, MS and MF. Transition rules are not mentioned in the procedures but implied by the synodic table. Since 3 cycles amount to a little less than 1 revolution, σ' is a subtractive displacement to which one must add 6,0° in order to obtain Mercury's total displacement. The σ'_j are identical with the net displacements $dB_j(3)$ derived from σ, but one zonal boundary, b_2 = 20;37,30° Cnc, is displaced by a tiny amount from the corresponding boundary for σ, 20° Cnc. The step function for σ and that for σ' are therefore mutually incompatible in the sense that they no longer exactly satisfy Eq. (2.41). The reason for this subtle anomaly, which is confirmed by the synodic table and thus deliberate, is unclear.[260]

j	b_j	α_j	σ'_j	r_j	δ_j	v_j	Π'	Z'	P'	$\overline{\sigma}'$
1	30° Ari	1,20;37,30°	−20°	0;48	0;1,2,30°	1,17,24	6,37,43	19,12	20;42,51,52,30	−17;22,45,..
2	20;37,30° Cnc	39;22,30°	−16°	1;3,16,52,30	0;0,50°	47,15				
3	30° Leo	4,0°	−16;52,30°	1;11,6,40	0;0,52,44,3,45°	4,33,4				

Table 3.14: Mercury system A_3, EL (and ES, MS, MF?): parameters of the step function for σ', periods of B and mean value $\overline{\sigma}'$.

3.4.4.3 Net displacements for other intervals

No. **5** P4 contains a list of net displacements for EL and MF for 63 = 21 · 3 cycles. The exact values, shown in Table 3.15, are obtained by computing the net displacement for 21 repetitions of σ'; they also underlie the synodic table. However, the numbers in No. **5** P4 differ slightly, presumably due to rounding or truncation. No. **5** P2 and P5 provide net displacements of EF, $dB_j(s)$, for s = 3 and 63 cycles, for six zodiacal regions (Table 3.15).[261] As already mentioned, at least some of these must have been rounded or truncated with respect to the unknown exact values. This is also apparent from the ratios $dB_j(3)/dB_j(63)$, which should be the same for all j (cf. Eq. 2.41 on p. 53), but they are not.

	s	t	dB_1	dB_2	dB_3	dB_4	dB_5	dB_6	dB_j/δ_j
EL, ES, MS, MF	21	1	−5;42,42,30°	−4;34,10°	−4;49,9,36,33,45°				5,29
EF	3	1^{yr}	−20°	−28°	−23°	−18°	−14°	−16°	?
	63	20^{yr}	−5;43°	−7;58°	−6;34°	−5;8°	−4°	−4;34°	?

Table 3.15: Mercury system A_3. Top: net displacements of EL, ES, MS, MF for s = 21 repetitions of σ' = 63 synodic cycles. Bottom: net displacements of EF for 3 and 63 synodic cycles.

---

257 The lower values of σ_j in Table 3.13b were obtained from $dB_j(3)$, the higher ones from $dB_j(63)$. Swerdlow (1998), pp. 133–134, used only $dB_j(3)$; cf. his $w_{i(1)}$ in Table 2.10C on p. 211.

258 This is also suggested by the similarity between the graph of the approximate σ (Fig. 3.10, right) and the graph of σ for EF as obtained in system A_1 (Fig. 3.6, right). This similarity is particularly significant because the σ_j and the zonal boundaries are very different in system A_1.

259 In Aaboe et al (1991) this step function is called system $A_{3.1}$. Note the erroneous value −20° for zone 2 in Aaboe et al. (1991), p. 38, a typographical error without consequence.

260 Cf. Swerdlow (1998), p. 131, for a possible explanation.

261 The boundary b_5 is assumed to be at 30° Sco, as mentioned in P2, but P5 suggests 30° Sgr.

3.4.4.4 Updating T with the synodic time

No. **5** does not provide an algorithm for updating T, nor does the synodic table contain a column T from which that algorithm could be reconstructed. It may be assumed that T is updated as usual by adding the synodic time τ (Eq. 3.9), using a value for c_{τ} similar to Eq. (3.25).

3.4.4.5 Subdivision of the synodic cycle

Procedures for the subdivision of the synodic cycle are not extant for system A_3, nor can we learn anything about this topic from the mentioned synodic table. If, as suggested by the procedures, the zodiacal positions of EL, ES, MS, and probably also MF and EF, were updated with step functions for σ, then the pushes between these phenomena are completely determined by the subdivision of the initial cycle, so that there would appear to be little need for such procedures.

3.4.5 Unidentified computational systems

For the remaining procedures that (may) concern Mercury (Table 3.16) the computational system could not be identified or fully reconstructed. The newly discovered No. **43** P3',P4' concern an inner planet, very likely Mercury. The former seems to mention 6 zones as for EF in system A_3, but the preserved values of σ_j are not the same. No. **6** P1',P2' concern Mercury's net displacement for 46^{yr}, but are otherwise difficult to understand, and nothing points to the known computational systems. Mention must also be made of two synodic tables for Mercury written on the same tablet as No. **45** and two daily motion tables.[262]

list of control values of the synodic arc	No. **43** P3'
list of control values of a temporal push?	No. **43** P4'
subdivision of the synodic cycle	No. **2** P4',P5'?,P6'?
net displacement for 46^{yr}	No. **6** P1'?,P2'

Table 3.16: Mercury: procedures for unidentified or incompletely known computational systems.

3.5 Venus

Venus is represented by at least six computational systems. Two of them, A_0 and A_3, are true type-A systems. The other systems, A_1, A_2, C_3, and C_4, are not really of type A because the algorithm for the synodic arc does not include transition rules. I classify them as type-C or pseudo type-A systems, the latter in order to maintain the well-established names A_1 and A_2. To the previously known procedure texts this study adds three further candidates (Nos. **2**, **49**, **51**).

3.5.1 Synodic cycle

Venus performs one revolution around the Sun in 224.7^d, which results in a mean synodic time $\overline{\tau} \approx 583.92^d \approx 9,53;12^{\overline{\tau}}$, and a mean total synodic arc $\overline{\Sigma} \approx 9,35;32$, the largest of all planets. Venus has several other unique features, some of which are reflected in the computational systems. Its motion has only a weak dependence on the zodiacal position, i.e. the zodiacal anomaly is not important.[263] Secondly, the approximation underlying the usual algorithm for the synodic time τ (Eq. 3.9) does not work well for Venus.[264] Indeed that algorithm is not used in systems A_1/A_2.

3.5.2 System A_0

System A_0, discovered by Neugebauer (*ACT*), is represented by one synodic table (*ACT* 400)[265] and, perhaps, one procedure text (No. **46**). The synodic arc of system A_0 is also used in system A_3 (§3.5.3) for updating the position of EL. None of the tablets with system A_0 mentions a date, but *ACT* 400 was written by the well-known astronomer Anu-aba-utēr from

[262] A 3425 (*ACT* 310; *HAMA*, pp. 418–419) and U 157 (*ACT* 1013), which concerns Mercury or Venus.

[263] This is because its orbit has a very small eccentricity.

[264] This was demonstrated e.g. by Britton (2001), pp. 544–545.

[265] The positions in *ACT* 401 may also have been computed with system A_0, but this is very uncertain; cf. Volume II.

Uruk, which implies a date within the range SE 110–124. In the absence of textual evidence nothing is known about the subdivision of the synodic cycle.[266]

3.5.2.1 Updating B with the synodic arc

System A_0 incorporates the simplest possible algorithm for updating B, namely a constant σ (Table 3.17a). This amounts to the assumption that Venus moves at its mean velocity. From the synodic tables it is inferred that this algorithm is used for EF and ML. No. **46** P28' and P29' contain the same algorithm, but formulated in terms of the *total* synodic arc $\Sigma = \sigma + 6{,}0 = 9{,}35;30°$. According to P28' this algorithm applies to ML, MF and MS; according to P29' it also applies to EL, EF and ES. Hence system A_0 appears to have been used for all of Venus's synodic phenomena.

j	α_j	σ_j	Π	Z	Y	P
1	6,0°	3,35;30°	12,0	7,11	19,11^{yr}	1;40,13,55,...

Table 3.17a: Venus system A_0: parameters of the step function for σ.

s	t	y	dB_1	dB_1/δ_1
5	3	8^{yr}	−2;30°	−5
142	85	227^{yr}	1°	2

Table 3.17b: Venus system A_0: net displacements for various intervals of s synodic cycles.

3.5.2.2 Updating T with the synodic time

Procedure texts with an algorithm for updating T are not preserved for system A_0. In the synodic table *ACT* 400 column T was computed in the usual manner (cf. §3.3.2.1) with

$$\tau = 9{,}53;10^{\bar{\tau}}. \tag{3.27}$$

From this one derives

$$c_{\bar{\tau}} = 6{,}17;40^{\bar{\tau}}, \tag{3.28}$$

or $12^{\overline{m}} + 17;40^{\bar{\tau}}$. This value is reproduced by Eq. (3.13) if one inserts $m = 1$, $\overline{\sigma} = 3{,}35;30°$, $e = 11;4^{\bar{\tau}}$, and if one rounds the exact result $6{,}17;41{,}28{,}40^{\bar{\tau}}$ to $6;17;40^{\bar{\tau}}$.

3.5.2.3 Net displacements for various intervals

Procedure texts concerning net displacements are not extant for system A_0. Apart from the 8-yr period (s = 5) which results in a net displacement $dB(5) = -2;30°$ and the exact return after $Y = 1151^{yr}$, system A_0 features only one other close return, namely 227^{yr} (Table 3.17b). Recall that for Venus the whole number of years corresponding to s cycles is $y = s + t$.

3.5.3 System A_3

System A_3 is represented by *ACT* 1050, which in its original state contained a synodic table for EF, EL, MF and ML followed by several brief procedures (No. **10**). The first successful attempt to understand this text was due to N.T. Hamilton (published posthumously in Hamilton & Aaboe 1998); a complete reconstruction was achieved by Britton (2001). Since the synodic table contains only zodiacal positions, nothing is known about the algorithm for the times.

3.5.3.1 Updating B with the synodic arc

The synodic table and the procedures imply that the positions of EL, MF, ML and EF are updated with (true) step functions for the synodic arc. Their defining parameters are different for each of these phenomena (Table 3.18), but all share the same period P, which coincides with that of system A_0 (Table 3.17a). For EL the synodic arc is constant as in system A_0.

[266] A rough impression of the angular and temporal push between ML and EF may, in principle, be inferred from the synodic tables (cf. Volume II).

	j	b_j	α_j	σ_j	r_j	δ_j	v_j
MF	1	12° Cnc	1,24°	3,35;30°	1;4	0;30°	2,48
	2	6° Lib	1,36°	3,49;52°	0;56,15	0;32°	3,0
	3	12° Cap	1,36°	3,35;30°	0;58,30,20,...	0;30°	3,12
	4	18° Ari	1,24°	3,30;8°	1;1,31,56,...	0;28°	3,0
ML	1	12° Cnc	2,54°	3,35;30°	0;53,20	0;30°	5,48
	2	6° Cap	1,22;40°	3,11;33,20°	1;15	0;26,40°	3,6
	3	28;40° Psc	1,43;20°	3,59;26,40°	0;54	0;33,20°	3,6
EF	1	12° Cnc	3,23;20°	3,59;26,40°	0;48	0;33,20°	3,6
	2	25;20° Lib	1,22;40°	3,11;33,20°	1;7,30	0;26,40°	3,6
	3	18° Cap	2,54°	3,35;30°	1;6,40	0;30°	5,48

Table 3.18: Venus system A_3: parameters of the step functions for σ for MF, ML and EF as reconstructed by Hamilton & Aaboe (1998) and Britton (2001). For EL cf. system A_0.

3.5.3.2 Net displacements for various intervals

No. **10** P1–P4 concern the net displacement of Venus for 5 cycles ('8^{yr}'). Values of $dB(s)$ for several intervals s are listed in Table 3.19. Recall that for Venus the corresponding whole number of years is $y = s + t$.

			MF			ML			EF			
s	t	y	$dB_1 = dB_3$	dB_2	dB_4	dB_1	dB_2	dB_3	dB_1	dB_2	dB_3	dB_j/δ_j
5	3	8^{yr}	−2;30°	−2;40°	−2;20°	−2;30°	−2;13,20°	−2;46,40°	−2;46,40°	−2;13,20°	−2;30°	−5
142	85	227^{yr}	1°	1;4°	0;56°	1°	0;52,20°	1;6,40°	1;6,40°	0;52,20°	1°	2

Table 3.19: Venus system A_3: net displacements for various intervals s (for EL cf. system A_0).

3.5.4 Systems A_1 and A_2

Systems A_1 and A_2, discovered by Kugler,[267] are represented by at most two procedure texts: No. **7**, which accompanies a synodic table (*ACT* 420), perhaps No. **8**, which also accompanies a synodic table (*ACT* 421a), and six further synodic tables.[268] The synodic table *ACT* 430 may have included procedures for systems A_1 and A_2, but they are now lost. Some of the unidentified procedure texts (§3.5.7) might also belong to systems A_1/A_2. Systems A_1 and A_2 are not truly of type A, because the algorithm for updating B does not incorporate transition rules. Often both systems are used in one and the same synodic table, such that certain phenomena are computed with system A_1 and others with A_2. Since a computational system is defined in this study as the complete set of algorithms by which the columns of a single synodic table are computed, systems A_1 and A_2 may be viewed as elements of a single computational system. System A_1 is thus far attested for MF, ML, EF and ES, and system A_2 for MS, ES and EL; hence ES is attested with both systems (cf. Table 3.20).

A_1	MF,ML: updating B with the synodic arc MF,ML: updating T with the synodic time	No. **7** P7,P11 No. **7** P8,P12
A_2	EF,ES,EL,MS: updating B with the synodic arc EF,ES,EL,MS: updating T with the synodic time	No. **7** P1,P3,P5,P9 No. **7** P2,P4,P6,P10
A_1,A_2?	updating B or T?	No. **8** P1',P2'

	Π	Z	Y	P	$\overline{\sigma}$
A_1	12,0	7,11	$19{,}11^{yr}$	1;40,13,55,...	3,35;30°
A_2	11,15	6,44	$17{,}59^{yr}$	1;40,14,51,...	3,35;28°

Table 3.20: Venus systems A_1, A_2: procedures, periods and mean synodic arc.

3.5.4.1 Updating B with the synodic arc

The procedure texts and the synodic tables imply that the algorithm for updating B for any of the synodic phenomena involves 5 different values of σ_j (Table 3.21). From the synodic tables it is inferred that

$$B_i = B_{i-1} + \sigma_j, \tag{3.29}$$

[267] Kugler (1907), p. 203.

[268] *ACT* 410–412, 421, 421a, 430.

where the σ_j are applied cyclically according to the fixed sequence j = 1, 4, 2, 5, 3. According to No. **7**, each σ_j is linked to a zodiacal region comprising 2 or 3 whole signs. Since no transition rules are applied when Venus crosses into another region, the σ_j not define a true step function. Furthermore, the connection between the σ_j and the zodiacal regions suggested in No. **7** is not fully consistent with the mentioned sequence, because Venus does not necessarily proceed by three regions if B is updated in this manner. The regions can therefore be meaningful only if interpreted in some approximate sense, or if they apply only at the initial time.[269]

The 5 values of σ_j add up to two different totals: either 17,57;30° = 3 revolutions −2;30°, which defines system A_1, or 17;57;20° = 3 full revolutions −2;40°, which defines system A_2 (Fig. 3.11). For ES both systems are attested in the synodic tables; they differ only with regard to σ_3. Again, if σ_j would be chosen on the basis of the zodiacal region in which Venus is located, then the same σ_j would occasionally repeat within 5 cycles, which would cause $dB(5)$ to differ from the canonical value −2;30° or −2;40°. As far as known this never happens in the synodic tables. Furthermore, with each sequence of 5 cycles Venus is displaced by $dB(5)$, leading to a drift of the σ_j with respect to the regions. After 60 cycles this amounts to a full zodiacal sign (30° in system A_1, 32° in system A_2). Hence any initial connection between the σ_j and the mentioned regions disappears with time.[270] The period of B (Table 3.20) is obtained as $P = \Pi/Z = 6{,}0/\overline{\sigma}$, where $\overline{\sigma}$ is the average of the 5 synodic arcs, $\overline{\sigma} = (\sigma_1 + \sigma_2 + \cdots + \sigma_5)/5$.[271] For system A_1 P is the same as in system A_0.

	j zone	α_j	σ_j	τ_j		j zone	α_j	σ_j	τ_j
MF	1 Ari–Gem	1,30°	3,39;50°	*9,48*$^{\tau}$	EF	1 ?	?	*3,44;10°*	*10,0*$^{\tau}$
	2 Cnc–Leo	1,0°	3,41;10°	*9,52*$^{\tau}$		2 ?	?	*3,30;30°*, 3,30;20°?	*9,23*$^{\tau}$
	3 Vir–Sco	1,30°	3,30;40°	*9,59*$^{\tau}$		3 ?	?	*3,32°*	*9,45*$^{\tau}$
	4 Sgr–Cap	1,0°	*3,25;30°*	9,55$^{\tau}$		4 ?	?	*3,36;20°*, 3,35;30°	*10,11*$^{\tau}$
	5 Aqr–Psc	1,0°	*3,40;20°*	9,52$^{\tau}$		5 ?	?	*3,34;30°*	*10,7*$^{\tau}$
total	A_1		17,57;30°	*49,26*$^{\tau}$	total	A_1		*17,57;30°*	*49,26*$^{\tau}$
MS	1 Ari–Tau	1,0°	3,40;30°	*9,49*$^{\tau}$	ES	1 Ari–Gem	1,30°	*3,37;30°*	10,1$^{\tau}$
	2 Gem–Leo	1,30°	3,42;50°	*9,31*$^{\tau}$		2 Cnc–Leo	1,0°	*3,38;30°*	9,59$^{\tau}$
	3 Vir–Lib	1,0°	3,30;10°	*9,57*$^{\tau}$		3 Vir–Sco	1,30°	*3,29;20°*, *3,29;30°*	9,46$^{\tau}$
	4 Sco–Cap	1,30°	3,24;40°?	*10,14*$^{\tau}$		4 Sgr–Cap	1,0°	*3,28;30°*	9,49$^{\tau}$
	5 Aqr–Psc	1,0°	3,39;10°	*9,55*$^{\tau}$		5 Aqr–Psc	1,0°	*3,43;30°*	9,51$^{\tau}$
total	A_2		17,57;20°	*49,26*$^{\tau}$	total	A_2/A_1		*17,57;20°*, *17,57;30°*	49,26$^{\tau}$
ML	1 Psc–Ari	1,0°	3,45;20°	10,14$^{\tau}$	EL	1 Ari–Gem	1,30°	3,36°	9,58$^{\tau}$
	2 Tau–Cnc	1,30°	3,29;40°	*9,45*$^{\tau}$		2 Cnc–Leo	1,0°	3,38°	10,4$^{\tau}$
	3 Leo–Vir	1,0°	3,28;50°	*9,31*$^{\tau}$		3 Vir–Sco	1,30°	3,32°	10,1$^{\tau}$
	4 Lib–Sgr	1,30°	3,38;30°	*9,57*$^{\tau}$		4 Sgr–Cap	1,0°	3,29;30°	9,40$^{\tau}$
	5 Cap–Aqr	1,0°	3,35;10°	*9,59*$^{\tau}$		5 Aqr–Psc	1,0°	3,41;50°	9,43$^{\tau}$
total	A_1		17,57;30°	*49,26*$^{\tau}$	total	A_2		17,57;20°	49,26$^{\tau}$

Table 3.21: Venus systems A_1/A_2: parameters for the synodic arc (σ) and the synodic time (τ) mentioned in procedure texts or (in italics) reconstructed from synodic tables (*ACT*, p. 301; *HAMA*, p. 462). The two values of σ_j preserved in No. **7** P1 (EF, right number) differ slightly from Neugebauer's reconstructions.

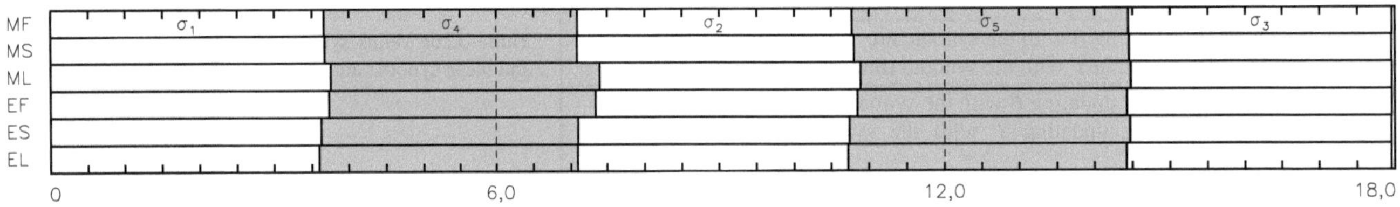

Fig. 3.11: Venus systems A_1 and A_2: synodic arcs σ_j for EF, ES, EL, MF, MS and ML laid out in the sequence j = 1, 4, 2, 5, 3 as implemented in the synodic tables. In this diagram the difference between systems A_1 and A_2 is not noticeable.

3.5.4.2 Net displacements for various intervals

Procedures for net displacements for intervals longer than 1 synodic cycle are not attested. Apart from the 8-yr period (s = 5, t = 3) and the exact return after Π cycles, systems A_1 and A_2 do not feature any other close returns.[272]

[269] Cf. also *HAMA*, pp. 462–463.

[270] This was pointed out by van der Waerden (1974), p. 278.

[271] Π can also be derived from the net displacements for 5 cycles; cf. §3.5.4.2.

[272] This is because the ratio $6{,}0°/dB(5)$ is a whole number in both systems: 2,15 (A_1) and 2,24 (A_2). After these whole multiples of 5 cycles Venus returns to exactly the same position. Hence $\Pi = 5 \cdot 2{,}15 = 11{,}15{,}5$ in system A_1, $5 \cdot 2{,}24 = 12{,}0$ in system A_2.

3.5.4.3 Updating T with the synodic time

From the synodic tables and the procedures (No. **7** P2,P4,P6,P8,P10) it is apparent that T is computed by a similar method as B. For each synodic phenomenon there are 5 different values of the synodic time τ_j (Table 3.21), from which T_i is obtained as

$$T_i = T_{i-1} + \tau_j, \tag{3.30}$$

where j follows the fixed periodic sequence 1, 4, 2, 5, 3. Unlike the synodic arcs, the 5 values of τ_j for each phenomenon always add up to exactly the same total $49{,}26^{\overline{\tau}} = 99^{\overline{m}} - 4^{\overline{\tau}}$ (Fig. 3.12). The resulting mean synodic time $\overline{\tau} = 49{,}26/5 = 9{,}53;12^{\overline{\tau}}$ is very similar to the system A_0 value $9{,}53;10^{\overline{\tau}}$. There is no constant difference in systems A_1 and A_2 between τ and σ, i.e. no unique value of c_τ (Eq. 3.9), which reflects the empirical behaviour of Venus (§3.5.1).

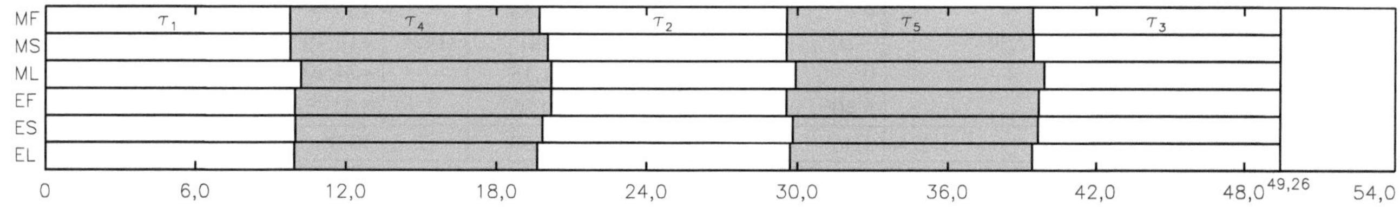

Fig. 3.12: Venus systems A_1 and A_2: the synodic times τ_j for EF, ES, EL, MF, MS and ML are laid out in the sequence j = 1, 4, 2, 5, 3 which is implemented in the synodic tables.

3.5.4.4 Subdivision of the synodic cycle

Procedures concerning the subdivision of the synodic cycle are not attested for systems A_1 and A_2. Since all 6 synodic phenomena are primary phenomena, there is little need for such procedures. Once the subdivision of the initial cycle is established, all subsequent pushes are fully determined by the algorithms for σ and τ. As pointed out by Neugebauer (*ACT*) and van der Waerden (1957), the resulting angular pushes exhibit certain strange artefacts. For instance, in some cycles the push between ES and MS turns out to be prograde, which constitutes a serious deviation from the empirical behaviour.

3.5.5 System C_3

System C_3 is represented only by No. **9**, which is early or pre-Seleucid. The present study includes a new translation of this text with minor changes compared to the edition of Britton & Walker (1991). All procedures concern the subdivision of the synodic cycle, but they also imply values for the synodic arc σ and the synodic time τ.

3.5.5.1 Updating B and T

Since no synodic tables are preserved, it is not known how system C_3 was implemented in practice. It is reasonable to assume that the scheme for the subdivision of the synodic cycle was applied repeatedly in order to compute successive cycles. This would result in a synodic arc $\sigma = \Sigma - 6{,}0^\circ = 3{,}36^\circ$ for all synodic phenomena. The resulting period of B is $P = 6{,}0/\sigma = \Pi/Z$ (Table 3.22), which makes system C_3 the only computational system in which the 8-yr period produces an exact return. Apart from multiples of that period system C_3 does not feature any other intervals that produce exact or close returns.

3.5.5.2 Subdivision of the synodic cycle

The parameters of the subdivision of the synodic cycle contained in or implied by No. **9** are compiled in Table 3.22. In the segments MS to ML (i) and EF to ES (ii) the daily displacement v is computed by means of linear interpolation, in accordance with the algorithm explained in §3.3.3.4, i being the day number measured from the beginning of the segment. The synodic time $\tau = 9{,}44^{\mathrm{d}} = 584^{\mathrm{d}}$ mentioned in the text is the correct total of the temporal pushes (1–8),[273] but instead of the correct total synodic arc $\Sigma = 9{,}36^\circ$ the text mentions $9{,}44^\circ$. The scheme also includes a list of control values for $\delta\tau$

[273] This number is close to the empirical value of τ measured in days (§3.5.1), which proves that the scheme is formulated in terms of days and not mean tithis.

(EL to MF), the duration of invisibility near IC, i.e. a subinterval of interval 7. The values are compiled in Table 3.22. For a detailed discussion cf. Britton & Walker (1991).

system C_3						system C_4				
	push	$\delta\tau$	v [°/d]	$\delta\Sigma$			push	$\delta\tau$	v [°/$\bar{\tau}$]	$\delta\Sigma$
1	MS to ML (i)	$2,8^d$	$0;45+i\cdot 0;0,14,3,45$	2,8°		1	EL to MS	$44^{\bar{\tau}}$	+0;15!	11°!
2	(ii)	$1,46^d$	1;15	2,12;30°		2	MS to ML (i)	$2,4^{\bar{\tau}}$	1	2,4°
	1–2 total	$3,54^d$		4,20;30°		3	(ii)	$1,43^{\bar{\tau}}$	1;12	2,3;45°!
3	ML to EF	$1,0^d$	1;15	1,15°			2–3 total	$3,47^{\bar{\tau}}$		4,7;45°
4	EF to ES (i)	$1,46^d$	1;15	2,12;30°		4	ML to EF	$1,10^{\bar{\tau}}$	1;12	1,24°
5	(ii)	$2,8^d$	$1;15-i\cdot 0;0,14,3,45$	2,8°		5	EF to ES (i)	$1,43^{\bar{\tau}}$	1;12	2,3;35°!
	4–5 total	$3,54^d$		4,20;30°		6	(ii)	$2,0^{\bar{\tau}}$	1	2,0°
6	ES	$12-x^d$	0	0°			5–6 total	$3,43^{\bar{\tau}}$		4,3;35°
7	ES to MS	32^d	−0;37,30	−20°		7	ES to EL	$34^{\bar{\tau}}$	+0;24!	+13;40°!
8	MS	$12+x^d$	0	0°			7+1 total	$1,18^{\bar{\tau}}$		+24;40°!
	2–4 total	$4,32^d$	1;15	5,40°			3–5 total	$4,36^{\bar{\tau}}$	1;12	5,31;20°
	1–8 total: τ =	$9,44^d$	Σ =	9,36°			1–7 total: τ =	$9,58^{\bar{\tau}}$	Σ =	10,0°

Π	Z	Y	P	$\bar{\sigma}$		Π	Z	Y	P	$\bar{\sigma}$
5	3	8^{yr}	1;40	3,36°		3	2	5^{yr}	1;30	4,0°

system C_3	Cap–Psc	Ari	Tau	Gem	Cnc–Vir	Lib–Sco	Sgr
$\delta\tau$ (EL to MF)	1^d	4^d	8^d	12^d	15^d	6^d	2^d

Table 3.22: Top, left; bottom: System C_3: subdivision of the synodic cycle. **Top, right**: System C_4: subdivision of the synodic cycle as reconstructed by Neugebauer & Sachs (1967), with the modifications of Britton & Walker (1991).

3.5.6 System C_4

System C_4 is represented by the probably early Seleucid synodic table BM 36301.[274] An unusual feature of that tablet is that it also contains a synodic table for Mars and an auxiliary table for the Moon.

3.5.6.1 Updating *B* and *T*

The synodic table implies that for EL $\sigma = 4{,}0°$, which is a much less accurate value than in any other system. For the resulting period of the zodiacal positions, $P = 6{,}0/\sigma = \Pi/Z$, cf. Table 3.22.

3.5.6.2 Subdivision of the synodic cycle

The first tentative reconstruction of the synodic cycle by Neugebauer & Sachs (1967) was later modified by Britton & Walker (1991), whose results are compiled in Table 3.22. The synodic time τ = 9,58, corresponding to the sum of all temporal pushes between ML and the next occurrence of ML, is closest to the empirical value (cf. §3.5.1) if interpreted as being measured in mean tithis. System C_4 has several odd features, most notably a forward motion between ES and MS instead of the expected retrograde motion. Also note that in the segments MS to ML (ii) and EF to ES (i) the usual relation $\delta\Sigma = \delta\tau \cdot v$ is not exactly satisfied, since $1{,}43 \cdot 1;12 = 2{,}3;36°$.

3.5.7 Unidentified computational systems

The remaining procedures for Venus, for which the computational system has not been identified, are listed in Table 3.23. The badly legible No. **49** contains at least one procedure for the synodic arc, but no value of σ_j is intact. Most other procedures concern the subdivision of the synodic cycle. Since they do not provide enough information for deriving the synodic arc they could not be classified in terms of a computational system.

3.5.7.1 Subdivision of the synodic cycle

The largest collection of procedures for the subdivision of the synodic cycle to be mentioned here comprises No. **46** P11'–P24', which contain schemes X.S_1 and X.S_2. The same tablet contains two other, incomplete schemes, one of which

[274] Neugebauer & Sachs (1967), Text C.

updating B with the synodic arc	No. **49** P3
subdivision of the synodic cycle scheme $X.S_1$	No. **46** P11',P12' ($\delta\Sigma$); No. **46** P17',P18' ($\delta\tau$)
scheme $X.S_2$	No. **46** P12a'–P16' ($\delta\Sigma$); No. **46** P19'–P24' ($\delta\tau$)
scheme $X.S_3$	No. **46** P27'
other	No. **2** P5',P6'; No. **46** P25',P26',P30'; No. **49** P1,P5'–P7'?; No. **51** P1'?
computations of unknown significance	No. **8** P1',P2'; No. **11** P2'

Table 3.23: Venus, unidentified computational systems: procedures.

is provisionally labeled $X.S_3$. Another fragmentarily known scheme, $X.S_4$, is represented by the synodic table BM 37151 (Aaboe & Huber 1977).

Schemes $X.S_1$ and $X.S_2$

Each of the procedures No. **46** P11'–P24' is devoted to one angular or temporal push. In their original state they provided $\delta\Sigma$ and $\delta\tau$ for two complete sequences of pushes between MF and the next occurrence of MF (see Table 3.24). The first, coarse sequence (scheme $X.S_1$) divides the synodic cycle into only two pushes, namely from MF to EF and from EF to MF. The second, finer sequence (scheme $X.S_2$) provides rules for all constituent phenomena of the synodic cycle. For some reason MS is initially skipped, so that the first push concerns MF to ML, but this is compensated by the inclusion of the push from MF to MS at the end of the sequence. The angular push from MS to ML, which appears to be omitted, may be obtained by subtracting $\delta\Sigma$ (MF to MS) from $\delta\Sigma$ (MF to ML). No. **46** P13' preserves two control values of $\delta\Sigma$ (EF to ES) for scheme $X.S_2$, 4,14° and 4,21°. It is not clear in which zodiacal signs they apply.[275]

	$\delta\Sigma$				$\delta\tau$						
	scheme $X.S_1$	scheme $X.S_2$			scheme $X.S_1$		scheme $X.S_2$				
	MF to EF	ES to EL	EL to MF	MF to MS	MF to EF	EF to MF	MF to ML	ML to EF	EF to ES	ES to EL	EL to MF
Ari	5,...°	?	?	?	5,23;30$^{\bar\tau}$	4,30$^{\bar\tau}$	4,22$^{\bar\tau}$	1,0$^{\bar\tau}$	4,6$^{\bar\tau}$	?	?
Tau	5,...°	?	?	?	5,...$^{\bar\tau}$	4,32$^{\bar\tau}$	4,19$^{\bar\tau}$	1,0$^{\bar\tau}$	4,8$^{\bar\tau}$	?	?
Gem	5,29°	?	?	?	?	4,34$^{\bar\tau}$	4,16$^{\bar\tau}$	58$^{\bar\tau}$	4,10$^{\bar\tau}$	19$^{\bar\tau}$	?
Cnc	5,27°	?	?	?	5,26$^{\bar\tau}$	4,...$^{\bar\tau}$	4,...$^{\bar\tau}$	56$^{\bar\tau}$	4,12$^{\bar\tau}$	17$^{\bar\tau}$	15$^{\bar\tau}$
Leo	5,28;30°	−6°	?	?	?	4,34$^{\bar\tau}$	4,17$^{\bar\tau}$	56$^{\bar\tau}$	4,7$^{\bar\tau}$	17$^{\bar\tau}$	15$^{\bar\tau}$
Vir	5,30°	−6°	?	?	?	4,...$^{\bar\tau}$	4,21$^{\bar\tau}$	56$^{\bar\tau}$	4,2$^{\bar\tau}$	17$^{\bar\tau}$	?
Lib	5,31;30°	?	?	?	5,23;30$^{\bar\tau}$	4,30$^{\bar\tau}$	4,25$^{\bar\tau}$	56$^{\bar\tau}$	?	?	?
Sco	5,...°	?	?	?	5,...$^{\bar\tau}$	4,30$^{\bar\tau}$	4,29$^{\bar\tau}$	56$^{\bar\tau}$	?	?	?
Sgr	5,...°	?	−2°	−10°	?	4,...$^{\bar\tau}$	4,...$^{\bar\tau}$	1,0$^{\bar\tau}$	3,58$^{\bar\tau}$	22;30$^{\bar\tau}$	4;30$^{\bar\tau}$
Cap	5,27°	?	−1°	−9°	5,26$^{\bar\tau}$	4,...$^{\bar\tau}$	4,...$^{\bar\tau}$	1,2$^{\bar\tau}$	3,58$^{\bar\tau}$	23$^{\bar\tau}$	1$^{\bar\tau}$
Aqr	5,28°	?	−1°	−9°	?	4,...$^{\bar\tau}$	4,29$^{\bar\tau}$	1,2$^{\bar\tau}$	4,1$^{\bar\tau}$	23$^{\bar\tau}$	1$^{\bar\tau}$
Psc	5,30°	?	−1°	?	?	4,30$^{\bar\tau}$	4,25$^{\bar\tau}$	1,2$^{\bar\tau}$	4,3$^{\bar\tau}$	22$^{\bar\tau}$	2$^{\bar\tau}$

Table 3.24: Venus: schemes $X.S_1$ and $X.S_2$ for the subdivision of the synodic cycle (No. **46** P11'–P24'). For EF to MF, MF to ML, ML to EF, and EF to ES $\delta\Sigma$ is not preserved.

Both sequences are worked out first for the angular pushes and then for the temporal pushes. Each procedure contains an introductory statement identifying the push, followed by a list of 12 control values, one for each zodiacal sign. The only possible indication about the location to which they pertain is provided in P19', which seems to imply 12°, but this interpretation is uncertain, and 12° is otherwise not attested as a control point for pushes. None of the control values form a zigzag function, or exhibit any other simple dependence on the zodiacal position. Schemes $X.S_1$ and $X.S_2$ are not attested in synodic tables, and the procedures do not inform us about the way in which the control values are used. It may be assumed that this involves linear interpolation with the zodiacal position of the parent phenomenon as the argument, e.g. $\delta\Sigma$ (MF to EF) is computed from B_{MF}. A strict mutual consistency between both schemes appears to be ruled out.

Scheme $X.S_3$

The incompletely known scheme $X.S_3$, attested in No. **46** P27', provides an algorithm by which B is computed for EL, ES, ML and MS from the position of the parent phenomenon. For EL and ES the parent phenomenon is EF; for ML and MS it is MF. The value of $\delta\Sigma$ is obtained in two steps. First an initial value $\delta\Sigma_0$ is computed under the assumption that the parent phenomenon occurs in a certain zodiacal sign. Then a correction, say C, is applied to $\delta\Sigma_0$ if the parent phenomenon is in a different zodiacal sign. Part **a** concerns the push from EF to EL, for which $\delta\Sigma_0 = 4{,}13°$, i.e.

$$\delta\Sigma\,(\text{EF to EL}) = 4{,}13^\circ + C_1. \tag{3.31}$$

[275] Neugebauer assigned them to Sco and Aqr, respectively (*ACT*, p. 397).

The uncorrected value is said to apply if EF occurs in Vir, i.e. $\delta\Sigma_0$ (EF to ES) $= 4{,}19;15^\circ$ and $\delta\Sigma_0$ (ES to EL) $= -6;15^\circ$. C_1 is provided as a list of 12 control values, one for each zodiacal sign (Table 3.25a). It is not clear whether they apply in the whole sign, or whether C_1 was obtained through linear interpolation between the control values.[276] As remarked by Neugebauer (*ACT*, p. 401), this approach reflects the empirical behaviour of Venus, because the push from ML to EF is nearly constant, while that from EL to MF varies rather strongly from cycle to cycle, and the latter variation can be modeled as a function of the zodiacal position. It therefore makes good sense to use EF as a parent phenomenon and model $\delta\Sigma$ (EF to EL) as a function of the zodiacal position. In the procedure, this is achieved by using the uncorrected zodiacal position of EL, $B_{\mathrm{EF}} + 4{,}13^\circ$, as the argument of C_1.

The procedure also mentions that ES can be dealt with in the same manner. Neugebauer takes this to be a reference to an alternative method for obtaining the zodiacal position of EL, namely by using ES as the parent phenomenon instead of EF. That makes sense if one assumes that successive positions of ES are computed from an algorithm for the synodic arc. By adding the retrograde angular push,

$$\delta\Sigma\,(\text{ES to EL}) = -6;15^\circ + C_2, \tag{3.32}$$

to B_{ES} one obtains B_{EL}. One would then expect the argument of C_2 to be the uncorrected zodiacal position of ES, i.e. $B_{\mathrm{EF}} + 4{,}19;15^\circ$. Since no additional table of coefficients is provided, it is possible that C_1 also plays the role of C_2, although the former has a different argument, namely the uncorrected position of EL, $B_{\mathrm{EF}} + 4{,}13^\circ$. However, both positions are not too far apart, so that the resulting error is small. Alternatively the statement may refer not to the computation of B_{EL} from B_{ES}, but to the computation of B_{ES} from B_{EF}, i.e.

$$\delta\Sigma\,(\text{EF to ES}) = 4{,}19;15^\circ + C_2. \tag{3.33}$$

Also then one expects the argument of C_2 to be the uncorrected position of ES. This interpretation, unlike the previous one, would imply that the retrograde arc from ES to EL is constant and equal to $-6;15^\circ$.

The procedure continues in part **b** with an analogous list of corrections for computing the zodiacal positions of ML and MS (Table 3.25a). Compared to the previous case, some important parameters that are necessary for applying the algorithm are lacking here. The parent phenomenon is not mentioned, but in analogy to the previous case, it can be assumed to be MF. Also the uncorrected values of the angular pushes from MF to MS, MF to ML, and MS to ML are not mentioned.

		$\delta\Sigma$			
	C_1	EF to ES	ES to EL	EF to EL	C_2
Ari	−2°	4,17;15°	−6;15°	4,11°	3°
Tau	−1°	4,18;15°	”	4,12°	1°
Gem	1°	4,20;15°	”	4,14°	−1°
Cnc	3°	4,22;15°	”	4,16°	−3°
Leo	3;30°	4,22;45°	”	4,16;30°	−3;30°
Vir	4°	4,23;15°	”	4,17°	−3;30°
Lib	2°	4,21;15°	”	4,15°	−2°
Sco	1°	4,20;15°	”	4,14°	−1°
Sgr	−2°	4,18;15°	”	4,11°	0°?
Cap	−3;30°	4,15;45°	”	4,9;30°	3;30°
Aqr	−3;30°	4,15;45°	”	4,9;30°	3;30°
Psc	−3°	4,16;15°	”	4,10°	3;30°

Table 3.25a: Venus: scheme X.S_3 for the subdivision of the synodic cycle.

push	v [°/$\bar{\tau}$]	$\delta\tau_1$	$\delta\Sigma_1$	$\delta\tau_5$	$\delta\Sigma_5$
MS to ML (1)	0;37,30	$40^{\bar{\tau}}$	25°	$40^{\bar{\tau}}$	25°
(2)	1;7,30	$1{,}17^{\bar{\tau}}$	1,26;37,30°	$1{,}17^{\bar{\tau}}$	1,26;37,30°
(3)	1;15	$1{,}59^{\bar{\tau}}$	2,28;45°	$2{,}2^{\bar{\tau}}$	2,32;30°
total		$3{,}56^{\bar{\tau}}$	4,20;22,30°	$3{,}59^{\bar{\tau}}$	4,24;7,30°
ML to EF	1;15	$57^{\bar{\tau}}$?		$1{,}0^{\bar{\tau}}$	
EF to ES (1)	1;15	$1{,}44^{\bar{\tau}}$	2,10°	$1{,}42^{\bar{\tau}}$	2,7;30°
(2)	1;8	$1{,}30^{\bar{\tau}}$	1,42°	$1{,}30^{\bar{\tau}}$	1,42°
(3)	0;40	$33^{\bar{\tau}}$	22°	$32^{\bar{\tau}}$	21;20°
(4)	0;10	$8^{\bar{\tau}}$	1;20°	$10^{\bar{\tau}}$	1;40°
total		$3{,}55^{\bar{\tau}}$	4,15;20°	$3{,}59^{\bar{\tau}}$	
ES to EL	−0;20	$20^{\bar{\tau}}$	−6;40°	$12^{\bar{\tau}}$	−4°
EL to MF	−0;20	$10^{\bar{\tau}}$	−3;20°	$16^{\bar{\tau}}$	−5;20°
MF to MS	−0;20	$23^{\bar{\tau}}$	−7;40°	$21^{\bar{\tau}}$	−7°
ES to MS	−0;20	$53^{\bar{\tau}}$	−17;40°	$49^{\bar{\tau}}$	−16;20°
MS to MS		$9{,}41^{\bar{\tau}}$		$9{,}42^{\bar{\tau}}$	

Table 3.25b: Venus: scheme X.S_4 for the subdivision of the synodic cycle.

Scheme X.S_4

In scheme X.S_4, reconstructed from the synodic table BM 37151 by Aaboe & Huber (1977), the subdivision of the synodic cycle varies from cycle to cycle. The precise nature of this variation is incompletely known due to the bad state of preservation of the text; perhaps it is based on a periodic sequence of five different schemes, similar to systems A_1/A_2. Values of $\delta\tau$ and $\delta\Sigma$ for sequences 1 and 5 reconstructed by Aaboe and Huber are compiled in Table 3.25b. For a discussion I refer to Aaboe & Huber (1977).

[276] Since the value 4,13° is said to apply if EF occurs in Vir, which implies that EL occurs in Tau or Gem, one expects $C_1 = 0$ in that region. Indeed C_1 changes from −1 to +1 between these signs.

Other schemes

Table 3.23 lists several other fragmentary procedures mentioning isolated parameters for the subdivision of the synodic cycle, whose relation to the other schemes is not clear. This includes the similarly structured No. **46** P25' and P30', which preserve values of $\delta\Sigma$ (P25') and $\delta\tau$ (P30') for several successive pushes. In P25' $\delta\Sigma$ (EF to EL) = 4,14;13° and $\delta\Sigma$ (ES to EL) = −6;20° are preserved; in P30' $\delta\tau$ (MF to ML) = 4,22;15$^{\overline{\tau}}$, $\delta\tau$ (ML to EF) = 45$^{\overline{\tau}}$, $\delta\tau$ (ES to EL) = 18$^{\overline{\tau}}$, $\delta\tau$ (EL to MF$^{?}$) = 21$^{\overline{\tau}}$ and $\delta\tau$ (MF$^{?}$ to MS) = 18$^{\overline{\tau}}$. As far as can be established, these values are roughly comparable with mean values implied by schemes X.S$_1$ and X.S$_2$.

3.6 Mars

The planet Mars is represented by systems A and B and several procedures for the subdivision of the synodic cycle and, perhaps, distance to the ecliptic.[277] Daily motion tables are not attested for Mars.

3.6.1 Synodic cycle

Mars performs one revolution around the Sun in 687.0^{d}, which results in a mean synodic time $\overline{\tau} \approx 779.96^{d} \approx 13{,}12;21^{\overline{\tau}}$, and a mean total synodic arc $\overline{\Sigma} \approx 6{,}48;43°$.

3.6.2 System A

System A, discovered by Schnabel (1924), is represented by eight procedure texts (Table 3.26a), seven synodic tables, four template tables, and three auxiliary tables.[278] No. **15** accompanies a synodic table (*ACT* 501a), and No. **12** accompanies an auxiliary table (*ACT* 803). Nos. **43** and BM 35857, an additional fragment of No. **16**, are published here for the first time. The only tablet for Mars system A preserving a dated colophon is No. **15**, which was written in SE 186. No. **12** was written by the well-known astronomer Šamaš-ēṭir from Uruk, which implies a date from the period SE 110–118. A further indication of the approximate time when system A was in use is provided by the initial dates of the synodic tables, which cover the period SE 69–172.

updating B with the synodic arc	No. **14** P5'?; No. **15** P1.a,b; No. **43** P6'
updating T with the synodic time	No. **13** P3'.b; No. **14** P3'?,P4'
period relations for B	No. **13** P11'.a
list of net displacements	No. **44** P3'
subdivision of the synodic cycle	cf. Table 3.27
distance to the ecliptic?	No. **14** P6'

Table 3.26a: Mars system A: procedures.

j	b_j	α_j	σ_j	r_j	δ_j	v_j
1	30° Ari	1,0°	45°	0;40	2;30°	24
2	30° Gem	1,0°	30°	1;20	1;40°	36
3	30° Leo	1,0°	40°	1;30	2;13,20°	27
4	30° Lib	1,0°	1,0°	1;30	3;20°	18
5	30° Sgr	1,0°	1,30°	0;45	5°	12
6	30° Aqr	1,0°	1,7;30°	0;40	3;45°	16

Π	Z	$Z+\Pi$	Y	P	$\overline{\sigma}$
2,13	18	2,31	4,44^{yr}	7;23,20	48;43,18,29,46,...°

Table 3.26b: Mars system A: parameters of the step function for the synodic arc, periods of B and mean synodic arc.

3.6.2.1 Updating B with the synodic arc

The synodic tables and the procedure texts imply that FA, S1 and LA are primary phenomena, while AR and S2 are satellites of S1.[279] The formulation of the algorithm in the newly discovered No. **43** P6', and in No. **15** P1.a,b, is based on template STEP.A.3 (§2.6.10.1), which includes transition rules involving coefficients r_j. No. **14** P5' may also contain this algorithm, but no parameters are preserved. The step function for σ is based on 6 zones (Table 3.26b; Fig. 3.13). Most σ_j

[277] The latter are discussed in the framework of system A, but are not necessarily associated with that system. The partly preserved, badly understood procedures for Mars contained in MNB 1856 (Neugebauer & Sachs 1969, Text H) probably belong to NMAT.

[278] ST: *ACT* 500–503, 501a, 501b, Neugebauer & Sachs (1967) Text C (which does not fully comply with system A); AT: *ACT* 803, 804, Steele (2005b); TT: Aaboe & Sachs (1966) Texts G (+) H, I, J, K.

[279] As pointed out by Aaboe (1987), p. 11, this arrangement makes good sense for Mars. If the positions of AR and S2 were computed from the same σ as those of FA, S1 and LA this would yield retrograde angular pushes from S1 to AR and from AR to S2 that are proportional to $\delta_j = \sigma_j/Z$. That would be a bad approximation of the empirical behaviour, which exhibits an entirely different dependence on the zodiacal position.

are now attested in procedure texts after the discovery of No. **43** P6', which complements the badly damaged No. **15** P1.a. In 3 zones $\sigma_j \geq \alpha_j$, which causes σ to differ considerably from the σ_j.[280] The resulting values of Π, Z and $Y = Z + 2\Pi$ are mentioned in No. **13** P11', which also provides a derivation of the accurate estimate $\overline{\sigma} \approx 48;43,18,30°$.

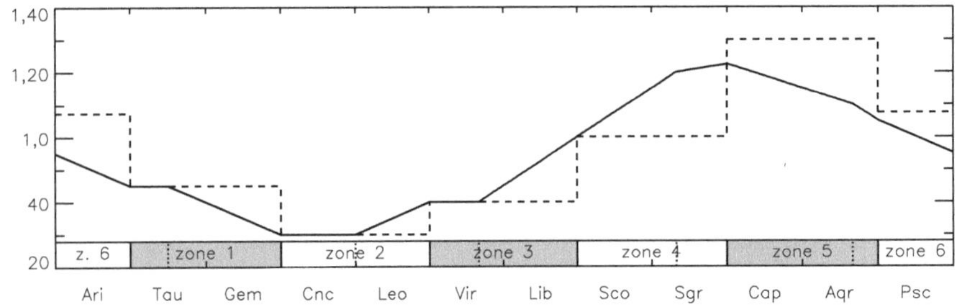

Fig. 3.13: Mars system A: synodic arc σ for FA, S1 and LA as a function of B_{i-1}. The dashed lines represent the preliminary values σ_j.

3.6.2.2 Updating T with the synodic time

No. **13** P3'.b contains the usual algorithm for updating T with the synodic time in accordance with Eqs. (3.8) and (3.9), where the constant c_τ equals

$$c_\tau = 12,23;37,52^{\tau}. \quad (3.34)$$

This algorithm is confirmed by the synodic tables. The constant c_τ can be written as 24^{m} + $23;37,52^{\tau}$. The latter number appears in No. **13** P3'.b, while in No. **14** P4' the constant is rounded to 2^{yr} (to be interpreted as 24^{m}) + 23^{τ}. In order to determine whether c_τ can be reproduced by Eq. (3.13), with $m = 2$ for Mars (Table 3.4), one may as usual set out by assuming e = $11;4^{\tau}$. This would require $\overline{\sigma}$ = 48;43,22,...°, close enough to the system A value (Table 3.26b, bottom) to suggest that c_τ was indeed computed from Eq. (3.13) using $e = 11;4$ and $\overline{\sigma}$ = 48;43,20°, or a value very close to it, and that the result $12,23;37,51,55,33,20^{\tau}$ was rounded to $12,23;37,52^{\tau}$.

3.6.2.3 Subdivision of the synodic cycle

The subdivision of the synodic cycle is rather well documented in the procedure texts (Table 3.27). As mentioned, the zodiacal positions of the primary phenomena FA, S1 and LA are obtained from one and the same step function for σ, while S2 and AR are treated as satellites of S1. The procedures fall into three categories: (i) those concerning pushes between the primary phenomena FA, S1 and LA; (ii) those concerning retrograde pushes (S1 to AR and AR to S2); (iii) those concerning daily motion in pushes or subintervals thereof.

i	$\delta\Sigma$ (FA to S1, S1 to LA, LA to FA)	scheme S_1	No. **14** P1',P2',P11'?
	$\delta\tau$ (idem)	scheme S_2	No. **13** P3'.c,P4'–P9'
ii	$\delta\Sigma$ (S1 to AR)	scheme S	No. **12** T1',P1'
		scheme T	No. **13** P1'
		scheme V	No. **16** P5'?
	$\delta\Sigma$ (AR to S2?)	scheme T?	No. **13** P2'
	$\delta\Sigma$ (S1 to S2)		No. **16** P4'?; No. **43** P8'
	$\delta\tau$ (S1 to S2)		No. **14** P7',P8'?
	subdivision of the synodic cycle		No. **15** P1.c; No. **16** P2',P3'
iii	zigzag function v_1		No. **13** P10'; No. **14** P9'
	zigzag function v_2		No. **43** P7'
	scheme S_3		No. **13** P10'.b; No. **47** P4'

Table 3.27: Mars system A: procedures for the subdivision of the synodic cycle.

ia. Angular pushes between LA, FA and S1 (schemes S_1 and S_2)

Two slightly different schemes for the angular pushes between the primary phenomena LA, FA and S1 are attested or implied in the procedure texts No. **14** (S_1) and No. **13** (S_2).[281] Most, if not all, synodic tables are based on S_1. Since the zodiacal positions of LA, FA and S1 are updated with σ, the angular pushes between them are fully determined once the initial positions are chosen. Since one and the same step function is used for LA, FA and S1 the algorithms for $\delta\Sigma$ can be fully consistent with the positions obtained from σ if certain conditions (§3.3.3.3) are satisfied, which they are. The

[280] For an explanation of this effect cf. §2.6.10.1 and Appendix C.

[281] For these schemes cf. also Aaboe & Sachs (1966), pp. 9–11; *HAMA*, p. 425.

$\delta\Sigma_j$ can therefore be viewed as preliminary values from which $\delta\Sigma$ is computed by applying transition rules (§2.6.10), thus ensuring that $\delta\Sigma$ always contains the same number of elementary steps, irrespective of the zone j. The total number of steps is 33 + 1,3 + 55 = 33 + 1,0 + 58 = 2,31 = $Z+\Pi=N$ (Eq. 3.3).[282] In other words, one obtains exactly the same result if one updates the positions of LA, FA or S1 with σ, or if one updates only one of them with σ and computes the other two from pushes. The purpose of these procedures for the subdivision of the synodic arc is not entirely clear. Perhaps they serve to determine the initial positions of LA, FA and S1, or to verify positions of LA, FA and S1 that were computed from the synodic arc.

	schemes S_1,S_2	scheme S_1		scheme S_2	
j	LA to FA	FA to S1	S1 to LA	FA to S1	S1 to LA
1	1,22;30°	2,37;30°	2,17;30°	2,30°	2,25°
2	55°	1,45°	1,31;40°	1,40°	1,36;40°
3	1,13;20°	2,20°	2,2;13,20°	2,13;20°	2,8;53,20°
4	1,50°	3,30°	3,3;20°	3,20°	3,13;20°
5	2,45°	5,15°	4,35°	5,0°	4,50°
6	2,3;45°	3,56;15°	3,26;15°	3,45°	3,37;30°
δN	33	1,3	55	1,0	58

Table 3.28: Mars system A: angular pushes $\delta\Sigma_j$ between the primary phenomena LA, FA and S1 according to schemes S_1 and S_2. The pushes from LA to FA are the same in both schemes. The bottom row gives the number of elementary steps in each push, $\delta N = \delta\Sigma_j/\delta_j$.

ib. Temporal pushes between LA, FA and S1 (schemes A.S_1 and A.S_2)

The temporal pushes between LA, FA and S1 are dealt with in No. **13** P3'–P9', and they can be inferred from several tabular texts. No. **13** deals with the subdivision of τ in accordance with scheme A.S_2. This remarkable text has been extensively discussed in the literature.[283] In what follows I summarise the results and add a few comments. Some aspects of the algorithm are still not fully understood; cf. the commentary. Recall that one can compute τ from the total synodic arc $\Sigma = \sigma + 6{,}0$ as $\tau = \Sigma + c'_\tau$ (Eq. 3.14), where the modified constant $c'_\tau = c_\tau - 6{,}0$ equals 6,23;37,52$^\tau$. This algorithm is now split up into three separate algorithms, one for each temporal push $\delta\tau_p$ between the primary phenomena (p = 1,2,3):[284]

$$\delta\tau = \delta\Sigma + c'_{\delta\tau}. \tag{3.35}$$

This can be used for computing the time of subsequent parent phenomena within one synodic cycle in accordance with

$$\begin{aligned} T_{\text{S1}} &= T_{\text{FA}} + \delta\tau_1 \\ T_{\text{LA}} &= T_{\text{S1}} + \delta\tau_2 \\ T_{\text{FA}} &= T_{\text{LA}} + \delta\tau_3. \end{aligned} \tag{3.36}$$

No. **13** provides a derivation of $c'_{\delta\tau}$ that reflects a deep understanding of how the subdivision of the synodic arc affects that of the synodic time. The starting point is Eq. (3.15), which for Mars ($m-n=1$) assumes the form $c'_\tau = 0;0{,}10\cdot e\cdot \overline{\Sigma} + 6{,}0\cdot(1+0;0{,}10e)$. The values of $c'_{\delta\tau}$ for the three pushes are obtained by an ingenious construction equivalent to

$$c'_{\delta\tau,p} = 0;0{,}10\cdot e\cdot \overline{\delta\Sigma}_p + r_p\cdot(1+0;0{,}10\cdot e). \tag{3.37}$$

As will become apparent, both terms of this expression are computed in such a way that the three $c'_{\delta\tau,p}$ add up to c'_τ, which in its turn guarantees that the $\overline{\delta\tau}_p$ add up to $\overline{\tau}_p$. Hence the subdivision of the synodic time is fully consistent with Eq. (3.14).[285]

The division of the first term, $0;0{,}10e\cdot\overline{\Sigma}$, called '*uddazallû*-coefficient of the position' in No. **13** P6', is a consequence of the division of $\overline{\Sigma}$ into mean angular pushes $\overline{\delta\Sigma}_p$. The latter is achieved by assigning to each mean push a fraction $\delta N_p/N$ of $\overline{\Sigma}$ (Table 3.28), i.e. $\overline{\delta\Sigma}_p = \overline{\Sigma}\cdot\delta N_p/N$, where $N = Z+\Pi$ is the number of elementary steps contained in Σ. The values of $0;0{,}10e\cdot\overline{\delta\Sigma}_p$ mentioned in No. **13** P6' (Table 3.29) imply that they were actually obtained from the equivalent expression $\overline{\delta\Sigma}_p = \overline{\sigma}\cdot\delta N_p/Z$,[286] with $\overline{\sigma}$ = 48;43,18,30°, and $0;0{,}10e$ = 0;1,50,40 (Table 3.5). The division of $6{,}0\cdot(1+0;0{,}10e)$ over

[282] This consistency is not immediately obvious from the values of $\delta\Sigma_j$ and Σ_j, since the former do not have to add up to Σ_j.

[283] *ACT*, pp. 382–388; van der Waerden (1957); *HAMA*, pp. 408–412; Swerdlow (1998), pp. 147–152.

[284] This interpretation goes back to van der Waerden (1957), p. 51. In *ACT* Neugebauer erroneously assumed that P3'–P4a' concern only the mean values $\overline{\delta\tau} = \overline{\delta\Sigma} + c'_{\delta\tau}$. In *HAMA*, pp. 406–412, certain notational changes suggest that this interpretation is abandoned, but his Eq. (24) for $\overline{\delta\tau}$ ($\Delta\tau_i$ in his notation) implies that they are still considered to be mean pushes, since $c'_{\delta\tau}$ is added to the mean angular push, $s_i\overline{\delta}$ in his notation. Note that his Eq. (24) should have the form $\Delta\tau_i = s_i\overline{\delta} + \gamma_i + (6{,}0+e)/6{,}0$, the equivalent of Eq. (3.35) in the present volume, but γ_i is missing.

[285] As with c_τ, the expression for $c'_{\delta\tau}$ is an approximation in the sense that Σ_p was replaced by $\overline{\Sigma}_p$. From a modern perspective this is justified for the same reason as with c_τ, because $0;0{,}10\cdot e = 0;1{,}50{,}40$ is small compared to 1.

[286] The equivalence of both expressions, $\overline{\Sigma}/N = \overline{\sigma}/Z$, is a direct consequence of Eq. (3.2). If one sets out from the original expression and inserts $\overline{\Sigma}$ = 6,48;43,18,30° one obtains $\overline{\delta\Sigma}_1$ = 1,29;19,23,54,38,..., $\overline{\delta\Sigma}_2$ = 2;42,24,21,39,20,..., and $\overline{\delta\Sigma}_3$ = 2,36;59,32,56,1,.... which, if rounded or truncated, leads to slightly different numbers compared to those in the text, except perhaps $\overline{\delta\Sigma}_1$. The advantage of the alternative expression is that Z = 18 has a finite-digit reciprocal, unlike N = 2,31.

p	push	δN_p	$\overline{\delta\Sigma}_p$	$0;0,10e \cdot \overline{\delta\Sigma}_p$	r_p	$r_p \cdot (1+0;0,10e)$	$c'_{\delta\tau,p}$
1	LA to FA	33	1,29;19,23,55°	2;44,45,6,46,46,40	30°	30;55,18,53,13,13,20	$33;40,4^{\overline{\tau}}$
2	FA to S1	1,0	2,42;24,21,40°	4;59,32,55,57,46,40	1,45°	1,48;13,40,4,2,13,20	$1,53;13,13^{\overline{\tau}}$
3	S1 to LA	58	2,36;59,32,56,40°	4;49,33,50,5,51,6,40	3,45°	3,51;55,1,9,54,8,53,20	$3,56;44,33,50^{\overline{\tau}}$
total	LA to LA	2,31	6,48;43,18,31,40°	12;33,51,52,47,20	6,0°	6,11;4	$6,23;37,52^{\overline{\tau}}$

Table 3.29: Mars system A: parameters for the subdivision of the synodic time (scheme A.S_2).

the three pushes (No. **13** P7'–P9') is achieved by assigning to each push a portion $r_p \cdot (1+0;0,10e)$, where r_p are angular distances satisfying the constraint $r_1 + r_2 + r_3 = 6,0°$. As explained in §3.3.2.1, $6,0 \cdot (1+0;0,10e)$ is the time that it takes for the elongation between Mars and the Sun to change by 6,0° corresponding to 1 synodic cycle. From this, van der Waerden (1957) correctly concluded that r_p is the change in elongation between Mars and the Sun associated with push p. He also made the plausible suggestion that the resulting configuration of Mars and the Sun is such that FA and LA occur at elongations of −15°, +15°, respectively, which implies that Mars' elongation at S1 is 2,0° (Fig. 3.14a). The parameters r_p can thus be viewed as yet another manifestation of the solar-distance principle, albeit one that thus far appears to be unique to Mars system A. The extent to which these sophisticated algorithms for the temporal pushes are applied in the synodic tables will be discussed in Volume II. Although the algorithms in No. **13** belong to scheme S_2, while the extant synodic tables follow S_1, they could be easily adapted for S_1 by repeating the derivation with modified values of δN_p.

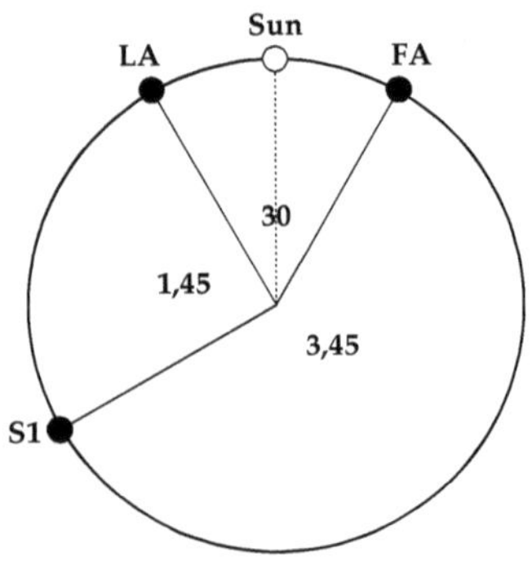

Figure 3.14a: Mars system A: relative positions of the primary phenomena with respect to the Sun as defined by the angles r_p.

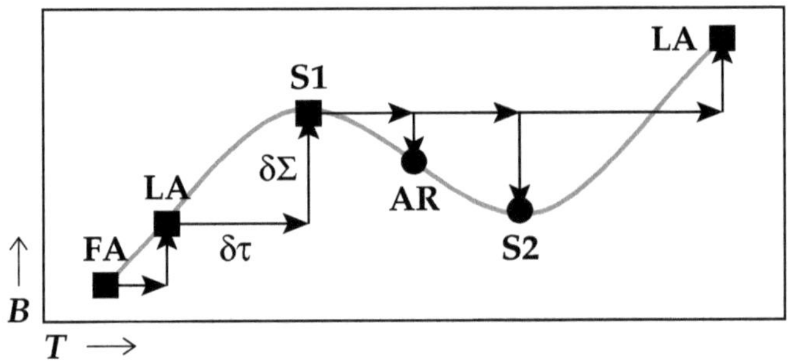

Figure 3.14b: Mars system A: subdivision of the synodic cycle.

ii. Retrograde push between S1 and S2

The procedure texts, auxiliary tables and synodic tables contain evidence of five different schemes for the retrograde arc known in the literature as R, S, T, U and Q (Table 3.30). Several arguments suggest that they are part of system A. First, in system A AR and S2 are not primary phenomena. Second, scheme S is attested in synodic tables of system A. Third, scheme T occurs in No. **13** together with other procedures for system A.

To begin with $\delta\Sigma$(S1 to AR): in schemes S, U, Q, and perhaps T, linear interpolation is used for computing its value at intermediate positions between the control points (c is the interpolation coefficient). Schemes R and S share the same control values, but in scheme R $\delta\Sigma$ is constant in both signs of each zone, while in scheme S interpolation is applied in the second sign of each zone (cf. Fig. 3.15). No. **12** P1' contains a list of values that may pertain to R or S, but the presence of an auxiliary table for scheme S on the same tablet suggests that they actually belong to scheme S. Most synodic tables seem to follow scheme S. Scheme T is attested in two procedures, each listing one control value for each zone. Neugebauer assumes that, analogous to scheme R, no interpolation is applied, so that $\delta\Sigma$ is constant within each zone, but that is not certain. Scheme U is attested in an auxiliary table.[287] Scheme Q was reconstructed by Steele (2005b) from the control values of $\delta\Sigma$ (AR to S2) contained in an auxiliary table[288] under the assumption that the ratio between them is the same as in schemes R and S. However, the assumption of a fixed ratio between these pushes is violated in schemes T and U, as will be argued, so that this reconstruction may turn out to be incorrect. Note that for schemes S, U and perhaps T, the control values are anchored at 0° of the zodiacal signs, but for scheme Q they pertain to 1°. Scheme V (not shown in Fig. 3.15) has been tentatively reconstructed from No. **16** P5'. The interpolation method for this scheme is not known.

---

[287] U 97+180(26) (*ACT* 804).

[288] W 20030,111, from the library of the Rēš, which suggests a date in the range 200–150 BC.

	$\delta\Sigma$(S1 to AR)					$\delta\Sigma$(AR to S2)				$\delta\Sigma$(S1 to S2)		
	R,S	T	U	Q?	V	R,S	T?	U	Q	R,S,T?	U	No. **43** P8'
Tau	6;48°	7°	6;45°	*6;36°*	*8°*	10;12°	10°	11;32,30°	9;54°	17°	18;17,30°	
Gem	”	”	7°	*6;48°*	”	”	”	11;45°	10;12°	”	18;45°	
Cnc	7;12°	7;30°	7;15°	*7°*	*7°*	10;48°	*10;30°*	11;57,30°	10;30°	18°	19;12,30°	19;30°
Leo	”	”	7;30°	*7;12°*	”	”	”	12;10°	10;48°	”	19;40°	”
Vir	6;48°	7°	7;15°	*7°*	6°	10;12°	*10°*	11;57,30°	10;30°	17°	19;12,30°	
Lib	”	”	7	*6;48°*	”	”	”	11;45°	10;12°	”	18;45°	
Sco	6;24°	6;30°	6;45	*6;36°*	5°	9;36°	9;30°	11;32,30°	9;54°	16°	18;17,30°	
Sgr	”	”	6;30°	*6;24°*	”	”	”	11;20°	9;36°	”	17;50°	
Cap	6°	6°	6;15	*6;12°*	6°	9°	*9°* (10;45°)	11;7,30°	9;18°	15°	17;22,30°	
Aqr	”	”	6°	*6°*	”	”	”	10;55°	9°	”	16;55°	
Psc	6;24°	6;30°	6;15°	*6;12°*	7°	9;36°	9;30°	11;7,30°	9;18°	16°	17;22,30°	
Ari	”	”	6;30°	*6;24°*	”	”	”	11;20°	9;36°	”	17;50°	
c	0;0,48	0?	0;0,30	*0;0,24*	?	0;1,12	0?	0;0,25	0;0,36	0;2	0;0,55	

Table 3.30: Mars system A: control values and interpolation coefficients c for $\delta\Sigma$ from S1 to AR, AR to S2 and S1 to S2 (uncertain reconstructions in italics). In scheme R the values of $\delta\Sigma$ apply to whole zodiacal signs ($c = 0$); in schemes S and U, perhaps also T, they pertain to 0° of the signs; in scheme Q to 1°. For scheme S interpolation is used in every second sign, setting out from Gem, while $c = 0$ in the other signs. Scheme T for AR to S2 has been completed on the assumption that the value $\delta\Sigma$ (Cap–Aqr) = 10;45° mentioned in No. **13** P2' is an error for 9°. Scheme V has been tentatively reconstructed from No. **16** P5'.

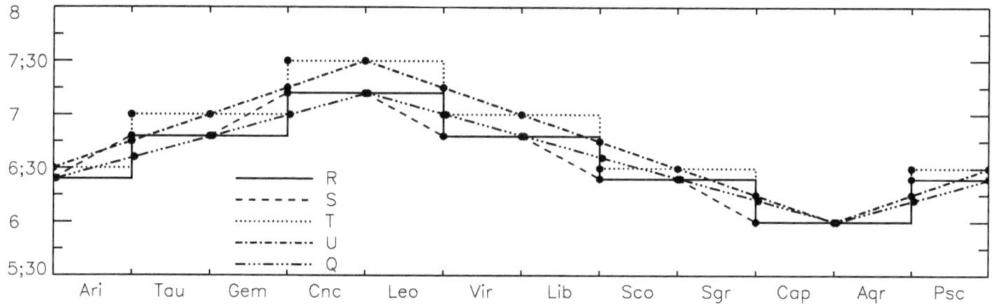

Fig. 3.15: Mars system A: angular push $\delta\Sigma$ (S1 to AR) for schemes R, S, T, U and Q as a function of the zodiacal position of S1. The control values are indicated by dots.

Schemes R, S, U, Q and perhaps T also provide values of $\delta\Sigma$ for the push from AR to S2 (Fig. 3.16). Neugebauer reconstructed control values for scheme R from the synodic table *ACT* 503. They coincide with the values for scheme S, which is attested in several synodic tables.[289] As pointed out by Neugebauer, they satisfy the relation

$$\delta\Sigma\,(\text{AR to S2}) = 1;30 \cdot \delta\Sigma\,(\text{S1 to AR}). \tag{3.38}$$

Scheme U was reconstructed by Neugebauer from the auxiliary table on the reverse of *ACT* 804, which contains control values at intervals of 1° (!) for several zodiacal signs. I assume that they represent $\delta\Sigma$ (AR to S1), because the values of $\delta\Sigma$ (S1 to AR) on the obverse of the same tablet belong to that scheme. Neugebauer did not draw that conclusion, nor did he give any other interpretation for these numbers. The reason for his hesitation is presumably that the numbers deviate from the other schemes by about 2°.[290] Moreover, they do not satisfy Eq. (3.38) or any other relation involving a fixed ratio.[291] However, Neugebauer & Sachs (1967) report that the zodiacal positions of S2 in the synodic table BM 36301,[292] which probably belongs to system A, imply values of $\delta\Sigma$ (S1 to S2) that are indeed about 2° larger than in scheme S, which supports my interpretation.

No. **13** P2' is a fragmentary procedure with several numbers probably representing $\delta\Sigma$ (AR to S2) for scheme T. That interpretation is preliminary, because it makes good sense only if one corrects the value for Cap–Aqr as indicated in Table 3.30. There is not a fixed ratio between these values and $\delta\Sigma$ (S1 to AR). The values of $\delta\Sigma$ (S1 to S2) are the same as in schemes R and S. Finally, scheme Q was reconstructed by Steele (2005b) from the mentioned auxiliary table.

By adding together both pushes one obtains the total angular push from S1 to S2, control values of which seem to be mentioned in a few fragmentary procedures. The newly discovered No. **43** P8' contains a value for the region Cnc–Leo that is similar to the value of scheme U. The badly understood No. **16** P4' mentions the number 14 for Vir–Lib, perhaps a variant of the same push.

[289] *ACT* 504; BM 36810+36947 = Aaboe & Sachs (1966), Text G.

[290] Cf. *ACT*, pp. 306, 375.

[291] Neugebauer invoked Eq. (3.38) in order to tentatively reconstruct $\delta\Sigma$ (AR to S2) for scheme U (*HAMA*, p. 459).

[292] Neugebauer & Sachs (1967), Text C.

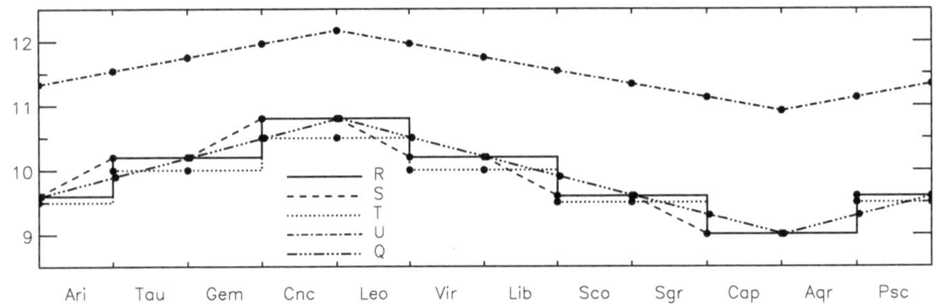

Fig. 3.16: Mars system A: $\delta\Sigma$ from AR to S2 for schemes R, S, T, U and Q as a function of the zodiacal position of AR. The control values are indicated by dots.

Next to nothing is known about $\delta\tau$ on the retrograde arc. A reconstruction of the corresponding algorithm from the tabular texts is still outstanding. In the only extant procedure dealing with that topic, No. **14** P7', not a single numerical value is preserved.

iii. Daily motion schemes: v_1, v_2 and scheme S_3

The most detailed description of the subdivision of the synodic cycle for Mars is found in several procedures providing values of the 'daily' displacement v for subintervals of pushes. Embedded in these procedures are two zigzag-type functions for v (Table 3.31) which I interpret as being expressed in degrees per mean tithi. No. **13** P10'.a and No. **14** P9' provide 12 control values of v_1, one for each zodiacal sign, forming a zigzag function of the zodiacal position with $m = 0;30°/\bar{\tau}$ in Cnc, $M = 0;50°/\bar{\tau}$ in Cap and $c = 0;3,20°/\bar{\tau}$ per zodiacal sign $= 0;0,6,40°/\bar{\tau}$ per °. The procedures do not explain how v_1 is determined in between the control values (e.g. piecewise constant in every sign or by means of linear interpolation). No. **13** P10' implies that v_1 describes Mars's motion near FA or LA.

Function v_2, attested in the newly discovered No. **43** P7', is defined with 6 control values, each pertaining to one zone of system A. They form a zigzag function with $m = 0;36°/\bar{\tau}$ in zone 2 (Cnc–Leo) and $M = 0;54°/\bar{\tau}$ in zone 5 (Cap–Aqr). No. **43** P7' does not explain how v_2 is determined in between the control values, and what is modeled by v_2, but the numerical values suggest that it describes daily motion between FA and LA.

		v_1 [°/$\bar{\tau}$]	v_2[°/$\bar{\tau}$]	scheme S_3	$\delta\tau$	v [°/$\bar{\tau}$]	$\delta\Sigma$
Cnc	$m =$	0;30	0;36	FA to S1 (1)	$3,10^{\bar{\tau}}$	0;40	2,6;40°
Leo		0;33,20	"	(2)	$30^{\bar{\tau}}$	0;36	18°
Vir		0;36,40	0;42	(3)	$30^{\bar{\tau}}$	0;24	12°
Lib		0;40	"	(4)	$30^{\bar{\tau}}$	0;12	6°
Sco		0;43,20	0;48	total	$4,40^{\bar{\tau}}$		2,42;40°
Sgr		0;46,40	"	S1 to S2	?	−0;12	?
Cap	$M =$	0;50	0;54	S2 to LA (1)	$30^{\bar{\tau}}$	0;12	6°
Aqr		0;46,40	"	(2)	$30^{\bar{\tau}}$	0;24	12°
Psc		0;43,20	0;48	(3)	$30^{\bar{\tau}}$	0;36	18°
Ari		0;40	"	(4)	$3,53^{\bar{\tau}}$	0;40	2,35;20°
Tau		0;36,40	0;42	total	$5,23^{\bar{\tau}}$		3,11;20°
Gem		0;33,20	"				

Table 3.31: Mars: control values of v_1 and v_2, the daily displacement near FA and LA and scheme S_3 for the subdivision of the synodic cycle (values of $\delta\Sigma$ reconstructed as $\delta\tau \cdot v$).

In No. **13** P10' v_1 is followed by scheme S_3 (Table 3.31), which concerns Mars's motion from FA to S1 and from S2 to LA. The resulting value of $\delta\Sigma$ (FA to S1) is nearly the same as that prescribed for zone 1 in scheme S_1 (Table 3.28). I have identified a version of the same scheme in the hitherto unexplained No. **47** P4'. That procedure includes a cryptic description of the motion between S1 and S2 which suggests that for a total of $\frac{1}{3}$ of the duration of that interval, presumably at the beginning and the end, Mars moves at $0;12°/\bar{\tau}$, and for $\frac{2}{3}$, presumably centred at the middle, it moves at $0;24°/\bar{\tau}$. Also, the interval S2 to LA (3) is here followed by 3 further intervals of $30^{\bar{\tau}}$ during which Mars moves at 0;40 (as in Table 3.31), 0;50 and again $0;40°/\bar{\tau}$; the rest of the interval from S2 to LA is omitted. The value of v near FA and LA coincides with v_1 in Ari and Lib, which suggests that v_1 describes Mars's daily motion when FA or LA occurs in these zodiacal signs. Since no daily motion tables have been found for Mars, scheme S_3 is not yet attested in tabular texts.

3.6.2.4 Net displacements for various intervals

No. **44** P3' lists net displacements $dB_j(s)$ for several intervals s longer than 1 synodic cycle (Table 3.32), namely for 47 and 79^{yr}. However, the values of dB_3 and dB_6 for 79^{yr} mentioned in the text (in brackets in Table 3.32) deviate from the exact values, probably due to rounding. No. **44** P3' also includes the correct statement that $4,44 = 284^{yr}$ (= Y) produces an exact return of the zodiacal position. As usual the net displacements are identified by the approximate number of years y, which for Mars equals $t+2s$ (Eq. 2.39), where t is the corresponding number of revolutions.

s	t	y	dB_1	dB_2	dB_3	dB_4	dB_5	dB_6	dB_j/δ_j
7	1	15^{yr}	−17;30°	−11;40°	−15;33,20°	−23;20°	−35°	−26;15°	−7
15	2	32^{yr}	10°	6;40°	8;53,20°	13;20°	20°	15°	4
22	3	47^{yr}	−7;30°	−5°	−6;40°	−10°	−15°	−11;15°	−3
37	5	79^{yr}	2;30°	1;40°	2;13,20 (2;15°)	3;20°	5°	3;45 (3;40°)	1

Table 3.32: Mars system A: net displacements for various intervals.

3.6.3 System B

System B, discovered by P. Huber, is represented by one procedure (No. **13** P11'.b) and one synodic table.[293]

3.6.3.1 Updating *B* and *T*

Procedures for updating B and T with the zigzag functions for σ and τ are not preserved. However, No. **13** P11'.b contains a derivation of the defining parameters of σ (cf. below). The synodic table preserves only columns σ and B for one unidentified phenomenon. Since P does not have a finite-digit reciprocal the mean synodic arc $\overline{\sigma} = 6{,}0/P$ = 48;43,18,29,46,...° is not a finite sexagesimal number, but it is well approximated by μ. There is no textual evidence for the zigzag function for τ, but I have reconstructed its defining parameters (Table 3.33) assuming that τ is related to σ through Eq. (3.9), and c_τ is the same as in system A (Eq. 3.34).

	d	m	M	Δ	μ	Π	Z	Y	P
σ	17°	17;19,8,30°	1,20;7,28,30°	1,2;48,20°	48;43,18,30°	2,13	18	$4{,}44^{yr}$	7;23,20
τ	$17^{\overline{\tau}}$	$12{,}40;57{,}0{,}30^{\overline{\tau}}$	$13{,}43;45{,}20{,}30^{\overline{\tau}}$	$1{,}2;48{,}20^{\overline{\tau}}$	$13{,}12;21{,}10{,}30^{\overline{\tau}}$	"	"	"	"

Table 3.33: Mars system B: parameters of the zigzag functions for the synodic arc (σ) and the synodic time (τ).

Derivation of the defining parameters of σ

In No. **13** P11'.b the defining parameters of σ are derived, setting out from the period relation for system A (values of Π and Z). The difference d = 17° is given, from which 2Δ is computed as $\Pi d/Z$ (Eq. 2.16 on p. 44). The mean synodic arc $\overline{\sigma}$ is then computed as $6{,}0° \cdot Z/\Pi$ (Eq. 3.5 on p. 60). Since Π does not have a finite-sexagesimal reciprocal, the result is rounded to 48;43,18,30°, which is adopted as μ of the zigzag function for σ. By adding $\Delta/2$ to μ and subtracting it from μ the extrema M and m can be calculated, but these operations are not preserved.

3.7 Jupiter

Jupiter is represented by the largest variety of computational systems of all planets: A, A_1, A_2, A', A", A"', A_6, B, B' and B". There is also much variation among the schemes for the subdivision of the synodic cycle. Some of these are not clearly associated with any of the computational systems.

3.7.1 Synodic cycle

Jupiter performs one revolution around the Sun in 11.87^{yr}, which results in a mean synodic time $\overline{\tau} = 398.88^d \approx 6{,}38;53^d \approx 6{,}45;13^{\overline{\tau}}$, and a mean (total) synodic arc $\overline{\Sigma} = \overline{\sigma} = 33;7°$.

[293] U 106 = *ACT* 510 ('system X'). Huber's results are reported in Aaboe (1958), p. 246; cf. also *HAMA*, p. 457.

3.7.2 System A

System A, discovered by Kugler,[294] is represented by at least 12 (at most 16) procedure texts (Table 3.34), 10 synodic tables and 2 template tables.[295] None of the procedure texts can be securely dated. The initial dates of the synodic tables cover the years SE 113–217, and the synodic table *ACT* 601 was written in SE 118. This provides a rough indication of the minimal period when system A was in use.

updating B with σ	No. **17** P1; No. **18** P2.a; No. **19** P1; No. **21** P2'; No. **23** P2'.a; No. **25** P2'.a; No. **26** P2'; No. **28** P1
zonal boundaries for σ	No. **18** P10.a
updating T with τ	No. **17** P1.b; No. **28** P1.b
list of net displacements $dB_j(s)$	No. **18** P1.b',P20; No. **23** P1'.b'; No. **25** P1'.b; No. **26** P1'.c; No. **46** P10'
setting time at LA, rising time at FA	No. **18** P1.c',P9b; No. **22** P1.b?; No. **23** P1'.c'; No. **25** P1'.c; No. **26** P1'.d
subdivision synodic cycle: scheme S_1	No. **18** P2.b,c;P10.b?; No. **19** P2?, P5 (variant); No. **23** P2'.b'; No. **25** P2'.b; No. **27** P2',P3'; No. **30** P1'.a'
scheme S_2	No. **18** P1.a'; No. **23** P1'.a'; No. **25** P1'.a'; No. **26** P1'.a
scheme S_3	No. **18** P23,P30,P31
scheme S_4	No. **18** P24
scheme S_5	No. **18** P11.a,b
zigzag function v	No. **18** P4,P27; No. **23** P2'.d; No. **25** P2'.d; No. **29** P1'.c'; No. **30** P1'.c'

Table 3.34: Jupiter system A: procedures.

3.7.2.1 Composition of the tablets

Only few procedure texts for Jupiter system A exhibit systematic features in terms of the selection and order of the procedures. In particular there are 7 procedures that often appear together on one tablet, always in the same sequence (Table 3.35). On several tablets some of them are merged into composite procedures. Most of the procedures of the sequence belong to system A, but this is not obvious for the procedure concerning Jupiter's distance to the ecliptic (cf. §3.7.10).

procedure	No. 18	No. 23	No. 25	No. 26	No. 29	No. 30
subdivision synodic cycle, scheme A.S_2 (PUSH.C)	P1.a'	P1'.a'	P1'.a	P1'.a'		
list of net displacements zone 1 (DIFF.A)	P1.b'	P1'.b'	P1'.b	P1'.c'		
rising time at FA, setting time at LA	P1.c'	P1'.c'	P1'.c	P1'.d'		
updating B with σ (STEP.E)	P2.a	P2'.a	P2'.a	P2'		
subdivision synodic cycle, scheme A.S_1 (PUSH.D)	P2.b,c	P2'.b	P2'.b		P1'.a'	P1'.a'
distance to the ecliptic at the stations	P3	P2'.c	P2'.c		P1'.b'	P1'.b'
control values of zigzag function for v	P4	P2'.d	P2'.d		P1'.c'	P1'.c'

Table 3.35: Jupiter system A: tablets with duplicates of the same sequence of procedures.

3.7.2.2 Updating B with the synodic arc

The algorithm for updating B with the step function for σ (Table 3.36, Fig. 3.17) is preserved in 8 procedures. No. **28** P1 and perhaps No. **17** P1 are based on template STEP.A.1 (§2.6.10). The other six, all part of the sequence of procedures shown in Table 3.35, are based on STEP.E, which lacks explicit transition rules. No. **18** P10.a provides only the zonal boundaries (STEP.F). The synodic tables imply that the same step function is used for all 6 synodic phenomena. Hence in system A all synodic phenomena are primary phenomena.

i	b_j	α_j	σ_j	r_j	δ_j	v_j
1	25° Gem	2,35°	30°	1;12	0;50°	3,6
2	30° Sco	3,25°	36°	0;50	1°	3,25

Π	Z	Y	P	$\overline{\sigma}$
6,31	36	$7{,}7^{\mathrm{yr}}$	10;51,40	33;8,44,48,...°

Table 3.36: Jupiter system A: parameters of the step function for σ, periods of B and mean synodic arc.

[294] Cf. the appendix to Kugler (1900), pp. 207–211.

[295] ST: *ACT* 600–608, 604a; TT: Aaboe & Sachs (1966), Texts D, F.

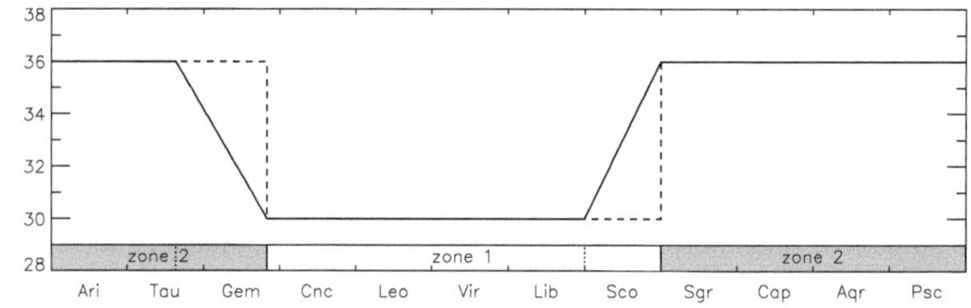

Fig. 3.17: Jupiter system A: graphical representation of σ as a function of B_{i-1}. The dashed lines represent the preliminary values σ_j. The dotted tick marks near the horizontal axis indicate transitions at which the influence of a subsequent zone becomes noticeable.

3.7.2.3 Updating T with the synodic time

Two procedures contain traces of the algorithm for updating T (Eqs. 3.8–3.9), very likely formulated in accordance with template T.1 or T.2. The synodic tables confirm that T is computed in that manner, with

$$c_{\tau} = 6{,}12;5{,}10^{\tau}, \tag{3.39}$$

which amounts to 12^{m} + $12;5{,}10^{\tau}$. This parameter is very close to the empirical value (§3.7.1) and can be reproduced in the usual manner from Eq. (3.13) with $m = 1$, if one assumes e = $11;4^{\tau}$ and $\overline{\sigma} \approx 33;8{,}45°$ (Table 3.36), and if one rounds the outcome $6{,}12;5{,}8{,}8{,}20^{\tau}$ to $6{,}12;5{,}10^{\tau}$.[296]

3.7.2.4 Net displacements for various intervals

Six procedures provide net displacements for intervals longer than 1 synodic cycle. All employ template DIFF.A. The parameters are compiled in Table 3.37. Recall that the closest whole number of years y equals $s+t$ for Jupiter (Eq. 2.39). For some reason the procedures in Table 3.35, as far as they are preserved, provide only values of dB for zone 1; this is certain for No. **18** P1.b'. Values for zone 2 are listed in No. **18** P20 and No. **46** P10'.[297] Some procedures also correctly mention that Jupiter returns to exactly the same position after 7,7 = 427^{yr} (= Y).

			A, A',A''		A',A''	
s	t	y	dB_1	dB_2	$dB_2 = dB_4$	dB_j/δ_j
11	1	12^{yr}	4;10°	5°	4;41,15°	5
1,5	6	$1{,}11^{\mathrm{yr}}$	−5°	−6°	−5;37,30°	−6
1,16	7	$1{,}23^{\mathrm{yr}}$	−0;50°	−1°	−0;56,15°	−1
1,27	8	$1{,}35^{\mathrm{yr}}$	3;20°	4°	3;45°	4
2,32	14	$2{,}46^{\mathrm{yr}}$	−1;40°	−2°	−1;52,30°	−2
3,59	22	$4{,}21^{\mathrm{yr}}$	1;40°	2°	1;52,30°	2

Table 3.37: Jupiter systems A, A' and A'': net displacements for various intervals. The values of dB_3 for systems A' and A'' coincide with the values of dB_2 for systems A, A' and A''.

3.7.2.5 Rising time at FA and setting time at LA

Several procedures, all embedded in the mentioned sequence (Table 3.35), are concerned with what I interpret as the time between sunset and the setting of Jupiter at LA (T_{set}), and the time between the rising of Jupiter and sunrise at FA (T_{rise}). For a graphical representation of these intervals cf. Fig. 3.3 (§3.1.2). Since these procedures had hitherto defied explanation, partly because of erroneous transliterations, a close investigation is warranted, for which reason the better preserved ones are presented here as a score:[298]

No. **18** Oi6'–7'	*ina* **ka**$_2$-*tu*$_2$ *ša*$_2$ 30 12.⌜30$^?$⌝/**be igi**	9 x$^?$ **be šu**$_2$	*ina*	**ka**$_2$-*tu*$_2$	*ša*$_2$	36	15 **be**	**igi** 7 ⌜**be šu**$_2$⌝	
No. **18** Oii20'	*ina* **ka**$_2$-*tu*$_2$ *ša*$_2$ 30 11 **be šu**$_2$	10 **be igi**	⌜*ina*	**ka**$_2$⌝-[*tu*$_2$	*ša*$_2$	36	xx **be**	**šu**$_2$ x]	
No. **25** O13'	[...		*ina*	**ka**$_2$-*tu*$_2$]	⌜*ša*$_2$⌝	36	15 **be**	**igi**	
No. **26** X8'	[...		*ina*	k]**a**$_2$-*tu*$_2$	*ša*$_2$	36 :	15 **be**	[**igi** ...]	

A hitherto overlooked but obvious interpretation is that 30 = σ_1, and 36 = σ_2. This implies that the procedures belong to system A, and that the underlying template can be represented as

[296] The unrounded value is attested for system B (Eq. 3.43).

[297] Uniquely in No. **18** P20 the preposition *ana* in *ana y* **mu.meš**, 'for y years', is written **a.ra**$_2$.

[298] Omitting No. **22** P1.b and No. **23** P1'.

ina bābtu(**ka$_2$**-*tu$_2$*) *ša$_2$* σ_j T_{rise} *šumma*(**be**) *innamer*(**igi**) T_{set} **be** *irabbi*(**šu$_2$**)	In the 'ward' of σ_j: T_{rise} if it appears (FA), T_{set} if it sets (LA).

Among the possible meanings of *bābtu*, 'ward' is suitable if one interprets it as a technical term for zodiacal zone.[299] For an astronomical interpretation we may explore the properties of T_{rise} and T_{set}. In zone 1 T_{rise} = 12;30 or 10 and T_{set} = 9 or 11; in zone 2 T_{rise} 15 and T_{set} = 7. Since two different pairs of T_{rise} and T_{set} are attested for zone 1, we conclude that there is no strict correspondence between a zone and the values of T_{rise} or T_{rise}. An interpretation of T_{rise} and T_{set} as the time between the rising of Jupiter at FA and sunrise, and that between the setting of Jupiter at LA and sunset, measured in time degrees, is roughly consistent with the empirical values of T_{rise} and T_{set}. Modern computations suggest that both are typically 11–14° (44–56 min.). It remains to be explained why they are different, and why their sum $T_{\text{rise}} + T_{\text{set}}$ is in the narrow range 21–22°.

3.7.2.6 Subdivision of the synodic cycle

The synodic tables and the procedures imply that all synodic phenomena are primary phenomena, whose positions are computed with one and the same step function for σ (Fig. 3.18). Hence the subdivision of the synodic cycle is fully determined by the subdivision of the initial cycle, suggesting that the purpose of these procedures is twofold. First, some may be used for establishing the initial subdivision of a synodic table, or verifying subsequent zodiacal positions that were updated with σ. Second, those schemes that provide or imply reasonably short finite sexagesimal values of v may be used for computing daily positions.

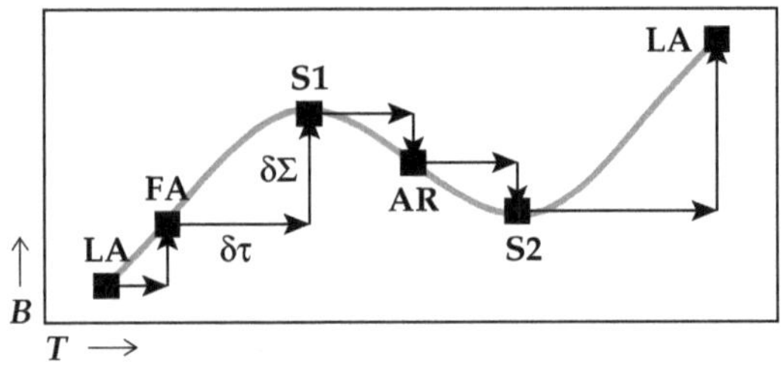

Fig. 3.18: Jupiter system A: subdivision of the synodic cycle.

Scheme A.S$_1$

In scheme A.S$_1$ (Table 3.38a), attested in numerous procedure texts, the angular pushes $\delta\Sigma_j$ constituting 1 synodic cycle add up to the (total) synodic arcs $\Sigma_j = \sigma_j$ in each zone j. As explained in §3.3.3.3, this is a precondition for the subdivision to be fully consistent with the positions obtained from the step function for σ. Transition rules, required for full consistency, are not explicitly mentioned in any of the procedure texts, but may be considered to be implied. The following templates are attested. PUSH.D (which provides only $\delta\Sigma$) is used in the procedures that belong to the sequence of Table 3.35. PUSH.B (which provides v and $\delta\tau$) is used in No. **18** P10.b; PUSH.A.1 (v, $\delta\tau$ and $\delta\Sigma$) in No. **27** P2',P3'. No. **19** P2, whose connection with system A is not certain, covers only the push from LA to FA using templates PUSH.B.3 and PUSH.C.1.

Minor variants of scheme A.S$_1$

Several procedure texts and tabular texts contain evidence of variants of scheme A.S$_1$. The differences mainly concern the retrograde motion between S1 and S2, a feature also known from Mars system A. One variant, partly preserved in No. **19** P5' (not shown in Table 3.38b), mentions a value $\delta\Sigma$ = 5° perhaps pertaining to LA to FA in zone 1. Variants attested only in tabular texts will be dealt with in Volume II and are not included in Table 3.38b.

[299] This is confirmed by a similar procedure for Jupiter system A''', where *qaqqar*(**ki**), 'place' is used instead of *bābtu* (No. **21** R9'–11'), and by No. **18** P8.b, where the same phrase T_{set} **be šu$_2$** T_{rise} **be igi** is preceded by 'from 10 Cap until [...]'.

scheme A.S$_1$						
push		$\delta\tau$	v_1 [°/$\bar{\tau}$]	$\delta\Sigma_1$	v_2 [°/$\bar{\tau}$]	$\delta\Sigma_2$
LA to FA		$30^{\bar{\tau}}$	0;12	6°	0;14,24	7;12°
FA to S1	(1)	$30^{\bar{\tau}}$	0;12,30	6;15°	0;15	7;30°
	(2)	$1,30^{\bar{\tau}}$	0;6,40	10°	0;8	12°
	total	$2,0^{\bar{\tau}}$		16;15°		19;30°
S1 to AR	(1)	$30^{\bar{\tau}}$	−0;2	−1°	−0;2,24	−1;12°
	(2)	$30^{\bar{\tau}}$	−0;6	−3°	−0;7,12	−3;36°
	total	$1,0^{\bar{\tau}}$		−4°		−4;48°
AR to S2	(1)	$30^{\bar{\tau}}$	−0;8	−4°	−0;9,36	−4;48°
	(2)	$30^{\bar{\tau}}$	−0;4	−2°	−0;4,48	−2;24°
	total	$1,0^{\bar{\tau}}$		−6°		−7;12°
S1 to S2		$2,0^{\bar{\tau}}$		−10°		−12°
S2 to LA	(1)	$30^{\bar{\tau}}$	0;4,30	2;15°	0;5,24	2;42°
	(2)	$38^{\bar{\tau}}$	0;7,30	4;45°	0;9	5;42°
	or (2)	$30^{\bar{\tau}}$	0;9,30	4;45°	0;11,24	5;42°
	(3)	*1,4*$^{\bar{\tau}}$	*0;10,45*?	10;45°	0;12,54?	12;54°
	total	$2,12^{\bar{\tau}}$?		17;45°		21;18°
LA to LA		$6,42^{\bar{\tau}}$?		30°		36°

scheme A.S$_2$						
LA to FA		*30*$^{\bar{\tau}}$, $28^{\bar{\tau}}$?	*0;12*	6°		
FA to S1	total	*2,0*$^{\bar{\tau}}$		16;45°		
S1 to AR	(1)	*30*$^{\bar{\tau}}$	*0*	*0°*		
	(2)	*30*$^{\bar{\tau}}$	−0;9	*−4;30°*		
	total	*1,0*$^{\bar{\tau}}$		*−4;30°*		
AR to S2		*1,4*$^{\bar{\tau}}$		−6°		
S2 to LA	(1)	*30*$^{\bar{\tau}}$	0	0°		
	(2)	*30*$^{\bar{\tau}}$	0;10	5°		
	(3)	*34*$^{\bar{\tau}}$	0;10	*5;40°*		
	(4)	*34*$^{\bar{\tau}}$	0;12,30	*7;5°*		
	total	*2,8*$^{\bar{\tau}}$		17;45°		
LA to LA		$6,42^{\bar{\tau}}$		30°		

Table 3.38a: Jupiter: subdivision of the synodic cycle, schemes A.S$_1$ and A.S$_2$. In italics: uncertain reconstructions.

scheme A.S$_3$						
push	$\delta\tau_1$	v_1 [°/$\bar{\tau}$]	$\delta\Sigma_1$	$\delta\tau_2$	v_2 [°/$\bar{\tau}$]	$\delta\Sigma_2$
LA to FA			6;15°	$29^{\bar{\tau}}$	*0;15*	7;30°
FA to S1			16;15°	$2,4^{\bar{\tau}}$	*0;8,45*	17;30°
S1 to AR			−5;10°	$58^{\bar{\tau}}$	*−0;4*	−4°
AR to S2			*−3;15°*	$1,4^{\bar{\tau}}$	*−0;4*	−4°
S2 to LA			15°	$2,10^{\bar{\tau}}$	*0;8,45*	17;30°
LA to LA						*34;30°*

scheme A.S$_4$						
LA to FA						7;30°
FA to S1						18°
S1 to AR						−4°
AR to S2						−5°
S2 to LA						19°
LA to LA						*35;30°*

scheme A.S$_5$						
LA to FA	$27^{\bar{\tau}}$	*0;12,13,20*	5;30°	$32^{\bar{\tau}}$	0;10,18,45	5;30°
FA to S1			15;37,30°			*18;45°*
S1 to AR			−5°			−6°
AR to S2	$54^{\bar{\tau}}$	−0;5,33,20	−5°			−6°
S2 to LA			18;20°	$2,15^{\bar{\tau}}$	0;9,46,40	22°
LA to LA			29;27,30°			*35;15°*

Table 3.38b: Schemes A.S$_3$, A.S$_4$ and A.S$_5$.

Scheme A.S$_2$

Scheme A.S$_2$ is attested in four procedures, all part of the sequence of Table 3.35, but the full scheme is not preserved. The following score edition may clarify some of the difficulties. It begins at the point where No. **25** P1' can at least be partly understood.[300]

1	No. **25** O6'	[xxxxxxx ina 1.4$^?$] **me** 6 **ki** *ana ar*$_2$-*šu*$_2$ **la**$_2$-*ma* **uš** ⸢**ta**⸣ [**uš** xxx]
		[... in 1,4$^?$] 'days' the position recedes backwards 6, then it is stationary (S2). From [the second station ...]
2	No. **25** O7'	[xxxxxxxxx 30] ⸢**me**⸣ 10 **ninda zi**-*šu*$_2$ **nu tuk** *u*$_3$ 10 **ninda** ⸢GAR⸣.[GAR-*ma* 10 **ninda a.ra**$_2$ 30 **du**-*ma* 5]
	No. **26** X1'	[...] ⸢x⸣ GAR.GAR ⸢x⸣ [...]
		[... 30] ⸢'days'⸣ its displacement is 10^{nin}. Nothing and 10^{nin} ⸢you accumulate⸣, [it is 10^{nin}. You multiply it by 30, it is 5.]
3	No. **25** O8'	[34] ⸢**me**⸣ 12.30 **zi**-*šu*$_2$ 10 *u*$_3$ 12.⸢30$^?$⸣ [GAR.GAR-*ma* 22.30 **a.ra**$_2$ 34 **du**-*ma* 12.45]
	No. **26** X2'	[... 10] ⸢*u*$_3$⸣ 12.30 ⸢GAR.GAR-*ma* 2⸣[2.30 ...]
		[34] 'days' its displacement is 0;12,30. 0;10 and 0;12,⸢30⸣ [you accumulate, it is 0;22,30, you multiply it by 34, it is 12;45.]
4	No. **25** O9'	[... 5 *u*$_3$ 12].⸢4⸣5 GAR.GAR-*ma* 17.45 **pap.pap ta** [...]
		[... 5 and 12];45 you accumulate, it is 17;45, the total. From [...]
5	No. **25** O10'	[...xx] ⸢6⸣ **ki** *ina* **nim du igi ta igi** *ana* **igi** *ina* ⸢x⸣ [...]
	No. **18** O1'–2'	[xxxxxxxxx] ⸢**ki** *ina* **nim**$^?$⸣ **du igi**
	No. **26** X3'	[...] ⸢x **du**$^?$⸣ **igi ta igi** [*ana* ...
	No. **23** O4'	⸢x⸣ 28 **me** 6 **ki du**-*ma* [...]
		[...] in 28 'days' the position proceeds 6 in the east, then it appears (FA). From appearance to appearance in [...]

[300] For the preceding lines cf. No. **25** O1'–5'.

6	No. **18** O2’	[xxxx] ⌜3? 36 1 10,40?⌝ **ki** *ana* **igi**-*šu*₂ **du**
		[...] ... the position proceeds forward 10;40.
7	No. **18** O2’–3’	*ina lib*₃-*bi* [xx] ⌜x⌝ **la**₂-*is-ma u*₃ **igi** *e-lat* 6 **me zal** *ša*₂ **mu**
	No. **25** O11’	[...x] 6 **me zal** *ša*₂ **mu**
	No. **26** X4’	[... **zal**] ⌜*ša*₂ **mu**⌝
	No. **23** O4’	6 **me zal**
		In ... [...] it recedes, and it appears (FA). (In) more than 6,0 ‘days’, the passing of the year,
8	No. **18** O4’	[*ina*] ⌜6.42?⌝ **me.meš ta igi** *ana* **igi** 30 **ki du**
	No. **25** O11’	*ina* 6.42 **me.meš ta** ⌜**igi**?⌝ [*ana* **igi** x xx]
	No. **26** X4’	*ina* 6.42 **me.**[**meš ta igi** *ana* **igi** x xx]
	No. **23** O4’	*ina* 6.42 **me.meš ta igi** [*ana* **igi** x xx]
		from appearance to appearance in 6,42 ‘days’ the position proceeds by 30.
9	No. **18** O4’–5’	**igi** *ša*₂ *lu-maš* / *gab-bi ki-i* **ka me** *u* **zi šar**₂.**šar**₂
	No. **25** O12’	x x xx xx xx x] ⌜**me**⌝ *u* **zi šar**₂.**šar**₂
	No. **26** X5’	x x xx xx xx x x x x **šar**₂].**šar**₂
		You compute the appearance for every zodiacal sign according to the ‘day’ and the displacement.

Several elements of scheme A.S$_2$ have been tentatively reconstructed (shown in italics in Table 3.38a). The ‘daily’ motion within the pushes is formulated in accordance with templates PUSH.C.1–2, which provide only $\delta\tau$ and $\delta\Sigma$. It appears that the push from S2 to LA is divided into four subintervals, two of duration $30^{\bar{\tau}}$ and two of duration $34^{\bar{\tau}}$, each characterised by a value of v. The computation of the total displacement $\delta\Sigma$ (S2 to LA) = 17;45 can be interpreted such that the values of v are first added for each pair of subintervals of equal duration, and then multiplied by the duration. A similar arrangement seems to underlie the push from S1 to AR, and perhaps also AR to S2 (not preserved apart from $\delta\Sigma$). A full synodic period from FA to FA equals $6{,}42^{\bar{\tau}} = 13^{\mathrm{m}}\ 12^{\bar{\tau}}$. This implies $\delta\tau$ (LA to FA) = $30^{\bar{\tau}}$, but the text mentions $28^{\bar{\tau}}$. The forward motion of 10;40° (line 3) might be interpreted as the total displacement ($\delta\Sigma$) for the intervals S2 to LA (2) and (3).

Scheme A.S$_3$

Scheme A.S$_3$ is attested in No. **18** P23,P30,P31. P23 consists of two parts, each pertaining to a different zone and based on a different template (PUSH.D and E, respectively). P30 and P31, also based on PUSH.D and E, respectively, are both concerned with zone 2. Assuming that $\delta\Sigma$(AR to S2) = −13;15° (No. **18** P23,P30) is an error for −3;15°, three significant differences with scheme A.S$_1$ remain. First, the pushes do not add up to σ_j in either zone. Secondly, the angular pushes in zone 1 do not have a constant ratio to the corresponding ones in zone 2. Thirdly, the angular and temporal pushes are not connected through the relation $\delta\Sigma = v \cdot \delta\tau$ (Eq. 3.19), because this would result in values of v that cannot be represented as finite sexagesimal numbers or by a reasonable number of digits. The reconstructed values of v (Table 3.38b) have been computed by replacing the exact value of $\delta\tau$ by the nearest round number, e.g. $29^{\bar{\tau}}$ by $30^{\bar{\tau}}$, and applying Eq. (3.19). Their significance (i.e. whether they might have been used for computing daily positions) is unclear.

Scheme A.S$_4$

Scheme A.S$_4$ is attested in No. **18** P24, which provides only values of $\delta\Sigma$ (template PUSH.D) for zone 2. As in scheme A.S$_3$, the pushes do not add up to σ_2 = 36°; in this case one gets 35;30°. Scheme A.S$_4$ is similar to scheme A.S$_5$. The values of $\delta\tau$ for LA to FA and AR to S2 in zone 2 are the same, but other parameters are different.

Scheme A.S$_5$

Scheme A.S$_5$ is attested in No. **18** P11.a,b. It is provisionally assigned to system A because it features two zones. Some parameters of scheme A.S$_5$ are the same as in scheme A’.S$_2$. The values of $\delta\Sigma$ for one full cycle do not appear to add up to σ_j = 30° (zone 1) and 36° (zone 2).

Zigzag functions for the daily displacement

Several procedures list control values (in at least two versions) forming a zigzag function of the zodiacal position (Table 3.39). On five tablets the procedure is embedded in the sequence shown in Table 3.35, for which reason they are discussed here in the framework of system A, but there is little else that connects these functions with any particular system. Version 1 is attested in No. **23** P2’.d; version 2, which is shifted by one zodiacal sign, in No. **18** P4 and No. **25** P2’.d. The control values preserved in No. **30** P1’.c are compatible with both versions. The procedures do not mention at which positions within the zodiacal signs the control values are anchored. The zigzag functions have thus far defied interpretation, partly because of apparent inconsistencies.[301] However, it has been overlooked that the smallest and the largest of the tabulated values each repeat in adjacent zodiacal signs. This implies that the extrema of the underlying zigzag function are not among the tabulated values. They can be reconstructed as $m = 0;10,48°/\bar{\tau}$ and $M = 0;15,36°/\bar{\tau}$ (variant 1) or $m = 0;11,12°/\bar{\tau}$ and $M = 0;16,24°/\bar{\tau}$ (variant 2). For both variants the interpolation coefficient is $c = 0;0,1,36$ per degree, equivalent to 0;0,48 per zodiacal sign. Comparison with scheme A.S_1 (Table 3.38a) suggests that these functions represent Jupiter's 'daily' displacement between LA to FA. A similar zigzag function, v_3, perhaps representing the same quantity, is partly preserved in No. **18** P27. It has a smaller interpolation coefficient, $c = 0;0,30$ per zodiacal sign. The few preserved control values differ slightly from those of v_1 and v_2. A tentative reconstruction (Table 3.39) yields $m = 0;12,15$ (between Cnc and Leo) and $M = 0;15,5$ (between Cap and Aqr).

variant	Ari	Tau	Gem	Cnc	Leo	Vir	Lib	Sco	Sgr	Cap	Aqr	Psc
v_1	0;14,24	0;13,36	0;12,48	*0;12*	*0;11,12*	0;11,12	0;12	0;12,48	0;13,36	*0;14,24*	0;15,12	0;15,12
v_2	*0;16*	*0;15,12*	*0;14,24*	*0;13,36*	*0;12,48*	0;12	0;12	0;12,48	0;13,36	*0;14,24*	0;15,12	*0;16*
v_3	*0;14*	0;13,30	0;13	*0;12,30*	*0;12,20*	*0;12,50*	*0;13,20*	*0;13,50*	0;14,20	0;14,50	0;15	*0;14,30*

Table 3.39: Jupiter: control values of three zigzag functions for the daily displacement (in italics: tentative reconstructed values).

3.7.3 Systems A_1 and A_2

Systems A_1 and A_2 are represented by No. **18** P11.c–f, which are actually concerned with the subdivision of the synodic cycle.[302] Tabular texts are not extant.

3.7.3.1 Updating *B* with the synodic arc

No. **18** P11.c–f mention two sets of zonal boundaries b_j and transition coefficients r_j (Table 3.40a). Since the r_j are the same as in system A it is likely that the underlying σ_j are also the same as in system A. This is sufficient to reconstruct a step function for the synodic arc for each of the two sets of coefficients. However, since these step functions are not directly attested in procedure texts or tabular texts it is not certain that they were ever used for computing synodic tables. For system A_2 the resulting lengths of the zones, α_j, coincide with those of system A, but for system A_1 they differ by 3°.

	j	b_j	α_j	σ_j	r_j	δ_j	v_j
A_1	1	27° Gem	2,38°	30°	1;12	0;20°	7,54
	2	5° Sgr	3,22°	36°	0;50	0;24°	8,25
A_2	1	0;45° Cnc	2,35°	30°	1;12	0;50°	3,6
	2	5;45° Sgr	3,25°	36°	0;50	1°	3,25

	Π	Z	Y	P	$\overline{\sigma}$
A_1	16,19	1,30	17,49^{yr}	10;52,40	33;5,41,59,...°
A_2	6,31	36	7,7^{yr}	10;51,40	33;8,44,48,...°

Table 3.40a: Jupiter systems A_1 and A_2: defining parameters of the step functions for σ, periods of *B* and mean synodic arcs.

	push	$\delta\tau_1$	v_1 [$°/\bar{\tau}$]	$\delta\Sigma_1$	$\delta\tau_2$	v_2 [$°/\bar{\tau}$]	$\delta\Sigma_2$
A_1	LA to FA	?	?	5;37,40°	?	?	*6;45,12°*
A_2	FA to S1	*2,0$^{\bar{\tau}}$*	*0;7,30*	*15°*	2,0$^{\bar{\tau}}$	0;9	18°
	S1 to AR	1,0$^{\bar{\tau}}$	0;5	−5°	1,0$^{\bar{\tau}}$	0;6	−6°
	AR to S2	?	?	−5°?	?	?	−6°?

Table 3.40b: Subdivision of the synodic cycle. In italics: uncertain reconstructions.

3.7.3.2 Subdivision of the synodic cycle

No. **18** P11.c–f are incomplete procedures, since some subintervals of the synodic cycle are lacking (Table 3.40b). A unique feature of these procedures is that the algorithms for the angular pushes, $\delta\Sigma$, incorporate the same transition rules known from system A. This made it possible to reconstruct some missing values of $\delta\Sigma$ by assuming $\delta\Sigma_j/\delta\Sigma_{j+1} = \sigma_j/\sigma_{j+1}$. The procedures also provide some temporal pushes $\delta\tau$ for zones 1 and 2. A possible interpretation is that, for a given push, $\delta\tau$ is always the same in both zones. The daily displacements v_j were reconstructed as $v_j = \delta\Sigma_j/\delta\tau_j$. For the push from LA to FA zones 1 and 2 are dealt with in this order, but for the retrograde pushes S1 to AR and AR to S2 the order is reversed.

[301] Cf. *ACT*, pp. 405, 424.

[302] The names system A_1 and system A_2 were introduced by Swerdlow (1998), pp. 93–94.

3.7.4 System A'

System A', discovered by Kugler,[303] is represented by 9, perhaps 10 tablets with procedures and 10 synodic tables.[304] A small newly found fragment adds some text to No. **18** P17,P18. Three procedures accompany a synodic table: No. **20** (*ACT* 613aa), No. **34**, and No. **35** (*ACT* 611). None of the tablets preserves a colophon with a date. A rough indication of the time when system A' was in use is implied by the initial dates of the synodic tables, which cover the years SE 61–239.

updating B with the synodic arc	No. **18** P14–P16; No. **20** P1.b; No. **23** P2'.g; No. **29** P5'?; No. **31** P2'.b; No. **32** P1; No. **34** P1'.c; No. **35** P1.a
zonal boundaries for σ (STEP.F)	No. **18** P17,P25; No. **31** P2'.a; No. **32** P2; No. **46** P3'
transition rules for σ (STEP.D)	No. **29** P7'?; No. **46** P6'
updating T with the synodic time τ	No. **18** P14–P16; No. **35** P1.b
subdivision of the synodic cycle, scheme S_1	No. **18** P9a.c,d,P18; No. **23** P3',P4'; No. **32** P3–P6; No. **34** P1'.a,b,P2'; No. **46** P4',P5'; No. **95** P1'?
scheme S_2	No. **18** P9a.a,b

Table 3.41: Jupiter system A': procedures.

3.7.4.1 Updating B with the synodic arc

The algorithm for updating B with the synodic arc is attested in 9 procedures. No. **20** P1.b, No. **32** P1 and No. **35** P1.a, are based on template STEP.A.1; No. **23** P2'.g on STEP.E; No. **31** P2' combines STEP.F and STEP.C; No. **46** P6' is based on STEP.D. No. **18** P14–P16 are actually concerned with the synodic time, but they include instructions for updating B based on STEP.E. The templates underlying No. **29** P5' and P7' could not be established. Other procedures listed in Table 3.41 mention only the zonal boundaries of the step function for σ in accordance with STEP.F. The procedure texts and the synodic tables imply that the same step function (Table 3.42) is used for all synodic phenomena. Compared to the step function of system A there are two additional zones, which are transitional between the zones of system A. They cause σ to have a smoother appearance than in system A (Fig. 3.19).

j	b_j	α_j	σ_j	r_j	δ_j	v_j
1	9° Cnc	2,0°	30°	1;7,30	0;50°	2,24
2	9° Sco	53°	33;45°	1;4	0;56,15°	56;32
3	2° Cap	2,15°	36°	0;56,15	1°	2,15
4	17° Tau	52°	33;45°	0;53,20	0;56,15°	55;28

Π	Z	Y	P	$\bar{\sigma}$
6,31	36	7,7^{yr}	10;51,40	33;8,44,48,...°

Table 3.42: Jupiter system A': defining parameters of the step function for σ, periods of B and mean synodic arc.

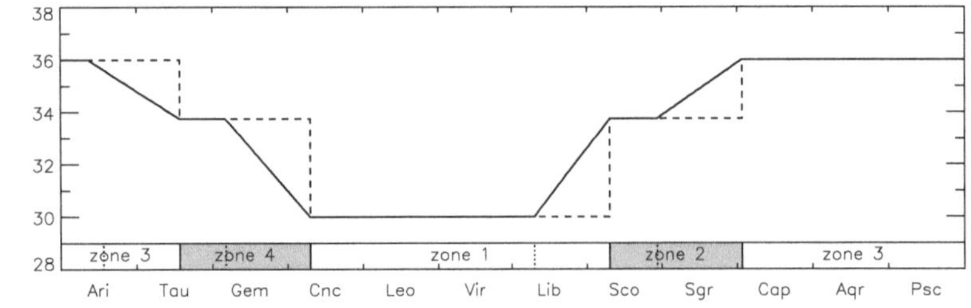

Fig. 3.19: Jupiter system A': graphical representation of σ as a function of B_{i-1}. The dashed lines represent the preliminary values σ_j. The step functions for σ in systems A'' and A''' have nearly the same appearance.

3.7.4.2 Updating T with the synodic time

Algorithms for updating T with the synodic time (τ) are partly preserved on two tablets. The badly damaged No. **35** P1.b appears to follow template T.2 (§3.3.2.1), but c_τ is not preserved. The synodic tables imply that τ is computed by the usual algorithm, i.e. $\tau_j = \sigma_j + c_\tau$, with c_τ = 6,12;5,10$^{\bar{\tau}}$ as in system A (variant 1 in Table 3.43).

[303] Kugler (1900), pp. 207–211 ('System B'); (1907), pp. 133–136 ('Zweites System').

[304] ST: *ACT* 609–614, 613a, 613aa, 613ab, 818.

A more elaborate algorithm for τ

No. **18** P14–P16 preserve a more sophisticated algorithm for the synodic time (τ) involving a different value of c_τ in each zone, say $c_{\tau,j}$, an approach that is reminiscent of the treatment of the subdivision of the synodic time in Mars system A. The attested values of $c_{\tau,j}$ are reproduced by the usual algorithm (Eq. 3.13) if one inserts $m = 1$ (Jupiter) and $e = 11;3,20^{\bar{\tau}}$ (which differs from the usual value $11;4^{\bar{\tau}}$), approximates $0;0,10 \cdot e = 0;1,50,33,20$ by 0;1,50, and if one replaces $\overline{\sigma}$ by the preliminary synodic arc σ_j for that zone, i.e.

$$c_{\tau,j} = 6,0 + e + 0;0,10 \cdot e\,\sigma_j = 6,11;3,20 + 0;1,50 \cdot \sigma_j. \tag{3.40}$$

The resulting values of $c_{\tau,j}$, in excess of $12^{\overline{m}} = 6,0^{\bar{\tau}}$, are compiled in Table 3.43 ('variant 2'). There is no evidence yet for the application of this algorithm in synodic tables.

	variant 1		variant 2	
j	$\tau_j - 12^{\overline{m}}$	$c_\tau - 12^{\overline{m}}$	$\tau_j - 12^{\overline{m}}$	$c_{\tau,j} - 12^{\overline{m}}$
1	$42;5,10^{\bar{\tau}}$	$12;5,10^{\bar{\tau}}$	$41;58,20^{\bar{\tau}}$	$11;58,20^{\bar{\tau}}$
2,4	$45;50,10^{\bar{\tau}}$	"	$45;50,12,30^{\bar{\tau}}$	$12;5,12,30^{\bar{\tau}}$
3	$48;5,10^{\bar{\tau}}$	"	$48;9,20^{\bar{\tau}}$	$12;9,20^{\bar{\tau}}$

Table 3.43: System A': parameters of the algorithms for τ_j in No. **18** P14–P16.

3.7.4.3 Subdivision of the synodic cycle

Since all synodic phenomena are primary phenomena, as in system A, the subdivision of the synodic cycle is fully determined by the subdivision of the initial cycle. This suggests that the purpose of these procedures is to establish the initial subdivision, or to verify zodiacal positions computed from σ. In both schemes A'.S$_1$ and A'.S$_2$ (Table 3.44) the constituent angular pushes add up to σ_j in each zone. Hence these schemes are, in principle, fully consistent with the zodiacal positions obtained with σ.

Scheme A'.S$_1$

Scheme A'.S$_1$ is best attested in the procedure texts. No. **32** P3–P6 and No. **18** P18 are formulated in accordance with template PUSH.B, which provides only $\delta\tau$ and v. In No. **18** P9a.c,d, No. **46** P4',P5' and No. **34** P1'.a,b PUSH.B is followed by PUSH.D, which provides only $\delta\Sigma$. It may be assumed that the same arrangement existed in No. **23** P3' (only the part based on PUSH.D preserved) and P4' (only the part based on PUSH.B preserved). As far as the angular pushes are concerned scheme A'.S$_1$ also underlies most of the synodic tables for system A', but the temporal pushes usually deviate from the schematic values in Table 3.44. The latter result in $\tau = 6,30^{\bar{\tau}}$, whereas the actual value of τ in system A' is variable and has a mean value close to $6,45^{\bar{\tau}}$ (Table 3.43). With regard to the zodiacal positions most synodic tables follow scheme A'.S$_1$, but some tables employ different values for $\delta\Sigma$ (S2 to LA).[305] As shown by P. Huber,[306] some values of v in scheme A'.S$_1$ are attested in daily motion tables from Uruk, e.g. for the intervals FA to S1 (2), S1 to S2, and S2 to LA in zones 1 and 3, but the durations $\delta\tau$ of these intervals, and hence also $\delta\Sigma$, usually deviate from scheme A'.S$_1$.

Scheme A'.S$_2$

Scheme A'.S$_2$ is attested in No. **18** P9a.a,b, which only covers zone 1. The parameters for zones 2–4 have been reconstructed on the reasonable assumption that $\delta\Sigma_j/\delta\Sigma_{j+1} = \sigma_j/\sigma_{j+1}$ (§3.3.3.3). For some subintervals scheme S$_2$ coincides with scheme S$_1$; (only) the deviations from scheme S$_1$ are compiled in Table 3.44. There is no evidence yet for the application of scheme S$_2$ in tabular texts.

3.7.4.4 Net displacements for various intervals

Procedures concerned with net displacements of Jupiter for intervals longer than 1 synodic cycle are not yet attested for system A'. Values of $dB_j(s)$ for several intervals s are listed in Table 3.37. Note that dB_1 and dB_3 are the same as dB_1 and dB_2 in system A. Since B has the same period in system A' as in system A, all net displacements contain the same number of elementary steps as in system A.

[305] *ACT*, p. 313.

[306] Huber (1957), pp. 273–6.

scheme A'.S_1							
push	$\delta\tau$	v_1 [°/$\bar{\tau}$]	$\delta\Sigma_1$	$v_2 = v_4$ [°/$\bar{\tau}$]	$\delta\Sigma_2 = \delta\Sigma_4$	v_3 [°/$\bar{\tau}$]	$\delta\Sigma_3$
LA to FA	$1^{\overline{m}}$	0;12,30	6;15°	0;14,3,45	7;1,52,30°	0;15	7;30°
FA to S1 (1)	$1^{\overline{m}}$	0;12,30	6;15°	0;14,3,45	7;1,52,30°	0;15	7;30°
(2)	$3^{\overline{m}}$	0;6,40	10°	0;7,30	11;15°	0;8	12°
total	$4^{\overline{m}}$		16;15°		18;16,52,30°		19;30°
S1 to S2	$4^{\overline{m}}$	−0;4,10	−8;20°	−0;4,41,15	−9;22,30°	−0;5	−10°
S2 to LA (1)	$3^{\overline{m}}$	0;6,23,20	9;35°	0;7,11,15	10;46,52,30°	0;7,40	11;30°
(2)	$1^{\overline{m}}$	0;12,30	6;15°	0;14,3,45	7;1,52,30°	0;15	7;30°
total	$4^{\overline{m}}$		15;50°		17;48,45°		19°
LA to LA	$13^{\overline{m}}$		30°		33;45°		36°
scheme A'.S_2 (where different from S_1)							
FA to S1 (2)	$3^{\overline{m}}$	0;6,15	9;22,30°	0;7,1,52,30	10;32,48,45°	0;7,30	11;15°
total	$4^{\overline{m}}$		15;37,30°		17;34,41,15°		18;45°
S1 to S2	$4^{\overline{m}}$	−0;5	−10°	−0;5,37,30	11;15°	−0;6	−12°
S2 to LA (1)	$3^{\overline{m}}$	0;7,55	11;52,30°	0;8,54,22,30	13;21,33,45°	0;9,30	14;15°
total	$4^{\overline{m}}$		18;7,30°		20;23,26,15°		21;45°
LA to LA	$13^{\overline{m}}$		30°		33;45°		36°

Table 3.44: Jupiter: schemes A'.S_1 and A'.S_2 for the subdivision of the synodic cycle.

3.7.5 System A"

System A", discovered by Neugebauer (*ACT*), is represented by three procedures, each written on a different tablet (No. **18** P7, No. **29** P4' and No. **30** P2'.d), and one template table with zodiacal positions of a single unspecified synodic phenomenon.[307] Neither the procedure texts nor the template table could be dated. All three procedures concern the updating of B with σ. Hence nothing is known about the subdivision of the synodic cycle. Furthermore, the procedures do not inform us about the algorithm for τ, although it is likely to be the same as in the other type-A systems for Jupiter.

3.7.5.1 Updating B with the synodic arc

The three mentioned procedures (partly) preserve the algorithm for updating B with σ. No. **18** P7 and No. **29** P4' are based on template STEP.E, which lacks explicit transition rules. The template of No. **30** P2'.d, a close duplicate of No. **29**, cannot be established due to the bad state of preservation, but it is presumably STEP.E. None of the procedures preserves a statement indicating for which synodic phenomena the step function is used. The parameters of the step function (cf. Table 3.45) are related to those of system A'. The zonal boundaries are slightly different, but the period relations are the same. Hence the elementary steps δ_j and the net displacements $dB_j(s)$ are also the same as in system A' (Table 3.37). As pointed out by Aaboe (1964a), this step function may have been constructed in order to ensure that α_j contains a whole number of elementary steps (v_j), a feature absent from system A', where v_2 and v_4 are fractional.

j	b_j	α_j	σ_j	r_j	δ_j	v_j
1	5;45° Cnc	2,0°	30°	1;7,30	0;50°	2,24
2	5;45° Sco	56;15°	33;45°	1;4	0;56,15°	1,0
3	2° Cap	2,15°	36°	0;56,15	1°	2,15
4	17° Tau	49;45°	33;45°	0;53,20	0;56,15°	52

Π	Z	Y	P	$\overline{\sigma}$
6,31	36	$7,7^{yr}$	10;51,40	33;8,44,48,...°

Table 3.45: Jupiter system A": defining parameters of the step function for σ, periods of B and mean synodic arc.

3.7.6 System A'''

System A''' is represented by two procedures, both (partly) preserved on two tablets. No. **18** P8.a and No. **21** P3'.a are concerned with updating B with the synodic arc; No. **18** P8.b and No. **21** P3'.b? with the setting time at LA and the rising time at FA. Other topics such as the synodic time, net displacements or the subdivision of the synodic cycle are not attested in the procedure texts. Tabular texts are not extant for system A'''.

[307] Aaboe & Sachs (1966), Text E.

3.7.6.1 Updating B with the synodic arc

The algorithm for updating B with the step function for the synodic arc is preserved in No. **18** P8.a and No. **21** P3'.a. Both are based on template STEP.E, which lacks explicit transition rules. The parameters of the step function (Table 3.46) are similar to those of systems A' and A". As suggested by Swerdlow (1998), pp. 98–99, system A"' may have been derived from system A' by shifting the zonal boundaries such that α_1, α_3 and $\alpha_2 + \alpha_4$ become multiples of 10°.[308]

j	b_j	α_j	σ_j	r_j	δ_j	v_j
1	9° Cnc	2,0°	30°	1;7,30	1;6,40°	1,48
2	9° Sco	48°	33;45°	1;4	1;15°	38;24
3	27° Sgr	2,20°	36°	0;56,15	1;20°	1,45
4	17° Tau	52°	33;45°	0;53,20	1;15°	41;36

Π	Z	Y	P	$\overline{\sigma}$
4,53	27	5,20	10;51,6,40	33;10,26,37,...°

Table 3.46: Jupiter system A"': defining parameters of the step function for σ, periods of B and mean synodic arc.

3.7.6.2 Setting time at LA and rising time at FA

System A"' is the only other computational system apart from system A for which there are procedures concerning Jupiter's setting time at LA and rising time at FA (No. **18** P8.b, and probably No. **21** P3'.b). The values of T_{set} and T_{rise} are preserved only in No. **18** P8.b: $T_{\mathrm{rise}} = 12$ and $T_{\mathrm{set}} = 10$. They differ from the system-A values, but their sum, 22, is very similar. No. **18** P8.b is assumed to belong to system A"' because this is true for P8.a and because in No. **21** P3'.b T_{set} and T_{rise} appear to be associated with the zones of system A"'. However, the zodiacal regions with which they are associated in No. **18** P8.b do not coincide with the zones of system A"' or other type-A system for Jupiter.

3.7.7 System A_6

System A_6, discovered by Neugebauer (*ACT*), is represented by one (composite) procedure, No. **44** P1, which deals only with the synodic arc (P1.a) and net displacements (P1.b).[309]

3.7.7.1 Updating B with the synodic arc

The partly preserved No. **44** P1.a contains rules for updating B with the synodic arc. Their formulation is based on template PSTEP (§2.6.10.1). Each zodiacal region consists of a whole number of zodiacal signs, and transition rules are lacking. Not all values of σ_j are preserved, and instructions on how to apply them seem to be absent. The present reconstruction (Table 3.47) differs in some details from that of Neugebauer. In the absence of synodic tables it is impossible to determine the precise algorithm by which B is updated with the σ_j. Two features suggest that σ is not a true step function. First, some of the ratios σ_{j+1}/σ_j are not finite sexagesimal numbers. This implies that the algorithm for σ does not incorporate transition rules (cf. §2.6.10.1). Second, the zodiacal regions for the net displacements dB_j do not coincide with those for σ, while they should if σ is a true step function (§2.6.10.4).

j	region	α_j	σ_j	region	$dB_j(11)$	region	$dB_j(65)$
1	Cnc–Lib?	2,0°?	30°	Gem–Sgr	?	?–Gem	−8
2	Sco?	30°?	30+x°	Cap–Tau	4;30	Cnc–Lib	−4;40°
3	Sgr	30°	34°			Sgr	−7;40°
4	Cap–Ari	2,0°	36°?			Cap	−9;40°
5	Tau	30°	34;30°?			Aqr–Psc	?
6	Gem	30°	33°			Ari–Gem	−8;30°

Table 3.47: Jupiter system A_6: synodic arcs and net displacements for 12^{yr} ($s = 11$) and 71^{yr} ($s = 65$).

3.7.7.2 Net displacements for various intervals

No. **44** P1.b contains a list of net displacements for several intervals including 12^{yr} and, probably, 65^{yr}. For every interval several values of dB are mentioned, each associated with a zodiacal region. Remarkably, these regions do not coincide

[308] However, the advantage of this arrangement is not obvious, and it comes at the cost of an inaccurate P.

[309] The name system A_6, introduced by Swerdlow (1998), p. 99, is maintained, although system A_6 is probably not a true type-A system.

with those of σ, an indication that system A_6 is not a true type-A system. The interpretation of the numbers $dB_j(65)$ is undertain; cf. also the commentary.

3.7.8 System B

System B, discovered by Kugler,[310] is represented by 6 procedure texts and 16 synodic tables.[311] Procedures concerning the subdivision of the synodic cycle[312] or net differences of T or B for intervals longer than 1 synodic cycle are not attested (see Table 3.48).

parameters of the zigzag function for τ (ZZ.A.3)	No. **18** P12.b,P21.b; No. **46** P1'.b
updating T with the zigzag function for τ (ZZ.B.1)	No. **24** P1.a; No. **36** P1.a; No. **37** P1.a
parameters of the zigzag function for σ (ZZ.A.3)	No. **18** P22.a; No. **31** P1'?
updating B with the zigzag function for σ (ZZ.B.1)	No. **24** P1.b; No. **36** P1.b; No. **37** P1.b
period relations and $\overline{\sigma}$	No. **18** P21.a
computing τ from σ, updating T with τ	No. **18** P13; No. **46** P2'
algorithm involving τ and Π	No. **18** P12.a', No. **46** P1'.a
computations involving σ	No. **31** P1'

Table 3.48: Procedures for Jupiter system B.

3.7.8.1 Updating *B* with the synodic arc

Five procedures contain rules for updating B with the zigzag function for σ (Table 3.49). No. **18** P22.a, and probably No. **31** P1', are based on template ZZ.A.3, which provides only the extrema and the difference d. No. **24** P1.b, No. **36** P1.b and No. **37** P1.b are based on template ZZ.B.1, which includes the reflection rules (Eqs. 2.14–2.15). No. **18** P22.a also mentions zodiacal positions for the extrema. In reality the extrema of a zigzag function of the event number do not occur at fixed positions, but they do tend to cluster around an average position, which is how the procedure must be understood. No. **37** P1.b mentions that the zigzag function for σ applies to FA, S1, AR and S2, omitting only LA; the synodic tables imply that it is used for all five synodic phenomena.

	d	m	B_m	M	B_M	Δ	μ	Π	Z	Y	P
σ	1;48°	28;15,30°	15° Vir	38;2°	15° Psc	9;46,30°	33;8,45°	6,31	36	7,7^{yr}	10;51,40
τ	1;48$^{\overline{\tau}}$	40;20,45$^{\overline{\tau}}$	15° Vir	50;7,15$^{\overline{\tau}}$	15° Psc	9;46,30$^{\overline{\tau}}$	45;14$^{\overline{\tau}}$	"	"	"	"

Table 3.49: Jupiter system B: parameters of the zigzag functions for the synodic arc (σ) and the synodic time (τ).

3.7.8.2 Updating *T* with the synodic time

Six procedures are concerned with the zigzag function for the synodic time (τ). No. **18** P12.b',P21.b and No. **46** P1'.b (template ZZ.A.3) provide only the defining parameters (Table 3.49). No. **24** P1.a, No. **36** P1.a and No. **37** P1.a are based on template ZZ.B.1, which includes reflection rules. The synodic tables imply that the zigzag functions for σ and τ are exactly in phase, so that the difference between them is constant (independent of the event number i) and given by

$$\tau - \sigma = 12;5,15^{\overline{\tau}}. \tag{3.41}$$

This constant represents the excess of c_τ over $12^{\overline{\text{m}}}$, i.e. $c_\tau - 12^{\overline{\text{m}}}$ (cf. Eq. 3.13). Hence the time of a synodic phenomenon is updated as

$$T_i = T_{i-1} + \tau + 12^{\overline{\text{m}}}, \tag{3.42}$$

a slight modification of Eq. (3.8). This algorithm is contained in the duplicates No. **18** P13 and No. **46** P2', which employ a formulation based on template T.1 (§3.3.2.1).

Algorithm for computing τ from σ

The duplicates No. **18** P13 and No. **46** P2' contain a derivation of τ from σ equivalent to Eq. (3.9) in which

[310] Cf. 'System C' in the appendix to Kugler (1900), and 'System III' in Kugler (1907).

[311] ST: *ACT* 620–629, 620a (+ Steele 2010, Text M), 620b, 621a, 622a (+ Steele 2010, Text N), 625a.

[312] The schemes for the subdivision of the synodic cycle that are implemented in the synodic tables remain to be reconstructed.

$$c_{\tau} = 6{,}12;5{,}8{,}8{,}20^{\bar{\tau}}, \quad (3.43)$$

i.e. the excess of c_{τ} over $12^{\bar{m}}$ is $12;5{,}8{,}8{,}20^{\bar{\tau}}$.[313] There is no evidence from synodic tables that this value of c_{τ} was used in practice.

3.7.8.3 Algorithm involving τ and Π

No. **18** P12.a' and No. **46** P1'.a are duplicates of a procedure that has hitherto defied interpretation.

No. **46** Oi1'	[xxx *mi-nu-u*$_2$ **a.ra**$_2$ 6].⌜31⌝ *lu*-**du**-*ma*
	[...What] should I multiply by [6],⌜31⌝

No. **46** Oi2'–3'	[*lu* 1.28.1.30.56.46 x] ⌜13.30.27⌝.46 **a.ra**$_2$ 6.31 **du**-*ma* / [1.28.1.30.56.46 x]
No. **18** Ri2'	[xx] 13.20.⌜2⌝7.46 ⌜**a.ra**$_2$⌝ [6.31 **du**-*ma* 1.28.1.30.56.46 x]
	[so that it is 1.28.1.30.56.46?] You multiply 13.30.27.46 by 6,31, it is 1.28.1.30.56.46.

No. **46** Oi3'	13.30.27.46 **murub**$_4$-*u*$_2$ [xx xxx] **gar**-*an*
No. **18** Ri2'–3'	[13.30]./⌜27⌝.46 **murub**$_4$-*u*$_2$ *a-na* 45.14 [**tab**? xx]
	13.30.27.46, the middle one, [you add?] to 45;14 [...] and put down.

The quantity sought in the first line is the number 13.30.27.46, which is called the 'middle one'. The number 6,31 can be identified as Π, and $45;14^{\bar{\tau}}$ as μ of τ. Since μ is measured in mean tithis, and 13.30.27.46 (correcting 13.20.27.46 in No. **18** Ri2') appears to be added to it, this number may also represent a time interval expressed in mean tithis, but its absolute value and interpretation remain unclear.

3.7.9 System B'

System B', discovered by Neugebauer (*ACT*), is represented by two procedure texts (Table 3.50).[314] The latter one follows a synodic table for system B' on the same tablet (*ACT* 640), which was written in SE 119 by the astronomer Anu-aba-utēr from Uruk. Other information about the date of system B' is not available.

parameters of the zigzag function for σ	No. **18** P22.b; No. **33** P1.b
parameters of the zigzag function for τ	No. **18** P21.c; No. **33** P1.a
net differences of T and B	No. **33** P1.c

Table 3.50: Jupiter system B': procedures.

3.7.9.1 Updating B with the synodic arc, T with the synodic time

B and T are updated with zigzag functions for σ and τ, respectively (Table 3.51). No. **33** P1.a,b are formulated in accordance with template ZZ.A.2, No. **18** P21.c and P22.b with template ZZ.A.3. Both templates provide the extrema and the difference d, but not the reflection rules. The zigzag functions for σ and τ are mutually inconsistent, because their amplitudes Δ (and hence the periods P) are not identical. As suggested by Swerdlow (1998), p. 85, this may be an artefact of the rounding of m and M with respect to system B. As a result the difference $\tau - \sigma$ varies (in system B it is constant and equal to $c_{\tau} - 12^{\bar{m}}$). Nothing is known about the subdivision of the synodic cycle, since this topic is not discussed in the extant procedures, and the mentioned synodic table contains only a single synodic phenomenon (S2).

	d	m	B_m	M	B_M	Δ	μ	Π	Z	Y	P
σ	1;46,40°	28;19,10°	15° Vir	37;58,20°	15° Psc	9;39,10°	33;8,45°	11,35	1,4	$12{,}39^{\text{yr}}$	10;51,33,45
τ	$1;46{,}40^{\bar{\tau}}$	$40;24{,}15^{\bar{\tau}}$	15° Vir	$50;3{,}31^{\bar{\tau}}$	15° Psc	$9;39{,}16^{\bar{\tau}}$	$45;13{,}53^{\bar{\tau}}$	2,24,49	13,20	$2{,}38{,}9^{\text{yr}}$	10;51,40,30

Table 3.51: Jupiter system B': parameters of the zigzag functions for σ and τ. The positions B_m and B_M represent approximate locations of the extrema.

[313] For the origin of this parameter cf. §3.7.2.2.

[314] Parts of No. **18** had been studied previously by Kugler (1907), but the procedures for system B' are on fragments that were joined only later.

3.7.9.2 Net differences for various intervals

No. **33** P1 mentions net differences of B and T for '12^{yr}' ($s = 11, t = 1$). Values of $dB(s)$ and $dT(s)$ for several intervals s are listed in Table 3.52.

s	t	y	$dT(s)$	$dB(s)$
11	1	12^{yr}	0;14,48$^{\bar{\tau}}$	0;15°
1,5	6	1,11^{yr}	−0;17,53$^{\bar{\tau}}$	−0;16,40°
1,16	7	1,23^{yr}	−0;3,4$^{\bar{\tau}}$	−0;1,40°
1,27	8	1,35^{yr}	0;11,44$^{\bar{\tau}}$	0;13,20°
2,32	14	2,46^{yr}	−0;6,8$^{\bar{\tau}}$	−0;3,20°
3,59	22	4,21^{yr}	0;5,36$^{\bar{\tau}}$	0;10°

Table 3.52: Jupiter system B': net differences of T and B for various intervals.

3.7.10 Unidentified computational systems

A considerable number of procedures, mainly concerned with the subdivision of the synodic cycle or Jupiter's distance to the ecliptic, could not be identified with a particular computational system (Table 3.53).

subdivision of synodic cycle, scheme X.S$_1$	No. **21** P1; No. **22** P1.a,P2?; No. **26** P1'.b'
scheme X.S$_2$	No. **29** P3', No. **30** P2'.c?
scheme X.S$_3$	No. **39** P7a',b'
scheme X.S$_4$	No. **29** P2'.b; No. **30** P2'.b; No. **38** P1'.c
scheme X.S$_5$	No. **23** P2'.f; No. **38** P1'.b,P2'.a
a scheme with linear interpolation for v	No. **38** P3'
other schemes	No. **18** P6,P26,P28; No. **23** P2'.e; No. **39** P1'–P6',P8'?; No. **46** P7',P8'
zigzag function for $\delta\Sigma$ (LA to FA)?	No. **18** P29; No. **29** P2'.a; No. **30** P2'.a; No. **38** P1'.a
zigzag function for v	cf. system A
distance to the ecliptic in general	No. **40** P2'.a; No. **47** P1
distance to the ecliptic at the stations	No. **18** P3; No. **23** P2'.c; No. **25** P2'.c'; No. **29** P1'.b'; No. **30** P1'.b'; No. **40** P2'.b
computations involving a trapezoid	No. **18** P5; No. **38** P4'; No. **40** P1'
control values of an unidentified quantity	No. **40** P2'.c
a Goal-Year type procedure?	No. **38** P2'.b; No. **46** P9'

Table 3.53: Jupiter, unidentified computational systems: procedures

3.7.10.1 Subdivision of the synodic cycle

Most of the procedures in Table 3.53 are concerned with the subdivision of the synodic cycle. Since the schemes, as far as preserved, do not distinguish between different zodiacal regions, they could not be associated with any particular type-A system.

Sequence of the procedures

Some of the mentioned procedures for the subdivision of the synodic cycle occur in a similar order on different tablets (Table 3.55). On Nos. **29** and **30** this sequence immediately follows the sequence of procedures for system A shown in Table 3.35 (p. 90).

Scheme X.S$_1$

Scheme X.S$_1$ is incompletely preserved in three procedures, one of which (No. **22** P1.a) is published here for the first time. Their formulation is based on the unique template PUSH.B.2. The interpretation of scheme X.S$_1$ remained completely unclear until P. Huber[315] made the crucial discovery that 1 **me** stands for 1,0 = 60 'days' (mean tithis). The following score is intended to shed some further light on scheme X.S$_1$:

1 | No. **21** O1–2 **mul$_2$.babbar me igi** 12 **zi**-*šu*$_2$ ⸢**en**⸣ [1 **me** 10.45 10.45 **a.ra$_2$**] / 1 **me du**-*ma* 10.45
| No. **22** O1–2 **mul$_2$.babbar me igi** 12 ⸢**zi**⸣-[*šu*$_2$ **en** xxxx 10.45 **a.ra$_2$**] / 1 **me du**-*ma* 10.4⸢5⸣
Jupiter. (On) the day when it appears its (daily) displacement is 0;12. Until 1,0 'days' 0;10,45$^?$.
You multiply 0;10,45 by 1,0 'days', it is 10,45.

[315] Huber (1957), p. 287.

scheme X.S$_1$			
push	$\delta\tau$	v [°/$^{\bar{\tau}}$]	$\delta\Sigma$
LA to FA	?	0;12	
FA to S1 (1)	1,0$^{\bar{\tau}}$	*0;10,45*	10;45°
(2)	1,0$^{\bar{\tau}}$	0;5,30	5;30°
total	2,0$^{\bar{\tau}}$		16;15°
S1 to S2 (1)	*30*$^{\bar{\tau}}$	±0;0,30?	±*0;15°*
(2)	1,0$^{\bar{\tau}}$	−0;4	−4°
total (1–2)	*1,30*$^{\bar{\tau}}$		−3;45°?
(3)	30$^{\bar{\tau}}$	−0;4,30	−2;15°
total	*2,0*$^{\bar{\tau}}$		−6° or −*6;30°*?
S2 to LA (1)	1,0$^{\bar{\tau}}$	0;5	5°
(2)	1,8$^{\bar{\tau}}$	0;11,15	12;45°
total	2,8$^{\bar{\tau}}$		17;45°

Table 3.54a: Jupiter, scheme X.S$_1$ for the subdivision of the synodic cycle. In italics: uncertain reconstructions.

scheme X.S$_2$			
push	$\delta\tau$	v [°/$^{\bar{\tau}}$]	$\delta\Sigma$
FA to S1 (1)	1,30$^{\bar{\tau}}$	0;10	15°
(2)	30	0;6,40	3;20°
total	2,0$^{\bar{\tau}}$		18;20°
S2 to LA (1)	?	0;3,20	?

scheme X.S$_3$			
push	$\delta\tau$	v [°/$^{\bar{\tau}}$]	$\delta\Sigma$
AR to S2? (1)	20$^{\bar{\tau}}$	−0;4,12	−1;24°
(2)	*30*$^{\bar{\tau}}$	*−0;5,20*	*−2;40°*
(3)	30$^{\bar{\tau}}$	−0;6,48	−3;24°
total	*1,20*$^{\bar{\tau}}$		−7;28°
S2 to LA? (1)	30$^{\bar{\tau}}$	0;7,30	3;45°
(2)	30$^{\bar{\tau}}$	0;8,30	4;15°
(3)	*1,0*$^{\bar{\tau}}$	*0;9,45*	*9;45°*
total	2,0$^{\bar{\tau}}$		17;45°

Table 3.54b: Schemes X.S$_2$ and X.S$_3$. Uncertain reconstructions are in italics.

procedure	No. **18**	No. **29**	No. **30**	No. **38**
zigzag function for $\delta\Sigma$ (LA to FA)	P29	P2'.a	P2'.a	P1'.a
scheme X.S$_4$ for $\delta\tau$ (LA to FA)		P2'.b	P2'.b	P1'.c
scheme X.S$_5$				P1'.b, P2'.a

Table 3.55: Jupiter: tablets with duplicates of the same sequence of procedures for the subdivision of the synodic cycle (unidentified system).

2 | No. **21** O2–3 | **ta** 1 **me** ⌜**gi**⌝ [**en** 1 **me**]/5.30 5.30 **a.ra**$_2$ 1 **me du**-*m*[*a* 5.30]
| No. **22** O2–3 | [**ta** 1 **me gi en** 1 **me** 5.30 5.30] /**a.ra**$_2$ 1 **me du**-*ma* ⌜5.30⌝ [...]
After completing 1,0 'days' until 1,0 days 0;5,30. You multiply 0;5,30 by 1,0 'days', it is 5;30.

3 | No. **21** O4–5 | **ta uš igi** 30 **ninda zi**-*šu*$_2$ ⌜x⌝[xxxx 4 **a.ra**$_2$] / 1 **me du**-*ma* 4 *ina* 1 **me** ⌜4$^?$⌝ [*ana ar*$_2$-*šu*$_2$ **la**$_2$]
| No. **22** O4–5 | **ta uš igi** 30 **ninda zi**-*šu*$_2$ [x xxxx 4 **a.ra**$_2$] / 1 **me du**-*ma* 4 *ina* 1 **me** 4 [xxxx]
After the first station its displacement is 0;30^{nin}. [...] You multiply [0;4 by] 1,0 'days', it is 4. In 1,0 'days' [it recedes backwards] 4.

4 | No. **21** O6–7 | **ta** 30 **me gi en** ⌜30⌝ [**me** 4.30 4.30 **a.ra**$_2$] /30 **du**-*ma* 2.15
| No. **22** O6 | **ta** 30 **me gi en** 30 **me** [xxxxxx] x xx xxx]
After completing 30 (error for 1,0?) 'days' until 30 'days' 0;4,30. You multiply 0;4,30 by 30, it is 2;15.

5 | No. **21** O7–8 | ⌜3⌝.[45 x xxx xxx x x] / *ana ar*$_2$-*šu*$_2$ **la**$_2$-*ma* **uš**
| No. **22** O7 | 3.45 *u*$_3$ 2.15 GAR.GAR-⌜*ma*⌝ [6 6 x xx xx x]
3;45 and 2;15 you accumulate, it is 6. 6 it recedes backwards, then it becomes stationary (S2).

6 | No. **21** O8–9 | **ta** [**uš** *ar*$_2$ 1 **me** 5] / 5 **a.ra**$_2$ 1 **me du**-*ma* [5 xxx]
From [the second station 1,0 'days' 0;5.] You multiply 0;5 by 1,0 'days', it is [5. ...]

7 | No. **22** O8 | **zi nu tuk en** 1 **me** 10 ⌜**ninda**⌝ [...]
The displacement does not exist. Until 1,0 'days' 10^{nin} (or 0;10^{nin}?) [...]

8 | No. **21** O10 | **du**-*ma* 11.15 **a.ra**$_2$ [1.8 **du**-*ma* 12.45]
| No. **22** O9 | [**ta**] 1 **me gi en** 1.8 **me** 1⌜2$^?$⌝.[x 11.15 **a.ra**$_2$ 1.8 **du**-*ma* 12.45]
| No. **26** X6' | [11].15 **a.ra**$_2$ 1.8 12.4[5]
After completing 1,0 'days' until 1,8 'days' 0;11,15. You multiply 0;11,15 by 1,8, it is 12;45.

9 | No. **21** O10–11 | [5 *u*$_3$ 12.45 GAR.GAR-*ma* / 17.45 **ki du š**[**u**$_2$ xxx]
| No. **22** O9–10 | [x x xxxx GAR.GAR-*ma* xx] xx] / **ki du šu**$_2$ [xxxxxxxx]
| No. **26** X7' | [17.4]⌜5⌝ **ki du**-*ma* **šu**$_2$
5 and 12;45 you accumulate, it is 17;45. The position proceeds by 17;45, then it sets (LA).

10 | No. **21** O12 | *ina lib*$_3$-*bi* 10 **ki** ⌜*ana*$^?$⌝ [*ar*$_2$-*šu*$_2$ **la**$_2$ xxxxx]
| No. **22** O10–11 | **ta me šu**$_2$ [xxxxxxxxxx x] 10 **ki** *ana ar*$_2$-*šu*$_2$ **la**$_2$ ⌜x⌝ [xxxxxxxxxxx]
After the day when it sets [...] within it$^?$ the position recedes backwards 10 ... [...]

11 | No. **22** O12–14 *lu-maš gab-bi* [xxxxxxxxxxxxxx] / GAR.GAR-*ma* 22.⌜30⌝ [xxxxxxxxxxxxx] / **du$_3$.du$_3$ ta** ⌜**uš**⌝ [xxxxxxxxxxxxx]
All zodiacal signs [...] / you accumulate, 22;30 [...] / procedure; from the [...] station [...]

The parameters of scheme X.S$_1$, as far as they could be reconstructed, are compiled in Table 3.54a. Some are the same as in scheme A.S$_1$ (Table 3.38a). A proximity to system A is also suggested by No. **26**, where scheme A.S$_2$ (P1'.a) is followed by a segment of scheme X.S$_1$ (P1'.b) introduced as a 'second opinion'. Not all subintervals of the synodic cycle are preserved. The motion during interval FA to S1 (1) has been reconstructed on the assumption of a constant v = 0;10,45°/d, but it is also possible that v in this interval decreases linearly from 0;12 (the value at FA) to 0;9,30 (not preserved), since the resulting $\delta\Sigma$ is the same. The instructions for the interval from S1 to S2 are problematic, as noted by Huber (1957), p. 288. One expects $\delta\Sigma$ (S1 to S2) to be close to $-10°$ (perhaps mentioned in 10) and not $-6°$ as implied by the text, perhaps suggesting that a segment is missing. Moreover, the $30^{\overline{\tau}}$ mentioned in 4 appears to be an error for $1{,}0^{\overline{\tau}}$, the duration of the preceding interval S1 to S2 (2). If one does accept that $\delta\Sigma$ (S1 to S2) = $-6°$, and that 30 *nindanu* in 3 represents $0;30^{\text{nin}}/\overline{\tau}$ = $0;0{,}30°/\overline{\tau}$, and $\delta\Sigma = -3;45°$ is the sum of the intervals S1 to S2 (1–2), then $\delta\Sigma$ for the interval S1 to S2 (1) must equal $30 \cdot 0;0{,}30 = +0;15°$, in the prograde direction, which is also odd. Finally, the interpretation of the interval of $1{,}0^{\overline{\tau}}$ during which v = 0;10 or $0;0{,}10°/\overline{\tau}$ (7) is also not clear. A comprehensive and coherent interpretation of scheme X.S$_1$ remains to be found.

Scheme X.S$_2$

Scheme X.S$_2$ is partly preserved in No. **29** P3' and, perhaps, No. **30** P2'.c. Both are based on template PUSH.B.3. The data are compiled in Table 3.54b.

Scheme X.S$_3$

Scheme X.S$_3$ is attested only in No. **39** P7a' and P7b', which are based on template PUSH.A.1. The data are compiled in Table 3.54b. The identification of the intervals is provisional.

Scheme X.S$_4$

Scheme X.S$_4$, partly preserved in No. **29** P2'.b, No. **30** P2' and No. **38** P1'.c, is concerned with Jupiter's interval of invisibility, $\delta\tau$ (LA to FA). Several values are provided, each associated with a segment of the zodiac: $27^{\overline{\tau}}$ (Cnc–Leo), x;$40^{\overline{\tau}}$ (Vir–Sco), $32^{\overline{\tau}}$ (Sgr–Cap) and $31;50^{\overline{\tau}}$ (Aqr). On three tablets scheme X.S$_4$ follows a procedure that provides parameters of what is probably a zigzag function for the corresponding arc $\delta\Sigma$ (LA to FA); cf. below. The values $27^{\overline{\tau}}$ and $32^{\overline{\tau}}$ also appear in scheme A.S$_5$ (No. **18** P11), suggesting a proximity between both schemes.

Scheme X.S$_5$

The unique scheme X.S$_5$, which had thus far partly defied interpretation, is incompletely preserved in the duplicates No. **23** P2'.f and No. **38** P1'.b, and in No. **38** P2'.a, which employs a different formulation. The following score edition aims to shed some light on this scheme:

| No. **38** P1'.b [x x] ⌜**a.ra$_2$**⌝ 20 *ana* **e me a.ra$_2$** 30 *ana* **uš** *ar$_2$-t*[*u$_2$* **a.ra$_2$** 40 x]
| No. **23** P2'.f *ana* **uš** GAM 20 *ana* **me** *a* GAM 30 *ana* **uš** 40 [xxxx]
To the station (S1): times 0;20. To daylight rising (AR): times 0;30. To the [second] station (S2): [times] 0;40. ...

| No. **38** P2'.a *ina šal-šu$_2$ ša$_2$* **na**-*šu$_2$* **uš igi**-*tu$_4$ ina mi-šil* **na**-⌜*šu$_2$*⌝ [**e me** xxxxxx]
At a third of its 'standing' the first station (S1). At half of its 'standing' [daylight rising (AR) ...]

A tentative interpretation, suggested by Sachs for No. **38** P2'.a,[316] is that the numbers represent schematic relative times of the synodic phenomena, measured from FA and expressed as a fraction of $\delta\tau$ (FA to LA), the time interval during which Jupiter is visible. The various schemes for the subdivision of the synodic cycle support this. If one considers, for instance, scheme A.S$_1$, where $\delta\tau$ (FA to LA) = $6{,}0^{\overline{\tau}}$ then S1 occurs $2{,}0^{\overline{\tau}} = 0;20 \cdot \delta\tau$ (FA to LA) after FA, AR $3{,}0^{\overline{\tau}} = 0;30 \cdot \delta\tau$ (FA to LA) after FA, and S2 $4{,}0^{\overline{\tau}} = 0;40 \cdot \delta\tau$ (FA to LA) after FA, in agreement with scheme X.S$_5$. No. **38** P2'.a is thus far the only procedure in which *manzāzu*(**na**), 'standing', is a technical term for $\delta\tau$ (FA to LA).[317]

[316] *ACT*, p. 430.

[317] This word usually denotes the time interval between sunset and the setting of the Moon or the planet.

Zigzag function for δΣ (LA to FA)?

Several procedures contain a list of coefficients forming a zigzag function of the zodiacal position with a minimum m = 6;10° in Gem, maximum M = 6;20° in Cap, mean value μ = 6;15° in Ari and Lib, difference d = 0;1,40° per zodiacal sign = 0;0,3,20°/°. It is not clear to which degree within the zodiacal signs the control values pertain. Various interpretations have been suggested for this function. According to Neugebauer[318] it represents $\delta\Sigma$ for the interval of $30^{\bar{\tau}}$ immediately preceding LA. This interpretation is suggested by the agreement between μ and this angular push in zone 1 of system A' (Table 3.44). According to Steele (2005a) it represents $\delta\Sigma$ (LA to FA), which makes better sense because in No. **38**, No. **29** and No. **30** the immediately following procedure is concerned with $\delta\tau$ (LA to FA).

Other schemes

Several procedures listed in Table 3.53 contain other schemes for the subdivision of the synodic cycle, often very incompletely preserved, for which I refer to the commentaries.

3.7.10.2 Distance to the ecliptic

Jupiter's distance to the ecliptic is only rarely mentioned in procedure texts, and thus far not attested in any tabular text. At the time when *ACT* was written these procedures were unknown. Neugebauer and Sachs (1967) published a first tablet (No. **47**), and Steele (2003, 2005a, 2007a) has discovered several tablets with similar procedures, to which the present study adds another one (No. **40**). The tablets contain two different procedures, both still not fully understood. The first one, partly preserved in No. **40** P2'.a and No. **47** P1, is as follows:

1	No. **47** O1–2	[xxx] ⸢**mul$_2$.babbar**⸣ $\frac{2}{3}$ **kuš$_3$** *e* **dur murub$_4$ nim** / [xx] **sig dur murub$_4$ sig**
	No. **40** X9'	[**ni**]**m** *u* **sig** *ša$_2$* **mul$_2$.babbar** $\frac{2}{3}$ **kuš$_3$ an.ta dur** ⸢**murub$_4$**⸣ **nim** $\frac{2}{3}$ ⸢**kuš$_3$**⸣ [xxxx]
		'Height and depth' of Jupiter. $\frac{2}{3}$ of a cubit above the middle band it is high, $\frac{2}{3}$ of a cubit below the middle band it is low.
2	No. **47** O3–4	*ki-i ina* ^{mul_2}**gir$_2$.tab nim** *ina zib*.**me bur$_3$** / *a-na* 1 **danna** 8 *u* **nim** *u* **sig**
	No. **40** X10'	[*ki*]-*i ina* **gir$_2$.tab nim** *ina zib*.**me sig** *a-na* 1 **danna**.⸢**am$_3$** 8? **nim**? 8? **sig**?⸣ [...]
		If it is high in Sco, it is low in Psc. For 1^b it ascends and descends 8^f.

In this procedure Jupiter is said to reach a maximum distance of $\frac{2}{3}$ cubit = 16^f (= 1;20°) above the ecliptic in Sco, and the same distance below the ecliptic in Psc. There is no mention of synodic phenomena. Furthermore, Jupiter is said to move up or down by 8^f per 30° of distance along the ecliptic, which amounts to $0;16^f$ per degree (= 0;1,20° per degree). As proposed by Neugebauer & Sachs (1967), these statements can be reconciled if one assumes that they describe a zigzag-type function of the zodiacal position which is constant in two regions centred at the extrema, each comprising 60° (Fig. 3.20).[319] The procedure does not explain where precisely the flattened regions begin (here assumed to be at 0° of the zodiacal signs). No. **47** P1 describes a similar latitudinal motion for Saturn, which is said to be 'high' in Vir–Sco, 'low' in the opposite three signs, but the numerical parameters are not preserved.

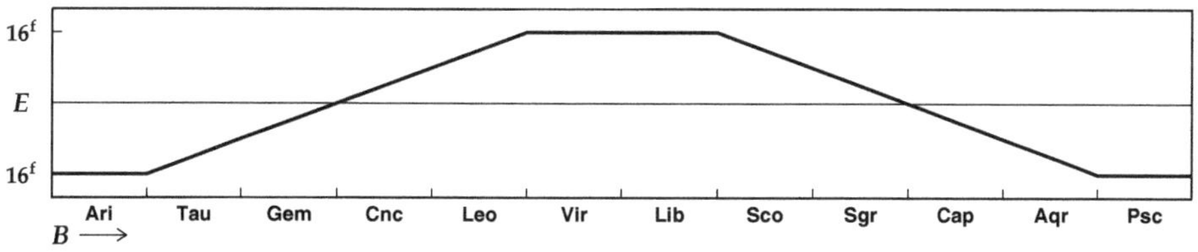

Fig. 3.20: Graphical representation of Jupiter's distance to the ecliptic as reconstructed from No. **40** P2'.a and No. **47** P1.

318 Cf. the commentaries to *ACT* 813, 817 and *HAMA*, p. 452.

319 In this interpretation the procedure also confirms that $1^c = 24^f$.

The other procedure concerns Jupiter's distance to the ecliptic at the stations. On most tablets it is embedded in the sequence of system A procedures shown in Table 3.35 (p. 90), but nothing else suggests a connection with system A. The following score edition includes the newly discovered No. **40** P2'.b and some new readings resulting from collations.[320]

1	No. **29** O2'	*ma-lak* *ša*$_2$ ⸢QU$^?$⸣
	No. **25** O17'	[**nim** *u* **sig** *ša*$_2$ **mu**]**l.babbar** *kal*$_3$ **mu** *ina* **dagal** *ma-l*[*ak* ...]
	No. **30** X2'	[...x] ⸢*ma*⸣-*lak* ⸢*ša*$_2$ QU$^?$⸣
	No. **23** O9'	[xxxxxxxxxx x x x x xx *ša*$_2$] ⸢QU$^?$⸣
		['Height and depth' of] Jupiter for the entire year. In the width of the path of ...

2	No. **18** Oi16'–17'	[*ina* **u**]**š** **igi**-*tu*$_4$ $\frac{1}{2}$ **kuš**$_3$ / [xxx *ina* **uš** *ar*$_2$-*tu*$_2$] ⸢$\frac{1}{2}$⸣ **kuš**$_3$ *i-šap-pil*
	No. **29** O2'	*ina* ⸢**uš** **igi**$^?$⸣ [x x xx x x x x x...]
	No. **25** O18'	[...x] $\frac{1}{2}$ **kuš**$_3$ *i-šap-pil*
	No. **30** X2'	*ina* ⸢**uš**⸣ **igi**-*tu*$_2$ $\frac{1}{2}$ **kuš**$_3$ [...]
	No. **23** O9'	*ina* **uš** **igi**-*tu*$_4$ $\frac{1}{2}$ **kuš**$_3$ *ša*$_2$-*qa* *ina* *ar*$_2$-*tu*$_4$ $\frac{1}{2}$ **kuš**$_3$ **sig**
	No. **40** X11'	[*ina* **uš** **igi**]-*tu*$_4$ $\frac{1}{2}$ **kuš**$_3$ *al-la* **murub**$_4$ **nim** *ina* **uš** [*ar*$_2$-*tu*$_4$ $\frac{1}{2}$ **kuš**$_3$ ...]
		At the first station it is $\frac{1}{2}$ cubit 'high' [above the middle], at the second station it is $\frac{1}{2}$ cubit 'low'.

Jupiter is said to be 'high' at S1, 'low' at S2. Modern data indicate that Jupiter is higher at S1 than at S2 in the region Lib–Psc, while the reverse is true in the other half of the zodiac. However, empirically the change of Jupiter's latitude between S1 and the immediately following S2 is at most about 0.3°, significantly smaller than $1^c = 2°$. Moreover, the procedure suggests that Jupiter is at opposite sides of the ecliptic at S1 and S2 (most clearly in No. **40** X11'). In reality this can happen only if Jupiter is very close to the ecliptic, which occurs only twice in 12^{yr}. In order to make sense of the procedure, 'high' and 'low' may therefore have to be interpreted in a different manner, as pointed out by J. Steele.

3.7.10.3 Procedures involving trapezoids

Some tablets with procedures for Jupiter include an apparently mathematical procedure involving a trapezoid. Two of them (No. **18** P5, No. **38** P4') were published by Neugebauer in *ACT*, a third one (No. **40** P1') has been newly discovered in the British Museum, and is published here for the first time. The connection between these difficult procedures, which still largely defy interpretation, and the other procedures on these tablets is not obvious. It may not be a coincidence that on two tablets the trapezoid procedure is preceded or followed by a procedure for Jupiter's distance to the ecliptic (No. **18** P3, No. **40** P2'), but the implications of this are unclear. For details cf. the commentaries.

3.8 Saturn

The planet Saturn is represented by systems A, A', B, B', B" and (at least) one incompletely known type-A system attested in two unusual, fragmentary synodic tables.[321] No. **47** P2' is concerned with Saturn's distance to the ecliptic and cannot be uniquely assigned to a computational system (cf. the commentary).

3.8.1 Synodic cycle

Saturn completes one revolution around the Sun in 29.46^{yr}. This makes it the slowest planet known to the Mesopotamians, which may explain its Akkadian name *Kajjamānu*, 'The steady one'. The mean synodic time of Saturn is $\overline{\tau} = 378.09^d \approx 6{,}18;5^d \approx 6{,}24;6^{\overline{\tau}}$, and the mean total synodic arc is $\overline{\Sigma} = \overline{\sigma} \approx 12;39°$, the smallest among the known planets.

3.8.2 System A

System A, discovered by Kugler,[322] is represented by four, perhaps five procedure texts (Table 3.56), one synodic table and two template tables.[323]

[320] Cf. Steele (2005a), pp. 48–49.

[321] BM 42878 and BM 45807; cf. Steele (2010) for a tentative reconstruction of the synodic arcs and subdivision of the synodic cycle.

[322] Kugler (1924) = *SSB* II, Anhang I, pp. 577–580.

[323] ST: BM 36948 (Steele 2002a); TT: Aaboe & Sachs (1966), Texts A, B. For Text A cf. also Aaboe (2001), p. 48.

zonal boundaries for the synodic arc	No. **41** P1; No. **42** P3
subdivision synodic cycle, scheme A.S_1, zone 1	No. **41** P2; No. **42** P4
zone 2	No. **41** P3; No. **42** P5
period relations	No. **41** P4; No. **42** P6; No. **45** P1'
net displacements	No. **44** P2 (?); No. **95** P2'?

Table 3.56: Procedures for Saturn system A.

3.8.2.1 Updating B with the synodic arc

Procedures with a complete set of rules for updating B with the step function for σ are not preserved. No. **41** P1 and No. **42** P3 mention only the zonal boundaries b_j for the 'small one' (zone 1) and the 'large one' (zone 2) in accordance with template STEP.F (§2.6.10); the other parameters were reconstructed from the tabular texts; cf. Table 3.57a and Fig. 3.21. The period relation for B, which is common to systems A and B, is mentioned in three procedures. A remarkable feature is that the v_j have fractional values.[324] The extant procedure texts do not reveal which are the primary synodic phenomena, but the template tables imply that they are FA, S1, S2 and LA. In analogy to systems A and A' of Jupiter we may therefore assume that all 5 synodic phenomena are primary phenomena.

j	b_j	α_j	σ_j	r_j	δ_j	v_j
1	10° Leo	3,20°	11;43,7,30°	1;12	1;18,7,30°	2,33;36
2	30° Aqr	2,40°	14;3,45°	0;50	1;33,45°	1,42;24

Π	Z	Y	P	$\overline{\sigma}$
4,16	9	4,25^{yr}	28;26,40	12;39,22,30°

Table 3.57a: Saturn system A: parameters of the step function for σ, periods of B and mean synodic arc.

s	t	y	dB_1	dB_2	dB_j/δ_j
28	1	29^{yr}	−5;12,30°	−6;15°	−4
29	1	30^{yr}	6;30,37,30°	7;48,45°	5
57	2	59^{yr}	1;18,7,30°	1;33,45°	1
85	3	88^{yr}	−3;54,22,30°	−4;41,15°	−3

Table 3.57b: Saturn system A: net displacements for various intervals.

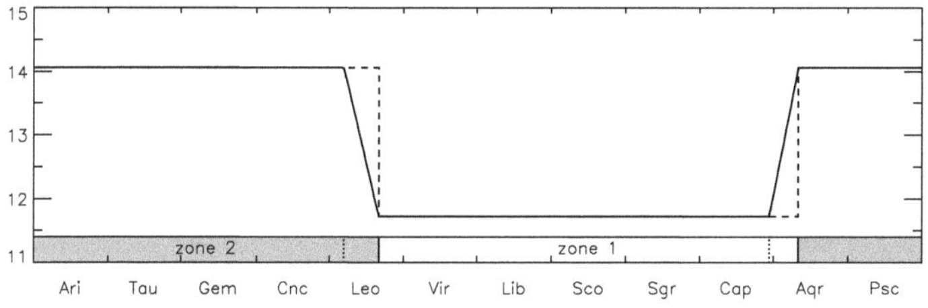

Fig. 3.21: Saturn, system A: graphical representation of σ as a function of B_{i-1}. The dashed lines represent the preliminary values σ_j.

3.8.2.2 Updating T with the synodic time

Procedures for updating T are not preserved for system A. Only one synodic table (Steele 2002a) preserves scant traces of a column T. They suggest that τ is obtained in the usual manner from Eq. (3.9), with c_τ close to

$$c_\tau = 6{,}11;27{,}20{,}37{,}20^{\overline{\tau}}. \tag{3.44}$$

This number, which amounts to $12^{\overline{m}}$ + $11;27{,}20{,}37{,}20^{\overline{\tau}}$, is reproduced by Eq. (3.13) if one sets $m = 1$, $e = 11;4^{\overline{\tau}}$ and $\overline{\sigma} \approx$ 12;39,22,30° (Table 3.57a).

3.8.2.3 Net displacements for various intervals

No. **44** P2 lists net displacements dB_j for 29^{yr}, probably also for 30^{yr}, using the templates DIFF.A.1 and DIFF.A.2. The values in the text are not identical with the ones in Table 3.57b: they are either rounded, or they pertain to an otherwise unknown variant of system A.[325] The numbers preserved in the badly damaged No. **95** P2' might also be interpreted as net displacements for system A.

[324] As pointed out by Swerdlow (1998), system A' may have been constructed in order to eliminate that feature.

[325] In *ACT*, p. 380, the net displacements for 29^{yr} are erroneously defined with plus signs.

push	$\delta\tau$	v_1 [°/$\bar{\tau}$]	$\delta\Sigma_1$	v_2 [°/$\bar{\tau}$]	$\delta\Sigma_2$
LA to FA	(40$^{\bar{\tau}}$?)	0;5	3;20°?	0;6	4°?
FA to S1 (1)	30$^{\bar{\tau}}$	0;5	2;30°	0;6	3°
(2)	3$^{\overline{m}}$	0;3,20	5°	0;4	6°
total	4$^{\overline{m}}$		7;30°		9°
S1 to AR	52;30$^{\bar{\tau}}$	−0;4,13,40	−3;41,40°	−0;5,4,24	−4;26,21°
AR to S2	1,0$^{\bar{\tau}}$	−0;3,20	−3;20°	−0;4	−4°
S2 to LA (1)	3$^{\overline{m}}$	0;3,35,30	5;23,15°	0;4,18,40	6;28°
(2)	30$^{\bar{\tau}}$	0;5	2;30°	0;6	3°
total	4$^{\overline{m}}$		7;53,15°		9;28°
LA to LA	6,32;30$^{\bar{\tau}}$?		11;41,35°?		14;1,39°?

Table 3.58a: Saturn: scheme A.S$_1$ for the subdivision of the synodic cycle.

push	$\delta\Sigma_1$	$\delta\Sigma_2$
LA to FA	3;20°	4°
FA to S1	7;30°	9°
S1 to S2	−6;40°	−8°
S2 to LA	7;33,7,30°	9;3,45°
LA to LA	11;43,7,30°	14;3,45°

Table 3.58b: Scheme A.S$_2$.

3.8.2.4 Subdivision of the synodic cycle

Procedure texts and template tables for system A contain evidence of two schemes for the subdivision of the synodic cycle. Scheme A.S$_1$ (Table 3.58a) is preserved in No. **41** P2,P3 and No. **42** P4,P5.[326] In scheme A.S$_2$ (Table 3.58b), reconstructed by Aaboe & Sachs (1966) from the template tables, some values of $\delta\Sigma$ are slightly different. Presumably by mistake, the values of $\delta\Sigma$ (S2 to LA) for this scheme are mentioned in No. **41** P2,P3 as if they describe retrograde motion in scheme A.S$_1$.

3.8.3 System A'

System A' is represented by one template table[327] containing only zodiacal positions for a single, unidentified synodic phenomenon, from which Aaboe & Sachs (1966) reconstructed the step function for σ (Table 3.59). Its period P deviates considerably from the system A value (and the empirical value). This may suggest that the table was computed from a faulty step function,[328] but according to Swerdlow (1998) system A' may also have been a deliberate creation whose purpose is to eliminate the fractional values of v_j in system A.

j	b_j	α_j	σ_j	r_j	δ_j	v_j
1	20° Leo	2,50°	11;43,7,30°	1;12	0;3,7,30°	54,24
2	10° Aqr	3,10°	14;3,45°	0;50	0;3,45°	50,40

Π	Z	Y	P	$\bar{\sigma}$
1,45,4	3,45	1,48,49^{yr}	28;1,4	12;50,56,20,...°

Table 3.59: Saturn system A': parameters of the step function for the synodic arc and periods of B.

3.8.4 Systems B, B' and B"

System B and its variants B' and B", discovered by Kugler,[329] are represented by 3 procedure texts and 13 synodic tables.[330] The extant procedure texts (Table 3.60) are concerned only with period relations and the zigzag functions for σ and τ. For the subdivision of the synodic cycle and other features inferred from the tabular texts cf. Volume II.

system B	parameters of zigzag function σ	No. **41** P6, No. **42** P8
system B, A	period relations	No. **45** P1'
system B"	parameters of zigzag function τ	No. **41** P5, No. **42** P7

Table 3.60: Procedures for Saturn, systems B and B".

[326] Swerdlow (1998), Table 3.5, lists an erroneous value −3;41,57,30° for $\delta\Sigma_1$ (S1 to AR).

[327] BM 78080 = Aaboe & Sachs (1966), Text C, first mentioned in *ACT*, p. 314.

[328] Neugebauer (*ACT*, p. 314) points out that if one shifts b_2 from 10° Aqr to 30° Aqr one obtains a modified step function resulting in a much better period $P = 28;18,8$.

[329] Cf. the appendix to Kugler (1900), pp. 207–211, and Kugler (1907), pp. 176–178.

[330] ST: *ACT* 700–709, 704a+705a (which have been joined), *SpTU* 4 170. Steele (2000c) found that the Saturn data in A 3405 = *ADRT* V 74 were also computed with system B.

3.8.4.1 Updating B and T

Procedures with complete instructions for updating B and T with the zigzag functions for σ and τ are not preserved. No. **41** P5,P6 and their duplicates No. **42** P7,P8 mention only the defining parameters (Table 3.61). While No. **41** P6 and No. **42** P8 (σ) comply with system B, the parameters in No. **41** P5 and No. **42** P7 (τ) are abbreviated (system B"). In most synodic tables columns B and T comply with system B; system B' is attested in the synodic tables *ACT* 704 (T) and 707 (B).[331] In system B' the extrema are rounded differently than in system B", but in such a way that Δ and P are the same as in system B.

		d	m	M	Δ	μ	Π	Z	Y	P	$\overline{\sigma}$
system B	σ	0;12°	11;14,2,30°	14;4,42,30°	2;50,40°	12;39,22,30°	4,16	9	4,25^{yr}	28;26,40	12;39,22,30°
system B'	σ	"	11;14,5°	14;4,45°	"	12;39,25°	"	"	"	"	"
system B	τ	0;12$^{\overline{\tau}}$	22;41,23,7,30$^{\overline{\tau}}$	25;32,3,7,30$^{\overline{\tau}}$	2;50,40$^{\overline{\tau}}$	24;6,43,7,30$^{\overline{\tau}}$	"	"	"	"	
system B'	τ	"	22;41,25$^{\overline{\tau}}$	25;32,5$^{\overline{\tau}}$	"	24;6,45$^{\overline{\tau}}$	"	"	"	"	
system B"	τ	"	"	25;24,5$^{\overline{\tau}}$	2;42,40$^{\overline{\tau}}$	24;2,45$^{\overline{\tau}}$	4,4	"	4,13^{yr}	27;6,40	

Table 3.61: Saturn systems B, B' and B": parameters of the zigzag functions for σ and τ, period relations and the mean synodic arc $\overline{\sigma} = 6{,}0/P$.

[331] Given the bad state of preservation of *ACT* 707 the reconstruction of system B' is provisional; cf. Volume II.

Chapter 4
Moon

About 48% of the procedure texts and 59% of the tabular texts are concerned with the Moon (Akk.: *Sîn*), making it the dominant topic of mathematical astronomy. §4.1 introduces the lunar phenomena and the terminology of the lunar texts. §4.2 contains a brief introduction to lunar systems K, A and B. A detailed discussion of the algorithms is found in §4.3–§4.5.[332] Editions of the procedure texts are found in §5.3.[333] The lunar functions are represented using the conventional symbols, with a few minor modifications. Functions pertaining to New Moon or Full Moon are provided with a superscript nm or fm, e.g. F^{nm} and F^{fm}; the lunation, day or tithi number is indicated by a subscript i, e.g. F_i.[334] Variants of a function are marked with accents, e.g. F', or superscripts in parentheses, e.g. $F^{(3)}$. Functions evaluated at successive days or tithis are, as usual, marked with a star, e.g. F^*. Sometimes a parameter or difference is provided with a subscript indicating the function, e.g. μ_F is the mean value of F, and d_F its monthly difference.

4.1 Lunar phenomena

4.1.1 Apparent motion

Like the Sun and the planets, the Moon moves along the zodiac from west to east, passing through the zodiacal signs in the sequence Ari, Tau, Gem, etc. (Fig. 4.1). Due to the inclination of its orbit, the Moon is not confined to the ecliptic, but can reach a maximum distance of about 5° above or below it. The points where the lunar orbit intersects the ecliptic are known as the ascending node (modern symbol: ☊), where the Moon crosses the ecliptic from below, and the descending node (☋), where it crosses it from above. The nodes perform a constant motion of about 19.4°/yr in the retrograde direction, opposite to the Moon's own motion (Fig. 4.1), resulting in 1 revolution per 18.60^{yr}. Due to the nodal motion, the draconitic month, $m_{\mathrm{drac}} \approx 27.212^{\mathrm{d}} = 27;12,44^{\mathrm{d}}$, defined as the time in which the Moon returns to the same node, is slightly shorter than the sidereal month, $m_{\mathrm{sid}} \approx 27.322^{\mathrm{d}} = 27;19,18^{\mathrm{d}}$, the time in which the Moon performs one revolution with respect to the stars.

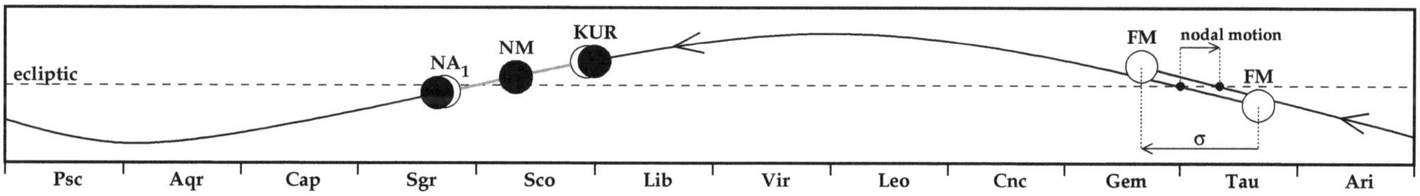

Fig. 4.1: View of the sky showing schematically, from right to left, the motion of the Moon between two Full Moons. The grey segment denotes the interval around New Moon, between KUR and NA_1, when the Moon is invisible. The synodic arc σ is the *net* displacement of the Moon (and the Sun) along the zodiac between successive lunations. The nodal motion is exaggerated in this figure.

[332] For some algorithms, especially those of lunar system B, the tabular texts are the only available source of information. Since a new investigation of these texts is outstanding (cf. Volume II), the results are to some extent preliminary.

[333] While some procedures deviate from the main systems A, B and K, none of them amounts to a coherent and complete computational system. For this reason they are discussed together with systems A and B or in the textual commentaries.

[334] In *ACT*, *HAMA* and elsewhere: subscript 1 = New Moon; 2 = Full Moon; $F(i) = F_i$.

4.1.2 Lunations

The fundamental synodic phenomena of the Moon are the lunations, i.e. New Moon, when the Moon and the Sun are in conjunction, and Full Moon, when they are in opposition.[335] The zodiacal position of the Moon (B) at lunation implies that of the Sun: at New Moon they coincide, while at Full Moon they are exactly opposite. In the Babylonian terminology, functions evaluated at (or near) New Moon are usually referred to as 'for day 1' (*ša₂* **u₄**.1.**kam**),[336] even though 'day 1' also denotes the first day of the month, which occurs several days after New Moon. By the same token, functions evaluated at (or near) Full Moon are referred to as 'for day 15'. For the Babylonian astronomers lunations were the starting point for predicting eclipses and Lunar Six intervals.

Synodic arc

All lunar systems include an algorithm for the synodic arc (σ), the net displacement of the Moon and the Sun along the zodiac between successive lunations. In lunar system A (§4.4.4) σ is a step function of the zodiacal position, and in system B (§4.5.3) a zigzag function of the lunation number. System K's algorithm for σ is still unknown. The Moon's *total* synodic arc Σ exceeds σ by a full revolution:[337]

$$\Sigma = \sigma + 6{,}0^\circ. \tag{4.1}$$

Since the Moon's elongation (= angular distance along the zodiac) from the Sun is exactly 0° at New Moon, and 180° at Full Moon, lunations satisfy a strict solar-distance principle (§3.3.1.3). Hence the Sun's synodic arc is exactly the same as that of the Moon,

$$\sigma_\odot = \sigma. \tag{4.2}$$

Period relations

The Moon's zodiacal position at successive lunations satisfies the usual period relation of a Babylonian function (§2.6.6), meaning that there is an exact return after Π months, corresponding to Z revolutions of the lunation. For each lunation the Moon itself covers one additional full revolution, so that Π lunations result in Π extra revolutions, which yields the following fundamental period relation:[338]

$$\Pi \text{ lunations } \sim Z \text{ revolutions of the lunation } \sim Z + \Pi \text{ revolutions of the Moon,} \tag{4.3}$$

where $\sim$ expresses exact correspondence (strictly speaking only nearly exact in system B, as explained in §3.3.1.4).

Zodiacal and lunar variation

Since the lunar texts are mainly concerned with lunations and associated phenomena, their algorithms reflect the lunar as well as the solar motion. One of the fundamental and innovative principles underlying the lunar systems is that seemingly irregularly varying quantities are modeled as the sum of different elementary periodic contributions. Most elementary lunar functions follow either the zodiacal ('solar') variation or the lunar variation, each of which has a distinct period. The only other elementary periodic motion represented in the lunar systems is the nodal motion. The zodiacal variation concerns quantities that are periodic on the zodiac, i.e. that assume the same value when the lunation returns to the same zodiacal position (B). Since the Sun performs one revolution around the zodiac in 1^{yr}, the period of these functions is close to $12;22,8^{\mathrm{m}}$, a close and commonly used approximation of the year.[339] Apart from functions computed from B, zigzag functions of the lunation number are also used for modeling the zodiacal variation. In order for these functions to satisfy $P \approx 12;22,8$ the parameters d and Δ must be suitably chosen.

The lunar variation concerns quantities that vary with the same period as the Moon's daily displacement (velocity). The period of this variation, known in modern astronomy as the lunar anomaly, is the anomalistic month, $m_{\mathrm{an}} = 27.555^{\mathrm{d}} = 27;33,16^{\mathrm{d}}$. Because the Moon's orbit is not fixed with respect to the stars, functions controlled by the lunar variation do not assume the same value if the Moon returns to the same zodiacal position. Hence they cannot be represented as

335 The term New Moon is used here in the conventional technical sense as synonymous with the conjunction and should not be confused with the first appearance of the crescent 1–3^{d} after the conjunction.

336 Occasionally 'for day 28', e.g. in *ACT* 155, a template table for lunar system B.

337 I.e. $n = 1$ in Eq. (3.2), p. 59.

338 This is a special case of the general period relation for a planet (Eq. 3.7), as can be seen by setting $m = 0$, $n = 1$, and replacing 'year' by 'synodic month' as the time needed for an exact return.

339 Proof: $12;22,8\ \overline{m}_{\mathrm{syn}} \approx 12;22,8 \cdot 29;31,50^{\mathrm{d}} \approx 6,5;15^{\mathrm{d}} = 365.25^{\mathrm{d}}$ (for a more accurate value of $\overline{m}_{\mathrm{syn}} \approx$ cf. Table B.1, p. 525). For the zodiacal variation and the implied year lengths in Babylonian astronomy cf. Britton (2002; 2007b).

periodic functions of B, and are modeled as zigzag functions of the lunation number. Since the anomalistic month is shorter than the synodic month ($m_{an}/\overline{m}_{syn} = 0.9333$), functions controlled by the lunar variation appear to vary with a period $P = p/(1-p) \approx 13.99^m$ if evaluated at successive lunations (for this effect cf. §2.6.8.2).

Time of the lunations

All lunar systems include algorithms for computing the synodic month, the time between successive lunations, which is the lunar equivalent of the synodic time, τ. The synodic month is not constant but varies rather irregularly between 29.27^d and 29.83^d, resulting in a mean value $\overline{m}_{syn} = 29.531^d = 29;31,50^d$. This is about 2 days longer than the sidereal month, because the Moon has to cover the extra distance σ in addition to 1 full revolution. Unlike the planetary systems, where τ is computed with a single algorithm, the more sophisticated algorithms of lunar systems A and B take into account two periodic contributions to the synodic month, reflecting the lunar variation (G) and the zodiacal variation (J). Various different conventions are used for expressing the time of the lunation: with respect to the preceding or following sunset or sunrise (M in systems A, B and K), or midnight (L in system B).

4.1.3 Eclipses

If a lunation occurs sufficiently close to a node a lunar eclipse may occur at Full Moon, or a solar eclipse at New Moon.[340] In order to predict eclipses Babylonian astronomers computed the Moon's distance to the ecliptic (E) and related functions representing the magnitude of eclipses (e.g. Ψ), such that an eclipse was considered possible if the value of this function was within a certain range. In the synodic tables, eclipse possibilities are often marked by the sign ḪAB next to the value of the eclipse function.[341] If an eclipse was considered impossible this is often marked by the sign BE or BAR. Similar conventions are known from the observational texts (Steele 2001/2). For lunar systems A and B there are also eclipse tables, which are basically synodic tables in which consecutive lines correspond to consecutive eclipse possibilities separated by 5 or 6 months.

4.1.4 Lunar Six intervals

Arguably the most remarkable achievement of Babylonian mathematical astronomy concerns the computation of the Lunar Six, which are time intervals between the rising or setting of the Moon and that of the Sun around New Moon or Full Moon.[342] Their Babylonian logographic names, **kur**, **na** (in two distinct meanings), **šu₂**, **me** and **gi₆**, are conventionally adopted as technical terms in modern translations.[343] The interval **na** that occurs shortly after New Moon is rendered here as NA_1, where the index 1 ('day 1') serves to distinguish it from NA near Full Moon.[344] For the definition of each interval cf. Table 4.1. They can be visualised as arcs parallel to the celestial equator, such that the arc length measured in degrees equals the duration in time degrees (Fig. 4.2). For KUR and NA_1 this arc connects the Moon and the Sun, for ŠU_2, NA, ME and GI_6 it connects the Moon and the 'anti-Sun', the point diametrically opposite to the Sun.

	time between	and	near
KUR	last *visible* moonrise before sunrise	sunrise	New Moon
NA_1	sunset	first *visible* moonset after sunset	New Moon
ŠU_2	last moonset before sunrise	sunrise	Full Moon
NA	sunrise	first moonset after sunrise	Full Moon
ME	last moonrise before sunset	sunset	Full Moon
GI_6	sunset	first moonrise after sunset	Full Moon

Table 4.1: Definition of the Lunar Six intervals.

The precise value of a Lunar Six interval depends on the choice of the reference points on the lunar and solar disks by which the horizon crossings are defined.[345] In the case of NA_1 and KUR simple observational considerations are sufficient to conclude that they are necessarily defined in terms of the lower rim of the Moon and the upper rim of the Sun. This is because the lunar crescent is on the Moon's lower rim at both events, and the conditions for seeing it are optimal when

[340] For the phenomenology and patterns of lunar and solar eclipses cf. Steele (2000b), pp. 9–15; Williams (2005), pp. 3–18.

[341] The Akkadian reading of ḪAB is unknown; cf. the Glossary.

[342] The modern term Lunar Six was introduced by Sachs (1948).

[343] In this capacity written with capitals. The Akkadian reading of the logograms is uncertain (cf. *ADRT* I, p. 21; *ACT*, p. 229), with the exception of **na** = *manzāzu*, 'standpoint' (lit.); cf. the Glossary.

[344] In the literature NA_1 is also called NA_N.

[345] In Fig. 4.2 this is ignored and the disk centres are used.

the Sun's upper rim is at the horizon. Similar considerations do not apply to the Full Moon, which is always easily visible as long as it is above the horizon — even during day time.

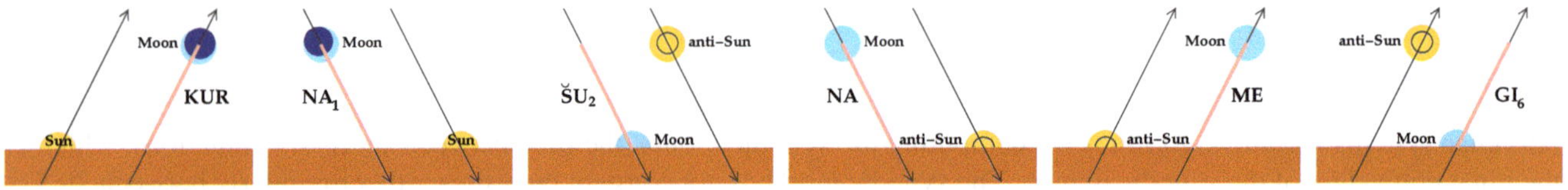

Fig. 4.2: Schematic views of the sky near the moment of sunset or sunrise showing the Moon, the Sun, or the 'anti-Sun', the imaginary disk diametrically opposite to the Sun, and the direction of setting or rising parallel to the celestial equator (arrows) for each of the Lunar Six intervals (thick bars), provisionally defined here in terms of disk centres.

The New Moon intervals NA_1 and KUR

The astronomical significance of the Lunar Six intervals is readily explained for NA_1 (Fig. 4.2), because the first appearance of the new crescent after sunset marks the beginning of the month (day 1) in the Babylonian calendar. Since the synodic month varies between 29.27 and 29.83^d (on average 29.53^d), this ideally occurs at the end of the 29th or the 30th day of the month.[346] Astronomically speaking, the importance of NA_1 stems from the fact that it is roughly proportional to the Moon's elongation from the Sun, and thus also to the width of the lunar crescent, which determines how bright the Moon stands out against the background sky. Hence there is a threshold value for NA_1 (and KUR) above which the crescent is sufficiently bright to be observed. This value is not explicitly mentioned in the available texts, but it can be estimated by analysing the values in column NA_1 and the month lengths (29 or 30^d) derived from them.[347] The interval KUR is the counterpart of NA_1 for the last appearance of the lunar crescent, which occurs about 1–3^d before the conjunction. It is of no use for the calendar, nor does it serve any other identifiable purpose within the framework of mathematical astronomy (cf. below).

The Full Moon intervals $ŠU_2$, NA, ME and GI_6

The intervals $ŠU_2$, NA, ME and GI_6 do not have to exceed some minimum value, because the Full Moon is always visible as long as it is above the horizon and not obstructed by clouds. Furthermore, the observational considerations from which it was inferred that NA_1 and KUR are delimited by the moments when the lower rim of the Moon and the upper rim of the Sun cross the horizon do not hold, since both rims are visible during Full Moon. The four intervals near Full Moon form two pairs, since $ŠU_2$ and NA differ only with regard to the order in which the Sun and the Moon cross the horizon, and the same holds for ME and GI_6 (Table 4.1).[348] If the Moon is not too far from the ecliptic $ŠU_2$ and ME occur before Full Moon and NA and GI_6 after it. Depending on whether Full Moon happens during the day or at night, their order is as shown on the left or the right of Fig. 4.3, respectively. If the Moon is far above or below the ecliptic the order can be different; the possible configurations are shown in Appendix E.4. Like KUR, the four intervals near Full Moon are of no use for the calendar, nor do they seem to have any other purpose within the framework of mathematical astronomy.

Fig. 4.3: Most common sequence of the Full Moon Lunar Six intervals for a Full Moon during day time (left) and night time (right). Thick horizontal bars indicate night time. Arrows indicate the rising (upward) or setting (downward) of the Sun and the Moon (those for the Sun coincide with the beginning or end of night time).

A satisfying, comprehensive explanation of why Lunar Six intervals are computed in mathematical astronomy remains to be found. This may be achieved by exploring the relations between mathematical astronomy and the observational texts, e.g. astronomical diaries and Lunar Six tablets, since they contain the earliest observations as well as predictions of Lunar Six intervals.[349] These predictions were made by means of the Goal-Year method,[350] which appears to have been fully operational as early as the 6th c. BC (Huber 2007). While in mathematical astronomy Lunar Six intervals are computed

[346] Months of 31^d are not attested, i.e. day 1 of the new month was declared no later than at the end of day 30, even if the lunar crescent was not observed.

[347] Cf. also §4.4.19.11; *ACT*, pp. 66–67, 83–84; *HAMA*, pp. 539–540. A new analysis of these data will appear in Volume II.

[348] Also near New Moon one could define intervals complementary to KUR and NA_1 such that the Sun and the Moon cross the horizon in the reverse order. However, since the lunar crescent cannot be seen during daylight, these were ignored by the Babylonian astronomers.

[349] These issues will be explored in a separate investigation.

[350] Brack-Bernsen (1997, 2002); Brack-Bernsen & Hunger (2002, 2010).

independently from one another (cf. §4.4.19), the Goal-Year method for predicting NA_1 proceeds from recorded values of NA_1, $ŠU_2$ and NA, and similarly that of KUR from records of KUR, ME and GI_6. Since NA_1 was important for calendaric purposes, it is only KUR whose practical significance remains unclear within the framework of the observational texts.

4.2 Lunar systems K, A and B

Nearly all lunar texts belong to systems A or B.[351] System K is introduced here as the name of the computational system underlying No. **52** P1–P4. Several tabular texts cannot be clearly assigned to lunar systems A or B (or K) because they, in the words of Neugebauer, contain a 'strange combination of elements of both systems'.[352] Significant deviations from these systems are also apparent among the procedures in §5.3.4 and in a few procedures on tablets otherwise devoted to system A or K. They are discussed in the textual commentaries or in the framework of systems A (§4.4) and B (§4.5). The earliest synodic tables from the late Achaemenid period contain evidence of functions and parameters related to those of systems A and B, but reflecting an earlier stage of development.[353] A new investigation of these tablets will appear in Volume II.

Some of the obvious distinctions between systems K, A and B are compiled in Table 4.2. The significance of the different conventions for the vernal equinox, if there is any, is not clear.[354] The newly discovered convention underlying system K whereby the vernal equinox occurs when the Sun is at 12° Ari was hitherto unknown from Babylonia.[355] System A exhibits a high degree of internal coherence that is lacking from system B.[356] System K deviates considerably from systems A and B, and reflects an older stage of development. In some respects its algorithms can be considered as intermediate between the Goal-Year methods and systems A and B.

	system K	system A	system B
algorithm for the synodic arc	?	step function of *B*	zigzag function of the lunation number
vernal equinox	12° Ari	10° Ari	8° Ari
time of New Moon defined with respect to	preceding sunset (M^{nm})	following sunset (M^{nm})	preceding midnight (L^{nm})
time of Full Moon defined with respect to	?	preceding sunset (M^{fm})	preceding midnight (L^{fm})

Table 4.2: Distinctive features of lunar systems K, A and B.

A few synodic tables from the Parthian era include a colophon with the phrase 'computed table of PN' (*tērsītu ša* PN), where PN is either Nabû-rēmanni (*ACT* 18: system A)[357] or Kidinnu (*ACT* 122, 123a: system B).[358] They are identified with Naburianos and Kidenas, hellenised names of two Babylonian astronomers mentioned by the Greek historian Strabo (ca. 64 BC – AD 24).[359] The suggestion has been made that these individuals, who are not the scribe or 'owner' of the mentioned tablets, are the creators of lunar systems A and B, respectively, but the available evidence is inconclusive. Only two instances of the phrase are securely dated, namely to 103 BC (*ACT* 122) and 49 BC (*ACT* 18), at least 250 years after lunar systems A and B were created. Since neither Kidinnu nor Nabû-rēmanni is directly attested as an astronomer, apart from these late attributions, their historical veracity is difficult to assess. During the Parthian era, Babylonian astronomers apparently followed the ancient Mesopotamian custom of ascribing scholarly texts to sages of a remote past.

351 Kugler (1900) called lunar system A 'System II', and lunar system B 'System I', a terminology originally also used by Neugebauer (e.g. 1938a–d).

352 *ACT* 90–93 (pp. 120–123).

353 Aaboe & Sachs (1969), Texts A–D; Aaboe et al. (1991), Text E.

354 During the first half of the 20th century the different conventions for the vernal equinox in systems A and B were often attributed to precession such that system B was considered to have been invented later than system A. Schnabel (1923, 1927) inferred from this that Babylonian astronomers had discovered the precession of the equinoxes before Hipparchos, an interpretation that was shown to be erroneous by Neugebauer (1950); cf. also *HAMA*, pp. 368–369.

355 Neither is this convention attested in other ancient sources, as far as known, except for a reference in the *Introduction to Aratus' Phenomena*, a work attributed to Achilles Tatius, a Roman-era Greek-writing scholar from Alexandria (quoted *HAMA*, p. 598).

356 This point was stressed by Britton (1999, 2007a).

357 Written ^{md}**ak**-*re*$_x$(**sipa**)-*man-nu*.

358 In *ACT* 122 presumably the same individual is named ^{m}*Ki-din*-DIŠ, likely a defective form of ^{m}*Ki-din*-^{d}DIŠ, which suggests that *Kidinnu* is an abbreviated form of *Kidin-Anu*. If so then Kidinnu probably originates from Uruk. For this name type cf. Tallquist (1905), p. 88.

359 *Geographica*, Book 16, Ch. 1:19: 'The mathematicians make mention of some individuals among them (= the Chaldaean astronomers), as Kidenas, Naburianus, and Sudines.' Sudines has not been identified in cuneiform texts. The identity of Kidinnu and Kidenas was first suggested by Schiaparelli in 1908 (cf. Fotheringham 1933, p. 33), that of Naburianos and Nabû-rēmanni by Schnabel (1923).

Chronology

The oldest lunar tables computed fully in accordance with lunar system A date to years 6–7 of the reign of Philip Arrhidaeus (318 –316 BC),[360] while the oldest known tablet with lunar system B is about 60 years later.[361] Near the beginning of the Seleucid era, perhaps a few decades before that, system A had thus reached its final form, system B at the latest by SE 54. There is only very little evidence to suggest that either system underwent any further change after that.[362] Several lunar tables from the preceding Achaemenid era have an affinity with system A, but none demonstrably with system B.[363] Furthermore, some parameters of system B are closer to the empirical value than in system A, while there appear to be no examples of the converse situation.[364] Most investigators therefore consider system A to be older than system B. Neugebauer cautioned against this conclusion,[365] because there are only few datable lunar tablets from the period before SE 54, after which both systems were used simultaneously. I would add that accuracy with regard to empirical data is a problematic criterion for establishing a chronology, since it presupposes an evolutionary development leading to a more accurate description of the predicted phenomena. The accuracy of an algorithm is to some extent subjective because it depends on which features are judged to be important – periods, mean values or extrema. Moreover, an algorithm may be more accurate not in a quantitative but a qualitative sense, because its mathematical structure is better capable of describing the various effects controlling the phenomena. For instance, Kugler argued that the step function underlying the synodic arc in system A is a more primitive and therefore older algorithm than system B's zigzag function, but it is now well accepted that the step function is, in many ways, the more flexible and powerful algorithm of the two. Most of the available evidence nevertheless points to system A being older than system B, and there can also be little doubt that system K is older than both systems A and B.

4.3 System K

Our only source of infomation about lunar system K is No. **52**, which does not contain procedures for the Full Moon case and lacks several important algorithms that must have been part of it, e.g. for the zodiacal position and the Moon's distance to the ecliptic. Therefore, the complete set of algorithms constituting system K is currently not available. Also synodic tables, which can be assumed to be the purpose of these algorithms, are not preserved. However, the algorithms for the Lunar Six intervals NA_1 and KUR, the main topic of No. **52**, are now relatively complete and well understood. The present study offers several improvements to their interpretation, and a first reconstruction of the algorithms for $ŠU_2$, NA, ME and GI_6. The resulting flow chart for system K, which is still incomplete for the mentioned reasons, is shown in Fig. 4.4.

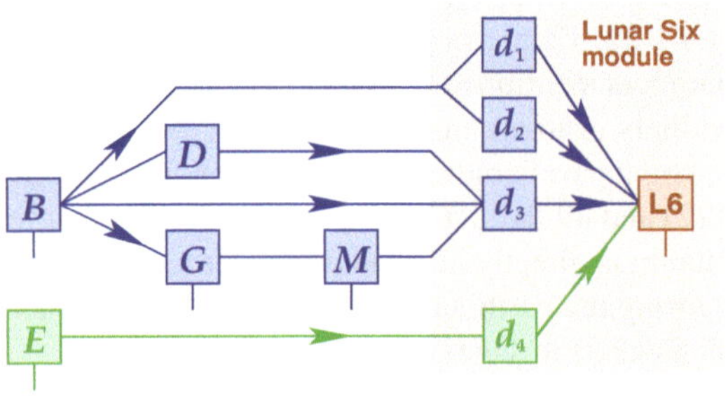

Fig. 4.4: Flow chart for lunar system K. Some functions depend on an initial value, which is indicated by a vertical bar below the box.

4.3.1 Zodiacal position of the Moon (B)

terminology	*qaqqaru*(**ki**) = 'position'
unit	zodiacal sign and degrees within it [0–30]
astronomical meaning	zodiacal position of the Moon

System K certainly included an algorithm for the Moon's zodiacal position (*B*), because several other algorithms depend on *B* (Fig. 4.4). It can be assumed that *B* is updated from one to the next lunation by adding the synodic arc, i.e.

[360] BM 40094+45662 (Aaboe & Hamilton 1979) and BM 36890 (Steele 2002b), both lunar system A.

[361] The template table BM 34162 (*ACT* 149) with data for the years SE 54–67 (258–245 BC).

[362] Suggested innovations concern the refinement of certain parameters of lunar system B in the Arsacid era; cf. p. 498 and Britton (2007b).

[363] Aaboe & Sachs (1969), Texts A–D; presumably also the undated BM 45861 (Steele 2002c).

[364] *ACT*, pp. 41–42.

[365] *ACT*, p. 42; *HAMA*, p. 475.

$B_i = B_{i-1} + \sigma$, but nothing is known about σ. There is a certain likelihood that σ is computed from B_{i-1} by means of linear interpolation, since that approach appears to be typical for system K (cf. the Lunar Six module).

4.3.2 Duration of the night (*D*)

terminology	*maṣṣarti*(**en.nun**) *mūši*(**gi$_6$**) = 'watch of the night'
unit	time degrees
astronomical meaning	duration of the night

The existence of this function is implied by step 3 of the Lunar Six module, which depends on it. Unlike lunar systems A and B, where *D* is called 'duration of the night' (LA$_2$ *ša$_2$ mūši*), its name in system K contains the term 'watch' (*maṣṣartu*). In the astronomical diaries this denotes one third of the night, but the meaning 'duration' (of the entire night) implied here is also attested in non-mathematical astronomical texts from Babylonia as well as Neo-Assyrian astrological texts.[366] An algorithm for *D* is not preserved in No. **52**. Since several coefficients of the Lunar Six module of system K are computed by linear interpolation with respect to control values pertaining to 12° of the zodiacal signs, it may be assumed that the algorithm for *D* is based on the same approach. Hence the equinoxes, when $D = 3{,}0°$, should correspond to the solar positions $B_{\odot}$ = 12° Ari or 12° Lib.[367]

4.3.3 The Moon's distance to the ecliptic? (*E*)

terminology	*šūqu*(**nim**) = 'height' ($E_{\uparrow}$); *šuplu*(**sig**) = 'depth' ($E_{\downarrow}$)
unit	?
astronomical meaning	the Moon's distance to the ecliptic?

The existence of a function representing the Moon's 'height and depth' (*E*) is implied by step 4 of the Lunar Six module (No. **52** P3,P4.d). The algorithm for *E* is not preserved, nor are there sufficient numerical examples in No. **52** P3,P4.d from which it might be reconstructed (only one value is mentioned in P4.d, without the unit of measurement; cf. the commentary). Also, there is no definite proof that 'height' and 'depth' are defined with respect to the Moon's ecliptic, as in lunar systems A and B, although the identical terminology makes this rather likely, and the interpretation of step 4 does not exclude this.

4.3.4 Duration of the synodic month (*G*)

terminology	*simānu* = 'duration'
unit	time degrees
astronomical meaning	duration of the synodic month -29^d
procedures:	
computing *G* from *B*	No. **52** P4.b

Function *G* represents the duration of the synodic month -29^d, such that G_i belongs to the interval between lunations $i-1$ and *i*. From *G* is computed the time of the lunation (*M*). The value of G_i is obtained by updating G_{i-1} with a difference which is computed from the Moon's zodiacal position at lunation. In the only extant procedure for *G*, which concerns New Moons, the difference is a subtractive coefficient, represented here as ZI_G after its Babylonian name *nisḫu*(**zi**) = 'subtraction',

$$G_i = G_{i-1} - \mathrm{ZI}_G(B_{i-1}). \tag{4.4}$$

ZI_G is computed from B_{i-1} by means of interpolation. There are probably 12 control values of ZI_G, each pertaining to 12° of a zodiacal sign, but only 2 of them could be reconstructed from No. **52** P4.b, namely 30° in 12° Psc and 45° in 12° Ari. It can be assumed that Eq. (4.4) was also used for Full Moons, but elementary considerations are sufficient to determine that this algorithm is not complete. The point is that G_i cannot always be smaller than G_{i-1}. There must be cases, unmentioned in No. **52** P4.b, where an additive difference is applied to G_{i-1}, say

$$G_i = G_{i-1} + \mathrm{TAB}_G(B_{i-1}). \tag{4.5}$$

Nothing is known about the magnitude of TAB_G, or the conditions under which it is applied.

[366] E.g. Mul.Apin II.i.9–21.

[367] It is therefore odd that the single exemplary value $D = 2{,}52;20°$ in P4 (NA$_1$ step 3), which should pertain to 1° Ari, cannot be readily explained by linear interpolation between 12° Psc and 12° Ari (cf. the commentary to No. **52**). Consequently one cannot use this to derive the value of *D* at 12° Psc.

4.3.5 *Time of the lunation (M)*

terminology	**gi₆ du** = '(time by which) the night has progressed' (M^{nm})
unit	time degrees
astronomical meaning	time of lunation with respect to sunset
procedures:	
M^{nm}	No. **52** P4.b

Function M represents the time of the lunation. The only known application of M occurs in step 3 of the Lunar Six module. The single numerical example of the algorithm for M^{nm} in No. **52** P4.b implies that M^{nm} is defined with respect to the preceding sunset. This convention differs from that of system A, where M^{nm} is defined with respect to the following sunset, but it agrees with system A's convention for M^{fm} (cf. Eq. 4.73).[368] The rule for updating M_i^{nm} implied by No. **52** P4.b (Fig. 4.5) can be represented as

$$M_i^{\mathrm{nm}} = \begin{cases} M_{i-1}^{\mathrm{nm}} + G_i^{\mathrm{nm}} & \text{if } M_{i-1}^{\mathrm{nm}} + G_i^{\mathrm{nm}} \leq 6{,}0\text{: date boundary not transgressed} \\ M_{i-1}^{\mathrm{nm}} + G_i^{\mathrm{nm}} - 6{,}0 & \text{if } M_{i-1}^{\mathrm{nm}} + G_i^{\mathrm{nm}} > 6{,}0\text{: date boundary transgressed.} \end{cases} \tag{4.6}$$

If the addition of G_i^{nm} results in a value in excess of 6,0° (=1^{d}) the date boundary (sunset) is transgressed. In that case 6,0° is subtracted from M_i^{nm} and 1^{d} is added to the date of the lunation. It is instructive to compare this algorithm with that of lunar system A (§4.4.18). Notably absent is a correction for the changing length of daylight (C').

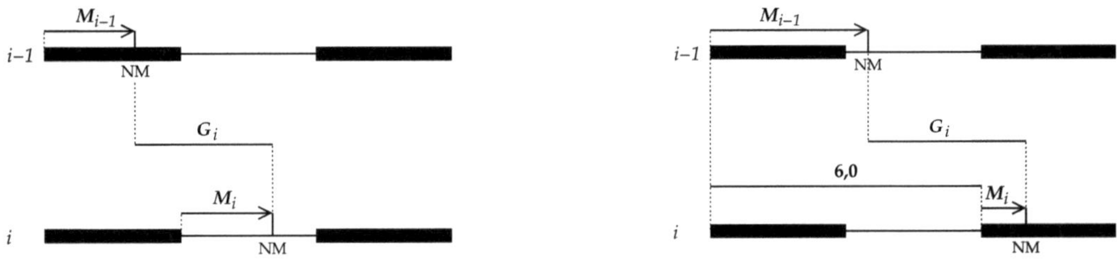

Fig. 4.5: System K: graphical representation of the computation of M^{nm}. Thick horizontal bars indicate night time. The arrow indicates the direction in which M is measured (from sunset). **Left**: date boundary not transgressed. **Right**: date boundary transgressed.

Since No. **52** does not contain an equivalent set of procedures for Full Moon, it is unknown how M^{fm} is defined, e.g. with respect to the previous sunset as for New Moon, or with respect to the following sunset, in which case

$$M_i^{\mathrm{fm}} = \begin{cases} M_{i-1}^{\mathrm{fm}} - G_i^{\mathrm{fm}} & \text{if } G_i^{\mathrm{fm}} < M_{i-1}^{\mathrm{fm}}\text{: date boundary not transgressed} \\ M_{i-1}^{\mathrm{fm}} + 6{,}0 - G_i^{\mathrm{fm}} & \text{if } G_i^{\mathrm{fm}} > M_{i-1}^{\mathrm{fm}}\text{: date boundary transgressed.} \end{cases} \tag{4.7}$$

If the subtraction of G^{fm} is impossible because $G^{\mathrm{fm}} > M^{\mathrm{fm}}$, then 6,0° is added before the subtraction, and the date of the lunation is modified accordingly by subtracting 1^{d}.

After computing M_i the corresponding date of lunation i is established by adding 29^{d} to the date of lunation $i-1$, taking into account the subtraction or addition of 6,0°. If that addition or subtraction occurred, the date boundary was transgressed, and the difference becomes 30^{d} instead of 29^{d}.

[368] The significance of this is not clear. The system A convention is probably motivated by the assumed dependence of the Lunar Six intervals on the Moon's elongation from the Sun, which is proportional to the time difference between the lunation and the Lunar Six event (N). In system K no such dependence is assumed.

4.3.6 Lunar Six intervals

terminology	*manzāzu*(**na**) [**u$_4$.1.kam**] = 'NA for day 1'; **u$_4$.1.kam** = 'day 1' = NA$_1$ **kur**; **u$_4$.na$_2$.a** ('day of disappearance') = KUR **me** = ME; **gi$_6$** = GI$_6$; **šu$_2$** = ŠU$_2$; *manzāzu*(**na**) = NA
unit	time degrees
astronomical meaning	cf. Table 4.1 (p. 113)
procedures and auxiliary tables:	
NA$_1$	No. **52** T1; P1,P2,P4
KUR	No. **52** T2; P3,P4
ŠU$_2$, NA	No. **52** T3
ME, GI$_6$	No. **52** T4

The Lunar Six intervals are the only known final quantities of lunar system K.[369] In their edition of No. **52**, Neugebauer and Sachs made a first attempt to reconstruct the algorithms for NA$_1$ and KUR. I here present several improvements to these algorithms as well as a reconstruction of the algorithms for ŠU$_2$, NA, ME and GI$_6$. While the Lunar Six modules of systems A (§4.4.19) and B (§4.5.18) both employ essentially the same sophisticated method involving a decomposition of the Lunar Six intervals into different geometric contributions, system K proceeds from the simpler assumption that the Lunar Six intervals can be updated from month to month by applying differences computed from the Moon's zodiacal position. The Lunar Six intervals (L6) for month i are obtained from those in month $i-1$ by applying a total difference consisting of at most 4 contributions:

$$\mathrm{L6}_i = \mathrm{L6}_{i-1} + d_1\,[-d_2] + d_3 + d_4. \tag{4.8}$$

The coefficients d_1, d_2 and d_3 are computed from the Moon's old zodiacal position, B_{i-1}, by means of linear interpolation. Some of the control values for the interpolation are provided in the auxiliary tables No. **52** T1–T4. A new analysis of the interpolation algorithm has revealed that all control values are anchored at 12° of the zodiacal signs.[370] Correction d_2 is placed in brackets because it is applied only under certain conditions (cf. below).

4.3.6.1 Step 1. Preliminary monthly difference d_1

Step 1 concerns an additive contribution d_1. In order to understand this contribution recall that the synodic month lasts on average 29.53^d. Therefore, the Moon's elongation from the Sun does not assume the same value if one proceeds from the previous Lunar Six event by a whole number of days (29 or 30). This results in a change of the Lunar Six interval with respect to the previous value that can be additive or subtractive, depending on the timing of the Lunar Six event with respect to the lunation. For Lunar Sixes that (ideally) occur before the lunation (KUR, ŠU$_2$ and ME) a 29^d shift leads to an increase, a 30^d shift to a decrease; for Lunar Sixes that (ideally) occur after the lunation (NA$_1$, NA and GI$_6$), the change is in the opposite direction. The additive nature of d_1 therefore implies that it pertains to 29^d in the case of KUR, ŠU$_2$ and ME, and to 30^d in the case of NA$_1$, NA and GI$_6$. The examples of step 1 for NA$_1$ and KUR in P1–P4 confirm this interpretation.

The value of d_1 is computed from B_{i-1} by means of linear interpolation with respect to 12 control values, each anchored at 12° of a zodiacal sign. Four different versions of d_1, TAB$_1$,...,TAB$_4$, are tabulated in No. **52** T1–T4 (Fig. 4.6). From the procedures it is clear that TAB$_1$ and TAB$_2$ pertain to NA$_1$ and KUR, respectively. The newly added fragment of No. **52** proves that TAB$_3$ and TAB$_4$ each pertain to one pair of Lunar Sixes near Full Moon: TAB$_3$ is used for ŠU$_2$ and NA; TAB$_4$ for ME and GI$_6$. This results in

$$d_1 = \begin{cases} \mathrm{TAB}_1 & \text{for NA}_1 \quad \text{date } +30^d \\ \mathrm{TAB}_2 & \text{for KUR} \quad \text{date } +29^d \\ \mathrm{TAB}_3 & \text{for ŠU}_2 \quad \text{date } +29^d \\ \mathrm{TAB}_3 & \text{for NA} \quad \text{date } +30^d \\ \mathrm{TAB}_4 & \text{for ME} \quad \text{date } +29^d \\ \mathrm{TAB}_4 & \text{for GI}_6 \quad \text{date } +30^d. \end{cases} \tag{4.9}$$

As mentioned, the shift of the date (29 or 30^d) follows from the time order of the Lunar Six event with respect to the lunation. In the case of NA$_1$ this shift has a special significance, because it represents the length of the month in whole days (hollow or full, respectively). Hence step 1 implies that the previous month is full, but this is a preliminary result that

[369] ŠU$_2$ and NA are not preserved in No. **52**. The other four are referred to by the usual logograms, but occasionally KUR is called **u$_4$.na$_2$.a**, 'day of disappearance', and NA$_1$ **u$_4$.1.kam**, 'day 1', terms also known from No. **53** P17',P18'.

[370] Cf. the commentary. Neugebauer & Sachs (1969) considered the interpolation scheme to be erroneous because they wrongly assumed that the control values pertain to 1° of the zodiacal signs.

can be modified in step 2. The use of one and the same coefficient, TAB_3, for both $ŠU_2$ and NA, and TAB_4 for both ME and GI_6, appears to be motivated by the fact that the mean synodic month, 29.53^d, is very close to the midpoint between 29^d and 30^d. The time interval between moonset and sunrise may therefore have been considered to be the same, on average, for the sunrise before the opposition ($ŠU_2$) and the sunrise after it (NA), and analogously for the time interval between moonrise and sunset before (ME) and after the opposition (GI_6).[371]

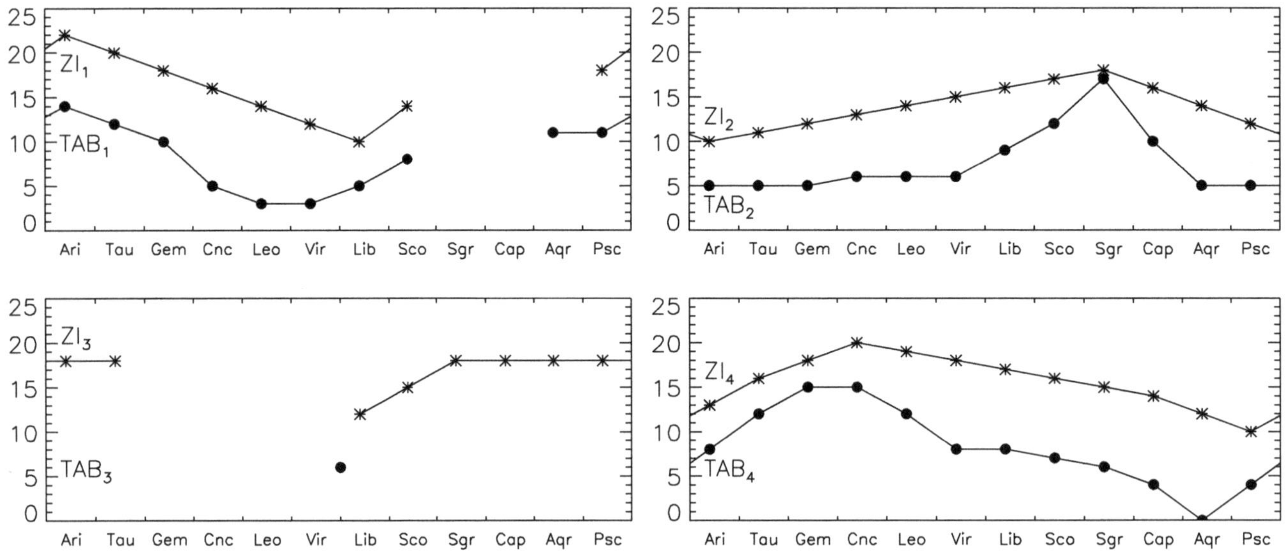

Fig. 4.6: Contributions $TAB_1,...,TAB_4$ ($= d_1$) and $ZI_1,...,ZI_4$ ($= d_2$) to the monthly difference of the Lunar Sixes measured in time degrees, plotted as functions of the Moon's zodiacal position. Dots and crosses correspond to the control values preserved in No. **52** T1–T4, and a few additional ones reconstructed from No. **52** P1–P4. The drawn lines represent linear interpolation between the control points, as formulated in the procedures.

4.3.6.2 Step 2. Optional subtractive correction d_2 corresponding to a shift by 1^d

Step 2 involves a subtractive contribution d_2 representing the effect of a shift of the Lunar Six event by 1^d with respect to what was assumed in step 1. If step 2 is executed then the total shift of the date becomes 30^d instead of 29^d, or vice versa. From the examples in No. **52** P1–P4 and from the tables No. **52** T1–T4 the following general rule can be inferred:

$$d_2 = \begin{cases} ZI_1 & \text{for } NA_1 & \text{date} - 1^d \\ ZI_2 & \text{for KUR} & \text{date} + 1^d \\ ZI_3 & \text{for } ŠU_2 & \text{date} + 1^d \\ ZI_3 & \text{for NA} & \text{date} - 1^d \\ ZI_4 & \text{for ME} & \text{date} + 1^d \\ ZI_4 & \text{for } GI_6 & \text{date} - 1^d. \end{cases} \tag{4.10}$$

All examples of step 2 in No. **52** P1–P4 concern NA_1 and KUR. The reconstruction of d_2 for the other Lunar Six intervals presented here has benefited from the newly added fragment proving that ZI_3 and ZI_4 pertain to the same pairs of Lunar Sixes as TAB_3 and TAB_4. Just like $TAB_1,...,TAB_4$, the corrections $ZI_1,...,ZI_4$ are computed from the Moon's zodiacal position at lunation by linear extrapolation with respect to control values pertaining to the 12° of the zodiacal signs, which are tabulated in No. **52** T1–T4. The use of one and the same coefficient, ZI_3, for both $ŠU_2$ and NA, and ZI_4 for both ME and GI_6, is analogous to the approach underlying step 1. The corresponding shifts of the date follow from the time order of the Lunar Six event with respect to the lunation, as explained in connection with step 1. In the case of NA_1 the execution of step 2 implies that the previous month is hollow (29^d) instead of full (30^d).

Since d_2 is subtractive, it leads to a smaller Lunar Six value. As explained in §4.1, the correct Lunar Six value is always the smallest one above the appropriate threshold value (Table 4.1, p. 113). NA_1 and KUR must exceed a minimum value for the lunar crescent to be visible, but as in lunar systems A and B, the precise condition that NA_1 and KUR must satisfy is not preserved. $ŠU_2$, NA, ME and GI_6 must only be computable in the Babylonian sense, i.e. larger than zero.

[371] This assumption only roughly accounts for the empirical behaviour, and it fails to capture all of the variations.

4.3.6.3 Step 3. Correction d_3 involving M and D

The third step involves the time of the lunation. Its astronomical interpretation remains unclear. All instances in No. **52** concern the New Moon intervals KUR and NA_1. Most of them employ a general formulation, which can be represented as follows:[372]

$$d_3 = \begin{cases} +(M^{nm} - D^{nm}) \cdot ZI & \text{if } M^{nm} > D^{nm} \\ -(D^{nm} - M^{nm}) \cdot ZI & \text{if } M^{nm} < D^{nm}. \end{cases} \tag{4.11}$$

In this expression the minus sign in the second line corresponds to a subtractive value in the Babylonian formulation. A corresponding rule for $ŠU_2$, NA, ME and GI_6 is not preserved. The coefficient ZI is called *nisḫu*(**zi**) in Akkadian. An interpretation as 'subtraction' is inappropriate, because d_3 can be additive or subtractive. Hence the proper translation is 'displacement', which implies that it cannot be identified with any of the 'subtractions' ZI_1,...,ZI_4 tabulated in T1–T4. In the procedures ZI is qualified as being 'for that month', which signifies that it is computed from the Moon's zodiacal position at lunation. Only in P4 (Ri34) a numerical value is assigned to ZI, which must be interpreted as 0;0,13,12 (cf. the commentary). Since it is multiplied by a time interval measured in degrees, the unit of ZI is time degrees per time degree. Hence 6,0 · ZI = 1;19,12° is the 'displacement' for 1^d. As proposed by Neugebauer and Sachs, this might represent the rising or setting time of the zodiacal arc traveled by the Sun during 1^d. The value 1;19,12 would be approximately suitable for the setting Sun if the Sun is in Cap–Gem, or for the rising Sun if the Sun is in the opposite half of the zodiac.[373] The astronomical interpretation should also explain why d_3 vanishes if New Moon coincides with sunrise ($M^{nm} = D^{nm}$), is additive if it occurs after sunrise ($M^{nm} > D^{nm}$), and subtractive if it occurs before sunrise ($M^{nm} < D^{nm}$). It therefore seems that d_3 is a correction that accounts for the setting or rising time of the tiny arc traveled by the Sun between New Moon and sunrise. However, one would expect there to be a separate version of d_3 for NA_1 that vanishes at sunset, but it appears that the same version is used for both KUR and NA_1.

4.3.6.4 Step 4. Correction d_4 involving E

The fourth contribution, d_4, involves the Moon's 'height and depth' (E). For KUR and NA_1 the algorithm for d_4 can be represented as

$$d_4 = \begin{cases} +0;24 \cdot E_{\uparrow} \\ -0;24 \cdot E_{\downarrow} \end{cases} \tag{4.12}$$

The terms 'height' ($E_{\uparrow}$) and 'depth' ($E_{\downarrow}$) usually denote the Moon's distance above or below the ecliptic, respectively, but it is not certain that they have that meaning here. Only in P4 (Ri36) a numerical value, 34.55.15, is assigned to $E_{\downarrow}$. If this represents degrees then probably it is 0;34,55,15°, in which case the coefficient is 0;24; if barleycorns then it is either $0;34,55,15^{še}$ or $34;55,15^{še}$. In the former case the coefficient is 0;24, in the latter case 0;0,24. In all instances of step 4 in No. **52** the Moon is in or near Ari. It is therefore not clear whether 0;24 applies only there or in the entire zodiac. If d_4 is the contribution to the Lunar Six interval resulting from the Moon's distance to the ecliptic, corresponding to R in the Lunar Six modules of systems A and B, then 0;24 should correspond to the coefficient r. However, this obvious interpretation is not without problems. First, the text implies that d_4 is the same for KUR and NA_1, whereas r is different if the Moon is rising (KUR) or setting (NA_1). Secondly, 0;24 should not be a constant, because r depends on the Moon's zodiacal position at lunation. The value 0;24 would be appropriate only if the Moon is near Ari and setting, or near Lib and rising.[374] Since the Moon is assumed to be in or near Ari, this explanation may work for NA_1, but it fails for KUR, when both the Moon and the Sun are rising.

4.4 System A

About 60% of the tabular texts and 90% of the procedure texts for the Moon belong to system A, making this the best documented computational system of Babylonian mathematical astronomy. The oldest evidence for elements of lunar system A ('proto-A') is contained in several tabular texts dating from the end of the fifth c. BC,[375] and an undated tabular text containing values of an early version of function Φ.[376] The earliest lunar tables computed fully in accordance with

[372] There is one apparent exception in P3 (KUR, Oii13–15; cf. below), perhaps due to a scribal error.

[373] Cf. the modern version of the coefficient q_{set} (Fig. E.1, p. 534).

[374] Cf. r_{set} in Fig. E.1 (right). In system A r_{set} is modeled as a zigzag function of the Moon's zodiacal position at lunation with a minimum in Ari which is either 0 or 0;18 (Fig. 4.35).

[375] Aaboe & Sachs (1969), Texts B–D, dated to about 400 BC by Britton (2010).

[376] BM 45861 (Steele 2002c).

lunar system A as we know it from the Seleucid era date to years 6–7 of the reign of Philip Arrhidaeus (318–316 BC).[377] The task of interpreting and reconstructing the algorithms of lunar system A was initiated and vigorously pursued by F. Kugler, whose *Babylonische Mondrechung* (1900) contains essentially correct representations of the algorithms for Φ, B, C, C', E, Ψ, F, G, J, K and M (the astronomical interpretation of Φ was established only much later). Kugler made extensive use of the procedure text No. 53, whose main fragment, 76-11-17,2418, was available to him. Other milestones were two papers by O. Neugebauer (1938b,c) mainly concerned with E and eclipse magnitude, Neugebauer's analysis of the procedures for the Lunar Six intervals (*ACT* 201 = No. **61.A**) and A. Aaboe's reconstruction of the algorithms for G, W, Λ, Y and Z.[378] Functions X_s and X_m (p. 124), each attested only once, were reconstructed and interpreted by Britton (2009).

4.4.1 Composition of the tablets

Less than half of the tablets contain a single procedure. These include a coherent group of 12 tablets with close duplicates of the interpolation scheme for computing G from Φ (§4.4.11), as well as three fragments, perhaps belonging to a single tablet, with related interpolation algorithms for computing W from G (§4.4.12).

	size	Φ	B	C	Ψ, Ψ'	F	E	$v_\odot$	$G+J$	K	M	G	NA_1	KUR	ME
No. **53**	A	P1,P2	P5'	P4'	P6'	P7'	P8'	P10',P11'.a	P14'	P15'.a	P15'.b	P16'	P17'	P18'	
No. **55**	F?	P1	P2	P3'											
No. **56**	F	P1	P2	P3'											
No. **58**	?		P1?				P2'								
No. **59**	?				P1'		P2'								
No. **60**	C?					P2'	P1'								P3'
No. **62**	?	P1',P2'	P3'												

Table 4.3: Moon system A: tablets with a collection of procedures for computing or updating various functions.

Most other tablets are collections of procedures. One group, whose main representative is No. **53**, consists of tablets with procedures for computing successive columns of a synodic table (Table 4.3). Most of the procedures on these tablets are directly connected with the production of synodic tables, and their order largely agrees with that of the corresponding columns in the synodic tables. However, No. **53** also includes procedures that do not strictly belong to lunar system A (P13': solstices and equinoxes) or whose purpose is not connected with the tabular texts or is unclear (P9', P11'.b). A second group consists of tablets with procedures concerned with different versions or aspects of a single function, or several closely related functions (Table 4.4). For instance, No. **65** contains a collection of procedures for computing Ψ from E and vice versa. No. **82** is a unique tablet with elaborate computations involving function Φ without any obvious direct connection to the production of synodic tables. Arguably the most remarkable tablets of the second group are Nos. **61.A–G**, which contain algorithms for the Lunar Six intervals ŠU_2, NA, ME and GI_6.

	size	content
Nos. **61.A–G**	C (2); F (5)	computing ŠU_2, NA, ME and GI_6
No. **65**	F	computing Ψ from E and vice versa; computing F from Φ
No. **69**	F?	computing G from Φ; differences of G
No. **79**	A,M?	computing Λ from Φ
No. **82**	D	computations involving Φ and G
No. **83**	E	computations involving F and variants

Table 4.4: Tablets with a collection of procedures for lunar system A concerned with different versions or aspects of a single function, or several closely related functions.

On some tablets procedures are combined with (significant) tabular content in the form of auxiliary tables. The combination of procedures with synodic tables, common among the planetary texts, is not attested among the lunar texts (Table 4.5).[379] The main function of the auxiliary tables is to provide coefficients that are needed in the accompanying procedures on the same tablet. A slightly different relation between auxiliary table and procedure can be identified in No. **79** which contains rules for computing Λ from Φ. While the procedure provides the interpolation algorithms for Λ in fully verbal form, the auxiliary table offers the same information in tabular form and largely devoid of verbs.

[377] BM 40094+45662 (Aaboe & Hamilton 1979) and BM 36890 (Steele 2002b).

[378] Aaboe (1968, 1969, 1971); Aaboe & Hamilton (1979).

[379] Strictly speaking Nos. **61.A–G** and several other procedure texts also fall in this category, but the tables represent only a small portion of these tablets, thus I ignore them here.

	size	auxiliary tables	procedures
No. **79**	A	Φ_k, Λ_k, c_k	computing G from Φ; Λ from Φ
No. **81**	?	Φ_k, Λ_k, c_k	unclear

Table 4.5: Moon system A: tablets with procedures and auxiliary tables.

4.4.2 Algorithms for lunar system A

The algorithms for lunar system A fall into three categories: (i) algorithms for computing and updating the functions in a synodic table, template table or eclipse table, (ii) the same for a daily motion table, (iii) other algorithms.

4.4.2.1 Algorithms for the synodic tables, template tables and eclipse tables

Lunar synodic tables are unusually elongated tablets with, in the case of system A, up to 16 columns. The modern designations of the columns were established by Neugebauer (1936), who modified the notation of Kugler in order to make it more consistent.[380] The leftmost column, T, contains the month names and, in month I, also the year number. Usually each side contains data for one calendar year, on the obverse for successive New Moons, on the reverse for successive Full Moons.[381] The New Moon data pertain to the New Moon occurring at the end of the month named in column *T*, while the Full Moon data pertain to the Full Moon in the middle of that month. The complete sequence of algorithms for computing a synodic table can be conveniently represented as a flow chart (Fig. 4.7a), in which each square represents an algorithm for computing or updating a certain function.[382] The arrows indicate how the outcome of an algorithm is passed on to other algorithms (cf. §2.6.4.4). By executing the flow chart all functions are updated to the next lunation. Most functions are auxiliary in the sense that they serve as input for other functions. The purpose of the synodic tables is defined by the 'final' functions, located at the end of a branch of the flow chart: Ψ (eclipse magnitude) and the Lunar Six intervals.[383]

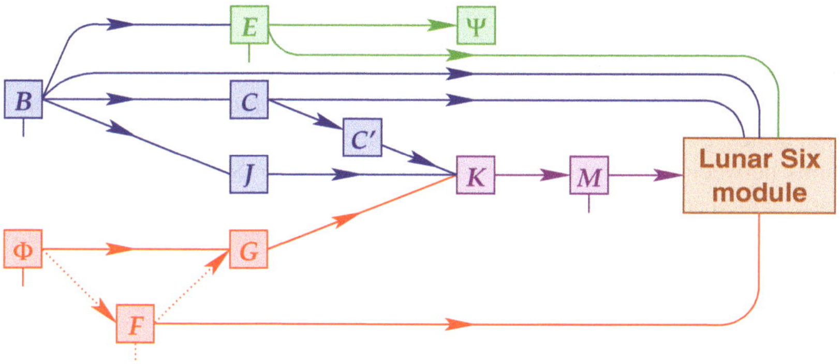

Figure 4.7a: Flow chart for the computation of a synodic table for lunar system A. Each box represents an algorithm for updating a function (from a previous value) or, in the case of a nested function (§2.6.5), computing it from other functions. The dotted lines indicate that G may be computed either from Φ or from F. Some functions depend on an initial value, which is indicated by a vertical line below the box.

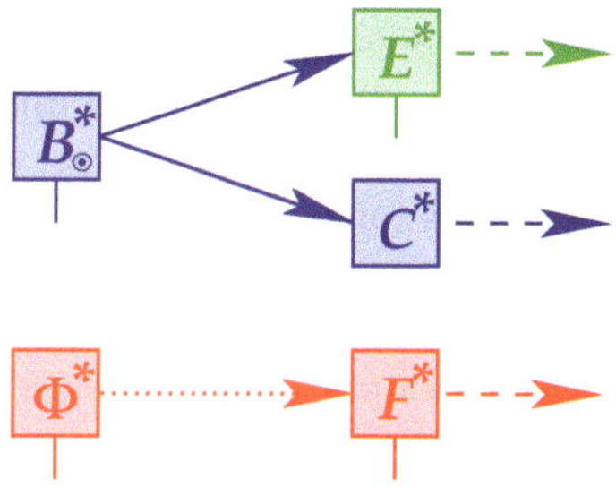

Figure 4.7b: Flow chart for the computation of a daily motion table for lunar system A. The dotted line indicates that F^* may be computed either from Φ^* or from its own zigzag function.

380 The usual sequence of the columns is T, Φ, B, C, E, Ψ, F, G, J, K, L, M, followed by NA_1 and KUR (on the New Moon side), $ŠU_2$, NA, ME and GI_6 (on the Full Moon side). Neugebauer's notation agrees with that of Kugler for E, K, L and M, but differs for the other columns: A (Kugler) = T (Neugebauer), B = Φ, C = B, D = C, F = Ψ, G = F, H = G, and I = J. Van der Waerden (e.g. in *Science Awakening* II) continued to use Kugler's notation. Neugebauer also used the symbols $P_1 = NA_1$, P_3 = KUR, $P_{2,1} = ŠU_2$, $P_{2,2}$ = NA, $P_{2,3}$ = ME, $P_{2,4} = GI_6$.

381 In order to provide necessary initial values for some columns the first line usually pertains to the last month (XII or XII_2) of the preceding year, so that there are 1+12 (normal year) or 1+13 (intercalary year) months on each side.

382 For a precursor of this flow chart cf. Aaboe (2001), p. 60.

383 Some synodic tables lack one or more of the auxiliary quantities, or the Lunar Six intervals (e.g. *ACT* 1). In the latter case the time of the lunation (M) is the final quantity.

Template tables and eclipse tables

Compared to lunar system B there are few template tables for system A. Most of them contain only columns Φ and G,[384] but one has only columns K and M.[385] They are computed by the same algorithms as the synodic tables. Eclipse tables[386] are essentially excerpts of synodic tables in which successive lines correspond to successive eclipse possibilities separated by 5 or 6 months. They are computed by the same algorithms as other synodic tables; some include a column W.

Functions Λ, Y, W, X_s and X_m

In some synodic tables columns G (§4.4.11) and J (§4.4.14), which represent the duration of 1^m, are replaced by the analogous columns W (§4.4.12) and Z (yet unattested; cf. §4.4.15), which represent the duration of 6^m, or Λ (§4.4.13) and Y (§4.4.16), which represent the duration of 12^m. The corresponding flow chart is obtained by carrying out the appropriate substitutions in Fig. 4.7a. For functions X_s and X_m, each attested in one tabular text, cf. Britton (2009).[387]

4.4.2.2 Algorithms for the daily motion tables

Two daily motion tables are extant for lunar system A (*ACT* 80, 81). Both are damaged, so that only columns T^*, Φ^*, $B^*_{\odot}$, C^*, E^* and F^* are (partly) preserved, but they may have contained further columns.

The tithi as a subdivision of the event frame

Successive values in the column of a daily motion table pertain to successive *real* tithis, a constructed unit defined as 1/30 of the *actual* synodic month. The tithi, called 'day' (*ūmu*) in Akkadian, is therefore not a constant unit of time, but a subdivision of the event frame (§2.5.1). Its real duration varies from lunation to lunation, like the duration of the synodic month. In order to distinguish this usage of the tithi from the mean tithi known from the planetary systems (§2.5.2.3), I reserve the symbol $^{\tau}$ for the real tithi, and $^{\bar{\tau}}$ for the mean tithi.[388] The mathematical proof that the starred functions pertain to real tithis and not mean tithis or days is that the net difference for 30^{τ} of a starred function, $df^*(30)$, equals the monthly difference d of the corresponding unstarred function whenever we can compute $df^*(30)$.

Purpose of the daily motion tables

The columns of the daily motion tables for lunar system A have the same astronomical interpretation as the corresponding ones in the synodic tables, except that the first column after Φ^* contains $B^*_{\odot}$, the zodiacal position of the Sun. A column for $B^*_{☽}$ is not preserved in the extant daily motion tables. Several procedure texts are concerned with the starred functions. A preliminary flow chart illustrating the relations between these functions is shown in Fig. 4.7b. Since no daily motion table is preserved to its right edge, we do not know which columns they may have contained beyond F^*. Whereas the daily motion tables for the planets and those for lunar system B have a clearly identifiable purpose, because they contain only times and zodiacal positions, that of the daily motion tables for lunar system A is not really understood. The final quantities in these tables are obviously not Lunar Six intervals or eclipse magnitude, since they are astronomically meaningful only near the lunations and not at intermediate tithis.

4.4.2.3 Other algorithms for lunar system A

The procedure texts for lunar system A contain numerous other algorithms. They can be divided into four categories, of which the first three serve purposes that are closely connected with the production of tabular texts: (i) procedures for net differences, $df(s)$, including net displacements, $dB(s)$, which can be used for verification; (ii) procedures for computing function values pertaining to Full Moon from function values pertaining to New Moon, or vice versa; (iii) period relations.

[384] Aaboe (1968) Texts A–D; Steele (2006x?), Text B (which also has column T).

[385] Steele (2006a), Text B.

[386] *ACT* 50–61, 51a, 61a; 'proto system A': Aaboe & Sachs (1969) Texts A,C,D.

[387] X_s, attested in BM 36311+ (Aaboe 1968, Text E; Britton 2009, pp. 375–386), is interpreted by Britton as the lunar variation of the Moon's net displacement for 1 saros, $dB(223)$; X_m, attested in BM 36699+ (Aaboe 1971, Text A; Britton 2009, pp. 375–386), as the lunar variation of the Moon's synodic arc, $\sigma = dB(1)$.

[388] These two different manifestations of the tithi in Babylonian mathematical astronomy were obviously known to Neugebauer, which makes it difficult to understand why he states (Neugebauer 1963: p. 532) that 'this clear and convenient definition [of the mean tithi] was perverted in later Indian astronomy and astrology to the use of thirtieths of true lunar months, i.e. to units variable in a very complicated fashion from month to month. In other words, the Hindus reintroduced into the definition of *tithis* (as they are still used in India today) exactly the complication which it was the purpose of the Babylonian invention to avoid'. In fact, the use of real tithis in lunar system A corresponds precisely to the 'perverted' practice of Indian astronomy.

(iv) The fourth category consists of procedures that do not appear to be immediately connected with the tabular texts (e.g. No. **82**). Some of them contain elaborate and sophisticated computations that reveal profound knowledge about the relations between various functions and parameters.

Introduction to the following sections

The following sections (§§4.4.3–4.4.19) are concerned with the functions of lunar system A. Information about the Akkadian terminology, units of measurement, astronomical interpretation, position of the corresponding column in the tabular texts and a list of procedures are provided at the beginning of each section. The first paragraph explains the algorithm by which the function is computed. Subsequent paragraphs are concerned with other algorithms, e.g. for net differences, the 'daily' ('starred') version of the function, period relations or other properties of the function.[389]

4.4.3 Duration of 223 synodic months (Φ)

terminology	2,13;20
unit	time degrees
astronomical meaning	contribution to the duration of $223^m - 6585^d$ resulting from the lunar variation
column	ii
procedures:	
updating Φ as a zigzag function	No. **53** P1.a,b; No. **55** P1.a; No. **56** P1; No. **62** P2'.a; No. **63** P1'; No. **65** P6
net differences	No. **55** P1.b; No. **63** P2'; No. **82** P1,P2.a,d,P12
relations between Φ^{fm} and Φ^{nm}	No. **53** P2; No. **62** P1'.a,b; No. **65** P6.a,b
checking rules for $\Phi_i + \Phi_{i+6}$	No. **62** P2'.b
computations involving Φ_{tr}	No. **82** P1,P3,P6
other computations involving Φ	No. **53** P3; No. **55** P1.a; No. **82** P1–P14; No. **100** P3'?

Function Φ is an approximation of the duration of 223^m, the interval commonly known as the saros.[390] To be precise, Φ_i is the amount by which the time between lunations i and $i+223$ exceeds 6585^d, such that only the lunar variation is taken into account. At the practical level pertaining to the production of synodic tables the only purpose of Φ is to serve as a source function for G (§4.4.11), and the astronomical significance of Φ is not obvious. Most of the procedures concerning Φ are for computing or verifying column Φ, but the extraordinary 'Saros Text' (No. **82**) is more theoretical in nature.

Updating Φ as a zigzag function of the lunation number

Numerous procedures contain the algorithm for updating Φ as a zigzag function of the lunation number (Fig. 4.8). For this algorithm cf. §2.6.8. No. **56** P1 and No. **55** P1.a are based on template ZZ.B.1, No. **53** P1.a follows template ZZ.B.2 and No. **62** P2', No. **65** P6 and perhaps No. **53** P1.b are based on template ZZ.B.3. The damaged No. **53** P2 and No. **63** P1'.a' may also contain this algorithm, but their template could not be established. Virtually all instances of Φ in the tabular texts are accounted for by the parameters in Table 4.6.

d	m	M	μ	Δ	δ	Π	Z	P	p
2;45,55,33,20°	1,57;47,57,46,40°	2,17;4,48,53,20°	2,7;26,23,20°	19;16,51,6,40°	0;0,22,13,20°	1,44,7	7,28	13;56,39,6,...	0;55,59,6,13,...

Table 4.6: Parameters of the zigzag function for Φ.

Properties of Φ

The correct astronomical interpretation of Φ as the duration of $223^m - 6585^d$ remained unknown until the 1950s, long after the algorithm for Φ was reconstructed by Kugler.[391] The issue was gradually resolved after Neugebauer (1957b) published the 'Saros Text' (No. **82**) in which a connection is made between Φ and the saros. The nature of the connection remained unclear until van der Waerden (1965) proposed that Φ represents the duration of the saros in excess of 6585^d.

[389] Some tablets include procedures concerning topics that are not an integral part of mathematical astronomy, e.g. solstices and equinoxes (No. **53** P13'). They are discussed only in the commentaries.

[390] For this pseudo-Babylonian term cf. *HAMA*, p. 497.

[391] Kugler (1900), pp. 116–128 ('Columne B'). Kugler interpreted Φ as the Moon's apparent diameter.

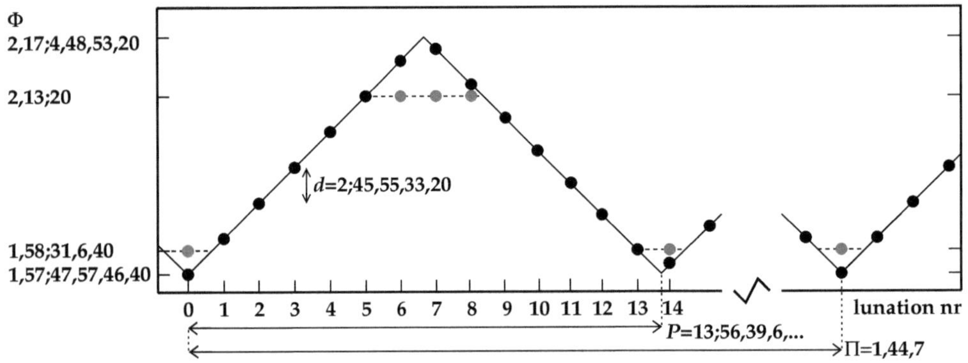

Fig. 4.8: Graphical representation of Φ as a function of the lunation number i. Also shown is the truncated version Φ_{tr} (dashed lines).

Definite proof of that interpretation came with Aaboe's (1968) discovery of the relations between Φ, G and Λ (§4.4.11). The empirical mean value of the excess is about $\frac{1}{3}^{\text{d}}$, which explains why the mean value of Φ is close to 2,0°. As pointed out already by Kugler (1900), the period P is the same as that of $F^{(1)}$, the Moon's daily displacement along the zodiac (§4.4.9), which defines the lunar variation. In other words, the zodiacal contribution to the saros is ignored in Φ.[392] This property carries over to G, which is computed from Φ. Two other features of Φ are inferred from the synodic tables.[393] First, nearly all instances of column Φ, even if they are centuries apart, are connectible in the sense that they derive from the same initial value (§2.6.8). All can thus be considered as segments of a single parent column Φ. Since Φ repeats only after $\Pi = 6447^{\text{m}}$, this can be exploited for dating fragments of synodic tables, even if they contain only a few damaged values of Φ. Secondly, Φ values pertaining to New Moon never coincide with values pertaining to Full Moon, because their grids are shifted by 0;0,11,6,40° = $\delta/2$, where δ (Table 4.6) is the smallest possible difference between two arbitrary values of Φ^{nm}, or two arbitrary values of Φ^{fm} (Eq. 2.21). It is therefore always possible to determine from a numerical value of Φ whether it pertains to New Moon or Full Moon. Furthermore, if one thinks of Φ^{nm} and Φ^{fm} as being sampled from the same function, then Φ^{nm} is shifted by half a synodic month with respect to Φ^{fm}.

Net differences for various intervals

Several procedures are concerned with net differences of Φ for intervals longer than 1 synodic month (Table 4.7a). For a general discussion of this topic cf. §2.6.8.4. Recall that t is the closest whole number of oscillations of the zigzag function. The significance of the net differences may be grasped by inspecting Fig. 4.8, which shows that 12^{m} amount to 1 oscillation of the zigzag function and a subtractive correction $d\Phi(12)$, while 14^{m} result in an additive correction $d\Phi(14)$. Both are mentioned in procedure texts: the former in No. **63** P2', the latter in No. **55** P1. The net difference for 223^{m} is mentioned in No. **82** P1,P2, and P12.[394] As will become apparent in §4.4.11, the parameter

$$\phi = 0;17,46,40°, \tag{4.13}$$

which equals $-d\Phi(223)$, plays an important role in the interpolation scheme by which G is computed from Φ. No. **82** P2.a mentions the following relation between ϕ and the monthly difference of Φ:[395]

$$d = 9;20 \cdot \phi. \tag{4.14}$$

s	t	$d\Phi(s)$	s	t	$d\Phi^*(s)$
6^{m}	0	16;35,33,20°	14^{τ}	0	19;17,9,37,46,40°
7^{m}	0	19;21,28,53,20°	15^{τ}	0	20;39,48,53,20°
12^{m}	1	−5;22,35,33,20°	28^{τ}	1	0;0,37,2,13,20°
14^{m}	1	0;9,15,33,20°	29^{τ}	1	1;23,16,17,46,40°
223^{m}	16	−0;17,46,40°			

Table 4.7a: Net differences of Φ and Φ^* for various intervals.

d	Π	Z	P
1;22,39,15,33,20°	10,24,42	22,19	27;59,33,6,51,..

Table 4.7b: Defining parameters and periods of Φ^* (as far as different from those of Φ).

392 For the lunar and zodiacal variation cf. §4.1.2.

393 *HAMA*, pp. 484–485.

394 The subtractive nature of $d\Phi(223)$ is not indicated in these procedures.

395 For the significance of this relation cf. *HAMA*, pp. 497–499; Britton (2009), pp. 361–368.

Checking rules for Φ values separated by 6^m and located on opposite branches

No. **62** P2'.b partly preserves a checking rule for the sum of two values of Φ separated by 6 months and located on opposite branches of the zigzag function. This rule is an application of the general principle explained in §2.6.8.5 (Eq. 2.24). If Φ_i and Φ_{i+6} are on opposite branches then $\Phi_i + \Phi_{i+6} = 2M - d\Phi(6) = 4,17;34,4,26,40°$ if Φ_i is increasing, or $\Phi_i + \Phi_{i+6} = 2m + d\Phi(6) = 4,12;11,28,53,20°$ if Φ_i is decreasing.

Truncated zigzag function Φ_{tr}

Some procedures are concerned with a modified zigzag function, Φ_{tr}, which is obtained by truncating Φ at $m' = 1,58;31,6,40°$ and $M' = 2,13;20°$ (Fig. 4.8). Unlike M, the latter is close to the empirical maximum of the saros interval.[396] The purpose of Φ_{tr} was clarified when Aaboe (1968) discovered how the interpolation schemes for G, W and Λ are constructed from two shifted versions of Φ_{tr} (§4.4.11). Instructions for truncating Φ are not preserved, but No. **82** P1,3,6 deal with intricate aspects of Φ_{tr}. The maximum 2,13;20° functions as the Babylonian technical term for Φ, presumably because it appears in the first line of the ΦG interpolation tables.

Empirical aspects

The fact that Φ represents only the lunar variation of the saros has important implications for the scenario by which Φ was derived from empirical data. If one considers the empirical variations in the length of the saros then the dominant contribution is the zodiacal variation and not the lunar variation (Brack-Bernsen 1997). Consequently the length of the actual, empirical saros varies with a slighly different period than Φ. Hence Φ cannot have been derived from measurements of the length of the saros, but only from some quantity, or a combination of quantities, for which the zodiacal variation is sufficiently suppressed with respect to the lunar variation. Two different scenarios have been proposed, by Brack-Bernsen and Britton, respectively. Brack-Bernsen (1990, 1997) discovered that the sum of the Lunar Six intervals near Full Moon, ŠU$_2$+NA+ME+GI$_6$, has the desired property, and is also exactly in phase with Φ. Britton (1990, 2007a, 2009) demonstrated that the zodiacal variation can be eliminated by measuring not the saros but the duration of 235^m.

Φ evaluated at consecutive tithis (Φ^*)

The daily motion tables (*ACT* 80, 81) include a column Φ^* with values of Φ evaluated at consecutive real tithis. Φ^* is modeled as a zigzag function of the tithi number, which can be viewed as the more rapidly varying zigzag function underlying Φ (cf. §2.6.8.2). This algorithm is not yet attested in procedure texts. Its defining parameters, as far as they differ from those of Φ, are compiled in Table 4.7b.

Net differences for various intervals

The net difference of Φ^* for 15^τ (Table 4.7a) may be restored in No. **62** P1'.a and No. **65** P6.a, two procedures concerned with the computation of Φ^{fm} from Φ^{nm} (cf. below). The net difference for 30^τ, $d\Phi^*(30)$, equals the monthly difference d, which confirms that the values of Φ^* pertain to real tithis and not to mean tithis or days.

Relations between Φ^{nm} and Φ^{fm}

Several procedures are concerned with the relation between Φ^{nm} and Φ^{fm}. First, No. **53** P2, No. **62** P1'.a and No. **65** P6.a contain a rule for computing Φ^{fm} from Φ^{nm}. This is achieved by updating Φ^{nm} with $d\Phi^*(15)$, the net difference of Φ^* for 15^τ (half a synodic month), and applying the appropriate reflection rule if the resulting preliminary value of Φ^{fm} falls outside the range $[m, M]$. These procedures employ essentially the same formulation as the procedures for updating the zigzag function, with the monthly difference d replaced by $d\Phi^*(15)$. No. **65** P6.a and No. **62** P1'.a are based on template ZZ.B.3; the template underlying No. **53** P2 could not be determined.

Some procedures preserve rules connecting Φ^{nm} and Φ^{fm} in a different but related manner. Since Φ_i^{nm} and Φ_i^{fm} can be viewed as instances of the same function Φ^* separated by 15^τ, they usually lie on opposite branches of the zigzag function for Φ^*. This implies that their sum is constant, which amounts to a checking rule (§2.6.8.5). In the present case we set out from an initial value Φ_0^*, representing for instance New Moon, and update this to Φ_{15}^* pertaining to the following Full Moon, by adding or subtracting $d\Phi^*(15)$. If we assume that one extremum is passed between Φ_0^* and Φ_{15}^*, so that both values are on opposite branches of the zigzag function,[397] and apply the appropriate reflection rule (Eqs. 2.14–2.15), then

[396] Britton (2009), p. 364.

[397] Since $d\Phi^*(15)$ is only slightly larger than Δ this is usually the case, but if Φ_0^* is very close to an extremum two extrema may occur between Φ_0^* and Φ_{15}^*, so that both are on the same branch.

$$\Phi_0^* + \Phi_{15}^* = \begin{cases} 2M - d\Phi^*(15) = 4,13;29,48,53,20° & (\Phi_0^* \text{ increasing}) \\ 2m + d\Phi^*(15) = 4,16;15,44,26,40° & (\Phi_0^* \text{ decreasing}). \end{cases} \quad (4.15)$$

Hence $\Phi_0^* + \Phi_{15}^*$ depends only on whether Φ_0^* is increasing or decreasing and not on the value of Φ_0^*. Now recall that Φ_i^{nm} pertains to the New Moon at the *end* of month *i*, while Φ_i^{fm} pertains to the Full Moon in the middle of that month (§4.4.2.1). Therefore, if we take Φ_0^* to be Φ_i^{nm} then Φ_{15}^* corresponds to Φ_{i+1}^{fm}, but if we take Φ_0^* to be Φ_i^{fm} then Φ_{15}^* corresponds to Φ_i^{fm}. These rules connecting Φ_i^{nm} to Φ_i^{fm} or Φ_{i+1}^{fm} are (partly) preserved in No. **62** P1' and No. **65** P6.

4.4.4 Zodiacal position of the Moon (B)

terminology	*qaqqaru*(**ki**) = 'position'; *lumāšu* = 'zodiacal sign'
unit	zodiacal sign and degrees within it [0–30]
astronomical meaning	zodiacal position of the Moon
column	iii
procedures:	
updating *B* with step funct. for σ	New Moon: No. **55** P2.a, No. **56** P2.a, No. **62** P3'.a; Full Moon: No. **53** P5'
net displacements	No. **55** P2.b, No. **56** P2.b?, No. **62** P3'.b, No. **65** P7.b
relation between B^{fm} and B^{nm}	No. **65** P7.a
other computations involving σ	No. **58** P1?

Function *B* represents the zodiacal position of the Moon. In synodic tables *B* implies the zodiacal position of the Sun ($B_\odot$): at New Moon $B_\odot$ coincides with *B*, while at Full Moon they are diametrically opposite. The value of *B* is expressed as a zodiacal sign and a number of degrees [0–30] measured from the beginning of the sign. Apart from the common term 'position' (*qaqqaru*), *B* is occasionally called *lumāšu*, 'zodiacal sign'.[398] For the coordinate system in general cf. §2.5.3.1.

Updating B with the step function for the synodic arc

B is updated from one to the next lunation by adding the synodic arc, $B_i = B_{i-1} + \sigma$, where σ is a step function of B_{i-1}. For this algorithm, reconstructed by Kugler,[399] cf. §2.6.10. It is preserved in four procedures, three of which concern New Moon, and one Full Moon. No. **53** P5', No. **62** P3' and No. **55** P2.a are formulated in accordance with template STEP.A.1; No. **56** P2.a follows STEP.A.2. Nearly all instances of column *B* in the synodic tables are accounted for by variant 1 (Tables 4.8a and 4.8b). A graphical representation of the algorithm for the New Moon case is shown in Fig. 4.9. For the Full Moon case the argument of σ is shifted by 6 zodiacal signs; compare b_j^{nm} and b_j^{fm} in Table 4.8a. Hence in both cases the actual argument of σ is the zodiacal position of the Sun ($B_\odot$).

var.	j	b_j^{nm}	b_j^{fm}	α_j	σ_j	r_j	δ_j	v_j
1	1	13° Vir	13° Psc	3,14°	30°	0;56,15	0;8°	24,15
	2	27° Psc	27° Vir	2,46°	28;7,30°	1;4	0;7,30°	22,8
2	1	29° Vir	29° Psc	2,44;30°	30°	0;56,40	0;30°	5,29
	2	13;30° Psc	13;30° Vir	3,15;30°	28;20°	1;3,31,...	0;28,20°	6,54

Table 4.8a: Parameters of the step functions for the synodic arc σ.

var.	Π	Z	P	$\overline{\sigma}$
1	46,23	3,45	12;22,8^{m}	29;6,19,0,...°
2	12,23	1,0	12;23^{m}	29;4,16,47,...°

Table 4.8b: Periods and mean synodic arc.

Variant 2 is attested only in one synodic table for 'proto system A' dating from the Achaemenid period.[400] It has two peculiar features. First, the zones are basically mirrored with respect to those of variant 1 (cf. α_j), apart from a small shift. Secondly, the transition coefficient $r_2 = 18/17$ is not a finite sexagesimal number.[401]

Properties of σ and B

Since σ is strictly periodic on the zodiac, this function represents only the zodiacal contribution to the synodic arc. In other words, the lunar variation is ignored in the algorithm for σ, which amounts to the assumption that the Moon moves with a mean velocity. A lunar correction to σ does not exist in lunar system A.[402] The adequacy of this approximation was

398 E.g. No. **53** Oi26'.

399 Kugler (1900), pp. 55–67 ('Columne C').

400 Aaboe & Sachs (1969), Text A.

401 In spite of that, the step function algorithm for *B* incorporates transition rules, which is made possible by a special choice of the initial value, as explained by Aaboe (1966).

402 However, Britton (2009) has argued that function X_m (cf. also p. 124), attested only in one tabular text, represents the lunar variation of the synodic arc.

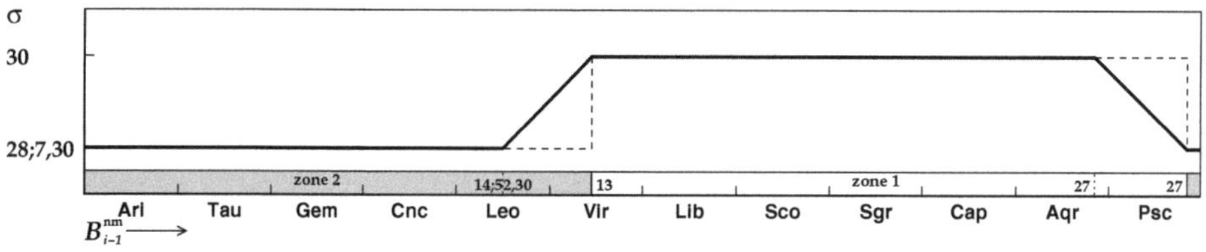

Fig. 4.9: The synodic arc σ as a function of B^{nm}_{i-1} (New Moon case; variant 1). The dashed lines represent the preliminary values σ_j.

demonstrated by Maeyama (1978, 1981) and Brack-Bernsen (1997). The period P (Table 4.8b) represents the fractional number of synodic months after which the lunation returns to exactly the same position for the first time, i.e. the length of the year expressed in synodic months. The number period Π is the smallest whole number of lunations after which the Moon and the Sun return to the same position, $Z = 225$ being the corresponding number of years. Since σ depends only on B it likewise returns to exactly the same value when B does so.

Net displacements for various intervals

Several procedures provide net displacements $dB_j(s)$ for intervals s longer than 1 synodic month (Table 4.9; for this topic cf. also §2.6.10.4). The formulation of No. **55** P1.b and No. **62** P3'.b is based on template DIFF.A.3; that of No. **65** P7.b on DIFF.A.4 (§2.6.10.4). They either pertain to close returns to the same zodiacal position (s = 12, 223, or 235^{m}) or to the opposite position ($s = 6^{\mathrm{m}}$). No. **65** P7 mentions $dB_j(s)$ for $s = 6^{\mathrm{m}}$, 12^{m} and 223^{m} (the saros). The 12^{m} interval is also mentioned in No. **62** P3'.b and No. **55** P2.b. The subtractive nature of dB_j is not always apparent in the procedures. Each net displacement is usually identified by the approximate number of years t, e.g. '18 years' for $dB_j(223)$.

s	t	dB_1	dB_2	dB_j/δ_j
6^{m}	0^{yr}	3,0°	2,48;45°	22,30
12^{m}	1^{yr}	−11;4°	−10;22,30°	−1,23
223^{m}	18^{yr}	10;48°	10;7,30°	1,21
235^{m}	19^{yr}	−0;16°	−0;15°	−2

Table 4.9: Net displacements of the Moon (and the Sun) for various intervals.

Relations between B^{nm} and B^{fm}

No. **65** P7, a procedure concerned with net displacements, begins with an algorithm by which the Moon's position at Full Moon, B^{fm}, can be computed from its position at the immediately preceding New Moon, B^{nm}, by adding a displacement amounting to half the Moon's total synodic arc, i.e. $\Sigma_1/2 = (6{,}0+\sigma_1)/2 = 3{,}15°$ (zone 1) and $\Sigma_1/2 = (6{,}0+\sigma_2)/2 = 3{,}14;3{,}45°$ (zone 2). This procedure could be used for verifying whether the numbers in columns B^{fm} and B^{nm} are correct, or for computing an initial value of B^{fm} from a given B^{nm}.

Alternative representation $\sigma(B_i)$ for backward iteration

There are situations in which it is desirable to compute the zodiacal position of the Moon backwards in time. In order to achieve this it is advantageous to express the synodic arc by which B_i was reached as a function of B_i itself, i.e. $\sigma(B_i)$. This alternative representation (Fig. 4.10) is readily obtained from the previous one, $\sigma(B_{i-1})$, by computing for each value of $\sigma(B_{i-1})$ the corresponding updated position $B_i = B_{i-1} + \sigma$. Comparison with $\sigma(B_{i-1})$ reveals that the transition regions are now located between b_j and $b_j + \sigma_j$, on the other side of the zonal boundaries, and the transition coefficients are the reciprocals of the previous ones, r_j, in agreement with the backward motion (from right to left in the figure). In the transition region between 13 Vir and 13 Lib the slope is thus $1 - r_1 = 0;3,45$; in the region between 27 Psc and 25;7,30 Ari it is $1 - r_2 = -0;4$. The old position is then obtained from the new one as

$$B_{i-1} = B_i - \sigma(B_i). \tag{4.16}$$

Although there is no direct textual evidence that the Babylonian astronomers made use of this approach, they must have done so if they wished to obtain the initial position of a synodic table from some later position. It is also useful for deriving the algorithms for Z (§4.4.15) and Y (§4.4.16).

'Daily' zodiacal position of the Sun ($B^*_\odot$)

The second column of the daily motion tables contains $B^*_\odot$, the zodiacal position of the Sun evaluated at intervals of 1^τ.[403] The purpose of $B^*_\odot$ is explained by the next column C^*, which is computed from $B^*_\odot$. The value of $B^*_\odot$ at tithi i is obtained

[403] *ACT* 80, 81. Neugebauer calls this B^* (*ACT*, p. 47).

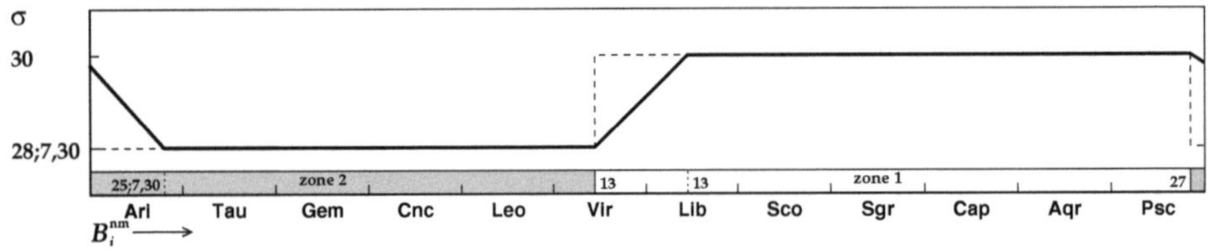

Fig. 4.10: Alternative representation of the synodic arc as a function of the updated position, B_i^{nm} (New Moon case; variant 1). The dashed lines represent the preliminary values σ_j.

from the previous value by adding the Sun's zodiacal displacement per tithi in accordance with

$$B^*_{\odot i} = B^*_{\odot i-1} + \sigma^*_{\odot}. \tag{4.17}$$

Here $\sigma^*_{\odot}$ is a step function of the Sun's zodiacal position (Table 4.10), whose preliminary values $\sigma^*_{\odot j}$ are related to the σ_j according to

$$\sigma^*_{\odot j} = \sigma_j/30. \tag{4.18}$$

Again, this implies that $B^*_{\odot}$ pertains to (real) tithis. The updating of $B^*_{\odot}$ with $\sigma^*_{\odot}$ involves transition rules based on the same r_j known from σ. It can be shown that this causes the net displacements for 30^{τ} to satisfy $dB^*_{\odot j}(30) = \sigma_j$, so that there is a strict consistency between the algorithm for updating $B^*_{\odot}$ and the one for updating B. The period P = 6,11;4$^{\tau}$ is the number of tithis after which the Sun carries out one revolution, i.e. 1^{yr} expressed in tithis. As expected, this equals 30 $\cdot P_B$.

j	b_j	α_j	σ^*_j	r_j	δ_j	v_j	$dB^*_{\odot,j}(15)$
1	13° Vir	3,14°	1°	0;56,15	0;4°	48,30	15°
2	27° Psc	2,46°	0;56,15°	1;4	0;3,45°	44,16	14;3,45°

Π	Z	P	$\overline{\sigma^*}$
1,32,46	15	6,11;4$^{\tau}$	0;58,12,38,1,51,...°

Table 4.10: Parameters of the step function for σ^*, the Sun's net displacement for 15^{τ} and periods of $B^*_{\odot}$.

'Daily' zodiacal position of the Moon ($B^*_{☽}$)

The existence of a function $B^*_{☽}$ representing the zodiacal position of the Moon evaluated at intervals of 1^{τ} is implied by references in several procedure texts to F^* (cf. §4.4.9), but a corresponding column is not yet attested in daily motion tables. The updating of $B^*_{☽}$ proceeds by adding the 'daily' displacement F^* in accordance with

$$B^*_{☽ i} = B^*_{☽ i-1} + F^*_i. \tag{4.19}$$

4.4.5 *Duration of day (C) and night (D)*

terminology	LA$_2$ *ša$_2$ ūmi*(**me**) = 'duration of daylight' (*C*) LA$_2$ *ša$_2$ mūši*(**gi$_6$**) = 'duration of the night' (*D*)
unit	time degrees
astronomical meaning	duration of daylight (*C*) and night (*D*)
column	*C*: iv; *C'*: ix or x
procedures:	
computing *C* from *B*	No. **53** P4', No. **55** P3', No. **56** P3
net differences of *C* for 12^{m}	No. **56** P4
net differences of *C'*	No. **79** P3'

Function *C* represents the length of daylight. From *C* is derived function *C'*, which is needed for computing *M*. A second application of *C* occurs in step 1 of the Lunar Six module, where the time of the Lunar Six event with respect to sunset or sunrise (*N*) is computed. The element LA$_2$ in the names of *C* and *D* certainly means duration, but the Akkadian reading has not been established.

Computing *C* from *B*

Function *C* is computed from the zodiacal position of the Sun ($B_{\odot}$) by means of linear extrapolation involving 12 control points located at 10° of the zodiacal signs (Table 4.11; Fig. 4.11).[404] Corresponding instructions are preserved in the three

[404] This algorithm was reconstructed by Kugler (1900), pp. 75–79 and 194 ('Columne D').

procedures listed above. They employ slightly different but equivalent formulations based on template IP.A, as can be illustrated for the control point in Psc:

10 zib 2.40 *ša₂ al* 10 *zib* **diri a.ra₂** 40 **du ki** 2.40 **tab**	10 Psc: 2,40. (The amount) by which it exceeds 10 Psc you multiply by 0;40, add with 2,40.	No. **53** P4'
[10 *zib*.**m**]**e** 2.40 **me** *ša₂ al-la* 10 *zib*.**me d**[**iri**] / [**a.ra₂** 40 **du k**]**i** 2.40 **tab**-*ma* **gar**-*a*[*n*]	10 Psc: 2,40 daylight. (The amount) by which it exceeds 10 Psc you multiply by 0;40, add with 2,40 and put down.	No. **55** P3'
ana tar-ṣa 10 *zib*.**me** 2.40 **me** *ša₂ al*-[*la* 10 *zib*.**me diri** GAM 40 **du**-*ma*] / **ki** 2.⌜40⌝ [*zib*.**me tab**]	Opposite 10 Psc: 2,40, the daylight. (The amount) by which it exceeds 10 Psc you multiply by 0;40 and add with 2,40 of Psc.	No. **56** P3

	Ari	Tau	Gem	Cnc	Leo	Vir	Lib	Sco	Sgr	Cap	Aqr	Psc
C_k	3,0°	3,20°	3,32°	3,36°	3,32°	3,20°	3,0°	2,40°	2,28°	2,24°	2,28°	2,40°
c_k	0;30	0;24	0;8	−0;8	−0;24	−0;40	−0;40	−0;24	−0;8	0;8	0;24	0;40
$dC_k(1)$	18;45°	11;15°	3;45°	−3;45°	−11;15°	−18;45°	−20°	−12°	−4°	4°	12°	20°
$dC_k(12)$	−6;55°	−4;9°	−1;23°	1;23°	4;9°	6;55°	7;22,40°	4;25,36°	1;28,32°	−1;28,32°	−4;25,36°	−7;22,40°
$-c_k/2$	−0;15	−0;12	−0;4	0;4	0;12	0;20	0;20	0;12	0;4	−0;4	−0;12	−0;20

Table 4.11: Control values C_k and interpolation coefficients c_k for computing C^{nm} from $B_\odot$, net differences $dC_k(1)$ and $dC_k(12)$, and interpolation coefficients $-c_k/2$ for computing net differences of C' (cf. text). All control values pertain to 10° of the zodiacal signs.

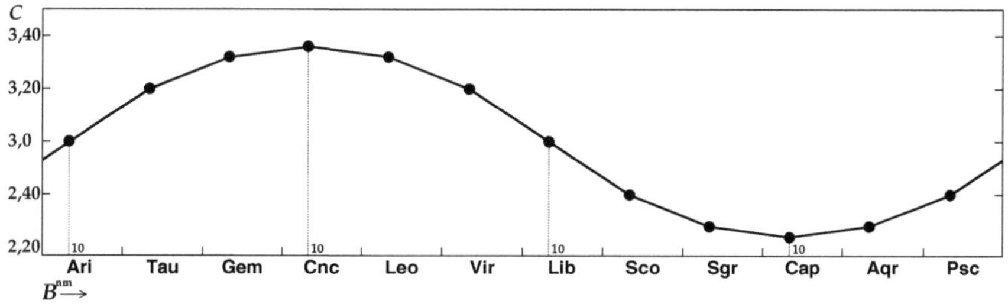

Fig. 4.11: Graphical representation of the computation of C from B^{nm} (= $B_\odot$). The dots are the control values for the interpolation scheme; the dotted lines denote solstices and equinoxes.

Net differences of C for various intervals

No. **56** P4 contains a list of 12 net differences of C for 12^{m}, $dC_k(12)$, one for each control point k. As usual the $dC_k(12)$ equal actual net differences of C only under certain conditions, namely if the Moon returns to the same zodiacal zone and the same interpolation interval. Under these conditions they are obtained by multiplying c_k (Table 4.11) by the Moon's net displacement for 12^{m} (Table 4.9), i.e. $dC_k(12) = c_k \cdot dB_j(12)$, where for each control point k one has to identify the appropriate zone j.

Duration of the night (D)

Function D represents the duration of the night,

$$D_i = 6{,}0 - C_i. \tag{4.20}$$

Instructions with a separate interpolation algorithm for D are not preserved. Neither is the relation between C and D explicitly mentioned in the procedure texts. In the synodic tables of system A there is no column D, but the computation of some Lunar Six intervals (step 1 of the Lunar Six modules) does require D.

Functions C', $C'^{(6)}$ and $\tilde{C}'$

Most synodic tables include a column C', which is computed from C by the following algorithm:

$$C'_i = -\frac{1}{2}(C_i - C_{i-1}). \tag{4.21}$$

The purpose of C' is to serve as a correction to the time of the lunation, M (§4.4.18). Since M is defined with respect to sunset, it is affected by the changing length of daylight. To be precise, C'_i is the amount by which sunset is shifted with respect to midday between lunations $i-1$ and i, such that a positive (additive) C' corresponds to a shift towards earlier times.[405]

Function $C'^{(6)}$

The existence of function $C'^{(6)}$ is implied by W, the duration of 6 synodic months (§4.4.12). It is computed from C by the following algorithm:

$$C'^{(6)}_i = -\frac{1}{2}(C_i - C_{i-6}). \quad (4.22)$$

The purpose of $C'^{(6)}$ is to serve as a correction, analogous to C', in the algorithm for updating the time of the lunation, $M^{(6)}$, over intervals of 6^{m} (§4.4.18).

Function $\tilde{C}'$

One synodic table[406] includes a column $\tilde{C}'$, which follows column Y and is computed from C by the following algorithm, reconstructed by Aaboe (1969):

$$\tilde{C}'_i = -\frac{1}{2}(C_i - C_{i-12}). \quad (4.23)$$

The purpose of $\tilde{C}'$ is to serve as a correction, analogous to C', in the algorithm for updating the time of the lunation, $\tilde{M}$, over intervals of 12^{m} (§4.4.18).

Net differences of C' *for various intervals*

The newly discovered procedure No. **79** P3' partly preserves a list of interpolation coefficients $-c_k/2$, one for every interpolation interval of C, which can be used for computing net differences of C' (or $\tilde{C}'$) from given net differences of B in accordance with

$$dC'_k(s) = -\frac{1}{2}c_k \cdot dB_j(s), \quad (4.24)$$

where s denotes the number of months. As usual, the $dC'_k(s)$ correspond to actual net differences of C' only if the Moon returns to the same interpolation interval k and the same zone j. They could be used for verifying column C' or $\tilde{C}'$, presumably over intervals for which the mentioned conditions can be satisfied, e.g. 1^{m} or 12^{m}.

'Daily' values of the duration of daylight (C^*)

The daily motion tables contain a column C^* with values of C pertaining to successive real tithis. It is not clear what the purpose of C^* may be, since the applications of C do not carry over to C^*. Procedures for C^* are not preserved. The numbers in column C^* are reproduced by the same interpolation algorithm as for C, such that C^* is computed from the Sun's 'daily' position $B^*_{\odot}$.

[405] The algorithm for C' and its astronomical meaning were established by Kugler (1900), pp. 188–191 ('Columne K').
[406] BM 40094+45662 (Aaboe 1969; Aaboe & Hamilton 1979).

4.4.6 Distance to the ecliptic (*E*)

terminology	*šūqu*(**nim**, **la₂**) *ša₂ Sîn* = "height' of the Moon' ($E_\uparrow$) *šuplu*(**sig**, **bur₃**) *ša₂ Sîn* = "depth' of the Moon' ($E_\downarrow$)
units	barleycorn (**še**)
astronomical meaning	the Moon's distance above or below the ecliptic
column	v
procedures:	
updating *E*	No. **53** P8'.a–c; No. **54** P1?; No. **56** P5; No. **57** P1; No. **58** P2'; No. **59** P2'; No. **60** P1'.a–e
alternative algorithm	No. **53** P8'.e; No. **60** P1'.f
relation between E^{nm} and E^{fm}	No. **65** P8.a
net differences	No. **60** P1'.f; No. **65** P8.b
computing *E* from Ψ	cf. §4.4.7
other computations involving *E*	No. **53** P8'.d,f; No. **54** P1?

Function *E* represents the distance of the Moon's centre above or below the ecliptic.[407] The value of *E* is expressed in barleycorns (**še** = *uṭṭatu*) or, in the Lunar Six procedures, degrees. Column *E*, usually situated between columns *C* and Ψ, contains the numerical value of *E* and an indication of its change, 'ascending' (**nim**, **la₂**) or 'descending' (**sig**, **bur₃**). In the procedure texts the term 'height and depth' (*šūqu u šuplu*) is used in the sense of an unspecified distance above or below the ecliptic. Two quantities are computed from *E*: (i) eclipse magnitude (Ψ, Ψ'); (ii) the contribution *R* to the Lunar Six intervals (step 6 of the Lunar Six module).

Updating *E*

The algorithm for updating *E*, reconstructed by Kugler,[408] can be represented in several different ways. We shall start with the formulation found most often in the procedure texts.

Step 1. Computing the preliminary monthly difference w as a generalised step function of B_i

In step 1 a preliminary value for the monthly difference of *E* is computed as a step function of the zodiacal position. This is the first of several occasions in lunar system A where the step function algorithm is used for modeling a quantity that is not a synodic arc (σ; cf. §2.6.10), other examples being $\Delta\Psi'$ (§4.4.8) and *J* (§4.4.14). Since these 'generalised' step functions differ essentially from the step functions for σ, they will be referred to by the distinct symbol *w*. Unlike σ which is computed from the old position B_{i-1}, *w* is computed from the *new* position B_i. In the procedure texts *w* is always an integral part of the algorithm for updating *E*. The most common template of this algorithm is as follows:[409]

ta b_j **en** b_{j+1} w_j **tab** *u* **la₂** *ša₂ al-la* b_{j+1} **diri a.ra₂**|GAM $\lvert c'_j\rvert$ **du ki** w_j **tab**[-*ma* **gar**-*an*]	From b_j to b_{j+1} you add and subtract w_j. (The amount by) which it exceeds b_{j+1} you multiply by $\lvert c'_j\rvert$, add to w_j [and put down].	(ESTEP.A.1)
ta b_j **en** b_{j+1} w_j **tab** *u* **la₂** *ša₂ al-la* b_{j+1} **diri a.ra₂**|GAM $\lvert c'_j\rvert$ **du ta** w_j **la₂**[-*ma* **gar**-*an*]	From b_j to b_{j+1} you add and subtract w_j. (The amount by) which it exceeds b_{j+1} you multiply by $\lvert c'_j\rvert$, subtract from w_j [and put down].	(ESTEP.A.2)

Here w_j is the preliminary value of *w* in zone *j*. Note that *w* is itself preliminary in the sense that the final monthly difference of *E* is computed from *w* in step 2. A comparison with template STEP.A for σ is revealing, because it explains some peculiarities. The first instruction, 'From b_j to b_{j+1} you add and subtract w_j', is essentially the same, with σ_j replaced by w_j. Also the zonal boundaries b_j are the same as for σ. As in STEP.A, the unnamed quantity that is implicitly compared with b_j and b_{j+1} is the old position B_{i-1}, which is assumed to be in zone *j*. Unlike STEP.A, the unnamed quantity to which w_j is applied is E_{i-1} and the subject of '(The amount by) which it exceeds b_{j+1} ...', is the *new* zodiacal position, B_i.[410] The second instruction, '(The amount by) which it exceeds...', concerns the transition rule to be applied when B_i is in the next zone ($j+1$). The resulting algorithm for *w* can be formulated as follows:

[407] Kugler (1900), pp. 132, 139, realised that *E* ('Columne E') is a measure of the Moon's distance to the ecliptic, but he was unsure about its precise astronomical definition, which was clarified by Neugebauer (1938b,c) and Pannekoek (1941). For the coordinate system of mathematical astronomy in general cf. §2.5.3.2.

[408] Kugler (1900), pp. 128–137, 142–147.

[409] No. **56** P5; No. **59** P2'?; No. **60** P1'.c.

[410] In the equivalent phrase in STEP.A the subject is the *preliminary* new position, $B_{i-1}+\sigma_j$.

$$w = \begin{cases} w_j & \text{if } b_j+\sigma_j \le B_i \le b_{j+1} \quad (B_i \text{ in zone } j) \\ w_j + c'_j \cdot (B_i - b_{j+1}) & \text{if } b_{j+1} \le B_i \le b_{j+1}+\sigma_{j+1} \quad (B_i \text{ in zone } j+1). \end{cases} \tag{4.25}$$

The corresponding flow chart is shown in Fig. 4.12. The transition coefficients c'_j turn out to be related to w_j and σ_j through the relation

$$c'_j = \frac{w_{j+1} - w_j}{\sigma_{j+1}}, \tag{4.26}$$

which amounts to linear interpolation between w_j and w_{j+1} in the 'transition region', which extends from b_{j+1} to $b_{j+1} + \sigma_{j+1}$; elsewhere w is constant. By the same token, $b_j + \sigma_j$ marks the end of the transition region between zones $j-1$ and j. There are specific reasons why c'_j is defined in this way, as will be explained further below.

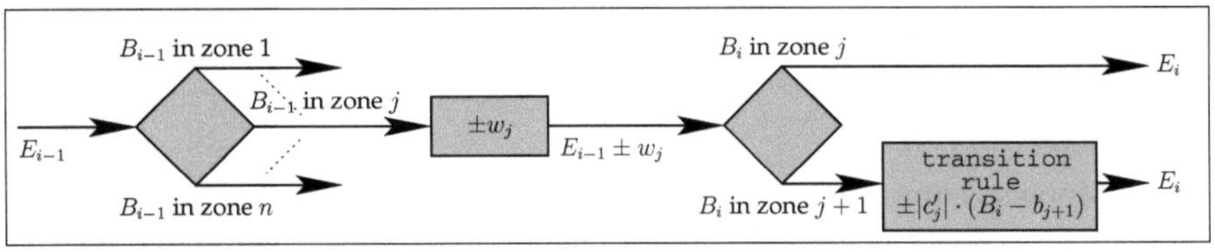

Fig. 4.12: Flow chart for step 1: updating E with the generalised step function for w.

j	w_j	c'_j
1	2,6;15,42^{še}	−0;16
2	1,58;45,42^{še}	0;15

Table 4.12: Parameters of the step function for w, the (preliminary) monthly difference of E. For the zonal boundaries cf. §4.4.4.

Inserting the defining parameters for the two zones (Table 4.12) we obtain four cases. For New Moons they can be represented as follows:

$$w^{\text{nm}} = \begin{cases} 1{,}58;45{,}42 + 0;15 \cdot (B_i^{\text{nm}} - 13^\circ\ \text{Vir}) & (13^\circ\ \text{Vir} \le B_i^{\text{nm}} \le 13^\circ\ \text{Lib}) \quad (1a) \\ 2{,}6;15{,}42 & (13^\circ\ \text{Lib} \le B_i^{\text{nm}} \le 27^\circ\ \text{Psc}) \quad (1b) \\ 2{,}6;15{,}42 - 0;16 \cdot (B_i^{\text{nm}} - 27^\circ\ \text{Psc}) & (27^\circ\ \text{Psc} \le B_i^{\text{nm}} \le 25;7{,}30^\circ\ \text{Ari}) \quad (2a) \\ 1{,}58;45{,}42 & (25;7{,}30^\circ\ \text{Ari} \le B_i^{\text{nm}} \le 13^\circ\ \text{Vir}) \quad (2b). \end{cases} \tag{4.27}$$

For a graphical representation cf. Fig. 4.13. The two templates ESTEP.A.1 and ESTEP.A.2 reflect that $c'_j \cdot (B_i - b_{j+1})$ can be additive or subtractive, depending on which of w_j and w_{j+1} is larger. We shall now consider some alternative

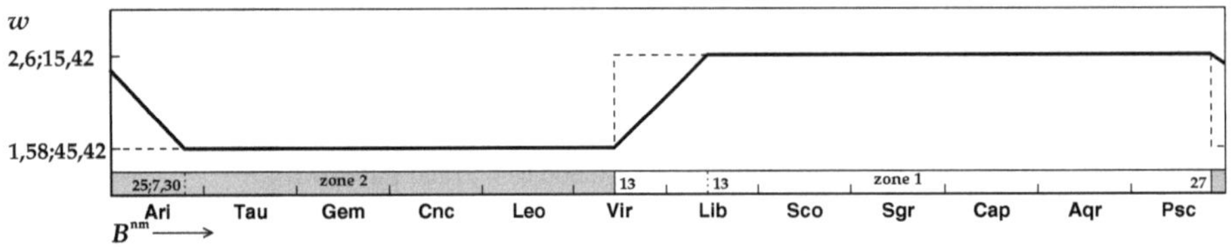

Fig. 4.13: Graphical representation of w, the monthly difference of E, as a function of the updated zodiacal position, B_i (New Moon case). Dashed lines represent the preliminary values w_j.

but equivalent templates.[411] The following one, attested in No. **57** P1, is very similar to ESTEP.A except that it contains explicit subtractions $B_i - b_{j+1}$:

ta b_j **en** b_{j+1} **ab₂** *ana* **ab₂** w_j **tab** *u* **la₂** *ša₂ al-la* b_{j+1} **diri** b_{j+1} **ta šа₃**-*šu₂* **nim** *ša₂* **tag₄** GAM|**a.ra₂** $\|c'_j\|$ **du** {**ki** w_j **tab**-*ma* | **ta** w_j **la₂**-*ma*} **gar**-*an*	From b_j to b_{j+1} month by month you add and subtract w_j. That which exceeds b_{j+1}: you deduct b_{j+1} from it, what remains you multiply by $\|c'_j\|$, {add it to w_j | subtract it from w_j} and put it down.	(ESTEP.B)

Another equivalent template (No. **53** P8'.b) is as follows:

ta b_j **en** b_{j+1} **ab₂** *ana* **ab₂** w_j **tab** *u* **la₂** *lib₃-bu-u₂ ša₂ qablītu*(**murub₄**-*tu₂*) **kur**-*ad₂* [...] *ana* 1 UŠ **ki** *ša₂ al* b_{j+1} **diri** GAM $\|c'_j\|$ **du ki** w_j **tab** *ana* **nim** *lu ana* **sig gar**	From b_j to b_{j+1} month by month you add and subtract w_j, whereby you reach the central part. [...] For (every) 1° by which the position exceeds b_{j+1} you multiply by $\|c'_j\|$, add it to w_j, and you put it down as the 'height' or as the 'depth'.	(ESTEP.C)

[411] The template of No. **54** P1 could not be determined.

The corresponding case where $(B_i - b_{j+1})|c'_j|$ is subtracted from w_j is not preserved for this template. The final statement is unexpected, because what is actually computed is w rather than E ('height and depth'). Yet another template is attested in No. **58** P2'.b:

ta b_j **en** b_{j+1} **iti** *ana* **iti** w_j **tab** *u* **la₂**-[*ka*]	From b_j to b_{j+1} month by month [your] addition and subtraction is w_j.	(ESTEP.D)
ta b_{j+1} **en** $b_{j+1}+\sigma_j$ σ_j σ_j GAM $|c'_j|$ **du**-*ma* $\sigma_j|c'_j|$ {**ki** w_j **tab** | **ta** w_j **nim**} w_{j+1}	From b_{j+1} until $b_{j+1}+\sigma_j$ is (a distance of) σ_j. You multiply σ_j times $|c'_j|$, it is $\sigma_j|c'_j|$, you {add it with | deduct it from} w_j, it is w_{j+1}.	

Like the other templates, ESTEP.D first provides the boundaries of zone j and the value of w_j. It does not continue with a general transition rule but exemplifies this rule for the special position $B_i^{\text{nm}} = b_{j+1}+\sigma_j$ at the end of the transition region (1a and 2a in Eq. 4.27). For $j=1$ this amounts to 25;7,30° Ari, for $j=2$ to 13° Lib. This approach is reminiscent of templates STEP.C and STEP.D for the synodic arc (§2.6.10).

Additional instruction for updating E

The templates ESTEP.A–D are incomplete in the sense that they do not state when w is added and when it is subtracted. Some procedure texts include an additional instruction for that:[412]

ki-i **tab tab** *ki-i* **la₂ la₂**	If it is increasing you add, if it is decreasing you subtract.

The formulaic representation is

$$E_i = E_{i-1} \pm w, \tag{4.28}$$

such that w is added if E_{i-1} is increasing (Moon above the ecliptic and ascending, or below the ecliptic and descending), and subtracted if it is decreasing (opposite configurations). The direction of change of E is stored in column E which contains the numerical value of E followed by two signs, the first of which denotes whether the Moon is 'high' (**la₂**) = above the ecliptic, or 'low' (**bur₃**) = below the ecliptic, the second whether the Moon is ascending (**la₂**) or descending (**bur₃**).[413]

Step 2. Modifying w if the Moon is in the nodal region

In step 2 w is modified depending on whether the Moon is inside or outside the nodal region, a band extending to a distance

$$\kappa = 2{,}24^{\text{še}} = 2^\circ \tag{4.29}$$

above and below the ecliptic. In No. **53** P8'.a,b the nodal region is called 'central part' (*qablītu*) or 'area of the node' (*qaqqar kiṣari*); in No. **60** P1' 'eclipse house' (**e₂** AN.MI = *bīt attalê*). If the Moon enters or leaves the nodal region, w is split up into a part w_{int} inside the nodal region and a part $w_{\text{ext}} = w - w_{\text{int}}$ outside the nodal region, and w_{int} is doubled, so that the final difference becomes

$$d = w_{\text{ext}} + 2w_{\text{int}} = w + w_{\text{int}}. \tag{4.30}$$

As indicated by the second equality, this is equivalent to w being augmented by w_{int}. The procedure for leaving the nodal region is analogous. If the previous value, E_{i-1}, was in the nodal region then the preliminary difference becomes $2w$. If addition or subtraction of $2w$ produces a preliminary new value $E_{i-1} \pm 2w$ outside the nodal region, then $2w$ is split up as $2w_{\text{int}} + 2w_{\text{ext}}$, where $2w_{\text{int}}$ is the part inside the nodal region. Now $2w_{\text{ext}}$ is halved, which leads to the same expression. Hence the formulaic representation of the updating of E becomes

$$E_i = E_{i-1} \pm d, \tag{4.31}$$

which replaces Eq. (4.28). Since Babylonian calculus does not know negative numbers, special attention is necessary when the Moon approaches the ecliptic. If $E_{i-1} > d$ then the Moon remains on the same side of the ecliptic, but if $E_{i-1} < d$ then the Moon crosses the ecliptic, and the new position becomes $E_i = d - E_{i-1}$. In view of No. **60** P1'.a,b,d,e we shall

[412] No. **56** P5; No. **57** P1.

[413] In column E **la₂** and **bur₃** are used instead of **nim** and **sig**, probably because the former occupy less space. Both pairs represent the same Akkadian words *šaqû*, 'to be high', and *šapālu*, 'to be low', respectively, and their derivations. Occasional phonetic writings imply that in first position, **nim**|**la₂** and **sig**|**bur₃** represent the statives *šaqi*, 'it is high', and *šapil*, 'it is low', respectively; in second position usually the present tense, *išaqqi*, 'it is ascending', and *išappil*, 'it is descending', rarely also the stative. In the procedure texts the scribes prefer **nim** and **sig**, although occasionally **la₂** and **bur₃** are used, sometimes in the same sentence (e.g. No. **61.G**).

work out both expressions for the case when the Moon is initially outside the nodal region, so that $w_{\text{ext}} = E_{i-1} - \kappa$ and $d = w_{\text{ext}} + 2w_{\text{int}} = 2w + \kappa - E_{i-1}$. Hence the Moon reaches the ecliptic ($E_i = 0$) if $E_{i-1} - w = \kappa/2 = 1{,}12^{\text{še}}$, so that

$$E_i = \begin{cases} 2\cdot[E_{i-1} - w - \kappa/2] & \text{if } E_{i-1} - w > \kappa/2 \quad \text{(Moon on same side of the ecliptic)} \\ 2\cdot[\kappa/2 - (E_{i-1} - w)] & \text{if } E_{i-1} - w < \kappa/2 \quad \text{(Moon crosses the ecliptic).} \end{cases} \tag{4.32}$$

Both cases are considered in No. **60** P1'.a,b,d,e.

Step 3. Reflection rules at the extrema of E

In step 3 E is confined to values below $M = 7{,}12^{\text{še}} = 6°$ by applying the usual reflection rule if addition of d produces a value $E_{i-1} + d > M$ (Eq. 2.14). In the procedure texts this is formulated as follows:

28' *ša*₂ *al* 7.12 **diri** 29' [*ina* 7.1]2 **e**₁₁-*ma* [*ana* **nim** *lu ana* **sig**] **gar**	28' (The amount) by which it exceeds 7,12 29' you subtract [from 7,1]2 and put down [as the 'height and depth'].	No. **53** P8'
9' ⌈*ša*₂⌉ *al-la* 7.12 **n**[**im diri**] 10' **ta** 14.⌈24⌉ **e**₁₁-*ma* **gar**-*an*	9' That which [exceeds] 7,12, the 'height', 10' you subtract from 14,[24] and put down.	No. **60** P1'

The formulation in No. **53** P8' is similar to the reflection rule in template ZZ.B.3 for the zigzag function; that in No. **60** P1' is similar to ZZ.B.1. With each reflection a reversal of the direction of motion is also implied.

Alternative algorithm for updating E across the nodal region

No. **53** P8'.e offers an alternative but equivalent algorithm for updating E when the Moon crosses the nodal region. Suppose that the Moon is at a distance E_0 above the ecliptic, just above the nodal region and descending. Then $w_{\text{ext}} = E_0 - \kappa$, so that $d = w_{\text{ext}} + 2w_{\text{int}} = 2w + \kappa - E_0$. If we momentarily allow E in this modern representation to assume negative values if the Moon is below the ecliptic, then Eq. (4.32) yields

$$E_1 = E_0 - d_0 = 2(E_0 - w_0) - \kappa. \tag{4.33}$$

(By exception all indices, including those of w, denote the lunation number here.) Updating once more and assuming that E_2 is below the nodal region we obtain

$$E_2 = E_1 - d_1 = (E_1 - \kappa)/2 - w_1 = E_0 - (w_0 + w_1 + \kappa). \tag{4.34}$$

The novel feature is that E_2 is expressed directly in terms of E_0, without reference to the intermediate value E_1. Similarly, if E_2 is still inside, but E_3 outside the nodal region, then $E_2 = 2(E_0 - w_0 - w_1) - \kappa$, leading to

$$E_3 = E_2 - d_2 = (E_2 - \kappa)/2 - w_2 = E_0 - (w_0 + w_1 + w_2 + \kappa). \tag{4.35}$$

These expressions can be generalised as follows: if it takes n lunations for the Moon to cross the nodal region, then the total change of E equals $w_0 + w_1 + w_2 + \cdots + w_n + \kappa$. Since the segment of $w_0 + w_1 + \cdots + w_n$ that lies within the nodal region is doubled, the total additional distance covered by the Moon when it has left the nodal region is κ. When comparing these expressions with actual procedures we need to remember that subtractions are possible in the Babylonian sense only if the result is positive. Since the Moon is on the other side of the ecliptic after crossing the nodal region, the value of E expressed as a positive distance above or below the ecliptic is $w_0 + w_1 + \cdots + \kappa - |E_0|$.

Alternative representation of d based on the nodal motion

Aaboe & Henderson (1975), setting out from a discovery by van der Waerden (cf. below), proposed the following elegant alternative representation of the monthly difference d:

$$d = \begin{cases} 4\cdot d_\eta & \text{(outside nodal region)} \\ 8\cdot d_\eta & \text{(inside nodal region).} \end{cases} \tag{4.36}$$

Here η is the Moon's elongation from the ascending node (☊),

$$\eta = B - B_{☊}, \tag{4.37}$$

d_η is the monthly difference of η,

$$d_\eta = \sigma - d_\Omega, \tag{4.38}$$

and

$$d_\Omega = -1;33,55,30^\circ \tag{4.39}$$

is the monthly displacement of the ascending node, which is retrograde (eastward) in agreement with empirical facts (cf. Fig. 4.1 on p. 111). The factors 4 and 8 represent the slope of the Moon's ascending or descending motion outside and

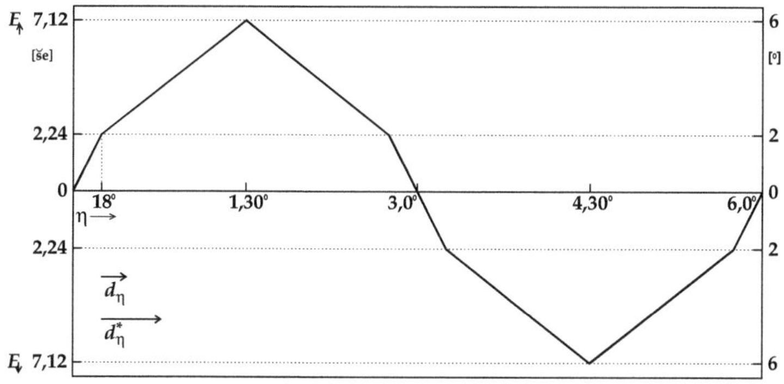

Fig. 4.14: Graphical representation of E as a function of η, the Moon's elongation from the ascending node. Also included are two arrows indicating the (average) differences of η for 1 synodic month (d_η) and for 1^τ (d^*_η).

inside the nodal region, taking into account that distances along the ecliptic are expressed in degrees while E is expressed in barleycorns (^{še}).[414] This representation makes the connection between E and η shown in Fig. 4.14 fully transparent.[415] It also allows us to understand the updating of E in terms of the updating of η with a generalised step function for d_η,

$$\eta_i = \eta_{i-1} + d_\eta(B_i). \tag{4.40}$$

Even though η is neither attested in the tabular texts nor explicitly mentioned in any procedure text, there can be no doubt that it underlies the algorithm for E. The reconstruction of this representation began with van der Waerden (1965), who discovered that, outside the nodal region, the preliminary monthly differences of E can be expressed as

$$w_j = 4 \cdot (\sigma_j - d_\Omega) = 4 \cdot w_{\eta j}, \tag{4.41}$$

where $w_{\eta j}$ is the preliminary monthly difference of η in zone j (Table 4.13). In order to prove Eq. (4.36), this equation

j	$w_{\eta j}$	$c'_{\eta j}$
1	31;33,55,30°	−0;4
2	29;41,25,30°	0;3,45

Table 4.13: Parameters of the step function for w_η, the monthly difference of η. For the zonal boundaries cf. §4.4.4.

must be shown to hold also if the Moon crosses a zonal boundary and w assumes a value in between w_1 and w_2. In other words, $w_{\eta j}$ must define a generalised step function with transition coefficients $c'_{\eta j}$ related to $w_{\eta j}$ and σ_j in the manner of Eq. (4.26). As we shall see, the c'_j of w_E are defined in such a way that this is the case. First note that $B_i - b_{j+1} = r_j \cdot y$, where $y = B_{i-1} + \sigma_j - b_{j+1}$ (Eq. 2.32 on p. 48).[416] Hence $c'_j \cdot (B_i - b_{j+1})$ in Eq. (4.25) can be rewritten as $r_j c'_j y$. Inserting Eqs. (4.26) and (4.41) this becomes

$$r_j c'_j y = r_j y \frac{w_{j+1} - w_j}{\sigma_{j+1}} = 4 r_j y \frac{\sigma_{j+1} - \sigma_j}{\sigma_{j+1}} = 4 y \frac{\sigma_{j+1} - \sigma_j}{\sigma_j} = 4 c_j y. \tag{4.42}$$

Consequently $w = w_j + c'_j \cdot (B_i - b_{j+1})$ can be written as $w = 4 \cdot (\sigma_j - d_\Omega + c_j y) = 4 \cdot (\sigma - d_\Omega)$, which proves Eq. (4.36).[417]

Periods of E, η and B_Ω

The time after which the ascending or descending node returns to the same zodiacal position is

$$P_\Omega = \frac{6,0}{|d_\Omega|} = \frac{\Pi}{Z} = \frac{4,0,0,0}{1,2,37} \approx 229.971^{\text{m}}, \tag{4.43}$$

[414] If one expresses E in degrees then the equivalent slopes are 4·0;0,50 = 0;3,20 degrees per degree and 8·0;0,50 = 0;6,20 degrees per degree.

[415] The same graph of E was drawn by Neugebauer (1938c, p. 262; *ACT* p. 49) without explicitly using the concept of elongation or nodal motion.

[416] This is essentially the definition of the transition rule. It can be proven by inserting $\sigma = \sigma_j + y \cdot c_j$ (Eq. 2.29) and $r_j = 1 + c_j$ (Eq. 2.31).

[417] For a less general version of this proof cf. Aaboe & Henderson (1975), pp. 206–7.

which amounts to 18.59^{yr}, very close to the modern empirical value. The number period of E, say Π_E, can be defined as $\Pi_E = \Pi_{\Omega} \cdot \Pi_B = 46{,}23 \cdot 4{,}0{,}0{,}0 = 3{,}5{,}32{,}0{,}0{,}0^{\text{m}}$, which amounts to approximately $1.94 \cdot 10^{8\text{yr}}$, for all practical purposes an infinitely long period.[418] However, Π_E is not a true number period, because a given value of E may occur repeatedly among Π_E successive lunations. It is therefore usually impossible to assign a unique date to a single value of E.

Net differences of η and E for various intervals

No. **65** P8.b and No. **60** P1'.f provide net differences of E for 12^{m}, $dE_j(12)$, for both zones. As usual they correspond to the actual net difference only under certain restrictions, namely if the Moon is in the same zodiacal zone and outside the nodal region initially and after 12^{m}, and if E is sufficiently far away from the extrema. These and several other net differences for positions outside the nodal region are listed in Table 4.14. As is true in general for net differences, it is not clear how the Babylonian astronomers computed them. If one adopts the formalism based on the nodal motion they are easily obtained as

$$dE_j(s) = \begin{cases} 4 \cdot [d\eta_j(s) \mod 3{,}0^\circ] & \text{(outside the nodal region)} \\ 8 \cdot [d\eta_j(s) \mod 3{,}0^\circ] & \text{(inside the nodal region),} \end{cases} \tag{4.44}$$

where

$$d\eta_j(s) = dB_j(s) - dB_{\Omega}(s) \tag{4.45}$$

are net differences of the nodal elongation, $dB_j(s)$ are the net displacements of the Moon listed in Table 4.9 and

$$dB_{\Omega}(s) = s \cdot d_{\Omega} + t' \cdot 6{,}0^\circ \tag{4.46}$$

is the net displacement of the ascending node for s months (i.e. $dB_{\Omega}(1) = d_{\Omega}$).[419] In this expression t' is the closest whole number of revolutions performed by the ascending node. Eq. (4.44) implies that the (preliminary) net differences of E inside the nodal region, thus far not attested in procedure texts, are just twice the values outside the nodal region (Table 4.14).

s	t'	t	dB_{Ω}	$d\eta_1$	$d\eta_2$	dE_1	dE_2
5^{m}	0	0	−7;49,37,30°	3,0−22;10,22,30°	3,0−31;32,52,30°	−1,28;41,30^{še}	−2,6;11,30^{še}
6^{m}	0	0	−9;23,33°	3,0 +9;23,33°	3,0 −1;51,27°	37;34,12^{še}	−7;25,48^{še}
12^{m}	0	1^{yr}	−18;47,6°	7;43,6°	8;24,36°	30;52,24^{še}	33;38,24^{še}
223^{m}	1	18^{yr}	10;54,43,30°	−0;6,43,30°	−0;47,13,30°	−0;26,54^{še}	−3;8,54^{še}
235^{m}	1	19^{yr}	−7;52,22,30°	7;36,22,30°	7;37,22,30°	30;25,30^{še}	30;29,30^{še}

Table 4.14: Net displacements of the ascending node and net differences of η and E (outside the nodal region) for various intervals.

'Daily' values of E and η (E^*, η^*)

The daily motion table *ACT* 81 includes a column E^* with values of E pertaining to successive (real) tithis. Procedures for E^* are not preserved, but the (generalised) step function for updating E^* is easily derived by considering the associated nodal elongations η^*.

Updating E^* and η^*

First, recall that E and η vary with the draconitic month, $\overline{m}_{\text{drac}} \approx 27.21^{\text{d}}$, which is about 2.3^{d} shorter than the synodic month ($\overline{m}_{\text{syn}} = 29.53^{\text{d}}$), the interval at which E is tabulated in the synodic tables. Hence with each lunation the underlying 'daily' functions E^* and η^* perform 1 extra oscillation compared to the monthly differences d_η and d_E. The difference of η^* for 1^τ is therefore 1/30 of the total change of η accumulated over 1 synodic month,

$$d^*_\eta = \frac{d_\eta + 6{,}0^\circ}{30}. \tag{4.47}$$

This defines a generalised step function for updating η^* with preliminary values $w^*_{\eta j} = (w_{\eta j} + 6{,}0^\circ)/30$ and transition coefficients $c^{*}_{\eta j}{}' = c'_{\eta j}/30$ (Table 4.15) in accordance with $\eta^*_i = \eta^*_{i-1} + d^*_\eta$, where i is the tithi number and the argument of d^*_η is $B^*_{\odot i}$. Second, the difference of E^* for 1^τ, d^*_E, is obtained from d^*_η by the usual transformation,

[418] For this and other periods of E cf. Aaboe & Henderson (1975), pp. 200–203.

[419] These expressions were first presented by Aaboe & Henderson (1975), pp. 199–200, using a slightly different notation. Neugebauer mentions $dE_1(12)$ and $dE_2(12)$ in his discussion of *ACT* 200i (p. 224) and *ACT* 204 (p. 250) without offering a derivation.

$$d_E^* = \begin{cases} 4 \cdot d_\eta^* & \text{(outside the nodal region)} \\ 8 \cdot d_\eta^* & \text{(inside the nodal region).} \end{cases} \tag{4.48}$$

This generalised step function for updating E^* has preliminary values $w_{Ej}^* = 4 \cdot w_{\eta j}^*$ (outside the nodal region) and transition coefficients ${c_{Ej}^*}' = c'_{Ej}/30$ (Table 4.15).[420] The mentioned daily motion table confirms this reconstruction.

Net differences of η^* and E^*

Net differences of E^* for intervals longer than 1^τ are obtained in the same fashion, leading to

$$dE^*(s) = \begin{cases} 4 \cdot [d\eta^*(s) \mod 3{,}0^\circ] & \text{(outside the nodal region)} \\ 8 \cdot [d\eta^*(s) \mod 3{,}0^\circ] & \text{(inside the nodal region).} \end{cases} \tag{4.49}$$

In order for the step functions d_η^* and d_η to be mutually consistent, the net differences of η^* and E^* for 30^τ must satisfy $d\eta_j^*(30) = w_{\eta j}$ and $dE_j^*(30) = w_{Ej}$ (outside the nodal region), which is indeed the case.[421]

j	$w_{\eta j}^*$	c'_j	w_{Ej}^*	c'_j	$d\eta_j^*(14)$	$d\eta_j^*(15)$	$dE_j^*(14)$	$dE_j^*(15)$
1	13;3,7,51°	−0;0,8	52;12,31,24^{še}	−0;0,32	3,2;43,49,54°	3,15;46,57,45°	10;55,19,36^{še}	1,3;7,51^{še}
2	12;59,22,51°	0;0,7,30	51;57,31,24^{še}	0;0,30	3,1;51,19,54°	3,14;50,42,45°	7;25,19,36^{še}	59;22,51^{še}

Table 4.15: Parameters of the step functions w_η^* and w_E^* and net differences of η^* and E^* for 14^τ and 15^τ (outside the nodal region).

Relation between E^{fm} and E^{nm}

No. **65** P8.a preserves elements of a hitherto unrecognised algorithm for computing E^{fm} from E^{nm}. Like the analogous algorithm for Φ (§4.4.3), this algorithm amounts to the addition of $dE_j^*(15)$, the net difference of E^* across half a synodic month (cf. Table 4.15). The procedure may have been used for verifying column E^{fm}, or computing an initial value for E^{fm} from an initial value for E^{nm}.

4.4.7 Eclipse magnitude (Ψ)

terminology	ḪAB-*rat* = 'disk'
unit	fingers (^{f})
astronomical meaning	eclipse magnitude
column	vi
procedures:	
computing Ψ from E	No. **65** P1,P2,P4
computing E from Ψ	No. **53** P11'.c; No. **65** P2–P4

Function Ψ quantifies the magnitude of eclipses, as was recognised already by Kugler (1900). Eclipse magnitude is the conventional, pragmatic name of this function, of which a fully satisfying astronomical interpretation has not yet been given. In the synodic tables Ψ is tabulated in column vi, after column E. For the alternative eclipse functions Ψ' and Ψ_3 cf. §4.4.8.

Computing Ψ from E

The procedure texts (No. **65** P1,P2,P4) and the synodic tables imply that Ψ is computed from E by means of linear interpolation in accordance with

$$\Psi = \begin{cases} 0;10 \cdot E - c_\Psi & (E \text{ decreasing from } \kappa = 2{,}24^{\text{še}} \text{ to } 1{,}44;24^{\text{še}}) \\ c_\Psi - 0;10 \cdot E & (E \text{ decreasing from } 1{,}44;24^{\text{še}} \text{ to } 0) \\ c_\Psi + 0;10 \cdot E & (E \text{ increasing from } 0 \text{ to } \kappa), \end{cases} \tag{4.50}$$

where the parameter $c_\Psi = 17;24^{\text{f}}$ represents the eclipse magnitude for a central eclipse ($E = 0$). For a graphical representation cf. Fig. 4.15. While Kugler (1900) succeeded in deriving some elements of this algorithm,[422] a complete

[420] The preliminary values w_{Ej}^* were first reconstructed by Neugebauer (*ACT*, pp. 54–55).

[421] Proof: $d\eta_1^*(30) = 2 \cdot d\eta_1^*(15) - 6{,}0° = 31;33,55,30°$, $d\eta_2^*(30) = 2 \cdot d\eta_2^*(15) - 6{,}0° = 29;41,25,30°$, $dE_1^*(30) = 2 \cdot dE_1^*(15) = 2{,}6;15,42^{\text{še}}$, $dE_2^*(30) = 2 \cdot dE_2^*(15) = 1{,}58;45,42^{\text{še}}$.

[422] Kugler (1900), pp. 147–157 ('Columne F' = Ψ, Ψ').

reconstruction was achieved by Neugebauer (1938c). Although Ψ is clearly connected with the prediction of eclipses, its precise astronomical interpretation continues to be a matter of debate.[423] Ψ is defined only in the nodal region ($E \leq \kappa$); if the Moon is further away from the ecliptic an eclipse was considered impossible, and column Ψ was left blank. The multiplication by 0;10 amounts to a conversion from barleycorns (^{še}) to fingers (^{f}). Note that Ψ is always positive,[424] and that column Ψ does not contain markers indicating whether the numbers pertain to positions above or below the ecliptic. Hence Ψ is fully determined by the magnitude of E and its direction of change (increasing or decreasing), so that opposite configurations (e.g. Moon above the ecliptic and ascending, or below the ecliptic and descending) produce the same value of Ψ. The constant c_Ψ causes the zeros of Ψ to be displaced with respect to those of E by an elongation $c_\Psi/1;20 = 13;3°$

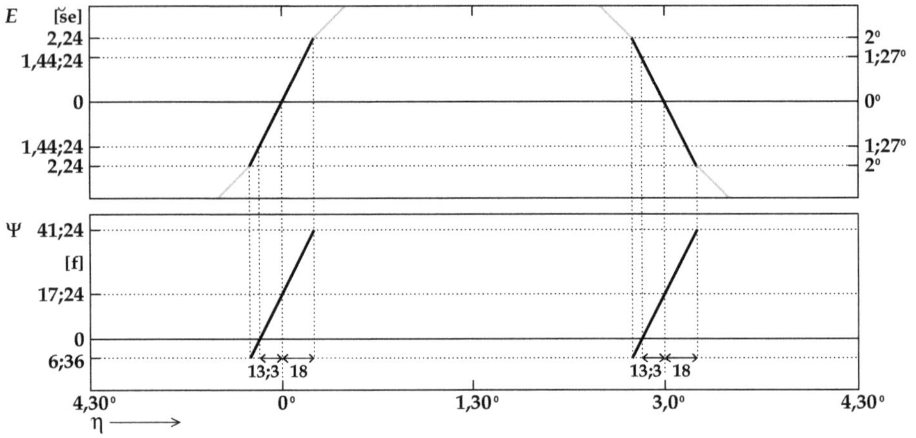

Fig. 4.15: Graphical representation of the computation of Ψ from E if the Moon is in the nodal region and ascending (left) or descending (right). **Top**: E as a function of η. **Bottom**: the corresponding values of Ψ.

(Fig. 4.15). When the Moon enters the nodal region from either side Ψ first decreases from 6;36^{f} ($E = \kappa$) to 0 ($E =$ 1,44;24^{še}); this segment is shown below $\Psi = 0$ in Fig. 4.15. In the tabular texts these values of Ψ are often accompanied by the sign BE, presumably indicating that an eclipse was considered impossible. In the next segment Ψ increases from 0 to 41;24^{f} at the other end of the nodal region. In the tabular texts these values of Ψ are sometimes accompanied by the sign ḪAB, which indicates the possibility of an eclipse.[425] Some procedures employ other control points for the interpolation than $E = 0$ and $\Psi = c_\Psi$, say E_0 and Ψ_0, resulting in

$$\Psi = \begin{cases} \left.\begin{array}{ll} \Psi_0 + 0;10\cdot(E - E_0) & \text{if } E \geq E_0 \\ \Psi_0 - 0;10\cdot(E_0 - E) & \text{if } E < E_0 \end{array}\right\} & E \text{ decreasing from } \kappa \text{ to } 1,44;24^{\text{še}} \text{ or increasing from 0 to } \kappa, \\ \left.\begin{array}{ll} \Psi_0 - 0;10\cdot(E - E_0) & \text{if } E \geq E_0 \\ \Psi_0 + 0;10\cdot(E_0 - E) & \text{if } E < E_0 \end{array}\right\} & E \text{ decreasing from } 1,44;24^{\text{še}} \text{ to 0,} \end{cases} \tag{4.51}$$

where E_0 and Ψ_0 are chosen such that the outcome coincides with Eq. (4.50). The formulation of these procedures is based on template IP.B.

Net differences of Ψ

Since successive eclipse possibilities are separated by 6, occasionally 5 months, the numbers in column Ψ are separated by 5 or 6 empty lines. The net differences of Ψ for these and other intervals coincide with the net differences of Ψ' (Table 4.17). They are not yet attested in procedure texts.

Computing E from Ψ

In No. **53** P11'.c and No. **65** P2–P4, E is computed from Ψ using an algorithm that is the inverse of Eq. (4.51):

$$E = \begin{cases} \left.\begin{array}{ll} E_0 + 6\cdot(\Psi - \Psi_0) & \text{if } \Psi \geq \Psi_0 \\ E_0 - 6\cdot(\Psi_0 - \Psi) & \text{if } \Psi < \Psi_0 \end{array}\right\} & E \text{ decreasing from } \kappa \text{ to } 1,44;24^{\text{še}} \text{ or increasing from 0 to } \kappa \\ \left.\begin{array}{ll} E_0 - 6\cdot(\Psi - \Psi_0) & \text{if } \Psi \geq \Psi_0 \\ E_0 + 6\cdot(\Psi_0 - \Psi) & \text{if } \Psi < \Psi_0 \end{array}\right\} & E \text{ decreasing from } 1,44;24^{\text{še}} \text{ to 0.} \end{cases} \tag{4.52}$$

[423] Cf. Neugebauer (1945); *HAMA*, pp. 521–523; Williams (2005), pp. 77–82. Neugebauer (1938c) originally considered Ψ^{fm} to be the distance, measured from a point at the rim of the lunar disk, by which the Moon is immersed into the Earth's shadow, but he later abandoned this interpretation after Pannekoek (1941) pointed out that this leads to inconsistencies.

[424] Unlike what is suggested in *ACT* (e.g. p. 55).

[425] For examples of this usage of BE and ḪAB in column Ψ cf. *ACT* 9, 10 and 60.

Since points at equal distance to the ecliptic but on opposite branches of E (ascending or descending) produce the same Ψ, the value of E computed with this algorithm can be either above the ecliptic on one branch or below the ecliptic on the opposite branch. These procedures were therefore presumably used only for verifying whether Ψ had been computed correctly from a given E.

4.4.8 Eclipse magnitude (Ψ')

terminology	ḪAB-*rat* = 'disk'
unit	fingers (^{f})
astronomical meaning	eclipse magnitude
column	vi
procedures:	
computing Ψ' from E	No. **53** P6'?
computations involving $\Psi^{(3)}$	No. **59** P1'; No. **80** P1

Some synodic tables[426] contain a variant of the eclipse magnitude, Ψ', which is also defined if the Moon is outside the nodal region, while coinciding numerically with Ψ everywhere inside the nodal region. Column Ψ' contains no markers indicating whether Ψ' pertains to a position above or below the ecliptic. There are different, mathematically equivalent representations of the algorithm for Ψ'. In one of the few extant procedures concerned with Ψ' it is computed from E by means of interpolation. Alternatively Ψ' can be updated month by month with a (generalised) step function for $\Delta\Psi'$, without recourse to E. I tentatively interpret No. **59** P1' and No. **80** P1 as procedures for another eclipse function, say $\Psi^{(3)}$.

Computing Ψ' from E

The only extant procedure that may be concerned with Ψ', No. **53** P6', contains at best an incomplete representation of the algorithm for computing Ψ' from E. The complete algorithm, reconstructed by Neugebauer from synodic tables,[427] is nevertheless well understood. It can be viewed as being constructed from the algorithm for Ψ (Eq. 4.50) through several modifications, the purpose of which is to obtain Ψ' as a zigzag function of η with a uniform slope. First, both branches of Ψ are continued beyond the boundaries of the nodal region. Accordingly, the doubling of the interpolation coefficient from 0;10 inside the nodal region to 0;20 outside of it compensates the halving of the slope of E (Fig. 4.16). Second, the increasing and decreasing branches are joined at the extremum $M = 2{,}0^{f}$. This is equivalent to a reflection rule, i.e. if the preliminary new value of Ψ' exceeds M then the final value is obtained by subtracting the excess, $\Psi' - M$, from M.[428] The resulting algorithm is as follows,

$$\Psi' = \begin{cases} 0;10\cdot(2E-\kappa)-c_{\Psi} & (E \text{ decreasing} \quad \text{to } \kappa = 2{,}24^{\check{s}e}) \\ 0;10\cdot E - c_{\Psi} & (E \text{ increasing from } \kappa \quad \text{to } 1{,}44;24^{\check{s}e}) \\ c_{\Psi} - 0;10\cdot E & (E \text{ increasing from } 1{,}44;24^{\check{s}e} \text{ to } 0) \\ c_{\Psi} + 0;10\cdot E & (E \text{ increasing from } 0 \quad \text{to } \kappa) \\ c_{\Psi} + 0;10\cdot(2E-\kappa) & (E \text{ increasing beyond } \kappa). \end{cases} \tag{4.53}$$

This amounts to a zigzag function $\Psi'(\eta)$ with a uniform slope 4,0/3,0 = 1;20^{f}/° featuring exactly two oscillations across 6,0° of elongation.[429] The constant c_{Ψ} causes the zeros and extrema of Ψ' to be displaced with respect to those of E by an elongation $c_{\Psi}/1;20 = 13;3°$.

Updating Ψ' with a generalised step function for $\Delta\Psi'$

In principle Ψ' can also be obtained without recourse to E by updating it from month to month with a (generalised) step function for the monthly difference $\Delta\Psi'$, to be computed from B_i as in step 1 of the algorithm for updating E. The essential steps leading to this alternative, mathematically equivalent representation which is not (yet) attested in procedure texts were made by Neugebauer (*ACT*, p. 56), who derived the preliminary values, w_j, and the transition coefficients, c'_j (Table 4.16). They are easily obtained by noting that Ψ' changes by 1;20 $\cdot d_\eta$ when η changes by the amount d_η, where 1;20^{f}/° is the slope of Ψ' in Fig. 4.16,

$$\Delta\Psi' = 1;20\cdot d_\eta. \tag{4.54}$$

[426] E.g. *ACT* 4a+, 6, 6aa+, 15, 16, 18, Steele (2002b) Text E.

[427] *ACT*, pp. 55–57.

[428] This affects Ψ' only in the small segments where E increases from 6,19;48$^{\check{s}e}$ to 7,12$^{\check{s}e}$.

[429] In Fig. 4.16 there appears to be a single oscillation because Ψ' is represented here alternatingly above and below the line $\Psi' = 0$.

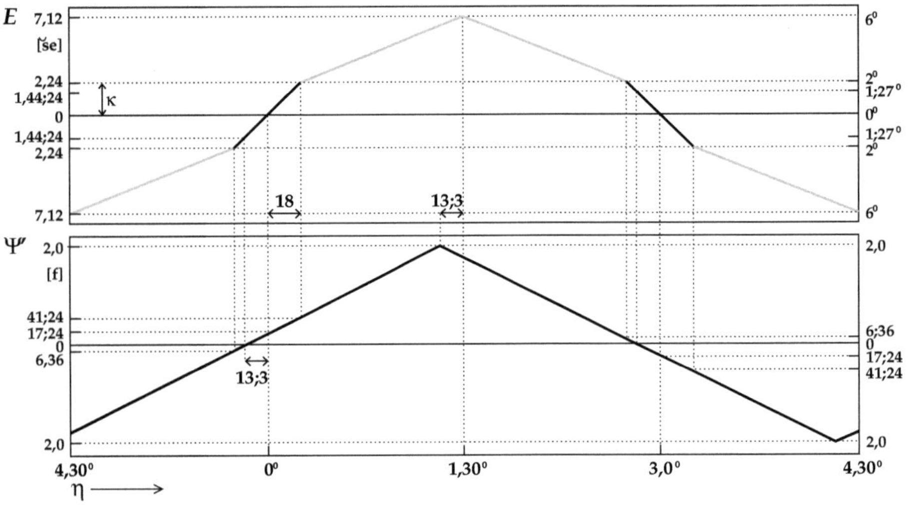

Fig. 4.16: Graphical representation of the computation of Ψ' from E. **Top**: $E(\eta)$. **Bottom**: the corresponding graph of $\Psi'(\eta)$.

This relation also holds for the preliminary values w_j (Table 4.16). From these the transition coefficients c'_j are obtained in the usual manner (Eq. 4.26). They guarantee that Eq. (4.54) remains valid if the transition rule is invoked. Also note that for positions inside the nodal region $\Delta\Psi' = 0;10 \cdot d_E$, for positions outside of it $\Delta\Psi' = 0;20 \cdot d_E$.

j	w_j	c'_j
1	42;5,14^{f}	−0;5;20
2	39;35,14^{f}	0;5

Table 4.16: Parameters of the step function for $\Delta\Psi'$. For the zonal boundaries cf. §4.4.4.

Net differences of Ψ' for various intervals

Net differences of Ψ' for various intervals are listed in Table 4.17. They are also easily derived by adopting the formalism based on the nodal motion, which results in

$$d\Psi'_j(s) = 1;20 \cdot [d\eta_j(s) \mod 3,0°]. \tag{4.55}$$

Since Ψ' coincides with Ψ everywhere in the nodal region, $d\Psi'_j(s)$ also describes the net differences of Ψ. The relation between $d\Psi'_j(s)$ and the net differences of E is as follows: for positions inside the nodal region $d\Psi'_j(s) = 0;10 \cdot dE_j(s)$, for positions outside of it $d\Psi'_j(s) = 0;20 \cdot dE_j(s)$.

s	t	$d\Psi'_1$	$d\Psi'_2$
5^{m}	0	−29;33,43,20^{f}	−42;3,50^{f}
6^{m}	0	12;31,32^{f}	−2;28,36^{f}
12^{m}	1^{yr}	10;17,28^{f}	11;12,48^{f}
223^{m}	18^{yr}	−0;8;58^{f}	−1;2,58^{f}
235^{m}	19^{yr}	10;8,30^{f}	10;9,50^{f}

Table 4.17: Net differences of Ψ' for various intervals.

Evidence for another eclipse function, $\Psi^{(3)}$

The badly understood procedures No. **59** P1' and No. **80** P1 contain computations involving another function for eclipse magnitude. The parameter 44;31,56,40^{f}, associated with the region Psc–Vir in both procedures, is considerably higher than $\Delta\Psi'$ in system A, but close to the minimum m of $\Delta\Psi'$ in system B (§4.5.9). Hence this may be the minimum m of the monthly difference of a variant, say $\Delta\Psi^{(3)}$. The number 46,20,33,20^{f}, mentioned in No. **59** P1, may be another value (μ?) of the same function, in which case the maximum can be reconstructed as M = 48;9,10^{f}. Since they are close to the corresponding parameters of $\Delta\Psi'$ in system B, the algorithm for updating $\Psi^{(3)}$ must incorporate a construction similar to step 3 of the algorithm for updating Ψ' in system B, because the period P cannot be expected to have the correct value without it. However, in No. **80** P1 the value 46;26,40^{f} appears to be associated with the region Vir–Psc, suggesting that this is the maximum of $\Delta\Psi^{(3)}$. A satisfying interpretation of these procedures and a consistent reconstruction of the algorithm for $\Psi^{(3)}$ remains to be found. In spite of the affinity with system B, $\Psi^{(3)}$ is provisionally assigned to system A, because No. **59** P2' is a system A procedure.

4.4.9 The Moon's daily displacement along the zodiac (F)

terminology	*nisḫu*(**zi**) *ša*$_2$ *ūmi*(**me**) *ša*$_2$ *Sîn* = 'daily displacement of the Moon'
unit	degrees per day
astronomical meaning	the Moon's daily displacement along the zodiac
column	vii
procedures:	
updating F as a zigzag function	$F^{(1)}$: No. **53** P7'.a; $F^{(2)}$: No. **60** P2'; $F^{(4)}$: No. **83** P5'.a
computing F from Φ	$F^{(1)}$: No. **53** P7'.b, No. **65** P5, No. **83** P6'; other: No. **92** P1?
net differences of F	$F^{(1)}$: No. **81** P6'?, No. **83** P2'.b; $F^{(4)}$: No. **83** P5'.c
truncating F	No. **83** P1',P2'.a,P5'b, No. **86** P2'
other computations involving F_{tr}	No. **83** P1',P3'–P6'
computing G from F	cf. §4.4.11
other computations involving F	$F^{(4)}$: No. **83** P4'; other: No. **52** P6.c; No. **81** P5'?,P6'?; No. **92** P2
computations involving y	No. **81** P5',P6'

Function F represents the Moon's daily displacement along the zodiac. The main application of F occurs in step 2 of the Lunar Six module (§4.4.19.2). Occasionally F takes over the role of Φ as a source function in the interpolation scheme for computing G (§4.4.11).

Updating F as a zigzag function

F is computed as a zigzag function of the lunation number.[430] The extant procedures for updating F are based on templates ZZ.B.3 (No. **53** P7'), ZZ.B.1 (No. **60** P2') and ZZ.A.4 (No. **83** P5'). The known variants are compiled in Table 4.18. $F^{(1)}$, also known as the 'unabbreviated' version of F, is attested in procedure texts, but rarely used in synodic tables.[431] The closely related $F^{(3)}$ underlies column F in the anomalous synodic table *ACT* 92.[432] The extrema of $F^{(3)}$ are lowered by 0;4,11,15 with respect to $F^{(1)}$, so that both functions have the same amplitude and period. Nearly all instances of column F are based on $F^{(2)}$, whose extrema are rounded at the second digit. Variant $F^{(4)}$, attested only in No. **83**, is unique in that it has the same period as $F^{(1)}$ in system B. Thus far there is no evidence for its use in synodic tables. Variant $F^{(5)}$ is attested in an early Seleucid synodic table.[433] A function y having the same period as $F^{(5)}$ but of unknown astronomical meaning is attested in the same synodic table and in No. **81** P5',P6'. Variant $F^{(6)}$ appears in a late Achaemenid synodic table classified as 'proto system A'.[434] Finally, at least one, perhaps two, incompletely known variants of F, whose connection with system A is not clear, are attested in No. **52** P6.b and c (cf. the commentary).

var.	d [°/d]	m [°/d]	M [°/d]	μ [°/d]	Δ [°/d]	Π	Z	P	p
1	0;42	11;4,4,41,15	15;56,54,22,30	13;30,29,31,52,30	4;52,49,41,15	1,44,7	7,28	13;56,39,6,...	0;55,59,6,13,...
2	0;42	11;4	15;57	13;30,30	4;53	4,53	21	13;57,8,...	0;55,59,14,8,...
3	0;42	10;59,53,26,15	15;52,43,7,30	13;26,18,16,52,30	4;52,49,41,15	1,44,7	7,28	13;56,39,6,...	0;55,59,6,13,...
4	0;42,11,15	10;50,23,26,15	15;44,31,52,30	13;17,27,39,22,30	4;54,8,26,15	4,11	18	13;56,40	0;55,59,6,28,...
5	0;42	11;3	15;57	13;30	4;54	14	1	14	0;56
6	0;42	*11;4*	15;56,50	*13;30,25*	*4;52,50*	*4,11*	*18*	*13;56,40*	*0;55,59,6,28,...*

Table 4.18: Parameters of the zigzag functions for F.

The relation between F and Φ

Function F, in particular $F^{(1)}$, is closely linked with Φ: both have the same period (P), and in the synodic tables the maximum of $F^{(1)}$ always coincides with that of Φ, so that both functions also have the same phase.[435] The relation between $F^{(1)}$ and Φ is the subject of several procedures in which $F^{(1)}$ is computed from Φ by means of linear interpolation. No. **65** P5 is based on template IP.D; No. **53** P7' and perhaps No. **83** P6' are partly based on IP.A (§2.6.7). The algorithm can be represented as

[430] This algorithm, and the astronomical meaning of F, were discovered by Kugler (1900), pp. 157–166 ('Columne G').

[431] The only example is *ACT* 18.

[432] The columns of this table are similar to those of system A, but they are based on parameters not attested elsewhere.

[433] MNB 1856 = Neugebauer & Sachs (1969), Text H.

[434] BM 36822 = Aaboe & Sachs (1969) Text A. Apart from M, the other defining parameters of $F^{(6)}$ are tentative reconstructions proposed by Britton (2009), p. 414.

[435] For a possible explanation cf. Britton (2009), pp. 410–411.

$$F^{(1)} = \begin{cases} m_{F^{(1)}} + 0;15,11,15 \cdot (\Phi - m_\Phi) & \text{if } \Phi \text{ increasing} \\ M_{F^{(1)}} - 0;15,11,15 \cdot (M_\Phi - \Phi) & \text{if } \Phi \text{ decreasing,} \end{cases} \quad (4.56)$$

where $0;15,11,15 = d_F/d_\Phi$ is the interpolation coefficient. One possible purpose of the algorithm is to compute $F^{(1)}$ from Φ rather than from its own zigzag function. Another purpose may be to verify whether $F^{(1)}$ was computed correctly from its own zigzag function. The same relation with Φ applies to $F^{(3)}$, but the other variants of F are not strictly connected to Φ, because their periods are slightly different.

Other properties of F

The period P, the fractional number of months after which F returns to the same value, represents, in modern terms, the number of anomalistic months contained in 1^{yr}. Since the anomalistic month is shorter than the mean synodic month (§4.1.2), there is a more rapidly varying zigzag function F^* underlying F that describes the Moon's daily displacement from tithi to tithi (cf. below). The periods of $F^{(2)}$ and some other variants of F differ slightly from that of $F^{(1)}$ and Φ. In the course of time this results in a notable drift of their phase away from $F^{(1)}$ and Φ. However, in the relatively short time span of a synodic table the effect remains small, and the resulting deviations that may occur in step 2 of the Lunar Six module are insignificant. This explains why the computationally more convenient $F^{(2)}$ is the preferred variant in the synodic tables.

Net differences for various intervals

No. **83** P2' and P5' mention net differences of $F^{(1)}$ and $F^{(4)}$ for intervals longer than 1^{m}, including 12^{m} ('1 year'), 14^{m}, 111^{m} ('9 years') and 223^{m} ('18 years'). These and several other values of $dF(s)$ for variants 1–5 are listed in Table 4.19. The text also correctly mentions that $F^{(4)}$ returns to the same value after $\Pi = 4,11^{\text{m}}$.

s	t	$dF^{(1)} = dF^{(3)}$	$dF^{(2)}$	$dF^{(4)}$	$dF^{(5)}$
6^{m}	0	4;12	4;12	4;13,7,30	4;12
12^{m}	1	−1;21,39,22,30	−1;22	−1;22,1,52,30	−1;24
14^{m}	1	0;2,20,37,30	0;2	0;2,20,37,30	0
111^{m}	8	−0;23,15	−0;26	−0;23,26,15	
223^{m}	16	−0;4,30	−0;10	−0;4,41,45	

Table 4.19: Net differences of F [°/d] for variants 1–5.

Truncated zigzag function F_{tr}

Several procedure texts and two early tabular texts contain evidence of a truncated zigzag function F_{tr}.[436] It is constructed from F by replacing values of F smaller than $m' = 11;15$°/d by m', and values larger than $M' = 15$°/d by M'. An explicit rule for the truncation is formulated e.g. in No. **83** P5', which concerns $F^{(4)}$:

[5]... *gab-bi ša₂ al-la* 15 **diri** 15 **e**-*šu₂ mim-ma ša₂ al* 11.15 [6][**la₂**]-*u₂* 11.15 **e**-*šu₂*

[5]... Whatever exceeds 15 you call 15. Whatever [6][is less] than 11;15 you call 11;15.

The likely purpose of the truncation can be inferred from the analogous truncation of Φ (cf. Fig. 4.8, p. 126), which occurs in the framework of the derivation of ΦG interpolation scheme 1 from two shifted instances of Φ_{tr} (§4.4.11). No. **77** confirms that $F_{\text{tr}}^{(1)}$ was likewise used for constructing an $F^{(1)}G$ scheme, mathematically equivalent to ΦG scheme 1, from two shifted instances of $F_{\text{tr}}^{(1)}$. Accordingly, the extrema of $F_{\text{tr}}^{(1)}$ and Φ_{tr} are connected by Eq. (4.56). An interesting feature of F_{tr} is that $m' = 11;15$ and $M' = 15$ are simpler numbers than the corresponding extrema of Φ_{tr}.[437]

'Daily' version of the Moon's daily displacement along the zodiac (F^*)

F^* represents the same quantity as F, but evaluated at successive tithis. Procedures concerning F^* are not preserved. F^* is used for updating the Moon's zodiacal position in the two extant daily motion tables (§4.4.4), one of which (*ACT* 81) preserves a column with F^*. From these tables it is inferred that F^* is a zigzag function of the tithi number, whose extrema coincide with those of $F^{(2)}$ (Table 4.20). 'Daily' versions of other variants of F are not attested. The net difference for

436 Synodic tables: the late Achaemenid tablet BM 36822 (Aaboe & Sachs 1969, Text A; B–D, both belonging to 'proto system A'; BM 36961 (Steele 2006a, Text G).

437 As shown by Britton (2007a), p. 108, and (2009), pp. 411–412, $m' = 11;15$ and $M' = 15$ may have been obtained from suitably averaged sums of empirical values of ŠU₂+NA+ME+GI₆, and an awareness that this sum corresponds to $2F$. On this topic cf. also Brack-Bernsen (1997).

30^τ is $dF^*(30)$ = 0;42°/d, which coincides with the monthly difference of F. This proves that successive F_i^* pertain to successive (real) tithis, and that the algorithm for F^* is fully consistent with F.

d	m	M	μ	Δ	Π	Z	P
0;20,56°/d	11;4°/d	15;57°/d	13;30,30°/d	4,53°/d	1,13,15	2,37	27;59,37,4,12,...$^\tau$

Table 4.20: Parameters of the zigzag function for F^*.

4.4.10 The Sun's daily displacement along the zodiac ($v_\odot$)

terminology	*nisḫu*(**zi**) *ša₂ ūmi ša₂ Šamaš₂* = 'daily displacement of the Sun'
unit	degrees per day
astronomical meaning	daily displacement of the Sun
procedures:	
zigzag function of $B_\odot$	No. **53** P10', P11'.b
unclear	No. **53** P9'; No. **101** P1?

Function $v_\odot$ is the Sun's daily displacement along the zodiac.[438] The only known application of $v_\odot$ occurs in step 2 of the Lunar Six module. In the extant procedures for the Full Moon Lunar Sixes $v_\odot$ is assumed to be constant and equal to 1°/d. Most of the Lunar Six columns in the synodic tables appear to be consistent with that assumption. However, No. **53** P10' provides parameters of a zigzag function of the zodiacal position, $v_\odot(B_\odot)$, with m = 0;55,32°/d, M = 1;2,44°/d, i.e. μ = 0;59,8°/d, Δ = 0;7,12°/d and interpolation coefficient c = 0;1,12°/d per degree. An exemplary value of that function appears in step 2 of the procedure for KUR (P18') on the same tablet. A related zigzag function for the Sun's monthly displacement along the zodiac is mentioned in No. **53** P11.a. Its extrema are 27;36° in Gem and 30;36° in Sgr. Since the ratios of the extrema of both functions are not the same, 27;36/0;55,32 = 29;49,..., and 30;36/1;2,44 = 29;16,..., they are not equivalent functions.

4.4.11 Duration of the synodic month (G)

terminology	*simānu* = 'duration'
unit	time degrees
astronomical meaning	duration of 1 synodic month -29^d (only the lunar variation)
column	viii
procedures:	
computing G from Φ (scheme 1)	cf. Table 4.21
computing G from Φ (other)	No. **66** P4'; No. **72** P1; No. **75** P1'; No. **82** P15; No. **85** P1',P2'?
computing G from $F^{(1)}$ (sch. 1)	No. **77** P1'
computing G from F (other)	No. **52** P6.b
monthly difference d	No. **52** P7.a; No. **69** P2; No. **70** P1; No. **72** P1.c; No. **82** P2,P13
net differences	No. **69** P2, No. **70** P2
other computations involving G	No. **66** P2', No. **72** P2?, No. **82** P9?, No. **84** P2'?

Function G is (an approximation of) the duration of the synodic month, such that G_i is the amount, measured in time degrees, by which the time between lunations $i-1$ and i exceeds 29^d. In the synodic tables column G is usually placed between columns F and J. The time of the lunation (M) is computed from G and two corrections, J and C' (Fig. 4.7a). Most procedure texts for G are concerned with the interpolation rules by which G is computed from Φ. Since G is computed from Φ, it inherits the same period, which is that of the lunar variation. Hence G does not incorporate a zodiacal variation, which amounts to the assumption that the synodic arc is constant. The zodiacal contribution to the synodic month is accounted for by the correction J (§4.4.14). Since $J = 0$ in zone 1 (except for the transition region 1a), this constant is equal to $\sigma_1 = 30°$. Apart from the most common interpolation scheme, several others are fragmentarily preserved in the procedure texts and the tabular texts. Moreover, there are indications of different methods for computing G, including one based on a zigzag function.[439]

[438] The conventional symbol v is maintained but the term 'velocity' is avoided, since v is not an instantaneous velocity in the modern sense, but a displacement (Akk.: *nisḫu*).

[439] *ACT* 92; cf. also *ACT*, p. 60; Aaboe & Hamilton (1979), p. 5.

procedure	intervals	interpol. coeff.	m	formulation
No. **52** P7.b	21–24	$c_k\phi$		MAL
No. **53** P16'	1–24	c_k	absent	MAL, IP.B
No. **64** P1	1–9	c_k		MAL
No. **67** P1	1–7	c_k		IP.C (!)
No. **68** P1'	13; 24–35	c_k	present	IP.B, MAL
No. **69** P1	1–12	c_k		MAL, IP.C
No. **70** P1	1–13; 35	c_k	?	MAL, IP.C, IP.B
No. **71** P1'	10–13	c_k	?	IP.C
No. **73** P1'	25–26?; 21?; 12–13?	$c_k\phi$	?	PHIG.A
No. **74** P1'	15–25	$c_k\phi$		PHIG.A
No. **76** P1'	12–15; 23–35	c_k	present	IP.B, MAL
No. **79** P1',P2'	10; 35	$c_k\phi$		PHIG.A
No. **84** P3'	13	$c_k\phi$	?	PHIG.A
No. **86** P1'?	1?	?		?
No. **77** P1'	12–16	c'_k	present	IP.E (G from $F^{(1)}$)

Table 4.21: Procedures for computing G from Φ (scheme 1). MAL denotes the *malûššu*-formulation (cf. below).

Interpolation scheme 1 for computing G from Φ

A remarkable and unique feature of lunar system A is that G is computed from Φ. It cannot be stressed enough that there is no obvious and compelling astronomical or mathematical reason for computing the duration of 1^{m} from the duration of 223^{m}, and not by a more straightforward, self-contained algorithm as in lunar system B. The highly intricate arguments that underlie the algorithm for G testify to a surprising level of mathematical skill and ingenuity on the part of the creators of lunar system A. Interpolation scheme 1, reconstructed by Kugler,[440] underlies nearly all instances of column G in the synodic tables, and is best represented in the procedure texts. It comprises 35 interpolation rules corresponding to one full oscillation of the zigzag function Φ. Each interpolation rule is defined by Φ_k, the control value of the source function, G_k, that of the goal function, and c_k, the interpolation coefficient between control points k and $k+1$ (Table 4.22).[441] If the control point k for which Φ is between Φ_k and Φ_{k+1} has been identified, then the corresponding G is obtained as

$$G = \begin{cases} G_k \pm c_k \cdot (\Phi - \Phi_k) & \text{if } \Phi \text{ on increasing branch} \\ G_k \pm c_k \cdot (\Phi_k - \Phi) & \text{if } \Phi \text{ on decreasing branch,} \end{cases} \tag{4.57}$$

where the plus (minus) sign applies if G is on the increasing (decreasing) branch. Although no single tablet or fragment contains the complete scheme, nearly all rules are preserved on at least one fragment (Table 4.21).

k	Φ_k [°]	G_k [°]	$\lvert\Phi_{k+1} - \Phi_k\rvert/\phi$	c_k	$c_k\phi$ [°]	$c_k \cdot \lvert\Phi_{k+1} - \Phi_k\rvert$ [°]
1	2,13;20 ↓	2,40	1	1	0;17,46,40	
2	2,13;2,13,20 ↓	2,40;17,46,40	1	2	0;35,33,20	
3	2,12;44,26,40 ↓	2,40;53,20	1	3	0;53,20	
4	2,12;26,40 ↓	2,41;46,40	1	4	1;11,6,40	
5	2,12;8,53,20 ↓	2,42;57,46,40	1	5	1;28,53,20	
6	2,11;51,6,40 ↓	2,44;26,40	1	6	1;46,40	
7	2,11;33,20 ↓	2,46;13,20	1	7	2;4,26,40	
8	2,11;15,33,20 ↓	2,48;17,46,40	1	8	2;22,13,20	
9	2,10;57,46,40 ↓	2,50;40	1	9	2;40	
10	2,10;40 ↓	2,53;20	41	9;20	2;45,55,33,20	1,53;22,57,46,40
11	1,58;31,6,40 ↓	4,46;42,57,46,40	1	8;20	2;28,8,53,20	
12	1,58;13,20 ↓	4,49;11,6,40	1	7;20	2;10,22,13,20	
13a	1,57;55,33,20 ↓	4,51;21,28,53,20	0;25,37,30 } 1	6;20	1;52,35,33,20	0;48,5,11,6,40
13b	1,57;47,57,46,40 $= m$	4,52;9,34,4,26,40	0;34,22,30	6;20	1;52,35,33,20	1;4,30,22,13,20
14	1,57;58,8,53,20 ↑	4,53;14,4,26,40	1	5;20	1;34,48,53,20	
15	1,58;15,55,33,20 ↑	4,54;48,53,20	1	4	1;11,6,40	
16	1,58;33,42,13,20 ↑	4,56	0;11,15	0	0	
17	1,58;37,2,13,20 ↑	4,56	1	2	0;35,33,20	
18	1,58;54,48,53,20 ↑	4,56;35,33,20	1	0	0	
19	1,59;12,35,33,20 ↑	4,56;35,33,20	1	2	0;35,33,20	
20	1,59;30,22,13,20 ↑	4,56	1	4	1;11,6,40	
21	1,59;48,8,53,20 ↑	4,54;48,53,20	1	5;20	1;34,48,53,20	
22	2,0;5,55,33,20 ↑	4,53;14,4,26,40	1	6;20	1;52,35,33,20	
23	2,0;23,42,13,20 ↑	4,51;21,28,53,20	1	7;20	2;10,22,13,20	

[440] Kugler (1900), pp. 167–179 ('Columne H').

[441] Note that in *ACT*, p. 60, Φ_k is tabulated in the same row as c_{k+1}.

k	Φ_k [°]	G_k [°]	$\lvert\Phi_{k+1}-\Phi_k\rvert/\phi$	c_k	$c_k\phi$ [°]	$c_k\cdot\lvert\Phi_{k+1}-\Phi_k\rvert$ [°]
24	2,0;41,28,53,20 ↑	4,49;11,6,40	1	8;20	2;28,8,53,20	
25	2,0;59,15,33,20 ↑	4,46;42,57,46,40	41	9;20	2;45,55,33,20	1,53;22,57,46,40
26	2,13;8,8,53,20 ↑	2,53;20	1	9	2;40	
27	2,13;25,55,33,20 ↑	2,50;40	1	8	2;22,13,20	
28	2,13;43,42,13,20 ↑	2,48;17,46,40	1	7	2;4,26,40	
29	2,14;1,28,53,20 ↑	2,46;13,20	1	6	1;46,40	
30	2,14;19,15,33,20 ↑	2,44;26,40	1	5	1;28,53,20	
31	2,14;37,2,13,20 ↑	2,42;57,46,40	1	4	1;11,6,40	
32	2,14;54,48,53,20 ↑	2,41;46,40	1	3	0;53,20	
33	2,15;12,35,33,20 ↑	2,40;53,20	1	2	0;35,33,20	
34	2,15;30,22,13,20 ↑	2,40;17,46,40	1	1	0;17,46,40	
35a	2,15;48,8,53,20 ↑	2,40	4;18,45 } 16;57,30	0	0	
35b	2,17;4,48,53,20 = M	2,40	12;38,45 }	0	0	

Table 4.22: ΦG interpolation scheme 1. Φ_k and G_k are the control values, c_k is the interpolation coefficient and $c_k\cdot|\Phi_{k+1}-\Phi_k|$ is the difference of G across interval k (included only where it differs from $c_k\phi$).

The resulting graph of G (Fig. 4.17) is nearly in antiphase with Φ except for a small phase difference. Most Φ_k are separated from Φ_{k+1} by a distance ϕ = 0;17,46,40° = $-d\Phi(223)$ (Eq. 4.13). These and other features of scheme 1 were

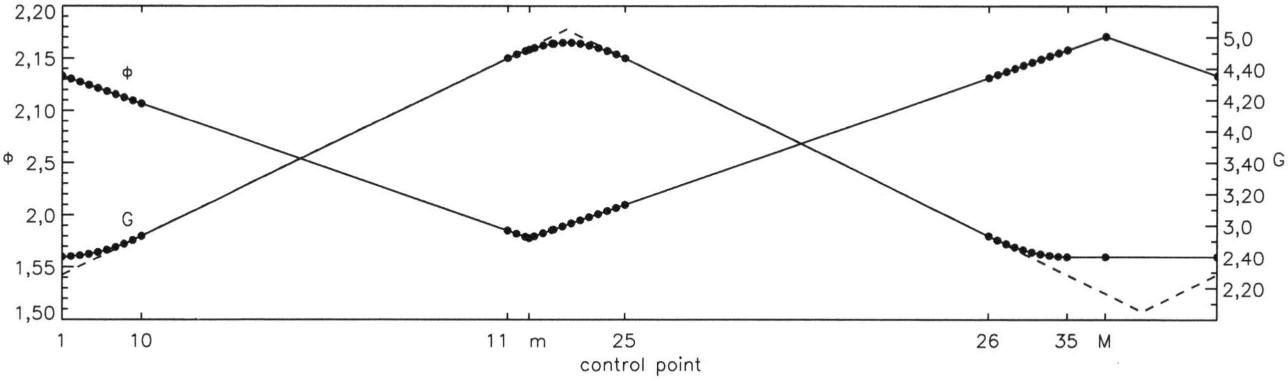

Fig. 4.17: Graphical representation of interpolation scheme 1 for computing G from Φ. Shown are Φ, G and $\hat{G}$ (dashed line). The dots denote the control points 1–35 and the extrema of Φ.

explained by A. Aaboe (1968, 1971), who discovered the intricate construction underlying the interpolation schemes for G, W and Λ. Each of these schemes is derived from two truncated versions of Φ shifted by 1^m (G), 6^m (W) or 12^m (Λ), thus explaining the existence of Φ_{tr}. Since the derivation of these schemes is accessible mainly through auxiliary tables[442] and is not discussed in the extant procedure texts, this topic is postponed to Volume II, apart from the following remarks. The number of basic intervals ϕ contained in one full oscillation of Φ equals $2\Delta_\Phi/\phi$ = 2,10;8,45, i.e. 130 whole intervals plus a fractional remnant 0;8,45ϕ. In order to obtain a consistent interpolation scheme, Φ_k and G_k must both return to their starting point after a whole number of intervals, which implies that this remnant must be incorporated as an anomalous interval. As can be seen in Table 4.22, the Babylonian astronomers chose to divide it over two intervals: nr. 16, of length 0;11,15ϕ, and nr. 35, of length 16;57,30ϕ = (17 + 0;8,45 - 0;11,15)ϕ, so that the sum of all intervals equals 17;8,45ϕ +31ϕ +2·41ϕ = 2,10;8,45ϕ, as required. The reason for inserting interval 16 is that this leads to a conveniently symmetric arrangement of the G_k values with respect to the maximum $G_{18} = G_{19}$. All adjacent intervals having the same interpolation coefficient were merged, resulting in 35 intervals.

Apart from the usual interpolation template IP.B (§2.6.7, p. 40) the following variants are worth mentioning. In some procedures the interpolation is formulated in terms of coefficients $c_k\phi$ instead of c_k. Accordingly, the difference $\Phi-\Phi_k$ or $\Phi_k-\Phi$ is first multiplied by $1/\phi$ = 3;22,30 before being multiplied by $c_k\phi$. Hence every interpolation involves an extra multiplication, but the result for G is the same. For the case when Φ and G are both increasing this template can be represented as follows:

[442] For the $\Phi\Lambda$ scheme: Aaboe (1968), Text E (BM 36311+36593), to which Britton (2009), p. 375, has added two small unnumbered fragments; for the ΦG scheme: Aaboe (1971) Text A (+) B, to which Britton (2009), p. 375, has added another fragment.

ana tar-ṣi Φ_k **tab**-*u*$_2$ G_k **gar**-*an* [*mim-ma*] *ša*$_2$ *al-la* Φ_k **tab**-*u*$_2$ **diri en** Φ_{k+1} **tab**-*u*$_2$ **a.ra**$_2$ 3.22.30 **du**-*ma* *ša*$_2$ **e**$_{11}$-*ka* **a.ra**$_2$ $c_k\phi$ **du**-*ma* **ki** G_k **tab**-*ma* **gar**-*an*	Opposite Φ_k, increasing, you put down G_k. [Whatever] (the amount) by which it exceeds Φ_k, increasing, until Φ_{k+1}, [increasing,] you multiply by 3;22,30, and what comes out for you, you multiply by $c_k\phi$, and add to G_k, and put down.	(PHIG.A)

The equivalent template for the case when Φ and G are on opposite branches need not be spelled out. The representation with $c_k\phi$ also underlies the small fragment No. **75**, where each entry appears to end with the computation of the difference for G across interval k, $c_k|\Phi_{k+1} - \Phi_k|$ (Eq. 4.57):

[...] **a.ra**$_2$ 3.22.30 **du**-*ma* $\lvert\Phi_{k+1} - \Phi_k\rvert/\phi$ **a.ra**$_2$ $c_k\phi$ **du**-*ma* $c_k\lvert\Phi_{k+1} - \Phi_k\rvert$ *bi-rit si-man*	[...] you multiply it by 3;22,30, it is $\lvert\Phi_{k+1} - \Phi_k\rvert/\phi$, you multiply it by $c_k\phi$, it is $c_k\lvert\Phi_{k+1} - \Phi_k\rvert$, the distance for the duration.	(PHIG.B)

The minimum m_Φ effectively represents an additional control point for the interpolation that splits interval 13 into two segments (Fig. 4.18). Not all procedure texts (Table 4.21) provide separate rules for both segments, but it can be assumed that this was always implied. No special treatment is required near the maximum M_Φ in interval 35, since G is constant there.[443]

Intervals 16, 18 and 35 near the extrema of G

Near the extrema of G (intervals 1–9, 16, 18 and 26–34) templates IP.B–D and PHIG.A are not used. In intervals 16, 18 and 35, where G is constant, there is no interpolation ($c_k = 0$). All instances of the instruction for interval 35 happen to be damaged at the crucial spot, but those for 16 and 18 are preserved in a few tablets. Both begin as usual with the phrase 'Opposite Φ_k, increasing, you put down G_k', which is followed by:

en Φ_{k+1} **tab**-*u*$_2$ *gab-bi* G_k **gar**-*an*	Until Φ_{k+1}, increasing, you put down G_k (for) all.	No. **53** P16'
en Φ_{k+1} **tab tab** *u* **la**$_2$ **nu tuk**	Until Φ_{k+1}, increasing, the addition and subtraction does not exist.	No. **74** P1'

The latter instruction contains an explicit representation of the number zero (§2.1).

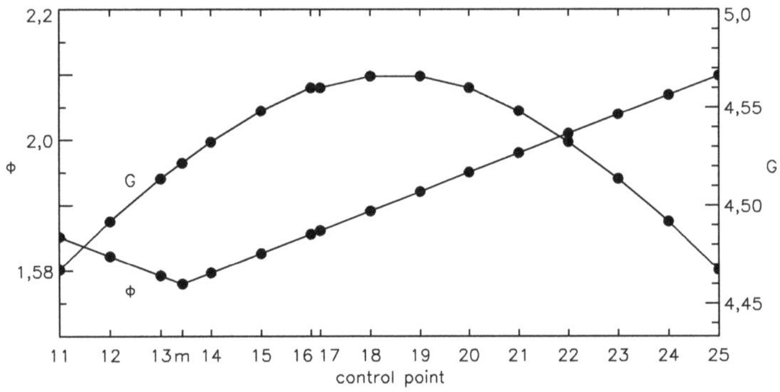

Fig. 4.18: Close-up of Φ and G in intervals 11–24 of interpolation scheme 1.

The 'malûššu-rule' for intervals 1–9 and 26–34

Also in the intervals 1–9 and 26–34 near the minimum of G the formulation usually differs from IP.B–D and PHIG.A. In intervals 1–9, where Φ decreases linearly with a constant difference ϕ from 2,13;20 to 2,10;40, G increases from its minimum 2,40 to 2,53;20 with a difference that increases linearly from ϕ to 9ϕ (Fig. 4.19, left). Only in No. **67** P1 are the interpolation rules for these intervals formulated in accordance with IP.C. In all other texts this is replaced by a single, seemingly opaque instruction that will be referred to as the '*malûššu*-rule', which warrants a close investigation. All known examples are presented in the following score:

No. **53** Ri3	*ana tar-ṣi*	2.13.20	**la**$_2$-*u*$_2$	2.40		**gar**-*an*	*mim-ma ša*$_2$ *al-la*	2.13.20	**la**$_2$-[*u*$_2$	**la**$_2$-*u*$_2$
No. **64** O1–2	*ana* ⸢*tar*⸣-*ṣa*	2.13.20	**la**$_2$-[*u*$_2$	2.40		**gar**	*ša*$_2$ *al* /	2.13.2]0 **la**$_2$		**la**$_2$-⸢*u*$_2$⸣
No. **70** O1–2	[*ana tar-ṣi*	2.13.20	**la**$_2$	2].⸢40 *si*⸣-*man*		⸢**gar**⸣-*an*	*ša*$_2$ *al*-[*la* /	2.13.20 **la**$_2$		**la**$_2$]-*u*$_2$
No. **69** O1	*ana tar-ṣa*	2.13.20	**la**$_2$	2.40		**gar**-[*an*	*ša*$_2$ *al*	2.13.20 **la**$_2$		**la**$_2$]

[443] Two ΦG texts contain evidence for the Akkadian reading of **la**$_2$. No. **53** P16' implies that in the position immediately after Φ_k, where it means 'decreasing', **la**$_2$ = *maṭû* (3 sg. stative G of *maṭû*). No. **67** P1 implies that in the comparison 'it is less' **la**$_2$ = *imaṭṭû* (3 sg. ipf. subj.).

No. **53** Ri3–4	[en 2.10.40] / **la**$_2$-*u*$_2$ **a.ra**$_2$ 3.22.30 **du** *ma-lu-uš-šu*$_2$ 17.46.40 **gar**-*an*
No. **64** O2–3	⌜**en**⌝ 2.10.40 **la**$_2$-⌜*u*$_2^?$⌝ [x xx **a.ra**$_2$] / [3.22.30] **du** *ma-lu-uš* 17.46.40
No. **70** O2–3	*ina* 2.13.20 **nim** *ša*$_2$ **e**$_{11}$-*ka* GAM 3.22.30 **du** / [*ma-lu-uš-šu*$_2$ 17.46.40
No. **69** O1–2	[*ina* 2.13.20 **nim**] / *ša*$_2$ **tag**$_4$ **a.ra**$_2$ 3.22.30 **du** *ma-lu-uš-š*[*u*$_2$ 17.46.40 **gar**-*an*]

No. **53** Ri4–5	*ana muḫ-ḫi a*-[*ḫa-miš*$_2$ GAR.GAR-*ma*] / **ki** 2.40 **tab**-*ma* **gar**-*an*
No. **64** O3	GA[R.GAR x xx xx xx]
No. **70** O3	GAR.GAR *ša*$_2$] ⌜**e**$_{11}$⌝-*ka* **ki** 2.40 **tab**-*ma* **gar**-*an*
No. **69** O2	[GAR.GAR-*ma* **ki** 2.40 **tab**-*ma* **gar**-*an*]

No. **53** Ri3	Opposite 2,13;20, decreasing, you put down 2,40. Whatever (the amount) by which it is less than 2,13;20,
No. **64** O1–2	Opposite 2,13;20, decreasing, you put down 2,40. That which is less than / 2,13;20,
No. **70** O1–2	Opposite 2,13;20, decreasing, you put down 2,40, the duration. That which is less than / 2,13;20,
No. **69** O1	Opposite 2,13;20, decreasing, you put down 2,40. That which is less than 2,13;20,

No. **53** Ri4	decreasing, until 2,10;40, decreasing, you multiply it by 3;22,30;
No. **64** O2–3	decreasing, until 2,10;40, decreasing, you subtract from 2,13;20, you multiply it by 3;22,30;
No. **70** O2	decreasing, you subtract from 2,13;20, what comes out for you, you multiply it by 3;22,30;
No. **69** O1–2	decreasing, you subtract from 2,13;20, / what remains you multiply it by 3;22,30;

No. **53** Ri4–5	while it fills up you put down 0;17,46,40, you accumulate it together, and / you add it to 2,40, and you put it down.
No. **64** O3	while it fills up you accumulate 0;17,46,40 [...]
No. **70** O3	while it fills up you accumulate 0;17,46,40, what comes out for you you add to 2,40, and you put it down.
No. **69** O2	while it fills up you put down 0;17,46,40, you accumulate it, and you add it to 2,40, and you put it down.

This instruction is comprehensible only because the algorithm that it represents has been well understood since it was reconstructed by Kugler.[444] First $G_1 = 2{,}40$ is associated with $\Phi_1 = 2{,}13;20$, after which $\Phi_1 - \Phi$ is multiplied by 3;22,30 $= 1/\phi$, yielding the fractional number of intervals ϕ contained in $\Phi_1 - \Phi$ (6;20 in the example shown in Fig. 4.19). The instruction 'you accumulate 0;17,46,40 (= ϕ) while it fills up' (*malûššu*) refers to the summation of the differences for G between control point 1 and the fractional position $(\Phi_1 - \Phi)/\phi$, where the difference for G in each interval is itself a cumulative sum over ϕ.[445] In Fig. 4.19 this amounts to $(1+2+3+4+5+6+0;20\cdot 7)\cdot\phi$. The 'filling up' describes how for each added term the distance to Φ_1 increases by ϕ, until the endpoint Φ is reached (6;20ϕ in Fig. 4.19). Finally, G is obtained by adding the sum of the G differences to G_1.

In intervals 26–34, where G approaches its minimum, the computation proceeds analogously, with the differences of G now linearly decreasing from 9ϕ (interval 26) to ϕ (interval 34), so that the 'filling up' proceeds in the reverse order from control point 35, as indicated in Fig. 4.19 (right).[446] Although an intact instruction for intervals 26–34 is not preserved, it can be reliably reconstructed on the basis of No. **68** and No. **76** and the instruction for intervals 1–9:

No. **68** R5'	*ana tar-ṣa* 2.13.8.8.53.2[0 **tab**-*u*$_2$ 2.53.20 **gar**-*an*]
No. **76** R6'	*ana tar-ṣa* 2.13.8.8.53.20 **tab** 2.53.20 [**gar**-*an*
	Opposite 2,13;8,8,53,20, increasing, you put down 2,53;20.

No. **68** R6'–7'	*ša*$_2$ *al-la* 2.13.8.8.53.[20 **tab**-*u*$_2$ **diri en** 2.15.48.8.53.20 **tab**-*u*$_2$] / *ina* 2.15.48.8.53.20 **t**[**ab**-*u*$_2$ **e**$_{11}$-*ma*
No. **76** R6'–7'	[*ša*$_2$ *al* 2.13.8.8.53.20 **tab** **diri en** 2.15.48.8.53.20 **tab**] / [ta] ⌜2⌝.15.48.8.53.20 **nim**
	That which exceeds 2,13;8,8,53,20, increasing, until 2,15;48,8,53,20, increasing, you deduct from 2,15;48,8,53,20, [increasing], and

No. **68** R7'–8'	**a.ra**$_2$ 3.22.30 **du**-*ma*] / *malûššu*(**diri**-*šu*$_2$) 17.46.40 GAR.GAR-*m*[*a* **ki** 2.40 **tab**-*ma* **gar**-*an*]
No. **76** R7'	*ša*$_2$ **tag**$_4$ [GAM 3.22.30 **du**-*ma* *malûššu*(**diri**-*šu*$_2$) 17.46.40 GAR.GAR-*ma* **ki** 2.40 **tab**-*ma* **gar**-*an*]
	[what remains] you multiply by 3;22,30, and while it fills up you accumulate 0;17,46,40, and you add it with 2,40, and you put it down.

In principle the same formulation could be used for intervals 10–14 and 21–25, where c_k also changes by the amount ϕ from interval to interval. This is attested only in the unusual procedure No. **52** P7.b, which is concerned with intervals 21–24 (cf. the commentary).

444 Kugler (1900), pp. 176–177.

445 *Malûššu* = 'in its filling up', inf. G *malû* + loc. + poss. suff. 3 sg. m.; cf. *GAG*, §66e; for the apocopated form *malûš* cf. §66f. The suggestion to connect *ma*- with *lu* (and not with the preceding **du** as in *ACT*) is due to H. Hunger, as reported by Britton (2009), p. 369.

446 This was overlooked by Neugebauer in his analysis of No. **68** (*ACT* 206).

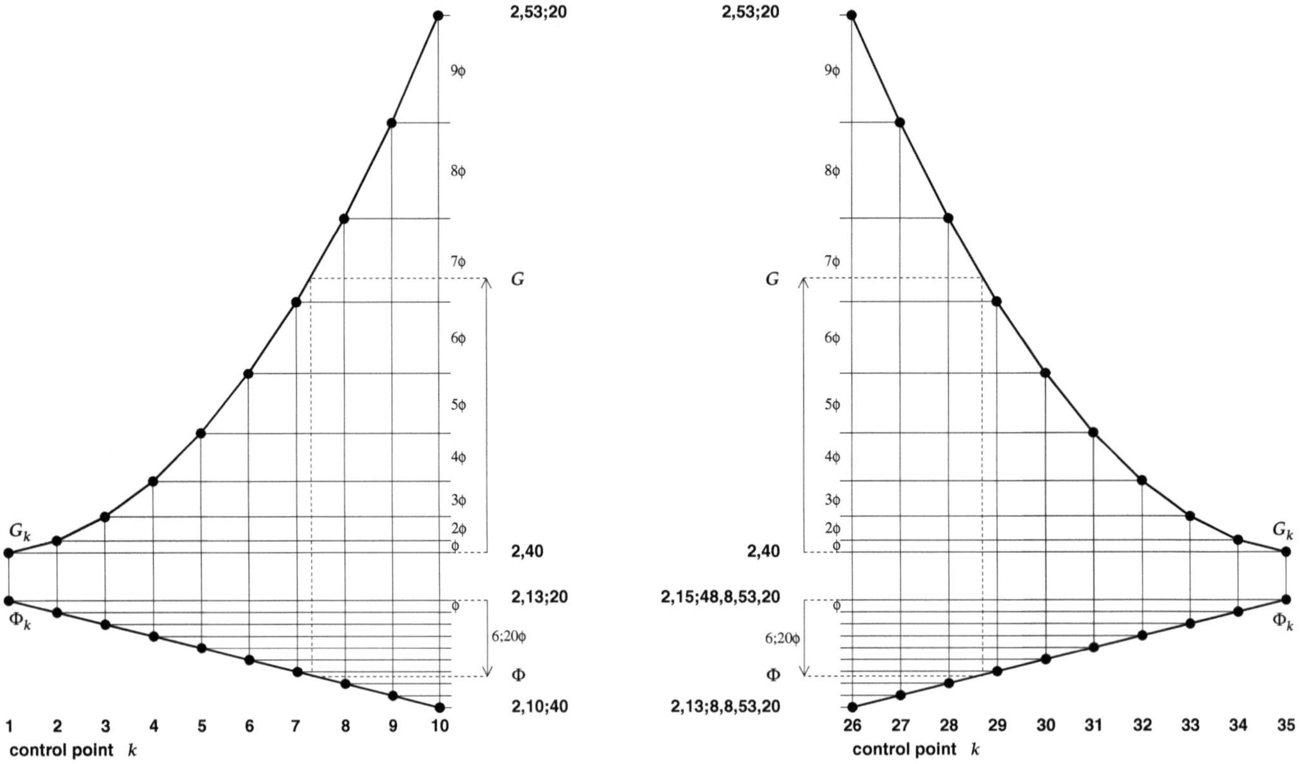

Fig. 4.19: Graphical representation of the computation of G (top) from Φ (bottom) in intervals 1–9 and 26–34 (scheme 1). The dashed line marks an exemplary value of Φ and the associated value of G; the arrows indicate the order in which the differences of Φ and G are summed up according to the *malûššu*-rule (see text).

Interpolation scheme 1 for computing G from F

No. **77** contains a version of interpolation scheme 1 in which Φ is replaced by $F^{(1)}$ (Table 4.23). The interpolation rules can be represented as

$$G = \begin{cases} G_k \pm c'_k \cdot (F - F_k) & (F \text{ on increasing branch}) \\ G_k \pm c'_k \cdot (F_k - F) & (F \text{ on decreasing branch}). \end{cases} \quad (4.58)$$

In order to obtain the same result as Eq. (4.57) the interpolation coefficients satisfy

$$c'_k = \frac{c_k}{0;15,11,15} = c_k \cdot 3;57,2,13,20, \quad (4.59)$$

where the factor 0;15,11,15 equals d_F/d_Φ, and the control points F_k are related to Φ_k through Eq. (4.56). It is difficult to say whether this alternative formulation was actually used for computing column G. The singularity of No. **77** contrasts with the large number of texts concerned with the ΦG scheme. This suggests that G was usually computed from Φ. Also the existence of column Φ, whose only identifiable purpose lies in the computation of G, points in that direction.[447] Analogous to the ΦG scheme, FG scheme 1 can also be construed from two truncated, shifted versions of its source function $F^{(1)}$, thus explaining the existence of F_{tr} (§4.4.9).[448]

k	F_k [°/d]	G_k	c'_k	k	F_k [°/d]	G_k	c'_k
1	15	2,40°	3;57,2,13,20	19	11,25,30	4,56;35,33,20°	7;54,4,26,40
2	14;55,30	2,40;17,46,40°	7;54,4,26,40	20	11;30	4,56°	15;48,8,53,20
3	14;51	2,40;53,20°	11;51,6,40	21	11;34,30	4,54;48,53,20°	21;4,11,51,6,40
4	14;46,30	2,41;46,40°	15;48,8,53,20	22	11;39	4,53;14,4,26,40°	25;1,14,4,26,40
5	14;42	2,42;57,46,40°	19;45,11,6,40	23	11;43,30	4,51;21,28,53,20°	28;58,16,17,46,40

[447] Moreover, column Φ is present in the synodic table *ACT* 18, whose column F contains $F^{(1)}$. On the other hand, the usual position of column F (vii) immediately before column G (viii) may be a fossilised feature of an earlier stage when G was computed from F. The fact that the extrema of the truncated F are more convenient numbers than the extrema of the truncated Φ may also support this.

[448] The basic interval between successive control values of $F^{(1)}$ is 0;15,11,15 ϕ = 0;4,30 = $-dF^{(1)}(223)$ (Table 4.19 on p. 144), analogous to the basic interval $\phi = -d\Phi(223)$ in the ΦG schemes.

k	F_k [°/d]	G_k	c'_k	k	F_k [°/d]	G_k	c'_k
6	14;37,30	2,44;26,40°	23;42,13,20	24	11;48	4,49;11,6,40°	32;55,18,31,6,40
7	14;33	2,46;13,20°	27;39,15,33,20	25	11;52,30	4,46;42,57,46,40°	36;52,20,44,26,40
8	14;28,30	2,48;17,46,40°	31;36,17,46,40	26	14;57	2,53;20°	35;33,20
9	14;24	2,50;40°	35;33,20	27	15;1,30	2,50;40°	31;36,17,46,40
10	14;19,30	2,53;20°	36;52,20,44,26,40	28	15;6	2,48;17,46,40°	27;39,15,33,20
11	11;15	4,46;42,57,46,40°	32;55,18,31,6,40	29	15;10,30	2,46;13,20°	23;42,13,20
12	11;10,30	4,49;11,6,40°	28;58,16,17,46,40	30	15;15	2,44;26,40°	19;45,11,6,40
13a	11;6	4,51;21,28,53,20°	25;1,14,4,26,40	31	15;19,30	2,42;57,46,40°	15;48,8,53,20
13b	11;4,4,41,15 $= m$	4,52;9,34,4,26,40°	25;1,14,4,26,40	32	15;24	2,41;46,40°	11;51,6,40
14	11;6,39,22,30	4,53;14,4,26,40°	21;4,11,51,6,40	33	15;28,30	2,40;53,20°	7;54,4,26,40
15	11;11,9,22,30	4,54;48,53,20°	15;48,8,53,20	34	15;33	2,40;17,46,40°	3;57,2,13,20
16	11;15,39,22,30	4,56°	0	35a	15;37,30	2,40°	0
17	11;16,30	4,56°	7;54,4,26,40	35b	15;56,54,22,30 $= M$	2,40°	0
18	11;21	4,56;35,33,20°	0				

Table 4.23: Interpolation scheme 1 for computing G from F.

Zigzag function $\hat{G}$

If one linearly extrapolates G beyond the anomalously long intervals 10 and 25, which belong to opposite branches of G, the result is a true zigzag function known as $\hat{G}$, a graph of which is included in Fig. 4.17 (p. 147). Since $\hat{G}$ has exactly the same period as Φ, it is effectively a scaled and shifted version of Φ. Its monthly difference (Table 4.24) satisfies

$$d_{\hat{G}} = 9;20 \cdot d_{\Phi}, \tag{4.60}$$

where $9;20 = c_{10} = c_{25}$. As far as known no instance of column G is computed from $\hat{G}$, but some of its defining parameters are mentioned in procedure texts. First, several procedures mention the monthly difference d, called the 'difference' (*tašpiltu*), or 'addition and subtraction' (*ṭīpu u mīṭu*) 'of the duration' (*ša simānu*).[449] Second, a hitherto misunderstood passage in No. **52** P7.a turns out to mention the extrema of variant 2 (Table 4.24), which is attested only there. Both are shifted upwards with respect to those of variant 1 by 0;1,58,31,6,40°; hence the period is the same.[450] No. **82** P15 mentions two numbers that may be the extrema of another variant (3), which is uniformly shifted upwards by 10;31,21,24,26,40°. Finally, No. **52** P6.b mentions the extrema of a variant 4, whose connection with lunar system A is even less clear. Its extrema differ considerably from those of the other variants of $\hat{G}$. However, they are very close to the extrema of the zigzag function underlying column G in the unusual synodic table *ACT* 92, which belongs to a variant of system A.

var.	d	m	M	Δ	Π	Z	P	p
1	25;48,38,31,6,40°	2,4;59,45,11,6,40°	5,4;57,2,13,20°	2,59;57,17,2,13,20°	1,44,7	7,28	13;56,39,6,...	0;55,59,6,13,..
2	"	2,5;1,43,42,13,20°	5,4;59,0,44,26,40°	"	"	"	"	"
3	"	2,15;31,6,35,33,20°	5,15;28,23,37,46,40°	"	"	"	"	"
4	23;20°	1,49;40°	4,32;20°	2;42;40°	8,8	35	13;56,34,17,...	0;55,59,4,55,...

Table 4.24: Parameters of the zigzag functions for $\hat{G}$.

Net differences of G for various intervals

No. **70** P2 and No. **69** P2 list net differences $d\hat{G}(s)$ for $s = 12^{\mathrm{m}}$ and 14^{m} (Table 4.25). They correspond to net differences of G in the regions where G coincides with $\hat{G}$.

s	t	$d\hat{G}(s)$
6^{m}	0	2,34;51,51,6,40°
7^{m}	0	3,0;40,29,37,46,40°
12^{m}	1	−50;10,51,51,6,40°
14^{m}	1	1;26,25,11,6,40°
223^{m}	16	−2;45,55,33,20°

Table 4.25: Net differences of $\hat{G}$ for various intervals.

449 No. **52** Rii27, No. **66** X2', No. **69** R1', No. **70** R5', No. **72** R11', No. **82** O19'.

450 The shift amounts to 1/60 of the minimum of the truncated version of Φ. The significance of this is wholly unclear.

Other interpolation schemes for computing *G* from *Φ* or *F*

Several procedure texts and tabular texts contain evidence of deviating schemes, none well enough preserved to allow a full reconstruction. At least one variant, differing from *ΦG* scheme 1 mainly near the extrema of *G*, is attested in three fragmentary auxiliary tables and template tables.[451]

k	Φ_k	G_k	$\lvert\Phi_{k+1}-\Phi_k\rvert/\phi$	c_k	$c_k \cdot \lvert\Phi_{k+1}-\Phi_k\rvert$
1	2,13;20° ↓	2,51;29,22,57,46,40°	*41;40*	*9;20*	*1,55;13,34,48,53,20°*
2	*2,0;59,15,33,20°* ↓	*4,46;42,57,46,40°*	*4*	*6;50*	*8;5,55,33,20°*
3	*1,59;48,8,53,20°* ↓	*4,54;48,53,20°*	*6;45,37,30*	?	?
4	1,57;47,57,46,40° = m_Φ	5,16$^?$;xx,41,42,13,20°	6;45,37,30	?	?
5	1,59;48,8,53,20° ↑	4,54;48,53,20°	4	6;50	8;5,55,33,20°
6	2,0;59,15,33,20° ↑	4,46;42,57,46,40°	41;40	9;20	1,55;13,34,48,53,20°
7	2,13;20° ↑	2,51;29,22,57,46,40°	*12;38,45*	x.20	?
8	*2,17;4,48,53,20°* = M_Φ	?	*12;38,45*	*x.20*	?

Table 4.26: Interpolation scheme 2 for computing *G* from *Φ*. In italics: uncertain reconstructions.

No. **72** P1 is the main witness for the more significantly deviating scheme 2. Since this tablet appears to be marred by inconsistencies, the reconstruction (Table 4.26) is provisional and incomplete; for details cf. the commentary. Eight intervals forming one complete oscillation of the zigzag function for *Φ* have been tentatively reconstructed, but the total number of intervals is probably larger. It appears that the maximum of *G* coincides with the minimum of *Φ*, and vice versa. Interval 5 corresponds to intervals 21–24 of scheme 1, but the intermediate values of *G* differ, since in scheme 1 c_k is not constant across intervals 21–24. In the adjacent interval 6 the interpolation coefficient is 9;20, the same as c_{25} in scheme 1. The length of this interval, if correctly reconstructed, is also nearly the same in both schemes (41;40ϕ versus 41ϕ). Elements of scheme 2, or similar schemes, are attested in No. **66** P4', No. **75** P1' and No. **85** P1'. No. **82** P15 contains hints of a third scheme; cf. the commentary. Another algorithm for computing a *G*-type function from an *F*-type function is partly preserved in No. **52** P6.b; cf. the commentary.

4.4.12 Duration of 6 synodic months (W)

terminology	[LA₂ *ša₂* 6 **iti** = 'duration of 6 months'?]
unit	time degrees
astronomical meaning	duration of 6 synodic months -176^d, 177^d or 178^d (only the lunar variation)
column	vi or viii
procedures:	
computing *W* from *Φ*	No. **88** P1', No. **89** P1', No. **90** P1'
auxiliary tables:	
computing *W* from *Φ*	Aaboe (1971), Text D; Steele (2010), Text F

Function *W* models the duration of 6 synodic months, such that W_i is the amount by which the time between lunations $i-6$ and i exceeds the next smaller whole number of days, which can be 176, 177 or 178. Several synodic tables and eclipse tables contain a column *W*.[452] From these tables it is inferred that the purpose of *W*, analogous to that of *G*, is to update the time of the lunation, $M^{(6)}$, over intervals of 6^m.

Properties of W

Since *W* is computed from *Φ* it acquires the same period, i.e. that of the lunar variation. From this Aaboe (1971) inferred the existence of a zodiacal correction to *W*, which he called *Z* (§4.4.15).

Computing W from Φ

The computation of *W* is achieved by means of an interpolation scheme comprising 124 rules, each pertaining to a segment of one full oscillation of the zigzag function for *Φ*,

$$W = \begin{cases} W_k \pm c_k \cdot (\Phi - \Phi_k) & \text{if } \Phi \text{ on increasing branch} \\ W_k \pm c_k \cdot (\Phi_k - \Phi) & \text{if } \Phi \text{ on decreasing branch,} \end{cases} \tag{4.61}$$

where Φ_k and W_k are the control values and c_k is the interpolation coefficient. The plus (minus) sign applies if *W* is on the increasing (decreasing) branch. For the complete scheme, as reconstructed by Aaboe & Hamilton (1979), cf.

[451] Aaboe (1968), Texts B–D. For a discussion cf. Volume II.

[452] *ACT* 54, 55, 60, 61a: cf. Aaboe & Hamilton (1979).

Fig. 4.20 and Appendix D.1 (p. 529).[453] Some of the interpolation rules are contained on three fragmentary procedure texts (Table 4.27), probably belonging to a single tablet. None of them preserves a complete rule for any interval. Their formulation is probably based on templates IP.B or IP.C. Two fragments of auxiliary tables also contain some control values of the interpolation scheme.[454]

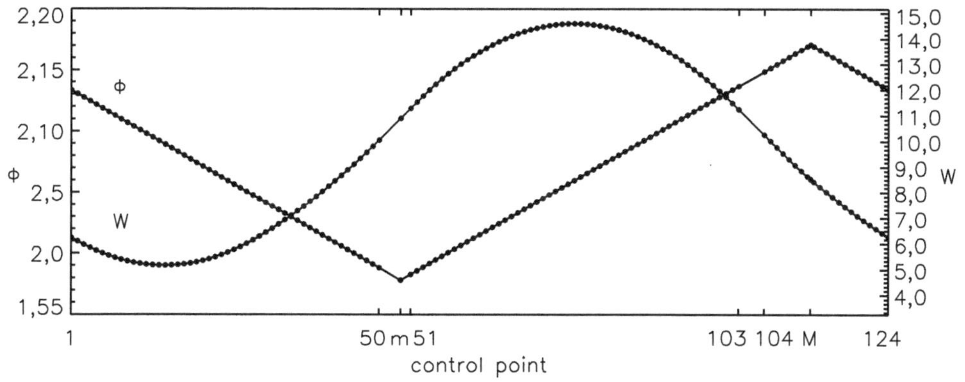

Fig. 4.20: Graphical representation of the interpolation scheme for computing W from Φ. Each control value is represented by a dot. In this representation 176^d+W is the duration of 6^m.

Subsequent values of Φ_k are usually separated by ϕ. This and other features were explained by Aaboe & Hamilton (1979), who showed how the ΦW scheme was derived from two shifted, truncated versions of Φ, similar to the construction underlying ΦG scheme 1.[455] Recall that there are 2,10;8,45 intervals ϕ within one oscillation of Φ. In order to obtain a consistent scheme in which Φ_k and W_k return to their starting point after a whole number of interpolation intervals, the fractional remnant 0;8,45ϕ was included as an anomalous 124th interval (Appendix D.1). Hence the 121 intervals of length ϕ, 1 interval of length 5ϕ (nr. 50), 1 of length 4ϕ (nr. 103), and interval 124, add up to the required 2,10;8,45ϕ. All adjacent intervals having the same interpolation coefficient were merged, resulting in 124 intervals.

	intervals
No. **88** P1'	15–27
No. **89** P1'	73–80
No. **90** P1'	113–121
AT: BM 45930 = Aaboe (1971), Text D	36–47
AT: BM 43024 = Steele (2010), Text F	(50), 51–55

Table 4.27: Procedure texts and auxiliary tables (AT) with interpolation rules for computing W from Φ.

4.4.13 Duration of 12 synodic months (Λ)

terminology	LA₂ *ša₂ šatti*(**mu**) = 'duration of the year'
unit	time degrees
astronomical meaning	duration of 12 synodic months -354^d (only the lunar variation)
column	viii or xii
procedures:	
computing Λ from Φ	No. **79** P4'–P6'
auxiliary tables:	
computing Λ from Φ	No. **79** T1'; No. **81** T1; Aaboe (1968), Text E

Function Λ models the duration of 12 synodic months, such that Λ_i is the amount by which the interval between lunations $i-12$ and i exceeds 354^d. Λ was discovered by Neugebauer (*ACT* 207d–e), but its astronomical meaning was established only later by Aaboe (1968). An Akkadian term for Λ, LA₂ *ša šatti* = 'duration of the year', is mentioned in No. **79** (Rii13). Apart from procedure texts and auxiliary tables, Λ is attested in two synodic tables, including the oldest known synodic table for lunar system A in its fully developed final stage.[456] Λ is used for updating $\tilde{M}$ (§4.4.18), a version

[453] Note that all W_k in Fig. 4.20 and Appendix D.1 represent the excess of 6^m over 176^d, because they have not been reduced to values in the range 0–6,0° corresponding to at most 1^d.

[454] For these tablets cf. Volume II. In BM 43024, the line with Φ_{51} and W_{51} is preceded by a line with $\Phi_{51}-\phi$ and the corresponding value of W instead of Φ_{50} and W_{50}, suggesting that this tablet is concerned with the derivation of the interpolation scheme.

[455] This topic is not covered by procedure texts and will be addressed in Volume II.

[456] BM 40094+45662 (Aaboe 1969; Aaboe & Hamilton 1979), dated to years 6–7 of Philip Arrhidaeus; *ACT* 55 (Aaboe & Hamilton 1979).

of the time of the lunation, over intervals of 12^m, analogous to the updating of M with G. One auxiliary table[457] offers a variant $\Lambda' = \Lambda + 3;24,7,...^\circ$, of unknown significance.

Properties of Λ

Since Λ is computed from Φ it acquires the same period, i.e. that of the lunar variation. The zodiacal correction to the duration of 12^m is modeled by Y (§4.4.16). As pointed out by Aaboe (1969), p. 11, Λ_i deviates from the quantity it is meant to represent, $\Sigma_{n=0}^{11} G_{i-n}$, by a constant amount $\max\Lambda - \max\Sigma_{n=0}^{11} G_{i-n} \approx 3{,}55;33{,}20 - 9{,}32;21{,}43 = -5{,}36;48{,}23^\circ$. Approximately the same constant, with the opposite sign, is added in the algorithm for Y (cf. Eq. 4.67), so that both constants (virtually) cancel out in $\Lambda_i + Y_i$, making this a good approximation of $\Sigma_{n=0}^{11}(G_{i-n} + J_{i-n})$.

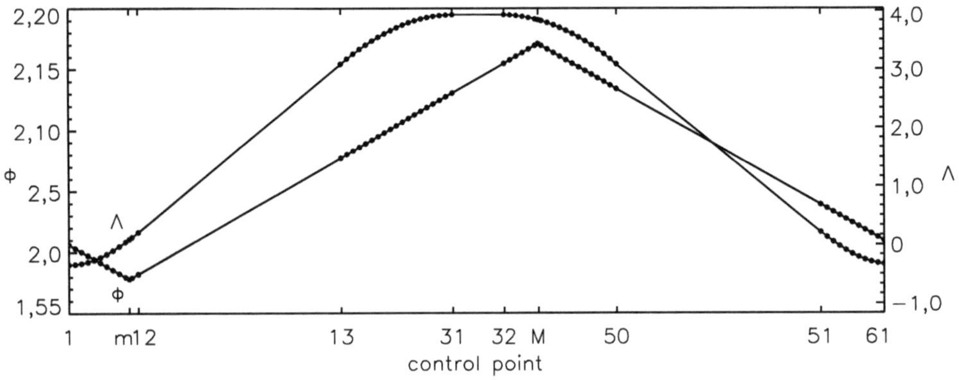

Fig. 4.21: Graphical representation of the interpolation scheme for computing Λ from Φ. Control points are represented as dots.

Computing Λ from Φ

Λ is computed from Φ by means of an interpolation scheme comprising 61 rules, each pertaining to one segment of a full oscillation of the zigzag function for Φ,

$$\Lambda = \begin{cases} \Lambda_k \pm c_k \cdot (\Phi - \Phi_k) & \text{if } \Phi \text{ on increasing branch} \\ \Lambda_k \pm c_k \cdot (\Phi_k - \Phi) & \text{if } \Phi \text{ on decreasing branch,} \end{cases} \tag{4.62}$$

where Φ_k and Λ_k are the control values, and c_k is the interpolation coefficient. The plus (minus) sign applies if Λ is on the increasing (decreasing) branch. For the complete scheme, reconstructed by Aaboe (1968), cf. Fig. 4.21 and Appendix D.2 (p. 531). Subsequent values of Φ_k are usually separated by the same distance ϕ known from ΦG scheme 1. The curves of Λ and Φ are nearly in phase, such that the maximum of G slightly precedes that of Φ. These and other features can all be explained in the framework of Aaboe's reconstruction. In order to obtain a consistent scheme in which Φ_k and Λ_k return to their starting point after a whole number of interpolation intervals, the fractional remnant $0;8{,}45\phi$ was incorporated as an anomalous interval. In this case it was attached to interval 31, where Λ attains its maximum, resulting in a total length of $8;8{,}45\phi$ for this interval. Hence the 58 intervals of length ϕ, the 2 intervals of length 32ϕ (nrs. 12 and 50) and interval 31 add up to the required $2{,}10;8{,}45\phi$. Adjacent intervals having the same interpolation coefficient were merged, resulting in 61 intervals.

	intervals
No. **79** P4'	61, 1–8, 9–?
No. **79** P6'	31–51
AT: No. **79** T1'	[...], 52–61, 1–12, (13–31)
AT: No. **81** T1'	1–12, (13–18, [...])
AT: BM 36311+ (Aaboe 1968, Text E)	13–59

Table 4.28: Procedures and auxiliary tables with interpolation rules for computing Λ from Φ.

Many of the interpolation rules are preserved in procedure texts and auxiliary tables (Table 4.28).[458] The formulation of the interpolation rules in No. **79** P4' and P6' is based on the usual template IP.B (p. 41). Even though Λ is additive or subtractive, whereas the analogously constructed function G is always additive, the additive or subtractive nature of Λ is ignored in this formulation. However, it is mentioned in the auxiliary tables No. **79** T1' and No. **81** T1' (in the latter only

[457] BM 36793 = Aaboe (1971), Text C. For this tablet cf. Volume II.

[458] For the auxiliary table BM 36311+, which is concerned with the derivation of the interpolation scheme from two shifted instances of Φ, cf. also Britton (2009), p. 375, and Volume II.

at the control values where Λ_k changes from additive to subtractive or vice versa). No. **79** T1' and No. **81** T1' are auxiliary tables sharing the unusual feature that the first part provides the control values (Φ_k, Λ_k) and the interpolation coefficients c_k up to interval 12, while the second part provides Φ_k (lacking in No. **81** T1'), Λ_k and another Φ value that is shifted with respect to the first one by ϕ. The latter Φ values are not control values of the interpolation scheme; they are needed only for deriving the scheme from two shifted instances of Φ, in accordance with Aaboe's reconstruction.

4.4.14 Zodiacal correction to G (J)

terminology	?
unit	time degrees
astronomical meaning	zodiacal correction to the synodic month G
column	ix
procedures:	
computing J^{nm} from B^{nm}	No. **53** P14'

Function J is a correction to the duration of the synodic month (G) that accounts for the zodiacal variation, i.e. the effect of the changing synodic arc of the Moon and the Sun. To be precise, $G_i + J_i$ represents the corrected duration, in excess of 29^{d}, of the interval between lunations $i-1$ and i. In the synodic tables, column J is usually placed between columns G and C'.[459]

Algorithm for computing J from B

The synodic tables imply that J is computed from the updated zodiacal position, B_i. For the New Moon case this algorithm can be represented as follows:

$$J_i^{\text{nm}} = \begin{cases} -57;3,45^\circ + 1;54,7,30\cdot(B_i^{\text{nm}} - 13^\circ\ \text{Vir}) & (13^\circ\ \text{Vir} \le B_i^{\text{nm}} \le 13^\circ\ \text{Lib}) \quad (1\text{a}) \\ 0^\circ & (13^\circ\ \text{Lib} \le B_i^{\text{nm}} \le 27^\circ\ \text{Psc}) \quad (1\text{b}) \\ -2;1,44\cdot(B_i^{\text{nm}} - 27^\circ\ \text{Psc}) & (27^\circ\ \text{Psc} \le B_i^{\text{nm}} \le 25;7,30^\circ\ \text{Ari}) \quad (2\text{a}) \\ -57;3,45^\circ & (25;7,30^\circ\ \text{Ari} \le B_i^{\text{nm}} \le 13^\circ\ \text{Vir}). \quad (2\text{b}) \end{cases} \qquad (4.63)$$

No. **53** P14' contains a verbal representation of each rule, with an incorrect version of rule 1b (cf. the commentary). For a graphical representation cf. Fig. 4.22. The negative values of J correspond to subtractive numbers in the Babylonian representation. For instance, the transition coefficient $-2;1,44$ is represented as '2;1,44, subtractive' in the procedure (cf. §2.4). Corresponding procedures for the Full Moon case are not preserved, but the algorithm for J^{fm}, as implied by the synodic tables, is readily obtained by inserting $B^{\text{nm}} = B^{\text{fm}} + 3,0^\circ$ into Eq. (4.63). Hence the actual argument of Z is the zodiacal position of the Sun. For a possible explanation of the value $J = -57;3,45^\circ$ cf. Britton (2003).

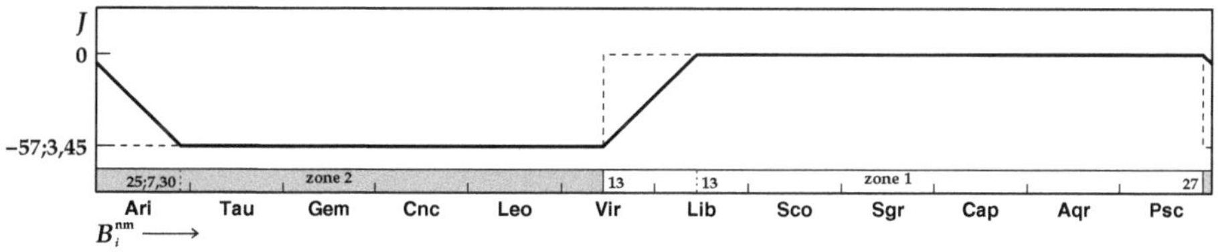

Fig. 4.22: Graphical representation of J_i as a function of B_i (New Moon case). The dashed lines represent the preliminary values w_j.

Representing J as a generalised step function of B

The algorithm for J can be construed as a generalised step function of the updated position in accordance with Eq. (4.25), p. 134, such that the preliminary values w_j and transition coefficients c'_j (Table 4.29) satisfy the same relation known from the step function for w_E (Eq. 4.26). The zonal boundaries, b_j^{nm}, are the same as for the synodic arc (§4.4.4). The algorithm for J^{fm} is likewise reproduced by Eq. (4.25) if one uses b_j^{fm} (Table 4.8a, p. 128) as the zonal boundaries.

j	w_j	c'_j
1	0°	−2;1,44
2	−57;3,45°	1;54,7,30

Table 4.29: Defining parameters of the generalised step function for J.

[459] The algorithm for J and its astronomical meaning were derived essentially correctly by Kugler (1900), pp. 179–181 ('Columne I').

4.4.15 Zodiacal correction to W (Z)

terminology	?
unit	time degrees
astronomical meaning	zodiacal correction to W

Function Z is a correction to the duration of 6 synodic months (W) that accounts for the zodiacal variation. To be precise, $W_i + Z_i$ represents the corrected duration, in excess of the nearest whole number of days (176, 177 or 178) of the interval between lunations $i-6$ and i. Textual evidence for Z is still lacking. Its existence was inferred by Aaboe (1971) from the analogous corrections J (for 1 month) and Y (for 12 months).

Algorithm for computing Z from B

Aaboe (1971) reconstructed the following expression for Z as a function of B_i (New Moon case):

$$Z_i^{\mathrm{nm}} = \begin{cases} -5{,}36;47{,}44^\circ & (13^\circ \text{ Vir} \leq B_i^{\mathrm{nm}} \leq 15;56^\circ \text{ Vir}) \quad (1a) \\ -5{,}36;47{,}44^\circ + 1;54{,}7{,}30 \cdot (B_i^{\mathrm{nm}} - 15;56^\circ \text{ Vir}) & (15;56^\circ \text{ Vir} \leq B_i^{\mathrm{nm}} \leq 13^\circ \text{ Psc}) \quad (1b) \\ 0^\circ & (13^\circ \text{ Psc} \leq B_i^{\mathrm{nm}} \leq 27^\circ \text{ Psc}) \quad (1c) \\ -2;1{,}44 \cdot (B_i^{\mathrm{nm}} - 27^\circ \text{ Psc}) & (27^\circ \text{ Psc} \leq B_i^{\mathrm{nm}} \leq 13^\circ \text{ Vir}). \quad (2) \end{cases} \tag{4.64}$$

For a graphical representation cf. Fig. 4.23 (top). The Full Moon version (Z^{fm}) is obtained by inserting $B^{\mathrm{nm}} = B^{\mathrm{fm}} + 3{,}0^\circ$. Hence in both cases the actual argument of Z is the zodiacal position of the Sun.[460]

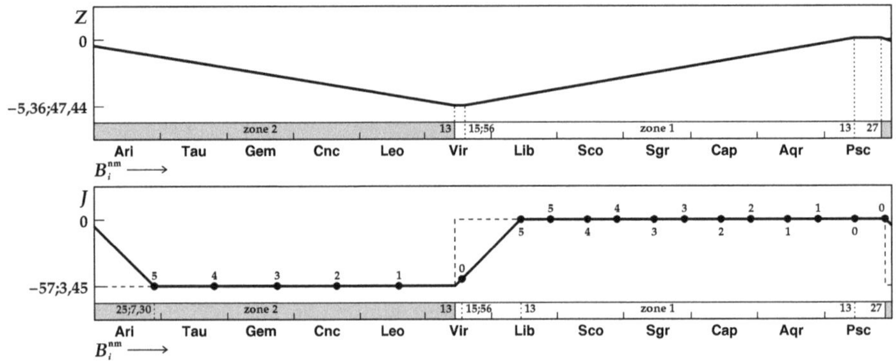

Fig. 4.23: Top: graphical representation of Z^{nm} as function of B_i^{nm} (New Moon case). **Bottom**: derivation of Z^{nm} from J^{nm} for three exemplary positions B_i^{nm} (labeled 0).

Derivation of the algorithm for Z

Aaboe's reconstruction is based on the assumption that Z_i represents the cumulative effect of J for lunations $i-5$ until i,

$$Z_i = \sum_{n=0}^{5} J_{i-n}, \tag{4.65}$$

where $J_{i-n} = J(B_{i-n})$. The circumstance that $W_i + Z_i \approx \sum_{n=0}^{5}(G_{i-n} + J_{i-n})$ and $W_i \approx \sum_{n=0}^{5} G_{i-n}$ implies that this expression is at least approximately satisfied.[461] In fact the analogy with Y_i (§4.4.16) suggests that it is exactly satisfied. If not then Z will be slightly different from Eq. (4.64). Aaboe (1971) did not provide the actual derivation of Eq. (4.64), which proceeds as follows. In order to find Z for a given B_i (the points labeled 0 in Fig. 4.23, bottom) we add the 6 values of J labeled 0–5. The positions of points 1–5 are obtained by repeatedly subtracting σ computed with the 'backward' algorithm (Eq. 4.16, p. 129), setting out from point 0. If B_i = 27 Psc then all 6 values of J are zero, so that $Z = 0$. If we shift B_i to the right, each time recomputing points 1–5, then points 0–5 one by one slide down the transition region between 27 Psc and 25;7,30 Ari, so that Z changes linearly, until B_i reaches 13 Vir. If we shift B_i up to 15;56 Vir, which corresponds to point 5 reaching 25;7,30 Ari,[462] the increase of J at point 0 is compensated by an equal decrease of J at point 5,[463] so that Z remains

[460] It will be obvious from Fig. 4.23 (top) that there is little point in attempting to formulate the algorithm for Z in terms of a generalised step function with preliminary values w_j and transition coefficients c'_j as is readily done for J and Y.

[461] Cf. Aaboe (1971), pp. 14–16. Note that $W_i = \sum_{n=0}^{5} G_{i-n}$ is exactly satisfied for the control values W_k and G_k but not for arbitrary intermediate values of W computed from Φ with the ΦW interpolation scheme.

[462] Proof: the distance between points 5 and 0 is 15;56 Vir − 25;7,30 Ari = 2,20;48,30° = 4 · 28;7,30° + 28;18,30°, where 28;18,30° = 28;7,30° + (15;56 − 13) · 0;3,45 = σ(15;56 Vir), the synodic arc at point 0.

[463] The essential property causing the slope to be constant in this segment is that the distance between adjacent points, σ_j, equals the width of the transition region in zone j, so that there is always at most one point in each transition region.

constant and equal to $6 \cdot (-57;3,45) + 1;54,7,30 \cdot (15;56 - 13) = -5,36;47,44°$. If B_i moves beyond 15;56 Vir, points 0–5 one by one slide up the transition region between 13 Vir and 13 Lib, resulting in a linear increase of Z, until point 5 is at 13 Lib, i.e. B_i = 13 Psc. Between there and 27 Psc $J = 0$.

4.4.16 Zodiacal correction to Λ (Y)

terminology	?
unit	time degrees
astronomical meaning	zodiacal correction to Λ
column	xiii

Function Y is a correction to the duration of 12 synodic months (Λ) that accounts for the zodiacal variation. To be precise, $\Lambda_i + Y_i$ represents the corrected duration, in excess of 354^d, of the interval between lunations $i-12$ and i. Column Y is attested in one synodic table,[464] from which Aaboe (1969) derived the algorithm for Y and its astronomical interpretation.

Algorithm for computing Y from B

The synodic table implies that Y is obtained from B by the following algorithm (New Moon case):

$$Y_i^{\rm nm} = \begin{cases} 0° & (13° \text{ Vir} \le B_i^{\rm nm} \le 15;56° \text{ Psc}) & \text{(1a)} \\ 1;54,7,30° \cdot (B_i^{\rm nm} - 15;56° \text{ Psc}) & (15;56° \text{ Psc} \le B_i^{\rm nm} \le 27° \text{ Psc}) & \text{(1b)} \\ 21;2,59° & (27° \text{ Psc} \le B_i^{\rm nm} \le 2;37,30° \text{ Vir}) & \text{(2a)} \\ 21;2,59° - 2;1,44 \cdot (B_i^{\rm nm} - 2;37,30° \text{ Vir}) & (2;37,30° \text{ Vir} \le B_i^{\rm nm} \le 13° \text{ Vir}). & \text{(2b)} \end{cases} \tag{4.66}$$

For a graphical representation cf. Fig. 4.24. The Full Moon version, $Y^{\rm fm}$, is obtained by inserting $B^{\rm nm} = B^{\rm fm} + 3{,}0°$.

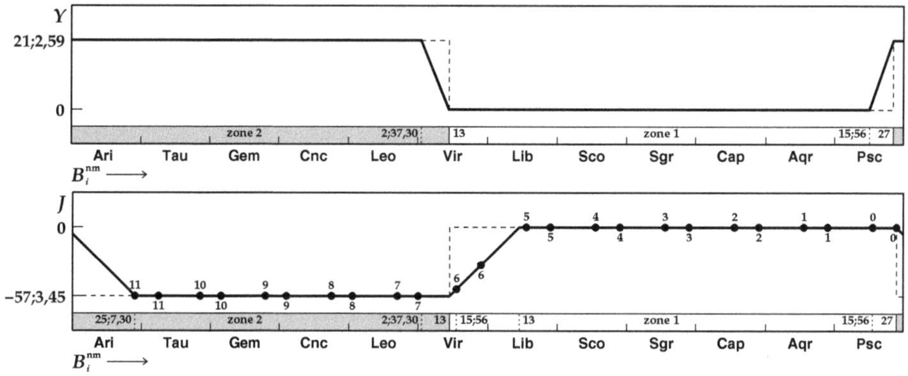

Fig. 4.24: Top: graphical representation of $Y^{\rm nm}$ as a function of $B_i^{\rm nm}$ (New Moon case). The dashed lines represent the preliminary values w_j. **Bottom**: derivation of $Y^{\rm nm}$ for two exemplary positions $B_i^{\rm nm}$ (labeled 0).

Derivation of the algorithm for Y

The following derivation of Eq. (4.68) sets out from the assumption that Y is the cumulative effect of J for the 12 lunations between $i-11$ and i, modified by a constant whose purpose will become apparent:[465]

$$Y_i = 5{,}36;47{,}44° + \sum_{n=0}^{11} J_{i-n}. \tag{4.67}$$

The value of Y at B_i (the points labeled 0 in Fig. 4.24, bottom) is obtained by summing over the 12 values of J labeled 0–11 and adding the constant. Points 1–11 are found by repeatedly subtracting σ computed with the 'backward' representation (Fig. 4.10), setting out from point 0. If B_i = 15;56 Psc then $Y = 5{,}36;47{,}4° + 6 \cdot (-57;3{,}45°) + (15;56 - 13) \cdot 1;54{,}7{,}30° = 5{,}36;47{,}44° - 5{,}36;47{,}44° = 0°$. If we shift B_i from 15;56 Psc to 27 Psc and recompute points 1–11 then point 6 slides up the transition region between 13 Vir and 13 Lib while all other points remain constant, resulting in a linearly increasing Y. If B_i = 27 Psc then $J = (27 - 13) \cdot 1;54{,}7{,}30° = -30;26°$ at point 6, so that $Z = 5{,}36;47{,}44° + 5 \cdot -57;3{,}45° - 30;26° = 21;2{,}59°$. If we shift B_i to the right of 27 Psc then one point slides down while another slides up, resulting in a constant Z. The next transition occurs at B_i = 2;37,30 Vir when point 11 reaches 13 Lib, where J is constant. Between there and

[464] BM 40094+45662 (Aaboe 1969; Aaboe & Hamilton 1979). The fragment BM 45662, which does not preserve column Y, was first published as *ACT* 128.

[465] For a somewhat different derivation cf. Aaboe (1969), who does not mention Eq. (4.67).

B_i = 13 Vir the decrease of J at point 5 (not shown in the figure) is not compensated so that Y decreases linearly to 0. It will be obvious that the constant 5,36;47,44° was added in order to obtain a function Y that vanishes in region 1a. Since approximately the same constant was subtracted from Λ, the sum $\Lambda + Y$ is barely affected.[466]

Representing Y as a generalised step function of B

Function Y can also be formulated as a generalised step function of B_i in accordance with Eq. (4.25) on p. 134. The zonal boundaries (b_j^{nm}) are the same as for the synodic arc (§4.4.4). The width of the transition regions, where Y changes linearly with B_i, equals $|dB_j(12)|$ = 11;4° in zone 1 and 10;22,30° in zone 2 (Table 4.9, p. 129). This is a consequence of the distance between points 11 and 0 being $\sigma_j + |dB_j(12)|$. Hence the transition coefficients, c'_j (Table 4.30), are related to the preliminary values of the step function, w_j, in accordance with

$$c'_j = \frac{w_{j+1} - w_j}{dB_{j+1}(12)}. \tag{4.68}$$

This is the equivalent of Eq. (4.26), p. 134, with the synodic arc σ_{j+1} replaced by $dB_{j+1}(12)$. The Full Moon version, Y^{fm}, can likewise be derived from Eq. (4.25) by using b_j^{fm} (Table 4.8a, p. 128) as the zonal boundaries.

j	w_j	c'_j
1	0°	−2;1,44
2	21;2,59°	1;54,7,30

Table 4.30: Defining parameters of the generalised step function for Y.

4.4.17 Monthly, 6-monthly and 12-monthly differences of the time of the lunation (K, $K^{(6)}$, $\tilde{K}$)

terminology	*simānu qatû* = 'final duration' (K)
unit	time degrees
astronomical meaning	K: monthly difference of the time of the lunation with respect to sunset (M)
	$K^{(6)}$: 6-monthly difference of the time of the lunation with respect to sunset ($M^{(6)}$)
	$\tilde{K}$: 12-monthly difference of the time of the lunation with respect to sunset ($\tilde{M}$)
column	K: xi; $\tilde{K}$: xiv
procedures:	
computing K	No. **53** P15'.a

Function K is the monthly difference of M, the time of the lunation with respect to sunset. In the synodic tables column K is usually placed between columns C' and M. Function $K^{(6)}$, the 6-monthly version of K, is the difference of $M^{(6)}$. The existence of this function is implied by W, but a corresponding column is not yet attested in the tabular texts. Function $\tilde{K}$, the 12-monthly version of K, is the difference of $\tilde{M}$. Column $\tilde{K}$ is attested in one synodic table, where it follows column Y.[467]

Algorithm for K

No. **53** P15'.a and the synodic tables imply that K_i is obtained as

$$K_i = G_i + J_i + C'_i. \tag{4.69}$$

This algorithm, reconstructed by Kugler,[468] is valid for both New Moons and Full Moons. $G_i + J_i$ represents the duration of the synodic month -29^{d}, and C'_i is a correction that accounts for the changing time of sunset. To be precise, C'_i is the amount by which sunset is shifted with respect to midday between lunations $i-1$ and i, such that additive values correspond to a shift towards earlier times (Fig. 4.25).

Algorithm for $K^{(6)}$

The synodic table implies that $K_i^{(6)}$ is obtained as

$$K_i^{(6)} = W_i + Z_i + C'^{(6)}_i. \tag{4.70}$$

[466] Cf. Aaboe (1969), p. 11, who derived that $\max \Lambda - \max \sum_{n=0}^{11} G_{i-n} \approx$ 3,55;33,20 − 9,32;21,43 = −5,36;48,23°.
[467] BM 40094+45662 (*ACT* 128): cf. Aaboe (1969); Aaboe & Hamilton (1979).
[468] Kugler (1900), pp. 184–192 ('Columne L').

The algorithm is valid for both New Moons and Full Moons. $W_i + Z_i$ represents the duration of 6 synodic months -176^d, 177^d or 178^d, and $C_i'^{(6)}$ (§4.4.5) is the amount by which sunset is shifted with respect to midday between lunations $i-6$ and i.

Algorithm for $\tilde{K}$

The algorithm for $\tilde{K}$ and its astronomical significance were established by Aaboe (1969). The mentioned synodic table implies that $\tilde{K}_i$ is obtained as

$$\tilde{K}_i = \Lambda_i + Y_i + \tilde{C}_i'. \tag{4.71}$$

The algorithm is valid for both New Moons and Full Moons. $\Lambda_i + Y_i$ represents the duration of 12 synodic months -354^d, and $\tilde{C}_i'$ (§4.4.5) is the amount by which sunset is shifted with respect to midday between lunations $i-12$ and i.

4.4.18 Time of the lunation (M, $M^{(6)}$, $\tilde{M}$)

terminology	*ūmu*(**me**)|*simānu ana rabê*(**šu₂**) *Šamaš₂* = 'time until sunset' (M^{nm}) *ūmu*(**me**)|*simānu ša₂ mūšu*(**gi₆**) *ittalak*(**du**) = 'time by which the night has progressed' (M^{fm})
unit	time degrees date: month and day number
astronomical meaning	time of lunation with respect to preceding (M^{fm}) or following (M^{nm}) sunset
column	M: xii; $\tilde{M}$: xv
procedures:	
updating M^{nm}	No. **53** P15'.b
date of New Moon	No. **53** P17'

Function M is the time interval between the lunation and the immediately preceding or following sunset. In the synodic tables, M is usually tabulated between column K and the Lunar Six columns. M is always a number in the range 0–6,0° corresponding to an interval of at most 1^d. Column M also includes the date (day number) of the lunation, which was established by a separate set of rules. Sometimes column M^{fm} includes the same indication about the length of the current month in whole days contained in column NA_1, such that '30' stands for a 29^d month, and '1' for a 30^d month. An Akkadian term for M^{nm} is mentioned in No. **53** P15' (Ri1) and P18' (Rii17); for M^{fm} in No. **61** (step 1). Apart from M which is updated from month to month, there is a version $M^{(6)}$ which is updated over intervals of 6^m, and another version, $\tilde{M}$, which is updated over intervals of 12^m. Only M and $\tilde{M}$ are attested in synodic tables.

Algorithm for updating M

From the synodic tables it is inferred that M^{nm} is the interval between New Moon and the *following* sunset, while M^{fm} is the interval between Full Moon and the *preceding* sunset. Accordingly, the monthly difference K_i (§4.4.17) is subtracted from M_{i-1}^{nm}, but added to M_{i-1}^{fm}.[469] In order to obtain a value of M in the range 0–6,0° it is occasionally necessary to add or subtract 6,0°. In the New Moon case 6,0° is added if $K_i^{nm} > M_{i-1}^{nm}$, because otherwise the subtraction of K_i^{nm} from M_{i-1}^{nm} is impossible in the Babylonian sense. In the Full Moon case the addition of K_i^{fm} may cause M_i^{fm} to exceed 6,0°, in which case 6,0° is subtracted. Both modifications signify that the date boundary (sunset) is transgressed by applying K_i. This results in the following expressions:

$$M_i^{nm} = \begin{cases} M_{i-1}^{nm} - K_i^{nm} & \text{if } K_i^{nm} < M_{i-1}^{nm}\text{: date boundary not transgressed} \\ M_{i-1}^{nm} + 6{,}0 - K_i^{nm} & \text{if } K_i^{nm} > M_{i-1}^{nm}\text{: date boundary transgressed,} \end{cases} \tag{4.72}$$

$$M_i^{fm} = \begin{cases} M_{i-1}^{fm} + K_i^{fm} & \text{if } M_{i-1}^{fm} + K_i^{fm} < 6{,}0\text{: date boundary not transgressed} \\ M_{i-1}^{fm} + K_i^{fm} - 6{,}0 & \text{if } M_{i-1}^{fm} + K_i^{fm} > 6{,}0\text{: date boundary transgressed.} \end{cases} \tag{4.73}$$

For a graphical representation cf. Fig. 4.25. No. **53** P15'.b contains a verbal representation of the algorithm for M^{fm}.

Establishing the date of the lunation

The corresponding day number of the lunation, say d_i, is determined as follows. Since the monthly difference K_i is defined as an excess over 29^d, these 29^d must be added to the date of the previous lunation. Second, if the date boundary

[469] The algorithm for M^{fm} was reconstructed essentially correctly by Kugler (1900), p. 184 ('Columne L').

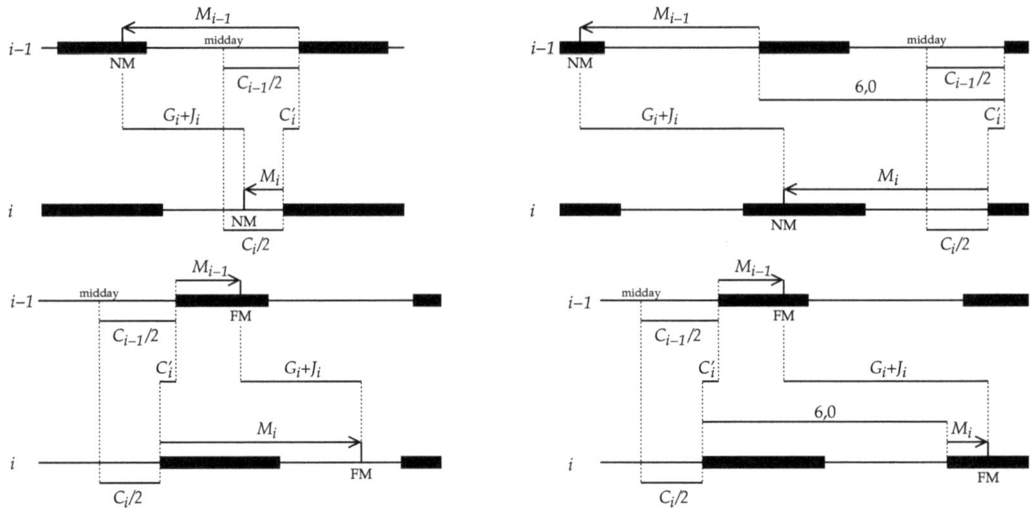

Fig. 4.25: Graphical representation of the computation of M. The arrow indicates the direction in which M is measured. Thick horizontal bars indicate night time. In all cases shown here, C'_i is additive. **Top, left**: New Moon (NM), date boundary not transgressed: $M_i^{\rm nm} + K_i^{\rm nm} = M_{i-1}^{\rm nm}$. **Top, right**: New Moon, date boundary transgressed: $M_i^{\rm nm} + K_i^{\rm nm} = M_{i-1}^{\rm nm} + 6{,}0$. **Bottom, left**: Full Moon (FM), date boundary not transgressed: $M_i^{\rm fm} = M_{i-1}^{\rm fm} + K_i^{\rm fm}$. **Bottom, right**: Full Moon, date boundary transgressed: $M_i^{\rm fm} + 6{,}0 = M_{i-1}^{\rm fm} + K_i^{\rm fm}$.

was transgressed during the computation of M_i an extra day is added, resulting in $30^{\rm d}$. Third, d_i also depends on whether the previous month $(i-1)$ was hollow $(29^{\rm d})$ or full $(30^{\rm d})$. This results in the following four rules, which apply equally to New Moon and Full Moon:

$$\mathrm{d}_i = \mathrm{d}_{i-1} \begin{cases} +0 & \text{previous month hollow,} & \text{date boundary not transgressed} \\ +1 & \text{previous month hollow,} & \text{date boundary transgressed} \\ -1 & \text{previous month full,} & \text{date boundary not transgressed} \\ +0 & \text{previous month full,} & \text{date boundary transgressed.} \end{cases} \tag{4.74}$$

These rules are formulated in No. **53** P17', and they are consistent with the dates mentioned in column M.

Algorithm for updating $M^{(6)}$

The existence of $M^{(6)}$ is implied by W (§4.4.12), but a corresponding column is not yet attested in the synodic tables. The analogy with M implies that $M^{(6)\rm nm}$ is the interval between New Moon and the following sunset and $M^{(6)\rm fm}$ the interval between Full Moon and the preceding sunset. The same considerations mentioned above in connection with M result in the following expressions for $M^{(6)}$:

$$M_i^{(6)\rm nm} = \begin{cases} M_{i-6}^{(6)\rm nm} - K_i^{(6)\rm nm} & \text{if } K_i^{(6)\rm nm} < M_{i-6}^{(6)\rm nm}\text{: date boundary not transgressed} \\ M_{i-6}^{(6)\rm nm} + 6{,}0 - K_i^{(6)\rm nm} & \text{if } K_i^{(6)\rm nm} > M_{i-6}^{(6)\rm nm}\text{: date boundary transgressed,} \end{cases} \tag{4.75}$$

$$M_i^{(6)\rm fm} = \begin{cases} M_{i-6}^{(6)\rm fm} + K_i^{(6)\rm fm} & \text{if } M_{i-6}^{(6)\rm fm} + K_i^{(6)\rm fm} < 6{,}0\text{: date boundary not transgressed} \\ M_{i-6}^{(6)\rm fm} + K_i^{(6)\rm fm} - 6{,}0 & \text{if } M_{i-6}^{(6)\rm fm} + K_i^{(6)\rm fm} > 6{,}0\text{: date boundary transgressed.} \end{cases} \tag{4.76}$$

The corresponding rules for the date of the lunation are the same as Eq. (4.74).

Algorithm for updating $\tilde{M}$

Function $\tilde{M}$, attested in one synodic table,[470] is defined in the same way as M, i.e. $\tilde{M}^{\rm nm}$ is the interval between New Moon and the following sunset, and $\tilde{M}^{\rm fm}$ the interval between Full Moon and the preceding sunset. The same considerations mentioned above in connection with M therefore result in the following expressions for $\tilde{M}$:

$$\tilde{M}_i^{\rm nm} = \begin{cases} \tilde{M}_{i-12}^{\rm nm} - \tilde{K}_i^{\rm nm} & \text{if } \tilde{K}_i^{\rm nm} < \tilde{M}_{i-12}^{\rm nm}\text{: date boundary not transgressed} \\ \tilde{M}_{i-12}^{\rm nm} + 6{,}0 - \tilde{K}_i^{\rm nm} & \text{if } \tilde{K}_i^{\rm nm} > \tilde{M}_{i-12}^{\rm nm}\text{: date boundary transgressed,} \end{cases} \tag{4.77}$$

$$\tilde{M}_i^{\rm fm} = \begin{cases} \tilde{M}_{i-12}^{\rm fm} + \tilde{K}_i^{\rm fm} & \text{if } \tilde{M}_{i-12}^{\rm fm} + \tilde{K}_i^{\rm fm} < 6{,}0\text{: date boundary not transgressed} \\ \tilde{M}_{i-12}^{\rm fm} + \tilde{K}_i^{\rm fm} - 6{,}0 & \text{if } \tilde{M}_{i-12}^{\rm fm} + \tilde{K}_i^{\rm fm} > 6{,}0\text{: date boundary transgressed.} \end{cases} \tag{4.78}$$

[470] BM 40094+45662 (*ACT* 128; Aaboe 1969; Aaboe & Hamilton 1979).

The corresponding rules for the date of the lunation are the same as Eq. (4.74).

4.4.19 Lunar Six intervals

terminology	*manzāzu*(**na**) [*ša*$_2$ **u$_4$.1.kam**] = 'NA [for day 1]' = NA$_1$ **kur**; **u$_4$.na$_2$.a** (= 'day of disappearance') = KUR **šu$_2$** = ŠU$_2$; **na** = NA; **me** = ME; **gi$_6$** = GI$_6$
unit	time degrees
astronomical meaning	cf. Table 4.1 (p. 113)
columns	NA$_1$: xiii; KUR: xiv; ŠU$_2$: xiii; NA: xiv; ME: xv; GI$_6$: xvi
procedures:	
computing NA$_1$	No. **53** P17'
computing KUR	No. **53** P18'
computing ŠU$_2$	No. **61.ADFG** P1
computing NA	No. **61.A–F** P2
computing ME	No. **61.ABCG** P3
computing GI$_6$	No. **61.ABCG** P4
auxiliary tables:	
q_{set}, r_{set}	No. **53** P17'; No. **61.ACEG** T1.a,b
q_{rise}, r_{rise}	No. **61.ACG** T2.a,b

The final columns of a synodic table for the Moon are usually occupied by the Lunar Six, which are time intervals between the setting or rising of the Moon and that of the Sun near New Moon (KUR and NA$_1$) or Full Moon (ŠU$_2$, NA, ME, GI$_6$). For their definition and astronomical significance cf. §4.1. KUR and NA$_1$ are written in adjacent columns after column M^{nm} (i.e. usually on the obverse); ŠU$_2$, NA, ME and GI$_6$ in adjacent columns after M^{fm} (usually on the reverse).[471]

The Lunar Six modules

The Lunar Six modules are among the most complex algorithms of Babylonian mathematical astronomy. The essential steps of these modules were reconstructed by Schaumberger (1935), whose results are actually based on a synodic table for lunar system B. Schaumberger also formulated a convincing geometric interpretation of the Lunar Six modules, which was adapted to lunar system A and refined by Neugebauer in his extensive commentary to No. **61.A** (*ACT* 201).[472] Neugebauer's results remain largely valid, but for some steps (1, 5, 6, 7, 10 and 12) I offer a slightly different or more complete reconstruction, or a different interpretation. The six different versions of the module, one for every Lunar Six interval, can be summarised in two flow charts (Fig. 4.26): one for the New Moon intervals KUR and NA$_1$, and one for the Full Moon intervals ŠU$_2$, NA, ME and GI$_6$. Each module has been divided into 13 subalgorithms ('steps'), numbered in accordance with their usual order in the procedure texts for the Full Moon intervals (No. **61**).[473] The steps do not correspond to visually distinct units of text, but most of them (2–6, 10–13) are dedicated to a single quantity or a few closely related quantities.[474] Some steps are introduced by a statement of purpose (4, 6). Note that the Lunar Six values in the synodic tables are usually not exactly reproduced by executing to the letter the steps as reconstructed here, perhaps because slightly different algorithms are used, or the outcome of some steps is rounded before being passed on to the next one.[475]

Brief astronomical explanation of the Lunar Six modules

The astronomical reasoning behind the Lunar Six modules can be summarised as follows. If the moment of lunation coincided with sunrise or sunset, and if the Moon were located at the ecliptic then the Sun and the Moon would cross the horizon simultaneously at the lunation.[476] This almost never happens for two reasons: first, lunations occur at any time of the day; second, the Moon is not confined to the ecliptic. Therefore, when the Moon crosses the horizon it is usually displaced from being in exact conjunction (New Moon) or opposition (Full Moon) with the Sun by a distance

[471] In No. **53** KUR is sometimes called **u$_4$.na$_2$.a** = *ūm bubbuli*, 'day of disappearance', and NA$_1$ **u$_4$.1.kam**, 'day 1'. Uniquely in No. **61.C**, ME is called **ki.na$_2$.a** (Akkadian reading unknown), 'place of disappearance'. This term makes little sense and is probably a scribal error.

[472] *ACT*, pp. 226–240; cf. also *HAMA*, pp. 535–540.

[473] In some Mss. of No. **61** the order is different for some steps, e.g. in No. **61.AB** P3,P4 step 10 follows step 11, and in No. **61.G** step 6 appears to follow step 7. In No. **53** P17',P18' the steps are partly garbled, and sometimes clearly inconsistent with the order in which they are meant to be executed.

[474] Neugebauer introduced 5 steps labeled a–f, which are related to the 13 steps adopted here according to a = 1, b = 2–3, c = 4–5, d = 6–7, e = 8–13.

[475] For an analysis of these deviations cf. Volume II.

[476] Moreover, a solar eclipse would occur at New Moon, and a lunar eclipse at Full Moon.

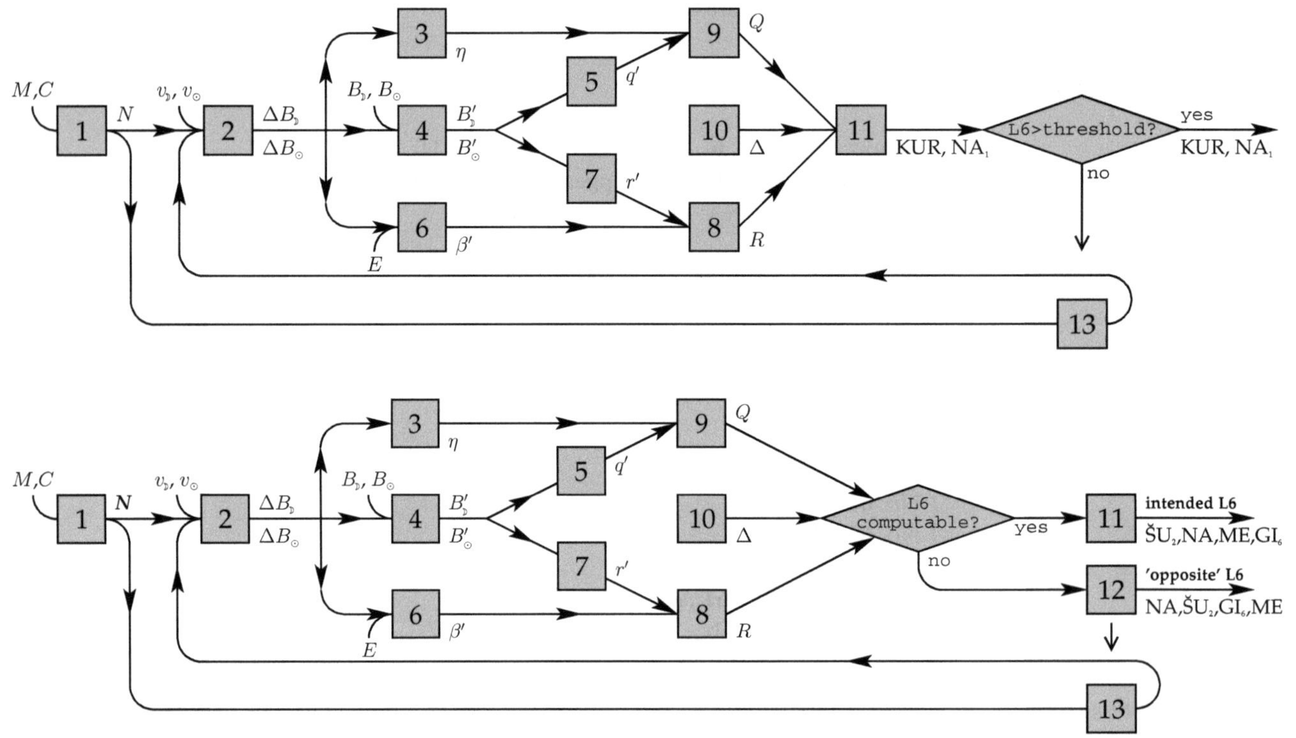

Fig. 4.26: Flow charts for the Lunar Six modules for KUR and NA$_1$ (**top**) and ŠU$_2$, NA, ME and GI$_6$ (**bottom**). Both flow charts are essentially identical up to step 10, although the algorithm for a given step may be different for each Lunar Six. The value of NA$_1$ or KUR obtained after the first execution of step 11 may be too small for the lunar crescent to be visible. In that case the final value is obtained by executing step 13 and repeating steps 1–11 for the following (NA$_1$) or previous (KUR) day. With ŠU$_2$, NA, ME or GI$_6$ the originally intended Lunar Six may turn out to be incomputable after the first execution of steps 1–10. In that case steps 12 and 13 are performed and steps 1–11 are repeated for the previous or following day, depending on the Lunar Six.

along the ecliptic, and a distance perpendicular to it (for a visualisation cf. Fig. 4.36, p. 176). If these displacements are known it is possible to compute the time interval between the lunation and the horizon crossing. The displacement along the ecliptic is obtained by multiplying the time between the Lunar Six event and the lunation by the rate of change of the Moon's elongation from the Sun (steps 1–3). The Moon's distance from the ecliptic is computed from its value at lunation by applying a correction that accounts for the change between the lunation and the horizon crossing (step 6). Both displacements are converted into differences in rising and setting time (steps 4–5, 7–9). These time intervals are added or subtracted, which yields the Lunar Six interval (step 11). Some Lunar Six intervals are subjected to a 'disk correction' (step 10) which, among other things, accounts for the fact that the horizon crossings are defined in terms of the upper or lower rim and not disk centre.[477] This summarizes the core of the Lunar Six modules (steps 1–11).

Near the end of the module the Lunar Six interval is evaluated in order to determine whether the computed value is final. This evaluation assumes a different form for KUR and NA$_1$ on the one hand, and ŠU$_2$, NA, ME and GI$_6$ on the other. The thin lunar crescent (KUR, NA$_1$) is visible only if it is sufficiently bright compared to the background sky, which implies that KUR and NA$_1$ must exceed a threshold value larger than zero. If NA$_1$ or KUR is smaller than the threshold, then steps 1–11 are repeated for the previous (KUR) or next (NA$_1$) day (step 13). Once the final value of NA$_1$ has been determined, the length, expressed in whole days (29 or 30), of the month that ended with NA$_1$ is known. With that information the date of the lunation in the next month, to be written in the next row of column M, can be established (§4.4.18).

For ŠU$_2$, NA, ME and GI$_6$ there is no threshold value, because the Full Moon is always visible as long as it is above the horizon. However, these intervals must still be computable in the Babylonian sense, i.e. larger than zero. If they are not, this signifies that the order in which the Sun and the Moon cross the horizon is the opposite from what was initially assumed. In that case one ends up with the 'opposite' Lunar Six interval (step 12), and the initially assumed Lunar Six interval is obtained by repeating steps 1–11 for the previous or the next day (step 13). Consequently the temporal sequence of the Lunar Six intervals differs from the most common one (Fig. 4.3 on p. 114). This issue does not emerge for KUR and NA$_1$ because their 'opposite' Lunar Six intervals are not observable and therefore not defined. Hence step 12 is absent

[477] In Figs. 4.2 and 4.36 this is ignored and the disk centres are used.

from the procedures for NA_1 and KUR. If step 12 is invoked one may end up with two different values of $ŠU_2$, NA, ME or GI_6, only one of which is valid. As will be argued in §4.4.19.12, the value obtained in step 12 is always the correct one.

Statement of purpose

Each Lunar Six module begins with a statement of purpose 'In order for you to construct L6' (L6 *ana* **du₃**-*ka*), where L6 stands for the name of the Lunar Six interval. As explained above, this purpose is provisional in the case of the Full Moon intervals because what has been announced as $ŠU_2$ might turn out to be NA in step 12, and similarly for NA, ME and GI_6.[478] In the case of KUR and NA_1 this cannot occur, and the statement of purpose is final.

4.4.19.1 Step 1. Time interval between lunation and Lunar Six event (N)

terminology	*ūmu*(**me**) *ana rabê*(**šu₂**) = 'time until sunset' = N_{NA_1}, N_{GI_6}
	me nim.a = 'time from sunrise' = N_{KUR}, $N_{\mathrm{ŠU}_2}$
	me *ana* **kur** = 'time until sunrise' = N_{NA}
	gi₆ du = '(time by which) the night has progressed' = N_{ME}
unit	time degrees
astronomical meaning	time interval between lunation and Lunar Six event
procedures:	
computing N_{KUR}	No. **53** P18'
computing $N_{\mathrm{ŠU}_2}$	No. **61.AD** P1
computing N_{NA}	No. **61.ADFG** P2
computing N_{GI_6}	No. **61.ABG** P4

Step 1 concerns the time of the Lunar Six event and the corresponding date. Both are expressed in relation to the nearest lunation (New Moon or Full Moon). The time is quantified by the interval N between the lunation and the Lunar Six event, i.e. sunrise (KUR, $ŠU_2$ and NA) or sunset (NA_1, ME and GI_6). N is always counted positive, irrespective of whether the Lunar Six event occurs before the lunation (KUR, and ideally $ŠU_2$ and ME) or after it (NA_1, and ideally NA and GI_6). For the formulation of step 1 in the procedure texts, which is different for every Lunar Six interval, I refer to the editions of No. **61** and No. **53**.[479]

NA_1

NA_1 occurs at one of the first instances of sunset after New Moon (Fig. 4.27). For consecutively later instances of sunset labeled $n = 0, 1, \ldots$ the time difference with New Moon N_{NA_1} is

$$N_{\mathrm{NA}_1} = M^{\mathrm{nm}} + n \cdot 6{,}0. \quad (4.79)$$

The earliest theoretical possibility of NA_1 ($n = 0$) occurs on the first sunset after New Moon (i.e. the day after New Moon) but this value of NA_1 is usually insufficient to guarantee visibility of the lunar crescent.

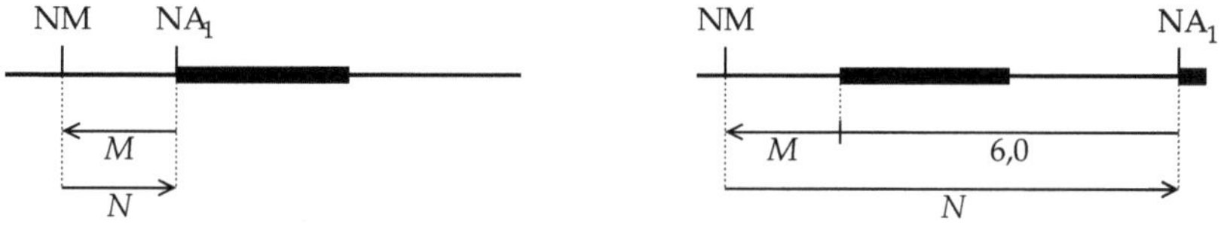

Fig. 4.27: Step 1. Definition of N_{NA_1}. **Left**: first sunset after New Moon ($n = 0$). **Right**: second sunset after New Moon ($n = 1$).

KUR

KUR occurs at one of the last instances of sunrise before New Moon. Two possibilities arise, depending on whether New Moon occurs during day time or during the night (Fig. 4.28). For a New Moon during day time, one obtains

$$N_{\mathrm{KUR}} = C^{\mathrm{nm}} - M^{\mathrm{nm}} + n \cdot 6{,}0, \quad (4.80)$$

[478] In No. **53** P18' (NA_1) the purpose is formulated using the alternative term 'day 1'.

[479] The Akkadian term for N_{KUR} is mentioned in No. **53** P18'; the others are not preserved but can be guessed from the terms for M.

where $n = 0, 1, \ldots$ labels progressively earlier possibilities and C is the length of daylight (§4.4.5). The latest possible KUR ($n = 0$) occurs on the same day as New Moon. If New Moon occurs during the night then

$$N_{\text{KUR}} = C^{\text{nm}} - M^{\text{nm}} + (n+1) \cdot 6{,}0, \tag{4.81}$$

and the latest possible KUR ($n = 0$) occurs on the day before New Moon. Usually this value of KUR is insufficient to guarantee visibility of the lunar crescent.

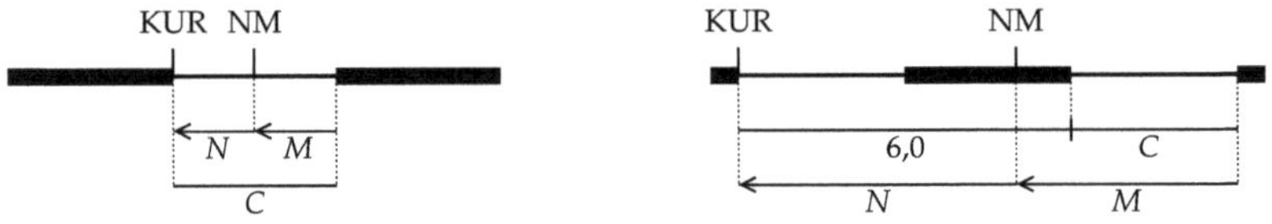

Fig. 4.28: Definition of N_{KUR} for the latest possible KUR ($n = 0$). **Left**: New Moon during day time. **Right**: New Moon during the night.

ŠU₂

For the Lunar Sixes near Full Moon, step 1 is initially executed on the assumption that the most common sequence is valid (Fig. 4.3, p. 114). The correctness of this assumption is established a posteriori after step 10. If it turns out to be wrong, N is modified in step 12 by adding 6,0°, which corresponds to a shift of 1^{d}, and with this new value steps 1–11 are repeated. ŠU_2 is initially assumed to occur at the last sunrise before Full Moon. The computation of N involves two cases (Fig. 4.29). If Full Moon occurs during day time ($M > D$), then ŠU_2 falls on the same day, i.e. $\text{d}_{\text{ŠU}_2} = \text{d}_{\text{FM}}$, and the time from sunrise to Full Moon is

$$N_{\text{ŠU}_2} = M^{\text{fm}} - D^{\text{fm}}, \tag{4.82}$$

where D is the length of the night. The computation of D as $6{,}0° - C$ is not mentioned explicitly in the procedures. If Full Moon falls during the night ($M < D$), ŠU_2 occurs 1^{d} earlier than Full Moon, i.e. $\text{d}_{\text{ŠU}_2} = \text{d}_{\text{FM}} - 1$ and 6,0° is added to $N_{\text{ŠU}_2}$, resulting in[480]

$$N_{\text{ŠU}_2} = M^{\text{fm}} + C^{\text{fm}}. \tag{4.83}$$

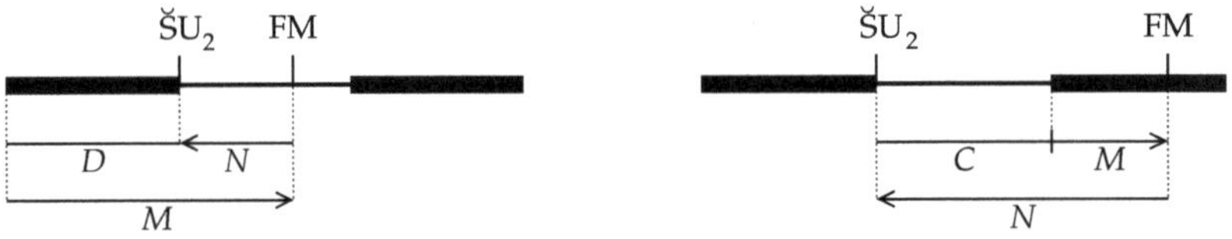

Fig. 4.29: $N_{\text{ŠU}_2}$. **Left**: Full Moon during day time. **Right**: Full Moon during the night.

NA

NA is initially assumed to occur at the first sunrise after Full Moon. If the latter occurs during day time ($M > D$), this is 1^{d} after Full Moon ($\text{d}_{\text{NA}} = \text{d}_{\text{FM}} + 1$), and the time from Full Moon to sunrise is (Fig. 4.30, left)

$$N_{\text{NA}} = 6{,}0 - (M^{\text{fm}} - D^{\text{fm}}). \tag{4.84}$$

If Full Moon is during the night ($M < D$), NA occurs on the same date ($\text{d}_{\text{NA}} = \text{d}_{\text{FM}}$), and

$$N_{\text{NA}} = D^{\text{fm}} - M^{\text{fm}}, \tag{4.85}$$

as shown in Fig. 4.30 (right).

[480] This expression for $N_{\text{ŠU}_2}$ replaces the one in *ACT*, p. 231; cf. the commentary to No. **61**.

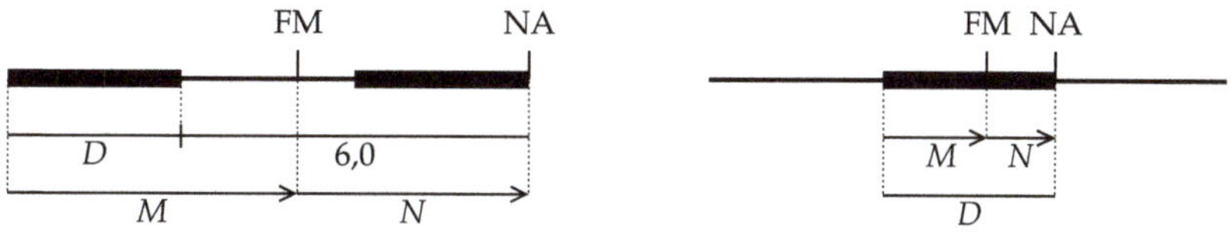

Fig. 4.30: N_{NA}. **Left**: Full Moon during day time. **Right**: Full Moon during the night.

ME and GI₆

Since ME and GI_6 occur at sunset, which is the moment of reference for M^{fm}, there is no ambiguity with regard to their timing (see Fig. 4.31). For ME, the time from sunset to Full Moon is initially taken to be

$$N_{ME} = M^{fm}. \tag{4.86}$$

The date coincides with the day of Full Moon, irrespective of whether the latter occurs during day or night. Hence no operations are involved for ME, and step 1 is absent. For GI_6 the time from Full Moon to sunset is initially set to

$$N_{GI_6} = 6{,}0 - M^{fm}, \tag{4.87}$$

the date of GI_6 always being 1^d after that of Full Moon, i.e. $d_{GI_6} = d_{FM} + 1$.

Fig. 4.31: Left: N_{ME}. **Right**: N_{GI_6}.

4.4.19.2 Step 2. Zodiacal displacement of the Moon ($\Delta B_☽$) and the Sun ($\Delta B_☉$)

terminology	*nisiḫ*(**zi**) *Sîn* = 'displacement of the Moon' = $\Delta B_☽$
	nisiḫ(**zi**) *Šamaš₂* = 'displacement of the Sun' = $\Delta B_☉$
unit	degrees
astronomical meaning	zodiacal displacement of the Moon ($\Delta B_☽$) and the Sun ($\Delta B_☉$)
procedures	No. **53** P18'; No. **61.ABCFG** P1–P4

Step 2 is concerned with the angular distances $\Delta B_☽$ and $\Delta B_☉$ by which the Moon and the Sun are displaced along the zodiac during the interval N. They are needed in steps 3 and 4. Like N these angular distances are additive for all Lunar Sixes (Fig. 4.32), both the ones that occur before, and those that occur after the lunation. The formulation of step 2 in the extant procedure texts is different for ŠU₂, NA, ME and GI₆ on the one hand, and NA₁ and KUR on the other.

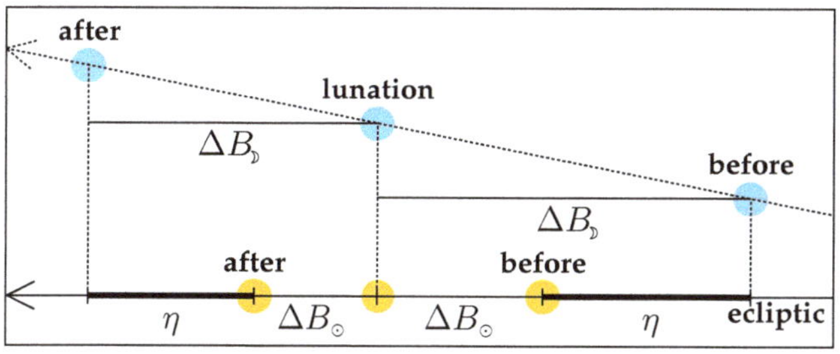

Fig. 4.32: Schematic view of the sky showing the positions of the Moon (ascending) and the Sun before, during and after a lunation. The disk at the equator represents the Sun during New Moon, or the anti-Sun (the disk diametrically opposite to the Sun) during Full Moon.

ŠU₂, NA, ME and GI₆

In No. **61** step 2 is formulated as follows:

[*ša*$_2$ **tag**$_4$] **a.ra**$_2$|GAM|GAM$_0$ 10 **du** [*ša*$_2$ **e**$_{11}$-*ka*] [*šu-u*$_2$ **zi** *šamaš*$_2$ **gar**-*an*]	[What remains (*N*)] you multiply by 0;0,10 ($v_{\odot}^{\mathrm{fm}}$), [what comes out for you,] [this you put down as the displacement of the Sun ($\Delta B_{\odot}$)].
a.ra$_2$|GAM|GAM$_0$ **zi** *sin ša*$_2$ **iti**-*šu*$_2$ **du**-*ma šu-u*$_2$ **zi** *sin* **gar**-*an*	You multiply it (*N*) by the displacement of the Moon for that month ($v_{☽}^{\mathrm{fm}}$), and this you put down as the displacement of the Moon ($\Delta B_{☽}$).

Hence $\Delta B_{☽}$ and $\Delta B_{\odot}$ are obtained by multiplying N by the corresponding displacement per time degree (v),

$$\Delta B_{\odot} = N \cdot v_{\odot}^{\mathrm{fm}}, \tag{4.88}$$

$$\Delta B_{☽} = N \cdot v_{☽}^{\mathrm{fm}}, \tag{4.89}$$

where $v_{\odot}^{\mathrm{fm}}$ is a constant 0;0,10° per time degree = 1° per day. By contrast, $v_{☽}^{\mathrm{fm}}$ varies from month to month, as indicated by the term 'displacement of the Moon for that month'. This corresponds to function F^{fm}, except that F is expressed in degrees per day. The conversion involves a division by 6,0, i.e. a multiplication by 0;0,10, an operation not mentioned explicitly in No. **61**.

KUR and NA$_1$

In No. **53** P18' (KUR) step 2 is formulated entirely in terms of sparse numerical examples. The following reconstruction of the underlying general algorithm, first proposed by Neugebauer (*ACT* p. 209) is therefore only tentative. What is certain is that the Moon's daily displacement at the Lunar Six event is assumed to deviate from the value at lunation. This variation is modeled by a zigzag function $v_{☽}$ whose daily difference is 0;18°.[481] The parameter $v_{\odot}$ = 0;57,56° employed in the procedure may be a particular value of the zigzag function whose extrema are mentioned in No. **53** P10' (cf. §4.4.10). The total displacement of the Moon and the Sun between New Moon and the Lunar Six event can thus be expressed as

$$\Delta B_{☽} = v_{☽}^{\mathrm{nm}} \cdot N_0 + 6{,}0 \cdot \sum_{k=1}^{n} v_{☽ k}, \tag{4.90}$$

$$\Delta B_{\odot} = v_{\odot}^{\mathrm{nm}} \cdot N. \tag{4.91}$$

Here n has the same meaning as in step 1, and N_0 represents the case $n = 0$ corresponding to the latest (KUR) or earliest (NA$_1$) possible occasion of the Lunar Six event. For each subsequently earlier (KUR) or later (NA$_1$) day labeled by k, up to day n, the lunar displacement is increased by 6,0 $v_{☽ k}$, the Moon's daily displacement on day k. Analogous to the case of ŠU$_2$, NA, ME and GI$_6$, the value of this function during lunation, $v_{☽}^{\mathrm{nm}}$, corresponds to F^{nm}, except that the latter is expressed in degrees per day, i.e. $v_{☽}^{\mathrm{nm}} = 0;0,10 \cdot F^{\mathrm{nm}}$. This operation is explicitly mentioned in No. **53** P18'. If one desires so, this more elaborate algorithm can be rationalised, because the interval N between New Moon and KUR or NA$_1$ can be as long as 3^{d}, during which $v_{☽}$ can vary considerably, whereas in the case of ŠU$_2$, NA, ME and GI$_6$ N is always less than 1^{d}. It is less clear why $v_{\odot}$ is modeled as a function of the zodiacal position, if that is the correct interpretation.

4.4.19.3 Step 3. Elongation of the Moon (η)

terminology	BE, perhaps to be read *bi*$_2$ = *birītu* = 'distance'?
unit	degrees
astronomical meaning	near New Moon: elongation between the Moon and the Sun near Full Moon: elongation between the Moon and the anti-Sun
procedures	No. **53** P18'; No. **61.ABD–G** P1–P4

In step 3 the elongation (angular distance along the ecliptic) between the Moon and the Sun or the anti-Sun at the Lunar Six event is computed by subtracting the distance covered by the Sun during the interval N from the distance covered by the Moon during the same interval. The outcome of step 3 is used in step 9. An instruction for step 3 is preserved only in No. **61**:

zi *šamaš*$_2$ **ta zi** *sin* **nim**-*ma* BE **gar**-*an*	You deduct the displacement of the Sun ($\Delta B_{\odot}$) from the displacement of the Moon ($\Delta B_{☽}$), and you put down the elongation (η).

[481] A similar zigzag function is known from lunar system B; cf. §4.5.10.

BE, 'elongation', is probably a phonetic abbreviation of *birītu*, 'distance'.[482] The formulaic representation is

$$\eta = \Delta B_{☽} - \Delta B_{\odot}. \tag{4.92}$$

Since the Moon moves faster than the Sun ($\Delta B_{☽} > \Delta B_{\odot}$), the value of η thus obtained is always positive, i.e. computable in the Babylonian sense. The astronomical interpretation of η depends on the Lunar Six event. For NA_1 and KUR η is the elongation between the Moon and the Sun; for $ŠU_2$, NA, ME and GI_6 it is the elongation between the Moon and the anti-Sun (Fig. 4.32). For Lunar Sixes that occur before the lunation (KUR, and ideally $ŠU_2$ and ME), η is the distance by which the Moon is behind the Sun (KUR) or the anti-Sun ($ŠU_2$, ME); for the other Lunar Sixes it is the distance by which the Moon is in front of the Sun (NA_1) or the anti-Sun (NA, GI_6).

4.4.19.4 Step 4. Zodiacal position of the Moon ($B'_{☽}$) and the Sun ($B'_{\odot}$) at the Lunar Six event

terminology	*qaqqar*(**ki**) *Sîn* = 'position of the Moon' = $B'_{☽}$ *qaqqar*(**ki**) *Šamaš* = 'position of the Sun' = $B'_{\odot}$
unit	zodiacal sign and degrees [0–30] within it
astronomical meaning	zodiacal position of the Moon ($B'_{☽}$) and the Sun ($B'_{\odot}$) at the Lunar Six event
procedures	No. **53** P18'; No. **61.ABD–G** P1–P4

In step 4 the zodiacal positions of the Sun and the Moon at the Lunar Six event, $B'_{☽}$ and $B'_{\odot}$, are computed from their respective positions at the lunation, $B_{☽}$ and $B_{\odot}$. The corresponding instructions in No. **61.ABCDE** are

zi *ša₂ šamaš₂|sin* **zi ki ki** *šamaš₂|sin* **tab**	The displacement by which the Sun|Moon moved you add with the position of the Sun|Moon.	(NA, GI_6)
zi *ša₂ šamaš₂|sin* **zi ta ki** *šamaš₂|sin* **nim|la₂**	The displacement by which the Sun|Moon moved you deduct|subtract from the position of the Sun|Moon.	($ŠU_2$, ME)

In No. **53** P18' (KUR) step 4 is formulated in terms of a numerical example. The formulaic representation of these algorithms, including the reconstructed one for NA_1, is

$$\left.\begin{aligned} B'_{☽} &= B_{☽} + \Delta B_{☽} \\ B'_{\odot} &= B_{\odot} + \Delta B_{\odot} \end{aligned}\right\} \text{NA}_1, \text{NA}, \text{GI}_6 \tag{4.93}$$

$$\left.\begin{aligned} B'_{☽} &= B_{☽} - \Delta B_{☽} \\ B'_{\odot} &= B_{\odot} - \Delta B_{\odot} \end{aligned}\right\} \text{KUR}, \text{ŠU}_2, \text{ME}. \tag{4.94}$$

The corrections $\Delta B_{☽}$ and $\Delta B_{\odot}$ are, as expected, added to the positions at lunation if the Lunar Six event occurs after the lunation, subtracted if it occurs before the lunation (cf. Fig. 4.32).[483] The outcome is used for computing 2 values of the coefficient q (step 5), one for the Moon and one for the Sun, and similarly for r (step 7). In No. **61.F** step 4 is not preserved, and in No. **61.G** it is badly damaged, if present at all (cf. the critical commentary). As will be argued in §4.4.19.5 and §4.4.19.7, one might indeed question the necessity of step 4, since the effect of the corrections $\Delta B_{☽}$ and $\Delta B_{\odot}$ on the values of q and r is not really sigificant. From a modern viewpoint it would suffice to compute q and r from the zodiacal positions at the lunation, $B_{☽}$ and $B_{\odot}$. This would have the additional advantage that the values of q for the Moon and the Sun are the same. Hence only one of them would have to be computed, and there would be no need for the averaging in steps 5 and 7. All that can be said is that the Babylonian astronomers saw a need for step 4.[484]

[482] Uniquely in No. **53** Rii39 (P18') the term 'stroke for the zodiac' (*mi-ḫi-il-tu₂ ša₂ lu-maš*) denotes elongation, even though BE is also used in the same procedure.

[483] If the new position lies across the boundary of a zodiacal sign the position in that sign is obtained by subtracting or adding 30°, but this is as usual not mentioned explicitly.

[484] In some synodic tables the Lunar Six columns may have been computed without step 4; perhaps some of the deviations between these columns and the values obtained by executing the complete Lunar Six modules to the letter can be explained in this manner. A detailed analysis of this issue will be presented in Volume II.

4.4.19.5 Step 5. Coefficient 'for the zodiac' (q')

terminology	*ṣiliptu*(**bar.nun**) *ša*$_2$ *lumāši* = '*ṣiliptu*-coefficient for the zodiac' (q') *ṣiliptu ša lumāši ša*$_2$ *elî*(**nim**) = '*ṣiliptu*-coefficient for the zodiac for rising' ($q_{\rm rise}$) *ṣiliptu ša lumāši ša*$_2$ *rabê*(**šu**$_2$) = '*ṣiliptu*-coefficient for the zodiac for setting' ($q_{\rm set}$)
unit	dimensionless
astronomical meaning	ratio between arc along ecliptic and associated rising ($q_{\rm rise}$) or setting time ($q_{\rm set}$)
procedures	No. **53** P17',P18' (traces?); No. **61.A–EG** P1–P4
auxiliary tables	$q_{\rm rise}$: No. **61.ADG** T2.a $q_{\rm set}$: No. **53** P17'; No. **61.ADEG** T1.a

The purpose of step 5 is to compute the '*ṣiliptu*-coefficient for the zodiac', q', to be used in step 9 for computing a contribution to the Lunar Six interval. The formulation of step 5 is mirrored in step 7, which concerns the similar coefficent r'. In most procedure texts q' is defined as the average of 2 values: one for the Moon ($q_{☽}$) and one for the Sun ($q_{\odot}$). They are computed from the zodiacal positions of the Moon and the Sun by an extrapolation algorithm. In the procedure texts this is formulated as follows:

ki *sin u* **ki** *šamaš*$_2$ *ša*$_2$ *al-la* 10|**me**|*u*$_4$*-me lu-maš* **diri** GAM$_0$ 16 **du**	(The amount) by which the position of the Moon and the position of the Sun exceed 10 (°) of the zodiacal sign you multiply by 0;0,16.	No. **61.A–E** P1–P4
ki bar.nun *ša*$_2$ *lu-maš ša*$_2$*-šu*$_2$ *ki* **tab tab** *ki* **la**$_2$ **la**$_2$	You add it with the corresponding *ṣiliptu*-coefficient for the zodiac if it is increasing, subtract if it is decreasing.	
mim-ma ša$_2$ *al-la* 10 **ḫun diri** *ša*$_2$ $^{\rm d}$*sin*|$^{\rm d}$**utu a.ra**$_2$ 16 **du ta** 1.24 **bar.nun** *ša*$_2$ **ḫun e**$_{11}$	(The amount) by which it exceeds 10 Ari for the Moon|Sun you multiply by 0;0,16, deduct from 1;24, the *ṣiliptu*-coefficient for Ari.	No. **53** P17'
ta ḫun en absin$_0$ **ta bar.nun la**$_2$ **ta rin**$_2$ **en** *zib* **ki bar.nun tab**	From Ari until Vir you subtract it from the *ṣiliptu*-coefficient; from Lib until Psc you add it with the *ṣiliptu*-coefficient.	

The control values, say q_{10}, are anchored at 10° of the zodiacal signs.[485] The distance to the control point, $B' - B_{10}$, is multiplied by the interpolation coefficient 0;0,16, and the result is added to or subtracted from the corresponding q_{10},

$$q = q_{10} \pm (B' - B_{10}) \cdot 0;0,16. \tag{4.95}$$

In No. **53** P17' this algorithm is exemplified for a single control point. In No. **61.G** step 5 is present in P2, but it is badly preserved. It appears that the formulation is similar to that of No. **61.A–E**, except that the subtraction $B' - B_{10}$ is explicit. No. **61** includes auxiliary tables (T1.a, T2.a), each containing 12 values q_{10} forming a zigzag function (Fig. 4.33). Superscripts imply that they pertain to 'settings' (T1.a) and 'risings' (T2.a).[486] I propose that this reflects the empirical

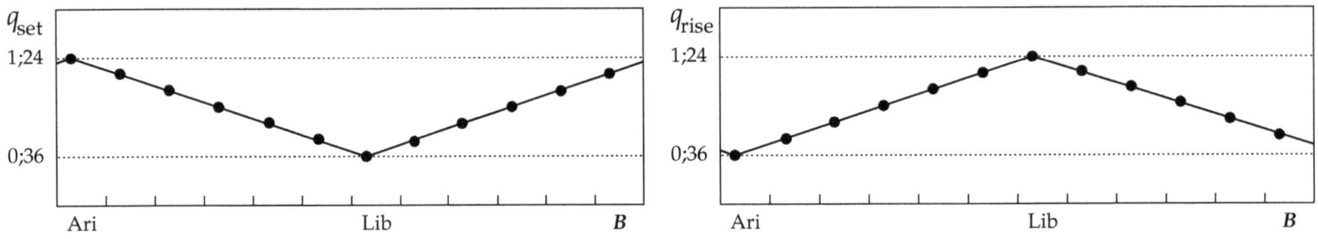

Fig. 4.33: Coefficients 'for the zodiac for setting' ($q_{\rm set}$) and 'rising' ($q_{\rm rise}$) against the zodiacal position of the Moon or the Sun. The dots are the control points q_{10}.

fact that $q_{☽}$ depends only on whether the Moon is setting or rising, and on the Moon's zodiacal position. By the same token $q_{\odot}$ depends only on whether the Sun is setting or rising, and on the Sun's zodiacal position.[487] Hence the coefficient for rising ($q_{\rm rise}$) is used for the rising Moon (KUR, ME, GI$_6$) and the rising Sun (KUR, ŠU$_2$, NA); the coefficient for setting

[485] This is a clear indication that No. **61** belongs to system A. The use of *ūmu*, lit. 'day', as a term for 10° in some Mss. remains unexplained.

[486] In *ACT* Neugebauer translated **nim** and **šu**$_2$ in the names of $q_{\rm rise}$ and $q_{\rm set}$ as 'sunrise' and 'sunset', thus assuming that these coefficients pertain to morning and evening, respectively. For the Full Moon intervals that interpretation is consistent only if one assumes, as Neugebauer does, that $q_{\odot}$ is computed from the position of the anti-Sun and not from the position of the Sun. However, the formulation of step 5 in No. **61** leaves no doubt that $q_{\odot}$ is computed from the position of the Sun ($B'_{\odot}$), and the same is implied by step 4. Moreover, in Neugebauer's interpretation the position of table T1.a ($q_{\rm set}$) after P1,P2 (setting Moon, rising Sun) and that of T2.a ($q_{\rm rise}$) after P4,P5 (rising Moon, setting Sun) would be the opposite of what one expects. In *HAMA* (pp. 535–7) Neugebauer appears to have abandoned his previous interpretation, without offering an improved translation.

[487] This suggests that the term 'corresponding *ṣiliptu*-coefficient' in the instruction for step 5 should be interpreted as a reference to the appropriate version of q, i.e. $q_{\rm rise}$ or $q_{\rm set}$.

(q_{set}) for the setting Moon (NA_1, $ŠU_2$, NA) and the setting Sun (NA_1, ME, GI_6).[488] No. **53** P17' includes a table with the same 12 values q_{10} pertaining to q_{set} – precisely the coefficient needed for NA_1. There may have been a corresponding table for q_{rise} in No. **53** P18' (KUR), but it is not preserved. The tabulated values of q_{10} imply an interpolation coefficient 0;48/3,0 = 0;0,16, in accordance with the procedure. The fact that q_{set} and q_{rise} are identical except for a shift of 6 zodiacal signs, and other qualitative features, are consistent with the astronomical interpretation of q_{set} and q_{rise} as conversion factors between zodiacal arcs and the associated setting or rising time. This interpretation is confirmed by a modern computation (Appendix E.1). In most procedure texts step 5 continues with an instruction to average $q_{☽}$ and $q_{☉}$:

ana muḫ-ḫi a-ḫa-miš₂ GAR.GAR-*ma* $\frac{1}{2}$-*šu₂-nu* GIŠ-*ma ana* **bar.nun gar**-*an*	You accumulate (them) together, compute half of it and put it down as the *ṣiliptu*-coefficient.	No. **61**.**A–F**
ana **ugu** *a-ḫa-miš₂ ša₂ sin u šam*[*aš₂* GAR].GAR-*ma* $\frac{1}{2}$-*šu₂* GIŠ	You accumulate (them) together for the Moon and the Sun and you compute half of it.	No. **53** P17'

Hence the final value of the coefficient 'for the zodiac' is

$$q' = (q_{☽} + q_{☉}) \cdot \frac{1}{2}. \tag{4.96}$$

The use of 'accumulating together' (*aḫāmiš kamāru*) reflects the symmetry of the operation (one value of q for the Moon and one for the Sun) and the loss of identity of the summands (§2.2). Even though the astronomical interpretation of $q_{☽}$ and $q_{☉}$, and the way in which q' is applied in step 9, imply that q' is the conversion factor between η and the corresponding rising or setting time, the averaging still requires additional clarification. In reality $q_{☽}$ and $q_{☉}$ are always very similar when they are averaged, because $q_{☽}$ and $q_{☉}$ are exactly the same at the moment of lunation due to the antisymmetry between q_{rise} and q_{set}.[489] At the Lunar Six events, the Moon and the Sun are slightly displaced from that fully symmetric situation, so that $q_{☽}$ and $q_{☉}$ are no longer exactly the same, but the difference is usually small. The averaging may be rationalised if one assumes that the 'coefficient for the zodiac' was considered to be changing continuously from $q_{☽}$ to $q_{☉}$ (or vice versa) in the course of the Lunar Six interval. In that interpretation, the averaging can be viewed as an attempt to define the optimal value of q' to be used in step 9 as the one pertaining to the middle of the arc η.[490]

4.4.19.6 Step 6. Moon's distance to the ecliptic at the Lunar Six event (β')

terminology	*šūqu*(**nim**) *u šuplu*(**sig**) *ša₂ Sîn* = "height and depth' of the Moon'
unit	degrees
astronomical meaning	distance of the Moon to the ecliptic at the Lunar Six event
procedures	No. **61**.**A–G** P1–P4

In step 6 the Moon's distance to the ecliptic at the Lunar Six event, say β', is computed from the value at the lunation, E.[491] This is used in step 8 for computing a correction to the Lunar Six interval. In some of the procedure texts step 6 begins with the instruction

nim *u* **sig** *ša₂ sin* **gar**-*an*	You put down the Moon's 'height and depth'.

This refers to E, the Moon's distance to the ecliptic at the lunation (§4.4.6), which is assumed to be known. However, β' turns out to be measured in degrees and not in barleycorns, the unit of E. The computation of β' therefore begins with a conversion, which is not mentioned in the procedures:

$$\beta = 0;0,50 \cdot E. \tag{4.97}$$

488 Apparently T1.a (q_{set}) follows P2 because the Moon is setting in P1 ($ŠU_2$) and P2 (NA); similarly T2.a (q_{rise}) follows P4 because the Moon is rising in P3 (ME) and P4 (GI_6).

489 This is because the New Moon rises or sets together with the Sun, and the Full Moon rises while the Sun sets, and vice versa.

490 These considerations also suggest that q' could be computed by a simpler algorithm than the one implied by the procedure texts, namely by evaluating $q_{☽}$ (or $q_{☉}$) at the lunation. According to a modern analysis the formula $R = q' \cdot \eta$ is a linear approximation involving the assumption that η is a small parameter, while the effect of evaluating $q_{☽}$ or $q_{☉}$ at the Lunar Six event instead of the lunation is quadratic in η. Hence the improvement resulting from the latter approach is formally negligible within the framework of this approximation (cf. also Appendix E.1). If both would be evaluated at the lunation then $q' = q_{☽} = q_{☉}$. The advantages are that (1) only one coefficient has to be determined, (2) there is no need for averaging and (3) step 4 becomes superfluous, because $q_{☽}$ is computed from $B_{☽}$. The procedure texts contain hints, but no unequivocal evidence for this approach. In No. **61**.**FG** the averaging appears to be omitted, but No. **61**.**G** P2 contains traces of step 4, and the averaging may have been mentioned in one of the gaps; cf. the critical commentary. Irrespective of that, the Lunar Six columns in some synodic tables may have been computed by the simplified method. This topic will be addressed in Volume II.

491 If necessary, distance above the ecliptic will be denoted as $\beta_{\uparrow}$, distance below the ecliptic as $\beta_{\downarrow}$.

The following instruction concerns a correction to β that accounts for the Moon's latitudinal displacement during the interval N:

zi *sin* GAM$_0$ 4 **du ki nim** *u* **sig** *ša*$_2$ *sin ki* **tab tab**|**la**$_2$ *ki* **la**$_2$ **la**$_2$|**tab**	You multiply the displacement of the Moon by 0;4 and you add|subtract it to|from the Moon's 'height and depth' if it is increasing, subtract|add if it is decreasing.

It is assumed that the Moon moves up or down at the rate 0;4° for every degree of displacement along the zodiac,[492] i.e.

$$\beta' = \beta \pm \Delta B_{☽} \cdot 0;4. \tag{4.98}$$

The additive or subtractive sense of the correction depends on two factors (Fig 4.34). First, it depends on whether β (i.e. E) is currently increasing or decreasing. This information is provided in column E by the terms 'ascending' (**la**$_2$) or 'descending' (**bur**$_3$): E increases if the Moon is above the ecliptic ('high') and ascending, or below the ecliptic ('low') and descending; E decreases if the opposite holds. Second, it depends on whether the Lunar Six event occurs before or after the lunation. This is not clearly expressed in the procedures, but is obvious from the following considerations: if the Lunar Six event occurs after the lunation (NA$_1$, and ideally NA and GI$_6$) the displacement must be added to β, but if it occurs before the lunation (KUR, and ideally ŠU$_2$ and ME), then the correction is oriented backwards in time, so that it must be subtracted. Stated more carefully in line with Akkadian usage, one should say that in the former case the sense of the displacement remains unchanged, while in the latter it is reversed, i.e. subtractive becomes additive, and vice versa.[493]

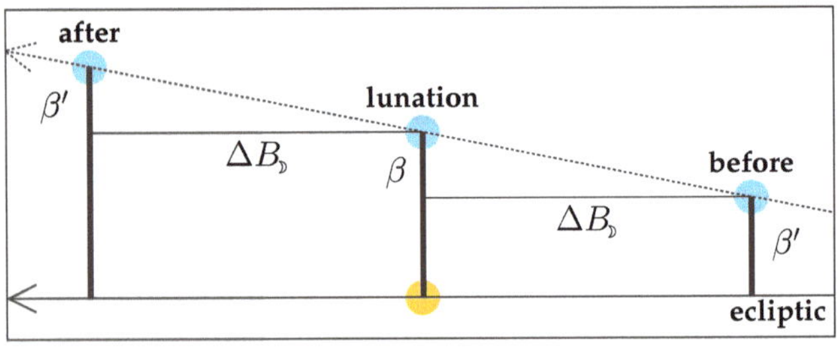

Fig. 4.34: Step 6. Schematic view of the sky showing the Moon (ascending and above the ecliptic) and the Sun, before, during and after a lunation. The disk at the ecliptic represents the Sun (New Moon case) or the anti-Sun, the disk diametrically opposite to the Sun (Full Moon case).

Compared to the algorithm for E (§4.4.6), the correction $\Delta B_{☽} \cdot 0;4$ amounts to a simplified treatment of the latitudinal motion. First, no distinction is made between the nodal region ($\beta \leq 2°$), where the algorithm for E implies a change of 0;6,40° per degree of nodal elongation, and outside of it, where this value is 0;3,20°. Furthermore, the motion of the nodes is ignored. Apparently this was considered sufficiently accurate for estimating latitudinal motion during the short interval N. The total latitudinal displacement over 90° of nodal elongation is $90 \cdot 0;4 = 6°$, which coincides with the maximum of E. In the procedure texts no consideration is given to the case when β' exceeds 6°, but it can be assumed that the usual reflection rule was applied in order to confine β' to values between 6° above and below the ecliptic.

4.4.19.7 Step 7. Coefficient 'for height and depth' (r')

terminology	*ṣiliptu*(**bar.nun**) *ša*$_2$ *šūqi*(**nim**) *u* *šupli*(**sig**) = '*ṣiliptu*-coefficient for 'height and depth'' = r, r' *ṣiliptu ša*$_2$ *šūqi u šupli ša*$_2$ *elî*(**nim**) = '*ṣiliptu*-coefficient for 'height and depth' for rising' = r_{rise} *ṣiliptu ša*$_2$ *šūqi u šupli ša*$_2$ *rabî*(**šu**$_2$) = '*ṣiliptu*-coefficient for 'height and depth' for setting' = r_{set}
unit	dimensionless
astronomical meaning	ratio between distance to ecliptic and associated rising (r_{rise}) or setting time (r_{set})
procedures	No. **53** P17',P18' (only traces); No. **61**.**A–G** P1–P4
auxiliary tables	r_{rise}: No. **61**.**ADG** P1–P4 r_{set}: No. **53** P17'; No. **61**.**ADEG** P1–P4

In step 7 the '*ṣiliptu*-coefficient for 'height and depth'', r', is computed from the positions of the Moon and the Sun. This coefficient is used in step 8 for computing a correction to the Lunar Six interval. Step 7 is formulated in exactly the same manner as step 5. In No. **61** P1–P4 step 7 sets out with the computation of $r_{☽}$ and $r_{\odot}$ using the same extrapolation

[492] The same uniform latitudinal motion is attested in the NMAT procedure text BM 41004 O8–9 (Brack-Bernsen & Hunger 2005/6, p. 103).

[493] This sign reversal of the latitudinal correction for KUR, ŠU$_2$ and ME was overlooked by Neugebauer in *ACT*, pp. 233–234.

rules as for $q_{☽}$ and $q_{\odot}$, except that the interpolation coefficient is 0;0,8. No. **61** includes the auxiliary tables T1.b and T2.b, each containing 12 control values r_{10} that form a zigzag function with extrema 0;18 and 0;42 (Fig. 4.35, top). Hence the interpolation coefficient is 0;24/3,0 = 0;0,8 as stated in the procedures. The version r_{set} (T1.b) applies to the setting Moon (NA_1, $ŠU_2$, NA) and the setting Sun (NA_1, ME, GI_6); the version r_{rise} applies to the rising Moon (KUR, ME, GI_6) and the rising Sun (KUR, $ŠU_2$, NA). Also No. **53** P17' contains a table with 12 values r_{10} forming a zigzag function for r_{set}. This function has the same mean value 0;30 as the one for $ŠU_2$, NA, ME and GI_6, but the extrema are different (0 and 1), so that the interpolation coefficient is 1/3,0 = 0;0,20 instead of 0;0,8. A similar table for r_{rise} is not preserved in No. **53** P17',P18' but is easily reconstructed by invoking the symmetry between r_{rise} and r_{set} (Fig. 4.35, bottom). Hence the complete set of interpolation rules can be represented as follows:

$$r = \begin{cases} r_{10} \pm (B' - B_{10}) \cdot 0;0,8 & (\text{ŠU}_2, \text{NA, ME, GI}_6) \\ r_{10} \pm (B' - B_{10}) \cdot 0;0,20 & (\text{NA}_1, \text{KUR}). \end{cases} \tag{4.99}$$

The qualitative features of r_{rise} and r_{set} are consistent with their astronomical interpretation as conversion factors between distance to the ecliptic and the associated rising (r_{rise}) or setting (r_{set}) time. A modern computation confirms this interpretation (Appendix E.1). At the lunation $r_{☽}$ is always equal to $r_{\odot}$, but at a Lunar Six event both are slightly different. Also r'

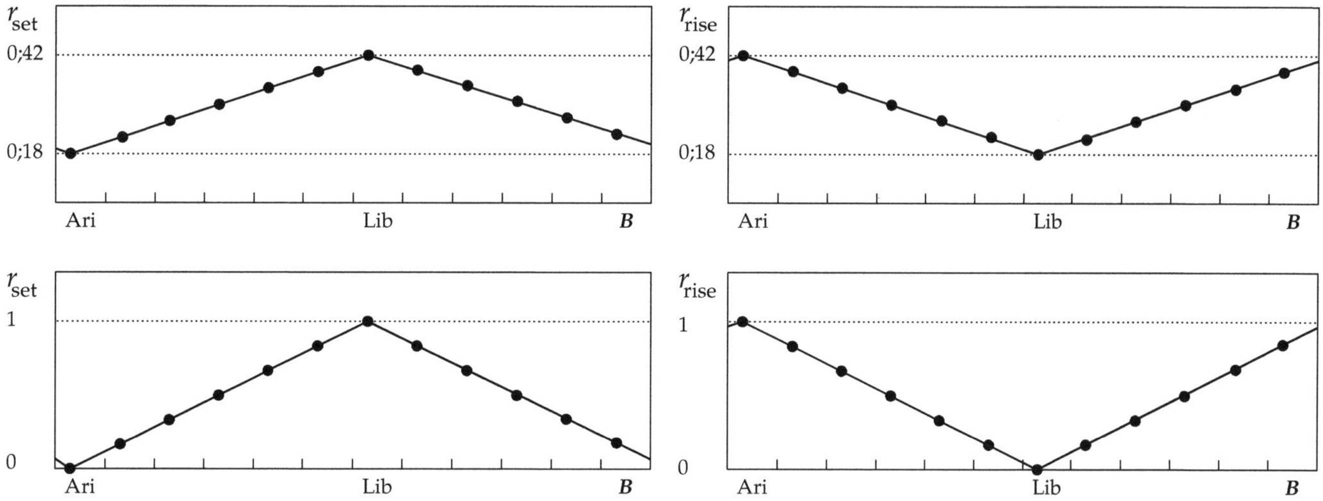

Fig. 4.35: Coefficients 'for height and depth for setting' (r_{set}) and 'for rising' (r_{rise}) against the zodiacal position of the Moon or the Sun. The dots are the control points. **Top**: for $ŠU_2$, NA, ME and GI_6. **Bottom**: for NA_1 and KUR.

is defined as the average of the values for the Moon and the Sun:

$$r' = (r_{☽} + r_{\odot}) \cdot \frac{1}{2}. \tag{4.100}$$

As I have argued in the case of q', the averaging of $r_{☽}$ and $r_{\odot}$ suggests that the Babylonian astronomers considered the 'coefficient for height and depth' to change continuously in the course of the Lunar Six interval. In that interpretation the averaging is an attempt to define the optimal value of r' to be used in step 8 as the one pertaining to the middle of the Lunar Six interval.[494] In No. **61.FG** the averaging is not preserved, but here both $r_{☽}$ and $r_{\odot}$ are computed, which implies that r' is defined as their average.[495]

[494] Neugebauer considered the averaging of $r_{☽}$ and $r_{\odot}$ to be meaningless (*ACT*, p. 234) because, in his view, it implies that the Sun's centre is not confined to the ecliptic. The latter would be inconsistent with other algorithms that clearly imply the Sun to be located at the ecliptic. For instance, in step 8 the Sun's distance to the ecliptic, if it existed, ought to be subtracted from $\beta'_{☽}$ and the difference multiplied by r', which is not done. However, the mere averaging of $r_{☽}$ and $r_{\odot}$ does not imply that the Sun is displaced from the ecliptic.

[495] Hence the extant procedures for step 7, like those for step 5, contain no evidence for the seemingly attractive approximation whereby the positions at the Lunar Six event would be replaced by those at the lunation, resulting in $r' = r_{☽} = r_{\odot}$.

4.4.19.8 Step 8. Contribution to Lunar Six resulting from Moon's distance to ecliptic (R)

terminology	DIŠ.UD *ša*$_2$ *šūqi*(**nim**) *u šupli*(**sig**) = 'DIŠ.UD of 'height and depth'' (only at FM; cf. below)
unit	time degrees
astronomical meaning	contribution to the Lunar Six resulting from the Moon's distance to the ecliptic
procedures	No. **53** P18'; No. **61.A–G** P1–P4

Step 8 is concerned with the contribution to the Lunar Six interval resulting from the Moon's distance to the ecliptic, R. The outcome is used in step 11. Three different formulations of step 8 are attested in the procedure texts:

<table>
<tr><td>a.ra$_2$|GAM$_0$|GAM DIŠ.UD ša$_2$ nim u sig ša$_2$ taš-kun tallak(du)</td><td>You multiply it (r') by the DIŠ.UD of 'height and depth' (β') which you had put down.</td><td>No. 61.A–F P1–P4</td></tr>
<tr><td>ša$_2$ t[ag$_4^?$ šu-u$_2$ a.ra$_2$] bar.nun egir-tu$_2$ ša$_2$ taš-kun du-ak</td><td>What rem[ains?, this] (β') you multiply [by] the last ṣiliptu-coefficient which you had put down.</td><td>No. 61.G P1</td></tr>
<tr><td>bar.nun GAM nim u sig [du ...]</td><td>[You multiply] the ṣiliptu-coefficient by the 'height and depth' [...].</td><td>No. 53 P18'</td></tr>
</table>

In No. **61.A–F** P1–P4 step 8 occurs immediately after step 7, which implies that the unnamed quantity ('it') is r' and 'DIŠ.UD of height and depth'[496] must here be interpreted as β'. This is peculiar, because elsewhere β' is called 'height and depth' and 'DIŠ.UD of height and depth' denotes R (cf. below). In No. **61.G** P1 'what remains' must be β'. Hence the computation of β' (step 6) comes immediately before step 8 here. Indeed, since r' is something 'which you had put down', it is not the outcome of the immediately preceding step. These considerations suggest that in No. **61.G** steps 6–8 are executed in the order 7,6,8.[497] In spite of the terminological difficulty with DIŠ.UD in No. **61.A–F** P1–P4, the algorithmic interpretation of step 8 is clear, and confirmed by the Lunar Six columns in the synodic tables. All variants express the same operation, which can be represented as

$$R = r' \cdot \beta'. \tag{4.101}$$

If necessary this relation can be specified for the case when the Moon is above the ecliptic or below it, i.e. $R_\uparrow = r' \cdot \beta'_\uparrow$ and $R_\downarrow = r' \cdot \beta'_\downarrow$. The astronomical interpretation of R is the setting or rising time of the arc β' (Fig. 4.36). This is fully consistent with the properties of r', as shown by a modern analysis (Appendix E.1). In the procedure texts R is not named within the body of step 8, but for an Akkadian term we may look to steps 11 and 12. In step 11 of No. **61.G** R is referred to as the 'height and depth which you had multiplied by the last *ṣiliptu*-coefficient'. In all instances of step 12 in No. **61.A–G** R is called the 'DIŠ.UD of height and depth', i.e. $R_\uparrow$ is the 'DIŠ.UD of height', and $R_\downarrow$ is the 'DIŠ.UD of depth'.[498]

4.4.19.9 Step 9. Contribution to the Lunar Six resulting from the lunar elongation (Q)

terminology	BE *u ṣiliptu*(**bar.nun**) = 'elongation 'and' the *ṣiliptu*-coefficient'
unit	time degrees
astronomical meaning	contribution to the Lunar Six interval resulting from the lunar elongation
procedures	No. **53** P17',18'; No. **61.A–G** P1–P4

Step 9 is concerned with the contribution to the Lunar Six interval resulting from the Moon's elongation from the (anti-)Sun. In the procedure texts this is formulated as follows:

<table>
<tr><td>BE GAM$_0$|GAM|a.ra$_2$ bar.nun du</td><td>You multiply the elongation (η) by the ṣiliptu-coefficient (q').</td><td>No. 61.A–G</td></tr>
<tr><td>a.ra$_2$ BE du</td><td>You multiply it (q') by the elongation (η).</td><td>No. 53 P17'</td></tr>
<tr><td>⌜nim⌝ 10 lu-maš gab-bi a.ra$_2$ / mi-ḫi-il-tu$_2$ ša$_2$ ⌜lu⌝-maš du</td><td>You multiply the (coefficient for) rising for 10 (°) for all zodiacal signs (q'?) by / the 'stroke' for the zodiac (η?).</td><td>No. 53 P18' (KUR)</td></tr>
</table>

The third variant deviates from the others. Its literal meaning is not fully understood,[499] but all seem to represent the same algorithm,

$$Q = q' \cdot \eta. \tag{4.102}$$

---

[496] The term DIŠ.UD is not used in No. **53** P17',P18'. For a possible Akkadian reading cf. below.

[497] Unfortunately the relevant lines are damaged in No. **61.G** P1,P2, and steps 7 and 8 are skipped in P3,P4, so this cannot be directly verified.

[498] The Akkadian reading of DIŠ.UD might be *ana ūmi*, literally 'to the daylight'. As a noun it has the female gender (cf. step 12 in No. **61** P1–P4). In the planetary texts the related terms *ana ūmi*(**me**) E|A, 'to rise to daylight', denote the acronychal rising (§3.1.2). It may therefore not be a coincidence that DIŠ.UD is used only in connection with the Lunar Six at Full Moon, when the Sun is close to opposition as is true for AR.

[499] The '(coefficient for) rising' may be an appropriate translation, since at KUR the Moon and the Sun are both rising.

The outcome Q is not explicitly named within step 9.[500] In steps 11–12 Q is usually referred to as the 'elongation and the *ṣiliptu*-coefficient' (BE *u* **bar.nun**),[501] sometimes simply 'elongation' (BE). Both can be viewed as abbreviations of 'the elongation which you had multiplied by the *ṣiliptu*-coefficient' (No. **61.G**). A geometric interpretation of Q is shown in Fig. 4.36.

4.4.19.10 Step 10. Disk correction (Δ)

terminology	2 [*ša*₂] ḪAB-*rat* = '2 for the disk'
unit	time degrees
astronomical meaning	correction to the Lunar Six interval (cf. below)
procedures:	
for NA_1	No. **53** P17'
for ME	No. **61.ABDG** P3
for GI_6	No. **61.ADG** P4

Step 10 concerns the 'disk correction' Δ which is applied to some Lunar Six intervals.[502] In the procedure texts this is formulated as follows:

2 ḪAB-*rat ana*|*itti*(**ki**) BE *u* **bar.nun tab**	You add the 2 for the disk to|with the elongation and the *ṣiliptu*-coefficient.	No. **61.ABDG** P3
2 ḪAB-*rat ina*(**ta**) BE *u* **bar.nun** *tanaḫḫis*(**la$_2$**)	You subtract the 2 for the disk from the elongation and the *ṣiliptu*-coefficient.	No. **61.ADG** P4
2 *ša$_2$* ḪAB-*rat* **ta na u$_4$**.1.**kam** *tanassaḫ*(**zi**)	You 'tear out' the 2 for the disk from NA of day 1 (= NA_1).	No. **53** P17'

Step 10 is absent from all duplicates of No. **61** P1 ($ŠU_2$) and P2 (NA), and from No. **53** P18' (KUR). The procedure texts therefore suggest that Δ is defined as follows:

$$\Delta = \begin{cases} -2 & (NA_1, GI_6) \\ +2 & (ME) \\ 0 & (KUR, ŠU_2, NA). \end{cases} \tag{4.103}$$

The Akkadian reading of ḪAB-*rat*, which probably means 'disk', is unknown.[503] Other than the qualification 'of the disk' the procedure texts do not contain any explicit references to the purpose of the correction. One likely goal, as suggested by the word 'disk', is to compensate for the fact that the Lunar Six intervals are defined in terms of the rims of the lunar and solar disks, and not their centres.[504] In the case of NA_1 and KUR simple observational considerations are sufficient to conclude that both are necessarily delimited by the moments when the lower rim of the Moon and the upper rim of the Sun cross the horizon (§4.1; Fig. 4.36). One would therefore expect identical subtractive corrections for NA_1 and KUR. While Δ is indeed subtractive for NA_1, the correction for KUR is zero. Moreover, 2° is much larger than what one expects if the purpose is to correct for the rim effect.[505] Lunar Six records in the observational texts (astronomical diaries and Lunar Six tablets) imply that they were measured to an accuracy of about 0;10–0;30°, which makes it difficult to understand how the rising time of the lunar or solar disk could be overestimated by such a large amount. Hence neither the absence of a disk correction for KUR, nor the magnitude of the disk correction for NA_1 can be explained in terms of rim definitions alone.

This also applies to Δ for $ŠU_2$, NA, ME and GI_6. Unlike KUR and NA_1 one cannot argue on observational grounds by which rims of the lunar and solar disks they were defined (§4.1). For $ŠU_2$, NA, ME and GI_6 one can always find a particular combination of rims for which Δ becomes additive, subtractive or zero, in accordance with the value attested for a given Lunar Six. If these rims conform with a single rule this suggests that they are the ones that were actually

[500] In No. **53** P17' (Rii11) Q is called NA_1, perhaps in anticipation of the final result $NA_1 = Q \pm R$.

[501] Hence the particle *u* 'and' replaces multiplication here.

[502] The position of step 10 is not the same in all texts. Between steps 9 and 11: No. **61.DG** P3,P4; after step 11: No. **61.AB** P3. A peculiar case is No. **61.A** P4, where the disk correction is mentioned only after step 13. In No. **61.C** P3 step 10 is omitted, probably by mistake (this Ms. contains numerous other scribal errors).

[503] Cf. *ACT*, pp. 197–198, to which one can add BM 41004 O9: *a-na* 1 **danna** 2-*ta* ḪAB-*ra-a-ta ki-i* **nim** *u ki i-šap-⌈pi⌉*, 'for 1 *bēru* (= 30° along the zodiac) 2 disks when it ascends and descends'. As pointed out by Neugebauer & Sachs (1967), p. 204, and Brack-Bernsen & Hunger (2005/6), the passage 6 **kuš$_3$ dagal** *ma-lak*, '6^c is the width of the path', in O8 implies that the Moon goes up and down by 6^c = 12° while proceeding by 180° along the zodiac, i.e. 1^c = 2° per 30°, so that 1 'disk' corresponds to 0;30^c = 1°. Accordingly, O11 mentions that the Moon's size is 12^f (= 0;30^c).

[504] Cf. *ACT*, pp. 206, 237–239.

[505] The correct magnitude of the rim effect is given by the rising time (or setting time) of the sum of the lunar and solar radii (0.25° each), which is about 0.7° (Appendix E.3). Rounded to the nearest degree this would be 1°. Even if one assumes that the Babylonian value of these radii is 0.5°, the correction would be 1.4°, considerably less than 2°.

used by the Babylonian astronomers. However, such a rule does not appear to exist, neither if one classifies the rims with respect to the horizon (upper/lower rim), nor if one does so with respect to the daily rotation of the sky (leading/trailing rim). For a detailed analysis of this issue cf. Appendix E.3.

It therefore appears that the Babylonian astronomers did not formulate the disk correction on the basis of geometric principles,[506] and that it must incorporate other factors besides the rim effect. More likely they derived it in a heuristic manner by evaluating deviations between predicted Lunar Sixes and empirical data. These data were readily available in astronomical diaries and Lunar Six tablets, records of Lunar Six values excerpted from the diaries.[507] On that assumption Δ is a composite correction incorporating all effects that contribute to these deviations. A preliminary analysis (Appendix E.3) indicates that the mentioned rim definitions for NA_1 and KUR, if combined with the effect of parallax,[508] result in a total correction for NA_1 with the correct order of magnitude and subtractive sense (about $-2°$). The resulting correction for KUR is only $-1°$, which might explain why the Babylonian astronomers seem to have ignored it. Also for $ŠU_2$, NA, ME and GI_6 there does not appear to be a combination of rim definitions and other effects that exactly reproduces the attested values of Δ to the nearest degree. Closest agreement is obtained if one uses the front rim of the Moon and the back rim of the Sun with respect to the daily rotation of the sky. It might be significant that this choice of rims for $ŠU_2$, NA, ME and GI_6 coincides with the one for NA_1, but not with the one for KUR.[509] However, the use of a single value $|\Delta| = 2°$ (apart from 0) may suggest that analogies between different Lunar Six intervals or other non-heuristic considerations influenced the formulation of the disk correction.[510]

4.4.19.11 Step 11. Assembling the Lunar Six interval

procedures | No. **53** P17',P18'; No. **61.A–G** P1–P4

In step 11 the Lunar Six interval is assembled from the contributions Q (step 9), R (step 8), and a possible disk correction Δ (step 10). In the procedures for KUR, NA_1, NA and ME this is formulated as follows:

[*ki*] **nim** ⌜**ki** BE **tab** *ki*⌝ **sig ta** BE **zi**-*aḫ*	[If] (the Moon) is 'high' you add it ($R_\uparrow$) with the elongation (Q). If it is 'low' you 'tear' it ($R_\downarrow$) out of the elongation.	No. **53** P17'
ki-i **nim**|**la₂** *ana*|**ki** BE [*u* **bar.nun**] **tab**	If (the Moon) is 'high' you add it to the elongation [and the *ṣiliptu*-coefficient].	No. **61.A–F** P2,P3
ki-i **sig** *ana*|*ina*(**ta**) BE [*u* **bar.nun**] *tumaṭṭa*(**la₂**)|*tanaḫḫis*(**la₂**)	If (the Moon) is 'low' you diminish the elongation [and the *ṣiliptu*-coefficient] by it ($R_\downarrow$)	
ki-i **nim nim** *u lu-u₂* **nim sig** *šu-u₂* **ki** BE **bar.nun** *u* 2 ḪAB-*rat* **tab**	If (the Moon) is 'high', ascending, or 'high', descending, you add this to the elongation, the *ṣiliptu*-coefficient and the 2 for the disk.	No. **61.G** P3
ki-i **bur₃ bur₃** *u lu-u₂* **bur₃ la₂** *šu-u₂* **ta** BE **bar.nun** *u* 2 ḪAB-*rat* **la₂**	If (the Moon) is 'low', descending, or 'low', ascending, you subtract this from the elongation, the *ṣiliptu*-coefficient and the 2 for the disk.	

The corresponding instructions in the procedures for $ŠU_2$ and GI_6 are:

[506] This point was already stressed by Pannekoek (1941), p. 5.

[507] For critical editions of the astronomical diaries cf. *ADRT* I–III; for the Lunar Six tablets cf. *ADRT* V.

[508] Refraction has no effect on NA_1 and KUR since the Moon and the Sun, being close to the conjunction, are subject to nearly identical vertical displacements of their apparent positions.

[509] NA_1 and KUR are both defined by the lower rim of the Moon and the upper rim of the Sun, but for NA_1 the lower rim corresponds to the front rim of the Moon and the back rim of the Sun; for KUR the reverse holds.

[510] Perhaps the 2° was initially established for $ŠU_2$, NA, ME and GI_6, since the Full Moon is more easily observable, and then adopted for the observationally more difficult NA_1. However, in *ACT*, p. 198, Neugebauer proposes the opposite, namely that the disk corrections for $ŠU_2$, NA, ME and GI_6 were constructed in analogy to that for NA_1. Since refraction does not affect NA_1 he concludes that it was also ignored in the disk corrections for $ŠU_2$, NA, ME and GI_6.

ki-i **nim** *ana*|**ta** BE [*u* **bar.nun**] **la**$_2$ *ki-i* **sig** *ana* BE [*u* **bar.nun**] **tab**	If (the Moon) is 'high' you diminish the elongation [and the *ṣiliptu*-coefficient] by it ($R_\uparrow$). If it is 'low' you add it ($R_\downarrow$) to the elongation [and the *ṣiliptu*-coefficient].	No. **61.ADF** P1, No. **61.AD** P4
ki-i **nim nim** *u lu-u*$_2$ **nim sig** *šu-u*$_2$ **ta** BE {*ša*$_2$ **a.ra**$_2$ **bar.nun** *tal-lik*|**bar.nun** *u* 2 ḪAB-*rat*} **la**$_2$	If (the Moon) is 'high', ascending, or 'high', descending, you subtract this from the elongation {which you had multiplied by the *ṣiliptu*-coefficient | the *ṣiliptu*-coefficient and the 2 for the disk}.	No. **61.G** P1,P4
ki-i **bur**$_3$ **bur**$_3$ *u lu-u*$_2$ **bur**$_3$ **la**$_2$ *šu-u*$_2$ **ki** BE {*ša*$_2$ **a.ra**$_2$ **bar.nun** *tal-lik*|**bar.nun** *u* 2 ḪAB-*rat*} **tab**	If it is 'low', descending, or 'low', ascending, you add this to the elongation {which you had multiplied by the *ṣiliptu*-coefficient | the *ṣiliptu*-coefficient and the 2 for the disk}.	

The formulaic representation of these instructions is as follows:

$$\text{KUR, NA}_1\text{, NA, ME} = \begin{cases} Q + R_\uparrow + \Delta & \text{(Moon above the ecliptic)} \\ Q - R_\downarrow + \Delta & \text{(Moon below the ecliptic)} \end{cases} \tag{4.104}$$

$$\text{ŠU}_2\text{, GI}_6 = \begin{cases} Q - R_\uparrow + \Delta & \text{(Moon above the ecliptic)} \\ Q + R_\downarrow + \Delta & \text{(Moon below the ecliptic).} \end{cases} \tag{4.105}$$

Depending on whether the Moon is above or below the ecliptic, R is added to Q or subtracted from it, as can be readily understood from the geometric interpretation (Fig. 4.36), which goes back to Schaumberger (1935) and Neugebauer (*ACT*). No. **53** P18' (KUR) contains at best an abbreviated representation of the complete instruction. Not only do we have to interpret 'height and depth' as R instead of the Moon's distance to the ecliptic, but one expects $R_\uparrow$ to be added to Q and $R_\downarrow$ to be subtracted and not both to be subtracted. In No. **53** P17' (NA_1) step 11 is alluded to by several opaque references; cf. the commentary. In No. **61.G** the conditions for addition and subtraction of R are formulated in an elaborate, but equivalent manner.[511]

In case step 11 involves a subtraction of R this is factually preceded by a comparison of R and $Q+\Delta$, because subtractions are possible in the Babylonian sense only if the result is positive, which requires $R < Q+\Delta$. This is not made explicit in the procedures,[512] but for the Full Moon Lunar Sixes the opposite case is considered in step 12. At the end of step 11 the result is named (L6 stands for the name of a Lunar Six interval):

ana L6 **me**|*e-bi*	You call it L6.	No. **61.A–F** P1
ana L6 **gar**-*an*	You put it down as L6	No. **53** P17', No. **61.G** P1

Thresholds for NA_1 and KUR

The values of NA_1 and KUR obtained in step 11 are final if they exceed a threshold above which the lunar crescent is considered to be visible. These thresholds are not mentioned in any text. Neugebauer inferred from the synodic tables that the actual conditions that NA_1 and KUR must satisfy also involve the elongation η, such that[513]

$$\text{NA}_1 + \eta > m_{\text{NA}_1} \approx 21^\circ, \tag{4.106}$$

$$\text{KUR} + \eta > m_{\text{KUR}} \approx 18\text{–}22^\circ. \tag{4.107}$$

If NA_1 or KUR does not satisfy the relevant condition then step 13 is executed, and the Lunar Six module is repeated for the next day (NA_1) or the previous day (KUR). When the final value NA_1 has been established, the length of the month whose end is signified by NA_1 is known. With that information the date of the next lunation, to be written in the next row of column M, is determined by the rules laid out in §4.4.18.

4.4.19.12 Step 12. Optional renaming of $ŠU_2$, NA, ME or GI_6

procedures | No. **61.A–G** P1–P4

Step 12 deals with the case when the subtraction of R in step 11 is impossible because $R > Q+\Delta$. As can be seen in Fig. 4.36, this signifies that the Moon is far enough below (KUR, NA_1, NA, ME) or above ($ŠU_2$, GI_6) the ecliptic for

[511] Perhaps the scribe of No. **61.G** intended to stay as close as possible to the terminology used in column E, where the numerical value of E is followed by the same pair 'high/low' and 'ascending/descending', even though the latter play no role within step 11.

[512] One would expect a conditional clause similar to 'if the elongation and the *ṣiliptu*-coefficient [and the 2 for the disk] exceeds the DIŠ.UD of 'height and depth'... ($Q+\Delta > R$).

[513] *ACT*, p. 67; *HAMA*, pp. 539–540. A new analysis of m_{NA_1} and m_{KUR} will appear in Volume II.

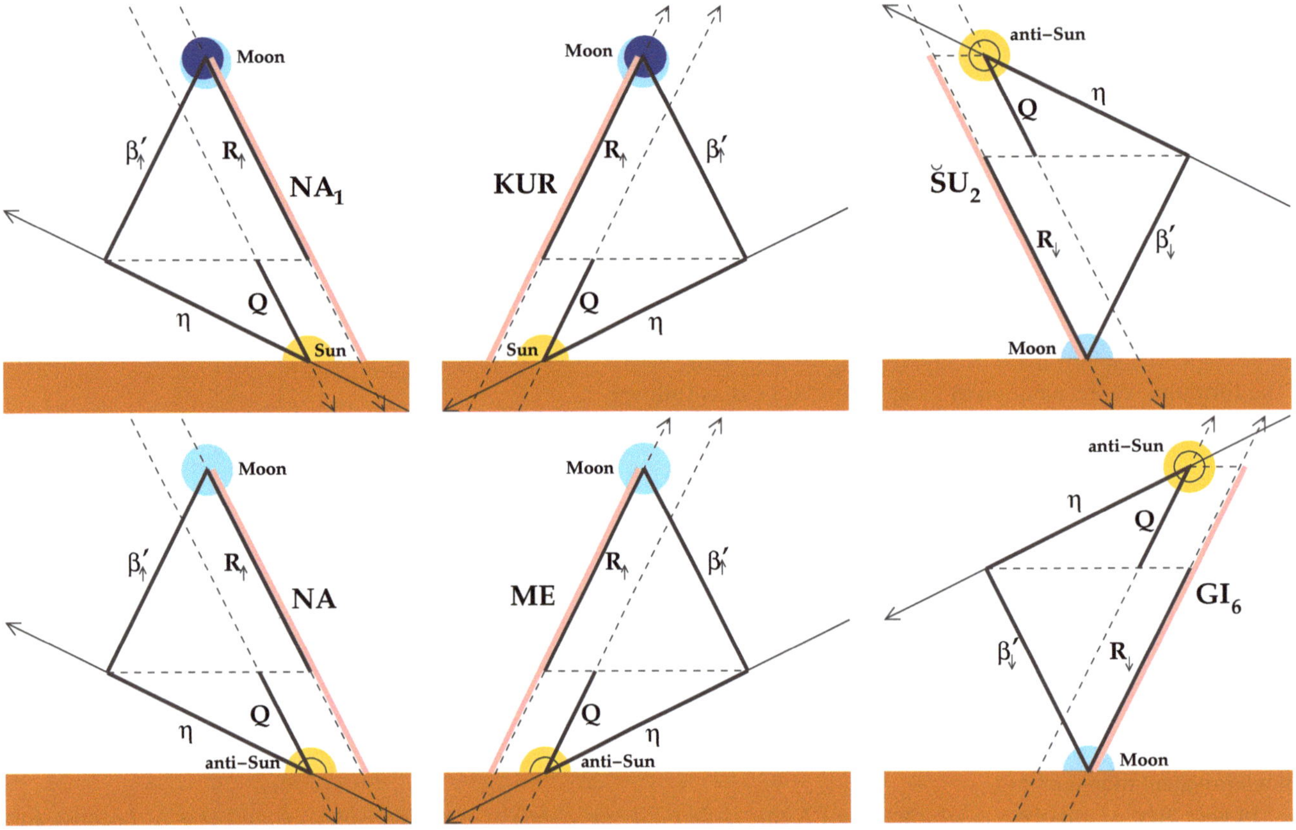

Fig. 4.36: Geometric interpretation of the computation of the Lunar Six intervals. The full arrow represents the ecliptic and the direction of lunar and solar motion along it; dashed arrows indicate the direction of setting or rising parallel to the celestial equator; horizontal dashed lines are parallel to the horizon. In all cases shown here the Lunar Six intervals are provisionally defined in terms of disk centres, and they are obtained as the sum of Q and R. The alternative cases where they are obtained as $Q-R$ correspond to the Moon being on the other side of the ecliptic.

the order in which the Moon and the Sun cross the horizon to be the reverse of what was initially assumed. The Full Moon intervals form two pairs, such that $\check{S}U_2$ and NA occur at consecutive sunrises with the opposite order of sunrise and moonset, and ME and GI_6 occur at consecutive sunsets with the opposite order of sunset and moonrise. Hence $R > Q + \Delta$ signifies that what was initially assumed to be NA is actually $\check{S}U_2$, or what was assumed to be ME is GI_6, or vice versa. In the case of KUR and NA_1 step 12 is absent, because their 'opposite' Lunar Six events are not defined (cf. §4.1). Moreover, for the marginal situations where a reversal might occur when computing KUR or NA_1, the Moon's elongation from the Sun (η) is necessarily small, so that the lunar crescent is too narrow to be visible anyway. On the day of KUR or NA_1 η is therefore always large enough for the initially assumed order of the horizon crossings to remain valid throughout the computation. In the procedures for $\check{S}U_2$, NA, ME and GI_6 step 12 begins with the mentioned condition $R > Q + \Delta$:

ki-i DIŠ.UD *ša*₂ **nim** *al-la* BE *u* **bar.nun** [*u* 2 ḪAB-*rat*] **diri**-*at* ...	If the DIŠ.UD of 'height' exceeds the elongation and the *ṣiliptu*-coefficient [and the 2 for the disk] ...	No. **61.ADFG** P1, **AD** P2!,P4, **B** P3!
ki-i DIŠ.UD *ša*₂ **sig|bur**₃ *al-la* BE *u* **bar.nun** *u* 2 ḪAB-*rat* **diri**-*at* ...	If the DIŠ.UD of 'depth' exceeds the elongation and the *ṣiliptu*-coefficient and the 2 for the disk ...	No. **61.ADG** P3

(The algorithmic interpretation implies that 'height' in No. **61.AD** P2 and **B** P3 is an error for 'depth'.) These conditions are followed by instructions that need not be quoted here, in which the subtraction is carried out in the opposite sense compared to step 11, i.e. $R-(Q+\Delta)$. The outcome is then renamed from $\check{S}U_2$, NA, ME or GI_6 (say $L6_1$) to the corresponding opposite interval NA, $\check{S}U_2$, GI_6 or ME (say $L6_2$):

ana L6₂ *tašakkan*(**gar**-*an*)	You put it down as L6₂.	No. **61.A** P1
[L6₁] *ana* L6₂ *taqabbi*(**e**-*bi*)	You call [L6₁] L6₂.	No. **61.ABDG** P1–P4
L6₁ *ana* L6₂ *itâr*(**gur**)	L6₁ becomes L6₂	No. **61.DG** P1–P4

The underlying algorithms can be represented as follows:

$$
\begin{aligned}
&R_\uparrow > Q &&\Rightarrow \text{NA} &&= R_\uparrow - Q &&\text{(Moon above the ecliptic)}\\
&R_\downarrow > Q &&\Rightarrow \check{\text{S}}\text{U}_2 &&= R_\downarrow - Q &&\text{(Moon below the ecliptic)}\\
&R_\downarrow > Q+2 &&\Rightarrow \text{GI}_6 &&= R_\downarrow - (Q+2) &&\text{(Moon below the ecliptic)}\\
&R_\uparrow > Q-2 &&\Rightarrow \text{ME} &&= R_\uparrow - (Q-2) &&\text{(Moon above the ecliptic).}
\end{aligned}
\tag{4.108}
$$

Finally the date of the Lunar Six event is established by the trivial instruction

ūma(**me**) *lā*(**nu**) *tušanni*(**man**-*ni*)	You do not change the day.	No. **61**.**ADG**
ūmu(**me**) *ina ūmišu*(**me**-*šu*$_2$)	The day (falls) on the same day.	No. **61**.**B**

This merely affirms that the renamed Lunar Six occurs on the same day established in step 1 for the initially assumed event ($d_{\check{S}U_2} = d_{NA}$ etc.).[514]

Possible complications with GI$_6$

When evaluating the condition for renaming GI$_6$ to ME, $R_\uparrow > Q-2$, a complication arises if $Q < 2$, again because Babylonian calculus does not know the concept of negative numbers. No special consideration appears to be given to that case in No. **61**.**DG**, but uniquely in No. **61**.**A** it is taken care of in a separate instruction (step 12') where $2-Q+R_\uparrow$ or $2-Q-R_\downarrow$ is computed, depending on whether the Moon is above or below the ecliptic, and the outcome is called ME. In the latter case, where $R_\downarrow$ must be subtracted from $2-Q$, another complication arises if $R_\downarrow > 2-Q$. This signifies that the Moon and the Sun cross the horizon in the originally assumed and correct order for GI$_6$ after all. The appropriate procedure would be to subtract $2-Q$ from $R_\downarrow$ and name the outcome GI$_6$, but this is ignored even in No. **61**.**A**.[515]

What if the Lunar Six value is too large?

Apart from the 'normal' case, and the case where subtraction of R is impossible, there is a third case, namely when addition of $Q+\Delta$ and R yields a Lunar Six value that is too large, in the sense that the correct date is 1^d earlier (NA, GI$_6$) or later (ŠU$_2$, ME) than initially assumed. This may happen if the Moon is far above (NA, ME) or below (ŠU$_2$, GI$_6$) the ecliptic. At first sight the Lunar Six modules do not appear to take this into consideration. Indeed no mention is made of it when R and $Q+\Delta$ are added, but it is in fact covered by the case when R is subtracted from $Q+\Delta$, because a shift of 1^d in the mentioned direction produces a date that coincides with the one initially assumed for the opposite Lunar Six event. For instance, NA is advanced by 1^d to the date originally assumed for ŠU$_2$, as illustrated in Figs. E.3 and E.4 (p. 538), second row on the left. In other words, even though this possibility remains unnoticed within the procedure for the initially assumed Lunar Six, it is taken care of in step 12 of the procedure for the opposite Lunar Six. The texts do not mention which of both computations then produces the correct result, but the Babylonian astronomers were obviously not confused. They knew that whenever step 12 is invoked, that computation yields the correct Lunar Six value, because the associated date is the latest (ŠU$_2$, ME) or the earliest (NA, GI$_6$), in agreement with the definition of the Lunar Six interval (Table 4.1).

This has implications for the Lunar Six values obtained by executing the four modules in the sequence ŠU$_2$, NA, ME and GI$_6$ (No. **61** P1–P4). If step 12 is invoked in P1 (ŠU$_2$) or P3 (ME), which results in a value for NA or GI$_6$, then P2 (NA) or P4 (GI$_6$) can be skipped. If step 12 is invoked in P2 (NA) or P4 (GI$_6$), which results in a value for ŠU$_2$ or ME, then P1 or P3 has already been carried out, but its outcome is discarded.

Implications for the temporal order of ŠU$_2$, NA, ME and GI$_6$

The mutations of ŠU$_2$, NA, ME and GI$_6$ in step 12 are treated as if they are independent from one another, except that ŠU$_2$ and NA always occur at consecutive sunrises and ME and GI$_6$ at consecutive sunsets. However, there are hidden constraints that reduce the set of possible configurations to the ones shown in Figs. E.3 and E.4 (p. 538). One results from the fact that a Lunar Six event is renamed only if it occurs sufficiently close to Full Moon, so that η is small enough for R to dominate over Q. It therefore happens only rarely that two Lunar Six events are renamed near the same Full Moon, and some combinations are prevented altogether. Secondly, the Moon's distance to the ecliptic changes by at most a few degrees in the course of 2–3^d. For that reason configurations of Lunar Sixes that would require the Moon to be on opposite sides of the ecliptic near the same Full Moon are also virtually excluded.

[514] The translation 'to change' for *šanû* D was suggested by H. Hunger (*ACT*: you do not count an (extra) day).

[515] It seems unlikely that the gap of about 4 signs at the end of No. **61**.**A**.R24 would have contained the relevant instructions.

4.4.19.13 Step 13. Optional shift of the Lunar Six event by 1^d

procedures | No. **53** P17'; No. **61.A–G** P1–P4

Modifying N

The shift of 1^d in the mentioned direction always corresponds to an increase of N by 6,0°, irrespective of the Lunar Six (cf. the definition of N in step 1). After modifying N, steps 1–11 must be repeated, which is sometimes explicitly stated in the form of an instruction 'you do as before'. In No. **61.G** the backward shift of the date of $ŠU_2$ is uniquely accompanied by the statement '$ŠU_2$ turns backward' (**šu₂** *ana arkišu inaḫḫis*). The condition for step 13 assumes a different form for the New Moon intervals and the Full Moon intervals.

NA_1 and KUR

No. **53** P17', which concerns NA_1, states that 'if the Moon is insufficient for being visible' one has to 'add the Moon's displacement for 1^d to it ($B_☽$) and declare (the immediately preceding month) full'. The fullness of the month is thus inferred from the insufficiency of NA_1. There is no instruction to execute steps 1–11 again for the next day in order to compute the new value of NA_1 and confirm that it exceeds m_{NA_1}. This contrasts with column NA_1 in the synodic tables which sometimes includes two values of NA_1 corresponding to successive sunsets.

$ŠU_2$, NA, ME and GI_6

For the Full Moon Lunar Six intervals the execution of step 12 implies that the Lunar Six event that was originally assumed to occur is advanced ($ŠU_2$, ME) or delayed (NA, GI_6) by 1^d in the direction away from Full Moon. Since η changes by about 12° per day it is certain that η is large enough for the Moon and the Sun to cross the horizon in the expected order on that day. In No. **61** the new Lunar Six value is once more provided with the name of the Lunar Six that was announced before step 1. The procedure texts successfully cope with all resulting configurations of Full Moon Lunar Sixes (Appendix E.4, p. 537), and they do so in a surprisingly brief and efficient manner.

4.5 System B

About 40% of the tabular texts and 10% of the procedure texts concerning the Moon belong to system B. Lunar system B is therefore only slightly less well represented than lunar system A by tabular texts, but more sparsely by procedure texts which, moreover, do not cover all algorithms of lunar system B. The reconstruction of lunar system B has therefore always relied more strongly on the tabular texts than is the case with system A. The synodic tables have a similar arrangement of columns as in system A, with some notable differences. Several functions of the Lunar Six module that are intermediate between M, the time of the lunation, and the Lunar Six intervals, are represented by columns, a feature unknown from system A. Second, the tablets from Uruk include numerous template tables, which are synodic tables containing only a selection of columns.

The reconstruction of lunar system B began with Epping & Strassmaier (1881), the very first publication about Babylonian mathematical astronomy, in which J. Epping correctly interpreted columns K and L. The algorithms for F, G, H and J were dealt with by Epping in two subsequent publications (1889, 1890).[516] Kugler successfully tackled most of the preceding columns in his *Babylonische Mondrechung* (1900). Several decades later Kugler (1924) also established the astronomical meaning of columns M, N, O, Q and R, and reconstructed some of the underlying algorithms. Essential features of the Lunar Six module, the most challenging part of lunar system B, were reconstructed and interpreted by Schaumberger (1935). Some aspects of lunar system B were (more fully) clarified by Schnabel, Neugebauer, Britton and others.

4.5.1 *Composition of the tablets*

Lunar system B is represented by five procedure texts.[517] Most tablets contain a collection of procedures, but they have little in common in terms of subject matter. No. **93** contains largely practically oriented algorithms for computing or updating various functions including C (or D?), R, the date of the lunation, and M or N. No. **97** contains algorithms

[516] Epping's results are entirely based on fragments of the synodic tables *ACT* 120 and 122.

[517] As far as their original sizes can be estimated they are of types C or F.

concerning F, A, G, E (?), H and J, many of which are not fully understood. No. **95** mainly deals with period relations, but also with functions A, B, and eclipse magnitude. No. **96** is more narrowly focussed on eclipse magnitude, apart from one procedure for computing G from F and vice versa. On two tablets, procedures are combined with tables: No. **98** (perhaps belonging to system B) accompanies a synodic table, and No. **94** contains a daily motion table with traces of a corresponding procedure.

4.5.2 Algorithms for lunar system B

The algorithms for lunar system B fall into three categories: (i) algorithms for computing or updating the columns of a synodic table, template table or eclipse table; (ii) the same for a daily motion table; (iii) other algorithms.

4.5.2.1 Algorithms for synodic tables, template tables and eclipse tables

The synodic tables for lunar system B contain up to 21 columns, each providing successive values of a certain function, sometimes also other information pertaining to that function. Since the content of these tables is less standardised than the ones for system A, the column number of a given function often varies from table to table, but their sequence is usually the same.[518] The leftmost column, T, mentions the year and the month name. All data in one line (row) pertain to the same month (lunation). Each tablet usually covers all lunations of one calendar year, such that New Moons are placed on the obverse, and Full Moons on the reverse.[519] For the corresponding flow chart cf. Fig. 4.37. By executing the flow chart, all functions are updated to the next lunation. Most functions are auxiliary and serve as input for other functions. The purpose of the synodic tables is defined by the 'final' functions, which are those located at the end of a branch of the flow chart: Ψ' or Ψ'' (eclipse magnitude), M (time of the lunation) and the Lunar Six intervals. For system B there are numerous template tables; they contain a selection of the columns of a synodic table. Eclipse tables are excerpts from synodic tables in which successive lines are separated by 5 or 6 months corresponding to eclipse possibilities.

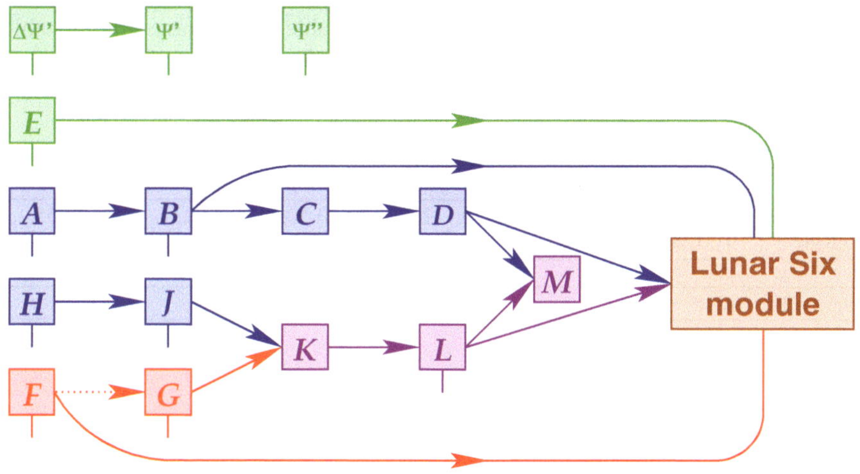

Fig. 4.37: Flow chart for computing a synodic table in accordance with lunar system B. Each box represents an algorithm for updating a function (from a previous value) or computing a function (from other functions). The dotted lines indicate that G may be computed either from its own zigzag function, or from F. Some functions depend on an initial value, which is indicated by a vertical line.

4.5.2.2 Algorithms for the daily motion tables

In the daily motion tables of lunar system B successive lines pertain to successive real days, not tithis as in system A. They also have a much simpler structure than in system A. Those for the Moon[520] contain at most three columns: T^* (the date), F^* (daily displacement along the zodiac) and $B^*_{☽}$ (zodiacal position); those for the Sun[521] contain only columns T^* and $B^*_{\odot}$. The main reason for assigning these tables to system B is that the mean values (μ) of F^* agree with mean values of F known from system B. There is no evidence that other functions of lunar system B were evaluated at successive days.

---

[518] The most complete known sequence of columns, preserved in *ACT* 120, a synodic table for New Moon, is T, A, B, C, D, $\Delta\Psi'$, Ψ', F' F, G, H, J, K, L, M, O, Q, R, NA_1, KUR.

[519] For technical reasons the tablets usually begin with the last month (XII or XII_2) of the preceding year, so that there are 1+12 (normal year) or 1+13 (intercalary year) months on each side.

[520] *ACT* 190–196, 194b, 194c.

[521] *ACT* 185–187.

4.5.2.3 Other algorithms

The procedure texts contain several other algorithms, which can be grouped together in the following categories: (i) procedures for net differences, $df(s)$, including net displacements $dB(s)$; (ii) procedures relating function values pertaining to New Moon to function values pertaining to Full Moon. Categories i and ii could be used for verification. (iii) Procedures concerned with the period relation for a certain function. (iv) Procedures that do not appear to be directly connected with the production of tables, or whose purpose is unclear. Some contain elaborate and sophisticated computations.

4.5.3 Synodic arc of the Moon and the Sun (A)

terminology	?
unit	degrees
astronomical meaning	synodic arc of the Moon and the Sun
column	ii (Babylon)
procedures:	
parameters of zigzag function	$A^{(1)}$: No. **95** P5'; $A^{(2)}$: No. **97** P4'.b; $A^{(3)}$: No. **97** P3'.b,4'.a
unclear	No. **95** P4'

Function A models the synodic arc (σ) of the Moon and the Sun, i.e. their common net displacement along the zodiac between successive lunations.[522] The synodic arc for the current month, A_i, represents the net displacement between the previous lunation ($i-1$) and the current one. In the tabular texts from Babylon, column A is placed immediately after column T. There is no column A in the synodic tables from Uruk, but it is present in some template tables from that city.

Updating A as a zigzag function of the lunation number

Procedures with a complete set of rules for computing A are not preserved. The synodic tables imply that A is computed as a zigzag function of the lunation number (§2.6.8.1) — one of the main features distinguishing system B from system A. For a graphical representation cf. Fig. 4.38. Two variants of the zigzag function are attested (Table 4.31). Some synodic tables and template tables[523] are based on variant 1, whose parameters are mentioned in No. **95** P5'. Other synodic tables[524] are based on variant 2, whose extrema are rounded compared to variant 1. The parameters of this variant are mentioned in No. **97** P4'.b.[525]

var.	d	m	M	μ	Δ	Π	Z	P	p	$\bar{A}$
1	0;18°	28;10,39,40°	30;1,59	29;6,19,20°	1;51,19,20°	2,46,59	13,30	12;22,8,53,20	0;55,30,43,23,...	29;6,16,55,...°
2	0;18°	28;10,40°	30;2	29;6,20°	1;51,20°	5,34	27	12;22,13,20	0;55,30,44,52,...	29;6,6,28,...°
	c	m	M	μ	Δ					
3	0;0;36°	28;12,19,20°	30;0,19,20°	29;6,19,20°	1;48°					

Table 4.31: Parameters of the zigzag functions for A.

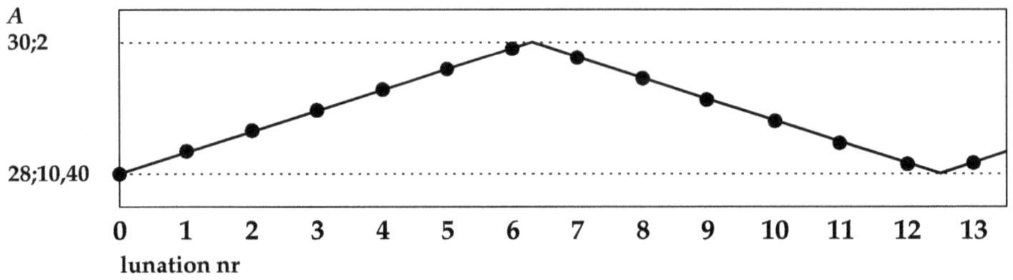

Fig. 4.38: Graphical representation of $A^{(2)}$ as a function of the lunation number.

As is true for any zigzag function of the lunation number, the extrema of A do not correspond to strictly fixed zodiacal positions. However, the period of A is such that they do cluster in certain zodiacal regions. In the tabular texts the minimum occurs in Gem–Cnc, the maximum in Sgr–Cap, roughly in agreement with empirical facts. No. **95** P5' and No. **97** P4'.b

[522] An Akkadian term for A is not attested. No. **97** P3'.b refers to A as *qaqqar*(**ki**) *Sîn* **ab₂** *ana* **ab₂** = 'the Moon's position (B) month by month'.

[523] E.g. *ACT* 102, 124, 141.

[524] *ACT* 123, 123a.

[525] The astronomical interpretation of A, the algorithm by which it is computed, and the defining parameters of $A^{(1)}$ were established by Kugler (1900), pp. 10, 107. $A^{(2)}$ was reconstructed by Schnabel (1923), p. 216.

state that the extrema are associated with the signs Cnc and Cap, respectively, which must therefore be interpreted in an approximative sense.

Properties of A

The period P, the fractional number of synodic months after which A returns to the same value, is a close approximation of the year. This implies that A describes only the zodiacal variation of the synodic arc, i.e. the lunar variation is ignored, as in system A. The mean synodic arc $\overline{A}$ (Table 4.31), obtained in the usual manner as 6,0/P, is not a finite sexagesimal number, but the mean value of the zigzag function, μ, is a good approximation of $\overline{A}$.[526]

Computing A as a zigzag function of the zodiacal position ($A^{(3)}$)

Britton, Horowitz & Steele (2007) discovered that No. **97** P3'.b and P4'.a mention a variant, say $A^{(3)}$, which is computed as a zigzag function of the zodiacal position.[527] The latter is suggested by the phrase 'for 1° the difference is 0;0,36', which describes an interpolation coefficient (c). Since the mean value (μ) is the same as for $A^{(1)}$, it appears that the starting point for the construction of $A^{(3)}$ was $A^{(1)}$, whose extrema were modified in such a way that, for the given c, one oscillation of the zigzag function fits exactly in the zodiac, i.e. $3{,}0 \cdot c = \Delta$. The procedure also mentions that the extrema occur in Cap (M; restored) and Cnc (m). Function $A^{(3)}$ is not yet attested in the tabular texts.

4.5.4 Zodiacal position of the Moon (B) and the Sun ($B_\odot$)

terminology	'position' = *qaqqaru*(**ki**)
unit	zodiacal sign and degrees within it [0–30]
astronomical meaning	zodiacal position of the Moon (B)
column	iii (Babylon); ii (Uruk)
procedures:	
unclear	No. **95** P4'

Function B represents the Moon's zodiacal position, expressed in degrees from the beginning of the zodiacal sign. In the synodic tables, where B is evaluated at successive lunations, B implies the zodiacal position of the Sun: at New Moon $B_\odot = B^{\mathrm{nm}}$, while at Full Moon $B_\odot = B^{\mathrm{fm}} + 3{,}0°$. In the tabular texts from Babylon column B is placed immediately after column A; in Uruk it usually follows column T, and column A is lacking.

Algorithm for updating B

Procedures with an algorithm for B are not preserved. The synodic tables imply that B is updated with the synodic arc A, such that A_i pertains to the arc between the previous lunation ($i-1$) and the current one (i),[528]

$$B_i = B_{i-1} + A_i. \tag{4.109}$$

Properties of B

Since the mean value of the monthly difference of B is μ_A, the period of B equals $P = 6{,}0/\mu_A$ = 12;22,7,51,53,...^{m} for variant 1, and 12;22,7,34,53,...^{m} for variant 2. Both are, as expected, very close to 12;22,8^{m}, the period of A, and good approximations of the year.[529]

Daily zodiacal position of the Moon (B^*)

B^* represents the zodiacal position of the Moon evaluated at intervals of 1^{d}.[530] Procedures for B^* are not preserved. The daily motion tables for the Moon,[531] where B^* usually occupies column ii, imply that B^* is updated from day to day by adding the Moon's daily displacement F^* (§4.5.10),

---

526 Since it is μ that determines the period of B, the latter is slightly different from the period of A; for this issue cf. also §3.3.1.4.

527 Function $A_B(\lambda)$ in Britton, Horowitz & Steele (2007), p. 49.

528 The algorithm for B was reconstructed by Kugler (1900), p. 107.

529 Since A and B have slightly different periods, the values of Π and Z for A (Table 4.31) do not strictly apply to B. Furthermore, a function like B whose difference is a zigzag function of the lunation number need not return to *exactly* the same value after a fixed number of lunations. For this issue cf. *HAMA*, pp. 382–386.

530 Neugebauer calls this ΣF^* (*ACT*, p. 77).

531 *ACT* 191–196.

$$B_i^* = B_{i-1}^* + F_i^*. \tag{4.110}$$

Daily zodiacal position of the Sun ($B^*_\odot$)

$B^*_\odot$ represents the zodiacal position of the Sun evaluated at intervals of 1^d.[532] In the daily motion tables for the Sun $B^*_\odot$ occupies column ii.[533] Procedures for $B^*_\odot$ are not preserved. The daily motion tables imply that $B^*_\odot$ is updated from day to day by adding the Sun's daily displacement $v_\odot$ (§4.5.11),

$$B^*_{\odot i} = B^*_{\odot i-1} + v_\odot, \tag{4.111}$$

where $v_\odot$ is constant and equal to 0;59,9°/d. Other values of $v_\odot$ or schemes with a varying $v_\odot$ are not attested in the daily motion tables.

4.5.5 Duration of daylight (*C*) and night (*D*)

terminology	[LA₂ *ša₂ ūmi*(**me**) = 'duration of daylight' (*C*)?] *mūšu*(**gi₆**) = 'night' (*D*) *mišil* **gi₆**= 'half the night' (*D*′)
unit	time degrees
astronomical meaning	duration of daylight (*C*) and night (*D*)
columns	*C*: iv; *D*: ≈iv; *D*′: ≈ v
procedures:	
computing *C* from *B*	No. **93** P1'?

C and *D* are complementary functions representing the duration of daylight and the night,[534]

$$D_i = 6{,}0 - C_i. \tag{4.112}$$

Instead of *D* some synodic tables contain a column with $D' = D/2$, appropriately called 'half the night' (*mišil mūši*) in No. **93** P4'. Columns *C* and *D*′ are attested mainly in synodic tables from Babylon, where they usually occupy two adjacent columns after *B*. Column *D* is known mainly from Uruk, where it follows *F*′ in the synodic tables. It usually includes the word 'night' (**gi₆** = *mūšu*), obviously an (abbreviated) term for *D*. Functions *C*, *D* and *D*′ are invoked at two separate instances: 1) when computing *M*, the time of the lunation with respect to sunrise or sunset; 2) in the Lunar Six module (step 1).

Computing C and D from B

The synodic tables imply that *C* is computed from the Sun's zodiacal position at lunation (B_i^{nm} or $B_i^{\mathrm{fm}} + 3{,}0$) by means of linear interpolation, as in system A. The badly damaged No. **93** P1' probably preserves a segment of this scheme, or the related scheme for *D*. The entire scheme, as reconstructed by Kugler,[535] consists of 12 control points, each pertaining to 8° of a zodiacal sign (Table 4.32; Fig. 4.39). The extrema of *C* are the same as in system A. The numbers in columns *C*, *D* and *D*′ are usually rounded at the third sexagesimal digit, so that they may deviate slightly from the exact values obtained by executing the reconstructed algorithm to full precision.

	Ari	Tau	Gem	Cnc	Leo	Vir	Lib	Sco	Sgr	Cap	Aqr	Psc
C	3,0°	3,18°	3,30°	3,36°	3,30°	3,18°	3,0°	2,42°	2,30°	2,24°	2,30°	2,42°
c	0;36	0;24	0;12	0;12	0;24	0;36	0;36	0;24	0;12	0;12	0;24	0;36

Table 4.32: Control values and interpolation coefficients (*c*) for computing *C* from $B_\odot$.

[532] *ACT* and *HAMA*: B^*.
[533] *ACT* 185–187.
[534] The terminology adopted here differs slightly from that of Kugler and Neugebauer, who used *D*′ for *D*, and vice versa.
[535] Kugler (1900), pp. 11, 108 ('Columne C').

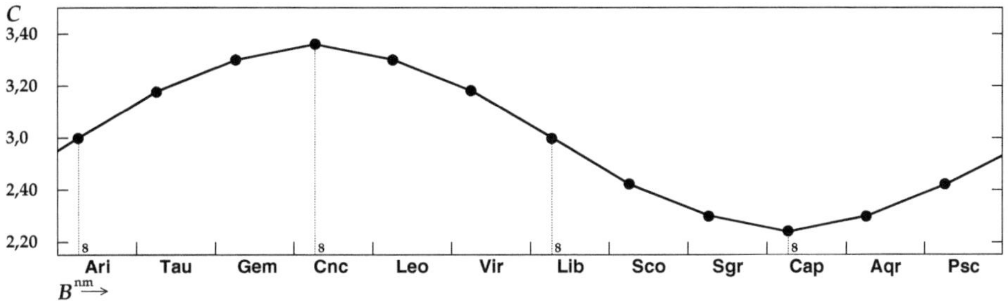

Fig. 4.39: Graphical representation of C^{nm} as a function of B^{nm}. The dots are control values of the interpolation scheme. The dashed lines mark equinoxes and solstices.

4.5.6 The Moon's distance to the ecliptic (E)

terminology	*šūqu*(**nim**) = 'height' ($E_{\uparrow}$)
	šuplu(**sig**) = 'depth' ($E_{\downarrow}$)
units	barleycorn or degrees?
astronomical meaning	the Moon's distance above or below the ecliptic
procedures:	
computations involving E	No. **97** P8'?

The existence of an algorithm for E is suggested by the Lunar Six module (steps 6 and 8), but the extant tabular texts for lunar system B do not include a corresponding column.[536] Moreover, eclipse magnitude is not computed from E, unlike in lunar system A, and the Lunar Six module of system B is not understood well enough to derive from column R, which is presumably computed from E, the underlying algorithm for E. The only available procedure with a possible reference to E (No. **97** P8') is badly preserved and difficult to interpret. For these reasons almost nothing is known about E in system B.

4.5.7 Eclipse magnitude (Ψ'')

terminology	ḪAB-*rat* = 'disk' or *kippatu*(**gam**) = 'circumference; circle'?
units	fingers (^{f}), degrees or units of $c_{\Psi} = 18^{\mathrm{f}}$
astronomical meaning	eclipse magnitude
column	$\approx$ V
procedures:	
computations involving Ψ''	No. **96** P1'?

Function Ψ'' measures the magnitude of eclipses.[537] At New Moon Ψ'' describes solar eclipses, at Full Moon lunar eclipses. Usually Ψ'' is expressed in fingers (^{f}) or degrees, but in some synodic tables (*ACT* 123, 125) in units of the parameter

$$c_{\Psi} = 18^{\mathrm{f}} = 1;30^{\circ}, \tag{4.113}$$

which represents the magnitude of a central eclipse. Eclipses were considered possible in the region where Ψ'' increases from 0 to $2c_{\Psi}$, shown in black in Fig. 4.40. In the synodic tables these values of Ψ'' are followed by the sign BAR.[538] Otherwise each value of Ψ'' is usually followed by **nim** or **la₂** if the Moon is above the ecliptic, **sig** if the Moon is below the ecliptic, a terminology known from column E of lunar system A. The Akkadian terms for Ψ'' are probably the same as those for Ψ'. A few tablets (e.g. *ACT* 123) contain both columns Ψ'' and Ψ', but usually they are mutually exclusive. Some template tables (*ACT* 150) contain only columns T and Ψ''. In some synodic tables the values of Ψ'' are rounded at the second sexagesimal digit.

Updating Ψ'' as a modified zigzag function of the lunation number

The algorithm underlying column Ψ'' is essentially the same as that for Ψ', except that the updating in step 1 is based on a constant difference d. Apart from isolated parameters, the algorithm is not preserved in procedure texts. All known instances of column Ψ'' are based on variant 1 (Table 4.33). Some features of the algorithm, especially step 3, still await a fully satisfying astronomical interpretation.

[536] Kugler and initially also Neugebauer thought that the Moon's distance to the ecliptic is modeled by Ψ'' and Ψ'; cf. below.

[537] This interpretation, established by Neugebauer (1945), replaces an earlier one as the Moon's distance to the ecliptic. The latter was proposed by Kugler (1900) and initially followed by Neugebauer (1938c).

[538] This usage of BAR differs slightly from that in the astronomical diaries and related observational texts (Steele 2001/2).

var.	d			m	M			Π	Z	P	p
1	$46;30^{f}$	$3;52,30°$	$2;35\cdot c_{\Psi}$	0	$1,58;27^{f}$	$9;52,15°$	$6;34,50\cdot c_{\Psi}$	1,30,58	7,45	11;44,15,29,...	0;55,17,22,22,...
2	?	?	?	0	$1,58;28^{f}$	$9;52,20°$	?	?	?	?	?

Table 4.33: Parameters of the zigzag functions for Ψ''.

Step 1. Preliminary update of Ψ''

A preliminary value of Ψ''_i is obtained by applying d to the previous value,

$$\Psi''_i = \Psi''_{i-1} \pm d, \tag{4.114}$$

where the plus sign applies if the magnitude of Ψ'' is increasing, the minus sign if it is decreasing. The value of d coincides with $\mu_{\Delta\Psi'}$, the mean value of the zigzag function for $\Delta\Psi'$ (§4.5.8).

Step 2. Reflection rule at extrema of Ψ''

Step 2 involves the usual reflection rule at the extremum M, which causes Ψ'' to assume the appearance of a zigzag function.[539] The incompletely known variant 2 is attested only through a slightly deviating value of M mentioned in No. **97** P9', a procedure whose purpose is not clear.

Step 3. Subtracting $2c_{\Psi}$ *at the nodes*

Step 3 involves the following, seemingly artificial, operation: whenever the magnitude of Ψ'' exceeds $2c_{\Psi} = 36^{f} = 3°$, for the first time after having reached 0, then $2c_{\Psi}$ is subtracted from Ψ''.[540] Subsequently the zigzag function proceeds its normal course on the increasing branch. Fig. 4.40 shows a modern graphical representation in which values of Ψ'' pertaining to positions below the ecliptic are placed below the horizontal axis.[541] The astronomical reasoning underlying step 3 is still unclear.

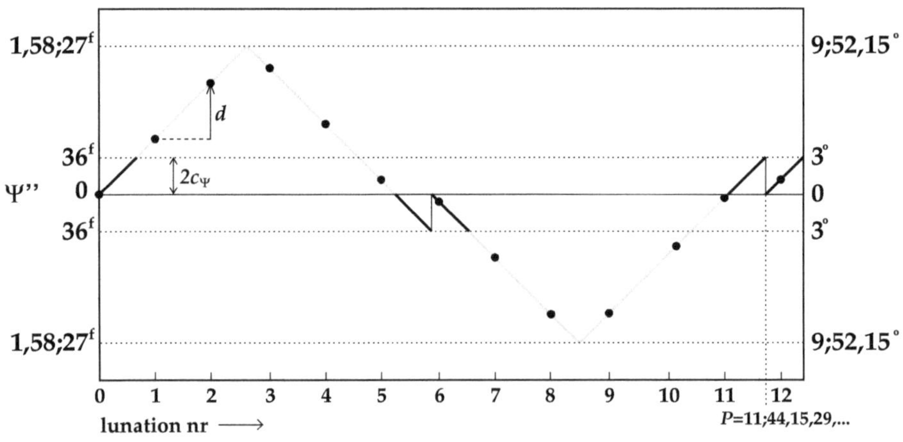

Fig. 4.40: Graphical representation of Ψ'' as a function of the lunation number. Eclipses were considered possible only in the black segments.

Period of Ψ''

As illustrated in Fig. 4.40, step 3 has the effect of extending the period of Ψ''. Since Ψ'' is not an ordinary zigzag function, P does not satisfy Eq. (2.16) on p. 44. If we set out from the definition of P as the fractional number of synodic months after which a function returns to the same value, then

$$P = \frac{4(M+c_{\Psi})}{d} = \frac{45;29°}{3;52,30°} = \frac{1,30,58}{7,45} = 11;44,15,29,..., \tag{4.115}$$

where $4(M+c_{\Psi})$ is the total change of Ψ'' during one oscillation. P is a good approximation of the draconitic year, which is shorter than the (sidereal) year due to the retrograde motion of the nodes. The underlying period $p = P/(P+1) =$

[539] For the reflection rules cf. §2.6.8.

[540] Step 3 was essentially reconstructed by Kugler (1900), p. 109 ('Columne E'), but the exact condition that triggers it was established by Neugebauer (1938c), p. 301.

[541] Unlike what is stated in *ACT*, p. 73, *HAMA*, pp. 521–524, Ψ'' is always a positive quantity.

0;55,17,22,... is the corresponding ratio of the draconitic month to the mean synodic month.[542] Since Ψ'' varies with the correct period, the lengthening of P caused by step 3 is compensated by a larger value of d compared to the mean value of $\Delta\Psi'$ in lunar system A, whose algorithm for Ψ' lacks an operation equivalent to step 3 (cf. the values of w_j in Table 4.16, p. 142).[543] Since d coincides with the mean value (μ) of $\Delta\Psi'$ (cf. below), and M is also the same, Ψ'' has exactly the same period as Ψ'.

Net differences for various intervals

Procedures concerning net differences of Ψ'' for intervals longer than 1 synodic month are not yet attested. Since Ψ'' is not an ordinary zigzag function, the usual expression for the net difference of a zigzag function (Eq. 2.22, p. 46) must be modified in order to take into account step 3. Each application of step 3 reduces the net difference by $2c_\Psi$, which occurs twice in the course of every oscillation, so that the correct expression is

$$d\Psi''(s) = sd - 4t(M + c_\Psi), \tag{4.116}$$

where, as usual, s is the number of synodic months. The number t, defined as the closest whole number of oscillations of Ψ'', is also allowed to represent *half* oscillations of Ψ'', e.g. t = 9.5, in order to cover close approaches to the opposite node.[544] Selected values are shown in Table 4.34.

s	t	$d\Psi''(s)$		
5^m	0.5	$-40;24^f$	$-3;22°$	$-2;14,40\cdot c_\Psi$
6^m	0.5	$6;6^f$	$0;30,30°$	$0;20,20\cdot c_\Psi$
11^m	1	$-34;18^f$	$-2;51,30°$	$-1;54,20\cdot c_\Psi$
12^m	1	$12;12^f$	$1;1°$	$0;40,40\cdot c_\Psi$
111^m	9.5	$-23;36^f$	$-1;58°$	$-1;18,40\cdot c_\Psi$
223^m	16	$-0;42^f$	$-0;3,30°$	$-0;2,20\cdot c_\Psi$

Table 4.34: Net differences $d\Psi''(s)$ for various intervals of s months.

4.5.8 Monthly difference of eclipse magnitude ($\Delta\Psi'$)

terminology	*ṭīpu*(**tab**) [*u mīṭu*(**la₂**)] *ša₂* ḪAB-*rat* = 'addition [and subtraction] for the disk'
units	fingers (^{f})
astronomical meaning	monthly difference of the eclipse magnitude Ψ'
column	≈ vi (Babylon)
procedures:	
parameters of $\Delta\Psi'$	$\Delta\Psi'^{(1)}$: No. **96** P2',P7'; $\Delta\Psi'^{(2)}$:
computations involving $\Delta\Psi'^{(3)}$	No. **95** P7'?; No. **96** P2',P7'
computations involving $\Delta\Psi'$	No. **96** P2',P6'?

$\Delta\Psi'$ is the monthly difference of the eclipse magnitude Ψ'. Column $\Delta\Psi'$ is mainly attested in tabular texts from Babylon, where it usually follows column D'. In Uruk Ψ'' appears to have been the preferred eclipse function, but a few template texts from that city (*ACT* 151, 152) do contain $\Delta\Psi'$ and Ψ'. In the extant texts, $\Delta\Psi'$ and Ψ' are measured in fingers. The terms *ṭīpu*(**tab**) *ša₂* ḪAB-*rat*, 'addition for the disk', mentioned in No. **96** P6', and *ṭīpu*(**tab**) *u mīṭu*(**la₂**), 'addition and subtraction' (No. **96** P7'), are probably technical terms for $\Delta\Psi'$.

Updating $\Delta\Psi'$ as a zigzag function of the lunation number

Procedures with a complete algorithm for $\Delta\Psi'$ are not preserved. The tabular texts imply that $\Delta\Psi'$ is modeled as a zigzag function of the lunation number (Fig. 4.41), as demonstrated by Kugler.[545] Two slightly different versions of the zigzag function, $\Delta\Psi'^{(1)}$ and $\Delta\Psi'^{(2)}$ (Table 4.35), are known from the tabular texts.[546] The extrema of variant 2 are rounded compared to those of variant 1, but both have the same mean value μ. Parameters of variant 1 are also mentioned in

542 For the nodal motion, the draconitic month and the draconitic year cf. §4.1. Neugebauer (*HAMA*, pp. 310, 523) mentions that the implied period relation Π = 1,30,58 synodic months = $\Pi + Z$ = 1,38,43 draconitic months was known to Hipparchos.

543 Without step 3, the period of Ψ'' would be $4M/d$ = 39;29/3;52,30 = 10;11,21,...^{m}, which does not have any obvious astronomical significance.

544 Note that the ratio s/t approaches P as s and t increase.

545 Kugler (1900), p. 110 ('Columne Δ'), which also mentions the defining parameters of $\Delta\Psi'^{(2)}$.

546 Variant 1: *ACT* 120, 121, 149, 151, 152; variant 2: *ACT* 123, 125c, 125d.

several procedures. In No. **96** the maximum is said to occur in the zodiacal sign Cap. The incompletely known variants 3a and 3b are reconstructed from procedure texts (cf. below).

var.	d	m	M	μ	Δ	P	p	Π	Z
1	0;33,20^f	44;46,55,33,20^f	48;13,4,26,40^f	46;30^f	3;26,8,53,20^f	12;22,8	0;55,30,43,5,...	46,23	3,45
2	”	44;46,55,30^f	48;13,4,30^f	”	3;26,9^f	12;22,8,24	0;55,30,43,13,...	3,26,9	16,40
1	0;2,46,40°	3;43,54,37,46,40°	4;1,5,22,13,20°	3;52,30°	0;17,10,44,26,40°				
2	”	3;43,54,37,30°	4;1,5,22,30°	”	0;17,10,45°				
	c	m	M	μ	Δ				
3a	0;1,6,40^f/°	*45;0,51,40^f*	*48;20,51,40^f*	46;40,51,40	*3;20^f*	6,0			
3b	*0;1,6,40^f/°*	*44;50^f*	*48;10^f*	*46;30^f*	*3;20^f*	”			

Table 4.35: Defining parameters of the zigzag functions for $\Delta\Psi'$. Uncertain reconstructions are in italics.

Properties of $\Delta\Psi'$

The period $P = 12;22,8^m$, the fractional number of synodic months after which $\Delta\Psi'$ returns to the same value, essentially coincides with the length of the year implied by function A. From this it follows that $\Delta\Psi'$, like A, models only the contribution resulting from the zodiacal variation. Note that μ is considerably higher than the mean value of $\Delta\Psi'$ in lunar system A (compare w_j in Table 4.16, p. 142). The reason for this is discussed in §4.5.9 (step 3).

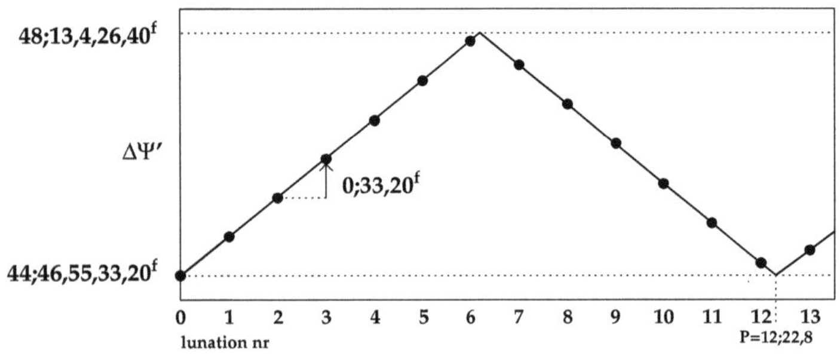

Fig. 4.41: Graphical representation of $\Delta\Psi'^{(1)}$ as a function of the lunation number.

Computing $\Delta\Psi'^{(3)}$ as a zigzag function of the zodiacal position?

In the extant procedure texts concerning $\Delta\Psi'$, all of which are damaged and difficult to interpret, $\Delta\Psi'$ appears to be modeled as a zigzag function of the zodiacal position at lunation, say $\Delta\Psi'^{(3)}(B)$, and not of the lunation number. By itself the existence of this variant is not surprising because, as argued above, the other variants of $\Delta\Psi'$ also model only the zodiacal variation. The complete algorithm for $\Delta\Psi'^{(3)}$ is not preserved and can be only partly reconstructed by making some reasonable assumptions. No. **96** P7' mentions only the interpolation coefficient c and (probably) the mean value μ (variant 3a in Table 4.35). Since $c = d/30$, it appears that c derives from the assumption that $\Delta\Psi'^{(3)}$ changes by d over one zodiacal sign. Since one oscillation of the zigzag function should fit exactly in the zodiac, the unknown Δ must satisfy $6,0 \cdot c = \Delta$, from which we obtain Δ, and hence m and M. An alternative assumption would be that μ is the same as for variants 1 and 2, which would imply slightly different values of m and M (variant 3b). Since μ is, in any case, similar to that of the other variants, the algorithm for $\Psi'^{(3)}$ must incorporate a correction equivalent to step 3 of the algorithms for Ψ'' and Ψ', because without it the period P cannot be expected to come out right.

4.5.9 Eclipse magnitude (Ψ')

terminology	ḪAB-*rat* = 'disk'; *kippatu*(**gam**) = 'circumference; circle'
units	fingers (^{f}), degrees or units of c_Ψ
astronomical meaning	eclipse magnitude
column	≈ vii (Babylon)
procedures:	
computations involving Ψ'	No. **95** P7'?; No. **96** P1'?,P5',P7'

Function Ψ' is an alternative measure of eclipse magnitude. It is mainly attested in tabular texts from Babylon, where it usually follows column D', and in a few template texts from Uruk (*ACT* 151, 152). Unlike column Ψ'', column Ψ' does not include an indication whether Ψ' pertains to a lunar position above (**nim**) or below (**sig**) the ecliptic. Eclipses were considered possible in the region where Ψ' increases from 0 to $2c_\Psi$, shown in black in Fig. 4.42. In some tables (*ACT* 151) these values are accompanied by the sign ḪAB, as in column Ψ''. Occasionally the sign BE indicates that an eclipse is excluded. The Akkadian terms for Ψ', ḪAB-*rat*, 'disk', and *kippatu*, 'circle', presumably reflect the geometric shape of the lunar and solar disk.

Some eclipse tables[547] include a column Ψ which is non-empty only if the Moon is in the nodal region, similar to column Ψ of lunar system A. Its algorithm remains to be reconstructed, but is probably similar to that for Ψ'.[548] In these tables the same logograms **nim** and **sig** which in column Ψ'' indicate the Moon's position with respect to the ecliptic now express its direction of motion (**nim** = ascending, **sig** = descending). The possibility of an eclipse is as usual indicated by ḪAB, its exclusion by BE (*ACT* 135).

Updating Ψ' as a modified zigzag function of the lunation number

The algorithm for Ψ' was reconstructed from the tabular texts by Kugler (1900) and Neugebauer (1938c), along with that for Ψ''.[549] It is not fully preserved in procedure texts. The algorithm consists of the same three steps as that for Ψ'', the only significant change being that the constant difference d is replaced by the variable difference $\Delta\Psi'$ (step 1). Some elements of the algorithm, especially step 3, still await a satisfying astronomical interpretation.

Step 1. Preliminary update of Ψ'

A preliminary value of Ψ'_i is obtained by applying $\Delta\Psi'$ to the previous value,

$$\Psi'_i = \Psi'_{i-1} \pm \Delta\Psi'_i, \tag{4.117}$$

where the plus sign applies if the magnitude of Ψ' is increasing, the minus sign if it is decreasing.

Step 2. Reflection rule at extrema of Ψ'

Step 2 involves the reflection rule at the extremum M (§2.6.8), which causes Ψ' to assume the appearance of a zigzag function. As far as known all instances of column Ψ' are based on variant 1 (Table 4.36).

var.	c_Ψ		m	M			Π	Z	P	p
1	18^f	1;30°	0	1,58;27^f	9;52,15°	6;34,50·c_Ψ	1,30,58	7,45	11;44,15,29,...	0;55,17,22,22,...
2	17;40^f	1;28,20°								

Table 4.36: Defining parameters and periods of Ψ'.

Step 3. Subtracting $2c_\Psi$ at the nodes

Step 3 proceeds in exactly the same manner as for Ψ''. Fig. 4.42 shows a modern graphical representation of Ψ' with values pertaining to positions below the ecliptic placed below the horizontal axis. An alternative value of c_Ψ (variant 2), attested only in No. **96** P5', is closer to the system-A value 17;24^f (§4.4.7). Its significance is not clear, and there is no evidence that it was applied in synodic tables.

Period of Ψ'

As with Ψ'', step 3 has the effect of extending the period of Ψ' to the value $P = 4(M + c_\Psi)/\mu_{\Delta\Psi'}$ = 45;29°/3;52,30° = 1,30,58/7,45 = 11;44,15,29,... Since $\mu_{\Delta\Psi'}$ coincides with d of Ψ'', this period is identical to that of Ψ''.[550]

---

[547] *ACT* 130, 135, 136.

[548] New results for Ψ will be presented in Volume II; cf. also *ACT*, p. 72.

[549] Kugler (1900), p. 110 (Columne E"). The precise condition triggering step 3 was established by Neugebauer (1938c).

[550] Since Ψ' is obtained as a summation over varying differences $\Delta\Psi'$, the implied 'number period' Π = 1,30,58 does not truly deserve that name, since Ψ' need not return to *exactly* the same value after Π lunations, even though P does represent an accurate average value of the period of Ψ'. For this issue cf. also Neugebauer (1948b) and *ACT*, pp. 36–37.

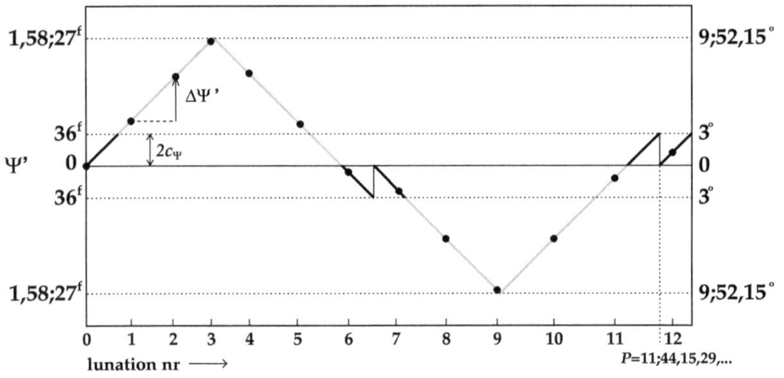

Fig. 4.42: Graphical representation of Ψ' as a function of the lunation number. Eclipses were considered possible only in the black segments.

4.5.10 The Moon's displacement along the zodiac per day (F) or time degree (F')

terminology	*nisḫu*(**zi**) *ša*₂ *Sîn ša*₂ *ūmi*(**me**) = 'the Moon's daily displacement' (F)
unit	degrees per day (F) or per time degree (F')
astronomical meaning	the Moon's displacement along the zodiac
column	F: $\approx$ vii (Babylon); F': $\approx$ iii (Uruk)
procedures:	
parameters of zigzag function	$F^{(1)}$, $F^{(3)}$: No. **97** P2'.c; $F^{(1)}$': No. **97** P3'.a; $F^{(1)*}$: No. **94** P1?
net differences	$F^{(1)}$: No. **97** P2'.b; $F^{(1)*}$: No. **94** P1
period relations	No. **95** P3'
computing $F^{(1)}$ from G	cf. §4.5.12

Function F represents the Moon's daily displacement along the zodiac. In the synodic tables from Babylon column F usually follows column Ψ'. In the synodic tables from Uruk F is replaced by F' which represents the same quantity expressed in degrees per time degree. In some tables from Babylon F' and F are in adjacent columns.[551] F and F' are used only in the Lunar Six module (step 2). The astronomical meaning of F, its algorithm and the defining parameters of $F^{(1)}$ were established by J. Epping.[552]

Updating F as a zigzag function of the lunation number

F is computed as a zigzag function of the lunation number. The variant F' models the same quantity but expressed in degrees per time degree, i.e.

$$F' = 0;0,10 \cdot F. \tag{4.118}$$

A complete procedure for updating F is not preserved, but several procedures mention defining parameters of $F^{(1)}$, $F^{(1)\prime}$ and $F^{(3)}$ (Table 4.37). All known instances of F and F' in the synodic tables are based on variants 1 and 2.[553] Variant $F^{(3)}$ is mentioned only in No. **97** P2'.c.

	d	m	M	μ	Δ	Π	Z	P	p
$F^{(1)}$	0;36	11;5,5	15;16,5	13;10,35	4;11	4,11	18	13;56,40	0;55,59,6,28,...
$F^{(2)}$	"	11;5	15;16	13;10,30	"	"	"	"	"
$F^{(3)}$	"	11;6,35,14,20	15;14,34,27,40	13;10,34,51	4;7,59,13,20	37,11,53	2,42,0	13;46,37,24,26,40	0;55,56,22,44,...
$F^{(1)\prime}$	0;0,6	0;1,50,50,50	0;2,32,40,50	0;2,11,45,50	0;0,41,50				
$F^{(2)\prime}$	"	0;1,50,50	0;2,32,40	0;2,11,45	"				
$F^{(3)\prime}$	"	0;1,51,5,52,23,20	0;2,32,25,44,36,40	0;2,11,45,48,30	0;0,41,19,52,13,20				
$F^{(1)*}$	0;18	11;6,35	15;14,35	13;10,35	4;8	4,8	9	27;33,20	
$F^{(3)*}$	"	11;6,35,14,20	15;14,34,27,40	13;10,34,51	4;7,59,13,20	37,11,53	1,21,0	27;33,14,48,53,20	

Table 4.37: Defining parameters of the zigzag functions for F, F' and F^*. The parameters d, m, M, μ and Δ are measured in °/d (F and F^*) or ° per time degree (F'). The periods (Π, Z, P and p) of F' are the same as for F.

[551] E.g. *ACT* 120, 121.
[552] Epping (1890), pp. 239–240 ('Columne a').
[553] $F^{(1)}$: *ACT* 120, 121; $F^{(2)}$: *ACT* 123, 123a.; $F^{(1)\prime}$: e.g. *ACT* 100, 101, 104, 120, 121.

Net differences for various intervals

No. **97** P2' seems to be concerned with the net difference of $F^{(1)}$ for 223^{m}, $dF^{(1)}(223)$, but the value itself is not preserved. Net differences for several intervals of s months, corresponding to t oscillations of the zigzag function, are compiled in Table 4.38.

s	t	$dF^{(1)}$	$dF^{(3)}$	s	t	$dF^{(1)*}$	$dF^{(3)*}$
12^{m}	1	−1;10	−1;3,58,26,40	28^{d}	1	0;8	0;8,1,33,20
14^{m}	1	0;2	0;8,1,33,20	29^{d}	1	0;26	0;26,1,33,20
111^{m}	8	−0;20	0;28,12,26,40	30^{d}	1	0;44	0;44,1,33,20
223^{m}	16	−0;4	1;32,24,53,20				

Table 4.38: Left: net differences $dF(s)$ [°/d] for various intervals of s months [$dF^{(2)}(s) = dF^{(1)}(s)$]. **Right**: net differences $dF^*(s)$ [°/d] for various intervals of s days.

Properties of F

The basic period underlying function F is the anomalistic month, $m_{\mathrm{an}} \approx 27.56^{\mathrm{d}}$, which defines the lunar variation (cf. also §4.1.2). Accordingly, the period P represents the number of anomalistic months contained in 1^{yr}. Since m_{an} is shorter than the synodic month, there is an underlying period p representing the ratio of the anomalistic month to the mean synodic month. As pointed out by Jones (1983), p. 10, μ_F was probably derived from the ratio of the mean total synodic arc implied by $A^{(1)}$, $\overline{\Sigma} = \mu_A + 6{,}0 = 6{,}29;6{,}19{,}20°$, to the mean synodic month implied by G, $\overline{m}_{\mathrm{syn}} = 29;31{,}50{,}8{,}20^{\mathrm{d}}$ (§4.5.12), which yields 13;10,34,51,...°/d, and rounding this number at the third sexagesimal. While the periods of variants 1 and 2 are accurate, that of variant 3 deviates more strongly from the empirical value.

The Moon's daily displacement at intervals of 1^{d} (F^*)

Function F^* represents the Moon's daily displacement along the zodiac at intervals of 1^{d}. It is used for computing daily positions of the Moon ($B^*_{☽}$). In the daily motion tables[554] F^* and $B^*_{☽}$ are usually tabulated in adjacent columns.

Updating F^* as a zigzag function of the day number

In the extant daily motion tables F^* is modeled as a zigzag function of the day number (variant 1 in Table 4.37). Procedures with a complete algorithm for updating F^* are not preserved. No. **97** P2' mentions some of the defining parameters of variant 3, thus named because its mean value coincides with that of $F^{(3)}$. The period P represents the anomalistic month expressed in days. The main reason for assigning the daily motion tables to system B is that the mean values of F^* agree with the mean values of F.[555]

Net differences for various intervals

No. **94** P1', a badly preserved procedure accompanying a daily motion table for the Moon, mentions a parameter that can be identified as $dF^{(1)*}(30)$, the net difference of $F^{(1)*}$ for 30^{d} (Table 4.38). Note that the monthly difference $d = 0;36°/\mathrm{d}$ is in between $dF^{(1)*}(29)$ and $dF^{(1)*}(30)$. This confirms that $F^{(1)*}$ pertains to successive days and not tithis, since the latter would require $dF^*(30) = d_F$.

4.5.11 The Sun's daily displacement along the zodiac ($v_{\odot}$)

terminology	*nisḫu*(**zi**) *ša*₂ *Šamaš*₂ *ša*₂ *ūmi*(**me**) = 'the Sun's daily displacement'
unit	degrees per day
astronomical meaning	the Sun's daily displacement along the zodiac
procedures:	
numerical value	No. **97** P4'

The existence of a function (or a constant) $v_{\odot}$ representing the Sun's daily displacement along the zodiac is implied by the Lunar Six module, but a corresponding column is not attested in the synodic tables. Secondly, the daily motion tables for the Sun[556] contain a column $v^*_{\odot}$ representing the same quantity evaluated from day to day, from which are computed the daily positions in the adjacent column $B^*_{\odot}$ (§4.5.4). The only value of $v^*_{\odot}$ attested in these tables is a constant

[554] *ACT* 190–196; cf. also Jones (1983), pp. 1–11.
[555] Cf. also *HAMA*, p. 480; Jones (1983), p. 10.
[556] *ACT* 185–187.

0;59,9°/d.[557] No. **97** P4' mentions the more accurate value $v_\odot$ = 0;59,8,9,48,40°/d.[558] More sophisticated models of $v_\odot$ or $v^*_\odot$ incorporating the zodiacal variation as they exist for lunar system A are not preserved. From the (mean) value of $v_\odot$ one obtains the length of the (sidereal) year by computing 6,0°/$v_\odot$. For $v_\odot$ = 0;59,8,9,48,40°/d this results in 1^{yr} = 6,5;15,33,45,39,..^{d} = 365.25938^d.

4.5.12 Duration of the synodic month (G)

terminology	*simānu* = 'duration'
unit	time degrees
astronomical meaning	duration of 1 synodic month -29^d (only the lunar variation)
column	≈ viii
procedures:	
computing G from $F^{(1)}$ and v.v.	No. **96** P3', No. **97** P6'
computations involving μ	No. **95** P3', No. **96** P8'

Function G is an approximation of the duration of the synodic month in excess of 29^d, such that G_i pertains to the interval between lunations $i-1$ and i. In the synodic tables from Babylon column G usually follows column F; most synodic tables from Uruk lack a column G.

Updating G as a zigzag function of the lunation number

The synodic tables imply that G is updated as a zigzag function of the lunation number (cf. §2.6.8). The algorithm is not preserved in procedure texts, but some of the defining parameters of the zigzag function are.[559]

d	m	M	μ	Δ
22;30°	1,52;34,35°	4,29;27,5°	3,11;0,50°	2,36;52,30°

Π	Z	P	p
4,11	18	13;56,40	0;55,59,6,28,6,...

s	12^m	14^m	111^m	223^m
t	1	1	8	16
dG	−43;45°	1;15°	−12;30°	−2;30°

Table 4.39: Defining parameters and periods of the zigzag function for G, and net differences of G for intervals of s synodic months with respect to t whole oscillations of the zigzag function.

Computing G from F, and vice versa

An alternative algorithm for G is preserved in No. **97** P6' and No. **96** P3', which contain interpolation rules for computing G from F according to

$$G = \begin{cases} m_G + 37;30 \cdot (F - m_F) & \text{if } F \text{ on increasing branch} \\ M_G - 37;30 \cdot (M_F - F) & \text{if } F \text{ on decreasing branch,} \end{cases} \quad (4.119)$$

where $37;30 = d_G/d_F$ plays the role of an interpolation coefficient. This algorithm may have been used for computing G from F instead of from its own zigzag function, or for verifying G. It implies that the minimum of F coincides with the maximum of G, and vice versa, and that the periods of F and G are identical. Both properties are satisfied in most synodic tables (only the reconstructed $F^{(3)}$, which is not attested in synodic tables, has a slightly different period). The inverse rule, which allows one to compute $F^{(1)}$ from G, is obtained by exchanging F and G, and replacing the interpolation coefficient by $d_F/d_G = 1/37;30 = 0;1,36$.

Properties of G

Since the period P is the same as that of $F^{(1)}$, function G models only the lunar contribution to the duration of the synodic month, as is true in system A. The zodiacal (solar) contribution is modeled by J, whose mean value is zero (cf. below). Hence the mean value of G, μ = 3,11;0,50° = 0;31,50,8,20^d, implies a mean synodic month $\overline{m}_{syn}$ = 29;31,50,8,20^d, a parameter well known from Hipparchos and Ptolemy. The same number appears as the 'month of the Moon' in No. **95** P3', and it also underlies the usage of mean tithis in the planetary systems (§2.5.2.3).

[557] Established by Schnabel (1927), p. 35.

[558] For a possible explanation of the empirical origin of this parameter cf. Britton, Horowitz & Steele (2007), pp. 49–50.

[559] The algorithm for G and its defining parameters were reconstructed by Epping (1890), pp. 227–229.

Net differences for various intervals

Net differences of G for intervals longer than 1 synodic month (Table 4.39) are not yet attested in the procedure texts.

4.5.13 Monthly difference of J (H)

terminology	?
unit	time degrees
astronomical meaning	monthly difference of J
column	variable
procedures:	
computations involving H	$H^{(2)}$: No. **97** P9'.a; $H^{(3)}$: No. **97** P10'?

Function H is the monthly difference of J. In the synodic tables from Babylon column H is usually placed between G and J. The synodic tables from Uruk lack a column H, even if they contain J.[560] Numerous template tables, mainly from Uruk, contain only columns T, H and J,[561] a few only T and H.[562]

Algorithm for updating H

Procedures with a complete algorithm for H are not preserved. Epping inferred from tabular texts that H is modeled as a zigzag function of the lunation number (Fig. 4.43).[563] All instances of column H are based on variant 1 (Table 4.40). No. **97** P9'.a contains badly understood computations involving a variant $H^{(2)}$, and No. **97** P10' mentions parameters of an incompletely known variant $H^{(3)}$. The period P is readily identifiable as 1/2 year expressed in synodic months (variant 1: $2P = 12;22,5,9,...^{m}$; variant 3: $2P = \mathit{12;18,45,...}^{m}$), which implies that H, and thus also J, is controlled by the zodiacal variation. A fully satisfying interpretation of the algorithm for H remains to be found; cf. the discussion in §4.5.14.

var.	d	m	μ	M	Π	Z	P
1	6;47,30°	0°	10;30°	21°	16,48	2,43	6;11,2,34,36,...
2	6;47,28,...°	"	*10;30°*	*21°*	?	?	?
3	6;49,20°	"	*10;30°*	*21°*	*31,30*	*5,7*	*6;9,22,51,59,...*

Table 4.40: Defining parameters of the zigzag functions for H. In italics: tentative reconstructions.

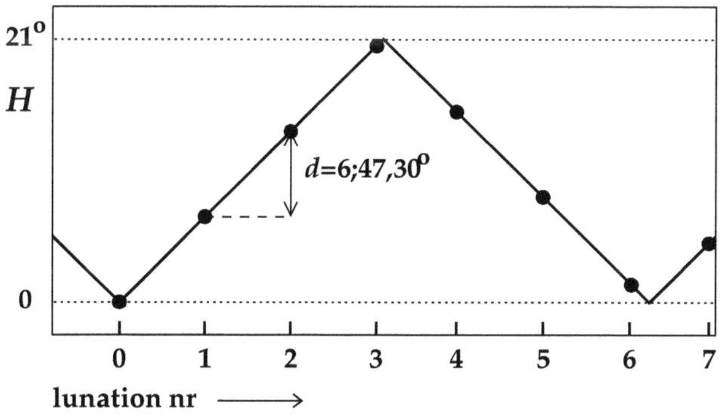

Fig. 4.43: Graphical representation of $H^{(1)}$ as a function of the lunation number.

[560] Cf. *ACT* 101 and the template tables *ACT* 170, 173.

[561] *ACT* 163–167.

[562] From Babylon: *ACT* 162; from Uruk: *ACT* 160, 161.

[563] Epping (1890), pp. 236–238 ('Columne c_1'), which also mentions the defining parameters of $H^{(1)}$.

4.5.14 Correction to the synodic month (J)

terminology	?
unit	time degrees
astronomical meaning	zodiacal correction to the synodic month G
column	variable
procedures:	
computations involving $J^{(2)}$	No. **97** P9'.b

Function J is a correction to the duration of the synodic month (G) that accounts for the zodiacal variation. The value of J can be additive or subtractive. In the tabular texts from Babylon column J follows column H. In the few synodic tables from Uruk that have a column J the position of this column varies strongly.

Algorithm for updating J

Procedure texts with a complete algorithm for updating J are not preserved. The synodic tables imply that J is updated as a zigzag function of the lunation number with H as the monthly difference.[564] The computation of J proceeds in two steps. A provisional value J_i is obtained from J_{i-1} by adding or subtracting H_i,

$$J_i = J_{i-1} \pm H_i, \qquad (4.120)$$

where the plus sign applies if J is increasing in magnitude, the minus sign if it is decreasing.[565] The second step involves reflection at M and 0. If J_{i-1} is increasing (in magnitude) and $J_{i-1} + H_i > M$ (additive or subtractive) then the reflection rule for the maximum is applied (Eq. 2.14). After reflection at M the additive or subtractive sense of H is reversed (but not that of J). If J_{i-1} is decreasing and $H_i > J_{i-1}$ then $J_i = H_i - J_{i-1}$, which amounts to reflection at 0. After reaching 0 the additive or subtractive sense of both H and J are reversed. In Fig. 4.44 additive values of J are shown above the horizontal axis, subtractive values below it.

var.	M	Π	Z	P
1	32;28,6°	46,23	3,45	12;22,8^{m}
1a	32;28°	1,4,56	5,15	12;22,5,42,51,...^{m}
2	32;28,5,15°	14,25,49	1,10,0	12;22,7,42,51,...^{m}

Table 4.41: Defining parameters and periods of J.

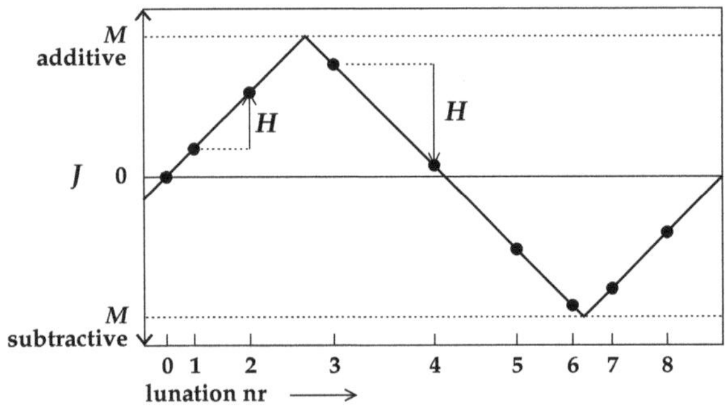

Fig. 4.44: Graphical representation of J as a function of the lunation number. Additive values are shown above, subtractive values below the horizontal axis.

Two different values of M, variants 1 and 1a, are attested in the tabular texts (Table 4.41). Most instances of column J are based on variant 1. Variant 1a, which is rounded at the second digit, underlies column J in the synodic table *ACT* 122. Variant 2 is mentioned only in No. **97** P9'.b. Since the monthly difference of J is itself a zigzag function, the period of J is obtained as

$$P = \frac{4M}{\mu_H}, \qquad (4.121)$$

from which, as usual, Π and Z are obtained as the smallest whole numbers satisfying $\Pi/Z = P$. All attested values of P (variants 1, 1a and 2 in Table 4.41) are equal or close to 12;22,8^{m}, the length of the year underlying all functions that model a zodiacal variation (e.g. A, B, $\Delta\Psi'$, H). Another consequence of the difference of J being variable is that J does not generally return to exactly the same value after Π lunations, so that Π is not a true number period.[566] In the synodic

[564] The algorithm for J, its basic astronomical meaning and the defining parameters of J and H were reconstructed by Epping (1890), pp. 236–238.

[565] Recall that in Babylonian calculus there are no negative numbers, and addition or subtraction is always defined in terms of the magnitude (§2.4).

[566] For this aspect of J cf. *HAMA*, pp. 493–494.

tables the minimum of H always nearly coincides with an extremum of J, but as pointed out by Neugebauer (*HAMA*, p. 496), the slight difference between their periods (taking $2P$ as the period of H) causes both functions to slowly drift out of phase.

Even though the general astronomical meaning of J has been well understood since Epping (1890), the elaborate nature of the algorithm has justifiably puzzled scholars.[567] In particular, since the general purpose of J is to account for the zodiacal variation of the synodic month, which in system A is modeled as a step function, an ordinary zigzag function of the lunation number with a constant monthly difference $d = \mu_H = 10;30°$ would appear to be an adequate model for J in system B, given that the same approach is used for several other functions representing zodiacal variations (A, $\Delta\Psi'$). A possible astronomical justification for modeling J with a variable monthly difference (H) was put forward by Britton (2003).

4.5.15 Corrected duration of the synodic month (K)

terminology	[*simānu qatû* = 'final duration'?]
unit	time degrees
astronomical meaning	corrected duration of the synodic month -29^d
column	variable

Function K is the corrected duration of the synodic month -29^d, which serves as the monthly difference of L, the time of the lunation. In the synodic tables from Babylon K usually follows column J. There is no column K in the tabular texts from Uruk. An Akkadian term for K is not attested, but it may be 'final duration' (*simānu qatû*) as in lunar system A.

Algorithm for K

Procedures with the algorithm for K are not preserved. Epping[568] correctly inferred from the tabular texts that

$$K_i = G_i + J_i. \qquad (4.122)$$

Recall that in lunar system A K incorporates a correction C' which accounts for the changing time of sunset or sunrise. This is not necessary in system B, because L is defined with respect to midnight.

4.5.16 Time of the lunation with respect to the preceding midnight (L)

terminology	*simānu* = 'duration'
unit	time degrees
astronomical meaning	time of the lunation with respect to the preceding midnight
column	variable
procedures:	
date of preceding midnight	No. **93** P3'

Function L represents the time of the lunation with respect to the immediately preceding midnight. In the synodic tables from Babylon column L follows column K; in the ones from Uruk it usually follows column Ψ''. L is always in the range 0–6,0°, corresponding to an interval of at most 1^d. Often column L includes a date, which I interpret as the date of the midnight preceding the lunation.[569] Usually column L also contains an indication about the length of the *preceding* month in whole days, i.e. the same number 30 or 1 contained in the previous line of column NA_1: 30 if the preceding month had 29^d, 1 if it had 30^d.[570] This information is needed for establishing the date of midnight.

Algorithm for updating L

Procedures for L are not preserved. From the synodic tables it is inferred that, for New Moons as well as Full Moons, L_i is obtained by adding the monthly difference K_i to the previous value, L_{i-1}:

[567] Schnabel (1927), p. 55; van der Waerden (1974), p. 241; Aaboe (2002). Neugebauer (*ACT*, p. 493) considered the use of the variable difference H to be a 'purely mathematical device'.

[568] Epping (1889) ('Columne d').

[569] Neugebauer (*ACT*, pp. 79–80), following Epping (1889), p. 93 and Kugler (1900), p. 31, interprets this date as pertaining to the lunation itself. This presupposes a convention whereby the day begins at midnight, which would be highly anomalous, since in Babylonia the day always began at sunset, also in astronomical contexts.

[570] E.g. *ACT* 101, 102, 104; absent e.g. in *ACT* 121a, 122.

$$L_i = \begin{cases} L_{i-1} + K_i & (L_{i-1} + K_i < 6{,}0) \\ L_{i-1} + K_i - 6{,}0 & (L_{i-1} + K_i > 6{,}0). \end{cases} \quad (4.123)$$

The conditional subtraction of 6,0° ensures that L_i is in the range 0–6,0°. This algorithm was reconstructed and understood essentially correctly by J. Epping.[571]

Rules for updating the date in column L

No. **93** P3' contains rules for establishing the day number, say d_i, of the midnight preceding the lunation, which is usually included in column L. Since K_i represents the excess over 29^{d}, these 29^{d} must be added to the previous date. If 6,0° were subtracted from L this must be compensated with a delay of 1^{d}. Thirdly, the date also depends on whether the previous month was hollow (29^{d}) or full (30^{d}). Altogether this results in the following rules, which are the same as in lunar system A (§4.4.18):

$$\mathrm{d}_i = \mathrm{d}_{i-1} \begin{cases} +0 & \text{previous month hollow,} & \text{date boundary not transgressed} \\ +1 & \text{previous month hollow,} & \text{date boundary transgressed} \\ -1 & \text{previous month full,} & \text{date boundary not transgressed} \\ +0 & \text{previous month full,} & \text{date boundary transgressed.} \end{cases} \quad (4.124)$$

4.5.17 *Time of the lunation with respect to sunrise or sunset (M)*

terminology	a: **kur** [*Šamaš₂*] = '(time until) [sun]rise'
	b: [**me**] **nim** = '[time] (since sun)rise'
	c: **šu₂** [*Šamaš₂*] = '(time until) [sun]set'
	d: **gi₆** [**du**] = '(time by which) the night [has progressed]'; **du** = abbrev. of the same?
unit	time degrees
astronomical meaning	time of the lunation with respect to sunrise or sunset
column	variable
procedures:	
computing M	No. **93** P4'

Function M is the time interval between the lunation and the preceding or following sunrise or sunset.[572] Column M usually follows column L. Apart from the value of M it includes the date of the lunation. No other functions are computed from M (cf. Fig. 4.37). Its purpose may be to enable a comparison of the time of the lunation with observations, given that L, which represents the same quantity with respect to midnight, is observationally ill defined.

Algorithm for M

From the synodic tables it is inferred that four different cases are distinguished when computing M. They correspond to a division of the range of L, [0–6,0°], into subintervals a,b,c and d (Fig. 4.45). The terms for M reflect these different definitions. The algorithms for M implied by the synodic tables can be represented as follows:

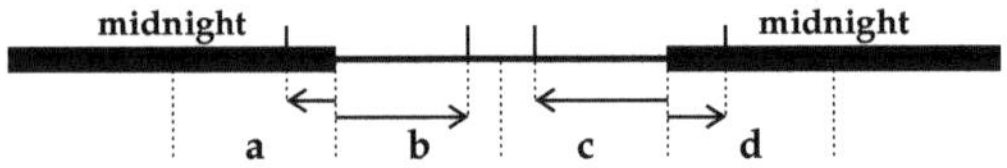

Fig. 4.45: Definition of M for cases a–d. The arrows point to the lunation.

$$M_i = \begin{cases} \frac{1}{2}D_i - L_i & \textbf{a}\text{: lunation between midnight and sunrise} & (0 \le L_i < \frac{1}{2}D_i) \\ L_i - \frac{1}{2}D_i & \textbf{b}\text{: lunation between sunrise and midday} & (\frac{1}{2}D_i \le L_i < 3{,}0) \\ 3{,}0 + \frac{1}{2}C_i - L_i & \textbf{c}\text{: lunation between midday and sunset} & (3{,}0 \le L_i < 3{,}0 + \frac{1}{2}C_i) \\ L_i - (3{,}0 + \frac{1}{2}C_i) & \textbf{d}\text{: lunation between sunset and midnight} & (3{,}0 + \frac{1}{2}C_i \le L_i < 6{,}0). \end{cases} \quad (4.125)$$

No. **93** P4'.a,b partly preserve the algorithms for cases a and b, respectively. For a graphical representation cf. Fig. 4.46.

[571] Epping & Strassmaier (1881), pp. 284–285 ('Columne B'); Epping (1889).

[572] This was established by Epping (1889), pp. 99–100 ('Columne f').

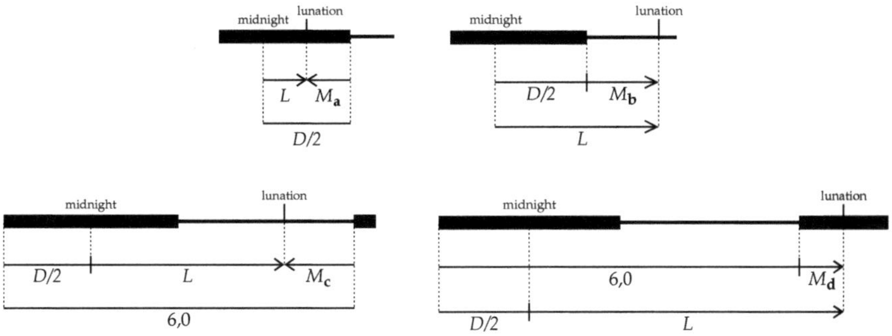

Fig. 4.46: Graphical representation of the computation of M for cases **a–d**.

Establishing the date of the lunation

The date of the lunation mentioned in column M is governed by the following rules. If the lunation occurs before sunset (a–c) it coincides with the date of the preceding midnight mentioned in column L. If the lunation occurs after sunset (d) it is 1^d after the date of the preceding midnight mentioned in column L.[573]

4.5.18 *Lunar Six intervals*

terminology	near New Moon: KUR, NA_1; near Full Moon: ŠU$_2$, NA, ME, GI$_6$
unit	time degrees
astronomical meaning	cf. Table 4.1 (p. 113)
columns	final 1–4 columns
procedures:	
step 1	No. **93** P4'?
step 11	No. **93** P2'?

The final columns of a synodic table usually contain Lunar Six intervals. The algorithms by which they are computed ('Lunar Six modules') have been divided into the same 13 steps as in system A.[574] Only two procedures for lunar system B may preserve some parts of these modules. This lack of procedures is partly compensated by the presence of columns N, O, Q and R in synodic tables, which contain intermediate functions of the Lunar Six modules. The reconstruction of these modules therefore relies mainly on the synodic tables and on the assumed parallels with system A, where they are well represented in procedure texts. Since N, O, Q and R are different for each Lunar Six, each Lunar Six column should ideally be preceded by a complete set of these columns, but in reality that is not the case. The general astronomical meaning of N, O and the Lunar Six intervals was established by Epping, that of Q and R by Kugler.[575] Schaumberger (1935) reconstructed the essential features of the algorithms underlying these functions and various intermediate steps not represented in the tables, and gave a precise geometric interpretation for each of them.[576] Through further analysis of the tabular texts Neugebauer[577] succeeded in refining some elements of these algorithms, but the sparse information provided by columns O, Q and R has not yet allowed a complete reconstruction. In particular, very little is known about the algorithms underlying steps 3, 4, 6, 7, 8 and 10.[578]

573 In fact it is inferred from these rules that the date in column L pertains to the midnight immediately preceding the lunation.

574 For the astronomical reasoning underlying the Lunar Six modules cf. §4.4.19.

575 Epping (1889), p. 93 ff; Kugler (1924), pp. 592–597.

576 Cf. also the review of Schaumberger (1935) in Neugebauer (1936).

577 *ACT*, pp. 81–85.

578 These issues will be addressed in a subsequent investigation of the tabular texts (Volume II).

4.5.18.1 Step 1. Time interval between lunation and Lunar Six event (N)

terminology	*napḫaru*(**pap**) = total?
unit	time degrees
astronomical meaning	time interval between lunation and Lunar Six event
column	variable
procedures:	
computing N	No. **93** P4'?

Function N represents the time between the lunation and sunrise (KUR, ŠU$_2$ and NA) or sunset (NA$_1$, ME and GI$_6$).[579] Since N is different for each Lunar Six interval, the synodic tables ideally contain a separate column N preceding each Lunar Six column.[580] In column N the numerical value of N is usually preceded by the sign **pap**, perhaps representing *napḫaru* = 'total'.[581] The more specific terms for N known from the system A procedures are not attested for system B, but can be assumed to apply here too. The algorithms for N, reconstructed by Schaumberger (1935), differ from the ones for system A only in that both systems employ different reference moments for the time of the lunation. Some of the algorithms may be partly preserved in No. **93** P4'. Column N includes a day number, which (ideally) represents the date of the sunset (NA$_1$, ME, GI$_6$) or sunrise (KUR, ŠU$_2$ and NA) of the Lunar Six event. The date in column N_{NA_1} is always 1 less than the date in column NA$_1$, i.e. 29 if the latter is 30, 30 if the latter is 1.[582] In other words, the sunset of NA$_1$ is assigned to the end of the day, while NA$_1$ itself is assigned to the beginning of the next day. The date in column N_{NA_1} is thus equivalent to the length of the month in whole days. There are only very few examples of column N for the other Lunar Six intervals; it appears that the dates in these columns do not always agree with the mentioned interpretation.[583]

NA$_1$

The earliest hypothetical possibility of NA$_1$ is the first sunset after New Moon, as shown in Fig. 4.47. If New Moon occurs between midnight and sunset (cases a–c of function M, i.e. $L < 6{,}0 - D/2$) then $N_{\mathrm{NA}_1} = 6{,}0 - \frac{1}{2}D^{\mathrm{nm}} - L^{\mathrm{nm}}$. By generalising this expression to the case when NA$_1$ occurs n days later one obtains

$$N_{\mathrm{NA}_1} = 6{,}0 - \tfrac{1}{2}D^{\mathrm{nm}} - L^{\mathrm{nm}} + n \cdot 6{,}0. \tag{4.126}$$

Note that if $n = 0$ this reduces to M (case c). If New Moon occurs between sunset and midnight (case d, i.e. $L > 6{,}0 - D/2$) the expression becomes

$$N_{\mathrm{NA}_1} = 12{,}0 - \tfrac{1}{2}D^{\mathrm{nm}} - L^{\mathrm{nm}} + n \cdot 6{,}0. \tag{4.127}$$

In either case the earliest theoretical possibility $n = 0$ falls on the day after New Moon. The value of NA$_1$ for that day is usually insufficient to guarantee visibility of the first crescent.

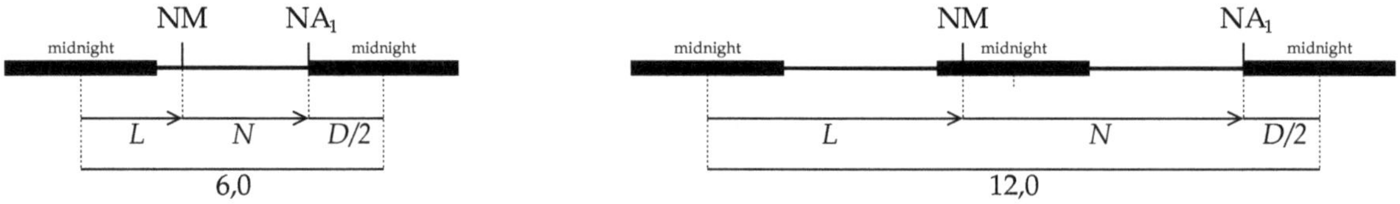

Fig. 4.47: N_{NA_1} for the earliest possible NA$_1$ ($n = 0$). **Left**: New Moon between midnight and sunset. **Right**: New Moon between sunset and midnight.

KUR

The latest hypothetical possibility of KUR is the last sunrise before New Moon (Fig. 4.48). If New Moon occurs between midnight and sunrise (case a, $L < D/2$), then $N_{\mathrm{KUR}} = 6{,}0 + L^{\mathrm{nm}} - \frac{1}{2}D^{\mathrm{nm}}$. By generalising this expression to the case when KUR occurs n days earlier one obtains

$$N_{\mathrm{KUR}} = (1+n) \cdot 6{,}0 + L^{\mathrm{nm}} - \tfrac{1}{2}D^{\mathrm{nm}}. \tag{4.128}$$

If New Moon occurs between sunrise and midnight (cases b–d, $L > D/2$) the corresponding expression is

[579] This was discovered by Epping (1889), pp. 93–94 ('Columne g').

[580] This is the case in *ACT* 100 (NA$_1$, KUR)

[581] Alternative reading: **kur$_2$**, a form of *nakāru*, 'to be different'?

[582] E.g. *ACT* 100, 101, 102, 120, 122.

[583] *ACT* 100, 102. For a new investigation of these columns cf. Volume II.

$$N_{\mathrm{KUR}} = L^{\mathrm{nm}} - \tfrac{1}{2}D^{\mathrm{nm}} + n \cdot 6{,}0. \tag{4.129}$$

If $n = 0$ this reduces to M (case b). These rules may be partly preserved in No. **93** P4'. The date of KUR depends on whether New Moon occurs during day time or during the night. If it occurs during the day (b,c) and $n = 0$ then KUR falls on the same day, i.e. in general $\mathrm{d}_{\mathrm{KUR}} = \mathrm{d}_{\mathrm{FM}} - n$. If New Moon falls in the night (a,d) and $n = 0$ then KUR occurs 1^{d} before New Moon, i.e. in general $\mathrm{d}_{\mathrm{KUR}} = \mathrm{d}_{\mathrm{FM}} - n - 1$. The value of KUR for the latest theoretical possibility $n = 0$ is usually insufficient to guarantee visibility of the lunar crescent.

Fig. 4.48: N_{KUR} for the latest possible KUR ($n = 0$). **Left**: New Moon between midnight and sunrise. **Right**: New Moon between sunrise and midnight.

ŠU₂

$\check{\mathrm{S}}\mathrm{U}_2$ is initially assumed to occur at the last sunrise before Full Moon. If Full Moon falls between midnight and sunrise (case a, i.e. $L < D/2$) then the time from sunrise to Full Moon is

$$N_{\check{\mathrm{S}}\mathrm{U}_2} = 6{,}0 + L^{\mathrm{fm}} - \tfrac{1}{2}D^{\mathrm{fm}}, \tag{4.130}$$

as shown in Fig. 4.49 (left). If Full Moon occurs between sunrise and midnight (cases b–d, i.e. $L > D/2$), then

$$N_{\check{\mathrm{S}}\mathrm{U}_2} = L^{\mathrm{fm}} - \tfrac{1}{2}D^{\mathrm{fm}}, \tag{4.131}$$

as shown in Fig. 4.49 (right). This is the same expression as M (case b). These rules may be partly preserved in No. **93** P4'. The date of $\check{\mathrm{S}}\mathrm{U}_2$ depends on whether Full Moon occurs during day time (b,c) or during the night (a,d). In the former case $\check{\mathrm{S}}\mathrm{U}_2$ falls on the same day, i.e. $\mathrm{d}_{\check{\mathrm{S}}\mathrm{U}_2} = \mathrm{d}_{\mathrm{FM}}$; in the latter case $\check{\mathrm{S}}\mathrm{U}_2$ occurs 1^{d} earlier than Full Moon, i.e. $\mathrm{d}_{\check{\mathrm{S}}\mathrm{U}_2} = \mathrm{d}_{\mathrm{FM}} - 1$.

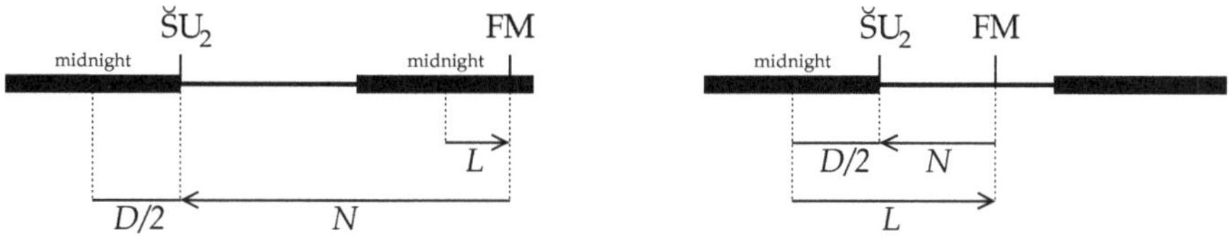

Fig. 4.49: $N_{\check{\mathrm{S}}\mathrm{U}_2}$. **Left**: Full Moon between midnight and sunrise. **Right**: Full Moon between sunrise and midnight. The times of Full Moon and Lunar Six event are indicated by vertical bars.

NA

NA is initially assumed to occur at the first sunrise after Full Moon. If the latter occurs between midnight and sunrise (M case a, i.e. $L < D/2$), then the time from Full Moon to sunrise is

$$N_{\mathrm{NA}} = \tfrac{1}{2}D^{\mathrm{fm}} - L^{\mathrm{fm}}, \tag{4.132}$$

as shown in Fig. 4.50 (left). If Full Moon occurs between sunrise and midnight (cases b–d, i.e. $L > D/2$), then

$$N_{\mathrm{NA}} = 6{,}0 + \tfrac{1}{2}D^{\mathrm{fm}} - L^{\mathrm{fm}}, \tag{4.133}$$

as shown in Fig. 4.50 (right). These rules may be partly preserved in No. **93** P4'. The date of NA depends on whether Full Moon occurs during day time or during the night. If it falls during the night then NA occurs on the same date, i.e. $\mathrm{d}_{\mathrm{NA}} = \mathrm{d}_{\mathrm{FM}}$. If Full Moon falls during day time, then the date of NA is one day later, i.e. $\mathrm{d}_{\mathrm{NA}} = \mathrm{d}_{\mathrm{FM}} + 1$.

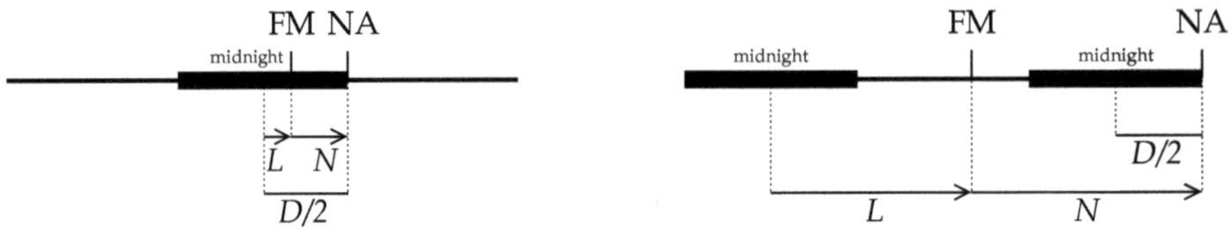

Fig. 4.50: N_{NA}. **Left**: Full Moon between midnight and sunrise. **Right**: Full Moon between sunrise and midnight.

ME

ME is initially assumed to occur at the last sunset before Full Moon. If the latter occurs between midnight and sunset (cases a–c of function M, i.e. $L < 6{,}0 - D/2$), then the time from sunset to Full Moon is

$$N_{\mathrm{ME}} = \tfrac{1}{2}D^{\mathrm{fm}} + L^{\mathrm{fm}}, \tag{4.134}$$

as sketched in Fig. 4.51 (left). If Full Moon occurs between sunset and midnight (M case d, i.e. $L > 6{,}0 - D/2$), then

$$N_{\mathrm{ME}} = \tfrac{1}{2}D^{\mathrm{fm}} + L^{\mathrm{fm}} - 6{,}0, \tag{4.135}$$

as sketched in Fig. 4.51 (right). The date of ME always coincides with the date of Full Moon, i.e. $\mathrm{d}_{\mathrm{ME}} = \mathrm{d}_{\mathrm{FM}}$.

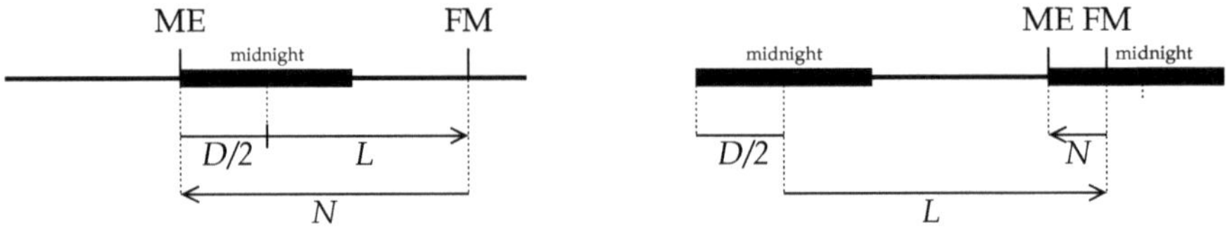

Fig. 4.51: N_{ME}. **Left**: Full Moon between midnight and sunset. **Right**: Full Moon between sunset and midnight.

GI$_6$

GI_6 is initially assumed to occur at the first sunset after Full Moon. If the latter occurs between midnight and sunset (cases a–c of M, i.e. $L < 6{,}0 - D/2$), then the time from Full Moon to sunset is

$$N_{\mathrm{GI}_6} = 6{,}0 - \tfrac{1}{2}D^{\mathrm{fm}} - L^{\mathrm{fm}}, \tag{4.136}$$

as sketched in Fig. 4.52 (left). This is the same expression as M (case c). If Full Moon occurs between sunset and midnight (M case d, i.e. $L < 6{,}0 - D/2$), then

$$N_{\mathrm{GI}_6} = 12{,}0 - \tfrac{1}{2}D^{\mathrm{fm}} - L^{\mathrm{fm}}, \tag{4.137}$$

as sketched in Fig. 4.52 (right). The date of GI_6 is always one day later than the date of Full Moon, i.e. $\mathrm{d}_{\mathrm{GI}_6} = \mathrm{d}_{\mathrm{FM}} + 1$.

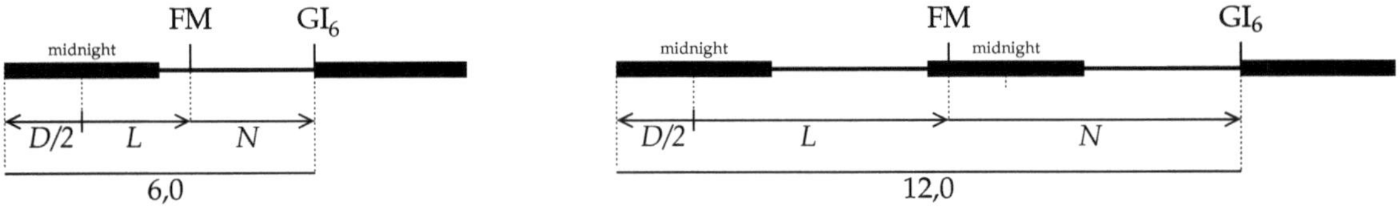

Fig. 4.52: N_{GI_6}. **Left**: Full Moon between midnight and sunset. **Right**: Full Moon between sunset and midnight.

4.5.18.2 Step 2. Zodiacal displacement of the Moon ($\Delta B_{\rightmoon}$) and the Sun ($\Delta B_{\odot}$)

terminology	*nisiḫ*(**zi**) *Sîn* = 'displacement of the Moon' ($\Delta B_{\rightmoon}$) *nisiḫ*(**zi**) *Šamaš* = 'displacement of the Sun' ($\Delta B_{\odot}$)
unit	degrees
astronomical meaning	zodiacal displacement of the Moon ($\Delta B_{\rightmoon}$) and the Sun ($\Delta B_{\rightmoon}$)

Step 2 is concerned with the zodiacal displacements $\Delta B_{\rightmoon}$ and $\Delta B_{\odot}$ covered by the Moon and the Sun during the interval N. Corresponding procedures are not preserved, and $\Delta B_{\rightmoon}$ and $\Delta B_{\odot}$ are not represented by a column in the tabular texts.

The existence of step 2 is implied by step 3 and by the fact that most synodic tables include a column F' (Eq. 4.118), whose only known application occurs in step 2. Schaumberger (1935) proposed that

$$\Delta B_{☽} = N \cdot F'_i, \tag{4.138}$$
$$\Delta B_{\odot} = N \cdot v_{\odot}. \tag{4.139}$$

In this expression $v_{\odot}$ is probably a constant, as in system A. The value is not exactly known, but it must be close to 0;0,10° per time degree corresponding to 1°/d.[584]

4.5.18.3 Step 3. Corrected elongation of the Moon (O)

terminology	BI, BE = *birītu*, 'distance'?
unit	degrees
astronomical meaning	corrected elongation between Moon and Sun (O^{nm}) or anti-Sun (O^{fm})
column	variable

Step 3 is concerned with function O, which is a measure of the Moon's elongation from the Sun (New Moon case) or the anti-Sun (Full Moon case) at the Lunar Six event.[585] Procedures for O are not preserved, but several synodic tables contain a column O, in which the numerical value of O is usually preceded by the sign BI or BE, a technical term for 'elongation' known from lunar system A.[586] Since the elongation is different for each Lunar Six interval one would expect the synodic tables to contain separate columns O preceding each Lunar Six column, but usually only one or two are present. The numbers in these columns imply that O is computed in two steps. First the true elongation is obtained by subtracting the distance covered by the Sun during the interval N from the distance covered by the Moon during the same interval,

$$\eta = \Delta B_{☽} - \Delta B_{\odot}. \tag{4.140}$$

Some synodic tables imply that η is approximated by $\Delta B_{☽}$, so that the effect of $\Delta B_{\odot}$ is neglected.[587] This can be justified because $\Delta B_{\odot}$ is about 12 times smaller than $\Delta B_{☽}$. The values in column O are close to, but not identical with η. Schaumberger[588] determined that O is usually smaller than η by a few degrees. Neugebauer (*ACT*, p. 82) proposed to represent O as

$$O = \eta - \varepsilon, \tag{4.141}$$

where ε is a small correction varying between −0;30° and +2°, depending on the Lunar Six interval. A detailed investigation of the algorithm underlying ε and its astronomical significance is outstanding (cf. Volume II).[589]

4.5.18.4 Step 4. Zodiacal position of the Moon ($B'_{☽}$) and the Sun ($B'_{\odot}$) at the Lunar Six event

terminology	[*qaqqar*(**ki**) *Sîn* = 'position of the Moon' = $B'_{☽}$] [*qaqqar*(**ki**) *Šamaš* = 'position of the Sun' = $B'_{\odot}$]
unit	zodiacal sign and degrees [0–30] within it
astronomical meaning	zodiacal position of the Moon ($B_{☽}$) and the Sun ($B_{\odot}$) at the Lunar Six event

In step 4 the zodiacal positions of the Sun ($B'_{\odot}$) and the Moon ($B'_{☽}$) are computed for the Lunar Six event by adding the displacements computed in step 2 to, or subtracting them from, the respective positions at lunation (cf. Eqs. 4.93–4.94 on p. 167). From these modified positions the coefficients q (step 5) and r (step 7) can be computed. However, q and r are insufficiently understood to determine whether they are computed from the zodiacal positions of the Moon and the Sun at the Lunar Six event, or from those pertaining to the lunation.[590] In the latter case step 4 is not part of system B's Lunar Six module. In the absence of relevant procedure texts this issue can only be resolved through a new analysis of the tabular texts (cf. Volume II).

---

[584] In the daily motion tables for system B $v_{\odot}$ = 0;0,9,51,30°/ time degree = 0;59,9°/d (§4.5.11).

[585] That O is closely related to the Moon's elongation from the (anti-)Sun was discovered by Epping (1889), pp. 94–97 ('Columne g').

[586] E.g. No. **53** P17', P18'. They are probably phonetic abbreviations of *birītu*, 'distance'.

[587] *ACT* 102; cf. *ACT*, p. 82 and *HAMA*, p. 539.

[588] Schaumberger (1935), pp. 385–386.

[589] Schaumberger (1935) proposed that ε models the effect of twilight. Alternatively it may play a similar role as the disk correction in lunar system A (step 10).

[590] Cf. the discussion in §4.4.19.5.

4.5.18.5 Step 5. Coefficient 'for the zodiac' (q)

terminology	cf. §4.4.19.5 (p. 168)
unit	dimensionless
astronomical meaning	ratio between zodiacal arc and associated rising (q_{rise}) or setting time (q_{set})

The Lunar Six module certainly incorporates an algorithm for a coefficient q by which differences in zodiacal position are converted into differences in rising and setting times as in system A. However, in system B the product $Q = q \cdot O$ (step 9) is added to O, and this yields the contribution to the Lunar Six interval corresponding to Q in lunar system A. Hence the coefficient q in system B is comparable to $q - 1$ in system A. In the absence of procedure texts, the reconstruction of step 5 has to proceed from the tabular texts. By analysing the ratio between O and Q Neugebauer was able to derive tentative reconstructions of q as zigzag functions or similar elementary functions of the zodiacal position of the Moon and the Sun with properties that are roughly consistent with $q - 1$ in system A.[591] The argument of these functions appears to vary between different tables: in some (*ACT* 101, 102) it is the position at lunation (B_i), for others Neugebauer suggests that it is $B_i + O_i/2$, the approximate midpoint of the arc between the Moon and the (anti-)Sun (*ACT* 100, 120). Since none of these functions exactly reproduces the attested values of Q there is still room for slightly different reconstructions. Perhaps the approach known from system A whereby q is computed separately for the positions of the Moon and the Sun at the Lunar Six event also underlies some instances of column Q in system B. If so, then q would be slightly different for the Moon ($q_{☽}$) and for the Sun ($q_{☉}$), which should have prompted the use of an average coefficient $q' = (q_{☽} + q_{☉})/2$ (cf. §4.4.19.5). These and other features of step 5 will be investigated in Volume II.

4.5.18.6 Step 6. Moon's distance to the ecliptic (β')

terminology	[*šūqu*(**nim**) *u šuplu*(**sig**) *ša*₂ *Sîn* = "height and depth' of the Moon']
unit	degrees?
astronomical meaning	the Moon's distance to the ecliptic at the Lunar Six event

The Lunar Six module may involve computing the Moon's distance to the ecliptic (β') at the Lunar Six event from the value at lunation (compare §4.4.19.6). This quantity may be used in step 8, but it is also conceivable that step 6 is omitted, and β' is approximated by β, the value at lunation. As in system A, the value of β, which is assumed to be expressed in degrees, could be obtained by converting E, which is presumably expressed in barleycorns. Since the extant synodic tables do not contain a column E, and this function is not attested in the procedure texts, our knowledge of step 6 is currently very limited. A new investigation of the tabular texts may shed light on these issues (cf. Volume II).

4.5.18.7 Step 7. Coefficient 'for height and depth' (r)

terminology	cf. §4.4.19.7)
unit	dimensionless
astronomical meaning	ratio between distance to ecliptic and rising (r_{rise}) or setting time (r_{set})

Column R (step 8) implies that the Lunar Six modules incorporate an algorithm for a coefficient r analogous to the one in system A (cf. §4.4.19.7 and the modern expression in Appendix E.1). This coefficient should represent the ratio between a function representing the Moon's distance to the ecliptic (β'), or a closely related function (cf. below), and the associated rising or setting time (R). Since the involved function representing the Moon's distance to the ecliptic is unknown, and in the absence of procedures or auxiliary tables for r as they exist in system A, almost nothing concrete is known about r. For instance, it is not clear whether r is computed as the average of a value for the Moon ($r_{☽}$) and one for the Sun ($r_{☉}$), each evaluated at the Lunar Six event (cf. step 4), or whether these are evaluated at the lunation, in which case $r = r_{☽} = r_{☉}$, so that the averaging is not necessary.

[591] Cf. *ACT* p. 83 and the commentaries to *ACT* 100–102, 120.

4.5.18.8 Step 8. Contribution to Lunar Six resulting from Moon's distance to ecliptic (R)

terminology	*ša₂ šūqi*(**la₂**)-*šu₂* = 'that of its 'height'' ($R_\uparrow$) *ša₂ šupli*(**sig**|**bur₃**)-*šu₂* = 'that of its 'depth'' ($R_\downarrow$)
unit	time degrees
astronomical meaning	contribution to Lunar Six interval reflecting the Moon's distance to the ecliptic
column	variable

Function R is an additive or subtractive correction to the Lunar Six interval representing the effect of the Moon's distance to the ecliptic, an interpretation first suggested by Kugler.[592] In column R, usually placed between columns Q and the first Lunar Six column, the value of R is often accompanied by the term 'that of its 'height'' or 'that of its 'depth', the former signifying that the Moon is above the ecliptic ($R_\uparrow$), the latter that it is below the ecliptic ($R_\downarrow$). Schaumberger and Neugebauer carried out detailed investigations of column R,[593] but it has thus far proven impossible to fully reconstruct the underlying algorithm. Schaumberger established that R is applied to $O+Q$ in accordance with Eqs. (4.143–4.144), which confirms the geometrical interpretation of the Lunar Six module shown in Fig. 4.36 (p. 176). R is presumably computed as in system A as the product of r and β', the Moon's distance to the ecliptic measured in degrees (step 6), or the equivalent function E measured in barleycorns (§4.5.6), or some other closely related function.[594]

4.5.18.9 Step 9. Contribution to Lunar Six resulting from the Moon's elongation (Q)

terminology	*ša₂ lu-maš* = '(that) for the zodiac'
unit	time degrees
astronomical meaning	contribution to the Lunar Six interval resulting from lunar elongation
column	variable

Function Q represents the contribution to the Lunar Six interval resulting from the Moon's elongation (distance along the zodiac), O, an interpretation first proposed by Epping,[595] and proven by Schaumberger.[596] Column Q is usually placed between columns O and R. It often includes the term '(that) for the zodiac', obviously a reference to O from which it is computed. Q is probably obtained by an algorithm similar to the one in system A, i.e.

$$Q = q \cdot O. \tag{4.142}$$

On that assumption Neugebauer (*ACT*, p. 83) succeeded in deriving rough estimates of q for several synodic tables (cf. step 5).

4.5.18.10 Step 11. Assembling the Lunar Six interval

column	variable
procedures:	
NA₁ or KUR?	No. **93** P2'

In step 11 the Lunar Six intervals are assembled from the contributions O (step 3), Q (step 9) and R (step 8). Step 10 is absent, but cf. below. The algorithms, reconstructed by Schaumberger (1935) from synodic tables, can be represented as follows:

$$\text{KUR, NA}_1\text{, NA, ME} = \begin{cases} O+Q+R_\uparrow & \text{(Moon above the ecliptic)} \\ O+Q-R_\downarrow & \text{(Moon below the ecliptic)} \end{cases} \tag{4.143}$$

$$\text{ŠU}_2\text{, GI}_6 = \begin{cases} O+Q-R_\uparrow & \text{(Moon above the ecliptic)} \\ O+Q+R_\downarrow & \text{(Moon below the ecliptic).} \end{cases} \tag{4.144}$$

These expressions are consistent with the geometric interpretation of the Lunar Six modules discussed in §4.4.19.11. Depending on whether the Moon is above or below the ecliptic, R is added to $O+Q$ or subtracted from it. There is no

[592] Kugler (1924), pp. 593–595 ('Columne n').

[593] Schaumberger (1935), pp. 381–383 ('Hilfskolumne R=n'); *ACT* p. 83.

[594] It seems worthwhile to reconsider the assumption of Schaumberger (1935), initially adopted by Neugebauer (1936) but later apparently abandoned, that R is computed from the eclipse magnitude Ψ'', which was still interpreted as the Moon's distance to the ecliptic in Neugebauer (1936).

[595] Epping (1889), p. 99 ('Columne l'). Cf. also Kugler (1924), pp. 592–593 ('Columne l'). Sidersky (1919) attempted to interpret Q through a comparison with the algorithms for the first appearance of the crescent in Maimonides's *Sanctification of the Moon*, for which Sidersky, unsuccessfully, claimed a Babylonian origin.

[596] Schaumberger (1935); cf. also Neugebauer (1936), pp. 285–286.

evidence of an explicit disk correction along the lines of step 10 in the Lunar Six module for system A (§4.4.19.10). However, the correction ε which is absorbed in O and whose precise purpose is unknown might play a similar role.[597] The only known reference to step 11 in the procedure texts is found in No. **93** P2'. All that can be understood is the instruction 'if it is high you add, if it is low you subtract', very likely pertaining to the computation of NA_1 or KUR.

Thresholds for NA_1 and KUR

If the values of NA_1 and KUR obtained in step 11 exceed a certain threshold then the lunar crescent is considered to be visible. If the threshold is not reached step 13 is executed, and the Lunar Six module must be repeated for the next day (NA_1) or the previous day (KUR). When the final value NA_1 is known the date of the immediately following Full Moon and New Moon can be determined with the help of the rules explained in connection with L (§4.5.16). As in system A, the thresholds for NA_1 and KUR are not mentioned in any text. Neugebauer analysed the available Lunar Six columns in synodic tables and eclipse tables,[598] but he did not discover a consistent set of criteria. In some tables NA_1 and KUR seem to satisfy similar criteria as in system A (Eqs. 4.106–4.107 on p. 175); in other tables NA_1 appears to have been subjected to the simpler condition

$$\mathrm{NA}_1 > m_{\mathrm{NA}_1} \approx 11\text{–}12^\circ. \tag{4.145}$$

4.5.18.11 Step 12. Optional renaming of $ŠU_2$, NA, ME or GI_6

The Lunar Six modules for $ŠU_2$, NA, ME and GI_6 must incorporate a set of rules for dealing with the case when the subtraction of R in step 11 is impossible because $R > O+Q$. From the analogous considerations for system A (Eq. 4.108 on p. 177), omitting the disk correction and replacing Q by $O+Q$, it is inferred that the following algorithms for the renamed Lunar Six values should apply in system B:

$$\begin{array}{llll} R_\uparrow > O+Q & \Rightarrow & \mathrm{NA} = R_\uparrow - (O+Q) & \text{(Moon above the ecliptic)} \\ R_\downarrow > O+Q & \Rightarrow & \check{\mathrm{S}}\mathrm{U}_2 = R_\downarrow - (O+Q) & \text{(Moon below the ecliptic)} \\ R_\downarrow > O+Q & \Rightarrow & \mathrm{GI}_6 = R_\downarrow - (O+Q) & \text{(Moon below the ecliptic)} \\ R_\uparrow > O+Q & \Rightarrow & \mathrm{ME} = R_\uparrow - (O+Q) & \text{(Moon above the ecliptic).} \end{array} \tag{4.146}$$

Other aspects of step 12 are the same as in system A (§4.4.19.12).

4.5.18.12 Step 13. Optional shift of the Lunar Six event by 1^d

Step 13 concerns the modification of N resulting from the shift of the Lunar Six event by 1^d. With the modified value of N steps 2–11 can be executed for a second time in order to obtain the final value of the Lunar Six interval. The rules governing step 13 are not preserved in the procedure texts, but should be exactly the same as in system A (§4.4.19.13).

[597] For a new investigation of this issue cf. Volume II.

[598] *ACT*, p. 67; *HAMA*, pp. 539–540. A new investigations of this issue will appear in Volume II.

Chapter 5
Critical editions

5.1 Conventions and notation

Each edition begins with a header providing the museum number(s) of the tablet or fragment(s), provenience (Babylon or Uruk), date (if known), measures, arrangement of obverse and reverse (O/R: turn around horizontal axis; O–R: turn around vertical axis), published photographs and copies, and a short description of the content. If the tablet is not preserved to full width, height or thickness, the incompletely preserved dimension is always followed by a reconstructed value of the original dimension in round brackets, e.g. 10(12), by a star (*) if this could not be estimated, e.g. 10(*).

5.1.1 Transliterations

Each transliteration includes markers for the structure of the tablet as defined by its sides, columns, sections (procedures and tables) and lines. The columns of a procedure text are ordered from left to right on the obverse, and usually from right to left on the reverse. Those of a tabular text are usually ordered from left to right on both sides. The division of a tablet into procedures (P1, P2,...) is inferred from the horizontal lines, sometimes from the empty space between successive lines. In the translations there is an additional level of structure for subprocedures (a,b,c, etc.), but these are not marked in the transliterations. Primed numbers are used if there may be missing columns, procedures, subprocedures, tables or lines before the first one that is preserved on the fragment. An effort is made to estimate the number of missing lines on each side of the tablet, if necessary in relation to the number for the other side. The vertical alignment of the signs on a tablet is not reproduced in the transliterations.[599]

notation:

O	obverse
R	reverse
UE	upper edge
LE	lower edge
X,Y	sides of unknown orientation
i,ii,...	columns
1,2,...	lines
P1,P2,...	procedures
P1.a,P1.b,...	subprocedures
T1,T2,...	tables

Signs and words

Cuneiform signs are transliterated in accordance with *Das akkadische Syllabar* (von Soden & Röllig 1991) and a few additional conventions introduced by Neugebauer.[600] Both the older 9 and the LB version of this number are transliterated as 9. The latter sign is also used in multiplications as an equivalent of **a.ra$_2$** and GAM, 'times', in which capacity it is

[599] In score editions (e.g. No. **61**) missing signs (x) are vertically aligned with the corresponding signs in the duplicates if it is plausible that they were present on the tablet. If this is not the case then the x's are distributed equidistantly.

[600] *ACT*, pp. 1–4, 38; cf. also Neugebauer (1947).

transliterated as GAM$_0$.[601] The sign MAN (20) is transliterated as *šamaš*$_2$ when it denotes the Sun;[602] 30 as *sin* when it denotes the Moon. The sign GAM is also used as a separator between numbers or words, in which capacity it is transliterated as :. For its usage in sexagesimal numbers cf. below. Logograms with an established (pseudo-) Sumerian reading are written in lower case boldface. If the Sumerian reading of the sign is unknown, not well established or not commonly used, the sign is transliterated with the default reading in capitals. For instance, GIŠ has a reading *našû*, 'to compute', which can be traced back to Sumerian **guru**$_{17}$(GIŠ-*tenû*). One might therefore transliterate **guru**$_x$, taking the sign to be a variant of GIŠ-*tenû*. Since this has not been adopted elsewhere in the literature I maintain the conventional GIŠ. Rare phonetic values encountered in mathematical astronomy are *ar*$_5$(GAM) [No. **94** R13'], *re*$_x$(SIPA) [col. of *ACT* 18] and *suḫ*$_x$(ŠAḪ) [No. **66** X4']. The latter two are not mentioned in *Das akkadische Syllabar*.

Babylonian astronomical texts contain numerous abbreviations, whose origin can be logographic, phonetic or unknown. Logographic abbreviations are set in lower case boldface like ordinary logograms. I make an exception for some abbreviated logograms whose proper Sumerian reading is not commonly used, e.g. I maintain the conventional transliteration ZIZ$_2$ of the logogram for month XII (*Šabāṭu*), even though the proper Sumerian reading is **ud**$_2$ (p. xxv). Phonetic abbreviations are set in lower case italics, like other Akkadian signs. Some phonetic abbreviations are pseudo-logograms in the sense that they are not declined and may take the plural determinative **meš** or **me**. In these cases, the phonetic component is set in lower case italics, e.g. *si-man*.**meš** = *simānū*, 'times; durations', and *zib*.**me** = *zibbātu*, 'Pisces' (lit. 'tails'). Unidentified abbreviations are written in capitals.

The transliterations are kept to a minimum. Normalised Akkadian words are not provided, e.g. I do not transliterate **e**-*bi*, 'you call it', as *taqabbi*(**e**-*bi*). Sometimes the normalised Akkadian word or underlying verb is provided in the philological notes, otherwise in the Glossary. Missing signs are enclosed by square brackets []. If no plausible reconstruction could be suggested each missing sign is represented by an x, such that the number of x's provides a rough indication of the number of missing signs. The x is also used for unidentified intact signs. Breaks of unknown length are rendered as [...].

notation:

šamaš$_2$	20
GAM	
0	GAM as digit zero
:	GAM as separator
ar$_5$	GAM
GAM$_0$	
re$_x$	SIPA
suḫ$_x$	ŠAḪ
⸢SIGNS⸣	signs damaged
x	sign unidentified
[xx]	break with space for approximately 2 signs
[...]	break of unknown length
⟨SIGNS⟩	signs omitted
{SIGNS}	signs included erroneously
?	suggested reading uncertain

Numbers

Sexagesimal numbers are represented with adjacent digits separated by a dot (period). I consider this convention, which goes back to Thureau-Dangin, to have significant advantages over other commonly used ones involving commas or empty spaces. The arguments were recently summed up by Proust (2009a). First, dots are the obvious choice since they are also used for separating signs within Sumerian words or logograms. Second, the use of commas could be confusing, because they also separate digits in translated, fully interpreted sexagesimal numbers. The use of dots has the advantage that it creates a notational distinction between translated, fully interpreted numbers and transliterated numbers (provided they contain at least two digits), so that the latter are immediately recognisable as having only a relative value. Empty spaces as separators are less satisfactory than dots, because sexagesimal numbers become more difficult to read in the absence of a visible marker between the digits. Moreover, nothing is gained by using empty spaces, since this involves the same interpretative steps required for placing dots or commas.

At the level of transliteration, sexagesimal numbers are interpreted only by segmenting them into digits in the most plausible way suggested by orthographic features and context, and placing a dot between adjacent digits. The first step

[601] The Akkadian reading of GAM$_0$, like that of **a.ra**$_2$ and GAM, is not clear (§2.2.4).

[602] In some texts MAN (20) is written with an extra horizontal wedge below it; this is usually mentioned in the critical notes.

of this process involves combining adjacent numerals into digits 0·59. Here one may encounter ambiguities resulting from the occurrence of multiples of 10 followed by a numeral 1·9 (e.g. 20 9), since they can be either combined into a single digit (29), or treated as separate digits (20.9). Usually the context and sometimes also the orthography help in making this an obvious choice. In some texts from Babylon, numerals are separated by a somewhat larger space if they are to be interpreted as separate digits. In texts from Uruk, the sign GAM is often inserted between such numerals in order to prevent them from being wrongly interpreted as a single digit. In the transliterations, these instances of GAM are represented by a zero, e.g. 50.09 (for 50 GAM 9).

5.1.2 Translations

Each translated procedure or table is marked by a label in the left margin (P1, P2, ..., T1, T2, etc.), followed by a header (in brackets) with a brief description of the content. I have introduced an additional level of structure for subprocedures, which are labeled P1.a, P1.b, etc. (they are not marked in the transliterations).[603] Within each procedure, the lines of the tablet are broken up or merged into meaningful units (subprocedures, paragraphs and sentences). Line numbers are placed in a raised position before the first word.[604] All logographically written verbs in main clauses are interpreted as second person of the present tense. There is no evidence for the use of imperatives in astronomical procedure texts.[605] The main principles underlying the translations (cf. also §1.6.1) can be summarised as follows:

(1) modern technical terms are avoided as much as possible;
(2) literal translations are preferred and important features of the Akkadian formulation are preserved;
(3) each semantically distinct meaning of an Akkadian word is translated by the same English word or phrase;
(4) translations are rendered in grammatically correct English.

There is obviously a tension between principles 2 and 4, and between 3 and 4, so that they cannot be applied in a dogmatic way.

Technical terms

Compared to the translations of Neugebauer and others, the effective meaning of a term is no longer always immediately obvious. For instance, 'height' replaces Neugebauer's 'positive latitude' as a translation of *šūqu* in contexts where it denotes distance above the ecliptic. In order to aid the reader I have included the corresponding modern symbol or a descriptive term in brackets after the first occurrence of the word in the procedure. In the case of *šūqu* the translation becomes 'height ($E_{\uparrow}$)'. A special case is the number 2.13.20, which is the name of function Φ (§4.4.3); hence I translate this as '2,13;20 (Φ)'. For the synodic phenomena of the planets the modern abbreviations (Tables 3.1–3.2, p. 57) are included, so that **igi** *ša₂* **kur** is translated as 'eastern appearance (MF)'. Guided by the main principles (1–4), I have adopted the following solutions for various classes of technical terms.

1. The names of planets, zodiacal signs and months

For the names of the planets, the Moon, the Sun, the zodiacal signs and the months I make an exception to principles 2 and 3, in that I translate them with modern equivalents (p. xxv).

2. Common Akkadian words

Common Akkadian words that acquire a technical meaning in a certain context are translated as follows:

(a) by the English common word enclosed in quotation marks. For example *ūmu*, 'day', can have a technical meaning '(mean) tithi' in certain contexts, in which case I translate *ūmu* as "day" (in quotation marks). If the literal meaning 'day' is implied I translate 'day' (without quotation marks). Other examples are ḪAB-*rat* = 'disk' (= eclipse magnitude Ψ, Ψ', etc., in lunar systems A and B), or *šūqu*(**nim**) *u šuplu*(**sig**) = 'height and depth', the term for distance to the ecliptic (E). The presence of quotation marks in a translation always signifies that the effective (functional) meaning of the word differs from the quoted, literal translation.
(b) by a corresponding English technical term. This option is chosen if (a) is impractical. For instance, the word *birītu*, 'distance', has various technical meanings including 'time interval' and 'elongation' (distance between two points

[603] In *ACT* a similar notation is also used on very few occasions, e.g. P2a, but these are not subprocedures.
[604] Usually before the first translated word of the line. Due to differences in word order between Akkadian and English, some words may appear on the wrong side of the line number.
[605] Imperatives are attested in OB mathematical problem texts.

measured along the zodiac). In the latter case the word is rendered by the abbreviation bi_2(BE) or *bi*, which probably represents *birītu*. Using the same translation 'distance' for all these instances would ignore the distinction between 'distance' and 'elongation' that is made at the orthographic level. For this reason bi_2 is translated as 'elongation'.

3. Technical terms whose literal meaning is irrelevant or unknown

In some cases where the effective meaning of a term is clear from the context the exact literal meaning is irrelevant or semantically far removed from the effective meaning, so that it does not lend itself as a translation. In these situations I have chosen to combine the Akkadian word with a descriptive English term representing the category to which the technical term belongs. An example is *uddazallû*, a loan word from Sumerian **u_4.da.zal.la_2**, whose original, literal meaning may be 'time that has passed'. However, in mathematical astronomy *uddazallû* effectively means 'coefficient', hence I translate it as '*uddazallû*-coefficient'. Other examples are '*ṣiliptu*-coefficient' and '*nappaltu*-coefficient'. It may also happen that the Akkadian reading of a logogram is unknown, so that we do not know the literal translation. If the effective meaning inferred from the context can be rendered in English by a single word or a concise phrase, this is adopted as a conventional translation. An example is LA_2, 'duration', in the phrase LA_2 *$ša_2$* **me|gi_6** = 'duration of daylight|night' (lunar system A).

4. Logograms with a complex technical meaning

The Lunar Six intervals NA_1, KUR, $ŠU_2$, NA, ME, GI_6 (§4.1.4) and some other terms (e.g. DIŠ.UD, also in the Lunar Six module) with a complex technical meaning cannot be rendered by a single English word or simple phrase. It is common practice to adopt the transliterated logograms (written in capitals) as pragmatic translations of these terms.

Numbers

In translations, sexagesimal numbers are represented by their absolute value whenever it is known, using the conventional notation, i.e. with commas between all digits except for a semicolon between the digit pertaining to $1 = 60^0$ and the one pertaining to 60^{-1} (cf. §2.1). Only in rare cases where the absolute magnitude of a number could not be established from the context is the relative notation based on dots used in the translation.

Formulaic representations

The translations of a few complex procedures (No. **61** P1–P4 and No. **53** P17',P18') are accompanied by formulaic representations which are explained in §4.4.19.

5.1.3 Critical and philological notes and commentaries

The critical and philological notes include remarks about the state of preservation, orthography and alternative readings of the cuneiform signs. Virtually all tablets were photographed and collated; if not, this is mentioned in the commentary. If a text has been published before, and a sign is read differently from what is stated in the latest edition, this is also mentioned. Changes with respect to editions preceding the last edition are usually not reported. Also trivial modifications affecting the placement of brackets around broken or damaged signs are not mentioned. The philological notes are kept to a minimum, since the Akkadian readings of the logograms can be found in the Glossary, and important linguistic and terminological issues are dealt with in the preceding chapters.

Each commentary begins with remarks about the physical appearance and the state of preservation of the tablet, archaeological and archival aspects, notable orthographic and linguistic features, the structure of the tablet, the nature of the procedures, a discussion of the colophon (if present) and general remarks about the astronomical content. This is followed by sections devoted to the individual procedures in which their state of preservation, typology, formulation and purpose are described. Many procedures are specific implementations of the templates defined in previous chapters. In that case the astronomical commentary only provides references to the appropriate templates and the relevant sections of Chapters 2–4, and remarks about any deviations from the templates. A complete astronomical commentary is provided only if the procedure does not follow a known template. This concerns mainly procedures whose purpose is not directly connected with the tabular texts, and procedures that reflect an older stage of development compared to the bulk of the corpus.

5.2 Planets

5.2.1 Mercury

No. 1

museum numbers	BM 45980 (81–7–6,423+434)
provenience	Babylon
date	data in synodic table: SE 133–153
measures	15.1 × 18.5 × 2.0–2.7 cm (type i)
arrangement	O/R
cuneiform text	photo: Fig. F.1 (R: procedures); *ACT*, Pl. 244 (R)
bibliography	Kugler (1907), pp. 178–186, 197–200 (partial edition of 81–7–6,423 R); *ACT* 820a, 301
contents	Mercury system A$_1$: synodic table (*T*, *B* for EF, EL, MF, ML) and corresponding procedures

Reverse

P1 1 *e-pe-šu$_2$ ša$_2$* **igi.meš** *an-ne$_2$-e-tu$_4$* **igi** *ša$_2$* **kur** *ana* **igi** *ša$_2$* **kur ta** ⸢1 **a en** 16 **maš$_2$** 1.46 **tab**⸣ *ša$_2$* ⸢*al*⸣ 16 ⸢**maš$_2$**⸣ **d**[**iri** GAM 1.20 **du ta** 16 **maš$_2$ en** 30 **mul$_2$.mul$_2$**]

2 2.21.20 **tab** *ša$_2$ al* 30 **mul$_2$.mul$_2$ diri** GAM 40 **du ta** 30 **mul$_2$**.⸢**mul$_2$ en**⸣ [1 **a** 1.34.13].⸢20⸣ [**tab** *ša$_2$ al* 1 **a diri** GAM 1.7.30 **du**]

3 *bi-rit* **igi** *ana* **igi** GIŠ.A 3.30.39 **ki**-*šu$_2$* **tab**-*ma* **ki** *u$_4$-mu* ⸢**tab**?⸣ [xxxxx **igi** *ša$_2$* **šu$_2$** *ana* **igi** *ša$_2$* **šu$_2$ ta** 6 **alla en** 26 **rin$_2$**]

4 ⸢2⸣.40 **tab** *ša$_2$ al* 26 **rin$_2$ diri** GAM 40 **du ta** 2⸢6⸣ **rin$_2$ en** 10 *zib* 1.⸢4⸣6.[40 **tab** *ša$_2$ al* 10 *zib* **diri** GAM 54 **du ta** 10 *zib* **en** 6 **alla** 1.36 **tab**]

5 [*ša$_2$ al*] ⸢6 **alla**⸣ **diri** GAM 1.40 **du u$_4$.meš gin$_7$ igi**-*u$_2$* **kimin ta igi** *ša$_2$* ⸢**kur en šu$_2$**⸣ [*ša$_2$* **kur** 12 **ḫun** 14 **mul$_2$.mul$_2$** 18 **maš.maš** 22 **alla**]

6 [26 **a** 30 **absin**] ⸢3⸣4 **rin$_2$** 44 **gir$_2$ pa** 42 **maš$_2$** 30 **gu** 24 *zib ina* **šu$_2$** *ina* [xx 36 **ḫun** 42 **mul$_2$.mul$_2$** 45 **maš.maš** 42 **alla** 36 **a** 22 **absin**]

7 [14 **rin$_2$ gir$_2$** 16 **pa**] ⸢20⸣ **maš$_2$** 22 **gu** ⸢*zib*⸣ **u$_4$.meš** *ina* **kur** *ina* 14 *ana* **igi**-*šu$_2$* **tur** [xxxxxxxxxxxxxxxxxxxx]

8 [xxxxx] ⸢**u$_4$.meš** xxx⸣ [xxxx] ⸢xx⸣ *ina* ⸢3⸣6 **u$_4$.meš** *ina* ⸢**šu$_2$**⸣ [xxxxxxxxxxxxxxxxxx]

Reverse

P1 **(Mercury system A$_1$)**

a **(MF: updating *B* with the synodic arc)**

[1]Procedure for these appearances. Eastern appearance (MF) to eastern appearance.
From 1 Leo until 16 Cap you add 1,46. (The amount) by which it exceeds 16 Cap [you multiply by 1;20.
From 16 Cap until 30 Tau] [2]you add 2,21;20. (The amount) by which it exceeds 30 Tau you multiply by 0;40.
From 30 Tau until [1 Leo you add 1,34;13,20. (The amount) by which it exceeds 1 Leo you multiply by 1;7,30.]

b **(MF: updating *T* with the synodic time)**

[3]The distance from appearance to appearance you compute, 3;30,39 you add to it, and to the ‘day’ of ⸢appearance?⸣ [you add it ...

c **(EF: updating *B* with the synodic arc)**

Western appearance (EF) to western appearance.
From 6 Cnc until 26 Lib] [4]you add 2,40. (The amount) by which it exceeds 26 Lib you multiply by 0;40.
From 26 Lib until 10 Psc [you add] 1,46;[40. (The amount) by which it exceeds 10 Psc you multiply by 0;54.
From 10 Psc until 6 Cnc you add 1,36.] [5][(The amount) by which it] exceeds 6 Cnc you multiply by 1;40.

d **(EF: updating *T* with the synodic time)**

The ‘days’: as before.

e **(Angular pushes from MF to ML)**

From eastern appearance (MF) to [eastern] setting (ML): [12 Ari, 14 Tau, 18 Gem, 22 Cnc, [6]26 Leo, 30 Vir], 34 Lib, 44 Sco (and) Sgr, 42 Cap, 30 Aqr, 24 Psc.

f **(Angular pushes from EF to EL)**

In the west (= EF to EL): [36 Ari, 42 Tau, 45 Gem, 42 Cnc, 36 Leo, 22 Vir, [7]14 Lib (and) Sco, 16 Sgr], 20 Cap, 22 Aqr (and) Psc.

g **(Temporal pushes from MF to ML)**

The ‘days’, in the east: in 14 forward, the minimum [...] [8][...]

h **(Temporal pushes from EF to EL)**

[...] the ‘days’ ... [...]... 36 ‘days’ in the west [...]

Critical and philological notes

Reverse

1 ⌜16 **maš$_2$** 1.46 **tab**⌝: only the upper parts are preserved.
Here and elsewhere ‘(The amount) by which it exceeds ...’: replaces ‘Whatever exceeds ...’ (*ACT*).

2 [1.34.13].⌜20⌝: only the upper half of 20 is preserved. It might also be a 30, but the preceding gap is too small to consider it as belonging to 1.7.30.

7 **tur** = *ṣeḫertu*, ‘smallest’ one. This confirms Neugebauer’s suggestion that the minimum of a time interval is mentioned here.

8 After ⌜**u$_4$.meš**⌝ there are very faint illegible traces. Before *ina* 36 there are traces including a vertical wedge, probably the end of a sign.

Commentary

Three physically joined fragments form the bulk of this large tablet. The obverse and most of the reverse are occupied by a synodic table with columns *T* and *B* for EF, EL, MF and ML, computed in accordance with system A_1, for which I refer to *ACT* 301 and Volume II. The data cover the years SE 133–153, suggesting that the tablet was written near SE 133 (179/8 BC). The investigation of this tablet began with Kugler (1907), who reconstructed the algorithm for σ from the synodic table on the main fragment 81–7–6,423. The procedure is preserved on the fragment 81–7–6,434 which was joined to it later (*ACT*). The end of the reverse contains a single procedure (P1). The right half of P1 is missing, but the textual restorations imply that it was written across the entire width of the tablet. A colophon is not preserved; it may have been present in the missing right half of the last line.

P1 Mercury system A_1: updating B and T (MF, EF); pushes from MF to ML and from EF to EL

P1 contains 8 subprocedures (a–h). Only a and b start at the beginning of a line. The introductory phrase ‘Procedure for these appearances’ points to the first appearances (MF, EF) dealt with in a–d. Parts e–h are concerned with ML and EL. Hence all of the procedures are directly connected with the synodic table. For Mercury system A_1 cf. §3.4.2. Part **a** (R1–2) is a procedure for updating the zodiacal position (*B*) of MF with the step function for σ. It is formulated in accordance with template STEP.A.1 (§2.6.10). Part **b** (R3) is an instruction for updating the time (*T*) of MF in accordance with Eq. (3.9). It is formulated in accordance with template T.1 (§3.3.2.1). Part **c** (R3–5) is an instruction for updating *B* of EF with the step function for σ also based on template STEP.A.1. Part **d** (R5) is an instruction for updating the time (*T*) of EF. The algorithm is not provided, but stated to be the same as that for MF. Part **e** (R5–7) is a list of angular pushes ($\delta\Sigma$) from MF to ML, part **f** (R6–7) for EF to EL (§3.4.2.3). Parts **g–h** (R7–8) are concerned with the corresponding temporal pushes ($\delta\tau$). As suggested by Neugebauer, $14^{\bar{\tau}}$ (R7) may be the minimum of $\delta\tau$ (MF to ML), which occurs in 15° Ari, and $36^{\bar{\tau}}$ (R8) may be $\delta\tau$ (EF to EL) at the same position (cf. Table 3.8, p. 70).

No. 2

museum number	BM 36436 (80–6–17,163)
provenience	Babylon
measures	5.5(*) × 6.2(*) × 3.0 cm; column width 2 cm
arrangement	O/R
cuneiform text	photo: Fig. F.4
contents	Mercury systems A_1, A_2, unknown system; a planet: various procedures

Side X
Column i'
(about m lines missing)
P0' 1' [xxxx] ⸢x⸣
2' [xxxx] ⸢x⸣
(about n+7 lines missing)
Column ii'
P1' (m lines missing)
1' [xxxx] ⸢**absin**$_0^?$⸣
2' [xxxx] **gir**$_2$
3' [xx] 22 **maš**$_2$
4' [x$^?$] ⸢2$^?$4⸣ **gu** *zib*

P2' 5' **šu**$_2$ *ša*$_2$ **šu**$_2$
6' *ana* **šu**$_2$ *ša*$_2$ **šu**$_2$
7' [t]**a** 30 **maš**
8' [**en** 30 **pa**] ⸢1⸣.48.30
9' [xxxx] ⸢x⸣
(n lines missing)
Column iii'
P3' (m lines missing)
1' 14 ⸢**gir**$_2^?$⸣ [x$^?$ **pa**]
2' 9 (erasure?) **maš**$_2$
3' 14 **gu**
4' 20 *zib*

P4' 5' **gu**$_4$ *ina* **igi šu**$_2$-[*šu*$_2^?$]
6' *ina* 35 **me** ⸢x⸣ [x]
7' *ina* **igi**-*at*$_2$ [x]
8' *ina* 15 **me** ⸢x⸣ [x]
9' **ta maš en** ⸢x⸣ [x]
10' 5 ⸢**ta**$^?$ **gir**$_2^?$⸣ [**en** x]
(n−1 lines missing)
Side Y
Column i'
P5' (about n+1 lines missing)
2' [xxxxx *ina*] ⸢8$^?$⸣ **me**
3' [x] ⸢6$^?$.1⸣5 ⸢**absin**$_0$ *ina*⸣ 14 **me**
4' [x] 6 **rin**$_2$ *ina* 14 **me**
5' [x] ⸢7$^?$⸣.15 **gir**$_2$ *ina* 21 **me**
6' [x] ⸢xx⸣2.20 **pa** *ina* 21 **me**
7' [x] ⸢x 30⸣ **maš**$_2$ *ina* ⸢x **me**⸣
(about m+1 lines missing)

Side X column i'

P0' (Unclear)

(about m lines missing)
1'[...] ... 2'[...] ...
(about n+7 lines missing)

Column ii'

P1' (Mercury system A_1: temporal push EF to EL?)

(m lines missing)
1'[...] ⸢Vir$^?$⸣,
2'[...] Sco,
3'[...] 22 Cap,
4'⸢2$^?$4⸣ Aqr, Psc.

P2' (Mercury system A_2: synodic arc for EL)

5'Evening setting (EL) 6'to evening setting: 7'[F]rom 30 Gem 8'[until 30 Sgr] 1,48;30 9'[...] ...
(n lines missing)

Column iii'

P3' (Mercury system A_2: temporal push EL to MF?)

(m lines missing)
1'14 ⸢Sco⸣ [... Sgr]
2'9 *(erasure?)* Cap
3'14 Aqr
4'20 Psc

P4' (Mercury: subdivision synodic cycle)

5'Mercury. Before [its] setting 6'in 35 'days' [...]
7'Before [its ...] 8'in 15 'days' ... [...]
9'From Gem until ... [...] 10'5.
⸢From Sco$^?$⸣ [until ...]
(n−1 lines missing)
Side Y column i'

P5' (A planet: subdivision synodic cycle?)

(about n+1 lines missing)
2'[... in] ⸢8$^?$⸣ 'days'.
3'[...] ⸢6;1⸣5 ⸢Vir in⸣ 14 'days'.
4'[...] 6 Lib in 14 'days'.
5'[...] ⸢7⸣;15 Sco in 21 'days'.
6'[...] ⸢2$^?$⸣2.30 Sgr in 21 'days'.
7'[...] ⸢...;30⸣ Cap in ... 'days'.
(about m+1 lines missing)

		Column ii'	Column ii'
P6'			**P6' (A planet: subdivision synodic cycle?)**
		(about n lines missing)	*(about n lines missing)*
	1'	⌜**ḫun** 1⌝[xxx]	1'⌜Ari⌝ 1 [...]
	2'	**mul₂** 2[xxx]	2'Tau 2[...]
	3'	**maš** 3 ⌜x⌝ [xxx]	3'Gem 3 [...]
	4'	**alla** ⌜4?⌝ [xxx]	4'Cnc 4 [...]
	5'	**a** 5.⌜20?⌝ [xxx]	5'Leo 5;⌜20⌝ [...]
	6'	**absin₀** 6 [xxx]	6'Vir 6 [...]
	7'	**ri**[**n₂** xxxx]	7'Li[b ...]
	8'	⌜**gir₂**⌝ [xxxx]	8'⌜Sco⌝ [...]
		(about m+1 lines missing)	*(about m+1 lines missing)*

Critical and philological notes

Side X' column i'
1' Traces of a single sign are visible.
2' Traces of a single sign are visible.
Column ii'
4' 24: of the 20 only traces of a winkelhaken are visible.
9' The ending of a sign similar to TAB is visible.
Column iii' The width of the preceding column suggests that not much clay is missing from column iii'. Also the open space before the second sign in iii'.3' and 4' suggests that these lines are completely preserved.
1' The zodiacal sign **gir₂** (Sco) must have been followed by **pa** (Sgr).
2' 9 is followed by traces of a small winkelhaken and a vertical wedge, probably the left side of an erased **maš₂**.
6' After **me** the beginning of two horizontal wedges are visible.
7' *ina* **igi**-*at₂* = *ina pānāt*, 'before' (temporal): after this one expects a synodic phenomenon.
8' After **me** a winkelhaken, perhaps the beginning of IGI, is visible.
9' Only an initial horizontal wedge remains of the zodiacal sign which is written after **en**, perhaps **alla** = Cnc or **gir₂** = Sco. It is not clear whether the sentence continues in 10'.
10' ⌜**ta**⌝: or ⌜**uš**⌝, 'degrees'. In the latter case one obtains '5°'. After ⌜**ta**⌝ the upper ends of three wedges are visible. One expects the name of a zodiacal sign, perhaps **gir₂**.
Side Y column i'
5' 7: or 8.
6' 2: preceded by traces of 2 signs, the latter perhaps 20.
7' The damaged sign at the beginning contains the head of a vertical wedge, presumably a number 1–8.
Column ii'
3' After 3 a small winkelhaken is visible.
5' ⌜20⌝: or ⌜30⌝.

Commentary

No original edges of the tablet are preserved on this newly discovered fragment. The handwriting is notably larger than usual. On both sides the text is divided into at least three unusually narrow columns which are further divided into sections by horizontal lines. I have numbered the columns on both sides from left to right, but the correct sequence of columns may be the opposite on the reverse side, which remains unidentified. On side X, 2 lines, each containing traces of one sign, remain of column i'. On side Y there are three columns, of which the first two (ia' and ib') are assumed to form a single procedure (P5'). Column ii' is a separate procedure (P6') because its lines are displaced with respect to those in column i'. All four partly preserved procedures on side X concern Mercury. P1' and P2',P3' are tentatively assigned to system A_1 and A_2, respectively. This makes P2',P3' the first known procedures for system A_2. The computational system of P4' has not been identified. The procedures on the reverse (P5',P6') also concern a planet, presumably Mercury or Venus.

P1' Mercury system A_1?: temporal push EF to EL

The beginning of P1' is lost. The 4 partly preserved lines mention zodiacal signs and numbers associated with them. No numbers are preserved for Vir and Sco (Xii'1'–2'); in Cap we have 22, in Aqr and Psc 24. They may be values of $\delta\tau$ (EF to EL), measured in mean tithis, for system A_1 (Table 3.8, p. 70).

P2' Mercury system A_2: synodic arc for EL

As stated at the beginning, this procedure concerns the synodic arc for EL. The sparse remains are fully consistent with Mercury system A_2: 1,48;30° = σ_1 (Table 3.10b, p. 71). Values for zones 2–4 were probably written in the missing lines below 8'.

P3' Mercury system A_2: temporal push EL to MF?

The beginning of P3' is lost. The 4 partly preserved lines mention zodiacal signs and numbers associated with them, perhaps values of $\delta\tau$ (EL to MF) for system A_2 (Table 3.11, p. 72). This makes P3' the only extant procedure concerned with the subdivision of the synodic time for system A_2.

P4' Mercury, unknown system: subdivision of the synodic cycle

The first 6 lines of this procedure are partly preserved. They may be concerned with the subdivision of the synodic cycle. An interval of $35^{\bar{\tau}}$ (or days?) preceding a 'setting', EL or ML, is mentioned in Xiii'6'. Presumably the corresponding value of v was mentioned in the gap at the end of that line. Xiii'7'–8' mention an interval of $15^{\bar{\tau}}$ preceding a synodic phenomenon, the name of which is missing. Xiii'9' mentions a zodiacal region between Gem and, perhaps, Cnc or Sco. The connection with the preceding lines is unclear.

P5' A planet: subdivision of the synodic cycle?

The beginning of this procedure is missing. Column ia' contains a number and a zodiacal sign, column ib' a time interval measured in 'days', presumably representing mean tithis. Since only Vir (Yi'3') – Cap (Yi'7') are preserved, Yi'3' may be preceded by Ari – Leo, which suggests 5 missing lines, and Yi'7' may be followed by Aqr – Psc, which suggests 2 additional lines. The numbers may represent angular pushes and the associated temporal pushes for an unidentified planet (Mercury?).

P6' A planet: subdivision of the synodic cycle?

This procedure also mentions zodiacal signs and numbers, perhaps representing pushes. The planet and the involved synodic phenomena remain unidentified.

No. 3

museum number	unnumbered fragment 2.4 (BM)
provenience	Babylon
measures	3.0(*) × 3.0(*) × ?(*) cm
cuneiform text	photo: Fig. F.2
contents	Mercury system A_2?: updating *B* with the synodic arc?

Side X

P1' (unknown number of lines missing)

1' [...] ⸢x⸣ **en** 30 **absin**$_0$ ⸢**ki**$^?$⸣ [...]

2' [...] **ḫun en** 30 ⸢**maš**⸣ [...]

P2' 3' [...] ⸢**igi**⸣ *ša*$_2$ **šu**$_2$ *ana* **igi** *ša*$_2$ [**šu**$_2$ ...]

4' [...] ⸢x **igi**$^?$ xx⸣ [...]

(unknown number of lines missing)

Side X

P1' **(ML?: updating *B* with the synodic arc?)**

(unknown number of lines missing)

1'[...] ... until 30 Vir ⸢the position$^?$⸣ [...] 2'[...] Ari until 30 ⸢Gem⸣ [...]

P2' **(EF: updating *B* with the synodic arc?)**

3'[...] western ⸢appearance⸣ (EF) to [western] appearance [...] 4'[...] ⸢... appearance$^?$ ...⸣ [...]

(unknown number of lines missing)

Critical and philological notes

Side X

1' Before **en** there are traces of a winkelhaken.

Commentary

This small fragment in the British Museum has not received a registration number.[606] It does not contain any edges of the tablet. Perhaps it belongs to the same tablet as No. **4**, but they do not physically join. Side X preserves remains of two procedures; the other side is destroyed. They may concern Mercury system A_2, but Venus cannot be ruled out.

P1' Mercury system A_2?, ML?: updating B with the synodic arc

The beginning and end of this procedure are missing. Nothing is known about the original length of the lines. The zonal boundaries mentioned in X1'–2' match those of the step function for the synodic arc for ML in Mercury system A_2.

P2' Inner planet, unknown system, EF: updating B with the synodic arc

Apart from a few signs in X3' and traces of X4' nothing remains of this procedure. Line X3' refers to the synodic arc or time for EF, which implies an inner planet, perhaps Mercury. The mention of EF excludes Mercury system A_2, where EF is a satellite of ML.

[606] Numerous small unnumbered fragments thought to have an astronomical content are kept in five plastic boxes (as of 2004). This fragment was kept in box 2.

No. 4

museum number	unnumbered fragment 2.8 (BM)
provenience	Babylon
measures	4.7(*) × 3.3(*) × ?(*) cm
cuneiform text	photo: Fig. F.3
contents	Mercury system A_2: various procedures

Side X

P1' (unknown number of lines missing)
1' [...] ⸢30?⸣ [xxxx...]
2' [...] 1.4⸢8.30 **tab** *ša*$_2$?⸣ [...]
3' [... **en** 30] ⸢**maš**⸣ 2.15.37.30 ⸢**tab**?⸣ [...]

P2' 4' [...] ⸢**du**?⸣ **ta** ⸢3?⸣ **lu** ⸢x⸣ [x...]
5' [...] ⸢x **en**?⸣ 15 [xx...]
(unknown number of lines missing)

Side X

P1' (EL: updating *B* with the synodic arc)

(unknown number of lines missing)
1'[...] ⸢30?⸣ [...]
2'[...] you add 1,4⸢8;30. (The amount) by which⸣ [it exceeds ...]
3'[... until 30] ⸢Gem you add?⸣ 2,15;37,30. [...]

P2' (Subdivision synodic cycle?)

4'[...] ⸢...⸣ From ⸢3?⸣ Ari? [...]
5'[...] ⸢... until?⸣ 15 [...]
(unknown number of lines missing)

Critical and philological notes

Side X

4' **du**: uncertain. A reading ⸢**tur**-*tu*$_2$⸣ cannot be excluded.
⸢3?⸣: only the first vertical wedge is fully preserved.
lu: perhaps Ari. If interpreted as **dib** = *etēqu* this might be a reference to omitted phenomena (cf. the commentary).

5' **en**: uncertain (one winkelhaken appears to be lacking).

Commentary

This small fragment in the British Museum has not received a registration number.[607] It does not contain any edges of the tablet. Only one side is preserved. Perhaps it belongs to the same tablet as No. **3**, but they do not physically join. The 5 partly preserved lines of text are divided over two sections.

P1' Mercury system A_2, EL: updating B with the synodic arc

Only the final 2 lines of this procedure are partly preserved. The numbers 1,48;30° (X2') and 2,15;37,30° (X3') are the synodic arcs σ_3 and σ_4 for EL of Mercury according to system A_2 (§3.4.3, p. 71). The traces of what appears to be 30 in line X1' may therefore belong to $\sigma_2 = 2{,}0;33{,}20°$. Also traces of the instruction to add the synodic arcs to the old position are preserved, which suggests that the formulation is based on one of the known templates (cf. §2.6.10.1, p. 48).

P2' Mercury, subdivision of the synodic cycle?

Too little remains of this procedure to determine its topic with certainty. If LU is interpreted as Ari then the position 3 Ari may be mentioned, of unknown significance; if interpreted as **dib** = *etēqu* it might be a reference to Mercury's omitted phenomena, known from procedures for system A_1 (§3.4.2.4, p. 70). The number 15 (X5'), if interpreted as 15°, might be a zodiacal position as they appear in the subdivision of the synodic cycle for Mercury systems A_1 and A_2.

[607] Numerous small unnumbered fragments thought to have an astronomical content are kept in five plastic boxes (as of 2004). This fragment was kept in box 2.

No. 5

museum number	BM 36321 (80–6–17,47)
provenience	Babylon
date	early Seleucid
measures	7.4 × 6.1 × 1.2–1.4 cm (type G)
arrangement	O/R
cuneiform text	photo: Fig. F.5; *ACT*, Pl. 253a (erroneously labeled 819d)
bibliography	*ACT* 816, Zu; *HAMA*, pp. 467–471; Aaboe et al. (1991), pp. 38–39; Swerdlow (1998), pp. 127–134
contents	Mercury system A$_3$: updating *B*; net displacements for various intervals

Obverse

P1
1 [xx t]**a** 30 **ḫun en** 30 **a** 16.52.30 [**la**$_2$]
2 [**ta**] ⌜30 **a**⌝ **en** 20.37.30 **alla** 16 [**la**$_2$]
3 [**t**]**a** 20.37.30 **alla en** 30 **ḫun** 20 [**la**$_2$]
4 [**m**]**u** *ana* **mu šu**$_2$ *ša*$_2$ **šu**$_2$ *u* **kur** *ša*$_2$ **nim**

P2
5 ⌜**igi**⌝ *ša*$_2$ **šu**$_2$ **ta** 30 **maš.maš en** 30 **maš**$_2$ 16 ⌜**la**$_2$⌝
6 **ta** 30 **maš**$_2$ **en** 30 **gir**$_2$**.tab** 14 ⌜**la**$_2$⌝
7 **ta** 30 ⌜**gir**$_2$⌝**.tab en** 30 **rin**$_2$ 18 **la**$_2$
8 **ta** 30 **rin**$_2$ **en** 30 **absin** 23 **la**$_2$
9 ⌜**ta**⌝ 30 **absin en** 30 **a** 28 **la**$_2$
10 ⌜**ta**⌝ 30 **a en** 30 **maš.maš** 20 **la**$_2$

P3
11 ⌜**uš**⌝ *ana* **uš** *ša*$_2$ **gu**$_4$**.ud** *ina* **ki** 16.52.30 **la**$_2$
12 1.50.56.15 ⌜**ki** *ana*? **igi**?⌝-*šu*$_2$ **du uš**

Reverse

1 *ina* **ki** 4.16 **la**$_2$ 1.45.11.6.40 *ana* **igi**-*šu*$_2$ **du** ⌜**uš**⌝
2 ⌜*ina*⌝ **ki** 5.20 **la**$_2$ 2.11.28.53.20 *ana* **igi**-*šu*$_2$ ⌜**du**⌝ [**uš**]

P4
3 [*ana*] *na-bal-kat*$_2$*-tu*$_4$ *ša*$_2$ 20 ⌜**mu**⌝**.an.na.meš šu**$_2$ *ša*$_2$ ⌜**šu**$_2$⌝
4 ⌜*u*⌝ **kur** *ša*$_2$ **nim ta** 30 ⌜**ḫun**⌝ **en** 30 **a** 4.49.17 **la**$_2$
5 **ta** 30 **a en** 20.30.07.30 **alla** 4.34.15 **la**$_2$
6 **ta** 20.37.30 **alla en** 30 **ḫun** 5.32.51.30 **la**$_2$

P5
7 ⌜*ana na*⌝*-bal-kat*$_2$*-tu*$_4$ *ša*$_2$ 20 **mu.an.na.meš igi** *ša*$_2$ **šu**$_2$
8 [*ḫi*]-⌜*pi*$_2$⌝ *eš-šu*$_2$ **ta maš en maš**$_2$ 4.34 **la**$_2$
9 [**t**]**a maš**$_2$ **en pa** 4 **la**$_2$
10 [**ta**] **pa en rin**$_2$ 5.8 **la**$_2$
11 [**ta r**]**in**$_2$ **en absin** 6.34 **la**$_2$

Obverse

P1 (EL, MF: net displacements for 3 cycles)

[1][...]
⌜From⌝ 30 Ari until 30 Leo [you subtract] 16;52,30.
[2][From] 30 Leo until 20;37,30 Cnc [you subtract] 16.
[3][Fr]om 20;37,30 Cnc until 30 Ari [you subtract] 20.
[4][Ye]ar by year, western setting (EL) and eastern rising (MF).

P2 (EF: net displacements for 3 cycles)

[5]Western ⌜appearance⌝ (EF).
From 30 Gem until 30 Cap ⌜you subtract⌝ 16.
[6]From 30 Cap until 30 Sco ⌜you subtract⌝ 14.
[7]From 30 Sco until 30 Lib you subtract 16.
[8]From 30 Lib until 30 Vir you subtract 23.
[9]From 30 Vir until 30 Leo you subtract 28.
[10]⌜From⌝ 30 Leo until 30 Gem you subtract 20.

P3 (ES, MS: net displacements for 3 cycles; synodic arc)

[11]⌜Station⌝ to station of Mercury. From the position you subtract 16;52,30. [12]1,50;56,15 ⌜the position⌝ proceeds ⌜forward⌝, it becomes stationary.

Reverse

[1]From the position you subtract 4,16. By 1,45;11,6,40 it proceeds forward, it ⌜becomes stationary⌝.
[2]⌜From⌝ the position you subtract 5,20. By 2,11;28,53,20 it ⌜proceeds⌝ forward, it [becomes stationary.]

P4 (EL, MF: net displacements for 63 cycles)

[3][For] the transgression of 20 years, ⌜western⌝ setting (EL) [4]and eastern rising (MF).
From 30 ⌜Ari⌝ until 30 Leo you subtract 4;49,17.
[5]From 30 Leo until 20;30,07,30 (error for 20;37,30) Cnc you subtract 4;34,15.
[6]From 20;37,30 Cnc until 30 Ari you subtract 5;32,51,30 (error for 5;42,51,30?).

P5 (EF: net displacements for 63 cycles)

[7]⌜For⌝ the transgression of 20 years, evening appearance (EF). [8]New ⌜break⌝.
From Gem until Cap you subtract 4;34.
[9][Fr]om Cap until Sgr you subtract 4.
[10][From] Sgr until Lib you subtract 5;8.
[11][From L]ib until Vir you subtract 6;34.

12 [ta abs]**in en a** 7.58 ⸢**la$_2$**⸣
13 [ta a] **en maš** 5.43 ⸢**la$_2$**⸣

Lower edge

Co 1 [xx] ^{m}**u.a.mu a-***šu$_2$* ^{m}**gi.u dumu** ^{m}*mu-š*[*e-zib*]

[12][From Vi]r until Leo ⸢you subtract⸣ 7;58.
[13][From Leo] until Gem ⸢you subtract⸣ 5;43.

Lower edge

Co **(Colophon)**

[1][Tablet of] Bēl-apla-iddin, son of Mušallim-Bēl, descendant of Mušē[zib].

Critical and philological notes

Obverse

1 In the gap at the beginning one might restore **gu$_4$.ud**, 'Mercury'. **la$_2$** is interpreted as *tanaḫḫis*, 'you subtract', but *inaḫḫis*, 'it recedes' cannot be ruled out; the same applies to the other instances of **la$_2$** on the tablet. However, the alternative interpretation **la$_2$** = *tumaṭṭa*, 'you diminish' (from *maṭû* D), can be ruled out for O11 and R1,2 due to the presence of *ina*, 'from'.

4 **kur** *ša$_2$* **nim** = *napāḫu*/*nipḫu ša elî*, 'morning rising': alternative term for MF.

11 *ina* **ki la$_2$** is interpreted as *ina qaqqari tanaḫḫis*, 'you subtract from the position', but *ina qaqqari inaḫḫis*, 'it recedes from the position' cannot be ruled out.

Reverse

1,2 At the end of these lines 1–2 signs are missing. The parallel phrase in O12 suggests **uš**, 'it becomes stationary'.

3 *kat$_2$*: replaces *kat* (*ACT*), also in R7.

5 20.30.07.30: replaces 20.30...17 (*ACT*). Between 30 and 07 there is an empty space, probably an erasure; perhaps the 0 is a remainder of the erased sign. As remarked by Neugebauer, one expects 20.37.30; cf. the commentary.

7 *ḫīpi eššu*, 'new break'.

Commentary

The tablet is nearly intact, except for some damage to the upper (lower) corners of the obverse (reverse). The colophon mentions Bēl-apla-iddin/Mušallim-Bēl/Mušēzib as the 'owner' of the tablet. He is attested in the colophons of three other astronomical tablets: No. **9** (Venus), *ADRT* V 60, which contains Jupiter observations, and *ADRT* I -321, an astronomical diary dated to year 2 of Philip Arrhidaeus (322/1 BC). R7 mentions 'new break', implying that the tablet was copied from an older original. The procedures on this tablet have been the subject of several investigations, starting with the first edition in *ACT*. All five are concerned with computing the zodiacal position of Mercury in accordance with system A$_3$ (§3.4.4). Apart from the synodic arcs they mention net displacements for 3 cycles ('1 yr') and 63 cycles ('20 yrs'). Not all synodic phenomena are covered. P1–P3 concern the net displacement of EL, MF, EF, MS and ES for 3 cycles, P4 and P5 that of EL, MF and EF for 63 cycles. The synodic arc is mentioned only in connection with the stations in P2; ML is not mentioned at all.

P1 EL, MF: net displacements for 3 cycles

P1 concerns the net displacement, $dB_j(3)$, of EL and MF for 3 synodic cycles ('1 yr'). For a discussion cf. §3.4.4.2. The zones are listed in the sequence 3,2,1, and for each zone the boundaries are given in the retrograde direction, in accordance with the subtractive nature of the displacements.

P2 EF: net displacements for 3 cycles

P2 concerns the net displacement of EF for 3 synodic cycles ('1 yr'). Values are provided for 6 zones, each comprising a whole number of zodiacal signs (Table 3.15, p. 74). In accordance with the subtractive nature of the displacements, the zones, and the boundaries of each zone, are listed in the retrograde direction, from zone 6 to zone 1. The net displacement of EF for 63 cycles ('20 yr') is dealt with in P5, which also mentions 6 zones, but they are not necessarily the same; cf. below.

P3 ES, MS: net displacement for 3 cycles; synodic arc

This procedure concerns Mercury's net displacement for 3 cycles at the stations. For a discussion cf. §3.4.4.1 and §3.4.4.2. The contents are garbled, and it is not stated what the numbers represent or in which part of the zodiac they apply. Three pairs of displacements are mentioned. The first number of each pair is always subtractive, the second one additive. The first number of the first pair is the net displacement for 3 cycles ('1 yr') for zone 3, $dB_3(3) = -16;52,30°$, which was already mentioned in P1, and the second one, 1,50;56,15° is the corresponding σ_3 (Table 3.13b). The other two pairs pertain to zones 2 and 1, in this order, but while the second number in each of these pairs represents σ_j, as expected, the first number is, strangely enough, the net displacement for 16^{yr}, computed as 16 times an approximate value of $dB_j(3)$, the net difference for 3 cycles ('1 yr'), i.e. $16 \cdot 16 = 4{,}16$ (zone 2) and $16 \cdot 20 = 5{,}20$ (zone 1). The reason for this mix up is not clear.

P4 *EL, MF: net displacements for 63 cycles ('20 yrs')*

This procedure contains the net displacement of EL or MF for 63 cycles ('20 years'). For a discussion cf. §3.4.4.3. Those for EL, listed in the sequence 3,2,1, differ only slightly from the reconstructed exact values (Table 3.15). If one assumes, with Neugebauer, that 32 in the latter is a scribal error for 42, then the agreement is even closer.

P5 *EF: net displacements for 63 cycles ('20 yrs')*

This procedure is concerned with the net displacement of EF for 63 cycles ('20 yrs'). Again the 6 zones, and the boundaries of each zone, are listed in the retrograde direction, from zone 6 to zone 1, in accordance with the subtractive nature of $dB_j(63)$. Only the zodiacal signs are given, but the precise boundaries may be assumed to be located at 30° of these signs as in P2. This assumption results in nearly the same zonal boundaries as in P2, but that between zones 4 and 5 is at 30 Sgr instead of 30 Sco.[608] If taken seriously this implies that the lengths (α_j) of zones 4 (30° according to P2) and 5 (1,0° according to P2) are exchanged, but it is fair to assume that the zones are meant to be the same in P2 and P5, so that one of them must be wrong. It is currently impossible to say which one, for lack of duplicate procedures, relevant synodic tables or other arguments.

[608] This deviation seems to have escaped the previous editors of the text.

No. 6

museum number	U 136
provenience	Uruk
measures	9.5(16?) × 8.0(*) × ? cm
arrangement	O/R
cuneiform text	photo: Fig. F.6 = *ACT*, Pl. 241; copy: *ACT*, Pl. 225
bibliography	*ACT* 800
contents	Mercury, unknown system: various procedures

Obverse

P1' (m lines missing)

1' [xxxxxxxxxxxxxxx] ⌜xxxxx⌝ [xxxxxxxxx]

2' [xxxxxxxxxxxxxxx] ⌜**alla**$^{?}$ *u*$_3$ **a**⌝ [xxxxxxxxx]

3' [xxxxxxxxxxxxxxx] ⌜x *u*$_3$ xx⌝ [xxxxxxxxx]

4' [xxxxxxxxxxxxxxx] ⌜xxxx⌝ [xxxxxxxxx]

5' [xxxxxxxxxxxxxxx] ⌜xx *u*$_3^{?}$ x⌝ [xxxxxxxxx]

6' [xxxxxxxxxxxxxxx] ⌜xxx **du**$^{?}$ x⌝ [xxxxxxxxx]

7' [xxxxxxxxxxxxxxx] ⌜**mu.meš**$^{?}$ x⌝ [xxxxxxxxx]

8' [xxxxxxxxxxxxxxx] ⌜x 30 x⌝ [xxxxxxxxx]

(n lines missing)

Reverse

P2' (about n+4 lines missing)

1' [xxxxxxxxxx] ⌜**igi**⌝ *ša*$_2$ **kur** *ana* ⌜**igi** *ša*$_2$ **kur**⌝ [x$^{?}$] ⌜46 **mu.meš**⌝ [xxxxxxx]

2' [xxxxxxx] ⌜*mi-nu-u*$_2$⌝ GAM 2.2⌜4⌝ **igi.meš** *lu*-**du**-*ma lu* 4.3[6 xxxxx 1.55]

3' [GAM 2.24 **d**]**u**-*ma* 4.36 : 13.20 : 1.55 : 5.3[3.20 xxxxxx]

4' [xxxxx] ⌜20$^{?}$⌝ *lu* GAM 2.24 **igi.meš du**-*ma* 1.12 *mi*-[*nu-u*$_2$ GAM 46 *lu*-**du**-*ma*]

5' [*lu* 2.24 3.7.49].33.54.46 ⌜GAM 46 **mu.meš**⌝ [**du**-*ma* 2.24 xxx]

6' [xxxxx 9].⌜2⌝3.28.41.4[4.18 xxxxxxxxxxxxxxx]

7' [xxxxxxxxxx] ⌜xxx⌝ [xxxxxxxxxxxxxxxxx]

(about m−4 lines missing)

Obverse

P1' (Unclear)

(*m lines missing*)

1'[...] ⌜...⌝ [...] 2'[...] ⌜Cnc$^{?}$ and Leo⌝ [...] 3'[...] ⌜... and ...⌝ [...] 4'[...] ⌜...⌝ [...] 5'[...] ⌜...and ...⌝ [...] 6'[...] ⌜...⌝ [...] 7'[...] ⌜years ...⌝ [...] 8'[...] ⌜... 30 ...⌝ [...]

(*n lines missing*)

Reverse

P2' (Mercury system A$_1$?: computations involving MF)

(*about n+4 lines missing*)

1'[...] eastern appearance (MF) to eastern appearance. [...$^{?}$] 46 years [...] 2'[...]

What should I multiply by 2,2⌜4⌝ appearances (MF) so that it is 4,3[6,0? ... You multiply 1,55 3'by 2,24], it is 4,36,0.

13.20 : 1,55 : 5.3[3.20 ...] 4'[... 0;30] ... you multiply by 2,24 appearances, it is 1,12.

Wh[at should I multiply by 46 so that 5'it is 2,24? You multiply 3;7,49],33,54,46 by 46 years, [it is 2,24...]

6'[... 9.2]3,28,41,4[4,18 ...] 7'[...] ... [...]

(*about m−4 lines missing*)

Critical and philological notes

Obverse

1' Only traces are visible.

2' Lines 2' and 3' contain an identical sign, perhaps *u*$_3$ (*ACT*). Before *u*$_3$ there is a winkelhaken, perhaps the end of **alla** (Cnc)?

3' The sign following *u*$_3$ might be 2 or 20. After that something like RIN$_2$.

4' Only traces.

5' Only traces. Neugebauer sees another *u*$_3$ (*ACT*).

6' **du**: very uncertain.

Reverse

1' Before 46 **mu.meš** there is room for about 1 additional sign, but perhaps nothing was written here.
2' 2.24: the 4 is damaged and might also be a 5, but a 4 is implied by the procedure.
4' Somewhere at the beginning one expects a 30; cf. the commentary. 20: due to the cursive script this might also be 2. *lu*: perhaps to be read *taš* = *tašpiltu*, 'difference'? Or read [*mi*]-⌜*nu-u*$_2$⌝, 'what'?
In the gap at the end, Neugebauer restored a multiplication by 23, but the multiplication by 46 in 5' suggests that the same number is to be restored here, and hence 2.24 at the beginning of 5'.
6' 23: a trace of the second winkelhaken of 20 is visible.
7' Only traces of numbers.

Commentary

This fragment does not preserve any edges of the tablet. It was not collated. On the badly damaged obverse only isolated signs can be identified. The handwriting is similarly cursive as that of No. **41** and No. **67**, both also from Uruk. Based on the textual restorations in R4'–5' the original width may have been about 16 cm. At least 2 procedures are partly preserved, of which the second concerns Mercury, perhaps system A$_1$ (very uncertain).

P1' Unclear

The beginning and end of this procedure are missing. The 8 lines with faint traces do not warrant a translation. If O2' can be read as **alla** *u*$_3$ **a**, 'Cnc and Leo', this may represent a zodiacal region as they are defined for the synodic arc in type-A systems. Regions consisting of 2 zodiacal signs are attested for Venus and Mars.

P2' Mercury: computations involving MF

The beginning and end of this procedure are missing. The computations are concerned with MF of Mercury, but they are not readily identifiable in terms of a known computational system. Line R1' mentions 46^{yr}, corresponding to 145 synodic cycles, a known interval for close returns of Mercury to nearly the same zodiacal position (cf. for instance $dB_j(145)$ for system A$_1$ in Table 3.9, p. 71). The text continues with a question, a rhetorical device regularly encountered in mathematical problem texts, but only rarely in astronomical procedures (cf. §1.7, p. 15). However, the computation does not operate with 2,25 but with 2,24. Neugebauer considers this to be an error (*ACT*, pp. 363–364), but a more satisfying explanation may be that 2,25 does not have a finite sexagesimal reciprocal, while 1/2,24 = 0;0,25, so that the use of 2,24 may be motivated by practical considerations. The number 4,36,0 equals 46 $\times$ 6,0°, the approximate distance traveled by Mercury and the Sun during 46^{yr}. The product 0;0,25 $\times$ 4,36,0 = 1,55° is thus an estimate of the mean synodic arc $\overline{\sigma}$, which comes out about 0;48° larger than the empirical value 1,54;12,...° also known from systems A$_1$, A$_2$ and A$_3$.

R3' mentions two other numbers having a ratio 2,24. As suggested by Neugebauer, 13.20 might represent 0;13,20, which equals $dB_3(145)$, the net displacement of MF for 145 synodic events in zone 3 of system A$_1$ (Table 3.9), but this is higly uncertain, since the significance of the ratio 0;13,20/2,24 = 0;0,5,33,20 is unclear, and nothing else points to system A$_1$. In R4' something is multiplied by 2,24, the result being 1.12. If one interprets this as 1,12, then the missing number is 0;30. The significance of 1,12 cycles, which correspond to 23^{yr}, is not clear. In R4'–5' one is asked for the factor by which 46^{yr} must be multiplied in order to get 2,24 cycles, which amounts to a computation of the period $P = \Pi/Z$. Since 46 does not have a finite digit inverse, the mentioned result 3;7,49,33,54,46 is an approximation of the exact ratio P = 3;7,49,33,54,46,57,... In R5' this is multiplied by 46 in order to verify the correctness of the result.[609] As remarked by Neugebauer, 9;23,28,41,44,18 (R6') equals $3 \cdot P$. The reason for considering this quantity is not clear.

[609] Since the value of P is only approximate, the exact product is not 2,24 but 2,23;59,59,59,59,16. It is not clear which of these should be restored in the gap at the end of R5'.

5.2.2 *Venus*

No. 7

museum numbers	BM 35495 (Sp3,1) +40102 (81–2–1,67+77) +46176 (81–7–6,637)
provenience	Babylon
date	data in synodic table: SE 180–242
measures	18.5(19.5) × 9.0(9.5) × 2.0–2.3 cm (type ii)
arrangement	O/R
cuneiform text	photo: Fig. F.10 (R); *ACT*, Pl. 244a (R); copy BM 35495: *LBAT* 106
bibliography	Kugler *SSB* I, p. 203; Neugebauer (1954), p. 75; *ACT* 420, 821b, Zld
contents	Venus systems A_1, A_2: synodic table for EF,ES,EL,MF,MS,ML; corresponding procedures

Reverse

Co 1 [xxxx *dil*]-⸢*bat*⸣ *ša*$_2$ **ta** 3 **en** 4.1
⟨^{m}⟩^{d}**amar.utu.mu.mu a** *ša*$_2$ ^{md}⸢**en**⸣**.mu**-*na* **a**
^{md}⸢*e*$_4$-*gi*$_7$-*ba-ti-la*⸣ *ana tar-ṣa*
^{md}⸢**amar**$^?$⸣.[**utu**]-⸢**dub**⸣-**numun** ^{lu_2}**umbisag aš u**$_4$
an.na ^{d}**en.lil**$_2$**.la**$_2$ **a** *ša*$_2$ ^{m}⸢^{d}**en.tin**⸣-*su* ⟨**a**⟩
⸢^{md}**ak**$^?$⸣.**mu.a ša**$_3$.⸢**bal**$^?$⸣.[**bal** xxx]

Column i

P1 1 [xxxxxx]-⸢*at*$^?$⸣ **uš**
2 [**igi** *ša*$_2$ **šu**$_2$ *a*]-⸢*na*$^?$⸣ **igi** *ša*$_2$ **šu**$_2$
3 [xxxxxx] ⸢x **maš**$^?$ **alla**$^?$⸣
4 [xxxxxx] ⸢x⸣ 3.30.⸢20 **tab**⸣
5 [xxxxxx] ⸢3⸣.35.30 ⸢**tab**⸣
6 [xxxxxxxx] ⸢xx **tab**⸣

P2 7 [**u**$_4$**.meš** *ša*$_2$ **igi** *ša*$_2$ **šu**$_2$ *ana* **igi** *ša*$_2$ **šu**$_2$]
(5 lines missing)

Column ii

P3 1 ⸢**uš** *ša*$_2$ **šu**$_2$ *a-na* **uš** *ša*$_2$ **šu**$_2$⸣
2 ⸢**gu** *zib*⸣.[**me** 3.4]⸢3.30⸣ **tab**
3 **ḫ**[**un mul**$_2$ **maš** 3.3]⸢7.30 **tab**⸣
4 ⸢**alla a**⸣ [3.3]⸢8.30⸣ **tab**
5 ⸢**absin rin**$_2$ **gir**$_2$**.tab** 3⸣.29.20 **tab**
6 ⸢**pa maš**$_2$ 3.28.30 **tab**⸣

P4 7 ⸢**u**$_4$**.meš** *ša*$_2$ **uš** *ša*$_2$ **šu**$_2$ *ana* **uš** *ša*$_2$ **šu**$_2$⸣
8 ⸢**gu** *zib*.**me** 9.51⸣ **tab**
9 [**ḫun**] ⸢**mul**$_2$ **maš** 10.1 **tab**⸣
10 [**alla a**] ⸢9.59 **tab**⸣
11 [**absin rin**$_2$] ⸢**gir**$_2$**.tab** 9.46 **tab**⸣
12 [**pa**] **maš**$_2$ 9.49 **tab**

Reverse

Co **(Colophon)**

[1][... Ve]nus which are from (year) 3,0 until 4,1. Marduk-nādin-šumi (or -šuma-iddin?), son of ⸢Bēl⸣-iddina, descendant of Egibatila. Opposite Mar[duk$^?$]-šāpik-zēri, scribe of 'When Anu (and) Enlil', son of Bēl-bullissu, ⟨son⟩ of ⸢Nabû$^?$⸣-nādin-apli, descen[dant of ...]

Column i

P1 **(System A_2. EF: updating *B* with the synodic arc)**

[1][...] ... station.
[2][Western appearance] to western appearance (EF).
[3][...] ⸢... Gem Cnc$^?$⸣
[4][...] Lib: ⸢you add⸣ 3,30;20.
[5][...] ⸢you add⸣ 3,35;30.
[6][...] ⸢you add⸣ ...

P2 **(System A_2. EF: updating *T* with the synodic time)**

[7]['Days' for western appearance to western appearance.]
(5 lines missing)

Column ii

P3 **(System A_2. ES: updating *B* with the synodic arc)**

[1]⸢Western station (ES) to western station⸣.
[2]⸢Aqr, Psc⸣: you add [3,4]⸢3;30⸣.
[3]A[ri, Tau, Gem]: you add [3,3]⸢7;30⸣.
[4]⸢Cnc, Leo⸣: you add [3,3]⸢8;30⸣.
[5]⸢Vir, Lib, Sco⸣: you add ⸢3⸣,29;20.
[6]⸢Sgr, Cap: you add 3,28;30⸣.

P4 **(System A_2. ES: updating *T* with the synodic time)**

[7]⸢'Days' for western station to western station⸣.
[8]⸢Aqr, Psc⸣: you add ⸢9,51⸣.
[9][Ari], ⸢Tau, Gem: you add 10,1⸣.
[10][Cnc, Leo]: ⸢you add 9,59⸣.
[11][Vir, Lib], ⸢Sco: you add 9,46⸣.
[12][Sgr], Cap: you add 9,49.

Column iii

P5 1 **šu$_2$** ⸢*ša$_2$* **šu$_2$** *a$^?$-na$^?$* **šu$_2$** *ša$_2$* **šu$_2$**⸣
2 ⸢**ḫun**⸣ [**mul$_2$ maš** 3.3]⸢6$^?$⸣ **tab**
3 ⸢**alla**⸣ [**a**] ⸢3.38⸣ **tab**
4 ⸢**absin rin$_2$ gir$_2$.tab** 3.32⸣ **tab**
5 ⸢**pa maš$_2$** 3.29⸣.30 **tab**
6 ⸢**gu**⸣ [*zib*].⸢**me** 3.41.50⸣ **tab**

P6 7 ⸢**u$_4$**⸣.[**meš**] ⸢*ša$_2$*⸣ [**šu$_2$** *ša$_2$* **šu$_2$** *a-na* **šu$_2$** *ša$_2$* **šu$_2$**]
8 ⸢**ḫun mul$_2$ maš** 9.5⸣8 **tab**
9 **alla a** 10.0⸢4$^?$ **tab**⸣
10 ⸢**absin**⸣ **rin$_2$ gir$_2$.tab** 10.1 **tab**
11 ⸢**pa**⸣ **maš$_2$** 9.⸢40 **tab**⸣
12 ⸢**gu**⸣ *zib*.**me** ⸢9.4⸣3 **tab**

Column iv

P7 1 ⸢**igi** *ša$_2$* **kur**⸣ [*a-na* **igi** *ša$_2$*] ⸢**kur**⸣
2 ⸢**ḫun**⸣ **m**[**ul$_2$ maš** 3.39.50] ⸢**tab**⸣
3 ⸢**alla a**⸣ ⸢3⸣.[41].10 **tab**
4 ⸢**absin**⸣ [**rin$_2$ gir$_2$.tab**] ⸢3.30.40⸣ **tab**
5 ⸢**pa maš$_2$**⸣ [3.25.30] **tab**
6 ⸢**gu** *zib*.**me**⸣ [3.42.20 **tab**]

P8 7 ⸢**u$_4$.meš** *ša$_2$*⸣ [**igi** *ša$_2$* **kur** *a-na* **igi** *ša$_2$* **kur**]
8 ⸢**ḫun**⸣ [**mul$_2$ maš** 9.48 **tab**]
9 ⸢**alla**⸣ [**a** 9.52 **tab**]
10 ⸢**absin**⸣ [**rin$_2$ gir$_2$.tab** 9.59 **tab**]
11 ⸢**pa maš$_2$**⸣ [9.55 **tab**]
12 ⸢**gu** *zib*.**me**⸣ [9.5]⸢2⸣ [**tab**]

Column v

P9 1 ⸢**uš**⸣ [*ša$_2$* **kur**] ⸢*a*⸣*-na* **uš** *ša$_2$* **kur**
2 ⸢**ḫun mul$_2$** 3⸣.40.30 **tab**
3 **maš alla** [**a**] ⸢3.4⸣2.50 **tab**
4 **absin** ⸢**rin$_2$**⸣ ⸢3⸣.30.10 **tab**
5 ⸢**gir$_2$.tab pa maš$_2$**⸣ 3.24.40$^?$ **tab**
6 [**gu** *zib*.**me** 3].⸢39.10⸣ **tab**

P10 7 [**u$_4$.meš** *ša$_2$* **uš** *ša$_2$* **kur** *a-na*] ⸢**uš** *ša$_2$* **kur**⸣
8 [**ḫun mul$_2$** xxxxxx] ⸢**tab**⸣
(4 lines missing)

Column vi

P11 1 **šu$_2$** *ša$_2$* **kur** *a-na* **šu$_2$** *ša$_2$* **kur**
2 *zib*.**me ḫun** 3.45.20 **tab**
3 **mul$_2$ maš alla** 3.29.40 **tab**
4 **a absin** 3.28.50 **tab**
5 **rin$_2$ gir$_2$.tab pa** 3.38.30 **tab**
6 ⸢**maš$_2$ gu**⸣ 3.35.10 **tab**

Column iii

P5 (System A$_2$. EL: updating *B* with the synodic arc)

[1]⸢Western⸣ setting (EL) ⸢to western setting⸣.
[2]⸢Ari⸣, [Tau, Gem]: you add [3,3]⸢6$^?$⸣.
[3]⸢Cnc⸣, [Leo]: you add ⸢3,38⸣.
[4]⸢Vir, Lib, Sco⸣: you add ⸢3,32⸣.
[5]⸢Sgr, Cap⸣: you add ⸢3,29⸣;30.
[6]⸢Aqr, Psc⸣: you add ⸢3,41;50⸣.

P6 (System A$_2$. EL: updating *T* with the synodic time)

[7]⸢'Days' for⸣ [western setting (EL) to western setting.]
[8]⸢Ari, Tau, Gem⸣: you add ⸢9,5⸣8.
[9]Cnc, Leo: ⸢you add⸣ 10,4.
[10]⸢Vir⸣, Lib, Sco: you add 10,1.
[11]⸢Sgr⸣, Cap: ⸣you add⸣ 9,40.
[12]⸢Aqr⸣, Psc: you add ⸢9,4⸣3.

Column iv

P7 (System A$_1$. MF: updating *B* with the synodic arc)

[1]⸢Eastern appearance (MF)⸣ [to] ⸢eastern⸣ [appearance.]
[2]⸢Ari⸣, Tau, Gem: ⸢you add⸣ [3,39;50].
[3]⸢Cnc, Leo⸣: you add 3,[41];10.
[4]⸢Vir⸣, [Lib, Sco]: you add ⸢3,30;40⸣.
[5]⸢Sgr, Cap⸣: you add [3,25;40].
[6]⸢Aqr, Psc⸣: [you add 3,42;20.]

P8 (System A$_1$. MF: updating *T* with the synodic time)

[7]⸢'Days' for⸣ [eastern appearance (MF) to eastern appearance.]
[8]⸢Ari⸣, [Tau, Gem: you add 9,48.]
[9]⸢Cnc⸣, [Leo: you add 9,52.]
[10]⸢Vir⸣, [Lib, Sco: you add 9,59.]
[11]⸢Sgr, Cap⸣: [you add 9,55.]
[12]⸢Aqr, Psc⸣: [you add 9,5]2.

Column v

P9 (System A$_2$. MS: updating *B* with the synodic arc)

[1][Eastern] ⸢station (MS)⸣ to eastern station.
[2]⸢Ari, Tau⸣: you add 3,40;30.
[3]Gem, Cnc, [Leo]: you add ⸢3,4⸣2;50.
[4]Vir, ⸢Lib⸣: you add 3,30;10.
[5]⸢Sco, Sgr, Cap⸣: you add 3,24;40$^?$.
[6][Aqr, Psc]: you add [3],⸢39;10⸣.

P10 (System A$_2$. MS: updating *T* with the synodic time)

[7]['Days' for eastern station to] ⸢eastern station⸣.
[8][...] you add.
(*4 lines missing*)

Column vi

P11 (System A$_1$. ML: updating *B* with the synodic arc)

[1]Eastern setting (ML) to eastern setting.
[2]Psc, Ari: you add 3,45;20.
[3]Tau, Gem, Cnc: you add 3,29;40.
[4]Leo, Vir: you add 3,28;50.
[5]Lib, Sco, Sgr: you add 3,38;30.
[6]⸢Cap, Aqr⸣: you add 3,35;10.

P12 7 ⸢**u$_4$.meš** *ša$_2$* **šu$_2$**⸣ *ša$_2$* **kur** ⸢*ana* **šu$_2$** *ša$_2$* **kur**⸣
8 *zib*.**me ḫun** 10:14 [**tab**]
9 ⸢**mul$_2$**⸣ [**maš alla** 9.45 **tab**]
(3 lines missing)
Column vii

P13 1 ⸢xxx 2.50 **kin**⸣ [xxxxx]
2 ⸢2.52 **diri** x⸣[xxxx]
3 ⸢2.55⸣ [**diri** xxxxx]
4 ⸢2.58⸣ [**diri** xxxxx]
5 [3 **diri** xxxxx]
6 3.⸢3⸣ [**diri** xxxxx]
7 3.⸢6⸣ [**diri** xxxxx]
8 3.⸢9⸣ [**kin** xxxxx]

P14 9 ⸢xxx-*ri*?-*tu$_4$*?⸣ [xxxxx]

P12 (System A$_1$. ML: updating T with the synodic time)

7 ⸢'Days' for⸣ eastern ⸢setting to eastern setting⸣.
8 Psc, Ari: [you add] 10,14.
9 ⸢Tau⸣, [Gem, Cnc: you add 9,45.]
(3 lines missing)
Column vii

P13 (List of leap years)

1 ⸢... 2,50: VI$_2$⸣ [...]
2 ⸢2,52: XII$_2$⸣ [...]
3 ⸢2,55⸣: [XII$_2$ ...]
4 ⸢2,58⸣: [XII$_2$ ...]
5 [3,0: XII$_2$ ...]
6 3,⸢3⸣: [XII$_2$ ...]
7 3,⸢6⸣: [XII$_2$ ...]
8 3,⸢9⸣: [XII$_2$ ...]

P14 (Unclear)

9 [...] ⸢...⸣ [...]

Critical and philological notes

Reverse

1 *ana tarṣa*, 'opposite', perhaps meaning 'in the presence of'. For a parallel in mathematical astronomy cf. *ACT* 122 (Colophon Zo). The name following *ana tarṣa* is of the type ^{m}DN-**dub.numun** = DN-šāpik-zēri, probably ^{md}AMAR.UTU-**dub.numun**, Marduk-šāpik-zēri (replaces '... MU' in *ACT*). UTU is broken and only traces remain of AMAR.

Column i

1 The sign before **uš** was copied as NE by Pinches.
uš: perhaps 'station', but the meaning of the sentence is not clear.
2 This line is parallel to line 1 in each of the columns ii–vi.
3 **maš alla**: very uncertain.

Column v

5 3.24.40: thus Neugebauer, but 3.24.30 is also possible.

Column vi

8 10:14: the initial 10 is followed by a separation mark, presumably in order to prevent 10.14 from being wrongly interpreted as 24.
9 9.45: replaces 9.15 (*ACT*).

Column vii

1 Neugebauer took column vii to consist of two separate columns, the first of which inscribed only in line 1, the second containing the consecutive leap years (cf. the commentary). It seems more likely that they form a single column, with the text at the beginning of line 1 being an introductory statement or label. However, the signs are badly damaged and could not be deciphered.
9 The sign preceding *ri* might be DIRI (*ACT*).

Commentary

This tablet is nearly completely preserved on three fragments. Nothing remains of the lower (upper) edge of the obverse (reverse), but not much clay is missing. The bulk of the tablet is occupied by a synodic table (*ACT* 420) with positions and dates for all of Venus's synodic phenomena (MF, ML, EF, ES, EL, MS) for SE 180–242. For a new edition of this table cf. Volume II. Hence the tablet was probably written near SE 180 (132/1 BC), during the Parthian era. Below the synodic table on the reverse there is a colophon consisting of a single line mentioning two individuals. The first one, Marduk-nādin-šumi (or -šuma-iddin?), son of Bēl-iddina, of the Egibatila (= Egibi) clan, is presumably the scribe. After collation, the name of the second individual turns out to be of the type DN-šāpik-zēri, probably Marduk-šāpik-zēri. He is said to be a 'scribe of 'When Anu (and) Enlil'', i.e. astronomer. The filiation (son of Bēl-bullissu, grandson of ⸢Nabû?⸣-nādin-apli) suggests that he may be identified with the astronomer Marduk-šāpik-zēri/Bēl-bullissu mentioned in two Parthian-era protocols of the counsel of the Esagila temple.[610] The word 'opposite' preceding his name probably implies that the tablet was written under his supervision. Below the colophon there are seven columns with procedures, all of which are directly connected with the synodic table.

[610] *CT* 49, 144 (dated 15 X SE 193) and *CT* 49, 186. For these documents cf. McEwan (1981); van der Spek (1985); Rochberg (2000).

P1–P12 Venus systems A_1 and A_2: updating B and T for MF, ML, EF, ES, EL, and MS

Columns i–vi contain procedures for updating *B* and *T* for all six successive synodic phenomena between ES to ML. Each column contains 2 procedures, the top one for updating *B* (odd numbered), the bottom one for updating *T* (even numbered). Most are badly damaged, but Neugebauer succeeded in partly restoring the missing text with the help of the algorithms that he reconstructed from the synodic tables. For a general discussion of systems A_1 and A_2 cf. §3.5.4; for the parameters cf. Table 3.21 (p. 78). The procedures for MF and ML belong to system A_1, those for EF, ES, EL and MS to system A_2. Each procedure for updating *B* begins with the statement 'SN to SN', each procedure for updating *T* with "days' for SN to SN'. All employ the same formulation based on the 'pseudo step function' template PSTEP (§2.6.10.1). P1 has a different beginning from the other procedures, but it is badly damaged. The two values of σ_j that are preserved in P1 are read differently from *ACT* (cf. Table 3.21).

P13 Leap years

Column vii contains a list with the actual intercalations needed for deriving the dates in column *T* of the synodic table. The first entry in Rvii1 is preceded by what may be an introductory statement, which is too strongly damaged for an interpretation.

P14 Unclear

Below P13 there is a horizontal line followed by a single line of text, too badly damaged to determine its meaning.

No. 8

museum numbers	BM 42799 (81–7–1,553) +45777 (81–7–6,193) +fragment
provenience	Babylon
date	data in synodic table: SE 183–242
measures	10.0(19?) × 9.0(*) × ? cm (type ii?)
arrangement	O/R
cuneiform text	photo: Fig. F.7 (BM 42799+45777 R); copy BM 45777: Kugler *SSB* I, Pl. 21
bibliography	Kugler *SSB* I; *ACT* 421a
contents	Venus system A_2?, synodic table for EF, ES, EL, MF, [MS, ML]; corresponding procedures

Reverse

Co 7' [...] ⸢xxxx⸣ **me**? 23 ⸢2? xxxx⸣ [...]

Column i'

P1' 8' [xxxx] ⸢xxx⸣
9' [xxxx] ⸢x⸣ 3 ⸢x⸣
10' [xxxx] ⸢x⸣ 3 ⸢x⸣
11' [xxxxx] ⸢x⸣ [xxx]
(n lines missing)

Column ii'

P2' 8' ⸢**uš**? xxxx *ša₂*⸣ **šu₂**
9' [xxxxx] **tab**
10' [xxxxx] **tab**
11' [xxxxxxx]2 **tab**
12' [xxxxxxx] ⸢**tab**⸣
(n−1 lines missing)

Reverse

Co **(Colophon)**

7'[...] ⸢...⸣ ... 23 ⸢2? ...⸣ [...]

Column i'

P1' **(System A_1/A_2?, EF or ES: updating *B*)**

8'[...] ⸢...⸣
9'[...] ⸢...⸣ 3 ⸢...⸣
10'[...] ⸢...⸣ 3 ⸢...⸣
11'[...] ⸢...⸣ [...]
(n lines missing)

Column ii'

P2' **(System A_1/A_2?, ES or EL: updating *B*)**

8'⸢... station? ...⸣ western ⸢...⸣
9'[...] you add.
10'[...] you add.
11'[...]2 you add.
12'[...] you add.
(n−1 lines missing)

Critical and philological notes

Reverse
Colophon
7' ME: or 1 10?
Column ii'
8' The faint traces before *ša₂* are compatible with **šu₂** or ˘**š**, less likely **igi**.

Commentary

This tablet is preserved on two joined fragments. A third fragment, discovered by Sachs and transliterated separately by Neugebauer (*ACT*, pp. 332–333), could not be identified (its museum number is unknown). No original edges of the tablet are preserved. The obverse, and the top of the reverse, contain a synodic table for Venus whose columns *T* and *B* are partly preserved for EF (only *B*), ES, EL and MF. For this table cf. *ACT* 421a and Volume II. The data in the synodic table suggest that the tablet may have been written near SE 183 (129/8 BC), during the Parthian era. The underlying algorithms are roughly consistent with systems A_1/A_2 (§3.5.4). It can be assumed that column *T* for EF was present on the left side, and columns *T* and *B* for MS and ML on the right side. This implies that about 1.5 cm of clay is missing from the left side, and about 7.5 cm from the right side, so that the original width was about 19 cm. The reverse is badly worn, with only few legible signs. Below the synodic table there is a horizontal line followed by a single line of text, probably a colophon, but almost nothing can be read. This is followed by remains of (probably) two columns, each containing faint remains of a procedure. This arrangement of synodic table, colophon and procedures is similar as in No. **7**, also a tablet concerned with Venus systems A_1/A_2, where each column contains a procedure for the synodic arc (σ) followed by a procedure for the synodic time (τ). Neugebauer did not produce an edition of the procedures, which are presented here for the first time.

P1' Updating B for EF or ES

The 4 partly preserved lines are likely concerned with the algorithm for updating *B* for Venus systems A_1/A_2, probably for EF or ES. The repeatedly occurring digit 3 confirms this, since the synodic arc is about 3,30°. It is presumably formulated in accordance with the pseudo step function template PSTEP (§2.6.10.1).

P2' Updating B for ES or EL

The 4 partly preserved lines probably contain the algorithm for updating *B* for Venus systems A_1/A_2, probably for the next morning phenomenon after the one dealt with in P1', i.e. ES or EL. It is presumably formulated in accordance with the pseudo step function template PSTEP .

No. 9

museum number	BM 33552 (Rm4,108)
provenience	Babylon
date	probably early Seleucid
measures	5.7(6.5) × 5.4 × 1.2–1.7 cm (type G)
arrangement	O/R
cuneiform text	photo: Fig. F.12; copy: Britton & Walker (1991), p. 99
bibliography	Britton & Walker (1991)
contents	Venus, system C$_3$: subdivision of the synodic cycle

Obverse

P1 1 [*dil-b*]*at* **ta** *muḫ-ḫi* **uš** ⸢*ar*$_2$⸣-*tu*$_2$ *ina* 2.⸢8⸣ [**me**]
2 2.8 ⸢**ki**⸣ **du ta** *muḫ-ḫi* 45 **z**[**i**]
3 **en** 1.15 **zi me** 14.3.45 *taš-pil-t*[*u*$_4$ x]
4 *ina* 4.32 *u*$_4$-*mu* 5.40 **ki du me** 1.15 [**ki**$^?$ **du**$^?$]
5 **egir** *ina* 2.8 **me** 2.8 **ki du uš** [**igi**-*tu*$_2^?$]
6 **ta** *muḫ-ḫi* 1.15 **zi en** *muḫ-ḫi* 4⸢5⸣ [**zi** x]
7 **me** 14.3.45 *taš-pil-tu*$_4$ **kuš**$_3$ **gal la**$_2$-*is* [x]
8 24 **me** *nen-mud ina* 32 ⸢**me**⸣ 20 ⸢**ki**⸣-*šu*$_2$ *ana* **šu**$_2$ ⸢**la**$_2$⸣-[*is*$^?$ x]
9 PAP.PAP *ina* 9.44 **me pap** 9.44 ⸢**ki**⸣ **du** [x]

P2 10 **ta pa en** *zib ma-la ša*$_2$ *ina* **šu**$_2$ *i-ra*[*b-bi* x]
11 *ina* 1 *u*$_4$-*mu* **igi** *ina* **ḫun** {**ḫun**} 4 ⸢*u*$_4$⸣-[*me* x]

Reverse

12 [*in*]*a* **mul**$_2$ 8 **me** *ina* **maš** 12 [**me** x]
13 ⸢**ta alla en absin**⸣ 15 ⸢*u*$_4$⸣-[*me* x]
14 *ina* ⸢**rin**$_2$ **gir**$_2$.**tab** 6⸣ **me** *ina* **pa** 2 ⸢*u*$_4$⸣-[*me* x]

P3 15 **ta uš** *ar*$_2$ **en ki šu**$_2$-*šu*$_2$ 4.20.[30 **ki**$^?$ **du**$^?$]
16 *ina* 3.45 **me du**-*šu*$_2$ [x]
17 **zi** ⸢**gal**⸣ 1.15 **zi tur** 45 [x]
18 *ina* 8 **mu.meš šu**$_2$ *u*$_3$ **igi gi** [x]

Co 19 [**im**].⸢**gid**$_2$.**da**⸣ ^{m}**u.a.mu a**-*šu*$_2$ ^{m}**gi.u**
20 **dumu** ^{m}*mu-še-zib*

Obverse

P1 **(Subdivision of the synodic cycle)**

[1]Venus. From the second station (MS) in 2,8 [days] [2]the position proceeds 2,8. From 0;45, the displacement, [3]until 1;15, the displacement, per day 0;0,14,3,45 is the difference. [...?] [4]In 4,32 days the position proceeds 5,40; per day [the position proceeds] 1;15. [5]Afterwards, in 2,8 days the position proceeds 2,8, (then) the [first] station (ES). [6]From 1;15, the displacement, until 0;45, [the displacement], [7]per day 0;0,14,3,45 is the difference.

It recedes by a large cubit [...]. [8]24 days it is stationary. In 32 days its position recedes to the west by 20. [9]Total: in 9,44 days the position proceeds 9,44 (error for 9,36) ... [...]

P2 **(Temporal push from EL to MF)**

[10]From Sgr (error for Cap) until Psc, wherever it sets in the west (EL) [...] [11]in 1 day appearance (MF); in Ari {Ari} 4 days [...],

Reverse

[12]in Tau 8 days, in Gem 12 [days, ...] [13]from Cnc until Vir 15 days [...] [14]in Lib, Sco 6 days, in Sgr 2 days [...]

P3 **(Subdivision of the synodic cycle: MS to ML)**

[15]From the second station (MS) until the position of its setting (ML) by 4,20;[30 the position$^?$ proceeds.$^?$] [16]in 3,45 (error for 3,54?) days its progression [.... [17]The largest displacement, 1;15, the smallest displacement, 0;45 [...] [18]In 8 years the setting and the appearance become 'intact'. [...]

Co **(Colophon)**

---

[19]*Giṭṭû*-tablet of Bēl-apla-iddin, son of Mušallim-Bēl, [20]descendant of Mušēzib.

Critical and philological notes

Obverse

7 **kuš**$_3$ **gal** = *ammatu rabītu*, 'large cubit'. This might refer to a cubit of 2;30° (cf. §2.5.3.2, p. 34). **la**$_2$-*is*: either *inaḫḫis*, 'recedes', or *tanaḫḫis*, 'you subtract'. The entire phrase can, if translated as 'it recedes by a large cubit', be meaningfully interpreted as a reference to the net displacement of Venus for 5 synodic cycles.

10 **pa**, Sgr: error for **maš**$_2$, Cap.

Reverse

15 4.20: one expects 4.20.30; cf. the commentary. The signs **ki du** must have been rather crammed, if they are to be restored here.

16 3.45: probably an error for 3.54; cf. the commentary.

18 **gi** = *išallamū*, 'they become intact', G. pres. 3 m. pl. of *šalāmu* G, 'to be/become intact'.

19 *giṭṭû*(**im.gid**$_2$.**da**) designates one-columnar tablets with a scholarly content; cf. p. 11.

Commentary

This well-preserved tablet was written by Bēl-apla-iddin/Mušallim-Bēl/Mušēzib, probably the same individual who wrote the procedure text No. **5**, the astronomical diary *ADRT* I -321, which is dated to year 2 of Philip Arrhidaeus (322/1 BC), and *ADRT* V 60, a tablet with Jupiter observations for 387–345 BC. The procedures concern the subdivision of the synodic cycle for Venus in accordance with system C_3. For the algorithms and the parameters cf. §3.5.5; for a detailed commentary cf. Britton & Walker (1991).

***P1** Subdivision of the synodic cycle*

P1 contains a scheme for the subdivision of the synodic cycle setting out from MS, providing values of $\delta\Sigma$, $\delta\tau$ and v for some of the pushes. The formulation is partly based on template PUSH.C.1, but in some segments v is modeled as a linear function of the day number, using a formulation not attested elsewhere. For the parameters cf. Table 3.22 (p. 80).

***P2** Temporal push from EL to MF*

P2 is essentially a list of values of $\delta\tau$ (EL to MF), each associated with a whole number of zodiacal signs, for the duration of invisibility near the inferior conjunction. For the parameters cf. Table 3.22 (p. 80).

***P3** Subdivision of the synodic cycle: MS to ML*

P3 contains a complementary description of the motion from MS to ML which is consistent with P1. The formulation is partly based on template PUSH.C.1. In a segment of this push v is modeled as a linear function of the day number, using an abbreviated version of the formulation from P1.

No. 10

museum number	BM 32599 (76–11–17,2343)
provenience	Babylon
date	probably late Achaemenid (4th c. BC); data in synodic table: 419–189 BC
measures	12.0(25?) × 9.0 × 2.0–3.0 cm (type iv?)
arrangement	O/R
cuneiform text	photo: Fig. F.9 (R: procedures); *ACT*, Pl. 247
bibliography	*ACT* 824, 1050; Hamilton & Aaboe (1998); Britton (2001)
contents	Venus system A$_3$: synodic table; procedures concerning net displacements

Reverse
Column i

P1 15 [**igi ta** 12 **alla en** 18 **maš**$_2$ 2.30 **la**$_2$ 2.13.20]
16 [25.20 **rin**$_2$ 12 **alla** 2.46.40]

Column ii

P2 15 [**du ta** 12 **alla en** 28.40 *zib* 2.46.40 **la**$_2$ 2.13].⸢20⸣
16 [6 **maš**$_2$ 12 **alla** 2].⸢30⸣

Column iii

P3 15 **kur ta** 12 **maš**$_2$ **e**[n 6 **rin**$_2$] 2.40 **la**$_2$ 2.30
16 12 **all**[**a** 18 **ḫun** 2].20 2.30

Column iv

P4 15 **šu**$_2$ *gab-bi* 2.30 **la**$_2$

Reverse column i

P1 **(EF: net displacement for 5 cycles)**

[15][Appearance (EF). From 12 Cnc to 18 Cap you subtract 2;30, (then) 2;13,20,] [16][25;20 Lib; 12 Cnc, 2;46,40.]

Column ii

P2 **(ML: net displacement for 5 cycles)**

[15][DU (ML). From 12 Cnc to 28;40 Psc you subtract 2;46,40, (then) 2;13],20, [16][6 Cap; 12 Cnc, 2];30.

Column iii

P3 **(MF: net displacement for 5 cycles)**

[15]Rising (MF). From 12 Cap to [6 Lib] you subtract 2;40, (then) 2;30, [16]12 Cnc; [18 Ari, 2];20, (then) 2;30.

Column iv

P4 **(EL: net displacement for 5 cycles)**

[15]Setting (EL). All (zodiacal signs): you subtract 2;30.

Critical and philological notes

Reverse

15 **la**$_2$: the Akkadian reading is not clear. Apart from *tanaḫḫis*, 'you subtract', it may represent *inaḫḫis* 'it recedes', *maṭi*, 'it is lacking', or *tumaṭṭa*, 'you diminish'. The astronomical or mathematical meaning is not affected by this.

Column ii

15 DU: the context implies that this is a term for ML; Akkadian reading unclear.

Commentary

This fragment preserves the right half of a tablet. The bulk of the tablet is occupied by a synodic table with zodiacal positions of Venus, first published as *ACT* 1050,[611] for which I refer to Britton (2001) and Volume II. It is followed on the reverse by 4 brief procedures (P1–P4), first published as *ACT* 824. As argued by Britton, the original tablet contained 20 columns subdivided into 5 groups, implying a width of about 25 cm. The resulting shape of the tablet is unusual, although not without parallel among the tabular texts. All rows and columns are separated by lines, resulting in a checker board structure. While Britton suggests that the tablet may have been written near 189 BC, the final date of the predictions, other extant astronomical tablets with a checker board format date from the Achaemenid or early Seleucid period. It is therefore more likely that the tablet was, as usual, written near the beginning of the predictions, in the late Achaemenid period. The unusual terminology for the synodic phenomena (e.g. DU for ML) confirms this. Each group of 4 columns provides zodiacal positions of EL, MF, ML and EF, computed in accordance with system A$_3$ (§3.5.3). Successive lines are separated by 5 synodic cycles = 8^{yr}. For a detailed commentary cf. Britton (2001).

[611] The fragment was first mentioned by Epping & Strassmaier (1881).

P1–P4 *Net displacements for 5 cycles*

The positions in the synodic table are updated from line to line by applying the appropriate net displacement for 5 cycles, $dB_j(5)$, which are provided in P1–P4. For the algorithm and the parameters cf. Table 3.19 (p. 77). They are formulated in accordance with template DIFF.A.2 (§2.6.10.4).

No. 11

museum number	BM 34389 (Sp,508)
provenience	Babylon
measures	5.7(*) × 4.1(*) × 2.6(*) cm
cuneiform text	photo: Fig. F.8; *ACT*, Pl. 243; copy: *LBAT* 150
bibliography	*ACT* 815
contents	Venus, unknown system: various procedures

Side X

P1' (unknown number of lines missing)

1' ⸢xxxx⸣ [...]

P2' 2' ⸢**ta**$^?$ **alla**⸣ *ana* **rin**$_2$ 20 x [...]

3' 8 *ša*$_2$ *dil-bat* ⸢xx⸣ [...]

4' *a*-⸢*na*⸣ 3 [...]

5' *a-na* 3 [...]

(unknown number of lines missing)

Side X

P1' (Unclear)

(unknown number of lines missing)

1'⸢...⸣ [...]

P2' (Unclear)

2'⸢From$^?$ Cnc⸣ to Lib 20 ... [...] 3'The 8 of Venus ...⸣ [...] 4'For 3 [...] 5'For 3 [...]

(unknown number of lines missing)

Critical and philological notes

Side X

1' Traces of several vertical wedges.

2' **ta**: dubious.

3' *dil-bat*: followed by traces of a sign beginning with a vertical wedge.

Commentary

This small fragment does not contain any edges of the tablet. It is inscribed on one side (denoted X); the other side is destroyed. Five lines are partly preserved. Horizontal dividing lines below X1' and X5' imply that the tablet contains at least 3 procedures. On the left side there is a double vertical line marking the edge of a column or of the tablet. The fragment might belong to the same tablet as No. **46**, but they do not physically join.[612]

P2' *Venus, unknown system: unclear*

This brief procedure consists of 4 lines, of which only the first few signs are preserved. X3' mentions the familiar 8-year period (cf. §3.5.1). The digits 3 mentioned in X4'–5' are of unknown significance.[613]

---

[612] If it belongs to the same tablet as No. **46** then it is positioned in column ii on the reverse, below P31'.

[613] As suggested by Neugebauer, they might belong to 3,47 = 227 years, an interval after which Venus returns to nearly the same position (compare Table 3.19, p. 77).

5.2.3 *Mars*

No. 12

museum number	U 150
provenience	Uruk
date	SE 105 ± 15 (likely before VII SE 118)
measures	6.2(20–25?) × 8.0(15?) × ? cm (type iii)
arrangement	O/R
cuneiform text	photo: Fig. F.13; *ACT*, Pl. 248 (R); copy: Pl. 225 (R)
bibliography	*ACT* 803, N
contents	Mars system A: auxiliary table + procedure (subdivision of the synodic cycle)

Obverse

T1'		(about 12 lines missing)			
	0'	[13	6.14.24	13	6]
	1'	[14	6.13.36	14	6]
	2'	[15	6.12.4]8	15	6
	3'	[16	6.12]	16	6
	4'	[17	6.11].12	17	6
	5'	[18	6.10].24	18	6
	6'	[19	6.9.36]	19	6
	7'	[20	6.8.4]8	20	6
	8'	[21	6].8	21	6
	9'	[22	6].7.12	22	6
	10'	[23]	6.6.24	23	6
	11'	24	6.5.36	24	6
	12'	25	6.4.48	25	6
	13'	26	6.4	26	6
	14'	27	6.3.12	27	[6]
	15'	28	6.2.24	28	[6]
	16'	29	6.1.36	[29	6]
	17'	30	6.0.4[8	30	6]

Reverse

Column i'

P1' 1 [x] *an-na*$^?$ x *ša*$_2$ **an**
2 [x] **ta uš igi**-*tu*$_2$ *ana a* **me gar**$^?$
3 ⌜**alla**⌝ *u* **a** 7.12 **gur**
4 **absin** *u* **rin**$_2$ 6.48 **gur**
5 **gir$_2$.tab** *u* **pa** 6.24 **gur**
6 [**maš$_2$** *u* **gu**] ⌜6 **gur** x⌝
7 [*zib u* **ḫun**] 6.24 ⌜**gur**⌝
8 ⌜**mul$_2$**⌝ *u* **maš** 6.⌜4⌝8 **gur**
Co 9 [...xx] ⌜*qat*$_3^?$⌝ ^{md}**utu.sur a** *ša*$_2$ ^{m}*ina*-⌜*qi*$_2^?$⌝-*bit*-^{d}**diš**
10 [...xx ^{m}*e*$_2$-*k*]*ur-za-kir*
11 [...xxxxxxxxx] ⌜**lugal.meš**⌝
(about 14 lines missing)

Obverse

T1' (Angular push from S1 to AR, scheme S)

(about 12 lines missing)

0'	[13	6;14,24	13	6]
1'	[14	6;13,36	14	6]
2'	[15	6;12,4]8	15	6
3'	[16	6;12]	16	6
4'	[17	6;11],12	17	6
5'	[18	6;10],24	18	6
6'	[19	6;9,36]	19	6
7'	[20	6;8,4]8	20	6
8'	[21	6];8	21	6
9'	[22	6];7,12	22	6
10'	[23]	6;6,24	23	6
11'	24	6;5,36	24	6
12'	25	6;4,48	25	6
13'	26	6;4	26	6
14'	27	6;3,12	27	[6]
15'	28	6;2,24	28	[6]
16'	29	6;1,36	[29	6]
17'	30	6;0,4[8	30	6]

Reverse column i'

P1 (Angular push from S1 to AR, scheme R or S)

1 [...] thus ... of Mars.
2 [...] from the first station to daylight rising (AR) you put down$^?$.
3 ⌜Cnc⌝ and Leo: it turns back 7;12.
4 Vir and Lib: it turns back 6;48.
5 Sco and Sgr: it turns back 6;24.
6 [Cap and Aqr]: ⌜it turns back 6 ...⌝.
7 [Psc and Ari]: ⌜it turns back⌝ 6;24.
8 ⌜Tau⌝ and Gem: it turns back 6;⌜4⌝8.

Co (Colophon)

9 [...] ⌜hand$^?$⌝ of Šamaš-eṭir, son of Ina-qibīt-Anu, 10 [... descendant of E]kur-zākir. 11 [...] were kings.
(about 14 lines missing)

Critical and philological notes

Reverse column i'

1 In this line the signs are smaller and less deeply impressed than in the rest of the text. The sign following *na*$^?$ looks like ŠA$_3$.
2 **gar**: uncertain.

6 ⌜6 **gur**⌝ is followed by traces of a sign, which was tentatively identified as LU by Neugebauer.

9 The damaged sign preceding ^{md}**utu.sur** is probably *qat₃*, 'hand of', which introduces the scribe of the tablet. The 'owner' must have been mentioned to the left of the preserved text.

Commentary

This fragment belongs to the lower (upper) right corner of the obverse (reverse) of a tablet. The left edge and upper (lower) edges of the obverse (reverse) are broken off. The obverse contains an auxiliary table with values of $\delta\Sigma$ (S1 to AR) for every degree of the zodiacal signs. A full reconstruction of this table (cf. below) implies that the original size of the tablet was about 20–25 × 15 cm. The reverse ends with a colophon, of which the first two lines are partly preserved. They mention Šamaš-ēṭir/Ekur-zākir as the scribe (less likely the 'owner') of the tablet; the name of the 'owner' and the date are not preserved. Šamaš-ēṭir/Ekur-zākir is a well-known astronomer associated with the Rēš temple who was active in the years SE 110–119 (202–192 BC). If he is the scribe then the tablet predates 12 VII SE 118, his earliest attestation as 'owner' of scholarly tablets.[614] The reconstructions leave room for about 14 more lines, which may have contained the rest of the colophon or an additional procedure.

T1' Angular push from S1 to AR (scheme S)

Two pairs of columns of this auxiliary table are partly preserved. The first column of each pair originally contained the numbers 1 to 30 representing degrees within a zodiacal sign, the second column the corresponding value of $\delta\Sigma$ (S1 to AR) in accordance with scheme S (Table 3.30, p. 87). The partly preserved columns belong to Sgr and Cap. On this basis Neugebauer (*ACT*, Pl. 211) proposed a full reconstruction of T1' with 12 pairs of columns, one for each zodiacal sign, of which the first 7, starting with Cnc and ending with the partly preserved columns for Sgr and Cap, are on the obverse, while the other 5 are on the reverse.

P1' Angular push from S1 to AR (scheme S or R)

This procedure summarises the information contained in T1'. To the left of P1' there were presumably 5 further pairs of columns of the auxiliary table. P1' begins with a damaged introductory statement mentioning Mars and the angular push, $\delta\Sigma$, from S1 to AR. What follows are 6 pairs of zodiacal signs and the associated values of $\delta\Sigma$. Neugebauer proposes that they belong to scheme R (Table 3.30), which would imply an inconsistency with the auxiliary table, which contains scheme S. However, at the beginning of each zodiacal sign the values of $\delta\Sigma$ are the same in both schemes. Hence P1' may also belong to scheme S, if one accepts that it is incomplete, in the sense that the interpolations to be performed in the second zodiacal sign of each pair are not mentioned, but implied.

[614] *ACT* 600 = AO 6476 +U 104, which was written by Anu-aba-utēr/Sîn-lēqi-unninni (cf. Volume II).

No. 13

museum numbers	BM 34676 (Sp,163+304+396+491+895)
provenience	Babylon
measures	10.2 × 15.0(*) × 2.4 cm (type D)
arrangement	O/R
cuneiform text	photo: Fig. F.11; *ACT*, Pl. 246; copy: *LBAT* 138
bibliography	*ACT* 811a; van der Waerden (1957), pp. 49–55; van der Waerden (1965), pp. 190–192 (1974: pp. 265–271); *HAMA*, pp. 406–412, 458; Swerdlow (1998), pp. 147–152
contents	Mars systems A and B: various procedures

		Obverse?
P0'	1	⌜x⌝ [...]
	2	**šu$_2^?$**-*šu$_2^?$* ⌜2$^?$⌝ [...]
	3	1.⌜50$^?$⌝ [...]
	4	20 [...]
	5	**u$_4$** ⌜1$^?$⌝ [...]
	6	5 [...]
	7	5 [...]
	8	**en** [...]
	9	*ša$_2$* [...]
	10	**gar**-⌜*an*⌝ [...]
	11	**alla** [...]
	12	⌜**gir$_2^?$**⌝ [...]
	13	**ḫun** [...]
	14	**ta** [...]
	15	**šu$_2^?$** ⌜x⌝ [...]
		(about n+20 lines missing)
		Reverse?
P1'		(n lines missing)
	1'	[xxxxxxxxxxxx *zib* **ḫu**]**n** ⌜6.30 **mul$_2$ maš**⌝ 7 **gur** *an*-⌜*na*⌝-[*a* xx]
P2'	2'	[**alla a** xx **absin rin$_2$** xx **gir$_2$.tab pa**] ⌜9$^?$.30$^?$ **maš$_2$ gu**⌝ 10.45 *zib* **ḫun** 9.30 **mul$_2$ maš** 10 *an-na*-⌜*a*⌝ [x]
P3'	3'	[xxxxxxx] **ḫun**$^?$ **mul$_2$ zi u$_4$** *bi-rit* **igi** *ana* **igi** GIŠ-*ma* 23.37.52
	4'	[**ki**-*šu$_2$* **tab**-*m*]*a* **u$_4$.meš e**-*bi* **ta šu$_2$** *ana* **igi ki** *bi-rit* **igi** *ana* **igi** GIŠ-*ma*
	5'	[xxxx **e**-*bi* **t**]**a igi** *ana* **uš ki** *bi-rit* **igi** *ana* **igi** 1.53.13.13 **iti.**⌜**meš**⌝
	6'	[xxxx **e**-*b*]*i* **ta uš igi**-*tu$_2$* *ana* **šu$_2$ ki** *bi-rit* **igi** *ana* **igi** 3.56.44.3⌜5⌝
P4'	7'	[xxxxx GAM] ⌜1⌝.50.40 **du**-*ma* **pap.pap gar**-*ma* 33.40.4 *ša$_2$* **kur** *u* **šu$_2$ ki**-*šu$_2$* **tab**-*ma* **e**-*bi*
P4a'	8'	[xxxxxxxx]-*ma* **pap.pap gar**-*an* GAM 1.50.40 **du**-*ma* **ki pap.pap tab**-*ma* **ki** *u$_4$-mu* *ša$_2$* **igi tab**-*m*[*a*]
P5'	9'	[xxxxxxxx] **igi.du$_8$.a ta** *lib$_3$-bi* **e$_{11}$**-*ma* **pap gar**-*an* 2 **ki pap tab**-*ma*
	10'	[xxxxx **ki** *u$_4$-mu* *š*]*a$_2$* **igi tab**-*ma* *u$_4$-mu* **uš e**-*bi* **ta uš igi**-*tu$_2$* *ana* **šu$_2$ gar.gar**-⌜*ma*⌝
	11'	[xxxxxxxx 1$^?$]5 **uš** *ša$_2$* *la-pan* ^{d}**utu sud ta** *lib$_3$-bi* ⌜**e$_{11}$**-*ma*⌝
	12'	[xxx] ⌜GAM 1.50⌝.40 ⌜**du**⌝ **ki** *u$_4$-mu* **uš** ⌜**tab**⌝-*ma* *u$_4$-mu* **šu$_2^?$**-*u$_2$* **e**-⌜*bi*⌝
P6'	13'	[12.3]⌜3.5⌝1.52.47.20 **murub$_4$**-*u$_2$* *ša$_2$* **ta igi** *ana* ⌜**igi** 2⌝.44.⌜45.6⌝.46.[46.40 **murub$_4$**-*u$_2$*]
	14'	[*ša$_2$* **t**]**a šu$_2$** *ana* **igi** 4.59.22.55.57.46.40 **murub$_4$**-*u$_2$* **ta** ⌜**igi** *ana* **uš** 4.49.33.50⌝
	15'	⌜5⌝.51.6.40 **murub$_4$**-*u$_2$* *ša$_2$* **ta igi** *ana* **šu$_2$** *an-na-a* **u$_4$.da.zal.la$_2$** *ša$_2$* **ki.meš**
P7'	16'	[30] *ša$_2$* **kur** *u*$^?$ **šu$_2$** GAM 1.50.40 **du**-*ma* 55.20 **ki** 30 *ša$_2$* **kur šu$_2$ tab**-*ma* 30.55.20 1.44.45
	17'	[**murub$_4$**]-*u$_2$* *ša$_2$* **ta šu$_2$** *ana* **igi ki**-*šu$_2$* **tab**-*ma* 33.40 **ta šu$_2$** ⌜*ana*⌝ **igi ki** *bi-rit* **igi** *ana* **igi** 32 **tab**

Obverse?

P0' (Unclear)

[1]... [...] [2]its setting? ... [...] [3]1.⸢50⸣ [...] [4]20 [...] [5]day ...[...] [6]5 [...] [7]5 [...] [8]until [...] [9]... [...] [10]you put down [...] [11]Cnc [...] [12]⸢Sco?⸣ [...] [13]Ari [...] [14]from [...] [15]... [...]

(about n+20 lines missing)

Reverse?

P1' (System A: angular push S1 to AR, scheme T)

[1'][... Psc, Ari]: ⸢6;30; Tau, Gem⸣: it recedes 7. Thus [for the positions.]

P2' (System A: angular push AR to S2?, scheme T?)

[2'][Cnc, Leo: ... Vir, Lib: ... Sco, Sgr:] ⸢9?;30⸣; Cap, Aqr: 10;45; Psc, Ari: 9;30; Tau, Gem: 10. Thus [for the positions.]

P3' (System A: updating *T* with the synodic time; temporal pushes)

a (Unclear)

[3'][...] Ari?, Tau, you subtract?.

b (Updating *T* with the synodic time)

The 'day': the distance (from) appearance (FA) to appearance you compute, and 23;37,52, [4'][you add to it] ⸢and⸣ you predict it as the 'days'.

c (Subdivision of the synodic cycle, scheme S_2: $\delta\tau$)

From setting (LA) to appearance (FA), position: you compute the distance from appearance to appearance, and [5'][..., you predict the 'days'.]

From appearance to station (S1), position: the distance from appearance to appearance, 1,53;13,13 months (error for 'days') [6'][... you pre]dict [the 'days'].

From first station to setting, position: the distance from appearance to appearance, 3,56;44,35.

P4' (System A, subdivision of the synodic cycle, scheme S_2: $\delta\tau$)

[7'][...] You multiply [... by 0;1],50,40, and you put down the total, it is 33;40,4, for the 'rising and setting'(?). You add it to it, and you predict it.

P4a' (System A: synodic time or temporal pushes)

[8'][...] and you put down the total. You multiply it by 0;1,50,40, and you add it to the total, and you add it to the 'day' of the appearance, and [you predict it.]

P5' (System A, subdivision of the synodic cycle, scheme S_2: $\delta\tau$ from FA to S1 and from S1 to LA)

[9'][...] you subtract [the ...] of the appearance (FA) from it, and you put it down as the total, you add 2,0 to the total, and [10'][...] you add [... to the 'day'] of the appearance, and you predict it as the 'day' of the station (S1). From first station to setting (LA) you accumulate it, and [11'][... 1]5° by which it is in front of the Sun, you subtract from it, and [12']you multiply [...] by 0;1,50,40, [add] it to the 'day' of the station, and you predict it as the 'day' of the setting.

P6' (System A, subdivision of synodic cycle, scheme S_2: values of 0;0,10$e \cdot \overline{\delta\Sigma_p}$)

[13'][12;33],51,52,47,20, the middle one which is from appearance (FA) to appearance.
2;44,45,6,46,[46,40, the middle one] [14'][which is] from setting (LA) to appearance.
4;59,22,55,57,46,40 (error for 4;59,32,55,57,46,40), the middle one from appearance to [station (S1)].
4;49,33,50,[15'][5],51,6,40, the 'middle one' which is from appearance (error for station) to setting.
Thus are the *uddazallû*-coefficients for the positions.

P7' (System A, subdivision of synodic cycle, scheme S_2: $\delta\tau$ from LA to FA)

[16']You multiply [30] for the 'rising and setting' by 0;1,50,40, it is 0;55,20, you add it to 30 for the 'rising (and) setting', it is 30;55,20. 1;44,45 (error for 2;44,45), [17']the middle one from setting (LA) to appearance (FA), you add to it, it is 33;40. From setting to appearance you add 32 to the distance (from) appearance to appearance.

P8’ 18’ [1.45] ⸢GAM 1.50⸣.40 **du**-*ma* 3.13.40 **ki** 1.45 **tab**-*ma* 1.[48].⸢1⸣3.40 4.59.33 **murub$_4$**-*u$_2$*
19’ [*ša$_2$* **ta igi** *ana* **uš igi**]-⸢*tu$_2$*⸣ **ki** 1.48.13.40 **tab**-*ma* 1.⸢53⸣.[13.13] ⸢**ta**⸣ **igi** *ana* **uš ki** *bi-rit ana bi-rit*
20’ [xxxxx] ⸢**murub$_4$**$^?$⸣ **me**$^?$ **tab**-*ma* [xxx]

P9’ 21’ [3.45] ⸢GAM 1.50⸣.40 **d**[**u**-*ma* 6.55] ⸢**ki** 3⸣.[45 **tab**-*ma* 3.51.5]⸢5⸣ 4.49.33.50
22’ [**murub$_4$**]-⸢*u$_2$* *ša$_2$*⸣ **ta uš i**[**gi**-*tu$_2$* *ana* **šu$_2$** **ki** 3.51.55 **tab**-*ma* 3].⸢5⸣6.44.34
23’ [**ta**] **uš igi**-*tu$_2$* *ana* ⸢**šu$_2$**⸣ **ki** *bi-rit* ⸢xxxx *ša$_2$* xxx **tab**-*ma* *u$_4$-mu* **šu$_2$**-*u$_2$* **e**⸣-*bi*

P10’ 24’ [z]**i** *ša$_2$* **an** *ša$_2$* *kal* **mu**.⸢**an.na**⸣ *ana* **igi**-*ka* **be**-*ma* *ina* **alla igi**-*ma* GIŠ$^?$ **igi**-*u$_2$* 30 **zi** *ina* **a** 33.20 **zi**
25’ *ina* **absin$_0$** 36.40 **zi** *ina* **rin$_2$** 40 *ina* **gir$_2$.tab** 43.10 **zi** *ina* **pa** 46.40 **zi** *ina* **maš$_2$** 50 **zi** *ina* **gu** 46.40 **z**[**i** *ina zib*]
26’ ⸢4⸣3.20 **zi** *ina* **ḫun** ⸢40⸣ **zi** *ina* **mul$_2$** 36.40 **zi** *ina* **maš.maš** 33.20 **zi** *ina* **alla** 30 **zi** *a-na* 1 **uš**
27’ 6.40 *taš-pil*-⸢*tu$_2$*⸣ *ta-za-za* 1-*me* 1.30 **me** *ša$_2$* **me** 40 **zi igi**-*at* **uš**-*šu$_2$* **igi**-*tu$_2$* 30 **me** *ša$_2$* **me**
28’ 36 **zi** 30 **me** *ša$_2$* **me** 24 **zi** 30 **me** *ša$_2$* **me** 12 **zi**-*ma* **uš ta uš igi**-*tu$_2$* **en uš** *ar$_2$-tu$_2$*
29’ *ša$_2$* **me** 12 *ana* *ar$_2$-ki-šu$_2$* **la$_2$**-*is-ma* **uš ta uš** *ar$_2$-tu$_2$* *u$_4$-mu* *ša$_2$* **zi tab**-*u$_2$* 30 **me** *ša$_2$* **me**
30’ 12 **zi** 30 **me** *ša$_2$* **me** 24 **zi** 30 **me** *ša$_2$* **me** 36 **zi** 3.53 **u$_4$.meš** *ša$_2$* **me** 40 **zi šu$_2$**
31’ **be**-*ma* **alla** 30 **zi be**-*ma* **rin$_2$** 40 **zi be**-*ma* **maš$_2$** 50 **zi be**-*ma* **ḫun** 40 **zi**-*ma* **šu$_2$**
32’ **ta šu$_2$** *ana* **igi** *ki* **igi**-*im-ma* *ina* *lu-maš*.**meš du$_3$**-*uš*

P11’ 33’ *ša$_2$* **an** 4.44 **mu.meš** ⸢2.13 x$^?$⸣ **igi.meš** 2.31 **bal.meš** 15.6 **ki du** *mi-nu-u$_2$* **a.ra$_2$** 2.13
34’ **igi.meš** *lu*-**du**-*ma* *lu*-⸢*u$_2$*⸣ 15.6 : 6.48.43.18.30 GAM 2.13 **igi.meš du**-*ma*
35’ 15.6 : 6 *tu*-⸢*ub*$^?$-*ba*$^?$⸣-*al* 48.43.18.40 *ana* **murub$_4$**-*u$_2$* **gar**-*an* 17 *taš-pil-tu$_2$*
36’ GAM 2.13 **igi.meš** ⸢**du**⸣-[*ma* 3]7.41 *mi-nu-u$_2$* GAM 18 **bal.meš** *lu*-**du**-*ma* *lu* 37.41 [x$^?$]
37’ 2.5.36.[40 GAM 1]8 **bal.meš du**-*ma* 37.41 : 2.5.36.40 *ša$_2$* **ta** 41 ⸢x⸣ [xxx]

Critical and philological notes

Obverse?

1 Probably *ša$_2$*, 4, or 7.

Reverse?

2’ The traces before **maš$_2$ gu** are compatible with 9.30, which number one expects on symmetry grounds.
10.45: one would expect 9; cf. the commentary.

3’ **ḫun mul$_2$**: replaces **e$_{11}$** (*ACT*). Pinches’s copy is erroneous here.
u$_4$: thus also Pinches’s copy; replaces *ma* (*ACT*).

4’,6’ **ki**: probably ‘position’ here.

4’–6’ **igi** *ana* **igi** = FA to FA? This is not what one expects; cf. the commentary.

7’ **kur** *u* **šu$_2$**: meaning uncertain, but cf. the commentary. The same phrase occurs twice in 16’, but in one of them the copula *u* is absent. This proves that one should not read **kur igi** (Pinches).
e-*bi* = *taqabbi*, ‘you predict’: as remarked by Neugebauer, this verb has no object, presumably by mistake.

8’ The first sign **pap** of the second instance of **pap.pap** looks more like TAB.

11’ **sud** = *rêqu*, ‘to depart; recede’?

12’ *u$_4$-mu* **šu$_2$**$^?$-*u$_2$*: **šu$_2$** appears to be written over another sign, perhaps KI or ŠU$_2$. It can be read either *ūmu rabû*, ‘day of setting’, (ideally *ūm rabê*), or *ūmu irabbû*, ‘the day when it sets’ (ideally *ūm irabbû*).

13’–15’ **murub$_4$**-*u$_2$* = *qablû*, ‘middle one’ = mean value.

14’ 55: replaces 45 (*ACT*). In the critical notes (*ACT*, p. 385) Neugebauer quotes the correct 55.
22: error for 32 (cf. the commentary).
uš 4.49.33.50: only traces remain.

15’ **igi**: error for **uš**, ‘station’.
u$_4$.da.zal.la$_2$ = *uddazallû*, ‘*uddazallû*-coefficient’. Attested meanings in astronomical contexts include ‘period’, ‘time interval’.

16’ **kur** *u* šu$_2$: replaces **kur**-*šu$_2$* **šu$_2$** (*ACT*, p. 386); cf. the commentary to line 7’.
kur šu$_2$: probably the same term as **kur** *u* **šu$_2$**.
1.44.45: error for 2.44.55 (cf. the commentary).

17’ **igi** *ana* **igi**, ‘appearance to appearance’: one expects **šu$_2$** *ana* **igi**, ‘setting to appearance’.
32: this number is unidentified. One does not expect any number; cf. the commentary.

19’ *bi-rit* at the end is written on the right edge.

20’ **murub$_4$** = *qablu*, ‘middle’: uncertain.

23’ After *bi-rit* the lower parts of several wedges are visible.

24’ GIŠ$^?$ *maḫrû*(**igi**-*u$_2$*): meaning unclear.

25’ 43.10: replaces the expected 43.20 (*ACT*).

27’ *ta-za-za* = *tazzaza*, ‘it stands’: 3. f. sg. pres. of *izuzzu* G. For the 3. f. sg. pres. with *ta*- in NB/LB Akkadian cf. *GAG* §75h.

29’ **zi** = *nisḫu* or *nasāḫu*, ‘displacement; motion’.
tab-*u$_2$* = *ušarrû*, ‘it begins’, 3 c. sg. pres. subj. *šurrû*, ‘to begin’. This alternative reading of **tab** fits better than the previous interpretation *ṭepû*, ‘to add’.

32’ *ki* **igi**-*im-ma* = *kī maḫrîma*, ‘like before’.

33’ 20$^?$: or 30 (thus Pinches). One expects nothing here (cf. the commentary).

P8' (System A, subdivision of synodic cycle, scheme S_2: $\delta\tau$ from FA to S1)

18'You multiply [1,45 by] 0;1,50,40, it is 3;13,40; you add it to 1,45, it is 1,[48],13,40. 4;59,33, the middle one 19'[from appearance (FA) to] first [station (S1)], you add to 1,48;13,40, it is 1,5[3;13,13]. From appearance to station you add 20'[...] (from) distance to distance, you add [to the ...] per 'day', and [you predict it as the 'day' of the station.]

P9' (System A, subdivision of synodic cycle, scheme S_2: $\delta\tau$ from S1 to LA)

21'You mul[tiply 3,45 by] 0;1,50,40, [it is 6;55, you add it to 3,45, it is 3,51;5]5. 4;49,33,50, 22'you add [the middle one] from first station [to setting to 3,51;55, it is 3],56,44,34. 23'[From] first station to setting you add ... to the distance ..., and you predict it as the 'day' of setting.

P10' (Subdivision of the synodic cycle)

a **(Zigzag function for v_1)**

24'In order for you to see the displacement of Mars for the whole year.
If it appears (FA) in Cnc, and the first ... it moves 0;30; in Leo it moves 0;33,20; 25'in Vir it moves 0;36,40; in Lib it moves 0;40; in Sco it moves 0;43,10 (error for 0;43,20), in Sgr it moves 0;46,40, in Cap it moves 0;50, in Aqr it moves 0;46,40, [in Psc] 26'it moves 0;43,20, in Ari it moves 0;40, in Tau it moves 0;36,40, in Gem it moves 0;33,20, in Cnc it moves 0;30. For 1° 27'0;0,6,40 stands as the difference.

b **(Scheme S_3)**

100 (and) 1,30 'days' it moves 0;40 per 'day'.
30 'days' before its first station it moves 28'0;36 per 'day'.
30 'days' it moves 0;24 per 'day'.
30 'days' it moves 0;12 per 'day', then it becomes stationary (S1).
From the first station to the second station (S2) 29'it recedes backwards 0;12 per 'day', becomes stationary.
From the second station, the day when the motion begins, 30 'days' 30'it moves 0;12 per 'day'.
For 30 'days' it moves 0;24 per 'day'.
for 30 'days' it moves 0;36 per 'day'.
for 3,53 'days' it moves 0;40 per 'day', (then) it sets (LA).

c **(Zigzag function for v_1)**

31'When Cnc it moves 0;30; when Lib it moves 0;40; when Cap it moves 0;50; when Ari it moves by 0;40, (then) it sets (LA). From setting to appearance you do as before in the zodiac.

P11' (System A: period relations; mean synodic arc)

a 33'For Mars. 4,44 years, 2,[1]3 ...? appearances, 2,31 revolutions, the position proceeds 15,6,0. What should I multiply by 2,13 34'appearances so that it is 15,6,0? : You multiply 6,48;43,18,30 by 2,13 appearances, it is 35'15,6,0. You ⸢subtract?⸣ 6,0, (it is) 48;43,18,30, you put it down as the middle one.

b **(System B: deriving the parameters of the zigzag function for σ)**

17, the difference, you multiply 36'by 2,13 appearances, [it is 3]7,41. What should I multiply by 18 revolutions so that it is 37,41? [...] 37'2,5;36,40 you multiply by 18 revolutions, it is 37,41. : 2,5;36,40, which you [...] from 41 ... [...]

35' *tu-⸢ub?-ba?⸣-al* = *tubbal* = 2 pres. G *abālu*, 'to carry (away)'. One expects 'you subtract'. A subtractive meaning is not attested for this verb, but occasionally in OB mathematics for the variant *tabālu* G (Høyrup 2002, p. 20).
40 *ana*: replaces 30 **šu₂** (Pinches; *ACT*). 40 is probably an error for 30; cf. the commentary.

Commentary

This tablet consists of five joined fragments. The upper edge is not preserved, but the shape of the tablet suggests that not much clay is missing. One side is nearly completely destroyed. As remarked by Neugebauer, the well-preserved side is slightly convex, suggesting that it is the reverse. All of the preserved procedures are for Mars. P1'–P10' are concerned with the subdivision of the synodic cycle for system A. They form three groups: P1' and P2' deal with the retrograde angular pushes, P3'–P9' with temporal pushes between the parent phenomena LA, FA and S1, and P10' provides values of the daily displacement for subintervals of pushes, and two zigzag functions representing angular pushes. For the algorithms cf. §3.6.2.3 (p. 84). P11'.b stands out in that it belongs to system B.

***P1'** System A: angular push from S1 to AR (scheme T)*

Only the last line is preserved. Not much text may be missing from the beginning, so that P1' is probably the first procedure on this side of the tablet. The remains of line 1' contain 2 values of a coefficient, each associated with 2 adjacent zodiacal signs. This information and the numerical values are sufficient to identify them as values of the angular push ($\delta\Sigma$) from S1 to AR of scheme T (§3.6.2.3; Table 3.30).

***P2'** System A: angular push from AR to S2 (scheme T?)*

P2' also contains a list of angular pushes, perhaps from AR to S2.[615] This suggests that they belong to the same scheme T. On symmetry grounds one expects the same value for the regions Sco–Sgr and Psc–Ari, so that one can restore 9;30 after the former. For a discussion cf. §3.6.2.3 (p. 84).

***P3'–P9'** System A: temporal pushes between LA, FA and S1*

P3'–P9' form two groups: P3'–P4a' provide algorithms for T for the parent phenomena, and P5'–P9' are concerned with the derivation of the involved parameters. As pointed out in the discussion of these algorithms (§3.6.2.3), Neugebauer interpreted them as pertaining to the mean temporal pushes $\overline{\delta\tau_p} = \overline{\delta\Sigma}_p + c'_{\tau,p}$.[616] However, nothing in the formulation of P3'–P4' and P7'–P9' suggests that $c'_{\tau,p}$ is added to a mean angular distance, and in that interpretation the procedures would be of little practical use. A more likely interpretation, due to van der Waerden (1957), is that they concern actual temporal pushes, $\delta\tau_p = \delta\Sigma_p + c'_{\tau,p}$, in which case the procedures are relevant for computing synodic tables (cf. Eq. 3.35, p. 85). Some features remain unclear. For instance, in P3' and P7'–P9' $c'_{\tau,p}$ is supposedly added to the 'distance from FA to FA' (*birīt* **igi** *ana* **igi**), i.e. the synodic arc σ (or Σ), which makes little sense (cf. §3.6.2.3).

***P3'** System A: updating T with the synodic time; angular pushes (LA to FA, FA to S1, S1 to LA)*

Part **a** mentions the signs Ari(?) and Tau and perhaps a subtraction, of unknown significance. Since Ari and Tau belong to different zones of system A (6 and 1, respectively), this reading might be wrong. Part **b** contains the usual rule for updating T with the synodic time τ. It is formulated in accordance with template T.1 (§3.3.2.1). Only FA is mentioned here as the synodic phenomenon to which this algorithm applies. The number 23;37,52$^{\overline{\tau}}$ is the excess of c_τ (Eq. 3.34, p. 84) over 24$^{\overline{\mathrm{m}}}$. The addition of c_τ to σ must have been mentioned in the broken part of 4' (... 'you predict as the time').

Part **c** contains 3 partly preserved analogous rules for computing $\delta\tau_p$ between the primary phenomena by an algorithm equivalent to Eq. (3.35). All 3 pushes are mentioned, but $c'_{\tau,p}$ is preserved only for the pushes from FA to S1 and from S1 to LA (R5'–6'). The value of $c'_{\tau,p}$ for the third push, from LA to FA, is mentioned in P4'. The 'months' (R5') must be a scribal error for 'days', to be understood as mean tithis. While the overall interpretation of the rules is clear, the phrase *birīt* **igi** *ana* **igi** is unexpected, as mentioned above. The usual meaning 'distance from FA to FA', a term for the synodic arc (σ), makes no sense here, since $c'_{\tau,p}$ is added to $\delta\Sigma_p$ and not σ.

***P4'** Temporal push from LA to FA*

This procedure consists of a single line. The number 33;40,4 is the parameter $c'_{\tau,1}$ (Table 3.29, p. 86), which pertains to the push from LA to FA. The phrase in which it is mentioned differs from the ones in P3'. Here $c'_{\tau,1}$ is preceded by at least a partial derivation from more basic parameters along the lines of Eq. (3.37), 0;1,50,40 being the factor $e \cdot 0;0,10$. The parameter $c'_{\tau,1}$ is accompanied by the term **kur** *u* **šu₂**, perhaps to be translated literally as 'rising and setting', which in R16' (P7') accompanies r_1. It is probably not a coincidence that all three instances occur in connection with the push from LA to FA. Hence it may be a technical term for that push, although this requires some twisting of the literal meaning: 'rising and setting' > distance (angular or temporal) between setting (LA) and rising (FA).

***P4a'** Updating T with the synodic time*

This procedure appears to contain a general algorithm involving the addition of a time interval to the time of FA, which allows two interpretations. Either it concerns the updating of the time of FA with the synodic time, $T_i = T_{i-1} + \tau$ (Eq. 3.8, p. 61), or the computation of the time of S1 by adding a push, $T_{\mathrm{S1}} = T_{\mathrm{FA}} + \delta\tau_1$ (Eq. 3.36, p. 85).

---

[615] In *ACT*, p. 382, 'from Φ to Ψ' (S1 to S2) should be 'from Θ to Ψ' (AR to S2), a typographical error without consequence.

[616] *ACT*, p. 383; *HAMA*, pp. 406–411.

P5' Subdivision of synodic time: FA to S1, S1 to LA

P5' is partly difficult to understand. What is clear is that it deals with the temporal pushes from FA to S1, and from S1 to LA. The formulation has a more archaic touch than that of the preceding procedures (e.g. the logogram **igi.du₈.a** for FA, which is otherwise **igi**, and R11'). Neugebauer refrained from offering an astronomical interpretation. Van der Waerden (1957), p. 53, proposed that the number 2 in R9' represents 2,0° = Mars's elongation from the Sun at S1 (Fig. 3.14a). This number is added to another quantity which may be the zodiacal position of the Sun, so that the sum (**pap** = *napḫaru*, 'total') is the zodiacal position of Mars, but the text is damaged here. The suggestion that P5' concerns elongation is supported by R11', which mentions the distance by which Mars is 'in front of the Sun'. Only the final digit 5 is preserved, but following Neugebauer this may be restored as 15°, an appropriate value for FA (Fig. 3.14a). This number is subtracted from something, probably the zodiacal position of Mars, so that the outcome, to be restored at the beginning of R12', would be the zodiacal position of the Sun. In what remains of R12' the time of LA appears to be computed, but the steps are not clear, and neither is the connection with R11'. Hence it is not clear how to restore the missing signs.

P6' Subdivision of the synodic time, scheme S_2

P6' consists of 4 lines mentioning the values of $0;0,10e \cdot \overline{\Sigma}$ and its 3 subintervals $0;0,10e \cdot \overline{\delta\Sigma_p}$ (Table 3.29), which are contributions to c'_τ and $c'_{\tau,p}$, respectively. The closing statement implies that they are called '*uddazallû*-coefficients of the position'. All 4 numbers are exact products of $0;0,10e = 0;1,50,40$ and the mean (*qablû*, 'middle one') angular distances $\overline{\delta\Sigma_p}$ (Table 3.29), provided that one corrects 4;59,<u>22</u>,55,57,46,40, representing $0;0,10e \cdot \overline{\delta\Sigma_2}$, to 4;59,<u>32</u>,55,57,46,40. The rounded value 4;59,33 mentioned in P8' proves that this correction is justified.

P7'–P9' Subdivision of the synodic time, scheme S_2

P7, P8 and P9 each provide 5 parameters corresponding to successive steps in the computation of $c'_{\tau,p}$, namely r_p, $r_p \cdot 0;0,10e$, $r_p \cdot (1+0;0,10e)$, $0;0,10e \cdot \overline{\delta\Sigma_p}$, and the result $c'_{\tau,p}$. Values of $0;0,10e \cdot \overline{\delta\Sigma_p}$ were already provided in P6'; here they are rounded at the third or fourth digit. In P7' r_1, which pertains to the arc from LA to FA, is accompanied by the phrase 'for its rising (**kur**)'. Neugebauer's suggestion that **kur** denotes FA is confirmed by No. **10** Riii1, where it denotes MF of Venus. In the final step, which is fully preserved only in P7', $c'_{\tau,p}$ is added to a distance. One expects this to be the angular push $\delta\Sigma_p$, but the formulation suggests something different. P7' seems to refer to the synodic arc from FA to FA, after which it mentions the number 32, of unknown significance. In P8' (19'–20') $c'_{\tau,2}$ is added to the 'distance to distance ...' (*birīt ana birīt* ...). In P9' a similar phrase may be restored. After several illegible signs, P9' ends with the instruction 'you call it the time of setting (LA)'. This implies that P9' did not end with the algorithm for $\delta\tau_3$ (Eq. 3.35), but that it included the algorithm for T (Eq. 3.36).

P10' Subdivision of the synodic cycle: daily motion v_1

This composite procedure is concerned with other aspects of the subdivision of the synodic cycle. Part **a** (R24'–27') provides 12 values of the 'displacement of Mars', one for each sign of the zodiac, forming a zigzag function v_1 (Table 3.31). The phrase 'if it appears in Cnc' (R24') suggests that v_1 applies when Mars is near FA. The meaning of the following phrase is not clear.

In part **b** (R27'– 32') the motion from FA to S1, and from S2 to LA, are each divided into 4 subintervals. This scheme, attested only here, has been named S_3 (Table 3.31). For each subinterval $\delta\tau$ and v are provided. The value of v for the segment FA to S1 (1) coincides with the mean value of v_1 (part a). This suggests that scheme S_3 pertains to the signs Ari and Lib, and that the values of v for other positions are obtained by scaling in proportion to the zigzag function for v_1. In part **c** (R31'–32') the extrema and mean value of v_1 are repeated, and it is stated that this function applies near LA. R32' suggests that v_1 is valid throughout the push from LA to FA.

P11' Period relations, mean synodic arc (system A); derivation of zigzag function for σ (system B)

P11' consists of two parts. Part **a** begins with the period relation of system A, namely Π = 2,13 synodic cycles correspond to $Z+\Pi$ = 2,31 = 151 revolutions, or $Z+2\Pi$ = 4,44 = 284^{yr} (Table 3.26b, p. 83). The corresponding total displacement is 15,6,0° = 2,31 · 6,0°. One is asked to compute from this the mean total displacement per synodic cycle, $\overline{\Sigma}$ = 15,6,0°/2,13, the outcome of which is given as 6,48;43,18,30°. Since 2,13 does not have a finite-digit reciprocal, this is an approximation of the exact value 6,48;43,18,29,46,27,58,12,....°. The answer is verified by multiplying with 2,13, the result of which is given as 15,6,0° (R35'). This is not entirely accurate, since the exact product is 15,6,0;0,0,30°. In R35' the mean synodic arc $\overline{\sigma}$ is obtained as the excess of $\overline{\Sigma}$ - 6,0° = 48;43,18,30°.

Part **b** was clarified by P. Huber,[617] who discovered that it contains a (partial) derivation of the defining parameters of the zigzag function for σ for system B, setting out from the period relation and mean synodic arc of part **a**. The parameters mentioned here are $d = 17°$, $\Pi = 2{,}13$, $Z = 18$, $\Pi d = 37{,}41°$ and $2\Delta = 2{,}5;36{,}40°$. For a discussion cf. §3.6.3 (p. 89). The computation involving 41,... at the end of R37' remains unexplained.

[617] As reported in Aaboe (1958), p. 246. Cf. also *HAMA* p. 458.

No. 14

museum number	BM 37024 (80–6–17,768)
provenience	Babylon
measures	10.1(12?) × 8.3(*) × ? cm (type C or F?); column width 3.5–4.0 cm
arrangement	O/R
cuneiform text	photo: Fig. F.14; Aaboe (1987), p. 6
bibliography	Aaboe (1987)
contents	Mars system A: various procedures

Obverse
Column i
(about m−5 lines missing)

P1' −4' [**ki gar ta igi** *ana* **uš igi**]
−3' [1.45 **alla ur**]
−2' [2.20 **absin rin**$_2$]
−1' [3.30 **gir**$_2$ **pa**]
0' [5.15 **maš**$_2$ **gu**]
1' ⌜3.56.15 *zib* **ḫun**⌝
2' 2.37.30 **mul**$_2$ **maš**

P2' 3' **ki gar ta uš igi**-*tu*$_2$ *ana* **šu**$_2$
4' 1.31.40 **alla ur**
5' 2.2.13.20 **absin**$_0$ **rin**$_2$
6' 3.3.20 **gir**$_2$ **pa**
7' ⌜4.3⌝5 **maš**$_2$ **gu**
8' [3.26].⌜15⌝ *zib* **ḫun**
9' [2.17.30] **mul**$_2$ **maš**

P3' 10' [xxxxxx] **u**$_4$
11' [xxxxxx] ⌜x⌝
(6 lines missing)
Column ii

P4' (m lines missing)
0' ⌜*u*⌝-*da*-⌜*zal*⌝-[xxxxxx]
1' *ša*$_2$ 2-⌜*ta*⌝ [**mu.meš** xxx]
2' **tag**$_4$ *u*$_4$-*m*[*u* xxxxx]
3' *ša*$_2$ 30 **ki** *ša*$_2$ ⌜x⌝ [xxxxxx]
4' *lu* **igi**-*šu*$_2$ **dib**-*qa* ⌜x⌝ [xxxx]
5' 30 *u*$_3$ 23 [xxxxx]
6' **pap** 2-*ta* **mu.meš** *u*$_3$ ⌜53⌝
7' **u**$_4$ *ki-i* **diri.še gal**$_2$-*u*$_2$
8' 2-*ta* **mu.meš** *u*$_3$ 23

P5' 9' **šu**$_2$ *ana* **šu**$_2$ **igi** *ana* **igi** *ina* **alla**
10' ⌜**ib**$_2$⌝.**tag**$_4$ *lu-maš*.**meš**
11' [*ga*]*b-bi* **gin**$_7$ *il-ki*
12' [xx] *ana* **šu**$_2$
13' [xx] *ana* **igi**
14' [xxx] **ḫun**
15' [xxx] ⌜*an*⌝-*na-a*
16' [xxxx] ⌜x⌝
Column iii

P6' (about m+6 lines missing)
6' 1 **u**$_4$ ⌜x⌝ [xxxxxx]
7' *ša*$_2$-*niš* ⌜**igi**?⌝ [xxxxx]
8' *al* **muru**[**b**$_4$ xxxxxxx]
9' **ta e me** *ana* **u**[**š** *ar*$_2$ xx]

Obverse column i
(about m−5 lines missing)

P1' **(Angular push from FA to S1, scheme S$_1$)**

−4'[You put down the position. From appearance (FA) to first station,]
−3'[1,45: Cnc, Leo.]
−2'[2,20: Vir, Lib.]
−1'[3,30: Sco, Sgr.]
0'[5,15 Cap, Aqr.]
1'⌜3,56;15: Psc, Ari⌝.
2'2,37;30: Tau, Gem.

P2' **(Angular push from S1 to LA, scheme S$_1$)**

3'You put down the position. From first station to setting (LA),
4'1,31;40: Cnc, Leo.
5'2,2;13,20: Vir, Lib.
6'3,3;20: Sco, Sgr.
7'⌜4,3⌝5: Cap, Aqr.
8'[3,26];⌜15⌝ Psc, Ari.
9'[2,17;30] Tau, Gem.

P3' **(Dates?)**

10'[...] day 11'[...] ...
(6 lines missing)
Column ii

P4' **(Synodic time)**

(m lines missing)
0'the *uddazallû*-coefficient [...] 1'for 2 [years ...] 2'remains, day [...] 3'of 30, the position of ... [...] 4'... it passes its appearance [...] 5'30 and 23 [...] 6'the total is 2 years and 53 7''days'; if there is a month XII$_2$ 8'2 years and 23.

P5' **(Updating *B* with the synodic arc?)**

9'Setting (LA) to setting, appearance (FA) to appearance in Cnc, 10'all other zodiacal signs, 11'like ... 12'[...] to the setting. 13'[...] to the appearance. 14'[...] Ari 15'[...] thus 16'[...] ⌜...⌝
Column ii

P6' **(Distance to the ecliptic?; retrograde motion)**

(about m+6 lines missing)
6'1 day ... [...] 7'Secondly: ⌜appearance⌝ [...] 8'above the midd[le ...]
9'from daylight rising (AR) to the [second] stat[ion ...]

	10’	43 *u$_4$-me* [xxx]
	11’	**uš igi**-*tu$_4$* [xxx]
	12’	**absin rin$_2$** 34 ⌜x⌝ [xxx]
	13’	**la$_2$**-*ma* [xxxxx]
P7’	14’	**u$_4$.meš gar ta uš** [**igi**? xxx]
	15’	*ana* **u**[**š** *ar$_2$*? xxxxx]
	16’	⌜x⌝ [xxxxxxx]
		Reverse
		Column i
P8’	1	40.44.2⌜6⌝.[40 **absin rin$_2$**]
	2	4⌜1?⌝.⌜2⌝6.40 **g**[**ir$_2$ pa**]
	3	5⌜1?.31?⌝.6.40 ⌜**maš$_2$**?⌝ [**gu**]
	4	4⌜4?.xxx⌝.20 *zib* [**ḫun**]
	5	⌜40?.xxx⌝.40 **m**[**ul$_2$ maš**]
	6	[xxxxxxxxx]
	7	[xxx] ⌜20?⌝ [xxxxx]
	8	⌜x⌝ [xx] **maš** ⌜x **nim**? x⌝ [x]
	9	⌜xx⌝ **zal** [xxx]
	10	⌜x **an**? 30?⌝ [xxxxx]
	11	⌜xx⌝ [xxxxxxx]
	12	⌜x⌝ [xxxxxxxx]
		(about m+5 lines missing)
		Column ii
P9’	1	(xxxxx)
	2	(xxxxx)
	3	[xxx]⌜x **zal**? *ša$_2$*?⌝ **mu**
	4	⌜x 2? **zi** *ša$_2$*? *u$_2$* ⌜x **a**⌝
	5	30 ⌜**alla**⌝
	6	⌜33.20⌝ **a**
	7	⌜36⌝.40 ⌜**absin**⌝
	8	40 ⌜**rin$_2$**⌝
	9	4⌜3⌝.20 ⌜**gir$_2$**⌝
	10	⌜46⌝.40 ⌜**pa**⌝
	11	50 **maš$_2$**
	12	46.⌜40⌝ **gu**
	13	43.20 *zib*
	14	40 **ḫun**
	15	36.40 [**mul$_2$**]
	16	33.20 [**maš**]
		(about m lines missing)
		Column iii
P10’		(about 4 lines missing)
	5	[xxx] **ḫun**
	6	[xx].30 **mul$_2$**
	7	⌜x 4⌝8 **maš**
P11’	8	⌜**ki**⌝.**me** *ina* ⌜xx *ana* **du$_3$**?-*ka*?⌝
	9	**ta šu$_2$** *ana* **igi**
	10	55 **alla**
	11	⌜x⌝.20 **a**
	12	[xxx] **absin$_0$**
	13	[xxx] **rin$_2$**
	14	[xxx] **gir$_2$**
	15	[xxx] **pa**
	16	[xxx **maš$_2$**]
	17	[xxx **gu**]
	18	[xxx *zib*]

[10’]43 ‘days’ [...] [11’]the first station [...] [12’]Vir Lib 34 ⌜...⌝ [...] [13’]it recedes, then [...]

P7’ (Temporal push S1 to S2?)

[14’]You put down the ‘days’. From [first] station [...] [15’]to [second] ⌜station⌝ [...] [16’]⌜...⌝ [...]

P8’? (A coefficient)

(perhaps a column missing?)

[... Cnc Leo]

Reverse column i’

[1]40.44.26.[40 Vir Lib]
[2]41.26.40 [Sco Sgr]
[3]51.31.6.40 ⌜Cap⌝ [Aqr]
[4]44...20 Psc [Ari]
[5]40...40 ⌜Tau⌝ [Gem]
[6][...] [7][...] ⌜?⌝ [...] [8]... [...] Gem ... [...] [9][...] it passes? [...] [10]... [...] ... [11]... [...] [12]... [...]

(about m+5 lines missing)

Column ii

P9’ (Daily motion v_1)

[1][...] [2][...] [3][...] ... for the year. [4]... the displacement which ...
[5]0;30 Cnc
[6]0;33,20 Leo
[7]0;36,40 Vir
[8]0;40 Lib
[9]0;43,20 Sco
[10]0;46,40 Sgr
[11]0;50 Cap
[12]0;46,40 Aqr
[13]0;43,20 Psc
[14]0;40 Ari
[15]0;36,40 [Tau]
[16]0;33,20 [Gem]

(about m lines missing)

Column iii

P10’ (A coefficient)

(about 4 lines missing)
[5][...] Ari
[6][...].30 Tau
[7][...].⌜4⌝8 Gem.

P11’ (Angular push from LA to FA?)

[8]Positions in ... [9]from setting (LA) to appearance (FA).
[10]55 Cnc
[11]...20 Leo
[12][...] Vir
[13][...] Lib
[14][...] Sco
[15][...] Sgr
[16][... Cap]
[17][... Aqr]
[18][... Psc]

19	[xxx **ḫun**]	[19][... Ari]
20	[xxx **mul₂**]	[20][... Tau]
21	[xxx **maš**]	[21][... Gem]
	(about m−5 lines missing)	*(about m−5 lines missing)*

Critical and philological notes

Obverse column i

3' **gar**: the interpretation *tašakkan*, 'You put down', replaces *ša₂* (Aaboe).
4' **ur**: alternative abbreviation of **ur.a**, Leo (usually = **a**).
6' 3.3.20: looks more like 2.3.20.
11' The upper parts of about three vertical wedges are visible.

Column ii

0' This line omitted by Aaboe.
10' *lu-maš*.**meš**, 'zodiacal signs': not identified by Aaboe.
11' *ilku*: the usual meaning of this word, 'duty; corvee work', does not make sense. Perhaps a more literal meaning derived from the cognate verb *alāku*, 'to go; proceed', must be assumed?

Column iii

6' After $\mathbf{u_4}$ the left halves of two horizontal wedges are visible, perhaps **tab**.

Reverse column i

2 41: or 44.
3 31.6: the 30 might also be 40 or 50.
4 44: or 45–48. This is followed by 3–8 and traces of 2 more digits before 20.
6–12 These lines were not transliterated by Aaboe.
6 Only vague traces of signs.
7 20: very uncertain.
8 **nim**: very uncertain.
10 30: or 20, 40 or 50.

Column ii

2–3 Below line 2 there appears to be a horizontal line, suggesting that P9' begins with line 3'.
4 2: or A?
ša₂: or $\mathbf{la_2}$?
The damaged sign after u_2 is probably a number 5–8.

Column iii

8 **ki.me**: replaces **ki** *ša₂* (Aaboe). The sign following *ina* resembles AB_2.

Commentary

This fragment preserves sections of the left edge and the lower (upper) edge of the obverse (reverse) of the tablet. The identification of obverse and reverse is uncertain. On both sides the text is divided into three columns, which are further divided into sections by horizontal lines. If one assumes that the tablet had three columns then the original width was about 12 cm. Based on the textual restorations (cf. P1'), the original height may have been at least 14 cm. All 11 extant procedures deal with Mars. Most of them belong to system A, perhaps with the exception of P11'. P1'and P2' are concerned with the angular pushes between the parent phenomena LA, FA, and S1, P3' and P4' with the synodic time, and P5' with the synodic arc. Most of the other procedures deal with the subdivision of the synodic cycle. A new investigation of the fragment has resulted in several improved readings. Moreover, the previous edition (Aaboe 1987) did not include a comprehensive translation, which is provided here for the first time.

P1' Angular push from FA to S1

Only the final 2 lines of this procedure are partly preserved, enough to conclude that P1' provides values of the angular push from FA to S1 for all 6 zones of system A (Table 3.28, p. 85). The introductory statement can be restored from P2', which deals with the adjacent push from S1 to LA. It can be assumed that P1' was preceded by a similar procedure for the angular push from LA to FA, which would require about 7 lines of text before Oi-4'.

P2' Angular push from S1 to LA

P2' contains values of the angular push from S1 to LA for all 6 zones of system A (Table 3.28). It can be assumed that P1' (and presumably P0') are structured in the same way as P2'. For the initial statement cf. §2.3.1 (p. 27).

P3' Synodic time or temporal pushes?

Almost nothing remains of this procedure except the word 'day', suggesting a connection with the synodic time or temporal pushes ($\delta\tau$). Therefore, P3' may actually belong to the same procedure as P4'.

P4' Synodic time

P4' is concerned with the synodic time τ (cf. §3.6.2.2). Its formulation is unusual and differs from the templates T.1 and T.2 (§3.3.2.1). The beginning is lost, and the precise meaning of Oii0'–4' is not clear. The 2 years mentioned in Oii1' and 8' are the approximate whole number of years contained in the synodic cycle of Mars (Table 3.4, p. 60). Oii5'–8' provide an algorithm for τ, such that c_τ is split up as 2 calender years plus an excess. $23^{\overline{\tau}}$ is an approximation of the excess of c_τ over 24^{m} (Eq. 3.34, p. 84). As explained by Aaboe (1987), pp. 7–8, the addition of $53^{\overline{\tau}}$ (Oii6'0) in a normal year versus $23^{\overline{\tau}}$ in an intercalary year can be reconciled with Eq. (3.34) if one assumes that this constant plus the 2 calender years are added to $\sigma - 30$ and not to σ, the subtraction of 30° from σ being compensated by the addition of $30^{\overline{\tau}}$ to c_τ.

P5' Updating B with the synodic arc?

Little more than the introductory statement is preserved. From this it would appear that P5' contains the algorithm for updating *B* with the synodic arc, but no such rules are preserved. Moreover, one would expect S1 to be mentioned alongside LA and FA as the primary phenomena of system A. The formulation is unusual, and the word *ilku*, if read correctly, is not understood (cf. the critical notes).

P6' Mars's distance to the ecliptic(?); retrograde motion

Only the last 8 lines are partly preserved. While Aaboe refrained from suggesting an interpretation, Oiii8' is reminiscent of the procedures for Jupiter's distance to the ecliptic (§3.7.10.2, p. 105). Perhaps this is the first such procedure for Mars. However, the algorithm and the precise astronomical meaning are not clear. In Oiii10' 43 'days' = mean tithis is a suitable number for the time between AR and S1. Also the number 34 (Oiii12') may pertain to a time interval on the retrograde arc, since it is too large for $\delta\Sigma$.

P7' Temporal push S1 to S2?

According to its introduction, this procedure is concerned with $\delta\tau$ (S1 to S2). Since there is no room for a list of coefficients in this column, P7' must have continued in the next column, presumably column i on the reverse. Hence P8' may be the continuation of P7'. Alternatively this may have been written in an additional column iv, now broken away, in which case P8' is a separate procedure.

P8' A list of coefficients

P8' is a list of coefficients. It appears that there were 6 values, each pertaining to a zone of system A. Since there is no introductory statement in column i', and one expects the list to begin with zone 2 (Cnc–Leo), P8' probably began in the previous column. As pointed out by Aaboe, the numbers are not consistent with an interpretation as $\delta\tau$ (S1 to S2), which should be about $75^{\overline{\tau}}$. The meaning of the coefficients therefore remains unclear. The lower half of column i' is strongly damaged, so that it is not clear how much of it still belongs to P8'.

P9' Control values of v_1

This procedure, of which the first 2 lines are missing, contains a table with 12 values of a coefficient, each associated with a zodiacal sign. They form the zigzag function v_1 representing Mars's daily displacement; cf. §3.6.2.3 and Table 3.31 (p. 88).

P10' A coefficient

Only the final 3 lines of this procedure are partly preserved. Since Ari – Gem are usually placed at the end of a list, it can be assumed that at least 9 lines of text are missing above Riii5. Since there is room for only 4 missing lines (Riii1–4) in this column, the first 5 of these must have been written at the bottom of the preceding column (i.e. the column to the right of it). Too little remains of the numerical values for an interpretation. The endings are not consistent with any of the known schemes for the retrograde pushes (Table 3.30).

P11' *Angular push from LA to S1, non system A?*

The introductory statement of this badly preserved procedure suggests that P11' concerns $\delta\Sigma$ (LA to S1). The only preserved number, 55 in Cnc, is consistent with that interpretation (cf. Table 3.28, p. 85). However, as Aaboe remarks, it pertains only to Cnc, since the list seems to continue with a (separate) value for Leo and, presumably, for all other zodiacal signs. This suggests that P11' does not belong to system A.

No. 15

museum number	BM 55546 (82–7–4,130)
provenience	Babylon
date	SE 186 (data in synodic tables: SE 186–211, 170–187)
measures	6.2(21) × 5.8(*) × 2.8? cm
arrangement	O/R
cuneiform text	photo: Fig. F.15 (R, LE: procedures and col.); *ACT*, Pl. 244a (R, LE)
bibliography	*ACT* 821aa, 430, 501a, Zmab
contents	Venus system A$_1$(?): synodic table [+ procedure?] (O); Mars system A: synodic table + procedures (R, LE)

Reverse

Co 1 [xxxxxxxxxxxxxxxxxxxxxxxxxxxxxx **m]u**.3.6.**kam** ^{md}**en.tin**-*su ina* **šu**.2-*šu*$_2$

P1 2 [xxxxxxxxxx 30 **alla a** 40 **absin rin**$_2$ 1 **gir**$_2$.**tab pa** 1.30 **maš**$_2$ **gu** 1.7.30 *zib* **ḫun**] ⸢4⸣5 **mul**$_2$.**mul**$_2$ **maš.maš** *ša*$_2$ *al* 30 **a diri**

3 [GAM$_0$ 1.20 **du** *ša*$_2$ *al* 30 **rin**$_2$ **diri** GAM$_0$ 1.30 **du** *ša*$_2$ *al* 30 **pa diri** GAM$_0$ 1.30 **du** *ša*$_2$ *al* 30 **gu diri** GAM$_0$ 45 **du** *ša*$_2$ *al* 30 **ḫun di]ri** GAM$_0$ 40 **du** *ša*$_2$ *al* 30 **maš diri** GAM 40 ⸢**du**⸣

4 [xxxxxxxxxxxxxxxxxxxxxxxxxxxxxxxxxxxx **m]ul**$_2$ **maš** *an-na-a*

Lower edge

5 [xxxxxxxxxxxxxxxxxxxxxxxxxxxxxxxxxxxx **m]eš gar ta** ⸢**uš**⸣ [**igi**$^?$ xx]

6 [xxxxxxxxxxxxxxxxxxxxxxxxxxxxxxxxxxxx **m]aš.maš** *an-na-a* ⸢x⸣ [xx]

7 [xxxxxxxxxxxxxxxxxxxxxxxxxxxxxxxxxxxx]⸢x⸣ 1 22 **mul**$_2$ **maš** ⸢x⸣ [xx]

Reverse

Co **(Colophon)**

[1][... Ye]ar 3,6. Bēl-bullissu [wrote it] with his (own) hands.

P1 **(Mars system A)**

a **(Updating *B* with the synodic arc: zones and σ_j)**

[2][... 30: Cnc, Leo; 40: Vir, Lib; 1: Sco, Sgr; 1,30: Cap, Aqr; 1,7;30: Psc, Ari;] 45: Tau, Gem.

b **(Updating *B* with the synodic arc: transition rules)**

(The amount) by which it exceeds 30 Leo [3][you multiply by 1;20.
(The amount) by which it exceeds 30 Lib you multiply by 1;30.
(The amount) by which it exceeds 30 Sgr you multiply by 1;30.
(The amount) by which it exceeds 30 Aqr you multiply by 0;45.
(The amount) by which it exce]eds [30 Ari] you multiply by 0;40.
(The amount) by which it exceeds 30 Gem you multiply by 0;40.

c **(Subdivision of the synodic cycle)**

[4][... T]au, Gem; thus

Lower edge

[5][...] ... you put down. From the [first$^?$] station [6][...] Gem. Thus ... [...] [7][...] ... 1.22 Tau, Gem ... [...]

Critical and philological notes

Lower edge

5 Similar phrases in No. **16** O14' and No. **14** Oi3',iii14' suggest that this line may be partly reconstructed as **ki.meš/u$_4$.meš gar ta uš igi**-*tu*$_2$ *ana* ..., 'You put down the positions/times. From the first station to ...'.

7 Before 1.22 there is a vertical wedge, clearly separated from 1.22. After **maš** there is also a vertical wedge.

Commentary

This fragment belongs to the upper (lower) right edge of the obverse (reverse) of a tablet. Nothing remains of the left edge and the lower (upper) edge of the obverse (reverse). The obverse contains a synodic table for Venus (*ACT* 430; cf. Volume II); the reverse a synodic table for Mars system A (*ACT* 501a; cf. Volume II) of which the rightmost columns (AR: *B*; S2: *T* and *B*) are partly preserved. Below the synodic table for Mars, and separated from it by an empty space, there is a colophon consisting of a single line, of which only a small portion is preserved. The tablet was written in SE 186 (126/5 BC), a date in between the initial and final dates of the predictions. The scribe, Bēl-bullissu, might be the same individual whose son Marduk-šāpik-zēri wrote No. **7**. An astronomer by the name Bēl-bullissu is also mentioned in several Parthian era documents from the Esagila temple.[618] The identity of these individuals is uncertain since the name of Bēl-bullissu's father and his clan affiliation are not preserved. Below the colophon is a procedure for Mars that accompanies the synodic table. The textual restorations in R3 suggest that about 15 cm of clay is missing from the left side of the fragment, so that the original width of the tablet may have been about 21 cm. In the synodic tables several additional columns must therefore be assumed to the left of the preserved ones. Perhaps the obverse also contained procedures for Venus below the synodic table for that planet.

***P1** Mars system A: updating B with the synodic arc; subdivision of the synodic cycle?*

This procedure consists of 6 lines, of which only the final parts are preserved. The textual restorations imply that less than half of each line is preserved. Parts **a–b** contain the rules for updating B with the synodic arc (§3.6.2.1). First the values of σ_j are provided for all 6 zones (part **a**), after which the corresponding transition rules are given in terms of coefficients r_j (part **b**). If taken together for each zone, this amounts to template STEP.A.3 (§2.6.10.1). Very little remains of part **c**. Several zodiacal signs and a station are mentioned, probably in connection with the subdivision of the synodic cycle. The number 1.22 (R7), which pertains to zone 1 (Tau – Gem), may be interpreted as 1,22° = $\delta\Sigma$ (LA to FA), close to the system A value 1,22;30° (Table 3.28, p. 85).

[618] *CT* 49 144 (dated 15 X SE 193) and *CT* 49 186. For these documents cf. McEwan (1981); van der Spek (1985); Rochberg (2000).

No. 16

museum numbers	BM 34721 (Sp2,211) (+) BM 35857 (Sp3,388)
provenience	Babylon
measures	BM 34721: 6.8(*) × 8.3(*) × 3.0(*) cm; column width 10 cm BM 35857: 4.0(*) × 4.5(*) × 1.8(*) cm
cuneiform text	photo: Fig. F.19; *ACT*, Pl. 243 (BM 34721); copy BM 34721: *LBAT* 139
bibliography	*ACT* 811b (BM 34721)
contents	Mars, system A and other?: various procedures

Side X
Column i'

P1' (unknown number of lines missing)
1' [xxxxxxxxxxxxxxxxxxxxxxxxxx] ⸢5? x⸣
2' [xxxxxxxxxxxxxxxxxxxxxxxx] ⸢x 1 **tab**?⸣
3' [xxxxxxxxxxxxxxxxxxxxxxxxxx]
4' [xxxxxxxxxxxxxxxxxxxxxxxx] ⸢3?⸣
5' [xxxxxxxxxxxxxxxxxxxxxxx] ⸢5⸣ *zib*.**me**
6' [xxxxxxxxxxxxxxxx?] (blank)

P2' 7' [xxxxxxxxxxxxx] ⸢xx *ša*$_2$⸣ *la*-**igi** *šamaš*$_2$ **sud uš** 8 **ki** *ša*$_2$ *ana* **e** ⸢**me**⸣
8' [xxx]⸢20 **ta**? 3⸣.30 ⸢x⸣ [xxx] ⸢**me**⸣ **zal**.⸢**la**$_2$?⸣ 2 **iti ki** 40 ME ME? 44 **me du** *ana* **me e** [x]
9' [xx] ⸢**ta**⸣ **absin en** *zib*.**me** [xxxx?] (blank)

P3' 10' [x] ⸢*lu*⸣-*maš gab-bi* **du**$_3$-⸢*uš*?⸣ [xxxx *la*]-⸢**igi**⸣ *šamaš*$_2$ **sud uš ki** ⸢*a-ḫa*⸣-*miš*$_2$ **tab**-*ma* **ta** 2 [xx]
11' [x] ⸢x **zal**⸣ *ša*$_2$ **mu ki** [xxxxxxxx **t**]**a absin en** *zib*.**me zal nu** [x]
11a' **gar** ⸢x⸣

P4' 12' [x] ⸢x⸣ AN *la*-⸢**igi**?⸣ [xxx *ina* 15 **alla** *u* **a** 1]⸢6⸣ *ina* 15 **absin** *u* **rin**$_2$ 14 *ina* 15 [**gir**$_2$.**tab** *u* **pa**]
13' ⸢16?⸣ *ina* 15 **maš**$_2$ ⸢*u*⸣ [**gu** x *ina* 15 *zib*.**me** *u* **ḫun** x *ina* 1]⸢5⸣ **mul**$_2$ *u* **maš.maš** ⸢10⸣[x]

P5' 14' **ki gar ta uš igi** ⸢*ana*⸣ [**e me alla** *u* **a** 7 **absin** *u* **rin**$_2$] 6 **gir**$_2$.**tab** *u* **pa** 5 **maš**$_2$ *u* **gu** 6? [*zib*.**me** *u* **ḫun**]
15' 7 **mul**$_2$.**mul**$_2$ *u* **m**[**aš** 8 xxxxxxx?] (blank)

P6' 16' ⸢x⸣ [xxxxxxxxxxxxxxxxxxxxxx] ⸢x **uš**? x⸣
(unknown number of lines missing)

Critical and philological notes

Side X column i'

1' Traces of two signs, the first of which may be 5 or 8.
4' ⸢3⸣: thus also Pinches; might be a 6 (thus Neugebauer). May belong to the next column.
5' 5: thus also Pinches; Neugebauer: *ša*$_2$
7' *uš*: uncertain; only the left part remains.
⸢xx⸣: the first sign might be ME; the second ends with a vertical wedge. Neugebauer: 5? **uš**?.
la-**igi** = *lapān*, 'in front of'.
sud = *rêqu*, 'to depart; be remote', probably the stat. 3 m. sg. subj. *rēqu*.
8' **me zal.la**$_2$**?**: uncertain; **me** might be ⸢x⸣ *ina*, where ⸢x⸣ is a sign ending with vertical wedge. **la**$_2$: or *ša*$_2$? Neugebauer: 6? *ina* NI *ša*$_2$ **u**$_4$? 1.
ME ME?: meaning unclear.
11' **ki**: *itti*, 'with', or *qaqqaru*, 'position'.
11a' This is written over the dividing line separating lines 4' and 5'.
13' 16: or 15?
14' **gar** probably stands for *tašakkan*, 'you put down'. An interpretation as *ša*$_2$, resulting in a verbless relative clause 'The position which (Is) from ...' cannot be excluded, but seems less likely.
maš$_2$ *u* **gu**: *u* is written above the line.
16' A winkelhaken is visible, perhaps belonging to **ki**.
uš: only upper part preserved; might also be **ta**.

Side X column i'

P1' (Unclear)

(*unknown number of lines missing*)

1'[...] ... 2'[...] ... 3'[...] 4'[...] 3 5'[...] ... Psc 6'[...]

P2' (Mars?: subdivision of the synodic cycle)

7'[...] by which it is distant in front of the Sun. Station, 8, the position by which [...] to the daylight rising (AR) 8'[...]⌜20. From? 3⌝.30 ... [...] ... 2 months, the position? ... 44 days it proceeds, (then) it rises to daylight [...]. 9'[...] ⌜from⌝ Vir until Psc [...]

P3' (Mars?: computations involving the subdivision of the synodic cycle)

10'[...] all zodiacal signs you do? [... by which] it is distant in front of the Sun you accumulate together, from 2 [...] 11'[...] ... passing of the year ... [...] from Vir until Psc it passes, no [...] 11a'...

P4' (Mars system A: S1 to S2?)

12'[...] of Mars in ⌜front? of⌝ [...: in 15 Cnc and Leo 16]; in 15 Vir and Lib 14; in 15 [Sco and Sgr] 13'⌜16⌝; in 15 Cap ⌜and⌝ [Aqr ... in 15 Psc and Ari ... in 1]5 Tau and Gem 10⌜+x⌝.

P5' (Mars system A: S1 to AR, scheme V?)

14'You put down the position. From the first station to [daylight rising (AR): Cnc and Leo 7; Vir and Lib] 6; Sco and Sgr 5; Cap and Aqr 6?; [Psc and Ari] 15'7; Tau and G[em 7 ...]

P6' (Unclear)

16'[...] ...

(*unknown number of lines missing*)

Commentary

The two fragments BM 34721 (*ACT* 811b) and the previously unpublished BM 35857 are very likely part of the same tablet, but they do not physically join. Neither fragment preserves an edge of the tablet, but in X13'–16' no signs are missing on the left side of BM 35857. Both fragments are inscribed on one side, the other side being destroyed. The textual restorations in X13' imply that the width of the preserved column was about 10 cm. On the right side of BM 34721 a vertical dividing line is visible, suggesting the presence of another column. The text is divided into at least 6 procedures by horizontal dividing lines. P4' and P5' are concerned with angular pushes for Mars. P2' and P3' deal with an outer planet, probably also Mars.

P1' Unclear

In P1' the surface is smeared, so that only isolated signs can be recognised. X5' contains a number and the sign Psc, perhaps belonging to a list of coefficients or pushes (compare P4',P5').

P2' Mars?: subdivision of the synodic cycle

The mention of AR (X7') suggests that P2' concerns the subdivision of the synodic cycle of an outer planet, perhaps Mars. The phrase 'by which it is distant in front of the Sun', also attested in another procedure for Mars, No. **13** P5' (O11'), refers to the planet's elongation from the Sun, 'in front' being defined with respect to the normal, forward motion of the planets. However, it may be noted that Vir to Psc (X9') is reminiscent of the region 15° Vir – 15° Psc, where σ increases from its minimum to its maximum according to Jupiter system B (Table 3.49, p. 100). A comprehensive interpretation of P2' remains to be found.

P3' Outer planet; unclear

This procedure consisting of 2 lines contains the same phrase 'in front of the Sun' mentioned in P2'. The term 'all zodiacal signs' suggests a list of coefficients associated with zodiacal signs (pushes?). It appears that the planet's elongation from the Sun is added to another quantity, the name of which is lost. The 'passing (?) of the year' (X11') is attested elsewhere in connection with the subdivision of the synodic cycle of Jupiter according to system A (§3.7.2.6).

***P4'** Mars, system A?: δΣ for S1 to S2, variant scheme?*

P4' originally provided 6 values of a parameter pertaining to Mars, one for each zone of system A. The sign Psc mentioned in X12', if read correctly, is unexpected, since the list ends with Tau and Gem, so that it must have begun with Cnc and Leo; perhaps it is not part of the list. The division of the zodiac into adjacent pairs of signs is known from Mars system A. The numbers probably represent an angular push, the name of which is broken off, perhaps S1 to S2 (compare Table 3.30 on p. 87).

***P5'** Mars, system A?: δΣ for S1 to AR, scheme V*

Also in P5' originally 6 values of a parameter were listed, one for each pair of adjacent zodiacal signs. They represent the angular push from S1 to AR, very likely for Mars system A. The preserved numbers suggest that they form a zigzag function similar to schemes R, S, T, U and Q (compare Table 3.30 on p. 87).

***P6'** Unclear*

Virtually nothing remains of this procedure, so that its topic could not be established.

5.2.4 *Jupiter*

No. 17

museum number	BM 34621 (Sp2,101)
provenience	Babylon
date	data in synodic table: SE 130–205
measures	8.7(19?) × 10.2(12?) × 2.2–2.5 cm (type ii?)
arrangement	O/R
cuneiform text	photo: Fig. F.20 (R: procedures); *ACT*, Pl. 251 (R); copy: Kugler *SSB* I, Pl. 10; *LBAT* 112
bibliography	Kugler *SSB* I, pp. 118–123 (ST); *ACT* 820aa, 602
contents	Jupiter system A: synodic table for [FA, S1, AR], S2, LA; procedures for updating *B* and *T*

Reverse

P1 6' [...] ⸢x⸣ **tab** ⸢*ša*$_2$ *al-la* x⸣ [xx **diri** xxxxx]

7' [... *ša*$_2$ *al-la* 30 **gir**$_2$**.tab diri a.ra**$_2$ 1.12 **du ki** 30 **gir**$_2$].⸢**tab**⸣ **tab**-*ma* **gar**-⸢*an* **u**$_4$**.meš** *bi-rit*⸣ [xxxxxx]

(few lines missing)

Reverse

P1 **(Updating *B* and *T*)**

a **(Updating *B* with the synodic arc)**

6'[...] you add ⸢...⸣. (The amount) by which it [exceeds] ⸢...⸣ [...] 7'[... (The amount) by which it exceeds 30 Sco you multiply by 1;12, add with 30] ⸢Sco⸣ and you put it down.

b **(Updating *T* with the synodic time)**

The 'days': the distance [(from) appearance to appearance you compute, you add to it ...]

(*a few lines missing*)

Critical and philological notes

Reverse

6' The traces before **tab** look like a vertical wedge followed by a winkelhaken; one expects 30 or 36. After *al-la* there are traces of two winkelhakens compatible with the expected 25 or 30.

7' *bi-rit*: only traces are visible.

Commentary

This fragment contains a section of the right edge of the tablet, but no other edges. The obverse and most of the reverse are occupied by a synodic table for Jupiter (cf. *ACT* 602 and Volume II), of which columns *T* and *B* are partly preserved for S2 and LA. Traces are preserved of a preceding column containing the term for AR. On the reverse the table is followed by an accompanying procedure. Since they are usually placed at the end of the tablet, at most a few centimeters of clay are missing from the lower (upper) edge of the reverse (obverse). A colophon is not preserved, but may have been present below or to the left of the procedure. The data in the synodic table suggest that the tablet was written near SE 130 (182/1 BC). Since the columns with *T* and *B* for AR are missing from the synodic table, and corresponding columns for FA and S1 can also be assumed to have been present, about 15 cm of clay are missing to the left of the fragment, so that the original width must have been about 19 cm. The text is written in a careful hand. Unlike the vertical wedges, all horizontal wedges are notably slanted.

P1 System A: updating B and T

Only a few signs from the end of the first 2 lines of P1 are preserved. Part **a** contains the algorithm for updating *B* with the synodic arc for system A of Jupiter (Table 3.36, p. 90). The formulation is probably based on templates STEP.A.1 or A.2 (§2.6.10). Part **b** contains the algorithm for updating *T* with the synodic time. It is presumably based on templates T.1 or T.2 (§3.3.2.1). This implies that P1 must have continued for at least 1 more line of text below 7'.

No. 18

museum numbers	BM 34081 (Sp,179) +34622 (Sp2,102) +34846 (Sp2,349) +42816 (81–7–1,580) +45851 (81–7–6,279) +46135 (81–7–6,590)
provenience	Babylon
measures	22.6 × 13.1(*) × ? cm (type A?); column width 7.5 cm
arrangement	O/R
cuneiform text	photo: Figs. F.17, F.18; BM 34081+34622+34846+45851+46135: *ACT*, Pl. 253 copy BM 34081+34622, 34846: *LBAT* 146–148
bibliography	Kugler *SSB* I, p. 147 (partial translit.); *ACT* 813
contents	Jupiter systems A, A$_1$, A$_2$, A', A'', A''', B and B': various procedures

Obverse
Column i

P1 (m lines missing)
1' [xxxxxxxxx] ⌜**ki**⌝ *ina* ⌜**nim**⌝ **du igi**
2' [xxxx] ⌜3? 36? 1? 10.40 x⌝ **ki** *ana* **igi**-*šu*$_2$ **du** *ina lib*$_3$-*bi*
3' [xx] ⌜x⌝ **la**$_2$-*is-ma u*$_3$ **igi** *e-lat* 6 **me zal** *ša*$_2$ **mu**
4' [*ina*] ⌜6.42?⌝ **me.meš ta igi** *ana* **igi** 30 **ki du igi** *ša*$_2$ *lu-maš*
5' *gab-bi ki-i* **ka me** *u* **zi šar**$_2$**.šar**$_2$ *ina* 12 **mu.meš** 4.10 ⌜**tab**⌝
6' *ina* 1.11 5 **la**$_2$ *ina* 7.7 **ki** *ana* **ki**-*šu*$_2$ *ina* **ka**$_2$-*tu*$_2$ *ša*$_2$ 30 12.⌜30?⌝
7' BE **igi** 9 BE **šu**$_2$ *ina* **ka**$_2$-*tu*$_2$ *ša*$_2$ 36 15 BE **igi** 7 ⌜BE **šu**$_2$⌝

P2 8' **igi** *ana* **igi ta** 25 **maš.maš en** 30 **gir**$_2$**.tab** 30 **tab ta** 30 **gir**$_2$**.tab en** [25]
9' **maš.maš** 36 **tab ta** 25 **maš en** 30 **gir**$_2$**.tab** 16.15 **ki du** [**uš ta uš**]
10' 4 **ki** *ana ar*$_2$-*šu*$_2$ **la**$_2$-*is-ma ana* **me e** 6 **ki** *ana ar*$_2$-*šu*$_2$ **la**$_2$-*is*-⌜*ma*⌝
11' *ar*$_2$-*tu*$_2$ **uš ta uš** *ar*$_2$-*tu*$_2$ *ana* **šu**$_2$ 17.45 **ki du šu**$_2$ [**ta šu**$_2$]
12' [*ana* **igi**] ⌜6 **ki**⌝ **du igi ta** 30 **gir**$_2$**.tab en** 25 **maš.maš** 1[9.30]
13' [**ki du uš ta u**]**š** 4.48 **ki** *ana ar*$_2$-*šu*$_2$ **la**$_2$-*is*-⌜*ma ana* **me e**⌝
14' [7.12 **ki** *ana ar*$_2$-*šu*$_2$ **la**$_2$-*is-ma* **uš** *ar*$_2$-*tu*$_2$ **ta**] ⌜**uš**⌝ *ar*$_2$-*tu*$_2$
15' [*ana* **šu**$_2$ 21.18 **ki du šu**$_2$ **ta šu**$_2$ *ana* **igi** 7.1]⌜2⌝ **ki du igi**

P3 16' [xxxxxxx *ina* **u**]**š igi**-*tu*$_4$ $\frac{1}{2}$ **kuš**$_3$
17' [xxx *ina* **uš** *ar*$_2$-*tu*$_2$] ⌜$\frac{1}{2}$⌝ **kuš**$_3$ *i-šap-pil*

P4 18' [12 **alla** 11.12 **a** 11.12 **a**]**bsin**$_0$ **rin**$_2$ 12 **zi**-*šu*$_2$ 12.4⌜8 **gir**$_2$ 13⌝.36
19' [**pa** 14.24 **maš**$_2$ 15.12 **gu**] 15.12 *zib* 14.24 **ḫun** 13.⌜36⌝ **mul**$_2$ 12.48
19a' **maš.maš**

P5 20' [xxxxxxx] 1-*en* **sag.ki gu**$_4$ *ša*$_2$ 12 **sag**
21' [xxxxxxx] ⌜x⌝ **me**? 10.45 **a.ša**$_3$-*šu*$_2$
22' [xxxxxxxx] ⌜**ta**?⌝ 10.45 **e**$_{11}$
23' [xxxxxxxx **sa**]**g** *ṭe-ri-tu*$_2$
24' [xxxxxxxxxxxx] ⌜x⌝
(n missing lines)

Column ii

P6 (m missing lines)
1' ⌜**mul**$_2$?**.babbar**?⌝ [xxxxxxx] ⌜xx 3?⌝ [xxxxx]
2' 6.⌜50 *ar*$_2$⌝-*šu*$_2$ **ta uš** *ar*$_2$-[*tu*$_2$ xx **zi**]-*šu*$_2$ **ta uš** *ar*$_2$-*tu*$_2$ 30 **me** 5 **zi**-*šu*$_2$ **en š**[**u**$_2$?]
2a' 10 **zi**-*šu*$_2$

P7 3' ⌜**ta**⌝ 5.4⌜5 **alla**⌝ **en** 5.4[5 **gir**$_2$**.tab** 30 **tab**] **ta** 5.45 **gir**$_2$**.tab en** 2 **maš**$_2$ 33.45 **tab**
4' ⌜**ta** 2⌝ [**maš**$_2$] ⌜**en**⌝ 17 **mul**$_2$**.mul**$_2$ ⌜36⌝ [**tab ta** 1]7 **mul**$_2$**.mul**$_2$ **en** 5.45 **alla** 33.45 **tab**

P8 5' [**ta** 9 **alla en** 9 **gir**$_2$**.tab** 3]0 **tab ta** 9 **gir**$_2$**.tab en** 27 **pa** ⌜33.45⌝ **tab**
6' [**ta** 27 **pa en** 17 **mul**$_2$**.mul**$_2$ 36 **tab t**]**a** 17 **mul**$_2$**.mul**$_2$ **en** 9 **alla** ⌜33.45⌝ **tab**

Obverse column i

P1 **(System A)**

a' **(Subdivision of the synodic cycle, scheme A.S$_2$)**

(m lines missing)

1'[...] ⸢the position⸣ proceeds in ⸢the east⸣, (then) it appears (FA). 2'[...] ⸢3 36 1 10 40?⸣ the position proceeds forward. In 3'[...] ⸢40⸣ it recedes, and it appears. Beyond 6,0 'days', the passing of the year, 4'[in] ⸢6,42⸣ 'days' the position proceeds 30 from appearance (FA) to appearance. You compute the appearance for 5'all zodiacal signs in accordance with the day and the displacement.

b' **(Net displacements for zone 1)**

In 12 years you ⸢add⸣ 4;10, 6'in 1,11 you subtract 5. In 7,7 the position (returns) to its position.

c' **(Rising times at FA, setting times at LA)**

In the 'ward' of 30: 12;[30?] if it appears (FA), 7'9 if it sets (LA). In the 'ward' of 36: 15 if it appears, 7 if ⸢it sets⸣.

P2 **(System A)**

---

a **(Updating *B* with the synodic arc)**

8'Appearance (FA) to appearance. From 25 Gem until 30 Sco you add 30. From 30 Sco until [25] 9'Gem you add 36.

b **(Subdivision of the synodic cycle, scheme A.S$_1$, zone 1)**

From 25 Gem until 30 Sco the position proceeds 16;15, [(then) it becomes stationary (S1).
From the station] 10'the position recedes backwards 4, then it rises to daylight (AR).
The position recedes backwards 6, then 11'the second station (S2).
From the second station to the setting (LA) the position proceeds 17;45, (then) it sets.
[From the setting] 12'[to the appearance (FA)] the position proceeds 6, (then) it appears.

c **(Subdivision of the synodic cycle, scheme A.S$_1$, zone 2)**

From 30 Sco until 25 Gem 13'[the position proceeds] 1[9;30, (then) it becomes stationary (S1).
From] the station the position recedes 4;48, then it rises to daylight (AR).
14'[The position recedes backwards 7;12, then the second station.
From] the second station 15'[until setting the position proceeds 21;18, (then) it sets.
From setting to appearance] the position proceeds 7;12, (then) it appears.

P3 **(Distance to the ecliptic at the stations)**

16'[... at the] first station [it ascends] $\frac{1}{2}$ cubit, 17'[... at the second station] it descends ⸢$\frac{1}{2}$⸣ cubit.

P4 **(System A?: parameters of zigzag function v_2 for the daily displacement?)**

18'[0;12: Cnc; 0;11,12: Leo; 0;11,12: V]ir; Lib: 0;12 is its displacement; 0;12,4⸢8: Sco; 0;13⸣,36: 19'[Sgr; 0;14,24: Cap; 0;15,12: Aqr]; 0;15,12: Psc; 0;14,24: Ari; 0;13.⸢36⸣: Tau; 0;12,48: 19a'Gem.

P5 **(Computations involving a trapezoid)**

20'[...] one trapezoid of which the front side is 12, 21'[...] 1 from 10.45, its area, 22'[...] you subtract ⸢from⸣ 10.45. 23'[...] the pinched side. 24'[...] ...

(n lines missing)

Column ii

P6 **(Unknown system: subdivision synodic cycle)**

(m lines missing)

1'⸢Jupiter?⸣ [...] ... [...] 2'6;50 backwards. From the second station its [displacement is ...] From the second station for 30 'days' its displacement is 0;5. Until the ⸢setting⸣ 2a'its displacement is 0;10.

P7 **(System A'': updating *B* with the synodic arc)**

3'⸢From⸣ 5;4⸢5 Cnc⸣ until 5;4[5 Sco you add 30.] From 5;45 Sco until 2 Cap you add 33;45.
4'⸢From 2⸣ [Cap] ⸢until⸣ 17 Tau [you add] ⸢36⸣. [From 1]7 Tau until 5;45 Cnc you add 33;45.

P8 **(System A''')**

a **(Updating *B* with the synodic arc)**

5'[From 9 Cnc until 9 Sco] you add 30. From 9 Sco until 27 Sgr you add ⸢33;45⸣.
6'[From 27 Sgr until 17 Tau you add 36.] From 17 Tau until 9 Cnc you add 33;45.

7' [xxxxxxx] ⸢**šu**$_2$⸣ **ta** 10 **maš**$_2$ **en** ⸢xx 10$^?$ BE⸣ **šu**$_2$ ⸢12 BE **igi**⸣

P9a 8' [**mul**$_2$.**babbar** *ina* **tur**-*ti* **ki** *šamaš*$_2$ **t**]**a** 9 **alla en** 9 ⸢**gir**$_2$.**tab ta šu**$_2$⸣ **en igi**

9' [*ša*$_2$ **me** 12.3]0 ⸢**zi egir**⸣ **igi** [30 **me**] ⸢*ša*$_2$ **me** 12⸣.30 ⸢**zi**⸣ 3 **iti**.⸢**meš**⸣ [*ša*$_2$ **m**]**e**

10' 6.15 **zi uš** 4 ⸢**iti**⸣.**meš** *ša*$_2$ ⸢**me**⸣ 5 **zi gur**-*ma* **uš** 3 **iti.meš**

11' *ša*$_2$ **me** 7.55 **zi** 30 **me** *ina* **igi**-⸢*at* **šu**$_2$⸣-*šu*$_2$ *ša*$_2$ **me** 12.30 **zi ta šu**$_2$

12' *ana* **igi** 6.15 **ki du** *u*$_3$ **igi** 15.3⸢7⸣.[30 **k**]**i du**-*ma* **uš igi**-*tu*$_2$ 10 **la**$_2$ **uš**

13' 18.7.30 **ki du šu**$_2$ *ša*$_2$-*niš* **ta šu**$_2$ *ana* [**igi** *ša*$_2$ **me** 1]⸢2⸣.30 **zi egir igi** 30 **me**

14' 12.30 **zi**-*ḫi* 3 **iti.meš** *ša*$_2$ **me** 6.[40 **zi**-*m*]*a* **uš** 4 **iti.meš** *ša*$_2$ **me**

15' 4.10 **zi gur**-*ma* **uš** 3 **iti.meš** *ša*$_2$ [**me** 6].23.20 **zi**-*ma* **uš**

16' 30 **me igi**-*at* **šu**$_2$-*šu*$_2$ *ša*$_2$ **me** 12.30 ⸢**zi**-*ma*$^?$⸣ *u* **šu**$_2$ **ta šu**$_2$ *ana* **igi**

17' 6.15 **ki du igi** 16.15 **ki du uš** [8.20 **z**]**i**-*ma* **gur**-*ma* **u**[**š** x$^?$]

18' 15.50 **du šu**$_2$ **zi** *an-na-a* **en uš** [xxxxx]

P9b 19' **mul**$_2$.[**babbar**$^?$] ⸢x⸣ **u**$_4$ **igi u**$_4$ **šu**$_2^?$ **uš** *ar*$_2$-*tu*$_2$ *ša*$_2$ 4 ⸢**danna**⸣ [xxxxxxx]

20' *ina* **ka**$_2$-*tu*$_2$ *ša*$_2$ 30 11 BE **šu**$_2$ 10 BE **igi** ⸢*ina* **ka**$_2$⸣-[*tu*$_2$ *ša*$_2$ 36 xxxxx]

P10 21' *ša*$_2$-*niš* **ta** 25 **maš.maš en** 30 [**gir**$_2$.**tab tur**-*tu*$_2$ **ta** 30 **gir**$_2$.**tab en** 25 **maš.maš**]

22' **gal**-*tu*$_2$ *ina* **tur**-*tu*$_2$ **ki** ^{d}**utu** *š*[*a*$_2$ **me** xxxxxxxxxxxxx]

23' [1]⸢2⸣.30 **zi** 3 **iti.meš** [xxxxxxxxxxxx]

24' [xx **it**]**i**.⸢**meš**⸣ *ša*$_2$ **me** ⸢7⸣.[xxxxxxxxxxxxxx]

25' [xxxxx] ⸢x⸣ [xxxxxxxxxxxxx]

(n−1 lines missing)

Column iii

P11 (m−2 lines missing)

1' [xxx] ⸢x **ki la**$_2$-*is* *ana* **me e ta e me** *ina* 50$^?$ **me** 6 **ki la**$_2$-*is*⸣ [**uš** *ar*$_2$-*tu*$_2$]

2' [**ta u**]**š** *ar*$_2$-*tu*$_2$ *ina* 1-*me* 35 **me** 22 **ki du**-*ma* **šu**$_2$ **ta šu**$_2$ *ina* 32 **me** 5.⸢30 **ki**⸣ [**du igi**$^?$]

3' **ta me**$^?$ **en me**$^?$ *šu*$_2^?$-⸢*u*$_2^?$⸣ **me.meš** *u* **ki.meš kimin**-*ma* *ina* ⸢x⸣ [**me** x$^?$]

b (Setting time at LA, rising time at FA)

[7'][... if] ˹it sets˺. From 10 Cap until ˹...: 10 if˺ it sets (LA), ˹11 if it appears (FA)˺.

P9a (System A': subdivision of the synodic cycle)

a (Scheme A'.S_2, zone 1, $\delta\tau$ and v)

[8'][Jupiter in (the region of) the small one.
With the Sun] from 9 Cnc until 9 Sco from setting (LA) to appearance (FA) [9']it moves [0;12],30 per 'day'.
For 30 'days' after the appearance it moves 0;12,30 per 'day'.
For 3 months [10']it moves 0;6,15 per 'day', (then) it is stationary (S1).
For 4 months it moves backward 0;5 per 'day', then it is stationary (S2).
For 3 months [11']it moves 0;7,55 per 'day'.
For 30 'days' before its setting (LA) it moves 0;12,30 per 'day'.

b (Scheme A'.S_2, zone 1, $\delta\Sigma$)

From the setting [12']until the appearance the position proceeds 6;15, and (then) it appears (FA). The position proceeds 15;37,[30], and then the first station. It recedes 10, (then) it becomes stationary (S2). [13']The position proceeds 18;7,30, (then) it sets.

c (Scheme A'.S_1, zone 1, $\delta\tau$ and v)

Otherwise: from setting to [appearance] it moves 0;12,30 [per 'day'].
After the appearance for 30 'days' [14']it moves 0;12,30.
For 3 months it moves 0;6,40 per 'day', then it becomes stationary (S1).
For 4 months [15']it moves backwards 0;4,10 per 'day', then it becomes stationary (S2).
For 3 months it moves [0;6],23,20 per [day], {then it becomes stationary (error)}.
[16']For 30 'days' before its setting (LA) it moves 0;12,30 per 'day', and? then it sets.

d (Scheme A'.S_1, zone 1, $\delta\Sigma$)

From setting until appearance [17']the position proceeds 6;15, (then) it appears. The position proceeds 16;15, (then) the station. It moves backwards [8;20], then it is stationary [...?]. [18']It proceeds 15;50, (then) it sets. Thus the displacement until the station [...]

P9b (System A: various procedures)

---

a (Stations; unclear)

[19']˹Jupiter˺, first ˹station?˺ and second station, for 4 'miles' [...]

b (Setting time at LA, rising time at FA)

[20']In the 'ward' of 30: 11 if it sets (LA), 10 if it appears (FA). [In the 'ward' of 36: ... if it sets, ... if it appears.]

P10 (System A)

---

a (Boundaries of the zones for σ)

[21']Alternatively: from 25 Gem until 30 [Sco the small one, from 30 Sco until 25 Gem] [22']the large one.

b (Subdivision of the synodic cycle, scheme A.S_1,?, zone 1)

In (the region of) the small one with the Sun [it moves ...] per 'day' [... For ...] [23']it moves 0;12,30 [per 'day']. For 3 months [it moves ... per 'day'.] [24'][For ...] months [it moves] 7;[...] per 'day' [...] [25'][...] ... [...]
(n−1 lines missing)
Column iii

P11 (Various systems: subdivision of the synodic cycle)

a (Scheme A.S_5, zone 2)

(m−2 lines missing)
[1'][From the first station in ... 'days'] ˹the position recedes˺ [...], ˹(then) it rises to daylight (AR). From daylight rising in 50?+x 'days' the position recedes 6˺, [(then) the second station.] [2'][From] the second station in 135 'days' the position proceeds 22, then it sets (LA). From the setting in 32 'days' the position [proceeds] 5;˹30˺, [(then) it appears? (FA).] [3']From 'day' until 'day'? these? are the 'days' and the positions; ditto.

4’ 15.37.30 ki du-*ma* uš igi-*tu*$_2$ uš ta uš [igi-*tu*$_2$ *ina* x me]

5’ 5 ki la$_2$-*is ana* me e ta e me *ina* 54 me 5 ki ⸢la$_2$-*is*⸣

6’ uš ⸢*ar*$_2$?⸣-*tu*$_2$ uš ta ⸢uš *ar*$_2$-*tu*$_2$?⸣ [*ina* xxx] ⸢x⸣ 18.20 ⸢ki du-*ma*⸣ [šu$_2$]

7’ ta šu$_2$ *ina* 27 me ⸢5?.30⸣ [ki du-*ma* igi xxxx]

8’ ⸢u$_4$?⸣.meš *u* ki *ša*$_2$ du ⸢kimin⸣ : ta [šu$_2$ *ana* igi ta 27 maš.maš en 5 pa]

9’ [xx] 5.37.40 ⸢ki⸣ du-*ma* igi *mi*-[*im-ma ša*$_2$ *al-la* 5 pa diri]

10’ a.ra$_2$ 1.[1]2 du-*ma* ki 5 pa tab ta [5 pa en 27 maš.maš 6.45.12 ki]

11’ du-*ma* igi *mim-ma ša*$_2$ *al-la* 27 maš.maš diri [a.ra$_2$ 50 du-*ma*]

12’ ki 27 maš.maš tab ta igi *ana* uš ta 5.[45 pa en 45 alla xx]

13’ *ina* 2 me 18 ki du uš *mim-ma ša*$_2$ *al* ⸢45⸣ [alla diri a.ra$_2$ 50 du-*ma*]

14’ ki 45 alla tab ta 45 ⸢alla⸣ [en 5.45 pa 15 ki du x]

15’ *mim-ma ša*$_2$ *al* 5.⸢45 pa⸣ d[iri a.ra$_2$ 1.12 du-*ma* ki 5.45 pa tab]

16’ uš igi-⸢*tu*$_2$ uš⸣ ta uš igi-⸢*tu*$_2$ en e me⸣ [ta 5.45 pa en 45 alla]

17’ ⸢*ina* 1-*šu*⸣ me 6 ki la$_2$-*is-ma ana* me e *ša*$_2$ 6 ta ⸢x⸣ [xxx *mim-ma ša*$_2$ *al*]

18’ 5.45 pa la$_2$-*u*$_2$ *ša*$_2$ la$_2$-*u*$_2$ a.ra$_2$ 50 du ⸢ta?⸣ [5.45 pa la$_2$ xx]

19’ ⸢ta⸣ 45 alla en 5.45 pa *ina* ⸢1-*šu*⸣ me 5 ki la$_2$ *ana* me ⸢e⸣ [xxxxx]

20’ ⸢*ša*$_2$?⸣ *al* 45 alla la$_2$-*u*$_2$ *ša*$_2$ la$_2$-*u*$_2$ a.ra$_2$ 1.12 du *ina* ⸢45⸣ [alla la$_2$ xx]

21’ ⸢ta⸣ e me *ana* uš *ar*$_2$-*tu*$_2$ ta 5.45 pa en 4⸢5⸣ [alla xxxxx ki]

22’ [la$_2$]-⸢*is* uš⸣ *ar*$_2$-*tu*$_2$ uš *ša*$_2$ 6 ta? ša$_3$? ⸢zi?⸣ *ša*$_2$ ⸢*al*⸣ [5.45 pa la$_2$-*u*$_2$ *ša*$_2$ la$_2$-*u*$_2$]

23’ [a.ra$_2$] ⸢50⸣ du-*ma* ta 5.⸢45⸣ [pa la$_2$ xxxxxxx]

24’ [xxxxx] ⸢45 x⸣ [xxxxxxxxxxx]

(n−3? lines missing)

Reverse

Column i

P12 (about n lines missing)

1’ [xxxx] ⸢x⸣ [xxxxxxxxxxxxx]

2’ [xx] 13.20.⸢2⸣7.46 ⸢a.ra$_2$?⸣ [6.31 du-*ma* 1.28.1.30.56.46 x 13.30.]

3’ ⸢27⸣.46 murub$_4$-*u*$_2$ *a-na* 45.14 [xxxxxxxxxxx]

4’ igi.meš du-*ma* gin$_7$ *ša*$_2$ ki.meš du$_3$-⸢*uš*⸣ [15 pa 45.14 u$_4$.meš murub$_4$?]

5’ 15 *zib* 50.07.15 u$_4$.meš nim 15 ab[sin 40.20.45 u$_4$.meš sig]

P13 6’ *ša*$_2$-*niš* 33.8.45 murub$_4$-*u*$_2$ *ša*$_2$ ki.meš a.⸢ra$_2$?⸣ [1.50.40 du-*ma* 1.1.8.8.20]

7’ 11.4 ki-*šu*$_2$ tab-*ma* 12.5.8.8.20 ⸢*a-na*⸣ gaba.ri m[u.an.na gar-*an* xx]

8’ *bi-ri-šu*$_2$-*nu* gar-*an* 12.5.8.8.20 ki-*šu*$_2$ tab-*ma* u$_4$.meš e-*bi* ⸢xx⸣.[xxxx]

P14 9’ *šal-šiš* u$_4$.meš *ša*$_2$ šu$_2$ igi *u* uš.meš gaba.ri mu.an.na *ana* igi-*ka* mul$_2$.babbar t[a 9 alla en 9 gir$_2$.tab]

10’ 30 danna *u*$_3$ mu.an.na *ana* mu.an.na 11.⸢3⸣.20 u$_4$.meš la$_2$ *u*$_3$ 30 dan[na xxx]

11’ ki *ša*$_2$ mul$_2$.⸢babbar⸣ mu *ana* mu tab ^{d}utu *ina* 30.45 u$_4$.meš du-*ma* 30.45 u$_4$.⸢meš⸣ [*u* 11.3.20]

12’ u$_4$.meš *ana* ugu *a-ḫa-miš*$_2$ GAR.GAR-*ma* 41.48.20 u$_4$.meš gaba.ri mu.an.na gar-[*an*]

P15 13’ ta 9 alla en 2 maš$_2$ mu *ana* mu 33.45 tab *u* ^{d}utu? mu *ana* mu uš.meš la$_2$? 33.4[5 xx]

14’ a.ra$_2$ 1.50 du-*ma* 1.1.52.30 ki 33.45 tab-*ma* tab? 34.46.⸢5⸣2.[30]

15’ *u* 11.3.20 u$_4$.meš *ša*$_2$ ^{d}utu *ana* ugu *a-ḫa-miš*$_2$ tab-*ma* 45.⸢50⸣.[12.30 u$_4$.meš gaba.ri mu.an.na gar-*an*]

b **(Scheme A.S$_5$, zone 1)**

In ... [‘days’] 4’the position proceeds 15;37,30, then it becomes stationary at the first station (S1). From the [first] station [in ... ‘days’] 5’the position recedes 5, (then) it rises to daylight (AR). From daylight rising in 54 ‘days’ the position ⌜recedes⌝ 5, 6’(then) it becomes stationary at the second station (S2). From ⌜the second station the position proceeds⌝ 18;20 [in ... ‘days’, (then) it sets (LA).] 7’From the setting [the position proceeds] ⌜5?;30⌝ in 27 ‘days’, [then it appears (FA)...] 8’The ‘days’ and position by which it proceeds: ditto.

c **(System A$_1$: LA to FA)**

From [setting (LA) to appearance (FA). From 27 Gem until 5 Sgr] 9’the position proceeds [...] 5;37,40, then it appears. Whatever (the amount) by which [it exceeds 5 Sgr], 10’you multiply it by 1;[1]2, and add it to 5 Sgr. From [5 Sgr until 27 Gem the position] 11’proceeds [6;45,12], and it appears. Whatever (the amount) by which it exceeds 27 Gem, [you multiply it by 0;50, and] 12’add it to 27 Gem.

d **(System A$_2$: FA to S1)**

From appearance (FA) to station (S1). From 5;[45 Sgr until 0;45 Cnc] 13’the position proceeds 18 in 2,0 ‘days’. Whatever (the amount) by which it exceeds ⌜0;45⌝ [Cnc you multiply by 0;50, and] 14’add to 0;45 Cnc. From 0;45 ⌜Cnc⌝ [until 5;45 Sgr the position proceeds 15 ...] 15’Whatever (the amount) by which it [exceeds] 5;45 Sgr [you multiply by 1;12, and add to 5;45 Sgr,] 16’(then) it becomes stationary at the first station.

e **(System A$_2$: S1 to AR)**

From the first station until daylight rising (AR). [From 5;45 Sgr until 0;45 Cnc] 17’the position recedes 6 in 60 ‘days’, then it rises to daylight (AR). Concerning? the 6 from ... [... Whatever (the amount) by which] 18’it is less than 5;45 Sgr: (the amount) by which it is less you multiply by 0;50, [you subtract] ⌜from?⌝ [5;45 Sgr ...] 19’From 0;45 Cnc until 5;45 Sgr the position recedes 5 in 60 ‘days’, (then) it rises to daylight. [...] 20’which is less than 0;45 Cnc: (the amount) by which it is less you multiply by 1;12, [you subtract] from 0;45 [Cnc ...]

f **(System A$_2$: AR to S2)**

21’From daylight rising (AR) to the second station. From 5;45 Sgr until 0;45 [Cnc the position] 22’[rece]des [... in ...], (then) it becomes stationary at the second station. Concerning the 6 from? ... which [is less than 5;45 Sgr: (the amount) by which it is less] 23’you multiply [by 0;50, and [subtract] from 5;4[5 Sgr...] 24’45 [...]

(n−3? lines missing)

Reverse column i

P12 **(System B)**

a **(Computations involving τ and Π)**

(about n lines missing)

1’[...] ... [...] 2’[you multiply] 13.20.27.46 (error for 13.30.27.46) by [6,31, it is 1.28.1.30.56.46 ... You add 13.30.]3’27,46, the middle one, to 45;14 [... you multiply by the ...] 4’the appearances, and you do as for the positions.

b **(Parameters of the zigzag function for τ)**

[15 Sgr: 45;14 ‘days’, the middle one]; 5’15 Psc: 50;7,15 ‘days’, the high one; 15 Vir: [40;20,45 ‘days’, the low one.]

P13 **(System B: computing τ from σ)**

6’Alternatively. 33;8,45, the middle one for the positions, [you multiply] by [0;1,50,40, it is 1;1,8,8,20.] 7’You add 11;4 to it, it is 12;5,8,8,20, [you put it down] as the value of the [‘year’. ...] 8’between them you put down, you add 12;5,8,8,20 to it, and you predict the ‘days’ ... [...]

P14 **(System A’: computing τ from σ, zone 1)**

9’Thirdly. In order for you to see the ‘days’ for setting (LA), appearance (FA) and stations, the value of the ‘year’. Jupiter. The (region of the) small one: from [9 Cnc until 9 Sco] 10’30 (i.e. one) ‘mile’, and year by year you subtract 11;3,20 ‘days’, and 30 (i.e. one) ‘mile’ [...] 11’you add [to] the position of Jupiter, year by year. The Sun proceeds in 30;45 (error for 30;55) ‘days’, and 12’you accumulate 30;45 (error for 30;55) ‘days’ and [11;3,20] together, 41;48,20 (error for 41;58,20) ‘days’ you put down as the value of the ‘year’.

P15 **(System A’: computing τ from σ, zone 2)**

13’From 9 Cnc (error for Sco) until 2 Cap you add 33;45 year by year, and year by year the Sun ... degrees?. 14’You multiply 33;4[5] by 0;1,50, it is 1;1,52,30. You add it to 33;45 ... (it is) 34;46,52,[30], 15’and 11;3,20 ‘days’ of the Sun you accumulate together, 45;50,[12,30 ‘days’ you put down as the value of the ‘year’.]

P16 16' ta 2 maš$_2$ en 17 mul$_2$ mu *ana* mu 36 tab be-⸢*ma*⸣ [xxxxx]
17' egir-*šu*$_2$ *ana* mu.an.na 4$^?$ ⸢la$_2$⸣-*is ina* 21 gu-*šu*$_2$ ⸢ta$^?$⸣ [xxxxxxx]
18' dib-*šu*$_2$-*ma* 36 ⸢mu$^?$⸣ [xx] ⸢a⸣.ra$_2$ 2 du-⸢*ma*⸣ [1.12 xx 36 a.ra$_2$ 10 du-*ma* 6 x]
19' 6 ta 1.12 ⸢zi-*ma*⸣ *ša*$_2$ tag$_4$ 1.⸢6⸣ [ki 36 tab-*ma* 37.6 *u* 11.3.20 u$_4$.meš]
20' *ša*$_2$ ^{d}utu ki 37.6 tab-*ma* ⸢x⸣ [48.9.20] ⸢xxx u$_4$ x⸣ [x]

P17 21' ta 9 alla en 30 gir$_2$.tab tur-*tu*$_2$ ta 9 alla en 2 maš$_2$ murub$_4$-*tu*$_2$
22' ta 2 maš$_2$ en 17 mul$_2$.mul$_2$ gal-*tu*$_2$ ta 17 mul$_2$ en 9 alla murub$_4$-*tu*$_2$

P18 23' [x$^?$] ⸢ta šu$_2$ *ana* igi *ša*$_2$ me 12⸣.30 zi egir igi 30 me *ša*$_2$ me 12.30 ⸢zi⸣
24' [3 iti.meš *ša*$_2$ me 6.40 zi-*ma* uš 4 iti.meš] *ša*$_2$ me 4.10 zi-⸢*ma* gur⸣-[*ma* uš]
(about m−2 lines missing)

Column ii

P19 (about n−3 lines missing)
1' ⸢x⸣ [xxxxxxxxxxxxxxxxxxx]
2' *ša*$_2$ mul$_2$.babbar [xxxxxxxxxxxxxxxxxx]
3' dib ⸢x⸣ [xxxxxxxxxxxxxxxxxx]
4' gur ⸢x⸣ [xxxx] ⸢xx⸣ [xxxxxxxxxxx]
5' ⸢x⸣ [xxxx].meš *ša*$_2$ mul$_2$.babb[ar xxxxxxxxxx] ⸢x⸣

P20 6' ⸢ki$^?$⸣.[meš$^?$ a.ra$_2$ 12 mu.meš] ⸢*ina*$^?$⸣ gal-*tu*$_2$ 5 t[ab a.ra$_2$ 1.11 mu.meš 6] ⸢la$_2$⸣ a.ra$_2$ 1.23
7' [mu.meš 1 la$_2$] ⸢a⸣.ra$_2$ 1.35 m[u.meš 4 tab a.ra$_2$] 2.46 40 mu.meš
8' [xxx] ⸢4⸣.21 mu.meš ⸢2⸣ [tab *ina* 7.7] ⸢mu⸣.meš ki-*šu*$_2$ kur-*ad*$_2$

P21 9' [xx 7].⸢7⸣ mu.meš 6.31 i[gi.meš] ⸢36 bal⸣.meš 3.36 ki du 33.8.
10' [45 muru]b$_4$-*u*$_2$ *ša*$_2$ ki.meš 15 ⸢*zib*⸣ 50.07.15 u$_4$.meš la$_2$ 15 absin$_0$ 50.20.
11' [4]⸢5⸣ u$_4$.meš sig 1.48 ⸢*taš*⸣-*pil-tu*$_4$ igi *ana* igi *ša*$_2$ ka 2-*i* 15 *zib*
12' ⸢50⸣.3.31 u$_4$.meš la$_2$ 1⸢5 absin$_0$⸣ 50.24.15 u$_4$.meš sig 1.46.40 *taš*
12a' igi *ana* igi

P22 13' ⸢ki⸣.meš *ša*$_2$ ⸢mul$_2$.babbar$^?$ 15 *zib*⸣ [38.2] ⸢la$_2$ 15⸣ absin$_0$ 28.15.⸢30 sig⸣ 1.48 *taš*
14' igi *ana* igi *ša*$_2$ ka 2-⸢*i*⸣ [15 *zib* 37.58.20 la$_2$] ⸢15⸣ absin$_0$ 28.9.10 sig 1.46.40 *taš* igi *ana* igi

P23 15' mul$_2$.babbar *ina* tur-*tu*$_2$ ta š[u$_2$ *ana* igi 6].⸢15⸣ ki du igi ta igi *ana* uš 16.15 ki du
16' uš ta uš *a-n*[*a* e me] 5.10 ki gur ta e me *ana* uš *ar*$_2$-*tu*$_2$
17' 13.15 ki gur ta u[š *ar*$_2$-*tu*$_2$] ⸢*ana* šu$_2$⸣ 15 ki du-*ma* šu$_2$ mul$_2$.babbar *ina* gal-*tu*$_2$

P16 (System A': computing τ from σ, zone 3)

[16']From 2 Cap until 17 Tau you add 36 year by year. If [...] [17']after it, you subtract 5 from[?] the year. In 21, its ... ⌜from[?]⌝ [...] [18']it passes it, and you multiply 36 [...] by 0;2, it [is 1;12. You multiply 36 by 0;0,10, it is 0;6] [19']You subtract 0;6 from 1;12, and what remains, 1;6, [you add to 36, it is 37;6, and 11;3,20 'days'] [20']of the Sun you add to 37;6, ... [48;9,20] ⌜...⌝

P17 (System A', zones for the synodic arc)

[21']From 9 Cnc until 30 (error for 9) Sco the small one. From 9 Cnc (error for Sco) until 2 Cap the middle one. [22']From 2 Cap until 17 Tau the large one. From 17 Tau until 9 Cnc the middle one.

P18 (Subdivision of the synodic cycle, scheme A'.S_1, zone 1)

[23'][...?] ⌜From setting (LA) to appearance (FA)⌝ it moves 0;12,30 per 'day'. After the appearance for 30 'days' it moves 0;12,30 per 'day'. [24'][For 3 months it moves 0;6,40 per 'day', then it is stationary (S1). For 4 months] it moves backwards 0;4,10 per 'day', [then it is stationary (S2).]
(*about m−2 lines missing*)
Column ii

P19 (Unclear)

(*about n−3 lines missing*)
[1']⌜...⌝ [...] [2']of Jupiter [...] [3']it passes ⌜...⌝ [...] [4']it turns back ⌜...⌝ [...] ⌜...⌝ [...] [5']⌜...⌝ [...] of Jupiter [...] ⌜...⌝

P20 (System A: net displacements for zone 2)

[6']⌜Positions[?]⌝. [For 12 years] in (the region of) the large one you add 5. [For 1,11 years] you subtract [6]. For 1,23 [7'][years you subtract 1.] For 1,35 y[ears you add 4.] For 2,46 40[?] years [8'][you add 2. For] 4,21 years [you add] 2. [In 7,7] years it reaches its position.

P21 (Systems B and B')

a (System B: period relations and mean synodic arc)

[9'][... 7],7 years, 6,31 appearances, 36 revolutions, the position proceeds 3,36,0.
33;8,[10'][45 is the 'middl]e one' for the positions.

b (System B: parameters of the zigzag function for τ)

15 [Psc]: 50;7,15 'days', the 'height'; 15 Vir: 50;20,[11']45 (error for 40;20,45) 'days', the 'depth'; 1;48, the difference.

c (System B': parameters of the zigzag function for τ)

Appearance to appearance according to a second opinion: 15 Psc, [12']50;3,31 'days', the 'height'; 15 Vir, 50;24,15 (error for 40;24,15) 'days', the 'depth'; 1;46,40 the difference (from) appearance to appearance.

P22 (Systems B and B': parameters of the zigzag function for σ)

a (System B)

[13']⌜Positions of Jupiter. 15 Psc: [38;2], the 'height'; 15 Vir: 28;15,30; the 'depth'; 1;48, the difference [14'](from) appearance to appearance.

b (System B')

According to a second opinion: [15 Psc: 37;58,20, the 'height']; 15 Vir: 28;9,10, the 'depth'; 1;46,40, the difference (from) appearance to appearance.

P23 (Subdivision of the synodic cycle, scheme A.S_3)

a (Zone 1)

[15']Jupiter in the (region of the) small one.
From set[ting (LA) to appearance (FA) the position proceeds 6];15, (then) the appearance.
From appearance to station (S1) the position proceeds 16;15, [16'](then) the station.
From station to [daylight rising (AR)] the position turns back 5;10.
From daylight rising to second station (S2) [17']the position turns back 13;15 (error for 3;15?).
From [second] ⌜station⌝ to setting the position proceeds 15, then it sets.

18’ **ta šu$_2$** *ana* **igi** 29 **u$_4$.meš tab t**[**a igi** *ana* **u**]**š** 4 **iti** 4 **me tab ta uš** *ana* **e me**
19’ 58 **me tab ta e me** *ana* **uš** *a*[*r$_2$-t*]*u$_2$* 2 **iti** 4 **me tab ta uš** *ar$_2$-tu$_2$ ana* **šu$_2$** 4 **iti**
19a’ 10 **me tab**

P24 20’ **mul$_2$.babbar** *ina* **gal**-*tu$_2$* **ta šu$_2$** *ana* **igi** 7.30 **ki** ⸢**du**⸣ **igi ta igi** *ana* **uš** 18 ⸢**ki du**⸣ **uš ta uš**
21’ *ana* **e me** 4 **ki gur ta e me** *ana* ⸢**uš**⸣ *ar$_2$-tu$_2$* 5 **ki gur ta uš** *ar$_2$-tu$_2$ ana* **šu$_2$** 19 **ki**
21a’ **du**-*ma* **šu$_2$**

P25 22’ *ša$_2$* **mul$_2$.babbar ta** 9 **alla en** 9 ⸢**gir$_2$.tab**⸣ **tur**-*tu$_2$* **ta** 9 **gir$_2$.tab en** 2 **maš$_2$ murub$_4$**-*tu$_2$*
23’ **ta** 2 **maš$_2$** ⸢**en** 17⸣ **mul$_2$ gal**-[*tu$_2$* xxxxxxxxxxxx] ⸢x⸣
24’ ⸢xxx⸣ [xxxxxxxxxxxxxxxxxxxxxxx]
(about m−2 lines missing)
Column iii
P26 (about n−3 lines missing)
1’ [xxxxxxxxxxxxxxxxxxxx] ⸢x **igi**⸣
2’ [xxxxxxxxxxxxxxxxxxx **u**]**š igi**-*tu$_4$*
3’ [xxxxxxxxxxxxxxxxx] ⸢xxxx *ana*⸣ **ugu**
4’ [xxxxxxxxxxxxxxxxxx] ⸢5⸣ **du**-*ma* 10.⸢43⸣
5’ [xxxxxxxxxxxxxxxxxx]-*ma* **uš me** *ana* **me** 5
6’ [xxxxxxxxxxxxxxxxx **m**]**e** *nen-mud* **ta uš** ⸢*ar$_2$*⸣ *ana* **šu$_2$**
7’ [xxxxxxxxxxxxxxx] 5.30 *taš ina* 30 **me igi** *ina* ⸢3⸣ [x]
7a’ **zi** ⸢x⸣
P27 8’ [xxxxxxxxxxxxx] **mul$_2$.mul$_2$** 13.30 **maš.maš** 1⸢3$^?$⸣ [xx]
9’ [xxx] 12.⸢40$^?$⸣ [xxx 1]4.20 **maš$_2$** 14.50 **gu** 15 [*zib*]
P28 10’ [xxx-*m*]*a ina* 4 **iti** 4 ⸢**u$_4$ uš**⸣ [xx] *ina* 2 **iti** *ana* **me e** *ar$_2$* **uš** ⸢x⸣ [x]
11’ [xxx] **uš** *u$_3$ ana* **nim la$_2$**-*is-ma* **u$_4$** ⸢x **zi**⸣ *ina* 12 **zi** ⸢x⸣ [x]
12’ [xx] ⸢x⸣ 36 **me** *ana* **me** 37.30 **la$_2$** *ina* 12 ⸢**mu.meš** x 6$^?$ x⸣ [x]
13’ [xx] **me** *ana* **me** *ina* 12 **mu.meš** 5 **uš ki**-*šu$_2$ ana* **nim d**[**u** xxx]
14’ [x] ⸢1⸣.48 : 1.36 1.24 : 1.12 : 49 : 46 : 42
15’ [x 4]2 : 12 **gal** 12 **la$_2$**

P29 16’ *ša$_2$* **ka** 2-*i* **mul$_2$.babbar** *ina* **alla** 6.10 **du**-*šu$_2$ ina* **a** 6.11.40 *ina* **absin$_0$** 6.13.20
17’ *ina* **rin$_2$** 6.15 *ina* **gir$_2$.tab** 6.16.40 *ina* **pa** 6.18.20 *ina* **maš$_2$** 6.20
18’ ⸢*ina*⸣ **gu** 6.18.20 *ina zib*.**me** 6.16.40 *ina* **ḫun** 6.15 *ina* **mul$_2$** 6.13.20 *ina* **maš**
18a’ 6.11.40

P30 19’ ⸢**mul$_2$.babbar**⸣ *ina* **gal**-*ti* **ta šu$_2$** *ana* **igi** 29 **u$_4$.meš tab ta igi** *ana* **uš** 4 **iti** 4 ⸢**me tab**⸣
20’ ⸢**ta uš**⸣ *ana* **e me** 58 **me tab ta e me** *ana* **uš** *ar$_2$-tu$_2$* 2 **iti** 4 **me tab** ⸢**ta**⸣
20a’ **uš** *ar$_2$-tu$_2$ ana* **šu$_2$** 4 **iti** 10 **me tab**
P31 21’ [**mul$_2$.babbar** *ina* **gal**]-⸢*ti*⸣ **ta šu$_2$** *ana* **igi** 7.30 **ki du ta igi** *ana* **uš** 17.30 **ki du**
22’ [**ta uš** *ana* **e me**] ⸢4 **ki gur**⸣ **ta e me** *ana* **uš** *ar$_2$-tu$_2$* 4 **ki gur ta uš** *ar$_2$-tu$_2$*
22a’ *ana* **šu$_2$** 17.30 **ki du šu$_2$**

b (Zone 2)

Jupiter in the (region of the) large one.
18' From setting to appearance you add 29 'days'.
Fr[om appearance to sta]tion you add 4 months, 4 'days'.
From station to daylight rising 19' you add 58 'days'.
From daylight rising to ⸢second station⸣ you add 2 months, 4 'days'.
From second station to setting you add 4 months, 10 'days'.

P24 (Subdivision of the synodic cycle, scheme A.S$_4$, zone 2)

20' Jupiter in (the region of) the large one.
From setting to appearance the position proceeds 7;30.
From appearance to station the position proceeds 18, (then) it becomes stationary.
From station 21' to daylight rising it turns back 4.
From daylight rising to the second station it turns back 5.
From the second station to setting the position 21a' proceeds 19, then it sets.

P25 (System A': zones for the synodic arc)

22' For Jupiter. From 9 Cnc until 9 Sco the small one. From 9 Sco until 2 Cap the middle one. 23' From 2 Cap until 17 Tau the large one.
(about m−2 lines missing)
Column iii

P26 (Subdivision of the synodic cycle; unclear)

(about n−3 lines missing)
1' [...] ... appearance 2' [...] the first station 3' [...] ... to 4' [...] you multiply [... by ...] 5, it is 10.43 5' [...], and it becomes stationary. Day by 'day', 5 6' [...] 'days', it is stationary. From the second station to setting (LA) 7' [...] 5.30, the difference. In 30 'days' it appears (FA), it moves 13 [...]

P27 (Values of a zigzag function v_3 for Jupiter's daily displacement?)

8' [...] Tau 0;13,30; Gem 0;1⸢3?⸣,[...] 9' [...] 0;12,40? [... 0;1]4,20; Cap 0;14,40; Aqr 0;15 [... Psc].

P28 (Subdivision of the synodic cycle; unclear)

10' [...], and in 4 months, 4 'days' it becomes stationary (S1). [...After the station] in 2 months it rises to daylight (AR). After the station ... [...] 11' [...] station and it recedes to the east, and ... from 0;12, the displacement, ... [...] 12' [...] 36. 'Day' by 'day' you subtract 37.30. In 12 years ... [...] 13' [...] 'day' by 'day'. In 12 years its position moves 5° to the east [...] 14' 1.48 1.36 1.24 : 1.12 : 49 : 46 : 42 15' .. 42 12, the largest one, you subtract? 12.

P29 (Zigzag function for $\delta\Sigma$ from LA to FA?)

16' According to a second opinion, Jupiter: in Cnc its progression is 6;10; in Leo 6;11,40; in Vir 6;13,20; 17' in Lib 6;15; in Sco 6;16,40; in Sgr 6;18,20; in Cap 6;20; 18' in Aqr 6;16,20; in Psc 6;16,40; in Ari 6;15; in Tau 6;13,20; in Gem 18a' 6;11,40.

P30 (Subdivision of the synodic cycle, scheme A.S$_3$, zone 2)

19' Jupiter in (the region of) the large one.
From setting (LA) to appearance (FA) ⸢you add⸣ 29 'days'.
From appearance to station (S1) you add 4 months, 4 'days'.
20' From station to daylight rising (AR) you add 58 'days'.
From daylight rising to the second station (S2) you add 2 months, 4 'days'.
From 20a' the second station to setting (LA) you add 4 months, 10 'days'.

P31 (Subdivision of the synodic cycle, scheme A.S$_3$, zone 2)

21' [Jupiter in (the region of) the large one].
From setting to appearance the position proceeds 7;30.
From appearance to station the position proceeds 17;30.
22' [From station to daylight rising] the position turns back 4.
From daylight rising to the second station the position turns back 4.
From the second station 22a' to setting the position proceeds 17;30, (then) it sets.

P32 23' [xxxxxxxxx] ⌜x 4?.5?.32.26?.30⌝ **a.ra₂** 30
24' [xxxxxxxxxxxxxxxx] ⌜2?⌝ **iti**
25' [xxxxxxxxxxxxxxxxxx] ⌜x⌝
(about m−2 lines missing)

Critical and philological notes

Obverse column i

1' *ina* **nim**: *ina* omitted in *ACT*. Perhaps **nim**, 'east'.
2' 3 36 1: only the lower halves are visible.
10.40: or **nim**?
3' Before **la₂** there is a winkelhaken, perhaps part of a number 10–50 (*ACT*: 40).
la₂-*is* = *inaḫḫis*, 'it recedes'.
e-lat = *elat*, 'beyond; besides'.
4' Traces at the beginning compatible with 6.42, as proposed by Neugebauer (cf. the commentary).
5' **ka** = *pû*, 'mouth'; *kī pî*, 'according to'.
šar₂.šar₂ = *tuštabbal* = 2 m. sg. pres. of *abālu* Št, 'to compute'.
The damaged sign after 4.10 is compatible with **tab**.
6' **ka₂**-*tu₂* = *bābtu*, 'ward' (= zone): also in Oi7' and Oii20'.
7' 9 **be šu₂**: **be** is written over an erasure. **be** = *šumma*, 'if'.
12' 6: upper part is visible.
14' There are traces of about 4 signs in the middle of the line, but they cannot be uniquely identified with the restored text.
16'–17' $\frac{1}{2}$ **kuš₃**: erroneously read as *šu₂-u₂* in *ACT* as pointed out by Steele (2005a), p. 48.
18' For the reconstructed numbers: cf. the commentary.
gir₂: only traces remain.
19' There is more space available in the gap at the beginning than what is suggested in *ACT*.
19a' **maš.maš** is written below the end of 19'.
20' **sag.ki gu₄** = *pūt alpi*, 'trapezoid' (*CAD* P sub *pūtu* 6).
21' **me**: perhaps the initial vertical belongs to the previous sign, read 2 *ina*?
22' **e₁₁**, 'you subtract': or **ki du**, 'the position proceeds'?
23' *ṭe-ri-tu₂* = *ṭerītu*: replaces DI.RI-*tu₂* (*ACT*). Verb. adj. derived from *ṭerû* G, 'to squeeze' (*CAD* Ṭ sub *ṭerû* A); *pūtu ṭerītu*, 'the squeezed side', i.e. the short side of the trapezoid. In this interpretation *pūtu ṭerītu* is synonymous with *pūtu ṣeḫertu*(**tur**-*tu₂*), 'the small side'.
24' The final vertical wedge of the last sign is visible.

Column ii

1' Near the middle of the line the lower halves of two vertical wedges are visible, followed by the lower half of a 3 (or perhaps a 6).
4' **ta** 2: only traces remain
7' After **en** there is a number, perhaps 10, followed by traces of what might be **alla** (uncertain).
⌜10 **be**⌝ **šu₂** 12 **be igi**: '10 if it sets (LA), 12 if it appears (FA): replaces ... **šu₂** 12 ... (*ACT*, p. 406).
8' **ki** = *itti*, 'with'.
9' **egir** = *arki*, 'after': only the lower half is visible; not accurately copied by Pinches.
10',17' **zi**[-*ma*] **gur** = *inassaḫ*[*ma*] *itâr*, a hendiadys construction meaning 'it moves backwards'.
11' After 12.30 **zi** Neugebauer reads **šu₂**, 'it sets', which is what one expects, but it is not on the tablet.
13' The gap offers sufficient room for the expected *ša₂* **me**, 'per 'day'' (omitted in *ACT*).
14' Before 12.30 one expects *ša₂* **me**, 'per 'day'', but it is omitted.
zi-*ḫi*: either *inassaḫi*, with a superfluous final *i*, 'it moves', or *nisḫi*, 'displacement'. Hence there is no need to assume that uniquely here ḪI instead of the usual GAM functions as a separation mark (*ACT*).
15' At the end **uš**, 'station/it is stationary' is an error, since one expects nothing; cf. the commentary.
16' -*ma*: reading uncertain, only the final vertical wedge is visible.
u, 'and': one expects nothing here.
19' **uš**?: the traces look like DU, but one expects **uš**.
25' traces of two horizontal wedges are visible.

Column iii

1' Only the lower parts of the signs are preserved. At the beginning there are traces of two vertical wedges.
3' *šu₂-u₂* = *šū*, 'this'. At the end faint traces are visible.
kimin-*ma*, 'ditto': the meaning is not clear here.
6' Of the second *ar₂-tu₂* only the upper ends of the wedges are visible.
7' 5: or 2.
8' *ša₂* **du** = *ša illiku*, 'by which it proceeds', or *ša alāku*, 'of its motion'?
After **kimin** there is a GAM₀, which seems to function as a separation mark.
16' This line is very worn. **en**: only traces are visible.
17' *ša₂*, 'concerning': this interpretation of the sign GAR is not certain. Perhaps a form of *šakānu*, 'to put'?
After 6 **ta** there are faint traces. Oiii22' contains a parallel, which is unfortunately also damaged.

P32 (Unclear)

23'[... you multiply] ⌜4,5,32,26,30⌝ by 30 24'[...] 2 months 25'[...] ...
(about m−2 lines missing)

18' **ta**: this is what one expects, but the traces look more like KI.
22' **ta ša$_3$ zi**: reading uncertain; meaning unclear. **ta ša$_3$** = *ina libbi*, 'from it'?
23' Between 50 and **du** there is another damaged sign, perhaps 5, as indicated in Pinches's copy; it was ignored by Neugebauer. Perhaps an erasure?
24' After 45, the upper part of a vertical wedge is visible.

Reverse column i

1' A single winkelhaken is visible.
3' *ana* points to an addition, so that one may restore **tab**, 'you add', in the gap at the end.
4' **gin$_7$** = *kīma*, 'like'.
7' *bi-ri-šu$_2$-nu*: *biri*, 'between'. The parallel, No. **46** P1', suggests that the preceding gap contained **igi** *ana* **igi**, 'appearance to appearance'.
8' After **e**-*bi* Neugebauer restored 1.1.8.8.20, but this is not consistent with the traces.
9' **mul$_2$.babbar**: crammed.
10' 11.3.20: the 3 might also be a 2.
11' ^{d}**utu** = *šamaš*, 'the Sun': significance unclear.
30.45: one expects 30.55; cf. the commentary.
12' 41.48.20: one expects 41.58.20; cf. the commentary.
13' **alla**, Cnc: error for **gir$_2$.tab**, Sco.
u ^{d}**utu**, 'and the Sun'?: meaning unclear. Perhaps read differently, e.g. U **igi**-*tu2*?
uš.meš la$_2$: meaning unclear.
14' **tab**-*ma* **tab**: the second **tab** was read as a separation mark by Neugebauer; meaning unclear.
15' The text in the gap at the end has been restored on the basis of Ri12'
16' **be**-⌜*ma*⌝ = *šumma*, 'if'.
17' 4: less likely 5.
gu-*šu$_2$*: clearly written; meaning unclear.
18' **dib**-*šu$_2$* = *ittiqšu*, 'it passes it'? Meaning unclear.
mu$^?$: reading very uncertain.
20'–23' The first half of these lines are (partly) preserved on the newly joined fragment BM 42816.
20' **tab**-*ma*, 'you add': one expects *ana* **ugu** *a-ḫa-miš$_2$* **tab**-*ma*, 'you add together', or a similar phrase (compare 12' and 15'). This is followed by a damaged sign, perhaps *ina* or **tab**. For 48.9.20 cf. the commentary. At the end of the line there are faint traces of several signs. Here one expects the same phrase as in Ri12', **u$_4$.meš gaba.ri nu.an.na gar**-*an*, 'you put down as the value of the 'year'', but the traces may not support this.
21' 30: error for 9. This is most likely a copying error resulting from the similarity between the cuneiform signs for 30 and 9.
The second instance of **alla** in this line is an error for **gir$_2$.tab** = Sco (same error as in 13').
23' **ta šu$_2$** *ana* **igi** *ša$_2$* **me** 12: only the upper parts of these signs are preserved.
24' [-*ma* **uš**]: part of it must have been written on the edge of the tablet.

Column ii

1' Only traces visible. Neugebauer suggests that this may be the last line of a procedure.
3' After **dib** there is a sign similar to ŠE.
4' After **gur** there is a broken sign beginning with a winkelhaken in raised position. After the first gap there are 2 broken signs, the second of which looks like *mim*.
5' The first sign might be UD or the beginning of GUR. At the end there are traces of three vertical wedges.
6' The traces at the beginning might belong to **ki**. There is more space in the gap than what is suggested by Neugeabuer.
6',7' **a.ra$_2$**: does not express multiplication here but functions as a logogram for *ana*, 'to' (cf. also No. **46** P10').
7' 40: one expects nothing here (cf. the commentary).
8' 2: replaces 4 (*ACT*).
10' 50.20: error for 40.20.
12' 50.24.15: error for 40.24.15.
taš = *tašpiltu*, 'difference'.
17' 13.15: this number is unexpected (cf. the commentary), but clearly written.
21'–22' The final signs in these lines are (partly) preserved on the new fragment BM 42816; they confirm Neugebauer's restorations.
24' At the beginning of the line about 3 damaged signs are visible, the first of which may be MU.

Column iii

1' Before **igi** a vertical wedge is preserved.
3' Before *ana* **ugu** only faint traces are preserved.
6' *nen-mud* = *nenmud*, 'it is/has become stationary'; stat. 3 m. sg. N *emēdu*, 'to be/become stationary'. Other occurrences are No. **9** O8; No. **46** Oii12'; No. **39** O9',11'. *ana* **šu$_2$**: thus also copied by Pinches; replaces **en** (*ACT*).
7a' **zi** is followed by a broken sign; the top end of a vertical wedge is preserved.
8' After **maš.maš** only 11 is preserved, but this may be any digit 11–13.
9' 40: or 50?
10' After *ar$_2$* **uš** a winkelhaken is visible, perhaps part of **igi**?
11' *ana* **nim**, 'to the east', this suggests prograde motion, which would seem to contradict the 'receding' (**la$_2$**).
After **u$_4$** a winkelhaken and several horizontal wedges are visible. At the end there is a vertical wedge, perhaps part of **la$_2$**.
14',15' Most instances of GAM$_0$ in these lines are transliterated as a separator (:), but their meaning is not clear.
16' **du**-*šu$_2$* = *alākšu$_2$*, 'its progression', a reading proposed by Steele (2005a), p. 42.

19' **šu$_2$**: as Neugebauer suggests, the scribe may have first written **igi** and corrected this to **šu$_2$**.
22a' 30: crammed.
25' Three horizontal wedges are visible, similar to the end of *i* or *tu*$_4$.

Commentary

Six fragments form the bulk of this large tablet. The left and right edges are partly preserved, but nothing remains of the original upper and lower edges. To the five fragments known to Neugebauer the small fragment BM 42816 is joined at the bottom of the reverse, adding some text to P16, P17 (column i) and P24, P25 (column ii). If the original tablet had size A then a total of about 7 cm may still be missing from the upper and lower edges. The tablet has several 'firing holes'. On both sides it is divided into three columns containing a total of at least 32 procedures for Jupiter. Most are visually separated by horizontal lines. The handwriting is slanted and occasionally crammed. The scribe sometimes continued a line in the space between two regular lines (e.g. Oi19', ii2'), suggesting that he was trying to retain the same line division as the original. A colophon is not preserved.

***P1** System A: various procedures*

In *ACT* this procedure was not translated. It consists of 3 subprocedures belonging to the sequence shown in Table 3.35 (p. 90). Part **a'** (Oi1'–5') concerns the subdivision of the synodic cycle (scheme S_2). Part **b'** (Oi5'–6') is a list of net displacements $dB_1(s)$ for zone 1, for intervals of s synodic cycles corresponding to $y = s + t$ years (§3.7.2.4, p. 91). Part **c'** (Oi6'–7') concerns Jupiter's rising time at FA and setting time at LA (§3.7.2.5, p. 91).

***P2** System A: updating B with σ; scheme S_1 for the subdivision of the synodic cycle*

Part **a** contains an incomplete algorithm for updating B with the synodic arc for system A (§3.7.2.2), formulated in accordance with template STEP.E (§2.6.10). Parts **b,c** belong to scheme S_1 for the subdivision of the synodic cycle (§3.7.2.6). They successively deal with zones 1 (**b**) and 2 (**c**). The pushes are formulated using the template PUSH.D (which provides only $\delta\Sigma$).

***P3** Distance to the ecliptic at the stations*

This brief procedure concerns Jupiter's distance to the ecliptic at the stations. As is the case on other tablets, it is embedded in a sequence of procedures for system A (cf. Table 3.35, p. 90). For a commentary cf. §3.7.10.2 (p. 105).

***P4** System A?: zigzag function v_2 for the daily displacement near LA and FA*

This procedure consists of 3 lines, of which the initial signs are broken away. For a reconstruction and discussion of this function, which probably represents Jupiter's daily displacement near LA and FA, and for a list of duplicates, cf. §3.7.2.6 (p. 92). It has been assumed that the coefficient follows the name of the sign up to Lib (Oi18'), whereas it precedes the name for other signs; this resolves the inconsistencies mentioned in *ACT*, p. 404.

***P5** Computations involving a trapezoid*

This badly preserved procedure contains computations involving a trapezoid. Similar procedures are No. **38** P4' and the newly discovered No. **40** P1', both also embedded in procedures for Jupiter. The dimensions of the trapezoid are (nearly) the same in these procedures. As in No. **38** P4', the long side is 12 and the surface 10.45, which implies that the short ('pinched') side is 9.30. There is too little text to restore the rest of the procedure. The absolute values of the numbers and the unit in which they are expressed are unknown. The astronomical significance and purpose of the procedure remain unclear.

***P6** Subdivision of the synodic cycle (unknown system; variant scheme)*

A few lines belonging to the end of P6 are partly preserved. They deal with the subdivision of the synodic cycle, presumably for Jupiter. It mentions a retrograde motion of 6;50° (2') which precedes S2. This value probably concerns the push from AR to S2 (compare with schemes S_1, S_2 of system A in Table 3.38a, p. 93). The motion from S2 to LA appears to be divided into 3 subintervals. For the first one no parameters are preserved; the second one lasts $30^{\bar{\tau}}$, during which the daily motion is 0;5°; during the third one the daily displacement is 0;10, but the duration is not mentioned.

***P7** System A": updating B with the synodic arc*

This procedure contains instructions for updating B with σ in accordance with system A" (Table 3.36). The formulation is based on template STEP.E (§2.6.10.1), i.e. the transition rules are not spelled out.

***P8** System A'''*

This composite procedure for system A''' (§3.7.6, p. 98) consists of two parts. Part **a** contains rules for updating B with the step function for σ in accordance with template STEP.E (§2.6.10). For the involved parameters cf. Table 3.36. Part **b**, which was not previously understood, concerns Jupiter's setting time at LA and its rising time at FA. One would expect the zodiacal regions associated with the setting and rising times to be those of system A''', but they are different; cf. also §3.7.6.2.

***P9a** Subdvision of the synodic cycle, schemes A'.S_2 and A'.S_1*

This procedure deals with the subdivision of the synodic cycle for zone 1 of system A'. Parts **a,b** (Oii8'–13') belong to scheme S_2, a variant of scheme S_1 (Table 3.44, p. 98). Part **a** is formulated in accordance with template PUSH.B, which provides only $\delta\tau$ and v. Part **b** provides the corresponding angular pushes $\delta\Sigma$ (PUSH.D). Parts **c,d** (Oii13'–18') also concern the subdivision of the synodic cycle for zone 1, but now the parameters agree with scheme S_1 (Table 3.44). Parts **c,d** were not translated in *ACT*, where one finds only a table with parameters. The scribe made an error when he put a stationary point after the first segment of the motion from S2 to LA (Oii15'); comparison with part **a** (Oii11') confirms this. P18 is another procedure on the same tablet which is concerned with the subdivision of the synodic cycle for scheme S_1, zone 1.

***P9b** System A?: motion between S1 and S2; setting time at LA, rising time at FA*

A horizontal line below Oii18', also visible in Pinches's copy, implies that P9b is separate from P9a (they were treated as one procedure, §9, in *ACT*). P9b consists of 2 lines, the ends of which are broken away. They were not translated in *ACT*. Collation has resulted in several improved readings. Part **a** (Oii19') mentions both stations (S1 and S2). In systems A and A' it takes $2{,}0^{\tau}$ for Jupiter to traverse the arc between S1 and S2, during which the Sun moves by about $2{,}0° = 4$ 'miles' (*bēru*), presumably the distance mentioned here. Since the rest is broken away, the purpose of this is not clear. Part **b** is concerned with Jupiter's setting time at LA and its rising time at FA. The preserved value of T_{set} differs from the one mentioned in P1.c'. For a discussion cf. §3.7.2.5 (p. 91).

***P10** System A: zones for the synodic arc; subdivision of the synodic cycle, scheme S_1?*

Only the beginning is preserved. Part **a** (Oii21'–22'), introduced as an alternative procedure ('Secondly'), mentions the zonal boundaries for σ in accordance with template STEP.F (§2.6.10). Part **b** (Oii22'–25') concerns the subdivision of the synodic cycle. As far as can be determined the formulation is based on template PUSH.B ($\delta\tau$ and v). Except for the daily displacement $v = 0;7,...$ mentioned in Oii24', the other parameters point to scheme S_1, zone 1 (Table 3.38a, p. 93).

***P11** Subdivision of the synodic cycle: scheme A.S_5; systems A_1 and A_2*

This composite procedure contains three different schemes for the subdivision of the synodic cycle. An unknown amount of text is missing from the beginning. Parts **a,b** (Oiii1'–7') belong to scheme S_5, which is provisionally assigned to system A because there appear to be 2 zones. For a discussion cf. §3.7.2.6 (p. 92). Both are formulated in accordance with the template PUSH.D. Part **a** deals with zone 2. The preserved text concerns the pushes from S1 to AR, AR to S2, S2 to LA and LA to FA. It can therefore be assumed that Oiii1' was preceded by at least one line for the missing push from FA to S1. Part **b** (Oiii3'-7') provides the same pushes for zone 1.

Part **c** (Oiii8'–12') contains an algorithm $\delta\Sigma$ (LA to FA) which includes transition rules. From this the step function for σ of system A_1 has been reconstructed. Part **d** (Oiii13'–16') contains a slightly different algorithm for $\delta\Sigma$ (LA to FA), from which the step function for σ of system A_2 has been reconstructed. For a discussion cf. §3.7.3 (p. 95). Part **e** (Oiii16'–20') concerns the push from S1 to AR for system A_2; part **f** the push from AR to S2. It can be assumed that the procedure continued below Oiii24' with the push from S2 to LA, thus completing a full cycle. This would require a few additional lines of text.

***P12** System B: computations involving τ and Π; parameters of the zigzag function for τ*

The beginning of this procedure is missing. The 5 partly preserved lines contain 2 subprocedures, both concerned with Jupiter system B (§3.7.8). Part **a'** (Ri2'–3') contains computations involving $\Pi = 6{,}31$ and the mean value of τ ($\mu = 45;14^{\bar{\tau}}$). The meaning is unclear; for a discussion cf. §3.7.8.3 (p. 101). The duplicate No. **46** P1'.a implies that the unidentified number 13.20.27.46 must be corrected to 13.30.27.46. Part **b'** (R4'–5') mentions the defining parameters of the zigzag function for τ (Table 3.49) in accordance with template ZZ.A.3 (§2.6.8.1), except that the unit ('days' = mean tithis) is written after each extremum.

***P13** System B: computing τ from σ*

This procedure for system B deals with the computation of the synodic time (τ) from the synodic arc (σ), in general agreement with Eq. (3.12) on p. 63. In Ri6'–7' the constant c_τ (Eq. 3.13) is derived. To that purpose the mean synodic arc, $\mu = 33;8{,}45°$, is multiplied by $0;0{,}10 \cdot e$, which is here approximated by the usual value 0;1,50,40 (Table 3.5), resulting in $1;1{,}8{,}8{,}20^{\bar{\tau}}$. This is added to $e = 11;4^{\bar{\tau}}$, resulting in $c_\tau = 12;5{,}8{,}8{,}20^{\bar{\tau}}$, which is called the 'value for the year' (Ri7'). Ri8' contains a general instruction for computing τ from σ equivalent to Eq. (3.9) on p. 61.

***P14–P16** System A': computing τ_j from σ_j for each zone j*

The introduction of P14 implies that it continues with the subject of P13, namely to compute τ from σ. What follows turns out to be the first of a set of 3 procedures, P14–P16, concerned with the computation of τ for each zone of system A'. Each of these procedures begins with an instruction for updating B in accordance with template STEP.E. Subsequently τ is computed from σ, but the algorithm differs from the usual one. For a discussion cf. §3.7.4.2 (p. 96). The involved parameters (Table 3.43) also differ slightly from the usual ones, since $e = 11;3{,}20^{\bar{\tau}}$ instead of the usual $11;3^{\bar{\tau}}$, and 0;0,10 $\cdot e$ = 0;1,50,33,20 is approximated by 0;1,50 (compare Table 3.5). This value is used in all 3 procedures P14–P16.

***P14** System A': computing τ_1 from σ_1*

P14 deals with zone 1, but the scribe appears to commit an error in the computation of $0;1{,}50 \cdot \sigma_1 = 0;55$, which he replaces by 0;45. This leads to $\sigma_1 + 0;1{,}50 \cdot \sigma_1 = 30;45$ instead of the expected 30;55. This error propagates into the result for $\tau_1 - 12^{\overline{m}}$, which is given as $41;48{,}20^{\bar{\tau}}$ instead of the expected $41;58{,}20^{\bar{\tau}}$.[619]

***P15** System A': computing τ_2 from σ_2*

P15 begins with an instruction for updating B in zone 2 (and 4) in accordance with template STEP.E. Subsequently τ_2 is computed from σ_2.

***P16** System A': computing τ_3 from σ_3*

The newly joined fragment BM 42816 has added some text to P15 (Ri20'). The procedure begins with an instruction for updating B in zone 3 in accordance with template STEP.E (Ri16'). The next line (Ri17') is not understood, and the numbers 5 and 21 remain unidentified. Subsequently (Ri18'–20') τ_3 is computed from σ_3, but unlike P14 and P15 this is done in two steps, such that $0;0{,}10 \cdot e = 0;1{,}50$ is split up as $0;2 - 0;0{,}10$, after which the multiplication by $0;0{,}10 \cdot e$ is done separately for each contribution, and the products are subtracted (Ri19').

***P17** System A': zones for σ*

In this brief procedure consisting of 2 lines, preserved mainly on the newly joined fragment BM 42816, the 4 zones of system A' are defined in accordance with template STEP.F, as in P25. For the parameters cf. Table 3.36 (p. 90). It is striking that Ri21' contains two trivial errors (cf. the critical notes).

***P18** Subdivision of the synodic cycle, scheme A'.S_1, zone 1*

The 2 partly preserved initial lines of this procedure agree with scheme A'.S_1 for the subdivision of the synodic cycle in zone 1 (§3.7.4.3, p. 97). The formulation is based on template PUSH.B (which provides only $\delta\tau$ and v). Since the same scheme was dealt with in P9a.a–b, the missing remainder of P18 may well deviate from scheme S_1.

---

619 Although Neugebauer does remark that 30;45 is an error for 30;55 (*ACT*, p. 412), he ignores that consequently also $\tau_1 - 12^{\overline{m}} = 41;48{,}20^{\bar{\tau}}$ is erroneous. Also in *HAMA* (p. 445) this number is quoted as if correct.

P19 *Unclear*

Too little remains of this procedure to establish its meaning, beyond the fact that it deals with Jupiter's (retrograde) motion. The beginning is missing, and only isolated signs are legible in the remaining 5 lines.

P20 *System A: period relations and net differences for zone 2*

This damaged procedure has a close parallel in No. **46** P10', which made it possible to restore some of the missing text. It provides net displacements of Jupiter in zone 2 of system A for several time intervals (cf. Table 3.37, p. 91). The orthography is unusual in that each time interval is preceded by **a.ra**$_2$, which is here a logogram for the preposition *ana*, 'for'.

P21 *Systems B,B': period relations; parameters of the zigzag functions for* τ

P21 provides period relations and defining parameters of the zigzag functions for τ for systems B and B' (§3.7.8). Part **a** contains the period relations for system B and Jupiter's total displacement after $Z = 36$ revolutions, i.e. $36 \cdot 6{,}0° = 3{,}36{,}0°$. This is followed by the mean value (μ) of the zigzag function for σ: 33;8,45°. Parts **b** and **c** provide the defining parameters of the zigzag function for τ (both with a trivial error) for systems B (part **b**) and B' (part **c**), including the approximate zodiacal positions of the extrema, in accordance with template ZZ.A.3 (§2.6.8.1). For the parameters cf. Table 3.49 (p. 100) and Table 3.51 (p. 101).

P22 *Systems B,B': defining parameters of the zigzag functions for* σ

P22 provides the defining parameters of the zigzag functions for σ for systems B (part **a**) and B' (part **b**). They are formulated in accordance with template ZZ.A.3.

P23 *Subdivision of the synodic cycle, scheme A.S*$_3$

This procedure concerns scheme A.S$_3$ for the subdivision of the synodic cycle. For a discussion cf. §3.7.2.6; Table 3.38b. Part **a** lists subsequent angular pushes for zone 1, starting and ending with LA, using template PUSH.D (which provides only $\delta\Sigma$). This is, oddly, followed in part **b** by rules for the corresponding temporal pushes $\delta\tau_2$. Exactly the same values of $\delta\tau_2$ are mentioned in P30, while P31 provides the corresponding values of $\delta\Sigma_2$. Hence P23, P30 and P31 all belong to scheme A.S$_3$. Presumably because the angular pushes do not add up to σ_j in each zone (30° and 36°), Neugebauer considered P23 to be erroneous.[620]

P24 *Subdivision of the synodic cycle, scheme A.S*$_4$*, zone 2*

P24 is concerned with scheme S$_4$ for the subdivision of the synodic cycle, perhaps pertaining to zone 2 of system A. For a discussion cf. §3.7.2.6 and Table 3.38b. The formulation is based on PUSH.D, which provides only $\delta\Sigma$.

P25 *System A': zones for the synodic arc*

Only the first 3 lines of P25 are partly preserved. In Rii22'–23' the zodiacal zones of system A' are defined (Table 3.36) in accordance with template STEP.F, as in P17 (§2.6.10.1). Since it is unlikely that the same procedure is written twice on the tablet, P25 may have continued beyond Rii23' with another topic, perhaps the subdivision of the synodic cycle.

P26 *Subdivision of the synodic cycle*

The beginning of this procedure is missing. The remaining 7 badly preserved lines imply that it is concerned with the subdivision of the synodic cycle, but little more can be said.

P27 *List of control values of a zigzag function* v_3

P27 contains a list of function values and corresponding zodiacal signs. For a reconstruction and possible interpretation cf. §3.7.2.6 and Table 3.39 (p. 95).

[620] *HAMA*, p. 448, n. 6.

P28 *Subdivision of the synodic cycle; net displacements*

This badly understood procedure seems to be concerned with the subdivision of the synodic cycle and net displacements of Jupiter. Not all intervals of the synodic cycle are preserved, and a coherent interpretation remains to be found. The interval of $4^{\overline{m}}\ 4^{\overline{\tau}}$ mentioned after the small gap at the beginning is $\delta\tau$ (FA to S1); the same number also appears in P30 (cf. Table 3.38b). The subsequent push from S1 to AR is said to last $2^{\overline{m}}$. This agrees with most other schemes (in P30: 58 'days'). No other intervals appear to be mentioned. A retrograde motion is mentioned (from S1 to AR?), but this seems to conflict with the eastward direction also mentioned there (Riii12'). The meaning of the numbers 12 (Riii11') and 36 (Riii12') remains unclear. In Riii12' a number 37.30 has to be subtracted 'day by day'. An interpretation as v on the retrograde arc (S1 to S2) seems unlikely (one expects something of the order $-0;4°/\overline{\tau}$). If interpreted as $-0;0,37,30$ this might represent the daily decrease of v (on approaching a station?), in which case P28 is one of very few known procedures containing a linear interpolation scheme for v (see §3.3.3.4, p. 67).

Riii13' mentions that Jupiter's net displacement for 12^{yr} is 5° in the eastward (i.e. prograde) direction, which coincides with the value in zone 2 of system A or zone 3 of system A' (Table 3.37, p. 91). This is followed in Riii14'–15' by a list of numbers, ordered from large to small, whose meaning is not clear. Except two (49 and 46) all are multiples of 12, a basic period of Jupiter: the first four numbers in Riii14' are $9 \cdot 12, 8 \cdot 12, 7 \cdot 12, 6 \cdot 12$. It may be significant that they differ by 1 from the nearest intervals for a close return of Jupiter: 1,47 yrs, 1,35 yrs, 1,23 yrs and 1,11 yrs.

P29 *Unknown system: subdivision of the synodic cycle, $\delta\Sigma$ for LA to FA?*

This procedure, a 'second opinion' according to the introductory phrase, contains a list of 12 numbers forming a zigzag function which probably represents the angular push ($\delta\Sigma$) from LA to FA; cf. §3.7.10.1.

P30 *Subdivision of the synodic cycle, scheme A.S_3, zone 2: $\delta\tau$*

This procedure belongs to scheme A.S_3 for the subdivision of the synodic time in zone 2 (§3.7.2.6; Table 3.38b). The formulation is based on template PUSH.E, which provides only $\delta\tau$.

P31 *Subdivision of the synodic cycle, scheme A.S_3, zone 2: $\delta\Sigma$*

This procedure also concerns scheme A.S_3 for the subdivision of the synodic cycle. Angular pushes are provided in accordance with the template PUSH.D.

P32 *Unclear*

The few traces of this procedure are insufficient to establish its topic. The previous procedure and the multiplication in Riii23' suggest a connection with the subdivision of the synodic cycle ($\delta\Sigma = v \cdot \delta\tau$?).

No. 19

museum number	BM 35943 (Sp3,479)
provenience	Babylon
measures	3.9(8?) × 2.2(*) × 2.2 cm
arrangement	O/R
cuneiform text	photo: Fig. F.16, *ACT*, Pl. 253a; copy: *LBAT* 149
bibliography	*ACT* 813a
contents	Jupiter: system A, other (?): various procedures

Obverse

P1 1 [**igi** *ana* **igi ta** 25 **maš.maš**] **en** 30 **gir$_2$.tab** 30 **tab**
2 [**ta** 30 **gir$_2$.tab en** 2]⸢5⸣ **maš.maš** 36 **tab**

P2 3 [xxxxxxx] ⸢x **en**⸣ 30 **me** *ša$_2$* **me** 12 **zi**-⸢*šu$_2$*⸣
4 [xxxxxxx *ina*] ⸢30⸣ **me** 6 **ki du** (erasure) [x$^?$]

P3 5 [xxxxxxxxxxx] ⸢xx **du** x⸣ [xxx]
(m lines missing)

Reverse

P4' (about m lines missing)
1' [xxxxxxxxx] ⸢x **du**⸣-*ma* 3 ⸢*ina*$^?$⸣ [xxxx]

P5' 2' [xxxxxx *ina* 3]0 **me** 5 **ki** *ana* **igi**-*šu$_2$* **d**[**u** xxx]
3' [xxxx 30 GAM 12.30 **d**]**u**-*ma* 6.15 *ina* 30 ⸢**me** 6⸣.[15 **ki du**]

Lower edge

4' [xxxxxx 30 GA]M 15 **du**-*ma* 7.30 *ina* 30 **me**
5' [7.30 **ki du** xxxx] ⸢x⸣ PAP.PAP ⸢*ar$_2$* **uš**⸣ *ar$_2$*-*tu$_2$* 14 **ki**$^?$
6' [xxxxxxxxx] ⸢**zi**⸣ 30 GAM 15 **du**-*ma* 7.30
7' [xxxxxxxxxxxx] ⸢*ina* **nim**$^?$ xxx⸣

Obverse

P1 **(System A: updating *B* with the synodic arc)**

[1][Appearance to appearance. From 25 Gem] until 30 Sco you add 30. [2][From 30 Sco until 2]⸢5⸣ Gem you add 36.

P2 **(Subdivision synodic cycle, scheme A.S$_1$?)**

[3][...] ⸢... until⸣ 30 'days' its displacement is 0;12 per 'day'. [4][... In] 30 'days' the position proceeds 6, [(then) it appears (FA).]

P3 **(Unclear)**

[5][...] ⸢...⸣ [...]
(m lines missing)

Reverse

P4' **(Subdivision of the synodic cycle, S1 to AR?)**

(about m lines missing)
[1'][...] ⸢you multiply⸣ [by ...] it is 3. ... [...]

P5' **(Subdivision of the syn. cycle, variant of scheme A.S$_1$?)**

[2'][... in 3]0 'days' the position proceeds forward by 5. [...] [3'][... you] multiply [30 by 0;12,30], it is 6;15. In 30 'days' [it proceeds] ⸢6⸣;[15]

Lower edge

[4'][...] you multiply [30 by] 0;15, it is 7;30. In 30 'days' [5'][the position proceeds 7;30 ...] (In) total after the second station the position [proceeds] 14. [6'][...] you multiply by 30, it is 7;30 [7'][...] ...

Critical and philological notes

Obverse

3 **igi**: one expects something like **ta šu$_2$ en igi**, 'from LA to FA' (replaces *ar$_2$* **igi**, 'after FA', *ACT*, p. 420).
4 30: traces of two winkelhakens are visible.

Lower edge

5' PAP.PAP = 'total', perhaps to be read *napḫaru*. It is preceded by a damaged sign, perhaps **uš** or **du**.
7' **nim**: perhaps 'east'?

Commentary

This small fragment belongs to the upper (lower) right edge of the obverse (reverse) of a tablet. The left edge of the tablet is not preserved, but the textual restorations (P1) imply that the originial width (or that of one column) was about 8 cm. The procedures are separated by horizontal dividing lines. The first two (P1,P2) and the last one (P5') certainly deal with Jupiter; for the others this is very likely as well. Since the tablet ends with a procedure (P5'), it appears that it lacks a colophon.

P1 *System A: updating B with the synodic arc*

This procedure contains the usual algorithm for updating *B* with the step function for σ for Jupiter system A. It is formulated in accordance with template STEP.E (§2.6.10). For the algorithm cf. §3.7.2.2.

P2 *System A: subdivision of the synodic cycle, scheme S_1: LA to FA, zone 1*

This procedure consisting of 2 lines concerns the subdivision of the synodic cycle. The daily motion of 0;12° and the duration of 30 'days' (O3) are consistent with scheme A.S_1 for the push from LA to FA in zone 1 (Table 3.38a, p. 93). The formulation is based on templates PUSH.B.3 and PUSH.C.1 (§3.3.3.2).

P3 *Unclear*

Virtually nothing remains of this procedure, which probably also concerns the subdivision of the synodic cycle.

P4' *Subdivision of the synodic cycle, S1 to AR?*

Only the final line remains of this procedure. The multiplication resulting in 3 might be the computation of $\delta\Sigma$ for the interval S1 to AR in zone 1, where Jupiter moves in the retrograde direction by 0;6° per mean tithi for $1^{\overline{m}}$ according to scheme S_1 of system A (Table 3.38a).

P5' *Subdivision of the synodic cycle, variant of scheme A.S_1*

The first 6 lines of P5' are partly preserved. They are concerned with the subdivision of the synodic cycle. Again the algorithms are similar to scheme A.S_1 (Table 3.38a), except for R2', which mentions a forward motion of 5° between LA and FA, where one expects 6°. R3' mentions $\delta\Sigma = 6;15°$ and the corresponding $\delta\tau = 30^{\overline{\tau}}$, probably pertaining to FA to S1 (1), zone 1. R4' continues with the same interval for zone 2 ($\delta\Sigma = 7;30°$). As is apparent from the end of R5', the motion after S2 is addressed here. The 14° may represent $\delta\Sigma$, but it does not match any interval known from schemes A.S_1–S_5. LE6' ($\delta\tau = 30^{\overline{\tau}}$ and $v = 0;15$) seems to repeat the interval FA to S1 (1) already mentioned in LE4'

No. 20

museum number	BM 55620 (82–7–4,220)
provenience	Babylon
date	data in synodic table: SE 202–210
measures	5.4(25?) × 3.9(*) × 2.2–2.5 cm (type iv?)
cuneiform text	photo: Fig. F.21 (R); *ACT*, Pl. 244a (R)
bibliography	*ACT* 823a, 613aa, Zna
contents	Jupiter system A': synodic table with *T*, *B* (FA, S1); various procedures

Reverse

P1 1 [...xxxx] ⸢**u**$_4^?$ **u**$_4$⸣ *ša*$_2$ ⸢x⸣ *ša*$_2$ ⸢**maš**$_2^?$⸣ [...]
2 [...xxxx].⸢**a**⸣.**meš** AN ⸢x **dur**$^?$⸣ [...]
3 [... **ta** 9 **gir**$_2$**.tab en**] ⸢2⸣ **maš**$_2$ 33.⸢45 **tab**⸣ [*ša*$_2$ *al-la* 2 **maš**$_2$ **diri** GAM 1.4 **du ki** 2 **maš**$_2$ **tab**-*ma* **gar**]
4 [... *ša*$_2$ *al* 17 **mu**]**l**$_2$ **diri** GAM ⸢56.15 **du**⸣ [...]
5 [...xxx] ⸢xxx⸣ [...]
(unknown number of lines missing)

Reverse

P1 **(System A': various procedures)**

a **(Unclear)**

[1][...] ... for ⸢...⸣, for Cap [...] [2][...] ... [...]

b **(Updating *B* with the synodic arc)**

[3][... From 9 Sco until] ⸢2⸣ Cap you add 33;⸢45. (The amount) by which it⸣ [exceeds 2 Cap you multiply by 1;4 ...] [4][... (The amount) by which it] exceeds [17 Tau] you multiply by ⸢0;56,15⸣, [add to ...] [5][... C]ap ... [...]

Critical and philological notes

Reverse

1' The first **u**$_4$ is uncertain. The sign following **u**$_4$ *ša*$_2$ looks a bit like 50.

2' **a.meš**: perhaps part of **igi.du**$_8$**.a.meš** = *tāmarāti*, 'appearances' (in the sense 'synodic phenomena' or specifically FA)?
AN: or read GAM 1?
dur: or **rin**$_2$ **u**$_4$? Correct reading and meaning uncertain.

Commentary

This small fragment does not preserve any edge of the original tablet. The obverse is occupied by a synodic table for Jupiter system A', which probably continued on the reverse. Nothing can be read in the first partly preserved line on the reverse, but the empty space below it suggests that it is the last line of the synodic table. Only columns *B* for FA and *T* for S1 are partly preserved for the years SE 202–209. Probably not much clay is missing from the top of the obverse, so that the initial year of the table, which gives an indication of the date of the tablet, cannot be much earlier than SE 202 (110/109 BC). The first preserved column must have been preceded by a column *T* for FA, which implies that about 2 cm are missing from the left side. Beyond column *T* for S1 there must have been a column for *B*, and probably columns *T* and *B* for AR, S2 and LA. This suggests that about 20 cm of clay may be missing on the right side. On the reverse the synodic table is followed by a procedure.

P1 System A': various procedures

The first 5 lines of P1' are partly preserved. P1' is tentatively divided into two parts. Too little remains of part **a** (R1'–2') for an interpretation. Part **b** (R3'–4') is a procedure for updating *B* with the synodic arc of system A' (Table 3.42, p. 96). Preserved traces of the transition rules imply that it is formulated in accordance with template STEP.A.1 (§2.6.10). Since not much clay is missing from the left edge there may not be sufficient space in R2' for this procedure to begin with zone 1 in accordance with the usual arrangement. As remarked by Neugebauer, the procedure may begin with zone 2, as in No. **23** P2'.

No. 21

museum number	BM 36801 (80–6–17,539)
provenience	Babylon
measures	3.9(6?) × 6.5(*) × 2.0 cm
arrangement	O/R
cuneiform text	photo: Fig. F.22; *ACT*, Pl. 253a
bibliography	*ACT* 813b, Zt; Huber (1957), pp. 287–288
contents	Jupiter systems A, A''', unknown system: various procedures

Upper edge

1 *ina a-mat* ^{d}**en** ^{d}**gašan**-*ia*$_2$ ^{d}BE ^{d}**ak** *u* ^{d}**ni**[**saba**$^?$ ... *liš-lim*]

Obverse

P1 1 **mul$_2$.babbar me igi** 12 **zi**-*šu*$_2$ ⸢**en**⸣ [1 **me** 10.45 10.45 **a.ra$_2$**]

2 1 **me du**-*ma* 10.45 **ta** 1 **me** ⸢**gi**⸣ [**en** 1 **me**]

3 5.30 5.30 **a.ra$_2$** 1 **me du**-*m*[*a* 5.30]

4 **ta uš igi** 30 **ninda zi**-*šu*$_2$ ⸢x⸣ [xxxx 4 **a.ra$_2$**]

5 1 **me du**-*ma* 4 *ina* 1 **me** ⸢4$^?$⸣ [xxxxxxxx]

6 **ta** 30 **me gi en** ⸢30⸣ [**me** 4.30 4.30 **a.ra$_2$**]

7 30 **du**-*ma* 2.15 ⸢3⸣.[45 *u*$_3$ 2.15 GAR.GAR-*ma* 6 6]

8 *ana ar*$_2$-*šu*$_2$ **la$_2$**-*ma* **uš ta** [**uš** *ar*$_2$ 1 **me** 5]

9 5 **a.ra$_2$** 1 **me du**-*ma* [5 xxxxxx]

10 **du**-*ma* 11.15 **a.ra$_2$** [1.8 **du**-*ma* 12.45 5 *u*$_3$ 12.45 GAR.GAR-*ma* 17.45]

11 17.45 **ki du š**[**u$_2$** xxxxxxxx]

12 *ina lib*$_3$-*bi* 10 **ki** ⸢*ana*$^?$⸣ [*ar*$_2$-*šu*$_2$ **la$_2$** xxxxx]

13 ⸢xxxxx *u*$_3$⸣ [xxxxxxxxxxx]

(m lines missing)

Reverse

Co (about m−1 lines missing)

1' ⸢*a-na* **mul$_2$**⸣.[**babbar**$^?$ xxxx]

2' **dumu** ^{lu2}*mu*-⸢*še*⸣-[*zib* xxx]

P2' 3' **ta** 25 **maš en** [30 **gir$_2$** 30 **tab**]

4' **ta** 30 **gir$_2$ en** [25 **maš** 36 **tab**]

P3' 5' **ta** 27 **pa en** 1[7 **mul$_2$** 36 **tab**]

6' **ta** 17 **mul$_2$ en** 9 [**alla** 33.45 **tab**]

7' **ta** 9 **alla en** 9 [**gir$_2$** 30 **tab**]

8' **ta** 9 **gir$_2$ en** 2[7 **pa** 33.45 **tab**]

9' *ina* **ki** *ša*$_2$ 36 ⸢**šu$_2$**⸣ [xxxxxx]

10' *ina* **ki** *ša*$_2$ 33.4⸢5⸣ [xxxxxx]

11' *ina* **ki** *ša*$_2$ 30 [xxxxxxxx]

Lower edge

1 (blank) 2 ⸢**ki** x⸣ [xxxxxxx]

2 (blank) *al-la* ⸢**ki**⸣ [xxxxxxx]

Upper edge

[1]At the command of Bēl, Bēltija, Ea, Nabû and Ni[saba$^?$ may it remain intact.]

Obverse

P1 **(Subdivision of the synodic cycle, scheme X.S$_1$)**

[1]Jupiter. (On) the day when it appears (FA) its displacement is 0;12. Until [1,0 days 0;10,45.] [2]You multiply [0;10,45 by] 1,0 'days', it is 10;45. After completing 1,0 'days' until [1,0 'days'] [3]0;5,30. You multiply 0;5,30 by 1,0 days, it is [5;30.] [4]After the first station (S1) its displacement is 0;30^{nin}, ⸢...⸣ [...] [5]You multiply [0;4] by 1,0 'days', it is 4. In 1,0 'days' [it moves] 4. [6]After completing 30 'days' until ⸢30⸣ ['days' 0;4,30.] [7]You multiply [0;4,30] by 30, it is 2;15. ⸢3⸣;[45 and 2;15 you accumulate, it is 6. By 6] [8]it recedes backwards, then it becomes stationary (S2). From [the second station 1,0 days 0;5]. [9]You multiply 0;5 by 1,0 'days', [it is 5...] [10]you multiply by [...], it is 0;11,15, you multiply it by [1,8, it is 12;45. From the second station] [11]the position proceeds 17;45, (then) it sets (LA) [...]

[12]From 10, the position [recedes backwards ...] [13]⸢...⸣ [...]

(m lines missing)

Reverse

Co **(Colophon)**

(about m−1 lines missing)

[1']⸢For Jup⸣[iter ...] [2']descendant of Mušē[zib ...]

P2' **(System A: updating B with σ)**

[3']From 25 Gem until [30 Sco you add 30.]

[4']From 30 Sco until [25 Gem you add 36.]

P3' **(System A''')**

a **(Updating B with σ)**

[5']From 27 Sgr until 1[7 Tau you add 36.]

[6']From 17 Tau until 9 [Cnc you add 33;45.]

[7']From 9 Cnc until 9 [Sco you add 30.]

[8']From 9 Sco until 2[7 Sgr you add 33;45.]

b **(Setting time at LA and rising time at FA?)**

[9']In the region of 36: ⸢setting⸣ [...] [10']In the region of 33;4[5 ...] [11']In the region of 30 [...]

Lower edge

[1]2 ⸢...⸣ [...] [2]than ⸢...⸣ [...]

Critical and philological notes

Obverse

2 **gi** = *ušallim*, 3 c. sg. pret. D *šalāmu*, 'it completed'. Only the beginning of the horizontal wedge of **gi** is visible; the parallels (e.g. O6) suggest this reading.
4 After **zi**-*šu*$_2$ the beginning of a sign similar to GUR is visible.
6 **ta** 30 **me** = 'from 30 days': probably an error for **ta** 1 **me** = 'from 1,0 days' (the duplicate No. **22** O7 has the same error).
7 Thus restored by Huber (1957), p. 288.
12 After **ki** a vertical wedge is visible.
13 The upper halves of four wedges consistent with 11.11 are visible before *u*$_3$.

Commentary

This fragment belongs to the upper (lower) left corner of a tablet. The right edge and the lower (upper) edge of the obverse (reverse) are broken off. The textual restorations imply that the original width of the tablet (or column) was about 6 cm. The original height is unknown, but probably at most a few centimeters of clay are missing. There are at least 3 procedures, which are separated by horizontal lines. All are concerned with Jupiter: P2 belongs to system A, P3 to A'''. The colophon mentions an individual from the well-known Mušēzib clan. Unusually, the colophon is followed by two more procedures. The cuneiform signs are carefully written in what looks like an early Seleucid hand (compare e.g. No. **53**). Also the terminology, especially in P1, is suggestive of a date in the early Seleucid period. On the upper edge there is an invocation, which is more elaborate than usual.

***P1** Subdivision of the synodic cycle, scheme X.S$_1$*

This difficult procedure is concerned with scheme X.S$_1$ for the subdivision of the synodic cycle for Jupiter. The first 13 lines of P1 are partly preserved; the end is missing. For a score edition of the duplicates and a discussion of scheme X.S$_1$ cf. §3.7.10.1.

***P2'** System A: updating B with the synodic arc*

P2' consists of 2 lines with an algorithm for updating *B* with the synodic arc for system A (§3.7.2.2). It is formulated in accordance with template STEP.E, which lacks the transition rules. For a general discussion of this algorithm cf. §2.6.10.

***P3'** System A$'''$: updating B with the synodic arc; setting time at LA, rising time at FA?*

This final procedure contains either 7 or 9 lines, depending on whether the lower edge is considered to be part of it. Part **a** contains rules for updating *B* with the step function for the synodic arc for system A'''(§3.7.6). It is formulated in accordance with template STEP.E, which lacks transition rules. I tentatively interpret what remains of part **b** as a procedure for the setting time at LA and the rising time at FA (§3.7.6.2). However, the time intervals themselves are not preserved.

No. 22

museum number	BM 41043 (81–4–28,590)
provenience	Babylon
measures	3.2(8?) × 7.0(*) × 2.0–2.4 cm
arrangement	O/R
cuneiform text	photo: Fig. F.23
contents	Jupiter, unknown system, system A: various procedures

		Upper edge
	0	*ina a-mat* ^{d}**en** ^{d}**gašan**-*ia*$_2$ ^{d}BE [^{d}**ak** xxxxx *liš-lim*]
		Obverse
P1	1	**mul$_2$.babbar me igi** 12 ⸢**zi**⸣-[*šu*$_2$ **en** 1 **me** 10.45 10.45 **a.ra**$_2$]
	2	1 **me du**-*ma* 10.4⸢5⸣ [**ta** 1 **me gi en** 1 **me** 5.30 5.30]
	3	**a.ra**$_2$ 1 **me du**-*ma* ⸢5.30⸣ [xxxxxxxxxx]
	4	**ta uš igi** 30 **ninda zi**-*šu*$_2$ [30 **a.ra**$_2$ 30 **du**-*ma* 15 ... 4 **a.ra**$_2$]
	5	1 **me du**-*ma* 4 *ina* 1 **me** 4 [xxxxxxxxx]
	6	**ta** 30 **me gi en** 30 **me** [4.30 4.30 **a.ra**$_2$ 30 **du**-*ma* 2.15]
	7	3.45 *u*$_3$ 2.15 GAR.GAR-⸢*ma*⸣ [6 6 xxxx 30 **me**$^?$]
	8	**zi nu tuk en** 1 **me** 10 ⸢**ninda**⸣ [**zi**-*šu*$_2$ xxxxxxx **ta**]
	9	1 **me gi en** 1.8 **me** 1⸢2⸣ [xxxxxxxxxxx]
	10	**ki du šu**$_2$ **ta me šu**$_2$ [xxxxxxxxxxx]
	11	10 **ki** *ana ar*$_2$-*šu*$_2$ **la**$_2$ ⸢x⸣ [xxxxxxxxxxx]
	12	*lu-maš gab-bi* [xxxxxxxxxxxxxx]
	13	GAR.GAR-*ma* 22.⸢30⸣ [xxxxxxxxxxxxx]
	14	**du$_3$.du$_3$ ta** ⸢**uš**⸣ [xxxxxxxxxxxxx]
	15	*ša*$_2$ 30 : 12 [xxxxxxxxxxxxx]
P2	16	⸢**igi** *ša*$_2^?$ x⸣ [xxxxxxxxxxxxxxxxxxx]
		(m lines missing)
		Reverse
P3'		(about m lines missing)
	1'	⸢x⸣ [xxxxxxxxxxxxxxxxxxx]
	2'	**an** ⸢x⸣ [xxxxxxxxxxxxxxxxxxx]
	3'	**igi**-*tu*$_2$ ⸢**tab**⸣ [xxxxxxxxxxxxxxxx]
	4'	*šu*$_2$-*u*$_2$ [xxxxxxxxxxxxxxxxx]
P4'	5'	⸢x⸣ [xxxxxxxxxxxxxxxxx]
	6'	⸢x⸣ [xxxxxxxxxxxxxxxxx]
	7'	10 ⸢x⸣ [xxxxxxxxxxxxxxxx]
	8'	10 ⸢x⸣ [xxxxxxxxxxxxxxxx]
	9'	10 ⸢x⸣ [xxxxxxxxxxxxxxxx]

Critical and philological notes

Obverse

2 **gi** = *ušallim*, 3 c. sg. pret. D *šalāmu*, 'it completed'.

9 12: one expects 11 but this is excluded.

Reverse

1' Traces of a vertical wedge.

5' The beginning of a horizontal wedge is visible.

6' The beginning of a vertical wedge is visible.

Upper edge

[0]At the command of Bēl, Bēltija, Ea, [Nabû and ... may it remain intact.]

Obverse

P1.a (Scheme X.S$_1$ for the subdivision of the synodic cycle)

[1]Jupiter. (On) the day when it appears (FA) its displacement is 0;12. [Until 1,0 'days' 0;10,45?.]. You multiply [0;10,45 by] [2]1,0 'days', it is 10;45. [After completing 1,0 'days' until 1,0 'days' 0;5,30.] [3]You multiply [0;5,30] by 1,0, it is 5;30. [..] [4]After the first station (S1) its displacement is 0;30^{nin}. [You multiply 0;0,30 by 30, it is 0;15. ...] [5]You multiply [0;4] by 1,0 'days', it is 4. In 1,0 'days' [it recedes backwards...] 4. [6]After completing 30 days until 30 'days' [0;4,30. You multiply 0;4,30 by 30, it is 2;15.] [7]3;45 and 2;15 you accumulate, it is [6. 6 ... 30 'days'] [8]there is no displacement (S2). Until 1,0 'days' 10^{nin} [...] After [9]completing 1,0 'days' until 1,8 'days' 0;12? [... You multiply ... by 1,8, it is 12;45. 5 and 12;45 you accumulate, it is 17;45.] [10]The position proceeds 17;45. After the day when it sets (LA) [...] [11]it recedes backwards 10 ... [...] [12]all zodiacal signs [...] [13]you accumulate, it is 22.⸢30⸣ [...] [14]the total from the ⸢station⸣ [until ...]

P1.b (System A: rising time at FA, setting time at LA?)

[... In the 'ward'] [15]of 30: 12 [... if it appears (FA), ...]

P2 (Unclear)

[16]⸢Appearance ...⸣ [...]

(*m lines missing*)

Reverse

P3' (Unclear)

(*about m lines missing*)

[1']⸢...⸣ [...] [2']... ⸢...⸣ [...] [3']first ⸢...⸣ [...] [4']this [...]

P4' (Unclear)

[5']⸢...⸣ [...] [6']⸢...⸣ [...] [7']10 ⸢...⸣ [...] [8']10 ⸢...⸣ [...] [9']10 ⸢...⸣ [...]

Commentary

This neatly written fragment preserves a part of the left edge and the upper edge of the obverse of a tablet. The textual restorations in P1 imply that the original width was about 8 cm. Four procedures are partly preserved, but only P1 can be partly understood.

P1 Subdivision of the synodic cycle, scheme X.S$_1$; rising time at FA and setting time at LA?

All 15 lines of P1 are at least partly preserved. Below O11 they contain only few signs, sufficient to conclude with a certain probability that O15 and the missing end of O14 form a separate subprocedure concerned with Jupiter's rising time at FA and its setting time at LA. For that procedure, which is probably associated with system A, cf. §3.7.2.5. The preceding lines are concerned with scheme X.S$_1$ for the subdivision of the synodic cycle. For this difficult scheme cf. §3.7.10.1, which contains a score of all duplicates. The formulation is based on the rare template PUSH.A.2 (§3.3.3.2). An essential step towards a better understanding of this scheme was made by P. Huber (1957), who recognised that 1 **me**, which occurs repeatedly in the text, stands for $1,0^{\bar{\text{T}}}$. O12–14 are too fragmentary for an interpretation. It is not certain that they belong to P1.a. The number 22.30 (O13) remains unidentified.

P2–P4' Unclear

Too little remains of the other procedures for an interpretation. Only the initial 1–2 signs or traces of them are preserved in P4'. They are indented by about 1.5 cm with respect to P3'. This may suggest that P4' is not a procedure but a table or a colophon. Traces of vertical lines support the former.

No. 23

museum number	DT 183
provenience	Babylon
measures	6.5(13) × 6.0(*) × 2.7 cm (type C or F?)
arrangement	O/R
cuneiform text	photo: Fig. F.24; *ACT*, Pl. 244
bibliography	*ACT* 814
contents	Jupiter systems A, A': various procedures

Obverse?

P1' (m lines missing)

1' [xxxxxxxxxxxxxxxxxxxxxxx] ⸢x **ta ki**?⸣ [xxxxxxxx]

2' [xxxxxxxxxxxxxxxxxxxxxxx] ⸢x 4⸣5 5 10 ⸢*ina*⸣ [xxxxxxxx]

3' [xxxxxxxxxxxxxxxxxxxxxx] ⸢x⸣ 28 **me** 6 **ki du**-*ma* [**igi** xxxxxxx]

4' [xxxxxxxxxxxxxxxxxxx] ⸢6⸣ **me zal** *ina* 6.42 **me.meš ta igi** [*ana* **igi** xxxxxx]

5' [xxxxxxxxxxx *ina* 12 **mu.meš** 4.10 t]**ab** *ina* 1.11 5 **la**$_2$ *ina* 7.7 **ki** *ana* **ki** *ina* **k**[**a**$_2$-*tu*$_2$ *ša*$_2$ 30 x BE **igi** xxx]

P2' 6' [**igi** *ana* **igi ta** 25 **maš en** 30 **gir**$_2$ 30 **tab ta** 30 **gir**$_2$ **en** 25 **ma**]**š** 36 **tab ta** 25 **maš en** 30 **gir**$_2$ 16.[15 **ki du uš ta uš** 4]

7' [*ana ar*$_2$-*šu*$_2$ **la**$_2$-*ma ana* **me** *a* 6 *ana ar*$_2$-*šu*$_2$ **la**$_2$-*ma* **uš** *ar*$_2$-*tu*$_2$ **ta uš** *a*]*r*$_2$-*tu*$_2$ 17.45 **ki du šu**$_2$ **ta šu**$_2$-*u*$_2$ 6 **ki d**[**u igi ta** 30 **gir**$_2$ **en**]

8' [25 **maš** 19.30 **ki du uš ta uš** 4.48 *ana ar*$_2$-*šu*$_2$ **la**$_2$-*m*]*a ana* **me** *a* 7.12 *ana ar*$_2$-*šu*$_2$ **la**$_2$-*ma* **uš** 21.18 **ki du š**[**u**$_2$ xxxx]

9' [xxxxxxxxxxxxxxx *ša*$_2$]-⸢x⸣ *ina* **uš igi**-*tu*$_4$ ½ **kuš**$_3$ *ša*$_2$-*qa ina ar*$_2$-*tu*$_4$ ½ **kuš**$_3$ **sig mul**$_2$**.ba**[**bbar** xxx]

10' [12 **alla** 11.12 **a** 11.12 **absin**$_0$ **rin**$_2$ 12 **zi**-*šu*$_2$ 12.48 **gir**$_2$ 13.36 **pa** 14.24 **maš**$_2$ 15.12 **gu** 15.12 *zi*]*b* 14.24 **ḫun** 13.36 **mul**$_2$ 12.14 **maš.maš u**$_4$**.meš** 4[0? xxx]

11' [xxxxxxxxxxxxxxx **e**]**n** 12 **ḫun** 47.30 **en** 12 **mul**$_2$ 44 **en** 12 **maš** 42 **u**[**š**? xxxx]

12' [xxxxxxxxxxxxxx **ma**]**š**$_2$? 20 12 50? **u**$_4$? 20 *ana* **uš** GAM 20 *ana* **me** *a* GAM 30 *ana* **uš** 40 [xxxx]

13' [xxxxxxxxxxxxx **t**]**a uš igi**-*tu*$_2$ 1.36.40 **zi**-*šu*$_2$ **ta uš** *ar*$_2$-*t*[*u*$_4$ xxxxxx]

14' [xxxxxxxxx **ta** 9 **gir**$_2$] **en** 2 **maš**$_2$ 33.45 **tab ta** 2 **maš**$_2$ **e**[**n** 17 **mul**$_2$ 36 **tab**]

15' [**ta** 17 **mul**$_2$ **en** 9 **alla** 33.45 **tab ta**] 9 **alla en** 9 **gir**$_2$ 30 **tab ta** 9 **g**[**ir**$_2$ **en** 2 **maš**$_2$ xxxx]

16' [xxxxxxxxxxx] ⸢45? **tab**? x⸣ [xxxxxxxxxxx]

(n lines missing)

Reverse?

P3' (about n lines missing)

1' [xxxxxxxxxxxxxxxxxxxxxxxxx] ⸢xx 8 x⸣ [xxx]

2' [xxxxxxxxxxxxxxxxxxxxxxxxx] ⸢x **igi en**? **uš**?⸣ [xxx]

3' [xxxxxxxxxxxxxxxxxxx 16.15 **ki du**] ⸢**uš** 8.20 **gur uš**⸣ [xxxx]

4' [xxxxxxxxxxxxxxxxxxxxxxxx] ⸢xx 8 **u**$_4$? **tab**?⸣ [xxxxx]

P4' 5' [xxxxxxxxxxxxxxxxxxxxxx **ta uš**] *ar*$_2$-*tu*$_2$ 19 **du** [xxxxx]

6' [xxxxxxxxxxxxxxxxxxxxxxxx] ⸢3⸣ **iti.meš** *ša*$_2$ **me** 7.[40 **zi**-*ma* xxxx]

7' [xxxxxxxxxxxxxxxxxxxxx *ina* **igi**-*a*]*t* **šu**$_2$-*šu*$_2$ 14.3.[45 **zi** xxxx]

(about m lines missing)

Critical and philological notes

Obverse?

1' KI: uncertain.

2' 45 5 10: unidentified.

5' **ka**$_2$-*tu*$_2$: *bābtu* = 'ward' = zodiacal zone.

9' ½ **kuš**$_3$, '½ cubit': erroneously read as *šu*$_2$-*u*$_2$ by Neugebauer, as pointed out by Steele (2005a).
ša$_2$-*qa* = *šaqa* pro *šaqi*, 'it is high' (3 sg. m. stat. G *šaqû*). This implies that **sig** is a stative, *šapil*, 'it is low'.

10' 12.14: one expects 12.48.

12' Before 40 one would also expect GAM.

16' 45: perhaps part of 33.45 (cf. the commentary).

Reverse?

2' The traces before **igi** are similar to ŠU$_2$.

3' **uš**: replaces *ana* (*ACT*).

Obverse

P1' (System A)

a' (Subdivision of the synodic cycle, scheme A.S_2?)

(*m lines missing*)
1'[...] ... From ... [...] 2'[...] ⌜... 4⌝5 5.10 in ... [...] 3'[...] 28 'days' the position proceeds 6, and it [appears (FA) ...]
4'[...] 6,0 'days' it passes, in 6,42 'days' from appearance [to appearance ...]

b' (Net displacements)

5'[... in 12 years] you add [4;10]; in 1,11 you subtract 5; in 7,7 position to position.

c' (Rising time at FA, setting time at LA)

In the 'wa[rd' of 30: ... if it appears,...]

P2' (Systems A and A')

a (System A: updating *B* with the synodic arc)

6'[Appearance to appearance. From 25 Gem until 30 Sco you add 30; from 30 Sco until 25 Gem] you add 36.

b (Subdivision of the synodic cycle, scheme A.S_1)

From 25 Gem until 30 Sco [the position proceeds 16;15, (then) it becomes stationary (S1). From the station it recedes backwards 4.] 7'[... From] the second station the position proceeds 17;45, (then) it sets (LA). From the setting the position [proceeds 6, (then) it appears (FA) ...]

8'[From 30 Sco until 25 Gem the position proceeds 19;30, (then) it becomes stationary (S1). It recedes backwards 4;48], then it rises to daylight (AR). It recedes backwards 7;12, then it becomes stationary (S2). The position proceeds 21;18, (then) it sets (LA) [...]

c (System A: distance to the ecliptic at the stations)

9'[...] high?. At the first station it is high by $\frac{1}{2}$ cubit, at the second (station) it is low by $\frac{1}{2}$ cubit.

d (Control values of the zigzag function for v_1)

Jupiter [...] 10'[... 0;12 Cnc; 0;11,12 Leo; 0;11,12 Vir; Lib: its displacement is 0;12; 0;12,48 Sco; 0;13,36 Sgr; 0;14,24 Cap; 0;15,12 Aqr; 0;15,12 Ps]c; 0;14,24 Ari; 0;13,36 Tau; 0;12,14 (error for 0;12,48) Gem.

e (Subdivision of the synodic cycle: control values of a temporal push?)

The 'days': ... [...] 11'[... until] 12 Ari 47;30, until 12 Tau 44, until 12 Gem 42 ... [...] 12'[... Ca]p? 20 12 50 'days'?, 20

f (Subdivision of the synodic cycle: scheme X.S_5; a scheme involving stations)

To the station (S1): times 0;20; to daylight rising (AR): times 0;30; to the station (S2): 0;40 [to ...] 13'[...] from the first station its displacement is 1.36.40. From the second station [...]

g (System A': updating *B* with the synodic arc)

14'[... From 9 Sco] until 2 Cap you add 33;45. From 2 Cap until [17 Tau you add 36.]
15'[From 17 Tau until 9 Cnc you add 33;45. From] 9 Cnc until 9 Sco you add 30.
From 9 [Sco until ...] 16'[...] you add [33];45 ...
(*n lines missing*)
Reverse?
(*about n lines missing*)

P3' (Subdivision of the synodic cycle, scheme A'.S_1, zone 1?)

1'[...] ⌜... 8 ...⌝ [...] 2'[...] ⌜... appearance until station⌝ [...] 3'[... the position proceeds 16;15,] ⌜it becomes stationary, it turns back 8;20, then it becomes stationary⌝ [...] 4'[...] ⌜... 8 'days'? ...⌝ [...]

P4' (Subdivision of the synodic cycle, scheme A'.S_1, zones 3,4)

5'[...From] the second [station] it proceeds 19[...] 6'[...] For 3 months [it moves] 0;7,[40] per 'day'. [...]
7'[... before] its setting [it moves] 0;14,3,[45 per 'day'...]
(*about m lines missing*)

Commentary

This fragment does not contain any edges of the original tablet. The textual restorations in P2' imply that the width of the tablet (or column) was about 13 cm. On the obverse the preserved text is divided into 2 procedures by a horizontal line; on the reverse an empty space separates 2 further procedures. Probably all preserved procedures concern Jupiter.

P1' System A: various procedures

The beginning of the procedure is missing. The remaining 5 strongly damaged lines of text contain part of the sequence of procedures discussed in §3.7.2.1 (cf. Table 3.35). Part **a'** (O1'–4') may belong to scheme A.S_2 for the subdivision of the synodic cycle; for a discussion with a score of all duplicates cf. §3.7.2.6. Part **b'** (O5') concerns Jupiter's net displacement for several intervals expressed in years, for zone 1 of system A (Table 3.37). Part **c'** (O5') is concerned with the setting time at LA and the rising time at FA (§3.7.2.5).

P2' Systems A and A': various procedures

The end of this procedure is lost. The 11 partly preserved lines form a composite procedure with (at least) 7 subprocedures **a–g**. Parts **a,b** belong to system A, part **f** to system A'. Part **a** (O6') is a procedure for updating B with the synodic arc based on template STEP.E (§2.6.10). Part **b** (O6'–8') contains elements of scheme A.S_1 for the subdivision of the synodic cycle (§3.7.2.6). Its formulation is based on PUSH.D, which provides only $\delta\Sigma$. Part **c** (O9') concerns Jupiter's distance to the ecliptic at the stations (§3.7.10.2). Part **d** (O10') contains 12 control values of Jupiter's daily displacement v_1, each associated with a zodiacal sign. For a discussion cf. §3.7.2.6. Part **e** (O11'–12') also contains a list of control values, perhaps representing a temporal push ($\delta\tau$), each associated with a segment of 30° beginning at 12° of a zodiacal sign. It has thus far defied interpretation. The value 47;30 (mean tithis?) pertains to 12° Psc (?) — 12° Ari, 44 to 12° Ari – 12° Tau, and 42 for 12° Tau – 12° Gem. Part **f** (O12'–13') contains scheme X.S_5 for the subdivision of the synodic cycle. It has a parallel in No. **38** P1'. For a discussion cf. §3.7.10.1. Line O13' is without parallel. It mentions a displacement of 1.36.40 'from S1'; the interpretation is not clear. Part **g** (O14'–16') is a procedure for updating B with the synodic arc for system A'. The first part (O14'–15') is formulated in accordance with template STEP.E, which lacks the transition rules. At the end of O15' zone 1 is mentioned for a second time, shortly before the break. As suggested by Neugebauer, this may have introduced the transition rules.

P3' Subdivision of the synodic cycle, scheme A'.S_1, zone 1?

The beginning of this procedure is lost. Only little is preserved of the remaining 4 lines. The retrograde angular push mentioned in R3' is consistent with scheme A'.S_1 for zone 1 (Table 3.44). This suggests that P3' is formulated in accordance with PUSH.D, which provides only $\delta\Sigma$.

P4' Subdivision of the synodic cycle, scheme A'.S_1, zones 3,4

Only the first 3 lines of P4' are partly preserved. Like P3' it deals with the subdivision of the synodic cycle for system A'. R5' mentions $\delta\Sigma = 19°$ for the push from S2 to LA in zone 3; R6' the interval of 3 months after S2 during which $v =$ 0;7,40°, also in zone 3. R7' mentions a daily displacement v for the subsequent interval of $1^{\overline{m}}$ month between S2 and LA, but now for zone 2 (or 4).

No. 24

museum number	BM 34574 (Sp2,46)
provenience	Babylon
date	data in synodic table: SE 190–231
measures	11.5(26?) × 11 × 2.2–3.2 cm (type iv?)
arrangement	O/R
cuneiform text	photo: Fig. F.25 (R: procedures, col.); *ACT*, Pl. 247 (R); copy: *LBAT* 126; Kugler *SSB* I, Pls. 17–18; Pannekoek (1947), p. 426 (O)
bibliography	Kugler *SSB* I (ST); Pannekoek (1947) (ST); *ACT* 823, 622, Zma
contents	Jupiter system B: synodic table with τ, T, σ, B for [FA, S1], AR, S2, LA; procedures for updating B and T

Reverse

Co 1 [xxxxxxxxxx] ⌜x⌝ *ana* **til gar**-*un*
P1 2 [xxxxxxx 40.20.4]⌜5⌝ **tur**-*er* **ta** 1.20.41.30 ⌜**nim**-*ma* **gar**⌝-*an*
3 [xxxxxxxxx 2]8.15.30 ⌜**tur**-*er* **ta** 56⌝.31 **nim**-*ma* **gar**-*an*
4 [xxxxxxxxx?] 1? (blank) 1.19

Reverse

Co (Colophon)

[1][...] completed it.

P1.a (Updating T with the zigzag function for τ)

[2][... That which] is smaller than [40;20,4]5 you ⌜deduct from 1,20;41,30 and put down⌝.

P1.b (Updating B with the zigzag function for σ)

[3][... That which] ⌜is smaller⌝ than [2]8;15,30 you deduct ⌜from 56⌝;31 and put down.
[4]1? 1.19

Critical and philological notes

Reverse

4 1: this impression looks like a vertical wedge, but it is not certain that it represents a sign. The number 1.19 is written about 5 cm further to the right in the empty space below R3 in a small and shallow handwriting. The significance of this number is not clear.

Commentary

This tablet is well preserved, except for the left edge, and a triangular part missing from the lower (upper) right corner. The obverse and the beginning of the reverse are occupied by a synodic table (cf. *ACT* 622 and Volume II) of which columns τ, T, σ and B for S2 and LA for Jupiter system B are partly preserved, and for AR the rightmost column with the logographic abbreviation (**e**) of that phenomenon. As proposed by Pannekoek (1947), they were likely preceded by columns τ, T, σ and B for FA, S1 and (certainly) AR. This implies that about 15 cm of clay is missing from the left edge, so that the original width may have been about 26 cm. On the reverse the synodic table is followed by a colophon and 2 procedures whose purpose is directly connected with the synodic table. The procedure is written in a notably larger hand (line separation 0.50 cm) than the synodic table (0.36 cm). The data in the synodic table suggest that the tablet was written near SE 190 (122/1 BC).

P1 Updating T and B

This procedure consists of two parts. Part **a** contains the algorithm for updating T with the zigzag function for the synodic time τ. Part **b** contains the algorithm for updating B with the zigzag function for the synodic arc σ. The parameters of both zigzag functions are those of Jupiter system B (§3.7.8), in agreement with the synodic table on the same tablet. Both are formulated in accordance with template ZZ.B.1 (§2.6.8.1).

No. 25

museum number	BM 36680 (80–6–17,412)
provenience	Babylon
measures	5.0(*) × 8.0(*) × 2.8 cm
arrangement	O/R
cuneiform text	photo: Fig. F.29; Steele (2005a), Fig. 1
bibliography	Steele (2003), pp. 278–282; Steele (2005a), pp. 34–38
contents	Jupiter system A: various procedures

		Obverse
P1'		(m lines missing)
	1'	[xxxxxxxxxx] ⸢xxxxxxx GAR.GAR-*ma*?⸣ [xxxx]
	2'	[xxxxxxxxxx] ⸢5.35? *ša*$_2$?⸣ 1.16 ⸢8? 5? **gar** *ina* 2 **me**⸣ [xxxx]
	3'	[xxxxxxxxxx] ⸢30 **me**? x 2.30⸣ [xxxx] ⸢x *ša*$_2$ **a**⸣ [xxxx]
	4'	[xxxxxxxxxx] ⸢**me**?⸣ *a-na* 30 ⸢**me**? 7? xxx⸣ [xxxxxxx]
	5'	[xxxxxxxxxx] ⸢x⸣.30 GAR.GAR-*ma* 9 9 ⸢**a.ra**$_2$⸣ 30 ⸢**du**⸣ [4.30 xxxx]
	6'	[xxxxxxxxxx] **me** 6 **ki** *ana ar*$_2$*-šu*$_2$ **la**$_2$-*ma* **uš** ⸢**ta**⸣ [xxxx]
	7'	[xxxxxxxxxx] ⸢**me**⸣ 10 **ninda zi**-*šu*$_2$ **nu tuk** *u*$_3$ 10 **ninda** ⸢GAR⸣.[GAR-*ma* 10 xxxx]
	8'	[xxxxxxxxxx] ⸢**me**⸣ 12.30 **zi**-*šu*$_2$ 10 *u*$_3$ 12.⸢30?⸣ [GAR.GAR-*ma* 22.30]
	9'	[**a.ra**$_2$ 34 **du**-*ma* 12.45 5 *u*$_3$ 12].⸢4⸣5 GAR.GAR-*ma* 17.45 **pap.pap ta** [xxxx]
	10'	[xxxxxxxxx] ⸢6⸣ **ki** *ina* **nim du igi ta igi** *ana* **igi** *ina* ⸢x⸣ [xxxx]
	11'	[xxxxxxxx] 6 **me zal** *ša*$_2$ **mu** *ina* 6.42 **me.meš ta** ⸢**igi**?⸣ [*ana* **igi** xxx]
	12'	[x **igi** *ša*$_2$ *lu-maš gab-bi ki-i* **ka**] ⸢**me**⸣ *u* **zi šar**$_2$**.šar**$_2$ *ina* 12 4.10 **tab** *ina* 1.11 5 [**la**$_2$ xxxxxxx]
	13'	[xxxxxxxxxxx *ina* **ka**$_2$-*tu*$_2$] ⸢*ša*$_2$⸣ 36 15 BE **igi** [xxx]
P2'	14'	[xxxxxxxxx **ta** 30 **gi**]**r**$_2$ ⸢**en**⸣ 25 **maš.maš** 36 **tab ta** ⸢2⸣[5 **maš.maš en** 30 **gir**$_2$ 30 **tab** ...]
	15'	[xxxxxxxxxxxx] *ana ar*$_2$*-šu*$_2$ **la**$_2$-*ma* **uš ta uš** [...]
	16'	[xxxxxxxxxx 19].⸢30⸣ **ki du uš** 4.48 [*ana ar*$_2$*-šu*$_2$ **la**$_2$ ...]
	17'	[xxxxxxxxxx **mu**]**l.babbar** *kal*$_3$ **mu** *ina* **dagal** *ma-l*[*ak* ...]
	18'	[xxxxxxxxxx] $\frac{1}{2}$ **kuš**$_3$ *i-šap-pil* [...]
	19'	[xxxxxxxxxx **ma**]**š**$_2$ 16 **gu** 15.12 *zib* 1[4.24 ...]
P3'	20'	(traces)
		(n lines missing)
		Reverse
		(about n+9 lines missing)
	1'	[xxxxxxxxxx] **en**? [...] ^{m}**numun**?-**mu** x [xxxx]
		(about m+10 lines missing)

Critical and philological notes

Obverse

5' 30 GAR.GAR: preceded by traces of a vertical wedge.

7' **nu tuk**: replaces TAB ... (Steele 2003)
10 **ninda**: replaces 14.

8' After *u*$_3$ 12 there are remains of a winkelhaken, perhaps belonging to 30.

19' *zib* 15.12: replaces *zib* 14.24.

Commentary

This fragment does not contain any edges of the tablet. Almost nothing can be read on the reverse. All preserved procedures are concerned with Jupiter, and are more or less associated with system A. All are part of the sequence of procedures discussed in §3.7.2.1 (Table 3.35).

P1' Various procedures

The beginning of P1' is missing. The 13 partly preserved lines can be divided into three parts. Part **a** (O1'–12') contains scheme A.S$_2$ for the subdivision of the synodic cycle.

Obverse

P1' **(System A: various procedures)**

a **(Subdivision of the synodic cycle, scheme A.S_2)**

(*m lines missing*)
1'[...] ⌜... you accumulate, it is ...⌝ [...] 2'[...] ⌜5.35 ... 1.16 8? 5? ... in 2 'days'⌝ [...] 3'[...] ⌜30 'days' ... 2.30⌝ [...] ... [...] 4'[...] ⌜... for 30 'days' 7? ...⌝ [...] 5'[...]...30 you accumulate, it is 0;9. ⌜You multiply⌝ 0;9 by 30, [it is 4;30 ...] 6'[... in ...] 'days' the position recedes backwards 6, then it is stationary (S2). From [the second station...] 7'[30] ⌜'days'⌝ its displacement is 10^{nin}. Nothing and 10^{nin} ⌜you accumulate⌝, [it is 10^{nin}. You multiply it by 30, it is 5.] 8'[34] 'days' its displacement is 0;12,30. 0;10 and 0;12,⌜30⌝ [you accumulate, it is 0;22,30. You multiply it by 34,] 9'[it is 12;45. 5 and 12];45 you accumulate, the total is 17;45. From [...] 10'[...in 28 'days'] the position proceeds 6 in the east, (then) it appears (FA). From appearance to appearance in ... [...] 11'[... more then] 6,0 'days', the passing of the year. In 6,42 'days' from appearance [to appearance ...]
12'[... You compute the appearance for every zodiacal sign according to] ⌜the 'day'⌝ and the displacement.

b **(Net displacements for zone 1)**

In 12 you add 4;10, in 1,11 [you subtract] 5 [...]

c **(Setting times at LA, rising times at FA)**

13'[In the ward of 30 ...; in the 'ward'] of 36: 15 if it appears.

P2' **(System A: various procedures)**

a **(Updating *B* with the synodic arc)**

14'[... From 30 Sc]o until 25 Gem you add 36. From 2[5 Gem until 30 Sco you add 30...]

b **(Subdivision of the synodic cycle, scheme A.S_1)**

15'[... it recedes backwards, and becomes stationary. From the station (S2) [...]
16'[...From the appearance (FA)] the position proceeds [19];30, it becomes stationary (S1), 4;48 [it recedes backwards, ...]

c **(Distance to the ecliptic at the stations)**

[...] 17'Jupiter the entire year. In the width of the pa[th ...] 18'[...] it descends by 1/2 cubit [...].

d **(Control values of zigzag function *v*)**

19'[... Ca]p 0;16 Aqr 0;15,12 Psc 0;1[4,24 ...]

P3' **(Unclear)**

20'(*traces*)
(*about n lines missing*)
Reverse
(*about n+9 lines missing*)
1'[...] ... [...] ...
(*about m+10 lines missing*)

For a score edition of all duplicates and a discussion of this badly known scheme cf. §3.7.2.6. The parameters are compiled in Table 3.38a (p. 93). The difficult lines O1'–4' are badly damaged and not preserved in any of the duplicates. Part **b** (O12') is a list of net displacements for zone 1 of system A (§3.7.2.4). Part **c** (O13') is concerned with Jupiter's rising time at FA and its setting time at LA. Only the rule for zone 2 is partly preserved. For a discussion cf. §3.7.2.5.

***P2'** Various procedures*

This composite procedures is concerned with Jupiter. Part **a** contains rules for updating *B* with the synodic arc, formulated in accordance with template STEP.E (2.6.10). For a discussion of this algorithm in the context of Jupiter system A cf. §3.7.2.2. Part **b** is concerned with the subdivision of the synodic cycle. The sparse remains are consistent with scheme S_1 (§3.7.2.6). Its formulation is based on template PUSH.D. Part **c** concerns Jupiter's distance to the ecliptic at the stations (§3.7.10.2). Part **d** is a list of control values for a zigzag function for the daily displacement (v) between LA and FA. It appears that the values are shifted by one zodiacal sign with respect to v_2 (§3.7.2.6).

No. 26

museum number	BM 40661 (81–4–28,206)
provenience	Babylon
measures	4.1(*) × 4.4(*) × 1.9(*) cm
cuneiform text	photo: Fig. F.30; Steele (2005a), Fig. 6
bibliography	Steele (2005a), pp. 45–47
contents	Jupiter system A, various procedures

Side X

P1' (unknown number of lines missing)

1' [...] ⌜x⌝ GAR.GAR ⌜x⌝ [xx...]

2' [... 10] ⌜*u*$_3$?⌝ 12.30 ⌜GAR.GAR-*ma* 2⌝[2.30 x...]

3' [...] ⌜x **du**?⌝ **igi ta igi** [*ana* ...]

4' [... **zal**] ⌜*ša*$_2$ **mu**⌝ *ina* 6.42 **me**.[**meš ta igi** *ana* **igi** ...]

5' [... **šar**$_2$].**šar**$_2$ *ša*$_2$ **ka** 2-*ia* **ta** 1 [...]

6' [... 11].15 **a.ra**$_2$ 1.8 12.4[5 ...]

7' [... 17.4]⌜5⌝ **ki du**-*ma* **šu**$_2$: *ina* 12 [**mu.meš** 4.10 **tab** ...]

8' [... *ina* **k**]**a**$_2$-*tu*$_2$ *ša*$_2$ 36 : 15 BE [**igi** ...]

P2' 9' [... **ta** 25 **maš.maš en** 30 **gir**$_2$ 30] ⌜**tab ta** 30 **gir**$_2$ **en** 25⌝ [**maš.maš** 36 **tab** ...]

(unknown number of lines missing)

Side X

P1' **(Various procedures)**

a' **(Subdivision of the synodic cycle, scheme A.S$_2$)**

(unknown number of lines missing)

1'[...] ... you accumulate ... [...] 2'[... 0;10] ⌜and⌝ 0;12,30 ⌜you accumulate, it is 0;2⌝[2,30 ...] 3'[...] ..., first appearance. From first appearance [to first appearance ...] 4'[...] per year. In 6,42 'days' [...] 5'You compute [the appearance for all zodiacal signs in accordance with the 'day' and the displacement.]

b' **(Subdivision of the synodic cycle, scheme X.S$_1$)**

According to a second opinion: from ... [...] 6'[... you multiply 0;11],15 by 1;8, it is 12;4[5 ...] 7'[... 17;4]5 the position proceeds, and it sets (LA).

c' **(Net displacements for zone 1)**

In 12 [years you add 4;10 ...]

d' **(Rising time at FA, setting time at LA)**

8'[... In the] 'ward' of 36: 15 if [it appears (FA) ...]

P2' **(Updating *B* with the synodic arc)**

9'[... From 25 Gem until 30 Sco] ⌜you add⌝ [30]. ⌜From 30 Sco until 25⌝ [Gem you add 36...]

(unknown number of lines missing)

Critical and philological notes

Side X

2' *u*$_3$: only the final vertical wedge is preserved.

8' 36 : 15: the separation mark between 36 and 15 consists of two small winkelhakens (same sign as the digit zero).

Commentary

No original edges of the tablet are preserved on this small fragment. Only one side is inscribed, the other side is destroyed. All of the partly preserved procedures deal with Jupiter and are at least loosely associated with system A.

P1' Various procedures

Only the final 8 lines of this procedure are partly preserved. They form a composite procedure, of which parts **a'**, **c'** and **d'** belong to the sequence of system A procedures discussed in §3.7.2.1 (Table 3.35). Part **a'** concerns scheme A.S$_2$ for the subdivision of the synodic cycle (§3.7.2.6). Part **b'**, introduced as a 'second opinion', concerns the motion between S2 and LA. Contrary to my earlier suggestion (Steele 2005a, p. 47), it appears that the computation in X6' concerns $\delta\Sigma$ for the interval of $1{,}8^{\overline{\tau}}$ before LA. The parameters point to scheme X.S$_1$ (Table 3.54a). The partly preserved number in X7' is therefore probably $\delta\Sigma$ (AR to LA) = 17;45°. Part **c'** is a list of net displacements for zone 1 of system A (§3.7.2.4). Part **d'** concerns Jupiter's rising time at FA and its setting time at LA (§3.7.2.5).

P2' Updating B with the synodic arc

Only one line of P2' is partly preserved, sufficient to conclude that P2' contains the algorithm for updating *B* with the step function for σ for Jupiter system A (§3.7.2.2). It is probably formulated in accordance with template STEP.E (§2.6.10).

No. 27

museum number	BM 35241 (Sp2,808)
provenience	Babylon
measures	4.8(12?) × 4.4(*) × 2.0(*) cm (type C or F?)
cuneiform text	photo: Fig. F.27; *ACT*, Pl. 254; copy: *LBAT* 97
bibliography	*ACT* 819b
contents	Jupiter system A: subdivision of the synodic cycle, scheme A.S$_1$

Side X

P1' (unknown number of lines missing)

1' [... xxxxxxxxxxxxx] ⌜xxx⌝ [...]

P2' 2' [... *ina* 30 **me** *ša*$_2$ **me** 6 **zi**] ⌜30⌝ GAM 6 **du**-*ma* ⌜3⌝ [...]

3' [... xxxxxxx *ina* 1-*šu*] **me** 4 **ki** *ana ar*$_2$-*šu*$_2$ **la**$_2$-*ma ana* [**me** *a* ...]

4' [... xxxxx *ina* 30 **me** *ša*$_2$ **me** 8] ⌜**zi**⌝ 30 GAM 8 **du**-*ma* [4 ...]

5' [... xxxxxxx **uš**$^?$] *ar*$_2$ *ina* 30 **me** *ša*$_2$ **me** 4 **zi** [30 GAM 4 **du**-*ma* 2 ...]

6' [... xxxxxxx] ⌜x **pap**⌝ *ina* 1-*šu* **me** 6 **ki** *ana ar*$_2$-*šu*$_2$ **la**$_2$-[*ma* **uš** *ar*$_2$-*tu*$_2$ ...]

7' [... xxx **ta uš**] ⌜**igi**$^?$⌝ **en uš** *ar*$_2$-*tu*$_2$ *ina* 4 **iti** 10 **ki** *ana a*[*r*$_2$-*šu*$_2$ **la**$_2$ ...]

P3' 8' [... xxxxxx *ša*$_2$] ⌜**me**⌝ 4.30 **zi** 30 GAM 4.30 **du**-*ma* 2.⌜1⌝[5 ...]

9' [... xxxxxxxx] ⌜**uš**$^?$⌝ *ar*$_2$ *ina* 38 **me** *ša*$_2$ **me** 7.⌜30⌝ [...]

10' [... xxxxxxxx] ⌜x *ina* 38⌝ **me** 4.⌜45⌝ [...]

(unknown number of lines missing)

Side X

P1' (Unclear)

(unknown number of lines missing)

1'[...] ... [...]

P2' (Subdivision of the synodic cycle, scheme A.S$_1$: S1 to S2, zone 1)

2'[... In 30 'days' it moves 0;6 per 'day'.] You multiply 30 by 0;6, it is 3. [...] **3'**[... In 60] 'days' the position recedes backwards 4, then [it rises] to [daylight (AR)...] **4'**[... In 30 'days'] it moves 0;8 [per 'day']. You multiply 30 by 0;8, it is [4 ...] **5'**[...] the second [station]. In 30 'days' it moves 0;4 per 'day'. [You multiply 30 by 0;4, it is 2 ...] **6'**[... In total] in 60 'days' the position recedes backwards 6, [then the second station ...] **7'**[... From the first station] until the second station in 4 months the position [recedes] backwards 10.

P3' (Subdivision of the synodic cycle, scheme A.S$_1$: S2 to LA, zone 1)

8'[... In 30 'days'] it moves 0;4,30 [per] 'day'. You multiply 30 by 0;4,30, it is 2;[15 ...] **9'**[...] ... In 38 'days' [it moves] 0;7,30 per 'day'. [You multiply 38 by 0;7,30, it is 4;45.] **10'**[...] 4 In 38 'days' [it moves] 4;45 [...]
(unknown number of lines missing)

Critical and philological notes

Side X

10' At the beginning of this line the upper ends of three vertical wedges are visible, probably a digit 4–8.

Commentary

No original edges of the tablet are preserved. The textual restorations imply that the width of the tablet (or column) was about 12 cm. Only one side is inscribed; the other side is destroyed. The preserved text is divided into 3 procedures by horizontal lines. P2' and P3' contain parts of scheme A.S$_1$ for the subdivision of the synodic cycle for zone 1.

P1' Unclear

Virtually nothing remains of the first procedure apart from a few traces in X1'.

P2' *Subdivision of the synodic cycle, scheme A.S_1: S1 to S2, zone 1*

This procedure is concerned with Jupiter's motion between S1 and S2. The numerical parameters are consistent with scheme A.S_1 in zone 1 (cf. Table 3.38a, p. 93). The formulation is based on template PUSH.A.1 (§3.3.3.2), i.e. v and $\delta\tau$ are explicitly multiplied, the result for each interval being the total displacement $\delta\Sigma$. The preserved part of X2' concerns the interval S1 to AR (2); this line must have begun with the preceding subinterval of $30^{\overline{\tau}}$ where $v = -0;2°$. Hence a considerable amount of clay is broken away on the left side of the fragment. The remaining lines deal with the other subintervals of AR to S2, and the corresponding total displacements.

P3' *Subdivision of the synodic cycle, scheme A.S_1: S2 to LA, zone 1*

Only the first 3 lines of this procedure are partly preserved. They concern the adjacent push from S2 to LA (subintervals 1 and 2) in accordance with scheme A.S_1, zone 1. Also here the formulation is based on template PUSH.A.1.

No. 28

museum number	BM 34571 (Sp2,43)
provenience	Babylon
date	data in synodic table: SE 147–218
measures	9.0(21?) × 7.7(*) × 2.5–3.4 cm (type ii?)
arrangement	O/R
cuneiform text	photo: Fig. F.28 (R: procedures); *ACT*, Pl. 245 (R); copy: *LBAT* 118; Kugler *SSB* I, Pl. 11
bibliography	Kugler *SSB* I; *ACT* 821, 603, Zl
contents	Jupiter system A: synodic table with *T*, *B* for [FA?], S1, AR, S2, [LA]; procedures for updating *B* and *T*

Reverse

Co 1 [**mul$_2$.babbar**] ⌜**ta**⌝ 2.26 **en** 3.37 ^{md}**en-ad**.[xxxxxxxxxxxxxxxxxxxxxx]

2 [*ana* **til gar**]-⌜*un*?⌝ *pa-liḫ* ^{d}**en** *li-iṣ-ṣur* ⌜x⌝ [xxxxxxxxxxxxxxxxxxxxxx]

3 [xxx] ⌜**dumu**⌝ *bar-zip*^{ki} *la* ⌜*u$_2$-kal*⌝-[*lam* xxxxxxxxxxxxxxxxxxxxx]

P1 4 [**ta** 25 **maš.maš en** 30 **gir$_2$**].**tab** 30 **tab ta** 30 **gir$_2$.tab en** 25 **maš.maš** {36 **nim**?} 36 **tab** *ša$_2$* [*al-la* 25 **maš.maš diri a.ra$_2$** 50 **du ki** 25 **maš.maš tab**-*ma* **gar**-*an*]

5 [*ša$_2$ al-la* 30 **gir$_2$.tab diri a.ra$_2$** 1.1]2 **du ki** 30 **gir$_2$.tab tab**-*ma* **gar**-*an* **u$_4$.meš** *bi-rit* [xxxxxxxxxxxxxxxxxxxxx]

Reverse

Co (Colophon)

[1][Jupiter] from 2,26 until 3,37. Bēl-aba-[...] [2][comple]ted it. He who honors Bēl may protect it [...] [3][...] He should not reveal it [to] a man from Borsippa [...]

P1 (System A: updating *B* and *T*)

a (Updating *B* with the step function for σ)

[4][From 25 Gem until 30 S]co you add 30. From 30 Sco until 25 Gem {... 36} you add 36. (The amount) by which [it exceeds 25 Gem you multiply by 0;50, add to 25 Gem, and put down.] [5][(The amount) by which it exceeds 30 Sco] you multiply [by 1;1]2, add to 30 Sco, and put down.

b (Updating *T* with τ)

The 'days': the distance [(from) appearance to appearance you compute, you add to it ...]

Critical and philological notes

Reverse

4 36 **nim**?: scribal error. **nim** might also be **ki**.

5 After *bi-rit* = *birīt*, 'distance', one expects **igi** *ana* **igi** GIŠ, '(from) appearance to appearance you compute'.

Commentary

This fragment preserves a part from the upper (lower) edge of the obverse (reverse) of a tablet. Nothing remains of the left and right edges. The obverse and the top of the reverse are occupied by a synodic table for Jupiter system A with columns *T* and *B* for AR and S2 for SE 147–218 (see *ACT* 603 or Volume II); hence the tablet may have been written near SE 147 (165/4 BC). Traces are preserved of a preceding column for S1. On the reverse the table is followed by a colophon and a procedure. The textual restorations of the procedure suggest that not much clay is missing on the left side, perhaps some 2 cm, while a considerable amount, perhaps 10 cm, is missing on the right side. Hence the columns for S2 must have been followed by columns for LA, but there is no compelling reason to assume the presence of columns for FA on the left side. In the colophon the name of the scribe or 'owner' of the tablet is partly preserved as Bēl-aba-[...,], probably Bēl-aba-uṣur.[621] At least two astronomers from Babylon carry this name. The colophon contains an unusual reference to an inhabitant of Borsippa, probably as part of an admonition to secrecy. The procedure is written in a notably larger hand (line separation 0.64 cm) than the synodic table (0.34 cm).

P1 System A: updating B and T

Part **a** (R4–5) contains the algorithm for updating *B* with the step function for the synodic arc (§2.6.10) for Jupiter system A (§3.7.2.2). It is formulated in accordance with template STEP.A.1. Part **b** contains the algorithm for updating *T* with the synodic time (τ), in accordance with template T.1 (§3.3.2.1).

[621] Cf. Hunger, *BAK* 178.

No. 29

museum number	BM 36434 (80–6–17,161)
provenience	Babylon
measures	7.5(*) × 5.0(*) × 2.0–2.7 cm
arrangement	O/R
cuneiform text	photo: Fig. F.32; Steele (2005a), Pl. X
bibliography	Steele (2005a)
contents	Jupiter systems A, A' (?) and A": various procedures

Obverse

P1' (few lines missing)

1' ⸢*ar*$_2^?$⸣-*šu*$_2$ [xxxxxxxxxxxxxxxxxxx]

2' *ma-lak ša*$_2$ ⸢QU$^?$⸣ *ina* ⸢**uš igi**$^?$⸣ [xxxxxxxxxxxxxxxx]

3' **absin**$_0$ **rin**$_2$ 12 **zi**-*šu*$_2$ 12.48 ⸢**gir**$_2$ 13⸣.[36 **pa** 14.24 **maš**$_2$ xxxxxxxxxxxxxxx]

P2' 4' 6.10 **alla** 6.20 **maš**$_2$ 1.40 **tab**-*šu*$_2$ *a-na* **uš a.**⸢**ra**$_2$⸣ [xxxxxxx]

5' *ina* **alla** 27 **me zal igi** *ina* **maš**$_2$ 32 **me zal igi** [xxxxxxx]

P3' 6' **mul**$_2$**.babbar ta igi**-*šu*$_2$ 1.30 **me** 10 **zi**-*šu*$_2$ 30 **me** 6.40 **zi** ⸢x⸣ [xxxxxxx]

7' **ta uš** *ar*$_2$-*tu*$_2$ 3.20 **zi**-*šu*$_2$ **ta uš** *ar*$_2$-*tu*$_2$ 30 [xxxxxxx]

P4' 8' **ta** ⸢5.45 **alla en** 5.45 **gir**$_2^?$ 30 **ta** 5.45 **gir**$_2^?$⸣ [xxxxxxx]

(n lines missing)

Reverse

P5' (about n lines missing)

1' ⸢x⸣ [xxxxxxxxxxxxxxxxxxxxxxxxx]

2' **mu** *ana* **m**[**u**$^?$ xxxxxxxxxxxxxxxxxxxx]

P6' 3' **u**$_4$ **igi** *ana* **u**$_4$ **igi** *ša*$_2$ **mul**$_2$**.babbar** [xxxxxxxxxxxxxxxxxxxxxx]

4' 11.3.20 **ki tab ki** [xxxxxxxxxxxxxxxxxxxxxx]

5' 32 **ki** *ša*$_2$ **du** ⸢x⸣ [xxxxxxxxxxxxxxxxxxxxxx]

6' *u*$_4$*-me ša*$_2$ **igi**-*ka* ⸢**ta**⸣ [xxxxxxxxxxxxxxxxxxxxx]

P7' 7' ⸢**zi**⸣ *gab*-⸢*bi ša*$_2$⸣ [xxxxxxxxxxxxxxxxxxxxxxxxx]

(few lines missing)

Obverse

P1' **(System A: various procedures)**

a' **(Subdivision of the synodic cycle, scheme S$_1$)**

(few lines missing)

1'behind it [...]

b' **(Distance to the ecliptic at the stations)**

[... In the width of] 2'the path of .... At the first station [...]

c' **(Control values of zigzag function for v_2)**

3'Vir, Lib: 0;12 is its displacement. 0;12,48: Sco. 0;13,[36: Sgr. ...]

P2' **(Subdivision of the synodic cycle)**

a **(Parameters of zigzag function for $\delta\Sigma$, LA to FA)**

4'6;10 Cnc. 6;20 Cap. [You multiply] 0;1,40, its addition for (one) degree, by [...]

b **(Scheme X.S$_4$)**

5'in Cnc it stays away 27 'days', then it appears (FA); in Cap it stays away 32 'days', then it appears [...]

P3' **(Subdivision of the synodic cycle, scheme X.S$_2$)**

6'Jupiter: from its appearance (FA) 1,30 'days' its displacement is 0;10, 30 'days' [its] displacement is 0;6,40, [...] 7'From the second station its displacement is 0;3,20; from the second station 30 ['days' ...]

P4' **(System A": updating B with the synodic arc)**

8'From ⸢5;45 Cnc until 5;45 Sco 30. From 5;45 Sco$^?$⸣ [...]

(n lines missing)

Reverse

P5' **(Updating B with the synodic arc, system A' or A"?)**

(about n lines missing)

1'... [...] 2'(from) year to year [...]

P6' **(Derivation of the algorithm for updating T with τ)**

3'Time of appearance (FA) to time of appearance of Jupiter: [...] 4'11;3,20 you add ... [...] 5'32, the position which proceeds ... [...] 6'time of your appearance from [...]

P7' **(Transition rules of step function for σ, system A'?)**

7'All the displacements for [the small one ...]

(few lines missing)

Critical and philological notes

Obverse

2' QU$^{?}$: the sign looks like QU, but the correct reading is unknown. *ša₂-qu*, i.e. *šaqû*, 'to be high' (inf.) or 'it is high' (stat.), makes little sense.

5' **zal** = *aḫāru* D, 'to stay away', either *uḫḫur* (3 m. sg. stat.) or *uḫḫar* (3 c. sg. pres).

Commentary

This fragment preserves a section of the left edge of the tablet, but no other original edges. The text is divided into procedures by horizontal lines. Probably all of them are concerned with Jupiter.

P1' System A: various procedures

The beginning of P1' is lost. What is preserved is sufficient to conclude that P1' contains several procedures of the sequence shown in Table 3.35 (p. 90). The duplicates suggest that part **a'** concerns the subdivision of the synodic cycle in accordance with scheme S_1 (§3.7.2.6). Its formulation is based on template STEP.F. Part **b'** concerns Jupiter's distance to the ecliptic at the stations (§3.7.10.2). Part **c'** is a list of control values of a zigzag function probably representing Jupiter's daily displacement v (§3.7.2.6: Table 3.39).

P2' Subdivision of the synodic cycle: various schemes

This brief procedure consisting of 2 lines can be divided into two parts, both of which concern the subdivision of the synodic cycle. Part **a** contains several control values and other defining parameters of a zigzag function of the zodiacal position, probably representing Jupiter's arc of invisibility, $\delta\Sigma$ (LA to FA). Part **b** contains scheme X.S_4, which deals with the corresponding duration $\delta\tau$; for a discussion cf. §3.7.10.1.

P3' Subdivision of the synodic cycle, scheme X.S_2

This procedure concerns scheme X.S_2 for the subdivision of the synodic cycle. For a discussion cf. §3.7.10.1 and Table 3.54b. The formulation is based on template PUSH.B.3.

P4' System A": updating B with the synodic arc

Only the first line of P4' is partly preserved. The traces are sufficient to establish that P4' contains the algorithm for updating B with the synodic arc for Jupiter in accordance with system A" (§3.7.5.1).

P5' Updating B with the synodic arc, system A' or A"?

Virtually nothing remains of P5' except traces of the last two lines (R1'–2'). The phrase 'year by year' (R2') is otherwise attested for Jupiter only in several procedures for updating B with the synodic arc for system A' (No. **18** P14–P16). It is therefore also possible that P5' is actually the end of P4', in which case it belongs to system A".

P6' Derivation of the algorithm for updating T with the synodic time?

The introductory statement and the parameter $11;3,20^{\overline{\tau}} = e$ (R3') suggest that P6' contains a derivation of the algorithm for updating T with the synodic time (τ). For a discussion of this algorithm cf. §3.3.2.1. The parameter $11;3,20^{\overline{\tau}}$ differs from the usual value $11;4^{\overline{\tau}}$. Due to a lack of parallels, the missing text cannot be restored, and the significance of the number 32 (R4') is unclear (perhaps it is the synodic arc?).

P7' Transition rules for σ, system A'?

Virtually nothing remains of P7' except the first signs of the first line (R7'). A tentative identification of its topic is implied by the fact that these signs are duplicated in No. **46** P6', which is concerned with the transition rules for σ for Jupiter system A'.

No. 30

museum number	BM 40659 (81–4–28,204)
provenience	Babylon
measures	4.1(*) × 5.4(*) × 2.2(*) cm
cuneiform text	photo: Fig. F.31; Steele (2005a), Fig. 5
bibliography	Steele (2005a), pp. 44–51
contents	Jupiter system A, A", unidentified system: various procedures

Side X

P1' (unknown number of lines missing)

1' [...xx] ⸢x 21.18 **ki**⸣ **du** ⸢**šu**$_2$⸣ [...]

2' [... *ina* **dagal**] ⸢*ma*⸣-*lak* ⸢*ša*$_2$ QU$^?$ *ina* **uš**⸣ **igi**-*tu*$_2$ $\frac{1}{2}$ **kuš**$_3$ [...]

3' [...] **rin**$_2$ 12 **zi**-*šu*$_2$ 12.48 **gir**$_2$ 1⸢3⸣.[36 ...]

P2' 4' [...] **alla** 6.20 **maš**$_2$ 1.40 **tab**-*šu*$_2$ *ana* ⸢**uš**$^?$⸣ [...]

5' [...xxx] *ina* **maš**$_2$ 3⸢2⸣ [...]

6' [...xxx] ⸢x⸣ 30 **me** 10 **zi**-*šu*$_2$ 30 [...]

7' [...xxx] ⸢x⸣ 3.20 **zi**-*šu*$_2$ **ta** [...]

8' [... **ta** 5.45 **alla**] ⸢**en** 5.45 **gir**$_2$ 30⸣ [**tab** ...]

(unknown number of lines missing)

Side X

P1' **(Various procedures)**

a' **(Subdivision of the synodic cycle, scheme A.S$_1$)**

(unknown number of lines missing)

1'[...] the position proceeds 21;18, (then) it sets (LA) [...]

b' **(Distance to the ecliptic at the stations)**

2'[... In the width of] the path ... At the first station [it is high] by $\frac{1}{2}$ cubit [...]

c' **(Control values of zigzag function for v)**

3'[... Vir], Lib its displacement is 0;12; 0;12,48: Sco, 0;1⸢3⸣,[36: Sgr ...]

P2' **(Subdivision of the synodic cycle; system A")**

---

a **(Zigzag function for $\delta\Sigma$, LA to FA?)**

4'[... 6;10]: Cnc. 6;20: Cap. Its addition is 0;1,40; for (one) degree [...]

b **(Scheme X.S$_4$ for $\delta\tau$, LA to FA)**

5'[...] in Cap 32 ['days' it ...]

c **(Scheme X.S$_2$?)**

6'[...] 30 'days' its displacement is 0;10. 30 [you multiply by 0;10, it is 5...] 7'[... 'days'] its displacement is 0;3,20. From [...]

d **(System A": updating B with the synodic arc)**

8'[... From 5;45 Cnc] until 5;45 Sco [you add] 30 [...]

(unknown number of lines missing)

Critical and philological notes

Side X

2' QU$^?$: uncertain; the sign looks more like BI.

3' 13: only two wedges of the 3 are visible before the break, but one expects 13.

8' It is possible that the line began with **ta** 5.45, in which case not much clay is missing from the left side.

Commentary

This small fragment does not preserve any original edges of the tablet. It is inscribed on one side; the other side is destroyed. A horizontal line divides the text into two procedures, both concerned with Jupiter.

P1' System A: various procedures

This composite procedure covers part of the sequence of procedures shown in Table 3.35 (p. 90). Part **a'** belongs to scheme A.S_1 for the subdivision of the synodic cycle (§3.7.2.6). The only preserved parameter is 21;18° = $\delta\Sigma$ (S2 to LA) in zone 2. Its formulation is probably based on PUSH.D, which provides only $\delta\Sigma$. Part **b'** concerns Jupiter's distance to the ecliptic at the stations (§3.7.10.2). Part **c'** lists control values of a zigzag function for Jupiter's daily displacement v; cf. Table 3.39 (p. 95). The only preserved control values (Lib and Sco) are compatible with both v_1 and v_2.

P2' Unknown system and system A": various procedures

The first 5 lines of P2' are partly preserved. The text can be divided into four parts. Parts **a–c** concern the subdivision of the synodic cycle for an unknown system; for a discussion cf. §3.7.10.1. Part **a** mentions parameters of a zigzag function for Jupiter's arc of invisibility, $\delta\Sigma$ (LA and FA). The number 0;1,40 is the interpolation coefficient (c) across one zodiacal sign. Part **b** lists control values for the corresponding temporal push $\delta\tau$ in accordance with scheme X.S_4. The daily displacements v = 0;10°/$\overline{\tau}$ and 0;3,20°/$\overline{\tau}$ mentioned in part **c** are suggestive of scheme X.S_2, but this identification is tentative. The formulation is based on PUSH.B.3. Part **d** is a procedure for updating B with the step function for σ in accordance with system A" (§3.7.5.1). Its template could not be identified due to the bad state of preservation.

No. 31

museum number	U 180(10) (Warka X 56)
provenience	Uruk
measures	5.4(10?) × 5.2(*) × ? cm
cuneiform text	photo: Fig. F.33 = *ACT*, Pl. 248; copy: *ACT*, Pl. 225
bibliography	*ACT* 805
contents	Jupiter systems A', B: various procedures

Side X

P1' (unknown number of lines missing)

1' [xxxxx] ⸢xx 30$^?$ x⸣ [xxxxxxxxx]

2' [xxxx 1]⸢5⸣ *zib* 38.2 [xxxxxxxxx]

3' [xxxx **ta**] ⸢**pa**⸣ **en** *zib a-na* ⸢1-*en*$^?$⸣ [xxxxxxxxx]

4' [xxxxx] ⸢x 4⸣3.30 **la$_2$** [xxxxxxxxx]

P2' 5' [**ta** 9 **alla en**] ⸢9⸣ **gir$_2$.tab tur**-*tu$_2$* **ta** 9 [**gir$_2$.tab en** 2 **maš$_2$ murub$_4$**-*tu$_2$* **ta** 2 **maš$_2$ en** 17 **mul$_2$**]

6' [**gal**-*tu$_2$* **t**]**a** 17 **mul$_2$ en** 9 **alla murub$_4$**-*t*[*u$_2$* **tur**-*tu$_2$* 30 **tab** 9 **gir$_2$.tab dib**]

7' [7.30 **tab**] **murub$_4$**-*tu$_2$ ip-pal-ka* **murub$_4$**-⸢*tu$_2$*⸣ [33.45 **tab**]

8' [2] ⸢**maš$_2$**⸣ **dib** 0.4 **tab** ⸢**gal**⸣-*tu$_2$ ip-pa*[*l-ka* **gal**-*tu$_2$* 36 **tab**]

9' [x]⸢x⸣ 17 **mul$_2$.mul$_2$ dib** 3.45 **l**[**a$_2$ gal**-*tu$_2$ ip-pal-ka*]

10' [xxxxxx]-*ma* **tab** ⸢xx⸣ [xxxxxxxx]

11' [xxxxxx] ⸢x⸣ [xxxxxxxxx]

(unknown number of lines missing)

Side X

P1' (System B: computations involving the zigzag function for σ)

(*unknown number of lines missing*)

1'[...] ⸢... 30 ...⸣ [...] 2'[... 1]⸢5⸣ Psc: 38;2 [...] 3'[... From] ⸢Sgr⸣ until Psc for ⸢1$^?$⸣ [degree ...] 4'[...] ... you subtract 43.30 [...]

P2' (System A': updating *B* with the synodic arc)

---

a (Zones of the synodic arc)

5'[From 9 Cnc until] ⸢9⸣ Sco the small one. From 9 [Sco until 2 Cap the middle one. From 2 Cap until 17 Tau] 6'[the large one]. From 17 Tau until 9 Cnc the middle one.

b (Updating *B*)

[The small one: you add 30. It passes 9 Sco], 7'[you add 0;7,30], it gives you the middle one. The middle one: [you add 33;45.] 8'It passes [2] Cap, you add 0;4, it gives you the large one. [The large one: you add 36]. 9'[...] It passes 17 Tau, you subtract 0;3,45, [it gives you the large one.] 10'[...] and you add ... [...] 11'[...] ... [...]

(*unknown number of lines missing*)

Critical and philological notes

Side X

1' Traces of 4–5 signs are visible; only a 30 is reasonably clear.

7' *tu$_2$* at the end of the line is visible only in the copy in *ACT* (Pl. 225).

8' **dib**: a form of *etēqu*, 'to pass', probably the 3 c. sg. pres. G *ittiq*, 'it passes'. *ip-pal-ka*: visible only in the copy in *ACT* (Pl. 225). 3 c. sg. pres. G *apālu*, 'to answer'.

9' Before 17 there is room for 1–2 signs, the second of which ends with a vertical wedge. 45: visible only in the copy in *ACT* (Pl. 225).

Commentary

No original edges of the tablet are preserved on this fragment. It is inscribed on one side; the other side is destroyed. The 2 partly preserved procedures, both concerned with Jupiter, are separated by a horizontal dividing line. The textual restorations in P2' imply that not much clay is missing from the left side, perhaps 1 cm, and about 3–4 cm from the right side. An early photograph, made during the excavation (Fig. F.33, right), reveals that a fragment has broken off on the right side, which affects lines X5'–11'. The tablet was not collated.

P1' System B: computations involving the zigzag function for σ

Only few signs of the 4 final lines of P1' are preserved. The number 38.2 (X2') points to the algorithm for updating B with the zigzag function for σ for Jupiter system B (Table 3.49), since it can be identified as 38;2° = M, which the text associates with the position 15° Psc. This points to template ZZ.A.3 (§2.6.8.1). The rest of P1' appears to be concerned with interpolation ('for 1 degree ...'), but the purpose is not clear, and the subtractive number mentioned in X4' remains unidentified (it does not match a net difference $dB(s)$ for system B).

P2' System A': updating B with the synodic arc

P2', of which only the beginning is partly preserved, contains algorithms for updating B with the synodic arc (σ) for Jupiter system A' (Table 3.42, p. 96). Part **a** mentions the boundaries of the zones of σ in accordance with template STEP.F (§2.6.10). Part **b** contains the algorithm for updating B; it is formulated in accordance with STEP.C. Very little remains of X10'–11', so it is not clear whether they are the continuation of X9'.

No. 32

museum number	BM 33869 (Rm4,431)
provenience	Babylon
measures	10.2 × 6.6 × 1.1–2.3 cm (type E)
arrangement	O/R
cuneiform text	photo: Fig. F.35; *ACT*, Pl. 250; copy: Kugler (1907), Pl. 16
bibliography	Kugler (1907), pp. 136–146; *ACT* 810
contents	Jupiter system A': various procedures

		Obverse
P1	1	⸢**mul$_2$.babbar**⸣ **ta** 9 **alla en** 9 **gir$_2$.tab** 30 **tab** *ša$_2$ al-la* 9 **gir$_2$.tab diri a.**⸢**ra$_2$**⸣ 1.7.[30 **du**]
	2	⸢**ta**⸣ 9 **gir$_2$.tab en** 2 **maš$_2$** 33.⸢45⸣ **tab** *ša$_2$ al-la* 2 **maš$_2$ diri a.ra** 1.⸢4⸣ **du**
	3	**ta** 2 **maš$_2$ en** 17 **mul$_2$.mul$_2$** 36 **tab** ⸢*ša$_2$ al-la*⸣ 17 **mul$_2$.mul$_2$ diri a.ra$_2$** 56.15 **du**
	4	**ta** 17 **mul$_2$.mul$_2$ en** 9 ⸢**alla**⸣ 33.⸢45⸣ [**tab**] ⸢*ša$_2$ al*⸣*-la* 9 **alla diri a.ra$_2$** 53.20 **du**
P2	5	**ta** ⸢9 **alla**⸣ **en** 9 **gir$_2$.tab tur**-*ti* **ta** 9 **gir$_2$.tab** ⸢**en** 2⸣ **maš$_2$ murub$_4$**-*tu$_4$* **ta** 2 **maš$_2$ en** 1⸢7⸣ **mul$_2$ gal**-[*tu$_4$*]
	6	**ta** 17 **mul$_2$.mul$_2$ en** 9 **alla murub$_4$**-⸢*tu$_4$*⸣
P3	7	*ina* **tur**-*ti* **ki** *šamaš$_2$* **ta** 9 **alla en** 9 **gir$_2$.tab** *ša$_2$* **me** 12.[30 **z**]**i** *ar$_2$* ⸢**igi**⸣ 30 **me**
	8	*ša$_2$* **me** 12.30 **zi** 3 **iti.meš** *ša$_2$* **me** 6.40 **zi**-*ma* **uš** [4 **iti.me**]**š** *ša$_2$* ⸢**me**⸣ [4.10 **zi**-*ma* **gur**-*ma* **uš**]
	9	3 **iti.meš** *ša$_2$* **me** 6.23.20 ⸢**zi**⸣ 30 **me** *ina* ⸢**igi**-*at*⸣ **šu$_2$**-*šu$_2$* ⸢*ša$_2$* **me**⸣ [12.30 **zi**-*ma* **šu$_2$**]
P4	10	*ina* **murub$_4$**-*tu$_4$* **ki** *šamaš$_2$* **ta** 9 **gir$_2$.tab en** 2 ⸢**maš$_2$**⸣ *ša$_2$* ⸢**me** 14.3⸣.[45 **zi** *ar$_2$* **igi**]
	11	30 **me** *ša$_2$* **me** 14.3.45 **zi** 3 **iti.meš** *ša$_2$* **me** [7.30 **zi** 4 **iti.meš** *ša$_2$* **me**]
	12	4.41.1⸢5⸣ **zi**-*ma* **gur**-*ma* **uš** 3 **iti.meš** *ša$_2$* **me** [7.11.15 **zi** 30 **me** *ina* **igi**-*at* **šu$_2$**-*šu$_2$*]
	13	*ša$_2$* **me** 14.3.45 **zi šu$_2$**
		Reverse
P5	1	*ina* **gal**-*tu$_4$* **ki** *šamaš$_2$* **ta** 2 **maš$_2$** ⸢**en** 1⸣7 **mul$_2$** *ša$_2$* ⸢**me**⸣ 16.5[2.30 **zi** *ar$_2$* **igi** 30 **me**]
	2	*ša$_2$* **me** 9 **zi** 3 **iti.meš** *ša$_2$* **me** 5.37.30 **zi**-*ma* ⸢**uš**⸣ [4 **iti.meš** *ša$_2$* **me** 8.37.30]
	3	**zi**-*ma* **gur**-*ma* **uš** 3 ⸢**iti**⸣**.meš** *ša$_2$* **me** 16.52.30 ⸢**zi** 30 **me**⸣ [*ina* **igi**-*at* **šu$_2$**-*šu$_2$*]
	4	*ša$_2$* **me** 16.52.30 **zi šu$_2$**
P6	5	*ina* **murub$_4$**-*tu$_4$* **ki** *šamaš$_2$* **ta** 17 **mul$_2$ en** 9 **alla** *ša$_2$* **me** 14.3.45 ⟨**zi**⟩ *ar$_2$* **igi** 30 **me**
	6	*ša$_2$* **me** 14.⸢3⸣.45 **zi** 3 **iti.meš** *ša$_2$* **me** 7.30 **zi** 4 **iti.meš** *ša$_2$* **me**
	7	⸢4.41.15⸣ **zi**-*ma* **gur**-*ma* **uš** 3 **iti.meš** *ša$_2$* **me** 7.11.15 ⸢**zi**⸣
	8	*ša$_2$* ⸢10⸣.[4?6?.5?]2.30 **zi** 30 **me** *ina* **igi**-*at* **šu$_2$**-*šu$_2$* *ša$_2$* **me** 14.3.45 [**zi šu$_2$**]

Critical and philological notes

Reverse

8 *ša$_2$* 10.[46.52]: thus restored by Neugebauer. The signs between *ša$_2$* and 2.30 are wiped out by what appear to be traces of an insect. The meaning of *ša$_2$* is not clear here. One would expect *ša$_2$* 1.30 ME ...: 'for 1,30 'days' ...', but the traces seem to rule this out.

Obverse

P1 (Updating *B* with the synodic arc)

[1]⸢Jupiter⸣.
From 9 Cnc until 9 Sco you add 30. (The amount) by which it exceeds 9 Sco you [multiply] by 1;7,[30].
[2]From 9 Sco until 2 Cap you add 33;⸢45⸣. (The amount) by which it exceeds 2 Cap you multiply by 1;⸢4⸣.
[3]From 2 Cap until 17 Tau you add 36. (The amount) ⸢by which⸣ it exceeds 17 Tau you multiply by 0;56,15.
[4]From 17 Tau until 9 ⸢Cnc⸣ [you add] 33;⸢45⸣. (The amount) ⸢by which⸣ it exceeds 9 Cnc you multiply by 0;53,20.

P2 (Zones for the synodic arc)

[5]From ⸢9 Cnc⸣ until 9 Sco the small one. From 9 Sco ⸢until 2⸣ Cap the middle one.
From 2 Cap until 1⸢7⸣ Tau the large one. [6]From 17 Tau until 9 Cnc the middle one.

P3 (Subdivision of the synodic cycle: scheme A'.S_1, zone 1)

[7]In (the region of) the small one.
With the Sun from 9 Cnc until 9 Sco it moves 0;12,[30] per 'day'.
For 30 'days' after the ⸢appearance⸣ (FA) [8]it moves 0;12,30 per 'day'.
For 3 months it moves 0;6,40 per 'day', then it becomes stationary (S1).
[For 4 mon]ths [it moves backwards 0;4,10 per 'day', then it becomes stationary (S2)]
[9]For 3 months it ⸢moves⸣ 0;6,23,20 per 'day'.
For 30 'days' before its setting it moves 0;12,[30 per 'day', then it sets (LA)]

P4 (Subdivision of the synodic cycle: scheme A'.S_1, zone 2)

[10]In (the region of) the middle one.
With the Sun from 9 Sco until 2 Cap [it moves] ⸢0;14,3⸣,[45] per ⸢'day'⸣.
[11]For 30 'days' [after the appearance (FA)] it moves 0;14,3,45 per 'day'.
For 3 months [it moves 0;7,30] per 'day', then it becomes stationary (S1).
[For 4 months] [12]it moves backwards 0;4,41,1⸢5⸣ per 'day', then it becomes stationary (S2).
For 3 months [it moves 0;7,11,15] per 'day'.
[For 30 'days' before its setting] [13]it moves 0;14,3,45 per 'day', (then) it sets (LA).

Reverse

P5 (Subdivision of the synodic cycle: scheme A'.S_1, zone 3?)

[1]In (the region of) the large one.
With the Sun from 2 Cap ⸢until 1⸣7 Tau it moves 0;16,5[2,30] per 'day'.
[For 30 'days' after the appearance (FA)] [2]it moves 0;9 per 'day'.
For 3 months it moves 0;5,37,30 per 'day', then it ⸢becomes stationary⸣ (S1).
[For 4 months] [3]it moves backwards [0;8,37,30 per 'day'], then it becomes stationary (S2).
For 3 months it moves 0;16,52,30 per 'day'.
For 30 'days' [before its setting] [4]it moves 0;16,52,30 per 'day', (then) it sets (LA).

P6 (Subdivision of the synodic cycle: scheme A'.S_1, zone 4)

[5]In (the region of) the middle one.
With the Sun from 17 Tau until 9 Cnc ⟨it moves⟩ 0;14,3,45 per 'day'.
For 30 'days' after the appearance (FA) [6]it moves 0;14,3,45 per 'day'.
For 3 months it moves 0;7,30 per 'day'.
For 4 months [7]it moves backwards 0;4,41,15 per 'day', then it becomes stationary (S2).
For 3 months it moves 0;7,11,15 per 'day', [8]so that? it moves 10;[46,5]2,30.
For 30 'days' before its setting [it moves] 0;14,3,45 per 'day', [(then) it sets (LA)].

Commentary

This tablet is nearly intact, except for some missing clay in the lower (upper) right corner of the obverse (reverse). It has an unusual shape (type E), and the text is written along the long side.[622] The text is divided into 6 procedures by horizontal dividing lines. The final procedure is also followed by a horizontal line, below which the tablet is uninscribed. There is

[622]For other tablets of this type cf. §1.4.

no colophon, although there is ample room for it. All procedures belong to Jupiter system A'. The first 2 deal with the synodic arc; the remaining 4 with the subdivision of the synodic cycle for each of the 4 zones of system A'.

P1 *Updating B with the synodic arc*

This procedure for updating B is formulated in accordance with template STEP.A.1. For the parameters of the step function for σ cf. Table 3.42 (p. 96).

P2 *Zones for the synodic arc*

This short procedure does not provide any new parameters. All it does is assign names, in accordance with template STEP.F (§2.6.10), to the zones for the synodic arc as defined in P1.

P3–P6 *Subdivision of the synodic cycle, scheme A'.S_1*

These 4 procedures contain an essentially complete version of scheme A'.S_1 for the subdivision of the synodic cycle. For a discussion of this scheme cf. §3.7.4.3 and Table 3.44 (p. 98). Each procedure deals with one zone of system A'. For each push or subinterval thereof the duration $\delta\tau$ (in mean tithis) and the daily displacement v (in degrees per mean tithi) are provided, in accordance with the templates PUSH.B (for this template cf. §3.3.3.2). The corresponding angular pushes $\delta\Sigma$ are not mentioned. However, in P5 not all parameters are what one expects for zone 3, since v is systematically larger by a factor 1;7,30, as pointed out by Neugebauer (*ACT*, pp. 378–379).

No. 33

museum number	A 3426
provenience	Uruk
date	SE 119; data in synodic table: SE 131–161
measures	9.8(11?) × 7.9(*) × 1.3–2.1 cm
cuneiform text	photo: Fig. F.38 (R: procedures, col.); *ACT*, Pl. 248 (R); copy: *ACT*, Pl. 219 (col.)
bibliography	*ACT* 820, 640, Q
contents	Jupiter system B': synodic table with τ, T, σ, B for S2; various procedures

Reverse

P1 1 [**mul$_2$.babbar**$^?$ 50].⌜0⌝3.31 **la$_2$** : 40.24.15 **sig** : **igi** *ana* **igi** 1.46.40 *taš a*[*n-na-a ša$_2$* **me.meš**]
2 [**mul$_2$.babbar**$^?$ 37].⌜5⌝8.20 **la$_2$** : 28.19.10 **sig** : **igi** *ana* **igi** 1.46.40 *taš an-na-*⌜*a*⌝ [*ša$_2$* **ki.meš**]
3 [*taš ša$_2$*] ⌜**me**⌝.**meš** *a-na* 12 **mu.meš** 14.48 **du** : *taš ša$_2$* **ki**.⌜**meš** *a-na* 1⌝2 **mu.meš** ⌜15⌝ [**du** x]
Co 4 [**im** ^{md}**diš**]-⌜**en**⌝-*šu$_2$-nu* **a** *ša$_2$* ^{m}*ni-din-tu$_4$*-^{d}**diš** ^{lu_2}**gala** ^{d}**diš a** ^{md}*sin*-⌜**ti-er$_2$**⌝ *qat$_3$* ^{m}⌜^{d}**diš-ad**⌝-[**gur a**-*šu$_2$*]
5 [xxxxx] ⌜**mu**.1⌝-*me*.19.**kam** ^{m}*an-ti-'i-i-ku-su u* ^{m}*an-ti-'i-i-ku-su* **a**-⌜*šu$_2$*⌝ [**lugal.me**]

Reverse

P1 **(Various procedures)**

a **(Parameters of the zigzag function for τ)**

[1][Jupiter$^?$. 50];3,31 is the 'height'; 40;24,15 the 'depth'. From appearance to appearance the difference is 1;46,40. Th[us for the times.]

b **(Parameters of the zigzag function for σ)**

[2][Jupiter$^?$. 37];⌜5⌝8,20 is the 'height'; 28;19,10 the 'depth'. From appearance to appearance the difference is 1;46,40. Thus [for the positions.]

c **(Net differences of T and B for 12^{yr})**

[3][The difference for] ⌜the times⌝ for 12 years: 0;14,48 it proceeds.
The difference for the positions for 12 years: 0;1[5 it proceeds...]

Co **(Colophon)**

[4][Tablet of Anu]-⌜bēl⌝šunu, son of Nidintu-Anu, lamentation priest of Anu, descendant of Sin-⌜lēqi-unninni⌝.
Hand of Anu-aba-[utēr, his son.] [5][...] Year 119, Antiochos and Antiochos, his so[n, were kings.]

Critical and philological notes

Reverse

1,2 **mul$_2$.babbar**: the presence of these signs in the gaps is uncertain since there is little room.
3 **du**: probably *illik*, 'it proceeds'.

Commentary

This fragment belongs to the upper (lower) edge of the obverse (reverse) of a tablet. No other edges are preserved. The textual restorations imply that not much clay is missing on the left and right sides, perhaps 0.5 cm on either side, so that the original width was about 11 cm. The tablet was excavated in Uruk by local inhabitants, probably near the south-eastern gate of the Rēš. The obverse and the top of the reverse contain a synodic table for Jupiter system B' (*ACT* 640; for a new edition see Volume II), which is followed on the reverse by corresponding procedures and a colophon. The tablet was 'owned' by the well-known astronomer Anu-bēlšunu/Nidintu-Anu of the Sin-lēqi-unninni clan, and written by his son, the astronomer Anu-aba-utēr, in the year SE 119 (193/2 BC).

P1 Jupiter system B': various procedures

Part **a** (R1) mentions the defining paramaters of the zigzag function for τ, part **b** (R2) those for σ. Both are formulated in accordance with template ZZ.A.2 (§2.6.8.1). For a discussion cf. §3.7.9. Part **c** (R3) provides the net differences of T and B for 12^{yr} (§3.7.9.2).

No. 34

museum number	BM 35078 (Sp2,614)
provenience	Babylon
date	data in synodic table: SE [...] 61–62 [...]
measures	7.5(*) × 5.0(*) × 0.5(*) cm; column width 7 cm
cuneiform text	photo: Fig. F.26; *ACT*, Pl. 254; copy: *LBAT* 152
bibliography	*ACT* 818
contents	Jupiter system A': various procedures; synodic table

Side X
Column i'

P1' (m lines missing)
1' [xxxxxx] ⌜xx⌝ [xxxxxxx 3 **iti.meš**]
2' [*ša*$_2$ **me** 6.40 **zi**]-*ma* **uš** 4 **iti.meš** *ša*$_2$ **me** 4.10
3' [**zi**-*ma* **gur**-*m*]*a* **uš** 3 **iti.meš** *ša*$_2$ **me** 6.23.20
4' [**zi**-*ma* 30 **me**] *ina* **igi**-*at* **šu**$_2$-*šu*$_2$ *ša*$_2$ **me** 12.30 **zi šu**$_2$
5' [6.15 **ki du igi** 1]⌜6⌝.15 **ki du uš** 8.20 **gur**
6' [**uš** 15.50 **ki du šu**$_2$] ⌜**zi**⌝ *an-na* **ta uš nu tuk**
7' [xxxxxxx] ⌜x⌝ **uš** 30 **tab**

P2' 8' [xxxxxxxx] ⌜x⌝ **ki du**
9' [xxxxxxxxx] ⌜x **igi**$^?$⌝
(n lines missing)

Side X column i'

P1' **(System A': procedures for zone 1)**

a **(Subdivision of the synodic cycle, scheme A'.S$_1$: $\delta\tau$ and v)**

(*m lines missing*]

1'[...] ... [... For 3 months] 2'[it moves 0;6,40 per 'day'], then it becomes stationary (S1). For 4 months it moves backwards 0;4,10 by 'day', 3'then it becomes stationary (S2). For 3 months [it moves] by 0;6,23,20 per 'day', 4'[then for 30 'days'] before its setting (LA) it moves 0;12,30 per 'day', (then) it sets.

b **(Subdivision of the synodic cycle, scheme A'.S$_1$: $\delta\Sigma$)**

5'[The position proceeds 6;15, (then) it appears (FA). The position proceeds 1]6;15, (then) it becomes stationary (S1). It turns back 8;20, 6'[(then) it becomes stationary (S2). The position proceeds 15;50, then it sets (LA).] Thus the displacement, from the station, nothing,

c **(Updating B with the synodic arc)**

7'[...] ... station you add 30.

P2 **(Subdivision of the synodic cycle)**

8'[...] ... the position proceeds 9'[...] ... it appears.
(*n lines missing*)

Critical and philological notes

Side X column i'

1' The duplicate lines No. **18** Oii13'–14' imply that the text before 3 **iti.meš** can be tentatively reconstructed as **ta šu**$_2$ *ana* **igi** *ša*$_2$ **me** 12.30 **zi** *ar*$_2$ **igi** 30 **me** 12.30 **zi**, 'From setting to appearance it moves 0;12,30 per 'day'; for 30 'days' after the appearance it moves 0;12,30 per 'day'.'

6' *an-na* = *annâ*, 'thus'; one expects *an-na-a*, as in the duplicate No. **18** Oii18'.
nu tuk, 'nothing': this apparently refers to the station, where $v = 0$.

7' Before **uš** there is a single wedge, either a winkelhaken or a horizontal one.

9' Before **igi**$^?$ the heads of three vertical wedges are visible.

Commentary

This fragment does not contain any edge of the original tablet. Only one side is preserved. There are two columns of text, the first of which contains two procedures separated by horizontal lines. Probably both deal with Jupiter. The textual restorations suggest that the original width of this column was about 7 cm. The second column belongs to a synodic table for Jupiter, of which the segment for SE 61–62 (251–249 BC) is partly preserved; for a new edition of this table (*ACT* 818) cf. Volume II. An unusual feature of the tablet is that the synodic table is placed to the right of the procedures, perhaps an indication that it serves as a numerical illustration for the procedures. A similar arrangement applies in No. **98**.

P1' Subdivision of the synodic cycle: scheme A'.S_1, zone 1

The beginning of P1' is lost. The preserved lines belong to scheme A'.S_1 for the subdivision of the synodic cycle (zone 1). For this scheme cf. Table 3.44 (p. 98). Part **a** is formulated in accordance with template PUSH.B, which provides only $\delta\tau$ and v. Oi'2' deals with the second segment of the push from FA to S1; this was probably preceded in Xi'0' by the push from LA to FA, and in Xi'1' by the first segment of the push from FA to S1. Part **b** mentions the corresponding values of $\delta\Sigma$; it is formulated in accordance with template PUSH.D. Part **c** contains the algorithm for updating B with the synodic arc for zone 1 ($\sigma_1 = 30°$), but the remains are too sparse to determine the underlying template.

P2' Subdivision of the synodic cycle?

Only a few signs of the first 2 lines of P2' are preserved. They suggest that P2' is also concerned with the subdivision of the synodic cycle, presumably in accordance with scheme A'.S_1 for zone 2.

No. 35

museum numbers	BM 34570 (Sp2,42+68+107+574+876) +VAT 1753+1755
provenience	Babylon
date	data in synodic table: SE 180–252
measures	15.5(25?) × 14.0 × 1.7–3.3 cm (type iii?)
arrangement	O/R
cuneiform text	photo: Fig. F.34 (R: procedures; col.); *ACT*, Pl. 245 (R); copy BM 34570: *LBAT* 119; Kugler *SSB* I, Pls. 12,13
bibliography	Kugler (1907), pp. 126–132; *ACT* 822, 611, Zm
contents	Jupiter system A': synodic table for FA, S1, AR, S2, [LA]; procedures for updating *B* and *T*

Reverse

P1 1 ⸢**ta**⸣ 9 **alla en** 9 **gir$_2$.tab** 30 **tab** *ša$_2$ al-la* 9 **gir$_2$.tab diri** GAM ⸢1⸣.[7.30] ⸢**du**⸣ **ki** 9 **gir$_2$.tab tab**-*ma* **gar**-*an* **ta** 9 **gir$_2$.tab** ⸢**en**⸣ [2 **maš$_2$** 33.45 **tab** *ša$_2$ al-la* 2 **maš$_2$ diri** GAM 1.4 **du ki** 2 **maš$_2$ tab**-*ma* **gar**]

2 ⸢**ta**⸣ 2 **maš$_2$ en** 17 **mul$_2$** 3⸢6 **tab**⸣ *ša$_2$ al-la* 17 **mu**[**l$_2$ diri** GAM 56].⸢15⸣ **du ki** ⸢17⸣ **mul$_2$.mul$_2$ tab**-*ma* **gar ta** 17 **mul$_2$.mul$_2$** [**en** 9 **alla** 33.45 **tab** *ša$_2$ al-la* 9 **alla diri** GAM 53.20 **du ki** 9 **alla tab**-*ma* **gar**]

3 [**u$_4$**].**meš** *bi-rit* **igi.du$_8$.am$_3$** *a-na* **igi.du$_8$.am$_3$** [xxxxxxxx] ⸢x⸣ *šu$_2$-u$_2$* **ki**-*šu$_2$* **tab** *ana* **mu**.⸢**an.na**?⸣-*ka* ⸢x⸣ [xxxxxxxxxxxxxxxxxxxxx]

4 [**igi**].**du$_8$.am$_3$** *ša$_2$* **mul$_2$.babbar gar**-*ma pa-a-liḫ* [xxxxxxxxx ^{md}x].⸢**tin**-*su*⸣-**e a** ^{md}**nanna**.[**u$_3$.tu** xxxxxxxxxxxxxxxxxxx]

Reverse

P1 (System A')

a (Updating *B* with the synodic arc)

[1]⸢From⸣ 9 Cnc until 9 Sco you add 30. (The amount) by which it exceeds 9 Sco you multiply by 1;[7,30], add to 9 Sco, and put down.
From 9 Sco un[til 2 Cap you add 33;45. (The amount) by which it exceeds 2 Cap you multiply by 1;4, add to 2 Cap, and put down.]
[2]⸢From⸣ 2 Cap until 17 Tau you add 36. (The amount) by which it exceeds 17 Ta[u you multiply by 0;56],15, add to 17 Tau, and put down.
From 17 Tau [until 9 Cnc you add 33;45. (The amount) by which it exceeds 9 Cnc you multiply by 0;53,20, add to 9 Cnc, and put down.]

b (Updating *T* with the synodic time)

[3]The 'days': the distance from appearance to appearance [you compute, ...] ..., this you add to it, [you ...] to your 'year' [...] [4]you put down the appearance of Jupiter.

Co (Colophon)

He who honors [..., son of Ea]-balāssu-iqbi, descendant of Nanna-[utu ...]

Critical and philological notes

Reverse

3 **igi.du$_8$.am** = *tāmartu*, 'appearance', probably a general term for synodic phenomenon here.
tab *ana* **mu.an.na**-*ka*: replaces ... ME-*ka* (*ACT*).

Commentary

This fragment preserves parts of the left edge and the upper and lower edges of the tablet. Nothing remains of the right edge. The obverse and the top of the reverse are occupied by a synodic table for SE 180–252, computed in accordance with Jupiter system A'. For a new edition of this table (*ACT* 611) cf. Volume II. Columns *T* and *B* for FA, S1 and AR, and *T* for S2 are partly preserved. The latter column was followed by column *B* for S2, and probably *T* and *B* for LA. This suggests that the original width of the tablet was about 25 cm. The reverse concludes with a procedure (P1) whose purpose is directly connected with the synodic table, and a colophon. Both are written across the entire width of the tablet. The data in the synodic table suggest that the tablet was written near SE 180 (132/1 BC). The scribe, NN/...-balāssu-iqbi//Nanna-utu, might be the same individual as Marduk-zēra-ibni/Ea-balāssu-iqbi//Nanna-utu, who wrote several lamentations (cf. *CTMMA* II) during the years SE 177–178, consistent with the estimated date of the tablet.

P1 *System A': updating B and T*

The procedure consists of two parts. Part **a** (R1–2) contains the algorithm for updating *B* with the synodic arc. It is formulated in accordance with template STEP.A.1. For the parameters cf. Table 3.42 (p. 96). Part **b** (R3–4) contains the algorithm for updating *T* of FA (§3.7.4.2). Its formulation is probably similar to template T.2 (§3.3.2.1). The numerical value of c_τ must have been mentioned in the gap following the second instance of 'appearance' (**igi.du$_8$.am$_3$**) in R3.

No. 36

museum numbers	BM 34789 (Sp2,282 +Sp3,386) +35855 (81–7–6,293+297?)
provenience	Babylon
date	data in synodic table: SE 171–243
measures	14.0(30?) × 8.2(*) × 1.5–3.0 cm (type iv?)
arrangement	O/R
cuneiform text	photo: Fig. F.36 (R: col., procedures); *ACT*, Pl. 246 (R); copy: *LBAT* 124, 125
bibliography	*ACT* 821a, 620a, Zlb
contents	Jupiter system B: synodic table with τ, T, σ, B for [FA, S1], AR, S2, LA; procedures for updating T and B

Reverse

Co 1 [... xxxxxxxxxx **a**] ⸢*ša*₂ ᵐᵈ**en-a-uru**₃⸣-*šu*₂ **a** ᵐᵈ**a**₂**.gal**₂-ᵈ**tu.tu** *qat*₃-*šu*₂ *im-šuḫ-ma ana* **til gar**-*un pa*-⸢*liḫ*⸣ [xxxx]

2 [... xxxxxxxxxxx] ⸢xxxx⸣

P1 3 [... xxxxxxxxxxx] ⸢*ša*₂ *al*⸣-*la* 50.07.15 ⸢**gal**-*u*₂ **ta**⸣ 1.40.14.30 **la**₂ *ša*₂ *al-la* 40.20.45 **tur**-*er-ma* ⸢**ta**⸣ 1.2[0.41.30 **la**₂]

4 [... *ša*₂ *al-la* 38.2 **gal**-*u*₂ **ta** 1.16.4] ⸢**la**₂ *ša*₂ *al-la*⸣ 28.⸢15.30⸣ **tur**-*er-ma* **ta** 56.31 **la**₂

Reverse

Co **(Colophon)**

[9][... son] ⸢of Bēl-apla-uṣur⸣šu, descendant of Ile"i-Marduk. His hand 'measured' it, and he finished it. He who honors [...] [10][...] ...

P1 **(Updating T and B)**

a **(Updating T with the zigzag function for τ)**

[11][...] That which is larger than 50;7,15 you subtract from 1,40;14,30. That which is less than 40;24,45 [you subtract] from 1,20;[41,30].

b **(Updating B with the zigzag function for σ)**

[12][... That which is larger than 38;2] you subtract [from 1,16;4.] That which is less than 28;15,30 you subtract from 56;31.

Critical and philological notes

Reverse

1 ᵐᵈ**en-a-uru**₃-*šu*₂: replaces ᵐᵈ**en**...*šu*₂? (*ACT* Colophon Zlb). The signs **a** and *šu*₂ are clear, but only faint traces of **uru**₃ are preserved, thus this reading is not entirely certain.
imšuḫ: 3 sg. G pret. *mašāḫu* = 'to measure'. In the astronomical corpus this verb is mainly known from diaries and almanacs, *mešḫū* = 'measurements' being the technical term for (Normal Star) Almanacs.

2 Very faint illegible traces of 3–4 signs are visible, after which the line is empty.

3 *ša*₂ *alla* ... *rabû*(**gal**-*u*₂): the algorithm implies that this means 'that which is larger than ...', not '(the amount) by which it is larger than ...'.

Commentary

This fragment includes segments of the upper (lower) edge of the obverse (reverse) of a tablet. No other edges are preserved. The obverse and the top of the reverse are occupied by a synodic table for Jupiter system B. For a new edition of this table (*ACT* 620a) cf. Volume II. Columns τ, T, σ and B are partly preserved for AR, S2 and LA. The missing columns (on the left side τ for AR and probably columns τ, T, σ and B; on the right side B for LA) imply that a few centimeters of clay are missing from the right edge, and perhaps about 12 cm from the left side. The data in the synodic table suggest that the tablet was written near SE 171 (141/0 BC). On the reverse the synodic table is followed by a colophon and a procedure, both written in a larger handwriting than the synodic table. The filiation of the scribe is partly preserved but badly legible; he may be a son of Bēl-apla-uṣuršu, of the Ile"i-Marduk clan. A member of that clan is also mentioned in the colophon of *ACT* 704, a synodic table for Saturn system B for SE 155–243, roughly consistent with the estimated date of the tablet. As remarked by Neugebauer, 34 lines are missing from the synodic table at the bottom of the obverse and the top of the reverse. Dividing them equally over both sides the original height can be estimated as 12 cm.

P1 *Updating* T *and* B

This procedure consists of two parts. Part **a** contains the algorithm for updating T with the zigzag function for the synodic time τ, including reflection rules. Part **b** contains the algorithm for updating B with the zigzag function for the synodic arc σ. The parameters of both zigzag functions are those of Jupiter system B (§3.7.8), in agreement with the synodic table. Both are formulated in accordance with template ZZ.B.1 (§2.6.8.1).

No. 37

museum number	BM 54120 (82–5–22,241)
provenience	Babylon
measures	5.6 × 3.8 × 1.0–1.2 cm (type G)
arrangement	O/R
cuneiform text	photo: Fig. F.37; Hunger (2003/4), p. 345
bibliography	Hunger (2003/4)
contents	Jupiter system B: updating *B* and *T*

Obverse

P1 1 **igi.meš** *ša$_2$* **mul$_2$.babbar** *ša$_2$ taš-pil-tu$_2$ ana* **du$_3$**-*ka* **ki.meš**

2 *ana* **šu$_2$ igi uš e.a uš** *ar$_2$* 1.48 **tab** *u* **la$_2$ en** 50.07.15 **gal**

3 *ša$_2$ al-la* 50.07.15 **diri ta** 1.40.14.30 **e$_{11}$**-*ma* **gar**

4 **en** 40.20.45 **tur** *ša$_2$ al* 40.20.45 **tur**

5 **ta** 1.20.41.30 **e$_{11}$**-*ma* **gar ki.meš** *ana* **šu$_2$ igi ⸢uš⸣** [**e.a**]

6 **⸢uš⸣** *ar$_2$* 1.48 **tab** *u* **la$_2$ en** 38.2 **gal** *ša$_2$* ⸢*al*⸣-[*la*]

7 [3]⸢8⸣.2 **diri ta** 1.16.⸢4 **e$_{11}$**⸣-[*ma* **gar** x$^?$]

Reverse

1 **⸢en⸣** 28.15.30 **tur** *ša$_2$ al*-⸢*la*⸣ [28.15.30 **tur**]

2 **ta** 56.31 **e$_{11}$**-*ma* **gar**-*an* 50$^?$ ⸢xxxxx⸣

3 (erasure)

4 (erasure)

5 (erasure)

Obverse

P1 **(System B)**

a **(Updating *T* with the zigzag function for τ)**

[1]In order for you to 'construct' the appearances (FA) of Jupiter per difference; the positions (error for times): [2]for setting (LA), appearance, station (S1), (daylight) rising (AR), second station (S2), you add and subtract 1;48 until 50;7,15, the largest value. [3]That which exceeds 50;7,15 you deduct from 1,40;14,30 and put down. [4](Similarly) until 40;20,45, the smallest value. That which is less than 40;20,45 [5]you deduct from 1,20;41,30 and put down.

b **(Updating *B* with the zigzag function for σ)**

The positions: for setting, appearance, station, [(daylight) rising,] [6]second station, you add and subtract 1;48 until 38;2, the largest value. That which exceeds [7]3]8;2 you deduct from 1,16;4 and [put down.]

Reverse

[1](Similarly) until 28;15,30, the smallest value. That which [is less than 28;15,30] [2]you deduct from 56;31 and put down.

Year$^?$ ...

Critical and philological notes

Obverse

1 The inclusion of *ša$_2$ taš-pil-tu$_2$* in the introductory statement is without parallel. The syntactic function of *ša$_2$* is not clear. The preliminary translation 'per difference' is analogous to the common expression *x ša ūmi* – '*x* per day' which is used for daily displacements (v). I thus essentially follow the solution adopted by Hunger (2003/4), who translated 'by the difference', but the alternative option, to read **gar** = *tašakkan*, 'you put down' cannot be ruled out, even though it is not attractive either, since there is nothing yet to be put down.
du$_3$: looks more like UD.
ki.meš 'positions': an error for **u$_4$.meš**, 'times'; cf. the commentary.

2 **e.a**: since this term appears between S1 and S2 it must be AR. It is probably a logogram for *elû*, '(daylight) rising', the **a** most likely being a pseudo-Sumerian nominalisation marker. The usual term for AR (as a noun) is **e me** or *a* **me**, 'daylight rising'; cf. §3.1.2 or the Glossary.

Reverse

2 After **gar**-*an* there is a sign similar to 50 (or **mu**?), perhaps the beginning of a short colophon? Traces of several further signs are visible, probably a number.

Commentary

This tablet is nearly intact. A small amount of clay is missing from the lower (upper) right corner of the obverse (reverse). It contains a single procedure which occupies the obverse and the first 2 lines of the reverse. The procedure is followed by a short colophon at the end of R2. It appears that there were 3 more lines below R2 but they are erased. The rest of the reverse is uninscribed.

***P1** Updating T and B*

The procedure contains algorithms for updating T (part **a**) and B (part **b**) with the zigzag functions for the synodic time and the synodic arc for Jupiter system B (§3.7.8). Both are formulated in accordance with template ZZ.B.1 (§2.6.8.1). The introductory statement is without parallel and problematic; cf. the philological notes. Each of the subprocedures begins with a list of the synodic phenomena to which the algorithm applies. I follow Hunger's suggestion to read **e.a** (and not **diri**) in O2, taking it to mean AR, even though the usual term for AR is **e me** or *a* **me**, 'daylight rising'.

No. 38

museum number	BM 34757 (Sp2,248)
provenience	Babylon
date	early or pre-Seleucid?
measures	5.1(*) × 6.3(*) × 2.0–2.9 cm; column width 7 cm
arrangement	O/R
cuneiform text	photo: Fig. F.39; *ACT*, Pl. 254; copy: *LBAT* 151
bibliography	*ACT* 817
contents	Jupiter, unknown system: various procedures

Side X

P1' (m lines missing)

1' [*ina* **rin**$_2$ 6.15 *ina*] ⸢**gir**$_2$**.tab**⸣ 3.16.40 *ina* ⸢**pa** 6⸣.[18.20 *ina* **maš**$_2$ 6.20]

2' [*ina* **gu** 6].⸢18$^?$⸣.20 *ina* *zib*.**me** 6.16.40 *ina* **ḫ**[**un** 6.15 *ina* **mul**$_2$ 6.13.20]

3' *ina* **maš.maš** 6.11.40 *ina* **alla** 6.11 *ina* ⸢**a** 6⸣.[11.40 *ina* **absin** 6.13.20]

4' ⸢**a.ra**$_2$⸣ 20 *ana* **e me a.ra**$_2$ 30 *ana* **uš** *ar*$_2$*-t*[*u*$_2$ **a.ra**$_2$ 40 *ana* xx *ina* **alla**]

5' 27 **me** *ina* **an**-*e* **zal igi** 40 *ina* **absin**$_0$ **rin**$_2$ [xxxxxxx]

6' *u* 2-*ta* **šu**.2.**meš** *ša*$_2$ **me** *ina* **an**-*e* **zal** *in*[*a* xxxxxxx]

7' **zal igi** *ina* **gu** 31 **me** *u*$_3$ 5 **šu**.2.[**meš** *ša*$_2$ **me** *ina* **an**-*e* **zal igi**]

P2' 8' **mul**$_2$**.babbar** *ina šal-šu*$_2$ *ša*$_2$ **na**-*šu*$_2$ **uš igi**-*tu*$_4$ *ina mi-šil* **na**-⸢*šu*$_2$⸣ [xxxxxx]

9' **mul**$_2$**.babbar** 12 **mu.meš** *ana* **egir**-*ka* **gur**-*ma* 15 **me** ⸢**ki**$^?$⸣ [xxxxxxx]

P3' 10' **mul**$_2$**.babbar me igi** 12 **zi**-*šu*$_2$ **en ugu** ⸢**uš**⸣ [xxxxxxx]

11' **nu tuk** *a-na* 2 **šeš.meš** *ṭa*$^?$-*ṭe-e*-[xxxxxxx]

12' 6 6 *taš-pil-tu*$_4$ **me** *ana* **me** 2 ⸢x⸣ [xxxxxxx]

13' 2.30 30 ⸢*a-na*$^?$⸣ [xxxxxxxxxxxx]

14' *ana* **me e** ⸢6⸣ [xxxxxxxxxxxxx]

15' ⸢xx⸣ [xxxxxxxxxxxxxxxxx]

(n lines missing)

Side Y

P4' (about n lines missing)

1' ⸢xx **mul**$_2^?$⸣**.babbar** *ina* **šu ud** 50$^?$ [12 **sag**]

2' **gal**-*tu*$_2$ 9.30 **sag tur**-*tu*$_2$ **uš**-*šu*$_2$ **e**[**n.nam**$^?$]

3' 10.45 *ana* **ugu** KI IGI **tab**-*ma ina* **ša**$_3$ [xx]

4' **du**$_3$**.du**$_3$**.bi** *ša*$_2$ **sag.ki gu**$_4$ **sag.ki**$^?$.[**meš**$^?$]

5' *ana muḫ-ḫi a-ḫa-miš*$_2$ **tab**-*ma* $\frac{1}{2}$-*šu*$_2$ GIŠ-⸢*ma*⸣

6' 10.45 **a.ša**$_3$ 10.45 *ana* **ugu** KI [IGI **tab**-*ma* x]

7' *ṣal-pi ina* **kaskal**-*šu*$_2$ *ina* **ša**$_3$-*šu*$_2$ (erasure?) **se**$_3$**.am**$_3$ [xx]

8' *ana ṣal-pi* **dib**-*iq tam-mar ki-ma a*-⸢*mu*⸣-[x$^?$]

9' 10.45 *šu-u*$_2$ *tam-mar* **ki šu**$_2$ *šu-u*$_2$ 12 ⸢x⸣ [x$^?$]

10' 2.24 9.30 **sag** *ṭe-ri-tu*$_2$ **a.ra**$_2$ 9.3[0 **du**-*ma*]

11' **ta** 2.24 **zi**-*ma ša*$_2$ *re-ḫi* **a.ra**$_2$ KI [IGI$^?$]

12' ⸢*ša*$_2^?$ 10.45⸣ **du**-*ma* 26.52.30 ⸢9.30 xx⸣

(about m lines missing)

Critical and philological notes

Side X

1' 3: error for 6.

3' **alla** 6.11: one expects **alla** 6.10 (cf. commentary).

5' **zal** = *uḫḫur*, 3 m. sg. stat. D *uḫḫuru*, 'to be delayed; stay away'; cf. the Glossary.
40 *ina*: Pinches copied them together as 50.

6' 2-*ta* **šu**.2.**meš**, 'two thirds'.

8' **na** = *manzāzu*, 'standing', here denotes the interval $\delta\tau$ (FA to LA) during which Jupiter is visible.

10' **me igi** = 'the day when it appears/of its appearance': *ūm innammeru* (pres. N *amāru*) or *ūm nanmurti*.

11' **šeš** = *aḫu*, 'part'.
ṭa: reading uncertain (ID?). The context requires 'you divide' here. The only verb 'to divide' that is known to be used in combination with 'into ... parts' is *zâzu* (cf. the Glossary), but this must be a different verb. The sign *ṭe* may also be read *de*.

Side X

P1' (Subdivision of the synodic cycle)

a (Control values of zigzag function for $\delta\Sigma$, LA to FA)

(m lines missing)

1'[In Lib 6;15, in] ⸢Sco⸣ 3;16,40 (error for 6;16,40), in ⸢Sgr 6⸣;[18,20, in Cap 6;20] 2'[in Aqr 6;1]⸢8?⸣,20, in Psc 6;16,40, in Ar[i 6;15, in Tau 6;13,20] 3'in Gem 6;11,40, in Cnc 6;11 (error for 6;10), in ⸢Leo 6⸣;[11,40, in Vir 6;13,20]

b (Scheme X.S$_5$)

[... To the station (S1)]: 4'times 0;20; to daylight rising (AR): times 0;30; to the second stati[on (S2): times 0;40 ...]

c (Scheme X.S$_4$: $\delta\tau$, LA to FA)

[In Cnc] 5'27 'days' it stays away in heaven, (then) it appears (FA). ... In Vir, Lib [...] 6'and $\frac{2}{3}$ of a day it stays away in heaven, (then) it appears. In [...] 7'it stays away [... in heaven], (then) it appears. In Aqr 31 'days' and $\frac{5}{6}$ [of a day] it stays away [in heaven, (then) it appears.]

P2' (Subdivision of the synodic cycle; net difference of T for 12^{yr})

a (Scheme X.S$_5$ for the subdivision of the synodic cycle)

8'Jupiter: at a third of its 'standing' the first station (S1); at one half of its 'standing' [rising to daylight (AR); ...]

b (Goal-Year type procedure for net difference of T?)

9'Jupiter: you turn back 12 years behind you, and you [add] 15 'days' ⸢with?⸣ [...]

P3' (Subdivision of the synodic cycle with linear interpolation for v)

10'Jupiter, (on) the day when it appears (FA) its displacement is 0;12, until the station (S1) [...] 11'nothing. ⸢You divide?⸣ it into 2,0 parts, [... it is] 12'0;0,6. The difference is 0;0,6, day by day. 2,0 ... [...] 13'2.30 30 to [...] 14'it rises to daylight (AR), 6 [...] 15'... [...]

(n lines missing)

Side Y

P4' (Computations involving a trapezoid)

(about n lines missing)

1'... Jupiter? in ... 2'The large [side is 12], the small side is 9;30, w[hat] is its width? 3'You add 10;45 to the 'position of appearance?', and from [... you subtract it?.]

4'Its procedure: the side[s] of the trapezoid 5'you add together, and you compute half of it, 6'it is 10;45; the area, 10;45, [you add] to the 'position [of appearance?'.] 7'You place? the 'crossing' in its path, in its middle, [...] 8'It passes to the 'crossing', you see it, as ... 9'10;45, this you see. The position of setting?, this is 12 ...[...] 10'2,24. 9;30, the pinched side, [you multiply] by 9;30, [and] 11'you 'tear it out' from 2,24, and what remains you multiply by the 'position of [appearance?'] 12'⸢of 10;45⸣, it is 26;52,30. 9;30 [...] ...

(about m lines missing)

12' The final sign is broken off except for an initial vertical wedge.
13' *a-na*: or **a.ra$_2$**.
15' Only the heads of 2–3 vertical wedges are visible.

Side Y

1' Several horizontal wedges remain of the first 1–2 signs.
mul$_2$.babbar, 'Jupiter': this reading, also implied by Pinches's copy, replaces ... *tu$_2$* (*ACT*).
ina ŠU U$_4$ 50: the signs are clear (50 is not entirely certain), but the correct reading is unknown. Alternatives: *aš-šu* **u$_4$**, 'according to the day', or *ina šu-tam-ḫi-*[ri], 'by squaring'?
In the gap at the end, one expects 12 **sag** (**sag** = *pūtu*, 'side'); cf. the commentary.
2' **en.nam**: or *mi-nu-u$_2$*. Only the beginning of the first wedge of **en** or *mi* is visible.
3' KI IGI: This term occurs repeatedly in the text (Y6',11') but is not attested anywhere else. The proper reading and the meaning are unclear. Perhaps *qaqqar nanmurti*, 'position of appearance'?
ša$_3$: replaces KI (*ACT*). *ina libbi*(**ša$_3$**): either 'in' or 'from'.
4' The sign at the end read as **ki** (also *ACT*) looks more like ŠU (thus also Pinches).
7' *ṣal-pi*, 'crossing; slant': the precise meaning is unclear; replaces NI.PI (*ACT*).
ina ḫarrāni(**kaskal**)*-šu$_2$*, 'in its path': replaces **ka$_2$** (*ACT*). This phrase seems to confirm that the procedure has an astronomical significance, but its meaning is not clear.
The following *ina* looks more like NU. The sign following **ša$_3$** was read *bi* by Neugebauer, but it is clearly *šu$_2$*. After it there appears to be an erasure (illegible traces are still visible); replaces *qaq* (*ACT*).
se$_3$: wrongly copied by Pinches as *qar* (thus read in *ACT*). The translation 'you place' is inferred from the context, but the correct Akkadian reading is not certain; perhaps the 2 m. sg. pres. of *šakānu* G, 'to place', or *nadānu* G, 'to give'.

8' The final sign is probably **mu**, less likely 50 (not 40); replaces ŠE (Pinches); ŠA$_3$ (*ACT*).
9' 12: or is this the beginning of **zi**?
12' Only the top of the first sign is preserved. It could be 2 or *ša*$_2$.
At the end one might restore 9.37.48.45 = 10.45 · 53.45.

Commentary

This fragment belongs to the left side of a tablet. No other edges are preserved. The textual restorations imply that not much clay is missing from the right side (original width about 7 cm). The width of this column in relation to the thickness of the tablet suggests that there was another column to the right of the preserved one. The orthography and formulation of the procedures point to the early Seleucid period or before that. Four procedures are partly preserved. Those on side X (P1'–P3') concern Jupiter; this is not certain for P4' (side Y). All of the procedures are unusual in one way or another.

P1' Jupiter, subdivision of the synodic cycle: LA to FA

The beginning of P1' is broken, but the absence of any traces above X1' suggests that this may actually be the first line of P1'. P1' can be divided into three parts concerned with different aspects of the subdivision of the synodic cycle. None of the schemes appears to be associated with any of the known computational systems. All are discussed in §3.7.10.1. Part **a** (X1'–3') contains 12 control values of the zigzag function for $\delta\Sigma$ from LA to FA. Part **b** (X4') contains scheme X.S$_5$ for the subdivision of the synodic time between FA and LA. Part **c** (X5'–7') contains scheme X.S$_4$, which concerns $\delta\tau$ (LA to FA).

P2' Jupiter: subdivision of the synodic cycle; net differences of T

This procedure consists of 2 subprocedures, which do not appear to be connected. Part **a** contains the same scheme X.S$_5$ as P1'.b, but formulated differently; for a discussion cf. §3.7.10. Part **b** is without parallel. The 'turning back' by 12^{yr} is reminiscent of Goal-Year procedures. According to this method the date of a synodic phenomenon in a future year (the Goal Year) is obtained by considering the date of the same phenomenon in the year preceding it by the appropriate period (in this case 12^{yr}) and applying a correction. For the 12^{yr} period, this correction should be about +17^{d}, very close to the 15^{d} mentioned in X9'.[623] In conclusion, part **b** provides net differences of the time of the synodic phenomena for 12^{yr} (corresponding to 11 cycles), perhaps also for other intervals.

P3' Jupiter: subdivision of the synodic cycle with a linearly varying v

The first 6 lines of P3' are partly preserved, but the end is broken away. Also this procedure for the subdivision of the synodic cycle is without parallel. It had hitherto defied explanation, but one feature is now understood. X10' mentions a commonly used value of the daily displacement near FA, v = 0;12°/$\bar{\tau}$. In X11' this number is divided by 2,0, which suggests that v is modeled as a linearly decreasing function that vanishes after 2,0$^{\bar{\tau}}$, a commonly assumed value of $\delta\tau$ (FA to S1). Hence 'nothing' (X11') refers to v being zero at S1. Therefore P3' contains one of only few known schemes for the subdivision of the synodic cycle featuring a daily displacement v that is modeled as a linearly changing function of the mean tithi number. For these schemes in general cf. §3.3.3.4. The rest of the procedure is too fragmentarily preserved to allow an interpretation.

P4' Computations involving a trapezoid

This procedure contains computations involving a trapezoid, which still largely defy interpretation. Similar procedures are attested on two other tablets, both also embedded in procedures for Jupiter (§3.7.10.3). The beginning and end of the procedure are lost. Neugebauer assumed that the procedure is purely mathematical with no relation to Jupiter, but this planet is probably mentioned in Y1'. In Y2' the short and long sides of a trapezoid are given. I provisionally interpret these numbers as 9;30 and 12 (unit unknown). If one desires to connect these numbers with Jupiter, one might interpret them also as 0;9,30°/$\bar{\tau}$ and 0;12°/$\bar{\tau}$, known values of v near LA and FA (cf for instance Table 3.38a, p. 93), but this does not seem to help in understanding the procedure. One is asked to find the 'width' (*šiddu*) of the trapezoid. This presupposes that the area is known. If this is the 10;45 mentioned in Y6', then the 'width' (in fact the height of the trapezoid) is 1, since the mean of its 'sides' is 10;45 = 0;30 · (9;30 + 12). In Y3' 10;45 is added to a quantity called KI IGI. Perhaps the translation is 'position of appearance', but the interpretation of this term remains unclear. Y6' implies that 10;45 is an area. The outcome of the addition is probably subtracted from another quantity, the name of which is lost (Y3').

[623] For these corrections cf. Gray (2009), p. 86.

Y4' marks the beginning of a separate part introduced by the term 'procedure', written with a logogram commonly used in mathematical problem texts. Hence the following lines should contain the solution to a problem. Elements that were hinted at in Y1'–3' are now formulated in more detail. The mean side of the trapezoid is computed as 10;45 (Y5'). This is stated to be a surface, implying that the 'width' (= height) of the trapezoid is 1. Y6' contains an instruction to add this to the KI IGI ('position of appearance'?), repeating Y3'. According to the next instruction a 'crossing' is placed in the middle of 'its path' (that of Jupiter?), the meaning of which is unclear. Then something 'passes to the crossing' (Y8'), followed by 'you see it', which in mathematical texts often introduces the solution to a problem. Y9' contains a term that may be translated as 'position of setting', perhaps the counterpart to the 'position of appearance'. If this is correct they may have a connection with the 'appearance' (FA) and 'setting' (LA) of Jupiter, as also suggested by the numbers. In Y10'–12' the square of 9;30, the narrow side of the trapezoid, is subtracted from 2,24 = 12 · 12, a square with side 12, the broad side of the trapezoid. The numerical result 2,24 − 1,30;15 = 53;45 is not mentioned, but directly multiplied by the 'KI IGI (position of appearance?) of? 10;45'. Since the outcome is 26;52,30 = 0;30 · 53;45 (Y12'), the 'KI IGI of? 10;45' is apparently 0;30. Apart from the number 9;30, the rest of the procedure is broken away. The meaning of all this remains unclear.

No. 39

museum number	BM 46056 (81–7–6,503)
provenience	Babylon
measures	5.2(*) × 6.7(*) × 2.3–2.7 cm
arrangement	O/R
cuneiform text	photo: Fig. F.40; *ACT*, Pl. 254
bibliography	*ACT* 819a
contents	Jupiter, unknown system: subdivision of the synodic cycle, various schemes

Obverse

P1' (m lines missing)
1' [...xxxx] ⸢**me igi** 12$^?$ **zi**⸣ [xxx...]
2' [...xx 30 GAM] 12 **du**-*ma* 6 *ina*$^?$ ⸢x⸣ [x...]

P2' 3' [...x]-⸢*pi*$^{?}$⸣ *ša*$_2$-*nu-tu* 9.⸢30 **zi**⸣ [30 GAM 9.30 **du**-*ma* 4.45 ...]
4' [...] ⸢x⸣ 30 **me** 4.45 **ki** ⸢**du**⸣ **pap.p**[**ap**$^?$ ...]

P3' 5' [...] *ša*$_2$-*nu-tu* 7.30 **zi** 30 GAM 7.[30 **du**-*ma* 3.45 ...]
6' [... *ina* 30 **me**] 3.45 **ki du** *ar*$_2$ *ina* 30 **me** *š*[*a*$_2$ **me** 3.30 **zi** ...]
7' [... 30 GAM] 3.30 **du**-*ma* 1.45 *ina* 30 **me** 1.[45 **ki du** ...]
8' [... *ina* 1]-⸢*šu*⸣ **me** 5.30 **ki du pap.pap** ⸢*ina*$^?$⸣ [...]

P4' 9' [...] ⸢x⸣ **uš igi** *nen-mud* **zi** ⸢x⸣ [...]
10' [...x] ⸢x⸣ 20 GAM 3 **du**-*ma* 1 *ina* ⸢20 **me**⸣ [...]

P5' 11' [...xx **uš**] ⸢**igi** *nen-mud*⸣ [xxx...]
(n lines missing)

Reverse
(about n lines missing)

P6' 1' [...xx] ⸢x⸣ *ma*$^?$ 10$^?$ ⸢x⸣ [xxxxxx...]
2' [...x] ⸢**ki**⸣ **du pap** *ina* 1-*šu* **me** ⸢x⸣ [xxxxxx...]

P7a' 3' [... *ša*$_2$-*nu*]-⸢*tu*⸣ 7.30 **zi** ⸢30⸣ [GAM 7.30 **du**-*ma* 3.45 ...]
4' [... 8].⸢30$^?$⸣ **zi** 30 GAM 8.⸢30$^?$ **du**-*ma*⸣ [4.15 xxx...]

P7b' 5' [...] *ina* 4 **iti** 1⸢7$^?$.45⸣ **ki** *ana* **igi**-[*šu*$_2$ **du** xxxx...]
6' [...] ⸢*ša*$_2^?$⸣ **me** 4.12 **zi** 20 ⸢GAM 4⸣.[12 **du**-*ma* 1.24 ...]
7' [... *ša*$_2$] **me** 6.48 ⸢**zi**⸣ 30 GAM 6.[48 **du**-*ma* 3.24 ...]
8' [...] ⸢**uš igi**-*tu*$_2^?$ 7⸣.28 **ki** *ana* *ar*$_2$-[*šu*$_2$ **la**$_2$-*ma* ...]
9' [...x] ⸢xxx⸣ 9 **zi** ⸢20$^?$ xxx **la**$_2^?$⸣ [xxxxxx...]

P8' 10' [...xx *ša*$_2$ **me** 4] ⸢**zi**⸣ 35 GAM 4 **du** ⸢2⸣.[20 xxxxx...]
11' [...xxxx] ⸢2$^?$ **ki**⸣ *ana ar*$_2$-*šu*$_2$ **la**$_2$-*m*[*a* xxxxxx...]
12' [...xxx] ⸢*ana ar*$_2^?$⸣-*šu*$_2$ **la**$_2$-*ma* **uš** ⸢*ar*$_2^?$⸣ [xxxxxx...]
13' [...xxxx] ⸢x *ina*$^?$ 3$^?$ **zi** 30$^?$⸣ [xxxxxxx...]
(about m lines missing)

Critical and philological notes

Obverse

1' The transliteration of this line is based on an unpublished copy by Strassmaier. It is no longer visible on the tablet, apart from faint traces (of **igi** 12?). Also the photograph in *ACT* (Pl. 254) does not show these signs. Strassmaier's copy was not included in *LBAT*; it was found among the Neugebauer files currently (2011) kept at the Institute for the Study of the Ancient World (New York).

2' 12 **du**: thus Strassmaier's copy; no longer visible on the tablet.
The damaged sign following *ina* looks like the beginning of **nim**.

3' *pi*: perhaps to be restored as *ṭup-pi*, 'tablet'.
After **zi** the text has been restored based on the parallel phrase in 5'.

4' Before 30 the remains of two wedges are visible.

7' The second 30 is written over another sign.

9' Before **uš** a horizontal wedge is visible, perhaps *ina*, or part of *ma*.
uš igi *nen-mud* = *nēmetta maḫrīta nenmud*, 'it is stationary at the first station': *nenmud* = 3 m. sg. stat. N *emēdu*, preceded by paronomastically used cognate noun in the accusative.
After **zi** two horizontal wedges are visible, similar to the beginning of TAB or DU.

Obverse

P1' (Subdivision of the synodic cycle: scheme A.S_1, LA to FA?)

(m lines missing)

1'[...] ⌜the day that it appears (FA)⌝ [its] ⌜displacement is 0;12⌝ [...] 2'[...] you multiply [30 by] 0;12, it is 6. In ... [...]

P2' (Subdivision of the synodic cycle: scheme A.S_1, S2 to LA?)

3'[... According to] another [tablet]: it moves 0;9,30, [you multiply 30 by 0;9,30, it is 4;45 ...] 4'[...] ⌜in?⌝ 30 'days' the position proceeds 4;45. The total [...]

P3' (Subdivision of the synodic cycle: FA to S1?)

5'[... According to] another [tablet?]: it moves 0;7,30, [you multiply] 30 by 0;7,[30, it is 3;45 ...] 6'[... in 30 'days'] the position proceeds 3;45. Afterwards in 30 'days' [it moves 0;3,30 per 'day' ...] 7'[...] You multiply [30 by] 0;3,30, it is 1;45. In 30 'days' [the position proceeds] by 1;[45 ...] 8'[... In] 60 'days' the position proceeds 5;30. (In) total in? [...]

P4' (Subdivision of the synodic cycle: S1 to S2)

9'[...] it is stationary at the first station. The displacement ... [...] 10'[...] you multiply 20 by 0;3, it is 1. In 20 'days' [it moves ... 1 ...]

P5' (Subdivision of the synodic cycle: motion near S1)

11'[...] it is stationary at the first [station ...]

(n lines missing)

Reverse

(about n lines missing)

P6' (Subdivision of the synodic cycle: unclear)

1'[...] ... 10? [...] 2'[...] the position proceeds [...]. (In) total in 60 'days' ... [...]

P7a' (Subdivision of the synodic cycle: scheme X.S_3)

3'[... According to ano]ther opinion]:
it moves 0;7,30. [You multiply 30 by 0;7,30, it is 3;45 ...]
4'[...] it moves 0;[8],⌜30⌝. You multiply 30 by 0;8,30, it is [4;15 ...]

P7b' (Subdivision of the synodic cycle: scheme X.S_3)

5'[...] In 4 months the position [proceeds] forwards 17?;45 [...]
6'[...] it moves 0;4,12 per 'day'. [You multiply] 20 by 0;4,[12, it is 1;24 ...]
7'[...] it moves 0;6,48 [per] 'day'. [You multiply] 30 by 0;[6,48, it is 3;24 ...]
8'[...] the first station the position [recedes] backwards ..., [then ...] 9'[...] ... it moves 9 ... [...]

P8' (Subdivision of the synodic cycle: S1? to S2; unclear)

10'[...] it moves [0;4 per 'day'.] You multiply 35 by 0;4, it is 2;[20 ...] 11'[...] ... the position recedes backwards, and [...] 12'[...] recedes backwards, and then the second station [...] 13'[...] ... it moves ... [...]

(about m lines missing)

Reverse

2' After **me** three horizontal wedges are visible, similar to *ma*.
4' 8.30: both instances of 30 are damaged, but a reading 40 or 50 seems to be excluded.
5' 17.45: 18.45 is also possible.
6' 4.12: replaces 7.12 (*ACT*).
8' **uš igi**-*tu*$_2$: very uncertain.
9' At the beginning the heads of several vertical wedges are visible. 20 and **la**$_2$: very uncertain.
10' The damaged sign between 35 and 4 may be **la**$_2$, or 10 or 20. After **du** there are illegible traces.
13' 3: the heads of three vertical wedges are visible.

Commentary

This fragment does not contain any original edges of the tablet. The textual restorations suggest that not much clay is missing from the left and right sides, perhaps a few centimeters. All procedures on the fragment concern the subdivision of the synodic cycle. The name of the planet is not preserved, but the parameters in P1'–P3' and P7a',b' imply Jupiter. The fragment does not preserve a single, coherent scheme, but parts of what are probably different schemes. Some (P1', P2') are consistent with known schemes, but others are not attested anywhere else. P2', P3' and P7a' are explicitly introduced as alternative procedures ('Secondly:...').[624]

P1' *Subdivision of the synodic cycle: scheme A.S$_1$, LA to FA?*

Only the final 2 lines of P1' are partly preserved. The preserved parameters $v = 0;12°/\bar{\tau}$, $\tau = 30^{\bar{\tau}}$ and $\delta\Sigma = 6°$ are consistent with Jupiter's motion from LA to FA according to scheme A.S$_1$ in zone 1 (Table 3.38a, p. 93).

P2' *Subdivision of the synodic cycle: variant of scheme A.S$_1$, S2 to LA?*

The introductory statement implies that P2' contains an alternative rule. However, the preserved parameters, $\delta\tau = 30^{\bar{\tau}}$ and v is $0;9,30°/\bar{\tau}$, i.e. $\delta\Sigma = 4;45°$ rather point to the interval S2 to LA (2), for which scheme A.S$_1$ prescribes the same value of $\delta\Sigma$ (in zone 1). Perhaps P2' might therefore belong to a variant of scheme A.S$_1$.

P3' *Subdivision of the synodic cycle; FA to S1?*

P3' mentions two consecutive intervals of $30^{\bar{\tau}}$ during which the planet, very likely Jupiter, moves in the forward direction. The names of the synodic phenomena are not preserved, and the numbers do not match any of the known schemes for Jupiter (cf. §3.7.2.6, §3.7.4.3 and §3.7.10.1). In the first interval $v = 0;7,30°/\bar{\tau}$, in the second $v = 0;3,30°/\bar{\tau}$, resulting in a total displacement 3;45 + 1;45 = 5;30° (O8'). As suggested by Neugebauer, this may pertain to Jupiter's motion immediately preceding S1.

P4' *Subdivision of the synodic cycle; motion near S1 or S2*

This procedure concerns the subdivision of the synodic cycle near S1. An interval of $20^{\bar{\tau}}$ is mentioned during which $v = 0;3°/\bar{\tau}$, resulting in a total displacement of 1°, presumably in the retrograde direction. This value of v is not attested in any of the known schemes, but it is close to values of v that pertain to Jupiter's motion near S1 and S2 in scheme A.S$_1$.

P5' *Subdivision of the synodic cycle; motion near S1*

Only the first line of P5' is partly preserved. Like P4' it mentions the first station (S1).

P6' *Subdivision of the synodic cycle; unclear*

P6' deals with an unidentified interval during which the planet (Jupiter?) moves in the forward direction for 60 'days' (R2'). Hence this cannot pertain to the retrograde motion between S1 and S2.

P7a' *Subdivision of the synodic cycle: scheme X.S$_3$*

P7a' and P7b' were considered to form a single procedure in *ACT*, but traces of a horizontal line between R4' and R5' suggest that they are separate. Both deal with the subdivision of the synodic cycle. The scheme, X.S$_3$, is not attested anywhere else. For a discussion cf. §3.7.10.1 (p. 102); the (reconstructed) data are compiled in Table 3.54b. If the beginning of R3' is restored correctly, P7a' is introduced as an alternative algorithm. Two intervals of $30^{\bar{\tau}}$ are mentioned, during which $v = 0;7,30°/\bar{\tau}$ and $0;8,30°/\bar{\tau}$, respectively, resulting in a total displacement 3;45 + 4;15 = 8°.

P7b' *Subdivision of the synodic cycle: scheme X.S$_3$, S2 to LA?*

The interval of 4^{m} resulting in a total displacement $\delta\Sigma = 17;45°$ (R5') can only pertain to the motion from S2 to LA or that from FA to S1. This is followed in R6'–7' by two subintervals in which the planet is accelerating, which suggests S2 to LA, S1 to AR. R8' mentions a retrograde displacement $\delta\Sigma = -7;28°$, probably pertaining to the interval from AR to S2. For a discussion cf. §3.7.10.1.

[624] It cannot be excluded that this fragment belongs to the same tablet as No. **27**, as suggested by Neugebauer (*ACT*, p. 434), but they do not physically join, and they do not contain parts of the same procedures.

P8' Subdivision of the synodic cycle: S1? to S2?

Only the first 4 lines of P8' are partly preserved. The number 35 (R10') presumably represents $\delta\tau$, which is multiplied by a daily displacement $v = 0;4°/\bar{\tau}$. These values are not attested elsewhere. In R11'–12' retrograde motion and the second station (S2) are mentioned, but little more can be understood.

No. 40

museum numbers	BM 82824 (83–9–28,30) +99697 (84–2–1) +99742 (84–2–1)
provenience	Babylon
measures	8.3(*) × 5.3(*) × 1.2(*) cm
cuneiform text	photo: Fig. F.41
contents	Jupiter, unknown system: various procedures

Side X

P1' (unknown number of lines missing)

1' [...xxxxxxxxx] ⸢xxxx⸣ [...]

2' [...] ⸢xxx⸣ [xxxx] ⸢x **tab**$^?$ *u*⸣ **la**$_2$ *ana* **sag.ki** ⸢x⸣ [...]

3' [...]-*ḫi*$^?$ **sag.ki tur**-*tu*$_4$ ⸢*pi*$^?$ x⸣ [xxx] ⸢10 x⸣ ŠU$_2$ ZALAG$_2$ **gibil**-*ši*$^?$ 1 LA$_2$ KUR$^?$ [...]

4' [...] **ta u**$_4$**.meš.am**$_3$ **ki u**$_4$ ⸢x *ana* **mu**$^?$⸣ *ša*$_2$ **ki** ⸢*ṣal*⸣-*pi* **dib**-*qa tam-mar* 31 [...]

5' [...] ⸢x *re*$^?$⸣-*ḫi* 27.28 **ki** ⸢x⸣-*ka* **tab**-*ma u*$_4$-*me* ⸢**ki** *ṣal-pi*⸣ ḪAB$^?$ **ud** *tam-m*[*ar*$^?$ ...]

6' [...] ⸢*ki*⸣-*i* **tag**$_4$ 9.⸢30⸣ *ana* **ugu** 10.50 **tab**-*ma* $\frac{1}{2}$-*šu*$_2$ GIŠ-*ma* **a.ra**$_2^?$ 32 ⸢**du**⸣ [...]

7' [...] ⸢xxxx **an** 10⸣ *tu*$^?$-*šaḫ-ḫi-zu* PI ⸢*an*⸣-*nu-šu*$_2$ 50 *ana tar-ṣa* ⸢x⸣ [...]

8' [...] ⸢x *ša*$_2^?$⸣ **u**$_4$**.meš** GIŠ 3 ME E *ina* 2 **uš** *ina* 2.12 ⸢ḪAB$^?$⸣ DU$_6$ MUŠ AN ⸢x⸣ [...]

P2' 9' [... **ni**]**m** *u* **sig** *ša*$_2$ **mul**$_2$**.babbar** $\frac{2}{3}$ **kuš**$_3$ **an.ta dur** ⸢**murub**$_4^?$⸣ **nim** $\frac{2}{3}$ ⸢**kuš**$_3^?$⸣ [...]

10' [... *ki*]-*i ina* **gir**$_2$**.tab nim** *ina zib*.**me sig** *a-na* 1 **danna**.⸢**am**$_3$ 8$^?$ **nim**$^?$ 8$^?$ **sig**$^?$⸣ [...]

11' [... *ina* **uš igi**]-*tu*$_4$ $\frac{1}{2}$ **kuš**$_3$ *al-la* **murub**$_4$ **nim** *ina* **uš** [*ar*$_2$-*tu*$_4$ $\frac{1}{2}$ **kuš**$_3$ ...]

12' [... **z**]**i**$^?$ *ša*$_2$ **mul**$_2$**.babbar** *ina* **alla a absin**$_0$ **rin**$_2$ ⸢16$^?$⸣ [...]

13' [...] (traces) [...]

(unknown number of lines missing)

Side X

P1' (Computations involving a trapezoid)

(unknown number of lines missing)

1'[...] ... [...] 2'[...] ...⸢addition$^?$ and⸣ subtraction to the width ... [...] 3'[...] ... the small width ... [...] ... new ... [...] 4'[...] from the days, with day ... [...] ... you see 10;50 by which it passes the position of the 'crossing', 31[...] 5'[...] 27.28 remains, you add it to your ⸢...⸣ and the day, the position of the 'crossing' ... you se[e ...] 6'[...] when [...] remains you add 9;30 to 10;50, you compute half of it and you multiply it by 0;32 [...] 7'[...]... you let proceed ... these ... 50. Opposite ...[you put ...] 8'[...] ... of the days you compute$^?$, 3 ... 2.12 ... [...]

P2' (Distance to the ecliptic)

---

a (As a function of the zodiacal position)

9'[... 'H]eight' and depth' of Jupiter: $\frac{2}{3}$ of a cubit above the middle band it is high, [it descends] $\frac{2}{3}$ of ⸢a cubit⸣ [below the middle band it is low.] 10'[I]f it is high in Sco, it is low in Psc. For 1 'mile' ⸢it ascends 8 and descends 8⸣ [fingers...]

b (At the stations)

11'[... at the fi]rst station it is $\frac{1}{2}$ cubit high above the middle; at the [second] station [it is $\frac{1}{2}$ cubit low ...]

c (Control values of an unidentified quantity)

12'[... displa]cement$^?$ of Jupiter in Cnc, Leo, Vir, Lib, ⸢16?⸣ [...] 13'[...] ... [...]

(unknown number of lines missing)

Critical and philological notes

Side X

1' Traces of about 4 signs.

2’ Before the gap, the lower parts of about 3 signs are visible. Before **tab** there is a vertical wedge, the end of a sign?
tab: uncertain; perhaps **zal**?
At the end there is a broken sign containing (at least) two horizontal wedges.

3’ The meaning of this line is unclear. The sign after **tur**-*tu₄* might also be RIN_2, E, or TAG_4.
10: or 20 or 30.
$ŠU_2$: may be the end of a sign.
$ZALAG_2$ usually represents *namāru* = ‘to shine; be clear’ or *nūru* = ‘light’, but there is insufficient context to determine the appropriate meaning. The same applies to **gibil** = *eššu*, ‘new’.

4’ **ta** $\mathbf{u_4}$**.meš.am**$_3$: perhaps *ina ūmī*, ‘from the days’, meaning obscure.
$\mathbf{u_4}$ is followed by the upper end of a vertical wedge.
ṣalpu: verbal adj. of *ṣalāpu*, ‘to cross out (a written sign)’. In mathematical texts the meaning ‘diagonal’ is attested. Provisionally the translation ‘crossing’ has been adopted here.
qaqqar(**ki**) *ṣal-pi etēqa*(**dib**-*qa*) *tam-mar*, ‘you see it passing the position of the ‘crossing”: nearly the same phrase occurs in No. **38** P4’ (Y8’).
31: is this perhaps to be restored as 32, the number mentioned in X6’?

5’ The sign after *ṣal-pi* looks like ḪAB (less likely: UR, IB, LU or KU); meaning unclear.

6’ **a.ra**$_2$: or *a-na*, with the same meaning?

7’ The signs before AN 10 could not be identified. Immediately before AN there could be A or GAR.
tu-šaḫ-ḫi-zu = *tušaḫḫizu*, ‘(that/by which) you let proceed’?, 2 m. pret. subj. Š *aḫāzu*, ‘to proceed’. However, *tu* is not certain.
PI: the sign is clear, but the meaning is not.

8’ GIŠ = *tanašši*, ‘you compute’?
ME E: perhaps **me e**, ‘it rises (to) daylight’ (AR)?
uš: station or degrees?

9’ **dur** ⸢**murub**$_4$⸣ = *riksu qablû*, ‘the middle band’, apparently a term for Jupiter’s ecliptic.
In the gap at the end one may restore something like *šapal*(**ki.ta**) *markasi*(**dur**) *qabli*(**murub**$_4$) *išappil*(**sig**), ‘it descends below the middle band’.

10’ After **danna.am**$_3$ there is a number, probably 8; it is followed by about 3 broken signs. The first may be **si** = *ubānu*, ‘finger’, or **nim**, ‘it goes up’, then a number, perhaps 4–8 (perhaps again 8), then perhaps **sig**, ‘it goes down’.

12’ The broken first sign may be part of **zi** = *nisḫu*, ‘displacement’. Following **rin**$_2$ there are 2 damaged signs: a winkelhaken (10) and a digit 6–8.

13’ Only the upper parts of 6–7 signs are visible, among which are several numbers.

Commentary

This newly discovered fragment does not include any portion of the original edges of the tablet. Only one side is preserved; the other is completely destroyed. The textual restorations in P2’.a suggest that not much clay is missing from the left side, and perhaps a few centimeters from the right side. A colophon is not preserved. The badly understood P1’ contains computations involving what appears to be a trapezoid, as well as other unidentified computations. P2’ concerns Jupiter’s distance to the ecliptic. Hence this tablet is the third one to combine procedures for Jupiter with an apparently mathematical procedure (cf. §3.7.10.3).

P1’ Computations involving a trapezoid

An unknown number of lines are missing before X1’. The interpretation of P1’ poses many problems. X3’–6’ contain similar phrases as No. **18** P5 and No. **38** P4’, procedures concerned with a trapezoid. No satisfactory translation could be found for the text between X1’ and the middle of X4’. The second half of X4’ contains a phrase also found in No. **38** P4, but the meaning of the term ‘crossing’ (*ṣalpu*) is unclear, and it is not clear what is ‘passing’ the ‘crossing’. In mathematical problem texts ‘you see’ (X4’) usually has the outcome of a computation as its object. X5’ mentions 27.28 as the outcome of a computation, most likely a subtraction. The significance of this number is unclear. Subsequently it is added to something. The end of X5’ is also not understood. Somewhat firmer ground is encountered in X6’, which can be interpreted as the computation of the area of a trapezoid with large base 10;50, small base 9;30 and height 0;32: i.e. $0;30 \cdot (9;30+10;50) \cdot 0;32 = 10;10 \cdot 0;32 = 5,25;20$. A very similar procedure, with 10;50 replaced by 10;45, is found in No. **38** P4. The rest of X7’ is unclear. The preposition ‘opposite ...’ is known from interpolation schemes. X8’ could not be translated.

P2’ Jupiter’s distance to the ecliptic; control values of an unidentified quantity

The reconstruction of the introductory statement suggests that not much text is missing from the left edge of the fragment, perhaps only the sign **nim**, ‘Height’. Part **a** (X9’–11’) concerns Jupiter’s distance above and below the ecliptic in general; this procedure is duplicated in No. **47** P1. Part **b** (X11’–12’) is concerned with Jupiter’s distance to the ecliptic at the stations. For a discussion of these procedures cf. §3.7.10.2. Part **c** (X12’) contains a rule whereby the value of a quantity, perhaps a ‘displacement’ (an angular push $\delta\Sigma$ or a daily displacement v?), is defined for different regions of the zodiac, of which only the first one between Cnc and Lib is preserved. This region is similar to zone 1 of Jupiter system A.

5.2.5 Saturn

No. 41

museum number	A 3418
provenience	Uruk
date	approx. SE 121–140
measures	5.9(7?) × 8.2 × 1.3–1.6 cm (type F)
arrangement	O–R
cuneiform text	photo: Fig. F.43; *ACT*, Pl. 249
bibliography	*ACT* 802, Zd; van der Waerden (1965), p. 185; (1974), pp. 262–263; Aaboe & Sachs (1966), p. 4
contents	Saturn systems A, B, B": various procedures

Upper edge

1 [*ina a-mat* ^{d}**diš** *u an-tu*$_4$] ⌜*liš*⌝*-lim*

Obverse

P1 1 [*ša*$_2$ **genna ta** 10 **a**] **en** 30 **gu tur**-*tu*$_2$
2 [**ta** 30 **gu**] ⌜**en**⌝ 10 **a gal**-*tu*$_2$

P2 3 [*ina* **tur**-*tu*$_2$ **ki** *šamaš*$_2$ *ša*$_2$] ⌜**me**⌝ 5 **zi**-*šu*$_2$
4 [*ar*$_2$ **igi** 30 **me** *ša*$_2$] ⌜**me**⌝ 5 **zi**-*šu*$_2$
5 [3 **iti.me**] ⌜*ša*$_2$ **me**⌝ 3.20 **zi uš**
6 [52.30 **me**] ⌜*ša*$_2$?⌝ **me** 4.1⌜3.40⌝ **gur**-*ma ana* **me** *a*
7 [1-*šu* **me**] ⌜*ša*$_2$?⌝ **me** 3.20 **gur**-*ma* **uš** *ar*$_2$
8 [3 **iti.me**] *ša*$_2$ **me** 3.35.30 **zi**
9 [30 **me** *ina* **igi**]-⌜*at*⌝ **šu**$_2$-*šu*$_2$ *ša*$_2$ **me** 5 **zi**-*ma* **šu**$_2$
10 [*ana*] ⌜1 **uš**⌝ 7.33.7.30 **gur**

P3 11 [*ina* **gal**]-⌜*tu*$_2$?⌝ **ki** *šamaš*$_2$ *ša*$_2$ **me** 6 **zi**
12 [*ar*$_2$ **igi**] ⌜30 **me**⌝ *ša*$_2$ **me** 6 **zi**
13 [3 **iti**].⌜**me**⌝ *ša*$_2$ **me** 4 **zi uš**
14 [52.30] ⌜**me**⌝ *ša*$_2$ **me** 5.4.24 **gur**-*ma ana* **me** *a*
15 [1-*šu* **me**] *ša*$_2$ **me** 4 **zi**-*ma* **gur**-*ma* **uš** *ar*$_2$
16 [3 **iti.me**] *ša*$_2$ **me** 4.18.40 **zi** 30 **me** *ina* **igi šu**$_2$-*šu*$_2$ *ša*$_2$ **me** 6 **zi šu**$_2$
17 [*ana* 1 **uš**] 9.3.45 **ki gur**-*ma* **uš**

Upper edge

(1)[At the command of Anu and Antu] may it remain intact.

Obverse

P1 **(System A: zonal boundaries for the synodic arc)**

(1)[For Saturn. From 10 Leo] until 30 Aqr the small one; (2)[from 30 Aqr] ⌜until⌝ 10 Leo the large one.

P2 **(Subdivison synodic cycle, scheme A.S$_1$, zone 1)**

(3)[In (the region of) the small one: with the Sun] its displacement is 0;5 per 'day'. (4)[After the appearance (FA)] for 30 'days' its displacement is 0;5 per ⌜'day'⌝. (5)[For 3 months] it moves 0;3,20 ⌜per 'day'⌝, (then) it is stationary (S1). (6)[For 52;30 'days'] it turns back 0;4,1⌜3,40 per⌝ 'day', and (then) it rises to daylight (AR). (7)[For 60 'days'] it turns back 0;3,20 ⌜per⌝ 'day', then the second station (S2). (8)[For 3 months] it moves 0;3,35,30 per 'day'. (9)[For 30 'days' bef]⌜ore⌝ its setting (LA) it moves 0;5 per 'day', then it sets. (10)[For] ⌜1°⌝ it turns back 7;33,7,30.

P3 **(Subdivision synodic cycle, scheme A.S$_1$, zone 2)**

(11)[In (the region of) the large] one: with the Sun it moves 0;6 per 'day'. (12)[After the appearance (FA)] for ⌜30 'days'⌝ it moves 0;6 per 'day'. (13)[For 3 months] it moves 0;4 per 'day', (then) it is stationary (S1). (14)[For 52;30 'day]s' it turns back 0;5,4,24 per 'day', then it rises to daylight (AR). (15)[For 60 'days'] it moves backwards 0;4 per 'day', then the second station (S2). (16)[For 3 months] it moves 0;4,18,40 per 'day'. For 30 'days' before its setting (LA) it moves 0;6 per 'day', (then) it sets. (17)[For 1 degree] the position turns back 9;3,45, then it is stationary (S2).

Reverse

P4 1 *ša*$_2$ **genna** 4.25 **mu.**⸢**meš**⸣ [4.16 **igi.meš**]
2 9 **bal.meš** 5[4 **ki.meš**]

P5 3 *ša*$_2$ **genna** 25.24.5 **nim** ⸢2⸣[2.41.25 **sig**]
4 12 *taš an-na-a* [*ša*$_2$ **me.meš**]

P6 5 *ša*$_2$ **genna** 14.4.42.30 **la**$_2$ 11.14.2.⸢30⸣ [**sig**]
6 12 *taš an-na-a* **ki.**⸢**meš**⸣

Co 7 **zi** *ša*$_2$ ^{d}*ṣal-bat-a-ni*
8 **im** ^{md}**diš.ad.gur a** *ša*$_2$ ^{md}**diš.en**-*šu*$_2$-*nu* **a** *ša*$_2$ ^{m}[*ni-din-tu*$_4$-^{d}**diš**]
9 **a** ^{md}*sin*-**ti.er**$_2$ ^{lu_2}**gala** ^{d}**diš** *u an*-[*tu*$_4$ xxxxx]

Reverse

P4 **(Systems A and B: period relations)**

^{1}For Saturn: 4,25 years, [4,16 appearances], ^{2}9 revolutions, 5[4,0 positions].

P5 **(System B'': parameters of zigzag function for τ)**

^{3}For Saturn: the 'height' is 25;24,5, [the 'depth' is 22;41,25], ^{4}the difference is 0;12; thus [for the 'days'].

P6 **(System B: parameters of zigzag function for σ)**

^{5}For Saturn: the 'height' is 14;4,42,30, [the 'depth' is] 11;14,2,⸢30⸣, ^{6}the difference is 0;12; thus the positions.

Co **(Colophon)**

^{7}The displacement of Mars.

^{8}Tablet of Anu-aba-utēr, son of Anu-bēlšunu, son of [Nidinti-Anu], ^{9}descendant of Sîn-lēqi-unninni, lamentation priest of Anu and An[tu, ...]

Critical and philological notes

Upper edge

1 Invocation: omitted in *ACT*.

Obverse

3 **ki** *šamaš*$_2$ = *itti šamaš*, 'with the Sun'.

6 4.13.40: thus also read by Kugler *SSB* II, p. 579 n. 2. In *ACT* 13.40 was omitted; in *HAMA*, p. 440, Neugebauer erroneously restores 4.13.20.

8 **zi** is interpreted as the finite verb *inassaḫ*, 'it moves', because of the copula *ma*, cf. also O9,15.

9 *ina* **igi**-*at* = *ina pānāt*, 'before' (temporal).

10 This line and O17 may be erroneous insertions; cf. the commentary.

Reverse

3 **nim** = *šūqu* 'height' = maximum; **sig** = *šuplu* 'depth' = minimum. Instead of **nim** the duplicate No. **42** R15 has the synonymous **la**$_2$ = *šūqu*.

6 **ki.meš** 'positions': in the duplicate No. **41** R6 this is preceded by *ša*$_2$, 'for'.

9 In the gap at the end of the line there is only room for a short filiation, which suggests that the scribe may have been Anu-aba-utēr himself ('with his own hand', *ina qāt ramānišu*), or his son.

Commentary

This fragment preserves about half of a tablet. The left edge of the obverse and the right edge of the reverse are missing. The orientation (O–R) is unusual for a procedure text. On the basis of the duplicate No. **42** the missing text can be securely restored. The restorations imply that not much clay is missing from the left (right) of the obverse (reverse), so that the original width was about 7 cm. The colophon includes a catchline, 'displacement of Mars', which implies that the tablet is part of a series in which it is followed by a tablet for Mars (which remains unidentified). The catchline is lacking on the duplicate No. **42**. The name of the scribe and the date of the tablet are not preserved. The tablet's 'owner' is the well-known astronomer Anu-aba-utēr/Anu-bēlšunu from the Sîn-lēqi-unninni clan. This implies that the tablet was written after 14/X/121 SE (191 BC), the date of his latest attestation as scribe of scholarly tablets.[625] All procedures concern Saturn. P1–P3 belong to system A; P4 mentions period relations that are common to systems A and B, and P5–P6 belong to system B. Exactly the same sequence of procedures is preserved on No. **42** (P3–P8), also from the Rēš library. Several features suggest that one is a copy of the other, or that both are copies of the same tablet. Many lines are terminated after the same sign, and both tablets share certain irregular formulations, e.g. **zi**-*ma* **gur**-*ma*, 'it moves and turns back', is used in P2 and No. **42** P4, but not in P1 and No. **42** P3. However, No. **42** has the usual orientation O/R instead of O–R.

[625] Other astronomical tablets 'owned' by Anu-aba-utēr were written by his cousin Anu-balāssu-iqbi or by Anu-uballiṭ/Ekur-zākir. Since there is little room on the tablet for a long filiation, the scribe might have been Anu-balāssu-iqbi, if one assumes that he is referred to as 'Anu-balāssu-iqbi, the son of his brother'.

***P1** System A, zonal boundaries of step function for synodic arc*

This procedure, a duplicate of No. **42** P3, provides the zonal boundaries of the step function for the synodic arc. For a discussion cf. §3.8.2.1. The parameters are compiled in Table 3.57a (p. 107). The formulation is based on template STEP.F, which is discussed in §2.6.10.

***P2,P3** Subdivision of the synodic cycle, scheme A.S_1 (and elements of scheme A.S_2)*

P2 and P3 concern scheme A.S_1 for the subdivision of the synodic cycle (§3.8.2.4). The formulation is based on the templates PUSH.B.1–3 and 5, which provide only v and $\delta\tau$. Both procedures end with a problematic phrase (O10 and 17). If one accepts the current restorations, which are based on the duplicates No. **42** P4,P5, then 'For 1° it turns back 7.33.7.30' (O10) describes a retrograde motion. This parameter does not match any of the values of v in scheme A.S_1, but it coincides with $\delta\Sigma$ (S2 to LA) in zone 1 in scheme A.S_2 (Table 3.58b, p. 108), 9;3,45° being the corresponding value for zone 2. In scheme A.S_1 these angular distances are 7;53,15° (zone 1) and 9;28°(zone 2); cf. Table 3.58a. Hence the numerical values appear to have been taken from another scheme, but the retrograde direction is then also erroneous, as is the term 'for 1 degree'.

***P4** Period relations for systems A and B*

This procedure contains the period relation common to systems A and B (Tables 3.57a, 3.61). The number 54,0 is the corresponding total distance covered by Saturn, $Z \cdot 6,0°$.

***P5** System B": parameters of the zigzag functions for τ*

This procedure provides the defining parameters of the zigzag function for τ for system B" (Table 3.61). It is formulated in accordance with template ZZ.A.1 (§2.6.8.1).

***P6** System B: parameters of the zigzag functions for σ*

This procedure provides the defining parameters of the zigzag function for σ for system B (Table 3.61). It is formulated in accordance with template ZZ.A.1 (§2.6.8.1).

5.2.6 *Mixed content*

No. 42

museum number	AO 6477
provenience	Uruk
measures	10.6 × 7.2 × ? cm (type E)
arrangement	O/R
cuneiform text	copy: *TU* 30
bibliography	Kugler *SSB* II, pp. 577–580 (partial translit.); Schnabel (1924) (partial translit.); *ACT* 801; Aaboe & Sachs (1966), p. 4
contents	Mercury system A_1; Saturn system A

		Obverse
P1	1	⌜*ša*$_2$ ^{d}⌝**gu$_4$.ud igi** *ša*$_2$ **kur** *a-na* **igi** *ša*$_2$ **kur**
	2	**ta** 1 **a en** 16 **maš$_2$** 1.46 **tab**
	3	16 **maš$_2$ dib**-*iq* GAM 20 **du**-*ma* **tab**
	4	**ta** 16 **maš$_2$ en** 30 **mul$_2$.mul$_2$** 2.21.20 **tab**
	5	30 **mul$_2$.mul$_2$ dib**-*iq* GAM 20 **du**-*ma* **la$_2$**
	6	**ta** 30 **mul$_2$.mul$_2$ en** 1 **a** 1.34.13.20 **tab**
	7	1 **a dib**-*iq* GAM 7.30 **du**-*ma* **tab**
	8	*bi-rit* **igi** *ana* **igi** GIŠ.GIŠ.ḪI-*ma* 3.30.39.4.20 **ki**-*šu*$_2$ **tab**-*ma*
	9	**me.meš e**-*bi* **gu$_4$.ud** *ina* **kur igi**-*su* **ta** 10 **lu en** 20 **mul$_2$.mul$_2$**
	10	*ina* **kur šu$_2$**-*šu*$_2$ **dib**-*iq* **kimin**
P2	11	*ša*$_2$ **gu$_4$.ud igi** *ša*$_2$ **šu$_2$** *ana* **igi** *ša*$_2$ **šu$_2$ ta** 6 **alla en** 26 **rin$_2$** 2.40 **tab**
	12	26 **rin$_2$ dib**-*iq* GAM 20 **du la$_2$ ta** 26 **rin$_2$ en** 10 *zib* 1.46.40 **tab**
	13	10 *zib* **dib**-*iq* GAM 6 **du la$_2$ ta** 10 *zib* **en** 6 **alla** 1.36 **tab**
	14	6 **alla dib**-*iq* GAM 40 **du tab** *bi-rit* **igi** *ana* **igi** GIŠ.GIŠ.ḪI-*ma*
	15	3.30.39.4.20 **ki**-*šu*$_2$ **tab**-*ma* **me.meš e**-*bi*
	16	**gu$_4$.ud** *ina* **šu$_2$ igi**-*su* **dib**-*iq* **ta** 30 **absin en** 5 **gir$_2$.tab**
	17	*ina* **šu$_2$ šu$_2$**-*šu*$_2$ **dib**-*iq* **kimin**
P3	18	*ša*$_2$ **genna ta** 10 **a en** 30 **gu tur**-*tu*$_2$
	19	**ta** 30 **gu en** 10 **a gal**-*tu*$_2$
P4	20	⌜*ina*⌝ **tur**-*tu*$_2$ **ki** *šamaš*$_2$ *ša*$_2$ **me** 5 **zi**-*šu*$_2$
	21	[*ar*$_2$] **igi** 30 **me** *ša*$_2$ **me** 5 **zi**-*šu*$_2$
	22	[3 **iti.me**] *ša*$_2$ **me** 3.20 **zi**-*ma* **uš**
		Reverse
	1	[52].30 **me** *ša*$_2$ **me** 4.⌜13$^?$⌝.40 ⌜**gur**⌝-*ma* *ana* **me** *a*
	2	[1-*š*]*u* **me** *ša*$_2$ **me** 3.20 **gur**-*ma* **uš** {**uš**} *ar*$_2$
	3	3 **iti.me** *ša*$_2$ **me** 3.35.30 **zi**
	4	30 **me** *ina* **igi**-*at* **šu$_2$**-*šu*$_2$ *ša*$_2$ **me** 5 **zi**-*ma* **šu$_2$**
	5	*a-na* 1 **uš** 7.33.7.30 **gur**
P5	6	*ina* **gal ki** *šamaš*$_2$ *ša*$_2$ **me** 6 **zi**
	7	*ar*$_2$ **igi** 30 **me** *ša*$_2$ **me** 6 **zi**
	8	3 **iti.me** *ša*$_2$ **me** 4 **zi**-*ma* **uš igi**
	9	⌜52⌝.30 **me** *ša*$_2$ **me** 5.4.24 **gur**-*ma* *ana* **me** *a*
	10	⌜1-*šu*⌝ **me** *ša*$_2$ **me** 4 **zi**-*ma* **gur**-*ma* **uš** *ar*$_2$
	11	3 **iti.me** *ša*$_2$ **me** 4.18.40 **zi**
	12	30 **me** *ina* **igi**-*at* **šu$_2$**-*šu*$_2$ *ša*$_2$ **me** 6 **zi**-*ma* **šu$_2$**
	13	*ana* 1 **uš** 9.3.45 **uš gur**-*ma* **uš**
P6	14	*ša*$_2$ **genna** 4.25 **mu.meš** 4.16 **igi.meš** 9 **bal.meš** 54 **ki**.[**meš**]

Obverse

P1 (Mercury system A_1: MF and ML)

a (MF: updating B with the synodic arc)

[1]For Mercury, morning appearance (MF) to morning appearance.
[2]From 1 Leo until 16 Cap you add 1,46. [3](If) it passes 16 Cap you multiply by 0;20 and add.
[4]From 16 Cap until 30 Tau you add 2,21;20. [5](If) it passes 30 Tau you multiply by 0;20 and subtract.
[6]From 30 Tau until 1 Leo you add 1,34;13,20. [7](If) it passes 1 Leo you multiply by 0;7,30 and add.

b (MF: updating T with the synodic time)

[8]The distance from appearance to appearance you compute, and 3;30,39,4,20 you add to it, and [9]you predict the 'days'.

c (Rules for the omission of MF and ML)

Mercury passes by its appearance in the morning (MF) from 10 Ari until 20 Tau, [10](and) its setting in the morning (ML), ditto.

P2 (Mercury system A_1: EF and EL)

a (EF: updating B with the synodic arc)

[11]For Mercury, evening appearance (EF) to evening appearance.
From 6 Cnc until 26 Lib you add 2,40. [12](If) it passes 26 Lib, you multiply by 0;20, subtract.
From 26 Lib until 10 Psc you add 1,46;40. [13](If) it passes 10 Psc, you multiply by 0;6, subtract.
From 10 Psc until 6 Cnc you add 1,36. [14](If) it passes 6 Cnc, you multiply by 0;40, add.

b (EF: updating T with the synodic time)

The distance from appearance to appearance you compute, and [15]3;30,39,4,20 you add to it, and you predict the 'days'.

c (Rules for the omission of EF and EL)

[16]Mercury passes by its appearance in the evening (EF) from 30 Vir until 5 Sco, [17]it passes by its setting in the evening (EL), ditto.

P3 (Saturn: system A, zonal boundaries)

[18]For Saturn. From 10 Leo until 30 Aqr the small one; [19]from 30 Aqr until 10 Leo the large one.

P4 (Saturn: subdivision synodic cycle, scheme A.S_1, zone 1)

[20]⌜In⌝ (the region of) the small one: with the Sun its displacement is 0;5 per 'day'.
[21][After] the appearance (FA) 30 'days' its displacement is 0;5 per 'day'.
[22][3 months] it moves 0;3,20 per 'day', then it is stationary (S1).
Reverse
[1][52];30 'days' it turns back 0;4,⌜13?⌝,40 per 'day', then it rises to daylight (AR).
[2][6]0 'days' it turns back 0;3,20 per 'day', then the second station {station} (S2).
[3]3 months it moves 0;3,35,30 per 'day'.
[4]30 'days' before its setting (LA) it moves 0;5 per 'day', then it sets.
[5]For 1° it turns back 7;33,7,30.

P5 (Saturn: subdivision synodic cycle, scheme A.S_1, zone 2)

[6]In (the region of) the large one: with the Sun it moves 0;6 per 'day'.
[7]After the appearance (FA) for 30 'days' it moves 0;6 per 'day'.
[8]3 months it moves 0;4 per 'day', then the first station (S1).
[9]⌜52⌝;30 'days' it turns back 0;5,4,24 per 'day', then it rises to daylight (AR).
[10]⌜60⌝ 'days' it moves backwards 0;4 per 'day', then the second station (S2).
[11]3 months it moves 0;4,18,40 per 'day'.
[12]30 'days' before its setting (LA) it moves 0;6 per 'day', then it sets.
[13]For 1° it turns back 9;3,45, then it is stationary (S2).

P6 (Saturn systems A,B: period relations)

[14]For Saturn: 4,25 years, 4,16 appearances, 9 revolutions, 54,0 posit[ions].

P7	15	*ša*$_2$ **genna** 25.24.5 **la**$_2$ 22.41.25 **sig** 12 *taš an-na-a ša*$_2$ ⌜**me.meš**⌝
P8	16	*ša*$_2$ **genna** 14.4.42.30 **la**$_2$ 11.4.2.30 **sig** 12 *taš ša*$_2$ **ki.meš**

Critical and philological notes

Obverse

2–3 This sentence begins with what is probably an asyndetic conditional clause (*GAG* §160). Hence **dib**-*iq* is most likely a 3 c. sg. pres. G of *etēqu* = *ittiq*, 'it passes by'. The same construction underlies O4–5 and O6–7.

8 GIŠ.GIŠ.ḪI = 'you compute'. The Akkadian reading is unknown, but in some procedure texts GIŠ = *tanašši* 'you compute' replaces GIŠ.GIŠ.ḪI (cf. the Glossary).

9 The sentence is syntactically correct but O16 suggests that **dib**-*iq* was omitted after **igi**-*su*.
igi-*su* (also O16): the suffix -*su* implies that *nanmurtu*(**igi**) + poss. suff. 3 m. sg. is realised as *nanmurassu*, 'its appearance' (nom./acc.), instead of the expected *nanmurtašu*.

Reverse

1 4.13.40: this is the expected number; thus read by Kugler. Thureau-Dangin's copy (*TU* 30) has 4 *re-ḫi*, probably by mistake. The duplicate No. **41** O6 also has the expected 4.13.40.

2 {**uš**}: dittography.

11–12 On the duplicate No. **41** these lines form a single line.

15 **la**$_2$ = *šūqu*, 'height'. The duplicate No. **41** R3 has **nim**, which has the same Akkadian reading.

16 11.4.2.30: error for 11.14.2.30 (cf. the commentary).

Commentary

Except for small bits of missing clay near the bottom (top) left and right corners of the obverse (reverse), this tablet from Uruk is nearly intact. The text is divided into eight sections by horizontal dividing lines. The first two procedures are devoted to Mercury; the other six concern Saturn, and are duplicated in No. **41**. Although the tablet does not contain a colophon, it may well have been written or 'owned' by Anu-aba-utēr, who is also the 'owner' of No. **41**. The tablet was not collated.

P1 Mercury system A_1: various procedures for MF

P1 is concerned with the position and time of MF of Mercury in accordance with system A_1. For a discussion of the algorithms cf. §3.4.2. Part **a** (O1–7) contains the algorithm for updating the zodiacal position (B) with the step function for the synodic arc σ (§3.4.2.1). The formulation is based on template STEP.B (§2.6.10), which involves transition coefficients c_j. Part **b** (O8–9) contains the algorithm for updating the time T with the synodic time τ (Eq. 3.9) with $c_\tau = 3;30,39,4,20^{\bar{\tau}}$. Part **c** (O9–10) contains rules for the omission of MF and ML (§3.4.2.4).

P2 Mercury system A_1: various procedures for EF

P2 is concerned with the position and time of EF, also for system A_1. Part **a** (O11–14) contains the algorithm for updating B with the step function for σ, using the template STEP.B (§2.6.10). Part **b** (O14–15) contains the algorithm for updating T with τ in accordance with Eq. (3.9), as in P1.b. Part **c** (O16–17) contains rules for the omission of EF and EL.

P3 Saturn system A: zonal boundaries

This procedure mentions the zonal boundaries of the step function for σ in accordance with system A for Saturn. For this algorithm cf. §3.8.2.1; for the involved parameters cf. Table 3.57a. The formulation is based on template STEP.F (§2.6.10).

***P4,P5** Saturn: subdivision of the synodic cycle, scheme A.S_1*

P4 and P5 are concerned with scheme A.S_1 for the subdivision of the synodic cycle of Saturn. For a discussion of this scheme cf. §3.8.2.4; for the parameters cf. Table 3.58a. The formulation is based on templates PUSH.B.1 (O22, R3,6–8,11), PUSH.B.2 (R4,12), PUSH.B.3 (O20–21, R3–4), PUSH.B.4 (R10), PUSH.B.5 (R1,2,9), all of which provide only v and $\delta\tau$. For a discussion of these templates cf. §3.3.3.2. Both procedures end with the same problematic phrase as the duplicates No. **41** P2,P3.

P7 **(Saturn system B": parameters of zigzag function for τ)**

[15]For Saturn: the 'height' is 25;24,5, the 'depth' is 22;41,25, the difference is 0;12; thus for the ⌜'days'⌝.

P8 **(Saturn system B: parameters of zigzag function for σ)**

[16]For Saturn: the 'height' is 14;4,42,30, the 'depth' is 11;4,2,30 (error for 11;14,2,30), the difference is 0;12; thus for the positions.

***P6** Saturn: period relation*

This procedure, a duplicate of No. **41** P4, mentions the period relation common to system A and system B for Saturn (Table 3.57a). The number 54,0 is the corresponding total distance covered by Saturn, $Z \cdot 6{,}0°$.

***P7** Saturn system B": parameters of the zigzag function for the synodic time*

P7 mentions the parameters of the zigzag function for the synodic arc σ and the synodic time τ in accordance with system B" of Saturn (§3.8.4). The parameters are compiled in Table 3.61 (p. 109).

***P8** Saturn system B: parameters of the zigzag function for the synodic arc*

These procedures, which are duplicates of No. **41** P5,P6, contain parameters for the zigzag functions for the synodic arc σ and the synodic time τ for system B of Saturn (§3.8.4). The parameters are compiled in Table 3.61.

No. 43

museum number	BM 33048 (87–10–15,31)
provenience	Babylon
measures	5.1(*) × 7.6(*) × 2.3–2.8 cm; column width 2.0–2.5 cm
arrangement	O/R
cuneiform text	photo: Fig. F.44
contents	Mercury, unknown system; Mars system A: various procedures

Obverse?
Column i'

P1' (m+3 lines missing)
1' [...].⸢30$^{?}$⸣ **maš**
2' [...]⸢2$^{?}$⸣.30 **maš**

P2'
3' [... x]
4' [... x] ⸢x⸣
5' [...] ⸢xx⸣
6' [...x] ⸢1⸣
7' [...] ⸢**absin**$_0$ 1$^{?}$⸣
8' [...] **rin**$_2$ 2$^{?}$
9' [...] ⸢**gir**$_2$⸣ 2
10' [...] 2
11' [...] 2
12' [...] 2
(n lines missing)
Column ii'

P3' (m lines missing)
1' [xxx **ma**]**š**$_2$
2' [xx]⸢1⸣ **gu**
3' ⸢4$^{?}$⸣.16 *zib*
4' 2.5.40 **ḫun**
5' 1.58.10 **mul**$_2$
6' ⸢1$^{?}$⸣.39.50 **maš**

P4'
7' **u**$_4$**.meš gar ta ki**
8' *šamaš*$_2$ **gu**$_4$**.ud**
9' 28 **alla**
10' ⸢3$^{?}$8⸣.20 **a**
11' 54 **absin**$_0$
12' 11.20 **rin**$_2$
13' 56.20 ⸢**gir**$_2$⸣
14' 50 [**pa**]
15' 5⸢2$^{?}$⸣ [**maš**$_2$]
(n lines missing)
Column iii'

P5' (m−1 lines missing)
1' **du**-*ma* [...]
2' **ki du** [...]
3' **uš** 1⸢8⸣ [...]
4' **uš.meš** [...]
5' *mi-šil ša*$_2$ ⸢x⸣ [...]
6' **e**$_{11}$-*ma* [... **a.ra**$_2$]
7' *mi-ḫi-il-ta* **du** [...]
8' *ana* **me e** [...]
9' 18 ⸢x⸣ [x...]
10' ⸢**uš**$^{?}$⸣ [x...]
(n+5 lines missing)

Obverse? column i'

P1' **(Unclear)**

(m+3 lines missing)
1'[...].⸢30$^{?}$⸣ Gem.
2'[...]⸢2$^{?}$⸣.30 Gem.

P2' **(Control values of an unidentified quantity)**

3'[...] 4'[...] ... 5'[...] ... 6'[...]
7'[...] ⸢Vir 1$^{?}$⸣.
8'[...] Lib 2.
9'[...] ⸢Sco⸣ 2.
10'[...] 2.
11'[...] 2.
12'[...] 2.
(n lines missing)
Column ii'

P3' **(Mercury: control values of the synodic arc?)**

(m lines missing)
1'[...] ⸢Cap⸣.
2'[...]⸢1⸣ Aqr.
3'⸢4$^{?}$⸣;16 Psc.
4'2,5;40 Ari.
5'1,58;10 Tau.
6'⸢1$^{?}$⸣,39;50 Gem.

P4' **(Mercury: control values of a temporal push?)**

7'You put down the 'days'. From the position of 8'the Sun. Mercury.
9'28 Cnc.
10'⸢38⸣;20 Leo.
11'54 Vir.
12'11;20 Lib.
13'56;20 Sco.
14'50 [Sgr.]
15'5⸢2$^{?}$⸣[+x Cap.]
(n lines missing)
Column iii'

P5' **(Outer planet: subdivision of the synodic cycle)**

(m−1 lines missing)
1'you multiply, and [...] 2'the position proceeds [...] 3'station, 18 [...] 4'stations$^{?}$ [...] 5'half of ... [...] 6'you subtract, and [...] 7'you multiply [by] the 'stroke' [...] 8'it rises to daylight (AR) [...] 9'18 ... [...] 10'⸢it is stationary$^{?}$⸣ [...]
(n+5 lines missing)

Reverse?
Column i'

P6' (about n−3 lines missing)
0' [*ina* **mul$_2$** *u* **maš** 45 **tab**]
1' [*ša$_2$ al* 30 **maš diri**]
2' ⸢**a**⸣.[**ra$_2$** 40 **du**]
3' *ina* **alla** *u* [**a** 30 **tab**]
4' *ša$_2$ al-*⸢*la*⸣ [30 **a diri**]
5' **a.ra$_2$** 1.⸢20⸣ [**du**]
6' *ina* **absin$_0$ rin$_2$** 40 ⸢**tab**⸣
7' *ša$_2$ al* 30 **rin$_2$ diri**
8' **a.ra$_2$** 1.30 **du**
9' *ina* **gir$_2$** *u* **pa** 1 ⸢**tab**⸣
10' *ša$_2$ al* 30 **pa diri**
11' **a.ra$_2$** 1.30 **d**[**u**]
12' *ina* **maš$_2$** *u* **gu** 1.[30 **tab**]
13' *ša$_2$ al* 30 **g**[**u diri**]
14' **a.ra$_2$** 45 [**du**]
15' *ina zib u* **ḫun** 1.[7.30 **tab**]
16' [*ša$_2$*] ⸢*al*⸣ 30 [**ḫun diri**]
17' [**a.ra$_2$** 40 **du**]
(about m−1 lines missing)

Column ii'

P7' (about n lines missing)
0' [**alla a** 36]
1' [**absin ri**]**n$_2$** ⸢4⸣2
2' **gir$_2$ pa** 48
3' **maš$_2$ gu** 54
4' *zib* **ḫun** 48
5' **mul$_2$ maš** 42
6' 3 *ki-ṣir*
7' **zi** *ša$_2$* **an**

P8' 8' **ta uš igi**
9' ⸢*ana*⸣ **uš** *ar$_2$-tu$_2$*
10' [x?] **alla a**
11' [1]⸢9⸣.30 **la$_2$**-*is* **uš**
12' [x? **absin$_0$**] **rin$_2$**
(about m lines missing)

Reverse? column i'

P6' **(Mars system A: updating *B* with the synodic arc)**

(about n−3 lines missing)
0'[In Tau and Gem you add 45.] 1'[(The amount) by which it exceeds 30 Gem] 2'[you multiply] by [0;40.]
3'In Cnc and [Leo you add 30.] 4'(The amount) by which it exceeds [30 Leo] 5'[you multiply] by 1;⸢20⸣.
6'In Vir, Lib ⸢you add⸣ 40. 7'(The amount) by which it exceeds 30 Lib 8'you multiply by 1;30.
9'In Sco and Sgr ⸢you add⸣ 1,0 10'(The amount) by which it exceeds 30 Sgr 11'you multiply by 1;30.
12'In Cap and Aqr [you add] 1,[30.] 13'(The amount) by which it exceeds 30 Aq[r] 14'[you multiply] by 0;45.
15'In Psc and Ari [you add] 1,[7;30.] 16'[(The amount) by it which] exceeds 30 [Ari] 17'[you multiply by 0;40.]
(about m−1 lines missing)

Column ii'

P7' **(Mars: control values of zigzag function v_2)**

(about n lines missing)
0'[Cnc, Leo 0;36.]
1'[Vir, Lib] 0;⸢4⸣2.
2'Sco, Sgr 0;48.
3'Cap, Aqr 0;54.
4'Psc, Ari 0;48.
5'Tau, Gem 0;42.
6'3 knots 7'of the displacement of Mars.

P8' **(Mars system A: angular push S1 to S2)**

8'From the first station 9'⸢to⸣ the second station:
10'[...] Cnc, Leo: 11'it recedes [1]⸢9⸣;30, then it is stationary.
12'[... Vir,] Lib:
(about m lines missing)

Critical and philological notes

Obverse? column i'

2' ⸢2⸣: or any number 2–8.
4' Traces of two vertical wedges are visible.
5' Traces of a number, perhaps 5, followed by the upper end of 2–3 vertical wedges.

Column ii'

2' ⸢1⸣: only the lower half is visible; could be part of any number 1–8.
7'–8' This sentence is problematic. **ki** = *qaqqaru*, 'zodiacal position', or read *itti*, 'with'? One might also read **absin$_0$** = Vir, but this does not result in a more satisfactory interpretation.
8' The damaged wedges to the left of 20 are assumed to belong to column i'.
20: since the number 20 seems to make no sense I have provisonally read *šamaš$_2$*, 'Sun'.
15' 5⸢2⸣: the 2 might be another digit 2–8.

Column iii'

5' After *ša$_2$* a winkelhaken is visible.
7' *miḫiltu* 'stroke': occasionally attested in LB mathematical and astronomical texts with a meaning 'number', 'coefficient', perhaps also 'push'; cf. the Glossary. The meaning here is unclear.

Reverse? column ii'

6' *kiṣru* 'knot': cf. the commentary.
10' The small gap at the beginning may have contained *ina*, 'in'.
11' It can be assumed that the gap at the beginning contained a 10; cf. the commentary.
la$_2$-*is* = *inaḫḫis*, 'it recedes'.

Commentary

This fragment does not preserve any edges of the tablet. Its original measures are unknown. The obverse (identification uncertain), probably also the reverse, contains three narrow columns, two of which are partly preserved to their full width of 2.0–2.5 cm. Similarly narrow columns are known from a few other planetary procedure texts (§1.4). They have been numbered from left to right on what is believed to be the obverse, and from right to left on the other side. The columns are further divided into procedures, of which there are 5 on the obverse and 3 on the reverse. P4', perhaps also P3' concern Mercury; P5' concerns an outer planet. All procedures on the reverse concern Mars. The original tablet must have contained several further procedures.

P1' *Unclear*

Only traces of numbers and the sign Gem can be identified.

P2' *Control values of an unidentified quantity*

Very little remains of this procedure. Oi'8'–9' mention the adjacent zodiacal signs Vir, Lib and Sco, each followed by a number, which is either 1 or 2. Hence P2' is probably a table of control values, each associated with a zodiacal sign, but the meaning of the numbers is not clear.

P3' *Mercury, unknown system: values of the synodic arc?*

P3' contains a list of numbers, each associated with a zodiacal sign. The first entries are missing. Since the preserved lines mention all signs between Cap and Gem, the lines above Oii'1' may have contained Cnc – Sgr, which would require at least 6 lines. Comparison with systems A_1, A_2 or A_3 for Mercury suggests that the numbers might be synodic arcs (σ_j). There is not enough information for a complete reconstruction of the underlying algorithm.

P4' *Mercury, unknown system: control values of a temporal push?*

The first 9 lines of P4' are preserved. The introductory statement is not fully understood (cf. the critical notes), but seems to suggest that the procedure is concerned with a temporal push ($\delta\tau$) immediately following the conjunction with the Sun. Most of 7 preserved control values, each associated with a zodiacal sign, are roughly consistent with the values of $\delta\tau$ (ML to EF) in system A_2 (§3.4.3.3), except 11;20 in Lib. At least 5 more lines must be assumed below Oii'15' for the remaining zodiacal signs. The computational system remains unidentified.

P5' *An outer planet; unclear*

The beginning and end of P5' are lost. The 10 partly preserved lines mention stations (Oiii'3',4',10'), forward motion (Oiii'2') and an acronychal rising (Oiii'8'). Hence P5' deals with the subdivision of the synodic cycle for an outer planet, perhaps Mars. Oiii'5'–7' contain a computation of unknown significance, for which there are no parallels. An unidentified quantity is halved, subtracted and multiplied by the 'stroke' (*miḫiltu*).

P6' *Mars system A, updating B with the synodic arc*

The beginning and end of P6' are missing. The preserved lines contain rules for updating *B* with the step function for σ in accordance with Mars system A. For a discussion cf. §3.6.2.1. The formulation is based on template STEP.A.3. The complete reconstruction suggests that not much text is missing above Ri'0' and below Ri'17'.

P7' *Mars: control values of zigzag function* v_2

The beginning is missing. Rii'1'–5' contain control values of a zigzag function v_2 of the zodiacal position representing the daily displacement of Mars, as stated in Rii'7'. The minimum 0;36°/d occurs in Cnc – Leo (zone 2), the maximum 0;54°/d in Cap – Aqr (zone 5). For a discussion of this function cf. §3.6.2.3.iii (p. 88). The meaning of '3 knots' is not clear, but note that Mars system A features 6 zones, so that there are 3 of them between the minimum of v_2 (0;36) and the maximum (0;54).

P8' *Mars system A, angular push S1 to S2*

This procedure concerns the angular push from S1 to S2, as stated in Rii'8'–9'. For a discussion of the various schemes for this push cf. §3.6.2.3.ii, especially Table 3.30. Only the values of $\delta\Sigma$ for Cnc and Leo are partly preserved. Since one expects a number between 10° and 20°, an initial 10 must be restored. In Rii'12' the zodiacal sign Lib is preserved. It must have been preceded by a number and Vir, with which it forms the next pair of signs. Both pairs of signs coincide with zones of system A (Table 3.26b). If all 6 zones were listed on the original tablet, 9 additional lines must be assumed below Rii'12'.

No. 44

museum number	BM 33801 (Rm4,361)
provenience	Babylon
date	early Seleucid
measures	6.9 × 5.7(*) × 2.0–2.5 cm (type G or H?)
arrangement	O/R
cuneiform text	photo: Fig. F.47; *ACT*, Pl. 250
bibliography	*ACT* 811, Zs
contents	Jupiter system A$_6$; Saturn system A; Mars system A: various procedures

Obverse

P1 1 ⸢mul$_2$⸣.babbar ta alla en rin$_2$ 30 ta rin$_2$ en gir$_2$.tab 30⸢x⸣
2 *ina* pa 34 ⸢ta maš$_2$⸣ en ḫun 3⸢6?⸣ [xx]
3 *ina* mul$_2$.mul$_2$ 34.⸢30?⸣ [x] ⸢*ina*⸣ maš.maš ⸢33⸣ [xx]
4 *a-na* 12 mu.meš 4 ⸢uš 30 ninda? tab?⸣ ta maš.maš ⸢en⸣ [pa? xx]
5 ta maš$_2$ en mul$_2$.mul$_2$ 4.30 ki-⸢*šu$_2$ ana*?⸣ [šu$_2$ x] ⸢la$_2$?⸣ [x] ⸢mul$_2$.babbar?⸣ [xxx]
6 8 uš ki-*šu$_2$ ana* šu$_2$ la$_2$ ta alla en rin$_2$ ⸢4.40 *ana*?⸣ [xx]
7 *ina* pa 7.⸢40?⸣ *ina* maš$_2$ 9.30 *ina* gu *u* ⸢*zib*⸣ [xxx]
8 *ina* ḫun *ina* ⸢mul$_2$⸣.mul$_2$ *u* maš.maš 8.30 la$_2$ *ina* 1.35 mu.m[eš xxx]

P2 9 genna ta a en gu *ina* 29 mu.meš 6 [x la$_2$ xx]
10 ⸢6 tab ta *zib*.me⸣ en alla *ina* 29 ⸢mu⸣.[meš x la$_2$ xx]
(n lines missing)

Reverse

P3’ (about n lines missing)
1’ ⸢gir$_2$ pa⸣ 10 maš$_2$ ⸢gu⸣ [15 *zib*.me ḫun 11.15]
2’ mul$_2$.mul$_2$ maš.maš 7.30 ki-*šu$_2$ ana* šu$_2$ la$_2$ *ina* [1.19 mu.meš]
3’ alla a 1.40 tab absin$_0$ rin$_2$ 2.15 tab gir$_2$ p[a 3.20]
4’ maš$_2$ gu 5 *zib*.me ḫun 3.40 *ina* mul$_2$.mul$_2$ [maš.maš 2.30]
5’ ki-*šu$_2$* tab *ina* 4.44 mu.me ki-⸢*šu$_2$*⸣ [kur-*ad*]

Co 6’ *an-na-a* til-*šu$_2$* ta giš.⸢da *ša$_2$ mi-ḫi-x*⸣ [xxx]
7’ *ana* šu.2 nu e$_3$ *gi$_3$-ṭu* [md]šu$_2$.dub.numun a [m]⸢x⸣[xx]
8’ a [m]*mu-še-zib* šu.2 [m]mu.[d]en dumu-*šu$_2$*

Critical and philological notes

Obverse

1 30⸢x⸣: any number 32–38 is equally possible; the sign following 30 is damaged but also filled with dirt (2008).
2 36: or 35. The following break has room for about 2 signs.
3 20: or 30.
33: replaces 32 (*ACT*). The top of the third wedge of the number 3 is preserved.
4 32: replaces 33 (*ACT*. The upper parts of both wedges of the digit 2 are preserved, and the presence of a third one can be excluded. Following Neugebauer one can restore **pa** = Sgr at the end, because the next zone begins at Cap.
5 In O5–10 several signs at the end were omitted by Neugebauer.
šu$_2$ = *rabû* probably means ‘west’ here corresponding to the retrograde motion.
7 40: or any number 20–50.
10 6: replaces 5 (*ACT*). *ina*: omitted in *ACT*.

Reverse

1’ 15: here Neugebauer transliterates [1]5 **l**[**a$_2$**], but no traces of these signs are currently visible on the tablet. **la$_2$** (‘it recedes’) is unexpected, because the retrograde nature of these displacements is also expressed at the end of the list in R2’.
2’ **la$_2$**: *ana* **šu$_2$**, ‘to the east’ = in the retrograde direction, suggests that **la$_2$** is to be read *inaḫḫis*, ‘It recedes’.
6’ *mi-ḫi-*: perhaps *miḫiltu* = ‘stroke; sign’, resulting in *lē’u*(**giš.da**) *ša miḫilti* = ‘wooden writing(?) board’ (otherwise not attested); or phonetically written *miḫirtu*(**gaba.ri**) = ‘copy’?
7’ *ana qātē* **e$_3$**: a parallel occurs in Hunger *BAK* 182: [4]*pa-liḫ* [d]*Marduk u* [d]*Ṣar-pa-ni-tu$_4$* [5]*ana qātē*(**šu.2**) *lā*(**nu**) *ušeṣṣi*(**e$_3$**), ‘He who honors Marduk and Ṣarpanītu shall not let it go out to (other) hands’. Whereas in the parallel **e$_3$** can only be a Š-stem of *aṣû*, a G stem seems preferable in the present case because **e$_3$** does not have an explicit subject.
gi$_3$-ṭu: one-columnar tablet with scholarly content; cf. p. 11.

Obverse

P1 **(Jupiter system A_6)**

a **(Updating *B* with the synodic arc)**

[1]⸢Jupiter⸣. From Cnc until Lib 30; from Lib until Sco ⸢30+*x*⸣; [2]in Sgr 34; ⸢from Cap⸣ until Ari 3⸢6?⸣ [...]; [3]in Tau 34;⸢30?⸣; in Gem ⸢33⸣ [...].

b **(Net displacements for various intervals)**

[4]For 12 years ⸢you add⸣ 4° 30nin.

From Gem until [Sgr ...] [5]From Cap until Tau its position ⸢recedes⸣ 4;30 to [the west].

... ⸢Jupiter?⸣ [...] [6]8° its position recedes to the west. From Cnc until Lib 4;40 [...]; [7]in Sgr 7;40; in Cap 9;30; in Aqr and Psc [...]; [8]in Ari, in Tau and Gem it recedes 8;30. In 1,35 years [...].

P2 **(Saturn system A?: net displacements)**

[9]Saturn. From Leo until Aqr in 29 years [you subtract] 6;[...; in 30 years] [10]you add 6.

From Psc until Cnc in 29 years [you subtract ...; in 30 years you add ...]

(n lines missing)

Reverse

P3' **(Mars system A: net displacements)**

a **(For 47 years)**

(about n lines missing)

[1']Sco, Sgr: 10. Cap, ⸢Aqr: you subtract 15⸣. [Psc, Ari: 11;15.] [2']Tau, Gem: 7;30 its position recedes to the east.

b **(For 79 years)**

In [1,19 years:] [3']Cnc, Leo: you add 1;40. Vir, Lib: you add 2;15. Sco, Sgr: [3;20.] [4']Cap, Aqr: 5. Psc, Ari: 3;40. in Tau, [Gem] [5']you add [2;30] to it.

c **(For 284 years)**

In 4,44 years [it reaches] its position.

Co **(Colophon)**

[6']This is its end. From a wooden board ... [...] [7']It should not go out to (strange) hands. *Giṭṭû*-tablet of Marduk-šāpik-zēri, son of [...], [8']descendant of Mušēzib. Hand of Iddin-Bēl, his son.

Commentary

This fragment preserves the upper (lower) part of the obverse (reverse) and a part of the right edge of a tablet. It contains 3 procedures, which are separated by horizontal lines. Each of them deals with a different planet, namely Jupiter (P1), Saturn (P2) and Mars (P3'). The tablet concludes with a colophon mentioning Marduk-šāpik-zēri//Mušēzib as the 'owner', and his son Iddin-Bēl as the scribe (Hunger, *BAK* 143). Colophons of this type are well known from Uruk, but rare in Babylon. A date is not given, but the mentioned astronomers are known to have lived in the early Seleucid era. The colophon also states that the tablet concludes a series, and that it was copied from a wooden board (inlaid with wax).

P1 Jupiter system A_6: various procedures

This composite procedure concerns Jupiter system A_6 (§3.7.7). Part **a** (O1–3) lists 6 values of the synodic arc (σ_j), each pertaining to a zodiacal region consisting of 2 adjacent signs (Table 3.47), in accordance with template PSTEP. For a discussion cf. §3.7.7.1. Part **b** (O4–8) mentions net displacements $dB_j(s)$ for 12^{yr} ($s = 11$), and probably 71^{yr} ($s = 65$). The data are compiled in Table 3.47 (p. 99). Values for 12^{yr} are provided for two regions: Gem – Sgr (comprising zones 6, 1–3) and Cap – Tau (zones 4–5). Only the latter value is preserved: 4;30°. O6 mentions a retrograde net displacement of 8° (ignored in *ACT*), probably pertaining to 71^{yr} (not preserved); compare the values for system A in Table 3.37 (p. 91). The number 4;40 (O6), apparently associated with the region Cnc – Sgr (zones 1–3), remains unidentified; perhaps it also pertains to 71^{yr}, in which case it should be subtractive. In O7–8 further subtractive net displacements, probably also pertaining to 71^{yr}, are listed for several regions, $-7;40$ in Sgr (zone 3), $-9;30$ in Cap, another value in Aqr and Psc and $-8;30$ in Ari – Gem. Only one of these regions coincides with a region for σ. It is peculiar that some of the regions seem to overlap.

***P2** Saturn system A?: net displacements*

Only the first 2 lines of P2 are partly preserved. They concern net displacements of Saturn, probably for system A (§3.8.2), but the zonal boundaries are only given as zodiacal signs and not in degrees. The formulation is based on templates DIFF.A.1 and DIFF.A.2 (§2.6.10.4). The value of dB_1 for 29^{yr} ($s = 28$) is only partly preserved as 6, but the exact value in zone 1 of system A is $-5;12,30°$ ($-6;15°$ in zone 2), suggesting that the mentioned value is rounded (cf. §3.57b). The following value +6° (O10) also belongs to zone 1. It probably pertains to 30^{yr}, and is also rounded.

***P3'** Mars system A: net displacements*

P3' is translated here for the first time; in *ACT* the content of P3' was presented only in the form of a table (p. 381). P3' contains a list of net displacements of Mars for system A for various intervals expressed in years. For a discussion cf. §3.6.2.4. The beginning is lost. For each interval $dB_j(s)$ is provided for all 6 zones (Table 3.32). The preserved intervals are 47^{yr} (part **a**), 79^{yr} (part **b**) and 4,44 = 284^{yr} (part **c**). However, two of the net displacements for 79^{yr} differ from what one expects, perhaps as a result of truncation.

No. 45

museum number	BM 34765 (Sp2,258)
provenience	Babylon
measures	7.6(*) × 6.3(*) × 2.8–3.3 cm
arrangement	O/R
cuneiform text	photo: Fig. F.48 (R); *ACT*, Pl. 254; copy: *LBAT* 154
bibliography	*ACT* 819c
contents	synodic tables for one or more planets; procedure for Saturn

Reverse
Column ii'

Co 6' **im** 3 **du**$_3$-*ša*$_2$ *ša*$_2$ ⸢xxx⸣ [...]
7' **im.dub** *ša*$_2$ **egir**-*šu*$_2$ [x...]
P1' 8' *ša*$_2$ ^{d}⸢**genna**⸣ 4.2⸢5 **mu.meš**⸣ [4.16 **igi.meš** ...]
9' 9 **bal.meš** 54 **ki**.[**meš** xx...]

Reverse column ii'

Co **(Colophon)**

6'Tablet 3. Procedure for ... [...] 7'The tablet which is after it [...]

P1' **(Saturn: periods)**

8'For Saturn. 4,25 years, [4,16 appearances ...] 9'9 revolutions, 54,0 position[s ...]

Critical and philological notes

Reverse column ii'

6' Before the break there are traces of several vertical wedges and winkelhakens.

Commentary

This fragment preserves a segment of the upper (lower) edge of the obverse (reverse) of the tablet. Nothing remains of the left and right edges. The thickness of the fragment decreases towards the right, suggesting that it is situated closer to the right side than to the left side. The obverse contains two synodic tables which remain unexplained (cf. *ACT*, p. 435, and Volume II). The second synodic table continues on the reverse. It is followed by (at least) two columns of text. Virtually nothing remains of the first column. The second one begins with a colophon, identifying the tablet as the third in a series (Rii'6'). It also contains a reference to the fourth tablet, but the catchline is not preserved (Rii'7'). This is followed by the remains of a procedure.

P1' Saturn: period relations of systems A and B

P1' mentions the period relations for Saturn that are common to systems A and B (Table 3.57a, p. 107). The number 54,0 is the corresponding total distance covered by Saturn, $Z \cdot 6,0°$.

No. 46

museum numbers	BM 34221 (Sp,327) +34299 (Sp,410) +35119 (Sp2,664) +35206 (763) +35445 (1034) +45702 (81–7–6,107)
provenience	Babylon
measures	18.2 × 14.7(*) × 2.1–3.0 cm (type A); column width 9 cm
arrangement	O/R
cuneiform text	photo: Figs. F.45, F.46; *ACT*, Pls. 251–252; copy: *LBAT* 140–145
bibliography	*ACT* 812; *HAMA*, pp. 465–466; Swerdlow (1998), pp. 170–171
contents	Jupiter: systems A',B; Venus: unknown system and system A_0?

Obverse
Column i

P1' (about m+14 lines missing)
1' [xxx *mi-nu-u*$_2$ **a.ra**$_2$ 6].⸢31⸣ *lu*-**du**-*ma*
2' [*lu* 1.28.1.30.56.46 x$^?$] ⸢13.30.27⸣.46 **a.ra**$_2$ 6.31 **du**-*ma*
3' [1.28.1.30.56.46 x] 13.30.27.46 **murub**$_4$-*u*$_2$
4' [*a-na* 45.14 x] **gar**-*an* 1.48 **a.ra**$_2$ 6.31 **du**-*ma*
5' [11.43.48 **gin**$_7$ *ša*$_2$ **ki.meš du**$_3$]-*uš* 15 **pa** 45.14 **u**$_4$**.meš**
6' [15 *zib* 50.7.1]5 15 **absin** 40.20.45 ⸢**u**$_4$⸣**.meš**

P2' 7' [*ša*$_2$-*niš* 33.8].45 **murub**$_4$-*u*$_2$ *ša*$_2$ **ki.meš a.ra**$_2$ 1.50.⸢40⸣ **du**-*ma*
8' [1.1.8.8].20 11.4 **ki**-*šu*$_2$ **tab**-*ma* 12.5.8.8.20
9' [*a-na* **gaba**].**ri.meš mu.an.na gar**-*an* ⸢x⸣ [x] ⸢**igi**$^?$ *ana*⸣ **igi** *bi-ri-šu*$_2$-*nu* **gar**-*an*
10' [12.5.8].8.20 **ki**-*šu*$_2$ **tab**-*ma* [**u**$_4$].⸢**meš**⸣ **e**-*bi*

P3' 11' [xx] ⸢x⸣ **gar ta** 9 **alla en** ⸢9⸣ [xxx$^?$] **gir**$_2$**.tab**
12' [**ta** 9] **alla en** 9 **gir**$_2$**.tab** ⸢**tur**-*ti*⸣
13' [**ta** 9] **gir**$_2$**.tab en** 2 **maš**$_2$ **murub**$_4$-*tu*$_4$ **ta** 2 **maš**$_2$ **en** 17 **mul**$_2$ [**gal**-*ti*]
14' [**ta** 1]⸢7⸣ **mul**$_2$**.mul**$_2$ **en** 9 **alla murub**$_4$-[*tu*$_4$]

P4' 15' [*ina* **tur**]-*ti* **ki** ^{d}**utu** *ša*$_2$ **me** 12.30 [**zi**]
16' [*ar*$_2$ **ig**]**i** 30 **me** *ša*$_2$ **me** 12.30 [**zi**]
17' [3] ⸢**iti.meš**⸣ *ša*$_2$ **me** 6.40 **zi**-*m*[*a* **uš**]
18' [4 **iti.meš** *ša*$_2$ **me** 4].⸢10⸣ **zi**-*ma* **gur**-*ma* ⸢**uš**⸣

Column ii

P5' (m lines missing)
1' [xxxx] ⸢x⸣ **igi** 6.15 [xxxxx]

P6' 2' **zi** *gab-ba ša*$_2$ **tur**-*ti* **a.ra**$_2$ 7.30 **du**-*ma* [**ki**-*šu*$_2$ **tab**-*ma*]
3' **murub**$_4$-*ti ip-pal-ka* **a.ra**$_2$ 12 **du**-*ma* [**ki**-*šu*$_2$ **tab**-*ma*]
4' **gal**-*ti ip-pal-ka* **zi** *gab-ba* [*ša*$_2$ **gal**-*ti*]
5' **a.ra**$_2$ 3.45 **du**-*ma* **ta šа**$_3$ **la**$_2$-*ma* **murub**$_4$-*t*[*i ip-pal-ka*]
6' [**a.ra**$_2$ 6].⸢40⸣ **du**-*ma* **ta šа**$_3$ **la**$_2$-*ma* **tur**-*ti* [*ip-pal-ka*]

P7' 7' [xxx] ⸢**uš**⸣ *ar*$_2$ 20 **du**-*ma* ⸢**igi**$^?$ 3$^?$ **danna**⸣ [xx] ⸢xxx⸣
8' [xxxxxx]⸢x⸣ 20 **du** ⸢5$^?$⸣ [xxxx **tu**]**r**-*ti*
9' [xxxxxx] ⸢20 x⸣ [xxxxx] **šu**$_2$

P8' 10' [xxxxxxxxxxx]-⸢*ma*⸣ **uš**
11' [xxxxxxxxxx] ⸢x⸣ **a**
12' [xxxxxxxxxx] ⸢x⸣ **me** *nen-mud*
13' [xxxxxxxxx] *ša*$_2$ **uš igi** *u* **šu**$_2$

P9' 14' [xxxxxxxxx] ⸢*a*⸣-*na* **egir**-*ka* **gur**-*ma*
15' [xxxxxxx] ⸢x⸣ [x] 6 *ša*$_2$ *i-kap-pa-pi*
16' [xxxxxx] ⸢**mu**⸣-*ka* ⸢**gibil**⸣ **e**$_{11}$-*ma* **e**-*bi*

P10' 17' **ki.m**[**eš** *ša*$_2$ **mul**$_2$**.babbar** *ina*] ⸢**gal**⸣-*tu*$_2$ 5 **tab** *a-na* 1.⸢11⸣ **mu.meš** 6 **la**$_2$
18' *a-na* [1.23 **mu.meš**] ⸢1⸣ **la**$_2$ *a-na* 1.35 **mu.meš** 4 **tab**
19' *a-na* ⸢2⸣.[46 **mu.meš** 2] ⸢**la**$_2$⸣ *a-na* 4.21 **mu.meš** ⸢2$^?$⸣ **tab**
20' *a-na* [xxxx 7.7] ⸢**mu**⸣**.meš ki**-*šu*$_2$ **kur**-⸢*ad*$_2$⸣

Obverse column i

P1' **(Jupiter system B)**

a' **(Computations involving Π and τ)**

(*about m+14 lines missing*)
1'[...] what should I multiply [by 6],⌜31⌝ so that it is 2'[1.28.1.30.56.46?] You multiply 13.30.27.46 by 6,31, it is 3'[1.28.1.30.56.46 ...] 13.30.27.46, the middle one, 4'[you add? it to 45;14 ...] you put it down. You multiply 1;48 by 6,31, it is 5'[11,43;48]. You [do as for the positions.].

b' **(Defining parameters of the zigzag function for τ)**

15 Sgr: 45;14 'days'; 6'[15 Psc: 50;7,1]5; 15 Vir: 40;20,45 'days'.

P2' **(Jupiter system B: computing τ from σ)**

7'[Alternatively: 33;8],45, the middle one for the positions, you multiply by 0;1,50,40, it is 8'[1;1,8,8],20. You add 11;4 to it, it is 12;5,8,8,20, 9'you put it down [for the val]ues of the year. (From) appearance to appearance you put down the distance between them, 10'you add [12;5,8],8,20 to it, and you predict the 'days'.

P3' **(Jupiter system A': zones for the synodic arc)**

11'[...] you put down. From 9 Cnc until 9 [...] Sco. 12'[From 9] Cnc until 9 Sco the small one.
13'[From 9] Sco until 2 Cap the middle one. From 2 Cap until 17 Tau [the large one].
14'[From 1]7 Tau until 9 Cnc the middle one.

P4' **(Jupiter: subdivision of the synodic cycle, scheme A'.S$_1$, zone 1)**

15'[In the (zone of)] the small one with the Sun [it moves] 0;12,30 per 'day'.
16'[After the appear]ance (FA) 30 'days' it [moves] 0;12,30 per 'day'.
17'[3] months it moves 0;6,40 per 'day', then [it is stationary (S1)].
18'[4 months] it moves backwards 0;4,10 [per 'day'], then it is stationary (S2).

Column ii

P5' **(Jupiter: subdivision of the synodic cycle, scheme A'.S$_1$, zone 1)**

(*m lines missing*)
1'[...] ... appearance (FA) [it proceeds] 6;15 [...]

P6' **(Jupiter system A': transition rules for σ)**

2'All displacements for the small one you multiply by 0;7,30, and [you add it with it, and] 3'it gives you the middle one. You multiply it by 0;12, and [you add it with it, and] 4'it gives you the large one. All displacements [for the large one] you multiply 5'by 0;3,45, and you subtract it from it, and [it gives you] the middle one. 6'You multiply it [by 0;6],40, and you subtract it from it, and [it gives you] the small one.

P7' **(Jupiter: subdivision of the synodic cycle; unclear)**

7'[...] the second station (S2) it proceeds 20, and ⌜... 3 'miles'⌝ [...] ... 8'[...] it proceeds 20, 5 [...] ⌜the small one⌝ 9'[...] ⌜20 ...⌝ [...] it sets (LA).

P8' **(Jupiter: subdivision of the synodic cycle; unclear)**

10'[...] and it is stationary. 11'[...] ... 12'[...] ... 'day', it is stationary. 13'[...] of the station, appearance (FA) and setting (LA).

P9' **(Jupiter?: Goal-Year type procedure?)**

14'[...] you turn back, and 15'[...] ... [...] 6 by which it 'bends'? 16'[...] you deduct from your new year, and you call it.

P10' **(Jupiter system A: net displacements)**

17'Positions [of Jupiter. In] the (region of the) large one you add 5.
For 1,11 years you subtract 6. 18'For [1,23 years] you subtract 1.
For 1,35 years you add 4. 19'For 2,[46 years] you subtract [2].
For 4,21 years you add ⌜2⌝. 20'For [... 7,7] years it reaches its position.

P11'	21'	*igi-gub-u*$_2$ *ša*$_2$ ⸢x⸣ [xxx] *ana* **du$_3$**-*ka* **ki.meš ta igi** *ša*$_2$ ⸢**kur**⸣
	22'	*a-na* **igi** *ša*$_2$ **šu$_2$** [xxx 5.2]7 **alla** 5.28.30 **a** 5.⸢30⸣ **a**[**bsin**]
	23'	5.31.30 **rin$_2$** ⸢5⸣.[xx **gir$_2$.tab** 5.xx **pa**] ⸢5⸣.27 **maš$_2$** 5.28 ⸢**gu**⸣
	24'	5.30 *zib* 5.[xx **ḫun** xx **mul$_2$**] ⸢5⸣.29 **maš.m**[**aš**]
P12'	25'	**ki.meš ta** ⸢**igi**⸣ [*ša*$_2$ **šu$_2$** *a-na* **igi** *ša*$_2$ **kur**] ⸢xxxx⸣ [xx]
	26'	⸢4$^?$⸣.[xxxxxxxxxxxxxxxxxx]
		(about 1 line missing)
P12a'		(about 2 lines missing)
P12b'		(about 3 lines missing)
		Reverse
		Column i
P13'	1	[xxxxxxxxxxxxxxxxxxxxxx]
	2	⸢**ta**⸣ [xxxxxxxxxxxxxxxxxxxxxx]
	3	4.1⸢4$^?$⸣ [xxxxxxxxxxxxxxxxxxxxxx]
	4	4.21 [xxxxxxxxxxxxxxxxxxxxxx]
P14'	5	**ki.meš** ⸢**ta**⸣ [xxxxxxxxxxxxxxxxxxxx]
	6	6 **a** 6 ⸢**absin**⸣ [xxxxxxxxxxxxxxxxxxx]
P15'	7	**ta šu$_2$** *ša*$_2$ **šu$_2$** *a-*⸢*na*⸣ [**igi** *ša*$_2$ **kur** xxxxxxxxxxxxx]
	8	2 **pa** 1 **maš$_2$** 1 **gu** 1 [*zib* x **ḫun** xxxxxxxxxxxxxx]
P16'	9	**ta igi** *ša*$_2$ **kur** *a-na* [**uš** *ša*$_2$ **kur** xxxxxxxxxxxxxxxxxx]
	10	10 **pa** 9 **maš$_2$** 9 **gu** [x *zib* xxxxxxxxxxxxxxxxx]
P17'	11	**u$_4$.meš ta igi** *ša*$_2$ **kur** [*a-na* **igi** *ša*$_2$ **šu$_2$** 5.26 **alla** xxx **a** xxxx **absin**]
	12	5.23.30 **rin$_2$** 5.⸢20$^?$⸣[x **gir$_2$.tab** xxx **pa** xxx **maš$_2$** xxx **gu** xxx *zib*]
	13	5.23.30 **ḫun** ⸢5⸣.[xx **mul$_2$** xxx **maš** x 5].⸢2⸣6 ⸢**alla**⸣
P18'	14	**u$_4$.meš ta igi.meš** [*ša*$_2$ **šu$_2$** *a-na* **igi** *ša*$_2$ **kur** xxx] ⸢**alla**⸣ 4.34 **a** ⸢4⸣.[xx **absin**]
	15	4.30 **rin$_2$** 4.30 **gir$_2$**.[**tab** xxx **pa** xxx **maš$_2$** xxx] ⸢**gu**⸣ 4.30 *zib* 4.30 **ḫun** ⸢4.32 **mul$_2$** 4.34⸣ [**maš**]
P19'	16	*ina* **kur** *ina* 12 **alla igi** [*ina* 4.13 **u$_4$.meš** *ina* **kur**] ⸢**šu$_2$**⸣ 4.17 **a** 4.21 ⸢**absin**⸣ 4.25 **rin$_2$**
	17	4.29 **gir$_2$.tab** 4.⸢20$^?$⸣[x **pa** xxx **maš$_2$** 4].29 **gu** 4.25 *zib* 4.22 **ḫun** 4.19 **mul$_2$** 4.16 **maš**
P20'	18	*ina* **kur** *ina* **alla šu$_2$** [*ina* 5]⸢6⸣ **u$_4$.meš** *ina* **šu$_2$ igi** 56 **a** 56 ⸢**absin**⸣ 56 **rin$_2$**
	19	⸢56 **gir$_2$.tab** 1 **pa**⸣ [1].⸢2⸣ **maš$_2$** 1.2 **gu** 1.⸢2⸣ *zib* 1 ⸢**ḫun**⸣ 1 **mul$_2$** 58 **maš**
P21'	20	[xxxxx *a*]-⸢*na*⸣ **uš** *ša*$_2$ **šu$_2$** 4.12 ⸢**alla.meš**$^?$⸣ 4.7 **a** 4.2 **absin**
	21	[xxx **rin$_2$** xxx] ⸢**gir$_2$**⸣.**tab** 3.58 **pa** 3.5⸢8 **maš$_2$**⸣ 4.1 **gu** 4.3 *zib* 4.6 **ḫun** 4.8 **mul$_2$** 4.10 **maš**
P22'	22	[xxxx] ⸢*a-na* 17$^?$ **u$_4$.meš** *ina* **šu$_2$ šu$_2$**⸣ 1⸢7⸣ **alla** 17 **a** 17 **absin$_0$**
	23	[xx **rin$_2$** xx **gir$_2$**].⸢**tab**⸣ 22.30 **pa** 23 **maš$_2$** 23 **gu** 22 *zib*
	24	[xx **ḫun** xx] ⸢**mul$_2$**⸣ 19 **maš** 17 **alla**

P11' (Venus, scheme X.S$_1$: angular push from MF to EF)

[21']In order for you to construct the *igigubbû*-coefficients for ... [...] Positions from eastern appearance (MF) [22']to western appearance (EF):
[... 5,2]7 Cnc, 5,28;30 Leo, 5,30 [Vir], [23']5,31;30 Lib, 5,[... Sco, 5,... Sgr], 5,27 Cap, 5,28 ⌜Aqr⌝, [24']5,30 Psc, 5,[... Ari, ... Tau], 5,29 Gem.

P12' (Venus, scheme X.S$_1$: angular push from EF to MF)

[25']Positions from [western] appearance (EF) [to eastern appearance (MF)]
... [...] [26']4?.[...]
(*about 1 line missing*)

P12a' (Venus, scheme X.S$_2$: angular push from MF to ML?)

(*about 2 lines missing*)

P12b' (Venus, scheme X.S$_2$: angular push from ML to EF?)

(*about 3 lines missing*)
Reverse column i

P13' (Venus, scheme X.S$_2$: angular push from EF to ES?)

[1][...] [2]From [...] [3]4,1⌜4⌝ [...] [4]4,21 [...]

P14' (Venus, scheme X.S$_2$: angular push from ES to EL?)

[5]Positions from [... to ...] [6]6 Leo, 6 Vir [... it recedes...]

P15' (Venus, scheme X.S$_2$: angular push from EL to MF?)

[7]From western setting (EL) to [eastern appearance (MF) ...]: [8]2 Sgr, 1 Cap, 1 Aqr, 1 [Psc, ... Ari ... it recedes...]

P16' (Venus, scheme X.S$_2$: angular push from MF to MS?)

[9]From eastern appearance (MF) to [eastern station (MS) ...]: [10]10 Sgr, 9 Cap, 9 Aqr, [... Psc, ... it recedes...]

P17' (Venus, scheme X.S$_1$: temporal push from MF to EF)

[11]'Days' from eastern appearance (MF) [to western appearance (EF).]:
[5.26 Cnc, ... Leo, ... Vir] [12]5,23;30 Lib, 5,⌜20?⌝ [... Sco ... Sgr ... Cap ... Aqr ... Psc] [13]5,23;30 Ari, ⌜5⌝,[... Tau ... Gem ... 5],⌜26 Cnc⌝.

P18' (Venus, scheme X.S$_1$: temporal push from EF to MF)

[14]'Days' from [western] appearances (EF) [to eastern appearance (MF):
...] ⌜Cnc⌝, 4,34 Leo, 4,[... Vir] [15]4,30 Lib, 4,30 Sco, [... Sgr ... Cap ...] ⌜Aqr⌝, 4,30 Psc, 4,30 Ari, ⌜4,32 Tau, 4,34⌝ [Gem].

P19' (Venus, scheme X.S$_2$: temporal push from MF to ML)

[16]It appears in the east (MF) in 12 Cnc, it sets [in the east (ML) in 4,... 'days'];
4,17 Leo, 4,21 Vir, 4,25 Lib, [17]4,29 Sco, 4,⌜20?⌝ [... Sgr ... Cap, 4],29 Aqr, 4,25 Psc, 4,22 Ari, 4,19 Tau, 4,16 Gem.

P20' (Venus, scheme X.S$_2$: temporal push from ML to EF)

[18]It sets in the east (ML) in Cnc, [in 5]6 'days' it appears in the west (EF);
56 Leo, 56 Vir, [19]56 Sco, 1,0 [Sgr, 1],⌜2⌝ Cap, 1,2 Aqr, 1,2 Psc, 1,0 ⌜Ari⌝, 1,0 Tau, 58 Gem.

P21' (Venus, scheme X.S$_2$: temporal push from EF to ES)

[20][...] to western station (ES): 4,12 ⌜Cnc⌝, 4,7 Leo, 4,2 Vir, [21][... Lib ...] Sco, 3,58 Sgr, 3,5⌜8 Cap⌝, 4,1 Aqr, 4,3 Psc, 4,6 Ari, 4,8 Tau, 4,10 Gem.

P22' (Venus, scheme X.S$_2$: temporal push from ES to EL)

[22][...] ⌜for 17? 'days' setting in the east (EL)⌝: 17 Cnc, 17 Leo, 17 Vir, [23][... Lib ...] Sco, 22;30 Sgr, 23 Cap, 23 Aqr, 22 Psc, [24][... Ari ...] Tau, 19 Gem, 17 Cnc.

P23’ 25 [*ina* šu$_2$ *ina* **alla**] ⸢**šu$_2$**⸣ *ina* 15 **u$_4$.meš** *ina* **kur igi** 15 **alla** 15 **a**
26 [xx **absin** xx **rin$_2$** xx] **gir$_2$.tab** 4.30 **pa** 1 **maš$_2$** 1 **gu** 2 *zib*
27 [xx **ḫun** xx **mul$_2$** xx] ⸢**maš**⸣**.maš** 15 **alla**

P24’ 28 [xxxx] ⸢3$^?$⸣8.⸢1⸣4.5 **u$_4$**.⸢**meš**$^?$⸣ [xxxxxx] ⸢x⸣

P25’ 29 [xxx] *ša$_2$* ^{d}*dil-bat* **igi** *ša$_2$* **kur** *ana* **šu$_2$** *ša$_2$* ⸢**kur** x⸣ [xxxxxxx]
30 [x *a*]-*na* **šu$_2$** *ša$_2$* **šu$_2$** 4.14.13 : **igi** *ša$_2$* **šu$_2$** *a-na* **uš** [*ša$_2$* **šu$_2$** xxxx]
31 [xxx *a*]-⸢*na*⸣ **šu$_2$** *ša$_2$* **šu$_2$** 6.20 **gur šu$_2$** *ša$_2$* **šu$_2$** *a-na* **igi** [*ša$_2$* **kur** xxxx]
32 [xxxxxx] ⸢23$^?$ **gur**⸣ *an-na-a igi-g*[*ub-u$_2$* xxxx]

P26’ 33 (traces)
(about m lines missing)
Column ii

P27’ 1 ⸢*u$_4$*⸣-*mu dil-bat ina* **šu$_2$ igi** *lu ina* **kur šu$_2$**-*u$_2$* **u$_4$.meš** *u* **ki.meš** *ana igi-gub-u$_2$* [x]
2 *ina* **šu**.2-*ka tu-kal* **be**-*ma dil-bat ina* **šu$_2$** *ina* **absin igi** 4.19.15 **tab**-*ma ina* **šu$_2$** ⸢**uš**⸣
3 ⸢6$^?$⸣.15 **gur**-*ma* **šu$_2$ pa** 2 **ta e$_2$ šu$_2$**-*šu$_2$* **la$_2$ maš$_2$** 3.30 **la$_2$**
4 ⸢**gu**⸣ 3.30 **la$_2$** *zib* 3 **la$_2$ ḫun** 2 **la$_2$ mul$_2$** 1 **la$_2$ maš.maš** 1 **tab**
5 ⸢**alla**⸣ 3 **tab a** 3.30 **tab absin** 4 **tab rin$_2$** 2 **tab gir$_2$.tab** 1 **tab uš** *ša$_2$* **šu$_2$ kimin**
6 [**ki.meš**] *ana igi-gub-u$_2$* **igi** *ša$_2$* **kur du$_3$**-*uš šum-ma* ⸢**pa**⸣ *i-pu-lu-ka* **pa ki e$_2$ igi**-*šu$_2$*
7 [x$^?$] ⸢**maš$_2$**⸣ 3.30 **tab gu** 3.30 **tab** *zib* 3.30 **tab ḫun** 3 **tab mul$_2$** 1 **tab maš** 1 **la$_2$**
8 [**alla** 3] ⸢**la$_2^?$ a**⸣ 3.30 **la$_2$ absin** 3.30 **la$_2$ rin$_2$** 2 **la$_2$ gir$_2$.tab** 1 **la$_2$ uš** *ša$_2$* **kur** ⸢**kimin**⸣

P28’ 9 [**šu$_2$ igi** *u*] ⸢**uš**⸣ *ša$_2$* ⸢**kur**⸣ *ana* **šu$_2$ igi** *u* **uš** *ša$_2$* **kur** 9.35.30 **ki tab**
10 [xxxx] ⸢x **ki** x-*tu$_4^?$*⸣ **du$_3$**-*uš*

P29’ 11 [**šu$_2$ igi** *u* **uš**] *ša$_2$* **šu$_2$** *a-na* **šu$_2$ igi** *u* **uš** *ša$_2$* **šu$_2$** 9.3⸢5.30 **ki**$^?$ **tab**$^?$ **zi**$^?$⸣ *u-tar* **du$_3$**

P30’ 12 [xxxx] ⸢**u$_4^?$**⸣.**meš** *ana* **igi**-*ka* **igi** [*ša$_2$*] ⸢**kur**⸣ [*ana* **šu$_2$** *ša$_2$* **kur** 4].⸢2⸣2.15 **šu$_2$** *ša$_2$* **kur**
13 [*ana* **igi** *ša$_2$* **šu$_2$** xx] 45 **igi** *ša$_2$* **šu$_2$** *ana* ⸢x⸣ [xxxxxxx] ⸢x **uš**⸣ *ša$_2$* **šu$_2$**
14 [xxxxx] *ana* **šu$_2$** *ša$_2$* **šu$_2$** 18 **šu$_2$** ⸢*ša$_2$*⸣ [**šu$_2$** *ana* **igi** *ša$_2$* **kur** x **igi**] ⸢*ša$_2$* **kur**⸣ *ana* **uš** *ša$_2$* **kur**
14a 18

P31’ 15 [xxxxxx] ⸢x **šu$_2^?$**⸣ **igi uš** ⸢*ša$_2$*⸣ [xxxxxxxx] **tab**
16 [xxxxxxxx] ⸢*ana*⸣ **uš** [xxxxxxxx]-⸢*tu$_4$*⸣ **du$_3$**-*uš*
17 [xxxxxxxxxxxxxxxxxxxxxx]
18 [xxxxxxxxxxxxxxxxxxxxx] ⸢10⸣
19 [xxxxxxxxxxxxxxxxxxxxxx] ⸢x⸣
20 [xxxxxxxxxxxxxxxxxxxxx] ⸢x⸣
21 [xxxxxxxxxxxxxxxxxxxxx]-⸢*ka*⸣
22 [xxxxxxxxxxxxxxxxxxxxx] ⸢**la$_2$**⸣
(about m+14 lines missing)

Critical and philological notes

Obverse column i

1’ ⸢31⸣: traces of the lower parts.

4’ The parallel No. **18** P12 suggests that part of the gap at the beginning may be restored as *ana* 45.14, ‘[you add?] to 45;14’; cf. the commentary.

5’ The gap before -*uš* has been restored on the basis of the parallel No. **18** P12, but there is little room for it.

9’ The traces after **gar**-*an* are compatible with U$_4$.
igi *ana*: only the upper ends of these signs are visible.

11’ Before **gir$_2$.tab** there is space for about 3 signs, but one expects nothing.

Column ii

3’ *ip-pal-ka* = *ippalka*, lit. ‘it answers you’ (3 c. sg. pres. G *apālu*).

7’ At the end, traces of about 3 signs are visible, the second of which looks like GIŠ.

8’ 5: or 6.

9’ 20: or 30.

11’ Before A the end of a horizontal wedge is visible, perhaps *ina*.

12’ Before **me** traces of a winkelhaken are visible.

15’ *i-kap-pa-pi*: probably *ikappap* = 3 c. sg. pres. G *kapāpu*, ‘to bend’. This verb is otherwise not attested in astronomical texts. There is insufficient context to establish its meaning here.

16’ **mu**-*ka* = *šattika*, ‘you year’. **gibil**: replaces AM ... (*ACT*). It is not clear what the object of *taqabbi*(**e**-*bi*) is.

20’ Between 7.7 and **mu.meš** there is space for about 4 signs. One may perhaps restore the line as *a-na* 7.7 **mu.meš nu tuk** *ina* 7.7 **mu.meš ki**-*šu$_2$* **kur**-*ad$_2$*, ‘For 7,7 years: nothing. In 7.7 years it reaches its position.’

P23' (Venus, scheme $X.S_2$: temporal push from EL to MF)

[25]It sets [in the west in Cnc], in 15 'days' it appears in the east (MF):
15 Cnc, 15 Leo, [26][... Vir ... Lib ...] Sco, 4;30 Sgr, 1 Cap, 1 Aqr, 2 Psc, [27][... Ari ... Tau ...] Gem, 15 Cnc.

P24' (Venus, scheme $X.S_2$: temporal push from MF? to MS?)

[28][...] ⸢3⸣8.⸢1⸣4.5 ⸢'days'⸣ [...] ⸢...⸣

P25' (Venus: subdivision of the synodic cycle, $\delta\Sigma$)

[29][... positions] of Venus. Eastern appearance (MF) to eastern setting (ML) [...]
[30][Western appearance (EF)] to western setting (EL) 4,14;13.
Western appearance to [western] station (ES) [...]
[31][Western station] to western setting (EL) it turns back 6;20.
Western setting to [eastern] appearance [...]
[32][...] ⸢it turns back 23?⸣. Thus the coefficients for [...]

P26' (Unclear)

[33]......] ...
(about m lines missing)
Column ii

P27' (Venus, scheme $X.S_3$ for the subdivision of the synodic cycle)

a (Angular push from EF to ES; ES to EL)

[1]The day when Venus appears in the west (EF) or sets in the east (ML): [2]you hold the times and positions for the *igigubbû*-coefficients [...] in your hands. If Venus appears in the west in Vir you add 4,19;15, and it is stationary in the west (ES). [3]It turns back ⸢6⸣;15, and sets (EL). Sgr: you subtract 2 from the house of its setting; Cap: you subtract 3;30; [4]Aqr: you subtract 3;30; Psc you subtract 3; Ari: you subtract 2; Tau: you subtract 1; Gem: you add 1; [5]Cnc: you add 3; Leo: you add 3;30; Vir: you add 4; Lib: you add 2; Sco: you add 1.
Western station (ES): ditto.

b (Angular push from MF to MS; MS to ML)

[6]You construct [the positions] for the *igigubbû*-coefficients of the eastern appearance (MF). If it gives you Sgr: the house of its appearance is Sgr. [7]Cap: you add 3;30; Aqr: you add 3;30; Psc: you add 3;30; Ari: you add 3; Tau: you add 1; Gem: you subtract 1; [8][Cnc]: you subtract [3]; Leo: you subtract 3,30; Vir: you subtract 3,30; Lib: you subtract 3; Sco: you subtract 1.
Eastern station (MS): ditto.

P28' (Venus system A_0?: total synodic arc for ML, MF, MS)

[9][Setting (ML), appearance (MF) and] eastern station (MS) until setting, appearance and eastern station: the position increases? by 9,35;30, [10][...] you construct ...

P29' (Venus system A_0?: total synodic arc for EL, EF, ES)

[11][Setting (EL), appearance (EF) and] western [station] (ES) until setting, appearance and western station: the position increases? by 9,35;30. .... procedure?

P30' (Venus, subdivision of the synodic time)

[12][...] In order for you to see the ⸢times⸣[...
Eastern] appearance (MF) [to eastern setting? (ML) 4],22;15.
Eastern setting [13][to western appearance (EF)] 45.
Western appearance to [western station (ES) ...].
Western station [14][...] to western setting (EL) 18.
[Western] setting [to eastern appearance ...]
Eastern [appearance] to eastern station (MS) [14a]18.

P31' (Venus?)

[15][...] to [...] setting, appearance, station: you add [...] [16][...] to station [...] you compute. [17–22][...]
(about m+14 lines missing)

21' *igi-gub-u_2* = *igigubbû*, '*igigubbû*-coefficient': apparently a technical term for (angular) push. After *igi-gub-u_2 $ša_2$* there is a sign beginning with a horizontal wedge.

25' After the first gap the upper parts of several signs similar to 2 2 10 1 are visible.
26' 4: or 5–8.

Reverse column i

2 All line numbers in this column are 1 lower in *ACT*.
3 14: or any number 11–18
14 **igi.meš**: the plural determinative **meš** is probably an error.
15 This line continues on the right edge.
16 *ina* **kur** = 'in the east'
17 This line continues on the right edge.
19 1.⌜2⌝ *zib*: the 2, also copied by Pinches, is very clear. Replaces 1.1 (*ACT*).
20 **alla.meš**: this unusual writing for Cnc is not without parallel in the procedure texts (No. **53** Oii22').
21 **pa** 3.58: or **pa** 3.57. The line continues on the edge.
22 ⌜*a-na* 17⌝: or ⌜*a-na* 14⌝.
28 The damaged sign at the end might be **la**$_2$.
29 At the end the lower half of a vertical wedge is visible.
32 ⌜23?⌝: or 26.

Column ii

3 6: or 5.
e$_2$ = *bītu*, 'house': apparently a technical term for zodiacal position, used repeatedly in P27' (cf. the Glossary).
5 **kimin**, 'ditto', is written with a large empty space between KI and MIN.
8 At the beginning there are traces compatible with **la**$_2$ or **tab**.
9 **ki tab**: the Akkadian reading and the syntactic construction are not clear. One would expect **ki ki tab**, 'you add with the position'.
10 At the beginning the final winkelhaken of a sign is visible.
du$_3$: wrongly copied as LAL by Pinches. There is a large empty space with two 'firing holes' between *tu*$_4$ and **du**$_3$.
11 *u-tar* = *uttar*, 3 c. sg. pres. G *(w)atāru*, 'it exceeds'? Reading uncertain.
zi: or **gaba**.
12 Before **meš** traces of a vertical wedge are visible, perhaps belonging to **u**$_4$.
14a The number 18 is written in a smaller script to the right of column i, in between lines 14 and 15.
15 Before **tab** there is a 'firing hole'.
17 There is another 'firing hole' at the end of this line.

Commentary

This tablet is partly preserved on four joined fragments which include segments of the left edge and the lower (upper) edge of the obverse (reverse), but no other edges. On both sides the text is divided over two columns. Considerable portions of the tablet, especially from the top of column i on the obverse and the bottom of column ii on the reverse, are still missing. At least 31 procedures are partly preserved. Within each column consecutive procedures are usually separated by double horizontal lines (single horizontal lines are visible elsewhere on the tablet). The present edition offers the first complete translation of the tablet; in *ACT* P19'–P24' were not fully translated. P1'–P10' are concerned with Jupiter and were probably preceded by further procedures for that planet. P1' and P2' (system B) are close duplicates of No. **18** P12 and P13. P3'–P6' (system A') concern the synodic arc and its subdivision. Also P7' and P8' seem to deal with the subdivision of the synodic arc, but the scheme cannot be fully reconstructed. The topic of P9' is not clear. P11'–P31' are concerned with Venus, although this is not entirely certain for each procedure. Between P12' and P13' the presence of two additional procedures may be assumed.

P1' Jupiter system B: computations involving τ and Π

This procedure is a duplicate of No. **18** P12. Since the beginning is missing from both, it cannot be fully reconstructed. The astronomical significance of the involved numbers, except the mean value (μ) of the zigzag function for τ and the number period Π (Jupiter system B), is not clear. For a discussion cf. §3.7.8.3. Part **b** (Oi5'–6') mentions the defining parameters of the zigzag function for τ: the extrema, the mean value μ and the corresponding (approximate) zodiacal positions.

P2' Jupiter system B: derivation of τ; updating T with τ

This procedure contains a derivation of the synodic time τ from the synodic arc σ (Oi7–9), followed by the algorithm for updating T with τ. The formulation of the latter is based on template T.1 (§3.3.2.1).

P3' Jupiter system A': zones for the synodic arc

This procedure defines the zones for the synodic arc of system A'. For a discussion cf. §3.7.4.1; the parameters are compiled in Table 3.36. The formulation is based on template STEP.E. A strange feature, which is not present in the duplicates, is that zone 1 is mentioned twice (Oi11'–12').

***P4',P5'** Jupiter: subdivision of the synodic cycle, scheme A'.S_1, zone 1*

P4', and probably also P5', are concerned with scheme A'.S_1 for the subdivision of the synodic cycle in zone 1. The parameters of this scheme are compiled in Table 3.44 (p. 98). Only the first 4 lines concerning the pushes from LA to FA and FA to S1 are partly preserved; the procedure must have continued for several lines in column ii. The formulation is based on template PUSH.B (which provides only $\delta\tau$ and v). The first preserved line of column ii is the last line of a procedure here labeled P5'. In its original state it probably provided the values of $\delta\Sigma$ (using template PUSH.D) for the same scheme as in P4'. The only preserved value, 6;15°, can be identified with $\delta\Sigma$ (LA to FA) in zone 1.

***P6'** Jupiter system A': transition rules for σ*

This procedure is concerned with the algorithm for updating *B* with the synodic arc. Only the transition rules are provided; they are formulated in accordance with template STEP.D, which involves coefficients c_j (Table 3.42). The synodic arcs themselves (σ_j) are not mentioned. In deviation from usual practice transitions between zones 1 and 3 are also considered, which involves a transition coefficient $\sigma_3/\sigma_1 - 1$ = 0;12. This transition cannot occur between successive occurrences of a synodic phenomenon, but it might be relevant for computing net displacements of Jupiter for intervals longer than 1 synodic cycle.

***P7'** Jupiter: subdivision of the synodic cycle; unclear*

This badly damaged, brief procedure is probably concerned with the subdivision of the synodic cycle, presumably for Jupiter, but a coherent interpretation is not possible, since there appear to be no parallels. The number 20 is probably a displacement ($\delta\Sigma$). The 'mile' (*bēru*) = 30° mentioned in Oii7' is without parallel in procedures for the subdivision of the synodic cycle. In Oii8' there may be a reference to zone 1 of system A' (?).

***P8'** Jupiter: subdivision of the synodic cycle; unclear*

P8' is also concerned with the subdivision of the synodic cycle, since a station, an appearance (FA) and a setting (LA) are mentioned, but nothing more can be said.

***P9'** Jupiter: Goal-Year type procedure for net differences of B or T?*

This brief, damaged procedure is difficult to interpret. Lines Oii14' and 15' are reminiscent of the Goal-Year method, 'your new year' being a common term for Goal Year. A parallel to Oii14' may be No. **38** P2'.b, which is probably concerned with the net difference of *T* for 12^{yr}. The term 'it bends', if correctly translated, is without parallel; its astronomical or mathematical meaning is not clear.

***P10'** Jupiter system A: net displacements in zone 2*

This procedure concerns the net displacement of Jupiter for several time intervals in zone 2 of system A (Table 3.37).

***P11'–P17'** Venus: schemes X.S_1 and X.S_2 for the subdivision of the synodic arc*

With P11' the planet changes from Jupiter to Venus. Although P11'–P17' are strongly damaged they clearly form a coherent group of procedures for the subdivision of the synodic arc, followed by a corresponding group (P18'–P24') for the subdivision of the synodic time. Both groups contain two different schemes, X.S_1 and X.S_2, which are discussed in §3.5.7.1. In scheme X.S_1 (P11',P12' and P17',P18') the synodic cycle from MF to MF is divided into pushes from MF to EF and from EF to MF. In scheme X.S_2 (P12a'–P16' and P19'–P24') basically all intervals from MF to MF are considered separately. However, MS is initially skipped, since P12a' and P19 concern the push from MF to ML, but this is compensated by the inclusion of the push from MF to MS (P16' and P24') at the end of the sequence. In their original state, each procedure provided 12 control values of $\delta\Sigma$ or $\delta\tau$.[626] The numerical data, as far as preserved, are compiled in Table 3.24 (p. 81).

***P11'** Venus, scheme X.S_1: angular push from MF to EF*

P11' concerns the push from MF to EF, the damaged P12' probably the push from EF to MF, thus completing a full cycle from MF to MF. P11' begins with an introductory statement '(In order for) you to construct the *igigubbû*-coefficient of ...', followed by the phrase 'positions from eastern appearance (MF) to western appearance (EF)'. The term *igigubbû* is

[626] In *ACT*, Neugebauer restored several of the missing values, but it is not clear what the basis for these restorations is.

not otherwise attested as a technical term for pushes. This is followed by the control values of $\delta\Sigma$, one for each zodiacal sign. Not all values are preserved, but it is clear that they do not form a zigzag function of the zodiacal position.

P12' *Venus, scheme X.S$_1$: angular push from EF to MF*

Only the first 2 lines are partly preserved. The topic of P11' and the parallel sequence of pushes in P18'–28' and the partly preserved introductory statement (Oii25') imply that it deals with the angular push from EF to MF. Since it can be assumed that 2 further procedures were present between P12' and P14', P12' cannot have continued much beyond Oii26', perhaps for 1 line.

P12a',b' *Venus, scheme X.S$_2$: angular pushes from MF to ML, and from ML to EF*

Nothing remains of these procedures. As pointed out by Neugebauer (*ACT*, 397), their presence after P12' is suggested by the corresponding sequence of procedures for the temporal pushes (P20'–P21').

P13' *Venus, scheme X.S$_2$: angular push from EF to ES*

After a gap of about 6 lines below P12', P13' continues with the push from EF to ES. Only the initial signs of the last 3 lines are preserved. The place of P13' within the sequence of procedures and the preserved numbers, 4,14° and 4,21°, are consistent with an interpretation as $\delta\Sigma$ (EF to ES). It is not clear to which zodiacal signs they pertain (Neugebauer assigned them to Sco and Aqr, respectively).

P14' *Venus, scheme X.S$_2$: angular push from ES to EL*

P14' consists of 2 lines, of which only the first 3–4 signs are preserved. The place of P14' within the sequence of procedures and the numbers imply that it concerns the push from ES to EL.

P15' *Venus, scheme X.S$_2$: angular push from EL to MF*

The partly preserved introductory statement in Ri7 and the numbers in Ri8 imply that P15' concerns the push from EL to MF, the endpoint of which is the starting point for P16'.

P16' *Venus, scheme X.S$_2$: angular push from MF to MS*

P16' also consists of 2 lines, the first of which contains an introductory statement. This and the numbers in Ri10 imply that P16' concerns the push from MF to MS.

P17'–P24' *Venus: schemes X.S$_1$, X.S$_2$ for the subdivision of the synodic time*

Immediately following the procedures for $\delta\Sigma$ there are 8 corresponding procedures for the subdivision of the synodic time. P17' and P18' belong to scheme X.S$_1$, according to which τ is divided into intervals from MF to EF (P17') and from EF to MF (P18'). The following procedures (P19'–P24') belong to scheme X.S$_2$, whereby τ is divided into the intervals MF to ML (P19'), ML to EF (P20'), EF to ES (P21'), ES to EL (P22'), EL to MF (P23') and, probably, MF to MS (P24'), thus presenting a nearly complete sequence of pushes, except that the push from MS to ML is not mentioned. The numerical data, as far as preserved, are compiled in Table 3.24 (p. 81). In P19', which concerns MF to ML,[627] the values of $\delta\tau$ appear to be associated with the 12th degree of the zodiacal signs, if one interprets 12° Cnc as indicative of the arrangement of all control points. In P19', P20' and P23' the parent phenomenon (MF or ML) is referred to in a verbal phrase, 'it appears/sets in the east'. Almost nothing remains of P24', which probably concerns the push from MF to MS, but the single partly preserved number in Ri28 remains unexplained.

P25' *Venus: subdivision of the synodic cycle*

P25' contains a list of successive angular pushes $\delta\Sigma$, namely for MF to ML (Ri29; number broken away), 4,14;13° for EF to EL (Ri30), EF to ES (Ri30; number broken away), −6;20° for ES to EL (Ri31), EL to MF (Ri31–32; number broken away). For a brief discussion cf. §3.5.7.1.

[627] Neugebauer assumed that the introductory statement of P19' contains an error, and that P19' concerns MS to ML.

P26' *Unclear*

Virtually nothing remains of P26'. Since the surrounding procedures concern the subdivision of the synodic cycle, this is probably also true for P26'.

P27' *Venus: scheme X.S_3 for the subdivision of the synodic cycle*

P27' contains another scheme for the subdivision of the synodic cycle of Venus. Part **a** deals with $\delta\Sigma$ from EF to ES and from ES to EL; part **b** with $\delta\Sigma$ from MF to MS and from MS to ML. For a discussion cf. §3.5.7.1. The data are compiled in Table 3.25a.

P28'–P29' *Venus, system A_0?: total synodic arc*

P28' and P29' are concerned with the total synodic arc of Venus. Both procedures mention the same value $\Sigma = \sigma + 6{,}0$ = 9,35;30°, which belongs to system A_0 (§3.5.2.1; Table 3.17a). In P28' this value is said to apply to ML, MS and MF, in P29' to EL, ES and EF. Nothing suggests that it is conceived of as a mean value (the term 'middle one', *qablītu*, which usually qualifies mean values, is lacking).

P30' *Venus: subdivision of the synodic time*

This fragmentary procedure appears to have a similar structure as P25', providing values of $\delta\tau$ for several successive pushes setting out from what is probably $\delta\tau$ (MF to ML) = 4,22;15$^{\overline{\tau}}$ (restoring the initial 4 of this number in Rii12 by comparing with scheme X.S_2 in Table 3.24, p. 81), ML to EF (45$^{\overline{\tau}}$), EF to ES$^?$ (broken), ES to EL (18$^{\overline{\tau}}$), EL to MF$^?$ (21$^{\overline{\tau}}$) and MF$^?$ to MS (18$^{\overline{\tau}}$). Cf. also §3.5.7.1.

P31' *Venus: subdivision of the synodic cycle*

Very little remains of this procedure. The synodic phenomena (setting, appearance and station) mentioned in Rii15 and 16 suggest that it is concerned with the subdivision of the synodic cycle.

No. 47

museum number	BM 37266 (80–6–17,1022)
provenience	Babylon
date	early Seleucid
measures	6.3 × 4.2(*) × ? cm (type G?)
arrangement	O/R
cuneiform text	photo: Fig. F.50; Neugebauer & Sachs (1967), p. 215
bibliography	Neugebauer & Sachs (1967), Text F; Steele (2003), pp. 274–275 (P1, P2)
contents	Jupiter, Saturn and Mars: unidentified systems; various procedures

Obverse

P1 1 [**nim** *u* **sig** *ša*$_2$] ⸢**mul$_2$.babbar**⸣ $\frac{2}{3}$ **kuš$_3$** *e* **dur murub$_4$ nim**

2 [$\frac{2}{3}$ **kuš$_3$**] **sig dur murub$_4$ sig**

3 *ki-i ina* $^{\text{mul}_2}$**gir$_2$.tab nim** *ina zib*.**me bur$_3$**

4 *a-na* 1 **danna** 8 *u* **nim** *u* **sig**

P2 5 **nim** *u* **sig** *ša*$_2$ **genna absin rin$_2$ gir$_2$.tab nim dib**-*ma ina* 1-*en* ⸢**dur**⸣

6 **du** *zib* **ḫun mul$_2$ sig dib**-*ma ina* 1-*en* **d**[**ur du**]

7 **ta mul$_2$.mul$_2$ igi**-*šu*$_2$ ⸢*ana* **nim**⸣ **gar** ⸢x⸣ [xxxx]

8 [**ta**] ⸢**gir$_2$**⸣.[**tab**] ⸢**igi**-*šu*$_2$⸣ [*ana* **sig gar** xx]

(n lines missing)

Reverse

P3' (about n lines missing)

1' [xxxxxxxxxxx] ⸢*an*⸣-*na-a*

2' **uš igi** *i* **pap** *ḫi-pi*$_2$ ⸢xxx⸣ **uš**-*šu*$_2$ **igi**-*tu*$_4$

P4' 3' 1 **iti** 35 1 **iti** 24 **iti** 12 **zi** *u* **uš**

4' **en ugu uš egir**-*tu*$_4$ *ma-la u*$_4$*-mu ša*$_2$ **uš**-*u*$_2$

5' $\frac{2}{3}$ 24 **zi** *u* $\frac{1}{3}$ 12 **zi** *u* **uš egir uš egir**-*tu*$_4$

6' 1 **iti** 12 1 **iti** 24 1 **iti** ⸢36 **ninda**⸣ **iti** 40 **ninda**

7' 1 **iti** 50 **ninda** *u* **iti** 40 ⸢**ninda**⸣ **zi**

Co 8' [xx ^{md}]⸢**šu$_2$.dub**⸣.**numun a** ^{md}**en-a-mu dumu** ^{m}*mu-še-zib*

9' [xxxxxx š]**u.2** ^{m}**mu**-^{d}**en a**-*šu*$_2$

Obverse

P1 **(Jupiter: distance to the ecliptic)**

[1][‘Height and depth’ of] Jupiter: $\frac{2}{3}$ of a cubit above the middle band it is ‘high’, [2][$\frac{2}{3}$ cubits] below the middle band it is ‘low’. [3]If it ‘high’ in Sco it is ‘low’ in Psc. [4]For 1 ‘mile’ it ascends and descends 8 fingers.

P2 **(Saturn: distance to the ecliptic)**

[5]‘Height and depth’ of Saturn: Vir, Lib, Sco it takes the ‘height’, then it proceeds$^{?}$ in the ‘first band’, [6]Psc, Ari, Tau it takes the ‘depth’, then it [proceeds] in the ‘first band’. [7]From Tau it is directed upwards [...] [8][From] Sco it is directed [downwards ...]

(n lines missing)

Reverse

P3' **(An outer planet: stations; unclear)**

(about n lines missing)

[1'][...] thus [2']the first station ... broken ... its first station.

P4' **(Mars: subdivision of the synodic cycle, scheme S$_3$)**

[3']1 month 35 (^{nin}) (error for 36), 1 month 24, a month 12, then it is stationary (S1). [4']Until the second station (S2), whatever (number of) ‘days’ it is stationary, [5']$\frac{2}{3}$ it moves 24 (^{nin}) and $\frac{1}{3}$ it moves 12, then the second station. After the second station [6']1 month it moves 12 (^{nin}), 1 month 24, 1 month 36^{nin}, a month 40^{nin}, [7']1 month 50^{nin}, and a month 40^{nin}.

Co **(Colophon)**

[8'][Tablet of] Marduk-šāpik-zēri, son of Bēl-apla-iddin, descendant of Mušēzib, [9'][...] hand of Iddin-Bēl, his son.

Critical and philological notes

Obverse

1 Traces at the beginning compatible with **mul$_2$.babbar**.

5 1-*en* **dur** = *ištēn markasu*, ‘first band’: among the possible translations of *ištēn* listed in the *CAD*, ‘first’ seems more likely than ‘one’, if one assumes the ‘first band’ is synonymous with the ‘middle band’ mentioned in P1.
The reading **dib** = *ṣabātu* was suggested to me by H. Hunger. Cf. *CAD Ṣ*, *ṣabātu* 8: *ḫarrāna ṣabātu*, ‘to take the road’. Replaces *etēqu* = ‘to pass’ (*ACT*).

6 **du**: either *illak*, ‘it proceeds’, or read **gub** = *izzaz*, ‘it stands’.

Reverse

2' *i*: since this line was copied from a damaged original (*ḫe-pi*$_2$ = ‘broken’), the sign *i* might originally have been *tu*$_4$, resulting in **uš igi**-*tu*$_4$ = first station.
šu$_2$: or **šu$_2$** = ‘setting’ (LA)?

Commentary

This fragment preserves the upper (lower) half of a tablet. There is some damage to the upper (lower) left corner. The obverse contains one procedure for Jupiter and one for Saturn; the reverse one unidentified procedure and a procedure for Mars. The tablet concludes with a colophon mentioning the 'owner', Marduk-šāpik-zēri/Bēl-apla-iddin//Mušēzib, and the scribe, his son Iddin-Bēl. From other tablets they are known to have lived near the beginning of the Seleucid era.

***P1** Jupiter: distance to the ecliptic*

This procedure describes Jupiter's distance to the ecliptic as a zigzag-type function of the zodiacal position. No. **40** P2'.a contains a duplicate. For a discussion cf. §3.7.10.2.

***P2** Saturn: distance to the ecliptic*

This is the only known procedure concerned with Saturn's distance to the ecliptic. It describes a motion similar to the zigzag-type function for Jupiter's distance to the ecliptic in P1, with Saturn being 'high' in Vir, Lib and Sco, 'low' in the opposite three signs Psc, Ari, Tau, roughly in agreement with empirical facts. However, the numerical parameters of the function are not preserved.

***P3'** Stations of an outer planet; unclear*

Only the last 2 lines are partly preserved. They mention the 'first station' and the 'second station', which points to an outer planet, perhaps Mars.

***P4'** Mars: scheme S_3 for the subdivision of the synodic cycle*

This procedure, which had hitherto defied interpretation, very likely contains a partly corrupted variant of scheme S_3 for the subdivision of the synodic cycle of Mars (contra Neugebauer & Sachs 1967). For a discussion cf. §3.6.2.3; the parameters are compiled in Table 3.31. The first subinterval of the push from FA to S1 and the last subinterval of the push from S2 to LA appear to be missing. For each month the daily displacement is expressed in units of 1^{nin} (= 0;1°) per mean tithi. This unit is explicitly mentioned in R6'–7', but otherwise omitted.

5.2.7 Unidentified planets

No. 48

museum numbers	BM 32397 (76–11–17,2130) +32421+32454
provenience	Babylon
measures	14.5(15?) × 8.0(*) × 2.5 cm (type B?); column width 7.5 cm
arrangement	O/R
cuneiform text	photo: Fig. F.51
contents	outer planet, unidentified system: various procedures

Side X
Column i'?

P1' (m lines missing)
1' [...xxxx] ⸢25 x⸣ [x]
2' [...xx] (blank) **la$_2$**
3' [...xx] ⸢x⸣ 20 **me la$_2$**
4' [...x] ⸢2$^?$⸣5 **me**$^?$ **la$_2$**
5' [...] ⸢**uš**⸣ *ar$_2$-tu$_4$* **me** *ana* **me**
6' [...] (blank) ⸢1$^?$⸣7 **me la$_2$**

P2' 7' [...] *ša$_2$-niš*
8' [...] ⸢x⸣ **igi**-*at* **uš igi**-*tu$_4$* 7 **me la$_2$**
9' [...] ⸢x **u$_4^?$** xx⸣ 1 **me**$^?$
10' [...] ⸢xxxx⸣ 25 **me la$_2$**
(n lines missing)

Column ii'?

P3' (about m−4 lines missing)
1' [xxxxxxxxxx]
2' [xxxxxxxxxx]
3' ⸢x⸣ [xxxxxxxxxx]
4' ⸢xxx⸣ [xxxxxxxx]
5' ⸢xxx⸣ [xxxxxxxx]

P4' 6' *ana* ⸢x⸣ 7 **mu** ⸢xxxx 1$^?$⸣ 20 ⸢**me**$^?$⸣ **šu$_2$**
7' ⸢**mu**$^?$ *ana*$^?$ **mu** 10 xx *ana ar$_2$-šu$_2^?$* **la$_2$**
8' ⸢xx⸣ 9.7.⸢5.4$^?$4⸣ **me tab**
9' ⸢**mu**⸣ *ana* **mu** 6.42.53.20 *ana ar$_2$*-⸢*šu$_2^?$* **la$_2^?$**⸣
10' ⸢xx 9$^?$⸣ 2⸢7$^?$ xx⸣
(about n+1 lines missing)

Side Y
Column i'?

P5' (about n+1 lines missing)
1' [...x] ⸢xx⸣ [xxxxxxx]
2' [...] ⸢20 *ina* x 7$^?$⸣ [xxxxxxx]
3' [...x] ⸢x⸣ *ina* ⸢55$^?$⸣ [xxxxxxx]
4' [...x] *ina* ⸢1.30⸣ [xxxxxxx]
(about m+10 lines missing)

Column ii'?

P6' (about n+2 lines missing)
1' [xx] ⸢x⸣-*šu$_2$* **la$_2$** ⸢x⸣ [xxxxxxxxx]
2' [xx] **sig**$^?$ **rin$_2^?$** ⸢x⸣ [xxxxxxxxx]
3' [xx].⸢**meš**⸣ 7.⸢50$^?$⸣ [xxxxxxxx]
4' [xx] ⸢51$^?$.6$^?$.6$^?$.30 **la$_2$** xxx 30 **me la$_2$**⸣

Side X column i'?

P1' (Outer planet: differences of T?)

(about m lines missing)
1'[...] ⸢25 ...⸣ [...] **2'**[...] you subtract. **3'**[...]⸢...⸣ 20 'days' you subtract$^?$. **4'**[...]⸢2⸣5 'days' you subtract. **5'**[...] the second station (S2), day by day **6'**[...] 17 'days' you subtract.

P2' (Outer planet: subdivision of the synodic cycle?)

7'[...] secondly: **8'**[...] ⸢...⸣ before the first station (S1) you subtract 7 'days'. **9'**[...] ⸢...⸣ 1 day$^?$. **10'**[...] ⸢...⸣ you subtract 25 'days'.
(about n lines missing)

Column ii'?

P3' (Unclear)

(about m−4 lines missing)
1'[...] **2'**[...] **3'**⸢...⸣ [...] **4'**⸢...⸣ [...] **5'**⸢...⸣ [...]

P4' (Net differences of B?)

6'For ⸢...⸣ 7 years ⸢...⸣ 20 days$^?$, setting$^?$. **7'**⸢Year by year ⸢it recedes backwards 10 ...⸣. **8'**⸢...⸣ 9.7.⸢5.44⸣ days you add. **9'**⸢Year⸣ by year it ⸢recedes⸣ backwards 6.42.53.21. **10'**⸢...⸣ 2⸢7 ...⸣
(about n+1 lines missing)

Side Y column i'?

P5' (Unclear)

(about n+1 lines missing)
1'[...] ⸢...⸣ [...] **2'**[...] ⸢20 ...⸣ [...] **3'**[...] ⸢...⸣ in ⸢55⸣ [...] **4'**[...] in ⸢1.30⸣ [...]
(about m+10 lines missing)

Column ii'?

P6' (Unclear)

(about n+2 lines missing)
1'[...] ... [...] **2'**[...] ... Lib$^?$ ... [...] **3'**[...] ... 7.⸢50⸣ ... [...] **4'**[...] ⸢51$^?$.6$^?$.6$^?$.30 ... 30 days you subtract.⸣

P7' 5' [xxxxx] ⸢xxx⸣ **en šu$_2$**-*šu$_2$* 17 **me tab**
6' [xxxxx] ⸢xx 30$^?$⸣ *ša$_2$* **šu$_2$**
7' [xxxxxx] ⸢20$^?$⸣ **šu$_2$** ⸢**nim**$^?$⸣
8' [xxxxxx] ⸢x⸣ **la$_2$**
9' [xxxxxx] ⸢x **du**⸣
10' [xxxxxx] ⸢xx⸣ **la$_2$**
(about m lines missing)

P7' **(Subdivision of the synodic cycle?)**

5'[... From] ⸢...⸣ until its setting you add 17 days. 6'[...] ⸢... 30$^?$⸣ for setting. 7'[...] ⸢20$^?$⸣ ... 8'[...] ... you subtract 9'[...] ... 10'[...] ... you subtract.
(about m lines missing)

Critical and philological notes

Side X column i'?
1' After 25 a vertical wedge is visible, perhaps **me** = 'days'.
3' Before 20 traces of a sign, perhaps a winkelhaken.
4' 25: or 35.
6' 17: or 7?
7' *ša$_2$-niš* = *šanîš*, 'secondly': or *ša$_2$ šamaš$_2$*, 'of the Sun'?
8' Before **igi** there are traces similar to the end of **iti**.
la$_2$: an interpretation as *inaḫḫis*, 'it recedes', is unlikely, because this should describe motion before S1; hence 'you subtract' (either *tanaḫḫis* or *tumaṭṭa*).
9' **u$_4$**: preceded by faint traces, followed by traces of a vertical wedge and, perhaps, **la$_2$**.
me: or **la$_2$**?
10' At the beginning there are very faint illegible traces.
Column ii'? Only very faint traces of signs.
6' *ana*: or 1?
šu$_2$: 'setting' or 'west'?
7' **mu** *ana*: very uncertain.
40: or 50.
The sign before *šu$_2$* might be 40 (ḪI).
8' 44: or 54.
9' -*šu$_2$* **la$_2$**: only the upper parts are visible.
10' At the beginning, the upper parts of several vertical wedges are visible. After 27, there are several winkelhakens, perhaps part of a number or a sign similar to ŠE.
Side Y column i'?
1' Traces of 2 signs, the first of which is similar to **mu** or *zib*.
2' 20: it may be that the column begins with *ina*, and 20 belongs to another column. After *ina* there are very faint traces.
3' 5: or 8.
Column ii'?
1' Before *šu$_2$* there are very faint traces. Perhaps *ana ar$_2$-šu$_2$* **la$_2$**, 'it recedes backwards'?
4' After the first **la$_2$** there are faint traces; the first sign might be 11.
5' Before **en** the upper parts of about 3 signs are visible, the last of which may be ME or IGI.
6' Before 30 there are faint traces.
8' Before **la$_2$** there are faint traces.
9' Before **du** there is a vertical wedge, perhaps part of **ki**?
10' Before **la$_2$** there are faint traces.

Commentary

This tablet is partly preserved on three joined fragments. They include a segment of the right edge, but no other edges of the original tablet. The surface is strongly damaged, especially on side Y, where very little can be read. Both sides contained at least two columns of text, which are further divided into sections by horizontal lines. Since the orientation of the sides is not clear, the correct sequence of the columns may be different from the one assumed here (numbered from left to right on both sides). On the obverse, some lines of column ii' are preserved to its full width of 7.5 cm. Hence the original width of the tablet was probably at least about 15 cm. Beyond the fact that some of the procedures deal with an outer planet, almost nothing can be understood.

P1' Outer planet: net differences of T?

Only the final signs of the last 6 lines of P1' are preserved. Several time intervals in days are mentioned, all of them subtractive. Xi'5' contains a reference to daily motion near the second station, which implies that P1' deals with an outer planet.

P2' Outer planet: subdivision of the synodic cycle?

The first 4 lines of P2' are partly preserved. They probably deal with the subdivision of the synodic cycle of an outer planet. Again several time intervals of a subtractive nature are mentioned. Line Xi'8' is without parallel among the procedure texts.

P3' Unclear

Almost nothing remains of this procedure, thus its topic cannot be established.

P4' Net differences of B?

The first 5 lines of P4' are partly preserved. They mention time intervals from 1 to 7 years, and corresponding differences measured in days. It is not clear what they represent. The parameters 9.7.5.44 days (Xii'8') and the retrograde yearly displacement (?) of 6.42.53.20 (Xii'9') remain unidentified.

P5' Unclear

Almost nothing remains of this procedure, thus its topic cannot be established.

P6' Unclear

Very little remains of this procedure, of which the final 4 lines are partly preserved.

P7' Subdivision of the synodic cycle?

The first 6 lines of P7' are partly preserved. Some of the phrases (Yii'5', 6') suggest a connection with the subdivision of the synodic cycle.

No. 49

museum numbers	BM 32406 (76-11-17,2139) +32491 (76-11-17) +32505 (76-11-17) +4 unnumbered fragments
provenience	Babylon
measures	16.0 × 8.3(*) × 2.0–2.9 cm (type B?); column width 8 cm
arrangement	O/R
cuneiform text	photo: Fig. F.57
contents	Mercury or Venus, unidentified system(s): various procedures

Side X
Column i?

P1
1 (traces)
2 (traces)
3 [...]
4 [xxx] (traces)
5 [...]
6 [xx] ⸢x **u**$_4$ *ana* **u**$_4^{?}$⸣ [...]
7 ⸢**ta** x 2$^{?}$⸣5 **šu**$_2$-*šu*$_2$ ⸢**en**$^{?}$⸣ [...]
8 ⸢x⸣ **u**$_4$.⸢**meš** x⸣ [...]
9 **ta** ⸢x⸣ [...]
10 [...]
(n lines missing)

Column ii?

P2
1 [...]
2 [...]
3 [...]
4 [...] (traces)
5 [...] ⸢*ša*$_2^{?}$ **šu**$_2^{?}$⸣ 2 ⸢x⸣ [xxxx] **la**$_2$
6 [...] ⸢x **me** xxxxx 1$^{?}$⸣2 **me la**$_2$
7 [...] **u**$_4$**.meš** *u* ⸢xx⸣ 12 **me la**$_2$

P3
8 [...] ⸢**mu**$^{?}$⸣ **igi** *ša*$_2$ ⸢**šu**$_2^{?}$ *ana*$^{?}$ **igi** *ša*$_2$⸣ **šu**$_2$
9 [...] ⸢4⸣6.20 ⸢xx **šu**$_2^{?}$ x⸣
10 [...] 5.29 [xxxx]
11 [...] ⸢4$^{?}$⸣8 ⸢16$^{?}$⸣ [xxxxx]
12 [...] ⸢xx⸣ [xxxxxxx]
(n lines missing)

Side Y
Column i'?

P4'
(about n lines missing)
1' ⸢8 **me šu**$_2^{?}$ xxx⸣ [xxxxxx]
2' ⸢**en šu**$_2$-*šu*$_2$⸣ 19 ⸢x⸣ [xxxxx]

P5'
3' ⸢**kur**$^{?}$ *ša*$_2^{?}$ xxx⸣ [xxx] ⸢xxx⸣
4' ⸢**alla**$^{?}$ 16 xx⸣ [xxx] ⸢xxx⸣
5' ⸢xxx **tab**⸣ [xx] ⸢xxx⸣
6' ⸢x **u**$_4^{?}$ xx⸣ [x] ⸢xxx⸣ **la**$_2$
7' ⸢x **igi**$^{?}$ x⸣ [x] ⸢xxx⸣ **la**$_2$
8' **u**$_4$.⸢**meš**$^{?}$ *u* x⸣ [xx] ⸢xx⸣ **la**$_2$
9' [x] ⸢xx⸣ [xxxx] ⸢xxxx⸣
10' [xxxxxxxxxx] ⸢xx⸣

Side X column i?

P1 **(Subdivision of the synodic cycle?)**

[1](*traces*) [2](*traces*) [3][...] [4][...] (*traces*) [5][...] [6][...] ⸢... day by day$^{?}$⸣[...] [7]⸢From ...⸣5 its setting ⸢until$^{?}$⸣ [...] [8]⸢...⸣ days ... [...] [9]From ⸢...⸣ [...] [10][...]
(*n lines missing*)

Column ii?

P2 **(Net differences of *T*?)**

[1][...] [2][...] [3][...] [4](*traces*) [5][...] ⸢western⸣ [...] 2 ... [...] you subtract [6][...] ⸢... 1⸣2 days you subtract [7][...] days and ⸢...⸣ you subtract 12 days.

P3 **(Inner planet, EF: updating *B* with the synodic arc?)**

---

[8][...] ⸢year$^{?}$, western⸣ appearance (EF) ⸢to$^{?}$⸣ western ⸢appearance⸣. [9][...] 46.20 ... [10][...] 5.29 [...] [11][...] ⸢4⸣8 ⸢16⸣ [...] [12][...] ... [...]
(*n lines missing*)

Side Y column i?

P4' **(Subdivision of the synodic cycle?)**

(*about n lines missing*)
[1']⸢8 days, setting$^{?}$ ...⸣ [...] [2']⸢until its setting⸣ 19 ⸢...⸣ [...]

P5' **(Net differences of *T*?)**

[3'][...] ⸢rising$^{?}$ ...⸣ [...] ⸢...⸣ [4']⸢Cnc$^{?}$ 16 ...⸣ [...] ⸢...⸣ [5']⸢...you add⸣ [...] ⸢...⸣ [6']⸢...⸣ [...] ⸢...⸣ you subtract. [7']⸢...⸣ [...] ⸢...⸣ you subtract. [8']'Days' ⸢and ...⸣ [...] ...⸣ you subtract. [9'][...] ⸢...⸣ [...] ⸢...⸣

Column ii?

P6' (about n−3 lines missing)

1' [xxxx] ⸢30 **a**$^?$ x⸣ [xxxxx]
2' [xxx] 33.30 ⸢**la**$_2$⸣ [xxxx]
3' [xxx] 10.30 **la**$_2$ ⸢x⸣ [xxx]
4' [xxx] ⸢**uš**$^?$⸣ 2.15 **me** ⸢xx⸣
5' ⸢x *ša*$_2^?$ xxx⸣ 7 ⸢xx **šu**$_2$⸣ [x] ⸢x⸣
6' ⸢**u**$_4$**.meš**⸣ 37 [xxx] **la**$_2$

P7' 7' ⸢x⸣ **la**$_2$ *u* ⸢**igi**$^?$⸣-*šu*$_2$ 3 ⸢x⸣ [xxx]
8' [x] ⸢xx 20⸣ [xxxx]
9' ⸢x⸣ [xx] ⸢x 30 x⸣ 5 [x] 30 ⸢xx⸣ [x]
10' ⸢**šu**$_2^?$ xx⸣ 5.12$^?$ ⸢1$^?$8$^?$.7$^?$⸣.30 ⸢**me**$^?$⸣ [x$^?$]
11' [xx] **kur** ⸢**en** xxx 15 **ki**$^?$ x⸣ [x]
12' [xxx] 24 ⸢**u**$_4$⸣ 1⸢5$^?$ x⸣ [xxx]

Column ii?

P6' (Subdivision of the synodic cycle?)

(*about n−3 lines missing*)
1'[...] ⸢30 ...⸣ [...] 2'[...] 33.30 ⸢you subtract⸣ [...] 3'[...] 10.30 you subtract ⸢...⸣ [...] 4'[...] ⸢station$^?$⸣ 2.15 days ⸢...⸣ 5'⸢... of ...⸣ 7 ⸢...⸣ [...] ⸢...⸣ 6'⸢'Days'⸣ 37 [...] you subtract.

P7' (Subdivision of the synodic cycle?)

7'⸢...⸣ you subtract and its ⸢appearance$^?$⸣ 3 ⸢...⸣ [...] 8'[...] ⸢... 20⸣ [...] 9'⸢...⸣ [...] ⸢... 30 ...⸣5 [...] 30 ⸢...⸣ [...] 10'⸢setting$^?$ ...⸣ 5.12 ⸢18$^?$.7$^?$⸣.30 ⸢days$^?$⸣ [...] 11'Eastern [...] ⸢until ... 15 ...⸣ [...] 12'[...] 24 ⸢days$^?$ 15 ...⸣ [...]

Critical and philological notes

Side X column ii?

8 **mu**: or is this the end of **zi**?

Side Y column ii?

4' **uš**: uncertain; only the end is preserved. Or read .. 13.15 **me**?
5' Before 7 there are traces of a sign similar to KUR.
11' The sign after **en** might be **igi**.

Commentary

This tablet is preserved on three numbered fragments and several small unnumbered fragments, all joined together. The orientation of the sides is not clear. The tablet is partly preserved to full width. The lower (upper) edge of side X (Y) is missing. On both sides the surface is badly worn, such that most signs are reduced to illegible traces. The text is divided into two columns, which are further divided into procedures by horizontal lines. Some of the procedures clearly deal with the subdivision of the synodic cycle of a planet; P3 deals with an inner planet. The name of the planet is not preserved, and its identity could not be inferred from the algorithms (but cf. P7'). The tablet's size (type B?) is unusual for procedure texts.

P1 Subdivision of the synodic cycle?

Very little can be read in this procedure. The 'setting' (Xi7) and the word 'days' (Xi8) suggest a connection with the subdivision of the synodic cycle of a planet.

P2 Net differences of T?

Very little can be read. In Xii5 a setting is mentioned. The final 3 lines end with **la**$_2$, which either means 'it recedes', or 'you subtract'. This suggests a connection with net differences of the date (T) of a synodic phenomenon. A similar structure is found in P5', and in No. **48** P1'...P3'.

P3 Updating B with the synodic arc for EF of inner planet?

The first line of P3 (Xii8), if correctly interpreted, points to an algorithm for updating the zodiacal position of a planet (B) with the synodic arc for EF, which implies an inner planet (Mercury or Venus).

P4' Subdivision of the synodic cycle?

Only the final 2 lines are partly preserved. The mention of 'settings' suggests a connection with the subdivision of the synodic cycle.

P5' Net differences of T?

Very little can be read. As in P2, several lines (Yi6'–8') end with **la**$_2$, 'you subtract' (or 'it recedes').

P6',P7' Subdivision of the synodic cycle?

Very little can be read. The numbers remain unidentified. Some of the terms ('station', 'appearance', 'setting') suggest a connection with the subdivision of the synodic cycle. The number 18;7,30° might be identified as $\delta\Sigma$ (S2 to LA) in zone 1 of scheme A'.S$_2$ for the subdivision of the synodic cycle of Jupiter (Table 3.44, p. 98).

No. 50

museum number	BM 33733 (Rm4,291)
provenience	Babylon
measures	4.0(*) × 4.7(*) × 2.3(*) cm; column width 2 cm
cuneiform text	photo: Fig. F.49
contents	a planet

Side X
Column i'

P1 (m lines missing)

1' [...x] ⸢x 5$^?$⸣
2' [...] ⸢x E⸣ **mu**
3' [...] ⸢x 20$^?$⸣ 1.0.15
4' [...] ⸢x⸣ **be**-*ma* **igi**
5' [...x] ⸢x⸣-*ma*
6' [...xx] ⸢x⸣ **uš**
(n lines missing)

Column ii'

P2' (about m+1 lines missing)

2' ⸢x⸣ [xxxxx]
3' **u$_4$.meš gar** ⸢**ta**⸣ [**igi**$^?$]
4' *ana* **uš igi**-⸢*tu*$_4^?$⸣
5' **diš alla šu$_2$**-*ma*$^?$ *ina* ⸢7$^?$ x⸣
6' 4 **u$_4$ uš** 4.4
7' **alla** 4.41 **a**
8' 5.31 **absin**
9' 5.38 **rin$_2$** 5.40
10' [**gir$_2$**] ⸢5⸣.29 **pa** 5.⸢14$^?$⸣
11' [**maš$_2$** 4].57 ⸢**gu**⸣ [x]
(about n-5 lines missing)

Column iii'

P3' (about m+8 lines missing)

9' ⸢x⸣ [...]
(about n−7 lines missing)

Side X column i'

P1' (Subdivision of the synodic cycle?)

(m lines missing)
1'[...] ⸢... 5$^?$⸣ **2'**[...] ⸢...⸣ year. **3'**[...] ⸢... 20$^?$⸣ 1.0.15 **4'**[...] if it appears **5'**[...] ⸢...⸣, and **6'**[...] ⸢...⸣ it is stationary,
(n lines missing)

Column ii'

P2' (Outer planet: subdivision of the synodic time?)

(about m+1 lines missing)
2'... [...]
3'You put down the 'days'. ⸢From⸣ [appearance$^?$] **4'**to first station (S1): **5'**If Cnc: it sets$^?$, and in ⸢7$^?$ ...⸣, **6'**4 'days', station. 4.4 **7'**Cnc, 4.41 Leo, **8'**5.31 Vir, **9'**5.38 Lib, 5.40 **10'**[Sco], ⸢5⸣.29 Sgr, 5.⸢14$^?$⸣ [Cap] **11'**[4].57 ⸢Aqr⸣ [...]
(about n−5 lines missing)

Column iii'

P3' (Unclear)

(about m+8 lines missing)
9'... [...]
(about n−7 lines missing)

Critical and philological notes

Side X column i'

1' 5: 8 also possible. The sign before it might be KI.
2' E: preceded by traces of a sign similar to LA$_2$.
3' 20: uncertain; perhaps to be combined with the following 1 into U$_4$?

Column ii'

2' Traces of a vertical wedge.
5' 7: 8 also possible. The sign after it might be **me**.
ma: uncertain; might also be GIŠ, or read with the following *ina* as **gal**, 'maximum'?
6' 4 **u$_4$**: or read 4.0 *ana*?

Column iii'

9' Remains of a winkelhaken are visible. This is all that remains of column iii', which is separated from column ii' by a vertical line.

Commentary

This small fragment is inscribed on one side (X); the other side is destroyed. No original edges are preserved. The inscribed side contains remains of three unusually narrow (2 cm) columns, separated by vertical lines.

P1' Subdivision of the synodic cycle?

Too little remains of P1' to establish its topic, but the terms 'it appears' (Xi'4') and 'station' (Xii6') point to the subdivision of the synodic cycle of a planet. The significance of the number 21(?).0.15 is unclear. The phrase 'if it appears' (Xi'4') is otherwise known from procedures concerned with the rising time of Jupiter at FA (§3.7.2.5).

P2' Subdivision of the synodic time

The beginning and end of P2' are lost. The preserved text implies that P2' is concerned with a temporal push ($\delta\tau$), perhaps from the first appearance to the first station (Xii3'–4'). While the term 'first station' suggests an outer planet (S1), this term is occasionally also used for the Evening Station (ES) of Venus (cf. No. **9**). Indeed the control values mentioned in Xii6'–11', each of which is associated with a zodiacal sign, are reminiscent of $\delta\tau$ (EF to ES) as defined in scheme X.S_2 for Venus (Table 3.24, p. 81). Xii5' is difficult to understand. In Xii6' a station is mentioned again, and an interval of 4 (or 4,0?) 'days'. The maximum of the control values occurs in Sco. Presumably the text continued with values for Psc – Gem, which would require about 3 more lines.

P3' Unclear

Virtually nothing remains of this procedure.

No. 51

museum number	BM 37380 (80–6–17,1137)
provenience	Babylon
measures	3.7(*) × 3.5(*) × 1.7–2.3 cm
arrangement	O/R
cuneiform text	photo: Fig. F.42
content	a planet: various procedures

Side X

P1’ (n lines missing)

1’ [...xx] ⸢**ki** x⸣ [xx]

2’ [...] ⸢x⸣ 5 ⸢10?⸣ [x?]

3’ [...] ⸢x⸣ 36 **ki du** [x?]

4’ [...] ⸢x⸣ *ana ar*$_2$*-šu*$_2$ [**la**$_2^?$]

5’ [...x] **la**$_2$ 5

6’ [...x] ⸢**du**?⸣ 5.27

Side Y

P2’ 1 [...] **šu**$_2$*-ma*

2 [...] ⸢x⸣*-ma*

3 [... **d**]**u igi**

4 [... *i*]*na*? **an**-*e*

5 [...]-*ma*

P3’ 6 [...] ⸢x⸣ [xx]

(about n lines missing)

Side X

P1’ (Subdivision of the synodic cycle?)

(*n lines missing*)

[1’][...] ⸢the position ...⸣ [...] [2’][...] ⸢...⸣ 5 ⸢10?⸣ [...] [3’][...] ⸢...⸣ 36 the position proceeds [...] [4’][...] ⸢...⸣ [it recedes] backwards [5’][...] it recedes, 5 [6’][...] ⸢...⸣ 5.27.

Side Y

P2’ (Subdivision of the synodic cycle?)

[1][...] it sets, and [2][...] ⸢...⸣ and [3][... pro]ceeds, (then) it appears. [4][... it stays away i]n? heaven. [5][...] and

P3’ (Unclear)

---

[6][...] ⸢...⸣ [...]

(*about n lines missing*)

Critical and philological notes

Side X

1’ Between **ki** and the gap at the end there is a large empty space, which suggests that the gap contained a single sign, perhaps **du**, ‘it proceeds’.

2’ 10: or the initial winkelhaken of another sign.

3’ 36: the damaged sign preceding it might be a digit 4–8, in which case 36 is the final digit of a larger number.

Side Y

4 *ina šamê* = ‘in heaven’: this might pertain to the interval from LA to FA during which an outer planet is invisible (cf. the Glossary).

Commentary

This small fragment belongs to the lower (upper) right corner of side X (Y) of a tablet. Two partly preserved procedures are probably concerned with the subdivision of the synodic cycle of a planet.

P1’ Unknown planet, subdivision of the synodic cycle?

The beginning and end are missing, and only a few signs of each line are preserved. They suggest that P1’ is concerned with the subdivision of the synodic cycle. The forward motion of ...36° mentioned in X3’ might be identified with $\delta\Sigma$ (EF to EL) = 36°, a parameter of Mercury system A$_1$ (Table 3.8 on p. 70), or with a value of Jupiter’s synodic arc in system A, σ_2 = 36°, but this remains speculation.

P2’ Unknown planet, subdivision of the synodic cycle?

Only the last few signs of each line are preserved. Y3–4 suggest a connection with the subdivision of the synodic cycle. The phrase ‘[to stay away] in heaven’ points to the interval during which an outer planet is invisible (LA to FA).

5.3 Moon

5.3.1 System K

No. 52

museum numbers	BM 36722 (80–6–17,455) +37205 (958) +40082 (81–2–1,47)
provenience	Babylon
date	early Seleucid (reign of Philip Arrhidaeus?)
measures	12 × 19 × 3.3 cm (type D)
arrangement	O/R
cuneiform text	photo: Figs. F.52, F.53; BM 36722 (+) 40082: Neugebauer & Sachs (1969), p. 113
bibliography	BM 36722 (+) 40082: NS = Neugebauer & Sachs (1969), Text K; *HAMA*, p. 552
contents	system K: *G*, Lunar Six; system A: *Φ*, *G* (including variants)

Obverse
Column i

	T1					**T2**				
1	⌜**ḫun**⌝	14	**tab**	22	**zi**	**ḫun**	5	**t[ab**	10	**zi]**
2	⌜**mul$_2$**⌝	12	**tab**	20	**zi**	⌜**mul$_2$**⌝	5	**tab**	1[1	**zi]**
3	⌜**maš**⌝	10	**tab**	18	**zi**	**maš**	5	**tab**	12	**zi**
4	⌜**alla**⌝	5	**tab**	16	**zi**	**alla**	6	**tab**	13	**zi**
5	⌜**a**⌝	3	⌜**tab**	14⌝	**zi**	⌜**a**⌝	6	**tab**	14	**zi**
6	⌜**absin**⌝	3	**tab**	⌜12	**zi**⌝	[**absin**]	6	**tab**	15	**zi**
7	⌜**rin$_2$**⌝	5	⌜**tab**⌝	10	**zi**	⌜**rin$_2$**⌝	9	**tab**	16	**zi**
8	**gir$_2$**	8	**tab**	14	**zi**	⌜**gir$_2$**⌝	12	**tab**	17	**zi**
9	⌜**pa**⌝	[x	**tab**	x]	⌜**zi**⌝	**pa**	17	**tab**	18	**zi**
10	⌜**maš$_2$**⌝	[x	**tab**	x	**zi]**	**maš$_2$**	10	**tab**	16	**zi**
11	[**gu**]	⌜11?	**tab**	x	**zi**⌝	**gu**	5	**tab**	14	**zi**
12	[*zib*]	⌜11	**tab**	18	**zi**⌝	*zib*	5	**tab**	12	**zi**
13	⌜*an-na*⌝*-a ša$_2$* **u$_4$**.1.**kam**					*an-na-a ša$_2$* **u$_4$.na$_2$.a**				
	T3					**T4**				
14	⌜**rin$_2$**	6?	**tab**⌝	12	**zi**	**rin$_2$**	8	**tab**	17	**zi**
15	[**gir$_2$**]	⌜x	**tab**⌝	15	**zi**	**gir$_2$**	7	**tab**	16	**zi**
16	[**pa**	x]	⌜**tab**⌝	18	**zi**	**pa**	6	**tab**	15	**zi**
17	[**maš$_2$**	x	**tab**]	⌜1⌝8	**zi**	**maš$_2$**	4	**tab**	14	**zi**
18	[**gu**	x	**tab**	1]⌜8⌝	**zi**	**gu**	**nu tuk**	**tab**	12	**zi**
19	[*zib*	x	**tab**	1]⌜8⌝	**zi**	*zib*	4	**tab**	10	**zi**
20	[**ḫun**	x	**tab**	1]⌜8⌝	**zi**	**ḫun**	8	**tab**	13	**zi**
21	[**mul$_2$**	x	**tab**	1]⌜8?⌝	**zi**	**mul$_2$**	12	**tab**	16	**zi**
22	[**maš**	x	**tab**	x]	**zi**	⌜**maš**⌝	15	**tab**	18	**zi**
23	[**alla**	x	**tab**	x]	⌜**zi**⌝	**alla**	1⌜5	**tab**⌝	20	**zi**
24	[**a**	x	**tab**	x	**zi]**	⌜**a**	12	**tab**⌝	19	⌜**zi**⌝
25	[**absin**	x	**tab**	x	**zi]**	⌜**absin**⌝	8	**tab**	18	⌜**zi**⌝
26	[*an-na-a ša$_2$* **šu$_2$** *u* **na**					*an*]-⌜*na*⌝*-a ša$_2$* **me** *u* ⌜**gi$_6$**⌝				

P1 27 [xxxxxx **bar** x] ⌜20⌝ **na** 8 **ḫun** 12 *zib* 11 **tab**

28 [*bi-rit* 11 *u$_3$* 14] ⌜3⌝ 3 *a-na* 30 [**šeš.meš** SE$_3$*-ma*]

29 [**igi** 30 : 2 3 **a.ra$_2$** 2 6 6 **a.ra$_2$** 4 **du**-*ma*]

30 2⌜4⌝ [24 **ta** 14 **tab** *ša$_2$* **ḫun zi**-*ma* 13.36 13.36]

31 **ki** 20 ⌜**na** *ša$_2$* **bar**⌝ [**tab**-*ma*] ⌜33.3⌝[6 *zib* 18 **zi**]

32 **ḫun** 22 **zi** *bi-rit* 18 *u$_3$* 22 4

33 4 *a-na* 30 **šeš.meš** SE$_3$*-ma* **igi** 30 : 2

34 4 **a.ra$_2$** 2 8 ⌜8⌝ **a.ra$_2$** 4 *tam-ṭi-tu$_4$* **du**-*ma*

35 32 32 **ta** 22 **zi** *ša$_2$* **ḫun zi**-*ma* 21.28

Obverse column i

T1 (Coefficients for NA_1)

[1]⸢Ari⸣	14	addition	22	subtraction
[2]⸢Tau⸣	12	addition	20	subtraction
[3]⸢Gem⸣	10	addition	18	subtraction
[4]⸢Cnc⸣	5	addition	16	subtraction
[5]⸢Leo⸣	3	⸢addition	14⸣	subtraction
[6]⸢Vir⸣	3	addition	⸢12	subtraction⸣
[7]⸢Lib⸣	5	⸢addition⸣	10	subtraction
[8]Sco	8	addition	14	subtraction
[9]⸢Sgr⸣	[...	addition	...]	⸢subtraction⸣
[10]⸢Cap⸣	[...	addition	...	subtraction]
[11][Aqr]	⸢11?	addition	...	subtraction⸣
[12][Psc]	⸢11?	addition	18	subtraction⸣

[13]Thus for ‘day 1’.

T2 (Coefficients for KUR)

Ari	5	addition	[10	subtraction]
⸢Tau⸣	5	addition	1[1	subtraction]
Gem	5	addition	12	subtraction
Cnc	6	addition	13	subtraction
⸢Leo⸣	6	addition	14	subtraction
[Vir]	6	addition	15	subtraction
⸢Lib⸣	9	addition	16	subtraction
⸢Sco⸣	12	addition	17	subtraction
Sgr	17	addition	18	subtraction
Cap	10	addition	16	subtraction
Aqr	5	addition	14	subtraction
Psc	5	addition	12	subtraction

Thus for the ‘day of disappearance’.

T3 (Coefficients for ŠU$_2$ and NA)

[14]⸢Lib	6?	addition⸣	12	subtraction
[15][Sco]	⸢...	addition⸣	15	subtraction
[16][Sgr	...]	⸢addition⸣	18	subtraction
[17][Cap	...	addition]	⸢1⸣8	subtraction
[18][Aqr	...	addition	1]⸢8⸣	subtraction
[19][Psc	...	addition	1]⸢8⸣	subtraction
[20][Ari	...	addition	1]⸢8⸣	subtraction
[21][Tau	...	addition	1]⸢8?⸣	subtraction
[22][Gem	...	addition	...]	subtraction
[23][Cnc	...	addition	...]	⸢subtraction⸣
[24][Leo	...	addition	...	subtraction]
[25][Vir	...	addition	...	subtraction]

[26][Thus for ŠU$_2$ and NA.

T4 (Coefficients for ME and GI$_6$)

Lib	8	addition	17	subtraction
Sco	7	addition	16	subtraction
Sgr	6	addition	15	subtraction
Cap	4	addition	14	subtraction
Aqr	nothing	addition	12	subtraction
Psc	4	addition	10	subtraction
Ari	8	addition	13	subtraction
Tau	12	addition	16	subtraction
⸢Gem⸣	15	addition	18	subtraction
Cnc	1⸢5	addition⸣	20	subtraction
⸢Leo	12	addition⸣	19	⸢subtraction⸣
⸢Vir⸣	8	addition	18	⸢subtraction⸣

Th]us for ME and ⸢GI$_6$⸣.

P1 (NA_1 steps 1,2)

1 [27][In order for you to compute day 1. If month I (begins on day) ...], NA_1 is ⸢20⸣ (in) 8 Ari: the addition is 12 (error for 14); Psc: 11. [28][The difference between 11 and 14] is 3. You divide 3 into 30 [parts:] [29][the reciprocal of 30 is 0;2, 3 times 0;2 is 0;6. You multiply 0;6 by 4, it is] [30]0;2⸢4⸣, [you ‘tear out’ 0;24 from 14, the addition for Ari, it is 13;36.] [31][You add 13;36] with 20, NA_1 for month I, [it is] ⸢33;36.

2 Psc⸣: [the subtraction (‘tearing out’) is 18,] [32]Ari: the subtraction is 22, the distance between 18 and 22 is 4. [33]You divide 4 into 30 parts: the reciprocal of 30 is 0;2, [34]4 times 0;2 is 0;8. You multiply 0;8 by 4, the deficit, it is [35]0;32. You ‘tear out’ 0;32 from 22, the subtraction for Ari, it is 21;28.

36 21.28 **ta** 33.36 **na u$_4$.1.kam** *ša$_2$* **gu$_4$ zi**-*ma*
37 *re-ḫi* 12.8 12.8 **na u$_4$.1.kam** *ša$_2$* **gu$_4$ gar**-*an*
38 **iti.gu$_4$** 30 12.8 **na e**-*bi*

P2 39 **u$_4$.1.kam** *ana* **du$_3$**-*ka ki-i* **bar** 30 17 **na**
40 12 **ḫun** 14 **tab ki** 17 **na tab**-*ma* 31
41 **gu$_4$** 1 31 **na e**-*bi mi-nu-u$_2$* **gi$_6$**
42 **du en.nun gi$_6$ ta ša$_3$ zi**-*ma ša$_2$* **tag$_4$**
43 ⸢**a.ra$_2$ zi**⸣ *ša$_2$* **iti**-*šu$_2$* **du**-*ma* **ki** 31 **tab**
Column ii
1 [xxxxxxxxxx] ⸢1 30⸣ [x]
2 [xxxxxxxxxx] ⸢**na**?-*su*?⸣ [x]
3 ⸢*mi-nu-u$_2$*⸣ *ki-i ana* **kur** *šamaš$_2$* ⸢*re*⸣-*ḫi* **a.ra$_2$ zi**
4 *ša$_2$* **iti**-*šu$_2$* **du**-*ma* **ta na** *ša$_2$* **gu$_4$ zi** *ki-i*
5 **sig** *šu-u$_2$ mi-nu-u$_2$ ki-i* **sig**
6 **a.ra$_2$** 24 **du**-*ma* **ta na** *ša$_2$* **gu$_4$ zi**-*ma* **gar**-*an*

P3 7 **u$_4$.na$_2$.a** *ana* **du$_3$**-*ka ki-i* **še** 27 15 **kur** 12 **ḫun** 5 **tab**
8 **ki** 15 **tab**-*ma* 20 *ki-i* ⟨*kun-nu*⟩ **gu$_4$** 26 20 **kur du$_{11}$.ga**
9 *ki-i tur-ru* 27 20 **kur du$_{11}$.ga** *mi-nu-u$_2$* **gi$_6$** ⸢**du**⸣
10 ⸢**ta**⸣ **en.nun gi$_6$ zi**-*ma ša$_2$ i-re-ḫi* **a.ra$_2$ zi**
11 *ša$_2$* **iti**-*šu$_2$* **du**-*ma* **ta u$_4$.na$_2$.a** *ša$_2$* **gu$_4$ zi**
12 *mi-nu-u$_2$ ki* **nim a.ra$_2$** 24 **du**-*ma*
13 **ki u$_4$.na$_2$.a** *ša$_2$* **gu$_4$ tab** *ki-i* **gi$_6$ du** *a*[*l-la*]
14 **en.nun gi$_6$** *i-ṣi mi-nu-u$_2$ ki-*⸢*i*⸣ [*ana* **kur** *šamaš$_2$ re-ḫi*]
15 **a.ra$_2$ zi** *ša$_2$* **iti**-*šu$_2$* **du**-*ma* **ki u$_4$**.[**na$_2$.a** *ša$_2$* **gu$_4$ tab**]
16 *ki-i* ⸢**sig**⸣ *šu-u$_2$ mi-nu-u$_2$ ki-i* [**sig**]
17 **a.ra$_2$** 24 **du**-*ma* **ta u$_4$**.⸢**na$_2$**⸣.[**a** *ša$_2$* **gu$_4$**]
18 **zi**-*ma* ⸢*ša$_2$*⸣ *i-re-*⸢*ḫi*⸣ [xxx]

P4 19 2 ⸢3? 10 30 x⸣ [xxxxxxxxx]
20 1.20 [xxxxxxxxxx]
20 [x] ⸢xxx⸣ [xxxxxxxxxx]
21 [xxxxxxxxxxxxxx]
22 [xxxxxxxxxxxxxx]
23 [xxxxxxxxxxxxxx]
24 3.⸢7?⸣ [xxxxxxxxxxxx]
25 1.1.20[xxxxxxxxxxxxxx]
26 2.3.19.1⸢5.3⸣[3.20 xxxxxxxx *ša$_2$*]
27 *al-la* 1.5⸢8⸣.[xxxxxxxxxxxxxx]
28 4.48.8.53.[20 xxxxxxxxxxxxxx]
29 44 44 **ta** 4.⸢30⸣ [**zi**-*ma* 3.46 xxxxxx]
30 1 **ḫun** *mi-nu-u$_2$ ki* **zi** [xxxxxx *zib* 30 **zi**]
31 **ḫun** 45 **zi** *mi-nu-u$_2$ ki-*⸢*i*⸣ [45 *al-la* 30 **zi** *ša$_2$*]
32 *zib* **diri** 15 **diri** 15 *a-na* 30 **šeš**.[**meš** SE$_3$-*ma*]
33 **igi** 30 2 2 **a.ra$_2$** 15 : 30 [1 **ḫun**]
34 *mi-nu-u$_2$ ki-i al-la* 12 **ḫun** [**la$_2$**-*u$_2$* 11 **la$_2$**]
Reverse
Column i
1 1 **ḫun** *mi-*⸢*nu-u$_2$*⸣ *ki al* 12 **ḫun la$_2$**-*u$_2$* 11 **la$_2$**
2 11 **a.ra$_2$** 30 5.30 5.30 **ta** 45
3 **zi** *ša$_2$* **ḫun zi**-*ma* 39.30 **ta** 3.46
4 *ša$_2$ taš-ku-nu* **zi**-*ma* 3.6.30
5 **ki** 4.31 **gi$_6$ du tab**-*ma*
6 7.37.30 6 **me** *u* **gi$_6$ ta ša$_3$ zi**-*ma*
7 **tag$_4$** 1.37.30 **gi$_6$ du**

[36]You ‘tear out’ 21;28 from 33;36, NA of day 1 (NA_1) for month II, and [37]there remains 12;8. You put down 12;8 as NA of day 1 for month II. [38]You predict that month II (begins on day) 30 (of the previous month), NA_1 is 12;8.

P2 (NA_1 steps 1,3,4)

1 [39]In order for you to compute day 1. If month I (begins on day) 30 (of the previous month), NA_1 is 17 [40](in) 12 Ari: you add 14, the addition, with 17, NA_1, it is 31. [41]You predict that month II (begins on day) 1 (following a 30 day month), NA_1 is 31.

3 Whatever (the time by which) the night has progressed (*M*): [42]you subtract (‘tear out’) the ‘watch’ of the night (*D*) from it, and what remains [43]you multiply by the displacement for that month (ZI), and you add it with 31.

Column ii

[1][...] ... [...] [2][...] ⸢...⸣ [...] [3]⸢Whatever remains until sunset [4]you multiply by the displacement for that month, and you ‘tear it out’ from NA_1 of month II.

4 If [5]it is low you multiply whatever (the amount by which) it is low [6]by 0;24, you ‘tear it out’ from NA_1 of month II and you put it down.

P3 (KUR steps 1,3,4)

1 [7]In order for you to compute the day of last visibility. If (in) month XII KUR is 15 (in) 12 Ari: you add 5, the addition, [8]with 15, it is 20. If ⟨(the previous month was) full⟩ you predict that on (day) 26 of month II KUR is 20. [9]If (the previous month was) hollow you predict that on (day) 27 KUR is 20.

3 Whatever (the time by which) the night has progressed (*M*), [10]you ‘tear it out’ from the ‘watch’ of the night (*D*), and what remains you multiply by the displacement [11]for that month, and you ‘tear it out’ from the ‘day of disappearance’ (KUR) of month II.

4 [12]You multiply whatever (the amount by which) it is high by 0;24, and [13]you add it with the ‘day of disappearance’ (KUR) of month II.

3 If (the time by which) the night has progressed [14]is less than the ‘watch’ of the night: whatever [remains until sunrise] [15]you multiply by the displacement for that month, and [you add it] to the ‘day [of disappearance’ (KUR) of month II].

4 [16]If it is ⸢low⸣: you multiply whatever (the amount by which) it is [low] [17]by 0;24, and you ‘tear it out’ from the ‘day of disappearance’ (KUR) [of month II], [18]and what remains [...]

P4 (Systems A and K: various procedures)

a **(System A: computations involving Φ and *G*?)**

[19]2 ⸢3? 10 30 ...⸣ [...] [20][...] ⸢...⸣ [...] [21][...] [22][...] [23][...] [24]3.⸢7?⸣ [...] [25]1.1.20[...] [26]2,3;19,15,3[3,20 ... which] [27][exc]eeds 1,5⸢8⸣;[...] [28]4,48;8,53,[20 ...

b **(System K: computing *G* and *M*)**

... is] [29]44. [You ‘tear out’] 44 from 4,⸢30⸣, [it is 3,46 ...] [30]1 Ari. Whatever the subtraction [... Psc: the subtraction is 30;] [31]Ari: the subtraction is 45. Whatever (the amount) by which [45 exceeds the subtraction] [32]for Psc: it exceeds it by 15. [You divide] 15 into 30 parts: [33]the reciprocal of 30 is 0;2. 0;2 times 15 is 0;30. [1 Ari:] [34]whatever it [is less] than 12 Ari: [it is less by 11.]

Reverse column i

[1]1 Ari: whatever it is less than 12 Ari: it is less by 11. [2]11 times 0;30 is 5;30. You ‘tear out’ 5;30 from 45, [3]the subtraction for Ari, it is 39;30. You ‘tear it out’ from 3,46 [4]which you had put down, it is 3,6;30. [5]You add it with 4,31, (the time by which) the night has progressed, it is [6]7,37;30. You ‘tear out’ 6,0, day and night, from it, and [7]there remains 1,37;30, (the time by which) the night has progressed.

8 ⌜*ki*⌝-*i* 26 21 **kur** *ki-i an-na-a* **tab** 5 **tab** *ša*$_2$ *zib*
9 5 **tab** *ša*$_2$ **ḫun** *bi-rit* **tab** *ana* **tab** *ia-a-nu*
10 5 **tab ki** 26 21 **kur tab**-*ma* 26 **gub** *zib* 12 **zi**
11 **ḫun** 10 **zi** *bi-rit* 12 **zi** *ša*$_2$ *zib u*$_3$ 10
12 **zi** *ša*$_2$ **ḫun** 2 : 2 *a-na* 30 **šeš.meš** SE$_3$-*ma*
13 **igi** 30 2 : 2 **a.ra**$_2$ 2 4 1 **ḫun** *mi-nu-u*$_2$
14 *ki-i al* 12 **ḫun la**$_2$-*u*$_2$ ⌜11⌝ [**la**$_2$ 11 **a.ra**$_2$ 4]
15 **du**-*ma* 44 ⌜44⌝ [**ki** 10 **zi** *ša*$_2$ **ḫun tab**-*ma*]
16 [10].⌜44 : 10.4⌝[4 **ta** 26 **zi**-*ma*] ⌜**tag**$_4$ 15.1⌝[6 **gar**-*an*]
17 [27 15.16 **kur** *ša*$_2$ **bar** d]**u**$_{11}$ **diri.še** 30 ⌜2⌝[1 **na**]
18 [*ki-i an-na-a* **tab** 11 **tab** *š*]*a*$_2$ *zib* 14 **tab** *ša*$_2$ ⌜**ḫun** *bi-rit*⌝
19 [**tab** *ana* **tab** 3 : 3 *a-na* 30 **še**]**š.meš** SE$_3$-*ma* **igi** 30
20 [2 : 2 **a.ra**$_2$ 3 6 1] **ḫun** *mi-nu-u*$_2$ *al* 12
21 [**ḫun la**$_2$-*u*$_2$ 11 **la**$_2$ 11 **a**].**ra**$_2$ 6 **du**-*ma* 1.6
22 [1.6 **ta** 14 **tab** *ša*$_2$ **ḫun zi**]-⌜*ma*⌝ **tag**$_4$ 12.54
23 [12.54 **ki**] ⌜2⌝1 **tab**-*ma* 33.54 **gar**-*an*
24 [**bar** 1 33.54] ⌜**na**⌝ **gar**-*an* **du**$_{11}$.**ga gur**-*ma* **diri.še**
25 [*mi-nu-u*$_2$] ⌜**ta**⌝ **ša**$_3$ *lu-us-suḫ zib* 18 **zi**
26 ⌜**ḫun** 2⌝2 **zi** *bi-rit* ⌜**tab** *ana* **tab**⌝ 4 *a-na* 30 **šeš.meš**
27 SE$_3$-*ma* 8 8 **a.ra**$_2$ 11 *tam-ṭi-tu*$_4$ *ša*$_2$ *al-la*
28 12 **ḫun la**$_2$-*u*$_2$ **du**-*ma* 11 **a.ra**$_2$ 8 1.28
29 **ta** 22 **zi** *ša*$_2$ **ḫun** ⌜**zi**⌝-*ma re-ḫi* 20.32
30 **gar**-*an* 20.32 **ta bar** 1 33.54 **zi**-*ma*
31 **tag**$_4$ 13.22 **gar**-*an* **bar** 30 13.22 **gar**
32 ⌜1⌝.37.30 **gi**$_6$ ⌜**du**⌝ *mi-nu-u*$_2$ *ki-i ana* **kur** *šamaš*$_2$
33 [1].⌜3⌝7.30 **ta** 2.52.20 **en.nun gi**$_6$ **zi**-*ma*
34 ⌜**tag**$_4$⌝ 1.14.50 1.14.50 **a.ra**$_2$ 13.12 **zi**
35 **du**-*ma* 16 16 **ta** 13.22 **na zi**-*ma*
36 13.6 **gar**-*an* 34.55.15 **sig**
Edge
37 [a].**ra**$_2$ 24 **du**-*ma* 14 14 **ta** 13.12
38 **tag**$_4$ 13.8 **bar** 30 13.8 **na** ⌜**du**$_{11}$⌝

Column ii
P5 1 [x] ⌜x⌝ *ša*$_2$ *si-ma-nu ana* **du**$_3$-*ka* **be**-*ma ina* **du**$_6$ 14 ⌜x⌝ [x]
2 ⌜*ana* **šu**$_2$⌝ *šamaš*$_2$ *ina* 15 **ḫun** AN.MI **gar** *u*$_3$ 12 **zi** *ša*$_2$ *s*[*in*]
3 *u*$_3$ **apin** 30 **ta** 14 *ša*$_2$ **du**$_6$ *a-di* 12 *ša*$_2$ **apin**
4 27 **u**$_4$.**meš šid**-*ma a-na* 27 **u**$_4$.**meš** 1.30 [x$^?$]
5 **ki** *si-man* **tab**-*ma* 12 *ša*$_2$ **apin gar** 1.10 **gi**$_6$ **d**[**u** xxx]
6 ^{d}*sin* 15$^?$ *ša*$_2$ **ḫun kur** *u* **ki** AN.MI ⌜x⌝ [xxx]
7 12 *nis-ḫi ša*$_2$ **me** ⌜**tab**$^?$⌝ *ki-ma* **mu** *u*$_4$-*mu* ⌜x⌝ [xxxx]
8 *ina* 2 *u*$_4$-*mu u*$_3$ 5 **danna** *si*-[*man* xxxxx]
9 **ḫun** 12 **gar** 1.10 **gi**$_6$ **du** ⌜**ta**⌝ [xxxxxxxx]

P6 10 *ša*$_2$-*niš ana tar-ṣa* **zi** *ša*$_2$ ⌜x⌝ [xxxxxxxx]
11 ⌜x 20 **bi** x⌝ [xxxxxxxxx]
12 [xxxxxxxxxxxx]
13 [xxxxxxxxxxxx]
14 [xxxxxxxxxxxx]
15 [*mim*]-⌜*ma ša*$_2$ *al-la*⌝ [xxxxxx]
16 [a.r]**a**$_2$ 33.20 **du ki** 1.49.⌜40⌝ [**tab**]
17 *mim-ma ša*$_2$ *al-la* 11.25 **la**$_2$-*u*$_2$ [x]
18 **a.ra**$_2$ 33.20 **du ta** 4.32.20 ⌜**nim**⌝ [x$^?$]
19 42 **a.ra**$_2$ 33.20 23.20 *taš-p*[*il-tu*$_4$]

c **(KUR steps 1,2)**

1 [8]If on (day) 26 KUR is 21, thus is the addition: the addition for Psc is 5, [9]the addition for Ari is 5, the difference from addition to addition is nothing. [10]You add 5, the addition, with (day) 26, (when) KUR is 21, it is 26, (the month) is full.

2 Psc: the subtraction is 12;[11]Ari: the subtraction is 10. The difference between 12, the subtraction for Psc, and 10, [12]the subtraction for Ari, is 2. You divide 2 into 30 parts: [13]the reciprocal of 30 is 0;2. 0;2 times 2 is 0;4. 1 Ari: whatever (the amount) by which [14]it is less than 12 Ari: [it is less] ⸢by 11⸣. You multiply [11 by 0;4], it is [15]0;44. [You add] ⸢0;44⸣ [with 10, the subtraction for Ari, it is [16]10];⸢44⸣. [You 'tear out'] ⸢10;4⸣[4 from 26, there] ⸢remains 15;1⸣[6]. [17]You predict [15;16 for KUR of month I].

d **(NA$_1$ steps 1–4)**

1 Month XII$_2$, (which began on day) 30 (of the previous month), [NA$_1$ is] ⸢2⸣[1]. [18][Thus is the addition: the addition] for Psc is [11], the addition for ⸢Ari⸣ is 14, ⸢the difference⸣ [19][from addition to addition is 3.] You divide [3 into 30 pa]rts: the reciprocal of 30 [20][is 0;2 : 0;2 times 3 is 0;6. 1] Ari: whatever (the amount) by which [21][it is less] than 12: [it is less by 11.] You multiply [11] by 0;6, it is 1;6. [22][You 'tear out' 1;6 from 14, the addition for Ari], and there remains 12;54. [23]You add [12;54 with] ⸢2⸣1, and you put down 33;54. [24]You put down, predict that [month I (begins on) day 1 (following a 30 day month), NA$_1$ is 33;54.]

2 (In case) it is hollow, month XII$_2$: [25][what] should I 'tear out' from it? Psc: the subtraction is 18, [26]Ari: the subtraction is 22, the difference from addition to addition is 4, you divide it into 30 parts, [27]it is 0;8. You multiply 0;8 by 11, the deficit by which it [28]is less then 12 Ari, i.e. 11 times 0;8 is 1;28. [29]You 'tear it out' from 22, the subtraction for Ari, and there remains 20;32, [30]you put it down. You 'tear out' 20;32 from 33;54, NA$_1$ for month I (which begins on) day 1 (following a 30 day month), and [31]there remains 13;22, you put it down: month I, (which begins on day) 30 (of the previous month), NA$_1$ is 13;22.

3 [32](The time by which) the night has progressed is 1,37;30. Whatever (the amount that) remains to sunrise (M): [33]you 'tear out' [1],37;30 from 2,52;20, the 'watch' of the night (D), and [34]there remains 1,14;50. You multiply 1,14;50 by 0;0,13,12, the displacement (ZI), [35]it is 0;16. You 'tear out' 0;16 from 13;22, NA$_1$, it is [36]13;6, you put it down.

4 You multiply 0;34,55,15, the 'depth', [37]by 0;24, it is 0;14. 0;14 from 13;12 (error for 13;22), [38]there remains 13;8. You predict that month I (begins on day) 30 (of the previous month), NA$_1$ is 13;8.
Column ii

P5 **(Computations involving eclipses)**

[1]In order for you to compute the [...] ... of the duration: if in month VII, (day) 14 ⸢...⸣ [...] [2]until sunset, an eclipse takes place in 15 Ari, and the displacement of the Moon is 12, [3]and month VIII (begins on day) 30 (of the preceding month): from (day) 14 of month VII until (day) 12 of month VIII [4]you count 27 days, and for 27 days you add 1.30 [...] [5]with the duration, and you put down (day) 12 of month VIII. 1.10, (the time by which) the night has progressed [...] [6]the Moon 15$^{?}$ of Ari ... position of the eclipse ⸢...⸣ [...] [7]⸢you add⸣ 12, the displacement per day. Like$^{?}$ the year$^{?}$, the day ⸢...⸣ [...] [8]in 2 days and 5^{b}, the dura[tion, ...] [9]Ari 12 you put down. 1.10, (the time by which) the night has progressed, ⸢from⸣ [...]

P6 **(Computations involving variants of F, G and Φ?)**

a **(Subject unclear)**

[10]Secondly: opposite the displacement of ... [...] [11]⸢... 20 ...⸣ [...] [12][...] [13][...] [14][...]

b **(Computing G from F, variant scheme)**

[15][What]ever the amount by which [it exceeds 15;26$^{?}$:] [16]you multiply it by 33;20, [add it] to 1,49;40. [17]Whatever the amount by which it is less than 11;25 [...?]: [18]you deduct it from 4,32;20 [...?] [19]0;42 times 33;20 is 23;20, the difference.

20 8.41.26.40 **gal** 4.3.6.40 [**tur**$^?$]
21 *ana tar-ṣa* 2.13.20 : 2.9 : *ana tar-ṣa* 1.⸢10$^?$⸣[x]
22 15.25.54.22.30 **tab** 11.25.4.4[1.15 x]
23 5.27.35 *si-ma-nu* 9.47.27 ⸢*si*⸣-*man ki*-⸢*ṣa*⸣-[*ri*$^?$ x]
24 4.30.12.15 **ab$_2$** *ana* **ab$_2$** 25.12 : 2.40 : 5.20 : 2⸢6⸣ [x]
P7 25 *ša$_2$-niš* : 5.4.59.0.44.26.40 **gal**
26 2.5.1.43.42.13.20 ⟨**tur**⟩ **ab$_2$** *ana* **ab$_2$**
27 25.48.38.31.6.40 **tab** *u* **la$_2$** *ša$_2$ si-ma-nu*
28 **ta ugu** 1.59.48.8.53.20
29 **en** 2.0.59.15.33.20 *ša$_2$ al-la*
30 1.59.48.8.53.20 **tab diri**
31 : **a.ra$_2$** 3.22.30 **du ta** *muḫ-ḫi*
32 : 1.34.48.53.20 **en** 2.28.8.
33 : 53.20 *ina taš-pil-tu$_4$* GAR.GAR-*ma*
34 **ta** ⸢4⸣.54.48.53.20 **zi**
Edge
Co 1 [xxxxxxx s]**ag.me.gar** *u* ^{d}*ṣal-bat-a*-⸢*nu*⸣ [x]
2 [xxxxxxxx *l*]*a u$_2$-kal-lam* [xx]
Column i-ii
3 (empty) ^{m}**mu**-^{d}**en** ⸢**a**⸣ [xxxxxxxxxxx] ⸢*la*⸣-*bi-ri-šu$_2$* [x]
4 [*ina a*]-*mat* ^{d}**en** ^{d}**gašan**-*ia$_2$* ⸢**en**$^?$⸣.**meš dingir.meš** *ša$_2$* **an**-*e u* [**ki**-*tim* xxxxxxxxxxxxxx]

Critical and philological notes

Obverse columns i–ii

1–6 By 2008 small flakes had separated from the obverse in lines 1–10 of columns i and ii. The original place on the tablet could not be established for all flakes. Signs read by NS but no longer identifiable among the flakes were maintained in the transliteration.

Column i

1 10: restored on the basis of Ri11.
3 **zi** at the end: now broken away.
4 **tab** 13 **zi**: now on separate flake.
5 6 **tab**: now partly on separate flake.
7 The first **zi** is now broken away.
8 **gir$_2$** 8 **tab** 14 **zi**: now broken away.
9 The first **pa** is now broken away.
11 11: or 12–18? Very uncertain.
12 11: restored on the basis of Oi27.
18: restored on the basis of Ri25.
13 *an-na-a ša$_2$* **u$_4$.1.kam**: NS read an additional NA before *ša$_2$*, but this could not be confirmed and there is no space for it. **1.kam** is now broken away.
14 The initial **rin$_2$** is now broken away.
6: or 5.
15 After the inital **gir$_2$**: 4, 5 or 6.
23 **alla** 15 etc.: on the newly joined fragment BM 37205, which extends until Oi28.
26 The new fragment proves that NS had correctly restored this line as *an-na-a ša$_2$* ... *an-na-a ša$_2$*....
27 The parallel sections imply that this line may have begun with a phrase similar to **u$_4$**.1.**kam** *ana* **du$_3$**-*ka*, 'In order for you to compute day 1', followed by the date of NA$_1$ and an indication whether the previous month was hollow or full (cf. the commentary), i.e. *ki-i* **bar** 30 or *ki-i* **bar** 1, 'If in month I (which begins on day) 30/1 (following a 29/30 day month)'. The space appears to be sufficient for this. The month name **bar** (I) is implied by Oi36, where the next month is said to be **gu$_4$** (II).
12 *zib*, '12 Psc': one expects 14 *zib*; cf. the commentary.
27 **tab** = *teṭeppi*, 'you add': 2 m. sg. pres. G *ṭepû*. **gar**-*an* = *tašakkan* in Oi37 and other procedure texts suggest that instructions are not realised as imperatives in the procedure texts. NS translate them as imperatives, but it is not clear whether they assume that the Akkadian forms are imperatives.
28 **šeš.meš** = 'parts'; cf. the Glossary. SE$_3$: probably a 2 m. sg. pres. of *zâzu*, 'to divide'; cf. the Glossary.
29 For a justification of the restored text cf. the commentary.
Between **igi** 30, 'the reciprocal of 30', and the result (2), there is a separator (:), perhaps to prevent 30 2 from being interpreted as 32. Elsewhere (e.g. Oii32) the separator is not used in equivalent positions.
30 This is the first line partly preserved in column i of the obverse of BM 40082 corresponding to line 27 in the edition of NS (p. 96). Similar corrections of the line numbering were applied in the other columns (obverse and reverse).
34 *tam-ṭi-tu$_4$* = *tamṭītu* 'deficit, subtraction': cognate noun of *maṭû* G, 'to be lacking', or D 'to diminish'. The only other attestation in the astronomical corpus is Ri27 below.
39 **u$_4$.1.kam**: 'day 1', i.e. the day of the first appearance of the lunar crescent.

c (Computations involving variants of F, G and Φ?)

[20]The largest (value) is 8.41.26.40; the [smallest (value)] is 4.3.6.40.

[21]Opposite 2,13;20: 2,9. Opposite 1.10[...] [22]15;25,54,22,30 add 11;25,4,4[1,15 ...] [23]5.27.35 the time interval; 9.47.27 the time interval for the ⌜'knot'⌝? [...] [24]4.30.12.15, month by month, 25.12. 2.40. 5.20. 26

P7 (System A: computations involving Φ and G)

a (Parameters of the zigzag function $\hat{G}^{(2)}$)

[25]Secondly: 5,4;59,0,44,26,40 is the largest one; [26]2,5;1,43,42,13,20 is the ⟨smallest one⟩. Month by month [27]25;48,38,31,6,40 is the addition and subtraction for the duration.

b (System A: computing G from Φ, scheme 1, intervals 21–24)

[28]From 1,59;48,8,53,20 [29]until 2,0;59,15,33,28,8, (the amount) by which it exceeds [30]1,59;48,8,53,20, increasing, [31]you multiply by 3;22,30. From 1;34,48,53,20 until 2;28,8,[33]53,20 you accumulate it in the difference, and [34]you 'tear it out' from 4,54;48,53,20.

Edge

Co (Colophon)

[35][...] Jupiter and Mars [...] [36][... the wise] shall not inform [the ignorant. Tablet of] [37]Iddin-Bēl, son of [... according to] an older version [...] [38][At] the command of Bēl and Bēltija, the ⌜lords⌝? (and) the gods of heaven and [earth ...]

42 **en.nun** = *maṣṣartu*, 'watch', in the sense of 'duration' (of the entire night).
en.nun gi$_6$ ta ša$_3$ zi: this can only mean 'you 'tear out' the duration of the night from it' (not 'you subtract it from the duration of the night' as in NS). Hence New Moon occurs during day time; cf. the commentary.

43 The time interval computed in 42 is multiplied by a coefficient called **zi** = *nisḫu*, the result of which is *added* to 31. Hence *nisḫu* cannot mean 'subtraction' here so that the alternative meaning 'displacement' must be assumed. The phrase *ša arḫišu*, 'for that month', refers to the coefficient's dependence on the zodiacal position of the Moon (or the Sun) at lunation.

Column ii

1–2 The few signs that could still be read by NS are now completely broken away. In the gaps one expects a rule for the alternative case of New Moon during the night as in Oii13–15; cf. the commentary.

3 *re-ḫi* **a.ra$_2$ zi**: now broken away.

4 **zi** *ki-i*: now broken away.

6 *-an*: written on right edge.

8 After *ki-i* one expects *kun-nu* = *kunnu*, '(the previous month was) full': perhaps omitted by mistake.

11 In this line **u$_4$.na$_2$.a**, 'day of disappearance', functions as a technical term for the Lunar Six interval KUR, similar to the usage of **u$_4$.1.kam** for NA$_1$ in No. **53** Rii15.

12 *ma*: now partly broken away.

13 After *al-la* NS restored $\frac{1}{2}$, but this would make sense only if the time of New Moon is defined with respect to midnight, as in lunar system B. In system K another convention is followed; cf. the commentary.

14 The text in the gap has been restored in accordance with the immediately preceding conditional phrase. The available space in the gap at the end is rather small, but the text could hardly have been shorter. *re-ḫi*, 'there remains', was omitted by NS, but this would result in a defective sentence. However, one expects the opposite conditional phrase; cf. the commentary.

18 The break at the end may have contained a final statement similar to **u$_4$.na$_2$.a** *ša$_2$* **gu$_4$**, 'you call it KUR of month II'.

19 3 10 30: only the upper parts are visible; hence 3 might also be 6. 10 30 replaces 19 (NS).

20 1.20: preserved on the fragment BM 37205.

24 ⌜7⌝ or ⌜4⌝.

30 The gap may be restored as [*ša$_2$* **ḫun** *al* **zi** *ša$_2$ zib* **diri** *zib* 30 **zi**].

31 The text restored in the gap replaces 45 *al-la* **zi** in NS.

Reverse column i

1 The beginning of this line is repeated from the last line of the obverse.
12 Ari: the formulation of this interpolation rule implies that 12 Ari is the control point, not 1 Ari as assumed in NS.

16 **gar**-*an*: there is space for 1–2 signs in the gap, but this restoration is uncertain.

17 **kur**, 'KUR': one might also restore the alternative term **u$_4$.na$_2$.a** (NS).

24 **gur**-*ma* **diri.še** = *turru-ma Addaru 2*. The context implies that Addaru$_2$ is the subject of the preceding stative *turru*, 'it is hollow'; cf. the commentary. In this construction the particle *-ma* has an emphasising function: '(in case) it is *hollow*, Addaru$_2$'.

25 *lu-us-su-uḫ* = *lussuḫ*, prec. 1 sg. G *nasāḫu*, '(what) should I 'tear out'?'.

26 *bi-rit* **tab** *ana* **tab**, 'the difference from addition to addition': one would expect *bi-rit* **zi** *ana* **zi**, 'the difference from subtraction to subtraction'.

28 After the gap the lower half of the signs 30.36 **zib** are preserved: replaces 11 (NS).

32 *ki-i ana* **kur** *šamaš$_2$*: *ki-i* omitted by NS.

Column ii

1 The broken sign before *ša$_2$* ends with three horizontal wedges followed by a vertical wedge, or perhaps two vertical wedges on top of each other. Perhaps *taš*, 'difference'?
The broken sign after 14 might be **la$_2$**. The gap at the end may have contained the numerical value of the time 'until sunset' mentioned in Rii2.

5,9 1.10: comparison with **la$_2$** in Ri1,14,28 suggests that 1.10 is a more likely interpretation than **la$_2$**.

6 15: doubtful, perhaps 17 or 18.

KUR *u*: meaning unclear.
After AN.MI the beginning of a sign similar to **igi** or u_3 is visible.
7 **me**: not used elsewhere in this text as a logogram for *ūmu* 'day'.
ki-ma = *kīma*: 'like' or 'when'?
u_4-*mu* is followed by the upper half of a vertical wedge.
10 The damaged sign following **zi** *ša*$_2$ might be **iti** (NS).
15 After *al-la* one expects a number; NS restored 15.26 (cf. the commentary).
18 The broken sign at the end might be **nim**. If so, this would be the only instance in the text where **nim** (= *šūlû*?) is used for subtraction instead of **zi**(*nasāḫu*), 'to tear out'.
21 The winkelhaken near the right edge may belong to a 20, 30, 40 or 50.
24 4.30.12.15: the 12 replaces :2 (NS).
25 5.4.59.0.44.26.40: interpreted as two numbers by NS.
26 2.5.1.43.42.13.20: interpreted as two numbers by NS.
31–33 The reason for the separation mark (:) at the beginning of these lines is not clear; perhaps to provide a separation from column i.
32–33 2.28.8.53.20: written across two lines, interpreted as 2 numbers by NS.
Edge
2 The gap at the end may have contained **im** '*ṭuppi*', 'Tablet of', to be connected with the following line.
3 The gap following ^{m}**mu-**^{d}**en**, 'Iddin-Bēl', must have contained his filiation, but there is probably no space for a second name (that of the scribe).

Commentary

This nearly complete tablet consists of three fragments, two of which were published by Neugebauer & Sachs 1967 (= NS) as 'Text K'. The newly discovered small fragment BM 37205 (size 5.0 by 3.2 cm) is positioned in between the other fragments. As a result, the original size of the tablet is now accurately known. All fragments are well preserved, except for parts of column i on the obverse, where the surface is rather worn, and the reverse of BM 37205, which is destroyed. The obverse of BM 37205 contains 6 partly preserved lines of text belonging to Oi23–28 and Oii20. On the lower edge of the reverse there is an unusual colophon consisting of 4 lines. The first 2 are confined to column ii, while the remaining 2 are written across both columns. The colophon begins with a reference to Jupiter and Mars, perhaps as part of an invocation, and an admonition to secrecy. Line 3 mentions Iddin-Bēl as the 'owner' or scribe of the tablet. He can be identified as Iddin-Bēl/Mušēzib, who is attested on several other astronomical tablets, one of which dates to the reign of Philip Arrhidaeus (322–315 BC).[628] The colophon ends with another invocation that is only partly understood. A date is not preserved, but orthographic and notational features suggest a date in the early Seleucid or late Achaemenid period, consistent with the biography of Iddin-bēl/Mušēzib.

The bulk of the tablet is occupied by P1–P4 and the auxiliary tables T1–T4, which are concerned with Lunar Six intervals (§4.1). The underlying algorithms belong to lunar system K (§4.3), thus far attested only on this tablet and believed to belong to an older stage of development than lunar systems A and B. Since P6, P7 and perhaps P4.a contain elements of lunar system A, the tablet does not offer a coherent collection of procedures, and it must have been written when lunar system A was already in existence. It is therefore not a surprise that, according to the colophon, the text was copied from an older tablet.

T1–T4 Tables of coefficients

The tablet begins with four tables T1–T4, divided over two columns, ia (T1, T3) and ib (T2, T4). Each table contains 12 values of two different coefficients called 'addition' (**tab** = *ṭīpu*) and 'subtraction' (**zi** = *nisḫu*).[629] These coefficients are denoted as TAB_1,...,TAB_4 and ZI_1,...,ZI_4, graphs of which are shown in Fig. 4.6. As demonstrated below, the 12 values of each coefficient are anchored to the 12th degree of the zodiacal signs. The meaning of the coefficients is explained in §4.3.6. TAB_1, ZI_1, TAB_2 and ZI_2 find their application in the computation of NA_1 and KUR, respectively, in accordance with the labels 'for day 1' and 'for the day of disappearance', which are provided below the last entries of T1 and T2. The coefficients that are tabulated in T3 and T4 are not used in any of the procedures on the tablet. The label of T4, 'for ME and GI_6',[630] confirms the suggestion by NS that they are meant to be used in procedures for the Lunar Six intervals at Full Moon ($ŠU_2$, NA, ME and GI_6) similar to the way in which the coefficients of T1 and T2 are used in P1–P4.

P1 NA_1 steps 1,2

In P1 steps 1 and 2 of the Lunar Six module (§4.3.6.1, §4.3.6.2) are demonstrated by means of a coherent numerical example for NA_1. It can be assumed that the damaged first line (Oi27) began with an introduction mentioning the aim of

[628] MNB 1856 (Neugebauer & Sachs 1969, Text H), dated to year 5 of Philip Arrhidaeus.
[629] In NS the tables are numbered I (T1), II (T3), III (T2) and IV (T4).
[630] This label is preserved on the new fragment BM 37205.

the procedure ('In order for you to compute day 1'), the date of the previous NA_1, and the length of the previous month, as is the case in P2–P4 (cf. the critical notes). It is not clear why the length of the previous month is mentioned here. Its value determines the date of the next KUR, but plays no role in the procedure for NA_1.[631] The other initial data are NA_1 = 20 in month I (month name not mentioned here but in step 2), and the Moon is in 8 Ari.

NA_1 step 1 (Oi27–31)

The value of TAB_1 in Ari is given as 12, and that in Psc as 11. While the latter agrees with T1 (Oi1), 12 is inconsistent with T1 (and with the rest of P1): TAB_1 is actually 14 in Ari, hence 12 must be a scribal error. The following reconstruction of Oi27–31, in the course of which the control points for interpolation will be shown to be the 12th degrees of the zodiacal signs, confirms this conclusion. The result of step 1 is mentioned in Oi36: NA_1 = 33;36 in month II. This was obtained by adding TAB_1 = 13;36 to 20, the value of NA_1 for the previous month (Oi31). Hence Oi27–30 should contain the computation of 13;36, which is obtained by subtracting 0;24 (Oi27), from 14, the control value for TAB_1 in Ari. The rest of Oi27 has been restored accordingly. In order to interpolate between the control values, 14 − 11 = 3 is divided by 30, i.e. multiplied by 0;2, the result being the interpolation coefficient 0;6, which is mentioned in Oi28 (Oi29 is restored accordingly). Multiplication of 0;6 by the distance between the Moon and the control point in Ari should result in the mentioned 0;24, i.e. this distance must be 4°. Since Psc is behind the Moon (in terms of its motion along the zodiac), the other control point (Ari) must be in front of it, i.e. at 8 + 4 = 12 Ari. It can be assumed that the other coefficients in T1–T4 also apply to the 12th degree; this is confirmed by the rest of the text.

NA_1 step 2 (Oi32–38)

The value obtained in step 1, NA_1 = 33;36, corresponds to month I being full. The execution of step 2 implies that this is considered too large, i.e. the lunar crescent should have appeared 1 day earlier, when NA_1 was smaller by the amount ZI_1. Accordingly, ZI_1 is computed for 8 Ari by interpolating between 18 (12 Psc) and 22 (12 Ari) in the same manner as TAB_1, leading to ZI_1 = 21;28 (Oi35). This is subtracted from 33;36, the result being NA_1 = 12;8 in month II, and month I is declared hollow (Oi38).

***P2** NA_1 steps 1, 3,4*

P2 also deals with NA_1 using slightly different initial conditions: the current month (I) follows a hollow month, NA_1 = 17°, and the Moon is at 12 Ari (Oi39–40). It is not clear what motivated the inclusion of a second example for NA_1, since there is no essential difference with P1. Again step 1 is demonstrated by means of a numerical example, while steps 3–4 are formulated in a general fashion.

NA_1 step 1 (Oi39–41)

Since the initial position coincides with the control point where TAB_1 = 14, no interpolation is required, so that 14 can be added directly, resulting in NA_1 = 31 in month II. Accordingly, month I is declared full (Oi41). The alternative possibility of a hollow month I, which would require the subtraction of ZI_1 from NA_1 (step 2), is not considered.

NA_1 step 3 (Oi41–43)

Step 3 (§4.3.6.3) begins with an instruction to subtract the 'watch of the night' from '(the time by which) the night has progressed'. The former denotes the duration of the night (D), the latter M^{nm}. The subtraction of D^{nm} from M^{nm} implies that $M^{nm} > D^{nm}$, i.e. New Moon takes place after sunrise. The time interval is multiplied by a quantity called 'displacement' (**zi**), the value of which is not mentioned, and the product is added to the outcome of step 1. The two lines between Oi43 and Oii3 are strongly damaged. Oii3 deals with a similar correction, which is now subtracted from NA_1 (Oii3–4). This may correspond to the alternative case of New Moon before sunrise ($M^{nm} < D^{nm}$), the condition for which may be restored in Oii2–3 ('If (the time by which) the night has progressed is less than the 'watch' of the night'). The result assumes the form of a general expression which is combined with step 1, quoted by its numerical value (Oi43).

NA_1 step 4 (Oii5–6)

Step 4 for NA_1 (§4.3.6.4) is presented in abstract terms, i.e. no numerical values are assigned. If the Moon is below the ecliptic a correction is subtracted from NA_1 for month II. No correction is applied if the Moon is above the ecliptic.

[631] Accordingly, it is impossible to infer the length of the previous month from the other data in the procedure.

***P3** KUR steps 1–4*

P3 deals with KUR and is stuctured similarly to P2. It begins with a numerical example of step 1 (Oii7–9). This is followed by steps 3–4 (Oii9–18), which are partly formulated in general terms, as was the case in P2. The only reference to the preceding numerical example is found in the instruction to add the corrections to the value of KUR for month II.

KUR step 1 (Oii7–9)

The initial data are: KUR = 15 on day 27 of month XII and the Moon is in 12 Ari. This position coincides with the control point where $TAB_2 = 5$, which is added to 15, so that KUR = 20 in the next month, 29 days after the previous KUR event. If one assumes that the previous month is correctly stated to be month XII, then month II (Oii8) must be a mistake for month I. Vice versa, if one assumes that the new month is month II, then the initial month must be corrected to month I. If the initial month (say month I) was full (the word *kunnu*, 'full' is omitted by mistake), the addition of 29 days to the previous date results in day 26 of month II; if it was hollow, one obtains day 27, as mentioned in the text.

KUR step 3 (Oii9–11 and 13–15)

Step 3 begins with a general rule for the case of New Moon during the night ($M^{nm} < D^{nm}$). The resulting correction is subtracted from KUR for month II. One would expect the alternative situation of New Moon during day time ($M^{nm} > D^{nm}$) to be dealt with as well. Indeed lines Oii13–15 again concern step 3, and the resulting correction is added to KUR as expected. However, the preceding condition $M^{nm} < D^{nm}$ (Oii13–14) is inconsistent with this, since it corresponds to the opposite case; presumably there is a scribal error.

KUR step 4 (Oii12 and 16–18)

Step 4 is mentioned twice: in the first instance, which follows step 3 for New Moon during the night, the Moon is above the ecliptic (Oii12), in the second instance, which follows step 3 for what is presumably New Moon during day time, the Moon is below the ecliptic (Oii16–17). This formulation differs from P2, where the case of the Moon being below the ecliptic occurs in connection with New Moon during the night (Oii5–6), if the interpretation of P2 is correct. The break at the end of Oii18 may have contained a final statement similar to 'you call it KUR for month II'.

***P4** System A: Φ, G; system K: G; M; KUR; NA_1*

This procedure consists of four parts. Part **a** (Oii19–28), which is heavily damaged, appears to be concerned with functions Φ and G of lunar system A (cf. §4.4.3 and §4.4.11). Its connection with the system K procedures is not clear. The rest belongs to system K. Part **b** (Oii29–Ri7) deals with the time of New Moon (function M^{nm}), which is needed in step 3 of the procedures for KUR and NA_1. Parts **c** (Ri8–17) and **d** (Ri17–38) contain further examples of the computation of KUR and NA_1.

a. (Oii19–28) System A: computations involving Φ and G

The beginning of P4 is heavily damaged and difficult to interpret. The numbers in Oii19–28 might be values of the system A functions Φ and G, or variants thereof. The preserved signs in Oii27 suggest an interpolation scheme for computing G from Φ similar to scheme 1 (§4.4.11). The number 2,3;19,15,33,20 (Oii26) may be a Φ value. If one assumes that it is connectible to the parent function (cf. §4.4.3) then it pertains to Full Moon (Φ^{fm}), and can be dated to month VI of Artaxerxes III year 16 (descending branch of Φ), or SE 3,11 month XI (ascending branch). The significance of these dates is not clear. In ΦG interpolation scheme 1 the former corresponds to G^{fm} = 4,1;53,34,48,53,20°, the latter to G^{fm} = 4,24;56,17,46,40°, neither of which is preserved. In Oii28 one may restore 4,48;8,53,20, probably a value of G. In scheme 1 it is associated with Φ = 1,58;20,58°, which number may be restored in Oii27. The mentioned values of Φ and G are not control values of scheme 1, but they may belong to a similar interpolation scheme. There is obviously no room for a scheme as extensive as scheme 1.

b. System K: computing G^{nm} and M^{nm} (Oii28–Ri7)

Part **b** concerns the time of New Moon (M^{nm}). The beginning is lost (Oii28), and the connection with part **a** is not clear. For a discussion of the algorithms for G and M cf. §4.3.4 and §4.3.5, respectively.

The computation begins in what turns out to be month XII, but can be partly reconstructed back to month XI (Fig. 5.1). The number 44 (Oii29), which was presumably computed in Oii28, is the value of ZI_G for the unknown position of the New Moon in month XII. It is subtracted from G_{XII} = 4,30°, the result being $G_{XII.2}$ = 3,46° (Oii29 has been restored accordingly). From Ri5 we learn that $M^{nm}_{XII.2}$ = 4,31°, which implies M^{nm}_{XII} = 4,31° − 3,46° = 45° (not preserved). The New Moon's position in month XII_2 is 1° Ari (Oii30). In Oii30–Ri3 the value of ZI_G for that position is computed by interpolating between the

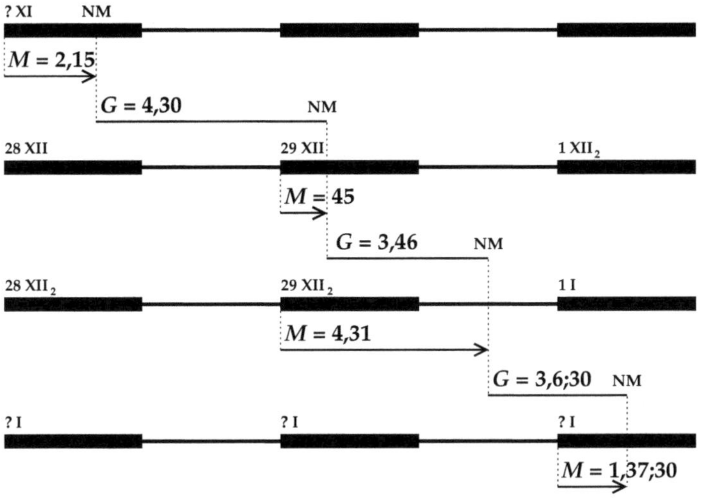

Fig. 5.1: Time of New Moon (NM) in months XI–I according to P4. Thick bars indicate night time.

control values 30° (12 Psc) and 45° (12 Ari), leading to ZI_G = 39;30° (Ri3). This is subtracted from $G_{XII.2}$ = 3,46°, so that G_I = 3,6;30°, which is added to $M^{nm}_{XII.2}$ = 4,31°, resulting in M^{nm}_I = 7,37;30°. Since this exceeds the date boundary, 6,0° is subtracted, leading to M^{nm}_I = 1,37;30°. This final result is applied later in step 3 for NA_1.

c. KUR: step 1 (Ri8–10)

This procedure proceeds from the same initial position of the Moon, i.e. 1 Ari. The other initial data are: KUR = 21° on the 26th, presumably of month XII_2 (a month name is not mentioned here but in Ri17). The value of TAB_2 is obtained by interpolating between the appropriate control values from T2, the result being 5, and this is added, so that KUR = 26° in month I. The corresponding date, 29 days after the previous KUR, is stated to be the 26th; this implies that month XII_2 was hollow. In Ri10 month I is declared to be full. It is unclear what the basis of this statement is, because the length of month I is established on the basis of NA_1 for month II, which is not available at this point.

Step 2 (Ri10–17)

The subsequent lines (Ri10–17) concern ZI_2, which implies that KUR = 26 was considered too large, and the correct KUR occurs 1 day later. The value of ZI_2 for 1 Ari is computed by means of interpolation, the result being 10;44°, and this is subtracted from 26°, so that KUR = 15;16° in month I.

d. NA_1: step 1 (Ri17–24)

Also here the Moon is assumed to be in 1 Ari. The other initial data are: NA_1 = 21° in month XII_2, which follows a hollow month XII (Ri17). The fact that the month is intercalary suggests that the data pertain to historical dates, although no year is mentioned, and the position 1 Ari may be idealised. The value of TAB_1 is computed by means of interpolation between the control values for Psc and Ari, the result being 12;54° (Ri22), and this is added to NA_1 = 21°, which results in NA_1 = 33;54° in month I. The corresponding date is 30 days after the previous event, i.e. month XII_2 is full.

Step 2 (Ri24–31)

The text continues with the case that the lunar crescent appeared 1 day earlier, which is here uniquely introduced by a question: 'If it is *hollow*, month XII_2, what should I 'tear out' from it?'. The correction ZI_1 is obtained by means of interpolation, the result being ZI_1 = 20;32°. This is subtracted from 33;54°, the outcome of step 1, which results in NA_1 = 13;22°. Accordingly, the previous month XII_2 is said to be hollow (Ri31).

Step 3 (Ri32–36)

The text continues with a numerical implementation of step 3 using M^{nm} = 1,37;30°, as computed in P4.b. The computation agrees with the general expression of d_3 for the case of a New Moon during the night ($M^{nm} < D^{nm}$). The time until sunrise, $D^{nm} - M^{nm}$ = 2,52;20 − 1,37;30 = 1,14;50°, is multiplied by the coefficient ZI = 0;0,13,12 (Ri34). This is the only instance of step 3 where numerical values are assigned to D (§4.3.2) and ZI. Both were presumably obtained by means of interpolation. If one makes the reasonable assumption that D = 3,0° if the New Moon is at the vernal equinox, which is located at 12 Ari in system K, then the current value of D differs by 3,0 − 2,52;20 = 7;40°. Assuming that the Moon is at 1 Ari, this difference should equal the product of the interpolation coefficient for Psc and the distance from 1 Ari to the control point, 12 − 1 = 11°. Hence 7;40/11 should be the value of the interpolation coefficient, but 7;40/11 is not a finite sexagesimal number, because 7;40 does not contain the factor 11. The solution may be that 2,52;20 was rounded to 3 digits, and that the exact value of 3,0$-D$ is not 7;40 but a number that does contain 11 as a factor.[632]

Alternatively we may abandon, for this algorithm, the assumption that the Moon is at 1 Ari and proceed from the fact that 7;40 can be factorised as 23 · 0;20. Hence the interpolation coefficient might be 0;20, and the Moon 23° away from the control point 12 Ari, i.e. either at 19 Psc or at 5 Tau. In the former case the control value in 12 Psc is 3,0 – 10 = 2,50°, in the latter case that in 12 Tau is 3,0 + 10 = 3,10°. This interpretation seems unattractive, since other parts of P4 imply that the Moon is at 1 Ari.

Step 4 (Ri36–38)

The text abruptly turns to step 4. The Moon's 'depth' (distance below the ecliptic?) is given as 34.55.15, perhaps to be interpreted as 0;34,55,15° (or barleycorns?). This is multiplied by 0;24, the result being 0;13,58,6, which is rounded to 0;14 and subtracted from NA_1. However, it is not subtracted from the value of NA_1 computed in step 3, but from the outcome of step 2 (in Ri37 13;12 is a trivial mistake for 13;22), the result being NA_1 = 13;8° in month I. The previous month (XII_2) is declared hollow, as had been done in step 2.

***P5** Computations involving a lunar eclipse*

P5 begins with the standard introduction, which is unfortunately damaged at the crucial point. The term 'duration' usually denotes G, the synodic month -29^d. Rii2 mentions a lunar eclipse that happened on day 14 (time indication broken away) of month VII, while the Moon was in 15 Ari. The 'displacement of the Moon' 12 (Rii2) can only be the daily displacement measured in degrees. There are still considerable difficulties with the rest of P5 that prevent a meaningful interpretation. In Rii4 27^d are added to day 14 of month VII, which results in day 12 of month VIII if one assumes that month VII was hollow (Rii3). The significance of this operation is not clear, but it may be noted that 27^d is close to the draconitic month (27.21^d), as well as the anomalistic month (27.56^d) and the sidereal month (27.32^d). In Rii5 something is added to the time (of the eclipse?). The number 1.10 (perhaps 1,10°), called '(the time by which) the night has progressed', is the time between sunset and the lunar eclipse. Since 1,10° is shorter than the duration of the night, the eclipse occurs during night time. Rii6–7 contain a reference to the zodiacal position of the eclipse, and the Moon's daily displacement of 12°. Rii8 mentions an interval of 2 days and 5 'miles' (*bēru*) = 2;25 days = 2 days + 2,30°. This might be an estimate of the synodic month -27 days. The meaning of Rii9 remains unclear.

***P6** Computations involving variants of F, G and Φ?*

P6 is concerned with functions F (the Moon's daily displacement) and G (duration of the synodic month -29^d). In the absence of any parallels or applications in synodic tables it is not clear to which computational system they belong. The first line (Rii10) begins with the adverb 'secondly', apparently a marker separating P6 and P7 from the system K procedures. Virtually nothing remains of part **a** (Rii10–14), except the term 'displacement', which could refer to F.

Part **b** (Rii15–19) contains rules for computing a G-type function, say $\hat{G}^{(4)}$, from an F-type function, say F_x. The number 33;20 (Rii18), attested only here, can be interpreted as the interpolation coefficient by which a difference in F_x is multiplied in order to get the corresponding difference for $\hat{G}^{(4)}$, i.e. $\Delta\hat{G}^{(4)} = 33;20 \cdot \Delta F_x$. In Rii19 this relation is used to connect the monthly differences of both functions, 23;20 = d of $\hat{G}(4)$ = 33;20 · 0;42. While 0;42 = d_{F_x} is identical to d_F of system A, 23;20 differs from the system A value d_G = 25;48,38,31,6,40. For a discussion of $\hat{G}^{(4)}$ cf. §4.4.11 (p. 151); the parameters are compiled in Table 4.24. If one divides the amplitude of $\hat{G}^{(4)}$ by 33;20 one obtains Δ = 4;52,48 for the amplitude of F_x, which is very close to 4;52,49,15,15 = Δ of $F^{(1)}$ in system A (§4.4.9). This supports the correctness of the reconstruction and interpretation of $\hat{G}^{(4)}$. The reconstruction of F_x is more problematic. By analogy with G of system A, $\hat{G}^{(4)}$ ought to be more or less out of phase with F_x, i.e. the maximum of the former should be close to the minimum of the latter and vice versa. Hence 11;25 (Rii17) should be near the minimum of F_x. Indeed this number is close to the minimum of $F^{(2)}$, m = 11;4 (§4.4.9), but it cannot be the minimum of F_x, since the text refers to values that are smaller (Rii17).

Part **c** (Rii20–24) is difficult to interpret. 8.41.26.40 is the maximum of a quantity, of which 4.3.6.40 might be the minimum. Rii21 mentions 2.13.20, which suggests an interpretion as 2,13;20, the maximum of the truncated version of Φ, which also functions as the name of Φ, but the rest of the line is unclear. The numbers 11;25,... and 15;25,54,22,30 (Rii21), might be values (extrema?) of a variant of F (compare §4.4.9). Lines Rii23–24 seem to concern G, but the meaning is unclear.

[632] For instance, if D = 5,52;20,1° the interpolation coefficient is 7;39,59/11 = 0;41,49. If D = 5,52;40 it is 7;20/11 = 0;40.

***P7** System A: computations involving Φ and G*

Part **a** (Rii25–27) provides parameters of variant 2 of the zigzag function $\hat{G}$ of lunar system A (Table 4.24 on p. 151). The formulation is based on template ZZ.A.5 (§2.6.8.1). Note that this interpretation differs from the one in NS; cf. the critical notes.

Part **b** (Rii28–34) consists of a single instruction for computing G from Φ by means of interpolation. A new interpretation leads to the surprising result that the algorithm is fully consistent with intervals 21–24 of ΦG scheme 1 of lunar system A (§4.4.11).[633] The numbers in Rii28–29 are the control points Φ_{21} = 1,59;48,8,53,20° (↑) and Φ_{25} = 2,0;59,15,33,20° (↑) of scheme 1. However, the interpolation is not formulated separately for the 4 intervals between Φ_{21} and Φ_{25} as in the templates IP.B–D or PHIG.A, but abbreviated using the *malûššu*-rule. If Φ is between Φ_{21} and Φ_{24} then $\Phi - \Phi_{21}$ is multiplied by $1/\phi$ = 3;22,30 (Rii31), where ϕ = 0;17,46,40° is the basic interpolation interval, the result being the distance by which Φ exceeds Φ_{21} measured in units of ϕ. For each further control point that is crossed, a contribution is added to ('accumulated in', Rii33) the difference of G. If $\Phi > \Phi_{22}$ then 1;34,48,53,20° = 5;20 ϕ is stored as the first contribution to the difference for G; if $\Phi > \Phi_{23}$ then 6;20 ϕ = 1;52,35,33,20° is added to it, etc., until $\Phi > \Phi_{25}$, in which case 8;20 ϕ = 2;28,8,53,20° is added to it. The contribution from the interval in which Φ is located, $[\Phi_k, \Phi_{k+1}]$ ($21 \leq k \leq 24$), is multiplied by the difference $\Phi - \Phi_k$ measured in units of ϕ. For instance, if $\Phi = \Phi_{24} + 0;10 \cdot \phi$ then the cumulative difference for G becomes (5;20 + 6;20 + 7;20 + 0;10 · 8;20)$\cdot\phi$. This is subtracted from the control value G_{21} = 4,54;48,53,20° (Rii33–34). The same result would be achieved by interpolating separately within each interval in accordance with ΦG scheme 1 as laid out in Table 4.22 (p. 147).

633 NS misunderstood this passage, partly because they did not recognise that 2;28,8,53,20° is written across lines Rii29–30. In *HAMA*, p. 511, where Neugebauer briefly revisits this procedure, he erroneously states that 4,54;48,53,20° = G_{21} applies to the entire segment between control points 21 and 26.

5.3.2 System A

No. 53

museum numbers	BM 32167 (76–11–17,1894) +32172 (1899) +32451 (2188) +32651 (2418) +32663 (2431) +32744 (2514) +32752 (2523)
provenience	Babylon
date	early Seleucid?
measures	20.5 × 20.4 × 2.5 cm (type A); column width 10.2 cm
arrangement	O/R
cuneiform text	photo: Figs. F.54, F.55; *ACT*, Pl. 239a (BM 32172); *ACT*, Pl. 234 (BM 32651) copy: Kugler *BMR*, Pls. 5–6 (BM 32651); *ACT*, Pls. 223–224 (BM 32651)
bibliography	Kugler (1900), pp. 66–192 (BM 32651); Neugebauer (1953), pp. 100–102 (P17'); *ACT* 200 (BM 32651); *ACT* 200aa, Zqb (BM 32172); Aaboe & Henderson (1975), pp. 208–211 (P8')
contents	Moon system A: various procedures

Upper edge

1 *ina a-mat* ^{d}**en** *u* ^{d}**gašan**-[*ia*$_2$ *liš-lim*]

Obverse

Column i

P1 1 **du**$_3$-*šu*$_2$ *ša*$_2$ 2.13.20 2.45.⸢5⸣[5].⸢33⸣.20 **ab**$_2$ *ana* **ab**$_2$ **tab** *u* ⸢**la**$_2$⸣ [*ša*$_2$ *al* 2.17.4.48.53.20 **diri**]

2 2.17.4.48.53.20 *ina lib*$_3$*-bi-šu*$_2$ **e**$_{11}$ *ša*$_2$ **t**[**ag**$_4$ *ina* 2.17.4.48.53.20 **e**$_{11}$]

3 *ša*$_2$ *al-la* 1.57.47.57.46.40 **la**$_2$-*u*$_2$ [*ina* 1.57.47.57.46.40 **e**$_{11}$ *ša*$_2$ **tag**$_4$]

4 **ki** 1.57.47.57.46.40 ⸢**tab**-*ma*⸣ **gar** ⸢5$^?$ x⸣ [xxxxxxxxx]

5 *ša*$_2$ *al* 2.17.4.48.53.20 **di**[**ri** xxxxxxxxxx *ša*$_2$ *al*]

6 1.57.47.57.46.40 **la**$_2$-⸢*u*$_2$⸣ [xxxxxxxxxxxxx]

P2 7 **du**$_3$-*šu*$_2$ *ša*$_2$ 2.13.20 *ša*$_2$ **u**$_4$.15.[**kam** xxxxxxxxxxxx]

8 2.17.4.48.53.20 [xxxxxxxxxxxxxxxxxx]

9 **ki** 1.57.47.57.[46.40 **tab** xxxxxxxxxxxx]

10 **tab**-*u*$_2$ *ša*$_2$ 14 [xxxxxxxxxxxxxxxxxxxx]

P3 11 **du**$_3$-*šu*$_2$ *ša*$_2$ ⸢2⸣.[13.20$^?$ xxxxxxxxxxxxxxxx]

(about 5 lines missing)

1' [xxxxxxxxxxx] ⸢*ki*⸣ 1.6 ⸢**nim**$^?$ x *ana* **ugu**$^?$⸣ [xxxxxxxxxxx]

Column ia

P4' 2' [10 **ḫun** 3 *ša*$_2$] ⸢*al* 10 **ḫun diri**⸣ **a.ra**$_2$ 40 **du ki** 3 **tab**

3' [10 **mul**$_2$ 3.2]0 *ša*$_2$ *al* 10 **mul**$_2$ **diri a.ra**$_2$ 24 **du ki** 3.20 **tab**

4' [10 **maš** 3].⸢3⸣2 *ša*$_2$ *al* 10 **maš diri a.ra**$_2$ 8 **du ki** 3.32 **tab**

5' [10] ⸢**alla**⸣ 3.36 *ša*$_2$ *al* 10 **alla diri a.ra**$_2$ 8 **du ta** 3.36 **la**$_2$

6' 10 **a** 3.32 *ša*$_2$ *al* 10 **a diri a.ra**$_2$ 24 **du ta** 3.32 **la**$_2$

7' 10 **absin**$_0$ 3.20 *ša*$_2$ *al* 10 **absin**$_0$ **diri a.ra**$_2$ 40 **du ta** 3.20 **la**$_2$

8' 10 **rin**$_2$ 3 *ša*$_2$ *al* 10 **rin**$_2$ **diri a.ra**$_2$ 40 **du ta** 3 **la**$_2$

9' 10 **gir**$_2$**.tab** 2.40 *ša*$_2$ *al* 10 **gir**$_2$**.tab diri a.ra**$_2$ 24 ⸢**du**⸣ **ta** 2.40 **la**$_2$

10' 10 **pa** 2.28 *ša*$_2$ *al* 10 **pa diri a.ra**$_2$ 8 **du ta** 2.28 **la**$_2$

11' 10 **maš**$_2$ 2.24 *ša*$_2$ *al* 10 **maš**$_2$ **diri a.ra**$_2$ 8 **du ki** 2.24 **tab**

12' 10 **gu** 2.28 *ša*$_2$ *al* 10 **gu diri a.ra**$_2$ 24 **du ki** 2.28 **tab**

13' 10 *zib* 2.40 *ša*$_2$ *al* 10 *zib* **diri a.ra**$_2$ 40 **du ki** 2.40 **tab**

Column ib

P5' 2' *lu*-⸢*maš*⸣.[**meš**$^?$ **ta** 13 *zib*]

3' **en** 27 **absin**$_0$ ⸢**ab**$_2$⸣ [*ana* **ab**$_2$]

4' 28.7.30 *ša*$_2$ *al* 13 ⸢*zib*⸣

5' **diri a.ra**$_2$ 1.4 **du ki**

6' 13 *zib* **tab ta** 27 **absin**

7' **en** 13 *zib* 30 **tab** *ša*$_2$ *al*

8' 27 **absin**$_0$ **diri a.ra**$_2$ 56.15

9' **du ki** 27 **absin tab**

Upper edge

[1]At the command of Bēl and Bēlti[ja may it remain intact.]

Obverse column i

P1 **(Φ)**

a **(Updating Φ as a zigzag function)**

[1]Procedure for 2,13;20 (= Φ). Month by month you add and subtract 2;45,[55],⌜33⌝,20. [That which exceeds 2,17;4,48, 53,20]: [2]you deduct 2,17;4,48,53,20 from it, what re[mains you deduct from 2,17;4,48,53,20]. [3]That which is less than 1,57;47,57,46,40 [you deduct from 1,57;47,57,46,40, what remains] [4]you add with 1,57;47,57,46, 40 ⌜and⌝ put down.

b **(Updating Φ as a zigzag function, alternative formulation?)**

⌜...⌝ [...] [5](The amount) by which it ⌜exceeds⌝ 2,17;4,48,53,20 [... (The amount) by which] [6]it is less than 1,57;47,57, 46,40 [...]

P2 **(Computing Φ^{fm} from Φ^{nm}?)**

[7]Procedure for 2,13;20 of day 15 (= Φ^{fm}) [...] [8]2,17;4,48,53,20 [...] [9][you add] to 1,57;47,57,[46,40 ...] [10]is the addition? for 14 [...]

P3 **(Computations involving Φ?)**

[11]Procedure for 2,[13;20 ...]

(*about 5 lines missing*)

[1'][...] if? ... is 1.6 ... [...]

Column ia

P4' **(Computing C^{nm} from B^{nm})**

[2'][10 Ari: 3,0. (The amount) by which it] ⌜exceeds 10 Ari⌝ you multiply by 0;40, add with 3,0.
[3'][10 Tau: 3,2]0. (The amount) by which it exceeds 10 Tau you multiply by 0;24, add with 3,20.
[4'][10 Gem: 3],⌜3⌝2. (The amount) by which it exceeds 10 Gem you multiply by 0;8, add with 3,32.
[5'][10] ⌜Cnc⌝: 3,36. (The amount) by which it exceeds 10 Cnc you multiply by 0;8, subtract from 3,36.
[6']10 Leo: 3,32. (The amount) by which it exceeds 10 Leo you multiply by 0;24, subtract from 3,32.
[7']10 Vir: 3,20. (The amount) by which it exceeds 10 Vir you multiply by 0;40, subtract from 3,20.
[8']10 Lib: 3,0. (The amount) by which it exceeds 10 Lib you multiply by 0;40, subtract from 3,0.
[9']10 Sco: 2,40. (The amount) by which it exceeds 10 Sco you ⌜multiply⌝ by 0;24, subtract from 2,40.
[10']10 Sgr: 2,28. (The amount) by which it exceeds 10 Sgr you multiply by 0;8, subtract from 2,28.
[11']10 Cap: 2,24. (The amount) by which it exceeds 10 Cap you multiply by 0;8, add with 2,24.
[12']10 Aqr: 2,28. (The amount) by which it exceeds 10 Aqr you multiply by 0;24, add with 2,28.
[13']10 Psc: 2,40. (The amount) by which it exceeds 10 Psc you multiply by 0;40, add with 2,40.

Column ib

P5' **(Updating B^{fm} with the step function for σ)**

[2']Zodiacal position[s. From 13 Psc] [3']until 27 Vir ⌜month by month⌝[4']28;7,30. (The amount) by which it exceeds 13 ⌜Psc⌝ [5']you multiply by 1;4, add with [6']13 Psc. From 27 Vir [7']until 13 Psc you add 30. (The amount) by which it exceeds [8']27 Vir you multiply by 0;56,15, [9']add with 27 Vir.

P6' 10' **nim** *u* **sig** *ša*$_2$ ḪAB AN.MI
11' **a.ra**$_2$ 20 **du** *ina* **nim**
12' **ki** 17.24 **tab** *ina* **sig**
13' **zi**
Column i

P7' 14' **du**$_3$-*uš ša*$_2$ **zi** *sin* **ab**$_2$ *ana* **ab**$_2$ 42 **tab** *u* **la**$_2$ *lib*$_3$*-bu-u*$_2$ *ša*$_2$ 15.56.54.22.30 **kur**-*ad*$_2$
15' *ša*$_2$ *al* 15.56.54.22.30 **diri ta** 15.56.54.22.30 **la**$_2$ *lib*$_3$*-bu-u*$_2$ *ša*$_2$ 11.4.4.41.15
16' **kur**-⸢*ad*$_2$⸣ *ša*$_2$ *al* 11.4.4.41.15 **la**$_2$-*u*$_2$ **ki** 11.4.4.41.15 **tab** : *ana tar-ṣi* 2.17.4.48.53.20
17' 15.56.54.22.30 **zi gar**-*an ana tar-ṣi* 1.57.47.57.46.40 11.4.4.41.15 **gar**
18' *ana tar-ṣa* 2.13.20 15 **zi** *ana tar-ṣa* 1.58.31.6.40 11.15 **zi** *ša*$_2$ *al* 2.13.20
19' ⸢**gal tur**⸣ **a.ra**$_2$ 15.11.15 **du** *ana* **zi tur** *u* **gal tab** *u* **la**$_2$

P8' 20' **du**$_3$-*šu*$_2$ *ša*$_2$ **nim** *u* **sig** *ša*$_2$ *sin* **ab**$_2$ *ana* **ab**$_2$ 12 **dagal** *ma-lak* ᵈ*sin* 2.24 **murub**$_4$-*ti qaq-qar ki-ṣa-ri*
21' **ta** 27 *zib* **en** 13 **absin**$_0$ **ab**$_2$ *ana* **ab**$_2$ 1.58.45.42 **tab** *u* **la**$_2$ *lib*$_3$*-bu-u*$_2$ *ša*$_2$ **murub**$_4$-*tu*$_2$ **kur**-*ad*$_2$
22' ⸢*šum*⸣-*ma* 3.52.11.39 **nim** 1.58.45.42 *ina* **ša**$_3$ **e**$_{11}$-*ma* 1.53.25.57 **tag**$_4$
23' ⸢*mi*⸣-*nu-u*$_2$ *ki-i al-la* 2.24 **la**$_2$ 30.34.3 **la**$_2$ 30.34.3 *ina* 1.52.25.57
24' **la**$_2$-*ma* 1.22.21.54 **nim** *ar*$_2$ 3.52.11.39 **nim gar**-*an* 30.34.3 *a-na* 1.58.45.42
25' **tab**-*ma* 2.29.19.45 **sig gar**-*an lib*$_3$*-bu-u*$_2$ *ša*$_2$ 13 **absin**$_0$ **kur**-*ad*$_2$ *ana* 1 **uš ki** *ša*$_2$ *al* 13
26' ⸢**absin**$_0$ **diri**⸣ GAM 15 **du ki** 1.58.45.42 **tab** *ana* **nim** *lu ana* **sig gar** *šum-ma lu-maš* 13 **rin**$_2$
27' ⸢**ta** 13 **absin**$_0$ **en**⸣ 13 **rin**$_2$ 30 1 **danna** 30 GAM 15 **du** 7.30 7.30 **ki** {**ki**} 1.58.45.42
28' [**tab**-*m*]*a* 2.6.15.42 ⸢*lu*⸣ 7.12 **nim nim** ⸢*lu* 7.12⸣ **sig** *šap-lu ša*$_2$ *al* 7.12 **diri**
29' [*ina* 7.1]2 **e**$_{11}$-*ma* [*ana* **nim** *lu ana* **sig**] **gar ta** *zib* **en absin** *ina* **murub**$_4$ *lu-maš gab-bi a-šar*
30' [x]-⸢*ti*? 2?⸣ [xxxx]-⸢*u*$_2$?⸣ 30 ⸢20 **ta**⸣ *zib* ⸢**en**⸣ **absin**$_0$ **gar**-*an* **an.ta murub**$_4$ *u* **ki.ta**
31' [**murub**$_4$ 3.52.11.39 **nim** *u* 2.29].⸢19.45 **sig**? **ki**⸣ *a-ḫa-miš*$_2$ **tab**-*ma* 6.21.31.24 1.58.45.42
32' [**ki** 1.58.45.42 **tab**-*ma* 3].⸢57.31⸣.24 2.24 **ki**-*šu*$_2$ **tab**-*ma* 6.21.31.24 **kur**-*ad*$_2$ **ta** 13 **absin**$_0$
33' [**en** 27 *zib* xxxxxxxxxxxx] ⸢*ša*$_2$ *ki-ṣir*⸣ **murub**$_4$ **kur gin**$_7$ **igi la**$_2$
Column ii

P9' (about 19 lines missing)
1' ⸢*ina* xx 22? **en**? xxxxxx⸣ [xxxxxxxxxx]
2' *ina* **pa** 1.2.44 **zi** 8 *ša*$_2$ **tur** *nap-pal-tu*$_2$ 8 *ina* 1.2.⸢4⸣[4 xxxxxxxxxx]
3' ⸢12? x 2?⸣ ᵈ**utu** *ina* **pa nim** *u* **sig** *nap-pal-tu*$_2$ **me** *šat-tuš ša*$_2$ ⸢*kal*⸣ *šat-tu*$_2$ [xxxxx]
4' *ša*$_2$ *kal šat-tu*$_2$ **e**$_{11}$-*ma ša*$_2$ **tag**$_4$ *šu*-⸢*u*$_2$⸣ **nim** *u* **sig** *ša*$_2$ *šamaš*$_2$ *ša*$_2$ *kal šat-tu*$_2$ [xxx]

P6' (Computing Ψ' from E)

[10']The 'height and depth' of the eclipse disk [11']you multiply by 0;20. When it is high [12']you add it with 17;24; when it is low [13']you 'tear it out'.

Column i

P7' ($F^{(1)}$)

a (Updating $F^{(1)}$ as a zigzag function)

[14']Procedure for the displacement of the Moon. Month by month you add and subtract 0;42, whereby you reach 15;56,54,22,30. [15'](The amount) by which it exceeds 15;56,54,22,30 you subtract from 15;56,54,22,30, whereby you [16']reach 11;4,4,41,15. (The amount) by which it is less than 11;4,4,41,15 you add with 11;4,4,41,15.

b (Computing $F^{(1)}$ from Φ)

Opposite 2,17;4,48,53,20 [17']you put 15;56,54,22,30 as the displacement. Opposite 1,57;47,57,46,40 you put 11;4,4, 41,15. [18']Opposite 2,13;20: 15, the displacement. Opposite 1,58;31,6,40: 11;15, the displacement.

(The amount) by which it (Φ) [19']exceeds or is less than 2,13;20 you multiply by 0;15,11,15, add to and subtract (from) the largest and the smallest displacement (F).

P8' (Updating E)

a [20']Procedure for the Moon's 'height and depth', month by month. The width of the path of the Moon is 12. The central part, the region of the node, is 2,24.

b (Step 1, zone 2, including a numerical example)

[21']From 27 Psc until 13 Vir you add and subtract 1,58;45,42 month by month, whereby you reach the central part. [22']If the 'height' is 3,52;11,39: you deduct 1,58;45,42 from it, and 1,53;25,57 remains. [23']⸢Whatever⸣ it is less than 2,24: it is less by 30;34,4, you subtract 30;34,4 from 1,52;25,57 (error for 1,53;25,57) [24']and you put down 1,22;21,54 (error for 1,22;51,54) as the 'height', after 3,52;11,39 (having been) the 'height'. You add 30;34,3 to 1,58;45,42 [25']and you put down 2,29;19,45 as the 'depth', whereby you reach 13 Vir.

For (every) 1° by which the zodiacal position exceeds 13 [26']Vir you multiply by 0;15, you add it with 1,58;45,42 and you put it down as the 'height' or as the 'depth'. If the zodiacal position is 13 Lib, [27']⸢from 13 Vir until⸣ 13 Lib is 30 (° i.e.) 1[b]. You multiply 30 by 0;15, it is 7;30 ([še]). [You add] 7;30 to ⟨⟨to⟩⟩ 1,58;45;42, it is 2,6;15,42.

c (Step 3: reflection at the extrema)

[28']⸢Whether⸣ it is 7,12 'high', ascending, ⸢or⸣ 7,12 'low', descending, (the amount) by which it exceeds 7,12 [29']you deduct [from 7,1]2 and put down [as the 'height and depth'].

d (Unidentified procedure for zone 2)

From Psc until Vir in the middle of all zodiacal signs, the place of [...] [30'][...] ⸢...⸣ [...] ⸢...⸣ 30.⸢20⸣ you put down from Psc ⸢until⸣ Vir.

e (Verifying E by invoking the alternative rule for crossing the nodal region)

Above the central part and below [31'][the central part] you add together [3,52;11,39, the 'height', and 2,29];19,45, the 'depth', it is 6,21;31,24. [You add] 1,58;45;42 [32'][to 1,58;45;42, it is 3],57;31,24; you add 2,24 to it, it reaches 6,21;31,24.

f (Unidentified procedure for zone 1)

From 13 Vir [33'][until 27 Psc ...] it reaches the [...] ⸢of the node⸣, the central part; you subtract it as before.

Column ii

P9' (Computations concerning solar motion)

a (Computation involving $v_{\odot}$; unclear)

(*about 19 lines missing*)

[1']⸢... 22? until? ...⸣ [...] [2']In Sgr the displacement is 1;2,44. The *nappaltu*-coefficient for the minimum is 0;8. You [subtract?] 0;8 from 1;2,4[4, it is ...] [3']⸢...⸣ the Sun in Sgr.

b (Computation involving 'height and depth' and C; unclear)

The 'height and depth': The *nappaltu*-coefficient for daylight in its year? for the entire year [...] [4']You subtract the [...] for the entire year, what remains is the Sun's 'height and depth' for the entire year [...].

P10’	5’	**zi** ^{d}**utu** *ina lu-maš*.**meš** *ina* **maš.maš** 55.32 *ina* **pa** 1.2.44 7.12 ⸢*bi-rit* **zi**$^{?}$ xxx⸣
P11’	6’	**ki** ^{d}**utu iti** *ana* **iti maš** 27.36 **pa** 30.36 **pa** 30.36 3 ⸢xxx **iti**$^{?}$ *ana*$^{?}$ **iti** 6⸣
	7’	36 **tab** *u* **la**$_{2}$ 1-*šu* **uš zi** *šamaš*$_{2}$ **šar**$_{2}$ 40 *ša*$_{2}$ *nap-pal-tu*$_{2}$ **me** *u* **gi**$_{6}$ ⸢20 xx **nim** *u*⸣ **sig**
	8’	*ša*$_{2}$ **e**-*u*$_{2}$ *u*$_{4}$-*me* 1$^{?}$ ^{d}**utu zi** *u*$_{4}$-*me* 40 *ša*$_{2}$ ⸢*nap-pal*$^{?}$⸣-*tu*$_{2}$ **me** *u*$_{4}$-*me* ⸢xxx **ta im**.1$^{?}$⸣
	9’	*a-na* **im**.2 **nim** *lu* **ta im**.2 *a-na* **im**.1 59.8 **zi** *šamaš*$_{2}$ ⸢**murub**$_{4}$⸣ [xx]
	10’	*nap-pal-tu*$_{2}$ **me** 36 6 **nim** *u* **sig** 24 *ina* 59.8 **nim e**$_{11}$-*ma* 38.⸢8$^{?}$ **tag**$_{4}^{?}$ xx⸣
	11’	7 **šu.si** ^{d}**utu nim** *u* **sig** :: *ana tar-ṣa* 1 : **šu.si** ḪAB-*rat ḫe-pit*$_{2}$-*tu*$_{2}$ 1.⸢38⸣.2⸢4⸣
	12’	**nim** *u* **sig gar**-*an ša*$_{2}$ *al* 1 ḪAB-*rat ḫe-pit*$_{2}$-*tu*$_{2}$ **la**$_{2}$ **ta** 1 **la**$_{2}$ *ša*$_{2}$ **ta**[**g**$_{4}$ **a.ra**$_{2}$ 6 **du**]
	13’	**ki** 1.38.24 **nim** *u* **sig tab**-*ma ana* **sig nim gar**-*an ša*$_{2}$ *al* 1 ḪAB-[*rat ḫe-pit*$_{2}$-*tu*$_{2}$ **diri**]
	14’	**a.ra**$_{2}$ 6 **du ta** 1.38.24 **nim** *u* **sig la**$_{2}$-*ma ana* **sig nim ga**[**r**-*an* **en** 17.24 ḪAB-*rat*]
	15’	**gin**$_{7}$ *an-nam* **du**$_{3}$ **nim** *u* [**sig**] **ta** ḪAB-*rat ana* **e**$_{3}$-*u*$_{2}$ [xxxxx]
P12’	16’	**zi gal**-*u*$_{2}$ *ana* **igi**-*ka mim-ma ša*$_{2}$ *e ki-ṣir u mim-ma* **ki.ta** *k*[*i-ṣir ana* **ugu** *a-ḫa-miš*$_{2}$]
	17’	GAR.GAR-*ma* $\frac{1}{2}$-*šu*$_{2}$-*nu* GIŠ.A *taš-pil-tu*$_{4}$ *ana* **ugu tab zi gal** *tam-ma*[*r* **zi tur** *ana* **igi**-*ka*]
	18’	*mim-ma ša*$_{2}$ *e ki-ṣir u* **sig** *ki-ṣir ana* **ugu** *a-ḫa-miš*$_{2}$ GAR.GAR-*ma* $\frac{1}{2}$-[*šu*$_{2}$-*nu* GIŠ.A *taš-pil-tu*$_{4}$]
	19’	**ta** *lib*$_{3}$-*bi* **e**$_{11}$-*ma* **zi tur**
P13’	20’	**gub.meš** *u* **lal**$_{2}$.**meš** *ša*$_{2}$ **mu** *ana* **mu** *ša*$_{2}$ ^{d}**utu** *ana* **du**$_{3}$-*ka* **mu** *ana* **mu** 1.36 [xxxxxx]
	21’	**me nim** *u*$_{3}$ **gi**$_{6}$-*ka* **tab** *ana* **šu**$_{2}$ *šamaš*$_{2}$ *ana* **kur** *šamaš*$_{2}$ **zi ta gub** *ana* **lal**$_{2}$ ⸢x⸣[xxxxxx]
	22’	*ana* **gaba.ri mu** 11 *ana* **me.meš** 3.36 **alla.meš** 2.24 **gi**$_{6}$ 3 ⸢**rin**$_{2}$ x⸣ [xxxxxx]
P14’	23’	**ta** 27 *zib* **en** 25.7.30 **ḫun** *a-na* 1 **uš** 2.1.⸢4⸣[4 **ta** *si-man* **la**$_{2}$ **ta** 25.7.30 **ḫun en** 13 **absin**$_{0}$]
	24’	57.3.45 **ta** *si-man* **la**$_{2}$ 2 **ta** 25.7.30 **ḫun en** 1[3 **absin** xxxxxx]
	25’	**ta** *si-man*.**meš la**$_{2}$-*is* **ta** 13 **absin**$_{0}$ **en** 27 *zib* **nu tuk ta** 27 *z*[*ib* **en** xxxxxx]
	26’	2.1.44 **la**$_{2}$ **a.ra**$_{2}^{!}$ 28.7.30 **ki** *ša*$_{2}$ *šamaš*$_{2}$ **du** 57.3.45 **la**$_{2}$ **en** 2[7$^{?}$ *zib*$^{?}$ x]
P15’	27’	*si-man qa-tu-u*$_{2}$ *ana* **du**$_{3}$-*ka a-šar* **tab ki** *si-man-ni-ka* **tab** *a-šar* **la**$_{2}$-*u*$_{2}$ **ta** *si-man*-[*ni-ka*]
	28’	**la**$_{2}$ *ana si-man qa-tu-u*$_{2}$ **gar** : *si-man ana* **šu**$_{2}$ *šamaš*$_{2}$ *ana* **du**$_{3}$-*ka si-man qa-tu-u*$_{2}$ **ta** *s*[*i-man*]
		Reverse
		Column i
	1	[*ana* **šu**$_{2}$ *šamaš*$_{2}$ **la**$_{2}$ *ki si*]-*man ana* **šu**$_{2}$ *šamaš*$_{2}$ *al-la* [*si-man qa-tu-u*$_{2}$] *i-ṣa* 6 6 **k**[**i** *si-man*]
	2	⸢*ana* **šu**$_{2}$ *šamaš*$_{2}$ **tab**⸣ *u si-man qa-tu-u*$_{2}$ **ta** ⸢**ša**$_{3}$⸣ (blank) [**la**$_{2}$]
P16’	3	*ana tar-ṣi* 2.13.20 **la**$_{2}$-*u*$_{2}$ 2.40 **gar**-*an mim-ma ša*$_{2}$ *al-la* 2.13.20 **la**$_{2}$-[*u*$_{2}$ **la**$_{2}$-*u*$_{2}$ **en** 2.10.40]
	4	**la**$_{2}$-*u*$_{2}$ **a.ra**$_{2}$ 3.22.30 **du** *ma-lu-uš-šu*$_{2}$ 17.46.40 **gar**-*an ana muḫ-ḫi a*-[*ḫa-miš*$_{2}$ GAR.GAR-*ma*]
	5	**ki** ⸢2.40 **tab**-*ma* **gar**⸣-*an ana tar-ṣi* 2.10.40 **la**$_{2}$-*u*$_{2}$ 2.⸢5⸣3.20 **gar**-*an ša*$_{2}$ *al* 2.1[0.40 **la**$_{2}$-*u*$_{2}$ **la**$_{2}$-*u*$_{2}$]

P10' (Parameters of the zigzag function for $v_{\odot}$)

[5']The displacement of the Sun in the zodiacal signs: in Gem 0;55,32, in Sgr 1;2,44; 0;7,12 is the ⸢distance between the displacements? ...⸣

P11' (Various procedures involving solar motion and E)

a **(Parameters of the zigzag function for the Sun's monthly displacement)**

[6']The position of the Sun, month by month: Gem 27;36, Sgr 30;36, Sgr 30;36. ⸢... month by month 6⸣.

b **(Computations involving 'height and depth' and a coefficient; unclear)**

[7']You add and subtract 36. 60 (^{nin}, i.e.) 1°, you compute as the displacement of the Sun. 0;40 is for the *nappaltu*-coefficient of day and night ⸢20 ... the 'height and⸣ depth' [8']which you have predicted. (Per) day the Sun moves by 1?. (Per) day 0;40, the *nappaltu*-coefficient of day [and night.] Whether it ascends ⸢from the south [9']to the north or ⟨it descends⟩ from the north to the south, the middle displacement of the Sun is 0;59,8 [...] [10']The *nappaltu*-coefficient of daylight: 36, it ascends and descends by 6. 0;24 you subtract from 0;59,8, above, it is 0;38,8. What ⸢comes out ...⸣ [11']The Sun ascends and descends by 7^{f}.

c **(Computing E from Ψ)**

Opposite 1^{f}, the 'broken disk', you put 1,3[8];24 [12']as the 'height and depth'. That which is less than 1, the 'broken disk', you deduct from 1. What rem[ains you multiply by 6] [13'](and) add to 1,38;24, the 'height and depth', and you put it down as the 'depth', ascending. (The amount) by which it [exceeds] 1, the ['broken disk'], [14']you multiply by 6 (and) deduct from 1,38;24, the 'height and depth', and you pu[t it down] as the 'depth', ascending. [Until the eclipse magnitude is 17;24] [15']you do like this. In order to produce the 'height and [depth'] from the 'disk' [...]

P12' (Computing the extrema of the zigzag function for F)

[16']In order for you to see the largest displacement: whatever is above the 'knot' and whatever is below the 'kn[ot'] you accumulate [together] [17']and you compute half of it, you add the difference to (it), you see the largest displacement.

[In order for you to see the smallest displacement:] [18']whatever is above the 'knot' and below the 'knot' you accumulate together and [you compute] half [of it], you subtract [the difference] [19']from it, it is the smallest displacement.

P13' (Computing solstices and equinoxes)

[20']In order for you to 'construct' the solstices and equinoxes of the Sun. Year by year 1,36 [...] [21']the time since sunrise and your night: you add; until sunset, until sunrise: you subtract ('tear out'). From solstice to equinox [...] [22']For the value of the year 11. For the daylight: 3,36 Cnc; 2,24 the night; 3,0 ⸢Lib ...⸣ [...]

P14' (Computing $G+J$)

[23']From 27 Psc until 25;7,30 Ari [you subtract] 2;1,44 for 1° [from the duration (G). From 25;7,30 Ari until 13 Vir] [24']you subtract 57;3,45 from the duration. 2: From 25;7,30 Ari until 1[3 Vir] [25']you subtract [...] from the durations. From 13 Vir until 27 Psc nothing. From 2[7 Psc until ...] [26']You multiply 2;1,44, subtractive, by 28;7,30, the position of the Sun, (it is) 57;3,45, subtractive. Until 2[7 Psc ...].

P15' (K, M)

a **(Computing K)**

[27']In order for you to 'construct' the final duration (K): where it is additive you add it (C') to your duration ($G+J$), where it is subtractive you [28']deduct it from [your] duration, you put it down as the final duration.

Reverse column i

b **(Updating M^{nm} with K^{nm})**

In order for you to 'construct' the time until sunset (M^{nm}): [1][you deduct] the final duration (K^{nm}) from [the time until sunset. If the ti]me until sunset is less than [the final duration]: 6,0; [2]⸢you add⸣ 6,0 t[o the time until sun]set and [you deduct] the final duration from ⸢it⸣.

P16' (Computing G from Φ)

1–9 [3]Opposite 2,13;20, decreasing, you put 2,40. Whatever (the amount) by which it is less than 2,13;20, decreas[ing, until 2,10;40], [4]decreasing, you multiply by 3;22,30; while it fills up you put down 0;17,46,40, you accumulate it together, and [5]you add it with 2,40, and you put it down.

6 **en** 1.⸢58⸣.31.6.40 **la**$_2$-*u*$_2$ **a.ra**$_2$ 9.20 **du**-*ma* **ki** 2.53.20 **tab**-*ma* ⸢**gar**-*an*⸣ [*ana tar-ṣi*]
7 1.58.31.6.40 **la**$_2$-*u*$_2$ 4.46.42.⸢57⸣.46.40 **gar**-*an ša*$_2$ *al-la* 1.58.31.⸢6⸣.[40]
8 **la**$_2$-*u*$_2$ **la**$_2$-*u*$_2$ **en** 1.58.13.20 *ma-ṭu-u*$_2$ **a.ra**$_2$ 8.20 **du ki** 4.46.42.57.4⸢6.40⸣
9 **tab**-*ma* **gar**-*an ana tar-ṣi* 1.58.13.20 **la**$_2$-*u*$_2$ 4.49.11.6.40 **gar**-*an mim-ma ša*$_2$ ⸢*al-la*⸣
10 1.58.13.20 **la**$_2$-*u*$_2$ *ma-ṭu-u*$_2$ **en** 1.57.55.33.20 **la**$_2$-*u*$_2$ **a.ra**$_2$ 7.⸢20 **du**⸣
11 **ki** 4.49.11.6.40 **tab**-*ma* **gar**-*an ana tar-ṣi* 1.⸢5⸣7.55.33.20 **la**$_2$-*u*$_2$
12 4.51.21.28.53.20 **gar**-*an ša*$_2$ *al-la* 1.57.55.33.10 **la**$_2$-*u*$_2$ **la**$_2$-*u*$_2$
13 **en** 1.57.58.8.53.20 **tab**-*u*$_2$ **a.ra**$_2$ 6.20 **du** *it-ti* 4.51.21.28.53.20
14 **tab**-*ma* **gar**-*an ana tar-ṣi* 1.57.58.8.53.20 **tab**-*u*$_2$ 4.53.14.4.26.40 **gar**-*an*
15 *ša*$_2$ *al-la* 1.57.58.8.53.20 **tab**-*u*$_2$ **diri en** 1.58.15.5⸢5⸣.33.20 **tab**-*u*$_2$
16 **a.ra**$_2$ 5.20 **du ki** 4.53.14.4.26.40 **tab**-*ma* **gar**-*an ana tar-*⸢*ṣi*⸣ 1.58.15.55.33.20
17 **tab**-*u*$_2$ 4.54.48.53.20 **gar**-*an ša*$_2$ *al-la* 1.58.15.⸢5⸣5.⸢33.20 **tab**-*u*$_2$ **diri**⸣
18 **en** 1.58.33.42.13.20 **tab**-*u*$_2$ **a.ra**$_2$ 4 **du ki** 4.⸢54.48.53⸣.[20 **tab**-*ma* **gar**-*an*]
19 *ana tar-ṣi* 1.58.33.42.13.20 4.56 **gar en** 1.58.37.2.13.20 ⸢**tab**-*u*$_2^?$⸣
20 *gab-bi* 4.56 **gar** *mim-ma ša*$_2$ *al-la* 1.58.37.2.13.20 **tab**-*u*$_2$ **diri en**
21 1.58.54.48.53.20 **tab**-*u*$_2$ **a.ra**$_2$ 2 **du ki** 4.56 **tab**-*ma* **gar** *ana tar-*⸢*ṣi*⸣
22 1.58.54.48.53.20 **tab**-*u*$_2$ 4.56.35.33.20 **gar**-*an* **en** 1.59.12.⸢35⸣.33.20
23 **tab**-*u*$_2$ *gab-bi* 4.56.35.33.20 **gar**-*an mim-ma ša*$_2$ *al-la* 1.59.12.35.33.20
24 **tab**-*u*$_2$ **diri en** 1.59.30.22.13.20 **tab**-*u*$_2$ **a.ra**$_2$ 2 **du ta** 4.56.35.33.⸢20⸣
25 **e**$_{11}$ **gar** *ša*$_2$ *al-la* 1.59.30.22.13.20 **tab**-*u*$_2$ **diri en** 1.59.⸢4⸣8.8.53.20
26 **tab**-*u*$_2$ **a.ra**$_2$ 4 **du ta** 4.56 **e**$_{11}$-*ma* **gar**-*an ana tar-ṣi* 1.59.48.8.53.⸢20⸣
27 **tab**-*u*$_2$ 4.54.48.53.20 ⸢**gar**⸣-*an ša*$_2$ ⸢*al*⸣-*la* 1.59.48.8.53.20 **tab**-⸢*u*$_2^?$⸣ [**diri**]
28 **en** 2.0.5.55.33.20 **tab**-*u*$_2$ **a.ra**$_2$ 5.20 **du ta** 4.54.⸢48.5⸣[3.20]
29 **e**$_{11}$-*ma* **gar**-*an ana tar-ṣi* 2.0.5.⸢55⸣.33.20 **tab**-*u*$_2$ ⸢4⸣.[53.14.4.26.40 **gar**-*an*]
30 ⸢*ša*$_2$ *al-la*⸣ 2.0.5.55.33.20 **tab**-*u*$_2$ **diri en** ⸢2⸣.[0.23.42.13.20 **tab**-*u*$_2$]
31 [**a.ra**$_2$ 6.20] ⸢**du**⸣ **ta** 4.53.14.⸢4⸣.[26.40 **e**$_{11}$-*ma* **gar**-*an ana tar-ṣi*]
32 [2.0.23.42.13.20 **tab**-*u*$_2$ 4.51.21.28.53.20 **gar**-*an ša*$_2$ *al-la* 2.0.23.42.13.20]
33 [**tab**]-⸢*u*$_2$ **diri**⸣ **en** 2.0.41.⸢2⸣[8.53.20 **tab**-*u*$_2$ **a.ra**$_2$ 7.20 **du ta**]
34 4.⸢5⸣1.⸢2⸣1.28.53.20 ⸢**e**$_{11}$⸣-[*ma* **gar**-*an ana tar-ṣi* 2.0.41.28.53.20 **tab**-*u*$_2$]
35 ⸢4.4⸣9.11.6.40 **gar**-*an* ⸢*ša*$_2^?$⸣ [*al-la* 2.0.41.28.53.20 **tab**-*u*$_2$ **diri en**]
36 2.⸢0⸣.59.15.3⸢3⸣.[20 **tab**-*u*$_2$ **a.ra**$_2$ 8.20 **du ta** 4.49.11.6.40 **e**$_{11}$-*ma* **gar**-*an*]
37 **ta** ⸢2⸣.0.⸢59⸣.[15.33.20 xxxxxxxxxxxxxxx]
38 **ta** [xxxxxxxxxxxxxxxxxxxxxxxxx]
39 [xxxxxxxxxxxxxxxxxxxxxxxxxx]
40 [xxxxxxxxxxxxxxxxxxxxxxxxxx]

Column ii

P17’ 1 [xxxxxxxxxxxxxxxxxxxxxxxxxx]
2 [xxxxxxxxxxxxxxxxxxxxxxxxxx]
3 [xxxxx] ⸢x⸣ **u**$_4$ ⸢x⸣ [xxxx] ⸢BE⸣ **ki** *ša*$_2$ ^{d}⸢*sin*⸣ [xxxxxxx]
4 [xxxxxx]⸢x⸣[xxx]⸢x⸣ **gar**-*an* BE **a.ra**$_2$ ⸢x⸣ [xxxxxxx]⸢x⸣
5 [xxxx t]**ab ki** ⸢x-*u*$_2^?$ x⸣ [xxxxx] *mim-ma ša*$_2$ ⸢*al*⸣-[*la*] ⸢10 **ḫun**⸣ **diri a.ra**$_2$ **bar**
6 [**nun**$^?$ **du** x *ki*] **nim** ⸢**ki** BE **tab** *ki*⸣ **sig ta** BE **zi**-*aḫ ar*$_2$ *lu* ⸢x⸣ **u**$_4$.28.**kam**
7 [xxx] *ša*$_2$ ^{d}*sin u* ^{d}**utu il**$_2$.**a** *mim-ma ša*$_2$ *al-la* 10 ⸢**ḫun**⸣ **diri** *ša*$_2$ ^{d}*sin* **a.ra**$_2$

10 Opposite 2,10;40, decreasing, you put 2,53;20. Whatever (the amount) by which it [is less than] 2,1[0;40, decreasing,] [6]until 1,58;31,6,40, decreasing, you multiply by 9;20, and add with 2,53;20, and put down.

11 [Opposite] [7]1,58;31,6,40, decreasing, you put 4,46;42,57,46,40. Whatever (the amount) by which it is less than 1,58;31,6, [40], [8]decreasing, until 1,58;13,20, decreasing, you multiply by 8;20, add with 4,46;42,57,46,40, and [9]put down.

12 Opposite 1,58;13,20, decreasing, you put 4,49;11,6,40. Whatever (the amount) by which it is less than [10]1,58;13,20, decreasing, until 1,57;55,33,20, decreasing, you multiply by 7;20, [11]add with 4,49;11,6,40, and put down.

13 Opposite 1,57;55,33,20, decreasing, [12]you put 4,51;21,28,53,20. Whatever (the amount) by which it is less than 1,57;55,33,20, decreasing, [13]until 1,57;58,8,53,20, increasing, you multiply by 6;20, add with 4,51;21,28,53,20, and [14]put down.

14 Opposite 1,57;58,8,53,20, increasing, you put 4,53;14,4,26,40. [15]Whatever (the amount) by which it exceeds 1,57;58,8,53,20, increasing, until 1,58;15,55,33,20, increasing, [16]you multiply by 5;20, add with 4,53;14,4,26,40, and put down.

15 Opposite 1,58;15,55,33,20, [17]increasing, you put 4,54;48,53,20. Whatever (the amount) by which it exceeds 1,58;15,55,33,20, increasing, [18]until 1,58;33,42,13,20, increasing, you multiply by 4, add with 4,54;48,53,20, and put down.

16–17 [19]Opposite 1,58;33,42,13,20, increasing, you put 4,56. Everything until 1,58;37,2,13,20, increasing: [20]you put down 4,56. Whatever (the amount) by which it exceeds 1,58;37,2,13,20, increasing, until [21]1,58;54,48,53,20, increasing, you multiply by 2, add with 4,56, and put down.

18–20 Opposite [22]1,58;54,48,53,20, increasing, you put 4,56;35,33,20. Everything until 1,59;12,35,33,20, [23]increasing: you put down 4,56;35,33,20. Whatever (the amount) by which it exceeds 1,59;12,35,33,20, [24]increasing, until 1,59;30,22,13,20, increasing, you multiply by 2, deduct from 4,56;35,33,20, [25]put down. Whatever (the amount) by which it exceeds 1,59;30,22,13,20, increasing, until 1,59;48,8,53,20, [26]increasing, you multiply by 4, deduct from 4,56, and put down.

21 Opposite 1,59;48,8,53,20, [27]increasing, you put 4,54;48,53,20. Whatever (the amount) by which it [exceeds] 1,59;48,8,53,20, increasing, [28]until 2,0;5,55,33,20, increasing, you multiply by 5;20, deduct from 4,54;48,5[3,20], [29]and put down.

22 Opposite 2,0;5,⌜55⌝,33,20, increasing, [you put] ⌜4⌝,[53;14,4,26,40]. [30]⌜Whatever (the amount) by which⌝ it exceeds 2,0;5,55,33,20, increasing, until ⌜2⌝,[0;23,42,13,20, increasing,] [31]⌜you multiply⌝ [by 6;20, deduct] from 4,53;14,⌜4⌝,[26, 40 and put down.]

23 [Opposite] [32][2,0;23,42,13,20, increasing, you put 4,51;21,28,53,20. Whatever (the amount) by which] ⌜it exceeds⌝ [2,0;23,42,13,20,] [33][increas]⌜ing⌝, until 2,0;41,⌜2⌝[8,53,20, increasing, you multiply by 7;20,] [34]⌜deduct⌝ [from] 4,51;21,28,53,20 [and put down.]

24 [Opposite 2,0;41,28,53,20, increasing,] [35][you put] 4,49;11,6,40. ⌜Whatever⌝ [(the amount) by which it exceeds 2,0;41,28,53,20, increasing, until] [36]2,0;59,15,33,[20, increasing, you multiply by 8;20, deduct from 4,49;11,6,40 and put down.]

25 [37]From 2,0;59,[15,33,20 ...] [38]from [...]

? [39][...] [40][...]

Column ii

P17' (Computing NA$_1$)

1–3?	[...] [2][...] [3][...] ... [...]	
4?	If? [...] the position of the ⌜Moon⌝ [and the Sun ...] [4][...] you put down.	$B'_☽, B'_⊙$?
9?	You [multiply] the elongation by [the *ṣiliptu*-coefficient ...]	$\eta \cdot q'$ [$\to Q$]?
?	[5][...] you add. The position [...].	
7?	(The amount) by which it exceeds ⌜10 Ari⌝ you [multiply] by the *ṣi*[*liptu*-coefficient?]. [6][...	$r = r_{10\text{Ari}} \pm (B' - 10\text{ Ari}) \cdot 0;0,20$?
11	... If] (the Moon) is high ⌜you add it with the elongation, if⌝ it is low you 'tear it out' from the elongation.	high $\Rightarrow Q + R_\uparrow$ [$\to$ NA$_1$] low $\Rightarrow Q - R_\downarrow$ [$\to$ NA$_1$]
?	Next ... [... of] day 28.	
4	[7]You compute [the position] of the Moon and the Sun.	$B'_☽, B'_⊙$

8 [16 du] **ta** 1.24 **bar.nun** *ša*$_2$ **ḫun e**$_{11}$ *mim-ma ša*$_2$ *al-la* [10 ḫu]**n diri** *ša*$_2$ ^{d}**utu a.ra**$_2$
9 [16 d]u$^?$ **ta** 1.24 **bar.nun** *ša*$_2$ **ḫun** *ša*$_2$ ^{d}**utu e**$_{11}$ **ta ḫun** [en] **absin**$_0$ **ta bar.nun la**$_2$
10 ⸢**ta**⸣ **rin**$_2$ **en** *zib* **ki bar.nun tab** *ana* **ugu** *a-ḫa-miš*$_2$ *ša*$_2$ *sin u šam*[*aš*$_2$ GAR].GAR-*ma* $\frac{1}{2}$-*šu*$_2$ GIŠ-*ma* **a.ra**$_2$
11 BE **du** *ša*$_2$ **tag**$_4$ *a-na* **u**$_4$.1.**kam** *ša*$_2$ ^{d}*sin* **gar**-*an ki-i sin* [*ma*]-*la a-ma-ri la ma-ṣu-u*$_2$
12 **zi** ^{d}*sin ša*$_2$ 1-*en u*$_4$-*mu* **ki**-*šu*$_2$ **tab**-*ma ana kun-nu* **e**-*bi* [**bar**].**nun** 1.24 **ḫun** 1.16 **mul**$_2$
13 1.8 **maš.maš** 1 **alla** 52 **a** 44 **absin**$_0$ 36 **rin**$_2$ 44 **gir**$_2$.**tab** 52 [**pa**] 1 **maš**$_2$ 1.8 **gu** 1.16 *zib*
14 **ḫun nu tuk mul**$_2$.**mul**$_2$ 10 20 **maš.maš** 30 **alla** 40 **a** 50 **absin**$_0$ 1 **rin**$_2$ [50] **gir**$_2$.**tab** 40 **pa** 30 **maš**$_2$ 20 **gu**
15 10 *zib* 2 *ša*$_2$ ḪAB-*rat* **ta na u**$_4$.1.**kam zi** *kun-nu u* [*nu*]-*bal-kut* **me** *ana* **me** *kun-nu*
16 *la nu-bal-kut* 1-*en* **me la**$_2$ *tur-ru la nu-bal-kut* **me** *ana* **me** *tur-ru u nu-bal-kut* 1-*en* **me tab**

P18’ 17 ⸢**u**$_4$.**na**$_2$⸣.**am**$_3$ *ana* **du**$_3$-*ka si-man ša*$_2$ **gub** *u* **gur il**$_2$.**a** *ina* **šu**.2 **dib** BE **iti.še u**$_4$.28.**kam**
18 1 **danna me** *ana* **šu**$_2$ *šamaš*$_2$ BE **il**$_2$ 6 **danna la**$_2$ *ša*$_2$ **me** 3 .$^?$ 6 **danna** 30 1 **danna**
19 *ina* **ša**$_3$ 6 **danna e**$_{11}$ 2.30 **tag**$_4$ 5 **danna me nim.a iti.še u**$_4$.2[8].**kam** 5 **danna**
20 **me nim.a** *sin u šamaš*$_2$ *ina* **ḫun lal**$_2$.**meš zi.meš** *ša*$_2$ *sin u* ^{d}**utu** *ina* ⸢**šu**.2 **dib**⸣ BE 28 12
21 ⸢**zi**⸣ *ša*$_2$ *sin* 57.56 **zi** *ša*$_2$ ^{d}**utu** *ina* **ḫun** 12 **zi** *ša*$_2$ *sin* **a.ra**$_2$ 10 **du**-*ma* 2
22 **a.ra**$_2$ 2.30 5 **danna** *u*$_4$-*me ša*$_2$ **nim.a du**-*ma* 5 *ina* 5 **danna** *si-man* 5 **uš**
23 ^{d}*sin ana* ^{d}**utu.šu**$_2$.**a** *tu-šaḫ-ḫa-az* 11.42 **zi** *ša*$_2$ ^{d}*sin ša*$_2$ **u**$_4$.27.**kam**
24 **ki** 5 **uš zi** *ša*$_2$ ^{d}*sin ša*$_2$ **u**$_4$.28.**kam tab**-*ma* 15.42 ^{d}*sin ana* ^{d}**utu.šu**$_2$.**a** *la*-**igi**
25 ^{d}**utu** BE ^{d}**utu** *ana* **im.mar.tu** *tu-šaḫ-ḫa-az* 57.56 **zi** ^{d}**utu** *ša*$_2$ **iti.še**
26 **u**$_4$.28.**kam** ⸢*ina* **ḫun**⸣ [**a.ra**$_2$ 10 **du**-*ma* 9.3]9.20 **a.ra**$_2$ 2.30 5 **danna du**-*ma* 24.8 **tur** *ina* 5 **danna**
27 ⸢*si*⸣-[*man* 24.8 **uš** ^{d}**utu** *ana* **im.mar.tu** *tu-šaḫ-ḫa-az*] 57.56 **zi** ^{d}**utu** *ša*$_2$ **u**$_4$.27.**kam**
28 [**ki** 24.8 **zi** *ša*$_2$ ^{d}**utu** *ša*$_2$ **u**$_4$.27.**kam tab**-*ma* 1.22.4 *ina* 17$^?$] **danna** *si-man ina* 16.42 **zi**
29 [xxxxxxxxxxxx]⸢x⸣-*tu ina* 10 **mul**$_2$ **bar.nun** GAM **nim** *u* **sig**
30 [xxxxxxxxxxxx] ⸢x⸣ ^{d}**utu** ⸢**gin**$_7$⸣ *ša*$_2$ ⸢**u**$_4$.1.**kam**$^?$ **du**$_3^?$ xx⸣
31 [xxxxxxxxxxxx] ⸢x⸣ **z**[**i** xxxxxxxx]
32 [xxxxxxxxxxxxxxxxxxxxxxxxxxx]
33 [xxxxxxxxxx **a.ra**$_2$] ⸢**me nim**$^?$⸣ ^{d}**utu** ⸢**du**⸣ **bar.nun** ⸢*ša*$_2$ *lu-maš*$^?$⸣
34 [xxxxxxxxxxxx] **ki** *šamaš*$_2$ *u* ^{d}*sin* **gin**$_7$ *ša*$_2$ **u**$_4$.1.**kam**

5	(The amount) by which it exceeds 10 ⌜Ari⌝ for the Moon you [multiply] by [8][0;0,16], deduct from 1;24, the *ṣiliptu*-coefficient for Ari. (The amount) by which it exceeds [10 Ar]i for the Sun you ⌜multiply⌝ by [9][0;0,16], deduct from 1;24, the *ṣiliptu*-coefficient for Ari for the Sun. From Ari [until] Vir you subtract it from the *ṣiliptu*-coefficient; [10]⌜from⌝ Lib until Psc you add it with the *ṣiliptu*-coefficient. You accumulate (them) together for the Moon and the Sun and you compute half of it,	$q_{☽} = 1;24 - (B'_{☽} - 10\ \text{Ari}) \cdot 0;0,16$ $q_{\odot} = 1;24 - (B'_{\odot} - 10\ \text{Ari}) \cdot 0;0,16$ $(q_{☽} + q_{\odot}) \cdot \frac{1}{2} [\to q']$
9	[11]and you multiply it by the elongation;	$q' \cdot \eta\ [\to Q]$
11	what comes out you put down as day 1 of the Moon.	$\to \text{NA}_1$ (!)
13	If the Moon is insufficient(ly bright) for being visible [12]you add with it the Moon's displacement for 1 day and you predict a full month.	$B'_{☽} + v_{☽} \to B'_{☽}$
5	The *ṣiliptu*-coefficient: 1;24 Ari, 1;16 Tau, [13]1;8 Gem, 1 Cnc, 0;52 Leo, 0;44 Vir, 0;36 Lib, 0;44 Sco, 0;52 [Sgr], 1 Cap, 1;8 Aqr, 1;16 Psc;	q_{set}
7	[14]Ari nothing, Tau 0;10, 0;20 Gem, 0;30 Cnc, 0;40 Leo, 0;50 Vir, 1 Lib, [0;50] Sco, 0;40 Sgr, 0;30 Cap, 0;20 Aqr, [15]0;10 Psc.	r_{set}
10	You subtract the 2 for the disk from NA_1.	$\text{NA}_1 - 2\ [\to \text{NA}_1]$
a	(The month) is full and (the day) is transgressed: the day (corresponds) to the (same) day. (The month) is full [16](and the day) is not transgressed: you subtract 1 day. (The month) is hollow and (the day) is not transgressed: the day (corresponds) to the (same) day. (The month) is hollow and (the day) is transgressed: you add 1 day.	date of lunation

P18' (Computing KUR)

	[17]In order for you to 'construct' the day of disappearance.	
1	You hold the computed duration (G) of the full month and the hollow month in (your) hands. If on the 28th of month XII [18]the time (from New Moon) to sunset is 1^b you compute the elongation. The length of daylight is 6^b. 6^b is 3,0 (°); you deduct 30 (°, i.e.) 1^b [19]from 6^b, 2,30 (°) remains, (i.e.) 5^b, the time from sunrise (to New Moon); on the 2[8]th of month XII [20]the time from sunrise (to New Moon) is 5^b, the Sun and the Moon being in conjunction in Ari.	$M = 1^b = 30°$ $N = C - M = 6^b - 1^b = 5^b = 2{,}30°$
2	You hold the displacements of the Moon and the Sun in (your) ⌜hands⌝; if on the 28th [21]the displacement of the Moon is 12 and the displacement of the Sun is 0;57,56 in Ari: you multiply 12, the displacement of the Moon, by 0;0,10, it is 2. [22]You multiply it by 2,30 (°, i.e.) 5^b, the time from sunrise (to New Moon), it is 5 (°); in 5^b time you let [23]the Moon proceed westwards by 5°. You add 11;42, the displacement of the Moon for the 27th, [24]to 5°, the displacement of the Moon for the 28th: the Moon is in front of the Sun by 15;42 (error for 16;42). [25]If you let the Sun proceed westwards you multiply 0;57,56, the displacement of the Sun for the 28th of month XII [26]i[n Ar]i [by 0;0,10: it is 0;0,9,3]9,20; you multiply it by 2,30 (°, i.e.) 5^b it is 0;24,8 (and) a little; in 5^b of [27]⌜time⌝ [you let the Sun proceed westwards by 0;24,8°. You add] 0;57,56, the displacement of the Sun on the 27th, [28][to 0;24,8, the displacement of the Sun on the 28th: it is 1;22,4].	$\Delta B_{☽}^{(28)} = v_{☽}^{(28)} \cdot N = 12 \cdot 0;0,10 \cdot 2,30 = 5°$ $\Delta B_{☽}^{(27)} = \Delta B_{☽}^{(28)} + v_{☽}^{(27)} = 5 + 11;42 = 16;42°$ $\Delta B_{\odot}^{(28)} = v_{\odot}^{(28)} \cdot N = 0;57,56 \cdot 0;0,10 \cdot 2,30 \approx 0;24,8°$ $\Delta B_{\odot}^{(27)} = \Delta B_{\odot}^{(28)} + v_{\odot}^{(27)} = 0;24,8 + 0;57,56 = 1;22,4°$
3	You 'tear out' [1;22,4 ...]^b of time, from 16;42, [29][it is 15;20 ...]	$\eta = 16;42 - 1;22,4 \approx 15;20°$
7?	[...] ⌜...⌝... in 10° of Tau.	r?
8	[You multiply] the *ṣiliptu*-coefficient by the 'height and depth'.	$r' \cdot \beta'\ [\to R]$
?	[30][...] the Sun ⌜you do as⌝ for ⌜day 1 ...⌝. [31][...] ... the displace[ment? ...] [32][...] [33][...] you multiply [by] ⌜...⌝ of the Sun.	
5?	The *ṣiliptu*-coefficient ⌜for the zodiac?⌝ [34][...]	q?
4	[35][You compute] [34]the position of the Sun and the Moon as for day 1.	$B'_{☽}, B'_{\odot}$

35 [**du**$_3$ xxxxxx]⌜x⌝.**a** [xxx]-⌜*ma*$^?$⌝ **u**$_4$.⌜28⌝.**kam** 9 **danna me** *ana* **šu**$_2$ *šamaš*$_2$ {10} 4.30
36 9 ⌜**danna** 6⌝ **danna la**$_2$ *ša*$_2$ **me** ⌜*ša*$_2$⌝ **iti.še u**$_4$.28.**kam** 3 6 **danna** *ina* 4.30 9 **danna**
37 **e**$_{11}$ 1.30 *ana* **kur** *šamaš*$_2$ *re-ḫi* 1.30 *ina* 6 12 **danna e**$_{11}$-*ma* 4.30 **ib**$_2$.**tag**$_4$
38 **u**$_4$.28.**kam** 4.30 ⌜**me**⌝ **nim.a** *ana* **kur e.a gar**-*an* ⌜**nim**⌝ 10 *lu-maš gab-bi* **a.ra**$_2$
39 *mi-ḫi-il-tu*$_2$ *ša*$_2$ ⌜*lu*⌝-*maš* **du nim** *u* **sig zi du**$_3$.**meš** *an-nu-tu*$_2$ *ša*$_2$ *kun-nu*
40 *u tur-ru u* **u**$_4$.**na**$_2$.**am**$_3$ (blank)

Co 41 **im.dub** *ša*$_2$ *ar*$_2$-*šu*$_2$ **u**$_4$ ⌜*ša*$_2$⌝ *šamaš*$_2$ *ina* 28.20 **alla gub**-*za* ⌜**nim**$^?$ *u*$^?$⌝ [x] ⌜x **igi**⌝ [xxxxxxxxxxxxxxxxxxxxxxxxxxxxx]
42 **ta** *muḫ-ḫi* ^{giš}**d**[a$^?$ sa]**r**$^?$-*ma* **igi.tab i**[**m.dub** xxxxxxxxxxxxxxxxxxxxxxxxxxxxxxxxxxx]
43 **a** *ša*$_2$ ⌜$^{md?}$⌝**pa**-*na-din*-**šeš a** ^{lu2}**sanga**.[xxxxxxxxxxxxxxxxxxxxxxxxxxxxxxxxxx]

Critical and philological notes

Obverse column i

1 *epūšu*(**du**$_3$-*šu*$_2$) = ‘procedure’.
2,13;20: name of Φ.
2.45.⌜5⌝[5...]: replaces *ša*$_2$ **u**$_4$.**2**.[**kam**...] (*ACT* p. 211).

4 5: or *ša*$_2$. This is followed by a winkelhaken and traces, perhaps belonging to *niš* (*ša*$_2$-*niš*, ‘secondly’) or **ab**$_2$? The gap at the end may have contained 2.45.55.33.20 **ab**$_2$ *ana* **ab**$_2$ **tab** *u* **la**$_2$, ‘Month by month you add and subtract 2;45,55,33,20’.

5 Perhaps the gap contained **ta** 2.17.4.48.53.20 **e**$_{11}$ *ša*$_2$ *al*, ‘you subtract from 2,17;4,48,53,20. (The amount) by which’

6 ⌜*u*$_2$⌝ [...]: replaces **di**[**ri**...] (*ACT*) The gap may have contained **ki** 1.57.47.57.46.40 **tab**-*ma* **gar**, ‘you add to 1,57;47,57,46,40 and put down’.

7 *ša*$_2$ 2.13.20 = ‘for Φ’: omitted in *ACT*.

1’ Line numbers of *ACT* are maintained, but primes have been added. Oi1’ is probably Oi17.
nim: or **ab2**...?

Column ia

2’–5’ Some of the signs described by Neugebauer as being broken off (*ACT*, p. 187) are actually on the tablet.

Column ib

2’ Between *lu-maš* and **ta** there is room for about 1 sign, perhaps **meš**.

4’ After 28.7.30 one expects **tab**, ‘you add’.

10’ ḪAB *attalê*(AN.MI), “disk’ of the eclipse’: probably a term for eclipse magnitude, or is the literal meaning intended?

Column i

14’ **ab**$_2$ *ana* **ab**$_2$, ‘month by month’, is taken to be part of the introductory phrase in analogy to Oi20, where this is certain, but it might also belong to the following sentence. The first **ab**$_2$ appears to have been written over an erasure.

18’–19’ 2.13.20 at the end of 18’ is understood to be the name of Φ, the following **gal** = *rabītu*, ‘large(st)’, i.e. the maximum of Φ, and **tur** = *ṣeḫertu*, ‘small(est)’, i.e. the minimum of Φ. In this interpretation, the sentence is lacking a final verb. Alternatively, the comparison might be expressed by **tur** = *iṣa*, ‘it is small(er)’, but then one would expect only the maximum of F and not also the minimum to be mentioned, since the former coincides with the maximum of Φ. A third option would be to interpret 2.13.20 as a numerical value of Φ (*ACT*, p. 190), but this is also problematic, because the associated value of F, 15, is not mentioned, and neither **tur** nor **gal** would appear to make sense. Although none of the options is fully satisfactory, the first one has been chosen in the translation because it seems to be the least problematic.

20’ *qaq-qar ki-ṣa-ri*: *kiṣ(a)ru* ‘knot, node’. **murub**$_4$-*ti* = *qablītu*, ‘middle’.

21’ **tab** *u* **la**$_2$: interpreted verbally here, i.e. *teṭeppi u tumaṭṭa*, ‘you add and subtract’.

23’ The first sign is definitely *mi*, as suggested by Pinches’s copy in *ACT*.
1.52.25.57: error for 1.53.25.57.

24’ 1.22.21.54: error for 1.22.51.54.

26’ *lu-maš* = *lumāšu*, ‘zodiacal sign’: here used in the sense of a zodiacal position.

28’ *lu*: replaces EN (*ACT*; Aaboe & Henderson 1975, pp. 208–9).
7.12 **sig** *šap-lu* = ‘7,12 it is low, descending’. One would rather expect ‘descending’ to be expressed by a pres. tense, but this rare phonetic writing proves that the stat. is used here.

30’ *u*$_2$: might also be GUR. 20: perhaps a separation mark (:)?

31’ Before **ki** there are traces of several signs, perhaps 19.45 **sig**$^?$, as restored by Neugebauer.

32’ ⌜5⌝[7]: traces of 50 are visible.

33’ Upper parts of several signs compatible with *ša*$_2$ *ki-ṣir* can be seen before **murub**$_4$.

Column ib

3’ After **absin**$_0$ traces compatible with ⌜**ab**$_2$⌝.

4’ After 28.7.30 one expects **tab**, ‘you add’, as in 7’.

Column ii

1’ This is approximately line Oii20 of the tablet.
22: or 20 **a**, ‘20 Leo’? The traces following 22 are compatible with **en** (or perhaps **tag**$_4$). It is followed by traces of two vertical wedges.

3’ The damaged signs at the beginning may be 11 **ḫun** 2
šat-tuš ša$_2$: if resolved in this manner, then *šattuš* is the apocopated form of *šattuššu*, ‘in its year’ (*GAG* §66f). Alternatively, *šattušša* might be a *sandhi* writing for *šattu ša*? Neither option yields a meaningful sentence.

4’ *šu*-⌜*u*$_2$⌝: replaces KI$^?$ PA$^?$ (*ACT*). This implies that **nim** *u* **sig** is the outcome of the subtraction; cf. the commentary.

5’ 7.12 *bi-rit*: replaces *ša*$_2$ 12 **danna** (*ACT*, p. 194). The signs near the edge might be **la**$_2$ 1 or 1 11.

6’ **pa** 30.36: dittography. 3 is followed by faint traces of about 3 signs that could not be identified. The first one looks like TAB.

1	[35][...]... [...] On the 28th the time (from New Moon) to sunset is 9^b, (i.e.) 4,30 (°) [36][...] the length of daylight for the 28th of month XII is 6^b. You deduct 3,0 (°, i.e.) 6^b from 4,30 (°, i.e.) 9^b, [37]1,30 remains until sunrise. You deduct 1,30 from 6,0 (°, i.e.) 12^b, 4,30 is the remainder [38]for day 28. You put down 4,30 as the time (from New Moon) to sunrise in order to predict KUR.	$M = 9^b = 4{,}30°$ $M - C = 4{,}30° - 3{,}0° = 1{,}30°$ $N_0 = 6{,}0 - (M - C) = 4{,}30°$
9	You multiply the (coefficient for) ⸢rising⸣ (for) 10 (°) of all zodiacal signs by [39]the 'stroke' for the zodiac.	$q' \cdot \eta$ $[\rightarrow Q]$
11	You 'tear out' the 'height and depth'.	error for $Q - R_{\downarrow}$ or $Q + R_{\uparrow}$ [→ KUR]
a	These procedures are for hollow [40]and full months and the day of disappearance.	

Co (Colophon)

[41]The tablet which is after it is 'The day when the Sun stands in 28;20 Cnc' ⸢...⸣ [...] ⸢...⸣ [...] [42]Copied from a wooden board and checked. Tab[let of ...], [43]son of Nabû-nādin-aḫi, descendant of Šangû-[DN ...]

7' 1-*šu* = *uššu*, '60'. Or *ku*, the poss. suff. 2 m. sg., which leads to 'your addition and subtraction'?
šar₂ = 'you compute'? Cf. **šar₂.šar₂** in the Glossary.
After **me** *u* **gi₆** there is a 20 followed by traces of several signs.

8' 1?: or *ana* (thus *ACT*) or **bar**?. *nap-pal*: very uncertain. **me** *u₄-me*: replaces **me** *u* **gi₆**. The traces at the end are compatible with **ta im.1**.

9' After **im.**1 one expects **sig** = *išappil*, 'it descends'.

10' **nim**: perhaps *šūqu*, 'height', but a meaning 'distance above the ecliptic?/equator?' does not seem to make sense. Nor is it an option to interpret **nim** as 'you deduct', since the immediately following **e₁₁**-*ma* already expresses subtraction.
After **tag₄**: traces of a number consisting of at least one winkelhaken followed by a vertical wedge and another winkelhaken. 38.8: replaces 28. the second 8 might also be a 5 or a 6.

11' ḪAB-*rat ḫe-piṭ₂-tu₂*: 'broken disk', apparently a term for eclipse magnitude.

13'–14' *ana* **sig nim gar**-*an*: the context implies that **sig** and **nim** both denote the change of the Moon's distance to the ecliptic ('descending ascending'). Note that in *ACT*, p. 196, Neugebauer wrongly translates 'increasing (and) decreasing (latitude)'. This corresponds to 'ascending (and) descending distance' only if one anachronistically adopts the modern convention whereby negative latitude increases if the Moon is below the ecliptic and ascending, and decreases if it is descending.

16' *ki-ṣir*: *kiṣ(a)ru*, 'wrist; joint; knot' here denotes the point where a tabulated zigzag function reverses its direction (increasing/decreasing); cf. the commentary.

21' After **lal₂** the head of a horizontal wedge is visible.

22' **alla.meš**: one expects **alla** = Cnc.
After **rin₂** the top of a vertical wedge is visible.

25' **ta** *si-man*.**meš la₂**-*is* = *ina* (or *ultu*) *simannī tanaḫḫis*, 'you subtract from the durations (of the month)'. The phonetic complement -*is* proves that in this phrase **la₂** is *naḫāsu* G. From NB administrative texts, where *naḫāsu* is written phonetically, it is known that this verb has a (transitive) meaning 'to deduct, subtract', apart from the more common intransitive meaning 'to retreat; become retrograde' (cf. also §2.2.3.2).

26' **a.ra₂**!: **ra₂** looks like UD (replaces *na* in *ACT*, p. 200).
ki *ša₂ šamaš₂*, 'the position of the Sun': the context implies that the distance along the zodiac between the Sun and 27 Psc is meant (cf. the commentary).

27' *a-šar* **la₂**-*u₂*: 'where it is subtractive' refers to columns C' and J, since in these columns the number is followed by the marker **tab** or **la₂** indicating whether it is 'additive' or 'subtractive' with respect to G.

Reverse column i

1 The final 6 is followed by remains of two winkelhakens on top of each other, perhaps the beginning of **ki**.

2 At the beginning of the line there are traces of *ana* **šu₂** *šamaš₂* **tab**, not included in the hand copy in *ACT* (Pl. 224). After **ta** there are traces of a winkelhaken followed by two vertical wedges, most likely **ša₃**. Following that is an empty space and a damaged sign at the end of the line, presumably **la₂**.

8 The phonetic writing *ma-ṭu-u₂* proves that **la₂** in the sense '(it is) decreasing' can have the Akkadian reading *maṭû*, the 3 sg. stat. G of *maṭû*.

10 The same phonetic writing that occurred in 8 proves that **la₂** = *maṭû* also when it represents 'it is less'.

15 1.58.15.55.33.20: the expected 55 replaces 52 (*ACT*).

16 *ṣi* is crammed and damaged.

17 At the end of the line the heads of several vertical wedges are visible.

33–38 These lines are partly preserved on the newly joined fragments (mainly on BM 37244, some signs on BM 32167+32451 +32663+32752).

37–38 These lines probably contain interpolation interval 25, but the formulation could not be reconstructed.

39–40 These 2 missing lines may have contained intervals 26–34; cf. the commentary.

Column ii

1 Since P16' probably continues until the end of Ri only 2 lines of P17' are missing before Rii3.

3 The damaged sign after **u₄** might be **tab** or *i*; the sign before **ki** is probably BE. The position of the Moon mentioned at the end of Rii3 implies a prior instruction for step 4 (cf. the commentary). Hence the gap at the end might have contained *u* ᵈ**utu il₂**, or something similar.

4 After **a.ra₂** a vertical wedge is visible, perhaps part of the expected **bar**.[**nun**] = *ṣiliptu*-coefficient.

5 **bar**: replaces ŠU$_2$ (*ACT*), perhaps to be connected with a missing **nun** at the beginning of Rii6, resulting in *ṣiliptu*(**bar.nun**), '*ṣiliptu*-coefficient'.
After **ki** one may restore *ša*$_2$ ^{d}*sin u* ^{d}**utu**, which results in 'The position of the Moon and the Sun'. This would then be the beginning of step 5. Or, if one assumes that **ki** is followed by *tanašši*(**il**$_2$) one gets 'you compute the position', an allusion to step 4.

6 The small gap at the beginning might contain a reference to step 8; perhaps **a.ra**$_2$ **nim** *u* **sig du**, 'you multiply it by the 'height and depth''.
ar$_2$ = *arki*, 'afterwards, next', as in Oi24'?
lu: perhaps to be restored as *lu-maš* in the sense 'zodiacal position', or read **dib**?
ta BE **zi**-*aḫ*: *ina/ultu* BE *tanassaḫ*, 'you 'tear it out' of the elongation'. The use of *nasāḫu*, 'to tear out' for subtraction suggests an early Seleucid or older date. Other logograms for subtraction in No. **53** are **la**$_2$ and **e**$_{11}$ (but in P17'–P18' **la**$_2$ is not used); **nim** is completely absent.

7 The gap at the beginning may have contained **ki(.meš)**, 'the position(s)', or **bar.nun** *ša*$_2$ *lu-maš*, '*ṣiliptu*-coefficient for the zodiac'; the former was chosen for the translation.
il$_2$**.a** = *tanašši*, 'you compute'.

11 **u**$_4$**.1.kam**, 'day 1', here has the technical meaning NA$_1$. The usual logogram **na** appears in Rii15.

12 **ki**-*šu*$_2$ = *ittišu*, 'to it': i.e. the zodiacal position of the Moon. Perhaps a mistake for **ki ki**-*šu*$_2$ = *itti qaqqarišu*, 'to its zodiacal position'?

15 *nu-bal-kut*: *nubalkut* = stat. 3 m. sg. of *nabalkutu*, '(a day) is transgressed'.

16 The translation of the final two cases in Rii16–17 as given in *ACT* (cases C and D) contains trivial errors.

17 **u**$_4$**.na**$_2$**.am**$_3$ = *ūm bubbuli*, 'day of disappearance', i.e. the day of KUR. Here it denotes KUR itself (compare Rii11; the term **kur** is used in Rii37).
si-man ša$_2$ **gub** *u* **gur il**$_2$**.a** = *siman ša kunni u turri našû*, 'the computed duration of the full month and the hollow month' (= G). For *kunnu* and *turru* (lit. 'confirmed and rejected') as terms for the full and hollow month cf. Beaulieu (1993).

18 3 is followed by a damaged sign containing a small winkelhaken in raised position, possibly a 0.

20 **lal**$_2$ = *šitqultu*, 'equinox'.
zi.meš *ša*$_2$ *sin u* ^{d}**utu** *ina* **šu.2 dib** = *nisḫī ša sin u šamaš ina qātē taṣabbat*, 'You hold the displacements of the Moon and the Sun in (your) hands', is used for quantities assumed to be known. Oddly enough, the displacements of the Moon and the Sun are not known at this point (cf. the commentary). Perhaps the phrase should be translated in the future tense, 'you shall hold...in your hand'.

23 *tu-šaḫ-ḫa-az* = *tušaḫḫaz*, 2 m. sg. pres. *aḫāzu* Š 'you let proceed'.
ana ^{d}**utu.šu**$_2$**.a** = *ana ereb šamši* 'to the west' (i.e. where the Sun sets) refers to the westward direction of the solar motion along the ecliptic when going *back* in time.

24 15.42: error for 16.42.
la-**igi** ^{d}**utu** = *lapān šamaš*, 'in front of Šamaš', refers to the Moon's position with respect to the Sun in terms of the daily rotation of the sky, i.e. it sets before the Sun.

25 BE: probably *bi*$_2$, 'elongation' (*ACT*: *šumma*, 'if').

28 **danna** *si-man*, 'miles (^{b}) of time': this term is not expected here. Perhaps the total time between KUR and New Moon is mentioned in the preceding gap for some reason, which would be $5^{b} + 1^{d} = 17^{b}$.

29 After **bar.nun** there are two small winkelhakens which are aligned vertically; they are transliterated as GAM. On this tablet GAM is used only twice for multiplication (Oi26',27') instead of **a.ra**$_2$.
Neugebauer correctly transliterates **nim** *u* **sig**, but *u* is mistakenly omitted from the copy.

30 Only the right part of the first sign after the break is preserved; this looks like BAR.
After collation it seems possible to read the end of the line as ⌜**gin**$_7$ *ša*$_2$ **u**$_4$**.1.kam du**$_3$ x⌝.

32 This single line is missing between BM 32651 and BM 32167+32451+32663+32752.

33 The reading of ⌜**me nim**⌝ is uncertain. Of the penultimate damaged sign only the lower half is preserved; this is consistent with a reading LU (or UR); the lower end of MAŠ is also preserved.

34 *šamaš*$_2$ *u* ^{d}*sin* = 'Sun and Moon': this order is unusual. It is also peculiar that only the Moon is provided with the determinative **dingir** (compare Rii18,20,21).

35 The traces before **a** are compatible with **il**$_2$**.a** = *našû*, 'computed' (verb. adj.). Before **u**$_4$.28.**kam** there are traces of what might be *ma*. Perhaps **iti.še** is mentioned in the preceding gap.

36 Before the 4 at the end there is a tiny spurious winkelhaken.

37 **ib**$_2$**.tag**$_4$ = *rīḫtu*, 'remainder'.

38 *ana* **kur e.a** = *ana* **kur** *qabê*, 'in order to predict KUR'.

38–39 *mi-ḫi-il-tu*$_2$ (= *miḫiṣtu*) = literally 'stroke': a coefficient (cf. the Glossary). It is multiplied by **nim** U *lu-maš*, and **nim** *u* **sig**, the 'height and depth', is subtracted from the product. This implies that (1) *miḫiltu* is the lunar elongation η (usually written BE); (2) 'height and depth' is R, and (3) **nim** U *lu-maš* is q_{rise} (cf. the commentary). If compared with the name of q_{rise} in No. **61**, *ṣiliptu*(**bar.nun**) *ša*$_2$ *lu-maš ša*$_2$ **nim**, '*ṣiliptu*-coefficient for the zodiac for rising', the main difference is the absence of *ṣiliptu*. The other elements are present but arranged differently. I assume that U represents 10 (degrees), since the control values of q_{rise} pertain to 10° of the zodiacal signs. Alternatively one might interpret **nim** U *lu-maš* as *elû u lumāšu* as an abbreviated writing for '(coefficient for) rising and the zodiac'.
du$_3$**.meš** = *epūšū*, 'procedures'. The sg. appears in Oi14 as **du**$_3$-*uš*.

41 *ṭuppu*(**im.dub**) *ša arkišu*, 'The tablet which is after it': cf. *BAK*, p. 2. A similar phrase occurs in No. **45**. R7: **im.dub** *ša*$_2$ *arkišu*(**egir**-*šu*$_2$) [...].
After **gub**-*za* one might restore **nim** *u* **sig**.

42 ⌜**sar**⌝-*ma* **igi.tab** = *šaṭirma bari*, 'written and verified' (Hunger *BAK*, pp. 3–4.)

43 *šangû*($^{\text{lu}_2}$**sanga**) = 'chief administrator of temple' (*CAD* Š I, *šangû* g3): reading suggested by C. Wunsch. The personal wedge (^{m}) appears to be lacking. Clan names of the type Šangû-DN are common in the NB and LB periods (Tallquist 1905, pp. 197–199).

Commentary

This large tablet contains instructions for nearly all the columns of a synodic table for lunar system A. The bulk of the tablet is formed by BM 32651 (*ACT* 200), to which are added three further fragments: BM 32172 (*ACT* 200aa), and the previously unpublished BM 32167+32451+32663+32752 and BM 32744.[634] All fragments belong to collection 76–11–17. BM 32651 has some damage to the lower (upper) left corner of the obverse (reverse), and the right edge. It is pitch black, presumably as a result of burning or secondary baking, whereas the other fragments have the usual brownish color. The new fragments preserve segments of the upper (lower) edge of the obverse (reverse), so that the tablet is now known to be an accurate square with sides 20.5 × 20.4 cm (type A). The obverse of BM 32167+32451+32663+32752 contains several additional lines of column i; the reverse contains most of the missing text of column ii, a part of the colophon and a few signs of column i. BM 32744, which is inscribed on one side, adds some text to column i of the reverse.

The text is divided into two columns. On the obverse, a small section of column i is split up into two subcolumns. The procedures are preceded by the standard invocation to Bēl and Bēltu near the upper edge. At the bottom of the reverse, about 40% of the colophon is preserved on the newly joined fragments.[635] It mentions a catchline, 'The day when the Sun stands in 28;20° of Cnc...', which is said to follow the present tablet, implying that both belong to a series.[636] The catchline suggests that the next tablet concerns solstices, because the Sun is near the quoted position at the summer solstice. However, no text with that catchline is known to me. The colophon also mentions that the tablet was copied from a wooden board and verified (R42). After a gap, R42 continues with the name of the scribe, [NN]/Nabû-nādin-aḫi//Šangû-[DN]. Perhaps the same name is partly preserved in the colophon of No. **61.A**: 'Nabû-nādin-aḫi/Nabû-nādin-aḫi; they might be the same individuals. Thus far no other astronomer from a clan Šangû-[DN] is attested.[637] The colophon does not preserve a date, but an exemplary value of *E* mentioned in P8' may suggest that the tablet was written near year 5 of Philip Arrhidaeus (319 BC), as proposed by Aaboe & Henderson (1975). This is consistent with orthographic features and the formulation of the procedures.

The tablet contains at least 18 visually marked procedures. On the obverse, additional procedures may have been present between Oi11 and Oi1', where 6 lines are missing, and in column ii, where about 21 lines are missing before Oii1'. No procedures are missing on the reverse. Near the end of the tablet there is a rare 'meta-instruction', according to which 'these procedures are for hollow and full (months) and for the day of disappearance', a reference to the computation of NA_1 and KUR in P17' and P18'. Most procedures are devoted to the computation of a single quantity. Their order on the tablet largely coincides with that of the columns of a synodic table for New Moon. However, the place of P16' (*G*) after P15' (*K*, *M*) is unexpected. Perhaps the scribe did not want P16' to be divided over both sides of the tablet. Besides procedures for computing or updating functions some contain checking rules (P11'.c, P12'), and others have no direct relevance for computing synodic tables (e.g. P9', P13'). Some procedures, especially P17' and P18', are rather sketchy. Although the tablet contains procedures for nearly all functions of a synodic table, it does not inform the user about such issues as the order of the procedures, the choice of the initial values and the flow of information between the procedures. All in all the tablet does not amount to a self-explanatory handbook for producing synodic tables.

P1 Procedures for Φ

P1 consists of two parts, both of which concern Φ. Part **a** (Oi2–4) is a procedure for updating Φ from month to month according to the zigzag function (§4.4.3). The monthly difference d = 2;45,55,33,20° (Oi1) and the reflection rules near M = 2,17;4,48,53,20° and m = 1,57;47,57,46,40° are preserved. The formulation is based on template ZZ.B.2 (§2.6.8.1). Part **b** (Oi5–6) is badly preserved. The extrema M and m suggest that this may be an alternative formulation of the zigzag function, perhaps based on template ZZ.B.3. In that case one would like to interpret the damaged signs at the end of Oi4 as *ša₂-niš*, 'secondly'.

[634] It was not previously recognised that BM 32172 and BM 32651 are part of the same tablet (Neugebauer believed that BM 32172 belongs to the same tablet as No. **62**). They do not physically join one another, but both join the fragment BM 32167+32451+32663+32752.

[635] The left corner of the colophon preserved on BM 32172 was edited in *ACT* as Colophon Zqb.

[636] This is only the second attestation of a catchline in the colophon of an astronomical procedure text. The other example is No. **45**, which concerns Mercury and Saturn, but there the catchline is broken off. The quoted zodiacal position is presumably generated by a solstice scheme. It also happens to be the position of the Full Moon in month X, SE 38 (and in the same month 225 years before or after that), when the Sun is in 28;20° Lib, which is near the winter solstice. The significance of this is not clear.

[637] Assuming that Šangû-[DN] is the clan name, which is not certain. A clan from Babylon with the similar name ^{lu_2}Šangû-parakki is attested twice in colophons of scholarly tablets (*BAK* 108). Well known are the scholars of the Šangû-Ninurta clan from Uruk, who were active in the Achaemenid and Seleucid periods.

P2 Computing Φ^{fm} from Φ^{nm}?

The introductory phrase suggests that P2 is concerned with the computation of Φ^{fm} from Φ^{nm}. For this algorithm cf. §4.4.3. Due to the damaged state, the template could not be established.

P3 Computations involving Φ?

If the number in Oi11 is correctly restored to 2,13;20, P3 is also concerned with Φ. Since virtually nothing else is preserved, little more can be said. It is not clear whether Oi1' belongs to the same procedure. The number 1.6, if read correctly, is unidentified.

P4' Computing C^{nm} from B^{nm}

P4' contains a complete set of interpolation rules for computing the length of daylight (C) from B^{nm} (§4.4.5). They are formulated in accordance with templates IP.A.1–2 (§2.6.7), but they lack the usual *ana tarṣa*... **gar**-*an*, 'Opposite... you put down...'.

P5' Updating B^{fm} with the step function for σ

This procedure for updating B^{fm} (§4.4.4) is formulated in accordance with template STEP.A.1, which includes transition rules. Since all other procedures on the tablet concern functions pertaining to New Moon, the presence of a procedure for Full Moon is unexpected, and presumably the result of a scribal error.

P6' Computing Ψ' from E

P6' contains a rule for computing a function for eclipse magnitude, say Ψ'', from the 'height and depth of the eclipse disk' (E). Strictly speaking the procedure implies the following algorithm:

$$\Psi'' = 17;24^{\mathrm{f}} \begin{cases} +0;20 \cdot E_{\uparrow} \\ -0;20 \cdot E_{\downarrow}. \end{cases} \tag{5.1}$$

This algorithm partly matches Ψ' outside the nodal region (§4.4.8, p. 141). In agreement with Ψ', a central eclipse ($E = 0$) results in $\Psi'' = 17;24^{\mathrm{f}} = c_{\Psi}$. However, P6' contains no indication that the algorithm applies only outside the nodal region. It therefore appears that P6' is an incomplete quotation from a procedure for Ψ', as proposed by Neugebauer.

P7' Procedures for $F^{(1)}$

P7' contains two procedures for the Moon's daily displacement F (§4.4.9). Part **a** (Oi14'–16') is a procedure for updating F as a zigzag function, using the parameters of $F^{(1)}$. It is formulated in accordance with the rare template ZZ.B.4 (§2.6.8.1). Part **b** (Oi16'–18') contains an interpolation algorithm for computing $F^{(1)}$ from Φ. The formulation deviates somewhat from the usual interpolation templates (§2.6.7). Values of $F^{(1)}$ are provided for 4 selected values of Φ — the extrema of the zigzag function and those of the truncated version. There is also a reference to the truncated version of F. The problematic phrase at the end of Oi18' is interpreted here as a general interpolation rule formulated in terms of the extrema, the factor 0;15,11,15 being the interpolation coefficient d_F/d_{Φ}.[638]

P8' Updating E

P8' is concerned with updating the Moon's distance to the ecliptic, E. For this algorithm cf. §4.4.6. Apart from Neugebauer's treatment in *ACT*, 190–193 ('Section 6'), P8' was analysed by Aaboe & Henderson (1975), pp. 208–211. About 19 lines of text are missing between the last preserved line of P8' (Oi33') and the first preserved line of P9' (Oii1'); it is unclear how much of that belongs to P8'. What remains consists of six parts. Generally formulated rules alternate with numerical examples in which the Moon is located at a specified position.

Part **a** (Oi20') contains an introductory phrase and some parameters: 12°, the 'width of the path of the Moon' (= $2M$; cf. step 3 in §4.4.6), and 2,24$^{\check{s}e}$ = κ, the boundary of the nodal region.

Part **b** (Oi21'–27') is concerned with steps 1 and 2 in zone 2. The formulation is partly based on template ESTEP.C (cf. §4.4.6), but this template is interrupted in Oi22'–25' by a numerical example for the case when the Moon is descending and about to enter the nodal region from E_0 = 3,52;11,39$^{\check{s}e}$ above the ecliptic. In accordance with step 1 the preliminary

[638] Cf. also the philological notes. This interpretation differs from Neugebauer's, cf. *ACT*, p. 190.

next position is obtained by subtracting w = 1,58;45,42$^{\check{s}e}$, resulting in E = 1,53;25,57$^{\check{s}e}$. This is followed by step 2 (Oi23'–24'), whereby (effectively) w_{int} is doubled (cf. Eq. 4.30 on p. 135). This is achieved by subtracting $\kappa - E_1$ = 30;34,3$^{\check{s}e}$ (Oi23'), from E_1, resulting in the final value E_1 = 1,22;51,54$^{\check{s}e}$, within the nodal region. The next position is said to be E_2 = 30;34,3$^{\check{s}e}$ + w = 2,29;19,45$^{\check{s}e}$ below the ecliptic, outside the nodal region. This can be understood as an application of Eq. (4.34) with $w_1 = w_0 = w$ = 1,58;45,42$^{\check{s}e}$, leading to $E_2 = |E_0 - (2w + \kappa)|$, so that $E_2 = 2w + \kappa - E_0$ = 2,29;19,45$^{\check{s}e}$ below the ecliptic. As pointed out by Aaboe & Henderson (1975), p. 210, E_2 can be dated to Philip Arrhidaeus year 5, month III, if one assumes that it is connectible to the parent synodic table of system A (for New Moons).[639] The general algorithm for step 1 continues in Oi25'–27' with a transition rule, formulated in accordance with template ESTEP.C, and numerically implemented for the position B_i = 13° Lib corresponding to the endpoint of rule 1a in Eq. (4.27) on p. 134.

Part **c** (Oi28'–29') concerns step 3, the reflection rule near the maximum of E.

Part **d** (Oi29'–30') contains an unidentified algorithm for zone 2 which is too damaged for a reconstruction.

Part **e** (Oi30'–33') contains a verification of the result obtained in part **b** which utilises the alternative algorithm for updating E across the nodal region (Eq. 4.35 on p. 136). First $E_0 + |E_2|$ is computed as 6,21;31,24$^{\check{s}e}$. This is compared with $2w + \kappa$, which leads to the same numerical result, as it should.

Part **f** (Oi32'–33') contains some algorithm for zone 1, but due to the bad state of preservation it could not be identified. Perhaps it continues in column ii.

P9' Various topics related to the Sun

An unknown number of lines (at most 19) are missing from the beginning of P9'. What remains of the procedure is only partly understood.

a. Computation involving $v_{\odot}$, unclear (Oii1'–3')

The number 1;2,44 (Oii2') is the maximum of the zigzag function for the Sun's daily displacement (§4.4.10). This function is also invoked further below in P10' and P18'. The significance of the coefficient 8 (Oii2'), called *nappaltu* (lit. 'compensation'), is unclear. The text implies that it is associated with the minimum of a quantity in Sgr. This quantity cannot be $v_{\odot}$, since its minimum occurs in Gem. The subtraction of 8 from 1;2,44 implies that 8 must be interpreted as 0;8 or 0;0,8. The former number coincides with the interpolation coefficient (c_k) for the length of daylight (C) in Sgr, where C attains its minimum 2,24°, and the interpolation coefficient its maximum (§4.4.5). However, it is not clear why c_k would be subtracted from $v_{\odot}$.

b. Computation involving the Sun's 'height and depth' and C (Oii3'–4')

It is also hard to make sense of part **b**, which concerns the **nim** *u* **sig** *ša Šamaš*, the 'height and depth of the Sun'. The usual meaning of this term, distance above and below the ecliptic, is unlikely to apply here. If the Sun were at a distance to the ecliptic, then the algorithms for eclipse magnitude and those for the Lunar Six intervals should depend on it, but they do not. It is therefore unlikely that there is such a concept within the framework of lunar system A. A more likely interpretation of 'height and depth' is suggested by the connection with the length of daylight (C) that the procedure appears to establish. The maximum of C occurs during the summer solstice, when the Sun reaches the largest distance (declination) above the celestial equator, the minimum during the winter solstice, when the Sun reaches the largest declination below the equator. Hence 'height and depth' may here denote declination (perhaps also in P11'.b). From Oii4' one infers that the 'height and depth of the Sun' is the outcome of a subtraction. Unfortunately the subtracted quantity cannot be identified, but the phrase 'for the entire year' implies that it is a function of the zodiacal position. It is also not clear from what it is subtracted. Hence the meaning of the computations remains unclear.

P10' Defining parameters of the zigzag function for $v_{\odot}$

P10' mentions the extrema of the zigzag function for $v_{\odot}$, the Sun's daily displacement along the zodiac: m = 0;55,32°/d in Gem and M = 1;2,44°/d in Sgr (§4.4.10). The maximum was also mentioned in P9' (Oii2'). After collation the end of the line is now read as '0;7,12, the distance of the displacements...', which can be identified as Δ.

P11' Various topics related to the Sun and eclipses

This procedure is concerned with various solar functions. Parts **a** and **c** are essentially understood, but part **b** continues to defy interpretation.

[639] The earliest attested synodic table of system A, *ACT* 128+, happens to be for the same year.

a. Defining parameters of a zigzag function for the Sun's monthly displacment (Oii6')

Oii6' mentions the extrema of a function describing the Sun's monthly displacement along the zodiac and their (approximate) zodiacal positions: 27;36° in Gem and 30;36° in Sgr. They can be interpreted in terms of a zigzag function; for a discussion cf. §4.4.10. The meaning of the monthly difference (?) 6 mentioned at the end of Oii6' is unclear.

b. Unidentified computations involving 'height and depth' and other parameters (Oii7'–11')

In Oii7' a monthly difference 36 is mentioned. It might be interpreted as 0;36 = d_F of lunar system B (§4.5.10), but this remains speculative. The rest of Oii7' also defies interpretation. As mentioned by Neugebauer, 0;40 is the amount by which the duration of daylight changes per degree of solar displacement along the zodiac in the daylight scheme of the astronomical text Mul.Apin (shortest day 2,0°, longest day 4,0°, resulting in an interpolation coefficient 2,0/3,0 = 0;40). In Oii8'–9' it is said that $v_\odot$ = 0;59,8° when the Sun 'ascends from the South to the North or ⟨descends⟩ from the North to the South'. This is the mean value of the zigzag function mentioned in P10'. As in P9', the upward and downward motion might be interpretated in terms of declination. In Oii10' the 'coefficient for daylight' is mentioned again. It is here associated with numbers of unknown significance. Perhaps 36 and 24 are the extrema of a function, and the number 6 by which the Sun (?) is said to ascend and descend (Oii10') is half the amplitude of the same function. Alternatively, 6 might be the maximum of E measured in degrees, in which case 'height and depth' would mean distance to the ecliptic, but one would then expect the subject to be the Moon. The number 20.8 (or 28?) is obtained by subtracting something from 0;59,8 = $v_\odot$, but the subtracted number, apparently 24 (or 20.4?), is inconsistent with the outcome, whether one reads the latter as 20.8 or as 28.[640] In Oii10' 0;59,8 appears to be called the 'height', which is unexpected. Also unexplained is the 7^f = 0;35° by which the Sun 'ascends and descends' (Oii11'). An interpretation in terms of declination seems to be ruled out because one would expect something similar to 23°.[641] In summary, part **b** largely defies explanation. Some elements seem to establish a connection between the Sun's motion along the zodiac and a motion which goes up and down, perhaps with respect to the celestial equator, but the nature of this connection remains unclear.

c. Computing E from Ψ (Oii11'–15')

As stated in Oii15, P11'.c is concerned with the computation of E from the eclipse magnitude (§4.4.7). This is achieved by means of an interpolation algorithm (Eq. 4.52, p. 140), which is implemented with control points $E_0 = 1{,}38;24^{\check{s}e}$ and $\Psi_0 = 1^f$. In Oii14' the outcome is assigned to **sig nim**, which is tentatively translated as 'descending (and) ascending (branch)', following a suggestion by Neugebauer (*ACT*, p. 196). It is understood that the former is associated with distance above the ecliptic and the latter with distance below the ecliptic, both being mentioned in this order in the previous clauses (cf. the philological notes). This correctly identifies among the 4 possible configurations of E (above/below the ecliptic, ascending/descending) the 2 that are associated with $\Psi = 1^f$. After the interpolation rule it is stated that one has to proceed in similar fashion until $\Psi = 17;24^f$ (Oii14'–15').

***P12'** Computing the extrema of the zigzag function for F*

The purpose of P12' is to compute the extrema of the 'displacement', most likely a reference to the zigzag function for the Moon's daily displacement, F (§4.4.9). However, the procedure is formulated without any numerical examples, so that it applies to any zigzag function. The term 'knot, wrist' (*kiṣru*) makes sense if one interprets it as a concrete position in the column of a synodic table rather than the mathematical concept of extremum, because the text refers to positions above and below the 'knot'.[642] Nevertheless, the procedure deviates from the correct computation of the extrema of F from the function values below and above the 'knot'. The correct expressions are

$$M = (F_{\text{above}} + F_{\text{below}} + d)/2, \qquad (5.2)$$

$$m = (F_{\text{above}} + F_{\text{below}} - d)/2, \qquad (5.3)$$

where d is the difference (*tašpītu*) of the zigzag function, but in the procedure d is added or subtracted after the multiplication by $1/2$, so that the outcome amounts to $M + d/2$ and $m - d/2$ instead of M and m.[643] One is therefore left with the unsatisfactory conclusion that there is a scribal error.

[640] The former, if interpreted as 0;20,8, would be obtained by subtracting 0;39, but this number is nowhere mentioned.

[641] Neugebauer translates **nim** *u* **sig** as 'increases and decreases' and interprets 7^f as the monthly change of $v_\odot$, i.e. 30 · $0;14^f$, the latter value being attested in *ACT* 135,220. However, **nim** *u* **sig** does not mean 'increase and decrease'.

[642] *ACT*, p. 199. Neugebauer's ad hoc translation, 'change' (of the slope of the zigzag function) hints at this. 'Knot' also has an astronomical meaning 'nodal region' known from the procedures for E, but this does not make sense here.

[643] This was ignored by Neugebauer in *ACT* I, p. 199. It seems impossible to stretch the grammatical interpretation far enough and reverse the order of the multiplication by $1/2$ and the addition or subtraction of d with respect to that in the present translation.

P13' Computations involving solstices and equinoxes

P13' is concerned with solstices and equinoxes, a topic that does not really belong to mathematical astronomy. The parameter 1,36° = 0;16^d (Oii20') is the length of the year -365^d. It is not associated with lunar system A, but known from the algorithm underlying BM 36731+, a list of solstices, equinoxes and dates of the heliacal rising of Sirius for the period between year 8 of Nabopolassar and year 25 of Nebuchadnezzar (618–580 BC).[644] In order to obtain the time and date of subsequent solstices or equinoxes this parameter is added if the time of the solstice or equinox is expressed in ° after sunset or sunrise, subtracted if it is expressed in ° before sunset or sunrise (Oii21'). The number 11 is (an approximation of) the yearly epact, the interval by which the year exceeds 12 synodic months. Also this number is not associated with lunar system A, but known from the so-called Uruk scheme for solstices and equinoxes, which prescribes a shift of their dates by 11^τ per year, except every 19th year, when it is 12^τ.[645] Both schemes for solstices and equinoxes predate lunar system A; they are also mutually inconsistent.

Oii22' mentions values of the length of daylight (C) and/or night (D) that are consistent with lunar system A (§4.4.5): C = 3,36° and D = 2,24° (summer solstice, Sun in Cnc), C = 3,0° (autumnal equinox, Sun in Lib), and presumably also C = 2,24° and D = 3,36° (winter solstice, Sun in Cap), and C = 3,0° (spring equinox, Sun in Ari). The dates when these values occur in lunar system A are not necessarily consistent with the mentioned schemes. In principle lunar system A only provides values of C and D at the lunations. Since solstices and equinoxes can occur in between lunations, one has to resort to interpolation in order to establish the date of the solstices and the equinoxes in lunar system A. The result is unlikely to agree with the schemes for solstices and equinoxes mentioned earlier, except in an average, approximative sense.

P14' Computing $G+J$

In P14' the final value of the duration of the synodic month ($G+J$) is obtained by adding to G the zodiacal correction J, which is computed as a generalised step function of B_i. For this algorithm cf. §4.4.14. However, the procedure is not entirely consistent with the reconstructed algorithm (Eq. 4.63). Oii23' contains rule 2a, and Oii23'–24' rule 2b. This is followed in Oii24' by the number 2 ('secondly'?) and another partly preserved rule for the same region — perhaps an alternative formulation. One would then expect rule 1a, but it is skipped. Instead Oii25' continues with rule 1b, except that the left boundary is at $b_1 = 13$ Vir. This deviation can be explained if one assumes that, unlike the other rules, this one is, presumably by mistake, formulated in terms of the *old* position (B_{i-1}) instead of the *new* position (B_i). In Oii26' the extrapolation is exemplified for the case $B_i - b_2 = 28;7,30°$ corresponding to the end of zone 2.[646] This is a rare case where an explicitly subtractive number is multiplied by a number, resulting in another subtractive number (§2.4).

P15' Computing K and M

Part **a** is concerned with the 'final duration', which is the technical term for $K = G+J+C'$ (§4.4.17). According to the procedure, an additive or subtractive correction is applied to the 'duration'. While Neugebauer interpreted the latter as G and the correction as $J+C'$, it is more likely that the 'duration' stands for $G+J$, since that quantity was computed in P14'. Hence the correction refers to C' (cf. Eq. 4.69), which is indeed additive or subtractive.

Part **b** is concerned with the 'time until sunset', i.e. M^{nm} (§4.4.18). The formulation of the procedure is fully consistent with the reconstructed algorithm (Eq. 4.72). Ri1–2 deal with the case when the subtraction of K results in a transgression of the date boundary.[647] The date of New Moon has to be corrected accordingly. This occurs at the end of P17', after the computation of NA_1 which determines the length of the previous month.

P16' Computing G from Φ: interpolation scheme 1

P16' contains interpolation rules for computing G from Φ in accordance with scheme 1 (§4.4.11). The place of this procedure on the tablet is unexpected, since the two preceding procedures (P14', P15') depend on G. Most entries are formulated in accordance with templates IP.B (§2.6.7). Lines Ri3–36 contain intervals 1–24 (Ri33–38 are partly preserved on the newly added fragments). The remaining intervals 25–35 would require about 7–9 lines: 2–3 for intervals 25 and 35 each, and about 3 for intervals 26–34 if they are formulated by means of the *malûššu*-rule (p. 149), as is the case with intervals 1–9 in Ri3–5. It therefore appears that some intervals must have been omitted.

---

[644] Neugebauer & Sachs (1967), Text A. For a more recent discussion cf. Britton (2002), pp. 29–30.

[645] Neugebauer (1948a); Britton (2002), pp. 43–44.

[646] I.e. at the point where, in Neugebauer's terminology, rule c begins, not rule b (*ACT*, p. 201).

[647] Neugebauer assumed that the text in the gap also deals with the associated subtraction of 1 day from the date of New Moon, but there is no space.

P17',P18' Computing NA$_1$ and KUR

The final two procedures are concerned with the Lunar Six intervals NA$_1$ (P17') and KUR (P18'). For the definition of the Lunar Six intervals and their astronomical significance cf. §4.1. A first, partial edition of P17' appeared in Neugebauer (1953), shortly before the one in *ACT* (pp. 226–240). Neugebauer again discussed these procedures in *HAMA*, pp. 535–540. In Kugler (1900), which contains the first edition of BM 32651, P17' and P18' are skipped. The newly joined fragments add a considerable amount of text to P18' (Rii33–40'). P17' and P18' each contain several steps of the Lunar Six module (§4.4.19), but they are by no means complete. There are notable differences with the procedures for ŠU$_2$, NA, ME and GI$_6$ (No. **61**) in terms of the formulation. Whereas No. **61** employs a fully general formalism, here we find a mixture of numerical examples and general expressions. Both P17' and P18' proceed from the assumption that New Moon occurs on the 28th of month XII, when the Sun is in Ari. It may be noted that synodic tables usually begin with that month. Some steps are merely alluded to in an opaque manner, or omitted altogether. In P18' step 1 is elaborately formulated in parallel units: time degrees (UŠ) and 'miles' (*bēru*). The order in which the steps are presented is not always compatible with the order in which they are meant to be executed.[648] P17' and P18' do not constitute a complete, self-explanatory set of procedures, and cannot have functioned as the sole source of instruction about KUR and NA$_1$.

Summary of P17'

P17' is formulated mainly in general terms. The first two lines (Rii1–2) are missing; they may have contained a statement of purpose and steps 1–3, the latter because step 4 can be restored in Rii3–4. Also step 9 appears to be mentioned here, even though the computation of the involved coefficient q' (step 5) occurs only in Rii7–10, after which step 9 is mentioned again (Rii11). This is followed in Rii5 by an unidentified addition, an instruction for the coefficient r (step 7), and step 10 in Rii6. The end of Rii6 is damaged and not understood; 'day 28' is presumably the date of New Moon. Rii7 begins with an instruction to compute the zodiacal positions (?) of the Moon and the Sun (step 4). Paradoxically, both were assumed to be known already in the previous lines, suggesting that the text beyond Rii7 (steps 4, 5, 9–12) is not fully harmonised with the preceding lines. From Rii12 it is clear that in P17' the month is assumed to be hollow, i.e. the lunar crescent becomes visible near sunset on day 29. The text continues with two unmarked tables of coefficients (q_{set} and r_{set}), an instruction for the disk correction, and a set of rules for establishing the date of New Moon (Rii14–16).

Summary of P18'

P18' consists of two parts, each dealing with a complementary initial condition: (**a**) New Moon during the day (Rii17–34?) and (**b**) during the night (Rii35?–40). The latter case is contained on the newly added fragments. The reason for this is connected to the computation of N (step 1) for KUR, which is different for the two conditions, whereas for NA$_1$ they are equivalent. P18' is formulated mainly in terms of numerical examples. After the introductory phrase (Rii17) there is an instruction to 'hold the computed duration for the full month and the hollow month in (your) hand'. I interpret this as a reference to G ('duration'), which the user is assumed to have computed. It is not clear why the full month and the hollow month are mentioned. Rii17–28 concern steps 1–3 for the case when New Moon occurs in Ari during day time on the 28th of month XII. The coherence and completeness of this part suggest that it may have been a separate procedure that was combined with the rest of P18'. Time intervals are elaborately expressed in double units ('miles' and time degrees). Beyond Rii28 the text is damaged. However, it is clear that there is not enough space for steps 4–13 in Rii29–32/33. Perhaps step 7 is alluded to in Rii29. After that step 8 appears in the form of a general rule, i.e. the numerical example has been abandoned here.

Part **b** probably begins in Rii35. Apart from step 1, which is complete, steps 8, 9(?) and 10 are alluded to, but the intermediate steps are absent. It therefore appears that the main purpose of part **b** is to exemplify step 1 for the case when New Moon occurs during the night. The final sentence, 'these procedures are for hollow and full months and the day of disappearance' (Rii39–40) is formally within the body of P18', but actually refers to the tablet as a whole.

Step 1. Time interval N between lunation and Lunar Six event (§4.4.19.1).

In both P17' and P18' New Moon is assumed to occur on the 28th of month XII.[649] The value of N_{NA_1} is not preserved in P17'; it might have been mentioned in Rii1–3.[650] In P18' both cases of N_{KUR} are considered. Rii17–20 concerns the case when New Moon occurs in Ari (the precise position is not specified) on the 28th of month XII, at $M^{\text{nm}} = 30°$ before sunset, which results in $N_{\text{KUR}} = 2{,}30°$. However, this is not the effective final value of N_{KUR}, because in step 2 the zodiacal displacements of the Moon and the Sun are modified by adding to each the displacement for 1 day. This implies that

[648] For example, at the point where step 10 is mentioned in P17', some of the involved quantities have not yet been provided, while in P18' step 10 is correctly placed near the end.

[649] In P17' a corresponding statement is not preserved, but month XII is consistent with the Sun being in Ari (Rii5), and day 28 is mentioned in Rii6. Furthermore, the fact that the first crescent appeared on day 30, i.e. after sunset on day 29 (Rii12), also points in that direction.

[650] If step 1 was formulated as a general instruction in P17' it must have been brief, since there is no room for anything similar to step 1 in P18'.

the case $n = 0$ is discarded without computing KUR, presumably because N_{KUR} is known in advance to be too short for producing a visible lunar crescent. The second case is considered in Rii35–38, where New Moon is assumed to occur at M^{nm} = 4,30° before sunset on the 28th of month XII, i.e. during the night of the 28th. This corresponds to the latest possible occasion of KUR before New Moon ($n = 0$).

Step 2. Zodiacal displacements $\Delta B_{☽}$ and $\Delta B_{\odot}$ (§4.4.19.2)

The computation of the displacements of the Moon ($\Delta B_{☽}$) and the Sun ($\Delta B_{\odot}$) between New Moon and the Lunar Six event is preserved only in P18'.a, where step 2 is a direct continuation of step 1. Here the Moon's displacement between the last sunrise before New Moon and New Moon is computed on the assumption that the daily motion (F) on the 28th is 12° (Rii20–23). After that, 1 day is subtracted from the date, and the daily displacement for the 27th, 11;42°, is added to the previous result (Rii23–24). The daily motion of the Sun is said to be 0;57,56° in Ari in agreement with the zigzag function defined in P10', since 0;57,56 = 0;55,32 + 2 · 0;1,12. In the example, $v_{\odot}$ is not assumed to vary from the 28th to the 27th (Rii27–30), unlike $v_{☽}$. The instruction to 'let the Moon/Sun proceed westwards', i.e. with the daily rotation, reflects the fact that these displacements point backwards in time, so that they are opposite to the forward motion along the ecliptic.[651] In P17' step 2 was probably contained in the damaged line Rii4. As explained in §4.4.19.2, $v_{☽}$ varies by 0;18° d^{-1} per day, and $v_{\odot}$ varies according to the same zigzag function known from P10'.

Step 3. Elongation of the Moon, η (§4.4.19.3)

In P17' step 3 might have been mentioned in the gap at the beginning of Rii4, because what follows appears to be step 9, which is the only step that depends on η. In P18' line Rii28 is preserved well enough to conclude that it contains a numerical example of step 3. Its outcome can be restored in the gap at the beginning of Rii29. The 'time' mentioned in Rii28 might be the corresponding time interval, which would be 1 day + 5^{b} = 17^{b}.

Step 4. Zodiacal positions of Moon, $B'_{☽}$, and Sun, $B'_{\odot}$ (§4.4.19.4)

In P17' step 4 seems to be alluded to in Rii3, but no algorithm is provided. Step 4 is in any case implied by P17', because in Rii7–10 q is computed separately for the Moon and the Sun, which makes sense only if their zodiacal positions are different due to the execution of step 4. In P18' step 4 is alluded to in Rii34–35, where one is merely instructed to compute the positions of the Sun and the Moon, in this unusual order, 'as (was done) for NA_1'.

Step 5. Coefficients for the zodiac q_{rise}, q_{set} (§4.4.19.5)

P17' includes an auxiliary table (Rii12–13) with 12 control values of the coefficient q_{set}, each pertaining to 10° of a zodiacal sign. P18' might contain an allusion to step 5 in Rii33, but the expected table for q_{rise} is not preserved (or omitted).[652] After evaluating the coefficient at the position of the Moon and the Sun, both are averaged, but this operation is preserved only in P17'.

The formulation of step 5 in P17' mixes numerical examples with general rules. While the extrapolation proceeds under the assumption that the positions of the Moon and the Sun are between 10° Ari and 10° Tau (Rii7–9), this is followed by a general interpolation rule (Rii9–10). Also the averaging of the coefficients for the Moon and the Sun is presented as a general operation, without numerical examples (Rii10).

Step 6. The Moon's distance to the ecliptic, β' (§4.4.19.6).

The computation of β' is not preserved, but this function is referred to in P17' (Rii6) and P18' (Rii29,39). However, there is no indication as to how it is obtained. In P17' step 6 might have been alluded to in the damaged lines preceding Rii6, but there is no room for a complete instruction along the lines of No. **61**. It can therefore not be excluded that step 6 is absent, i.e. no correction was applied for the Moon's displacement between New Moon and the Lunar Six event. If so, then β' in step 9 is replaced by β (the value at lunation). However, the gaps in P17',P18' prevent a definite conclusion about this issue.

Step 7. Coefficients for the Moon's 'height and depth', r_{rise} and r_{set} (§4.4.19.7)

P17' includes an auxiliary table (Rii14–15) with 12 control values of the zigzag function for r_{set}, each pertaining to 10° of a zodiacal sign (Fig. 4.35, top). A corresponding table for r_{rise}, which should equal r_{set} except for a shift of 6 zodiacal signs, is not preserved in P18'. The extrapolation alluded to in Rii5–6 must be connected to step 7 (for q it is mentioned in Rii7). This rule is specific in that the control point is in Ari, but general in the sense that it does not identify a position within Ari. The coefficient is presumably computed separately for the Moon and the Sun, and averaged. A corresponding

[651] Wrongly translated as 'before sunset' in *ACT*, p. 209.

[652] In principle there is enough room in the gap in Rii31–33. The values of q_{rise} should equal those of q_{set} except for a shift by 6 zodiacal signs.

instruction is not preserved.[653] There is no room in the gap in Rii6 for the averaging, but also in P18' the averaging is missing from step 7. It is unclear how to interpret this state of affairs. Probably r is meant to be computed for both the Moon and the Sun, in which case they must also have been averaged. But it cannot be completely ruled out that only the coefficient for the Moon was intended to be multiplied by the Moon's distance to the ecliptic in step 8.

Step 8. R: contribution to the Lunar Six from the Moon's distance to the ecliptic (§4.4.19.8)

Step 8 is readily identified in P18'.a (Rii29), less easily in P18'.b (Rii38–39). The latter passage is without parallel, but appears to be concerned with steps 8 and 9; cf. §4.4.19.8. In Rii39 (step 10) the 'height and depth' must also be a reference to R. In P17' R is alluded to in Rii6 (step 10).

Step 9. Q: contribution to the Lunar Six from lunar elongation (§4.4.19.9)

Step 9 (§4.4.19.9) appears to be mentioned twice in P17' (Rii4 and 10–11). The formulation of step 9 in P18' — if this interpretation is correct — is without parallel (Rii38–39).

Step 10. Disk correction Δ (§4.4.19.10)

The disk correction is mentioned only in P17' (Rii15), where 2° 'for the disk' are subtracted from NA_1.

Step 11. Computation of the Lunar Six interval (§4.4.19.11)

In P17' step 11 is represented by an incomplete instruction (Rii6) calling for an unnamed quantity, apparently R, to be subtracted from or added to the 'elongation'. This makes sense only if 'elongation' is interpreted as Q. This only foreshadows what is mentioned in subsequent lines, because most of the involved quantities have not yet been computed. No name is assigned to the outcome, but in Rii11 Q is called NA_1 (**na**). Since NA_1 later becomes $Q \pm R$, this only makes sense as a provisional result. However, the latter operation was already alluded to in Rii6, so that it appears that the scribe has mixed up the instructions.

In P18' step 11 is alluded to in Rii39. No reference is made to the preceding numerical example. However, the instruction 'you subtract the 'height and depth'' is at best an abbreviated representation of the correct and full procedure, since one expects $R_\uparrow$ to be added to Q and $R_\downarrow$ to be subtracted and not both to be subtracted.[654] Presumably, step 11 is also mentioned in the first part of P18', perhaps in the damaged lines Rii30–33.

As explained in §4.4.19.11, the sunrise immediately preceding New Moon is discarded as a possible moment for KUR in P18' (step 2), since the positions of the Moon and the Sun were updated by adding the displacements for 1 day before continuing with step 3.

Step 13. The Lunar Six event shifts by 1 day (§4.4.19.13)

In P17' step 13 is referred to in line Rii11, where it is stated that 'if the Moon is insufficient for being visible' one has to 'add the displacement of the Moon for 1 day to it and declare (the month) full'. Hence there is no instruction to repeat steps 1–10 for the next day in order to compute the actual value of NA_1 and confirm that it is sufficient.[655]

P17'.a: Establishing the date of New Moon

Part **a** at the end of P17' (Rii15–16) contains rules for establishing the date of New Moon (§4.4.18).[656]

[653] It might have been mentioned in the gap in Rii5 preceding the extrapolation, as was done in Rii7–8 (step 5).

[654] Both R and the Moon's distance to the ecliptic (β) are referred to as 'height and depth'. The term 'DIŠ.UD of the 'height and depth'' which denotes R in step 11 of No. **61** is absent.

[655] This appears to contrast with the practice attested in the synodic tables, since they always contain the actual value of NA_1 for day 1 of the new month, also if the previous month is hollow.

[656] Note that in *ACT*, pp. 206–207, Neugebauer has exchanged 'transgressed' and 'not transgressed' in his cases C and D (Rii14 in his line count).

No. 54

museum number	Rm 839
provenience	Babylon
measures	4.2(*) × 3.4(*) × 2.0 cm
cuneiform text	photo: Fig. F.56; *ACT*, Pl. 236
bibliography	*ACT* 200e
contents	Moon system A: computations involving *E*

Side X

P1
1 [... 1.5]8.45.42 **tab**
2 [...xxx] *ša*$_2$ **nim** *u* **sig** *ša*$_2$ **iti**
3 [...xxxx]⸢x⸣ **ki-***šu*$_2$ **kur**
4 [...xxxxx]⸢3? 44.46⸣
5 [...xxxxxx] 38
(unknown number of lines missing)

Side X

P1 **(Computations involving *E*; unclear)**

[1][...] you add [1,5]8;45,42. [2][...] for the 'height and depth', per month. [3][...] ... it reaches its position [4][...]⸢3? 44.46⸣ [5][...] 38
(*unknown number of lines missing*)

Critical and philological notes

Side X

3 Before **ki-***šu*$_2$ there is a winkelhaken and a vertical wedge, perhaps belonging to 11.
4 The initial 3 might also be a digit 4–8.
44.46: replaces 30,3,33 (ACT).
5 38: written on the right edge.

Commentary

This small fragment from the upper right corner of a tablet is inscribed on one side; the other side is destroyed. The size of the tablet could not be reconstructed. The handwriting is unusually large.

P1 Computations involving E; unclear

Only 5 lines of this procedure are partly preserved. The parameters and the term 'height and depth' point to the Moon's distance to the ecliptic, *E* (§4.4.6), but the purpose of the algorithm could not be established. The number in X1 can be restored as 1,58;45,42, the preliminary monthly difference of *E* in zone 2. The digits 44.46 are suggestive of $\Delta\Psi'$ of lunar system B, whose minimum m = 44;46,55,...[f], but this identification remains speculative.

No. 55

museum number	BM 34833 (Sp2,333)
provenience	Babylon
measures	6.0(7) × 5.5(10.5?) × 1.5–2.2 cm (type F?)
arrangement	O/R
cuneiform text	photo: Fig. F.58; *ACT*, Pl. 239; copy: *LBAT* 84
bibliography	*ACT* 200a; P2: Aaboe & Henderson (1975)
contents	Moon system A: procedures for Φ, B^{nm}, C^{nm}

		Obverse
P1	1	[xxxxxxxxxx 2.45.55.3]⸢3⸣.20 ⸢**tab**⸣ [*u* **la**$_2$]
	2	[xxxxxxx 1].⸢5⸣7.47.⸢5⸣[7.46.40 xx]
	3	[*ša*$_2$ *al-la* 2.17.4.48.53.20] ⸢**gal**⸣-*u*$_2$ **ta** 4.⸢3⸣[4.9.37.46.40]
	4	[**nim**-*ma* **gar**-*an* *ša*$_2$ *al-la* 1.57.47].⸢5⸣7.46.40 **tur**-⸢*er*⸣
	5	[**ta** 3.55.35.55.3]⸢3⸣.20 **nim**-*ma* **gar**-*an*
	6	[xxxx] ⸢x⸣ *ana* 15 **ab**$_2$ 9.15.33.20 **mu.du**
P2	7	[*lu-maš*.**meš ta**] 27 *zib*.**me en** 13 **absin** 28.7.30 **tab**
	8	[*ša*$_2$ *al-la* 13 **absin**] **diri a.ra**$_2$ 1.4 **du ki** 13 **absin tab**-*ma* **gar**-*an*
	9	[**ta** 13 **absin en** 27 *zib*].⸢**me ab**$_2$⸣ *ana* **ab**$_2$ 30 **tab** *ša*$_2$ *al-la* 27 *zib*.**me diri**
	10	[**a.ra**$_2$ 56.15 **du ki** 27 *zib*].⸢**me tab**⸣ **ta** 27 *zib*.**me en** 13 **absin**
	11	[*ana* 12 **ab**$_2$ 10.22.30 **ta** 13 **absin en** 27] ⸢*zib*.**me** 11⸣.4 **mu.du**
P3'		(at least 10 lines missing)
		Reverse
		(at least 10 lines missing)
	1'	[xxxxxx] ⸢xx 10⸣ [**gu** 2].⸢24⸣ [**me** xxx]
	2'	[*ša*$_2$ *al-la* 10 **gu**] ⸢**diri a**⸣.[**ra**$_2$] ⸢24 **du ki**⸣ 2.28 **t**[**ab**-*ma* **gar**-*an*]
	3'	[10 *zib*.**m**]**e** 2.40 **me** *ša*$_2$ *al-la* 10 *zib*.**me d**[**iri**]
	4'	[**a.ra**$_2$ 40 **du k**]**i** 2.40 **tab**-*ma* **gar**-*a*[*n*]

Critical and philological notes

Obverse

3 **gal**: replaces **tab** (*ACT*); only the final horizontal wedge is visible.
6 **mu.du**: written on the right edge.
9–11 The final 1–2 signs of these lines are written on the right edge.
11 The signs near the left broken edge are compatible with *zib*.**me**, and *ana* 12 **ab**$_2$ must be restored at the beginning of the line, not here as in *ACT*. Below the line a short section of a horizontal line is visible.

Reverse

2' **du**: copied as GIŠ by Pinches.

Commentary

This small fragment belongs to the right upper (lower) corner of a tablet. The textual restorations imply that the original width was about 7 cm. From the estimated number of lines missing from P3' it follows that the height of the tablet was at least about 10.5 cm. The sequence of procedures on the obverse suggests that no further procedures were present between P2 and P3', and that there was no additional column. Since the lower half of the reverse is not inscribed it appears that the tablet had no colophon.

P1 Updating Φ as zigzag function; net differences

Part **a** (O1–5) is a procedure for updating Φ as a zigzag function (cf. §4.4.3). The digits in O3 and O5 can be identified as belonging to $2M$ and $2m$, respectively, which points to template ZZ.B.1 (cf. §2.6.8.1). Part **b** (O6) is concerned with net differences of Φ. Only $d\Phi(14)$, the net difference for 14 months (Table 4.7a), is preserved (the text associates it with 15 months, presumably by mistake).

Obverse

P1 **(*Φ*)**

a **(Updating *Φ* as a zigzag function)**

1[... 2;45,55,3]3,20 you add [and subtract.] 2[... 1],57;47,57,4[6,40 ...]
3[That which] is larger [than 2,17;4,48,53,20 you deduct] from 4,[34;9,37,46,40 and put down.]
4[That which is smaller than 1,57;47],57,46,40 5you deduct [from 3,55;35,55,3]3,20 and put down.

b **(Net differences)**

6[...] for 15 (error for 14) months it proceeds 0;9,15,33,20.

P2 **(*B*)**

---

a **(Updating *B* with the step function for *σ*)**

7[Zodiacal positions. From] 27 Psc until 13 Vir you add 28;7,30. 8[(The amount) by which it] exceeds [13 Vir] you multiply by 1;4, add to 13 Vir and put down. 9[From 13 Vir until 27 Ps]c you add 30 month by month. (The amount) by which it exceeds 27 Psc 10[you multiply by 0;56,15], ⸢add⸣ [to 27]⸢Psc⸣.

b **(Net displacements for 12^m)**

From 27 Psc until 13 Vir 11[for 12 months 10;22,30, from 13 Vir until 27 Psc for 12] months 11;4 it proceeds.

P3' **(Computing C^{nm} from B^{nm})**

(*at least 10 lines missing*)

Reverse

(*at least 10 lines missing*)

1'[...] ... [...]

10 [Aqr]: ⸢2,28⸣ [is the daylight.] 2'[(The amount) by which it] exceeds [10 Aqr] you multiply by 0;24, add with 2,28 [and put down].

3'[10 Ps]c: 2,40 is the daylight. (The amount) by which it ex[ceeds] 10 Psc 4'[you multiply by 0;40], add with 2,40 and put down.

P2 Updating B^{nm} with the step function for σ; net displacements for 12^m

Part **a** is a procedure for updating B^{nm} with the step function for σ (§4.4.4). It is formulated in accordance with template STEP.A.1 (§2.6.10.1). Part **b** mentions net displacements of the Moon for 12^m, $dB_j(12)$, for both zones of system A. The formulation is based on template DIFF.A.3 (§2.6.10.4).

P3' Computing C^{nm} from B^{nm}

This procedure provides interpolation rules for computing the length of daylight (*C*) from B^{nm} (§4.4.5). The formulation agrees with templates IP.A.1–2 (§2.6.7). Some of the missing text has been restored accordingly. Only the final two entries (Aqr and Psc) are partly preserved. If one assumes that P3' contained entries for all 12 zodiacal signs then about 20 lines must be assumed missing between O11 and R1'. If divided equally between obverse and reverse, they would occupy about 3 cm of clay.

No. 56

museum number	BM 33631 (Rm4,187)
provenience	Babylon
measures	6.9(8) × 12.0(13?) × 2.3–2.8 cm (type F)
arrangement	O/R
cuneiform text	photo: Fig. F.59; *ACT*, Pl. 236
bibliography	*ACT* 200b
contents	Moon system A: Φ, B^{nm}, C^{nm}, E^{nm}

		Obverse
P1	0	[xxxxxxxxxxxxxxxxxxx]
	1	⸢2.13⸣.[20$^?$ xxxxxxxxxxxxxx]
	2	**en** 2.1⸢7⸣.[4.48.5.20 xxxxxxxxx *ša$_2$ al-la*]
	3	2.17.4.48.5⸢3⸣.[20 x **ta** 4.34.9.37.46.40 xxx]
	4	**en** 1.57.47.5[7.46.40 xx *ša$_2$ al-la* 1.57.47.57.46.40 x]
	5	**ta** 3.⸢55⸣.3⸢5⸣.[55.33.20 xxxxxxxxx]
	6	⸢xx 2$^?$ 10$^?$ 3 x 5 x⸣ [xxxxxxxxxxxx]
P2	7	⸢**te**$^?$⸣ *lu-maš ana* **du$_3$**-*ka* ⸢**ta**⸣ [27 *zib*.**me en** 13 **absin** 28.7.30 **tab**]
	8	*ša$_2$ al-la* ⸢13 **absin**⸣ **diri** [13 **absin ta ša$_3$**-*šu$_2$* **nim a.ra$_2$** 1.4 **du**]
	9	**ki** 13 ⸢**absin**⸣ **tab**-⸢*ma*⸣ **gar**-*an* ⸢**ta**⸣ [13 **absin en** 27 *zib*.**me** 30 **tab**]
	10	⸢*ša$_2$ al-la* 27 *zib*.**me diri** 27 *zib*.**me ta**⸣ [**ša$_3$**-*šu$_2$* **nim a.ra$_2$** 56.15 **du**]
	11	⸢**ki** 27 *zib*.**me tab**-*ma* **gar**-*an* xxx⸣ **ta** 27 *zib*.⸢**me**⸣ [**en** 13 **absin** xxxx]
P3	12	*ana tar-ṣa* (erasure?) ⸢10 **ḫun** 3 **me** *ša$_2$ al*⸣-[*la* 10 **ḫun diri** GAM 40 **du**-*ma*]
	13	[**ki**] ⸢3 **ḫun tab**⸣-[*ma*]
	14	*ana tar-ṣa* 10 **mul$_2$** ⸢3.20⸣ [**me** *ša$_2$ al*]-⸢*la* 10 **mul$_2$**$^?$⸣ [**diri** GAM 24 **du**-*ma*]
	15	⸢**ki**⸣ [x$^?$] ⸢3.20⸣ **mul$_2$** [**tab**-*ma*]
	16	*ana tar-ṣa* 10 **maš.maš** 3.⸢32⸣ [**me** *ša$_2$*] ⸢*al*⸣ 10 ⸢**maš**⸣.[**maš diri** GAM 8 **du**-*ma*]
	17	**ki** ⸢3⸣.[32 **maš.maš tab**-*ma*]
	18	*ana tar-ṣa* 10 **alla** 3.36 **me** *š*[*a$_2$ al-la* 10 **alla diri** GAM 8 **du**-*ma*]
	19	**ta** 3.[36 **alla nim**-*ma*]
	20	*ana tar-ṣa* 10 **a** 3.32 **me** *ša$_2$ al*-[*la* 10 **a diri** GAM 24 **du**-*ma*]
	21	**ta** 3.3⸢2⸣ [**a nim**-*ma*]
	22	*ana tar-ṣa* 10 **absin** 3.20 **me** *ša$_2$ al-la* [10 **absin diri** GAM 40 **du**-*ma*]
	23	**ta** 3.2[0 **absin nim**-*ma*]
	24	[*ana tar*]-*ṣa* 10 **rin$_2$** 3 **me** *ša$_2$ al-la* ⸢10 **rin$_2$**$^?$⸣ [**diri** GAM 40 **du**-*ma*]
	25	⸢**ta**⸣ 3 **rin$_2$** ⸢**nim**⸣-[*ma*]
		Reverse
	1	⸢*ana tar-ṣa* 10 **gir$_2$**$^?$.**tab** 2.40 **me** *ša$_2$ al-la* ⸢10 **gir$_2$**$^?$.**tab diri** GAM 2[4 **du**-*ma*]
	2	**ta** 2.40 **gir$_2$.tab nim**-*ma*
	3	*ana tar-ṣa* 10 **pa** 2.28 **me** *ša$_2$ al-la* 10 **pa diri** GAM [8 **du**-*ma*]
	4	**ta** 2.28 **pa nim**-*ma*
	5	*ana tar-ṣa* 10 **maš$_2$** 2.24 **me** *ša$_2$ al-la* 10 **maš$_2$** [**diri** GAM 8 **du**-*ma*]
	6	**ki** 2.24 **maš$_2$ tab**-*ma*
	7	*ana tar-ṣa* 10 **gu** 2.28 **me** *ša$_2$ al-la* 1[0 **gu diri** GAM 24 **du**-*ma*]
	8	**ki** 2.28 **gu tab**-*ma*
	9	*ana tar-ṣa* 10 *zib*.**me** 2.40 **me** *ša$_2$ al*-[*la* 10 *zib*.**me diri** GAM 40 **du**-*ma*]
	10	**ki** 2.⸢40⸣ [*zib*.**me tab**]
P4	11	⸢**mu** *ana*$^?$ **mu**$^?$⸣ [**ḫun**] ⸢*ana*⸣ [**ḫun**] 6.55 **m**[**ul$_2$** *ana* **mul$_2$** 4.9]
	12	⸢**maš** *ana* **maš**⸣ 1.23 **alla** *ana* **a**[**lla** 1.23]
	13	**a** *ana* **a** 4.9 **absin** *ana* **ab**[**sin** 6.55]
	14	**rin$_2$** *ana* **rin$_2$** 7.22.40 **g**[**ir$_2$.tab** *ana* **gir$_2$.tab** 4.25.36]
	15	⸢**pa**⸣ *ana* **pa** 1.28.3[2 **maš$_2$** *ana* **maš$_2$** 1.28.32]
	16	**gu** *ana* **gu** 4.25.36 *z*[*ib*.**me** *ana zib*.**me** 7.22.40]

Obverse

P1 (Updating Φ as a zigzag function)

[0][...] [1]⸢2,13⸣;[20? ... You add and subtract 2;45,55,33,20.]
[2]Until 2,1⸢7;[4,48,5,20 ... That which is less than] [3]2,17;4,48,5[3,20 you subtract from 4,34;9,37,46,40 ...]
[4]Until 1,57;47,5[7,46,40 ... That which exceeds 1,57;47,57,46,40 you subtract] [5]from 3;⸢55⸣,3⸢5⸣,[55,30,20...]
[6]⸢... 2? 10? 3 ... 5 ...⸣ [...]

P2 (B)

a (Updating B^{nm} with the step function for σ)

[7]In order for you to construct the ⸢approach?⸣ of a zodiacal sign.
⸢From⸣ [27 Psc until 13 Vir you add 28;7,30.] [8]That which exceeds ⸢13 Vir⸣: [you deduct 13 Vir from it, you multiply it by 1;4], [9]you add it with 13 ⸢Vir and⸣ you put it down. ⸢From⸣ [13 Vir until 27 Psc you add 30]. [10]That which exceeds 27 Psc: [you deduct] 27 Psc ⸢from⸣ [it, you multiply it by 0;56,15], [11]⸢you add it with 27 Psc⸣.

b (Net displacements?)

⸢...⸣ From 27 Ps[c until 13 Vir ...]

P3 (Computing C^{nm} from B^{nm})

[12]Opposite [...?] ⸢10 Ari 3,0, the daylight. (The amount) by which it [exceeds 10 Ari you multiply by 0;40 and] [13]⸢add⸣ [to] ⸢3,0 of Ari.

[14]Opposite 10 Tau ⸢3,20⸣, [the daylight. (The amount) by which it exceeds] ⸢10 Tau⸣ [you multiply by 0;24 and] [15][add to 3],20 of Tau.

[16]Opposite 10 Gem 3,⸢32⸣, [the daylight. (The amount) by which it exceeds] ⸢10 Gem⸣ [you multiply by 0;8 and] [17][add] to 3,[32 of Gem].

[18]Opposite 10 Cnc 3,36, the daylight. (The amount) by which it exceeds 10 Cnc you multiply by 0;8 and [19][deduct] from 3,[36 of Cnc.

[20]Opposite 10 Leo 3,32, the daylight. (The amount) by which it [exceeds 10 Leo you multiply by 0;24 and] [21][deduct] from 3,32 of [Leo].

[22]Opposite 10 Vir 3,20, the daylight. The amount by which it [exceeds 10 Vir you multiply by 0;40 and] [23][deduct] from 3,20 of [Vir].

[24]Opposite 10 Lib 3,0, the daylight. The amount by which it [exceeds] 10 Lib [you multiply by 0;40 and] [25]deduct from 3,0 of Lib.

Reverse

[1]Opposite 10 Sco 2,40, the daylight. The amount by which it exceeds 10 Sco you multiply by 0;2[4 and] [2]deduct from 2,40 of Sco.

[3]Opposite 10 Sgr 2,28, the daylight. The amount by which it exceeds 10 Sgr you multiply by [0;8 and] [4]deduct from 2,28 of Sgr.

[5]Opposite 10 Cap 2,24, the daylight. The amount by which it [exceeds 10 Cap you multiply by 0;8 and] [6]deduct from 2,24 of Cap.

[7]Opposite 10 Aqr 2,28, the daylight. The amount by which it [exceeds 10 Aqr you multiply by 0;24 and] [8]add to 2,28 of Aqr.

[9]Opposite 10 Psc 2,40, the daylight. The amount by which it [exceeds 10 Psc you multiply by 0;40 and] [10][add] to 2,40 of [Psc].

P4 (Net differences of C^{nm} for 12^{m})

[11]Year ⸢to year?⸣ [Ari] ⸢to⸣ [Ari] 6;55, Ta[u to Tau 4;9], [12]⸢Gem to Gem⸣ 1;23, Cnc to C[nc 1;23], [13]Leo to Leo 4;9, Vir to Vi[r 6;55], [14]Lib to Lib 7;22,40, Sc[o to Sco 4;25,36], [15]⸢Sgr⸣ to Sgr 1;28,3[2], [Cap to Cap 1;28,32], [16]Aqr to Aqr 4;25,36, P[sc to Psc 7;22,40].

P5 17 **nim** *u* **sig** *ana* **du$_3$**-*ka* **ta** 2[7 *zib*.**me en** 13 **absin**]
18 1.58.45.42 **tab** *u* **la$_2$** [*ša$_2$ al-la* 13 **absin diri**]
19 GAM 15 **du ki** 1.58.4[5.42 **tab**-*ma* **gar**-*an*]
20 *ki-i* **tab tab** *ki-i* **la$_2$ la$_2$ ta** 1[3 **absin en** 27 *zib*.**me**]
21 ⸢2.6.15.42⸣ **tab** *u* **la$_2$** ⸢*ša$_2$*⸣ [*al-la* 27 *zib*.**me diri**]
Lower edge
22 [GAM 16 **du ta** 2.6.1]⸢5⸣.42 [**la$_2$**-*ma* **gar**-*an*]
23 [*ki-i* **tab tab** *ki-i* **la$_2$ la$_2$** x$^?$] ⸢xx⸣ [xxxxxxx]

Critical and philological notes

Obverse

1 ⸢2.13⸣: probably the beginning of 2.13.20; replaces **a.ra$_2$** (*ACT*).

5 3.⸢55⸣.3⸢5⸣[55.33.20...]: replaces 2.45.55 (*ACT*).

6–11 These lines were not transliterated in *ACT*, except for isolated signs, presumably because they are unclear on the available photograph (*ACT*, Pl. 236).

6 This line is heavily worn. The first sign looks like EN or MU, the second like ḪUN. The sign after 13 looks like **gin$_7$**.

7 **te** = *ṭeḫû*, 'approach': replaces EN (*ACT*). Or read **kur** = *kašādu*, 'approach'?

11 Neugebauer erroneously restored m_Φ at the end of the line. The sign before **ta** looks like **nim** or **kin**.

12 Between *ana tar-ṣa* and 10 **ḫun** there is space for about 5 signs, but nothing is expected. The surface is damaged here; perhaps there is an erasure.

15 There is an empty space between **ki** and 3.20.

Reverse

11 At the beginning of this heavily damaged line there is a clear **mu** followed by traces compatible with **meš** (rather than **me**).

17 The 20 which follows **ta** contains a small horizontal wedge at the bottom, a common feature in LB astronomical texts.

Lower edge

23 Traces of a few signs are visible, perhaps including a winkelhaken.

Commentary

This fragment preserves sections of the left, upper and lower (upper) edges of the obverse (reverse) of a tablet. The right edge is completely broken off. Since the text on the reverse is a continuation of the obverse, there is no doubt that the tablet has only one column. The textual restorations imply that about 1 cm is missing on the right, so the original width was about 8 cm. Traces of what might be a colophon are preserved on the upper edge of the obverse. The five partly preserved procedures are concerned with Φ, B^{nm}, C^{nm} and E^{nm}. Their order agrees with that of the corresponding columns of a synodic table.

P1 Updating Φ as a zigzag function

The remains of this procedure point to Φ (§4.4.3). The number in O5' can be reconstructed as $2m$, where m is the minimum of Φ. This points to reflection rules that are formulated in terms of $2m$ and $2M$, suggesting that P1 is concerned with updating Φ as a zigzag function in accordance with template ZZ.B.1. The textual restorations are based on that interpretation. O6 is partly illegible and defies interpretation. Perhaps it mentions net differences of Φ.

P2 Procedures for B^{nm}

The topic of P2 remained unidentified in *ACT*, probably due to the low quality of the available photograph (Pl. 236). Part **a** is a procedure for updating B^{nm} with the step function for σ (§4.4.4). It is formulated in accordance with template STEP.A.2 (§2.6.10.1). If correctly read, the purpose of P2 is described in terms of 'approaching(?)' the zodiacal signs, a term not used elsewhere in this connection. The duplicates (cf. §4.4.4) suggest that part **b** (O11') is concerned with net displacements, perhaps for 12^m.

P3 Computing C^{nm} from B^{nm}

P3 contains interpolation rules for computing C from B^{nm} (§4.4.5). They are formulated in accordance with templates IP.A.1–2.

P5 (Updating E^{nm}: step 1)

[17]In order for you to construct the 'height and depth'.
From 2[7 Psc until 13 Vir] [18]you add and subtract 1,58;45,42. [(The amount) by which it exceeds 13 Vir] [19]you multiply by 0;15, add to 1,58;4[5,42 and put down.] [20]If it is increasing you add, if it is decreasing you subtract.
From [13 Vir until 27 Psc] [21]you add and subtract 2,⌜6;15,42⌝. (The amount) ⌜by which⌝ [it exceeds 27 Psc]
Lower edge
[22][you multiply by 0;16, subtract from 2,6;1]5,42 [and put down.] [23][If it is increasing you add, if it is decreasing you subtract. ...] ⌜...⌝ [...]

P4 *Net differences of C^{nm} for 12^m*

P4 is concerned with net differences of C^{nm} for 12^m (cf. $dC_k(12)$ in Table 4.11, p. 131); hence the phrase 'from year to year'. For an explanation of these numbers cf. §4.4.5. The additive or subtractive nature of the differences is not indicated. Hence P4 is of little practical use for updating C across 12^m. Most likely it is meant to be used for verifying the correctness of the values in column C.

P5 *Updating E^{nm}: step 1*

As stated in the introductory phrase, P5 concerns the Moon's distance to the ecliptic at New Moon (E^{nm}). In fact P5 contains only some parts of the algorithm for updating E^{nm}. Step 1 is provided for both zones. The formulation is based on template ESTEP.A (§4.4.6). Step 3 (reflection rule at the extrema) may have been mentioned in the break at the end of R23, but step 2 (concerning the nodal region) appears to be omitted.

No. 57

museum numbers	BM 34245 (Sp,352) +41608 (81–6–25,223)
provenience	Babylon
measures	6.8 × 5.0(*) × 2.0 cm (type G?)
arrangement	O/R
cuneiform text	photo: Fig. F.60; *ACT*, Pl. 235; copy: *LBAT* 85–86
bibliography	*ACT* 200c
contents	Moon system A: updating *E*

Upper edge

0 *ina a-mat* ^{d}**en** *u* ^{d}**gašan**-*ia*$_2$ *liš-lim*

Obverse

P1 1 [**nim**] *u* ⸢**sig** *ša*$_2$ **ab**$_2$?⸣ *ana* ⸢**ab**$_2$?⸣ *ana* ⸢**du**$_3$?⸣-*ka* ⸢**ta**⸣ 2⸢7 *zib*⸣.**me en**

2 13 **absin ab**$_2$ *ana* **ab**$_2$ 1.58.⸢4⸣[5].42 **tab** *u* ⸢**la**$_2$?⸣ *ša*$_2$ *al*

3 13 **absin diri** 13 **absin** [t]a **ša**$_3$-*šu*$_2$ **nim** *ša*$_2$ **tag**$_4$

4 GAM 15 **du ki** 1.58.[45].42 **tab** *ki-mu-u*$_2$

5 **igi**-*u*$_2$ *ki-i* **tab tab** *ki*-⸢*i* **la**$_2$?⸣ **la**$_2$ **ta** 13 **absin**

6 **en** 27 *zib*.**me ab**$_2$ *ana* **ab**$_2$ ⸢2⸣.6.15.42 **tab** *u* **la**$_2$

7 *ša*$_2$ *al* 27 *zib*.**me diri** 2⸢7 *zib*⸣.**me** ⸢**ta**⸣ **ša**$_3$-*šu*$_2$ **nim**

8 *ša*$_2$ **tag**$_4$ GAM 16 **du ta** 2.6.15.42 ⸢**la**$_2$?⸣

9 *ki-mu* **igi**-*u*$_2$ *ki* **tab tab** *ki* **la**$_2$ **la**$_2$ **en** ⸢7.12⸣ [**nim** x]

(n lines missing)

Reverse

(about n lines missing)

1' ⸢xx 3?7 x⸣ [xxxxxxxxxxxxx]

Obverse

[0]By the command of Bēl and Bēltija may it remain intact.

P1 **(Updating *E*)**

[1]In order for you to construct the ['height'] and ⸣'depth', month⸣ by ⸢month⸣.

a **(Zone 2)**

⸢From 27⸣ Psc until [2]13 Vir, month by month you add and subtract 1,58;⸢4⸣[5],42. That which exceeds [3]13 Vir: you deduct 13 Vir from it; what remains [4]you multiply by 0;15, add to 1,58;[45],42. As [5]before, you add if it is increasing, subtract if it is decreasing.

b **(Zone 1)**

From 13 Vir [6]until 27 Psc month by month you add and subtract ⸢2⸣,6;15,42. [7]That which exceeds 27 Psc: you deduct 27 Psc from it; [8]what remains you multiply by 0;16, you deduct from 2,6;15,42. [9]As before, you add if it is increasing, subtract if it is decreasing, until [it is high] by ⸢7,12⸣ [...]

(*n lines missing*)

Reverse

c **(Subject unclear)**

(*about n lines missing*)

[1']⸢... 3?7 ...⸣ [...]

Critical and philological notes

Obverse

6 ⸢2⸣.6.15.42: traces of the 2 are visible.

9 7.12: replaces 5... (*ACT*).

Reverse

1' The first sign could be 9 (= GAM$_0$) or GAM. This is followed by the lower half of a vertical wedge, perhaps 1.

Commentary

The tablet is preserved to full width. Nothing remains of the lower (upper) part of the obverse (reverse). On the reverse, a single line with traces of numbers is written near the top, below which the tablet is uninscribed. It therefore appears that there was no colophon. The obverse contains a single procedure. It is not clear whether the single line on the reverse belongs to the same procedure.

P1 Updating E

As stated in O1, the purpose of P1 is to update the Moon's distance to the ecliptic from month to month. For a discussion of the algorithm cf. §4.4.6. The formulation is based on template ESTEP.B. Parts **a** and **b** are concerned with zones 2 and

1, respectively. At the end of part **b** the beginning of step 3 is preserved. Step 2, which deals with the case when the Moon enters or leaves the nodal region, is not preserved; it may have been present below O9.

No. 58

museum number	BM 34148 (Sp,250)
provenience	Babylon
measures	4.9(10?) × 5.9(*) × 2.5–3.0 cm
arrangement	O/R
cuneiform text	photo: Fig. F.61; *ACT*, Pl. 236; copy: *LBAT* 87
bibliography	*ACT* 200d
content	Moon system A: procedures for *B* (?) and *E*

Obverse

P1

1 [xxxxx] ⌜**zi**⌝.**meš** 3.45 **a.ra**$_2$ 4 *igi-gub-*⌜*u*$_2$⌝ [x]
2 [xxx] ⌜15$^?$⌝ **la**$_2$ 15 16-*u*$_2$ 4 *igi-gub-e* ^{d}[xx]
3 ⌜*šu*$^?$⌝-*u*$_2$ **ta** *lib*$_3$*-bi* 27 *zib*.**me en** 27 [xxx]
4 *ša*$_2$ ⌜*u*$_4$⌝-*mu* ^{d}**utu** 15-*u*$_2$ *ša*$_2$ 1 **zi**-*šu*$_2$ ⌜x⌝ [xx]
5 [1]⌜5⌝-*u*$_2$ *ša*$_2$ ḪAB-*rat-ti-šu*$_2$ **ta** 13 **a**[**bsin** xx]
6 [**mu**] ⌜**nu**⌝ **zu**-*u*$_2$ 4 *igi-gub-e* ⌜**igi**$^?$ **ta**⌝ [xxx]
7 [xxx]⌜x⌝ **a.ra**$_2$ 3 6 *lu-maš* **d**[**u** xxx]
8 [xx]-⌜*uš*$^?$⌝ **gar** *ma-la* ^{d}**utu** [xxxx]
9 [xxx]-*ma* 13 **zi** ^{d}**u**[**tu** xxxxx]
10 [xxx] 11 **zi** ^{d}**utu** ⌜x⌝[xxxxxx]
11 [xxx **g**]**ar.gar**-*ma* 14 **z**[**i** xxxxxx]
12 [xxx] 16 *u*$_4$*-mu* [xxxxxxx]
13 [xxxx 2]7 *zib*.**m**[**e** xxxxxxxxx]

(n lines and perhaps 1 column missing)

Reverse

P2'

(n lines and perhaps 1 column missing)

1' [xxxxx].⌜51⌝ GAM 20 **du** *ana* 1⌜4$^?$⌝.10 ⌜x⌝ [xxxx]
2' [x] ⌜*ana* **iti**$^?$⌝ **nim**-*u*$_2$ *u* **sig e**$_{11}$-*ma re*-[*ḫi*]
3' [1.58].45.42 **ta** 27 *zib*.**me en** 1⌜3⌝ [**absin**]
4' [**iti** *ana*] **iti** 1.58.45.42 **tab** *u* ⌜**la**$_2$⌝
5' {⌜**ta**⌝ 13 **rin**$_2$ **ta** 13 **rin**$_2$} **ta** 13 ⌜**absin**⌝
6' [**e**]**n** 13 **rin**$_2$ 30 30 GAM 15 **du**-*ma* 7.3[0]
7' [**ki** 1.5]8.45.[42] **tab** 2.6.15.4[2 xx]
8' [**ta** 13 **r**]**in**$_2$ **en** 27 *zib*.**me iti** *ana* **i**[**ti**]
9' [2.6.15.4]2 **tab** *u* **la**$_2$-*ka* 25.[7.30 **ḫun**$^?$]
10' [**ta** 2]7 *zib*.**me en** 25.7.30 **ḫun** 2[8.7.30]
11' [GAM] 16 **du**-*ma* 7.30 **ta** 2.[6.15]
12' [4]⌜2⌝ **nim**-*ma re-ḫi* 1.58.45.[42 **ta**]
13' [25].7.30 **ḫun en** 13 **absin iti** *an*[*a* **iti**]
14' [1.58].45.42 **tab** *u* **la**$_2$

Obverse

P1 **(Computations involving *B* and eclipse magnitude?)**

[1][...you multiply the ... of] the displacements, 0;3,45, by 4, the *igigubbû*-coefficient of [...]. [2][... it is] ⌜15$^?$⌝ .... 15 16th, 4, the *igigubbû*-coefficient of [...], [3]⌜this⌝ from 27 Psc until 27 [...] [4]per$^?$ day the Sun [...] 1/15th of 1, its displacement ..., [...] [5][a 1]5th of its 'disk' from 13 V[ir until ...].

[6][Since] you do not know it: 4, the *igigubbû*-coefficient of ... [you ...] from [...]. [7][...] you mult[iply ...] by 3,0, (i.e.) 6 zodiacal signs, [...] [8][...] ... you put down. Whatever the Sun [...] [9][...] and 13, the displacement of the Su[n, ...] [10][...] 11, the displace[ment of ...] [11][...] you accumulate up, and it is 14, the displace[ment of ...] [12][...] 16 days [...] [13][... 2]7 Psc [...]

(n lines and perhaps 1 column missing)

Reverse

P2' **(*E*)**

a **(Subject unclear)**

(n lines and perhaps 1 column missing)

[1'][...] you multiply [...].51 by 20, to$^?$ 1⌜4⌝.10 ⌜...⌝ [...]

b **(Updating *E*)**

[2'][...] ⌜by month$^?$⌝ you deduct ... and the 'depth', there rema[ins] [3'][1,58];45,42.

From 27 Psc until 13 [Vir], [4'][month by] month the addition and subtr[action] is 1,58;45,42. [5']{⌜From⌝ 13 Lib until 13 Lib.} From 13 Vir [6']until 13 Lib is (a distance of) 30. You multiply 30 by 0;15, it is 7;⌜30⌝, [7']you add it [to 1,5]8;45,42, (it is) 2,6;15,4[2 ...].

[8'][From 13 Li]b until 27 Psc, month by m[onth] [9']your addition and subtraction is [2,6;15,4]2. 25;[7,30 Ari:] [10'][From 2]7 Psc until 25;7,30 Ari is (a distance of) 2[8;7,30]. [11']You multiply it [by] 0;16, it is 7;30, you [12']deduct it from 2,[6;15,42 ...], there remains 1,58;45,[42. From] [13'] [25]; 7,30 Ari until 13 Vir month by month [14']the addition and subtraction is [1,58];45,42.

Critical and philological notes

Obverse

1 3.45: not 1.1.45 (Pinches's copy). The outcome of the multiplication is 15, which can be restored at the end of 1 or the beginning of 2.

2 ^{d}[xx]: perhaps ^{d}**utu** (compare O4).
la$_2$: *tanaḫḫis* = 'you subtract', or *šūqu* = 'height' (maximum)?

3 The initial damaged sign is compatible with ⌈*šu*⌉.
4 After **zi**-*šu*$_2$ the upper half of a vertical wedge is visible.
6 **nu**: thus also in Pinches's copy.
7 After the break there are two winkelhakens, perhaps the end of **zi**, as suggested by Neugebauer.
10 After ^{d}**utu** there is a number of which the two leftmost wedges are visible, perhaps 4,5,...,8.

Reverse

1' *ana*: or 1.
4: or 7?
2' *ana* **iti**: very uncertain.
nim-*u*$_2$: the ending -*u*$_2$ excludes an interpretation as *šūqu*, 'height', suggesting that it must be read as a form of *elû* here (or its Š-stem, 'to subtract'?). The interpretation remains unclear.
5' The damaged sign at the beginning might be **ta**. As indicated in Pinches's copy, this is followed by 13 (*ACT*: 12).
9' **tab** *u* **la**$_2$-*ka* = 'your addition and subtraction': *ka* replaces BE-*ma* in Pinches's copy. The gap after 25 at the end probably contained 7.30 or 7.30 **ḫun**.
12' Before **nim** there are remains of a vertical wedge, perhaps belonging to the final digit of 2.6.15.42.

Commentary

Only the left edge and the upper (lower) edge of the obverse (reverse) of the tablet are preserved on this fragment. From the textual restorations it follows that not much clay is missing on the right side, so that the original width may have been about 5 cm. However, the thickness of the fragment increases from the left to the right. This suggests that about half of the tablet is missing, so that it must have had two columns, of which only one is partly preserved. Hence the original width may have been about 10 cm.

P1 Computations involving B and eclipse magnitude?

The first 13 lines of P1 are partly preserved. The interpretation of this procedure, which has no known parallel in the corpus, poses many problems. Some words and parameters (e.g. O10,11) suggest a connection with the synodic arc, e.g. the positions mentioned in O3,5,13, 27° Psc and 13° Vir, are zonal boundaries of the step function for σ (§4.4.4). Other terms suggest a connection with eclipse magnitude (O5). Due to the bad state of preservation, the computations could not be fully reconstructed. The number 3.45 (O1) might be identified as the number period Z = 3,45 of B. Interpreted as 0;3,45 this number equals 1/16, the reciprocal of 16. It is multiplied by 4, the '*igigubbû*-coefficient of ...'. The result of the multiplication, 15, is mentioned at the beginning of O2. It reappears in the same line in the ratio 15/16 = 0;56,15. This ratio equals the transition coefficient r_1 for the step function for σ (cf. §4.4.4). In O4 1/15 of the Sun's daily displacement, $v_{\odot}$ = 1°/d, is computed, and in O5 1/15 of the 'disk', a term for eclipse magnitude. The significance of these operations remains unclear.

P2' Procedure for E^{nm}

The beginning of P2' is missing. R1'–7' are translated here for the first time. The meaning of R1'–2' is unclear; the numbers in R1' remain unidentified. The partly preserved computation in R2' appears to result in w_2 = 1,58;45,42$^{\check{s}e}$, the preliminary monthly difference of E in zone 2 (§4.4.6).

Part **b** (R3'–14') is a procedure for updating E^{nm}. For a discussion of this algorithm cf. §4.4.6. Its structure is somewhat confusing, but the interpretation poses no problems. The formulation is based on template ESTEP.D, which is not attested anywhere else. Between 27° Psc and 13° Vir the monthly difference is said to be 1,58;45,42 = w_2. At the beginning of R5' there is a repetition, presumably by mistake. In R5'–7' w is computed by applying the transition rule for an exemplary position at 13 Lib (as is done in No. **53** P8'). This is the boundary between regions 1a and 1b (Eq. 4.27, p. 134), where w = 2,6;15,42$^{\check{s}e}$ = w_1. Since 13 Lib is 30° beyond 13° Vir, the result is $w_2 + 0;15 \cdot 30 = w_1$ (cf. also Fig. 4.13 on p. 134).

R8' contains a similar procedure for zone 1. One would expect the boundary b_1 = 13° Vir for the preliminary difference (w) to be mentioned here, but the text mentions 13° Lib, presumably by mistake. This is followed by an exemplary computation of w for 25;7,30° Ari, which involves the transition rule. Analogous to R3'–7' this position is the boundary between regions 2a and 2b.

No. 59

museum number	BM 45655 (81–7–6,49)
provenience	Babylon
measures	5.3(*) × 8.4(*) × 3.3–4.2 cm
arrangement	O/R
cuneiform text	photo: Fig. F.62; *ACT*, Pl. 237; copy: *LBAT* 88
bibliography	*ACT* 200f
contents	Moon system A: procedures for eclipse magnitude and updating *E*

Side X

P1' (m lines missing)

1' [...xxx] ⸢x 2? ḪAB?-*rat*?⸣ d⸢*sin*⸣ [xx...]

2' [...xx] ⸢**ta**? **ki** 4⸣5 *ki-ṣir* [x...]

3' [... **ta** *zib*.**me en**] **absin** ⸢*ana*?⸣ **ab**$_2$ 44.31.56.40 **tab** [x...]

4' [...] ⸢xx⸣ 20 *ana* 6 **ab**$_2$ 4.27.11.⸢40⸣ [x...]

5' [...] ⸢8?⸣.20 **tab** *ana* 18 **mu** 1 **tab** *ša*$_2$ [xx...]

6' [...x] ⸢2?⸣ **ki tab**-*ka* **tab ki** ḪAB-⸢*rat*⸣ [xx...]

7' [...xx] ⸢4⸣6.20.33.20 ⸢*ana*?⸣ [xx...]

8' [...xxx *ana*] ⸢6⸣ **ab**$_2$ 4.2⸢8?.3 xx⸣ [x...]

9' [...xxxx] ⸢xxxxx⸣ [x...]

(n lines missing)

Side Y

P2' (about n lines missing)

1' [...xxx] ⸢xxxx⸣ [...]

2' [... **ta** 13 *zib*].**me en** 27 **absin** [2.6.15.42 ...]

3' [... *ša*$_2$ *al* 27] ⸢*zib*⸣.**me diri** GAM$_0$ 16 [**du ta** 2.6.15.42 ...]

4' [... **ta** 27 **absin en** 13 *zib*.**me**] ⸢1⸣.58.45.42 [...]

5' [...xx] ⸢x⸣ **ki tab**-*ka* **tab** [...]

6' [...x] ⸢2?⸣ **du**-*ma* **tab.meš** *an-nu-*⸢*ut-ti*⸣

P3' 7' [...x] ⸢xx.21⸣ [...]

P4' 8' [...xxxx] ⸢5⸣.3⸢7? x⸣ [...]

(about m lines missing)

Side X

P1' (Computations involving $\Psi^{(3)}$)

(about m lines missing)

1'[...] ⸢...2, the 'disk?' of the Moon⸣ [...] 2'[...] ⸢from? the position, 4⸣5, the knot [...] 3'[... from Psc until] Vir ⸢for?⸣ (one) month you add 44;31,56,40 [...] 4'[...] ⸢...⸣ 20. For 6 months [you add] 4,27;11,40 [...] 5'[...] you add [...].8.20. For 18 years you add 1. That which [...] 6'[...] you add [...].2 to your addition. To the 'disk' [...] 7'[...] ⸢4⸣6;20,33,20 ... [...] 8'[... for] ⸢6⸣ months 4,2⸢8;3 ...⸣ [...] 9'[...] ⸢...⸣ [...]

(n lines missing)

Side Y

P2' (Updating *E*)

(about n lines missing)

1'[...] ... [...]

2'[... From 13 P]sc until 27 Vir [2,6;15,42 ...] 3'[... (the amount) by which] it exceeds [27] Psc [you multiply] by 0;16 [, deduct from 2,6;15,42 ...] 4'[... From 27 Vir until 13 Psc] 1,58;45,42 [...] 5'[...] you add to your addition. [...] 6'[...] you multiply by [...], and these are the additions.

P3' (Subject unclear)

7'[...] ⸢...21⸣ [...]

P4' (Subject unclear)

8'[...] ⸢5⸣.3⸢7 ...⸣ [...]

(about m lines missing)

Critical and philological notes

Side X

1' 2 ḪAB-*rat*: very uncertain.

2' **ta**: or ...UD?

3' **ab**$_2$: replaces 30 (*ACT*); correctly copied by Pinches. Before **ab**$_2$ there is a damaged sign not shown in Pinches's copy; perhaps 1 = *ana*.

4' Pinches's copy is inaccurate here. The damaged sign at the beginning looks like **meš** or a digit 30, 40 or 50. It is followed by faint traces of another sign.

5' *ana* **tab**: replaces *ša*$_2$ **tab** (*ACT*); correctly copied by Pinches.

6' Before **ki** there are traces of two vertical wedges.

9' The upper halves of several wedges are visible; reading uncertain.

Side Y

5' Before **ki** a vertical wedge is visible.

6' *ti*: written on edge.

7' The upper parts of several digits are visible. The sign before 21 is a digit 3–8.

Commentary

This fragment preserves a section of the right edge of what was probably a large tablet, judging by its thickness. No other edges are preserved. The orientation of the sides is not known. At most 4 procedures are partly preserved. As far as can be determined, they are concerned with the Moon's distance to the ecliptic (E) and eclipse magnitude.

P1' Computations involving $\Psi^{(3)}$

The beginning and end of P1' are missing. X2' mentions the 'knot', perhaps a reference to the Moon's nodal region. I interpret the number 44;31,56,40 (X3'), presumably measured in fingers, as the minimum of a hitherto unknown function representing the monthly difference of eclipse magnitude, $\Delta\Psi^{(3)}$. For a discussion cf. §4.4.8. In X3' this parameter is associated with the region Psc – Vir. X4' mentions the difference for 6^{m}, computed as $6 \cdot 44;31,56,40 = 4,27;11,40^{\text{f}}$. The number 46;20,33,20 is likely the mean value μ. The corresponding difference for 6^{m}, computed as $6 \cdot 46;20,33,20 = 4,38;3,20^{\text{f}}$, might be intended in X8', in which case the number 4,28;3 is an error for 4,38;3.

P2' Updating E

The beginning of this procedure is missing. The preserved part is concerned with updating E (§4.4.6). The formulation is based on template ESTEP.A. The number 1,58;45,42 (Y4') is w_2, and 0;16 is the transition coefficient $|c_1'|$ (Table 4.12, p. 134).

P3',P4' Subject unclear

Too little remains of P3' and P4' to establish even the topic of these procedures. P3' consists of a single line with traces of several digits. Only a single line with a number remains of P4'.

No. 60

museum number	BM 55530 (82–7–4,112)
provenience	Babylon
measures	6.7(13.5?) × 6.1(*) × 2.0–2.8 cm (type C?)
arrangement	O/R
cuneiform text	photo: Fig. F.63; *ACT*, Pl. 239a
bibliography	*ACT* 200i
contents	Moon system A: *E*, *F* and ME

Obverse

P1' (n lines missing)

1' [xxx] 2.24 gal-*u* ⟨*ki*⟩ *al*-⸢*la*⸣ [1.12 *mi-šil* e$_2$ an.mi diri 1.12]

2' ta ⸢ša$_3$ e$_{11}$⸣ *ša$_2$ re-ḫi* GAM$_0$ ⸢2 du⸣ [*ana* e$_2$ an.mi gar-*an mim-ma ša$_2$* GAM$_0$ 2 du]

3' ta 1.12 e$_{11}$ *u* tab *ša$_2$* ⸢iti⸣-*ka* ki-⸢*šu$_2$* tab ki.ta e$_2$⸣ [an.mi gar-*an*]

4' *ki al* 1.12 *mi-šil* e$_2$ an.mi la$_2$ ta 1.12 ⸢e$_{11}$⸣ [*ša$_2$ re-ḫi*]

5' GAM$_0$ 2 du *ana* e$_2$ an.mi gar *mim-ma ša$_2$* GAM$_0$ 2 du 1.12 ki-*šu$_2$* tab ⸢*u*⸣ [tab *ša$_2$* iti-*ka*]

6' ki-*šu$_2$* tab ki.ta-*nu* e$_2$ an.mi gar-*an ša$_2$ al* 13 absin$_0$ ⟨diri⟩ GAM$_0$ 15 [du]

7' ki 1.58.45.42 tab

8' ta 13 absin$_0$ en 27 *zib* iti *ana* iti 2.⸢6⸣.15.42 tab *u* la$_2$ *ša$_2$ al* 2[7 *zib* diri]

9' GAM$_0$ 16 ⟨du⟩ ⸢ta⸣ 2.6.15.42 la$_2$ en 7.⸢12 nim *ša$_2$*⸣ *al-la* 7.12 n[im diri]

10' ta 14.⸢24⸣ e$_{11}$-*ma* gar-*an* en 2.24 ⸢tur-*er* ta⸣ 2.24 [xx]

11' *ki* ⸢*al*⸣ [1].⸢12⸣ *mi-šil* e$_2$ an.mi ⸢diri⸣ 1.12 ta ⸢ša$_3$⸣-*šu$_2$* e$_{11}$ [*ša$_2$ re-ḫi*]

12' ⸢GAM$_0$ 2 du *ana* e$_2$ an.mi gar⸣-*an* ⸢*mim-ma ša$_2$* GAM$_0$⸣ 2 du-⸢*ka*⸣ ta 1.12 [e$_{11}$]

13' *ša$_2$* ⸢*re-ḫi*?⸣ *u* tab ⸢*ša$_2$* iti-*ka*⸣ ki-*šu$_2$* tab *šap-la-nu* e$_2$ ⸢an⸣.mi [gar-*an*]

14' *ki* ⸢*al*⸣ 1.1⸢2 la$_2$ ta 1.12 e$_{11}$⸣ *ša$_2$ re* GAM$_0$ 2 du *ana* e$_2$ an.[mi gar-*an*]

15' *mim-ma ša$_2$* GAM$_0$ 2 ⸢du-*ku*?⸣ 1.1⸢2⸣ [ki]-⸢*šu$_2$* tab⸣ *u* tab *ša$_2$* iti-*ka* ki-*šu$_2$* tab ⸢*šap*?⸣-[*la-nu*]

16' e$_2$ an.mi ⸢gar⸣ mu *ana* ⸢mu ta⸣ 27 *zib* en 13 absin 33.⸢3⸣[8.24 tab *u* la$_2$]

17' ta 13 absin$_0$ en 27 *zib* 30.⸢5⸣2.24 tab *u* ⸢la$_2$⸣

P2' 18' ⸢zi⸣ *ana* du$_3$-*ka* en 15.57 ⸢gal *ša$_2$*?⸣ *al* 15.⸢57 diri⸣ [xx]

19' [xx] ⸢x⸣ ⸢11⸣.4 tur *ša$_2$ al* 11.⸢4⸣ tur ta 22.8 [e$_{11}$ xx]

Reverse

P3' 1 ki *sin u* ki *šamaš$_2$* gar zi *ša$_2$ sin is-suḫ* ta ki *s*[*in* la$_2$]

2 *u* zi *ša$_2$ šamaš$_2$ is-suḫ* ta ki *šamaš$_2$* la$_2$ ki *sin u* ki *š*[*amaš$_2$*]

3 *ša$_2$ al* 10 *lu-maš* diri GAM$_0$ 16 du ki bar.nun *ša$_2$ lu-maš*

4 *ša$_2$-šu$_2$ ki* tab ⸢la$_2$⸣ *ki* la$_2$ tab *ana* ugu {*a* ugu} *a-ḫa-miš$_2$* GAR.GAR-*m*[*a*]

5 $\frac{1}{2}$-*šu$_2$-nu* GIŠ-*ma ana* bar.nun gar nim *u* sig *ša$_2$ sin* gar zi *sin*

6 GAM$_0$ 4 du ki nim *u* sig *ki* tab la$_2$ *ki* la$_2$ tab *ana muḫ-ḫi*

7 GAR.GAR-*ma* $\frac{1}{2}$-*šu$_2$-nu* GIŠ-*ma ana* bar.nun gar ki *sin u* ki *šam*[*aš$_2$*]

8 *ša$_2$ al* 10 *lu-maš* diri GAM$_0$ ⸢20?⸣ du ki bar.nun *ša$_2$ lu-m*[*aš*]

9 *ša$_2$-šu$_2$ ki* tab tab *ki* la$_2$ la$_2$ *ana* ugu *a-ḫa-miš$_2$* GAR.GAR-*ma*

10 $\frac{1}{2}$-*šu$_2$-nu* GIŠ-*ma* GAM$_0$ ⸢gi?⸣ *ša$_2$* nim *u* sig *ša$_2$* gar-*nu* du

11 BE GAM$_0$ bar.nun du *ki* nim ki BE tab *ki* sig ta BE [la$_2$]

12 *ki ma-la* igi *la ma-ṣu-u$_2$* 6 ki-*šu$_2$* tab *ana* {1} me [*e-bi*]

13 ⸢*ki*⸣-*ma* igi-*u$_2$* du$_3$-*uš ana* ki.na$_2$.a e.a ⸢gar⸣-[*an*]

P4' 14 ⸢36 ḫun?⸣ (blank) ⸢1 x⸣ [xx]

(about n lines missing)

Critical and philological notes

Obverse

3' e$_2$: replaces nu (*ACT*). ki.ta-*nu* = *šaplānu* 'below' (adv.).

3',5' *u* tab: there is no reason to assume an omitted la$_2$ before *u* tab (*ACT*, p. 222).

5' The end is restored on the basis of O2'.

9' 16 is followed by a crammed sign, which is probably ta, perhaps written over an erasure. Hence the expected du seems to be omitted.

10' tur-*er* = *iṣeḫḫer*, 'it is small(est)': i.e. if *E* is smaller than 2,24^{še} the Moon is in the nodal region.

Obverse

P1' **(Updating *E*)**

a **(Step 2: Moon remains on same side of ecliptic)**

(*n lines missing*)

1'[...] 2,24, the largest one. ⟨If⟩ [it] exceeds [1,12, half the 'eclipse house'], 2'you ⸢deduct⸣ [1,12] from it. That which remains [you multiply] by 2, [put down as the 'eclipse house'. Whatever you multiplied by 2] 3'you deduct from 1,12, and you add the addition for your month with it, [you put it down] below as [the 'eclipse house'.]

b **(Step 2: Moon crosses ecliptic)**

4'If it is less than 1,12, half the 'eclipse house', you deduct it from 1,12, [what remains] 5'you multiply by 2, put down as the 'eclipse house'. Whatever you multiplied by 2: you add 1,12 with it, [and the addition for your month] 6'you add with it, you put it down below as the 'eclipse house'.

c **(Steps 1 and 3: Moon outside nodal region)**

(The amount) by which it ⟨exceeds⟩ 13 Vir you [multiply] by 0;15, 7'add with 1,58;45,42. 8'From 13 Vir until 27 Psc you add and subtract 2,6;15,42 month by month. (The amount) by which it [exceeds] 2[7 Psc] 9'you ⟨multiply⟩ by 0;16, subtract ⸢from⸣ 2,6;15,42, until it is high by 7,12. That which [exceeds] 7,12 high 10'you deduct from 14,[24] and put down, until 2,24, (where) ⸢it is small(est).

d **(Step 2: Moon remains on same side of ecliptic)**

From⸣ 2,24, [...] 11'If it exceeds 1,12, half of the 'eclipse house', you deduct 1,12 from it, [what remains] 12'you multiply by 2, put down as the 'eclipse house'. That which you multiplied by 2 you [deduct] from 1,12. 13'That which ⸢remains⸣: and? the addition for your month you add with it, [you put it down] below as the 'eclipse house'.

e **(Step 2: Moon crosses ecliptic)**

14'If it is less than 1,12, you deduct it from 1,12, what remains you multiply by 2, [put down] as the 'eclipse house'. 15'That which you multiplied by 2: [you add] 1,12 [with] it, and the addition for your month you add with it, 16'you put it down ⸢below⸣ as the 'eclipse house'.

f **(Net differences of *E* for 12^{m})**

Year by year, from 27 Psc until 13 Vir [you add and subtract] 33;3[8,24], 17'from 13 Vir until 27 Psc you add and subtract 30;52,24.

P2' **(Updating $F^{(2)}$ as a zigzag function)**

18'In order for you to construct the displacement: until 15;57, the largest (value): that which exceeds 15;[57 you deduct from 31;44 ...] 19'[...] ... 11;4, the smallest; that which is smaller than 11;4 [you deduct] from 22;8 [...]

Reverse

P3' **(Lunar Six interval ME: cf. No. 61.C.)**

13' *ša$_2$* ⸢*re-ḫi*⸣ *u* **tab**: replaces *ša$_2$ r*[*i*] ⟨*ḫi*⟩ **la$_2$** *u* **tab** (*ACT*). *u* remains unexplained.
14' *re*: phonetic abbreviation of *reḫi*, 'there remains'.
15' ⸢**du**-*ka*?⸣ 1.1⸢2⸣ [**ki**]-⸢*šu$_2$* **tab**⸣: replaces **du ki** [**tab** *itti*(**ki**)]-*šu$_2$* **la$_2$** *u* **tab** (*ACT*) The sign *ka*, which is badly damaged, might also be *ku* or *ki*. Either would be suitable as a phonetic complement for *tallik*, 2 m. sg. pret. of *alāku*, 'you multiplied'.
17' After **la$_2$** the line appears to be empty.
18' **en** replaces 13 (*ACT*).
19' The sign at the beginning might be **tur**, but this is very uncertain. In the first occurrence of 11.4 **tur**, **tur** is assumed to represent *ṣeḫertu*, 'small(est)', in the second *iṣeḫḫer*, 'it is small(er)'.
Reverse Cf. No. **61.C**.

Commentary

This fragment from the lower (upper) left corner of the obverse (reverse) of a tablet contains 4 partly preserved procedures. An unknown number of lines are missing above O1'; about the same number are missing below R14. The curvature of the fragment implies that there must have been another column on the right side, which is completely broken away. Hence the original width may have been about 13.5 cm. The fact that only one of the expected four Lunar Six procedures (P3') is preserved supports this. Furthermore, P1' may have been preceded by procedures for Φ, *B* and *C*, and P2' followed by procedures for *G*, *J*, *K* and *M*. Some of the terminology of the tablet is peculiar; for instance, the term 'eclipse house' for the nodal region is found only here.

P1' Procedures for E

The beginning of P1' is missing. The 17 partly preserved lines are concerned with various steps of the algorithm for updating E (§4.4.6). They can be divided into six parts (**a–f**). Parts **a**, **b**, **d**, **e** are concerned with step 2; part **c** with steps 1 and 3; part **f** with net differences of E.

Part **a** deals with the case when the Moon is initially above the nodal region ($E_0 > \kappa = 2{,}24^{\check{s}e}$) and the updated position, E_1, is in the nodal region on the same side of the ecliptic. As explained in §4.4.6, this corresponds to $E_0 - w > \kappa/2$ (Eq. 4.32 on p. 136). In that case the subtraction $E_0 - d$ is possible, the result being $E_1 = 2(E_0 - w - \kappa/2)$. In O3' the next position, E_2, turns out to be below the nodal region. It is obtained by subtracting 'what you had multiplied by 2', which corresponds to $E_0 - w - \kappa/2$, from $\kappa/2$, and adding w, called the 'addition for your month' (replacing 'positive and negative difference', *ACT*, p. 223). The phrase 'for your month' reflects that w is computed from the Moon's zodiacal position at lunation, B (step 1). The result is $E_2 = \kappa/2 - (E_0 - w_0 - \kappa/2) + w_1 = \kappa + w_0 + w_1 - E_0$ below the ecliptic, in accordance with Eq. (4.34).

Part **b** (O4'–6') deals with the case when E_1 is on the opposite side of the ecliptic, which occurs when $E_0 - w_0 < 1{,}12^{\check{s}e}$, corresponding to the second case of Eq. (4.32). The subtraction of d from E_0 is reversed, leading to $E_1 = 2 \cdot [\kappa/2 - (E_0 - w_0)]$. The next position is also below the nodal region, so that $E_2 = \kappa/2 + [\kappa/2 - (E_0 - w_0)] + w_1 = \kappa + w_0 + w_1 - E_0$.

Part **c** (O6'–10') begins with step 1, i.e. the computation of the (preliminary) difference w as a step function of B_i. For this algorithm cf. §4.4.6. The formulation is based on the usual template ESTEP.A. The Moon is assumed to be initially outside the nodal region. Immediately following step 1 the text continues with the reflection rule at $7{,}12^{\check{s}e}$ (step 3). After the reflection E decreases (O9'–10'). From then on the differences are subtracted until the Moon returns to the nodal region at $E = 2{,}24^{\check{s}e}$ (O10').

In parts **d** and **e** (O11'–16') the rules for updating E in the nodal region are repeated. This suggests that the scribe intended to provide instructions for a complete cycle of the Moon's latitudinal motion.

Part **f** (O16'–17') provides net differences of E for 12^{m} ('1 year'); cf. Table 4.14 in §4.4.6.

P2' Updating $F^{(2)}$ as a zigzag function

P2' partly preserves the algorithm for updating $F^{(2)}$ as a zigzag function of the lunation number (§4.4.9). O18'–19' contain the reflection rules at the extrema, formulated in accordance with template ZZ.B.1 (§2.6.8.1).

P3' Lunar Six interval ME

For this procedure cf. No. **61**.**C**. The few signs preserved below P3' may belong to an auxiliary table with control values of the coefficient q_{rise}; cf. No. **61**.T2.a. The number 36, interpreted as 0;36, could be the control value for 10° Ari.

No. 61

A	museum number	BM 35399 (Sp2,981)
	measures	9.9 × 12.6 × 3.2 cm (type F)
	arrangement	O/R
	cuneiform text	photo: Figs. F.65, F.66; *ACT*, Pl. 238; copy: *ACT*, Pl. 226; *LBAT* 91
	bibliography	*ACT* 201
	contents	O: ŠU$_2$, NA, q_{set}, r_{set}; R: ME, GI$_6$, q_{rise}, r_{rise}
B	museum number	BM 47734 (81–11–3,439)
	measures	8.9 × 5.2(14) × 1.9–2.5 cm (type F)
	arrangement	O/R
	cuneiform text	photo: Fig. F.64
	contents	O: NA; R: ME, GI$_6$
C	museum number	BM 55530 (82–7–4,112)
	measures	6.7(13.5) × 6.1(*) × 2.0–2.8 cm (type C?)
	arrangement	O/R
	cuneiform text	photo: Fig. F.63; *ACT*, Pl. 239a
	bibliography	*ACT* 200i
	contents	O: *E*; R: ME
D	museum numbers	BM 33451 (Rm4,5) +47744 (81–11–3,449) +47924 (81–11–3,631)
	measures	6.0(8.0) × 10.8 × 2.0–2.8 cm (type F)
	arrangement	O/R
	cuneiform text	photo: Fig. F.68
	contents	O: ŠU$_2$, NA; R: q_{set}, r_{set}, ME, GI$_6$, q_{rise}, r_{rise}
E	museum number	BM 35076 (Sp2,612)
	measures	3.5(10.0) × 2.9(*) × 1.5–2.2 cm (type F?)
	arrangement	O/R
	cuneiform text	photo: Fig. F.71; *ACT*, Pl. 237; copy: *LBAT* 92
	bibliography	*ACT* 201a
	contents	O: NA; R: q_{set}, r_{set}
F	museum number	BM 35125 (Sp2,671)
	measures	5.5(9.0) × 5.2(*) × 2.0(*) cm (type F?)
	cuneiform text	photo: Fig. F.69; *ACT*, Pl. 240; copy: *LBAT* 93
	bibliography	*ACT* 201aa
	contents	O: ŠU$_2$, NA
G	museum number	MMA 86.11.363
	measures	7.9 × 8.0(*) × 1.5–2.6 cm (type F)
	arrangement	O/R
	cuneiform text	photo: *CTMMA* II, Pls. 105–106; copy: *CTMMA* II, Pl. 132
	bibliography	*CTMMA* II, No. 81
	contents	O: ŠU$_2$, NA; R: NA, q_{set}, r_{set}, ME, GI$_6$, q_{rise}, r_{rise}

About the transliteration and the translation

The transliteration is presented as a score. The translation incorporates all Mss. Raised sigla indicate that a word or phrase is present only in the mentioned Mss. Variants are surrounded by accolades {...} and separated by vertical dashes, |. If the variant is significantly different the entire step is presented separately. Variants judged to be erroneous are translated only in the critical notes, marked by an asterisk (∗). Formulaic representations provide a link to the astronomical commentary (§4.4.19). An attempt is made to preserve the order of the operations. Double arrows represent causal connections, i.e. A ⇒ B means 'if A then B'; $x \to$ QN means 'x is named QN'. The same in square brackets, [...], indicates that x is not named, usually because it is passed on immediately to the next step.

	A.O0	*ina a-mat* ^{d}**en** *u* ^{d}**gašan**-*ia*$_2$ *liš-lim*

P1

1	A.O1	**šu**$_2$ *ana* **du**$_3$-*ka ki si-man ša*$_2$ **gi**$_6$ **du** ⸢*al-la* LA$_2$ *ša*$_2$⸣ **gi**$_6$ **diri** ⸢LA$_2$ *ša*$_2$⸣ **g**[**i**$_6$ **ta gi**$_6$ **du nim**]
	D.O1–2	**šu**$_2$ *ana* **du**$_3$-*ku ki-i si-man ša*$_2$ **gi**$_6$ **du** ⸢*al-la* LA$_2$ *ša*$_2$⸣ [x x x x x x] / ⸢**gi**$_6$⸣ **du e**$_{11}$ →
	A.O2	**me** *ina* **me**-*šu*$_2$ *ki si-man ša*$_2$ **gi**$_6$ **du** *al-la* LA$_2$ *ša*$_2$ **gi**$_6$ *i-ṣa* **gi**$_6$ **du** ⸢*u*$_3$⸣ [x x x]
	D.O2–3	*ša*$_2$ **tag**$_4$ **me** *ina* **me**-*šu*$_2$ *ki-i si-man ša*$_2$ **gi**$_6$ **du** [xx x x x] / ⸢*i-ṣa*⸣ **gi**$_6$ **du** *u* LA$_2$ *ša*$_2$ ⸢**me**⸣ →
	A.O3	**ki** *a-ḫa-miš*$_2$ **tab**-*ma* 1-*en u*$_4$-*mu* **la**$_2$ →
	D.O3	**ki** *a-ḫa-miš*$_2$ **tab**-*ma* ⸢1-*en* **me** **la**$_2$⸣→
2	A.O3	*ša*$_2$ **tag**$_4$ GAM$_0$ 10 **du** *šu-u*$_2$ **zi** *šamaš*$_2$ **gar**-*an* ⸢GAM$_0$⸣ [**zi** *sin*]
	D.O3–4	⸢*ša*$_2$ **tag**$_4$⸣ [xx x x] / *ša*$_2$ ⸢**e**$_{11}$⸣-*ka šu-u*$_2$ **zi** *šamaš*$_2$ **gar**-*an* ⸢**a.ra**$_2$⸣ [x x] →
	A.O4	*ša*$_2$ **iti**-*šu*$_2$ **du**-*ma šu-u*$_2$ **zi** *sin* **gar**-*an* →
	D.O4–5	[x xx xx xx] / ⸢**zi** *sin* **gar**⸣-*an* →
3	A.O4	**zi** *šamaš*$_2$ **ta zi** *sin* **nim**-*ma* BE ⸢**gar**⸣-[*an*]
	D.O5	**zi** *šamaš*$_2$ **ta** ⸢**zi** *sin*⸣ [xx x xx] →
4	A.O5–6	*u*$^?$ **ki** *sin u* **ki** *šamaš*$_2$ **gar**-*an* **zi** *ša*$_2$ *šamaš*$_2$ **zi ta ki** *šamaš*$_2$ **nim zi** *ša*$_2$ *sin* **zi** [**ta**] / **ki** *sin* **nim** →
	D.O5–6	[x x x x x xx] / ⸢**zi** *ša*$_2$ *šamaš*$_2$ **zi**⸣ **ta ki** *sin*(sic) **nim** [x x x x x x x x] →
5	A.O6	**ki** *sin u* **ki** *šamaš*$_2$ *ša*$_2$ *al-la* 10 *lu-maš* **diri** GAM$_0$ 16 **du ki bar.nun** *ša*$_2$ *l*[*u-maš*]
	D.O6–7	[x x] / *u* ⸢**ki** *šamaš*$_2$ *ša*$_2$ *al*⸣-*la* 10 *lu-m*[*aš* x xx xx x x xx x xx] →
	A.O7	*ša*$_2$-*šu*$_2$ *ki* **tab tab** *ki* **la**$_2$ **la**$_2$ *ana muḫ-ḫi a-ḫa-miš*$_2$ GAR.GAR-*ma* →
	D.O7–8	[xx] / *ki-i* **tab tab** *ki-i* **la**$_2$ **l**[**a**$_2$ x xx xxx xxx] →
	A.O7	$\frac{1}{2}$-*šu*$_2$-*nu* GIŠ-*ma ana* **bar.nun** ⸢**gar**⸣-[*an*]
	D.O8	[xxx xx x xx xx]
?	G.O1'–3'	[xxxxxxxxxxxxxxxxx] GIŠ$^?$ **tab**$^?$ [xxxxxx] / [xxxxxxxxxxxxxxxxx] ⸢x⸣ **a.ra**$_2$ BE [xxxxx] / ⸢*tal*$^?$⸣-*lik* U ⸢x⸣→
6	A.O8	**nim** *u* **sig** *ša*$_2$ *sin* **gar**-*an* **zi** *sin* GAM$_0$ 4 **du** →
	D.O9	⸢**nim**$^?$ *u*$^?$ **sig**$^?$ *ša*$_2^?$⸣ *sin* **gar**-*an* [x x xx x x] →
	F.X1'	[x x x x x **gar**]-⸢*an* **zi** *sin*⸣ [x x] ⸢**du**⸣ →
	G.O3'–4'	[**bar.nun** *gab-bi ša*$_2$ **nim** *u* **sig**] ⸢**zi**⸣ [^{d}]⸢*sin šu*$_{10}$⸣-*a-ti*$_3$ **a.ra**$_2$ 4 **d**[*u-ak*$^?$] / *ša*$_2$ **nim**-*ka* →
	A.O8	**ki nim** *u* **sig** *ša*$_2$ *sin ki* **tab la**$_2$ *ki* **la**$_2$ [**tab**]
	D.O9	[x x xx x x xx x x xx x x]
	F.X1'–2'	[x x xx x x] / [x x x x x x] →
6?	G.O4'–5'	⸢**ki**⸣ [xxxxxx] ⸢*ša*$_2^?$⸣ *taš-kun ki-i* **tab** *šu-u*$_2$ **ki**-*šu*$_2$ **tab** *u* ⸢*ki-i*⸣ [**la**$_2$ *šu-u*$_2$] / **ta** *lib*$_3$-*bi-šu*$_2$ **la**$_2$ →
7	A.O9	**ki** *sin u* **ki** *šamaš*$_2$ *ša*$_2$ *al-la* 10 *lu-maš* **diri** GAM$_0$ 8 **du ki** ⸢**bar**⸣.**nun** *ša*$_2$ **nim** *u* **sig**⸢*ša*$_2$⸣-[*šu*$_2$]
	D.O10–11	[x x xx x x xx x xx x xx xx x x xx x] / [x xx xx] →
	F.X2'	[xx x x *qa*]*q-qar šamaš*$_2$ *ša*$_2$ 10 *lu-maš* [x x xx x xx x x xx xx]
	A.O10	*ki* **tab tab** *ki* **la**$_2$ **la**$_2$ *ana muḫ-ḫi a-ḫa-miš*$_2$ GAR.GAR-*ma* $\frac{1}{2}$-*šu*$_2$-*nu* GIŠ-*ma* →
	D.O11	[xx x x xx x x x xx xxx xxx xxx xx]
	F.X3'	[x x x x] ⸢**la**$_2$ **la**$_2$⸣ →
8	A.O10–11	GAM$_0$ DIŠ.UD *ša*$_2$ [x] *u* ⸢**sig**⸣ / *ša*$_2$ *taš-kun* **du** →
	D.O12	**a.ra**$_2$ DIŠ.⸢UD *ša*$_2$⸣ [x x x x xx x] →
	F.X3'	GAM DIŠ.UD *ša*$_2$ **nim** *u* [x x xx x] →
	G.O5'	*ša*$_2$ **t**[**ag**$_4^?$ *šu-u*$_2$ **a.ra**$_2$] **bar.nun egir**-*tu*$_2$ *ša*$_2$ *taš-kun* **du**-*ak* →

At the command of Bēl and Bēltija may it remain intact. **A**

P1 (ŠU₂)

In order for you to construct $ŠU_2$. **AD**

1 **AD**

If the time by which the night has progressed exceeds the ⸢length of⸣ night you subtract ⸢the length of night⸣ [from] (the time by which) ⸢the night has progressed⸣. [D]What remains:[D] the day (of $ŠU_2$) is on the same day. If the time by which the night has progressed is less than the length of night you add (the time by which) the night has progressed and the length of ⸢daylight⸣ together, and you subtract 1 day.

$M > D \Rightarrow M - D\ [\rightarrow N_{ŠU_2}]$
$d_{ŠU_2} = d_{FM}$
$M < D \Rightarrow M + C\ [\rightarrow N_{ŠU_2}]$
$d_{ŠU_2} = d_{FM} - 1$

2 **AD**

What remains you multiply by 0;0,10. [D]What comes out for you[D], this you put down as the displacement of the Sun. You multiply it by [the displacement of the Moon] for that month, this you put down as the displacement of the Moon.

$N_{ŠU_2} \cdot 0;0,10 \rightarrow \Delta B_{\odot}$
$N_{ŠU_2} \cdot v_{☽} \rightarrow \Delta B_{☽}$

3 **AD**

You subtract the displacement of the Sun from the displacement of the Moon and [you] ⸢put it down⸣ as the elongation.

$\Delta B_{☽} - \Delta B_{\odot} \rightarrow \eta$

4 **AD**

You put down the position of the Moon and the position of the Sun. The displacement by which the Sun moved you subtract from the position of the Sun. The displacement by which the Moon moved you subtract [from] the position of the Moon.

$B'_{\odot} = B^{fm}_{\odot} - \Delta B_{\odot}$
$B'_{☽} = B^{fm}_{☽} - \Delta B_{☽}$

5 **AD**

(The amount) by which the position of the Moon and the position of the Sun exceed 10 (°) of the zodiacal sign you multiply by 0;0,16. You add it with the corresponding *ṣiliptu*-coefficient for the zod[iac] if it is increasing, subtract if it is decreasing. You accumulate (them) together, compute half of it and ⸢put⸣ it down as the *ṣiliptu*-coefficient.

$q = q_{10} \pm (B' - B_{10}) \cdot 0;0,16$
$(q_{☽} + q_{\odot}) \cdot \frac{1}{2} \rightarrow q'$

? **G**

[...]... add [... which] you had multiplied by the elongation? [...] and ...

$\ldots q' \cdot \eta$?

6 **ADF**

You put down the Moon's 'height and depth'. The displacement of the Moon you multiply by 0;4. You subtract it from the Moon's 'height and depth' if it is increasing, [add] if it is decreasing.

$\beta' = \beta^{fm} \mp \Delta B_{☽} \cdot 0;4$

6 **G**

[All the *ṣiliptu*-coefficients for 'height and depth'.] The displacement of ⸢the Moon, this⸣ [you] mul[tiply] by 0;4, what comes out for you, [you ...] to [...] which you had put down. If it is increasing you add this to it, and ⸢if⸣ [it is decreasing] you subtract [this] from it.

$\beta' = \beta^{fm} \pm \Delta B_{☽} \cdot 0;4$
(error?)

7 **A[D]F**

(The amount) by which the position of the Moon and the position of the Sun exceed 10 (°) of the zodiacal sign you multiply by 0;0,8. You add it with the ⸢corresponding *ṣiliptu*-coefficient⸣ for 'height and depth' if it is increasing, subtract if it is decreasing. [A[D]]You accumulate (them) together, you compute half of it, and[A[D]]

$r = r_{10} \pm (B' - B_{10}) \cdot 0;0,8$
$(r_{☽} + r_{\odot}) \cdot \frac{1}{2}\ [\rightarrow r']$ **A[D]**

8 **ADF**

you multiply it by the DIŠ.UD of 'height and ⸢depth⸣' which you had put down.

$r' \cdot \beta'\ [\rightarrow R]$

8 **G**

What rem[ains?, this] you multiply [by] the last *ṣiliptu*-coefficient which you had put down.

$\beta' \cdot r'\ [\rightarrow R]$

9	**A**.O11	BE GAM$_0$ **bar.nun** **du** →
	D.O12	[x xx xx x] →
	F.X3'–4'	[x x] / [xx x] →
	G.O5'	BE **a.ra$_2$** ⸢**bar.nun**⸣ [xx]
11	**G**.O6'	*ki-i* **nim** *u* ⸢**sig** xx$^?$⸣ [xx$^?$ *ša$_2$* a].**ra$_2$** **bar.nun egir**-*tu$_2$ tal-lik* →
	A.O11	*ki* **nim** *ana* BE **la$_2$** →
	D.O12	[xx x x BE *u* **bar.nun** x]
	F.X4'	[x x x x x] →
	G.O6'–7'	*ki-i* **nim nim** *u lu*-⸢*u$_2$* **la$_2$ bur$_3$**⸣ *šu-u$_2$* / ⸢**ta**⸣ BE *ša$_2$* **a.ra$_2$ bar.n**[**un** *tal-lik* ⸢**la$_2$**⸣ →
	A.O11	*ki* [x] *ana* BE **tab** →
	D.O13	*ki-i* **sig** *ana* ⸢BE⸣ *u* **bar.nun** **tab** →
	F.X4'	[x x] ⸢*a-na* BE⸣ **tab** →
	G.O7'–8'	*ki-i* **bur$_3$ bur$_3$** *u lu-u$_2$* **bur$_3$ la$_2$** *šu-u$_2$* **ki** BE *ša$_2$* **a.ra$_2$** / **bar.nun** *tal-lik* **tab** →
	A.O11	*ana* **šu$_2$** ⸢**me** xxx⸣
	D.O13	⸢*ana* **šu$_2$** **e**⸣ →
	F.X4'	*a-na* **šu$_2$** **me** ⸢x⸣ [xxx]
	G.O8'	*ana* **š**[**u$_2$** xx] ⸢**gar**⸣-*an* →
12	**A**.O12	*ki* DIŠ.UD *ša$_2$* **nim** *al-la* BE *u* **bar.nun diri**-*at$_2$* BE$^?$ *u* **bar.nun ta** *lib$_3$-bi* ⸢**nim**⸣
	D.O13–14	⸢*ki-i* DIŠ⸣.[UD x x xx x x xx x] / BE *u* **bar.nun** ⸢**ta**⸣ **ša$_3$**-*šu$_2$* **la$_2$** →
	F.X5'	[x xx x x xx x x xx] ⸢**diri**$^?$-*at*$^?$ BE *u$_3$* **bar.nun**⸣ **ta** *lib$_3$-b*[*i* x] →
	G.O8'–9'	⸢*ki*⸣-*i* DIŠ.UD *ša$_2$* **nim** *al-la* BE *u* **bar.nun diri**-*at* / BE *u* **bar.nun ta** *lib$_3$*-⸢*bi-šu$_2$*⸣ **la$_2$** →
	A.O13	*ana* **na gar**-*an* →
	D.O14	*ana* **na e šu$_2$** *ana* [**na gur** x x xx] →
	F.X5'–6'	[x x x] / [x x xx] →
	G.O9'	*ana* **na** ⸢**e**⸣-*bi* **šu$_2$** *ana* **na gur** *u$_4$-mu* **nu man**-*ni* →
13	**A**.O13	6 **ki** *si-man* **igi**-*u$_2$* **tab** *ki* **igi**-*u$_2$* **du$_3$**-*uš* 1-*en u$_4$-mu* **la$_2$** *ana* **šu$_2$ e**-*bi*
	D.O14–15	[x x xx] / **igi**-*u$_2$* **tab** *ki*-⸢*ma*⸣ **igi**-*u$_2$* **du$_3$**-*uš ana* **šu$_2$** ⸢**e**⸣ [1-*en* **me** x]
	F.X6'	[x x xx **igi**]-*u$_2$* **tab** ⸢*ki* **igi**⸣-*u$_2$* ⸢**du$_3$**⸣-*uš* 1-*en* **me** **l**[**a$_2$** x x x]
	G.O9'–10'	6 **ki** *si-man*/ **igi**-*u$_2$* **tab gin$_7$** **igi**-*u$_2$* **du$_3$**-*uš ana* **šu$_2$** ⸢**e**⸣-*bi* 1-*en* **me la$_2$** ⸢**šu$_2$**⸣ *ana ar$_2$-šu$_2$* **la$_2$**-*is*

P2

1	**A**.O14	**na** *ana* **du$_3$**-*ka ki si-man* ⸢*ša$_2$* **gi$_6$ du**⸣ *al-la* LA$_2$ *ša$_2$* **gi$_6$** ⸢**diri**⸣ LA$_2$ *ša$_2$* **gi$_6$** ⸢**ta**⸣ **gi$_6$ du**
	D.O16	**na** *ana* **du$_3$**-*ka ki-i si-man ša$_2$* **gi$_6$ du** [xx x x x xx xx x x x x] →
	F.X7'	[x x xx x xx x xx *al-l*]*a* LA$_2$ *ša$_2$* **gi$_6$** *i-ṣi* **gi$_6$ du ta** LA$_2$ *ša$_2$* [x] →
	G.O11'	**na** ⸢*ana*⸣ **du$_3$**-*ka* ⸢*ki-i si*⸣-*man ša$_2$* **gi$_6$ du** ⸢*al*⸣-*la* LA$_2$ *ša$_2$* **gi$_6$** *i*-⸢*ṣi* **gi$_6$**⸣ **du ta** LA$_2$ *ša$_2$* **gi$_6$** →
	A.O15	**nim** *ša$_2$* **tag$_4$ ta** 6 ⸢**nim**⸣ [1-*e*]*n* **me tab** *ki si-man ša$_2$* **gi$_6$ du** *al-la* L[A$_2$ x g]**i$_6$** *i-ṣa*
	D.O16–17	[x] / ⸢*ša$_2$* **tag$_4$**⸣ **me** *ina* **me**-*šu$_2$ ki-i si-man ša$_2$* **gi$_6$** [**du** xx x x x x] →
	F.X7'–8'	[x] / [x x x x xx x xx x xx *al*]-*la* LA$_2$ *ša$_2$* **gi$_6$ diri** →
	G.O11'–12'	[x] / *ša$_2$* **tag$_4$** [x x **me**-*š*]*u$_2$* ⸢*ki*⸣-*i si-man* ⸢*ša$_2$*⸣ **gi$_6$ du** *al*-⸢*la*⸣ LA$_2$ *ša$_2$* **gi$_6$ diri**
	A.O16	**gi$_6$ du ta** LA$_2$ *ša$_2$* **gi$_6$ nim me** *ina* **me**-*šu$_2$* →
	D.O17–18	[LA$_2$ *ša$_2$* **gi$_6$ ta gi$_6$ du nim**] / [*ša$_2$* **tag$_4$ ta**] ⸢6 **la$_2$** 1-*en*⸣ **me tab** →
	F.X8'–9'	⸢LA$_2$ *ša$_2$* **gi$_6$ ta gi$_6$**⸣ [**du** x] / [x x x x x xx x x] →
	G.O13'	⸢LA$_2$ *ša$_2$*⸣ [x x x x x x] ⸢**tag$_4$**⸣ **ta** ⸢6$^?$⸣ **la$_2$** 1-*en* **me tab** →
2	**A**.O16–17	*ša$_2$* **tag$_4$** GAM$_0$ 10 **du** *ša$_2$* ⸢**e$_{11}$**-*ma šu*⸣-*u$_2$* / **zi** *šamaš$_2$* **gar**-*an* →
	D.O18–19	*ša$_2$* **tag$_4^?$** ⸢**a.ra$_2$** 10$^?$⸣ [x x xx xx x x xx] →
	F.X9'	[x x x x x x **e$_{11}$**-*m*]*a šu-u$_2$* **z**[**i** x xx] →
	G.O13'–14'	*ša$_2$* **tag$_4$** **a.ra$_2$** 10 **du** / *ša$_2$* ⸢**tag$_4$**⸣-[xx xxx] **zi** ^{d}[**utu**] →

9 **A[DF]G**

You multiply the elongation by the *ṣiliptu*-coefficient.

$\eta \cdot q' \; [\rightarrow Q]$

11 **ADF**

If (the Moon) is ‘high’ you diminish the elongation [D][‘and’ the *ṣiliptu*-coefficient][D] by it. If it is low you add it to the elongation [D]‘and’ the *ṣiliptu*-coefficient[D]. You call it $\check{S}U_2$. [AF]⌜...⌝[AF]

high $\Rightarrow Q - R_{\uparrow} \rightarrow \check{S}U_2$
low $\Rightarrow Q + R_{\downarrow} \rightarrow \check{S}U_2$

11 **G**

If the ‘height and depth’ which you had multiplied by the last *ṣiliptu*-coefficient: If (the Moon) is high, ascending, or ⌜high?, descending?⌝, you subtract this from the elongation which [you had multiplied] by the *ṣiliptu*-coefficient. If it is low, descending, or low, ascending, you add this with the elongation which you had multiplied by the *ṣiliptu*-coefficient. ⌜You put⌝ it down as ⌜$\check{S}U_2$⌝.

high $\Rightarrow Q - R_{\uparrow}$
low $\Rightarrow Q + R_{\downarrow}$
$\rightarrow \check{S}U_2$

12 **ADFG**

If the DIŠ.UD of ‘height’ exceeds the elongation ‘and’ the *ṣiliptu*-coefficient you {[A]deduct|[B[F]G]subtract} the elongation ‘and’ the *ṣiliptu*-coefficient from it. {[A]You put it down as NA.|[DFG]You call it NA.} [DG]$\check{S}U_2$ becomes NA.[DG] [DFG]You do not change the day[DFG].

$R_{\uparrow} > Q \Rightarrow R_{\uparrow} - Q \rightarrow \mathrm{NA}$
$\mathrm{d}_{\mathrm{NA}} = \mathrm{d}_{\check{S}U_2}$

13 **ADFG**

You add 6,0 to the previous time, you do as before. {[AF]You subtract 1 day. You call it $\check{S}U_2$.|[DG]You call it $\check{S}U_2$. You subtract 1 day.} [G]$\check{S}U_2$ turns backward.[G]

$N_{\check{S}U_2} + 6{,}0 \rightarrow N_{\check{S}U_2}$
$\mathrm{d}_{\check{S}U_2} - 1 \rightarrow \mathrm{d}_{\check{S}U_2}$

P2 (NA)

In order for you to construct NA. **AD[F]G**

1 **A**

If the time by which the night has progressed exceeds the length of night, you deduct the length of night from (the time by which) the night has progressed. What remains you subtract from 6,0. You add 1 day. If the time by which the night has progressed is less than the length [of] night, you deduct (the time by which) the night has progressed from the length of night. The day (of NA) is on that day.

$M > D \Rightarrow 6{,}0 - (M - D) \; [\rightarrow N_{\mathrm{NA}}]$
$\mathrm{d}_{\mathrm{NA}} = \mathrm{d}_{\mathrm{FM}} + 1$
$M < D \Rightarrow D - M \; [\rightarrow N_{\mathrm{NA}}]$
$\mathrm{d}_{\mathrm{NA}} = \mathrm{d}_{\mathrm{FM}}$

1 **DFG**

If the time by which the night has progressed is less than the length of night, you deduct (the time by which) the night has progressed from the length of night. What remains: the day (of NA) is on that day. If the time by which the night has progressed exceeds the length of night, you deduct the length of night from (the time by which) the night has progressed. What remains you subtract from 6,0. You add 1 day.

$M < D \Rightarrow D - M \; [\rightarrow N_{\mathrm{NA}}]$
$\mathrm{d}_{\mathrm{NA}} = \mathrm{d}_{\mathrm{FM}}$
$M > D \Rightarrow 6{,}0 - (M - D) \; [\rightarrow N_{\mathrm{NA}}]$
$\mathrm{d}_{\mathrm{NA}} = \mathrm{d}_{\mathrm{FM}} + 1$

2 **A[B]DFG**

What remains you multiply by 0;0,10. What ⌜comes out, this⌝ {[A[B]DF]you put down as|[G]is} the displacement of the Sun. You multiply {[A[B]DF]it|[G]this} by the Moon’s displacement for that month, and [G][what comes out,][G] this you put down as the displacement of the Moon.

$N_{\mathrm{NA}} \cdot 0;0{,}10 \rightarrow \Delta B_{\odot}$
$N_{\mathrm{NA}} \cdot v_{\rightmoon} \rightarrow \Delta B_{\rightmoon}$

	Source	Text
	A.O17	GAM$_0$ zi sin ša$_2$ iti-šu$_2$ du-ma šu-u$_2$ zi sin ⌜gar-an⌝
	B.O1’	[xx xx x x xx] →
	D.O18–19	[xx] / [x x x xx xx šu]-⌜u$_2$⌝ zi sin gar-an →
	F.X9’–10’	[x x x x xx xx] / [xx x x xx] →
	G.O14’–15’	šu$_{10}$-a-ti$_3$ a.ra$_2$ zi ^{d}sin / ša$_2$ it[i-šu$_2$ xx ša$_2$ e$_{11}$-ma šu]-u$_2$? ana z[i? x xx] →
3	A.O18	zi šamaš$_2$ ta zi sin nim-ma BE gar-an →
	B.O1’	[x] ⌜šamaš$_2$ ta⌝ zi sin n[im-ma x xx]
	D.O19	[x x x x x xx x xx]
	E.O1’	[xx x xx] →
	F.X10’	[x x x x x nim]-ma BE [xx] →
	G.O15’–16’	⌜zi? ^{d}utu⌝ [x z]i? [x] / [xx xxx ša$_2$ e$_{11}$-m]a šu-⌜u$_2$⌝ [x xx] →
4	A.O18	ki sin u ki šamaš$_2$ gar-an zi ⌜ša$_2$ šamaš$_2$⌝
	B.O2’	[x x x x šama]š$_2$ ⌜gar-an zi ša$_2$⌝ šamaš$_2$ →
	D.O20	[x x x x x xx x x x] →
	E.O1’	ki s[in? x x x xx x x x] →
	F.X10’?	[x x x x x xx x x x]
	G.O16’–18’	[x x x] / [x x xx] zi ^{d}utu [xxxxxxxx] / [xxxxxx] ⌜x⌝-ka →
	A.O19	⌜zi⌝ ki ki šamaš$_2$ tab zi ša$_2$ sin zi ki ki sin tab →
	B.O2’–3’	zi ki ki šamaš$_2$ tab zi ⌜ša$_2$ sin⌝ z[i x x x] / [x] →
	D.O20–21	[x x x x x x x x x x x x] / [x] →
	E.O1’–2’	[x x x x] / [x x x x z]i ki ki si[n? x] →
	G.O18’	ki ki ^{d}utu [x x xxxxx]
5	A.O19–20	ki sin u ki šamaš$_2$ ša$_2$ al-la 10 [xx] ⌜diri →
	B.O3’	⌜ki⌝ sin u ki šamaš$_2$ ša$_2$ al-la me lu-maš diri →
	D.O21	[x x x x x x xx x xx x] →
	E.O2’–3’	[x x x x x x xx x] / [xx x] →
	G.O19’	ša$_2$ [xx x xx x] →
	G.O19’–20’	10 lu-maš ta lib$_3$-[bi-šu$_2$ la$_2$ xxxxxxxx] / ša$_2$ [xx]
	A.O20	⌜GAM$_0$⌝ 16 ⌜du⌝ ki bar.nun ša$_2$ lu-⌜maš ša$_2$⌝-šu$_2$ ki tab tab ki la$_2$ la$_2$ →
	B.O3’–4’	GAM 16 du ki [xx] / ⌜ša$_2$ lu⌝-maš ša$_2$-šu$_2$ ki tab tab ki la$_2$ la$_2$ →
	D.O21	[xx xx x x bar.n]un ša$_2$ lu-maš ša$_2$-šu$_2$ ⌜ki⌝-[i x x xx x x]
	E.O3’	[GA]M 16 du ki bar.nun ša$_2$ lu-ma[š xx x x x x x x] →
	G.O20’	[x xx x x xx] ša$_2$ lu-maš ša$_2$-šu$_2$ ⌜ki⌝-[i x x xx x x]
	A.O20–21	ana ugu / [a-ḫa]-miš$_2$ GAR.⌜GAR-ma⌝ $\frac{1}{2}$-šu$_2$-nu GIŠ-ma ana bar.nun gar-an →
	B.O4’–5’	ana ugu a-ḫa-miš$_2$ GAR.⌜GAR-ma⌝ [xxx xx] / ana bar.nun gar-an →
	D.O22	[x x xxx xxx xxx GIŠ].⌜a⌝ ana bar.nun gar-an →
	E.O3’–4’	[x x] / [xxx GAR].GAR-ma $\frac{1}{2}$-⌜šu$_2$ GIŠ⌝-ma ana bar.nun gar-an →
	G	
?	G.O21’–22’	ša$_2$-niš ⌜zi? ^{d}utu⌝ šu$_{10}$-a-ti$_3$ ša$_2$ zi ⌜^{d}⌝[sin? xxxxxxxx] / ⌜x⌝] [x] te-e-ṣe-pi šu-u$_2$ ⌜ki⌝ [xxxxxxxx]
6	A.O21	nim u sig ša$_2$ sin gar-an zi sin
	B.O5’	nim u sig [x] sin gar-an zi sin →
	D.O22–23	ni[m x x x x] / [xx x x] →
	E.O4’	n[im x x x x xx x x]
	A.O22	[x] ⌜4⌝ du ⌜ki nim u⌝ sig ša$_2$ sin ki tab tab ki la$_2$ la$_2$ →
	B.O5’–6’	GAM 4 du ⌜ki⌝ [x x x] / ša$_2$ ⌜sin⌝ ki tab tab ki la$_2$ la$_2$ →
	D.O23	[xx x x x x x] ⌜sig⌝ ša$_2$ sin (eras.) ⌜ki-i⌝ tab tab ki-[i x x]
	E.O5’	[x] ⌜4⌝ [x x x x x x] ki tab tab ki la$_2$ ⌜la$_2$⌝ →

3 **AB[DE]FG**	
You subtract the displacement of the Sun from the displacement of the Moon, and {$^{\mathrm{AB[DE]F}}$you put it down|^{G}what comes out$^?$, this you put down} as the elongation.	$\Delta B_{\rightmoon} - \Delta B_{\odot} \rightarrow \eta$
4 **AB[D]E[F]**	
You put down the position of the Moon and the position of the Sun. The displacement by ⸢which⸣ the Sun moved you add to the position of the Sun. The displacement by which the Moon moved you add to the position of the Moon.	$B'_{\odot} = B^{\mathrm{fm}}_{\odot} + \Delta B_{\odot}$ $B'_{\rightmoon} = B^{\mathrm{fm}}_{\rightmoon} + \Delta B_{\rightmoon}$
4 **G**	
[...] The displacement of the Sun [...] to you, [you add] to the position of the Sun [...]	$B'_{\odot} = B^{\mathrm{fm}}_{\odot} + \Delta B_{\odot}$
5 **ABDE**	
(The amount) by which the position of the Moon and the position of the Sun exceed 10 (°) of the zodiacal sign you multiply by 0;0,16. You add it with the corresponding *ṣiliptu*-coefficient for the zodiac if it is increasing, subtract if it is decreasing. You accumulate (them) together, you compute half of it and you put it down as the *ṣiliptu*-coefficient.	$q = q_{10} \pm (B' - B_{10}) \cdot 0;0,16$ $(q_{\rightmoon} + q_{\odot}) \cdot \frac{1}{2} \rightarrow q'$
5 **G**	
[...] That which [exceeds 10 (°) of the zodiacal sign: you subtract] 10 (°) of the zodiacal sign from it, [...] what [comes out for you, you multiply by 0;0,16. You add it with] the corresponding *ṣiliptu*-coefficient for the zodiac [if it is increasing, subtract if it is decreasing.]	$q = q_{10} \pm (B' - B_{10}) \cdot 0;0,16$
? **G**	
Alternatively: ⸢the displacement$^?$ of the Sun⸣, this of$^?$ the displacement [...] ... [... which] you had doubled$^?$, this [...] to [...]	$...\Delta B_{\odot}..?$
6 **ABDE**	
You put down the Moon's 'height and depth'. The displacement of the Moon you multiply by 0;4. You add it with the Moon's 'height and depth' if it is increasing, subtract if it is decreasing.	$\beta' = \beta^{\mathrm{fm}} \pm \Delta B_{\rightmoon} \cdot 0;4$

7	**A**.O22	**ki** *sin u* **ki** *šamaš*$_2$ *ša*$_2$ *al-la* 10
	B.O6’	**ki** *sin u* **ki** *šamaš*$_2$ *ša*$_2$ ⸢*al-la* 10⸣ →
	D.O24	[x x x x x x] ⸢*al*⸣-*la* 10 →
	E.O5’	[xxxxxxxxxx]
	A.O23	[xx] **diri** GAM$_0$ 8 **du** **ki bar.nun** *ša*$_2$ **nim** *u* **sig** *ša*$_2$-*šu*$_2$ *ki* **tab tab** *ki* **la$_2$ la$_2$** →
	B.O6’–7’	*lu-maš* **d**[**iri** x x x] / **ki bar.nun** *ša*$_2$ **nim** *u* **sig** *ša*$_2$-*šu*$_2$ *ki* **tab tab** *ki* **la$_2$ la$_2$** →
	D.O24–25	*lu-maš* ⸢**diri**⸣ **a.ra$_2$** ⸢8 **du**⸣ [x xx x x x x xx] / [xx x x *k*]*i*-⸢*i*⸣ **la$_2$ la$_2$** →
	E?	
	G.O23’	[x x xx x] **nim** *u* **sig** *ša*$_2$-*šu*$_2$ *ki*-[*i* x x xx x x]
	A.O23–24	*ana muḫ-ḫi* / [*a-ḫa*]-*miš*$_2$ GAR.G[AR-*m*]*a* $\frac{1}{2}$-*šu*$_2$-*nu* GIŠ-*ma* →
	B.O7’–8’	*ana* **ugu** *a-ḫa-m*[*iš*$_2$] / GAR.GAR-*ma* $\frac{1}{2}$-*šu*$_2$-*nu* GIŠ-⸢*ma*⸣→
	D.O25	*ana* **ugu** *a-ḫa-miš*$_2$ GAR.GAR-⸢*ma*⸣ [xxx xx] →
	EG?	
?	**G**.O24’–25’	[xxxxx] ⸢d⸣*sin ša*$_2$ **iti b**[**i**? xxxxxxxxxx] / [xxxx] ⸢xx⸣ [xxxxxxxxxx]
8	**A**.O24	GAM$_0$ DIŠ.UD *ša*$_2$ **nim** *u* **sig** *ša*$_2$ *taš-kun* ⸢**du**⸣ →
	B.O8’	GAM DIŠ.UD *ša*$_2$ **nim** *u* **sig** *ša*$_2$ *taš*-⸢*kun*⸣ **du** →
	D.O25–26	[xx xx x x x x] / [x] ⸢**gar**-*un*⸣ **du** →
	E.O6’	[x xx x x x x] *ša*$_2$ **gar**-*un* ⸢**du**⸣ →
9	**A**.O24	⸢BE GAM$_0$⸣ [xx x]
	B.O8’–9’	BE [x xx] / **du** →
	D.O26	BE **a.ra$_2$** **bar.nun** **du** →
	E.O6’	[x x xx x]
11	**A**.O25	[x x x] BE **tab** *ki* **sig** *ana* BE **la$_2$** *ana* **na** **e**-*bi* →
	B.O9’	*ki* **nim** *ana* BE ⸢**tab** *ki*⸣ [x x x x] ⸢*ana* **na**⸣ [xx]
	D.O26	*ki-i* **la$_2$** *ana* BE *u* **bar.nun** [x xx x x BE *u* **bar.nun** x x x x]
	E.LE1	[x x x x x] ⸢**bar.nun** **tab** *ki*⸣ [x x xxx x x x x] →
12	**A**.O25–26	⸢*ki*⸣ DIŠ.UD *ša*$_2$ **nim** *al-l*[*a* x] / [x xx **dir**]**i** BE *u* **bar.nun** **ta** *lib*$_3$-*bi* **nim** →
	D.27	*ki-i* DIŠ.UD ⸢*ša*$_2$ **nim**⸣ [*al*]-*la* BE *u* **bar.nun** **diri**-*at* BE *u* [xx x xx x] →
	E.LE1–2	[x xx x x] / [xx x x xx] ⸢**diri**⸣-*at* BE *u* **bar.nun** **ta** *lib*$_3$-*bi* **nim** →
	G.R1’	[x xx xx x x **bar**].⸢**nun** **ta**⸣ *lib*$_3$-*bi*-⸢*šu*$_2$⸣ [x] →
	A.O26	⸢*ša*$_2$ **tag$_4$**⸣ *ana* **šu$_2$** **e**-*bi* **m**[**e nu man**-*ni*]
	D.O27–28	[x x x x x] / **na** *ana* **šu$_2$** **gur** ⸢**me**⸣ [x xx] →
	E.LE2–3	[x x x x x x x x x] / [x x xx] →
	G.R1’	[x x x x x x x x x x x xx]
13	**A**.O27	[6 **ki** *si*]-*man* **igi**-*u*$_2$ **tab** *ki* **igi**-*u*$_2$ **du$_3$**-*uš* 1-*en* **me** ⸢**tab** *ana* **na** **e**⸣-[*bi* xxxx?]
	D.O28	[x x] ⸢*si*⸣-*man* **igi**-*u*$_2$ **tab gin$_7$** [xx xx xx x x x x x]
	E.LE3	[x **k**]**i** *si-man* **igi**-*u*$_2$ **tab gin$_7$** **igi**-*u*$_2$ [xx xx x x x x x]
	G.R2’	⸢6 **ki**⸣ *si-man* **igi**-*u*$_2$ **tab gin$_7$** **igi**-*u*$_2$ **d**[**u$_3$**-*uš* xx x x x x xxxx]

T1

	A.O28	[xx x xx x š]**u$_2$**→
	D.Ri1	[**ba**]**r.nun** *ša*$_2$ *lu-maš ša*$_2$ **šu$_2$**
	E.Ri1	[xx x xx x x]
	G.Ri3’	**bar.nun** *ša*$_2$ *lu-maš ša*$_2$ **šu$_2$**

7 **ABD[E]G**

(The amount) by which the position of the Moon and the position of the Sun exceed 10 (°) of the zodiacal sign you multiply by 0;0,8. You add it with the corresponding *ṣiliptu*-coefficient for 'height and depth' if it is increasing, subtract if it is decreasing. $^{\mathbf{ABD[E]}}$You accumulate (them) together, you compute half of it, and$^{\mathbf{ABD[E]}}$

$r = r_{10} \pm (B' - B_{10}) \cdot 0;0,8$

$(r_{☽} + r_{☉}) \cdot \frac{1}{2}\ [\rightarrow r']$ **ABD[E]**

? **G**

[...] the Moon for that month [...] ... [...]

?

8 **ABDE**

you multiply it by the DIŠ.UD of 'height and depth' which you had put down.

$r' \cdot \beta'\ [\rightarrow R]$

9 **ABD**

You multiply the elongation by the *ṣiliptu*-coefficient.

$\eta \cdot q'\ [\rightarrow Q]$

11 **ABDE**

If (the Moon) is 'high' you add it to the elongation $^{\mathbf{DE}}$'and' the *ṣiliptu*-coefficient$^{\mathbf{DE}}$. If it is 'low' you diminish the elongation $^{\mathbf{DE}}$['and' the *ṣiliptu*-coefficient]$^{\mathbf{DE}}$ by it. You call it NA.

high $\Rightarrow Q + R_{\uparrow} \rightarrow$ NA
low $\Rightarrow Q - R_{\downarrow} \rightarrow$ NA

12 **ADEG**

$^{\mathbf{AD}}$If the DIŠ.UD of 'height' (error for 'depth')$^{\mathbf{AD}}$ exceeds the elongation 'and' the *ṣiliptu*-coefficient, you deduct the elongation 'and' the *ṣiliptu*-coefficient from it. ⌜What remains⌝ you call ŠU$_2$. $^{\mathbf{DEG}}$NA becomes ŠU$_2$.$^{\mathbf{DEG}}$ [You do not change] ⌜the day⌝.

$R_{\downarrow} > Q \Rightarrow R_{\downarrow} - Q \rightarrow$ ŠU$_2$!

$d_{ŠU_2} = d_{NA}$

13 **ADEG**

You add [6,0 to] the previous time, you do as before. You add 1 day. You call it NA.

$N_{NA} + 6,0 \rightarrow N_{NA}$;
$d_{NA} + 1 \rightarrow d_{NA}$

T1.a (Coefficient q_{set}) **ADEG**

Ṣiliptu-coefficient for the zodiac for setting.

	A.O28	⸢1.24 **ḫun** 1.16⸣ [x] ⸢1.8 **maš**⸣ [x x xx x xx x]
	D.Ria2–7	1.24 **ḫun** / [1].16 **mul**$_2$ / [1].8 **maš** / [x al]**la** / [xx] ⸢**a**⸣ / [4]4 ⸢**absin**⸣
	E.Ria2–7	[xx x / xxx x / xx x / x x / xx x / xx x]
	G.Ria4'–9'	1.24 **ḫun** / 1.16 **mul**$_2$ / 1.8 **maš** / 1 **alla** / 52 **a** / 44 **absin**

	A.O29	[xx x xx xx xx x x x xx x xxx xx]
	D.Ria8,ib2–7	36 **rin**$_2$ / 44 **gir**$_2$**.tab** / 52 **pa** / 1 **maš**$_2$ / 1.8 **gu** / 1.16 *zib*.**me** / 1.24 **ḫun**
	E.Ria8,ib2–6	[xx x] / [xx xx] / [xx xx] / [x] ⸢**maš**$_2$⸣ / [xx] ⸢**gu**⸣ / [xxx xx]
	G.Rib4'–9'	36 ⸢**rin**$_2$⸣ / 44 ⸢**gir**$_2$**.tab**⸣ / 52 ⸢**pa**⸣ / 1 **maš**$_2$ / 1.8 **gu** / 1.16 *zib*.**me**

	A.O30	[xx x x x x] ⸢*ša*$_2$ **šu**$_2$⸣ →
	D.Rii1	**bar.nun** *ša*$_2$ [x x x x x]
	E.Rii1	⸢**bar**⸣**.nun** *ša*$_2$ **nim** *u* **sig** *ša*$_2$ **šu**$_2$
	G.Riia3'	**b**[**ar.nun** x x x x x x]

	A.O30	18 [x xx x xx x x x xx x xx x]
	D.Riia2–7	18 **ḫun** / 22 **mul**$_2$ / 26 **maš** / 30 **alla** / 34 **a** / 38 **absin**
	E.Riia2–7	18 **ḫun** / 22 **mul**$_2$ / 26 **maš** / 30 **alla** / [xx] **a** / [xx **abs**]**in**
	G.Riia4'–9'	[xx x] / 2[2 x] / 2[6 x] / 3[0 x] / [xx x] / [xx x]

	A.O31	[xx x] 36 **gir**$_2$**.tab** 34 **pa** 30 ⸢**maš**$_2$ 26 **gu** 22 *zib*⸣.[**me**]
	D.Riib2–7	[xx x] / [xx xx] / [xx x] / [x x] / [xx x] / [xx xx]
	E.Riib2–7	42 **rin**$_2$ / 36 **gir**$_2$**.tab** / 34 **pa** / 30 **maš**$_2$ / 26 **gu** / 22 *zib*.**me**
	G.Riib4'–9'	[xx x] / [xx xx] / [xx x] / [x x] / [xx x] / [xx xx]

P3

	A.R1	[x x xx] →
	D.R9	**me** *ana* **du**$_3$-*ku* →
	G.R10'	**me** *ana* **du**$_3$-*ka* →

2	**A**.R1	[**gi**$_6$ **d**]**u** GAM$_0$ 10 **du** ⸢*šu-u*$_2$ **zi**⸣ *šamaš*$_2$ ⸢**gar**⸣-*an* ⸢**me** *ina* **me**-*šu*$_2$⸣ →
	D.R9	⸢**gi**$_6$⸣ **du a.ra**$_2$ 10 **du** **me** *ina* **me**-*šu*$_2$ **me** *ana* ⸢x⸣ →
	E.R8	[xx xx xx xx x]-*šu*$_2$ **me** *ana ar*$_2$-⸢*ki*-[xx] →
	G.R10'	**gi**$_6$ **du a.ra**$_2$ 10 **du**-*ku* **me** *ina* 1 **me** →

	A.R1–2	[GAM$_0$ **zi** *sin ša*$_2$ **iti**-*šu*$_2$] / [**du**-*ma šu-u*$_2$ **zi**] *sin* **gar**-*an* →
	D.R9	[xxxxxxxx]
	E.R8	[xxxxx]
	G.R10'	[xxxxxxxxxx]

3	**A**.R2	⸢**zi** *šamaš*$_2$ **ta**⸣ **zi** *sin* **nim**-[*ma* BE **gar**-*an*]
	DG	

4	**A**.R3	[x x x x x xx x] ⸢*ša*$_2$ *šamaš*$_2$⸣ [**zi ta k**]**i** *šamaš*$_2$ **nim** ⸢**zi** *ša*$_2$⸣ [*sin* **zi ta ki** *sin* **nim**
	C.R1–2	**ki** *sin u* **ki** *šamaš*$_2$ **gar zi** *ša*$_2$ *sin is-suḫ* **ta ki** *s*[*in* **la**$_2$] / *u* **zi** *ša*$_2$ *šamaš*$_2$ *is-suḫ* **ta ki** *šamaš*$_2$ **la**$_2$ -
	DG	

5	**A**.R4	[x x x x x x xx 10 *lu-ma*]*š* **diri** GAM$_0$ 16 ⸢**du ki bar.nun** *ša*$_2$ *lu*⸣-[*maš*] *ša*$_2$-[*šu*$_2$]
	C.R2–4	**ki** *sin u* **ki** *š*[*amaš*$_2$] / *ša*$_2$ *al* 10 *lu-maš* **diri** GAM$_0$ 16 **du ki bar.nun** *ša*$_2$ *lu-maš* / *ša*$_2$-*šu*$_2$ →
	DG	

	A.R5	[*ki* **tab tab** *ki* **la**$_2$ **l**]**a**$_2$ *ana* ⸢*muḫ-ḫi a-ḫa*⸣-*miš*$_2$ GAR.GAR-*ma* $\frac{1}{2}$-⸢*šu*$_2$⸣-*nu* **GIŠ.a** [x xx xx]
	C.R4–5	*ki* **tab** ⸢**la**$_2$⸣ *ki* **la**$_2$ **tab** *ana* **ugu** {*a* **ugu**} *a-ḫa-miš*$_2$ GAR.GAR-*m*[*a*] / $\frac{1}{2}$-*šu*$_2$-*nu* GIŠ-*ma ana* **bar.nun gar**
	DG	

1;24 Ari, 1;16 Tau, 1;8 Gem, 1 Cnc, 0;52 Leo, 0;44 Vir, 0;36 Lib, 0;44 Sco, 0;52 Sgr, 1 Cap, 1;8 Aqr, 1;16 Psc. | q_{10}

T1.b (Coefficient r_{set}) **ADEG**

Ṣiliptu-coefficient for 'height and depth' for setting.
0;18 Ari, 0;22 Tau, 0;26 Gem, 0;30 Cnc, 0;34 Leo, 0;38 Vir, 0;42 Lib, 0;36 Sco, 0;34 Sgr, 0;30 Cap, 0;26 Aqr, 0;22 Psc. | r_{10}

P3 (ME)

In order for you to construct ME. **[A]DG**

? **E**

[...] ... 1 day ... [...] | ?

2 **ADEG**

You multiply (the time by which) the night has progressed by 0;0,10. [A]⸢This you put down as the displacement⸣ of the Sun.[A] The day (of ME) is on {[ADE]that day. ME ...|[G]1? day. } [You multiply it by the displacement of the Moon for that month. [A]This] you put down as [the displacement] of the Moon.[A] | $M_{\text{ME}} \cdot 0;0,10 \rightarrow \Delta B_{\odot}$ | $\text{d}_{\text{ME}} = \text{d}_{\text{FM}}$ | $M_{\text{ME}} \cdot v_{☽} \rightarrow \Delta B_{☽}$

3 **A**

You subtract ⸢the displacement of the Sun from⸣ the displacement of the Moon [and you put it down as the elongation]. | $\Delta B_{☽} - \Delta B_{\odot} \rightarrow \eta$

4 **AC**

You put down the position of the Moon and the position of the Sun. The displacement by which the {[C]Moon|[A]Sun} moved you subtract from the position of the {[C]Moon|[A]Sun}. The displacement by which the {[C]Sun|[A]Moon} moved you subtract from the position of the {[C]Sun|[A]Moon}. | $B'_{\odot} = B^{\text{fm}}_{\odot} - \Delta B_{\odot}$ | $B'_{☽} = B^{\text{fm}}_{☽} - \Delta B_{☽}$

5 **AC**

(The amount) by which the position of the Moon and the position of the ⸢Sun⸣ exceed 10 (°) of the zodiacal sign you multiply by 0;0,16. [You add*] it with the corresponding *ṣiliptu*-coefficient for the zodiac if it is increasing, ⸢subtract*⸣ if it is decreasing. You accumulate (them) together*, you compute half of it, and you put it down as the *ṣiliptu*-coefficient. | $q = q_{10} \pm (B' - B_{10}) \cdot 0;0,16$ | $(q_{☽} + q_{\odot}) \cdot \frac{1}{2} \rightarrow q'$

6	**A**.R6	[x *u* **si**]**g** *ša*$_2$ *sin* **gar**-*an* **zi** *sin* GAM$_0$ 4 **du ki nim** *u* **sig** *ša*$_2$ *s*[*in* x x x] →
	C.R5–6	**nim** *u* **sig** *ša*$_2$ *sin* **gar zi** *sin* / GAM$_0$ 4 **du ki nim** *u* **sig** *ki* **tab la**$_2$ →
	D.R10	**bar.nun** *gab-bi ša*$_2$ ⸢**nim**⸣ ⟨*u* **sig**⟩ **zi** *sin* **a.ra**$_2$ 4 **du ki** ⸢**nim**⸣ [x x xx x x]
	G.R11'–12'	**bar.nun** *gab-bi ša*$_2$ **nim** *u* **sig zi** ^{d}*sin* **a.r**[**a**$_2$ x x x] ⸢**nim**⸣ [x x x xx] / *ki-i* **tab la**$_2$ →
	A.R6	[x x x]
	C.R6–7	*ki* **la**$_2$ **tab** *ana muḫ-ḫi* / GAR.GAR-*ma* $\frac{1}{2}$-*šu*$_2$-*nu* GIŠ-*ma ana* **bar.nun gar** →
	D.R11	*ki-i* **la**$_2$-*u*$_2$ **tab** →
	G.R12'	*ki-i* **la**$_2$ **tab** →
7	**A**.R7	[x x] *u* **ki** *šamaš*$_2$ *ša*$_2$ *al-la* 10 *lu-maš* **diri** [x] 8 **du ki bar.nun** *ša*$_2$ **nim** [*u* **sig** *ša*$_2$-*šu*$_2$]
	C.R7–9	**ki** *sin u* **ki** *šam*[*aš*$_2$] / *ša*$_2$ *al* 10 *lu-maš* **diri** GAM$_0$ ⸢20$^{?}$⸣ **du ki bar.nun** *ša*$_2$ *lu-m*[*aš*] / *ša*$_2$-*šu*$_2$ →
	DG	
	A.R8	[x x x] *ki* **la**$_2$ **la**$_2$ *ana* **ugu** *a-ḫa-miš*$_2$ GAR.GAR-*ma* $\frac{1}{2}$-*šu*$_2$-*nu* GIŠ-⸢*ma*⸣
	B.R1'	⸢*ana* **ugu** *a-ḫa*⸣-[*miš*$_2$ xxx xxx xx] →
	C.R9–10	*ki* **tab tab** *ki* **la**$_2$ **la**$_2$ *ana* **ugu** *a-ḫa-miš*$_2$ GAR.GAR-*ma* / $\frac{1}{2}$-*šu*$_2$-*nu* GIŠ-*ma* →
	DG	
8	**A**.R9	[GAM$_0$ DIŠ.UD x x x] **sig** *ša*$_2$ *taš-kun* **du** →
	B.R1'	[x xx x x x x x xx x]
	C.R10	GAM$_0$ ⸢GI$^{?}$⸣ *ša*$_2$ **nim** *u* **sig** *ša*$_2$ **gar**-*nu* **du**
	DG	
9	**A**.R9	BE GAM **ba**[**r.nun d**]**u** →
	B.R2'	BE GAM **bar.nun du** →
	C.R11	BE GAM$_0$ **bar.nun du** →
	D.R11	BE **a.ra**$_2$ **bar.nun du** →
	G.R12'	BE **a.ra**$_2$ **bar.nun d**[**u**-*m*]*a* →
10	**ABC**	
	D.R11	2 ḪAB-*ra*[*t* x x x xx x]
	G.R12'	⸢2⸣ Ḫ[AB-*r*]*at* **ki** BE *u* **bar**.[**nun tab**]
11	**G**.R13'	*ki-i* DIŠ.UD *ša*$_2$ **nim** *u* **sig** *ša*$_2$ **a.ra**$_2$ **bar.nun** *ar*$_2$-⸢*tu*$_2$⸣ *tal-lik* →
	A.R9	*ki* **nim** *ana* B[E x]→
	B.R2'	*ki* **nim** *ana* B[E x]→
	C.R11	*ki* **nim ki** BE **tab** →
	D.R12	⸢*ki-i* **nim** *ana*⸣ BE *u* **bar.nun tab** →
	G.R13'–14'	*ki-i* **nim nim** *u lu-u*$_2$ **nim sig** *š*[*u-u*$_2$] / **ki** BE **bar.nun** *u* 2 ḪAB-*rat* **tab** →
	A.R9–10	[x x x x x] / [x x xx] →
	B.R2'	[x x x x x x x xx] →
	C.R11	*ki* **sig ta** BE [**la**$_2$]
	D.R12	⸢*ki*⸣-*i* **sig** *ana* BE *u* **bar**.[**nun** x x x x] →
	G.R14'–15'	*u ki-i* ⸢**bur**$_3$ **bur**$_3$⸣ *u lu-u*$_2$ **bur**$_3$ **la**$_2$ *šu-u*$_2$ **ta** BE **bar.nun** *u* 2 ḪAB-*rat* **la**$_2$ *ana* **me e**-*bi* →
10	**A**.R10	[2] ḪAB-*rat ana* BE *u* **bar.nun tab** →
	B.R2'	[x xx x x x xx x]
	CDG	
12	**A**.R10	⸢*ki-i* DIŠ.UD *ša*$_2$ **sig** *al-la*⸣ [x xx x x xx x]
	B.R3'	*ki* DIŠ.UD *ša*$_2$ ⸢**nim**⸣ *al-la* BE *u* **bar.nun** *u*$_3$ 2 ḪAB-*rat* **diri**-*at*$_2$ →
	C.R12	*ki ma-la* **igi** *la ma-ṣu-u*$_2$
	D.R12–13	[xx xx] / *ša*$_2$ **bur**$_3$ *al-la* BE *u* **bar.nun** *u* ⸢2$^{?}$ ḪAB⸣-*rat* **diri**-[*at*] →
	G.R15'	*ki-i* DIŠ.UD *ša*$_2$ **sig** *al-l*[*a* B]E **bar.nun** *u* 2 ḪAB-*rat* **diri**-*at* →

6 **ACDG**	
{^{AC}You put down the Moon's 'height and depth'.|^{DG}All the *ṣiliptu*-coefficients for 'height and depth'.} You multiply the displacement of the Moon by 0;4 and you subtract it from the ^{AG}Moon's^{AG} 'height and depth' if it is increasing, add if it is decreasing.*	$\beta' = \beta^{\text{fm}} \mp \Delta B_{☽} \cdot 0;4$
7 **ABC**	
(The amount) by which the position of the Moon and the position of the Sun exceed 10 (°) of the zodiacal sign you multiply by 0;0,8.* You add it with the corresponding *ṣiliptu*-coefficient for 'height and depth'* if it is increasing, subtract it if it is decreasing.* You accumulate (them) together, you compute half of it, and	$r = r_{10} \pm (B' - B_{10}) \cdot 0;0,8$ $(r_{☽} + r_{\odot}) \cdot \frac{1}{2}\ [\rightarrow r']$
8 **A[B]C**	
you multiply it by the {$^{A[B]}$[DIŠ.UD]|^{C}⌜GI$^{?}$⌝} of 'height and depth' which you had put down.	$r' \cdot \beta'\ [\rightarrow R]$
9 **ABCDG**	
You multiply the elongation by the *ṣiliptu*-coefficient.	$\eta \cdot q'\ [\rightarrow Q]$
10 **DG**	
[You add] the 2 for the disk with the elongation 'and' the *ṣiliptu*-coeff[icient].	$Q+2$
11 **ABCD**	
If (the Moon) is 'high' you add it to the elongation ^{D}'and' the *ṣiliptu*-coefficient^{D}. If it is 'low' [you subtract] it from the elongation ^{D}'and' the *ṣiliptu*-coefficient^{D}. [You call it ME.]	high $\Rightarrow Q + R_{\uparrow}$ (**D**: $Q+2+R_{\uparrow}$) low $\Rightarrow Q - R_{\downarrow}$ (**D**: $Q+2-R_{\downarrow}$) $\rightarrow$ ME
10 **AB**	
You add the [2] for the disk to the elongation 'and' the *ṣiliptu*-coefficient.	ME + 2
11 **G**	
If the DIŠ.UD of 'height and depth' which you had multiplied by the last *ṣiliptu*-coefficient: If (the Moon) is high, ascending, or high, descending, you add ⌜this⌝ to the elongation, the *ṣiliptu*-coefficient 'and' the 2 for the disk. If it is 'low', descending, or 'low', ascending, you subtract this from the elongation, the *ṣiliptu*-coefficient 'and' the 2 for the disk. You call it ME.	high $\Rightarrow Q+2+R_{\uparrow}$ low $\Rightarrow Q+2-R_{\downarrow}$ $\rightarrow$ ME
12 **ABDG**	
If the DIŠ.UD of 'depth'* exceeds the elongation 'and' the *ṣiliptu*-coefficient 'and' the 2 for the disk you subtract the elongation 'and' the *ṣiliptu*-coefficient 'and' the 2 for the disk from it. You call {^{AB}ME GI_6|^{DG}it GI_6. ME becomes GI_6}. ⌜You do not change the day⌝.	$R_{\downarrow} > Q+2 \Rightarrow$ $R_{\downarrow} - Q - 2 \rightarrow GI_6$ $d_{GI_6} = d_{ME}$
12 **C**	
If it is insufficient for being visible:	

	A.R11	[x xx] u$_3$ 2 ḪAB-rat ⸢ta lib$_3$-bi⸣-šu$_2$ la$_2$ me ana gi$_6$ e-bi →
	B.R3'–4'	BE u bar.nun / u 2 ḪAB-rat ta lib$_3$-bi-šu$_2$ la$_2$ me ana gi$_6$ e-bi →
	C	
	D.R13–14	[x x xx x x xx x xx x] / ana gi$_6$ e me ana gi$_6$ gur$^?$ →
	G.R15–16'	BE bar.nun u 2 ḪAB-rat / ta lib$_3$-bi-šu$_2$ la$_2$ ana ⸢gi$_6$ e-bi⸣ [x x x gu]r→

	A.R11	⸢me nu$^?$ man$^?$-ni$^?$⸣
	B.R4'	me ina me-šu$_2$ →
	C	
	D.R14	[x x] ⸢man-ni⸣ →
	G.R16'	⸢me nu man-ni⸣ →

13	A.R12	⸢6⸣ [x s]i-man igi-u$_2$ tab ⸢gin$_7$⸣ igi-u$_2$ du$_3$-uš 1-en u$_4$-me la$_2$ ana me gar-an
	B.R4'–5'	6 ki si-man / igi-u$_2$ tab ki-ma igi-u$_2$ du$_3$-uš 1-en u$_4$-mu la$_2$ ana me gar-⸢an⸣
	C.R12–13	6 ki-šu$_2$ tab ana {1} me [e-bi] / ⸢ki⸣-ma igi-u$_2$ du$_3$-uš ana ki.na$_2$.a e.a ⸢gar⸣-[an]
	D.R14–15	[x x xx x] / ⸢ki-ma$^?$ igi-u$_2$ du$_3$⸣-[uš xx xx x x x x]
	G.R16–17'	6 ki si-man igi-u$_2$ tab / gin$_7$ igi-u$_2$ ⸢du$_3$-uš⸣ [xx xx x x x] e-bi

P4

1	A.R13	gi$_6$ ana du$_3$-ka gi$_6$ du ta 6 la$_2$ 1-en u$_4$-⸢mu tab⸣→
	B.R6'–7'	⸢gi$_6$⸣ ana du$_3$-ka gi$_6$ du ta 6 la$_2$ 1-en$^?$ me [x] →
	D.R16	⸢gi$_6$ ana du$_3$-ku⸣ [xx x x x xx x x] →
	G.R18'	gi$_6$ ana du$_3$-ka g[i$_6$ du x x x xx xx x] →

2	A.R13–14	⸢GAM$_0$⸣ 10 du-⸢ma$^?$ zi šamaš$_2$ gar-an⸣ / GAM$_0$ zi sin ša$_2$ iti-šu$_2$ du-ku zi sin gar-an
	B.R6'–7'	GAM 10 du ⸢šu⸣-u$_2$ / [x] ⸢šamaš$_2$ gar-an⸣ GAM zi sin ša$_2$ iti-šu$_2$ du šu-u$_2$ zi sin gar-an
	D.R16	[x x x x x x]
	G.R18'	⸢ana⸣ igi-šu$_2$ du gin$_7$ na du$_3$-uš

3	A.R14	zi šamaš$_2$ ta z[i sin nim-ma BE g]ar-[an]
	B.R8'	[x x x z]i ⸢sin nim-ma⸣ [x xx] →
	DG	

4	A.R15	ki sin u$_3$ ki šamaš$_2$ gar-an ⸢zi⸣ ša$_2$ šamaš$_2$ zi ⸢ki ki šamaš$_2$⸣ tab z[i x x x x x x] tab
	B.R8'–9'	[x] sin u ki šamaš$_2$ gar-an / [x x x x x x x x x ša$_2$ si]n zi ki ki sin tab
	DG	

5	A.R16	ki sin u$_3$ ki šamaš$_2$ ša$_2$ a[l-la] u$_4$-me lu-maš ⸢diri⸣ GAM$_0$ 1[6 du] →
	B.R10'	[x x x x x x xx x xx x x x d]u →
	DG	

	A.R16	[x] bar.nun ša$_2$ ⸢lu-⸣m[aš ša$_2$-šu$_2$ ki tab tab ki la$_2$] la$_2$
	B.R10'–11'	ki bar.nun / [x xx xx x x x x x x] →
	DG	

	A.R17	ana ugu a-ḫa-miš$_2$-šu$_2$ GAR.GAR-[ma] $\frac{1}{2}$-šu$_2$-nu GIŠ-[ma ana] bar.[nun gar-a]n →
	B.R11'–12'	[x x a-ḫ]a-miš$_2$ GAR.GAR-ma / [xxx xx x xx xx] →
	DG	

6	A.R17	nim u sig ša$_2$ sin [gar-an]
	B.R12'	[x x x x x xx] →
	D.R17	bar.nun gab-bi [ša$_2$ nim u$^?$ sig$^?$] →
	G.R19'	bar.nun gab-bi [x x x x] →

	A.R18	zi sin GAM$_0$ 4 du ki n[im] u sig ša$_2$ {zi} ⟨sin⟩ ki ta[b ta]b ki la$_2$ la$_2$ →
	B.R12'	[x] ⸢sin$^?$ GAM$^?$ 4$^?$⸣ [x]
	D.R17–18	[x x xx x x x x xx x x xx x x] / ki-i la$_2$-u$_2$ la$_2$ →
	G.R19'–20'	[x x a.r]a$_2$ 4 du-ku ki nim u sig ša$_2$ ^{d}sin / ki-i tab tab ki-[i x x] →

13 **ABDG**

You add 6,0 to the previous time, you do as before. You subtract 1 day. {^{AB}You put it down as ME.|^{G} You call it ME.}

$N_{\mathrm{ME}} + 6{,}0 \rightarrow N_{\mathrm{ME}}$
$\mathrm{d}_{\mathrm{ME}} - 1 \rightarrow \mathrm{d}_{\mathrm{ME}}$

13 **C**

You add 6,0 to it, [you call?] it ME, you do as before. You call it the 'place of disappearance', you put it down.

$N_{\mathrm{ME}} + 6{,}0 \rightarrow N_{\mathrm{ME}}$

P4 ($\mathbf{GI_6}$)

In order for you to construct GI_6. **ABDG**

1 **AB[D]G**

You subtract (the time by which) the night has progressed from 6,0. You ⌜add⌝ 1 day.

$6{,}0 - M\ [\rightarrow N_{\mathrm{GI}_6}]$;
$\mathrm{d}_{\mathrm{GI}_6} = \mathrm{d}_{\mathrm{FM}} + 1$

2 **AB**

You multiply it by 0;0,10 ⌜^{A}and^{A} you put {^{A}it|^{B}this} down as the displacement of the Sun⌝. You multiply it by the displacement of the Moon for that month. You put {^{A}it|^{B}this} down as the displacement of the Moon.

$N_{\mathrm{GI}_6} \cdot 0;0{,}10 \rightarrow \Delta B_{\odot}$
$N_{\mathrm{GI}_6} \cdot v_{☽} \rightarrow \Delta B_{☽}$

2 **G**

You go to the beginning and proceed as (with) NA.

3 **AB**

You subtract the displacement of the Sun from the displacement of ⌜the Moon, and⌝ you put it down as [the elongation].

$\Delta B_{☽} - \Delta B_{\odot} \rightarrow \eta$

4 **AB**

You put down the position of the Moon and the position of the Sun. The ⌜displacement⌝ by which the Sun moved you add ⌜to the position of the Sun⌝. The dis[placement by which the Mo]on moved you add to the position of the Moon.

$B'_{\odot} = B^{\mathrm{fm}}_{\odot} + \Delta B_{\odot}$
$B'_{☽} = B^{\mathrm{fm}}_{☽} + \Delta B_{☽}$

5 **AB**

(The amount) by which the position of the Moon and the position of the Sun ⌜exceed⌝ 10 (°) of the zodiacal sign [you multiply] by ⌜0;0,16⌝. [You add it] to the [corresponding] *ṣiliptu*-coefficient for ⌜the zodiac⌝ [if it is increasing], subtract [if it is decreasing]. You accumulate (them) together, you compute half of it [and] put it down [as] the ⌜*ṣiliptu*-coefficient⌝.

$q = q_{10} \pm (B' - B_{10}) \cdot 0;0{,}16$
$(q_{☽} + q_{\odot}) \cdot \frac{1}{2} \rightarrow q'$

6 **ABDG**

{^{AB}[You put down] the Moon's 'height and depth'.|^{DG}All the *ṣiliptu*-coefficients [for 'height and depth'.]} You multiply the displacement of the Moon by 0;4. You add it with the Moon's* 'height and depth' if it is increasing, subtract if it is decreasing.

$\beta' = \beta^{\mathrm{fm}} \pm \Delta B_{☽} \cdot 0;4$

7	**A**.R18–19 **DG**	**ki** *sin u* **k**[**i** *šamaš*$_2$ *ša*$_2$ *al-la* 10 *lu-maš*] **diri** / GAM$_0$ 8 **du ki bar.nun** *ša*$_2$ **nim** *u* [**sig** *ša*$_2$*-š*]*u*$_2$ →
	A.R19–20 **DG**	*ki* **tab tab** *ki* **la**$_2$ **la**$_2$ ⌜*ana* **ugu**⌝ *a-ḫa-*⌜*miš*$_2$ GAR⌝.[GAR-*ma*] / [$\frac{1}{2}$-*š*]*u*$_2$*-nu* GIŠ.**a**
8	**A**.R20 **DG**	GAM$_0$ DIŠ.UD [*ša*$_2$ **nim** *u*] **sig** *ša*$_2$ *taš-kun* **du** →
9	**A**.R20 **D**.R18 **G**.R20’	BE GAM **bar.nun du** → [x xx xx x] → [x xx xx **d**]**u**$^?$-*ku* →
10	**A** **D**.R18 **G**.R20’	 [x xx x] → 2 ḪAB-*rat* **ta** BE *u* **bar.nun la**$_2$
11	**G**.R21’	[*ki*]-*i* DIŠ.UD *ša*$_2$ **nim** [*u* **sig** *ša*$_2$ **a.ra**$_2$ **bar.nun** *ar*$_2$*-tu*$_2$ *t*]*al-lik* →
	A.R20 **D**.R18–19 **G**.R21’–22’	*ki* **nim** **t**[**a** x x$^?$ xx$^?$ x] [xx] / **nim** *ana* BE *u* **bar.nun** ⌜**la**$_2$$^?$⌝→ *ki-i* **nim nim** *u lu-u*$_2$ / [**nim si**]**g** *šu-u*$_2$ [**ta** BE **bar.nun** *u* 2 ḪAB-*rat* **la**$_2$] →
	A.R21 **D**.R19 **G**.R22’–23’	[*ki*] **sig** *ana* BE *u* **bar.n**[**un** **tab** *ana* **gi**$_6$ **e**] → [xx x x x x xx x x x x] → [*u ki-i*] **bur**$_3$ **bur**$_3$ *u lu-u*$_2$ **bur**$_3$ **la**$_2$ *šu-u*$_2$ / [**ki** BE] ⌜**bar.nun**⌝ [*u* 2 ḪAB-*rat* x x x **ga**]**r**-*an* →
12	**A**.R21 **D**.R19–20 **G**.R23’–24’	*ki* DIŠ.UD *ša*$_2$ **nim** *al-la* BE *u* **bar.nun** → [xx xx x x] / *al-la* BE *u* **bar.nun** *u* 2 ⌜ḪAB-*rat*⌝ → *ki-i* DIŠ.⌜UD⌝ *ša*$_2$ **nim** / [xx x x xx x$^?$ x$^?$ xx$^?$] →
	A.R21 **D**.R20 **G**.R24’	**diri**-⌜*at*$_2^?$⌝ BE *u* **bar.n**[**un** **ta** *lib*$_3$*-bi* **nim**] ⌜**diri**-*at*⌝ [x x xx *u* 2 ḪAB-*rat* x xx x] → [x x x xx x$^?$ x$^?$ xx$^?$ x xx x] →
	A.R22 **D**.R20–21 **G**.R24’–25’	[x$^?$] *ana* **me** ⌜**e** **me nu man**⌝-*nu*→ [x x x] / ⌜**gi**$_6$⌝ *ana* **me gur** **me nu man**-*ni* → [x x] ⌜**e**⌝-*bi* **me**$^?$ **la**$_2$$^?$ **gi**$_6$ *ana* **me gur** / [x x xx] →
13	**A**.R22 **D**.R21 **G**.R25’–26’	6 **ki** *si-man* **igi tab** *ki* **igi du**$_3$*-uš* 1-*e*[*n u*$_4$*-mu* **tab** *ana* **gi**$_6$ **e**]-*bi* 6 **ki** *si-man* **igi tab** *k*[*i*$^?$*-ma*$^?$ xx xx xx x x x x x] [x x xx x x x **ig**]**i du**$_3$*-uš* 1-*en* **me** **tab** / [xxxxxx x] **e**-*bi*
10	**A**.R23	[2] ḪAB-*rat* **ta** BE *u* **bar.nun la**$_2$→
12’	**A**.R23	*ki-i* 2 ḪAB-*rat al-la* BE *u* **bar.nun diri** [BE *u* **bar.nun ta** *lib*$_3$*-bi* **nim**]
	A.R24	[*ki*] **nim** *šu-u*$_2$ *ana ša*$_2$ **tag**$_4$ **tab** *ki* **sig** *šu-u*$_2$ *ana* ⟨*ša*$_2$⟩ **tag**$_4$ **la**$_2$ ⌜**gi**$_6$ *ana* **me**⌝ [**gur**$^?$ xxxx]-*an*$^?$

T2

	A.R25 **D**.Ri22 **G**.Ri27’	[**bar.n**]**un** *ša*$_2$ *lu-maš ša*$_2$ **nim** [**ba**]**r.nun** *ša*$_2$ *lu-maš ša*$_2$ **nim** [xx x xx x x]
	A.R25 **D**.Rib23–28	1.24 **rin**$_2$ 1.16 **gir**$_2$**.tab** 1.8 **pa** 1 **maš**$_2$ 5[2 x xx xx] 1.24 **rin**$_2$ / 1.16 **gir**$_2$**.tab** / 1.8 **pa** / 1 **maš**$_2$ / 52 **gu** / 44 *zib*.**me**

7 **A**

(The amount) by which the position of the Moon and the posi[tion of the Sun exceed 10 (°) of the zodiacal sign] you multiply by 0;0,8. You add it with the ⌜corresponding⌝ *ṣiliptu*-coefficient for 'height and [depth'] if it is increasing, subtract if it is decreasing. ⌜You accumulate (them) together⌝, you compute [half] of it,

$r = r_{10} \pm (B' - B_{10}) \cdot 0;0,8$

$(r_{☽} + r_{\odot}) \cdot \frac{1}{2}$ [$\rightarrow r'$]

8 **A**

you multiply it by the DIŠ.UD [of 'height and] depth' which you had put down.

$r' \cdot \beta'$ [$\rightarrow R$]

9 **A[D]G**

You multiply the elongation by the *ṣiliptu*-coefficient.

$\eta \cdot q'$ [$\rightarrow Q$]

10 **[D]G**

You subtract the 2 for the disk from the elongation 'and' the *ṣiliptu*-coefficient.

$Q - 2$

11 **AD**

If (the Moon) is high you subtract it from the elongation 'and' the *ṣiliptu*-coefficient. [If] it is low [you add] it to the elongation 'and' the ⌜*ṣiliptu*-coefficient⌝. [You call it GI_6].

high $\Rightarrow Q - R_{\uparrow}$(**D**: $Q - 2 - R_{\uparrow}$)
low $\Rightarrow Q + R_{\downarrow}$ (**D**: $Q - 2 + R_{\downarrow}$)
$\rightarrow GI_6$

11 **G**

⌜If⌝ the DIŠ.UD of 'height [and depth' which] you had multiplied [by the last *ṣiliptu*-coefficient]: If (the Moon) is high, ascending, or [high, descen]ding, [you subtract] this [from the elongation, the *ṣiliptu*-coefficient 'and' the 2 for the disk. If] it is low, descending, or low, ascending, [you add] this [to the elongation], the *ṣiliptu*-coefficient ['and' the 2 for the disk.] You put [it down as GI_6].

high $\Rightarrow Q - 2 - R_{\uparrow}$
low $\Rightarrow Q - 2 + R_{\downarrow}$
$\rightarrow GI_6$

12 **ADG**

If the DIŠ.UD of 'height' exceeds the elongation 'and' the *ṣiliptu*-coefficient **D[G?]**'and' the 2 for the disk**D[G?]** [you subtract] the elongation 'and' the *ṣiliptu*-coeffi[cient **[DG?]**'and' the 2 for the disk**[DG?]** from it]. You call it ME. **G**...**G** **DG**GI_6 becomes ME.**DG** You do not change the day.

$R_{\uparrow} > Q \Rightarrow R_{\uparrow} - Q \rightarrow$ ME **A**
$R_{\uparrow} > Q - 2 \Rightarrow R_{\uparrow} - Q + 2$
$\rightarrow$ ME **[DG]**
$d_{ME} = d_{FM}$

13 **ADG**

You add 6,0 to the previous time, you do as before. You add 1 day. You call it [GI_6].

$N_{GI_6} + 6,0 \rightarrow N_{GI_6}$;
$d_{GI_6} + 1 \rightarrow d_{GI_6}$

10 **A**

You subtract the [2] for the disk from the elongation 'and' the *ṣiliptu*-coefficient.

$Q - 2$

12' **A**

If the 2 for the disk exceeds the elongation 'and' the *ṣiliptu*-coefficient, you subtract [the elongation 'and' the *ṣiliptu*-coefficient from it; if] (the Moon) is high you add it to what remains, if it is low you subtract it from ⌜what⌝ remains. GI_6 [becomes] ⌜ME⌝ [...].

$2 > Q \Rightarrow 2 - Q$
high $\Rightarrow 2 - Q + R_{\uparrow} \rightarrow$ ME
low $\Rightarrow 2 - Q - R_{\downarrow} \rightarrow$ ME

T2.a (Coefficient q_{rise}) **ACDG**

Ṣiliptu-coefficient for the zodiac for rising.

1;24 Lib, 1;16 Sco, 1;8 Sgr, 1 Cap, 0;52 Aqr, 0;44 Psc, 0;36 Ari, 0;44 Tau, 0;52 Gem, 1 Cnc, 1;8 Leo, 1;16 Vir.

q_{10}

T2.b (Coefficient r_{rise}) **ADG**

Ṣiliptu-coefficient for 'height and depth' for rising.

A.R26	[3]6 **ḫun** 44 **mul$_2$** 52 **maš** 1 **alla** 1.8 **a** 1.16 **absin** [x$^?$]
C.R14	⌜36 **ḫun**$^?$⌝
D.Ria23–28	36 **ḫun** / 44 **mul$_2$** / 52 **maš** / 1 **alla** / 1.8 **a** / 1.16 **absin**

A.R27	[bar.nu]**n** *ša$_2$* **nim** *u* **sig** *ša$_2$* **nim**
D.Rii22	**bar.nun** *ša$_2$* **nim** [x x x x]
G.Rii27’	[xx x x x x x] ⌜**nim**⌝

A.R27	18 **rin$_2$** ⌜22⌝ **gir$_2$.tab** 26 **pa** 30 **maš$_2$** [xx x xx xx]
D.Riib23–28	[xx x] / [xx xx] / 2[6 x] / 30 [x] / 34 [gu] / 38 [*zib*.me]

A.R28	[4]2 [x xx x] 34 **maš** 30 **alla** 26 **a** 22 **ab**[sin x$^?$]
D.Riia23–28	42 **ḫun** / 38 **mul$_2$** / 34 **maš** / 30 **alla** / 26 **a** / 22 **absin**

(Colophon)

D.R29	**im.dub** ^{md}**ak-sum-šeš a** *ša$_2$* ^{md}**ak-sum-**⌜**šeš** x⌝ [xxx]

Critical and philological notes

Asterisks (∗) indicate variants not included in the main translation.

P1 (ŠU$_2$)

step 1

A.O1 *si-man ša$_2$* **gi$_6$ du** = ‘the time by which the night has progressed’; cf. the Glossary. This is often abbreviated to **gi$_6$ du**.

The sign *man* is written here with an additional horizontal wedge below the two winkelhakens. This peculiar orthography occurs in some but not all instances of MAN in Ms. **A**. Examples are O2,14,15 (*si-man*) and O19 (*šamaš$_2$*). On the same tablet the usual orthography is found e.g. in R12,22 (*si-man*).

In **A** the most frequent logogram for subtraction is **nim**, and after that **la$_2$**, so that either may be restored in the gap at the end.

A.02 Neugebauer restored the gap at the end as LA$_2$ *ša$_2$ mūši*(**gi$_6$**), ‘duration of the night’, which leads to an erroneous expression (*ACT*, p. 231). One expects LA$_2$ *ša$_2$ ūmi*(**me**), ‘duration of daylight’; cf. the commentary. The latter is now confirmed by **D**.O3; cf. below.

D.O3 *ša$_2$* **tag$_4$**, ‘What remains’, refers to the outcome of the preceding subtraction. This appears to be superfluous because no further operation is performed with the outcome, but most Mss. also have it in NA step 1, where the same objection holds.

LA$_2$ *ša$_2$* ⌜**me**⌝ [x$^?$]: the damaged sign read as **me** looks more like ŠU$_2$, but **gi$_6$** can be ruled out. Hence Neugebauer’s restoration of the break in the same passage in **A**.O2 can be discarded and replaced by LA$_2$ *ša$_2$* **me**; cf. the astronomical commentary. After ⌜**me**⌝ there is an erasure of about 1 sign.

step 2

A.O3 **du** *šu-u$_2$*: *šu* is written over an erasure.

D.O3 [x 10 **du**-*ma*]: in **D** multiplication is expressed by **a.ra$_2$**, so that one can restore this here and in other broken passages.

step 3

A.O4 BE = ‘elongation’, probably to be read *bi$_2$*, a phonetic abbreviation of *birītu*, ‘distance’; cf. the Glossary.

step 4

A.O5 The initial sign looks like an elongated *u*; one expects nothing.

zi *ša$_2$ šamaš$_2$* **zi**: the version of step 4 in **C**.R1–2 (P3) proves that this is to be read *nisḫu ša šamaš issuḫ*, ‘the displacement by which the Sun moved’. Although Neugebauer’s translation of **C** implies that he had the same understanding of **C**.R1–2, he did not apply this to the instances of step 4 in the other Mss.

D.O6 *sin*: scribal error for *šamaš$_2$*.

step 5

A.O7 *ša$_2$-šu$_2$* = *šâšu*, ‘that; the aforementioned’: the context implies that this is an attribute of *ṣiliptu*(**bar.nun**), ‘*ṣiliptu*-coefficient’. In *ACT*, Neugebauer did assume this in step 7 but not in step 5, where he considered *šâšu* to be an attribute of *lu-maš*, ‘that zodiacal sign’. However, also here the purpose of *šâšu* is to direct the user to the appropriate version of q (for rising or for setting). Since q has not been defined at this point (cf. P3,P6), ‘corresponding’ has been adopted as a more suitable translation than ‘that; aforementioned’.

step ?

G.O1’–3’ In the commentary to step 8 it is argued that in **G** step 7 precedes step 6, even though it seems difficult to assign the few legible signs in O1’–3’ to step 7. The meaning and identification in terms of steps therefore remain unclear.

G.O2’ The traces before **a.ra$_2$** are compatible with the end of *lik*, **nim**, *ka*, etc.

0;18 Lib, 0;22 Sco, 0;26 Sgr, 0;30 Cap, 0;34 Aqr, 0;38 Psc, 0;42 Ari, 0;38 Tau, 0;34 Gem, *r*$_{10}$
0;30 Cnc, 0;26 Leo, 0;22 Vir.

(Colophon) **D**

Tablet of Nabû-nādin-aḫi, son of Nabû-nādin-a[ḫi, descendant of ...]

step ?

G.O3' *tallik*, 'you have multiplied': the pret. implies that this phrase is not an instruction, but probably a relative clause, 'the ... [which] you had multiplied by the elongation(?) [...] and ...'. The meaning of BE, usually 'elongation', is uncertain here, because the expected multiplication of the elongation and the coefficient q is mentioned below in O5' (step 9).

step 6

A.O8 At the end of the line about 1 sign is missing, presumably **tab**. Neugebauer (*ACT*, 228; 234 n. 56) considered *ki* **tab la**$_2$ to be an error for *ki* **tab tab** and erroneously restored the gap as **la**$_2$.

D.O9–10 Since the lines on **D** are shorter than those on **A** by a few signs, it is likely that step 6 continues a few signs into **D**.O10.

G.O3' The last sign before the break is copied as BI. It is assumed to be a broken **du**, which is the expected sign. In the gap at the end there is room for about 1–2 more signs, perhaps only *ku* or *ak*.

step 6?

G.O4'–5' **ki**-*šu*$_2$: the copy has a clear *šu*$_2$, not U as transliterated in *CTMMA*. The parallel phrases in R11'–12' and R19'–20' suggest that these lines are concerned with step 6, but the interpretation is problematic because the phrases **ki**-*šu*$_2$ **tab** (O4') and **ta** *lib*$_3$*-bi-šu*$_2$ **la**$_2$ (O5') are in the reverse order compared to the other Mss.

step 7

F.X2' It is uncertain whether one should restore *qaq-qar sin*, 'position of the Moon', in the gap at the beginning of step 7. Perhaps only the position of the Sun is computed here; cf. the commentary.

F.X3' The initial two signs near the left edge are most likely **la**$_2$ **la**$_2$, not U BE (*LBAT*). Hence it appears that *ana muḫ-ḫi*... GIŠ-*ma* is lacking, as is the case in **G**; cf. the astronomical commentary. There is about enough space in the gap between 10 *lu-maš* and GAM DIŠ.UD to restore the rest of the phrase as it is preserved in **A**.

GIŠ = *tanašši* 'you compute', 2 m. sg. pres. G of *našû* (cf. the Glossary).

step 8

F.X3' DIŠ.UD: technical term for R (cf. §4.4.19.8). The stat. *atrat* in step 12 implies that it is a f. sg. noun.

step 11

A.O11 **sig** is in Pinches's copy (*LBAT* 91) but no longer on the tablet. After **me** there are about 3 damaged signs, the last of which might be *an* or ME, as indicated in Pinches's copy.

D.O12 BE *u* **bar.nun** = 'the elongation and the *ṣiliptu*-coefficient'. Cf. the Glossary and §4.4.19.9. The particle *u* 'and' stands for multiplication here (cf. §2.2.6).

F.X4' Collation has revealed that the sign copied as ME by Pinches and transliterated as such by Neugebauer is ŠU$_2$. Hence this section concerns ŠU$_2$. Since the next section concerns NA, the order of the sections on the tablet is the standard one. Just before the gap at the end there is a vertical wedge. The gap may contain about 3 signs.

G.O6' *ki-i* **nim** *u* ⸢**sig** xx$^?$⸣ [xx$^?$ *ša*$_2$ **a.r**]**a**$_2$ **bar.nun** *tal-lik*: Between **sig** and *ša*$_2$ **a.ra**$_2$ **bar.nun** there are about 2 damaged signs, and there is room for about 2 more signs in the gap, but as suggested by the parallels R14',21' nothing is expected here; probably there is an erasure.

Only in **G** does step 11 begin with this clause, which lacks a verbal predicate, since *tallik* belongs to the relative clause. Its syntactic relation with the conditional clause following it is not entirely obvious. At least two interpretations offer themselves: if *kī* is the conjunction 'if' then the clause might be proleptic with respect to the following conditional clause, which has been assumed in the translation. If *kī* is a preposition, neither the comparative ('like') nor the causal ('on account of') or any of the other meanings listed in the dictionaries appear to apply. One might tentatively translate *kī* as 'concerning', so that the clause functions as a casus pendens to the conditional clause. However, no other instances of this usage of *kī* are known to me. Note that in the procedures for ME and GI$_6$ the term 'height and depth' is replaced by 'the DIŠ.UD of height and depth' at the same location in step 10 (**G**.R13',21'), but not in the conditional clause following it, although it is clear from the context that it must mean the same here, i.e. distance to the ecliptic (for DIŠ.UD cf. also the Glossary and the commentary to ŠU$_2$ step 8). It is unclear how this affects the interpretation of the phrase. Whatever the correct grammatical interpretation, the algorithmic interpretation is clear.

ki-i **nim nim** *u lu*-⸢*u*$_2$ **la**$_2$ **bur**$_3$⸣: this reading was established after collation.

G.O7' **bur**$_3$ **bur**$_3$ *u lu-u*$_2$ **bur**$_3$ **la**$_2$: replaces 20 *u lu-u*$_2$ 10 **la**$_2$ (*CTMMA* II). Instead of **nim** and **sig** which are used in the first condition of step 10 and everywhere else as logograms for distance to the ecliptic and its change, the scribe of **G** curiously but consistently uses the alternative logograms **la**$_2$ and **bur**$_3$ in the second condition. Both pairs represent the same verb forms *šaqi*/*išaqqi*, 'it is high/ascending' and *šapil*/*išappil*, 'it is low/descending', respectively.

step 12

A.O12 *at*$_2$: omitted in *ACT*. The gap at the end of the line probably contained no more than the missing part of **nim**.

DG **šu**$_2$ *ana* **na** *itâr*(**gur**) = 'ŠU$_2$ becomes NA'. Equivalent phrases are used in step 11 for NA, ME and Gl$_6$; cf. the commentary.

step 13

ADFG **du**$_3$-*uš*: *teppuš*, 'you do; construct; compute'

D.O15 As suggested by C.B.F. Walker, the gap at the end after *ana* **šu**$_2$ **e**, 'you call it ŠU$_2$', probably contained the instruction to subtract 1 day from the date, which in the other Mss. precedes this phrase.

F.X6' About 9 signs are missing from the beginning of the line, so that one might restore [**me nu man**-*ni* 6 **ki** *si-man* **igi**]-*u*$_2$ here. The gap at the end might be restored to *ana* **šu**$_2$ **me**, 'you call it ŠU$_2$'.

G.O10' ⸢**šu**$_2^?$⸣ *ana arkišu*(*ar*$_2$-*šu*$_2$) *inaḫḫis*(**la**$_2$-*is*), 'ŠU$_2$ turns backward'. For the meaning of this phrase, cf. the commentary.

P2 (NA)

step 1

ADFG **na** = *manzāzu*, lit. 'standing', is the technical term for the Lunar Six interval NA.

A.O15 ⸢**nim**⸣ [1-*e*]*n* **me tab**: collation by C.B.F. Walker.

DG In **D**.O17, **G**.O17, and perhaps **F**.X8', there is the seemingly superfluous phrase *ša*$_2$ **tag**$_4$, 'What remains'; compare **D**.O2 (ŠU$_2$ step 1). In both positions it is lacking in **A**.

F.X7' Pinches's copy of the final 3 signs before the gap is not accurate, but in *ACT* Neugebauer correctly transliterated **ta** LA$_2$ *ša*$_2$ ....

step 2

AG **e**$_{11}$ = *illi*, 'it comes out', 3 c. sg. pres. G of *elû*.

B.O1' The broken beginning probably contained the end of step 2.

G.O14' *ša*$_2$ **tag**$_4$, 'what remains': replaces *ša*$_2$ **e**$_{11}$ (*CTMMA* II).

step 3

A.O18 At the end of the line there are traces of *ša*$_2$ between **zi** and ⸢*šamaš*$_2$⸣.

E.O1' The beginning is broken off. It probably contained the end of step 3.

G.O15'–16' The identification of these lines is tentative.

step 4

A.O19 For the unusual orthography of the sign *šamaš*$_2$ cf. the critical notes to **A**.O1.

E.O1' The damaged sign read as *sin* might also be a *šamaš*$_2$; therefore the identification of the passage is uncertain. Based on the length of the gaps the most likely solution is to identify **ki** with the beginning of step 4.

E.O2' The damaged sign read as *sin* might also be a *šamaš*$_2$; therefore the identification of the passage is uncertain. Based on the length of the gaps, **ki ki** has been identified with the end of step 4.

G.O18' **ki ki** $^{\mathbf{d}}$**utu** might belong to step 4. The identification of the preceding lines O16'–17' is uncertain, but they do suggest that step 4 is formulated differently than in the other Mss. Unfortunately, there is no parallel within **G**.

step 5

B.O3' *ūmu*(**me**) here denotes the 10th degree of the zodiacal sign, since the parallels have a 10. Cf. also **A**.R16 (Gl$_6$ step 5). Outside the present text, no examples are known to me for this usage of *ūmu*, the origin of which is unknown. The usual meaning 'day' makes no sense here.

E.O4' $\frac{1}{2}$-*šu*$_2$: perhaps an error for $\frac{1}{2}$-*šu*$_2$-*nu*.

G.O19' Beyond **la**$_2$ which has been restored in the gap at the end there is room for about 8 more signs, but one expects nothing.

G.O20' In the small gap at the beginning one might restore **tag**$_4$, 'remains', or **e**$_{11}$-*ka*, 'comes out for you'. No more than about 7 signs can be restored in the gap at the end. Accordingly, step 5 is assumed to end with **la**$_2$ **la**$_2$.

G.O21' *ša*$_2$-*niš* = 'Secondly': this suggests an alternative rule begins here, but its content and identification in terms of steps (5 or 7?) are unclear due to the bad state of preservation.

G.O22' *te-e-ṣe-pi*: probably *tēṣepi* < *tēṣep*, 2 m. sg. pret. of *eṣēpu*, 'to double' with a spurious final *i*; if one takes the *i* seriously this is *tēṣepī*, a 2 f. sg. pret. The pret. signifies that this is not an instruction. Perhaps it is part of a relative clause, '[... which] you had doubled (?)'. *Eṣēpu* is otherwise not attested in astronomical texts, but it does occur in the LB mathematical text W 23209-x (Oiv16,19), in the sense 'to paste, add'. Whether this meaning also applies here could not be established.

step 6

B.O6' *ki* **tab tab** *ki* **la**$_2$ **la**$_2$, 'you add if increasing, subtract if decreasing', is consistent only with NA or Gl$_6$. Since the final section on the reverse of **B** concerns Gl$_6$, this section concerns NA. The 10 is, perhaps, written over an erasure of **me** (C.B.F. Walker).

E.O5' The broken sign near the beginning is taken to be a 4, but this is uncertain.

step 7

E.O5' Between the end of step 6 in O5' and the beginning of O6', which belongs to step 8, only about 20 signs are missing, of which the final 7 or so contain the beginning of step 8. Hence only about 10 signs are available for step 7, which is clearly insufficient. Perhaps the averaging is omitted from step 7 as is the case in **FG**. However, it is present in step 5, and the text would still be too long to fit in the gap. Perhaps step 7 is omitted; this is attested also in **DG** for ME and Gl$_6$.

step ?

G.O24'–25' Only a few signs can be read in these lines. Since step 6 probably follows step 7 in **G**, they might belong to step 6, but the meaning is unclear. About 4 lines containing steps 8–10 are missing between O25' and R1', i.e. about 2 more lines follow O25' on the obverse.

step 9

A.O24 As remarked by C.B.F. Walker, there is no room to restore **bar.nun du** at the beginning of O25, but there is at the end of this line, which must have continued on the edge.

step 11

B.O9' Collation indicates that the damaged sign after BE is consistent with the expected **tab**. At the end, about 2 signs are missing, perhaps **gar**-*an*.

D.O26 **D** has **la**$_2$ = *šaqi*, 'it is high', while the other Mss. have **nim**, with the same Akkadian reading. There is room in the gap for restoring the expected **tab** *ki-i* **sig** *ana* BE *u* **bar.nun la**$_2$ *ana* **na e**-*bi*.

step 12

AD The beginning of step 12 is preserved only in **AD**, both of which have DIŠ.UD *ša*$_2$ **nim**, 'the DIŠ.UD of 'height". As remarked by Neugebauer, one expects DIŠ.UD *ša*$_2$ **sig**, 'the DIŠ.UD of 'depth" (cf. the astronomical commentary). If this is indeed a scribal error, then (this line of) **A** is copied from **D**, or vice versa, or both derive from a common source.

A.O26 Restoration **m**[**e nu man**-*ni*] proposed by C.B.F. Walker.

G.R1' About 2 more lines are missing between this line and the upper edge of the reverse. After *lib*$_3$-*bi*-⌜*šu*$_2$⌝ there is probably **la**$_2$ as in ŠU$_2$ step 11.

step 13

A.O27 Perhaps there was no text beyond **e**-*bi* in the gap at the end, even though there is room for a few signs.

T1 (Coefficients for setting)

A The restoration of O28–30 is due to C.B.F. Walker, who recognised from **D**.Ri–ii that they contain the coefficients q and r for the setting Moon or Sun. Due to the bad state of preservation and the absence of duplicates Neugebauer could not make sense of these lines. Whereas **A** orders the coefficients in rows of six, the other Mss. present them in columns of six.

P3 (ME)

step 2

C.R1 Steps 2–3 are lacking but may have been written at the end of column i of the reverse, if **C** had two columns.

D.R9 The meaning of **me** *ana* ⌜x⌝ is unclear. There is room for about 8 signs in the gap at the end, but one can only guess at what might be restored here; possibly **a.ra**$_2$ **zi** *sin* *ša*$_2$ **iti**-*šu*$_2$ **du**. The usual statement *šu-u*$_2$ **zi** *sin* **gar**-*an* by which the displacement of the Moon is named in **A**, or an equivalent phrase, is probably absent since it is also lacking for the Sun. Cf. the commentary.

G.R10' As is the case in **D**, the usual phrases for naming the results are probably lacking.

step 4

C.R1–2 The phonetically written *issuḫ* proves that **zi** in this position is the finite verb in a relative clause introduced by the preceding *ša*$_2$. The absence of the subjunctive ending is common for Late-Babylonian Akkadian (*GAG*, §83g).

step 5

C.R4 ∗ **C** has *ki* **tab la**$_2$ *ki* **la**$_2$ **tab**, 'you subtract...if it is increasing, add...if it is decreasing', instead of the expected *ki* **tab tab** *ki* **la**$_2$ **la**$_2$. Since the scribe of **C** has committed several obvious errors including a repetition of the signs *a* **ugu** in the same line, this anomaly can be rather safely dismissed as a scribal error.

step 6

A.R6 In *ACT*, p. 228, Neugebauer erroneously restored the gap as [...*ki* **tab tab** *ki* **la**$_2$ **la**$_2$].

C.R6–7 ∗ The other Mss. imply that the scribe of **C** has erroneously inserted the phrase 'you accumulate (them) together, you compute half of it and you put it down as the *ṣiliptu*-coefficient' at the end of step 6. The same phrase appears correctly in R5. Since nothing is computed in step 6 that can be 'accumulated together', this is another scribal error.

step 7

C.R8 ∗ 20: very uncertain. If this reading is correct 20 may be interpreted as 0;0,20. At the same position the other Mss. have 0;0,8. This might be an alternative version of r, which is also attested in No. **53** (cf. the astronomical commentary).

∗ **bar.nun** *ša*$_2$ *lu-maš*, '*ṣiliptu*-coefficient for the zodiac', is probably a scribal error, because this coefficient was dealt with already in step 5. The context and the parallels imply that the correct phrase is **bar.nun** *ša*$_2$ **nim** *u* **sig**, '*ṣiliptu*-coefficient for 'height and depth". In *ACT*, p. 226, Neugebauer attempted to resolve the discrepancy in **C**.R3,8 by assuming that **bar.nun** *ša*$_2$ *lu-maš* *ša*$_2$-*šu*$_2$ means '*ṣiliptu*-coefficient for the corresponding zodiacal sign', i.e. that 'zodiacal sign' is the argument of the coefficient, so that the phrase denotes q as well as r. This ad hoc solution is unsatisfactory, because the parallels imply that *šâšu* (steps 5,7) is an attribute to **bar.nun**, i.e. 'the corresponding *ṣiliptu*-coefficient'. Secondly, *lu-maš*, 'zodiacal sign', is not the argument of the coefficient, but a qualifier which identifies the coefficient as being q (not r).

step 8

C.R10 ⌜GI$^?$⌝: apparently an alternative logogram for DIŠ.UD, which is not attested anywhere else as far as I know. Unfortunately, this does not seem to bring us further in identifying the Akkadian reading of either GI or DIŠ.UD.

gar-*nu* = *taškunu*: replaces **gar**-*un* (*ACT*).

step 12

A.R10 There appears to be little room to restore the expected signs, including *u*$_3$ 2 ḪAB-*rat*, in the gap at the end.

A.R11 ⸢**ta** *lib*$_3$-*bi*⸣-*šu*$_2$ **la**$_2$: replaces **tab**... (*ACT*). At the end ⸢**me nu**$^?$ **man**$^?$-*ni*$^?$⸣ has been restored, as suggested by C.B.F. Walker.

B.R3' ∗ **nim** = 'height': this must be a scribal error, since all other Mss. have the expected **sig**, 'depth'.

C.R12 *kī mala amāri*(**igi**) *lā maṣû*, 'if it is insufficient for being seen' is unique for **C**. The same phrase with an explicit subject appears in No. **53** Rii9' as *kī sin mala amāri lā maṣû*, 'if the Moon is insufficient for being seen'.

step 13

A.R12 Reading ⸢**gin**$_7$⸣ established by C.B.F. Walker. The entire phrase ⸢**gin**$_7$⸣ **igi**-*u*$_2$ **du**$_3$-*uš* is copied more accurately by Strassmaier than by Pinches.

C.R12 There is a superfluous 1 after *ana*. In the gap after **me** one can restore either **gar**-*an* or **e**-*bi*, the former being somewhat more likely, given R13. The logogram **ki.na**$_2$**.a** is unique and its Akkadian reading is unknown; it is presumably an alternative name of ME, although this should probably not be given much weight, in view of the numerous errors and oddities of **C**.
e.a: most likely a 2 m. sg. pres. *taqabbi*, 'you call; predict'; less likely the verbal adjective *qabê*, 'predicted' (attribute to **ki.na**$_2$**.a**).

P4 (GI$_6$)

step 1

B.R6' 1-*en*$^?$ written over an erasure (C.B.F. Walker).

C.R14 It is unclear whether P4 was present in **C**. Only 1 line with a few traces remain of the section following P3, but yhey do not appear to be compatible with step 1; perhaps they belong to a table of coefficients.

step 2

A.R13 Possibly -*ma* is to be replaced (or followed) by *šu*-*u*$_2$, as suggested by C.B.F. Walker. The traces at the end of the line do not allow a unique reading, but are consistent with **zi** *šamaš*$_2$ **gar**-*an*.

A.R14 Reading **du**-*ku* established by C.B.F. Walker.

B.R7' **zi** *sin* **gar**-*an* written over an erasure (C.B.F. Walker).

step 5

A.R16 Instead of the usual 10 there is *u*$_4$-*me*; cf. the notes to NA step 5.

A.R17 *a*-*ḫa*-*miš*$_2$-*šu*$_2$ = *aḫāmiššu*, the adverb *aḫāmiš*, 'together', combined with a poss. suff. 3 m. sg.: an unknown construction that appears to be without parallel.

step 6

A.R18 **zi**, 'displacement': scribal error for *sin*, 'Moon'.

B.R12' Traces might be ⸢*sin* GAM 4⸣.

step 7

A.R19 **sig** is still visible on Strassmaier's copy.

step 9

D.R18 Since 2 ḪAB-*rat* is mentioned in R20 (step 11) one expects 2 ḪAB-*rat* **ta** BE *u* **bar.nun la**$_2$, but it is uncertain whether there is enough room.

step 11

A.R20 In the light of R21 one expects BE *u* **bar.nun** in the gap at the end, in which case the line must have continued on the edge.

A.R21 Neugebauer's reading *ana* BE **diri** could not be confirmed; collation suggests that BE is followed by *u* **bar.n**[un tab]. This leaves very little room to restore the expected *ana* **gi**$_6$ **e**(-*bi*) in the same gap. The damaged sign after **diri** might be *at*$_2$.

D.R19 Since in R20 (step 12) 2 ḪAB-*rat* is mentioned, the disk correction has been applied, so that one expects 2 ḪAB-*rat* also after both instances of BE *u* **bar.nun**. However, it appears to be missing. After the former there are traces that look like the beginning of **la**$_2$, but this is uncertain. Note that throughout P4 the copula *u*, 'and', in the phrase *u* 2 ḪAB-*rat* actually replaces a subtraction.

G.R22' In the gap a subtraction **ta**...**la**$_2$ has been restored on the basis of the parallels in **AD** and the other instances of step 11 in **G**.

step 12

A.R22 The gap at the beginning appears too small for restoring *ša*$_2$ **tag**$_4$. Since this phrase is not used consistently in parallel positions in step 12 in **A** (for NA, but not for ŠU$_2$ and ME), it is probably absent here.

D.R20 Since *u* 2 ḪAB-*rat* is present after the first instance of BE *u* **bar.nun** it can also be restored in the gap.

G.R24' It is possible that BE *u* **bar.nun** is followed by *u* 2 ḪAB-*rat* in the gap, but there is not much room.

step 12'

A.R24 Before the second **tag**$_4$ one expects *ša*$_2$.
la$_2$: the preposition *ana* implies that this is *tumaṭṭa*, 2 m. sg. pres. D *maṭû*, 'you reduce'.
The damaged signs at the end are consistent with **gi**$_6$ *ana* **me**, which suggests that the first sign in the gap at the end is **gur**. The remaining space of about 4 signs might be occupied by the phrase **me nu man**-*ni*, 'you do not change the day'. One might also assume an instruction for the case $R_\downarrow > 2 - Q > 0$ (cf. the commentary), but there is little space.

Commentary

This text contains 4 composite procedures, P1–P4, for computing the Lunar Six intervals ŠU$_2$, NA, ME and GI$_6$. Because of their unparallelled complexity, clear formulation and rigorous structure, these procedures are among the most astonishing achievements of Babylonian mathematical astronomy. For the astronomical definition of the Lunar Six intervals cf. §4.1, Table 4.1 (p. 113) and Fig. 4.2 (p. 114). For a detailed explanation of the algorithms cf. §4.4.19. P2 (NA) is followed by an auxiliary table, T1, with control values of the '*ṣiliptu*-coefficients for setting' (q_{set}, r_{set}). Similarly P4 (GI$_6$) is followed by T2 with control values of the '*ṣiliptu*-coefficients for rising' (q_{rise}, r_{rise}). The extent to which the individual steps of P1–P4 and the tables are preserved in the seven Mss. is shown in Fig. 5.2. In their original state, probably all Mss. contained procedures for all four intervals ŠU$_2$, NA, ME and GI$_6$. In Mss. **DG** some steps of P3 and P4 are omitted. This might also be the case in **EF**, which do not preserve P3 and P4.

The manuscripts

Four of the Mss. (**ACEF**) were previously edited by O. Neugebauer (*ACT*). Ms. **G**, which is kept in the Metropolitan Museum, was edited by C.B.F. Walker (*CTMMA* II). The remaining two are published here for the first time. The Mss. in the British Museum belong to four collections: Sp2 (**AEF**), 81–11–3 (**B**; two fragments of **D**), to 82–7–4 (**C**), and Rm4 (a fragment of **D**). Two types of tablets can be distinguished. **ABDG**, perhaps also **EF**, are exclusively concerned with ŠU$_2$, NA, ME and GI$_6$. As far as the original dimensions are preserved or can be reconstructed, most share the same elongated rectangular format of about 9 × 13 × 2.5 cm (type G), with only little variation (about 1 cm in width and length). Ms. **C**, which is not exclusively concerned with Lunar Six intervals, is probably a larger tablet (type C?).

The individual Mss. are briefly characterised as follows. **A** is nearly intact and offers the most complete text. There is some damage to the right edge, the upper (lower) right corner and the lower (upper) left corner of the obverse (reverse). All steps are fully written out for all Lunar Sixes, and it is the only tablet with a preserved standard invocation. Unlike the other Mss., the tables of coefficients (q and r) are not in columnar format. Some of the previously assumed scribal errors[657] could be eliminated; the remaining ones are few and mostly trivial.[658]

B is a fragment including sections of the left and right edges of the tablet. It is a close duplicate of **A**. As far as the text is preserved, no steps are omitted (except, as in all other Mss., step 10 for ŠU$_2$ and NA). The lines on **B** have nearly the same height as on **A**, but they are shorter by a few signs, so that the original height must have been a bit larger than that of **A**, perhaps 14 cm.

C is a fragment from the lower (upper) left corner of the obverse (reverse) of a tablet. Part of one column is preserved to full width. The obverse contains a procedure for E (cf. No. **60**); the reverse a procedure for ME. There must have been one additional column, now completely broken off, which contained the beginning of the procedure for ME, so that its original width was about 13.5 cm. The procedure for ME is followed by traces of what was probably a list of coefficients (T2.a). The missing part of the tablet may also have contained the corresponding procedures for ŠU$_2$, NA and GI$_6$. The scribe has committed several obvious errors.[659] Orthography and terminology sometimes deviate from that of the other Mss. (e.g. ME steps 4, 7–8, 11–12). There are some similarities to No. **53** (e.g. ME step 11).

D consists of several fragments joined by I. Finkel (May 1988). On both sides the surface is heavily damaged in several places. About 2–3 cm of the right edge are missing, so the original width was about 8 cm. It contains a slightly abbreviated duplicate of **A**. In the procedures for ME and GI$_6$ steps 3–5 and 7–8 are omitted, as in **G**. **D** is the only Ms. preserving a colophon. The tablet's 'owner' is Nabû-nādin-aḫi/Nabû-nādin-a[ḫi]. As far as known, he is not attested elsewhere.

E is a small fragment including a segment of the lower (upper) edge of the obverse (reverse) of the tablet. The textual restorations imply that about 1 cm is missing on the left side and about 5.5 cm on the right side, so the original width was about 10 cm. The preserved text is consistent with NA, GI$_6$ or NA$_1$. Neugebauer (*ACT*, pp. 240–241) thought that it concerns NA$_1$, but this is unlikely, because the coefficients preserved on the reverse (T1.b = r_{set}) point to the Full Moon Lunar Sixes. The values of these coefficients for NA$_1$, as provided in No. **53** P16', are different. Since the preserved procedure precedes this table it most likely concerns NA and not GI$_6$.[660]

F is a small fragment. No edges of the tablet are preserved. It contains the latter half of the procedure for ŠU$_2$[661] and the beginning of the procedure for NA. The reconstruction of O7'–8' suggests that about 2.5 cm of clay is missing on the

[657] O2,8; cf. *ACT*, pp. 226–240.

[658] The one significant error is **nim** instead of **sig** in O25 (NA step 11), which also occurs in **D**; trivial are e.g. **zi** instead of *sin* in R18, and a superfluous *ša*$_2$ in R24.

[659] In R6–7 (ME step 6) and R8 (step 7).

[660] After the table only a few signs are preserved of the next procedure; they should belong to the procedure for ME; cf. the critical and philological notes.

[661] Not ME as stated in *ACT*, pp. 241–242.

left side of the obverse, and about 1 cm on the right side, so the original width of the tablet (or one of its columns) was about 9 cm. **F** shares with **G** some peculiarities in steps 4,5 and 7; cf. the commentary.

G is a large fragment preserving sections of the left and right edges of the tablet. The surface of the obverse is smeared out here and there. The formulation of some steps (e.g. 6 and 11) deviates significantly from that of the other Mss. It shares certain features with **D**, others with **F**. The scribe of **G** appears to prefer elaborate formulations where other Mss. are concise. Some of the damaged phrases, labeled with question marks in the transliteration, could not be identified in terms of the reconstructed steps (e.g. O2'–3', 21'–22', 24'–25'). Near the bottom of the obverse (O23'), NA step 7 appears to be mentioned, followed by 2 lines of unknown meaning. Unlike other Mss., the procedure for NA continues on the reverse. As in **D**, step 8 is present for ŠU$_2$ but omitted for ME and GI$_6$, and steps 9,10 are present for ŠU$_2$, ME and GI$_6$. Hence steps 8–10 can be assumed to have been present in the procedure for NA in the gap between obverse and reverse. This would require about 4 lines. Some peculiarities of **G** affect the astronomical interpretation; cf. the commentary to steps 4,5,7. Steps 3–5,7,8 in the procedure for ME and steps 2–5,7,8 in the procedure for GI$_6$ are omitted. The same omissions occur in **D**, except GI$_6$ step 2.

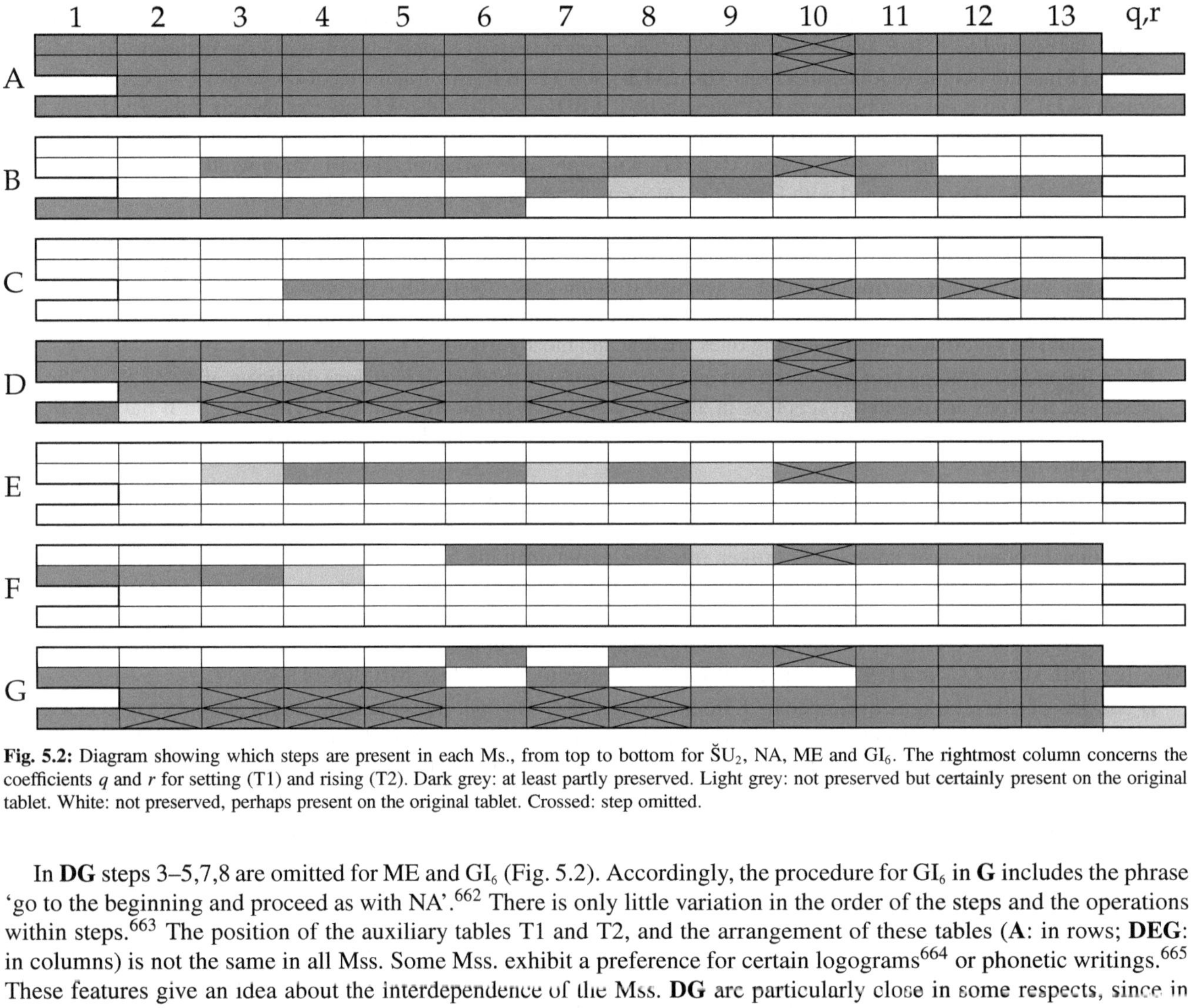

Fig. 5.2: Diagram showing which steps are present in each Ms., from top to bottom for ŠU$_2$, NA, ME and GI$_6$. The rightmost column concerns the coefficients q and r for setting (T1) and rising (T2). Dark grey: at least partly preserved. Light grey: not preserved but certainly present on the original tablet. White: not preserved, perhaps present on the original tablet. Crossed: step omitted.

In **DG** steps 3–5,7,8 are omitted for ME and GI$_6$ (Fig. 5.2). Accordingly, the procedure for GI$_6$ in **G** includes the phrase 'go to the beginning and proceed as with NA'.[662] There is only little variation in the order of the steps and the operations within steps.[663] The position of the auxiliary tables T1 and T2, and the arrangement of these tables (**A**: in rows; **DEG**: in columns) is not the same in all Mss. Some Mss. exhibit a preference for certain logograms[664] or phonetic writings.[665] These features give an idea about the interdependence of the Mss. **DG** are particularly close in some respects, since in both nearly the same steps of the modules for ME and GI$_6$ are omitted, step 6 has a different beginning and they share several orthographic features (**a.ra$_2$** instead of GAM, *ki-i* instead of *ki*). In other respects **G** deviates strongly from all

[662] **G**.R18' (GI$_6$ step 1).

[663] For instance in NA step 1, where **A** deals with the two cases in the reverse order compared to **DFG**. A second example concerns the placement of step 10 (the disk correction) in the module for ME.

[664] For instance, multiplication is done with **a.ra$_2$** in **DG**, GAM in **BF**, and GAM$_0$ in **AC**. In **G** the names of the Moon and the Sun are written ^{d}*sin* and ^{d}**utu**, respectively, whereas the other Mss. have *sin* and *šamaš$_2$*.

[665] E.g. *ki-i* in **DG** versus *ki* in the other Mss. for *kī*, 'if'. Other variations seem arbitrary. In **ABD** the logogram for subtraction varies between **nim**, **la$_2$**, and **e$_{11}$**. The same is true for *ūmu*, 'day', which is written **me**, *u$_4$-me* or *u$_4$-mu*, and *kīma*, 'like', which is written **gin$_7$** or *ki-ma*.

other Mss. including **D**. **FG** both omit the averaging from steps 5 and 7. **AD** share an error in NA step 11, but the correct version is not preserved in any Ms.

In the following textual commentary I only discuss issues that were not addressed in the main commentary (§4.4.19). Nearly all Mss. conform with the sequence of steps as shown in the flow chart (Fig. 4.26, bottom), with few exceptions. The absence of significant variation in this respect therefore points to a close relationship between the Mss. One exception is No. **61**.**G**, where step 7 is executed before step 6.

Step 1. Time and date of the Lunar Six event (§4.4.19.1)

The expression for $N_{\check{S}U_2}$ for the case when Full Moon occurs during day time (Eq. 4.83, p. 164) replaces the erroneous one in *ACT*, p. 231. No. **61**.**A** was the sole Ms. available to Neugebauer for his reconstruction of $N_{\check{S}U_2}$, but it is damaged at the crucial spot in O2. Neugebauer restored the gap as LA$_2$ *ša*$_2$ **gi**$_6$, 'duration of the night', instead of the expected LA$_2$ *ša*$_2$ **me**, 'duration of daylight'. Furthermore, he assumed that the instruction to 'subtract 1 day' concerns $N_{\check{S}U_2}$ itself, rather than the date of ŠU$_2$. The parallels (ŠU$_2$ step 13, NA steps 1,13, ME step 13, GI$_6$ step 1) leave no doubt that this instruction concerns the date.[666] The resulting expression for $N_{\check{S}U_2}$ in *ACT* is inconsistent, as Neugebauer noted himself. Due to the duplicate No. **61**.**B**, LA$_2$ *ša*$_2$ **me** can be restored in **A** with some degree of probability, thus removing the need to assume a scribal error.

Step 2. Zodiacal displacement of Moon and Sun (§4.4.19.2)

In **G** P2 the formulation of step 2 deviates slightly from the other Mss., but the underlying algorithm is probably the same. In **G** P4 step 2 is replaced by the phrase 'you go to the beginning and do as with NA'.

Step 3. Elongation (§4.4.19.3).

In **G** P2 the formulation of step 3 deviates from the other Mss., but the algorithmic interpretation is the same.

Step 4. Zodiacal position of Moon and Sun (§4.4.19.4)

The state of preservation of **G** is such that no complete version of step 4 is preserved. We therefore cannot be certain that its formulation is the same as in the other Mss. Moreover, steps 5 and 7, both of which use the outcome of step 4, are also incompletely preserved in **G**. In particular, it is not certain that in **G** the coefficients q and r are computed separately for the Moon and the Sun and averaged. Without these operations, step 4 is superfluous, because these coefficients become different for the Moon and the Sun only if step 4 is executed.

Step 5. Coefficient for the zodiac (§4.4.19.5)

In **FG** there is no intact version of step 5. In **F** there seems to be no averaging of $q_☽$ and $q_{\odot}$, and step 4 may be lacking as well. This suggests that q may be computed not for the Lunar Six event, but for the moment of Full Moon, when $q_☽ = q_{\odot}$, so that there is no need for averaging. In **G**, where step 5 is partly preserved only for NA, a somewhat different formulation is used, but the control values of q are the same. At the end of step 5 the result q' is explicitly named by the phrase 'you put it down as the *ṣiliptu*-coefficient'. One would expect a similar instruction at the end of step 7 (r'), but it is always lacking there. It therefore appears that the abbreviated name '*ṣiliptu*-coefficient' is reserved for q'. Also in steps 9–11 this term is used for q' but not for r'.

Step 6. Moon's distance to ecliptic (§4.4.19.6)

Two different introductory phrases are attested for step 6. 'You put down the Moon's 'height and depth'' (in all procedures of **ABCEF**, and in P1 (ŠU$_2$) and P2 (NA) of **D**) refers to the initial value E from which β' is computed. The second version, 'All the *ṣiliptu*-coefficients for 'height and depth''[667], is used in **G**, and in P3 (ME) and P4 (GI$_6$) of **D**. It is not entirely clear why this phrase appears here. The *ṣiliptu*-coefficient for 'height and depth' is the technical term for r; hence it might actually refer to step 7, where r is computed. The alternative pairs of Akkadian terms for β', **nim** and **sig** for 'height' and **la$_2$** and **bur$_3$** for 'depth', do not reflect any difference in meaning. For instance, the scribe of **G** uses **nim** and **sig** in one part of a sentence, but **la$_2$** and **bur$_3$** in another part of the same sentence (step 11). The formulation of step 6 in P1 (ŠU$_2$) of Ms. **G** deviates from the other Mss. in that the sign of the correction appears to be wrong, assuming that these lines indeed belong to step 6.

[666] Moreover, the subtraction of 1 day from a time interval would be formulated as 'you subtract 6,0 (°)'.

[667] 'All' refers to the 12 zodiacal signs of the coefficients listed in T1,T2.

Step 7. Coefficient for the Moon's distance to the ecliptic (§4.4.19.7)

In **FG** there is no averaging in the procedure for ŠU$_2$, and presumably neither for the other Lunar Sixes. Perhaps a single value of r is computed for the moment of Full Moon, when $r_{☽} = r_{\odot}$. The outcome of step 7 is passed on anonymously to step 8, except in **G**, where step 6 is probably placed in between steps 7 and 8. This prompts the scribe of **G** to refer to r in step 8 as the 'last *ṣiliptu*-coefficient' (*ṣiliptu arkītu*) in order to distinguish it from q.

Step 9. Contribution to the Lunar Six due to elongation (§4.4.19.9)

The outcome of step 9 is not explicitly named. In steps 10,11 it is usually referred to by the descriptive term 'elongation and the *ṣiliptu*-coefficient', which is sometimes abbreviated to 'elongation', while in **G** the elaborate version 'elongation which you had multiplied by the *ṣiliptu*-coefficient' is used in the introductory phrase of step 10, but nowhere else.

Step 10. Disk correction (§4.4.19.10)

The Mss. differ somewhat as to where exactly after step 9 the correction is introduced. In **AB** it is applied to ME at the end of step 10; in **DG** it is applied to ME and GI$_6$ at the end of step 9. A peculiar case is GI$_6$ in **A**, where the relevant instruction is not mentioned in step 10 or 11, but at the beginning of step 11', which deals with the special case $Q < 2$. It is therefore fair to conclude that the disk correction is also implied in steps 10 and 11 (as indicated in the formulaic representation).

Step 11. Computation of the Lunar Six interval (§4.4.19.11)

In **G** the conditions for addition and subtraction of R are formulated differently from the other Mss. Instead of 'high' comes 'high, ascending, or high, descending', which is obviously equivalent. Instead of 'low' **G** has 'low, descending, or low, ascending', which amounts to 'low'. One wonders why the scribe of **G** chose to formulate step 11 in a superfluously elaborate way. Perhaps the purpose was to stay close to the terminology used in column E, which contains the numerical value of E followed by the same pair of expressions 'high/low' and 'ascending/descending', even though the latter play no role in step 10.

Step 13. New timing for the Lunar Six event

In **G**, the backward shift of the date of ŠU$_2$ is uniquely accompanied by the statement 'ŠU$_2$ turns backward' (**šu$_2$** *ana arkišu inaḫḫis*).

No. 62

museum number	BM 32241 (76–11–17,1968)
provenience	Babylon
measures	5.2(10) × 5.7(*) × 2.2(*) cm (type E, F or G?)
cuneiform text	photo: Fig. F.70; *ACT*, Pl. 239a
bibliography	*ACT* 200aa (together with BM 32172)
contents	Moon system A: procedures for Φ and B^{nm}

Obverse

P1' (unknown number of lines missing)

1' [xxxxxxxxxxxxxxxxxxxxxxx].20 ⌜**e**$_{11}$⌝-[*ma* xxxx]

2' [xxxxxxxxxxxxxxxxxxx].⌜40⌝ **e**$_{11}$-*ma* **gar**-⌜*an*⌝ [xx]

3' [xxxxxxx 20.39.48.53.20 **ki**] 2.13.20 *ša*$_2$ **u**$_4$.1.**kam tab**-*ma ša*$_2$ ⌜*al-la*⌝

4' [2.17.4.48.53.20 **diri ta** 2.17.4.48.53].⌜20⌝ **la**$_2$ *ša*$_2$ *al* 1.57.47.57.4⌜6.40⌝

5' [**la**$_2$ **ki** 1.57.47.57.46.40 **tab** xx 4.16.15.4]4.26.40 **tab**-*u*$_2$ 4.13.29.48.53.20

P2' 6' [xxxxxxxxxxxxxx] ⌜**tab**$^?$⌝ *ša*$_2$ *al* 2.17.4.48.53.20 **diri ta** ⌜2⌝.

7' [17.4.48.53.20 **la**$_2$ *ša*$_2$ *al* 1.57.47.5]7.46.40 **la**$_2$-*u*$_2$ **ki** 1.57.47.57.4⌜6⌝.[40 **tab**]

8' [xxxxxxx 4.17.34.4.26].40 **tab** 4.12.11.28.53.20 **la**$_2$ [xxx]

P3' 9' [*lu-maš*.**meš ta** 27 *zib* **en** 13 **absin**$_0$ 28.7.30] **tab** *ša*$_2$ *al* 13 **absin**$_0$ **diri a.ra**$_2$ ⌜1⌝.[4 **du ki**]

10' [13 **absin**$_0$ **tab ta** 13 **absin**$_0$ **en** 27 *zib* 30 **tab** *ša*$_2$] ⌜*al*⌝ 27 *zib* **diri a.ra**$_2$ 5⌜6⌝.[15 **du ki** 27 *zib* **tab**]

11' [**ta** 27 *zib* **en** 13 **absin**$_0$ *ana* 12 **ab**$_2$] ⌜10⌝.22.30 **tab ta** 1⌜3⌝ [**absin**$_0$ **en** 27 *zib ana* 12 **ab**$_2$ 11.4 xx]

12' [xxxxxxxxxxxxxxx] ⌜xx **tab**$^?$ x⌝ [xxxxxxxx]

(unknown number of lines missing)

Obverse

P1' **(Relations between Φ^{fm} and Φ^{nm})**

a **(Computing Φ^{fm} from Φ^{nm})**

(*unknown number of lines missing*)

1'[...] you subtract [from ...],20 [and ...]

2'[...] you subtract [from ...],⌜40 and you put it down [...]

3'[... 20;39,48,53,20] you add [with] 2,13;20 (= Φ) for day 1, and (the amount) by which 4'[it exceeds 2,17;4,48,53,20] you subtract [from 2,17;4,48,53],20. (The amount) by which it [is less than] 1,⌜57;47,57,46,40⌝ [you add with 1,57;47,57,46,40.]

b **(Checking rules for $\Phi^{fm} + \Phi^{nm}$)**

5'[... decreasing: 4,16;15,4]4,26,40; increasing: 4,13;29,48,53,20.

P2' **(Procedures for Φ)**

a **(Updating Φ as a zigzag function)**

6'[...] .... (The amount) by which it exceeds 2,17;4,48,53,20 7'[you subtract] from ⌜2⌝,[17;4,48,53,20. (The amount) by which] it is less than 1,57;47,5]7,46,40 [you add] to 1,57;47,57,4⌜6⌝,[40] 8'[...]

b **(Checking rules for $\Phi_i + \Phi_{i+6}$)**

[... 6 months: 4,17;34,4,26],40: increasing; 4,12;11,28,53,20, decreasing [...]

P3' **(Procedures for B^{nm})**

a **(Updating B^{nm} with the step function for σ)**

9'[Zodiacal positions. From 27 Psc to 13 Vir] you add [28;7,30]. (The amount) by which it exceeds 13 Vir [you multiply] by ⌜1⌝;[4, add with 10'13 Vir. From 13 Vir to 27 Psc you add 30. (The amount) by which] it exceeds 27 Psc [you multiply] by 0;56,[15, add with 27 Psc].

b **(Net displacements)**

11'[From 27 Psc to 13 Vir for 12 months] you add (error for subtract) ⌜10⌝;22,30. From 1[3 Vir to 27 Psc for 12 months ... 11;4.] 12'[...] ⌜... you add$^?$ ...⌝ [...]

(*unknown number of lines missing*)

Critical and philological notes

Obverse

3' As remarked by Neugebauer, this might be the first line of a new section.
ša$_2$ *al*: replaces **gar**-*an* (*ACT*).

4' The damaged sign before **la**$_2$ is 20, not 40 (*ACT*). It is likely that no signs are missing after 1.57.47.57.4⌜6.40⌝.

5' It appears that nothing is missing at the end of the line.

6' The damaged sign at the beginning looks like **tab**, 'you add', but one expects [... **tab** *u*] **la**$_2$, '... you add and subtract'.
At the end there are traces of a vertical wedge, presumably belonging to 2. It is assumed that the other digits are written in the next line, but this is not certain.

8' Before 4.17.34.4.26.40 one might restore *ana* 6 **ab**$_2$, 'for 6 months'.

11' **tab**, 'you add': one expects **la**$_2$, 'you subtract'.

12' Before **tab** there is a digit 5–8.

Commentary

This fragment does not preserve any edge of the tablet except, perhaps, a small segment of the right edge. It is inscribed on one side, the other side being destroyed. The restored text in P2',P3' implies that about 5 cm of clay are missing from the left side, so the original width of the tablet was about 10 cm. Three sections, separated by horizontal dividing lines, are partly preserved. Apart from Φ (P1',P2'), P3' is concerned with *C*. Since these topics usually appear at the beginning of a tablet, the inscribed side is probably the obverse, and not much clay might be missing before O1'. Neugebauer erroneously assumed that BM 32241 and BM 32172 are part of the same tablet (*ACT* 200aa), although they do not physically join. BM 32172 and several other fragments are now joined with No. **53**, and BM 32241 is not part of that tablet.

P1' Computing Φ^{fm} from Φ^{nm}; checking rule for $\Phi^{fm} + \Phi^{nm}$

The beginning of P1' is missing, and the remaining lines are badly damaged. As suggested by Neugebauer (*ACT*, pp. 212–213), P1' is concerned with the relation between $\Phi^{\rm nm}$ and $\Phi^{\rm fm}$. Part **a** begins with two lines too damaged for an interpretation. This is followed by an algorithm for computing $\Phi^{\rm fm}$ from $\Phi^{\rm nm}$ (cf. §4.4.3). Its formulation is based on template ZZ.B.3. P1'.b partly preserves two numbers known from another algorithm connecting values of $\Phi^{\rm fm}$ and $\Phi^{\rm nm}$ that can be used for verification (cf. also §4.4.3).

P2' Updating Φ as a zigzag function; checking rule for $\Phi_i + \Phi_{i+6}$

This procedure for Φ consists of two parts. Part **a** (O6'–7') contains rules for updating Φ (§4.4.3) as a zigzag function of the lunation number. It is formulated in accordance with template ZZ.B.3 (§2.6.8.1). In the gap at the beginning of O6' the monthly difference $d = 2;45,55,33,20°$ should be mentioned, so that one would expect a phrase similar to 'Month by month you add and subtract 2;45,55,33,20', but cf. the critical notes. Part **b** partly preserves a checking rule for the sum of two Φ values separated by $6^{\rm m}$ and located on opposite branches of the zigzag function; for the explanation cf. §4.4.3. It is therefore also conceivable that in part **a** Φ is updated across $6^{\rm m}$, in which case one has to restore the missing difference as $d\Phi(6) = 16;35,33,20°$ (cf. Table 4.7a on p. 126) instead of d.

P3' Procedures for B^{nm}

Only the beginning of this procedure for $B^{\rm nm}$ (§4.4.4) is partly preserved. Part **a** (O9'–10') contains the algorithm for updating $B^{\rm nm}$ with the step function for σ. It is formulated in accordance with template STEP.A.1 (§2.6.10). Part **b** (O11'–12') concerns the net displacement of the Moon and the Sun for $12^{\rm m}$, $dB_j(12)$ (Table 4.9). However, the 10;22,30° (zone 2) ought to be subtractive. The corresponding net displacement for zone 1, $dB_2 = -11;4°$(i.e. subtractive), may be restored at the end of O11'. Too little is preserved of O12' for an interpretation. Perhaps this is another set of net differences.

No. 63

museum number	BM 35162 (Sp2,714)
provenience	Babylon
measures	5.5(*) × 5.0(*) × 1.7 cm
arrangement	O/R
cuneiform text	photo: Fig. F.72; *ACT*, Pl. 239; copy: *LBAT* 94
bibliography	*ACT* 203
contents	Moon system A: procedures involving Φ

Obverse

P1' (unknown number of lines missing)

1' [...xxxxxxxxx] ⌜x⌝
2' [...xxxxxxxxx].20 **kur**-*ad*$_2$
3' [...xxxxxxxx].⌜2?⌝2.20
4' [...xxxxxx].⌜20⌝ **e**$_{11}$-*ma*
5' [...xx 1.57].47.57.46.40

Edge

6' [... 1.57.47].⌜57.46⌝.40
7' [...x] ⌜xx 1.57⌝.47.57.46.40

Reverse

P2' 1 [...xxx] ⌜**gar**?⌝-*an* **tab u**$_4$ *ana* **u**$_4$
2 [...x 1.22.3]⌜9⌝.15.33.20 **tab** *u* **la**$_2$
3 [...xxxx *ana* 1]2 **iti** 5.22.35
4 [33.20 ...xxxxxxx] ⌜^{giš}⌝**da**
5 [...?]
(no further lines?)

Obverse

P1' **(Updating Φ as a zigzag function?)**

(*unknown number of lines missing*)
1'[...] ... 2'[... until] you reach [2,17;4,48,53],20.
3'[...].⌜2?⌝2.20
4'[...] you subtract [from 2,17;4,48,53],⌜20⌝.
5'[... 1,57];47,57,46,40.

Edge

6'[...],⌜47,46⌝,40 7'[...] ... ⌜1,57⌝;47,57,46,40.

Reverse

P2' **(Computations involving Φ^*; net differences of Φ)**

1[...] you ⌜put down⌝. The addition, day by day [...] 2[1;22,3]⌜9⌝,15,33,20 you add and subtract. 3[... for 1]2 months 5;22,35,4[33,20 ...]

Co **(Colophon)**

[... Copied from] a wooden board.
(*no further lines?*)

Critical and philological notes

Obverse

1' Traces of two vertical wedges and a sign similar to ŠU.
2' Before **kur** there is a 20, perhaps the final digit of 2,17;4,48,53,20 = M_Φ.
3' Before 2.20 there are traces of at least two winkelhakens.

Edge

7' The first sign ends with 3 vertical wedges; the second one looks a bit like KU.

Reverse

1 The first sign is either a digit in the range 4–8, or GAR preceded by traces of another vertical wedge. In the latter case one may read **gar**-*an*, 'you put down'.
2 Alternatively one might restore [... *ana* 14 **iti**] (9).15.33.20. In *ACT* Neugebauer actually restored [... *ana* 15 **iti**] (15 months) on account of No. **55** O6, which has 15 instead of the expected 14.
4 ^{giš}**da** = *lē'u*, 'wooden board', presumably belongs to a colophon. After **da** the line is blank.

Commentary

This small fragment belongs to the lower (upper) right corner of what is probably the obverse (reverse) of the tablet. The obverse is largely destroyed, with only a few signs near the edge remaining. No lines could be restored completely, so the original width of the tablet remains unknown. Horizontal dividing lines between the sections are not apparent. The last line on the reverse may belong to a colophon. If read correctly, the tablet was copied from a wooden board. The handwriting is larger than average. All preserved procedures are concerned with Φ (§4.4.3).

P1' Updating Φ as a zigzag function

The sparse traces in P1' suggest that it contains the rules for updating Φ as a zigzag function. The underlying template (§2.6.8.1) could not be determined. The minimum m is mentioned in E7', perhaps the maximum M in O2' and O4'. Since this procedure is usually placed at the beginning of a tablet, only a few lines of text may be missing above O1'.

P2' *Computations involving* Φ^*; *net differences of* Φ

Only the last 4 lines are partly preserved. This procedure may have begun on the obverse. Since R1 mentions a 'daily' difference, I have restored d = 1;22,39,15,33,20°, the 'daily' difference of Φ^*, in R2, instead of 0;9,15,33,20° = $d\Phi^*(14)$ (*ACT*), but the latter cannot be completely excluded. The net difference of Φ for 12 months, $d\Phi(12) = -5;22,35,33,20°$ (Table 4.7a, p. 126) is partly preserved in R3.

No. 64

museum number	BM 33480 (Rm4,34)
provenience	Babylon
measures	4.8(7–8?) × 2.8(*) × 2.2 cm (type G or H?)
cuneiform text	photo: Fig. F.73
contents	Moon system A: computing *G* from *Φ* (scheme 1?)

Obverse?

P1 1 [*si-man ana*] **du$_3$**-*ka ana* ⸢*tar*⸣-*ṣa* 2.13.20 **la$_2$**-[*u$_2$* 2.40 **gar**-*an ša$_2$ al*]

2 [2.13.2]0 **la$_2$ la$_2$**-⸢*u$_2$* **en**⸣ 2.10.40 **la$_2$**-⸢*u$_2$*?⸣ [**ta** 2.13.20 x **a.ra$_2$**]

3 [3.22.30] **du** *ma-lu-uš* 17.46.40 GA[R.GAR-*ma* **ki** 2.40 **tab** x]

4 [xxxxx] ⸢x⸣ GAM 3 4-*'u-u$_2$* ⸢GAM?⸣ [xxxxxx]

5 [xxxxx] ⸢GAM⸣ 7 8-⸢*u$_2$* GAM? 3?⸣ [xxxxxxx]

(n lines missing)

Reverse?

1' [xxxxx] ⸢xx 20 x⸣ [xxxxxxx]

2' [xxxxx] ⸢x 11? xx⸣ [xxxxxxx]

(approx. n+3 lines missing)

Obverse?

P1 **(Computing *G* from *Φ*)**

[1][In order] for you to construct [the duration].

1–9 Opposite 2,13;20, decreasing, [you put down 2,40. That which] [2]is less than [2,13;2]0, decreasing, until 2,10;40, decreasing, [you subtract from 2,13;20,] [3]multiply [by 3;22,30]. While it fills up you accu[mulate] 0;17,46,40 [and you add it with 2,40. ...]

[4][...] ... times 3/4 ... [...] [5][...] ⸢times⸣ 7/8 ⸢...⸣ [...]

(*n lines missing*)

Reverse?

[1'][...] ⸢... 20 ...⸣ [...] [2'][...] ⸢... 11? ...⸣ [...]

(*approx. n+3 lines missing*)

Critical and philological notes

Obverse?

2 In the gap at the end there is room for about 6 signs. This suggests that the difference $\Phi - \Phi_k$ is explicitly computed here (otherwise the gap should have contained only **a.ra$_2$**, 'times').

3 *ma-lu-uš* = *malûš*, an apocopated form of *malûššu* = inf. G *malû* + loc. + poss. suff. 3 sg. m. (*GAG*, §§66e,f; 150i), 'in its filling up'; cf. §4.4.11.

4 The first sign is a vertical wedge or perhaps ŠU$_2$.

Reverse?

1' Apart from 20 only traces of signs are visible.

2' Traces of signs, perhaps including a digit 11, are visible.

Commentary

This small fragment belongs to the upper edge of a tablet. Nothing is preserved of the other edges. The inscribed side is probably the obverse, since the preserved text contains the beginning of *ΦG* scheme 1. The other side is destroyed. The textual restorations in lines O1–2 imply that the original width of the tablet or column was about 7–8 cm.

P1 Computing G from Φ; unclear

P1 contains the usual interpolation algorithm for intervals 1–9 of *ΦG* scheme 1 (§4.4.11). The formulation is based on the *malûššu*-rule. The numbers 3/4 and 7/8 (O4–5) might be ratios of interpolation coefficients c_k. Indeed 3, 4, 7 and 8 are the values of c_k for intervals 3, 4, 7 and 8 of scheme 1, respectively, which were dealt with in the previous lines. The meaning of this is not clear. The few remaining signs on the reverse are insufficient for an interpretation.

No. 65

museum numbers	BM 34079 (Sp,177) +35152 (Sp2,704) +35324 (Sp2,896)
provenience	Babylon
measures	8.5 × 11.5(12?) × 1.8–2.6 cm (type F)
arrangement	O/R
cuneiform text	photo: Fig. F.74; *ACT*, Pl. 240; copy: *LBAT* 95–96 (BM 34079+35324)
bibliography	*ACT* 204; Aaboe (1968), p. 8 (remark); Aaboe & Henderson (1975), pp. 191–192 (P7)
contents	Moon system A: procedures for E, Ψ, Φ, F, and B

Obverse

P1 (probably no lines missing)

1’ [xx] ⸢xxxx⸣ [xxxxx] ⸢x⸣ [xxxx]

2’ [1.4]4.24 **nim gar**-*an* **nim** *u* **sig** *ša*$_2$ *al* ⸢1⸣.4⸢4⸣.[24 **diri**]

3’ [*ša*$_2$] *al* 1.44.24 **la**$_2$ GAM 10 **du** [*ana*] ḪAB-[*rat* **gar**-*an*]

4’ **en** 17.24 **gin**$_7$ *an-nam* **du**$_3$ ⸢**ta ša**$_3$ **nim**?⸣ [xx]

5’ ⸢**nim**⸣ *u* **sig** *ina* 1.44.24 **la**$_2$ *ša*$_2$ **tag**$_4$ ⸢*ina* 3⸣[4.48 xx]

6’ ⸢*a*⸣-*na* ḪAB-*rat* **gar**-*an* **nim** *u* ⸢**sig** *ša*$_2$ *al*⸣ 1.[44.24 **la**$_2$]

7’ *ina* **ša**$_3$ **la**$_2$ *ša*$_2$ **tag**$_4$ GAM 10 ⸢**du** *ana*? 3?⸣4.⸢48⸣ [**tab** x]

P2 8’ *ša*$_2$-*niš* **ta** 17.2⸢4 ḪAB-*rat*⸣ **nim** *u* **sig** *ša*$_2$ ⸢*al*⸣ [2.15.36 **la**$_2$]

9’ *ina* 2.15.36 **la**$_2$ *ša*$_2$ **tag**$_4$ GAM 10 **du t**[**a** 40 **la**$_2$ *ana* ḪAB-*rat* **gar**-*an*]

10’ *ana tar-ṣa* ḪAB-*rat ša*$_2$ ⸢**nim**⸣-*u*$_2$ **gar**-*an* ⸢*tu*?-x⸣ [x *ana*] ⸢*tar*⸣-[*ṣa* 1 ḪAB-*rat*]

11’ 1.38.24 **nim gar**-*an ana tar-ṣa* 40 [ḪAB-*rat* 2.15.36 **gar**-*an*]

12’ **nim** *u* ⸢**sig**⸣ GAM 10 **du** ḪAB-*rat* **kur**-*ad*$_2$ 17.⸢24 GAM⸣ [6 **du** 1.44.24]

13’ 16.14 GAM 6 **du** ⸢1.38.24⸣ **nim kur**-*ad*$_2$ ⸢x⸣ [xx]

14’ ⸢xx ḪAB?-*rat*? xx⸣ 36 GAM 6 **du**-*ma* ⸢3.36 x⸣ [xx]

P3 15’ [**nim** *u* **sig ta**] ⸢ḪAB-*rat*⸣.**meš** *ana šu-ṣu-u*$_2$ ⸢ḪAB? x⸣ [x] ⸢x⸣ ḪAB-*rat* [x]

16’ 17.[24 x] ⸢xxx⸣ ḪAB-*rat ša*$_2$ *al* 1.18.⸢2⸣4 **la**$_2$ *ina* **š**[**a**$_3$ **la**$_2$]

17’ *ša*$_2$ ⸢*i*?-*re*⸣-*ḫi* GAM ⸢6 **du**⸣ *a-na* **nim** *lu a-na* ⸢**sig**⸣ [**gar**-*an ša*$_2$ *al*]

18’ ⸢1⸣7.24 **diri**-*at*$_2$ 17.24 [*ina*] ⸢**ša**$_3$ **la**$_2$ *ša*$_2$ **tag**$_4$ GAM⸣ [6 **du** *ana* **nim**]

19’ [*lu*] *ana* **sig gar**-*an*

P4 20’ ⸢*ana*⸣ *tar-ṣa* 1 **šu.si** ḪAB-*rat ḫe-pi-ti* 1.⸢38.24⸣ [**gar**-*an*]

21’ *ša*$_2$ *al* 1 ḪAB-*rat* **la**$_2$-*at*$_2$ **ta** 1 **la**$_2$ *ša*$_2$ **tag**$_4$ GAM 6 **d**[**u**]

22’ **ki** 1.38.24 **nim** *u* **sig tab**-*ma a-na* **nim** *u* **sig gar**-*an* [*ša*$_2$ *al* 1 **diri**]

23’ ⸢GAM 6⸣ **du ta** 1.38.24 **nim** ⸢*u* **sig la**$_2$-*ma a-na*⸣ [**nim** *u* **sig gar**-*an*]

24’ [**en** 17].24 ḪAB-*rat* **gin**$_7$ *an-n*[*am* **du**$_3$ xxxxx]

25’ [xxxx] ⸢xxxxxxx⸣ [xx *ša*$_2$ *al*]

26’ [1.38.24] **diri** GAM 10 **du ta** 1 ḪAB-*rat* **l**[**a**$_2$-*ma a-na* ḪAB-*rat*]

27’ [**gar**-*an š*]*a*$_2$ *al* 1.38.24 **la**$_2$ GAM 10 **d**[**u ki** 1 **tab**-*ma ana* ḪAB-*rat* **gar**-*an*]

Reverse

P5 1 [**zi** *sin* **ta** 2.13].20 *ana šu-ṣu-u*$_2$ *ana* ⸢*tar-ṣa*⸣ 2.17.4.48.[53.20]

2 [15.56.54].22.30 **gar**-*an* ⸢*a-na tar-ṣa*⸣ 1.[57].47.57.[46.40]

3 [11.4.4.41.15 **gar**-*an ša*$_2$ *al* 2.17].⸢4.48.5⸣[3.20 **la**$_2$]

4 ⸢**ta** 2.17.4.4⸣8.53.20 ⸢**nim** *ša*$_2$ **tag**$_4$⸣ [GAM 15.11.15 **du**]

5 ⸢*ša*$_2$ **nim**-*ka*? **ta**⸣ 15.56.54.22.30 **n**[**im**-*ma* **gar**-*an*]

6 *ša*$_2$ *al* 1.57.47.57.46.40 **diri** 1.5[7.47.57].

7 46.40 *ina* **ša**$_3$ **la**$_2$ *ša*$_2$ **tag**$_4$ GAM 15.11.15 **du**

8 **ki** [11].4.4.41.15 **tab**-*ma* **gar**-*an*

P6 9 2.13.20 *ša*$_2$ **u**$_4$.1.**kam** *ana* **u**$_4$.14.**kam** *ana šu-ṣu-u*$_2$ 20.⸢3⸣[9.48.53.20]

10 ⸢**tab** *u*⸣ **la**$_2$ *ša*$_2$ *al* 2.17.⸢4⸣.48.53.20 **diri** [**ta** 2.17.4.48].

11 ⸢5⸣[3.20] **la**$_2$ *ša*$_2$ ⸢*al*⸣ 1.57.47.⸢5⸣[7.4]6.[40 **la**$_2$ **ki** 1.57].

12 4⸢7.5⸣[7.4]6.40 **tab**-*ma* **gar**-*an* [xxxxxx]

13 ⸢4?⸣.13.29.⸢48?.5?⸣[3.20 xxx] ⸢x **u**$_4$?⸣.14.⸢**kam**⸣ [xx]

Obverse

P1 (Computing E from Ψ and vice versa)

(*probably no lines missing*)

1'[...] ... [... Opposite nothing(?), the 'disk'], 2'you put down 1,44;24 ($^{\check{s}e}$) as the 'height'.

(The amount) by which the 'height and depth' [exceeds] 1,44;24 3'[or is less] than 1,44;24 you multiply by 0;10, you put it down as the 'di[sk']. 4'Until 17;24 (^{f}) you do like this.

From ⌜...⌝ [...] 5'you deduct the 'height and depth' from 1,44;24, what remains ⟨you multiply by 0;10⟩, [subtract] from 3[4;48 ...], 6'you put it down as the 'disk'. The 'height and depth' which is [less] than 1,[44;24] 7'you subtract from it, what remains you multiply by 0;10, you [add it] to 34;48.

P2 (Computing E from Ψ and vice versa)

8'Secondly, from 17;24, the 'disk'. The 'height and depth' which is [less] ⌜than⌝ [2,15;36] 9'you subtract from 2,15;36, what remains you multiply by 0;10, [you subtract] fr[om 40, you put it down as the 'disk'].

10'Opposite the 'disk' you put down ⌜the 'height and depth'?⌝ ... [Opposite 33;48, the 'disk'], 11'you put down 1,38;24 as the 'height'. Opposite 40, [the 'disk', you put down 2,15;36.]

12'You multiply the 'height and depth' by 0;10, you reach the 'disk'. You [multiply] 17;24 by [6, it is 1,44;24.] 13'You multiply 16;14 by 6, you reach 1,38;24, the 'height' ⌜...⌝ [...] 14'⌜.. the 'disk'? ...⌝ You multiply 0;36 by 6, it is ⌜3;36⌝ ...[...]

P3 (Computing E from Ψ)

15'In order to produce [the 'height and depth' from] ⌜the 'disks': the 'disk'? ...⌝ [...] 16'17;[24 ...] ⌜...⌝ The 'disk' which is less than 1,18;24 [you subtract] from it, 17'what remains you multiply by 6, [you put it down] as the 'height' and 'depth'. [That which] 18'exceeds 17;24: you subtract 17;24 from it, what remains [you multiply] by [6], 19'you put it down as [the 'height' and] 'depth'.

P4 (Computing E from Ψ and vice versa)

20'Opposite 1^{f} of the 'disk' [you put down] 1,38;24. 21'That which is less than 1, the 'disk', you subtract from 1, what remains you multiply by 6, 22'add with 1,38;24, the 'height and depth', and you put it down as the 'height and depth'. [(The amount) by which it exceeds 1] 23' you multiply by 6, deduct from 1,38;24, the 'height and depth', and [you put it down] as [the 'height and depth'.] 24'[Until] the 'disk' is [17];24 you [do] like this.

[...] 25'[...]... [... (The amount) by which it] 26'exceeds [1,38;24] you multiply by 0;10, sub[tract] from 1 and 27'[you put it down as the 'disk'.] (The amount) by which it is less than 1,38;24 you multiply by 0;10, [add with 1 and you put it down as the 'disk'.]

Reverse

P5 (Computing $F^{(1)}$ from Φ)

1In order to produce [the Moon's displacement from 2,13];20 (= Φ). ⌜Opposite⌝ 2,17;4,48,[53,20] 2you put [15;56,54],22,30. ⌜Opposite⌝ 1,[57];47,57, [46,40 3you put 11;4,4,41,15. That which is less than 2,17];⌜4,48,5⌝[3,20] 4⌜you deduct from 2,17;4,4⌝8,53,20. What remains [you multiply by 15;11,15], 5⌜what comes out for you you deduct from⌝ 15;56,54,22,30 [and put down.] That which exceeds 1,57;47,57,46,40: you subtract 1,5[7;47,57],46,40 from it, what remains you multiply by 15;11,15, add with [11];4,4,41,15 and put down.

P6 (Relations between Φ^{fm} and Φ^{nm})

a (Computing Φ^{fm} from Φ^{nm})

9In order to transform 2,13;20 for day 1 (= Φ^{nm}) into (that for) day 14 (= Φ^{fm}) 10you add and subtract 2[0;39,48,53,20]. (The amount) by which it exceeds 2,17;[4],48,53,20 you subtract [from 2,17;4,48],5[3,20]. 11(The amount) by which it [is less] than 1,57;47,[57,4]6,[40] 12you add [with 1,57];4[7,57,4]6,40 and put down. [...]

b (Checking rule for $\Phi^{\mathrm{fm}} + \Phi^{\mathrm{nm}}$)

13⌜4⌝,13;29,⌜48,5⌝[3,20 ...] ... ⌜day⌝ 14 [...].

P7 14 ⸢*lu*⸣-*maš ša*$_2$ **u$_4$**.1.**kam** *ana* **u$_4$**.14.**kam ta** *zib* **en** ⸢**absin**⸣ 3.1⸢4⸣.[3.45]
15 ⸢**ta**⸣ **absin en** *zib* ⸢3⸣.15 *lu-maš ša*$_2$ 6 **ab$_2$ ta** ⸢*zib*⸣ [**en absin**]
16 2.48.45 **tab ta absin en** *zib* 3.0 *lu-maš* ⸢*ša*$_2^?$ **mu**⸣ [**ta** *zib*]
17 **en absin** 10.22.30 **la$_2$** ⸢**ta absin en**⸣ *zib* 11.[4 **la$_2$** *lu-maš ša*$_2$ 18 **mu.meš**]
18 **ta** *zib* **en absin** 10.07.30 **tab ta** ⸢**absin en**⸣ *zib* [10.48 **tab**]

P8 19 **nim** *u* **sig** *ša*$_2$ **u$_4$**.1.⸢**kam**⸣ *ana* **u$_4$**.1⸢4.**kam ta** *zib*⸣ [**en absin** 59.22.51]
20 ⸢**ta absin en**⸣ *zib* 1.⸢3.7.5⸣1 ⸢xxx⸣ [x 12 **ab$_2$ ta** *zib* **en absin**]
21 3⸢3⸣.38.24 ⸢**tab**$^?$⸣ *u* ⸢**la$_2^?$ ta absin en**$^?$⸣ [*zib* 30.52.24 xxx]
22 ⸢xxxxxxx⸣ [xx] ⸢xxx⸣ 3 ⸢4$^?$⸣ [xxxxxxxx]
23 (traces)
24 (traces)
(probably no further lines)

Critical and philological notes

Obverse

1' At the end one expects *ana tar-ṣa*, followed by a term for zero (perhaps **nu tuk**), and ḪAB-*rat*.

2'–3' In the gap at the end of O2' I restore **diri**, so that the beginning of O3' is not an erroneous repetition as assumed in *ACT*, p. 245; cf. the commentary.
nim *u* **sig** = 'height and depth': here and elsewhere this translation replaces 'increasing and decreasing (latitude)' (*ACT*).

4' **ša$_3$** is followed by 1–2 damaged signs, probably only **nim**.

5' ⸢*ina* 30⸣: copied as AN U by Pinches. Only one winkelhaken remains of 30. In this reading, the expected multiplication by 10, the result of which should be subtracted from 34;48, is omitted (cf. the commentary). Alternatively, *ina* may be a damaged GAM, in which case one can read GAM 10 [**du**], '[you multiply] by 10', but this is doubtful after collation.

9' My restoration of the end of the line differs from that of Neugebauer (*ACT*, p. 246).

10' Between *ana tar-ṣa* and **gar**-*an* Pinches's copy is not accurate. Collation suggests ḪAB-*rat sha2* **nim**-*u*$_2^?$ **gar**-*an tu*$^?$- (*tu* is very uncertain). In *ACT* Neugebauer reads AL$^?$ ina$^?$ GAR$^?$ ... without providing a translation. This is followed by about 2 signs, the first of which might be MI or MU. In the gap at the end one can probably restore *ana tar-ṣa* 1 ḪAB-*rat*.

13' Pinches accurately copied the initial number as 16.14, but one expects 16.24, as was transliterated by Neugebauer without comment. The broken sign at the end of the line might be BI or TA.

14' ḪAB-*rat*: replaces GAM (Pinches's copy in *LBAT*). The sign after ⸢3.36⸣ begins with a winkelhaken.

15' After *šu-ṣu-u*$_2$ there is a damaged sign, perhaps ḪAB.

16' 1.18.24: one expects 17.24 (not 1.38.24 mentioned in *ACT*, p. 247, since this is a value of E, not of Ψ). The damaged signs before ḪAB-*rat* might be ⸢**gar**-*an*⸣.

17' After the initial *ša*$_2$ there is a sign that begins with two horizontal wedges, as noted by Neugebauer (not so on Pinches's copy), perhaps *i*.

25' Traces of about 8 signs.

27' In *ACT* Neugebauer omitted **ki** 1 **tab**-*ma* from the restoration in the gap at the end.

Reverse

1 There is not much room for **zi** *sin* **ta** = *nisiḫ Sîn ina*, 'The Moon's displacement from', in the gap at the beginning. If one omits **zi** *sin* **ta** then *šu-ṣu-u*$_2$ = *šuṣû* must be translated as 'to transform' instead of 'to produce'.

3 At the end of the line traces suggest the expected number (M_Φ). The gap following it must have contained **la$_2$**, 'is less' (replacing **diri**, 'exceeds', in *ACT*).

5 *ša*$_2$ **nim**-⸢*ka*$^?$⸣ = *ša illika*, 'what comes out for you': not transliterated in *ACT*.

9 2,13;20 is the Babylonian name of Φ; cf. the Glossary.
As proposed by Aaboe (1968), the gap after *ana šu-ṣu-u*$_2$ can be restored as 20.39.48.53.20.

14 *lu-maš* = 'zodiacal sign': the context implies the meaning 'zodiacal position'.

15 There does not appear to be writing in the empty space between 15 and *lu-maš*.

16 After the 3 there is a 0 or a separator; the former interpretation fits better here.

19 'Height and depth': replaces 'increasing and decreasing' (*ACT*). In the light of the modified reading of R20–21 the gap at the end has been restored differently than in *ACT*; cf. the commentary.

20 1.⸢3.7⸣.[5]1: replaces 2.6.15.[42] (*ACT*). The 7 was correctly copied by Pinches (*LBAT*).

22 Traces of signs are visible. Only the 3 is clearly legible.

23–24 Traces of signs.

P7 **(Procedures involving B)**

a **(Computing B^{fm} from B^{nm})**

^{14}Zodiacal position of day 1 (B^{nm}) to (that of) day 14 (B^{fm}).
From Psc until ⌜Vir⌝ 3,1⌜4⌝;[3,45]. ^{15}⌜From⌝ Vir until Psc ⌜3⌝,15.

b **(Net displacements for various intervals)**

Zodiacal position for 6 months: From P[sc until Vir] ^{16}you add 2,48;45, from Vir until Psc 3,0.
Zodiacal position ⌜for 1 year⌝: [From Psc] ^{17}until Vir you subtract 10;22,30, ⌜from Vir until⌝ Psc [you subtract] 11;[4. Zodiacal position for 18 years]: ^{18}From Psc until Vir you add 10;7,30, from ⌜Vir until⌝ Psc [you add 10;48].

P8 **(Procedures involving E)**

a **(Computing E^{nm} from E^{fm})**

^{19}'Height and depth' of day 1 (E^{nm}) into (that of) day 14 (E^{fm}).
From Psc [until Vir 59;22,51.] ^{20}From Vir until Psc 1,⌜3;7,5⌝1 ... [...]

b **(Net differences of E)**

[For 12 months: from Psc until Vir] ^{21}⌜you add?⌝ and ⌜subtract?⌝ 3⌜3⌝;38,24. ⌜From Vir until⌝ [Psc 30;52,24 ...]
^{22}... [...] ...3... [...] ^{23}*(traces)* ^{24}*(traces)*
(probably no further lines missing)

Commentary

The tablet is nearly completely preserved. The upper (lower) edge of the obverse (reverse) and the right edge are partly damaged. The first partly preserved line, which is probably the first line of the tablet, is nearly illegible. On the reverse, the last few lines contain only illegible traces; perhaps they belong to a colophon. The procedures on the obverse deal with the transformation between E and Ψ in both directions. The reverse contains procedures for computing F from Φ (P5), Φ^{fm} from Φ^{nm} (P6), net differences of B (P7), and net differences of E (P8). The procedures are not ordered in the same way as the corresponding columns of a synodic table. Most of them are concerned with verification.

P1–P4 contain rules for computing eclipse magnitude (Ψ) from E and vice versa by means of interpolation in accordance with Eqs. (4.51) and (4.52) on pp. 140–140. Recall that Ψ is defined in such a way that opposite configurations of the Moon in terms of its position and motion with respect to the ecliptic produce the same value of Ψ. If one computes E from Ψ this always results in two possible values, which is not mentioned in the procedures. This suggests that their purpose is only to verify whether Ψ has been computed correctly for a given value of E. The interpolation rules involve computing the distance to the control point of the goal function, E_0 or Ψ_0. The necessary subtractions are possible in the Babylonian sense only if the result is positive. If Ψ approaches $17;24^{\text{f}}$ then E approaches 0, so that the extrapolation cannot be continued further. This explains why the cases $\Psi < 17;24^{\text{f}}$ and $\Psi > 17;24^{\text{f}}$ are treated separately, and why different control points are used on either side of $\Psi = 17;24^{\text{f}}$. Except for P3, the transformation between E and Ψ is considered in both directions. P2 also includes numerical examples. It is not clear why the interpolations are presented for more than two different pairs Ψ_0 and E_0, since a single pair for each segment delimited by the points where $\Psi = 0$ or $E = 0$ would suffice.

P1 Computing E from Ψ and vice versa

P1 begins with the computation of E from Ψ (Oi1'–2') using $\Psi_0 = 0^{\text{f}}$ and $E_0 = 1,44;24^{\text{še}}$ as control points for the interpolation. This applies if the Moon is below the ecliptic and ascending or above it and descending (Fig. 5.3). The point $\Psi_0 = 0$ is not preserved but can be restored based on Oi2'–4', where Ψ is computed from E using the same control points. Lines O4'–7' deal with the same transformation in both directions but now using $\Psi_0 = 34;48^{\text{f}}$, which point is also associated with $E_0 = 1,44;24^{\text{še}}$. This applies if the Moon is above the ecliptic and ascending, or below the ecliptic and descending.

P2 Computing E from Ψ and vice versa

P2 begins with the statement that this section concerns eclipse magnitudes 'from 17;24 (onwards)'. However, Ψ is then computed from E using $\Psi_0 = 40^{\text{f}}$, which corresponds to $E_0 = 2,15;36^{\text{še}}$ (O8'–9'). This applies if the Moon is above the ecliptic and ascending, or below it and descending (Fig. 5.3). The meaning of the beginning of O10' is unclear. After that, $E = 1,38;24^{\text{še}}$ is associated with $\Psi = 1^{\text{f}}$, which applies if the Moon is below the ecliptic and ascending, or above it and descending. O12' appears to contain a computation of Ψ near $E_0 = 0$, but it is incomplete, because one expects $E \cdot 0;10$ to be added to $\Psi_0 = 17;24^{\text{f}}$, which is not mentioned. This is followed by numerical examples of the computation of E from

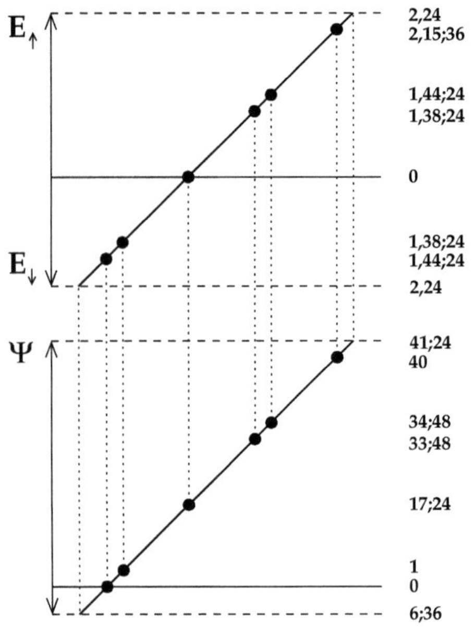

Fig. 5.3: Graphical representation of the interpolation rules for computing Ψ from E and vice versa in P1–P4. The dots are the control points (E_0,Ψ_0). The dashed lines are the boundaries of the nodal region. Only the ascending branch of E is shown; on the descending branch the same transformation applies with $E_\uparrow$ and $E_\downarrow$ exchanged. **Top**: E [barleycorns] vs. nodal elongation (η). **Bottom**: eclipse magnitude Ψ [fingers] vs. nodal elongation (η).

Ψ, all of which employ Ψ_0 = 17;24^{f} and E_0 = 0 as control points. At the end of O12', if the text is restored correctly, E is computed for $\Psi = 0$, resulting in E = 6 ·17;24 = 1,44;24$^{\check{s}\mathrm{e}}$, and O13' does the same for Ψ = 1^{f}, i.e. $E = 6 \cdot$ 16;24$^{!}$ = 1,38;24$^{\check{s}\mathrm{e}}$. After a few illegible signs O14' continues with the case $E = 6 \cdot$ 0;36 = 3;36$^{\check{s}\mathrm{e}}$, which corresponds to Ψ = 18^{f} (not preserved).

P3 Computing E from Ψ

In P3 E is computed from Ψ using Ψ_0 = 17;24^{f} and $E_0 = 0$. Although O15'–16' are damaged, they appear to deal with the case $\Psi <$ 17;24^{f},[668] since O17'–19' deal with the case $\Psi >$ 17;24^{f}. Perhaps O15'–16' mention that Ψ = 17;24^{f} is associated with $E = 0$. The inverse transformation is not mentioned.

P4 Computing E from Ψ and vice versa

P4 concerns the computation of E from Ψ and vice versa using Ψ_0 = 1^{f} and E_0 = 1,38;24$^{\check{s}\mathrm{e}}$. The rule for computing E is correctly said to apply until the point where Ψ = 17;24^{f}. After that the opposite transformation is mentioned (O25'–26').

P5 Computing $F^{(1)}$ from Φ

P5 is concerned with the computation of $F^{(1)}$ from Φ in accordance with the algorithm explained in §4.4.9.[669] It is formulated in accordance with the interpolation templates IP.D.1–2.

P6 Relations between Φ^{fm} and Φ^{nm}

Aaboe & Henderson (1975) established that P6 is concerned with the relation between Φ^{fm} and Φ^{nm}. In P6.a Φ^{fm} is computed from Φ^{nm} by adding the net difference of Φ^* for half a synodic month; cf. §4.4.3 for a general discussion. The formulation is based on template ZZ.B.3. P6.b contains a checking rule for $\Phi^{\mathrm{nm}} + \Phi^{\mathrm{fm}}$; again cf. §4.4.3 for a discussion.

P7 Computing B^{fm} from B^{nm}; net displacements for various intervals

Part **a** contains a rule for computing B^{fm} from B^{nm} by adding the appropriate displacement; for a discussion cf. §4.4.4. Part **b** is concerned with net displacements $dB(s)$ for s = 6^{m}, 12^{m} and 223^{m} (cf. §4.4.4). Lines R15–16 mention $dB_j(6)$ = 2,48;45° (zone 2) and 3,0° (zone 1); R16–17 mention $dB_j(12)$ = −10;22,30° (zone 2) and −11;4° (zone 1). It is now clear that R18 mentions $dB_j(223)$ = 10;7,30° (zone 2). In the gap at the end the corresponding number for zone 1, 10;48°, can therefore be safely restored.

[668] Assuming that 1,18;24 in O16' is to be corrected to 17;24.

[669] Several errors in Neugebauer's translation have been corrected; cf. the critical notes.

P8 *Computing E^{nm} from E^{fm}; net differences of E*

This damaged procedure consists of (at least) two parts. According to the introductory phrase, part **a** is concerned with computing E^{fm} ('for day 14') from E^{nm} ('for day 1'). For this algorihm cf. §4.4.6. Collation has established that O20 mentions the corresponding net difference of E^* for 15^{τ} for zone 2, $dE_2^*(15)$. Hence the gap at the end of O19 must have contained the corresponding number for zone 1.[670] Part **b** is concerned with net differences of E for intervals longer than 1 month. Also for this topic cf. §4.4.6, especially Table 4.14. Only the value for 12^{m} in zone 2, $dE_2(12)$, is (partly) preserved. This may have been followed in R22 by further net differences, but the traces are insufficient to determine which ones.

[670] Neugebauer assumed that O19–20 contained the preliminary monthly differences of E, $1{,}58;45{,}42^{\check{s}e}$ and $2{,}6;15{,}42^{\check{s}e}$.

No. 66

museum number	VAT 1762
provenience	Babylon
measures	4.6(*) × 4.2(*) × 2.2–3.0 cm
arrangement	O/R
cuneiform text	photo: Fig. F.67; *ACT*, Pl. 237
bibliography	*ACT* 204a
contents	Moon system A: *Φ*, *G*, perhaps *C′*

Side X

P1' (m lines missing)

1' [...xxxxx] ⸢xx⸣ [...]

P2' 2' [... 25.48].⸢3⸣8.31.⸢6⸣.40 *taš-pil-t*[*u*$_2$ ...]

3' [...x 2].45.55.33.20 **a.ra**$_2$ 9.⸢20⸣ [**du** ...]

4' [...xx]⸢2⸣.30 *ša*$_2$ *ina* ⸢*lib*$_3$⸣-*bi* ⸢*as-suḫ*$_x$⸣ 7.24 ⸢x⸣[...]

5' [...x 19].⸢1⸣7 **a.ra**$_2$ 30 **du**-*ma* 9.38.⸢30⸣ [...]

6' [...xx] ⸢2$^?$⸣4$^?$.22.8.⸢4⸣9.12.⸢9⸣.6.⸢20$^?$⸣ [...]

7' [...xxxx] ⸢1$^?$⸣.50.50.50 ⸢*mi*⸣-*nu-u*$_2$ *ki*-⸢x⸣ [x...]

8' [...xxx] ⸢1⸣.57.⸢47⸣.57.⸢46⸣.[40 xxx...]

(n lines missing)

Side Y

P3' (about n lines missing)

1' [...] ⸢GAM 6 **du**$^?$ x⸣ [x...]

2' [...] **du**-*ma mi*-⸢*nu*$^?$⸣-[*u*$_2$ ...]

3' [...] (uninscribed)

P4' 4' [... 2.51.2]9.22.57.46.[40 ...]

5' [...]-*ma* **a.ra**$_2$ 9.20 **d**[**u** ...]

6' [...x]⸢6$^?$⸣.52.⸢10$^?$⸣[...]

(about m lines missing)

Side X

P1' (Subject unclear)

(m lines missing)

1'[...] ⸢...⸣ [...]

P2' (Computations involving *Φ* and *G*; unclear)

2'[... 25;48],⸢3⸣8,31,⸢6⸣,40, the differen[ce ...]
3'[... you multiply 2];45,55,33,20 by 9;⸢20⸣, [it is 25;48, 38,31,6,40 ...]
4'[...]⸢2⸣.30, which ⸢I subtracted⸣ from it, 7.24 ...[...]
5'[...] you multiply [19];⸢1⸣7 by 0;30, it is 9;38,⸢30⸣ [...]
6'[...]⸢2$^?$4$^?$.22.8.⸢4⸣9.12.⸢9⸣.6.⸢20$^?$⸣ [...]
7'[...] ⸢1$^?$⸣.50.50.50. Whatever ... [...]
8'[...] ⸢1⸣,57;47,57,⸢46⸣,[40 ...]
(n lines missing)

Side Y

P3' (Subject unclear)

(about n lines missing)
1'[...] ⸢you multiply ... by 6⸣ [...] 2'[...] you multiply [... by ...] Whatever$^?$ [...] 3'[...] *(uninscribed)*

P4' (Computing *G* from *Φ*, variant scheme?)

4'[... 2;51,2]9,22,57,46,[40 ...] 5'[...] and ⸢you multiply⸣ it by 9;20, [it is ...] 6'[...]⸢6$^?$⸣.52.⸢10$^?$⸣.[...]
(about m lines missing)

Critical and philological notes

Side X

1' The lower halves of 2 signs are visible, the first of which is 3 or 6, the second 10–50.

4' ⸢2⸣.30: or ⸢3⸣.30.
suḫ$_x$: replaces E (*ACT*). *as-suḫ*$_x$ = *assuḫ*, 1 sg. pret. G of *nasāḫu*. The reading *suḫ*$_x$(ŠAḪ) is not mentioned in sign lists but is attested in late texts, as pointed out to me by H. Hunger.

5' The computation implies that the number at the beginning must be restored to 19.17, the outcome to 9.38.30.

6' 24: or 34. The 4 might also be **gar**. 49: or 59.
12: replaces 13 (*ACT*).

7' 1: could also be 2 or 3, or the final vertical wedge of another sign.
mi-nu-u$_2$ *ki*: established after collation. Traces of one wedge belonging to the following sign are visible; one expects *i* or *al*.

Side Y

2' *mi*-⸢*nu*$^?$-*u*$_2$$^?$⸣: replaces NIM (*ACT*).

5' 9.20: replaces 30 (*ACT*).

6' 6: or any digit 5–8.
10: or any other sign beginning with a winkelhaken.

Commentary

This small flake does not preserve any edges of the tablet. The measures of the original tablet could not be estimated. On both sides a horizontal dividing line is visible.

P1' Subject unclear

Only a few damaged signs of the last line of P1' remain on the fragment. The topic could not be established.

P2' Computations involving Φ, G and C'?

The first 8 lines of this procedure are partly preserved. The numbers and computations in X2'–3' point to Φ and G of lunar system A: X2' probably mentions the monthly difference $d_{\hat{G}}$, and X3' contains the familiar relation $d_{\Phi} \cdot 9;20 = d_{\hat{G}}$ (§4.4.11). The rest of P2' is more difficult to interpret. The numbers in X4' remain unidentified. The multiplication in X5' might be interpreted as the computation of an exemplary value of C' (§4.4.5) that can be dated to the Full Moon in month VI_2 of SE 151, assuming that it belongs to the parent synodic table of lunar system A. The difference between the values of C for months VI and VI_2 of SE 151 is 3,20;45 – 3,1;28 = 19;17, resulting in $C' = -9;38,30°$ (Eq. 4.21, p. 131). This identification is speculative, because the agreement might be accidental. The numbers in X6' remain unidentified. As pointed out by Neugebauer, the number in X7' may represent m_F = 0;1,50,50,50, the minimum of $F^{(1)\prime}$ in lunar system B (Table 4.37, p. 188). This interpretation is unlikely to be correct, since the rest of P2', including m_{Φ} = 1,57;47,57,46,40 in X8', points to lunar system A.

P3' Subject unclear

The beginning of this procedure is missing. A few signs of the 3 final lines are preserved. A multiplication by 6 is mentioned. The remains are insufficient for determining the topic of P3'.

P4' Computing G from Φ, variant scheme?

This procedure contains traces of what may be an alternative scheme for computing G from Φ (§4.4.11). The number in Y4' has been restored as 2,51;29,22,57,46,40, a control value of G in scheme 2 (cf. Table 4.26 on p. 152); this value does not occur in scheme 1. The number in Y6' might be another, unidentified control value of the same scheme. It might also be restored as $36;52,20,44,26,40 = d\hat{G}/dF$, as suggested by Neugebauer.

No. 67

museum number	MLC 2205
provenience	Uruk
measures	7.4(8?) × 9.1 × 2.1 cm (type H)
arrangement	O/R
cuneiform text	photo: Fig. F.75; *ACT*, Pl. 241
bibliography	*ACT* 205
contents	Moon system A: computing G from Φ (scheme 1)

Obverse

P1 1 [*a-na tar-ṣi*] ⸢2⸣.13.20 **la**$_2$-*u*$_2$ [2].40 **gar**-*an*
2 [*mim-ma ša*$_2$ *al*]-*la* {2.13.2.13} 2.13.20 **la**$_2$-*u*$_2$
3 [*i-maṭ-ṭ*]*u-u*$_2$ **en** 2.13.2.13.20 **la**$_2$-*u*$_2$ **ta** 2.13.20
4 [**e**$_{11}$]-*ma ša*$_2$ *re-ḫi* GAM 1 **du**-*ma* **ki** 2.40 **tab**-*ma* **gar**-*an*
5 [*a-na tar-ṣ*]*i* 2.13.2.13.20 **la**$_2$-*u*$_2$ 2.40.17.46.40 **gar**-*an*
6 [*mim-ma ša*$_2$ *al-l*]*a* 2.13.2.13.20 **la**$_2$-*u*$_2$ *i-maṭ-ṭu-u*$_2$ **en**
7 [2.12.44.26.40] **la**$_2$-*u*$_2$ **ta** 2.13.2.13.20 **e**$_{11}$-*ma ša*$_2$ *re-ḫi*
8 [GAM 2 **du**-*ma* **ki** 2].40.17.46.40 **tab**-*ma* **gar**-*an*
9 [*a-na tar-ṣi* 2.12].44.26.40 **la**$_2$-*u*$_2$ 2.40.53.20 **gar**-*an*
10 [*mim-ma ša*$_2$ *al-la* 2.1]2.44.26.40 **la**$_2$-*u*$_2$ *i-maṭ-ṭu-u*$_2$ **en**
11 [2.12.26.40 **la**$_2$-*u*$_2$] **ta** 2.12.44.26.40 **e**$_{11}$-*ma ša*$_2$ *re-ḫi* GAM 3
12 [**du**-*ma* **ki** 2.40].53.20 **tab**-*ma* **gar**-*an a-na tar-ṣi* 2.12.26.40
13 [**la**$_2$-*u*$_2$ 2.41.46].40 **gar**-*an mim-ma ša*$_2$ *al-la* 2.12.26.40 **la**$_2$-*u*$_2$
14 [*i-maṭ-ṭu-u*$_2$] **en** 2.12.8.53.20 **la**$_2$-*u*$_2$ **ta** 2.12.26.40
15 [**e**$_{11}$-*ma*] *ša*$_2$ *re-ḫi* GAM 4 **du**-*ma* **ki** 2.41.46.40 **tab**-*ma* **gar**-*an*
16 *a-na tar-ṣi* 2.12.8.53.20 **la**$_2$-*u*$_2$ 2.42.57.46.40 **gar**-*an*
17 [*m*]*im-ma ša*$_2$ *al-la* 2.12.8.53.20 **la**$_2$-*u*$_2$ *i-maṭ-ṭu-u*$_2$
18 [**e**]**n** 2.11.51.6.40 **la**$_2$-*u*$_2$ **ta** 2.12.8.53.20
19 ⸢**e**$_{11}$⸣-*ma ša*$_2$ *re-ḫi* GAM 5 **du**-*ma* **ki** 2.42.57.46.40
20 [**tab**]-*ma* **gar**-*an a-na tar-ṣi* 2.11.51.6.40 **la**$_2$-*u*$_2$
21 [2].44.26.40 **gar**-*an mim-ma* ⟨*ša*$_2$ *al-la*⟩ 2.11.51.6.40

Reverse

1 [**la**$_2$-*u*$_2$ *i-maṭ-ṭ*]*u-u*$_2$ **en** 2.11.33.20 **la**$_2$-*u*$_2$
2 [**ta** 2.11.51.6.40] ⸢**e**$_{11}$⸣-*ma ša*$_2$ *re-ḫi* GAM 6 **du**-*ma*
3 [**ki** 2.44.26.40 **tab**-*m*]*a* **gar**-*an*
4 [*a-na tar-ṣi* 2.11.33.20 **la**$_2$]-*u*$_2$ 2.46.13.20 **gar**-*an*
5 [*mim-ma ša*$_2$ *al-la* 2.11.33.20 **la**$_2$-*u*$_2$ *i*]-*maṭ-ṭu-u*$_2$
6 [**en** 2.11.15.33.20 **la**$_2$-*u*$_2$ **ta** 2.11].33.20 **e**$_{11}$-*ma*
7 [*ša*$_2$ *re-ḫi* GAM 7 **du**-*ma* **ki** 2.46.13].⸢20⸣ **tab**-*ma* **gar**-*an*
(rest uninscribed)

Critical and philological notes

Obverse

2 2.13.2.13: instead of Φ_1 the scribe wrote down Φ_2.

21 The scribe omitted *ša*$_2$ *al-la*.

Obverse

P1 (Computing G from Φ)

1 [1][Opposite] ⌜2⌝,13;20, decreasing, you put down 2,40. [2][That which is le]ss than {2,13;2,13} 2,13;20, decreasing, [3]until 2,13;2,13,20, decreasing, [4]you deduct from 2,13;20, and what remains you multiply by 1, and you add it with 2,40 and you put it down.

2 [5][Opposite] 2,13;2,13,20, decreasing, you put down 2,40;17,46,40. [6][That which is le]ss than 2,13;2,13,20, decreasing, until [7][2,12;44,26,40], decreasing, you deduct from 2,13;2,13,20, and what remains [8][you multiply by 2, and] you add it with 2,40;17,46,40 and you put it down.

3 [9][Opposite 2,12];44,26,40, decreasing, you put down 2,40;53,20. [10][That which is less than 2,1]2;44,26,40, decreasing, until [11][2,12;26,40, decreasing], you deduct from 2,12;44,26,40, and what remains [12][you multiply] by 3, and you add it [with 2,40];53,20 and you put it down.

4 Opposite 2,12;26,40, [13][decreasing], you put down 2,41;46,40. That which is less than 2,12;26,40, decreasing, [14]until 2,12;8,53,20, decreasing, you deduct from 2,12;26,40, [15]and what remains you multiply by 4, and you add it with 2,41;46,40 and you put it down.

5 [16]Opposite 2,12;8,53,20, decreasing, you put down 2,42;57,46,40. [17]That which is less than 2,12;8,53,20, decreasing, [18]until 2,11;51,6,40, decreasing, you deduct from 2,12;8,53,20, [19]and what remains you multiply by 5, and you [add] it with 2,42;57,46,40 and [20]you put it down.

6 Opposite 2,11;51,6,40, decreasing, [21]you put down [2],44;26,40. That which

Reverse

[1]is less ⟨than⟩ 2,11;51,6,40, decreasing, until 2,11;33,20, decreasing, [2]you deduct [from 2,11;51,6,40], and what remains you multiply by 6, and [3][you add it with 2,44;26,40 a]nd you put it down.

7 [4][Opposite 2,11;33,20, decreasing], you put down 2,46;13,20. [5][That which] is less [than 2,11;33,20, decreasing,] [6][until 2,11;15,33,20, decreasing,] you deduct [from 2,11];33,20, and [7][what remains you multiply by 7, and] you add it [with 2,46;13,20 a]nd you put it down.

(*rest uninscribed*)

Commentary

This tablet is unique among the procedure texts on account of its oval shape. On the obverse some clay is missing from the upper edge and from the left edge. The surface of the reverse is strongly damaged. In the upper half a few signs are preserved of lines R1–7. Below that the tablet is probably empty. The tablet is also one of the few known lunar procedure texts from Uruk,[671] and the only such text for lunar system A. A rough terminus ante quem is 150 SE (162/1 BC), since no scholarly cuneiform tablets from Uruk are attested after that date. The handwriting is highly slanted, as is true for two other procedure texts from Uruk: No. **6** (Mercury) and No. **41** (Mars, Saturn). The scribe uses certain rare phonetic writings, e.g. *i-maṭ-ṭu-*u_2 for $\mathbf{la}_2$.

P1 Computing G from Φ, scheme 1

P1 contains the interpolation rules for intervals 1–7 of ΦG scheme 1 (§4.4.11). Since the tablet appears to be empty below R7 the remaining intervals 8–35 are missing (at least two more tablets of the same size would be needed for the complete scheme). The formulation of intervals 1–7 is unique in that they are not abbreviated as usual by means of the *malûššu*-rule, but written out fully in accordance with template IP.C, which involves explicit subtractions $\Phi - \Phi_k$.

[671] Neugebauer does not mention a provenience in *ACT*, but cf. Oelsner *Materialien*, p. 181.

No. 68

museum number	BM 35253 (Sp2,820)
provenience	Babylon
measures	5.1(7–8) × 4.8(15?) × 2.3 cm (type F?)
arrangement	O/R
cuneiform text	photo: Fig. F.76; *ACT*, Pl. 239; copy: *LBAT* 97
bibliography	*ACT* 206
contents	Moon system A: computing G from Φ (scheme 1)

Obverse

P1' (at least about 14 lines missing)

1' [4.51].21.28.53.[20 xxxxxxxxxxxx]

2' [xxx 1].57.47.57.4[6.40 xxxxxxxxx]

3' [*ana tar*]-*ṣa* 1.57.47.57.4[6.40 xxxxxxxxxx]

4' [*ša*$_2$ *a*]*l-la* 1.57.47.57.[46.40 **diri** xxxxxxxxxx]

(about 18 lines missing)

Reverse

(about 11 lines missing)

0' [*a-na tar-ṣa* 2.0.41.28.53.20 **tab**-*u*$_2$ 4.49.11.6.40 **gar**-*an*]

1' [xxxxx **en** 2.0.59.1]5.33.20 [**tab**-*u*$_2$ **a.ra**$_2$ 8.20 **du ta** 4.49.11.6.40 **e**$_{11}$-*ma* **gar**-*an*]

2' [*ana tar-ṣa*] ⸢2⸣.0.59.15.33.20 ⸢**tab**⸣-[*u*$_2$ 4.46.42.57.46.40 **gar**-*an*]

3' [*ša*$_2$ *a*]*l-la* 2.0.59.15.33.⸢20⸣ [**tab**-*u*$_2$ **diri en** 2.13.8.8.53.20 **tab**-*u*$_2$]

4' [**a.r**]**a**$_2$ 9.20 **du ta** 4.46.4⸢2⸣.[57.46.40 **e**$_{11}$-*ma* **gar**-*an*]

5' *ana tar-ṣa* 2.13.8.8.53.2[0 **tab**-*u*$_2$ 2.53.20 **gar**-*an*]

6' *ša*$_2$ *al-la* 2.13.8.8.53.[20 **tab**-*u*$_2$ **diri en** 2.15.48.8.53.20 **tab**-*u*$_2$]

7' *ina* 2.15.48.8.53.20 **t**[**ab**-*u*$_2$ **e**$_{11}$-*ma* **a.ra**$_2$ 3.22.30 **du**-*ma*]

8' **diri**-*šu*$_2$ 17.46.40 GAR.GAR-*m*[*a* **ki** 2.40 **tab**-*ma* **gar**-*an*]

9' [*ana*] ⸢*tar*⸣-*ṣa* ⸢2.15.4⸣[8.8.53.20 **tab**-*u*$_2$ 2.40 **gar**-*an*]

(at least about 14 lines missing)

Critical and philological notes

Obverse

1'–4' Pinches's copy has several additional signs in the right upper corner of the obverse that were broken off by the time the photograph in *ACT* Pl. 239 was taken.

1' 53: thus Pinches's copy; no longer on the tablet. The restored number 4,51;21,28,53,20 is G_{13a}. However, it is not clear how to restore the text in O1'–2'.

2' The first 2 digits were wrongly copied by Pinches as 58.57: the tablet has the expected 57.47.
57.4[6...]: thus Pinches's copy; no longer on the tablet.

3' 47.57.4[6...]: thus Pinches's copy; no longer on the tablet.

4' 47.57: thus Pinches's copy; no longer on the tablet

Reverse

1' There is little space for the textual restorations in the gap at the end.

6' This and the following line are restored differently from *ACT* pp. 253–254; cf. the commentary for an explanation. In *ACT* 2.15.48.8.53.20 **tab**-*u*$_2$ was omitted, which implies a large empty space in this line, but since the sentence does not end until R8' one does not expect an empty space in R6'.

7' *ina*: very clear but omitted in *ACT*; Pinches's copy has 10.
e$_{11}$-*ma*: omitted in *ACT*.

8' **diri**-*šu*$_2$ = *malûššu* = inf. G *malû* + loc. + poss. suff. 3 m. sg., 'while it fills up'.
ki 2.40 **tab**-*ma*: replaces **ta** 2.53.20 **e**$_{11}$-*ma* (*ACT*).

Obverse

P1' **(Computing G from Φ)**

13a *(at least about 14 lines missing)*
1'[4,51];21,28,53,[20 ...] **2'**[... 1],57;47,57,4[6,40 ...]

13b **3'**[Oppos]ite 1,57;47,57,4[6,40, increasing, you put down 4,52;9,34,4,26,40.] **4'**[(The amount) by which] it exceeds 1,57;47,57,[46,40, increasing, until 1,57;58,8,53,20, increasing], **5'**[you multiply by 6;20, you add it with 4,52;9,34,4,26, 40 and you put it down.]
(about 18 lines missing)
Reverse
(about 11 lines missing)

24 **0'**[Opposite 2,0;41,28,53,20, increasing, you put down 4,49;11,6,40. **1'**(The amount) by which it exceeds 2,0;41,28, 53,20, increasing, until 2,0;59,1]5,33,20, [increasing, you multiply by 8;20, you subtract it from 4,49;11,6,40 and you put it down.]

25 **2'**[Opposite] 2,0;59,15,33,20, increasing, [you put down 4,46;42,57,46,40.] **3'**[(The amount) by which it exceeds] 2,0;59,15,33,20, [increasing, until 2,13;8,8,53,20, increasing,] **4'**you multiply ⌜by⌝ 9;20, [you subtract it] from 4,46;42,57,46,40 and you put it down.]

26–34 **5'**Opposite 2,13;8,8,53,2[0, increasing, you put down 2,53;20.] **6'**That which exceeds 2,13;8,8,53,[20, increasing, until 2,15;48,8,53,20, increasing,] **7'**[you subtract] from 2,15;48,8,53,20, in[creasing, and you multiply it by 3;22,30 and] **8'**while it fills up you accumulate 0;17,46,40, and [you add it with 2,40 and you put it down.]

35 **9'**[Oppo]site ⌜2,15;4⌝[8,8,53,20, increasing, you put down 2,40.]
(at least about 14 lines missing)

Commentary

This fragment includes a small segment of the left edge of the tablet, but no other edges. Both sides contain portions of interpolation scheme 1 for computing G from Φ, probably the only topic of the tablet. From the textual restorations it appears that the original width of the tablet was approximately 7–8 cm, and the height at least about 15 cm (cf. below).

P1' Computing G from Φ

The fragment preserves interpolation rules for intervals 13b (obverse) and 24–35 (reverse) of ΦG scheme 1 (§4.4.11). The formulation of intervals 13b and 24–25 is based on template IP.B (§2.6.7); that of intervals 26–35 is based on the *malûššu*-formulation (p. 148). Parts of intervals 13a (O1'–2') and 23 (R1') are also preserved, but their formulation appears to be anomalous, so that the full text could not be restored. The missing intervals 1–12 would take up about 14 lines of text above O1'. The same number of lines may be assumed below R9', but only a few lines are needed for completing interval 35, so that some other procedure, a colophon, or an empty space must have followed P1'. Intervals 14–23 would require about 30 lines of text, to be restored between O4' and R1' (about 18 on the obverse and 12 on the reverse). If the tablet contained a complete scheme, there were at least about 36 lines of text on both sides, corresponding to a height of about 15 cm.

No. 69

museum number	BM 46116 (81–7–6,567)
provenience	Babylon
measures	5.3(9?) × 4.0(12?) × 1.7–2.5 cm (type F?)
arrangement	O/R
cuneiform text	photo: Fig. F.77; *ACT*, Pl. 240
bibliography	*ACT* 207a
contents	Moon system A: computing G from Φ (scheme 1)

Obverse

0 *ina a-mat* ⸢^{d}**en**⸣ [*u* **gašan**-*ia*$_2$] ⸢*liš-lim*⸣

P1 1 *si-man*.**meš** *ana* **du**$_3$-*ka ana tar-ṣa* 2.13.20 **la**$_2$ 2.40 ⸢**gar**⸣-[*an ša*$_2$ *al* 2.13.20 **la**$_2$ **la**$_2$ *ina* 2.13.20 **nim**]

2 *ša*$_2$ **tag**$_4$ **a.ra**$_2$ 3.22.30 **du** *ma-lu-uš*-⸢*šu*$_2$⸣ [17.46.40 **gar**-*an* GAR.GAR-*ma* **ki** 2.40 **tab**-*ma* **gar**-*an*]

3 *ana tar-ṣa* 2.10.40 **la**$_2$ 2.53.20 **gar**-*an ša*$_2$ ⸢*al*⸣ [2.10.40 **la**$_2$ **la**$_2$ *ina* 2.10.40 **nim**]

4 *ša*$_2$ **tag**$_4$ GAM 9.20 **du ki** 2.⸢5⸣[3.20 **tab**-*ma* **gar**-*an*]

5 *ana tar-ṣa* 1.58.31.6.40 **la**$_2$ ⸢4⸣.[46.42.57.46.40 **gar**-*an ša*$_2$ *al* 1.58.31.6.40 **la**$_2$ **la**$_2$ *ina* 1.58.31.6.40 **nim** *ša*$_2$ **tag**$_4$]

6 **a**.⸢**ra**$_2$⸣ 8.20 **du ki** 4.46.[42.57.46.40 **tab**-*ma* **gar**-*an*]

7 *ana tar-ṣa* 1.⸢58.13⸣.[20 **la**$_2$ 4.49.11.6.40 **gar**-*an ša*$_2$ *al* 1.58.13.20 **la**$_2$ **la**$_2$ *ina* 1.58.13.20 **nim** *ša*$_2$ **tag**$_4$]

8 **a.r**[**a**$_2$ 7.20 **du ki** 4.49.11.6.40 **tab**-*ma* **gar**-*an*]

(about 15 lines missing)

Reverse

(about 16 lines missing)

P2 1' 25.[48.38.31.6.40 *taš-pil-tu*$_4$ ...]

2' 50.10.⸢5⸣[1.6.40 **mu.du** ...]

3' 1.26.25.[11.6.40 **mu.du** ...]

Obverse

^{0}At the command of Bēl [and Bēltija] may it remain intact.

P1 **(Computing *G* from *Φ*, scheme 1)**

^{1}In order for you to construct the durations.

1–9 Opposite 2,13;20, decreasing, [you] put 2,40. [That which is less than 2,13;20, decreasing, you deduct from 2,13;20]; ^{2}what remains you multiply by 3;22,30; while it fills up [you accumulate 0;17,46,40, and you add it with 2,40, and you put it down.]

10 ^{3}Opposite 2,10;40, decreasing, you put 2,53;20. That which [is less than 2,10;40, decreasing, you deduct from 1,10;40;] ^{4}what remains you multiply by 9;20, [add] with 2,5[3;20, and put down.]

11 ^{5}Opposite 1,58;31,6,40, decreasing, you put 4,[46;42,57,46,40. That which is less than 1,58;31,6,40, you deduct from 1,58;31,6,40, what remains] ^{6}you multiply by 8;20, [add] with 4,46;[42,57,46,40, and put down.]

12 ^{7}Opposite 1,58;13,[20, decreasing, you put 4,49;11,6,40. That which is less than 1,58;13,20, decreasing, you deduct from 1,58;13,20, what remains] ^{8}[you multiply] by [7;20, add with 4,49;11,6,40, and put down.]

(about 15 lines missing)

Reverse

(about 16 lines missing)

P2 **(Monthly difference and other net differences of *G*)**

$^{1'}$25;[48,38,31,6,40, the difference ...]

$^{2'}$50;10,5[1,51,6,40 it proceeds ...]

$^{3'}$1;26,25,[11,6,40 it proceeds ...]

Critical and philological notes

Obverse

2 **gar**-*an* GAR.GAR-*ma*: might also be restored as GAR.GAR-*ma*; cf. the score with all duplicates offering intervals 1–9 in §4.4.11.

4 GAM: the text employs both **a.ra**$_2$ (O2,6,8) and GAM for 'times'.

Commentary

This fragment from the upper (lower) left corner of the obverse (reverse) is inscribed on both sides. Based on the textual restorations on the obverse, the original width was about 9 cm, the height about 12 cm. The text of P2 is written with notably larger signs and a larger indentation than that of P1. The topics of P1 and P2 suggest that the tablet did not contain any other procedures.

P1 *Computing G from Φ (scheme 1)*

This procedure contains a part of interpolation scheme 1 for computing G from Φ (§4.4.11). On the obverse intervals 1–12 are partly preserved. The division of P1 into intervals is reflected in the layout of the tablet. The first line of each entry begins with a larger than average vertical wedge (*ana*), and the second line is slightly indented shorter, except intervals 1–9 in O1–2. If the tablet contained the entire interpolation scheme, the remaining intervals 13–35 would require about 30 lines of text, which may be restored between O8 and R1'. Accordingly, the original height was at least about 12 cm. The formulation of intervals 1–9 is based on the *malûššu*-rule (§4.4.11). The other intervals are formulated in accordance with template IP.C, which involves explicit subtractions $\Phi - \Phi_k$ and $\Phi_k - \Phi$.

P2 *Monthly difference and other net differences of $\hat{G}$*

P2 is a list with the monthly difference d and the net differences for 12^{m}, $d\hat{G}(12)$, and 14^{m}, $d\hat{G}(14)$. For this topic cf. §4.4.11.

No. 70

museum number	BM 34134 (Sp,236)
provenience	Babylon
measures	6.0(7.5?) × 6.5(*) × 1.9–2.5 cm (type F?)
arrangement	O/R
cuneiform text	photo: Fig. F.78; *ACT*, Pl. 240; copy: *LBAT* 98
bibliography	*ACT* 207b
contents	Moon system A: computing *G* from *Φ* (scheme 1)

Obverse

P1 1 [*si-man ana* **du**$_3$*-ka ana tar-ṣi* 2.13.20 **la**$_2$ 2].⸢40 *si*⸣*-man* ⸢**gar**⸣*-an ša*$_2$ *al-*[*la*]
2 [2.13.20 **la**$_2$ **la**$_2$]*-u*$_2$ *ina* 2.13.20 **nim** *ša*$_2$ **e**$_{11}$*-ka* GAM 3.22.30 **du**
3 [*ma-lu-uš-šu*$_2$ 17.46.40 GAR.GAR *ša*$_2$] ⸢**e**$_{11}$⸣*-ka* **ki** 2.40 **tab**-*ma* **gar**-*an*
4 [*ana tar-ṣi* 2.10.40 **la**$_2$ 2.53].20 **gar**-*an ša*$_2$ *al-la* 2.10.40 **la**$_2$ **la**$_2$*-u*$_2$
5 [**en** 1.58.31.6.40 **la**$_2$ *ina* 2.10].40 **nim.a** ⸢*ša*$_2$⸣ **tag**$_4$ GAM 9.20 ⸢**du**⸣
6 [**ki** 2.53.20 **tab**-*ma* **gar**-*an ana t*]*ar-ṣi* 1.58.31.6.40 ⸢**la**$_2$⸣
7 [4.46.42.57.46.40 **gar**-*an ša*$_2$ *a*]*l* 1.58.31.6.[40 **la**$_2$ **la**$_2$*-u*$_2$]
8 [**en** 1.58.13.20 **la**$_2$ GAM 8.20 **du ki** 4.]46.42.57.4[6.40 **tab**-*ma*] **gar**
9 [*ana tar-ṣi* 1.58.13.20 **la**$_2$ 4.49.11].6.40 **gar**-*an ša*$_2$ ⸢*al*⸣
10 [1.58.13.20 **la**$_2$ **la**$_2$*-u*$_2$ **en** 1.57.5]5.33.20 **la**$_2$ GAM 7.20 [**du**]
11 [**ki** 4.49.11.6.40 **tab**-*ma* **gar**-*an ana tar-ṣi* 1.57.5]5.33.20 ⸢**la**$_2$⸣
12 [4.51.21.28.53.20 **gar**-*an ša*$_2$ *al* 1.57].⸢55.33.20⸣ [**la**$_2$ **la**$_2$*-u*$_2$]
(n lines missing)
Reverse
(about n lines missing)
1' [xxxxxxxxxx] **tab** ⸢x⸣ [xxxxxx]
2' [xxxxxxxxx].⸢5⸣3.20 [xxxxxx]
3' [xxxxxxxx].⸢x3⸣.20 **tab** *gab-b*[*i* xxxxx]
4' [*ana tar-ṣi* 2.15.48.8].⸢5⸣3.20 **tab en** 2.1⸢3.20⸣ [**la**$_2$ 2.40 **gar**-*an*]
5' [xxxxx 25.48].38.31.6.40 *taš-pil-*⸢*tu*$_4$⸣

P2 6' [xx 50.10.51].⸢5⸣1.6.40 **mu.du**
7' [xx 1.26.25.1]1.6.40 **mu.du**

Critical and philological notes

Obverse

1 The gap has been restored from the duplicate No. **69**.
si-man immediately before **gar**-*an* replaces *ina* **mu** (*ACT*, p. 257).

3 **e**$_{11}$ (DUL.DU): before DU only the final vertical wedge of DUL is visible.

5 **nim.a**: **a** is omitted in Pinches's copy (and Neugebauer's transliteration).

8 **en** 1.58.13.20 **la**$_2$ GAM 8.20 **du**: there is insufficient space in the gap for restoring *ina* 1.58.31.6.40 **nim** *ša*$_2$ **tag**$_4$ before GAM, as done by Neugebauer. However, omitting **en** 1.58.13.20 **la**$_2$, as suggested by Neugebauer, renders the formulation defective, and does not yield enough space for his restoration either. The current restoration fits exactly; cf. the commentary to O10.

10 Neugebauer (*ACT*, p. 257) assumes that between **la**$_2$ and GAM the phrase *ša*$_2$ **tag**$_4$ *ina* 1,58;13,20 **nim**, 'what remains you deduct from 1,58;13,20' is missing. This is not necessarily the case, since it is often absent in the duplicates; e.g. systematically so in No. **53** P16'. This reflects the two possible interpretations of *ša*$_2$ *alla Q* **la**$_2$**/diri**; cf. §2.3.2.1 (p. 28).

13 Depending on whether or not interval 13 is split up at m_Φ = 1,57;47,57,46,40°, either this value and the corresponding G_{13b} = 4,52;9,34,4,26,40°, or Φ_{14} = 1,57;58,8,53,20° (increasing) and G_{14} = 4,53;14,4,26,40° must be restored here.

Reverse

1' After **tab** there are remains of a vertical wedge.

3' ⸢x⸣3: the 3 is preceded by traces of a winkelhaken belonging to a digit 10–50. The word *gabbi*, 'all', implies an interpolation interval where *G* is constant, i.e. 16, 18 or 35.

4' ⸢50⸣: remains of three winkelhakens can be seen, not two as copied by Pinches.

Obverse

P1 **(Computing** G **from** Φ**)**

[1][In order for you to construct the duration.

1–9 Opposite 2,13;20, decreasing, you put down 2],⌜40 as the duration⌝. That which is le[ss than] [2][2,13;20, decreas]ing, you deduct from 2,13;20; what comes out for you you multiply by 3;22,30, [3][while it fills up you accumulate 0;17,46,40; what] ⌜comes out⌝ for you you add with 2,40 and you put it down.

10 [4][Opposite 2,10;40, decreasing], you put down [2,53];20. That which is less than 2,10;40, decreasing, [5][until 1,58;31,6,40, decreasing], you deduct [from 2,10];40. ⌜What⌝ remains you ⌜multiply⌝ by 9;20, [6][add with 2,53;20 and put down.

11 Oppos]ite 1,58;31,6,40, ⌜decreasing⌝, [7][you put down 4,46;42,57,46,40. (The amount) by which it is less th]an 1,58;31,6,[40, decreasing], [8][until 1,58;13,20, decreasing, you multiply by 8;20, add with 4],46;42,57,4[6,40 and] you put it down.

12 [9][Opposite 1,58;13,20, decreasing, you put down 4,49;11],6,40. (The amount) by which it [10]is less ⌜than⌝ [1,58;13,20, decreasing, until 1,57;5]5,33,20, decreasing, [you multiply] by 7;20, [11][you add with 4,49;11,6,40 and you put it down.

13 Opposite 1,57;5]5,33,20, ⌜decreasing⌝, [12][you put down 4,51;21,28,53,20. (The amount) by which it is less than 1,57];⌜55,33,20⌝, [decreasing,] [13][until ..., you multiply by 6;20, you add with 4,51;21,28,53,20 and you put it down.]

(*n lines missing*)

Reverse

(*about n lines missing*)

[1'][... Opposite ...], increasing, ⌜...⌝.[...]

[2'][... Opposite ...],⌜5⌝3,20, [...]

[3'][... Opposite ...],⌜...3⌝,20, increasing, all [...]

35 [4'][Opposite 2,15;48,8],⌜5⌝3,20, increasing, until 2,13;20, [decreasing, you put down 2,40.]

a **(Monthly difference of** G**)**

[5'][... 25;48,38],31,6,40, the ⌜difference⌝.

P2 **(Net differences of** G**)**

[6'][... 50;10,51],51,6,40 it proceeds.

[7'][... 1;26,25,1]1,6,40 it proceeds.

Commentary

This fragment preserves a part of the upper (lower) edge of the obverse (reverse) and the right edge of a tablet. The textual restorations on the obverse imply that the original width was about 7.5 cm; the original height is unknown. It appears that the tablet does not have a colophon.

P1 Computing G from Φ

P1 provides rules for computing G from Φ in accordance with interpolation scheme 1 (§4.4.11). The obverse concerns intervals 1–13. The formulation undergoes a subtle change in O6 from template IP.C to IP.B. Only the first half of interval 13 is preserved (O11–12), so that it is not clear whether the minimum of Φ (control point 13a) is included or not. R3' might belong to intervals 16, 18 or 35, near an extremum of G, as suggested by the word *gabbi* = 'all' (for the formulation of these intervals cf. §4.4.11). Since R4' concludes with interval 35, if restored correctly, R1'–3' should concern the immediately preceding intervals, but the text could not be restored. If one assumes that the tablet contained a complete copy of scheme 1 then the original height must have been at least about 13 cm. Part **a** (O5') mentions the monthly difference of $\hat{G}$.

P2 Net differences of $\hat{G}$ *for various intervals*

This procedure mentions net differences of $\hat{G}$ for 12^{m} and 14^{m} (cf. §4.4.11; Table 4.25).

No. 71

museum number	BM 34737 (Sp2,228)
provenience	Babylon
measures	4.2(7–8?) × 6.3(*) × 2.2(*) cm
cuneiform text	photo: Fig. F.79; *ACT*, Pl. 243; copy: *LBAT* 99
bibliography	*ACT* 207c
contents	Moon system A: computing G from Φ (scheme 1)

Side X

P1' (unknown number of lines missing)

1' [xxxxxx] **la**$_2$ ⸢x⸣ [xxxxxxxxxxxxxx]
2' [*ana tar-ṣa* 2.10].⸢40⸣ **la**$_2$ 2.⸢53.20⸣ [**gar**-*an* *ša*$_2$ *al-la* 2.10.40 **la**$_2$ **la**$_2$]
3' [**en** 1.58].31.6.40 **la**$_2$ *ina* 2.⸢10.40⸣ [**nim** *ša*$_2$ **tag**$_4$ GAM 9.20]
4' [**du ki** 2].53.20 ⸢**tab**-*ma* **gar**⸣-[*an*]
5' [*ana tar-ṣa* 1].58.31.6.40 **la**$_2$ ⸢4.4⸣[6.42.57.46.40 **gar**-*an*]
6' [*ša*$_2$ *a*]*l-la* 1.58.31.6.40 **la**$_2$⸣ [**la**$_2$ *ina* 1.58.31.6.40 **nim** *ša*$_2$ **tag**$_4$]
7' [GAM 8.20] ⸢**du ki** 4.46.42⸣.[57.46.40 **tab**-*ma* **gar**-*an*]
8' [*ana tar-ṣa* 1].⸢58.13.20 **la**$_2$ 4.49⸣.[11.6.40 **gar**-*an*]
9' [*ša*$_2$ *al-la*] ⸢1.58.13.20 **la**$_2$-*u*$_2$⸣ [**la**$_2$ *ina* 1].⸢5⸣[8.13.20 **nim** *ša*$_2$ **tag**$_4$]
10' [GAM 7].20 **du** ⸢**ki** 4.49.11⸣.[6.40 **tab**-*ma* **gar**-*an*]
11' [*ana tar-ṣa* 1].⸢57⸣.[55.33.20 **la**$_2$ 4.51.21.28.53.20 **gar**-*an*]
(unknown number of lines missing)

Side X

P1' **(Computing** G **from** Φ**)**

(*unknown number of lines missing*)
1'[...] decreasing, ... [...]

10 2'[Opposite 2,10];⸢40⸣, decreasing, [you put down] 2,⸢53;20⸣. [That which is less than 2,10;40, decreasing,] 3'[until 1,58];31,6,40, decreasing, [you deduct] from 2,⸢10;40⸣. [What remains you multiply by 9;20,] 4'⸢add⸣ [with 2],53;20 ⸢and put down⸣.

11 5'[Opposite 1],58;31,6,40, decreasing, [you put down] ⸢4,4⸣[6;42,57,46,40.] 6'[That which is le]ss ⸢than 1,58;31,6,40, decreasing⸣, [you deduct from 1,58;31,6,40. What remains] 7'⸢you multiply⸣ [by 8;20], ⸢add with 4,46;42⸣,[57,46,40 and you put down.]

12 8'[Opposite 1],⸢58;13,20, decreasing⸣, [you put down] ⸢4,49⸣;[11,6,40.] 9'[That which is less than] ⸢1,58;13,20, decreasing⸣, [you deduct from 1],⸢5⸣[8;13,20. What remains] 10'you multiply [by 7];20, ⸢add with 4,49;11⸣,[6,40 and you put it down.]

13 11'[Opposite 1],⸢57⸣;[55,33,20, decreasing, you put down 4,51;21,28,53,20.]
(*unknown number of lines missing*)

Critical and philological notes

Side X

1' After **la**$_2$ there are badly worn traces, perhaps of two winkelhakens or the sign GAM.
3' *ina*, 'from', replaces the more common logographic writing **ta**.
4' Available space and traces between **tab** and **gar** imply that *ma* is written here.
6' 1.58: badly worn.
ša$_2$ **tag**$_4$: to be restored here or in X7'.
9' *ša*$_2$ **tag**$_4$: to be restored here or in X10'.

Commentary

This fragment is inscribed on one side, probably the obverse (cf. below); the other side is destroyed. No original edges are preserved. The curvature near the bottom of side X, if not caused by damage, suggests that the fragment is situated close to the lower edge of side X of the tablet. The textual restorations imply that not much clay is missing from the left edge, so that the original width of the tablet (or that of the column) must have been 7–8 cm.

P1' *Computing G from Φ (scheme 1)*

This procedure concerns the computation of G from Φ by means of interpolation scheme 1 (§4.4.11). X2'–11' contain the rules for intervals 10–13. The formulation is based on template IP.C. Presumably intervals 1–9 were dealt with in the lines preceding X1'. If their formulation is based on the *malûššu*-rule only a few lines are needed for this. Consequently, side X is likely the obverse of the tablet. Since little remains of interval 13, it is not clear whether the minimum of Φ is treated as a separate control point.

No. 72

museum numbers	BM 36004 (Sp3,547 +82–7–4,107) +40611 (81–4–28,156)
provenience	Babylon
date	early Seleucid?
measures	10.7(13) × 6.6(*) × 2.0–2.7 cm (type C?)
arrangement	O/R
cuneiform text	photo: Fig. F.82; *ACT*, Pl. 235 (BM 36004); copy: *LBAT* 100 (BM 36004)
bibliography	*ACT* 207ca
contents	Moon system A: computing G from Φ

Obverse

P1 1 [*ana tar-ṣi* 2.13.20$^?$ **la**$_2$]-⸢*u*$_2^?$⸣ 2.⸢5⸣1.29.22.57.46.⸢40 xx⸣ *mim-ma ša*$_2$ *al* 2.1⸢3.20 **la**$_2^?$⸣-[*u*$_2$ **la**$_2$ xxxxx **a.ra**$_2$]
2 [9.20 **du**]-⸢*ma*⸣ **ki** 2.51.29.22.57.4⸢6⸣.40 **tab** *ana tar-ṣi* 2.13.20 **la**$_2$-*u*$_2$ 2.⸢2⸣[5.32.50.22.13.20 **gar**-*an*]
3 [*mim-ma ša*$_2$] ⸢*al*⸣ 2.13.20 **la**$_2$ ⸢GIŠ-*ma* **a.ra**$_2$ 9.20 **du**⸣-*ma* **ki** 2.25.32.50.22.13.20 [**tab** xxxx]
4 [*ana tar*]-*ṣi* 1.59.48.8.53.20 4.⸢5⸣4.48.53.20 *si-man* **gar**-*an mim-ma ša*$_2$ *al* 1.⸢5⸣[9.48.8.53.20]
5 [**diri**] GIŠ-*ma* **a.ra**$_2$ 6.50 **du**-*ma* **ta** ⸢4.54⸣.48.53.20 **nim**-*ma ša*$_2$ **tag**$_4$ *a-na si-*[*man* **gar**-*an* xxx]
6 [*ana tar*]-*ṣi* 2.0.59.15.33.20 4.46.⸢4⸣2.57.46.40 *si-man* **gar**-*an mim-ma ša*$_2$ *al* 2.0.5[9.15.33.20]
7 [**diri**$^?$ GIŠ-*m*]*a* **a.ra**$_2$ 9.20 **du**-*ma* **ta** 4.46.⸢42⸣.57.46.40 **e**$_{11}$ *mim-ma ša*$_2$ *al* 2.13.⸢20⸣ [xxx **diri** GIŠ-*ma*]
8 [**a.ra**$_2$ x].⸢20⸣ **du**-*ma* **ta** 2.51.29.2[2.57.4]⸢6⸣.40 **nim**-*ma si-man* (blank) [**gar**-*an*]
9 [*mim-ma ša*$_2$ *a*]*l* 2.13.20 **la**$_2$-⸢*u*$_2^?$ GIŠ x⸣ [xxxx **ki** 2].⸢51⸣.29.22.⸢57.46.40⸣ [**tab** xxxxx]
10 [xxx *s*]*i-man* 2.⸢5⸣1.29.⸢22⸣.[57.46.40 xxx] ⸢x⸣ [xxxxxxxxxxxxxx]
11 [xxxx] ⸢x *ša*$_2$ *si-man* **gar** x⸣ [xxxxxxxxxxxxxxxxxxxxxx]
(n missing lines)

Reverse

(about n−2 missing lines)
1’ [*ana tar-ṣi* 2.0].⸢59.15.33.20⸣ [4.46.42.57.46.40 xxxxxx]
2’ [xx *mim-ma ša*$_2$ *al* 2].0.59.15.33.20 **diri** [GIŠ **a.ra**$_2$ 9.20 **du**-*ma* **ta** 4.46.42.57.46.40 xxxx]
3’ [xxxxx *mim*]-*ma ša*$_2$ *al* 2.13.20 **diri** [GIŠ **a.ra**$_2$ xxx **du**-*ma* **ta** xxxxxxxxxxxxx]
4’ [*ana tar-ṣi*] ⸢2$^?$⸣.13.20 2.5⸢1.29.22.5⸣[7.46.40 x] ⸢x⸣ [xxxxxxxxxxxxxxxxxxxx]
5’ [xxxx].⸢x⸣7.4.17.4.48.⸢53.20⸣ [xxxxx *si*]-*man* (blank) [xx]
6’ [*ana tar-ṣi*] ⸢1⸣.57.47.57.⸢46.40 5.16$^?$.xx.4$^?$1.4⸣2.13.20 ⸢*si*⸣-*man* [**gar**-*an*]
7’ [*mim-ma ša*$_2$ *al*] 1.59.48.8.⸢53.20 **en**$^?$ 2.1x.26.40 **diri**$^?$ GIŠ$^?$ **a.ra**$_2^?$⸣ 6.50 **d**[**u**$^?$-*ma*]
8’ [**ta** xxxx]8.8.53.20 **e**$_{11}$ *si-man* **gar**-⸢*an*⸣ [*mim-ma ša*$_2$ *al*] ⸢2.1x.26.40⸣ [xx 2].0.59.15.33.20 [xxx]
9’ [**ta** xxxx].⸢1$^?$2.4$^?$⸣5.33.20 **e**$_{11}$ *ša*$_2$ **tag**$_4$ 1.59.⸢4.46.40⸣ [xxx].⸢x.1$^?$⸣8.31.21.2⸢4⸣ [xxx]
10’ [xx **ta** 4].46.42.5⸢7.4⸣6.40 **nim**-*ma* **tag**$_4$ 4.[x].18.⸢10.37⸣.[xxx] *si-man* [xx]
11’ [xxx 2.45].55.33.20 **a.ra**$_2$ 9.20 **du**-*ma* ⸢25⸣.48.38.31.⸢6⸣.[40 *taš-pil*]-⸢*tu*$_4^?$⸣ *ša*$_2$ *si-man* ⸢**gar**⸣-[*an*]

P2 12’ [xxxx].⸢30⸣ **la**$_2$-*u*$_2$ 2.8.31.6.40 **tab**-*u*$_2$ ⸢**ta**⸣ *zib* **en absin** 3.24 [xxx]

Co 13’ [xxxxxxx] *šamaš*$_2$ *a-na la ma-*⸢*še-e ša*$_2^?$*-ṭir*$^?$⸣ ^{d}**sag.me.gar** *u* ⸢^{d}⸣[*ṣal-bat-a-nu*$^?$ xxx]
14’ [xxxxxxx]-⸢*u*$_2^?$⸣ **nu** *u*$_2$*-kal-lam* **im** ^{m}⸢^{d}⸣**šu**$_2$**.dub.numun** ⸢**a**⸣ [*ša*$_2$ xxxxx]

Critical and philological notes

Obverse

1 The broken sign before *mim-ma* ends with a horizontal wedge and two vertical wedges. One would expect *si-man* **gar**-*an*, ‘you put down as the duration’.
2.13.20 **la**$_2$-[*u*$_2$]: replaces 2.13.21.11 (*ACT*).
2 2.25...: restored on the basis of O3.
3 GIŠ = *tanašši*, ‘you compute’. The alternative would be to read GIŠ as the phonetic complement -*is* in **la**$_2$-*is* = *inaḫḫis*, ‘lags behind’, but this does not fit the context.
4 Probably nothing is missing before *ana tar-ṣi*.
5 The first sign after the gap is definitely GIŠ, which most likely represents *tanašši*, ‘you compute’, as in O3.
6 Probably nothing is missing before *ana tar-ṣi*.
9 After GIŠ there are remains of two vertical wedges.
10 After the gap in the middle a damaged sign is visible, perhaps **iti** or x 30.
11 The broken sign before *ša*$_2$ looks like DIRI or KAL. After **gar** (the beginning of) a sign similar to LA$_2$ is visible.

Obverse

P1 (Computing G from Φ)

1 [1][Opposite 2,13;20?, decreas]⸢ing⸣, [...] 2,⸢5⸣1;29,22,57,46,⸢40⸣. [You compute?] (the amount) by which it [is less than] 2,1⸢3;20, decreasing⸣, [...] [2][you multiply it by 9;20] ⸢and⸣ add it with 2,51;29,22,57,4⸢6⸣,40.

1' Opposite 2,13;20, decreasing, [you put down] 2,⸢2⸣[5;32,50,22,13,20.] [3][You compute (the amount) by which] it is less ⸢than⸣ 2,13;20, ⸢you multiply it by 9;20⸣ and [add] it with 2,25;32,50,22,13,20 [...]

5? [4][Oppo]site 1,59;48,8,53,20 you put down 4,⸢5⸣4;48,53,20 as the duration. You compute (the amount) by which it [exceeds] 1,⸢5⸣[9;48, 8,53,20], [5]you multiply it by 6;50, deduct it from ⸢4,54⸣;48,53,20, and what remains [you put down] as the dura[tion ...]

6 [6][Opposi]te 2,0;59,15,33,20 you put down 4,46;⸢4⸣2,57,46,40 as the duration. [You compute] (the amount) by which it [exceeds?] 2,0;5[9,15,33,20], [7]you multiply it by 9;20 and deduct it from 4,46;4[2],57,46,40.

7 [You compute] (the amount) by which it [exceeds] 2,13;20 [...], [8]you multiply it [by ...];20, deduct it from 2,51;29,2[2,57,46],40, and [put it down] as the duration.

1 [9]⸢You compute⸣ [(the amount) by which it is less th]an 2,13;20, ⸢decreasing⸣, [... you add it with 2],⸢51⸣;29,22,⸢57,46, 40⸣ [...] [10][... the dura]tion, 2,⸢5⸣1;29,⸢22⸣,[57,46,40 ...] ... [...] [11][...] ⸢... of the duration you put down ...⸣ [...]

(n missing lines)

Reverse

(about n−2 missing lines)

6 [1'][Opposite 2,0];⸢59,15,33,20⸣ [you put down 4,46;42,57,46,40 ...] [2'][... You compute (the amount) by which it] exceeds [2],0;59,15,33,20, [you multiply it by 9;20 and deduct it from 4,46;42,57,46,40 ...]

7 [3'][You compute (the amount) by wh]ich it exceeds 2,13;20, [you multiply it by ... and deduct it from ...]

1,7? [4'][Opposite] ⸢2?⸣,13;20 [you put down] 2,5⸢1;29,22,5⸣[7,46,40 ...] ⸢...⸣ [...] [5'][...].⸢...⸣7.4.17.4.48.⸢53.20⸣[... the dur]ation [...]

4 [6'][Opposite] ⸢1⸣,57;47,57,⸢46,40⸣ [you put down] ⸢5,16?⸣;[...]⸢5?,4?1⸣,4⸣2,13,20 as the duration.

a [7']⸢You compute?⸣ [(the amount) by which it] ⸢exceeds⸣ 1,59;48,8,⸢53,20 until? 2,10...,26,40 ..., you multiply it by⸣ 6;50 [and] [8']deduct it [from ...]8,8,53,20, (and) you put it down as the duration.

b [You compute (the amount) by which it excee]ds ⸢2,1x;26,40⸣ [... 2],0;59,15,33,20 [...] [9'][...] you deduct [from ...]⸢1?2⸣,4?5,33,20. That which remains, 1,59;⸢4,46,40⸣ [...],1?8,31,21,2⸢4⸣ [...] [10'][... you deduct from 4],46;42,5⸢7, 4⸣6,40, and there remains 4,[...];18,⸢10,37⸣,[...], the duration [...]

c **(Relation between the monthly differences of Φ and G)**

[11'][... 2;45],55,33,20 you multiply by 9;20, and it is ⸢25⸣;48,38,31,⸢6⸣,[40], you put it down as [the differ]ence of the duration.

P2 (Subject unclear)

[12'][...].⸢30⸣, decreasing, 2.8.31.6.40 increasing. From Psc until Vir 3.24 [...].

Co (Colophon)

[13'][...] Šamaš. ⸢Written?⸣ in order not to ⸢be forgotten?⸣. Jupiter and [Mars? ...] [14'][... May the wise] not show it [to the ignor]ant. Tablet of Marduk-šāpik-zēri, son [of ..., descendant of ...]

Reverse

2' This line is apparently a duplicate of O4–5.

4' ⸢2?⸣.13.20: only the final vertical wedge of the 2 is preserved; hence this restoration is uncertain.

5' 17.4.17.4: as suggested by Neugebauer, this might be an error for 2.17.4.

6' Between 5 and 42 there is room for about 3 digits, the first of which is most likely 16 (not 3 as suggested by Neugebauer), the second 15,25,...,55 or 18,28,...,58 (not 37) and the third 41 or 51.

7' After **diri** there are 2–3 damaged signs, then 6.50 (replacing 6.40 in *ACT*).

8' 2.1x.26: x might be 2,3 or 4.

9' The first number could be 12, 22 or 32; the second 45 or 55. The 18 after the second break might also be a 28 or 38.

10' Neugebauer's tentative identification of the first number as a Φ value turns out to be incorrect.

11' The new fragment reveals that Neugebauer's reconstruction of the beginning of the line was incorrect.

13' *ma*: looks more like *ba*. The signs after *ma-še-e* are damaged and crammed; *ša*₂ looks more like PA. Reading suggested by C. Wunsch. Or **gu₄.ud** ᵈ**sag.me.gar**? Parallels: *ana lā mašê šaṭir*: 'written in order not to be forgotten' (*CAD* M/1, p. 400).

14' **nu** *u*$_2$*-kal-lam* = *lā ukallam*, 'he shall not reveal it', presumably preceded by *la mu-du-u*$_2$ or **nu zu**-*u*$_2$ = *lā mūdû*, 'to the ignorant'. A possible parallel is *BAK* 533:5, *la mu-du-u mu-da-a li-kal-lim mu-du-u la mu-da-a la* [*u*$_2$*-kal*]*-lam*, 'may the ignorant reveal it to the wise, may the wise not reveal it to the ignorant'.

Commentary

This tablet is preserved on three joined fragments, two of which, forming BM 36004, were published as *ACT* 207ca. Fragment BM 40611 was mentioned in Aaboe (1968),[672] but it was never published. The joined fragments preserve a segment of the upper (lower) edge of the obverse (reverse), but no other edges. The textual restorations in O2–8 suggest that about 1 cm of clay is missing from the left and right sides, so the original width was about 13 cm. P1 contains rules for computing G from Φ and occupies the bulk of the tablet. P2 consists of a single line of text (R12'). Below P2 there is a colophon. The tablet's 'owner', Marduk-šāpik-zēri, might be the same individual as Marduk-šāpik-zēri/Bēl-apla-iddin//Mušēzib, who wrote No. **47**, No. **44** and the NMAT procedure text BM 41004 (Brack-Bernsen & Hunger 2005/6). None of these tablets carries a date. Content and orthography suggest a date early in the Seleucid era.

P1 Computing G from Φ

P1 contains interpolation rules for computing G from Φ (§4.4.11). As pointed out by Neugebauer, P1 appears to suffer from inconsistencies. In spite of the newly added fragment the interpretation continues to pose problems, so that a complete and fully consistent scheme could not be reconstructed. The results are compiled in Table 4.26 (p. 152). The order in which the intervals are dealt with is also peculiar, and some intervals appear to be mentioned twice or even three times. To make things worse, most of the preserved interpolation rules are based on an abbreviated and slightly modified version of the interpolation template IP.B (§2.6.7). First, for a given interval k the value of Φ_{k+1}, where interval $k+1$ begins, is not mentioned; second, the phrase concerning the computation of $\Phi - \Phi_k$ includes the word *tanašši*(GIŠ), you compute'.

Interval 1

P1 begins with an interpolation rule for what is probably interval 1 (O1–2). This rule appears to be repeated in O9–10 and once more in R4'.

Interval 1'

It appears that O3 contains an alternative rule for interval 1, since Φ_1 is here associated with another G value, 2,25;32,50,22,13,20°. As remarked by Neugebauer, it is smaller than the minimum of G in scheme 1 (m_G = 2,40). It is unclear how this rule fits in with scheme 2; for this reason it has not been included in the reconstruction (Table 4.26).

Interval 5

O4–5 partly preserve an interpolation rule provisionally identified as interval 5.

Interval 6

O6–7 contains an interpolation rule involving Φ = 2,0;59,15,33,20° and an interpolation coefficient 9;20, probably pertaining to interval 6. It might also concern interval 1 if one assumes that this Φ value is the *right* boundary of interval 2, Φ_2. Other examples from ΦG interpolation schemes where the interpolation proceeds from the right boundary are not known to me, but it cannot be excluded for this anomalous text.

Interval 7

O7–8 deal with interval 7, which begins at Φ_7 = 2,13;20 (↑). The other boundary, Φ_8, is not mentioned, but is provisionally assumed to be 2,17;4,48,53,20° = M_Φ.[673] The final digit of c_7 is 20.

Interval 4

After an unknown number of missing lines, the reverse of the tablet returns to the increasing branch of Φ. After what appear to be repetitions of intervals 6 and 7 (R3'–4'), R4' mentions 1,57;47,57,46,40° = m_Φ, the control value for interval

[672] Thereafter also in Aaboe (1971) and Aaboe & Hamilton (1979).

[673] The number in R5' was corrected to M_Φ = 2,17;4,48,53,20 by Neugebauer (*ACT*).

4. The corresponding G value, ⌜5,16$^{?}$⌝;[x]5$^{?}$,4$^{?}$1⌝,42,13,20, is presumably the maximum of G. It is considerably larger than in scheme 1.

Interval a

Part **a** (R7') continues with an interpolation rule involving Φ_3 or Φ_5. Its place within scheme 2 is not clear.

Interval b

Part **b** (R8'–10') contains further interpolation rules which are damaged and difficult to interpret. The computation at the beginning of R8' seems to deal with the decreasing branch of G, because something is subtracted from ...8,8,53,20, presumably a control value of G. At the end of R8' there is another interpolation rule pertaining to the decreasing branch of G. The other computations in R9'–10' are difficult to interpret. The number 4,46;42,57,46,40° (R10') is G_2 or G_6.

***c** Relation between the monthly differences of Φ and G*

No problems are posed by R11', which mentions the monthly differences of Φ and $\hat{G}$ and the relation between them, i.e. $d_\Phi = 2;45,55,33,20 = 9;20 \cdot 25;48,38,31,6,40 = 9;20 \cdot d_{\hat{G}}$ (§4.4.11).

***P2** Subject unclear; perhaps net differences of G*

P2 consists of a single line. Several numbers are mentioned, each associated with a zodiacal region. The region Psc – Vir resembles zone 2 of lunar system A (§4.4.4). Given the topic of P1, the numbers may be net differences of G, but their origin remains to be explained.

No. 73

museum number	BM 41990 (81–6–25,612)
provenience	Babylon
measures	4.0(*) × 5.0(*) × 2.6–3.5 cm
arrangement	O/R
cuneiform text	photo: Fig. F.81; *ACT*, Pl. 236; copy: *LBAT* 101
bibliography	*ACT* 207cb
contents	Moon system A: computing *G* from *Φ* (variant scheme?)

Obverse

P1’ (unknown number of lines missing)

1’ [...xxxxxxxxxxx] 1 [xxxx]
2’ [...xxxxx **en** 2.13].⸢8.8⸣.53.⸢20⸣ [**tab** xxx]
3’ [... *ana tar-ṣa*] ⸢2.13.8.8⸣.53.20 **tab** 2.⸢5⸣[3.20 **gar**]
4’ [...xx 4].⸢4⸣6.42.57.46.40 [xxx]
5’ [...xxxxx] ⸢x *si*⸣-*man* **tab** [x?]
6’ [...xxxxxx] ⸢x⸣ *ana tar-ṣa* 1.59.48.8.⸢5⸣[3.20]
7’ [...xxxxxx] ⸢xxxx 8? 10?⸣ [x]
(n lines missing)

Reverse

(about n lines missing)

1’ [...xx] ⸢xxx⸣ [xxxx]
2’ [... *ša₂ al-la* ...].13.20 **la₂ diri a.ra₂** 3.[22.30 **du** xx]
3’ [...x] *si-man ana tar-ṣa* 1.57.5⸢5⸣.[33.20 xx]
4’ [...x] **a.ra₂** 3.22.30 **du** 10 [xx]

Obverse

P1’ (Computing *G* from *Φ*)

(25) 1’[...] 1 [...] 2’[... until 2,13;8],⸢8⸣ 53,⸢20⸣, [increasing, ...] 3’[...

(26) Opposite] ⸢2,13;8,8⸣,53,20, increasing, [you put] 2,⸢5⸣[3;20 ...]

(11) 4’[... 4],⸢4⸣6;42,57,46,40 [...] 5’[...] ... the duration you add. [...] 6’[...] ...

(21) Opposite 1,59;48,8,⸢5⸣[3,20, decreasing, ..., you put down ...] 7’[...] ... [...]
(n lines missing)

Reverse

(about n lines missing)

(?) 1’[...] ⸢...⸣ [...] 2’[... (The amount) by which] it exceeds (error for is less than?) [...];13,20, decreasing, [you multiply] by 3;[22,30, ...] 3’[...] the duration.

(13) Opposite 1,57;5⸢5⸣,[33,20 ...]
4’[...] you multiply by 3;22,30, 10 [...]

Critical and philological notes

Obverse

3’ 20 **tab**: erroneously copied by Pinches as MI.
5’ Before ⸢*si*⸣-*man* the upper half of a vertical wedge is visible.
6’ The broken sign before *ana tar-ṣa* might be **du** or **uš**.
7’ Traces of about 4 signs, the first of which may be 30 or MEŠ, followed by 8 or 7 and at least one winkelhaken.

Reverse

1’ Traces of about 3 signs, the second of which might be **du**.
2’ **la₂ diri**, ‘it exceeds ..., decreasing’: one expects either **la₂ la₂**, ‘it is less than ..., decreasing’, or **tab diri**, ‘it exceeds ..., increasing’; cf. the commentary.
3’ The sign *man* resembles KUR, being written with an additional horizontal wedge below the 20, an orthography also attested in No. **53**.

Commentary

This small fragment preserves a segment of the right edge of a tablet. The last line of text on the reverse is followed by a horizontal line, below which the fragment is uninscribed. It therefore appears that the tablet did not have a colophon.

P1’ Computing G from Φ, variant scheme?

The fragment contains interpolation rules for computing *G* from *Φ* (§4.4.11), but the interpretation poses problems. In the translation all interpolation rules have been provisionally labeled in terms of the intervals of scheme 1. O2’–3’ seem to offer rules for intervals 25–26. O4’ mentions G_{11} (= G_{25}), O6’ Φ_{21}, and R3’–4’ Φ_{13}. The interpolation rule in R2’ is peculiar because it refers to values larger than the control value of *Φ*, while *Φ* is said to be decreasing. It therefore appears that this control value is the right boundary of the interpolation interval (Φ_{k+1}) instead of the usual left boundary (Φ_k). One would expect successive control values Φ_k to follow the zigzag function for Φ, which is not the case if the Φ values are interpreted in terms of scheme 1. Hence this is probably a variant scheme. The number 3;22,30 (R4' and perhaps R2') can be identified as $1/\phi$, which implies that the interpolations are formulated in terms of interpolation coefficients $c_k\phi$, as in template PHIG.A.

No. 74

museum numbers	BM 36438 (80–6–17,165) +37012 (756) +37026 (770) +37274 (1030) +37319 (1075)
provenience	Babylon
measures	10.0 × 10.2(20?) × 1.8–2.4 cm (type D)
arrangement	O/R
cuneiform text	photo: Fig. F.83; *ACT*, Pl. 239a (BM 36438); Aaboe & Hamilton (1979), pp. 20, 22
bibliography	*ACT* 207cc (BM 36438); Aaboe & Hamilton (1979), pp. 18–23
contents	Moon system A: computing *G* from *Φ* (scheme 1)

Obverse

P1 (at least about 20 lines missing)

-2' [ana tar-ṣa 1.57.58.8.53.20 tab 4.53.14.4.26.40 gar-an]
-1' [mim-ma ša$_2$ al 1.57.58.8.53.20 tab diri en 1.58.15.55.33.20 tab]
0' [a.ra$_2$ 3.22.30 du-ma ša$_2$ e$_{11}$-ka a.ra$_2$ 1.34.48.53.20 du-ma]
1' [ki 4.53.14.4.26.40 tab-ma gar]-an
2' [ana tar-ṣa 1.58.15.55.33.20 tab 4.54.48.53.20] ⌜gar⌝-an
3' [mim-ma ša$_2$ al 1.58.15.55.33.20] tab diri en 1.⌜58.33.42.13⌝.20 tab
4' ⌜a.ra$_2$⌝ [3.22.30 du-ma ša$_2$ e$_{11}$]-ka a.ra$_2$ 1.11.6.40 du-⌜ma⌝
5' ki 4.[54.48.53.20] tab-ma gar-an
6' ana tar-ṣa 1.58.33.⌜4⌝2.13.20 tab 4.56 gar-an
7' en 1.58.37.2.13.20 tab tab u la$_2$ nu tuk
8' ana tar-ṣa 1.58.37.2.13.20 tab 4.56 gar-an
9' ⌜mim-ma ša$_2$ al⌝ 1.58.37.2.13.20 tab diri en 1.58.54.48.53.20 tab
10' a.ra$_2$ 3.⌜22.30⌝ du-ma ša$_2$ e$_{11}$-ka a.ra$_2$ 35.33.20 du-ma
11' ki 4.56 tab-ma gar-an
12' ana tar-ṣa 1.58.54.48.53.20 tab 4.56.35.33.20 gar-an
13' en 1.59.12.35.3⌜3⌝.20 tab tab u la$_2$ nu tuk
14' ana tar-ṣa 1.59.12.3[5].33.20 tab 4.56.35.33.20 gar-an

Reverse

1 ⌜mim⌝-[ma ša$_2$] ⌜al 1.5⌝9.1⌜2⌝.[35].33.20 tab ⌜diri⌝ en 1.59.30.22.13.[20 tab]
2 ⌜a.ra$_2$⌝ 3.22.30 du-ma ša$_2$ e$_{11}$-ka [a].⌜ra$_2$⌝ 35.33.20 du-[ma]
3 ta 4.56.35.33.20 ⌜e$_{11}$⌝-ma gar-an
4 ana tar-ṣa 1.59.30.22.13.20 [tab] 4.56 gar-an
5 mim-ma ša$_2$ al 1.59.30.22.13.20 ⌜tab diri en⌝ 1.59.48.8.53.20 tab
6 a.ra$_2$ 3.22.30 du-ma ša$_2$ ⌜e$_{11}$-ka⌝ a.ra$_2$ 1.11.6.40 du-ma ta 4.56 e$_{11}$-ma gar-an
7 ana tar-ṣa 1.59.48.8.53.20 tab 4.54.48.53.20 gar-an mim-ma ša$_2$ al
8 1.59.48.8.53.20 tab diri en 2.0.5.55.33.20 a.ra$_2$ 3.22.30 du-⌜ma⌝
9 ⌜ša$_2$⌝ e$_{11}$-ka a.ra$_2$ 1.34.48.{8}.53.20 du ta 4.54.48.⌜53⌝.[20 e$_{11}$-ma gar-an]
10 ana tar-ṣa 2.0.5.55.33.20 tab 4.53.14.4.26.40 gar-an mim-[ma ša$_2$ al]
11 [2.0].⌜5.55.3⌝3.20 tab diri [en] 2.0.⌜5$^?$⌝3.42.13.20 a.⌜ra$_2$⌝ [3.22.30 du-ma ša$_2$]
12 [e$_{11}$-ka a].⌜ra$_2$⌝ 1.5⌜2⌝.[35].⌜3⌝3.20 du-ma ⌜ta 4.53⌝.[14.4.26.40 e$_{11}$ gar-an]
13 [ana tar-ṣa 2.0.2]⌜3.4⌝[2.13.20] tab 4.51.21.⌜28⌝.53.⌜20 gar-an mim-ma⌝ [ša$_2$ al]
14 [2.0.23.42.13.20 tab diri en] ⌜2.0⌝.41.28.53.20 a.ra$_2$ 3.22.30 ⌜du⌝-[ma]
15 [ša$_2$ e$_{11}$-ka a.ra$_2$ 2.10.22.1]⌜3⌝.20 du-ma ta 4.51.21.28.53.20 e$_{11}$ gar-an
16 [ana tar-ṣa 2.0.41.28.53.20] tab 4.49.11.6.40 gar-an mim-ma ša$_2$ al-la
17 [2.0.41.28.53.20 tab] ⌜diri⌝ en 2.0.59.15.⌜3⌝3.20 a.ra$_2$ 3.22.30 du-ma
18 [ša$_2$ e$_{11}$-ka a.ra$_2$ 2.28.8].⌜5⌝3.20 du-ma ta 4.49.11.6.40 e$_{11}$-ma gar
19 [ana tar-ṣa 2.0.59.15.3]⌜3⌝.20 tab 4.46.42.57.46.40 ⌜gar⌝-an
20 [mim-ma ša$_2$ al 2.0.59.15.3]⌜3⌝.20 tab diri en 2.13.8.8.53.20 tab
21 [a.ra$_2$ 3.22.30 du-ma] ⌜ša$_2$ e$_{11}$-ka a.ra$_2$⌝ 2.45.55.33.[20]
22 [du-ma ta 4.46.42.57.46.40] ⌜e$_{11}$⌝-[ma gar-an]

(at least about 20 lines missing)

Obverse

P1 **(Computing G from Φ, scheme 1)**

(at least about 20 lines missing)

14 $^{-2'}$[Opposite 1,57;58,8,53,20, increasing, you put 4,53;14,4,26,40]. $^{-1'}$[Whatever (the amount) by which it exceeds 1,57;58,8,53,20, increasing, until 1,58;15,55,33,20, increasing,] $^{0'}$[you multiply by 3;22,30, and what comes out for you you multiply by 1;34,48,53,20, and] $^{1'}$[add with 4,53;14,4,26,40 and put] down.

15 $^{2'}$[Opposite 1,58;15,55,33,20, increasing], you put [4,54;48,53,20]. $^{3'}$[Whatever (the amount) by which] it exceeds [1,58;15,55, 33,20], increasing, until 1,58;33,42,13,20, increasing, $^{4'}$you multiply by 3;22,30, and what comes out for you you multiply by 1;11,6,40, $^{5'}$add with 4,[54;48,53,20] and put down.

16 $^{6'}$Opposite 1,58;33,42,13,20, increasing, you put 4,56. $^{7'}$Until 1,58;37,2,13,20, increasing, the addition and subtraction do not exist.

17 $^{8'}$Opposite 1,58;37,2,13,20, increasing, you put 4,56. $^{9'}$Whatever (the amount) by which it exceeds 1,58;37,2,13,20, increasing, until 1,58;54,48,53,20, increasing, $^{10'}$you multiply by 3;22,30, and what comes out for you you multiply by 0;35,33,20, and $^{11'}$you add it with 4,56, and put it down.

18 $^{12'}$Opposite 1,58;54,48,53,20, increasing, you put 4,56;35,33,20. $^{13'}$Until 1,59;12,35,33,20, increasing, the addition and subtraction do not exist.

19 $^{14'}$Opposite 1,59;12,35,33,20, increasing you put 4,56;35,33,20.

Reverse

^{1}Whatever (the amount) by which it exceeds 1,59;12,35,33,20, increasing, until 1,59;30,22,13,20, [increasing], ^{2}you multiply by 3;22,30, and what comes out for you you multiply by 0;35,33,20, and you deduct it ^{3}from 4,56;35,33,20, and you put it down.

20 ^{4}Opposite 1,59;30,22,13,20, [increasing], you put 4,56. ^{5}Whatever (the amount) by which it exceeds 1,59;30,22,13, 20, increasing, until 1,59;48,8,53,20, increasing, ^{6}you multiply by 3;22,30, and what comes out for you you multiply by 1;11,6,40, you deduct it from 4,56, and you put it down.

21 ^{7}Opposite 1,59;48,8,53,20, increasing, you put 4,54;48,53,20. Whatever (the amount) by which it ^{8}exceeds 1,59;48,8, 53,20, increasing, until 2,0;5,55,33,20, increasing, you multiply by 3;22,30, and ^{9}what comes out for you you multiply by 1;34,48,{8,}53,20, deduct from 4,54;48,53,[20 and put down.]

22 ^{10}Opposite 2,0;5,55,33,20, increasing, you put 4,53;14,4,26,40. Whatever (the amount) [by which] ^{11}it exceeds [2,0];5,55,33,20, increasing, until 2,0;53,42,13,20 (error for 2,0;23,42,13,20), increasing, [you multiply by 3;22,30, and what] ^{12}[comes out for you] you multiply by 1;52,35,33,20, and deduct from 4,53;14,4,26,40 and put down.

23 ^{13}[Opposite 2,0;2]3,4[2,13,20], increasing, you put down 4,51;21,28,53,20. Whatever (the amount) by which it ^{14}[exceeds 2,0;23,42,13,20, increasing, until] ⌜2,0⌝;41,28,53,20 you multiply by 3;22,30, and ^{15}[what comes out for you you multiply by 2;10,22,1]3,20, deduct it from 4,51;21,28,53,20 and put down.

24 ^{16}[Opposite 2,0;41,28,53,20], increasing, you put down 4,49;11,6,40. Whatever (the amount) by which it ^{17}exceeds [2,0;41,28,53,20, increasing], until 2,0;59,15,33,20, you multiply by 3;22,30, and ^{18}[what comes out for you] you multiply by 2;28,8,53,20, and you deduct from 4,49;11,6,40 and put it down.

25 ^{19}[Opposite 2,0;59,15,3]3,20, increasing, you put down 4,46;42,57,46,40. ^{20}[Whatever (the amount) by which] it exceeds [2,0;59,15,3]3,20, increasing, until 2,13;8,8,53,20, increasing, ^{21}[you multiply by 3;22,30, and what] comes out for you [you multiply] by 2;45,55,33,[20], ^{22}[deduct from 4,46;42,57,46,40 and put down.]

(at least about 20 lines missing)

Critical and philological notes

Obverse

7',13' **tab** *u* **la$_2$ nu tuk**: 'the addition and subtraction do not exist'. Aaboe & Hamilton (1979), p. 18, assumed that **tab** and **la$_2$** are finite verbs (forms of *ṭepû* and *maṭû*, respectively), but since **nu tuk**, 'nothing', follows **tab** *u* **la$_2$** they must be interpreted as nouns, i.e. *ṭīpu u maṭû* = 'addition and subtraction', and **nu tuk** is the predicate (probably *ul ibašši*, 'do not exist').

Reverse

9 1.34.48.8.53.20: error for 1.34.48.53.20.

11 2.0.⌜5$^{?}$⌝3.42.13.20: the 53 is an error for 23; the 50 may have been partly erased.

Commentary

The five joined fragments include segments of the lower (upper) edge of the obverse (reverse), and the left and right edges of the tablet.

***P1** Computing G from Φ, scheme 1*

The procedure covers intervals 14–25 of interpolation scheme 1 for computing G from Φ (§4.4.11). The formulation of intervals 14–15, 17 and 19–25 complies with template PHIG.A, which involves interpolation coefficients $c_k\phi$. The missing intervals 1–13 would take up about 20 lines. A similar number of lines can be assumed missing below R22; they would suffice for intervals 26–35. This implies that the original height of the tablet was about 20 cm.

No. 75

museum number	BM 32942 (78–5–31,31)
provenience	Babylon
measures	3.3(*) × 4.7(*) × 1.4(*) cm
cuneiform text	photo: Fig. F.80
contents	Moon system A: computing G from Φ, variant scheme?

Side X

P1' (unknown number of lines missing)

1' [...] ⸢x **la**$_2^?$ 9$^?$ 2⸣[...]
2' [... **a.ra**$_2$ ...].⸢2⸣2.30 **du ta** ⸢4.x⸣ [...]
3' [... **a.ra**$_2$ ...] **du**-*ma* 6.50 ⸢x⸣ [...]
4' [... 8.5].⸢5$^?$⸣5.33.20 ⸢*bi*⸣-[*rit*$^?$ ...]
5' [...].⸢20⸣ *si-man ana tar-ṣa* 1.⸢x⸣[...]
6' [...**a.ra**$_2$ ...] ⸢**du**⸣-*ma* 1.11.6.40 ⸢*bi*$^?$⸣-[*rit*$^?$ ...]
7' [... *ana tar*]-⸢*ṣa*⸣ 1.58.54.48.5⸢3⸣.[20 ...]
8' [...].33.20 *bi-rit si*-⸢*man*⸣ [...]
9' [... *ana tar*]-*ṣa* 1.58.33.4⸢2⸣.[13.20 ...]
10' [...].⸢33$^?$⸣.20 ⸢*si*⸣-*man ana t*[*ar-ṣa* ...]
(unknown number of lines missing)

Side X

P1' **(Computing G from Φ, variant scheme?)**

(unknown number of lines missing)

1'[...] ⸢... 2 ...⸣[...]
2'[...] you multiply [... by ...].⸢2⸣2.30, [you subtract it] from ⸢4.x⸣[...]
3'[...] you multiply [... by ...], it is 6;50, ⸢...⸣[...]
4'[... 8;5],⸢5$^?$⸣5,33,20, ⸢the dist⸣[ance for the duration ...] **5'**[...],⸢20⸣, the duration.
Opposite 1.⸢...⸣[...] **6'**[...] ⸢you multiply⸣ [... by ...], it is 1;11,6,40, ⸢the dist⸣[ance for the duration...] **7'**[...]

(18?) [Opp]⸢osite⸣ 1,58;54,48,5⸢3⸣,[20 ...] **8'**[...],33,20, the distance for the ⸢duration⸣[...] **9'**[...]

(16?) [Opp]osite 1,58;33,4⸢2⸣,[13,20 ...] **10'**[...],⸢33$^?$⸣,20, the duration.
Oppos[ite ...]

(unknown number of lines missing)

Critical and philological notes

Side X

1' At the beginning of the line there are traces of two horizontal wedges.
3' The traces at the end are consistent with the beginning of BI or TAB.
5' At the end there are traces of two winkelhakens, probably belonging to a digit 20–50.

Commentary

This small flake does not preserve any edges of the original tablet. It is inscribed on one side; the other side is destroyed.

P1' Computing G from Φ, variant scheme?

The sparse remains of P1' especially X6',8', are sufficient to conclude that it is concerned with the computation of G or a related quantity (W or Λ) from Φ (§4.4.11). The numbers are suggestive of scheme 2, or a variant of scheme 1. The term 'distance for the duration' (*birīt siman*), which is not attested elsewhere, appears to denote $c_k|\Phi_{k+1} - \Phi_k|$, the difference of G across interpolation interval k (cf. template PHIG.B, p. 148). X3' mentions 6;50, which may be identified with the interpolation coefficient c_2 or c_5 in scheme 2 (Table 4.26, p. 152); it does not occur in scheme 1. It appears to be the outcome of a multiplication, perhaps 6;50 = 8;5,55,33,20 · 4ϕ. The number 1;11,6,40 = 4ϕ (X6') is also the result of a multiplication. It may have been obtained as $c_k|\Phi_{k+1} - \Phi_k|$ for an interval where $|\Phi_{k+1} - \Phi_k| = \phi$ and $c_k = 4$ (in scheme 1: intervals 4 and 20). The Φ values mentioned in X7',9' may be identified with control points 18 and 16 of scheme 1, respectively. While in scheme 1 they belong to the increasing branch of Φ, their order in the present text suggests that they belong to the decreasing branch. Hence the underlying interpolation scheme may be some variant of scheme 1 or 2.

No. 76

museum number	BM 77620 (84–2–11,363)
provenience	Babylon
measures	5.7(9) × 4.5(12) × 2.0–2.9 cm (type F?)
arrangement	O/R
cuneiform text	photo: Fig. F.84
contents	Moon system A: computing *G* from *Φ* (scheme 1)

Obverse

P1' (at least about 8 lines missing)

0' [*ana tar-ṣa* 1.58.13.20 **la₂** 4.49.11.6.40 **gar**-*an ša₂ al* 1.58.13.20 **la₂ la₂**]
1' [**en** 1.57.55.33.20 **la₂ a**].⌜**ra₂** 7.20⌝ **du ki** 4.4[9.11.6.40 **tab**-*ma* **gar**-*an*]
2' [*ana tar*]-*ṣa* 1.57.55.33.20 [**la₂** 4.51.21.28.53.20 **gar**-*an ša₂ al* 1.57.55.33.20 **la₂ la₂**]
3' [**en** 1.57].⌜47⌝.57.46.40 **la₂ a.ra₂** ⌜6⌝.[20 **du ki** 4.51.21.28.53.20 **tab**-*ma* **gar**-*an*]
4' *ana tar-ṣa* 1.57.47.57.46.⌜40⌝ [**tab** 4.52.9.34.4.26.40 **gar**-*an ša₂ al* 1.57.47.57.46.40 **tab diri**]
5' **en** 1.57.58.8.53.20 **ta**[**b a.ra₂** 6.20 **du ki** 4.52.9.34.4.26.40 **tab**-*ma* **gar**-*an*]
6' *ana tar-ṣa* 1.57.58.8.53.20 [**tab** 4.53.14.4.26.40 **gar**-*an ša₂ al* 1.57.58.8.53.20 **tab diri en** 1.58.15]
7' 55.33.20 **tab a.ra₂** 5.20 [**du ki** 4.53.14.4.26.40 **tab**-*ma* **gar**-*an*]
8' *ana tar-ṣa* 1.58.⌜15.55⌝.[33.20 **tab** 4.54.48.53.20 **gar**-*an ša₂ al* 1.58.15.55.33.20 **tab diri en**]
9' 1.58.⌜33⌝.[42.13.20 **tab a.ra₂** 4 **du ki** 4.54.48.53.20 **tab**-*ma* **gar**-*an*]
(about 6 lines missing)

Reverse

(about 6 lines missing)
0' [*ana tar-ṣa* 2.0.23.42.13.20 **tab** 4.51.21.28.53.20 **gar**-*an ša₂ al* 2.0.23.42.13.20 **tab diri en**]
1' ⌜2.0⌝.[41.28.53.20 **tab** GAM 7.20 **du ta** 4.51.21.28.53.20 **nim**-*ma* **gar**-*an*]
2' *ana tar*-⌜*ṣa*⌝ [2.0.41.28.53.20 **tab** 4.49.11.6.40 **gar**-*an ša₂ al* 2.0.41.28.53.20 **tab diri en** 2.0.59]
3' 15.33.⌜20⌝ **tab** GAM ⌜8.20⌝ [**du ta** 4.49.11.6.40 **nim**-*ma* **gar**-*an*]
4' *ana tar-ṣa* 2.0.59.15.33.20 **tab** ⌜4⌝.[46.42.57.46.40 **gar**-*an ša₂ al* 2.0.59.15.33.20 **tab diri**]
5' **en** 2.13.8.8.53?.20 **tab** GAM 9.20 ⌜**du**⌝ [**ta** 4.46.42.57.46.40 **nim**-*ma* **gar**-*an*]
6' *ana tar-ṣa* 2.13.8.8.53.20 **tab** 2.53.20 [**gar**-*an ša₂ al* 2.13.8.8.53.20 **tab diri en** 2.15.48.8.53.20 **tab**]
7' [**ta**] ⌜2⌝.15.48.8.53.20 **nim** *ša₂* **tag₄** [GAM 3.22.30 **du**-*ma* **diri**-*šu₂* 17.46.40 GAR.GAR-*ma* **ki** 2.40 **tab**-*ma* **gar**-*an*]
8' [*ana tar-ṣa* 2.15].⌜4⌝8.8.53.20 **tab** 2.40 ⌜**gar**⌝-[*an* xxxxxxxxxxxxxxxxxxxxxxxxxx]
9' [xxxxxx] ⌜x 2.17.4.48⌝.[53.20 xxxxxxxxxxxxxxxxxxxxxxx]
(at least about 7 lines missing)

Critical and philological notes

Obverse

1' Multiplication is expressed by **a.ra₂** in O1',3',5',7', by GAM in R3'and 5'.

Reverse

1' Here and elsewhere subtraction is assumed to be expressed by **nim** = *tušelli*, 'you deduct' in analogy to R7'.

9' Before 2.10 the upper part of a wedge is visible. The traces of 4 signs after 2.10 are compatible with the digits 5–8, 3–8, 20–50 and 5–8. Since the previous line contains the beginning of interval 35, the number must be 2,17;4,48,53,20 = M_{Φ}.

Commentary

This fragment includes a segment of the left edge of the tablet, but no other edges. The textual restorations imply that the original width was about 9 cm, and the height at least about 12 cm (cf. below).

Obverse

P1' (Computing G from Φ, scheme 1)

(*at least about 8 lines missing*)

12 $^{0'}$[Opposite 1,58;13,20, decreasing, you put down 4,49;11,6,40. (The amount) by which it is less than 1,58;13,20, decreasing,] $^{1'}$[until 1,57;55,33,20, decreasing], you multiply ⸢by 7;20⸣, you add it with 4,4[9;11,6,40 and you put it down.]

13a $^{2'}$[Opposi]te 1,57;55,33,20, [decreasing, you put down 4,51;21,28,53,20. (The amount) by which it is less than 1,57;55,33, 20, decreasing,] $^{3'}$[until 1,57];⸢47⸣,57,46,40, decreasing, [you multiply] by ⸢6⸣;[20, you add it with 4,51;21, 28,53,20 and you put it down.]

13b $^{4'}$Opposite 1,57;47,57,46,⸢40⸣, [increasing, you put down 4,52;9,34,4,26,40. (The amount) by which it exceeds 1,57;47, 57,46,40, increasing,] $^{5'}$until 1,57;58,8,53,20, increas[ing, you multiply by 6;20, you add it with 4,52;9,34,4, 26,40 and you put it down.]

14 $^{6'}$Opposite 1,57;58,8,53,20, [increasing, you put down 4,53;14,4,26,40. (The amount) by which it exceeds 1,57;58,8, 53,20, increasing, until 1,58;15,]$^{7'}$55,33,20, increasing, [you multiply] by 5;20, [you add it with 4,53;14,4,26,40 and you put it down.]

15 $^{8'}$Opposite 1,58;⸢15,55⸣,[33,20, increasing, you put down 4,54;48,53,20. (The amount) by which it exceeds 1,58;15, 55,33,20, increasing, until] $^{9'}$1,58;⸢33⸣,[42,13,20, increasing, you multiply by 4, you add it with 4,54;48,53,20 and you put it down.]

(*about 6 lines missing*)

Reverse

(*about 6 lines missing*)

23 $^{0'}$[Opposite 2,0;23,42,13,20, increasing, you put down 4,51;21,28,53,20. (The amount) by which it exceeds 2,0;23,42, 13,20, increasing, until] $^{1'}$⸢2,0⸣;[41,28,53,20, increasing, you multiply by 7;20, you deduct it from 4,51;21,28,53,20 and you put it down.]

24 $^{2'}$Opposi⸢te⸣ [2,0;41,28,53,20, increasing, you put down 4,49;11,6,40. (The amount) by which it exceeds 2,0;41,28, 53,20, increasing, until 2,0;59],$^{3'}$15,33,⸢20⸣, increasing, you [multiply] by 8;20⸣, [you deduct it from 4,49;11,6,40 and you put it down.]

25 $^{4'}$Opposite 2,0;59,15,33,20, increasing, [you put down] ⸢4⸣,[46;42,57,46,40. (The amount) by which it exceeds 2,0;59,15, 33,20, increasing,] $^{5'}$until 2,13;8,8,53$^{?}$,20, increasing, you ⸢multiply⸣ by 9;20, [you deduct it from 4,46;42, 57,46,40 and you put it down.]

26–34 $^{6'}$Opposite 2,13;8,8,53,20, increasing, [you put down] 2,53;20. [That which exceeds 2,13;8,8,53,20, increasing, until 2,15;48,8,53,20, increasing] $^{7'}$you deduct [from] ⸢2⸣,15;48,8,53,20. What remains [you multiply by 3;22,30 and while it fills it up you accumulate 0;17,46,40, and you add it with 2,40 and you put it down.]

35 $^{8'}$[Opposite 2,15];⸢4⸣8,8,53,20, increasing, [you] ⸢put down⸣ 2,40. [...]
$^{9'}$[...] ⸢... 2,17;4,48⸣,[53,20 ...]

(*at least about 7 lines missing*)

P1' Computing G from Φ scheme 1

The obverse contains intervals 12–15 of interpolation scheme 1 for computing G from Φ (§4.4.11); the reverse intervals 23–35 (only traces remain of 23). The beginning of each interval is visually marked by a larger than average initial vertical wedge (*ana*) and a slight indentation of the second line, a feature also known from No. **69** and other ΦG procedures. Entries 12–15, 23–25 and probably 35 are formulated in accordance with the template IP.B. Interval 13 is split up at m = 1,57;47,57,46,40°, i.e. the interpolation is formulated separately for interval 13a, where Φ is decreasing (O2'–3'), and interval 13b, where it is increasing (O4'–5'). This tablet is one of only two that partly preserve instructions for intervals 26–34. R6'–7' can be restored in accordance with the *malûššu* rule for these intervals (§4.4.11); presumably the same formulation was used for intervals 1–9, to be restored above O0'. Intervals 10–11 are also missing, which makes the total number of lines to be restored above O0' about 8. Comparison with No. **53** P14' suggests that about 12 lines are missing between O9' and R0', 6 of which belong to the obverse. Hence about 14 lines are to be restored on both sides, which results in a height of about 12 cm.

No. 77

museum number	CBS 1493
provenience	Babylon
measures	4.8(7) × 5.0(*) × 2.3(*)
cuneiform text	photo: Fig. F.85
bibliography	*ACT* 208
contents	Moon system A: computing *G* from *F* (scheme 1)

Side X

P1' (at least about 9 lines missing)

1' [**en** 11.6 **la**$_2$ **a.ra**$_2$ 28.58.1]7.46.40 **d**[**u**-*ma* **ki** 4.49.11.6.40 **tab**]
2' [*ana tar-ṣa* 11.6 **z**]**i** 4.5[1].21.28.53.20 **gar**-*a*[*n* **ta** 11.6 **la**$_2$]
3' [**en** 11.4.4.41.15] **la**$_2$ **a.ra**$_2$ 25.1.14.4.26.40 **du**-*m*[*a* **ki** 4.51.28.53.20 **tab**]
4' [*ana tar-ṣa*] 11.4.4.41.15 **zi** 4.52.9.3[4.4.26.40 **gar**-*an*]
5' [**ta** 11.4.4.41].15 **la**$_2$ **en** 11.6.39.22.30 **ta**[**b a.ra**$_2$ 25.1.14.4.26.40]
6' [**du**-*ma* **k**]**i** 4.52.9.34.4.26.40 **ta**[**b** *ana tar-ṣa* 11.6.39.22.30]
7' [**zi** 4.53].14.4.26.40 ⌜*si*⌝-*man* **gar**-*an* [**ta** 11.6.39.22.30 **tab**]
8' [**en** 11.11.9.22.30] **a.ra**$_2$ ⌜21⌝.[4.11.51.6.40 **du**-*ma* **ki** 4.53.14.]
9' [4.26.40 **tab** *ana*] ⌜*tar*⌝-*ṣa* 11.1[1.9.22.30 **zi** 4.54.48.53.20 **gar**-*an*]
10' [**ta** 11.11.9.22.30 **tab en**] ⌜1⌝1.15.3⌜9⌝.[22.30 **tab a.ra**$_2$ 15.48.8.53.20 **du**-*ma*]
11' [**ki** 4.54.48.53.20 **tab** *ana tar-ṣa* xxxx].⌜xx⌝[xxxxxxxx]
(perhaps about 30 lines missing)

Side X

P1' (Computing *G* from $F^{(1)}$, scheme 1)

12 (*at least about 9 lines missing*)
[...] 1'[until 11;6, decreasing, you multiply by 28;58,1]7,46,40, [and add with 4,49;11,6,40.]

13a 2'[Opposite 11;6, the displ]acement, you put down 4,51;21,28,53,20. [From 11;6, decreasing] 3'[until 11;4,4,41,15], decreasing, you multiply by 25;1,14,4,26,40, and [add it with 4,51;21,28,53,20.]

13b 4'[Opposite] 11;4,4,41,15, the displacement, [you put down 4,52;9,34,4,26,40.] 5'[From 11;4,4,41],15, decreasing, until 11;6,39,22,30, decreasing, [you multiply by 25;1,14,4,26,40] 6'[and add it with] 4,52;9,34,4,26,40.

14 Opposite 11;6,39,22,30, 7'[the displacement], you put down 4,53;14,4,26,40 as the duration. [From 11;6,39,22,30, increasing], 8'[until 11;11,9,22,30, increasing, you multiply] by 21;[4,11,51,6,40, and add it with 4,53;14],9'[4,26,40.]

15 Opposite 11;11,9,22,30, the displacement, you put down 4,54;48,53,20. 10'[From 11;11,9,22,30, increasing, until 1]1;15,39, [22,30, increasing, you multiply by 15;48,8,53,20, and] 11'[add it with 4,54;48,53,20.]

16 (*traces*)
(*perhaps about 30 lines missing*)

Critical and philological notes

Obverse

7' *si-man* = 'duration' (= *G*): overlooked in *ACT*.
11' Traces are visible of about 2 digits that must belong to interval 16.

Commentary

This fragment is inscribed on one side; the other side is damaged. No edges are preserved. The textual restorations imply that about 1 cm of clay is missing from both sides, so the original width was about 7 cm.

P1' Computing G from $F^{(1)}$, scheme 1

The fragment contains a single procedure with intervals 12–16 of interpolation scheme 1 for computing *G* from $F^{(1)}$ (§4.4.11). The formulation is based on the unique template IP.E (§2.6.7), which is not attested elsewhere. The missing intervals 1–11 and 17–35 would require about 9 and 30 additional lines, respectively.

No. 78

museum number	BM 36930 (80–6–17,671)
provenience	Babylon
measures	4.0(*) × 4.1(*) × 2.0–2.3 cm
cuneiform text	photo: Fig. F.86
contents	Moon system A: computing G, W or Λ from F or Φ?

Obverse

P1 1 [...xxx].⌜6.40⌝ **zi**
2 [...xx].6.40 **zi**
3 [...x] ⌜xx⌝.1.51.6.⌜40⌝ **zi**-*šu*$_2^?$
4 [...] ⌜x.5⌝4.48.53.20
5 [...] ⌜x⌝ **tab**-*šu*$_2$ *ana* ⌜**ugu**$^?$ xx⌝
6 [...]-⌜*tu*$^?$⌝ **tab**-*u*$_2$ 1-*me* 1.20 **u$_4$.meš**
(unknown number of lines missing)

Obverse

P1 **(An interpolation scheme?)**
1 [...].⌜.6.40⌝, the displacement.
2 [...].6.40, the displacement.
3 [...] ⌜... ⌝ 1 51.6.⌜40⌝, its$^?$ displacement.
4 [...] ⌜... 5⌝4.48.53.20
5 [...] ⌜...⌝ its addition to ⌜...⌝
6 [...] ... increasing, ... 1.20 days
(*unknown number of lines missing*)

Critical and philological notes

Obverse

3 Before 1.51 there are traces followed by what might be a 4.
zi-*šu*$_2$$^?$: written on right edge.

4 Before 54 there are traces of several vertical wedges.

Commentary

This small fragment preserves a segment of the upper edge of a tablet. One side of the fragment, assumed to be the reverse, is uninscribed.

P1 Computing G, W or Λ from F or Φ?

The first 6 lines of P1 are partly preserved. The digits in O1–4 probably belong to control values or interpolation coefficients of a scheme for computing G, or a related function (W, Λ), from F or Φ. A unique identification could not be made. The repeatedly occurring term 'displacement' suggests that the source function is F, but this remains speculative.

No. 79

museum numbers	BM 35564 (Sp3,71) +40081 (81–2–1,45) (+) 77256 (83–9–28,7)
provenience	Babylon
measures	13.9(23?) × 12.7(*) × 3.0–4.4 cm (type A or M?)
arrangement	O/R
cuneiform text	photo: Figs. F.90, F.91; *ACT*, Pl. 242 (BM 35564+40081)
bibliography	*ACT* 207d (BM 35564+40081)
contents	Moon system A: ΦG scheme; net differences of C'; $\Phi\Lambda$ scheme

Obverse
Column i'

P1' (about n+17 lines missing)

1' [xxxxxxxxxxxxxxxxxxxxxxxx] ⸢x 2$^?$ x **du**$^?$⸣
2' [xxxxxxxxxxxxxxxxxxx] ⸢x.43$^?$⸣.40
3' [xxxxxxxxxxxxxxxxx].⸢3.20$^?$ x⸣ **diri tab**$^?$ **gar**-*an*
4' [... *mim-ma* *ša*$_2$ *al* xxx **la**$_2$ **la**$_2$ **en** xxx **la**$_2$]-⸢*u*$_2$⸣ **a.ra**$_2$ 3.22.30 **du**
5' [**diri**-*šu*$_2$ 17.46.40 **gar**-*an ana*] ⸢**ugu** *a-ḫa*⸣-*miš*$_2$ GAR.⸢GAR⸣-[*ma ša*$_2$] ⸢**e**$_{11}$⸣ **ki** 2.53.20 **tab**-*ma* **gar**-*an*

P2' 6' [*ana tar* 2.15.]⸢48.8⸣.53.20 **tab** ⸢2.40⸣ **gar**-*an a-di* ⸢2⸣.[17.4.48.53.20 x 2].⸢40⸣ **gar**-*an a-di* 2.13.20 **la**$_2$ 2.40 **gar**-*an*

P3' 7' [...] ⸢x 2⸣-*u*$_2$ *ša*$_2$ *si-man*.**meš** *ša*$_2$ **iti ta** *lu-maš* 10$^?$ [xxxx] ⸢x⸣ **du**$_3$
8' [...] ⸢**ta**⸣ 10 $^{\text{mul}_2}$**alla en** 10 **maš**$_2$ **tab ta** ⸢10⸣ [**maš**$_2$ **en** 10 $^{\text{mul}_2}$]**alla la**$_2$
9' [... **ta** 10] ⸢$^{\text{mul}_2}$⸣**alla en** 10 **a** *lu-maš* **a.ra**$_2$ 4 ⸢x⸣ [xxx] ⸢LA$_2$⸣ *ša*$_2$ *u*$_4$-*me* **du**
10' [... **ta** 10] ⸢**a**⸣ **en** 10 **absin**$_0$ *lu-maš* **a.ra**$_2$ 1⸢2⸣ [xx] LA$_2$ *ša*$_2$ *u*$_4$-*me* **du**
11' [... **ta** 10] ⸢**absin**$_0$⸣ **en** 10 **gir**$_2$ *lu-maš* **a.ra**$_2$ ⸢20⸣ [x] LA$_2$ *ša*$_2$ *u*$_4$-*me* **du**

Column ii'

T1' (n lines missing)

1'	[2.3.35.55.33.20 **la**$_2$	7.45.33.20 **tab**]	⸢**a.ra**$_2$⸣ 1⸢6.8⸣.[45]
2'	[2.3.18.8.53.20 **la**$_2$	2.58.31.6].40 **tab**	**a.ra**$_2$ 15.8.[45]
3'	[2.3.0.22.13.20 **la**$_2$	1.30.44].⸢26⸣.40 **la**$_2$	**a.ra**$_2$ 14.8.4[5]
4'	[2.2.42.35.33.20 **la**$_2$	5.42.13.20] ⸢**la**$_2$⸣	[a].⸢**ra**$_2$⸣ [13 **du**]
5'	[2.2.24.48.53.20 **la**$_2$	9.33.20 **la**$_2$	**a.ra**$_2$ 11] ⸢**du**⸣
6'	[2.2.7.2.13.20 **la**$_2$	12.48.53.20 **la**$_2$	a].⸢**ra**$_2$ 9 **du**⸣
7'	[2.1.49.15.33.20 **la**$_2$	15.28.53].20 **la**$_2$	**a.ra**$_2$ 7 **du**
8'	[2.1.31.28.53.20 **la**$_2$	17].33.20 **la**$_2$	**a.ra**$_2$ 5 **du**
9'	[2.1.13.42.13.20] **la**$_2$	1⸢9⸣.2.13.20 **la**$_2$	**a.ra**$_2$ 3 **du**
10'	[2.0.55.55.3]3.20 **la**$_2$	1⸢9⸣.55.33.20 **la**$_2$	**a.ra**$_2$ 1 **du**
11'	[2.0.38.8].⸢5⸣3.20 **la**$_2$	20.13.20 **la**$_2$	**a.ra**$_2$ 1 **du**
12'	[2.0.20.2]2.13.20 **la**$_2$	1⸢9.5⸣5.33.20 **la**$_2$	**a.ra**$_2$ 3 **du**
13'	[2.0.2.3]5.33.20 **la**$_2$	1⸢9⸣.2.13.20 **la**$_2$	**a.ra**$_2$ 5
14'	[1.5]⸢9⸣.44.48.53.20 **la**$_2$	1[7].33.20 **la**$_2$	**a.ra**$_2$ 7
15'	⸢1⸣.59.27.2.13.20 **la**$_2$	15.[2]8.53.20 **la**$_2$	**a.ra**$_2$ 9
16'	1.59.9.15.33.20 **la**$_2$	12.⸢4⸣8.53.20 **la**$_2$	**a.ra**$_2$ 11
17'	1.58.51.28.53.20 **la**$_2$	9.⸢3⸣3.20 **la**$_2$	**a.ra**$_2$ 13
18'	1.58.33.42.13.20 **la**$_2$	5.42.13.20 **la**$_2$	**a.ra**$_2$ 14.8.45
19'	1.58.15.55.33.20 **la**$_2$	1.30.44.26.⸢40 **la**$_2$⸣	**a.ra**$_2$ 15.8.45
20'	1.57.58.8.53.20 **la**$_2$	2.58.31.6.40 **tab**	**a.ra**$_2$ 16.8.45
21'	1.57.55.33.20 **tab**	7.45.33.20 **tab**	**a.ra**$_2$ 17.8.[45]
22'	1.58.13.20 **tab**	12.50.22.13.20 **tab**	**a.ra**$_2$ 1⸢8.8⸣.[45]
23'	1.58.31.6.40 **tab**	18.12.57.46.40 **tab**	**a.ra**$_2$ [18.8.45]
24'	1.58.48.53.20 **tab**	23.35.33.20	⸢**ta** 1⸣.[58.48.53.20 **tab**]
25'	**en** 2.7.42.13.20 **tab**	2.13.4.48.⸢5⸣[3.20 **tab**	3.4.53.20]
26'	2.8	2.13.22.3[5.33.20 **tab**	3.10.13.20]
27'	2.8.17.⸢46.40 **tab**⸣	2.13.[40.22.13.20 **tab**	3.15.15.33.20]

Obverse column i'

P1' (Computing G from Φ, scheme 1?)

(about n+17 lines missing)

1'[...] ⌜...⌝ 2'[...] ⌜...43?⌝,40 3'[...] ⌜3.20?...⌝ exceeds ... you put it down.

10–14? 4'[(The amount) by which it is less than ..., decreasing, until ... decreasing], you multiply by 3;22,30. 5'[... You put down 0;17,46,40], you accumulate it together, and [what] ⌜comes⌝ out you add with 2,53;20, and you put it down.

P2' (Computing G from Φ, scheme 1?)

35 6'[Opposite 2,15];⌜48,8⌝,53,20 you put down ⌜2,40⌝. Until ⌜2⌝,[17;4,48,53,20 ...] you put down [2],⌜40⌝. Until 2,13;20, decreasing, you put down 2,40.

P3' (Net differences of C')

7'[...] ... of the durations of the month from the zodiacal position 10? [...] ⌜...⌝ you construct.
8'[...] ⌜from⌝ 10 Cnc until 10 Cap you add. From ⌜10⌝ [Cap until 10] Cnc you subtract.
9'[... From 10] Cnc until 10 Leo you multiply the zodiacal position by 0;4, [...] the length of daylight.
10'[... From 10] Leo until 10 Vir you multiply the zodiacal position by 0;12 [...] the length of daylight.
11'[... From 10] ⌜Vir⌝ until 10 Sco you multiply the zodiacal position by 0;20 [...] the length of daylight.

Column ii'

T1' (Computing Λ from Φ)

(n lines missing)

52	1'[2,3;35,55,33,20, decreasing,	7;45,33,20, additive],	times 1⌜6;8⌝,[45].
53	2'[2,3;18,8,53,20, decreasing,	2;58,31,6],40, additive,	times 15;8,[45].
54	3'[2,3;0,22,13,20, decreasing,	1;30,44],⌜26⌝,40, subtractive,	times 14;8,4[5].
55	4'[2,2;42,35,33,20, decreasing,	5;42,13,20], ⌜subtractive⌝,	[you multiply ti]mes [13].
56	5'[2,2;24,48,53,20, decreasing,	9;33,20, subtractive],	⌜you multiply⌝ [times 11].
57	6'[2,2;7,2,13,20, decreasing,	12;48,53,20, subtractive],	⌜you multiply⌝ [times 9].
58	7'[2,1;49,15,33,20, decreasing,	15;28,53],20, subtractive,	you multiply times 7.
59	8'[2,1;31,28,53,20, decreasing,	17];33,20, subtractive,	you multiply times 5.
60	9'[2,1;13,42,13,20], decreasing,	19;2,13,20, subtractive,	you multiply times 3.
61	10'[2,0;55,55,3]3,20, decreasing,	19;55,33,20, subtractive,	you multiply times 1.
1	11'[2,0;38,8],53,20, decreasing,	20;13,20, subtractive,	you multiply times 1.
2	12'[2,0;20,2]2,13,20, decreasing,	1⌜9;5⌝5,33,20, subtractive,	you multiply times 3.
3	13'[2,0;2,3]5,33,20, decreasing,	1⌜9⌝;2,13,20, subtractive,	times 5.
4	14'[1,5]9;44,48,53,20, decreasing,	1[7];33,20, subtractive,	times 7.
5	15'1,59;27,2,13,20, decreasing,	15;[2]8,53,20, subtractive,	times 9.
6	16'1,59;9,15,33,20, decreasing,	12;⌜4⌝8,53,20, subtractive,	times 11.
7	17'1,58;51,28,53,20, decreasing,	9;⌜3⌝3,20, subtractive,	times 13.
8	18'1,58;33,42,13,20, decreasing,	5;42,13,20, subtractive,	times 14;8,45.
9	19'1,58;15,55,33,20, decreasing,	1;30,44,26,⌜40, subtractive⌝,	times 15;8,45.
10	20'1,57;58,8,53,20, decreasing,	2;58,31,6,40, additive,	times 16;8,45.
11	21'1,57;55,33,20, increasing,	7;45,33,20, additive,	times 17;8,[45].
12	22'1,58;13,20, increasing,	12;50,22,13,20, additive,	times 18;8,[45].
	23'1,58;31,6,40, increasing,	18;12,57,46,40, additive,	times [18;8,45].
	24'1,58;48,53,20, increasing,	23;35,33,20.	From 1,[58;48,53,20, increasing,]
(13)	25'until 2,7;42,13,20, increasing,	2,13;4,48,⌜5⌝[3,20, increasing,	3,4;53,20.]
(14)	26'2,8	2,13;22,3[5,33,20, increasing,	3,10;13,20.]
(15)	27'2,8;17,⌜46,40, increasing⌝,	2,13;[40,22,13,20, increasing,	3,15;15,33,20.]

Reverse

Column i

1	2.8.35.33.20 **tab**	2.13.58.8.5⸢3⸣.[20 **tab**	3.20]
2	2.8.53.20 **tab**	2.14.15.55.33.20 **tab**	[3.24.26.40]
3	2.9.11.⸢6.40⸣ **tab**	2.14.33.42.13.20 **tab**	⸢3.2⸣[8.35.33.20]
4	2.9.28.53 **tab**	2.14.⸢5⸣1.28.53.20 **tab**	3.32.⸢2⸣[6.40]
5	2.9.46.40 **tab**	2.15.9.15.⸢33.20⸣ **tab**	3.3⸢6 x$^?$⸣
6	2.10.4.⸢26.40⸣ **tab**	2.15.2⸢7.2.13.20⸣ **tab**	3.⸢39.15⸣.[33.20]
7	2.10.22.13.20 **tab**	2.15.4⸢4⸣.[48].⸢5⸣3.⸢20⸣ **tab**	3.⸢42.13⸣.[20]
8	2.10.40 **tab**	2.16.2.3[5].⸢3⸣3.20 **tab**	3.⸢44.53⸣.[20]
9	2.10.⸢57⸣.46.40 **tab**	2.16.20.⸢22.13.20⸣ **tab**	3.⸢47⸣.15.33.⸢20⸣
10	2.11.15.3⸢3.20⸣ **tab**	2.16.38.8.53.20 **tab**	3.49.20
11	2.11.33.20 **tab**	2.16.55.55.33.20 **tab**	3.51.6.40
12	2.11.51.6.40 **tab**	2.17.13.42.13.20 **tab**	3.52.35.33.20
13	2.12.8.53.20 **tab**	2.17.31.28.53.20 **tab**	3.53.46.40
14	2.12.26.40 **tab**	2.17.49.15.33.20 **tab**	3.54.40
15	2.12.⸢44⸣.26.40 **tab**	2.18.7.2.13.20 **tab**	3.55.15.33.20
16	2.13.2.13.20 **tab**	2.18.⸢24⸣.48.53.20 **tab**	3.55.33.20
17	2.13.20 **tab**	2.18.⸢42.3⸣5.33.20 **tab**	3.55.33.20

Between columns i and ii, written perpendicularly

0 2$^?$5 ⸢xxx 3 xxxx⸣ **gar** *ana tar* 1.⸢57$^?$.57$^?$.58.x⸣

Column ii

P4’ 1 [*ša$_2$ al* 2.0.55.55.33.20 **la$_2$ la$_2$**] ⸢**en** 2⸣.0.3⸢8.8⸣.5⸢3⸣.[20 **la$_2$ a.ra$_2$** 1] ⸢**du**⸣ **ki** 19.55.33.20 ⸢**tab**-*ma* **gar**-*an*⸣

2 [*ana tar* 2.0.3]⸢8⸣.8.53.20 **la$_2$** 20.1[3.20 **gar** *ša$_2$ al* 2.0.38.8.53.20] ⸢**la$_2$**$^?$⸣ **la$_2$** *a-di* 2.0.20.22.13.20

3 [**la$_2$ a.ra$_2$** 1 **du**] ⸢**ta**⸣ 20.13.20 **nim** ⸢**gar** *ana*⸣ [*tar* 2.0.20.22].13.20 **la$_2$** 19.55.33.20 **gar**

4 [*ša$_2$ al* 2.0.20.22.13.20] ⸢**la$_2$**⸣ **la$_2$ en** 2.0.2.3⸢5⸣.[33.20 **la$_2$ a.ra$_2$** 3 **du**] ⸢**ta**⸣ 19.55.33.20 **nim gar**

5 [*ana tar* 2].0.2.35.33.20 **la$_2$** [19.2.13.20 **gar** *ša$_2$ al* 2.0.2.35.33.20 **la$_2$ la$_2$ en**] ⸢1.5⸣9.44.48.53.20 **la$_2$** [**a.ra$_2$** 5 **du ta** 19.2.13.20 **nim gar**]

6 [*ana tar* 1.59.44.48.53.20 **la$_2$** 17.33.20 **gar** *ša$_2$ al* 1.59.44.48.53 **la$_2$ la$_2$ en** 1.59.27.2.13.20 **la$_2$ a**].⸢**ra$_2$** 7⸣ **du ta** 17.33.20 **nim gar**

7 [*ana tar* 1.59.27.2.13.20 **la$_2$** 15.28.53.20 **gar** *ša$_2$ al* 1.59.27.2.13.20 **la$_2$ la$_2$**] ⸢**en**⸣ 1.59.9.15.33.20 **la$_2$**

8 [**a.ra$_2$** 9 **du ta** 15.28.53.20 **nim gar** *ana tar* 1.59.9.1]5.33.20 **la$_2$** 12.⸢48.5⸣3.20 **gar**

9 [*ša$_2$ al* 1.59.9.15.33.20 **la$_2$ la$_2$ en** 1.58.51.28.53.20 **la$_2$ a.ra$_2$** 11 **du**] ⸢**ta**⸣ 12.4⸢8.53⸣.20 **nim**-*ma* **gar**

10 [*ana tar* 1.58.51.28.53.20 **la$_2$** 9.33.20 **gar** *ša$_2$ al* 1.58.51.28.53.20 **la$_2$**] ⸢**la$_2$**⸣ **en** 1.⸢58.33.42.1⸣3.20 **la$_2$**

11 [**a.ra$_2$** 13 **du ta** 9.33.20 **nim gar** *ana tar* 1.58.33.42.13].⸢20⸣ **la$_2$** ⸢5.42.1⸣3.20 **gar**

12 [*ša$_2$ al* 1.58.33.42.13.20 **la$_2$ la$_2$ en** 1.58.15.55.33.20 **la$_2$ a.ra$_2$** 14.8.45] ⸢**du**⸣ **ta** 5.42.13.20 **nim gar**

13 [*ana tar* 1.58.15.55.33.20 **la$_2$** 1.30.44.26.40] ⸢LA$_2$⸣ *ša$_2$* **mu gar**-⸢*an*⸣

14 [xxxxxxxxxxxxxxxxxxxxxxxxxxxxxxxxxx] ⸢x 20$^?$ **la$_2$**$^?$ 10$^?$⸣ [xx]

15 [xxxxxxxxxxxxxxxxxxxxxxxxxxxxxxxxxx] ⸢50$^?$ **gar**-*an*$^?$⸣

16 [xxxxxxxxxxxxxxxxxxxxxxxxxxxxxxxxxx].⸢45$^?$⸣ **du**$^?$-*ma*

17 [xxxxxxxxxxxxxxxxxxxxxxxxxxxxxxxxxx] ⸢5$^?$ x⸣ **a**$^?$ ⸢x⸣

P5’ 18 [xxxxxxxxxxxxxxx].**meš** *ša$_2$* **mu** *a-na* **mu**

P6’ 19 [xxxxxxxxxxxxxx] ⸢x **du**⸣-*ma ana tar-ṣa* 2.13.2.13.20 **tab** 3.55.33.20 **gar**-*an*

Reverse column i

(16)	[1]2,8;35,33,20, increasing,	2,13;58,8,5⸢3⸣,[20, increasing	3,20]
(17)	[2]2,8;53,20, increasing	2,14;15,55,33,20, increasing	[3,24;26,40]
(18)	[3]2,9;11,⸢6,40⸣, increasing	2,14;33,42,13,20, increasing	⸢3,2⸣[8;35,33,20]
(19)	[4]2,9;28,53, increasing	2,14;⸢5⸣1,28,53,20, increasing	3,32;⸢2⸣[6,40]
(20)	[5]2,9;46,40, increasing	2,15;9,15,⸢33,20⸣, increasing	3,3⸢6 x?⸣
(21)	[6]2,10;4,⸢26,40⸣, increasing	2,15;2⸢7,2,13,20⸣, increasing	3,⸢39;15⸣,[33,20]
(22)	[7]2,10;22,13,20, increasing	2,15;4⸢4⸣,[48],⸢5⸣3,⸢20⸣, increasing	3,⸢42;13⸣,[20]
(23)	[8]2,10;40, increasing	2,16;2,3[5],⸢3⸣3,20, increasing	3,⸢44;53⸣,[20]
(24)	[9]2,10;⸢57⸣,46,40, increasing	2,16;20,⸢22,13,20⸣, increasing	3,⸢47⸣;15,33,⸢20⸣
(25)	[10]2,11;15,3⸢3,20⸣, increasing	2,16;38,8,53,20, increasing	3,49;20
(26)	[11]2,11;33,20, increasing	2,16;55,55,33,20, increasing	3,51;6,40
(27)	[12]2,11;51,6,40, increasing	2,17;13,42,13,20, increasing	3,52;35,33,20
(28)	[13]2,12;8,53,20, increasing	2,17;31,28,53,20, increasing	3,53;46,40
(29)	[14]2,12;26,40, increasing	2,17;49,15,33,20, increasing	3,54;40
(30)	[15]2,12;⸢44⸣,26,40, increasing	2,18;7,2,13,20, increasing	3,55;15,33,20
(31)	[16]2,13;2,13,20, increasing	2,18;⸢24⸣,48,53,20, increasing	3,55;33,20
	[17]2,13;20, increasing	2,18;⸢42,3⸣5,33,20, increasing	3,55;33,20

Between columns i and ii, written perpendicularly

? [0]2[?]5 ⸢... 3 ...⸣ you put down. Opposite 1,⸢57[?];57[?],58,...⸣

Column ii

P4' (Computing Λ from Φ)

61 [?][Opposite 2,0;55,55,33,20, decreasing, you put 19;55,33,20.] [1][(The amount) by which it is less than 2,0;55,55,33, 20, decreasing], until 2,0;38,8,53,[20, decreasing, you mul]tiply [times 1], add with 19;55,33,20, put down.

1 [2][Opposite 2,0;3]8,8,53,20, decreasing, [you put] 20;1[3,20. (The amount) by which it] is less than [2,0;38,8,53,20, decreasing], until 2,0;20,22,13,20, [3][decreasing, you multiply times 1], subtract from 20;13,20, put down.

2 Oppo[site 2,0;20,22],13,20, decreasing, you put 19;55,33,20. [4][(The amount) by which it] is less than [2,0;20,22,13, 20], decreasing, until 2,0;2,35,[33,20, decreasing, you multiply times 3], subtract from 19;55,33,20, put down.

3 [5][Opposite 2,0;2,35,33,20, decreasing, you put 19;2,13,20. (The amount) by which it is less than 2,0;2,35,33,20, decreasing, until 1,59;44,48,53,20, decreasing, you multiply times 5, subtract from 19;2,13,20, put down.]

4 [Opposite] 1,59;44,48,53,20, decreasing, [6][you put 17;33,20. (The amount) by which it is less than 1,59;44,48,53,20, decreasing, until 1,59;27,2,13,20, decreasing], you multiply times 7, subtract from 17;33,20, put down.

5 [7][Opposite 1,59;27,2,13,20, decreasing, you put 15;28,53,20. (The amount) by which it is less than 1,59;27,2,13,20, decreasing], until 1,59;9,15,33,20, decreasing, [8][you multiply times 9, subtract from 15;28,53,20, put down.]

6 [Opposite 1,59;9,15,33,20, decreasing, you put 12;48,53,20.] [9][(The amount) by which it is less than 1,59;9,15,33,20, decreasing, until 1,58;51,28,53,20, decreasing, you multiply times 11], subtract from 12;4⸢8,53⸣,20, put down.

7 [10][Opposite 1,58;51,28,53,20, decreasing, you put 9;33,20. (The amount) by which] it is less [than 1,58;51,28,53,20, decreasing], until 1,⸢58;33,42,1⸣3,20, decreasing, [11][you multiply times 13, subtract from 9;33,20, put down.

8 Opposite 1,58;33,42,13,20,] decreasing, you put 5;42,13,20. [12][(The amount) by which it is less than 1,58;33,42,13, 20, decreasing, until 1,58;15,55,33,20, decreasing], ⸢you multiply⸣ [times 14;8,45], subtract from 5;42,13,20, put down.

9–? [13][Opposite 1,58;15,55,33,20, decreasing,] you put down [1;30,44,26,40] as the duration of the year. [14][...] ⸢...⸣ [...] [15][...] ⸢...⸣ [16][...] ⸢45[?]⸣ you multiply, and [17][...] ⸢...⸣

P5' (Computations involving Λ?)

[18][...] ⸢...⸣ from year to year.

P6' (Computing Λ from Φ)

[19][...] ⸢... you multiply⸣, and

31 opposite 2,13;2,13,20, increasing, you put down 3,55;33,20. [20][...

20 [xxxxxxxxxxxxxxxx] ⸢xx⸣ 3.55.33.20 **gar** *ša*$_2$ *al* 2.15.27.2.13.20 **tab diri** GAM$_0$
21 [xxxxxxxxxxxxxxxxx] ⸢xxxxx⸣ **diri a.ra**$_2$ 3.22.30 ⸢**du**⸣
22 [xxxxxxxxxxxxxxxxxxxxx] ⸢x⸣ **u**$_4^?$**.meš** 1.⸢50 x **uš**$^?$⸣**.meš** ⸢1$^?$.50$^?$⸣
23 [xxxxxxxxxxxxxxxxxxxxxxxx] ⸢2$^?$.40⸣ **igi**-*u*$_2$ **nim** [xx]
24 [xxxxxxxxxxxxxxxxxxxxxxx]-*ma* **ta** 3.55.33.20 **n**[**im**] ⸢xx⸣ [xx]
25 [xxxxxxxxxxxxxxxxxxxxxx] *ana tar* 2.17.4.48.53.20 **tab**
26 [xxxxxxxxxxxxxxxxxxxxxxx] ⸢3⸣.55.33.20 **gar**-*an*
27 [*ša*$_2$ *al* xxxxxxxxxxxxxxxx **ta** 3.5]5.33.20 **nim** *ša*$_2$ **tag**$_4$
28 [xxxxxxxxxxxxxxxxxxxxx] *ana tar* 2.13.4.48.[53.20 **la**$_2$]
29 [xxxxxx **gar** xxxxxx *ša*$_2$ *al* 2.13.4.48.53.20 **la**$_2$] ⸢**la**$_2$ **en**⸣ 2.3.53.42.13.20
30 [**la**$_2$ GAM$_0$ 18.8.45 **du ta** xxxxxx **nim gar** *ana tar*]-⸢*ṣa*⸣ 2.3.5⸢3.42.13.20⸣ [**la**$_2$ x]
(about n lines missing)

Critical and philological notes

Obverse column i'

1' The signs are badly damaged.
2' 43: preceded by vague traces of two horizontal wedges.
3' 20: or 30 – 50. Between **diri** and **tab** there may be an erased sign.
7' Before 2-*u*$_2$ there is a sign similar to LA$_2$. 10: or any sign beginning with a winkelhaken. **du**$_3$: replaces LA$_2$(?) (*ACT*).
9'–11' LA$_2$ probably means 'length; duration' here (Akk. reading unclear; cf. the Glossary).

Column ii'

15' **a.ra**$_2$ 9: the 9 is written using the old form (elsewhere on the tablet the LB form is used).

Reverse column ii

1 One expects *ana tar* 2.0.55.55.33.20 **la**$_2$ 19.55.33.20 **gar** at the beginning of the line, but there is no place for it. Perhaps this is contained in the badly legible line 0 which is, unusually, written perpendicularly between columns i and ii?
tab-*ma*: *ma* omitted in *ACT*.
2 Before **la**$_2$ *a-di*: traces of a sign, perhaps **la**$_2$.
7 **en**: replaces **ta** (*ACT*).
14 The damaged sign before 20 looks like ŠU$_2$.
16 **du**-*ma*: or KA?
17 ⸢5⸣: or ⸢4⸣.
20 The signs before 3.55 might be ⸢**nu kur**⸣ (not 30.13 as suggested by Neugebauer).
21 3.22.30: replaces 3.22.20 (*ACT*).
22 ⸢x⸣ **u**$_4$**.meš**: or ⸢DU⸣**.meš**?
uš.meš: followed by traces similar to 1.50 or GAR 50.
23 2: or A or 5, less likely 8.
29 **la**$_2$ **en**: replaces *ša*$_2$ **tag**$_4$ (*ACT*).
30 *ṣa*: replaces 5 (*ACT*).

Commentary

This text is preserved on three fragments, which include segments of the lower (upper) of the obverse (reverse) and the right edge of the tablet. To the previously known fragments BM 35564+40081 (*ACT* 207d) a small fragment from the lower (upper) edge of the obverse (reverse) is added. It does not physically join the other fragments, but not much clay is missing between them. The tablet is unusually thick. The obverse and the upper half of the reverse are divided into two (main) columns. Column i is not preserved to full width. The textual restorations in P3' suggest that about 9 cm of clay are missing from the left side, so that the original width of the tablet was probably about 23 cm. Column ii of the obverse contains an auxiliary table, T1', which continues on the reverse in column i (the right column). On the reverse there are further procedures in column ii to the left of T1' (P3'), and below T1' (P4',P5'). It is unclear whether P4' and P5' are confined to column i or were written across the entire tablet.

32–37a Opposite 2,15;27,2,13,20, increasing,] you put down 3,55;33,20. (The amount) by which it exceeds 2,15;27,2,13, 20, increasing, [21][you multiply] by [...], you multiply it by 3;22,30, [22][...] ⌜...⌝ days? 1.⌜50 ... 1.50?⌝ [23][...] ⌜2.40⌝ previous ... [...] [24][...] and you subtract it from 3,55;33,20. ⌜...⌝ [...] [25][...]

37b–49? Opposite 2,17;4,48,53,20, increasing, [26][...] you put down 3,55;33,20. [27][(The amount) by which ...] you subtract [from 3;5]55,33, 20, what remains [28][...

50 Opposite 2,13;22,35,33,20, decreasing, you put down 3,4;53,20. ...] Opposite 2,13;4,48,[53,20, decreasing] [29][you put down ... (The amount) by which it] ⌜is less⌝ [than 2,13;4,48,53,20, decreasing], ⌜until⌝ 2,3;53,42,13,20, [30][decreasing, you multiply by 18;8,45, deduct from ...

51 Oppo]⌜site⌝ 2,3;5⌜3,42,13,20⌝, [decreasing, you put down 12;50,22,13,20. ...]
(about n lines missing)

P1' Computing G from Φ, scheme 1?

The remains of P1' are sufficient to establish that it is concerned with the computation of G from Φ (§4.4.11). Perhaps a preceding auxiliary table, now lost, contained a complete scheme 1 in tabular form analogous to the $\Phi\Lambda$ scheme in T1'. The multiplication by $0;3,22,30 = 1/\phi$ (O4') and the 'accumulation' mentioned in O5', on the new fragment, suggest that this is the *malûššu* rule (p. 148). If P1' belongs to scheme 1, then the addition of the difference for G to $G = 2,53;20$ in O5' implies that this concerns intervals 10–14. Parallels of the *malûššu* rule being used for these intervals are not attested. Since c_k changes by 1 from interval to interval across intervals 10–14, the *malûššu* formulation could be used, but it would require a modification, since interval 10 is anomalously long.

P2' Computing G from Φ, scheme 1, interval 35?

Below Oi'5' there is a dividing line followed by what is probably the interpolation rule for interval 35 of scheme 1.

P3' Net differences of C'

The newly discovered fragment adds a substantial amount of text to this procedure, which had hitherto defied interpretation. It is now clear that P3' contains an algorithm for computing net differences of C' (§4.4.5), say $dC'_k(s)$, in accordance with Eq. (4.24) on p. 132. In its original state P3' must have contained rules for all interpolation intervals of C (cf. Table 4.11, p. 131), but only 4 rules, pertaining to the signs Cnc–Sco, are preserved. The remaining 8 rules would require as many lines of text, which must have been written at the beginning of column ii'.

T1' Computing Λ from Φ

This auxiliary table preserves a section of the interpolation scheme for computing Λ from Φ (§4.4.13). Since the previous edition (*ACT*) predates the complete reconstruction of the $\Phi\Lambda$ scheme by Aaboe (1968), Neugebauer's commentary must be revised in many details. Neugebauer did incorporate Aaboe's results into *HAMA*, pp. 511–512, but he did not explore the consequences for the present text. It can be assumed that T1' contained interpolation rules for all intervals of the scheme. Since T1' ends with interval 31, intervals 32–50 are probably to be restored above Oii'0, which would require about 19 lines of text.

T1' consists of two parts. The first part, Oii'1'–22', contains a complete sequence of interpolation rules for intervals 52–61 and 1–12, providing the control values Φ_k, Λ_k, and the interpolation coefficient c_k. Separate instructions for subintervals 10a and 10b, which are located on either side of $\Phi_{10b} = m_\Phi$, are not provided. Control point 12 (Oii'22') marks the beginning of interval 12, which has a length 32ϕ. One would expect the table to jump from here to Φ_{13}, where c_k changes, but in fact the table continues to step through Φ at intervals of ϕ for 2 more lines (Oii'23'–24'). Oii'24'–5' partly preserve the phrase 'between Φ_{12} until Φ_{13}', apparently a reference to the fact that $c_{12} = 18;8,45$ throughout this anomalously long interval.

In Oii'25' the table's content changes abruptly: from here on each line contains 2 values of Φ followed by Λ in the third subcolumn. The interpolation coefficients c_k are omitted, perhaps due to a lack of space. This part of T1' was not fully understood in the time when *ACT* was written, but its interpretation no longer poses any problems.[674] The first column of Φ values has the same meaning as before. The second Φ column contains the Φ value shifted by ϕ which, together with the previous column, is used for deriving Λ_k in accordance with the algorithm reconstructed by Aaboe. The third column

[674] Cf. Aaboe (1968); *HAMA*, pp. 511–512. For this algorithm cf. Volume II.

contains the control value Λ_k, thus continuing the second column of the preceding part. The Φ value in Ri17 is located within interval 31, at a distance ϕ beyond Φ_{31}.

P4' Computing Λ from Φ, intervals 61, 1–9

P4' contains instructions for computing Λ from Φ in some of the same intervals dealt with in T1'. The new fragment adds some of the missing text in Rii1–5. As far as the instructions are preserved, their formulation is based on template IP.B.2 (§2.6.7). On that assumption the instructions for intervals 61 and 1–8 have been restored, but in some lines (Rii5–6) there is little room for this. Especially in Rii1 there is insufficient room for the expected initial phrase 'Opposite ... you put down ...'; perhaps it was written in the badly legible line written perpendicularly between columns i and ii? Rii13–14 may have contained interval 9. The 'duration of the year' mentioned in that line is probably the technical term for Λ. Rii15–17 may contain further interpolation rules, presumably for intervals 10–12, perhaps also 13–30, but the text is too damaged for a reconstruction. If present, the interpolation algorithms for these intervals must have been formulated in terms of a *malûššu*-type rule (cf. §4.4.11).

P5' Computations involving Λ?

This procedure consisting of a single line may also concern Λ, as suggested by the preserved text.

P6' Computing Λ from Φ, intervals 31–

P6' continues with interpolation rules for computing Λ from Φ. An unknown number of lines are missing from the end of P6'. Some of the preserved text is too damaged for a complete reconstruction. Rii19 deals with interval 31. Little remains of Rii20–24, but the preserved numbers, Φ_{32}, G_{32} and $1/\phi = 3;22,30$, and the subtraction in Rii24, are sufficient to conclude that this must be a *malûššu*-type rule for intervals 32–37a. This is followed in Rii25–27 by what is probably a *malûššu*-type rule for intervals 37b–49. Rii25 mentions $\Phi = 2,13;4,48,53,20(\downarrow)$, which lies at a distance of ϕ beyond Φ_{50}, in the anomalously long interval 50, where $c_{50} = 18;8,45$. Hence the gap at the beginning of this line probably contained the interpolation rule for the interval beginning at Φ_{50}.

No. 80

museum number	BM 36887 (80–6–17,628)
provenience	Babylon
measures	5.5(*) × 6.4 × 1.5 cm (type E or H)
arrangement	O/R
cuneiform text	photo: Fig. F.88
contents	Moon system A: computations involving eclipse magnitude

Obverse

P1 1 [...xx] 6.40 GAM$_0$ ⸢6 x **du**⸣
2 [...xx] 20 **ki** ⸢ḪAB? *ša*$_2$?⸣ AN.MI
3 [...xx] ⸢x⸣ *ma-lak* GAR.GAR-*ma*
4 [...x *mi*?]-⸢*nu*?⸣-*u*$_2$ **ki** *ša*$_2$ ḪAB? 40 ⸢x⸣
5 [...x A]N?.MI **e**$_{11}$ *ša*$_2$ **tag**$_4$
6 [...x] ⸢40⸣ **e**$_{11}$ 4 **ta** *lib*$_3$*-bi* **nim**
7 [...] ⸢**ta** *zib*⸣ **en absin**$_0$ 44.31.56.40
8 [...**ta absin**$_0$ **en**] ⸢*zib*⸣ 46.26.40
9 [...] ⸢x⸣ 10 *ki-i* **la**$_2$ **ta** *zib*
10 [**en absin**$_0$ ...] ⸢**ta**⸣ **absin**$_0$ **en** *zib*
11 [...xx]-*u*$_2$ 11.40 *u*$_3$
12 [...xx] 32 ⸢xxx⸣

Lower edge

13 [...] 5

Reverse

14 [...xxx] 40 **ta ša**$_3$ **nim**
15 [...xxx] ⸢x⸣ **mu** *ana* **mu ta** ⸢*zib*⸣
16 [**en absin**$_0$ ...x] ⸢x⸣.40 **tab ta absin**$_0$ **en** *zib*
17 [...] **tab** *mim-ma ša*$_2$ BE **u**$_2$? *u*$_3$? **gal**
18 [...] ⸢AN?.MI⸣ ḪAB **ki**? 32 **tab**

Obverse

P1 **(Computations involving eclipse magnitude?)**

[1][...] ⸢you multiply⸣ 6.40 by ⸢6...⸣ [2][...] 20 with ⸢the 'disk' of⸣ the eclipse. [3][...] ⸢...⸣ the path you accumulate, and [4][... What]ever the position? of the 'disk' 40 ⸢...⸣ [5][...] you deduct [from the ...] of the eclipse, what remains [6][...] you deduct [from ...].⸢40⸣, you subtract 4 from it. [7][...] ⸢from Psc⸣ until Vir 44;31,56,40. [8][... from Vir until] ⸢Psc⸣ 46;26,40. [9][...] ... 10 if decreasing, from Psc [10][...]⸢from⸣ Vir until Psc [11][...] ... 11.40 and [12][...] 32 ⸢...⸣

Lower edge

[13][...] 5

Reverse

[14][...] you deduct 40 from it; [15][...] ⸢...⸣ from year to year, from ⸢Psc⸣ [16][until Vir ...] you add [...].40, from Vir to Psc [17][...] you add. Whatever ... [18][...] ⸢eclipse⸣, you add the 'disk' with? 32.

Critical and philological notes

Obverse

1 The damaged sign before **du** looks like **ab**$_2$ = *arḫu*, 'month'.
2 ḪAB: or **gur**?
3 Before *ma-lak* the right half of a sign similar to ḪAB or **gur** is visible.
4 **ki**, 'position': or *ki* = *kī*, 'when'?
9 Before 10 there are remains of a winkelhaken.
12 After 32 there are traces of signs similar to 11.20.40.

Reverse

15 Before **mu** *ana* **mu** there are traces of a sign similar to **tab**.

Commentary

This fragment represents the right half of a small tablet. The left edge of the tablet is missing. The lower half of the reverse is uninscribed.

P1 Computations involving $\Psi^{(3)}$

The single procedure on this tablet is difficult to interpret due to the bad state of preservation and a lack of parallels. A comprehensive reconstruction of the algorithms has not been possible so far. What is clear is that the computations deal with eclipse magnitude. The number 44;31,56,40^f (O7), also attested in No. **59**, is tentatively interpreted as a value of the monthly difference of a hitherto unrecognised eclipse function $\Psi^{(3)}$; for a discussion cf. §4.4.7. The number 46;26,40^f (O8) is presumably also a value of that function.

No. 81

museum number	BE 15557
provenience	Babylon
measures	10(*) × 11(*) × ? cm
arrangement	O/R
cuneiform text	photo: Fig. F.89 = *ACT*, Pl. 243
bibliography	*ACT* 207e
content	Moon system A: computing Λ from Φ; unidentified procedures

		Side X	
		Columns i'-ii'	
T1'		(m lines missing)	
	0'	[2.0.55.55.33.20	19.55.33.20]
	1'	[2.0.38].8.53.20	2[0.13.20]
	2'	[2.0.20.2]2.13.20	19.5[5.33.20]
	3'	[2.0.2.3]5.33.20	19.2.[13.20]
	4'	[1.59.44.4]8.53.20	17.3[3].30
	5'	[1.59.27].2.13.20	15.28.53.[20]
	6'	[1.59.9.1]5.33.20	12.4[8.5]3.2[0]
	7'	[1.58.51.2]8.53.20	9.[3]3.20
	8'	[1.58.33.42].13.20	5.4[2].13.20
	9'	[1.58.15.55.33.20] **la**$_2$	1.30.44.26.40 **la**$_2$
	10'	[1.57.58.8.53.20] ⸢**la**$_2$⸣	2.58.31.6.40 **tab**
	11'	[1.57.55.33.2]0	7.45.33.20
	12'	[1.58.13.20]	12.50.22.13.20
	13'	[1.58.31.6.4]0	18.12.57.46.40
	14'	[1.58.48.53.2]0	23.35.33.20
	15'	[xxxxx]2.13.4.48.53.20	3.4.53.20
	16'	[2.13.22.3]5.13.20	3.10.13.20
	17'	[2.13.40.2]2.13.20	3.15.15.33.20
	18'	[2.13.58.8.53].20	3.20
	19'	[2.14.15.55.33.20]	[3.2]4.26.40
	20'	[2.14.33.42.13.20]	[3.28.3]5.33.20
	21'	[2.14.51.28.53.20]	[3.32.26.40]
		(n lines missing)	
		Column iii'	
P1'		(about m+8 lines missing)	
	1'	⸢x⸣ [...]	
	2'	2 [...]	
	3'	20 [...]	
	4'	2.10[...]	
	5'	**du** [...]	
	6'	30 [...]	
	7'	⸢x⸣ [...]	
	8'	⸢x⸣ [...]	
	9'	10$^?$ [...]	
		(about n+2 lines missing)	
		Side Y	
		Column i'	
P2'		(about n+6 lines missing)	
	7'	**igi** [...]	
	8'	**a** [...]	
	9'	**ki** [...]	
	10'	*ša*$_2$ 5 [...]	

T1' (Computing Λ from Φ)

Side X, columns i'–ii'

(m lines missing)

61	0'[2,0;55,55,33,20	19;55,33,20]
1	1'[2,0;38],8,53,20	2[0;13,20]
2	2'[2,0;20,2]2,13,20	19;5[5,33,20]
3	3'[2,0;2,3]5,33,20	19;2,[13,20]
4	4'[1,59;44,4]8,53,20	17;3[3],30 (error for 17;33,20)
5	5'[1,59;27],2,13,20	15;28,53,[20]
6	6'[1,59;9,1]5,33,20	12;4[8,5]3,2[0]
7	7'[1,58;51,2]8,53,20	9;[3]3,20
8	8'[1,58;33,42],13,20	5;4[2],13,20
9	9'[1,58;15,55,33,20], decreasing	1;30,44,26,40, subtractive
10	10'[1,57;58,8,53,20], ⸢decreasing⸣	2;58,31,6,40, additive
11	11'[1,57;55,33,2]0	7;45,33,20
12	12'[1,58;13,20]	12;50,22,13,20
	13'[1,58;31,6,4]0	18;12,57,46,40
	14'[1,58;48,53,2]0	23;35,33,20
(13)	15'[...] 2,13;4,48,53,20	3,4;53,20
(14)	16'[2,13;22,3]5,13,20	3,10;13,20
(15)	17'[2,13;40,2]2,13,20	3,15;15,33,20
(16)	18'[2,13;58,8,53],20	3,20
(17)	19'[2,14;15,55,33,20]	[3,2]4;26,40
(18)	20'[2,14;33,42,13,20]	[3,28;3]5,33,20
(19)	21'[2,14;51,28,53,20]	[3,32;26,40]

(n lines missing)

Column iii'

P1' (Subject unclear)

(about m+8 lines missing)

1'⸢...⸣ [...] 2'2 [...] 3'20 [...] 4'2.10[...] 5'... [...] 6'30 [...] 7'⸢...⸣ [...] 8'⸢...⸣ [...] 9'10? [...]

(about n+2 lines missing)

Side Y, column i'

P2' (Subject unclear)

(about n+6 lines missing)

7'... [...] 8'... [...] 9'... [...] 10'for 5 [...]

P3' 11' *ana* ⸢*tar*?⸣ [...]
12' 1 ⸢x⸣ [...]
13' *mim-ma* [...]
14' **du** [...]
Column ii'
P4' (about n lines missing)
1' [... 1]1.50 *ana* 6 [xxxxxxx]
2' [...] ⸢x⸣ 7.28.46.4[0 xxxxx]
3' [...] ⸢xx⸣ **iti.ne** 5 *e*? x [xxx]

P5' 4' [...xx] 2.7.19.5.23.20 **tab** x [xxx]
5' [...] ⸢5?⸣.10.12.18.53.20 **tab** 2.28 x *ana* 6 [xx]
6' [...]-*ka* 2.7.26.23.20 **la**$_2$-*u*$_2$ **en** 10 [xx]
7' [...] ⸢x⸣ 4.40.27.46.40 **la**$_2$-*u*$_2$ 11.44 x *ana* 6 ⸢**iti**⸣
8' [...] 8.27.46.40 **tab** *ana* 6 **iti** 7.6 **tab**
9' [...] 49.10.55.33.20 **tab** *ana* 6 **iti** 7.6 **tab**
10' [...] **gi**$_6$ **babbar sa**$_5$ **sig**$_7$
11' [...] 1 2 3 4

P6' 12' [...] ⸢4?⸣1.26.4.41.15 **tab** 7.6 **tab** *ana* 6 **iti**
13' [...] ⸢x⸣.2? 30.29.31.52.30 **tab** *ana* 6 **iti**
14' [...] 4.12 29.31.52.30 **tab** 2.28 **tab**
15' [...x] 45.54.22.30 7.6 **tab** *ana* 6 ⸢**iti**⸣
16' [...xx] 29.31.52.30 11.44 [xx]
17' [...xx 2]9.31.52.30 11.[44 xx]
(about m lines missing)

Critical and philological notes

Side X

4' 17.33.30: one expects 17.33.20.

15' 2.13.4.48.53.20 is displaced to the right. It is unclear what was written in the preceding gap; perhaps a remark about the anomalous interval between this and the preceding line (cf. the commentary)?

16' 22.13.20: one expects 22.13.30.

Side Y, column ii'

1'–17' The correct separation into numbers and digits is uncertain.

3' E: doubtful. The following sign cannot be recognised from the photo.

4'–9' 'increasing', or 'additive'?; 'decreasing', or 'subtractive'?

4' The sign following **tab** cannot be recognised from the photo.

5' The sign following 2.28 might be **en**.

7' The first sign cannot be recognised from the photo.

12' The number before 1 at the beginning of the line is 10, 20, 30, 40 or 50.

13' The number may be 13;30,29,31,52,30; cf. the commentary.

Commentary

This is the only known tablet with mathematical astronomy that was found during the German excavations in Babylon led by R. Koldewey. It was excavated in 1902 in the area Amran, near the city wall towards the Euphrates (§1.3.1). The current location of the tablet is unknown, so it could not be collated.[675] The tablet's width and height were estimated from the photographs; its thickness is unknown. No edges are preserved. The obverse contains an auxiliary table (T1') consisting of two columns, followed by traces of a procedure; the reverse contains three more procedures.

T1' Computing Λ from Φ

The beginning of this auxiliary table is missing. The two partly preserved columns contain parameters for the $\Phi\Lambda$ interpolation scheme (cf. §4.4.13, p. 153, and Appendix D.2, p. 531). Another column, now destroyed, might have contained the interpolation coefficients c_k. However, the few remaining signs of column iii' do not appear to be aligned with the rows of columns i'–ii', suggesting that they belong to a different procedure or table (cf. P1').

[675] It is probably kept in the Eski Şark Eserleri Müzesi (Istanbul) or in the Iraq Museum (Pedersén 2005).

P3' (Subject unclear)

11'⸢Opposite?⸣[...] 12'1 ⸢...⸣ [...] 13'Whatever [...] 14'... [...]

Column ii'

P4' (Subject unclear)

(*about n lines missing*)
1'[... 1]1.50 for 6 [months ...] 2'[...] ⸢...⸣ 7.28.46.4[0 ...] 3'[...] ⸢...⸣ month V 5 ... [...]

P5' (Computations involving a function *y*)

4'[...] 2.7.19.5.23.20, increasing?, ... [...]
5'[...] 5.10.12.18.53.20. 2.28 ... For 6 [months...]
6'[...] your [...] 2.7.26.23.20, decreasing?, until 10? [...]
7'[...] ⸢...⸣ 4.40.27.46.40, decreasing?, 11.44 .... For 6 ⸢months⸣.
8'[...] 8.27.46.40, increasing?. For 6 months you add 7.6.
9'[...] 49.10.55.33.20, increasing?. For 6 months you add 7.6.
10'[...] black white red green
11'[...] 1 2 3 4.

P6' (Computations involving *y* and *F*?)

12'[... 4]1.26.4.41.15, increasing. You add 7;6 for 6 months.
13'[...] 2.30.29.31.52.30 you add for 6 months.
14'[...] you add 4.12 29.31.52.30. You add 2;28 15'[... ...] 45.54.22.30 7;6 you add for 6 mo[nths]. 16'[...] 29.31.52.30 11;44 [...] 17'[... 2]9.31.52.30 11;[44 ...]
(*about m lines missing*)

As in No. **79**, the auxiliary table consists of two parts. The first part (X1'–12') contains Φ_k and Λ_k for intervals 61, 1–12. One would expect T1' to contain only parameters pertaining to the control points 1–61, but the anomalous interval 12, which has a length 32ϕ, is represented by two additional entries beyond Φ_{12}, located at distances ϕ and 2ϕ, respectively (X13'–14'), even though c does not change there. After this Φ jumps by 30ϕ to Φ_{13} (X15'). From here on the Φ values no longer coincide with the control values Φ_k as tabulated in Appendix D.2, since they are shifted by the amount ϕ, as in No. **79** T1'. Although it is possible that T1' differs from the known interpolation scheme, it is more likely that the shifted Φ values were used for deriving the interpolation scheme in accordance with the reconstruction by Aaboe (1968), as was argued for No. **79** T1'. This is also suggested by the inclusion of some Φ values where the interpolation coefficient (presumably) does not change.

***P1'–P4'** Subject unclear*

In these procedures only the first 1 or 2 signs can be read in some lines, which is insufficient to establish their subject. Perhaps P3' contains an interpolation scheme. The beginning of P4' is missing; Yii'1' appears to mention a difference or net displacement 'for 6 months', as in P5' and P6'. The numbers in P4' could not be identified.

***P5'** Computations involving a function y*

The subject of P5' remains largely unclear. Neugebauer identified the numbers 2.28 (Yii'5'), 7.6 (Yii'8',9') and 11.44 (Yii'11') as the minimum, mean value and maximum of a zigzag function, attested in the early Seleucid synodic table MNB 1856,[676] whose astronomical interpretation remains unknown. The period (P) of this function, 14^{m}, is the same as that of $F^{(5)}$, a zigzag function representing the Moon's daily displacement (§4.4.9) also represented by a column in MNB 1856. Hence function y represents the lunar variation (§4.1.2) of some quantity. The repeatedly occurring phrase 'for 6 months' points to net differences for 6^{m}. Yii'8' seems to imply that $\mu = 7.6$ is also the net difference for 6^{m}. The meaning of the other numbers is unclear, but the sequence of digits in Yii'9' is reminiscent of Φ, G and Λ. The four colours and the four numbers mentioned in Yii'10'–11' are enigmatic.

***P6'** Computations involving y and F?*

Only the first 6 lines of P6' are partly preserved. As in P5', several parameters of function y are mentioned. At least one of the other parameters points to $F^{(1)}$ (§4.4.9), since 30.29.31.52.30 (Yii'13') can be interpreted as the final digits of $\mu = 13;30,29,31,52,30$ (Table 4.18, p. 143). The number 4.12 (Yii'14') might be $4;12 = dF(6)$ (Table 4.19, p. 144). A satisfying interpretation of this procedure remains to be found.

[676] For a critical edition cf. Neugebauer & Sachs (1969), Text H, and Volume II.

No. 82

museum numbers	BM 36705 (80–6–17,437) +36725 (458) +37475 (1232) +37484 (1241)
provenience	Babylon
measures	9.5 × 18.0(18.5?) × 2.7 cm (type D)
arrangement	O/R
cuneiform text	photo: Figs. F.92, F.93; Neugebauer (1957b), Pls. I–II (BM 36705+36725); Aaboe (1968), Pl. V (BM 36705+36725+37484)
bibliography	BM 36705+36725: Neugebauer (1957b); Aaboe (1968), pp. 35–38
contents	Moon system A: computations involving *Φ* and *G*

		Obverse
P0		(perhaps 1 line missing)
	-1'	[xxxxxxxxxxxxxxxxx] ⸢*u*$_3$⸣ 20 [xx]
P1	0'	[xxxxxxxxxxx **a.ra**$_2$] 1.5.⸢4⸣.22.30 **du**-⸢*ma*⸣
	1'	[xxxxxxxxxxxx 2.1]⸢3⸣.20 *a-na* 1.58.31.6.40 **m**[**u.meš**]
	2'	[xxxxxx *ša*$_2$] ⸢**e**?-*ma*⸣ 14.48.53.20 **mu.m**[**eš** x] 17.46.40 *taš-pil-tu*$_2$
	3'	⸢*ša*$_2$ 18⸣ **mu.meš** *ša*$_2$ **e**-*ma* 13.39.35.36 ⸢x⸣ [xx]6.15 *šu-u*$_2$
P2	4'	17.46.40 **a.ra**$_2$ 9.20 **du**-*ma* 2.45.55.⸢3⸣[3.20 x] ⸢xx⸣ 13.46.38.15 **me**
	5'	**ta** *muḫ-ḫi* **zi** *sin* **gal en** *muḫ-ḫi* **zi** *sin* **tur** ⸢2?⸣ [xxxxxx] ⸢xx⸣
	6'	1.22.39.49.30 *si-man šu-u*$_2$ *ina* 18 **mu.meš** ⸢x⸣ [xxxxxx] ⸢x⸣
	7'	*sin a-*⸢*na*⸣ *šamaš*$_2$ **kur** *u*$_3$ *ina* 18 **mu.meš** ⸢1? 3?.16.20?⸣ [xxxx]
	8'	**ta** 3.13.21.4 **danna nim**-*ma re-ḫi* [xxxxx *mi-nu-u*$_2$ **a.ra**$_2$]
	9'	1.5.4.22.30 **du**-*ma lu* 1.22.39.⸢4⸣[9.30 1.16].⸢13.10?⸣.[11.24.36]
	10'	**a.ra**$_2$ 1.5.4.22.30 **du**-*ma* 1.22.3⸢9.49.30 x 20?⸣.13 [xxx]
	11'	**ta** *muḫ-ḫi* **zi gal en** *muḫ-ḫi* **zi tur** 1.16.13.10.⸢11⸣.2⸢4⸣.[36]
	12'	**a.ra**$_2$ 9.20 **du**-*ma* 11.51.22.55.6.29.36 **a.ra**$_2$ 10 [**du**-*ma*]
	13'	1.58.{39}.33.49.11.4.56 27.33.16.30 (blank) [xx?]
	14'	*u*$_3$ 1.58.33.49.11.4.56 *a-na muḫ-ḫi a-ḫa-miš*$_2$ GAR.GAR-[*ma*]
	15'	29.31.50.19.11.4.56 **šid ab**$_2$ **kur** *ina* **zi murub**$_4$ **ab**$_2$ *a-na* **ab**$_2$ [xx]
	16'	*ina* 29 **me** *ina* 3.11.1.⟨55⟩.6.29.36 *si-man sin a-na šamaš*$_2$ **kur** *si-man* **gar** *a-na tar-ṣi* [xx]
	17'	1.53.56.47.24.2.40 *si-man* **gar** (erasure) *a-na tar-ṣi* 1.58.3[0 xxxxxxxx]
	18'	4.28.5.25.45.9.30 2.45.55.33.20 *ša*$_2$ **ab**$_2$ **a.ra**$_2$ 9.20 **du**-*m*[*a*]
	19'	25.48.38.31.6.40 **tab** *u* **la**$_2$ *ša*$_2$ *si-man*.**meš** 2.45.55.33.20 [x?]
	20'	**a.ra**$_2$ 7 **du**-*ma* 19.21.28.53.20 19.16.51.6.40 **ta ša**$_3$ **nim**-*m*[*a*]
	21'	*re-ḫi* 4.37.46.40 $\frac{1}{2}$-*šu*$_2$ GIŠ-*ma* 2.18.53.20 2.18.53.20
	22'	**ta** 2.45.55.33.20 {**ta ša**$_3$} **nim**-*ma re-ḫi* 2.43.{42.43.}36.40
	23'	**a.ra**$_2$ 9.20 **du**-*ma* 25.27.2.13.20 (blank) 25.27.2.13.20
	24'	*u*$_3$ 51.37.17.2.13.20 GAR.GAR-*ma* 1.17.4.19.⸢15⸣.33.20
	25'	**ki** ⸢x⸣ 3.11.1.6.29.26 (blank) *ša*$_2$ ⸢*si*⸣-[*man*? **tab**?]
P3	26'	**ta zi gal en zi tur** 17.46.40 **a.ra**$_2$ 12.38.45 **du**-*ma* **zi** [x]
	27'	17.46.40 **a.ra**$_2$ 50 **du**-*ma* **ta** *muḫ-ḫi* **zi** *sin* **gal en** *muḫ-ḫi* **z**[**i** *sin* **tur**]
	28'	**ta tur en gal** 17.46.40 **a.ra**$_2$ 2.25.37.30 **du**-*ma* **zi tur** UL LA NI
P4	29'	[xxxx] ⸢x 3.1.3?⸣2 ⸢**ki du** *u*$_3$?⸣ **zi**-*šu*$_2$ BE **ta zi**-*šu*$_2$ **gal**
	30'	[xxx] ⸢x **ta**⸣ [xx] 4.55.46.40 **me** 1.0.46.1.30 **du**
	31'	[xxx] ⸢3?⸣5.32.35 **me** ⸢x⸣ [xx] 30 **du** *ina* 5.6.9.53.20 **me**
	32'	[xxx 1.0].⸢4⸣6.1.30 **du** ⸢xxx⸣ [xx].⸢4?⸣8.45 **me** 2.1.32 **ki du**
	33'	[xxxxxxxx] ⸢xxxx **zi**⸣ *ina* **ki zi** *bi*?-*rit* ⸢x⸣ [xx]
		Reverse
P5	1	[xx] ⸢x **nim**⸣ $\frac{1}{2}$ **uš** *ina* 30 ⸢x⸣ [xxx] 31.16.20 ⸢**zi**⸣ 9.6.26 ⸢2?⸣ [xxx]
P6	2	[xxxx] ⸢3?⸣2.56 *ina* **zi murub**$_4$ ⸢*ina*?⸣ [x? 1.16.13.10].11.24.36 **zi** *ina* **ki zi** ⸢x⸣ [xx]
	3	[xxxx] ⸢x 5?⸣5.21.53.4⸢2⸣ [xx z]**i gal** *ina* 1.3.30.58.29.30.30 **zi**
	4	[xxxx] ⸢3?.5?⸣3.48.1⸢2?⸣ [xxx 1]8 **mu.meš** *si-man ša*$_2$ **gal** 2.13.20

Obverse

P0 **(Subject unclear)**

(*perhaps 1 line missing*)
$^{-1'}$[...] ⌜and?⌝ 20 [...]

P1 **(Computations involving Φ and Φ_{tr}; net difference $d\Phi(223)$; partly unclear)**

$^{0'}$[...] you multiply [... by] 1,5;⌜4⌝,22,30 and $^{1'}$[... 2,13];20 to 1,58;31,6,40, th[is] $^{2'}$[... which] you have predicted?, it is 14;48,53,20. Thi[s ...] 0;17,46,40, the difference $^{3'}$for 18 years which you have predicted?. 13;39,35,36 ⌜...⌝ [...] this is [...].6.15.

P2 **(Computations involving Φ and G)**

a **(Relation between ϕ and d_{Φ})**

$^{4'}$You multiply 0;17,46,40 by 9;20, it is 2;45,55,3[3,20]

b **(Computing the mean synodic month from the anomalistic month)**

[...] ⌜the Moon?⌝ 13;46,38,15 days $^{5'}$from the Moon's largest displacement until the Moon's smallest displacement. ⌜2?⌝ [...] ⌜...⌝ $^{6'}$1,22,39;49,30, this is the duration. In 18 years ⌜...⌝ [...] $^{7'}$the Moon reaches the Sun and in 18 years ⌜1? 3?.16.20?⌝ [...] $^{8'}$you subtract from 3.13.21.4^{b}, there remains [...] [What] $^{9'}$should I multiply [by] 1,5;4,22,30 so that it is 1,22,39;4[9,30]? You multiply [1,16];⌜13,10⌝,[11,24,36] $^{10'}$by 1,5;4,22,30, it is 1,22,39;⌜49,30. ... 20?⌝,13 [...] $^{11'}$from the largest displacement until the smallest displacement. You multiply 1,16;13,10,⌜11⌝,2⌜4⌝,[36] $^{12'}$by 9;20, it is 11,51;22,55,6,29,36, [you multiply] it by 0;10, [it is] $^{13'}$1;58,{39},33,49,11,4,56. 27;33,16,30 [...] $^{14'}$and 1;58,33,49,11,4,56 you accumulate together, [it is] $^{15'}$29;31,50,19,11,4,56, the number (of days) of the month for (the Moon) reaching (the Sun) with the mean displacement, month by month [...]. $^{16'}$In 29 days in 3,11;1,⟨55⟩,6,29,36, the duration, the Moon reaches the Sun.

c **(Control values of Φ and G)**

You put down the duration. Opposite [...] $^{17'}$you put down 1.53.56.47.24.2.40 as the duration.
Opposite 1;58,3[3,49,11,4,56 ...] $^{18'}$4,28;5,25,45,9,30.

d **(Computations involving differences and net differences of Φ and G)**

You multiply 2;45,55,33,20 per month by 9;20, it is $^{19'}$25;48,38,31,6,40, the addition and subtraction for the durations. $^{20'}$You multiply 2;45,55,33,20 by 7, it is 19;21,28,53,20. You deduct 19;16,51,6,40 from it, $^{21'}$there remains 0;4,37,46,40. You compute half of it, it is 0;2,18,53,20. You deduct 0;2,18,53,20 $^{22'}$from 2;45,55,33,20 {from it}, there remains 2;43,{42.43.}36,40. $^{23'}$You multiply it by 9;20, it is 25;27,2,13,20.

e **(Computations involving G)**

$^{24'}$You accumulate 25;27,2,13,20 and 51;37,17,2,13,20, it is 1,17;4,19,15,33,20. $^{25'}$[You add it] with ⌜...⌝ 3,11;1,6,29,26 (error for 3,11;1,55,6,29,36) for the ⌜dur⌝[ation?].

P3 **(Computations involving Φ_{tr})**

$^{26'}$From the largest displacement until the smallest displacement, you multiply 0;17,46,40 by 12;38,45, it is the [...] displacement. $^{27'}$You multiply 0;17,46,40 by 50, and from the Moon's largest displacement until the Moon's [smallest displacement], $^{28'}$from the smallest until the largest, you multiply 0;17,46,40 by 2;25,37,30, it is the smallest displacement ...

P4 **(Computations involving Φ?)**

$^{29'}$[...] ⌜... 3.1.3?⌝2 ⌜the position proceeds and?⌝ its displacement ... from its largest displacement $^{30'}$[...] ⌜... from ...⌝ [In ...] 4.55.46.40 days it proceeds 1.0.46.1.30. $^{31'}$[...].⌜3?⌝5.32.35 days ⌜...⌝ it proceeds [...].30. In 5.6.9.53.20 days $^{32'}$it proceeds [...1.0].⌜4⌝6.1.30 ... ⌜...⌝.[...]⌜4?⌝8.45 days the position proceeds 2.1.32. $^{33'}$[...] ⌜... displacement⌝ subtract from the position, the distance between? ⌜...⌝ [...]

Reverse

P5 **(Subject unclear)**

^{1}[...] ⌜...you subtract?⌝; half a degree?; in 30 ⌜...⌝ [...] 31.16.20, ⌜the displacement⌝, 9.6.26 ⌜2?⌝ [...]

P6 **(Computations involving Φ_{tr}; partly unclear)**

^{2}[...] ⌜3⌝2.56 in the middle displacement ⌜in?⌝ [... 1,16;13,10],11,24,36, the displacement. In the position of the displacement? ⌜...⌝ [...] ^{3}[...] ⌜... 5⌝5.21.53.4⌜2⌝ [...] the largest displacement. In 1,3,30;58,29,30,30, the [...] displacement, ^{4}[...] ⌜3?.5?⌝3.48.1⌜2?⌝ [... 1]8 years, the duration for the maximum of 2,13;20 (= Φ).

P7	5	[xxx] ⸢xx⸣ [xx] ⸢35?⸣ [xx] 8 (blank) ⸢x⸣
P8	6	23.52.13.20 *ša*$_2$ 1⸢8 **mu**⸣.[**meš** xxxxx 1]4.4.26.40 (blank) 7 **ab**$_2$
P9	7	34.24.51.21.28.53.⸢20⸣[xxx *š*]*a*$_2$ *si-man*.**meš** *ša*$_2$ **ab**$_2$ *ana* **ab**$_2$
	8	[x]3.33.56.2.57.4⸢6.40?⸣[xx *š*]*a*$_2$ **ki** *si-man* **tab**
P10	9	⸢4⸣.53.47.29.40.44.26.40 *s*[*i-man* xxx] 18.16.5.19.15.33.20 *si-man*
P11	10	[1]⸢3⸣.46.38.15 **me ta zi gal en z**[**i tur**] **a.ra**$_2$ 6 **du**-*ma*
	11	1.22.39.49.30 **danna me** *a-na* 1.5.4.22.30 **šeš.meš bar**-*ma*
	12	1.16.13.10.11.24.36 (blank) ⸢*ana* **uš.meš**⸣
P12	13	17.46.40 **tab** *u* **la**$_2$ *ša*$_2$ 18 **mu.meš** [**a.ra**$_2$] 1.5.4.22.3[0 **du**-*ma* 19].16.51.6.40
	14	$\frac{1}{2}$-*šu*$_2$ GIŠ-*ma* **ki** 1.16.13.10.⸢11.24.36 **tab**-*ma*⸣ [1.25.51.35.44.4]4.36
	15	*ana tar-ṣi* 2.13.20 **zi gal** 1.4.5[4? xxxxxxxxx] ⸢**zi tur**⸣
P13	16	17.46.40 **tab** *u* **la**$_2$ *ša*$_2$ 18 **mu**.⸢**meš a.ra**$_2$⸣ [xxxxxxxxxx]
	17	2.45.55.33.20 [xxxxxxxxxxxx]
	18	25.48.38.31.6.[40 xxxxxxxxxxxxx]
	19	[x? **u**]**š** *sin ina* **ki zi** [x] ⸢x⸣ [xxxxxxxxxxxx]
	20	[xx] 1.30.36? [xxxxxxxxxxxxx]
	21	⸢x⸣-*u*$_2$ ⸢x⸣ [xxxxxxxxxxxxxxx]
	22	**ta** ⸢x⸣ [xxxxxxxxxxxxxxx]
	23	⸢**ta**?⸣ [xxxxxxxxxxxxxxx]
	24	*si-man* ⸢*ana*?⸣ [xxxxxxxxxxxxxxx]
P14	25	*ki-i* [xxxxxxxxxxxxxxxxxxxx] ⸢**zi**⸣ **tur**
	26	1.6.⸢41?.30⸣[xxxxxxxxxxxxxxxxx **z**]**i gal**
	27	1.25.42.⸢3?1⸣[xx] 44 [xxxxx] ⸢**zi**?⸣ (blank)
	28	1.6.42.[xxx] 48.4.30 [xxxx **z**]**i gal** 1.[xx].22.30
	29	$\frac{1}{2}$-*šu*$_2$ GIŠ-*ma* **a.ra**$_2$ 17.46.40 **du**-*ma* 9.22.33.13.4[0] **tab** *u* [**la**$_2$]
	30	1.25.58.17.38.40 *a-na tar-ṣi* (blank) **zi** [**tur**]
	31	1.6.41.26.32 *a-na tar-ṣi* **zi gal** 19.16.51.6.40 ⸢*bi*⸣-[*ri*]-⸢*šu*$_2$⸣-*nu*
P15	32	⸢*a*⸣-*na tar-ṣi* 2.17.4.48.53.20 2.15.31.6.35.33.20
	33	[*a-na t*]*ar-ṣi* 2.0.59.15.33.20 4.46.42.57.46.40
	34	[*a-na tar-ṣi* 1.57.47].57.46.40 5.15.28.23.37.46.40
	35	[xxxxxxxxx] (blank) *bi-ri-šu*$_2$-*nu*
	36	[xxxxxxxxxxxx] ⸢1.22.57⸣.[4]⸢6⸣.40 [xxxx]
		(perhaps 1 line missing)

Critical and philological notes

Obverse

-1' 20: perhaps to be restored as 30.

2' **mu.meš**: perhaps not to be read *šanāti*, 'years', but *šuāti*, 'this' (accus.).

2',3' **e**-*ma* is provisionally interpreted as *taqbû-ma*, '(which) you predicted'. Alternatively one might read *e-ma* = 'whenever; wherever'. There is insufficient context for a resolution.

3' If one reads *e-ma* then the preceding sign must be read as **gar** = *tašakkan*, 'you put down'. After 36 there is a small winkelhaken and a horizontal wedge, perhaps the beginning of MU.

4' Before 13.46.38.15 there are traces of GAM or a digit 30 (*sin*?) – 50.

5' After **tur** there is a vertical wedge. At the end of the line there are traces of about 2 signs.

6' After **mu.meš** there are 1–2 broken signs consisting of a 10 and the beginning of two wedges. At the end of the line one can see the end of four horizontal wedges, perhaps belonging to ⸢ŠU⸣.

10' After 1.22.39.49.30 there are traces of a sign similar to GAM.

15' **šid** = *minûtu*, 'number'.
ab$_2$ **kur**: perhaps *araḫ kašādi*, 'the month for (the Moon) reaching (the Sun)', which could be a technical term for the synodic month. Between *ina* and **murub**$_4$ there is **zi** = *nisḫu*, 'displacement' (omitted by Neugebauer).

16' *ina* 29 **me** is written on the left edge in a smaller script. **gar**: replaces *ša*$_2$ (Neugebauer 1957b).

17' **gar**: replaces *ša*$_2$ (Neugebauer 1957b). This is followed by a faint sign similar to GAR, probably an erasure (Neugebauer 1957b).

25' After **ki** there is a sign ending with a winkelhaken. 26: error for 36. The expected **tab**, 'you add', must have been written in the gap at the end.

26' It is not clear how to restore the gap at the end. Neugebauer assumed **gal**, 'largest', but this does not appear to make sense.

28' UL LA NI: confirmed by collation; meaning unknown.

P7 (Subject unclear)

[5][...] ⸢...⸣ [...] ⸢35?⸣ [...] 8 ⸢...⸣

P8 (Net differences?)

[6]23.52.13.20 is for 18 year[s ... 1]4.4.26.40: 7 months.

P9 (Computations involving *G*?)

[7]34;24,51,21,28,53,20 [...] for the durations, month by month [8][2?]3.33.56.2.57.46.40 [...] you add with the duration.

P10 (Subject unclear)

[9]4.53.47.29.40.44.26.40 the dur[ation ...] 18.16.5.19.15.33.20 the duration.

P11 (Computations involving the anomalistic month)

[10][1]⸢3⸣;46,38,15 days from the largest displacement until the [smallest] displacement you multiply by 6,0, it is [11]1,22,39; 49,30[b] (error for degrees?) of time. You divide it into 1,5;4,22,30 parts, it is [12]1,16;13,10,11,24,36 ⸢for the degrees⸣.

P12 (Computations involving *Φ* and the anomalistic month; partly unclear)

[13]0;17,46,40, the addition and subtraction for 18 years, [you multiply by] 1,5;4,22,3[0, it is 19];16,51,6,40. [14]You compute half of it and add it with 1,16;13,10,11,24,36, it is [1,25;51,35,44,4]4,36. [15]Opposite 2,13;20 the largest displacement, 1.4.5[4 ...] the smallest displacement.

P13 (Computations involving *Φ*; partly unclear)

[16]0;17,46,40, the addition and subtraction for 18 years, [you multiply] ⸢by⸣ [...] [17]2;45,55,33,20 [...] [18]25;48,38,31,6, [40 ...] [19][...] the Moon in the position of the [...] displacement [20][...] 1.30.36? [...] [21]... [...] [22]from ... [...] [23]⸢from?⸣ ... [...] [24]the duration ... [...]

P14 (Subject unclear)

[25]If [.. the largest displacement ...] the smallest ⸢displacement⸣. [26]1,6,41,30 [...], the largest displacement. [27]1,25,42,31 [...].44.[...] ⸢the displacement?⸣. [28]1,6,42,[...],48,4,30 [...], the largest [disp]lacement. 1,[...],22,30, [29]you compute half of it, multiply it by 0;17,46,40, it is 9;22,33,13,40, the addition and [subtraction]. [30]1,25;58,17,38,40: opposite the [smallest] displacement. [31]1,6;41,26,32: opposite the largest displacement. 19;16,51,6,40 ⸢between them⸣.

P15 (Computing *G* from *Φ*, variant scheme)

[32]Opposite 2,17;4,48,53,20: 2,15;31,6,35,33,20.
[33]Opposite 2,0;59,15,33,20: 4,46;42,57,46,40.
[34][Opposite 1,57;47],57,46,40: 5,15;28,23,37,46,40.
[35][...] between them. [36][...] ⸢1;22,57⸣,[4]⸢6⸣,40 [...]
(perhaps 1 line missing)

29' The broken sign before 3 might be ME, but this is highly uncertain. 32: or 52?. **du** is followed by a sign similar to LU or u_3.
31' The initial 35 might also be 55. 53: replaces 33 (Neugebauer 1957b).
32' 2.1.32: or 3.1.32? (thus Neugebauer 1957b). 48: or 58.
33' Before **zi** *ina* **ki** the upper parts of about 4 signs are visible, perhaps 2 1, another strongly damaged sign, and then perhaps a 2. **ki**: replaces DI (Neugebauer 1957b). *bi?-rit* ⸢x⸣ = *birīt*, 'distance': GAL 1-*en* (Neugebauer 1957b). The sign *bi* is crammed.

Reverse

1 The damaged sign before **nim** looks like IGI. After 9.6.26 the lower halves of two vertical wedges are visible.
2 After **ki zi** at the end of the line the beginnings of two small horizontal wedges positioned on top of one another are visible.
5 At the beginning only traces of signs are visible.
8 The initial sign was restored as 20 by Neugebauer, but there are no traces of it on the tablet.
11 **danna** = *bēru* ('mile') = 30°, but this does not make sense here. The meaning remains unclear. **me** has been interpreted as *ūmu*, 'time', but it might also be the plural determinative for **danna**.
bar = *zâzu*, 'to divide'; cf. the Glossary.
12 *ana*: replaces *ša*$_2$ (Neugebauer 1957b). **uš.meš**: perhaps 'degrees', since 'stations' makes no sense here.
19 **uš** is probably the first sign of the line.
22 **ta** is followed by a winkelhaken.
23 **ta** or **uš**.
24 After *si-man* there is a vertical wedge.
26 41: or 42?
27 After 1.25.42 there is perhaps 31.

31 Traces at the end are compatible with *bi-ri-šu$_2$-nu* = *birišunu*, 'in between them' (*biri* = in between). This form cannot be derived from *birītu*, 'distance, elongation', contra Aaboe (1968), p. 37. It is the predicate of the preceding number (Δ_Φ), which implies that 'them' refers to the minimum and the maximum of Φ.

Commentary

This nearly complete tablet, commonly known as the 'Saros Tablet', consists of four fragments, all of which belong to the BM collection 80–6–17. The bulk of the tablet (BM 36705+36725) was edited first by Neugebauer (1957b). The fragment BM 37484 was joined by Aaboe (1968); with the present edition a fourth fragment, partly filling the gap in the lower (upper) left corner of the obverse (reverse), is joined to the tablet. Perhaps one line of text is still missing from the top of the obverse, and one from the bottom of the reverse. The reverse is partly illegible because of damage to the surface. It appears that there was no colophon. The text consists of 16 procedures separated by horizontal lines, making this one of the largest known procedure texts. Many of the procedures are without parallel. Those that are (partly) understood concern intricate aspects of Φ and related functions, without any obvious practical application to the production of synodic tables.

P0 Subject unclear

Virtually nothing remains of the first procedure, which may have consisted of 2 lines.

P1 Computations involving Φ and Φ_{tr}; net difference $d\Phi(223)$; partly unclear

This procedure consisting of 4 lines is the first of several concerned with function Φ and its truncated version Φ_{tr} (§4.4.3). Neugebauer (1957b), van der Waerden (1965) and Aaboe (1968) succeeded in interpreting some of the computations. 1,5;4,22,30 (O0') can be identified as Δ_Φ/ϕ, the amplitude of Φ in units of ϕ = 0;17,46,40° = $-d\Phi(223)$, the parameter mentioned in O2' (cf. Table 4.7a, p. 126). This number is multiplied by something, perhaps ϕ, in which case the result Δ_Φ = 19;16,51,6,40° can be restored at the beginning of O1'. O1'–2' mention parameters of Φ_{tr}, namely 1,58;31,6,40° = m', 2,13;20° = M' and 14;48,53,20° = $M' - m'$. These parameters reappear in P3. In O2'–3' ϕ is called the 'difference for 18 years', i.e. the net difference of Φ for 223^{m} ('18 years'). From this statement van der Waerden (1965) inferred the astronomical meaning of Φ as the duration of 223^{m} in excess of 6585^{d} (cf. §4.4.3). The significance of the clause 'which you have predicted' is not clear to me. As pointed out by Aaboe (1968), the number 13.39.35.36 might be identified as 13;39,35,36^{d}, a good approximation of half the sidereal month ($m_{sid}/2$; cf. §4.1.1), but the role of this parameter within the procedure is wholly unclear.[677] The digits 6.15 (O3') remain unidentified.

P2 Computations involving Φ and G (partly unclear)

P2 was partly explained by Neugebauer (1957b), with a few modifications by Aaboe (1968). It can be divided into 5 subprocedures. Part **a** (O4') concerns the relation between d_Φ and $d\Phi(223)$ (Eq. 4.14, p. 126). The aim of part **b** (O4'–16') is to compute the mean synodic month, $\overline{m}_s$, from a given value of the anomalistic month, m_a. This computation is somewhat artificial, since $\overline{m}_s$ is a more fundamental and readily available parameter than m_a. The purpose of the procedure may rather be to verify that the assumed value of m_a is correct. O4' mentions the time interval 13;46,38,15^{d} 'from the Moon's largest displacement until the Moon's smallest displacement', which is identifiable as an approximation of $m_a/2$ = 27;33,16,30^{d}/2. This value of m_a is not attested elsewhere. It was presumably obtained from a given value of $\overline{m}_s$ by the inverse of the procedure followed here, namely by multiplying $p_\Phi = m_a/\overline{m}_s$ (Table 4.6, p. 125) by the common value $\overline{m}_s$ = 29;31,50,8,20^{d} (Eq. 2.4, p. 33) and rounding the 4th sexagesimal digit of the product, 31, to 30. In O6' m_a is expressed as 1,22,39;49,30° = 6,0 · 13;46,38,15; this number is the subject of further manipulations below O8'. The rest of O6' is unclear. In O7' the Moon is said to reach the Sun, obviously a reference to New Moon (conjunction). The number 3.13.21.4 (O8') is unidentified, and the unit 'mile' (*bēru*) is puzzling.[678] In O8'–16' the mean synodic month ($\overline{m}_s$) is derived from m_a through a series of manipulations equivalent to the following expression (Eq. 2.19, p. 45):

$$\overline{m}_s = \frac{m_a}{p_\Phi} = m_a + \frac{m_a}{P_\Phi}. \tag{5.4}$$

In order to understand this expression recall that in the synodic tables Φ is tabulated at successive lunations separated by a synodic month whose duration slightly exceeds the underlying period of Φ, the anomalistic month. Hence there is an underlying, more rapidly varying zigzag function Φ^* whose period is $p_\Phi = m_a/\overline{m}_s$. First m_a/P_Φ is computed. The first

[677] Interpreted as 13;39,35,36° this number is close to μ_F (§4.4.9).

[678] If correct, this number corresponds to 1.36.40.32°, but also in R11 a number measured in degrees is followed by the unit 'mile' (*bēru*) for unclear reasons.

step of this computation is phrased as a question: 'What should I multiply by 1,5;4,22,30 so that it is 1,22,39;49,30?'. In O9'-10' the answer is given as

$$1,16;13,10,11,24,36 \cdot 1,5;4,22,30 = 1,22,39;49,30. \tag{5.5}$$

The number 1,22,39;49,30 (O6') is $m_a/2$ measured in time degrees, and 1,5;4,22,30 = Δ_Φ/ϕ, the amplitude of Φ in units of ϕ; hence 1,16;13,10,11,24,36 = $\frac{1}{2}m_a\phi/\Delta_\Phi$. Since $m_a/2$ is the time that it takes for Φ^* to proceed from m to M, $\frac{1}{2}m_a\phi/\Delta_\Phi$ is the time that it takes for Φ^* to change by the amount ϕ. This is multiplied in O12' by 9;20 = d_Φ/ϕ, which is d_Φ in units of ϕ (cf. part **a**). Hence the product, 11,51;22,55,6,29,36 = $\frac{1}{2}m_a d_\Phi/\Delta_\Phi = m_a/P_\Phi$, is the time that it takes for Φ^* to change by the amount d_Φ. In O13' this is multiplied by 0;0,10, i.e. converted from time degrees to days, which yields 1;58,33,49,11,4,56^d. Each synodic month Φ^* performs one extra oscillation with respect to Φ, which takes 1 anomalistic month. Hence $\overline{m}_s$ is obtained by adding m_a/P_Φ to m_a (Eq. 5.4), resulting in

$$27;33,16,30^d + 1;58,33,49,11,4,56^d = 29;31,50,19,11,4,56^d. \tag{5.6}$$

This is stated to be 'the number (of days) of the month for approaching', an appropriate term for the synodic month. The term 'mean displacement' (O15') probably reflects that the synodic month thus computed is a mean value amounting to the assumption that the Moon (and the Sun) move with their mean velocities. The slight deviation from the aforementioned common value 29;31,50,8,20^d (Eq. 2.4, p. 33) can probably be attributed to rounding at intermediate steps. In O16' the excess of $\overline{m}_s$ over 29^d is converted to time degrees, i.e. 6,0 · 0;31,50,19,11,4,56 = 3,11;1,55,6,29,36°, but the digit 55 is omitted by the scribe. This number represents an estimate of the mean value of G (§4.4.11), say $\overline{G}$, which reappears in part **e** (O25'). One might ask whether the result could not have been obtained by a more straightforward procedure, e.g. by estimating $1/P_\Phi$ and multiplying this by m_a. Even though P_Φ = 1,44,7 cannot be inverted in the Babylonian sense, it would have been possible to find a reasonable approximation of $1/P_\Phi$ such as 0;0,0,34,34,36.

Part **c** (O16'–18') contains two statements reminiscent of ΦG scheme 1 (§4.4.11). The first rule involves a control value 1.53.56.47.24.2.40. The formulation suggests that this is a G value and not a Φ value as claimed by Neugebauer (1957b), and that the corresponding Φ value was written in the gap at the end of O16'. Since it is less than m_G, the initial digit may have to be corrected to 2, resulting in 2,53;56,47,24,2,40°. The second rule involves a control value Φ = 1,58;30... and G = 4,28;5,25,45,9,30. The origin of this number is unclear (it deviates from scheme 1), but if one corrects the final digit to 20, as suggested by Neugebauer, it is identical to the number computed in part **e**.

Part **d** (O18'–23') contains further manipulations involving parameters of Φ and G. O18' mentions the relation between d_G and d_Φ (Eq. 4.60, p. 151). In O19'–22' $d_\Phi - (7d_\Phi - \Delta_\Phi)/2 = d_\Phi - d\Phi(14)/4$ is computed; for $d\Phi(14)$ cf. Table 4.7a (p. 126). In O23' this quantity is transformed into the corresponding one for G by multiplying it by 9;20, the result being 25;27,2,13,20 = $d_G - d\hat{G}(14)/4$. For $d\hat{G}(14)$ cf. Table 4.25 (p. 151). It is not clear what these quantities are good for.[679]

In part **e** (O24'–25') the outcome of part **d** is subjected to further manipulations. In O24' 51;37,17,2,13,20 = $2d_G$ is added to 25;27,2,13,20, resulting in 1,17;4,19,15,33,20 = $(\Delta_{\hat{G}} - d_{\hat{G}})/2$. The reason for separating parts **d** and **e** is that this quantity could have been computed directly rather than from the complex outcome of **d**, so that each must serve a different purpose. In O25' $(\Delta_{\hat{G}} - d_{\hat{G}})/2$ is added to $\overline{G}$, the outcome of part **b** (O16'), the result being 4,28;5,25,45,9,20.[680] The significance of this quantity is not really clear. It appears to play a role as a particular value of G, since it was already mentioned in part **c**, if one corrects the final digit to 30.

P3 Computations involving Φ_{tr}

The computations in P3 were explained by Aaboe (1968), pp. 37–38. Three instructions call for the parameter 0;17,46,40 = ϕ to be multiplied by numbers which are differences between the extrema of Φ and those of its truncated version (cf. §4.4.3), measured in units of ϕ. Hence P3 converts these numbers from units of ϕ to time degrees, in accordance with

$$\begin{aligned} M_\Phi - M'_\Phi &= 12;38,45 \cdot \phi &&= 3;44,48,53,20° \\ M'_\Phi - m'_\Phi &= 50 \cdot \phi &&= 14;48,53,20° \\ m'_\Phi - m_\Phi &= 2;25,37,30 \cdot \phi &&= 0;43,8,53,20°. \end{aligned}$$

The outcome of the multiplications is not provided, but the second one was mentioned in P1. Note that the sum of the three numbers is Δ_Φ. The repeatedly occurring phrase 'From the largest displacement and the smallest displacement' is not fully understood. The term 'displacement' (usually) refers to F, so that it might reflect the fact that $F^{(1)}$ is exactly in phase with Φ, i.e. their extrema coincide (cf. §4.4.9). Also the instructions in O27'–28' are difficult to connect with the other computations. The meaning of UL LA NI (O28') is unclear.

---

679 The tentative interpretation offered by Neugebauer (1957b), p. 17, applies only under special conditions.

680 If the digit 55 were included in $\overline{G}$ the sum would be 4,28;6,14,22,2,56.

***P4** Lunar motion?*

The new fragment BM 37475 adds a considerable amount of text to P4. Unfortunately this has not resolved the difficulties of this procedure, which continues to defy interpretation, given the lack of parallels. Some of the numbers appear to represent displacements along the zodiac, others time intervals measured in days. Their origin is unclear.

***P5** Subject unclear*

This procedure consists of a single line (R1), of which a part is preserved on the new fragment. The meaning remains unclear, and the numbers could not be identified. It cannot be excluded that P5 is actually the last line of P4.

***P6** Computations involving Φ_{tr} (partly unclear)*

The new fragment adds some text to this procedure. As pointed out by Aaboe (1968), several quantities computed in P2 and P3 reappear in P6. P6 begins with a broken number and the term 'middle (= mean) displacement', the meaning of this term is unclear. The number 1,3,30;58,29,30,30 (R3) is the product of 1,16;13,10,11,24,36 = $\frac{1}{2}m_a\phi/\Delta_\Phi$ and 50 = $(M'_\Phi - m'_\Phi)/\phi$. Since $\frac{1}{2}m_a\phi/\Delta_\Phi$ is the time that it takes Φ^* to change by the amount ϕ, the product is the time that it takes Φ^* to proceed between m' and M'. One might expect P6 to also provide the product of $\frac{1}{2}m_a\phi/\Delta_\Phi$ with the other two intervals mentioned in P3, but the digits in R2–4 do not confirm this, and they remain unidentified.

***P7** Subject unclear*

P7 consists of a single heavily damaged line, the meaning of which is unclear.

***P8** Net differences?*

P8 consists of a single line mentioning (probably) two numbers, each associated with a time interval, namely the saros (223^m) and 7^m. The latter period may have a connection with the interval of 14^m implied in O18'–23'. Perhaps the numbers are net differences for these intervals, but they could not be identified.

***P9** Computations involving G; unclear*

P9 is concerned with G, as suggested by the term 'duration' and the number 34;24,51,21,28,53,20, which Neugebauer identified as 1;20 $\cdot d_G$ (§4.4.11). It is not clear what this number is for. The number in R8 remains unidentified.

***P10** Subject unclear*

The purpose of P10 is not clear. Again the word 'duration' suggests a connection with G, but none of the numbers has been identified. The first number looks like an exemplary value of G. It does not appear in the parent synodic table of lunar system A. The second number is probably incomplete, because the gap may have contained additional digits. If one assumes that the final digits of both numbers belong to the same power of 60 then their sum is 5.12.3.35.0.0.0.0. The vanishing of the final 4 digits suggests that this sum is a fundamental parameter, but its significance remains unclear.

***P11** Computations involving Φ and the anomalistic month*

This procedure lists several parameters that were computed in P2: $m_a/2$ = 13;46,38,15^d = 1,22,39;49,30°, 1,5;4,22,30 = Δ_Φ/ϕ, and 1,16;13,10,11,24,36 = $\frac{1}{2}m_a\phi/\Delta_\Phi$. Their mention may be viewed as a preparatory step for P12. It is not clear why 1,22,39;49,30° is followed by the unit 'miles' (*bēru*), since it is actually expressed in time degrees. The meaning of the phrase 'for the degrees' is not clear.

***P12** Computations involving Φ and the anomalistic month*

In R13 1,5;4,22,30 = Δ_Φ/ϕ is multiplied by ϕ, the result being Δ_Φ (the same multiplication was probably mentioned in O0'–1'). The outcome is halved and added to 1,16;13,10,11,24,36 = $\frac{1}{2}m_a\phi/\Delta_\Phi$, the result being 1,25;51,35,44,44,36. The purpose of computing the sum of half the amplitude of Φ and the time that it takes for Φ^* to change by the amount ϕ remains elusive. R15 is formulated like the entry of an interpolation scheme for computing G from Φ. It is not clear which quantity is associated with the mentioned Φ = 2,13;20, the maximum of Φ_{tr}. The word 'largest displacement' suggests F, the Moon's daily displacement along the zodiac, in which case the maximum of F_{tr} must be meant here. The significance of the number 1.4.54$^?$... is not clear.

P13 *Computations involving Φ and G*

P13 is strongly damaged. Most of the legible numbers are also mentioned elsewhere on the tablet. R16 mentions ϕ = 0;17,46,40, a repetition of a statement in P1. R17 mentions d_Φ, and R18 $d_{\hat{G}}$. In R19–24 only isolated words and digits can be read.

P14 *Computations involving Φ; unclear*

The beginning of P14 is strongly damaged. The numbers in R26–28 have not been identified. References to the 'smallest/largest displacement' (R25) suggest a connnection with F. In R28 one may follow Neugebauer and restore 1,5;4,22,30 = Δ_Φ/ϕ, the amplitude of Φ in units of ϕ. This number is halved and multiplied by ϕ, which should result in 9;38,25,33,20 = $\Delta_\Phi/2$, but the text has 9.22.33.13.40 (R29). This is called the 'addition and subtraction', the usual term for a difference, not for an amplitude. It is therefore unlikely that the number in R28 is to be restored as 1,5;4,22,30. If one assumes that the outcome of the multiplication is correct, then the number should be 1.3.17.14.17.15, which is incompatible with the preserved digits. The conclusion seems unavoidable that there is a scribal error. The numbers 1,25;58,17,38,40 (R30) and 1,6;41,26,32 (R31) are here associated with the smallest and the largest 'displacement', respectively (extrema of F?). As pointed out by Neugebauer, the difference between these numbers is 19;16,51,6,40° = Δ_Φ. Accordingly, they correspond to the extrema of Φ shifted by a constant which equals 51;6,31,14,40°. The significance of this is not clear.

P15 *Computing G from Φ, variant scheme*

This procedure is concerned with the computation of G from Φ, as explained by Neugebauer (1957b) and Aaboe (1968). It mentions only three pairs of control values, and no interpolation rules. The first control value for Φ is the maximum M = 2,17;4,48,53,20 (R32) which, unlike schemes 1 and 2 (cf. §4.4.11), is here associated with G = 2,15;31,6,35,33,20°. Hence this may be the minimum of a variant of G (variant 3 in Table 4.24 on p. 151). In R33 Φ = 2,0;59,15,33,20 is associated with G = 4,46;42,57,46,40, as in schemes 1 and 2. In R34 a Φ value is mentioned which, following Neugebauer, may be restored as m_Φ = 1,57;47,57,46,40. It is here associated with G = 5,15;28,23,37,46,40. This value is considerably larger than in scheme 1, but probably similar to the value associated with m_Φ in scheme 2. It probably represents the maximum of this variant of G, because the difference 5,15;28,23,37,46,40° − 2,15;31,6,35,33,20° = 2,59;57,17,2,13,20° coincides with the amplitude of (the common version of) $\hat{G}$ (cf. Table 4.24). The number restored in R36, 1;22,57,46,40, can be identified as 0;30 d_Φ.

No. 83

museum number	BM 36775 (80–6–17,512)
provenience	Babylon
measures	9.4(10.0) × 6.2(*) × 1.3–2.5 cm (type E, F or G)
arrangement	O/R
cuneiform text	photo: Fig. F.94; Aaboe (1968), Pl. IV
bibliography	Aaboe (1968), Text F
contents	Moon system A: computations involving *F*

Obverse

P1’ (n lines missing)
0’ [xxxxxxxxxxxxxxxxxxxxxxxxxxxxxxx] ⌜x 5 x⌝ [x]
1’ [xx 11].15 **la**$_2$ ⌜x⌝ [xxxxxxxxxxxx] ⌜1$^?$1⌝.15 **ta** [x]
2’ [xxx] ⌜3⌝.45 ⌜**zi**-*ma*⌝ [xxxx] ⌜**ki**$^?$ x⌝ [xxxxx] *ki-i al* 11.15
3’ [xxx] **tab**-*ma* 2.11$^?$.⌜x⌝ [xxxx] ⌜50$^?$⌝ [xxxxx].⌜2$^?$⌝.15 **gar**-*an*

P2’ 4’ [xxxxx] ⌜**ta**⌝ [xxxxx] ⌜x⌝ 51$^?$ [xxxx] ⌜x⌝.46.52.30
5’ [xxxxxxxxxxxx *ša*$_2$] ⌜*al*⌝ 15 **diri** 15 **e**-*šu*$_2$
6’ [xxxxxxxxxxxxxx 42 *ša*$_2$] ⌜**iti**⌝ 1.21.39.22.30
7’ [*ša*$_2$ **mu** 2.20.37.30 *ša*$_2$ 14 **iti** 23.15 *ša*$_2$ 9 **mu**].⌜**meš**⌝ 4.30 *ša*$_2$ 18 **mu.meš**
8’ [xxxxxxxxxxxxxxx] ⌜x⌝ **zi** (blank) **gal**
9’ [xxxxxxxxxxxxxxx] (blank)

P3’ 10’ [xxxxxxxxxxxxxxx] ⌜xx⌝ 11.15 **zi**
11’ [xxxxxxxxxxxxxxx] ⌜x⌝ **an.ta** 15 **ta**
12’ [xxxxxxxxxxxxx] ⌜15⌝ **gar**-*an ki-i al* 15 **diri**
13’ [xxxxxxxxxxxxx] ⌜**gar**⌝-*an ki-i al* 15 **la**$_2$ ⌜2$^?$.15$^?$⌝ [x]

Reverse

P4’ 1 [xxxxxx] 13.5.9.22.30 [xxxxx]
2 [xxxx] ⌜**du**$^?$⌝ 11.⌜15⌝ **la**$_2$ 42.11.15 ⌜x⌝ [xxx] ⌜x⌝

P5’ 3 ⌜**ta**⌝ 15.44.31.52.30 **en** 10.58.23.26.15 **iti**
4 [*ana* **iti**] 42.11.15 **tab** *u* **la**$_2$ *lib*$_3$*-bu-u*$_2$ *ša*$_2$ **iti.meš** *ta-ba-nu-u*$_2$
5 [x] ⌜x⌝ *gab-bi ša*$_2$ *al-la* 15 **diri** 15 **e**-*šu*$_2$ *mim-ma ša*$_2$ *al* 11.15
6 [**la**$_2$]-*u*$_2$ 11.15 **e**-*šu*$_2$ 42.11.15 *ša*$_2$ **iti** 1.22.1.52.30 *ša*$_2$ **mu**
7 [2].20.37.30 *ša*$_2$ 14 **iti** 23.26.⟨15⟩ *ša*$_2$ 9 **mu** 4.41.15 *ša*$_2$ 18
8 [**mu**] 4.11 *šal-miš*$_2$ **zi**-*šu*$_2$ **kur**-*ad*$_2$ **ta** [1]5.4⌜4.31.52.30⌝
9 [**en** 10].50.23.26.15 4.54.⌜8.26.1⌝[5 x] **bala.meš sar**

P6’ 10 [*ana tar-ṣa* 2.1]3.20 15 **zi** *ana tar-ṣa* ⌜1.5⌝[8.31.6.40 11.15 **zi**]
11 [**ta** 15.5]⌜6⌝.54.22.30 **en** [11.4.4.41.15 xxxxxxxx]
12 [xxxxxxx] (traces) [xxxxxxxxxx]
(about n lines missing)

Critical and philological notes

Obverse

1’ **la**$_2$ is followed by the beginning of two horizontal wedges, perhaps *u*$_2$ or *is*. In the former case one obtains **la**$_2$-*u*$_2$ = *maṭû*, ‘decreasing’, in the latter case **la**$_2$-*is* = *tanaḫḫis*, ‘you subtract’.

2’ 3.45: replaces 2.45 (Aaboe); **ki**$^?$: replaces TA. It is followed by traces of two vertical wedges, perhaps a number. **zi** = *nisḫu*, ‘displacement’, or *tanassaḫ*, ‘you subtract’; the latter seems more appropriate.

3’ 2.11: or 2.12–2.13. The following sign might be 40 or 50. 2.15: or 3.15.

4’ Before 46 there is a digit 5–9.

5’ Here and elsewhere **e**-*šu*$_2$ has been interpreted as *taqabbišu*, ‘you call’. Alternatively, it may be read as an infinitive *qabûšu*, ‘its prediction’.

6’ Aaboe restored the gap as [*mim-ma ša*$_2$ *al* 11.15 **la**$_2$-*u*$_2$ 11.15 **e**-*šu*$_2$], ‘Whatever is less than 11;15 you call 11;15’, but this rule for the minimum of the truncated *F* is probably dealt with in P1 already.

8’ Before **zi** a vertical wedge is visible.

Obverse

P1' **(Truncating *F*; partly unclear)**

(n lines missing)

0'[...] ⌜... 5 ...⌝ [...] 1'[...] is less than [11];15 ⌜...⌝ [...]⌜1?1⌝;15 from [...] 2'[...] ⌜3⌝;45 you subtract, it is [11;15...] ⌜...⌝ [...] If it [is less] than 11;15 3'[you call it 11;15 ...] you add, it is 2.11? ⌜...⌝ [...] ⌜50⌝ [...].⌜2?⌝.15 you put down.

P2' **(Various procedures for $F^{(1)}$)**

a **(Truncating *F*)**

4'[...] from [...].46.52.30. 5'[... Whatever] exceeds 15 you call 15. 6'[...]

b **(Net differences of $F^{(1)}$)**

[0;42 is for] the month. 1;21,39,22,30 7'[is for the year. 0;2,20,37,30 is for 14 months. 0;23,15 is for 9 month]s. 0;4,30 is for 18 years. 8'[...] the largest displacement. 9'[...]

P3' **(Truncating *F*; unclear)**

10'[...] ⌜of ...⌝ 11;15, the displacement?. 11'[...] ... above, 15, from 12'[...] you put down 15. (The amount) by which it exceeds 15 13'[...]. If it is less than 15 [...] 2,1⌜5?⌝ [...]

Reverse

P4' **($F^{(4)}$; truncation; unclear)**

1[...] 13;5,9,22,30 [...] 2[...] ⌜...⌝ is less than 11;⌜15⌝, 0;42,11,15 ⌜...⌝ [...]

P5' **($F^{(4)}$)**

a **(Updating $F^{(4)}$ as a zigzag function)**

3⌜From⌝ 15;44,31,52,30 until 10;58,23,26,15 4you add and subtract 0;42,11,15 month by month, by means of which you compute the months.

b **(Truncating $F^{(4)}$)**

5All ⌜...⌝. Whatever exceeds 15 you call 15. Whatever 6is less than 11;15 you call 11;15.

c **(Net differences of $F^{(4)}$)**

0;42,11,15 is per month.
1,22;1,52,30 is per year.
7[0;2],20,37,30 is per 14 months.
0;23,26,⟨15⟩ is per 9 years.
0;4,41,15 is per 18 8years.
(After) 4,11 (events) it completely reaches the same displacement.
From [1]5;44,31,52,30 9[until 10];50,23,26,15 4;54,⌜8,26,1⌝[5 ...] revolutions it begins.

P6' **(Computing $F^{(1)}$ from Φ)**

10[Opposite 2,1]3;20 15, the displacement.
Opposite 1,5[8;31,6,40 11;15, the displacement.]
11[From 15;56],54,22,30 until [11;4,41,15 ...] 12[...] ... [...]
(about n lines missing)

10' Before 11.15 traces of 2 signs are visible, perhaps ⌜4⌝ or ⌜*ša*$_2$⌝ followed by ⌜KUR⌝ or ⌜40⌝.
11' Before **an.ta** there are traces of a vertical wedge, perhaps part of *ša*$_2$.
13' 2.15?: the 5 may also be 4 or 6.

Reverse

2 **du**: or **nim**, or a similar sign.
After 42.11.15 there are traces of a sign beginning with two horizontal wedges; at the end of the line are traces of 2 vertical wedges.
4 *libbû ša*: a conj., probably to be translated 'by means of which' (cf. the Glossary).
iti.meš, 'months': this is probably a reference to the fact that the mentioned difference is a monthly difference.
tabannû = 2 m. sg. pres. G *banû*, 'you construct': not attested anywhere else in the corpus in the sense 'to compute', for which the commonly used verbs are *našû*(GIŠ) and *epēšu*(**du**$_3$).
5 The traces before *gab-bi* look like UR; is not ...**meš** or **lu.mash**.
7 23.26: error for 23.26.15 (cf. the commentary).
8 *šalmiš* = 'entirely, intactly'.
9 **bala.meš sar**: written on the edge; **bala.meš** = *palû*, 'regnal years' (?); **sar** is perhaps *ušarra*, 'it begins', a 3 c. sg. pres. *šurrû* (cf. *CAD* 17 Š.III 'to begin, originate', with numerous astronomical applications).

Commentary

The upper (lower) edge of the obverse (reverse) of this tablet is missing, and the surface of the obverse is strongly damaged. The text contains at least 5 procedures, which are separated by horizontal lines. All procedures deal with aspects of F, the Moon's daily displacement (§4.4.9). The tablet was first edited by Aaboe (1968), to which the present edition offers several corrections. A colophon is not preserved, but may have been present below P6'.

P1' Truncating F; other computations involving F_{tr}

The first procedure is strongly damaged, and several lines are missing from the beginning. Some of the numbers can be identified as parameters of F_{tr}: the minimum m'_F = 11;15 (O2' and perhaps O1') and 3;45 (O2'), which is perhaps $M' - m'$. The number 2.11... (O3') is close to 0;2,11,15°/°, the mean value of the truncated version of F'. It is unclear how this number can be understood as the outcome of the preceding addition. The number ...2$^?$.15.[...] (O3') remains unidentified.

P2' Various procedures for $F^{(1)}$ or close variant: truncation; differences

P2' is also concerned with various aspects of F. Part **a** (O4'–5') concerns the truncated version of $F^{(1)}$. The extrema were probably mentioned in O4'–5'. If one restores the broken number at the end of O4' as 0;1,50,40,46,52,30 = 0;0,10 · 11;4,4,41,15, then this is the minimum (m) of $F^{(1)'}$. O5' mentions 15°/day, which is the maximum m' of the truncated version of F. Part **b** (O6'–7') is concerned with net differences $dF(s)$ for various time intervals. As pointed out by Aaboe, they are compatible with $F^{(1)}$ or $F^{(3)}$; cf. Table 4.19 (p. 144). The number at the end of O6' (for '1 year') is $dF^{(1)}(12)$ = $-$1;21,39,22,30°/day, but its subtractive nature is not mentioned. Similarly, the number at the end of O7' is 0;4,30 = $-dF^{(1)}(223)$ ('18 years'). The parallel procedure P4' enabled Aaboe to fill the intermediate gap with further differences of $F^{(1)}$, namely d = 0;42, $dF^{(1)}(14)$ = 0;2,20,37,30 and $dF^{(1)}(111)$ = $-$0;23,15 (for '9 years'). Little can be said about the final 2 lines, except that O8' contains a reference to the maximum.

P3' F: truncation; partly unclear

P3' consists of 4 badly damaged lines, assuming that a new procedure begins on the reverse (P4'). It again deals with the truncation of F at m' = 11;15°/d and M' = 15°/d. The digits ⌜2.15$^?$⌝ (O13') are unidentified.

P4' $F^{(4)}$; truncation; partly unclear

This procedure and the following one deal with $F^{(4)}$, a variant of F attested only here (cf. §4.4.9). The numbers in R2 can be identified as 11;15 = m', and 0;42,11,15 = d. The number in R1, if interpreted as 13;5,9,22,30, is close to the mean value $(m' + M')/2$ = 13;7,30°/day; its significance is not clear.

P5' Various procedures for $F^{(4)}$

P5' continues with the topic of P4'. Part **a** (R3–4) contains an instruction for updating $F^{(4)}$ as a zigzag function, with M = 15;44,31,52,30, m = 10;58,23,26,15 and d = 0;42,11,15°/d. The formulation is based on template ZZ.A.4 (§2.6.8.1), which lacks the reflection rules. The phrase 'by means of which you construct the months' is not attested elsewhere. In part **b** (R5–6) F is truncated at the usual values m' = 11;15 and M' = 15 (§4.4.9). Part **c** (R6–8) concerns net differences $dF^{(4)}(s)$ for 1^m, 12^m ('year'), 14^m, 111^m ('9 years') and 223^m ('18 years' = the saros). The net difference for 111^{m}contains a scribal error (compare Table 4.19, p. 144). R8 mentions the number period Π = 4,11, which is correctly stated to result in an exact return of $F^{(4)}$. R9 repeats M and m, and it mentions Δ = 4;54,8,26,15°/d. The meaning of the phrase at the end of R9 is not clear.

P6' Computing $F^{(1)}$ from Φ

This procedure, of which only the first 2 lines are partly preserved, concerns the computation of $F^{(1)}$ from Φ in accordance with Eq. (4.56) on p. 144. The first control value of Φ (R10) and the associated control value of $F^{(1)}$ correspond to the point where the zigzag functions Φ_{tr} and F_{tr} are truncated. As suggested by Aaboe, the digits in R11 may belong to the maximum of $F^{(1)}$, which suggests that the following gap contained the minimum.

No. 84

museum number	BM 36698 (80–6–17,430)
provenience	Babylon
measures	5.0(*) × 4.5(*) × 2.8 cm
cuneiform text	photo: Fig. F.87
content	Moon system A: computations involving *G*

Side X

P1' (unknown number of lines missing)

1' ⌜**ta** xx⌝ [...]

P2' 2' *si-man ša*$_2$ **du**-*ku* [...]

P3' 3' *ana tar-ṣa* 1.57.⌜5⌝[8.8.53.20 **tab**-*u*$_2$ 4.53.14.4.26.40 **gar**-*an* ...]

4' **a.ra**$_2$ 3.2⌜2⌝.[30 **du**-*ma* ...]

5' **ta** 4.5⌜3.1⌝[4.4.26.40 ...]

6' (erasure?) GIŠ? 3.22.⌜30⌝ [...]

7' *ana tar-ṣa* 1.⌜57?⌝.[...]

8' ⌜xx⌝ [...]

(unknown number of lines missing)

Side X

P1' (Subject unclear)

(unknown number of lines missing)

1'⌜From ...⌝ [...]

P2' (Computations involving *G*?)

2'The duration which you multiplied? [...]

P3' (Computing *G* from *Φ*, scheme 1?)

13 3'Opposite 1,57;⌜5⌝[8,8,53,20, increasing, you put 4,53;14,4,26,40 ...] 4'[you multiply] by 3;2⌜2⌝,[30, ...] 5'[subtract] from 4,5⌜3;1⌝[4,4,26,40 ...]

? 6'*(erasure?)* you compute?, 3;22,⌜30⌝ [...]

7'Opposite 1,⌜57?⌝,[...] 8'⌜...⌝ [...]

(unknown number of lines missing)

Critical and philological notes

Side X

6' GIŠ (or: GUR?) is preceded by what is probably an erased sign.

Commentary

This small fragment from the left side of a tablet is inscribed on one side (X); the other side is destroyed. It might belong to the same tablet as No. **79**, but they do not physically join.

***P1',P2'** Subject unclear; computations involving G?*

The remains of P1' are insufficient to identify its topic. P2' contains a single line of text of which the first few signs are preserved. The word 'duration' suggests that it is concerned with *G* or a related function (*W*, *Λ*).

***P3'** Computing G from Φ, interpolation scheme 1?*

The first 6 lines of this procedure are sufficiently preserved to conclude that X3'–5' deal with interval 13 of *ΦG* interpolation scheme 1 (§4.4.11). Presumably X6'–8' also belong to this scheme, but the preserved signs are insufficient for a precise identification.

No. 85

museum number	BM 37114 (80–6–17,863)
provenience	Babylon
measures	3.0(*) × 5.2(*) × 3.0 cm
arrangement	O/R
cuneiform text	photo: Fig. F.95
contents	Moon system A: computing *G* from *Φ*

Side X

P1' (m lines missing)

1' [...] ⸢**gar**?-*an* xxxx⸣ [...]
2' [...] **du**-*ma* **ta** 4.46.[42.57.46.40 **nim**? ...]
3' [...] *ša₂ al* 2.13.20 ⸢xx⸣ [...]
4' [... **ta** 2.51.29.22].⸢5⸣7.46.40 **n**[**im**? ...]

P2' 5' [...] 6.25.⸢44?⸣ [...]
6' [...] 10 *mi*-⸢*nu-u₂*⸣ [...]
7' [...] ⸢x⸣-*ma* [...]
8' [...] ⸢x⸣ **du** [...]
(n lines missing)

Side Y

P3' (about n lines missing)

1' [...] ⸢xx⸣ [...]
2' [...] ⸢x⸣ 4.30 ⸢**zi**?⸣ [...]
3' [...] **gar**-*an ina* ⸢x⸣ [...]
4' [...].10 *bi*-⸢*rit*⸣ [...]
5' [...] ⸢x⸣ [...]
(about m+4 lines missing)

Side X

(m lines missing)

P1' (Computing *G* from *Φ*, scheme 2?)

1'[...] ⸢you put down? ...⸣ [...]

6 2'[...] you multiply [... by ...] and [you deduct? it] from 4,46;[42,57,46,40 ...]

7 3'[...] (The amount) by which it ⸢...⸣ 2,13;20 ⸢...⸣ [...] 4'[... you deduct from 2,51;29,22],⸢5⸣7,46,40 [...]

P2' (Subject unclear)

5'[...] 6.25.⸢44?⸣ [...] 6'[...] 10. Whatever [(the amount) ...] 7'[...] ⸢...⸣ and [...] 8'[...] ⸢..⸣ you multiply [...]
(n lines missing)

Side Y

P3' (Subject unclear)

(about n lines missing)
1'[...] ⸢...⸣ [...] 2'[...] ⸢...⸣ 4.30 ⸢the displacement?⸣ [...] 3'[...] you put down ... [...] 4'[...].10, ⸢the distance⸣ [...] 5'[...] ... [...]
(about m+4 lines missing)

Critical and philological notes

Side X

3' After 20 there is a sign beginning with a vertical wedge, perhaps **la₂**, followed by traces of another sign.
5' 4⸢4⸣: the 40 may also be 50, the 4 may also be 5–8.
7' Before *ma* there are remains of a horizontal wedge.
8' Before **du** there are remains of a vertical wedge.

Side Y

1' Remains of a winkelhaken and a vertical wedge are visible.
3' After *ina* there is a vertical wedge.
5' Two horizontal wedges and a vertical wedge are visible.

Commentary

This small flake does not preserve any edges of the original tablet. The horizontal dividing line on side X implies that the tablet contains several procedures.

P1' Computing G from Φ, scheme 2?

Just enough signs can be read to allow the conclusion that P1' concerns the computation of G from Φ (§4.4.11). X2' probably mentions $G = 4,46;42,57,46,40°$, which is associated with $\Phi = 2,0;59,15,33,20°$ (↑) in both schemes 1 and 2 (cf. Table 4.22, p. 147 and Table 4.26, p. 152). X3' suggests that this interval continues until $\Phi = 2,13;20°$ (↑), which points to scheme 2. The surviving digits in X4' are compatible with $G_7 = 2,51;29,22,57,46,40°$ (scheme 2), which is associated with $\Phi_7 = 2,13;20°$ (↑).

P2',P3' Subject unclear

The remains of these procedures are insufficient for determining their subject. Perhaps they provide interpolation rules. The numbers could not be identified.

No. 86

museum number	BM 42097 (81–6–25,719)
provenience	Babylon
measures	3.0(*) × 4.0(*) × 1.2(*) cm
cuneiform text	photo: Fig. F.96
contents	Moon system A: computations involving Φ, F and G

Side X

P1' (unknown number of lines missing)

1' [... *ša*₂] ⸢*al-la* 2⸣.1⸢3?⸣.[...]

2' [...] ⸢2⸣.13.20 **la**₂-*u*₂ 2.1⸢3⸣.[...]

3' [... 19.16].51.6.40 ⸢**zi**⸣[...]

4' [...] ⸢x **tab**⸣ [...]

P2' 5' [...] **ta ugu** 15 **zi** [...]

6' [...] ⸢x⸣ 5 42 *taš-pil-tu*₂ [...]

P3' 7' [...].⸢x⸣1.51.⸢6⸣.[...]

8' [...] ⸢x 30 x⸣ [...]

(unknown number of lines missing)

Side X

P1' (Computing G or a related function from Φ)

(unknown number of lines missing)

1'[... which] ⸢is less than 2⸣,1⸢3?⸣.[...] 2'[...] ⸢2⸣,13;20, decreasing, 2,1⸢3⸣;[20 ...] 3'[... 19;16],51,6,40, the displacem[ent ...] 4'[...] ⸢... you add⸣ [...]

P2' (Computing F from Φ?)

5'[...] you subtract from 15 [...] 6'[...] 5. 0;42, the difference [...]

P3' (Subject unclear)

7'[...]⸢...⸣1.51.6 [...] 8'[...] ⸢... 30 ...⸣ [...]

(unknown number of lines missing)

Critical and philological notes

Side X

1' 3: traces of two vertical wedges are preserved.

4' The large empty space before **tab** suggests that **tab** is the last sign, so that not much clay may be missing from the right edge. The sign before **tab** might be a digit 4–8.

7' ⸢x⸣1.51.6: at the beginning there are remains of two winkelhakens belonging to a digit 20–50.

8' Before and after 30 there are remains of a vertical wedge, probably digits.

Commentary

This small fragment is inscribed on one side (X); the other side is destroyed. No original edges are preserved, but probably not much clay is missing on the right side (cf. the critical notes to X4.). The surviving text is divided into three sections by horizontal lines.

P1' Computing G or a related function from Φ

Only a few signs of the last 4 lines of P1' are preserved. They are sufficient to conclude that P1' contains an interpolation scheme similar to ΦG scheme 1, but the goal function (G, W or Λ) could not be identified. The only preserved control value of Φ coincides with Φ_1 of ΦG scheme 1 (Table 4.22, p. 147). The digits in X3' may be restored as 19;16,51,6,40 = Δ_Φ, the amplitude of Φ (§4.4.3).

P2' Computing F from Φ?

The number 15 (X5') may be identified as the maximum of the truncated version of F, i.e. $M' = 15°$/day (§4.4.9). This identification is confirmed in X6', where the monthly difference $d = 0;42$ is mentioned. Something is subtracted from 15, which points to the interpolation algorithm for computing Φ from F or vice versa (§4.4.9). If this is true, the text may be restored as 'The amount by which it exceeds 15 you subtract from 15'.

P3' Unidentified computations

Too little is preserved of P3' to identify its topic with any degree of certainty.

No. 87

museum number	BM 77224 (83–6–30,4)
provenience	Babylon
measures	5.8(*) × 6.5(*) × 2.6(*) cm
cuneiform text	photo: Fig. F.97
contents	Moon system A: *ΦΛ* interpolation scheme

Side X

P1' (unknown number of lines missing)

1' [xxxxxx] ⸢xxx⸣ [xxxx *ana tar* 1.58.51.28.53.20 **la$_2$** 9.33.20 **gar**]
2' [*ša$_2$ al* 1.58.51.2]8.53.20 **la$_2$** [**la$_2$ en** 1.58.33.42.13.20 **la$_2$**]
3' [**a.ra$_2$** 13 **du ta** 9].⸢3⸣3.20 **nim gar** *ana tar* 1.5⸢8⸣.[33.42.13.20 **la$_2$** 5.42.13.20 **gar** *ša$_2$ al*]
4' [1.58.33.42.13.20 **la$_2$ la$_2$ en** 1].⸢58⸣.15.5⸢5⸣.33.20 **la$_2$** ⸢**a**⸣.[**ra$_2$** 14.8.45 **du ta** 5.42.13.20]
5' [**nim gar** *ana tar* 1.58.15].⸢5⸣5.33.20 **la$_2$** 1.30.[44.26.40 **gar** xxxxxx]

P2' 6' [xxxx 1.58.15.55.3]3.20 **la$_2$** 1.30.44.26.40 [xxxxxxxx]
7' [xxxxxx 1.58.15.5]⸢5$^{?}$⸣.33.20 **la$_2$** 1⸢7$^?$.50 xx⸣ [xxxxxxxx]

P3' 8' [xxxxxxxx]⸢x⸣.33.20 ⸢**la$_2$**⸣ **la$_2$** ⸢**en**⸣ 1.⸢5⸣[7.58.8.53.20 **la$_2$** xx]
9' [xxxxxxxxx] ⸢x⸣ **tab** ⸢*ša$_2$* **nu**$^?$ x⸣ : *ana tar-*⸢*ṣa*⸣ 1.⸢50$^?$.xx⸣ [xxxxxxxxxx]
10' [xxxxxxxx 1.5]⸢7$^?$⸣.55.33.20 ⸢xxx⸣ [xxxxxxxx]
11' [xx *ana tar* 1.57.55.3]⸢3⸣.20 **tab** 7.4$^?$5.33.20 ⸢**gar**⸣ [*ša$_2$ al*]
12' [1.57.55.33.20 **tab**] ⸢**diri**⸣ **en** 1.⸢58⸣.13.⸢20 **tab**⸣ [**a.ra$_2$** 17.8.45]
13' [**du ki** 7.45.33.20 **tab gar** xx] ⸢xxxx⸣ [xxxxx]
14' [xxxxxxxxxx] ⸢xx⸣ [xxxxxxxxx]
(unknown number of lines missing)

Side X

P1' **(Computing *Λ* from *Φ*)**

(*unknown number of lines missing*)
$^{1'}$[...] ⸢...⸣ [...

7 Opposite 1,58;51,28,53,20, decreasing, you put down 9;33,20.] $^{2'}$[(The amount) by which it is less than 1,58;51,2]8,53,20, decreasing, [until 1,58;33,42,13,20, decreasing,] $^{3'}$[you multiply by 13], deduct [from 9];33,20, put down.

8 Opposite 1,5⸢8⸣;[33,42,13,20, decreasing, you put down 5;42,13,20. (The amount) by which it is less than] $^{4'}$[1,58;33,42,13,20, decreasing, until 1,5]8;15,55,33,20, decreasing [you multiply] ⸢by⸣ [14;8,45, deduct from 5;42,13,20], $^{5'}$[put down.

9 Opposite 1,58,15];⸢5⸣5,33,20, decreasing, [you put down] 1;30,[44,26,40...]

P2' **(Computing *Λ* from *Φ*)**

9? $^{6'}$[...] it is less than [1,58;15,55,3]3,20, decreasing, 1;30,44,26,40 [...] $^{7'}$[... 1,58;15,5]5,33,20, decreasing, 1⸢7$^?$.50...⸣ [...]

P3' **(Computing *Λ* from *Φ*)**

10? $^{8'}$[... (The amount) by which it is less than ...]⸢...⸣,33,20, decreasing, until 1,5[...] $^{9'}$[...] you put down. Opposite 1,⸢50...⸣ [...] $^{10'}$[... 1,5],⸢7$^?$⸣;55,33,20 ⸢...⸣ [...]

11 $^{11'}$[Opposite 1,57;55,3]⸢3⸣,20, increasing, ⸢you put down⸣ 7;45,33,20. [(The amount) by which it exceeds] $^{12'}$[1,57;55,33,20, increasing], until 1,⸢58⸣;13,⸢20, increasing⸣, [you multiply by 17;8,45,] $^{13'}$[add with 7;45,55,33,20, put down. ...] $^{14'}$[...] [...]
(*unknown number of lines missing*)

Critical and philological notes

Side X

1' The lower halves of several digits are visible.
7' 7: or 8. 50: or 40.
9' After **nu** there are faint traces of a sign similar to ME. 50: followed by faint traces of several signs.
10' 20: followed by faint traces of several signs.

11' 45: or 55.
13' Traces of several digits.

Commentary

This fragment does not preserve any edges of the tablet.[681] It is inscribed on one side, the surface of the other side being destroyed. Three procedures are partly preserved, all of them concerned with the computation of Λ from Φ.

P1'–P3' Computing Λ from Φ

These procedures contain segments of the interpolation scheme for computing Λ from Φ. For this scheme cf. §4.4.13 and Appendix D.2. P1' partly preserves intervals 7–9. It seems that the interpolation rule for interval 9 continues in P2'. P3' concerns intervals 10–11. Since the full scheme comprises 61 intervals, side X is probably the obverse of the tablet.

[681] In Britton (2009), p. 385, it is suggested that this fragment was joined to No. **79**, but this is not the case, and they are unlikely to be part of the same tablet.

No. 88

museum number	BM 33593 (Rm4,149)
provenience	Babylon
measures	3.5(*) × 4.7(*) × 2.8(*) cm
cuneiform text	photo: Fig. F.98; *ACT*, Pl. 239
bibliography	*ACT* 207; Aaboe & Hamilton (1979), pp. 9–10
contents	Moon system A: computing *W* from *Φ*

Side X

P1' (unknown number of lines missing)

1' [*ana tar-ṣa* 2.9.11.6.40 **la**$_2$ 5.9.59.45.11].⌜6⌝.40 ⌜*ša*$_2$⌝ [*al-la* 2.9.11.6.40 **la**$_2$ **la**$_2$ GAM 42.30 **du ta** 5.9.59.45.11.6.40 **nim gar**]

2' [... 5.9.47.9].⌜37⌝.46.40 *ša*$_2$ [*al-la* ...]

3' [... 5.10.10].07.24.26.40 *ša*$_2$ [*al-la* ...]

4' [... 5.11.]8.38.31.6.40 *ša*$_2$ [*al-la* ...]

5' [... 5.12].⌜4⌝2.42.57.46.40 [*ša*$_2$ *al-la* ...]

6' [... 5.14].50.22.13.20 [*ša*$_2$ *al-la* ...]

7' [... 5.17.3]⌜5⌝.33.20 [*ša*$_2$ *al-la* ...]

8' [... 5.20.58].1.28.53.20 *ša*$_2$ [*al-la* ...]

9' [... 5.24.54].19.[15].33.20 *ša*$_2$ *a*[*l-la* ...]

10' [... 5.29.26.10.22].13.20 *ša*$_2$ *al-*[*la* ...]

11' [... 5.34.33.34.4]8.53.20 *ša*$_2$ *al-l*[*a* ...]

12' [... 5.40.16.32.35.33].20 *ša*$_2$ *al-l*[*a* ...]

13' [... 5.46.35.3.42.13.20] *ša*$_2$ *al-l*[*a* ...]

(unknown number of lines missing)

Side X

P1' **(Computing *W* from *Φ*)**

(unknown number of lines missing)

15 1'[Opposite 2,9;11,6,40, decreasing, 5,9;59,45,11],⌜6⌝,40. ⌜(The amount) by which⌝ [it is less than 2,9;11,6,40, decreasing, you multiply by 0;42,30, deduct from 5,9;59,45,11,6,40, put down.]

16 2'[... 5,9;47,9],37,46,40. (The amount) by which [...]

17 3'[... 5,10;10],7,24,26,40. (The amount) by which [...]

18 4'[... 5,11];8,38,31,6,40. (The amount) by which [...]

19 5'[... 5,12];42,42,57,46,40. (The amount) by which [...]

20 6'[... 5,14];50,22,13,20. (The amount) by which [...]

21 7'[... 5,17;3]5,33,20. (The amount) by which [...]

22 8'[... 5,20;58],1,28,53,20. (The amount) by which [...]

23 9'[... 5,24;54],19,[15],33,20. (The amount) by which [...]

24 10'[... 5,29;26,10,22],13,20 (The amount) by which [...]

25 11'[... 5,34;33,34,4]8,53,20. (The amount) by which [...]

26 12'[... 5,40;16,32,35,33],20. (The amount) by which [...]

27 13'[... 5,46;35,3,42,13,20.] (The amount) by which [...]

(unknown number of lines missing)

Commentary

This small fragment does not contain any edges of the original tablet. It is inscribed on one side; the other side is destroyed. Probably No. **89** belongs to the same tablet. P1' preserves segments of the interpolation rules for intervals 15–27 of the *ΦW* scheme (cf. §4.4.12 and Appendix D.1). For interval 15 I have included a tentative reconstruction of the complete rule, assuming that it is based on template IP.B (p. 41).

No. 89

museum number	BM 42747 (81–7–1,511)
provenience	Babylon
measures	3.5(*) × 4.7(*) × 2.0(*) cm
cuneiform text	photo: Fig. F.99
content	Moon system A: computing *W* from *Φ*

Side X

P1' (unknown number of lines missing)

1' [*ana tar-ṣa* 2.4.47.2.13.20 **tab** 2.31.0.29].⸢37.46?⸣.40 ⸢*ša*₂ *al*⸣-[*la* 2.4.47.2.13.20 **tab diri** GAM 7.2.30 **du ki** 2.31.0.29.37.46.40 **tab gar**]

2' [... 2.33.5].⸢40⸣.44.2⸢6.40⸣ *ša*₂ *al*-[*la* ...]

3' [... 2.34.37].2.13.20 *ša*₂ *al*-[*la* ...]

4' [... 2.35.27].46.40 *ša*₂ *al*-[*la* ...]

5' [... 2.35.42].⸢5⸣7.46.40 ⸢*ša*₂ *al*⸣-[*la* ...]

6' [... 2.35.2]⸢2⸣.35.33.20 ⸢*ša*₂ *al*⸣-[*la* ...]

7' [... 2.34.2]⸢6⸣.40 (blank) ⸢*ša*₂ *al*⸣-[*la* ...]

8' [... 2.32.51.5]⸢1.6.40 *ša*₂⸣ [*al-la* ...]

(unknown number of lines missing)

Side X

P1' **(Computing *W* from *Φ*)**

(*unknown number of lines missing*)

73 1'[Opposite 2,4;47,2,13,20, increasing, 2,31;0,29],⸢37,46?⸣,40. ⸢(The amount) by which⸣ [it exceeds 2,4;47,2,13,20, increasing, you multiply by 7;2,30, add with 2,31;0,29,37,46,40, put down.]

74 2'[... 2,33;5],⸢40⸣,44,2⸢6,40⸣. (The amount) by which [it exceeds ...]

75 3'[... 2,34;37],2,13,20. (The amount) by which [it exceeds ...]

76 4'[... 2,35;27],46,40. (The amount) by which [it exceeds ...]

77 5'[... 2,35;42],⸢5⸣7,46,40. ⸢(The amount) by which⸣ [it exceeds ...]

78 6'[... 2,35;2]⸢2⸣,35,33,20. ⸢(The amount) by which⸣ [it exceeds ...]

79 7'[... 2,34;2]⸢6⸣,40. ⸢(The amount) by which⸣ [it exceeds ...]

80 8'[... 2,32;51,5]⸢1,6,40. (The amount) by which⸣ [it exceeds ...]

(*unknown number of lines missing*)

Commentary

This small fragment does not contain any edges of the original tablet. It is inscribed on one side; the other side is destroyed. The formulation and handwriting suggest that it belongs to the same tablet as No. **88** and perhaps No. **90**, even though they do not physically join.

P1' Computing W from Φ

The preserved text contains segments of the interpolation rules for intervals 73–80 (§4.4.12). The tabulated values of *W* correspond to the ones in Appendix D.1 minus 12,0° (= 2^d), because the latter have not been reduced to values within the range 0–6,0°. The rule for interval 73 has been tentatively reconstructed on the basis of template IP.B (p. 41).

No. 90

museum number	BM 34497 (Sp,621)
provenience	Babylon
measures	3.4(*) × 4.3(*) × 2.2(*) cm
cuneiform text	photo: Fig. F.100; copy: *LBAT* 156
bibliography	*ACT* 1005; Aaboe & Hamilton (1979), p. 10
contents	Moon system A: computing *W* from *Φ*

Side X

P1' (unknown number of lines missing)

1' [... GAM 40.5]⸢1.15⸣ [**du** ...]
2' [... GAM] ⸢39⸣.51.15 [**du** ...]
3' [... GAM] ⸢3⸣8.51.15 ⸢**du**⸣ [...]
4' [...] GAM 37.48.20 ⸢**du**⸣ [...]
5' [...] ⸢GAM⸣ 36.51.15 ⸢**du**⸣ [...]
6' [...] GAM 35.51.15 [**du** ...]
7' [... GAM] ⸢3⸣4.51.15 [**du** ...]
8' [...] ⸢GAM⸣ 33.51.1⸢5⸣ [**du** ...]
9' [... GAM 32].⸢51⸣.[15 **du** ...]
(unknown number of lines missing)

Side X

P1' **(Computing *W* from *Φ*)**

(unknown number of lines missing)

113 1'[Opposite 2,16;38,8,53,20, decreasing, 2,13;13,34,48,53,20. (The amount) by which it is less than 2,16;38,8,53,20, decreasing, you multiply by 40;5]⸢1,15⸣, [deduct from 2,13;13,34,48,53,20, put down.]

114 2'[... you multiply by] ⸢39⸣;51,15 [...]

115 3'[...] ⸢you multiply⸣ [by] ⸢3⸣8;51,15 [...]

116 4'[...] ⸢you multiply⸣ by 37;48,20 [...]

117 5'[...] ⸢you multiply by⸣ 36;51,15 [...]

118 6'[... you multiply] by 35;51,15 [...]

119 7'[... you multiply by] ⸢3⸣4;51,15 [...]

120 8'[... you multiply] ⸢by⸣ 33;51,1⸢5⸣ [...]

121 9'[... you multiply by 32];⸢51⸣,[15 ...]

(unknown number of lines missing)

Critical and philological notes

Side X

9' ⸢51⸣: only traces of the upper parts are preserved.

Commentary

This fragment does not preserve any edges of the original tablet. Only one side is preserved; the other side is destroyed. Perhaps it belongs to the same tablet as No. **88** and No. **89**.

P1' Computing W from Φ

P1' partly preserves interpolation rules for intervals 113–121 of the *ΦW* scheme (§4.4.12). The rule for interval 113 has been tentatively reconstructed on the basis of template IP.B (p. 41).

No. 91

museum number	BM 42840 (81–6–25,604)
provenience	Babylon
measures	4.2(*) × 4.8(*) × 2.4(*) cm
cuneiform text	photo: Fig. F.106
contents	Moon system A: interpolation rules for a *ΦG*-type scheme?

Side X

P1' (unknown number of lines missing)

1' [... *ša*$_2$] ⸢*al-la*⸣-*nu*-⸢*šu*$_2$ **la**$_2^?$⸣ [...]
2' [...].6.40 ⸢*ša*$_2$ *al*⸣-*la-nu-šu*$_2$ ⸢**la**$_2^?$⸣ [...]
3' [...].⸢55$^?$.13⸣.20 *ša*$_2$ *al-la-nu-šu*$_2$ ⸢**la**$_2^?$⸣ [...]
4' [...].⸢55$^?$⸣.33.20 *ša*$_2$ *al-la-nu-šu*$_2$ ⸢**la**$_2^?$⸣ [...]
5' [...]⸢5$^?$.26$^?$.20 *ša*$_2^?$⸣ *al-la-nu-šu*$_2$ **la**$_2$ [...]
6' [...].⸢33$^?$.20 *ša*$_2$ *al-la-nu*⸣-*šu*$_2$ ⸢**la**$_2^?$⸣ [...]
(unknown number of lines missing)

Side X

P1' (Interpolation rules)

(unknown number of lines missing)
1' [... That which] ⸢is less than that⸣ [...]
2' [...].6.40 ⸢That which is less than that⸣ [...]
3' [...].⸢55$^?$.13⸣.20. That which ⸢is less⸣ than that [...]
4' [...].⸢55$^?$⸣.33.20. That which ⸢is less⸣ than that [...]
5' [...] ⸢5$^?$.26$^?$.20. That which⸣ is less than that [...]
6' [...].⸢33$^?$.20. That which is less than⸣ that [...]
(unknown number of lines missing)

Critical and philological notes

Side X

1' *al-la-nu-šu*$_2$ = *allānuššu*, 'beyond it'; cf. *CAD* A$_1$ sub *alla* d.
ša$_2$: there is insufficient context to determine whether the correct translation is 'That which ...', as has been assumed, or '(The amount) by which it ...'.
3' 55: or 56.
4' 55: or any digit 52–58.
5' 5.26: very uncertain.
6' 33: very uncertain.

Commentary

This small flake does not preserve any edges of the original tablet. It is inscribed on one side; the other side is destroyed.

P1' Interpolation rules for computing a G-type function from Φ?

The 6 partly preserved lines contain identically structured segments of an interpolation algorithm reminiscent of the *ΦG*, *ΦW* and *ΦΛ* schemes (§4.4.11–4.4.13). The preserved digits support this general conclusion, but they are insufficient for a more precise identification. The formulation with *allānuššu*, 'beyond it', is thus far unique in the astronomical corpus.

No. 92

museum numbers	BM 36744 (80–6–17,477) +37031 (775)
provenience	Babylon
measures	9.3 × 9.3(13-14?) × 1.7–3.3 cm (type F)
arrangement	O/R
cuneiform text	photo: Fig. F.101; Neugebauer & Sachs (1969), p. 112
bibliography	Neugebauer & Sachs (1969) = NS, Text J
contents	Moon system A or related: computations involving *F*, *Φ*, *E* (?) and eclipse magnitude

Obverse

P1
1 *ina* 18 **mu.meš** 2.⸢47⸣.23$^?$ **uš** *sin* **la**$_2$
2 *mim-ma ša*$_2$ *al* 1.57.47.57.46.40 **diri a.ra**$_2$
3 10.05.20 **du**-*ma* **ki** 11.36 **zi tab igi.gub**-*šu*$_2$
4 GIŠ-*ma* ⸢**a**⸣.**ra**$_2$ 2.47.22 **du a.ra**$_2$ 6 **du**-*ma*
5 ⸢**ta**⸣ 3.⸢16⸣.21 **e**$_{11}$-*ma si-man ša*$_2$ **gi**$_6$-*ka* **la**$_2$
6 1 **u**$_4$ 20$^?$ **zi** ⸢**me nim**$^?$⸣.**a gi**$_6$ **du tab**

P2
7 13.53.47.50 ⸢*ḫi*$^?$⸣-[*pi*$^?$] *eš-šu*$_2$ **zi** *ša*$_2$ **me tab** *u* **la**$_2$
8 **a.ra**$_2$ 2 **du ta** ⸢**zi** *ša*$_2^?$ **me nim**$^?$⸣ *u*$_4$-*me* **e**$_{11}$
9 [x] *ša*$_2$ **la**$_2$ 3.7.31 ⸢xx⸣ 4 **tab** 3.34 **tab**
10 [xx]⸢x⸣ **en u**$_4$ 3.24 ⸢**la**$_2$ xx *ki*$^?$-*i*$^?$ **zi**$^?$⸣
11 [xxx] ⸢x *ša*$_2^?$⸣ *al* 26.2.24 **la**$_2$ **tab**
12 [xxxxxx] **igi.gub**-*šu*$_2$ GIŠ-*ma* ḪAB$^?$ **nim**
13 [xxxxxxxxxxx] ⸢xxx⸣
(m lines missing)

Reverse

P3
(about m lines missing)
1’ [xxxxxxxxxx] ⸢x⸣ **a.ra**$_2$ 46 [x]
2’ [xxxxxx] ⸢*ša*$_2$⸣ *al* 1$^?$ ⸢*ki*⸣ **la**$_2$ **la**$_2$ ⸢*ki*⸣ **tab** ⸢**tab**⸣
3’ *ša*$_2$ **tag**$_4$ **ki** *mim-ma ša*$_2$ *al* 2 *ki* **la**$_2$ **tab**-*ma* ⸢*ki*$^?$⸣
4’ *mim-ma ša*$_2$ *al* 22.30 **diri a.ra**$_2$ ⸢20$^?$ xx⸣ [x]
5’ **ta** 12.15.20 ḪAB ⸢xxxx⸣ **igi.gub**$^?$ 11.4 **la**$_2^?$
6’ **a.ra**$_2$ 57.51 **du ta** 11.4 **zi** ḪAB **gar**

T1
1 ⸢xx⸣ 2.12 [x$^?$]
2 ⸢12 **ki**$^?$ **la**$_2$⸣ **igi** [xx]
3 3.59.1.50 [x]
4 4.17.35.51 [x]
5 4.41.14.8 **du**
6 4.33.12.49
7 7.31.22
8 0.26.5
9 12.1.59
10 17.54.41
11 1.4$^?$1.20.59
12 1.48.32.2.50
13 2.16.34.4.38.50.20
14 2.44.26.55.26
15 3.40 **du**

Obverse

P1 (Computations involving *Φ* and *F*?)

[1]In 18 years the Moon lacks$^?$ 2.⸢47⸣.23$^?$°. [2]Whatever (the amount) by which it exceeds 1,57;47,57,46,40: [3]you multiply it by 0;10,5,20 and add it with 11;36, the displacement.
[4]You compute its *igigubbû*-coefficient, and you multiply it by 2.47.22, you multiply it by 6, and [5]you deduct it from 3.16.21, and you subtract the duration of your night, [6]1 day of the Sun$^?$ you subtract; ⸢the time since sunrise⸣, (the time by which) the night has progressed, you add.

P2 (Computations involving *F*?)

[7]13;53,47,50 — recent ⸢break⸣ — the daily displacement you add and subtract, [8]you multiply by 2, you deduct it from ⸢the daily displacement ...[1]⸣[9][...] which is less, 3.7.31 ⸢...⸣ 4 you add 3.34 you add [10][...] ⸢...⸣ until 3.24 ⸢you subtract ...⸣[11][...] ⸢... that which⸣ is less than 26.2.24 you add [12][...] you compute its *igigubbû*-coefficient, and you deduct the ‘disk’$^?$ [13][...] ⸢...⸣
(m lines missing)

Reverse

P3 (Computations involving *E* and eclipse magnitude?)

(about m lines missing)
[1’][... you multiply] ⸢...⸣ by 46 [...] [2’][...] that which you subtract if increasing, add if increasing [3’]what remains you add with what exceeds 2 if decreasing, ⸢...⸣[4’]Whatever exceeds 22.30 you multiply by ⸢20 ...⸣ [...] [5’][you subtract] it from 12.15.20, the ‘disk’ ⸢...⸣ the *igigubbû*-coefficient$^?$, 11;4, ... [6’]you multiply it by 57.51, you subtract it from 11;4, you put it down as the ‘disk’.

T1 (Cf. Volume II)

Critical and philological notes

Obverse

1 2.47.23: 47 could be 57; 23: replaces 22 (NS). **la**$_2$: either *inaḫḫis*, ‘it lags behind’, or a form of *maṭû*: *imaṭṭi*, ‘it falls short’, or *maṭi*, ‘it is lacking’.

5 3.⸢16⸣.21: more likely than 3.13.21 (NS). *si-man ša*$_2$ **gi**$_6$-*ka*, ‘the time of your night’.

6 1 ⸢**u**$_4$ 20$^?$⸣ **zi** ⸢**me nim.a**⸣replaces *ana* ... 1,41$^?$ (NS). Meaning unclear.

7 13.53.47.50 ⸢*ḫe*$^?$⸣-[*pi*$^?$] *eš-šu*$_2^?$.

8 **zi**: replaces 12 (NS).
9 The final **tab** is written on the edge.
10 **u$_4$**: replaces KI. **zi**: written on edge; omitted in NS.
11 **tab**: replaces *u*$_2$ (NS).
12 ḪAB$^?$ **nim**?: replaces **a.ra$_2$** (NS). The reading **ra$_2$**(DU) remains a possibility.
Reverse
2' ⸢*ki*⸣ **la$_2$ la$_2$** ⸢*ki*⸣ **tab** ⸢**tab**⸣: replaces KI$^?$.LAL$^?$ **la$_2$ e$_{11}$**-*ma* (NS)
3' 2: replaces 13 (NS).
4' After **a.ra$_2$** there are two winkelhakens, perhaps part of a digit 20–50. The number 54$^?$ read by NS at the end of the line could not be confirmed.
5',6' ḪAB: might also be SI (thus NS)
5' **igi.gub** 11.4 **la$_2$**$^?$: replaces KI$^?$.LAL$^?$ DU$^?$ (NS)

Commentary

This fragment belongs to the upper (lower) part of the obverse (reverse) of a tablet. On account of the curvature, about 4–6 cm of clay may be missing from the lower (upper) part of the obverse (reverse). An improved reading of the colophon indicates that the tablet is a copy of an older, damaged exemplar. The main text is divided into (at least) three sections separated by horizontal lines. Remarkably, the tablet concludes with a numerical table written at a right angle to the procedures. The handwriting is unusually large and strongly slanted, the latter suggesting a date in the late Seleucid or Arsacid era. The procedures contain elements of lunar system A, but none is fully understood. For the numerical table (T1), whose significance remains to be established, cf. Volume II.

P1 Computations involving Φ and F?

P1 contains computations involving the saros ('18 years') and several functions known from lunar system A. The number 2.47.22° is said to be the amount by which the Moon 'is lacking' or 'lags behind' during 1 saros. Perhaps this refers to the net difference of G, the duration of the synodic month, across 223^{m}, i.e. $dG(223)$, which has the value −2;45,55,33,20 in lunar system A (cf. Table 4.25, p. 151). If so then the number must be interpreted as −2;47,22°.

O2–3 contain a single interpolation rule. The control value of the source function, 1,57;47,57,46,40, can be identified as the minimum of Φ (§4.4.3); that of the goal function, 11.36 (O3), can be interpreted as 11;36°/day, perhaps the minimum of a variant of F (§4.4.9). The interpolation coefficient is given as 10.05.20; comparison with the system A value d_F/d_Φ = 0;15,11,15 (Eq. 4.56) suggests that it should be interpreted as 0;10,5,20, or 0;15,20 if one assumes that the intermediate 0 is a scribal error. The reconstructed maximum of F is thus 11;36 + 0;10,5,20 $\cdot\Delta_\Phi$ = 14;50,31,20,59,15,33,20, or 11;36 + 0;15,20 $\cdot\Delta_\Phi$ = 11;36 + 4;55,38,23,42,13,20 = 16;31,38,23,42,13,20. These extrema deviate considerably from those of the known variants of F (compare Table 4.18, p. 143). In the absence of parallels the interpretation remains uncertain.

The instructions in O3–6 are more difficult to interpret. 'Its *igigubbû*-coefficient' (O3) is a term for the reciprocal, but in astronomical contexts the meaning may be different, and it is not clear what 'it' refers to. This is multiplied by 2.47.22 (= the number in O1?), and then by 6 (= 6,0?), i.e. by 16.44.12. The result is subtracted from 3.16.21. The 'time of your night' (O5) may correspond to the system A function M^{fm} (§4.4.18) or D (§4.4.5). This suggests that 3.16.21 is a time interval, perhaps 3,16;21°. The value of the 'time interval of your night' is not given.

P2 Computations involving F; unclear

The first 7 lines of P2 are partly preserved. The number in O7, if interpreted as 13;53,47,50, may be a value of F, as also suggested by the term 'displacement per day'. In the following instructions a difference ('addition and subtraction') is multiplied by 2, and the result is subtracted from F (?). The interpretation is not clear. The rest of the procedure also defies explanation.

P3 Computations involving E and eclipse magnitude?

The reverse preserves 6 more lines forming the end of a procedure, perhaps the continuation of P2. The phrase 'if decreasing you subtract, if increasing you add' (R2') suggests that a zigzag-type function, perhaps E = the Moon's distance to the ecliptic, is updated by adding or subtracting its (monthly) difference d. The threshold value 2 mentioned in R3' might represent 2° = κ, the boundary of the nodal region in the algorithm for E (§4.4.6). The rest of the procedure is difficult to interpret. The repeated mention of the 'disk' (R5',6'), points to eclipse magnitude. Perhaps this function is computed from E by an algorithm similar to that for Ψ (§4.4.7).

5.3.3 System B

No. 93

museum number	A 3413
provenience	Uruk (library of the Rēš?)
measures	7.2(*) × 6.9(*) × 1.4–1.6 cm (type F?)
arrangement	O/R
cuneiform text	photo: Fig. F.102; *ACT*, Pl. 238
bibliography	*ACT* 202
contents	Moon system B: various procedures

Obverse

P1' (m lines missing)

1' [xxxxxxxxx] ⸢xxx⸣ [xxxxxxxx]

2' [xxx **a.ra**$_2$ *bi-ti-ka*] ⸢**du**⸣-*ma* **b**[**ar.nun** LA$_2$-*ti* xxx]

3' [xxx **a.ra**$_2$ *bi-ti*]-⸢*ka*⸣ **du**-*ma* **bar.nun** ⸢LA$_2$-*ti*⸣ [xxx]

4' [xxx **a.ra**$_2$ *bi*]-⸢*ti*⸣-*ka* **du**-*ma* **bar.nun** LA$_2$-*ti* ⸢x⸣ [xxx]

5' [xxx **a**].⸢**ra**$_2$⸣ *bi-ti-ka* **du**-*ma* **bar.nun** LA$_2$-*ti* ⸢x⸣ [xxx]

6' [xxx **a**].**ra**$_2$ *bi-ti-ka* **du**-*ma* **bar.nun** LA$_2$-*ti* ⸢x⸣ [xxx]

7' [xxx] **zi**-*ḫi* *u*$_3$ *tar-di-tu*$_4$ ⸢x⸣ [xxx]

P2' 8' [xx] *ša*$_2$ *lu-maš*.**meš** (erasure) **a.ra**$_2$ 40 *igi-gub*-[xx]

9' [xx] **me** 20 **du**-*ma* *ki-i* **nim tab** *u*$_3$ *ki-i* **sig** ⸢**la**$_2$⸣

P3' 10' [*kun-nu*] ⸢*u*$_3$⸣ 6 **bal** *u*$_4$-*mu ana u*$_4$-*mu* : *kun-nu u*$_3$ 6 **nu bal** *u*$_4$-*me* **la**$_2$

11' [*tu-ru*] ⸢*u*$_3$⸣ 6 **bal** *u*$_4$-*mu ut-tar* : *tu-ru u*$_3$ 6 **nu bal** *u*$_4$-*me ana u*$_4$-*me*

P4' 12' [xx$^?$ *si*]-⸢*man*$^?$⸣ *a-na* **du**$_3$-*ka ki-i mi-šil* **gi**$_6$ GIŠ.GIŠ.ḪI *al-la si-man a-tar*

13' [*si-man* **ta** *mi-šil*] **gi**$_6$ **e**$_{11}$-*ma ša*$_2$ **tag**$_4$ *a-na* **kur** *šamaš*$_2$ **e**-*bi* : *ki-i mi-šil* **gi**$_6$ GIŠ.GIŠ.ḪI

14' [*al-la si-man i-ṣa mi*]-⸢*šil* **gi**$_6$⸣ **ta** *si-man* **e**$_{11}$-*ma ša*$_2$ **tag**$_4$ ⸢*a*$^?$⸣-[*na* **me**$^?$ **nim.a**$^?$ **e**-*bi*]

15' [xxxxxxxxxxxxxx *a*]*l-la si-man* [xxxxxx]

16' [xxxxxxxxxxxxxx **e**$_{11}$]-⸢*ma*⸣ *ša*$_2$ ⸢**tag**$_4$⸣ [xxxxxx]

(n lines missing)

Reverse

P5' (about n lines missing)

1' [xxxxxxxxxx] ⸢2⸣ [xxxxxxx]

2' [xxxxxxxxxxxxxxxxxxxx]

3' [xxxxxxx] ⸢24$^?$⸣ [xxxxxxx]

4' [xxxxxxxxxxxxxxxxxxxx]

5' [xxxxxxxxx] ⸢10⸣ [xxxxxx] ⸢**la**$_2$⸣

6' [xxxxxxxxxxxxxxxxxxxx]

7' [xxxxxxxx] ⸢x 20 x⸣ [xxxxx] ⸢**la**$_2$⸣

8' [xxxxxx] ⸢3.18⸣ [xxxxxxxx] ⸢**la**$_2$⸣

9' [xxxx] ⸢xxxx 21$^?$⸣ [xxx] ⸢**la**$_2$⸣

10' [xxxx] ⸢xx 10⸣ [xxxxxx]

11' [xxxx] ⸢xxxxx⸣ 12 ⸢x⸣ [xx] ⸢**la**$_2$⸣

12' [xxxx] ⸢3$^?$.36$^?$⸣ [xxxxxxxxxxx]

(about m lines missing)

Critical and philological notes

Obverse

1' Traces of signs similar to 2, a horizontal wedge, and a vertical wedge are visible

2' *bi-ti-ka*: probably a phonetic abbreviation of *birītika*, 'your distance'.
LA$_2$-*ti*: Akkadian reading unknown; I assume this is the same word as in LA$_2$ *ša ūmi*/*mūši*, 'duration of daylight/night'.

4'–6' The damaged sign at the end might be **du**.

7' *tardītu* = 'addition'.

8' About 2 signs are missing at the beginning; they were restored as **bar.nun** by Neugebauer, but this is uncertain (cf. the commentary). After *lu-maš*.**meš** about 4–5 signs were erased by the scribe.

Obverse

P1' (Computing *C* or *D* from *B*?)

(*m lines missing*)
1'[...] ⸢...⸣ [...]
2'[...] you multiply [... by your distance], it is the *ṣi*[*liptu*-coefficient for the duration? ...]
3'[...] you multiply [... by] ⸢your⸣ [distance], it is the *ṣiliptu*-coefficient for the duration? [...]
4'[...] you multiply [... by] your ⸢distance⸣, it is the *ṣiliptu*-coefficient for the duration? ⸢...⸣ [...]
5'[...] you multiply [...] ⸢by⸣ your distance, it is the *ṣiliptu*-coefficient for the duration? ⸢...⸣ [...]
6'[...] you multiply [...] ⸢by⸣ your distance, it is the *ṣiliptu*-coefficient for the duration? ⸢...⸣ [...]
7'[...] the subtraction and the addition ⸢...⸣ [...]

P2' (Computing Lunar Six intervals from *Q* and *R*?)

8'[...] for the zodiacal signs [you multiply] by 0;40, the *igigubbû*-coefficient [...] 9'[...] ... you multiply by 0;20, if it is 'high' you add and if it is 'low' ⸢you subtract⸣.

P3' (Establishing the date of the lunation or the preceding midnight)

10'[(The month) is full] and it transgresses 6,0: the day (corresponds) to the day.
It is full and it does not transgress 6,0: a day is lacking.
11'[It is hollow and] it transgresses 6,0: a day is in excess.
It is hollow and it does not transgress 6,0: the day (corresponds) to the day.

P4' (Computing *M* or *N*)

12'In order for you to compute the [dur]ation? [...].

a If half of the computed night exceeds the duration (from midnight to lunation) 13'you deduct [the duration from half of the computed night]; what remains you call (the time) until sunrise.

b If half of the computed night 14'[is less than the duration (from midnight to lunation)], you deduct ⸢half the night⸣ from the duration; what remains [you call the time (since) sunrise].
15'[...] than the duration [...] 16'[... you deduct a]nd what ⸢remains⸣ [...]
(*n lines missing*)
Reverse

P5' (Unclear)

(*about n lines missing*)
1'[...] ⸢2⸣ [...] 2'[...] 3'[...] ⸢24?⸣ [...] 4'[...] 5'[...] ⸢10⸣ [...] ⸢...⸣6'[...] 7'[...] ⸢... 20 ...⸣ [...] 8'[...] ⸢3.18⸣ [...] ⸢...⸣9'[...] ⸢... 21?⸣ [...] ⸢...⸣10'[...] ⸢... 10⸣[...] 11'[...] ⸢...⸣12 ⸢...⸣ [...] ⸢...⸣12'[...] ⸢3?.36?⸣ [...]
(*about m lines missing*)

9' **me**: meaning unclear. Perhaps read *me* = 100? Or read 20 as *šamaš*$_2$?
10' **la**$_2$: presumably *maṭi*, 3 m. sg. stat. G *maṭû*, or *imaṭṭi*, 3 c. sg. pres. G, 'it is lacking'. A reading *tumaṭṭa* (2 m. sg. pres. D), 'you diminish', is less likely, on account of *uttar* in the next line.
11' *ut-tar*: *uttar* = 3 c. sg. pres. G *watāru*, 'it exceeds'.
12' Before *a-na* a winkelhaken is visible, perhaps belonging to MAN.
GIŠ.GIŠ.ḪI: the correct logographic reading is unclear, but the meaning is certainly 'you compute'; cf. the Glossary.
14' In the gap at the end the time interval between sunrise and the moment of lunation must be mentioned, probably **me nim**(.**a**), 'time (since) sunrise'.

Commentary

This fragment belongs to the right edge of a tablet. No other edges are preserved. From the textual restorations on the obverse it follows that about 1.5 cm is missing on the left, so the original width of the tablet, or column, was about 11 cm. This fragment is one of the few extant lunar procedure texts from Uruk. It probably originates from the Rēš library (cf. p. 9). The preserved text on the obverse is divided into 4 procedures by horizontal lines. The reverse is badly damaged, so that only few signs can be read.

P1' Computing C or D from B?

An unknown number of lines are missing from the beginning of P1'. Each of the lines O2'–6' preserves an identical segment of an instruction but no numerical parameters, which makes a unique interpretation difficult. Possibly they belong to an interpolation scheme for the length of daylight, C, or night, D (§4.5.5).[682] Perhaps an interpolation coefficient (c_k) is multiplied by the distance to the control point, $B - B_k$, in accordance with Eq. (2.10) on p. 41 (§2.6.7). The term *ṣilipti*(**bar.nun**) LA$_2$-*ti*, provisionally translated as '*ṣiliptu*-coefficient for the duration', probably denotes the product $c_k \cdot (B - B_k)$, which is to be added to or subtracted from the control value C_k or D_k, 'duration' being a well-established technical term for C or D. The 'subtraction and addition' (O7') might refer to the subtractive or additive nature of this term.

P2' *Assembling the Lunar Six interval from Q and R?*

This short procedure consisting of 2 lines is almost intact. In O9' the result of a multiplication is added to another quantity if the Moon is above the ecliptic, subtracted if it is below the ecliptic. This points to the computation of the Lunar Six intervals from Q and R in step 11 of the Lunar Six module (§4.5.18.10). The digit 40, 'the *igigubbû*-coefficient', if interpreted as 0;40,[683] might be an exemplary value of q, and the quantity 'for the zodiacal signs' by which it is multiplied might be the elongation (η). By the same token, 20 may be an exemplary value of r if interpreted as 0;20,[684] which is presumably multiplied by β' in O9'.[685]

P3' *Establishing the date of the lunation or the preceding midnight*

P3' contains rules for establishing the date of the lunation as written in column M (§4.5.17), or the date of the midnight immediately preceding the lunation as written in column L (§4.5.16).

P4' *Computing M or N*

The first 5 lines of P4' are partly preserved. They concern the time of the lunation, M (§4.5.17), or that of the Lunar Six event, N (§4.5.18.1). The preserved text implies a Lunar Six event at sunrise, i.e. KUR, ŠU$_2$ or NA. Part **a** is consistent with case a of the algorithm for M (Eq. 4.125), or with the corresponding case of N_{KUR}, $N_{\text{ŠU}_2}$ or N_{NA} (§4.5.18.1); part **b** is consistent with case b. Perhaps the procedure continues in O15'–16' with the other cases (c, d).

P5' *Unclear*

After a break of unknown length, the tablet continues on the reverse with a badly damaged procedure or (auxiliary?) table. Only a few digits and some occurrences of the sign **la$_2$** ('subtractive', or 'you subtract'?) are preserved.

682 This interpretation differs from the one in *ACT*, p. 243.

683 This value would be approximately correct if the Moon is in Ari and rising.

684 This value is approximately correct if the Moon is in Ari and setting. Hence the two coefficients appear to be contradictory, because they apply in opposite circumstances.

685 Alternatively one might try to interpret P2' as step 6 of the Lunar Six module (§4.5.18.6), which is concerned with the computation of β'. That interpretation is less likely, because one would expect the elongation (η) to be multiplied by a coefficient of the order 0;4. The numbers in O8'–9' rule this out.

No. 94

museum numbers	BM 45818 (81–7–6,239) +45838 (262) +46192 (653)
provenience	Babylon
measures	7.5(*) × 9.5(*) × ? cm
arrangement	O/R
cuneiform text	photo: Fig. F.107 (R); *ACT*, Pl. 239 (R)
bibliography	*ACT* 221, 194a, Zkc; Neugebauer & Sachs (1967); Jones (1983) (partial transliteration)
contents	Moon system B: daily motion table (O,R), procedure (R)

Reverse
Column i

P1' (unknown number of lines missing)
1' ⸢**tur**-*er*⸣ [xxxx]
2' **nim**-⸢*ma* [xxxx]
3' **mu.du** [xxxx]
4' 44 [xxxx]
5' [xxxxx]
Co (about 5 lines missing)
11' [**mu**.1-*me*-*1*.19.**kam**]
12' [*ša*$_2$ *ši-i*] ⸢**mu**.2-*me*-*43*.**kam**⸣
13' [^{m}]⸢*ar*$_5$⸣-*ša*$_2$-*ka-a* **lugal**
14' [*u* ^{mi_2}]⸢*pi*⸣-*ri-us-ta-na-a*
15' [**dam**-*šu*$_2$] ⸢**gašan**⸣
(unknown number of lines missing)

Reverse column i

P1 (Computations involving F^*)

(unknown number of lines missing)
[1']⸢it is small(est)⸣ [...] [2']you deduct, ⸢and⸣ [...] [3']it proceeds [...] [4']0;44 [...] [5'][...]

Co (Colophon)

(about 5 lines missing)
[11'][Year 179] [12'][which is] ⸢year 243⸣, [13']⸢Ar⸣šakā was king [14'][and] Piriustanā, [15'][his spouse], ⸢was queen⸣.
(unknown number of lines missing)

Critical and philological notes

Reverse
13' *ar*$_5$: a rare phonetic reading of GAM.

Commentary

This fragment preserves a small segment of the left edge of a tablet, but no other edges. The obverse and reverse contain a daily motion table for the Moon (*ACT* 194a), computed in accordance with system B, for which I refer to Volume II. The leftmost column on the reverse preserves a procedure and part of a colophon.

P1' Computations involving F^* and its net difference for 30^d

The beginning of P1' is lost. Only the first few signs of 4 lines are preserved, sufficient to conclude that the procedure is concerned with the zigzag function for F^*, and $dF^*(30) = 0;44$, the net difference for 30^d (cf. §4.5.10).

No. 95

museum numbers	BM 55555 (82–7–4,143) +55562 (152)
provenience	Babylon
date	cf. the commentary
measures	10.0(12–13?) × 12.0(*) × 3.2–4.0 cm (type C?)
arrangement	O/R
cuneiform text	photo: Fig. F.103; *ACT*, Pl. 243a
bibliography	*ACT* 210; Steele (2006b); Britton (2007b), pp. 129–130 (P3')
contents	Jupiter?; Saturn system A?; Moon system B: periods; computations involving A, B and Ψ'

Obverse
Column i
(m lines missing)
P1' 1' [...xxx] ⸢x⸣ **ugu tab**-*ma*?
2' [...xx] ⸢xx⸣ GAM$_0$ **e**-*bi*
3' [...x] ⸢x⸣ 14.3.45
4' [...] ⸢x⸣7.30 **ki** 11.4
5' [...] **tab**-*ma* **u$_4$.meš** ⸢x⸣
6' [...xxxx] BE
(n lines missing)
Column ii
(about m−14 lines missing)
P2' 1' [xxxxxxx 29] ⸢**mu**?⸣**.meš** 5.12 [xxxxx]
2' [xxxxxx] 7.30 **tab** *a-na* 1.⸢28?⸣ [**mu.meš** xx]
3' [xxxxx] ⸢*a*⸣-*na* 4.25 **mu.meš** ⸢**ki**⸣-*šu*$_2$ ⸢x⸣-[xx]
4' [xxxxx] (blank) ⸢5?9 x⸣ [x]

P3' 5' [xxxx] ⸢x⸣ **bal**-*šu*$_2$ ^{d}*sin ana* **ki**-*šu*$_2$ [**gur** xx]
6' [29.31].50.08.20 **u$_4$.meš šid**-*ut* **i**[**ti** *ša*$_2$ ^{d}*sin*]
7' [5.54].⸢22⸣.1.40 **u$_4$.meš** *ša*$_2$ ⸢1⸣2 **iti.meš** *ša*$_2$ ^{d}⸢*sin*⸣
8' [5.27].⸢5⸣1.20 **u$_4$.meš šid** 12 **iti** *ša*$_2$ ^{d}*sin ana* **ki**-*šu*$_2$ ⸢**gur**⸣
9' [1.4]⸢9.4⸣4.31.20 **u$_4$.meš** *ša*$_2$ 18 **mu.meš** *ša*$_2$ ^{d}*sin ana* **ki** ⸢**gur**⸣
10' [1].⸢4⸣9.45.19.20 **u$_4$.meš** *ša*$_2$ 18 **mu** *ša*$_2$ ^{d}*sin*
11' [1].⸢4⸣9.34.⸢2⸣5.27.18 **u$_4$.meš** *ša*$_2$ 18 **mu** *ša*$_2$ ^{d}**utu**
12' [*ana*] ⸢x⸣ **gur** *ina* 18 **bal.meš**
13' [10.5]⸢3⸣.52.42 **u$_4$.meš** *ana* **gaba.ri mu.an.na** *ša*$_2$ ^{d}**utu tab**
14' [xx] ⸢1⸣1.4 ⸢*ana* **gaba**⸣**.ri mu.an.na** *ša*$_2$ ^{d}**utu** [**tab** x?]
15' [xxx]**.meš šid**-*ut* **mu.an.na** *ša*$_2$ ^{d}*sin*
16' [xxxxx] **ab$_2$.meš šid**-*ut* **mu.an.na** *ša*$_2$ ^{d}*sin ana* **k**[**i**?-*šu*$_2$ **gur**]
17' [xxxxx **ab$_2$.meš**] ⸢**šid**-*ut*⸣ **mu.an.na** *ša*$_2$ ^{d}**utu** ⸢*ana*⸣ [**ki**-*šu*$_2$ **gur**]
18' [xxxxxxxxxx] ⸢1⸣.45 *tur-ru* [1.58 *kun-nu* xxxx]
(about n+2 lines missing)
Reverse
Column i
(about n−5 lines missing)
P4' 1' **ki.meš** ⸢*ša*$_2$ xxx⸣ [xxxxxxxxx]
2' 30 BAR 29.51.42 ⸢x⸣ [xxxxxx]
3' 30 BAR 28.21.42 ⸢*taš*?⸣ [xxxxxx]

P5' 4' **ki.meš u$_4$**.14.**kam** *ana* **u$_4$**.14.**kam** ⸢**alla**?⸣ [28.10.39.40]
5' **maš$_2$** 30.01.59 **ab$_2$** *ana* **ab$_2$** [xxxxx]

P6' 6' ⸢*ina*⸣ 18 **mu.meš** 38 **iti an.mi** [*sin* xxxxx]
7' ⸢*ina* 18 **mu**⸣**.meš** 38 **iti.meš an.m**[**i** *šamaš*$_2$ xxxxx]
8' [xx] ⸢**u$_4$**⸣**.meš a.ra$_2$** 3.43 [xxxxxx]
9' [xxx] ⸢x⸣ **u$_4$.meš a.ra$_2$** 3.4⸢3⸣ [xxxxxx]
10' [xxx] ⸢x⸣ **u$_4$.meš a.ra$_2$** 3.43 [xxxxx]

Obverse column i

P1' **(Jupiter?)**

(m lines missing)
1'[...] ⸢...⸣ you add? to it, and 2'[...] ⸢...⸣ ... you call it 3'[...] ⸢...⸣ 0;14,3,45 4'[...] ⸢...⸣7.30 with 11;4 5'[...] you add, and the days ⸢...⸣ 6'[...] ...
(n lines missing)
Column ii

P2' **(Saturn system A: net displacements?)**

(about m−14 lines missing)
1'[... For 29] ⸢years?⸣ 5;12 [... you subtract ...]
2'[For ... years ...]7,30 you add.
For 1,⸢28?⸣ [years you ...] 3'[...]
For 4,25 years its position ⸢...⸣ [...]
4'[...] ⸢5?9 ...⸣ [...]

P3' **(Moon system B: lunar and solar periods)**

a 5'[... in?] ⸢...⸣ revolutions the Moon [returns] to its position [...].
6'[29;31],50,08,20 days is the number for the mo[nth of the Moon.]
7'[5,54];⸢22⸣,1,40 days is for ⸢1⸣2 months of the ⸢Moon⸣.
8'[5,27];⸢5⸣1,20 days is the number for 12 months in which the Moon ⸢returns⸣ to its position.
9'[1,4],⸢9,4⸣4;31,20 days for 18 years in which the Moon ⸢returns⸣ to its position.
10'[1],⸢4⸣9,45;19,20 days for 18 years of the Moon.

b 11'[1],⸢4⸣9,34,⸢2⸣5,27,18 days for 18 years in which the Sun 12'returns [to] the ⸢...⸣ in 18 revolutions.
13'[10;5]⸢3⸣,52,42 days you add to the value of the 'year' of the Sun.
14'[...] ⸢1⸣1;4 [you add] ⸢to the value⸣ of the 'year' of the Sun.
15'[... month]s is the number for the 'year' of the Moon.
16'[...] months is the number for the 'year' of the Moon in which [it returns] to [its] pos[ition.]
17'[... months] is the ⸢number⸣ for the 'year' of the Sun in which [it returns] to [its position.]
18'[...] 1,45 are hollow, [1,58 are full ...]
(about n+2 lines missing)
Reverse column i

P4' **(Moon system B: computations involving B?)**

(about n−5 lines missing)
1'Positions ⸢of ...⸣ [...]
2'... 29;51,42 ⸢...⸣ [...]
3'... 28;21,42, ⸢the difference?⸣ [...]

P5' **(Moon system B: parameters of $A^{(1)}$)**

4'Positions for day 14 until (the next) day 14. Cn[c?: 28;10,39,40], 5'Cap: 30;1,59, month by month [...]

P6' **(Moon system B?: period relations for eclipses)**

6'⸢In⸣ 18 years there are 38 months for a [lunar] eclipse [...]
7'⸢In 18⸣ years there are 38 months for a [solar] eclip[se ...]
8'[...] days times 3,43 [...] 9'[...] 20 days times 3,43 [...]
10'[...] days times 3,43 [...]

P7’ 11’ [xxx *kip*]-*pat ina igi-gub*-⸢*u*$_2$⸣ [xxxxxxxxx]
12’ [xxx]-*iš* 2.10.20 ⸢x⸣ [xxxxxxxxx]
13’ [xxx] ⸢x⸣ *taš* **ab**$_2$ *ana* **ab**$_2$ [xxxxxxxxxx]
14’ [xx **ta**] ⸢**absin**$_0^?$⸣ **en** *zib a-n*[*a* 1 **uš** xxxxxxxx]
15’ [xxx **ab**$_2$] *ana* ⸢**ab**$_2$⸣ [xxxxxxxxxxx]
(about m−4 lines missing)
Column ii
(about n−5 lines missing)
P8’ 1’ [...]-*tim*
2’ [...]
3’ [...]
4’ [...] ⸢x⸣-*ka*
(about m+9 lines missing)

Critical and philological notes

Obverse column i

1’ The sign before **ugu** might be a digit 30–50. **tab**-*ma*: or *šap*?
2’ Before GAM$_0$ (or 9?) remains of a vertical wedge are visible.
3’ The traces before 14 include a horizontal wedge, which rules out 3.14.3.45 (*ACT*).
4’ 7.30 is preceded by traces of two winkelhakens on top of each other, which rules out 28.7.30 (*ACT*).
5’ After **u**$_4$**.meš** there is a sign, perhaps 30 or KUR.

Column ii

1’ Before **meš** there are faint traces.
2’ 8: very uncertain.
4’ After 59 there are traces of a sign, perhaps a digit 3–6.
5’ **bal** is preceded by a sign containing a vertical wedge and traces of further wedges.
6’ **šid** = *minûtu*, ‘number’.
12’ Before **gur** only the upper part of a vertical wedge is visible. Britton (2007b) has proposed a restoration *ana* **gub**, ‘(it returns) to the solstice’, instead of *ana* **ki**-*šu*$_2$ (*ACT*). A comparison with other instances of *šu*$_2$ on the tablet confirms that the preserved wedge is more likely the final wedge of **gub** (or **ki**?) than part of *šu*$_2$.
13’ **mu.an.na** = *šattu*, ‘year’; here it means ‘period; interval’.
15’ **meš**: **ab**$_2$**.meš**, ‘months’, or **u**$_4$**.meš**, ‘days’; cf. the commentary.

Reverse column i

1’ After *ša*$_2$ there are traces of what appear to be numbers; perhaps 7.12.1 (very uncertain).
2’,3’ 30 BAR: meaning unclear. The interpretation **eš.bar** = *purussû*, ‘decision; prediction’, suggested in *ACT*, does not appear to result in a meaningful sentence.
2’ After 42 there are traces of a winkelhaken.
4’ **alla**: or *mim* plus traces.
8’ Before **u**$_4$**.meš** Neugebauer identified a damaged 20, but this could not be confirmed.
9’ 10: or 20, 30 (erroneously transliterated as line 8’ in *ACT*).
11’ *kip-pat*: *kippatu* = ‘circle, circumference’, a term for eclipse magnitude.
u$_2$: replaces *ba* (*ACT*).
12’ 20: followed by a sign that begins with a vertical wedge.
13’ *taš*: phonetic abbreviation of *tašpiltu*, ‘difference’.
14’ Before **en** there is a 1, perhaps the end of a sign now damaged.

Commentary

This fragment belongs to the right edge of a tablet; other edges are not preserved. Both sides contained two columns of text, but almost nothing remains of column i on the obverse and column ii on the reverse. If the columns were of equal width then the original width of the tablet must have been about 12–13 cm. The subject of P1’ and P2’ could not be established with certainty, but the parameters point to Jupiter (P1’) and Saturn (P2’). If correct, this is the only known procedure text containing both planetary and lunar procedures. As far as the other procedures are understood they belong to, or have an affinity with, lunar system B. None of these has a practical purpose related to the production of synodic tables. The tablet is similar to No. **96**, which belongs to the same BM collection 82–7–4. As pointed out by Steele (2006b) and Britton (2007b), there are also orthographic similarities with No. **97**. According to Steele (2006b) and Britton (2007b), citing Rawlins (1991), the tablet may have been written after 135 BC. This conclusion is based on the mention of a year length that is otherwise unattested in Babylonian astronomy and considered to have been imported from Greek astronomy after having been established by Hipparchos through observations in 135 BC (cf. P3.b).

P7' **(Moon system B?: computations involving eclipse magnitude)**

11'[... the 'cir]cle' from the coefficient [...] 12'[...]... 2.10.20 ⌜...⌝ [...] 13'[...] the difference from month to month [...] 14'[... from] ⌜Vir⌝ until Psc, for [1° you ...] 15'[... from month] to month [...]

(about m−4 lines missing)

Column ii

P8' **(Unclear)**

(about n−5 lines missing)

1'[...] ... 2'[...] 3'[...] 4'[... for] you to ⌜...⌝

(about m+9 lines missing)

P1' Procedure for Jupiter?

The beginning and end of this procedure are missing. Six lines of text are partly preserved, each containing only a few signs. Neugebauer's restorations of the digits 14.3.45 (Oi3') and 7.30 (Oi4') as parameters of the algorithm for B of lunar system A (§4.4.4) are inconsistent with the traces of the adjacent signs. Steele (2006b) proposed that P1' is concerned with Saturn, but I tentatively identify 14.3.45 as 0;14,3,45 = v, a value of Jupiter's 'daily' displacement known from scheme A'.S$_1$ for the subdivision of the synodic cycle (Table 3.44, p. 98). A connection with planets is also suggested by the number 11.4, which can be interpreted as $11;4^{\overline{\tau}} = e$, the yearly epact, an important parameter of the planetary systems (cf. §3.3.2.1).

P2' Saturn system A, net displacements?

As suggested by Steele (2006b), this procedure probably concerns Saturn. Only a few signs of the final 4 lines are preserved. The interval of 4,25 years (Oii3') is tentatively identified as the number of years $Y = \Pi + Z$ corresponding to an exact return of Saturn's position in systems A, B and B' (cf. Tables 3.57a; 3.61). The digits 5.12 in Oii1' might belong to $-5;12,30° = dB_1(28)$, Saturn's net displacement for 29^{yr} in zone 1 of system A (Table 3.57b, p. 107), the digits 7.30 (Oii2') to $dB_1(29)$ or $dB_1(57)$.

P3' Moon system B: lunar and solar periods

P3' contains a list of periods for the Moon (part **a**) and the Sun (part **b**). Part **a** (Oii5'–10') begins with $29;31,50,8,20^{\mathrm{d}} = \overline{m}_{\mathrm{syn}}$, a value of the mean synodic month compatible with the mean value of G (§4.5.12). In Oii7' the duration of 12^{m} is given as $5,54;22,1,40^{\mathrm{d}} = 12 \cdot \overline{m}_{\mathrm{syn}}$. Oii10' gives the duration of 223^{m} (1 saros) as $1,49,45;19,20^{\mathrm{d}} = 223 \cdot \overline{m}_{\mathrm{syn}}$. These intervals are said to produce a return of the Moon's position, but strictly speaking they only produce a close return to the same position in lunar system B. 5,27;51,20 (Oii8') is the duration of 12^{m} expressed in days. As explained by Neugebauer, it is reproduced by computing the time required for the Moon to perform 12 sidereal revolutions at a rate 13;10,35°/day (= μ of $F^{(1)}$; cf §4.5.10). The closest 5-digit approximation of 12 × 6,0 / 13;10,35 would be $5,27;51,33^{\mathrm{d}}$, but the text has $5,27;51,20^{\mathrm{d}}$. A closer approximation of that number would be obtained if one assumes that the computation is based on a slightly different mean value 13;10,35,30°/d, a variant of F not attested anywhere else. 1,49,44;31,20 (Oii9') can be explained in a similar fashion as the number of days required for the Moon to carry out 241 = 4,1 full revolutions, the closest number of sidereal revolutions contained in 1 saros. The nearest 5-digit approximation of 4,1 × 6,0 / 13;10,35 (using μ of $F^{(1)}$) is $1,49,44;30,18^{\mathrm{d}}$. Closer agreement is obtained if one employs $F^{(3)}$, whose mean value μ = 13;10,34,51°/d, which results in $1,49,44,31,33^{\mathrm{d}}$.

Part **b** continues with solar periods. Oii11'–12' mention that it takes $1,49,34;25,27,18^{\mathrm{d}}$ for the Sun to return to its position 18 times. As pointed out by Britton (2007b), pp. 129–130, the corresponding length of the year, 1,49,34;25,27,18 / 18 = $6,5;14,44,51^{\mathrm{d}}$, cannot be connected in any obvious way with parameters of lunar systems A or B, and might originate from Hipparchos. Some entries below Oii12' contain the word **mu.an.na**, 'year'. The context implies that this must be a technical term for 'period' or 'interval'. In Oii12'–13' $10;53,52,42^{\mathrm{d}}$ are added to the duration of 18 solar years in order to obtain the duration of 18 lunar years. The origin of this number is not clear. The 18 solar years are called the 'value of the 'year' (**mu.an.na**) of the Sun'. Elsewhere (§2.2.2.1) I have argued that the inclusion of the word *meḫertu*(**gaba.ri**), 'value; equivalent', in Oii13' can be explained by noting that the identity of the sum differs from that of the summands. Oii14' contains a similar statement by which 11;4 is added to the solar 'year'. This parameter can be identified as e, the excess of 1^{yr} beyond 12 (mean) synodic months (or, equivalently, $6,0^{\overline{\tau}}$; cf. Eq. 3.10 in §3.3.2.1). If this interpretation is correct then the 'year of the Sun' appears to have different meanings in different contexts. In Oii15' a "year' of the Moon' is mentioned, but its value is broken away. Oii16' provided a period ('year'), measured in months, after which the

Moon returns to the same position, but the numerical value is broken away. In Oii17' a similar period is mentioned for the Sun.[686] As explained by Neugebauer, the number 1,45 (Oii18') is the average number of hollow months in 1 saros if one sets out from the length of the saros mentioned in Oii10', i.e. $1,49,45^{d} = 1,45 \times 29^{d} + 1,58 \times 30^{d}$.

***P4'** Moon system B?: computations involving A and B?*

The sparse remains of P4' concern the Moon's zodiacal position. The meaning of 30 BAR which appears at the beginning of Ri2' and 3' is not clear. The numbers 29;51,42 and 28;21,42 might be values of the Moon's synodic arc, function A of lunar system B (they are in between the extrema of A; cf. §4.5.3). Their significance is unclear.

***P5'** Moon system B: parameters of A^{fm}*

This procedure consisting of 2 lines concerns the zodiacal position of Full Moon (B^{fm}), as mentioned in the introductory statement (Ri4'). It provides the extrema of $A^{(1)}$ (§4.5.3) and the corresponding approximate zodiacal positions.

***P6'** Moon system B?: computations involving the saros*

P6' deals with eclipse possibilities, of which there are 38 in 1 saros = $3,43^{m}$, as mentioned in Ri6'–7'. The saros also shows up in each of the following lines. Three times this number is multiplied by a time interval measured in days, but the outcome is not preserved, nor can it be reconstructed. Similar multiplications occur in No. **96** O3'–4' and No. **97** P2'.

***P7'** Moon system B?: computations involving eclipse magnitude*

From the introductory phrase it appears that P7' deals with a function for eclipse magnitude (cf. §4.5.7, §4.5.9), but its precise purpose is not clear. The number 2.10.20 (Ri12') remains unidentified. In Ri13' a monthly difference is mentioned. Ri14' contains a reference to an interpolation coefficient (c), perhaps for computing $\Delta\Psi'$ from B, similar to the algorithm for $\Delta\Psi^{(3)}$ (cf. §4.5.8). Unfortunately the value of c is not preserved.

***P8'** Unclear*

Too little remains of this procedure to determine its content.

686 Neugebauer restored the missing periods as $4,1 = 241^{m}$ and $3,43 = 223^{m}$, respectively.

No. 96

museum number	BM 55545 (82–7–4,129)
provenience	Babylon
measures	6.2(14–15?) × 7.3(*) × 2.5–3.0 cm (type C?)
arrangement	O/R
cuneiform text	photo: Fig. F.108; *ACT*, Pl. 243a
bibliography	*ACT* 211
contents	Moon system B: various procedures

Obverse

P1' (n lines missing)

1' ⸢9.52.15 GAM⸣ [xxxxxxxxxxxxxxxxxxxxxxxx 9.52.15]

2' GAM 12 **du**-*ma* 1.⸢58.27⸣ [xxxxxxxxxxxxxxxxxxxxxxxxx]

3' *ina* 18 **mu.meš** 38 **ab$_2$.meš** ⸢**an**⸣.[**mi** xxxxxxxxxxxxxxxxxxxxxxxxx]

4' 4.29 **u$_4$.meš** GAM 3.43 **du**-[*ma* 16.39.47 xxxxxxxxxxxxxxxxxxxx]

P2' 5' *kip-pat ina igi-gub-bu-u$_2$* ⸢ḪAB$^?$⸣-*r*[*at$^?$* xxxxxxxxxxxxxxxxxx **ta** *zib*.**me en absin** *ana* 1 **uš**]

6' 1.6.40 **tab ta absin en** *zib*.**me** *ana* 1 [**uš** 1.6.40 **la$_2$** xxxxxxxxxxxxxxxxxxxx]

7' *ina* **maš$_2$** ⸢44.46⸣.55.33.20 **sig** 30 [xxxxxxxxxxxxxxxxxxxxxxxxx]

8' *ana tar-ṣa* **an.ta** 31.22.30 *ina* 1$^?$ ⸢x⸣ [xxxxxxxxxxxxxxxxxxxxxxxx]

P3' 9' *si-man* **ta zi** GIŠ-*u$_2$* *mim-ma ša$_2$* [*al* 15.16.5 **la$_2$** GAM 37.30 **du ki** 1.52.34.35 **tab** *mim-ma ša$_2$ al* 11.5.5 **diri**]

10' GAM 37.30 **du ta** 4.29.27.5 [**la$_2$ zi ta** *si-man* GIŠ-*u$_2$* *mim-ma ša$_2$ al* 1.52.34.35 **diri** GAM 1.36 **du**]

11' **ta** 15.16.5 **la$_2$** *mim-ma ša$_2$* ⟨*al*⟩ 4.⸢29⸣.[27.5 **la$_2$** GAM 1.36 **du ki** 11.5.5 **tab** xxxxx]

11a' 5

P4' 12' *ša$_2$* **tag$_4^?$** ⸢KI⸣ RI ⸢MU$^?$⸣ *ana* ⸢*tar-ṣa*⸣ **nu ur** *ana tar*-[*ṣa* xxxxxxxxxxxxxxxxxxxx]

13' 14.26.40 ⸢**tab**⸣ **ab$_2$** *ana* **ab$_2$** 5 *taš ša$_2$* ⸢*ana*⸣ [xxxxxxxxxxxxxxxxxxxxxx]

14' GAN *ši-i* **igi**-*šu$_2$* *lu ana* **nim** *lu* ⸢*ana*⸣ [xxxxxxxxxxxxxxxxxxxxxxxx]

15' **be**-*ma* 2 **šu.si** 2.40 **e**-*bi* [xxxxxxxxxxxxxxxxxxxxxxxxxxx]

Reverse

P5' 1 *igi-gub-bu-u$_2$* GAM *ša$_2$* **ta** *si-man* GIŠ-*u$_2$* ⸢x⸣ [xxxxxxxxxxxxxxxxxxxxxxxxxx]

2 **be**-*ma* **ki** *kip-pat* **dib**-*iq* 35.20 35.20 ⸢**nim** x$^?$⸣ [xxxxxxxxxxxxxxxxxxxxxxxxxx]

3 *ša$_2$* **ta** 2.13.20 GIŠ-*u$_2$* *mim-ma ša$_2$ al* **murub$_4$ diri** ⸢GAM 1$^?$⸣ [xxxxxxxxxxxxxxxxxxxxxx]

4 GAM 3.14.24 **du la$_2$** *an-na-a ša$_2$* GAM *taš* [xxxxxxxxxxxxxxxxxxxxxxxxxxxxx]

P6' 5 **tab** *ša$_2$* ḪAB-*rat* **ta** *si-man* GIŠ-*u$_2$* *mim-ma ša$_2$ al* ⸢x⸣ [xxxxxxxxxxxxxxxxxxxxxxxx]

P7' 6 GAM *ana tar-ṣa* **uš** *ana* **uš** *ina* **ḫun** 46.40.51.40 *ina* **alla** [xxxxxxxxxxxxxxxxxxx **ta alla en maš$_2$**]

7 *ana* 1 **uš** 1.6.40 **la$_2$ ta maš$_2$ en alla** *ana* 1 **uš** 1.[6.40 **tab** xxxxxxxxxxxxxxxxxxxxxxx]

8 10.30 30 **pa** 10.15 GAM *ša$_2$* 18 *ana* 18 ⸢xx⸣ [xxxxxxxxxxxxxxxxxxxxxxxxx]

9 30 **pa** 1.24 **ta pa en maš** 1⸢1$^?$ xxx⸣ [xxxxxxxxxxxxxxxxxxxxxxxxxxxxx]

10 *ana tar-ṣa si-man* **tab** *u* **la$_2$** 1.12 ⸢20$^?$⸣ [xxxxxxxxxxxxxxxxxxxxxxxxxx]

11 *mim-ma* GAM **bal**-*šu$_2$* *lu*-**du** *lu* 6.⸢4⸣[7.30$^?$ xxxxxxxxxxxxxxxxxxxxxxx]

12 ⸢x⸣ 46.30 **tab** *u* **la$_2$** 30 **maš** 48.13.4.[26.40 **tab** *u* **la$_2$** xxxxxxxxxxxxxxxxxx]

P8' 13 [x].**da.meš** *ša$_2$* **me** 1.49.45.19.20 ⸢x⸣ [xxxxxxxxxxxxxxxxxxxxxxxxx]

14 [xx] ⸢30 x 20 30 **du**$^?$ xxx⸣ [xxxxxxxxxxxxxxxxxxxxxxxxxx]

(about n lines missing)

Critical and philological notes

Obverse

1' 9: replaces 10 (*ACT*).
15: also 12 or 18 possible. This is followed by faint traces compatible with GAM.

2' 58: replaces 48 (*ACT*).

5' *kippatu*(**gam**) = 'circle': a technical term for eclipse magnitude. The word is consistently written without an ending.
The sign after *u$_2$* may be ḪAB or *šu$_2$* (*ACT*: **nim**).

7' 44: or 45.

8' The traces after *ina* 1 belong to a sign similar to TAR.

9' *al* 15.16.5 **la$_2$**, 'is less than 15;5,6': replaces *al* 15.16.5 **diri**, 'exceeds 15;5,6'.

Obverse

P1' **(Computations involving Ψ'' or Ψ')**

(*n lines missing*)
[... you multiply] **1'**⸢9;52,15 by⸣ [...] **2'**You multiply [9;52,15] by 12, it is 1,⸢58;27⸣ [...]
3'In 18 years 38 months (with) an ⸢eclipse⸣ [...] **4'**You multiply 4,29 days by 3,43, [it is 16,39,47 ...]

P2' **(Computations involving $\Delta\Psi'$; partly unclear)**

5'The 'circle' from the *igigubbû*-coefficient ⸢of the 'disk'?⸣.
[... From Psc until Vir for 1°] **6'**you add 0;1,6,40.
From Vir until Psc for 1[° you subtract 0;1,6,40 ...]
7'In Cap ⸢44;46⸣,55,33,20, the 'depth', 30 [...]
8'Opposite the 'height?' 31.22.30, in 1? ⸢...⸣ [...]

P3' **(Computing G from $F^{(1)}$ and vice versa)**

a **9'**The duration (of the month) from the computed displacement.
(The amount) by which it [is less than 15;16,5 you multiply by 37;30, add to 1,52;34,35.
(The amount) by which it exceeds 11;5,5] **10'**you multiply by 37;30, [subtract] from 4,29;27,5.

b [The displacement from the computed duration.
(The amount) by which it exceeds 1,52;34,35 you multiply by 0;1,36], **11'**subtract from 15;16,5.
(The amount) by which it [is less than] 4,29;[27,5 you multiply by 0;1,36, add to 11;5,5.]

P4' **(Computations involving eclipse magnitude; unclear)**

12'What remains? ... the opposite does not exist. Oppos[ite...] **13'**⸢you add⸣ 14.26.40. Month to month the difference ⸢for⸣ [...] is 5. [...] **14'**... this; whether it [...] its face? upwards or [...] **15'**If it is 2 fingers you predict 2.40 [...]

Reverse

P5' **(Computations involving eclipse magnitude; unclear)**

1The *igigibbû*-coefficient for the 'circle' from the computed duration. ⸢...⸣ [...]
2If it passes the position of the 'circle', 35;20, you subtract 35;20. ... [...]
[...] **3**which you had computed from 2,13;20. (The amount) by which it exceeds the middle [you multiply] ⸢by 1⸣ [...]
4you multiply by 3.14.24, subtract. Thus for the 'circle'. The difference [...]

P6' **(Computations involving $\Delta\Psi'$)**

5The addition for the 'disk' from the computed duration. (The amount) by which it [exceeds] the midd[le ...]

P7' **(Computations involving $\Delta\Psi'$)**

6The 'circle'. Opposite (one) degree to (the next) degree, in Ari: 46;40,51,40; in Cnc: [...
From Cnc to Cap] **7**for 1° you subtract 0;1,6,40. From Cap until Cnc for 1° [you add] 0;1,[6,40 ...]
810.30 30 Sgr 10.15, the 'circle' for 18 to 18 ⸢...⸣ [...] **9**30 Sgr 1.24. From Sgr until Gem 1⸢1⸣ [...]
10Opposite the duration the addition and subtraction is 1.12 ⸢20⸣ [...] **11**What should I multiply by its period so that the difference is 6,⸢4⸣[7,30...] **12**⸢...⸣ the addition and subtraction is 46;30. 30 Gem: [...] it is 48;13,4,[...]

P8' **(Computations involving G and the saros)**

13[...] ⸢...⸣ ... of the day; 1,49,45;19,20,⸢...⸣ [...] **14**[...] ⸢30 ... 20 30 ...⸣ [...]
(*about n lines missing*)

10' **zi ta** *si-man* **giš**-u_2, 'displacement from the computed duration': this phrase, omitted by Neugebauer, introduces the inverse transformation. Its inclusion largely resolves the inconsistent numbers of signs in O9'–10' as restored in *ACT*.
11a' 5: or ZA; meaning unclear.
12' $\mathbf{tag_4}$: the identity of this sign is uncertain; *râḫu*($\mathbf{tag_4}$), 'to remain', does not result in a meaningful sentence.
⸢KI⸣ RI ⸢MU⸣: meaning unclear. **nu ur** = *ul ibašši*, 'does not exist'. Instead of **ur** one might read **tuk**-*ši* (less likely).
13' *taš* = *tašpiltu*, 'difference'.
14' GAN: meaning uncertain. A reading *i* seems excluded. **igi**-*šu*$_2$: perhaps *pānšu*$_2$, 'its face'?
Reverse
3 $\mathbf{murub_4}$ = *qablu*, 'middle'.
7 *ana* 1: read as 2 by Neugebauer, but the large distance between the wedges implies that they are separate signs.
8 Traces following 18 *ana* 18 compatible with **mu** *ana* or 51.
9 **maš** 10 followed by traces of a number 1–8 and about 3 other signs.
10 20: very uncertain.
12 The sign before 46.30 is similar to KU (**ḫun**?), KI or LU.
14 Only the upper parts of the signs are preserved. The first 30 is followed by a digit 2–8.

Commentary

This fragment belongs to the lower (upper) left corner of the obverse (reverse) of a tablet. Nothing remains of the right edge and the upper (lower) edge of the obverse (reverse). The textual restorations in O9'–11' imply that the original width was at least about 14 cm. Most of the 8 partly preserved procedures belong to lunar system B, but P5' seems to have a connection with lunar system A, while P4'–P6' are too badly understood for a definite identification. P1'–P3' and P7' have a close affinity with No. **95**.

P1' Computations involving Ψ'' or Ψ'

The 4 final lines of P1' are partly preserved. O1'–2' contain computations involving eclipse magnitude. The preserved parameter $M = 9;52,15°$ is compatible with Ψ' (§4.5.7) and Ψ'' (§4.5.9). The purpose of the procedure is not wholly clear. In O2' M is multiplied by 12, which amounts to a conversion from ° to ^{f}, the result being 1,58;27^{f} (O2'). O3'–4' mention that there are 38 eclipse possibilities in 1 saros of 3,43 = 223 months ('18 years'). This number of months is multiplied by 4,29 = 269^{d}, an interval of unknown significance.[687] Similar multiplications of the form 3,43 times an interval measured in days (?) are attested in No. **95** P6', but in that procedure the number of days is not preserved.

P2' Computations involving $\Delta\Psi'$; partly unclear

According to the initial statement, a quantity called 'circle' is computed from (?) the '*igigubbû*-coefficient for the disk'. The precise meaning is not clear, but 'circle' and 'disk' are both known as technical terms for eclipse magnitude or its monthly difference. The number 0;1;6,40^{f} (O6') can be identified as the interpolation coefficient c from the algorithm for $\Delta\Psi'^{(3)}$, a (reconstructed) variant of the monthly difference of Ψ' (Table 4.35, p. 186). The number 44;46,55,33,20^{f} (O7') is the minimum m of $\Delta\Psi'^{(1)}$, which the text associates with the zodiacal sign Cap. Line O8' so far defies explanation, and the number 31.22.30 remains unidentified.[688] Some phrases and numbers mentioned in P2' reappear in P7'.

P3' Computing G from $F^{(1)}$ and vice versa

In P3' the 'duration', a term for G, the duration of the synodic month in excess of 29^{d}, is computed from the Moon's daily displacement F (cf. §4.5.10). The algorithm is based on interpolation with the extrema of both zigzag functions as the only control points, such that the minimum of $F^{(1)}$ is associated with the maximum of G, and vice versa.[689]

P4' Computations involving eclipse magnitude; unclear

P4' appears to be concerned with eclipse magnitude, but not much else is understood. O12' continues to defy interpretation. The number 14.26.40 and the monthly difference 5 (O13') and the number 2.40 (O15') remain unidentified. The meaning of O14'–15' is not clear, but the unit fingers points to eclipse magnitude.

P5' Computations involving eclipse magnitude; unclear

P5' is assumed to be a separate procedure, but it might also be the continuation of P4'. The underlying algorithm appears to be without parallel and is not fully understood, but some elements have been clarified due to improved readings. The connection with lunar system B is not clear. The '*igigubbû*-coefficient for the circle' (R1) is a technical term connected with eclipse magnitude. According to the introductory statement it is computed from the 'duration', as appears to be the case in P5'. It is doubtful that this refers to G, since G is controlled by the lunar variation (cf. §4.5.12 while $\Delta\Psi'$ is controlled by the zodiacal variation, so that both have fundamentally different periods. Perhaps it refers to J (§4.5.14), the zodiacal contribution to the synodic month, which has the same period as $\Delta\Psi'$, but the nature of the connection between these functions remains unclear.

The number 35.20 (R2) might be interpreted as 35;20^{f}, an alternative value of $2c_{\Psi}$, i.e. c_{Ψ} = 17;40^{f} (cf. step 3 in the algorithms for Ψ' and Ψ''; §4.5.7 and §4.5.9).[690] If this is correct, the 'position of the circle' (R2) is presumably the point where the Moon's centre crosses the ecliptic. In R3 an unidentified quantity is computed from '2,13;20'. This number points to Φ of lunar system A (§4.4.3), which is surprising for a tablet containing procedures for system B. The computed

---

687 If 4,29 is interpreted as a number of months then it equals $\Pi + Z$ for F, the number of anomalistic months contained in Π = 4,11^{m}.

688 The number 31;22 might represent the Sun's displacement in 30^{d}, computed as 30 · 1;2,44, where 1;2,44 is the maximum of a zigzag function for the Sun's daily displacement $v_{\odot}$ thus far attested only in lunar system A (§4.4.10).

689 Neugebauer overlooked this when reconstructing P3' in *ACT*, p. 274.

690 In *ACT*, Neugebauer proposed an interpretation as 0;35,20°, a value of the lunar diameter known from Greek astronomy.

quantity remains unidentified. In R4 a quantity is multiplied by 3.14.24, the outcome of which has a connection with eclipse magnitude, but the algorithm is not understood.

***P6'** Computations involving $\Delta\Psi'$; unclear*

P6' consists of a single line concerning the computation of the 'addition for the 'disk' from the 'computed duration'. The algorithm itself is not preserved. The goal function, 'addition for the disk', is probably $\Delta\Psi'$. The source function is called 'duration'. As in P5', it is doubtful whether this is *G*; perhaps it refers to *J* (§4.5.14). Again the nature of the connection between these functions remains unclear.

***P7'** Computations involving $\Delta\Psi'$*

The introductory phrase, 'The 'circle' opposite (1) degree to (the next) degree', may be a reference to the interpolation coefficient c = 0;1,6,40 (R6–7) also mentioned in P2', which belongs to $\Delta\Psi'^{(3)}$ (Table 4.35, p. 186). What follows may be an instruction for computing $\Delta\Psi'^{(3)}$ as a zigzag function of the zodiacal position. In R6' the number 46;40,51,40^{f}, presumably a value of $\Delta\Psi'$, is connected with the sign Ari. This number is close to μ = 46;30^{f} of variants 1 and 2; perhaps it belongs to a separate variant 3a (Table 4.35). At the beginning of R8 several numbers and the sign Sgr are mentioned, but the meaning is not clear. The phrase 'circle (= eclipse magnitude) from (one) 18 to (the next) 18' refers to the net difference of the eclipse magnitude for 1 saros = 223 months; perhaps these are the numbers in R9–10, but they do not match any of the $d\Psi''(s)$ (Table 4.34, p. 185). It has hitherto been overlooked that the number in R11 can be restored as 6;47,30° = d of $H^{(1)}$ (§4.5.13). Unfortunately the rest of the algorithm is broken, so that the nature of the connection between H and $\Delta\Psi'$, if there is any, remains unclear. The number 46;30^{f} (R12) is the mean value (μ) of $\Delta\Psi'$ (variants 1, 2 and 3b), which is said to be associated with Ari (presumably at 30°), and the number 48.13.4..., said to pertain to 30° Gem, can be interpreted as M = 48;13,4,26,40^{f} or 48;13,4,30^{f}, maxima of variants 1 and 2.

***P8'** Computations involving G and the saros*

Very little remains of this procedure. Neugebauer recognised that the number in R13 can be restored as 1,49,15;19,20,58, 20^{d} = 3,43 · 29;31,50,8,20^{d}, the number of days in 1 saros, where 29;31,50,8,20^{d} is the mean synodic month computed as 29^{d} + μ_G/6,0 (§4.5.12).

No. 97

museum number	BCM A.1845–1982.2 + unnumbered fragments a,b
provenience	Babylon
measures	10.9(20?) × 12.1(*) × 2.8–3.6 cm (type A?)
arrangement	O/R
cuneiform text	photo: Figs. F.104, F.105; Horowitz (2000), Fig. 3
bibliography	Horowitz (2000); Britton, Horowitz & Steele (2007) = BHS
contents	Moon system B: various procedures

Obverse
Column i
P0’ (about 10 lines missing)
11’ [...] ⸢7$^?$⸣.58
12’ [...] ⸢4$^?$⸣8
13’ [...] ⸢x⸣-*šu*$_2$
14’ [...]
15’ [...] ⸢x⸣ *ša*$_2$ **me** *ana* **me**
16’ [...] ⸢x 4$^?$⸣
(about n+12 lines missing)
Column ii
P1’ (perhaps 1 line missing)
1’ [xxxx] ⸢x-*ma*$^?$ 1$^?$⸣ [xxxxxxxxxxxx]
2’ [xxx].22.13.42.⸢10$^?$⸣ [xxxxxxxxxxxx]
3’ [xx].⸢4⸣3.18.45.30 [xxxxxxxxxxxx]
4’ [xx].⸢5$^?$⸣7.35.40 **tab** *u* ⸢**la**$_2$⸣ [xxxxxxxxxxx]

P2’ 5’ ⸢x⸣ *ša*$_2$*-niš* **mu** *ana* **mu** 11.3.⸢5⸣[5$^?$ xxxxxxxxxxxx]
6’ ⸢x⸣ 6 **mu.meš** *šal-šu*$_2$ **u**$_4$**.meš** ⸢x⸣ [xxxxxxxxxxxx]
7’ ⸢3⸣6 GAM$_0$ 3.43 **du**-*ma* 2.1[3.48 xxxx 32 GAM$_0$ 4.11]
8’ ⸢**du**⸣-*ma* 2.13.52 : 15.1⸢4⸣.[34.27.40 **la**$_2$ xxxx]
9’ 13.10.34.51 **zi** ^{d}*sin* ⸢*ša*$_2$⸣ [xxxxxxxxx]
10’ 11.6.3$^?$5.14.20 **sig** 18 ⸢*taš*⸣ [xxxx] ⸢xxxx⸣
11’ 36 *taš* **ab**$_2$ *ana* **ab**$_2$ [x$^?$]

P3’ 12’ *šal-šiš* 2.32.40.50 **la**$_2$ 1.50.⸢50⸣.[50 **sig** 6] ⸢*taš*⸣ **ab**$_2$ *ana* **ab**$_2$
13’ 29.6.19.20 **ki** ^{d}*sin* **ab**$_2$ *ana* ⸢**ab**$_2$⸣ [**maš**$_2$ 30.0.1]9.20 ⸢**alla** 28.12.19⸣.20
14’ *a-na* 1 **uš** 36 *taš* **ib**$_2$**.tag**$_4$ [xxxx] ⸢x⸣ **du**$_3$ **ta alla** *ana* ⸢**maš**$_2$ **gar**⸣-*an*

P4’ 15’ 59.8.9.48.40 **zi** ⸢*ša*$_2$⸣ [*šamaš*$_2$ xx] ⸢29⸣.6.19.20 **ki** *šamaš*$_2$ **ab**$_2$ *ana* **ab**$_2$
16’ ⸢**maš**$_2$⸣ 30.0.19.20 **alla** 28.⸢1⸣[2.19.20 *a*]-⸢*na*⸣ 1 **uš** 36 *taš*-⸢*pil-tu*$_2$⸣
17’ [xx **maš**$_2$] ⸢30⸣.02 **alla** 28.[10.40 xxx] 18 *taš-pil-tu*$_4$

P5’ 18’ [xxxx] ⸢x 30$^?$ x⸣ [xx **la**$_2$ xx]⸢8⸣.40 **sig** 7$^?$.⸢20$^?$⸣ *taš ša*$_2$ 12 *lu-maš*
19’ [xxxxxxxxxxx]⸢4$^?$5.20$^?$ **sig**$^?$ 5$^?$.20$^?$⸣ *taš ša*$_2$ ⸢6$^?$⸣ *lu-maš*
20’ [xxxxxxxxxx].⸢2$^?$⸣6.40 *mi* **ab**$_2$ *ana* **ab**$_2$ ⸢xx⸣ [x]
21’ [xxxxxxxxx] ⸢*taš*⸣ **ab**$_2$ *ana* **ab**$_2$ *ša*$_2$*-niš* 6 ⸢x$^?$ *taš*$^?$ **la**$_2$⸣
22’ [xxxxxxxxx] ⸢*taš*⸣ **ab**$_2$ *ana* **ab**$_2$

P6’ 23’ [xxxxxxxx *ša*$_2$] *al-la* 15.16.5 ⸢**la**$_2$ GIŠ⸣-*ma*
24’ [GAM$_0$ 37.30 **du ki** 1].⸢5⸣2.34.35 **tab**-⸢*ma ša*$_2$ *al*⸣-*lu*
25’ [11.5.5 xxxxx] ⸢x⸣ **diri** GIŠ-*ma* GAM$_0$ 37.30 **du**-⸢*ma*⸣
26’ [**ta** 4.29.27.5 xx]-⸢*ma*$^?$⸣ [xx$^?$]

P7’ 27’ [xxxxxxxxx] ⸢x *ša*$_2$⸣ *al-la* 1.52.34.3⸢5⸣ [x$^?$]
28’ [xxxxxxxxxxx] ⸢*ana*$^?$ 3$^?$.11$^?$.0$^?$.50$^?$⸣
(n lines missing)

Obverse column i

P0' **(Unclear)**

(about 10 lines missing)
11'[...] ⌜7?⌝.58 12'[...] ⌜4?⌝8 13'[...] its ⌜...⌝ 14'[...] 15'[...] ⌜...⌝ from day to day 16'[...] ⌜...4?⌝
(about n+12 lines missing)

Column ii

P1' **(Unclear)**

(perhaps 1 line missing)
1'[...] ⌜... and ...⌝ [...] 2'[...].22.13.42.⌜10?⌝[...] 3'[...].⌜4⌝3.18.45.30 [...] 4'[...].⌜5?⌝7.35.40 you add and sub[tract ...]

P2' **(Computations involving the year, $F^{(1)}$, $F^{(3)}$ and $F^{(3)*}$)**

a **(Computations involving the year; partly unclear)**

5'⌜...⌝ Secondly: year by year 11;3,⌜5⌝[5? ...] 6'[...] 6 years, a third day, ⌜...⌝ [...]

b **(Net difference of $F^{(1)}$ for 223^m?)**

7'You multiply 0;⌜3⌝6 by 3,43, it is 2,1[3;48. ...] 8'You multiply [0;32 by 4,11], it is 2,13;52.

c **(Parameters of the zigzag functions for $F^{(3)}$ and $F^{(3)*}$)**

15;1⌜4⌝,[34,27,40, the maximum ...] 9'13;10,34,51, the [mean] displacement of the Moon [...] 10'11;6,35,14,20, the minimum. 0;18, the difference [day by day ...] 11'0;36, the difference month by month.

P3' **(Functions $F^{(1)}$' and $A^{(3)}$)**

a **(Parameters of the zigzag function for $F^{(1)}$')**

12'Thirdly: 0;2,32,40,50, the 'height', 0;1,50,⌜50⌝,[50, the 'depth', 0;0,6], the difference from month to ⌜month⌝.

b **(Computing $A^{(3)}$ as a zigzag function of the zodiacal position)**

13'29;6,19,20: the Moon's position from month to ⌜month⌝. [Cap: 30;0,1]9,20; Cnc: 28;12,19,20. 14'For 1° the difference is 0;0,36. The remainder [of ...]: you do ⌜..., you put it down⌝ from Cnc to ⌜Cap⌝.

P4' **(Parameters of solar motion, $A^{(3)}$ and $A^{(2)}$)**

a **($v_{\odot}$ and parameters of the zigzag function for $A^{(3)}$)**

15'0;59,8,9,48,40 is the displacement ⌜of⌝[the Sun ...] ⌜29⌝;6,19,20, the position of the Sun from month to month. 16'⌜Cap⌝: 30;0,19,20, Cnc: 28;⌜1⌝[2,19,20 f]or 1° the ⌜difference⌝ is 0;0,36.

b **(Parameters of the zigzag function for $A^{(2)}$)**

17'[Secondly: Cap]: ⌜30⌝;02, Cnc: 28;[10,40]; 0;18: the difference.

P5' **(Unclear)**

18'[...] ⌜... 30 ...⌝ [... the 'height', ...]⌜8⌝.40, the 'depth', 7?.⌜20?⌝, the difference for 12 zodiacal signs. 19'[...] ⌜45.20? the 'depth', 5?.20?⌝, the difference for ⌜6?⌝ zodiacal signs. 20'[...].⌜2⌝6.40. What ⌜...⌝ month by month? 21'[...] ⌜the difference⌝ month by month. Secondly: 6, ⌜... the difference? you subtract⌝. 22'[...] ⌜the difference⌝ month by month.

P6' **(Computing G from $F^{(1)}$)**

23'[...] ⌜You compute⌝ (the amount) by which it ⌜is less⌝ than 15;16,5, 24'[you multiply it by 37;30, and] add it [with 1,5]2;34,35. You compute (the amount) by which 25'it exceeds [11;5,5 ...] ⌜...⌝, you multiply it by 37;30, ⌜and⌝26'[subtract it from 4,29;27,5 ...], ⌜and?⌝ [...]

P7' **(Computing $F^{(1)}$ from G?)**

27'[...] ⌜...⌝ (the amount) by which [it exceeds] 1,52;34,3[5 ...] 28'[...] ⌜... 3,11;0,50⌝
(n lines missing)

Reverse
Column i

P8' (about n lines missing)
1' [xxxxxxxxxxxxxx] ⸢x⸣ [x$^?$]
2' [xxxxxxxxxxxxxxxx] ⸢xxx⸣ 4.23 ⸢xx⸣ [x]
3' [xxxxxxxxxxxxxx] GAM$_0$ 4.23 **du** : *taš* **du$_3^?$** *lu*-[x]
4' [xxxxxxxxxx 8].2⸢5.5⸣3.20 ⸢GAM$_0$⸣ 4.23 ⸢**du**⸣ [x]
5' [xxxxxxxxxxx] ⸢8⸣.2⸢5.5⸣3.20 *ša$_2$* **nim** *u* ⸢**sig**⸣ [x]
6' [xxxxxxxxxx]⸢7⸣.6.23.20 ^{mul_2}**alla** 3⸢7⸣.[xxxx].20 **tab**
7' [xxxx **maš$_2$** xxxx]⸢1$^?$⸣.20 **la$_2$ iti** *ana* ⸢**iti**⸣ 12 *taš* [x]

P9' 8' [xxxxxxxxx **en**] ⸢**ḫun**$^?$⸣ **la$_2$**-*u$_2$* 6 ⸢**ta.am$_3$** 4 **tab**⸣ [x]
9' [xxxxxxxxxx] ⸢x⸣ **nim** *u* **sig** *ša$_2$* ⸢**ta**$^?$ *zib*.**me**$^?$⸣ **en** ⸢**ḫun**$^?$⸣ **du**-⸢x⸣
10' [xxxxxx GAM$_0$ xx] ⸢x⸣ **du**-*ma* 6.8.12 *mi-nu*-⸢*u$_2^?$*⸣ [x]
11' [xxxxxxxxx *lu*$^?$]-*u$_2$* 6.8.12 : 6.47.2⸢8⸣.[xx]
12' [xxxxxxxxxx] ⸢6⸣.8.12 : 6.47.28.[x].⸢5$^?$ x⸣ [x]
13' [xxxxxxxxxxx] 21 **la$_2$** $\frac{1}{2}$ GIŠ-*ma* 10.30 : 10.⸢30⸣ *taš*
14' [xxxxxxxxxxx] ⸢x⸣ 16.40$^?$ *mi-nu*-⸢*u$_2$* GAM$_0$⸣ 4.23
15' [**du**-*ma lu* xxxxxxxxx].⸢30⸣ : 2.9.52.2[1] GAM$_0$ 4.23
16' [9.29.16.28.3 xxxx] ⸢xx-*šu$_2^?$*⸣ GIŠ-*ma* 32.28.5.15 **tab** *u* **la$_2$**
17' [xx **alla** 32.28.5].⸢15⸣ **la$_2$ maš$_2$** 32.28.5.15 **sig**

P10' 18' [xxxxxxxx] **ta** *zib*.**me en ḫun la$_2$**-*u$_2$* 4 **ta.am$_3$** *ša$_2$* **tab**
19' [xxxxxxxx] ⸢6$^?$⸣.49.20 *taš* **ab$_2$** *ana* [**ab$_2$**] ⸢*an*⸣-*na-a* ⸢*taš*⸣ *ša$_2$* *taš-šu$_2$*
20' [x 6.4]9.20 [x] ⸢x⸣ **tab** (blank) ⸢**maš$_2^?$**⸣ [x 6.4]⸢9⸣.[20$^?$ x **ab$_2$**] ⸢*ana* **ab$_2^?$**⸣

P11' 21' [xxx] ⸢x⸣ **mu gar**-*an* **u$_4$** ⸢x⸣ **du** GIŠ *ša$_2$* **me** 3⸢2$^?$⸣ [xxxxxx]
22' [xxx] ⸢**zi**$^?$ **kur**$^?$⸣ *ša$_2$* **me** 32 **tur ki** *taš* [xxxxxx]
23' [xxxx] ⸢xxxx 5$^?$⸣ [xxxxxx]

P12' 24' [xxxxx] *ša$_2$* **tag$_4$** ⸢xx⸣ GAM$_0$ 5⸢4$^?$⸣ [xxxxxxx]
25' [xxxxx] 39.2.5$^?$.30.⸢10$^?$⸣ [xxxxxxx]
26' [xxxxxx] 52.⸢50$^?$⸣ *mi*-[xxxxxxxx]
(about 1 line missing?)

Critical and philological notes

Obverse column i

11' 7: or 8.
12' 48: or 58.
13' The sign before *šu$_2$* has an ending similar to **tab**.
15' Before *ša$_2$* there is a sign with an ending similar to U$_4$.
16' 4: perhaps any digit 4–8. It is preceded by a digit 30–50.

Column ii

2' 10: or any digit 20–50.
5' The lower parts of two vertical wedges remain of the initial sign.
6' After **u$_4$.meš** the beginning of a sign compatible with MU is visible.
9' As suggested by BHS, the gap may have contained **murub$_4$** = *qablu*, 'middle'.
10' 35: less likely 45 or 55. **sig** = *šuplu*, 'depth', in the sense of minimum value.
12' **la$_2$** = *šūqu*, 'height': replaces 1.10 (BHS).
13' **alla** = Cnc: replaces **ki** (BHS).
14' **ib$_2$.tag$_4$** = *rīḫtu*, 'remainder': replaces -*tum ina lib$_3$* (BHS). **maš$_2$ gar**-*an*: replaces xx 6 **kuš$_3$** (BHS).
16' **maš$_2$** = Cap: replaces **ki** (BHS).
18' 30: followed by traces of a vertical wedge. 7.20: or 8.20.
19' 45: or 55. The second 20: or any digit 20–50. 6: or 5.
20' 26: the 20 might be any digit 20–50.
21' After 6 there are faint traces of a sign.
23' *al-la*: replaces **tab** *taš* **ki** (BHS).
24' ⸢*ma ša$_2$ al*⸣-*la*: replaces xx-ma ⸢x⸣ (BHS).
25' Before **diri** there is a vertical wedge.
27' ⸢x *ša$_2^?$*⸣ *al-la*: replaces DIŠ 4 TAB *taš* KI (BHS).

Reverse column i

2' 4: replaces 6 (BHS)
3' *taš* **du$_3$** *lu*-: replaces 3 BAR.NUN (BHS).

Reverse column i

P8' **(Computations involving E?; unclear)**

(*about n lines missing*)

1'[...] ⸢...⸣ [...] 2'[...] ⸢...⸣ 4,23 ⸢...⸣ [...] 3'[...] you multiply [...] by 4,23 : you construct? the difference ... [...] 4'[...] ⸢you multiply [... 8].25.5⸣3.20 ⸢by⸣ 4,23 [...] 5'[...] ⸢8⸣.2⸢5.5⸣3.20 of the 'height and ⸢depth⸣' [...] 6'[...]⸢7⸣.6.23.20, Cnc: 3⸢7⸣ [...].20 you add. 7'[...]⸢1⸣.20, the 'height'. Month by ⸢month⸣ the difference is 12 [...]

P9' **(Computations involving $H^{(2)}$ and $J^{(2)}$)**

a **(Computations involving $H^{(2)}$)**

8'[...] ⸢Ari?⸣, decreasing, 6 ⸢each, you add 4⸣ [...] 9'[...] ⸢...⸣ 'height and depth' which proceeds ⸢from Psc?⸣ until ⸢Ari? ...⸣ [...] 10'[...] you multiply [by ...]⸢...⸣, it is 6.8.12. What 11'[should I multiply by ...] so that it is 6.8.12? : 6;47,28 [...] 12'[...] 6.8.12 : 6;47,28 [...] ⸢5...⸣. 13'[...] 21, the maximum. You compute half (of it), it is 10;30. 10;⸢30⸣, the difference. 14'[...] ⸢...⸣ 16.40?.

b **(Computations involving $J^{(2)}$)**

What [should I multiply] ⸢by⸣ 4,23 15'[so that it is ...].⸢30⸣? [You multiply] 2,9;52,2[1] by 4,23, 16'[it is 9,29,16;28,3 ...] You compute ⸢its ...th⸣, it is 32;28,5,15, the addition and subtraction. 17'[... Cnc: 32;28,5],15, the 'height'; Cap: 32;28,5,15, the 'depth'.

P10' **(Computations involving $H^{(3)}$?)**

18'[...] from Psc until Ari, decreasing, 4 each for the addition 19'[...] ⸢6⸣;49,20, the difference month by [month.] Thus the ⸢difference⸣ of its difference. 20'[... 6;4]9,20 [...] ⸢...⸣ you add; Cap: [6;4]9,[20... month] ⸢by month⸣.

P11' **(Unclear)**

21'[...] ⸢...⸣ year you put down?. ... ⸢...⸣ ... per day 3⸢2⸣? [...] 22'[...] ⸢...⸣ per day 32, the small(est) one, with the difference [...] 23'[...] ⸢... 5?⸣ [...]

P12' **(Unclear)**

24'[...] what remains ⸢...⸣ times 5⸢4?⸣ [...] 25'[...] 39.2.5?.30.⸢10?⸣ [...] 26'[...] 52.⸢50?⸣ what [...]

(*about 1 line missing?*)

6' ^{mul_2}**alla**, 'Cnc': the determinative is unexpected; replaces TAB 11 *lib*$_3$ (BHS). 7.6.23.20: or 8.6.23.20. 20 **tab** is written on the right edge. 37: or 38 if one assumes that one wedge is broken off.

8' **ḫun**, 'Ari': or **absin**$_0$, 'Vir'? **la**$_2$-*u*$_2$ 6: replaces 2 43 5 (BHS).

9' ⸢**ta**? *zib*.**me**?⸣ **en** ⸢**ḫun**? **du**-⸢x⸣: this provisional reading, suggested by the assumed parallel Ri18', is uncertain. Replaces ^{d}UTU KI *lib*$_3$ DUL.DU-*ma* (BHS). The signs ^{d}, KI and *lib*$_3$(**ša**$_3$) can be ruled out, but the others remain possible.

11' *u*$_2$: replaces 3 (BHS).

13' Replaces 32 *taš* **alla**$_x$ 30 : 30 *taš* (BHS).

14' Before 16 there are traces of a sign similar to 9. The 40 might also be a 50. GAM$_0$, 'times': omitted in BHS.

15' 2[1]: replaces 20 (BHS).

18' **ḫun**, 'Ari', or **absin**$_0$, 'Vir'? Replaces TA ŠU 1-EN ŠU LAL 13 4 TA A x *ša*$_2$ AB$_2$ (BHS).

19' ⸢*an*⸣-*na-a* ⸢*taš*⸣ *ša*$_2$: replaces : DIRI1 2 GAR (BHS).

20' Before **tab** there are vague traces of several vertical signs. The text after **maš**$_2$ was omitted by BHS.

21' **gar**-*an* **u**$_4$: or *ša*$_2$ ^{d}**utu**, 'of the Sun'? The traces after **u**$_4$ could not be identified.

22' **zi kur**: very uncertain.

23' At the beginning of the line there is a sign similar to **tag**$_4$; reading very uncertain.

24' *ša*$_2$ **tag**$_4$ = *ša rēḫu*, 'what remains': replaces 4 MAŠ$_2$ (BHS).

Commentary

The tablet is partly preserved on four joined fragments, which include a segment of the right edge and the lower (upper) edge of the obverse (reverse). On each side there were two columns of text, but very little remains of column i on the obverse, and nothing of column ii on the reverse. If both columns were equally wide then the original width of the tablet was about 20 cm. The tablet is divided into at least 11 procedures by horizontal lines. All procedures, as far as they are understood, belong to lunar system B but, as pointed out by BHS, some of the parameters deviate from the usual system B values. BHS ascribe these deviations to deliberate corrections introduced in the Arsacid era. Similarities with No. **95**, and the use of GAM as a separator between multiples of 10 and digits 1–9, which is rare in Babylon, may also be signatures of a late date.

P0' *Unclear*

The scant traces of the first column on the obverse are provisionally assigned to a single procedure, P0'. Its topic is not clear, except that O15' mentions a daily function.

P1' *Unclear*

The beginning of P1' is missing. The 4 remaining lines preserve sequences of digits, none of which has been identified. The fact that P2' concerns F and is introduced by the phrase 'Secondly' may suggest that P1' also concerns F. The number in O4' is probably a monthly or daily difference.

P2' *Computations involving the year, $F^{(1)}$ and $F^{(3)*}$*

P2' consists of 7 lines concerned with solar and lunar motion. Only the first half of each line is preserved. Part **a** (Oii5'–6'), as far as it is understood, deals with the length of the year. The number 11.3.50... (Oii5') was restored by BHS as 11;3,55$^{\bar{\tau}}$, a value of the yearly epact e, i.e. the excess of 1 year over 12 synodic months (Table 3.5, p. 62). This value of e is not attested elsewhere but can be computed from the value of $v_\odot$ mentioned in O15' and the mean synodic month implied by G (§4.5.12), namely

$$e = \frac{6{,}0}{0;59{,}8{,}9{,}48{,}40} - 12 \cdot 29;31{,}50{,}8{,}20 = 10;53{,}32{,}5{,}39{,}...^{\text{d}}, \tag{5.7}$$

which can be converted into mean tithis by multiplying it by $1^{\bar{\tau}}/1^{\text{d}}$ = 1;0,57,13,26,40 (for this factor cf. §2.5.2.3), resulting in 11;3,55,23,35,...$^{\bar{\tau}}$. The interval of 6 years and 1/3 days (?) (Oii6') remains unexplained.

Part **b** (Oii7'–8') contains instructions for computing $dF^{(1)}(223) = 3{,}43d - 32\Delta = -0;4^\circ$ (Table 4.38, p. 189). However, only the products $3{,}43d$ = 2,13;48° and 32Δ = 2,13;52° are mentioned, the difference, $dF^{(1)}(223)$, is not. Part **c** (Oii8'–11') mentions parameters of $F^{(3)}$ and F (Table 4.37 in §4.5.10). The number restored as 15;14,34,27,40 by BHS at the end of Oii8' is the maximum M; 13;10,34,51 = μ (Oii9'), 11,6,35,14,20 = m (Oii10'), 0;18 = d_{F*}, the difference of the daily versions F^*, and 0;36 = d_F, the monthly difference of F (Oii11').

P3' *Zigzag functions for $F^{(1)}$ and $A^{(3)}$*

P3' consists of 6 nearly intact lines. Part **a** (Oii12'–13') mentions defining parameters of $F^{(1)}$, namely the maximum M = 0;2,32,40,50 and the minimum m = 0;1,50,50,50 (Table 4.37, p. 188). Part **b** (Oii13'–14') contains instructions for computing $A^{(3)}$ as a zigzag function of the zodiacal position (§4.5.3). In Oii13'–14' the mean value μ = 29;6,19,20°, maximum M = 30;0,19,20° and minimum m = 28;12,19,20° of this function are mentioned. The extrema are located in the zodiacal signs Cap (M) and Cnc (m), respectively. This is followed in Oii14' by a damaged instruction for computing $A^{(3)}$ in between the extrema, using c = 0;0,36°/° as the interpolation coefficient.

P4' *Parameters of the solar motion and the zigzag functions for $A^{(3)}$ and $A^{(2)}$*

P4' mentions several parameters pertaining to the solar and lunar motion. Part **a** (Oii15'–16') provides the daily displacement $v_\odot$ = 0;59,8,9,48,40°/d (§4.5.11) and the same parameters of $A^{(3)}$ known from P3'. Part **b** (Oii17') mentions parameters of $A^{(2)}$, namely the maximum M = 30;2°, miminum m = 28;10,40° and monthly difference d = 0;18° (§4.5.3).

P5' *Unclear*

This strongly damaged procedure consists of 5 lines. Its subject could not be identified. Several parameters and net differences are mentioned: two minima (Oii18',19'), differences for 6 (?) and 12 zodiacal signs (Oii18',19'), and for 1 month (Oii20',22'). Some of these differences are associated with distances along the zodiac, suggesting that they pertain to functions whose argument is the zodiacal position.

P6' *Computing G from $F^{(1)}$*

P6' is concerned with the computation of G from $F^{(1)}$ by means of interpolation (§4.5.12). The maximum of $F^{(1)}$, M = 15;16,5, the associated minimum of G, m = 1,52;34,35°, and the interpolation coefficient 37;30 are partly preserved. The formulation is based on a variant of templates IP.B.1 and IP.B.2 (§2.6.7).

P7' Computing $F^{(1)}$ from G?

Only the first 2 lines of P7' are partly preserved. The number 1,52;34,35° (Oii27') can be identified as the minimum (m) of G; the number 3,11;0,50° (28') as the mean value μ (§4.5.12). Perhaps $F^{(1)}$ is computed from G, i.e. P7' deals with the inverse of the algorithm in P6'.

P8' Computations involving E? Unclear

The beginning of P8' is missing. The remaining 7 lines of text preserve instructions that are difficult to interpret. The term 'height and depth' (Ri5') may refer to the Moon's distance to the ecliptic, E (§4.5.6), but the context is too damaged for an interpretation. The number 4,23 (Ri3',4') can be interpreted as the number of revolutions, $\Pi + Z$, performed by the Moon during Π synodic months (cf. Eq. 3.7 on p. 61 with $m = 0$, $n = 1$), where $\Pi = 4{,}11$ and $Z = 12$, periods known from $F^{(1)}$ and $F^{(2)}$ (§4.5.10). The second instance of 4,23 is multiplied by a number ending with 8.25.53.20, an unidentified quantity said to be 'for 'height' and 'depth''. If this number is complete then the product is 36.57.28.46.40. The numbers in Ri6', one of which is connected with the sign Cnc, are also difficult to interpret. Ri7' mentions a monthly difference 12, which could not be identified.

P9' Computations involving $H^{(2)}$ and $J^{(2)}$; unclear

P9' contains 10 lines, which concern $H^{(2)}$ (§4.5.13) and $J^{(2)}$ (§4.5.14), variants of H and J attested only here. The instructions are only partly understood. Part **a** is concerned with $H^{(2)}$. Ri8' is similar to Ri18' (P10'), which suggests that 'decreasing', was preceded by the zodiacal region 'from Psc until Ari (or: Vir?)'. This phrase is not attested in any other procedure text, and its meaning is unclear. Perhaps it pertains to the zigzag function for H. The number 4 might reflect that this zigzag function performs 2 oscillations in the course of 1 year, so that it goes up and down 4 times (cf. Fig. 4.43, p. 191). In Ri9' the region Psc–Ari (or Vir?) is associated with a quantity called 'height and depth'. BHS's reading, 'height and depth of the Sun', could not be confirmed after collation. The term 'height and depth' usually means distance to the ecliptic, but it might also refer to the extrema of a function — the zigzag function for H? As proposed by BHS, the number 6;47,28° (Ri11',12') is probably the monthly difference (d) of $H^{(2)}$. Ri10'–11' contain an instruction of the kind 'what should I multiply by x so that it is 6.8.12?'. However, x is not preserved, and the number 6.8.12 is unidentified, so that the purpose remains unclear.[691] We are on more solid ground in Ri13', where the maximum of H, $M = 21°$, is halved, resulting in 10;30°.

Part **b** continues with computations involving $J^{(2)}$. The number 2,9;52,21 (Ri15') equals $4M$ of $J^{(2)}$. Since $J^{(2)}$ goes up and down by M this is the total change of $J^{(2)}$ during one full oscillation of the zigzag function. This number is multiplied by 4,23, which yields the total variation of $J^{(2)}$ over 4,23 revolutions, $4{,}23 \cdot 4M = 9{,}29{,}16;28{,}3°$, probably to be restored at the beginning of Ri16'. This could be part of a computation of the net difference of $J^{(2)}$ for 223^{m}. The following computations involve $32;28{,}5{,}15° = M$; their purpose is unclear.

P10' Computations involving $H^{(3)}$?

P10' consists of 3 lines, of which the first one partly duplicates Ri8'. As proposed earlier, this statement may concern the zigzag function for H, but the precise meaning is unclear. The '4 each' may refer to 4 different zodiacal regions delimited by the extrema of H, but one would expect them to consist of 3 zodiacal signs (Fig. 4.43, p. 191). A satisfying interpretation remains to be found. The repeatedly occurring sequence of digits ...49.20 might be restored as 6;49,20°, the monthly difference of a hitherto unknown variant of H, say $H^{(3)}$ (cf. §4.5.13).

P11' Unclear

P11' contains 3 lines. Its subject could not be determined. The number 32 (Ri22'), perhaps the minimum of a function, could not be identified.

P12' Unclear

Only the first 3 lines of P12' are partly preserved. Its subject could not be determined, and the preserved digits could not be identified.

[691] Perhaps $x = 6;47{,}28°$, but this does not appear to yield a meaningful computation.

5.3.4 *Unidentified systems*

No. 98

museum number	BM 35203 (Sp2,760)
provenience	Babylon
date	data in synodic table: SE 60–61
measures	7.8(*) × 7.2(*) × 2.4(*) cm
cuneiform text	photo: Fig. F.109; *ACT*, Pl. 237; copy: *LBAT* 90
bibliography	*ACT* 200h
contents	Moon, unknown system: procedure, synodic table

Side X
Column i'

P1' 1 [...xxx] ⸢1$^{?}$8 *u*$_3^{?}$⸣
2 [...x] 30 **ki** ⸢AN.MI$^{?}$⸣ [x]
3 [...] ⸢2$^{?}$8$^{?}$⸣.10.30 **ki** 1.⸢13$^{?}$ **du**⸣
4 [...] ⸢x⸣ *ina* 54 **mu**.⸢**meš** *ši*⸣-*kin*-*ša*$_2$
5 [...] ⸢x-*et*$^{?}$⸣ *i*-*šaḫ*-*ḫaṭ*
6 [...x m]**u.meš ki** AN.MI
7 [...x] ⸢30$^{?}$⸣ **nim**$^{?}$ *ana* **ki pa gur**$^{?}$
8 [...xx] **nu gar**-*an*
9 [...xx]⸢x⸣ **du**-*kam*
10 [...xx]⸢1$^{?}$⸣6.**am**$_3$
11 [...xxx] ⸢x⸣-*u*$_2$
12 [...xxxxx]⸢x⸣ 18.**am**$_3$
13 [...xxxxxxx] ⸢xx⸣
14 [...xxxxxxx ⸢**gub**-*az*$^{?}$⸣
(n lines missing)

Side X column i'

P1' **(Computations involving eclipses)**

[1][...] ⸢18 and⸣ [2][...] 30 the position$^{?}$ of ⸢the eclipse$^{?}$⸣ [...] [3][...] ⸢28⸣.10.30, the position ⸢proceeds 1.13⸣ [4][...] ⸢...⸣ in 54 years its shape [5][...] ⸢...⸣ it jumps. [6][...] years the position of the eclipse [7][...] the moon$^{?}$ is high, it returns$^{?}$ to the position of Sgr. [8][...] you do not put down. [9][...] ⸢...⸣ goes. [10][...] 16. [11][...] ... [12][...] ⸢...⸣ 18. [13][...] ⸢...⸣ [14][...] ⸢it stands$^{?}$⸣

(n lines missing)

Critical and philological notes

Side X column i

1 18: or 28 etc.; replaces 46 (*ACT*).
2 **ki** ⸢AN.MI⸣: or **absin** ⸢x⸣ (copied as ŠU AN by Pinches).
3 28: or 25.
4 Before *ina* there are traces of two vertical wedges.
5 Before *et* the upper parts of three vertical wedges are visible.
7 My transliteration of this line differs substantially from that in *ACT*, but it agrees closely with Pinches's copy, except for **nim**, which he copied as UŠ. **gur** = *itâr*, 'it returns'.
8 **nu**: copied correctly by Pinches; replaces **tab** (*ACT*).
9 Before **du** there is a broken sign ending in a vertical wedge.
12 Before 18 the heads of three vertical wedges are visible, perhaps a number.

Commentary

This fragment preserves a segment of the upper edge of a tablet. It is inscribed on one side; the other side is destroyed. The text is divided into two columns containing a procedure (i') and a synodic table (ii'). For a new edition of the latter cf. Volume II. The data in the table pertain to SE 60 month V until SE 61 month IV (252/1 BC), which may provide an approximate date for the tablet.

P1' Computations involving eclipses; unclear

The end of this badly preserved procedure is lost. Perhaps column i' was preceded by another column with text belonging to P1', or to another procedure. Despite several improved readings, P1' remains difficult to interpret in the absence of parallels. Xi'2,4,6 and 12 mention terms or numbers that point to eclipses. 18^{yr}refers to the well-known saros cycle of 223^{m}; 54^{yr} (Xi'4) is an eclipse cycle comprising 3 saroi.[692] Xi'7 seems to mention a return of the eclipse to the sign Sgr. The numbers in Xi'2,3, and 10 could not be identified, and no algorithm could be reconstructed from this procedure.

[692] Known in Greek astronomy as the 'exeligmos'.

No. 99

museum number	BM 35739 (Sp3,261)
provenience	Babylon
measures	6.5(*) × 4.3(*) × ?(*) cm; column width 2.3 cm
cuneiform text	photo: Fig. F.110; *ACT*, Pl. 237; copy: *LBAT* 89
bibliography	*ACT* 200g
contents	Moon, unknown system: computations involving eclipses

Side X
Column i'

P1'
1 [...xx]
2 [...x] *ša*$_2$ ^{d}**utu**
3 [... **n**]**im kaskal** 19
4 [...] 52.35
5 [...] **zi**-*ma* **murub**$_4$
6 [...] ⸢AN⸣.MI **gar**-*an* 20
7 [...x] ⸢*al*⸣ **murub**$_4$ **nim**
8 [...xxxx] ⸢x⸣
(n lines missing)

Column ii'

1 *ana* **mu**-*ka* **me.a** ⸢**gar**-*an*?⸣
2 18 GAM 11.6.40 **du**
3 3.20 18 **zal mu** ⸢*ina* 20⸣ **ninda**
4 *šal-šu*$_2$ ḪAB-*rat sin al*
5 **murub**$_4$ **kaskal**.2 **nim** ^{d}**utu šu**$_2^?$
6 ḪAB-*rat al* **murub**$_4$ **kaskal**.2 ⸢**sig**?⸣
7 ^{d}*sin i-par-šid-ma šal-šu*$_2$
8 ḪAB-*rat sin al* ⸢**murub**$_4$⸣
9 ⸢x⸣ 20 *šal-šu*$_2$ ⸢xx⸣-[x]
(n−1 lines missing)

Column iii'

1 **a**.⸢**ra**$_2^?$ PI?⸣ **nu** ⸢x⸣ [...]
2 3⸢2 *ša*$_2^?$⸣ AZ ŠU? [...]
3 1-*en* **nim**? ⸢6?⸣ [...]
4 **nim**-⸢*ma*? x⸣ [x...]
5 ⸢**en**? *al*⸣ [xx...]
6 ⸢x **kaskal**.2?⸣ [xxx...]
7 1 ⸢x⸣ [xxxx...]
8 ⸢xx⸣ [xxxx...]
(n lines missing)

Side X column i'

P1' (Computations involving eclipses; unclear)

^{1}[...] ^{2}[...] of the Sun. ^{3}[... it is h]igh?, the path, 19 ^{4}[...] 52.35 ^{5}[...] you 'tear out'?, and the middle ^{6}[...] an ⸢eclipse⸣ occurs, 20 ^{7}[...] it is 'high' ⸢above⸣ the middle. ^{8}[...] ⸢...⸣

(*n lines missing*)

Column ii'

^{1}you predict for your year, ⸢put down⸣. ^{2}You multiply 18 by 11.6.40, ^{3}it is 3.20. 18 passings of the year. ⸢In 20⸣^{nin}, ^{4}a third of the disk the Moon ascends above ^{5}the middle of the path, covers the Sun. ^{6}The disk is 2 ⸢below⸣ the middle of the path, ^{7}the Moon 'escapes', and a third ^{8}of the disk the Moon [is above] the ⸢middle⸣. ^{9}... 0;20, a third ⸢of the disk⸣ [...]

(*n−1 lines missing*)

Column iii'

^{1}⸢times? ...⸣ not ⸢...⸣ [...] ^{2}3⸢2 ...⸣ [...] ^{3}one ... ⸢6⸣ [...] ^{4}⸢...⸣ [...] ^{5}⸢... beyond⸣ [...] ^{6}⸢... the path?⸣ [...] ^{7}1 ⸢...⸣ [...] ^{8}⸢...⸣ [...]

(*n lines missing*)

Critical and philological notes

Side X column i'

3 **nim**: either a form of *šaqû*, 'to be high' (= above the ecliptic), or *šūlû*, 'to deduct'. **kaskal** = *ḫarrānu*, 'path'.

5 **zi**: either *tanassaḫ*, 'you tear out', or *inassaḫ*, 'it moves'. **murub**$_4$ = *qablītu*, 'middle': the nodal region where eclipses occur?

6 **gar**-*an* 20: this reading seems preferable to *ša*$_2$ ^{d}*šamaš*$_2$, because in Xi'2 *Šamaš* is written ^{d}**utu**.

7 As shown in Pinches's copy, there is a vertical wedge (*ana*) before **murub**$_4$. It is preceded by a damaged sign, of which two vertical wedges are visible.

8 Traces of a sign similar to *taš* or *al* are visible.

Column ii'

1 **me.a**: probably *taqabbi*, 2 m. sg. pres. G *qabû*, 'you predict'. **gar**-*an*?: replaces ŠU$_2$ E? (*ACT*); Pinches also copied **gar**.

3 **zal**: perhaps a form of *nasāḫu*, 'to pass' (of time). *ina* 20: replaces ⸢EN⸣ (*ACT*); 20^{nin} = 0;20° corresponds to 1/3 of the 'disk' mentioned in the next line.

4 *sin*: or 30? In Xii'7 *Sîn* is written as ^{d}*sin*, which may suggest 30.

5 **kaskal**.2 = **kaskal** = *ḫarrānu*, 'path'. ^{d}**utu šu**$_2$: or **bal** DIŠ **šu**$_2$; replaces *tam*?-*mar* (*ACT*).

6 **sig**: omitted in *ACT*.

7 *i-par-šid-ma* (replaces *i-pir-rid-ma* in *ACT*) = *ipparšidma*: 3 c. sg. pret. *naparšudu*, 'to escape'. This verb is not attested elsewhere in the astronomical corpus. Perhaps it describes the end of an eclipse.

9 Before 20 the upper part of a wedge is visible. The traces after *šal-šu₂* are compatible with ḪAB- or *ša₂* ḪAB-, 'of the 'disk'.

Column iii'

1 Pinches's copy is not very accurate here. **ra₂**: or NA. PI: or GAL. The meaning is unclear.

2 *ša₂*: or 1?. AZ ŠU$^?$: Akkadian reading unclear.

3 1-*en*: thus also copied by Pinches; replaces $\frac{1}{2}$-*šu₂* (*ACT*). **nim**: or **du**?

5 Replaces ŠU$^?$ E$^?$ 15$^?$ (*ACT*).

6 The sign before **kaskal**.2$^?$ looks like DU₃.

Commentary

This fragment preserves a segment of the upper edge of a tablet. It is inscribed on one side; the other side is destroyed. There are three columns of text, which are provisionally assigned to a single procedure (certainly Xii' 1 is not the beginning of a new procedure). Only the second column is preserved to its full width of 2.3 cm. Orthography (e.g. ^{d}**utu**), terminology (e.g. the unusual verb *naparšudu* = 'to escape') and content point to an early date (early Seleucid or Achaemenid?). Collation has resulted in several corrections to the previous edition (*ACT*), but the text remains difficult to interpret for lack of parallels.

P1' Computations involving eclipses; unclear

The beginning and end of P1' are lost. Columns i' and especially ii' imply that this procedure is concerned with eclipses, but the algorithms could not be reconstructed. The number 19 (Xi'3) might refer to the 19^{yr} cycle known from the Babylonian calendar, but this remains speculative. The number 52.35 (Xi'4) remains unidentified. The repeatedly occurring word 'middle' (**murub₄** = *qablītu*) is known from lunar system A as a technical term for the nodal region, where eclipses may occur (§4.4.6). The phrase 'you predict for your year' (Xii'1) is reminiscent of Goal-Year procedures, 'your year' being the usual expression for the Goal Year. The factor 18 in the multiplication 18 · 11.6.40 = 3.20 (Xii'3) is most likely the saros period expressed in years, so that 11.6.40 must be the yearly difference of some function related to eclipses. In the following lines there is repeated mention of 'a third of the disk of the Moon', perhaps a reference to the fraction of the Moon's diameter that is eclipsed. Also the number 20^{nin} (Xii'3) may refer to this if one assumes that, as in lunar system A, the diameter of the lunar or solar 'disk' equals 1°. In Xii'7 the unusual expression 'it escapes' might refer to the final stages of an eclipse. Column iii' is too strongly damaged for an interpretation.

No. 100

museum number	BM 36665 (80–6–17,397)
provenience	Babylon
measures	6.0(*) × 6.2(*) × 3.0 cm
arrangement	O/R
cuneiform text	photo: Fig. F.111
contents	Moon, unknown system: computations involving eclipses

Obverse

P1' (m lines missing)
1' ⸢10⸣ [xxxxxxxxxxx...]
2' 3.20.11 ⸢x⸣ [x] ⸢xxx⸣ [...]

P2' 3' 6.40 **a.ra$_2$** 1⸢4⸣ 1.33.20 ⸢6⸣.[40? ...]
4' **ta u$_4$.1.kam en u$_4$.14.kam** 6.⸢40⸣ [...]
5' AN.MI **gar**-*an* **ta u$_4$**.⸢14⸣.**kam en u$_4$**.[x.**kam** ...]

P3' 6' [x] ⸢x **u$_4$**⸣.**da.zal.la$_2$** *tar*-⸢*di*⸣-*tu$_4$ ša$_2$* AN.⸢MI⸣ [...]
7' [xx] ⸢x⸣ **si**? 2.13.20 **a**.⸢**ra$_2$**⸣ 30 1.[6.40 ...]
8' [xx] ⸢xx⸣ 10 2.13.20 ⸢xxx⸣ [...]
9' [xxxx] ⸢x 1⸣8 **mu** ⸢x⸣ [x...]
10' [xxxxx] ⸢x⸣-*ma* ⸢x⸣ [xx...]
(n lines missing)

Reverse

P4' (about n lines missing)
1' [xxx].⸢**meš**⸣ *ana* 1-*en* **gur** *ša$_2$* ⸢x⸣ [...]

Co 2' [x] ⸢xx⸣ 2-*u$_2$* **nu al.til ta** [...]
3' **egir**-*šu$_2$* 1 2 ŠU$_2$ AN *qat$_3$*? *ša$_2$* ^{m}[x...]
4' *la* **la$_2$** *ina* **ugu zi** PAP ⸢x⸣ [...]
5' ⸢xxxxx⸣ [xxxx...]
(about m lines missing)

Obverse

P1' **(Unclear)**

(m lines missing)
1' ⸢10⸣ [...] 2' 3.20.11 ⸢...⸣ [...] ⸢...⸣ [...]

P2' **(Computations involving eclipses; unclear)**

3' 6.40 times 1⸢4⸣ is 1.33.20 ⸢6⸣.[40? ...] 4' From day 1 until day 14 6.⸢40⸣ [...] 5' an eclipse will occur. From day ⸢14⸣ until day [...]

P3' **(Computations involving eclipses; unclear)**

6' [...] the *uddazallû*-coefficient of the addition for the ⸢eclipse⸣ [...] 7' [...] ⸢...⸣ ... 2.13.20 times 30 is 1.[6.40 ...] 8' [...] ⸢...⸣ 10 2.13.20 ⸢...⸣[...] 9' [...] ⸢1⸣8 years [...] 10' [...] ⸢...⸣ and ⸢...⸣ [...]
(n lines missing)

Reverse

P4' **(Unclear)**

(about n lines missing)
1' [...] ⸢...⸣ becomes? 1 ... ⸢...⸣ [...]

Co **(Colophon)**

2' [...] ⸢...⸣ the second [...]. Not finished. [Copied] from [... The tablet which is] 3' after it: ... Hand? of [...] 4' not ... in ... ⸢...⸣ [...] 5' ⸢...⸣
(about m lines missing)

Critical and philological notes

Obverse

2' The broken sign after 3.20.11 may be MA or another sign beginning with three horizontal wedges. After the break a vertical wedge and a sign similar to ZU are visible.
5' **gar**-*an* = *iššakkan*, 3 c. sg. pres. N *šakānu*, 'to occur'.
6' **u$_4$.da.zal.la$_2$** = *uddazallû*, '*uddazzallû*-coefficient'. *tardītu* 'additon': cognate noun of *redû* D, 'to add'.
7' ⸢x⸣ SI = *ubānu*, 'finger': or read ⸢LA$_2$⸣ ḪAB? Meaning unclear.
8' 10 2.13.20: preceded by traces of a winkelhaken and a vertical wedge; followed by traces of several vertical wedges.

Reverse

1' After *ša$_2$* the beginning of a horizontal wedge is visible.
2' Before 2-*u$_2$* there are traces compatible with 2 GUR. **nu al.til** = *ul qati*, 'not finished': for this phrase cf. Hunger *BAK*, p. 2. **ta** = *ultu*, 'from', presumably followed by *lē'i*(^{giš}**da**) *šaṭirma bari*, 'copied from a wooden board and checked' or a similar phrase.
3' ŠU$_2$.AN might be read **en$_2$** = *šiptu*, 'incantation', but this does not appear to make sense.
4' The correct reading and meaning of this line are unclear. PAP: read **pap** = *napḫaru*, 'total', or **kur$_2$** = *nakāru*, 'to be different'? The sign following PAP is similar to ŠEŠ.
5' The heads of several wedges are visible.

Commentary

This fragment belongs to the left edge of a tablet. The main text is divided into at least 4 procedures by horizontal lines. P2' and P3' are concerned with eclipses. The underlying algorithms could not be reconstructed due to the bad state of preservation and a lack of parallels. On the reverse a colophon is partly preserved. R2' implies that the tablet is the second one in a series containing at least one more tablet, perhaps with incantations (R3'). The rest of the colophon is difficult to interpret.

P1' *Unclear*

Only a few digits are preserved; their meaning is not clear.

P2' *Computations involving eclipses*

This procedure consisting of 3 lines is concerned with eclipses. A number 6.40 is multiplied by 14 days, which implies that 6.40 is the daily difference of some function, whose astronomical meaning is not clear. In O4' the product 1.33.20, which should be a difference for 14^d, is followed by the phrase 'from day 1 until day 14'. Since there are 13^d between day 1 and day 14, 'day 1' and 'day 14' must perhaps be interpreted as technical terms for New Moon (conjunction) and Full Moon (opposition), respectively. Hence the purpose of P2' may be to compute the value of the function at New Moon from the value at the preceding Full Moon, or vice versa.

P3' *Computations involving eclipses and function Φ of lunar system A?*

The first 5 lines of P3' are partly preserved. Beyond the fact that it deals with eclipses very little is understood. Mention is made of a coefficient and an 'addition for the eclipse'. The number 2.13.20 (O7',8') may be interpreted as 2,13;20°, which points to function Φ of lunar system A (§4.4.3). It is multiplied by 30, perhaps to be interpreted as 0;30 or as 30, the number of tithis in a synodic month. The astronomical interpretation of this computation is not clear.

P4' *Unclear*

R1' appears to be the final line of another procedure (or of P3'). Too little remains for an interpretation.

No. 101

museum number	BM 37123 (80–6–17,872)
provenience	Babylon
measures	3.0(*) × 3.4(*) × 1.2–2.3 cm
cuneiform text	photo: Fig. F.112
contents	computations involving a solar quantity

Obverse

P1 1 ^{d}**utu ta** *zib* **en** [...]

2 ⸢**ta**⸣ **absin**$_0$ **en** ⸢*zib*⸣ [...]

3 **ta** 56.⸢40$^{?}$⸣.[...]

4 *ana* **me** 1.⸢20 x⸣ **mu** [...]

5 ⸢*taš*⸣-*pil-tu*$_4$ ⸢x⸣ [...]

6 ⸢x 30$^{?}$ x 5$^{?}$⸣ [...]

(unknown number of lines missing)

Obverse

P1 **(Daily displacement of the Sun?)**

[1]The Sun, from Psc until [Vir ...] [2]⸢from⸣ Vir until ⸢Psc⸣ [...] [3]from 0;56,⸢40$^{?}$⸣,[...] [4]for (each) day 1.⸢20 ...⸣ years [...] [5]the difference ⸢...⸣ [...] [6]⸢... 30$^{?}$ ... 5$^{?}$⸣ [...]

(*unknown number of lines missing*)

Critical and philological notes

Obverse

4 *ana* **me**, 'for (1) day', or 1 **me** = '1 day', or 1-*me* = 100?

Commentary

This small fragment belongs to the upper (lower) left corner of a tablet. It is inscribed on one side; the surface on the reverse is not inscribed though well-preserved.

P1 A function for the Sun's daily displacement?

The first 6 lines are partly preserved. Only little could be understood. The Sun is mentioned, followed by what appear to be instructions concerning the Sun's daily displacement, $v_{\odot}$ (§4.4.10). The mention of two zodiacal regions, from Psc to what must be Leo, and from Vir to what must be Aqr, suggests that $v_{\odot}$ is modeled as a function of the zodiacal position, perhaps a zigzag function or a step function. Each region contains 6 zodiacal signs and coincides more or less with a zone for the synodic arc in lunar system A (§4.4.4). The digits in O3, if read as 0;56,⸢40⸣,..., might represent the minimum of $v_{\odot}$, which occurs in Gem. As far as known this value is not attested elsewhere.

No. 102

museum number	BM 36712 (80–6–17,445+820+919)
provenience	Babylon
measures	8.4 × 6.0(*) × 1.2–1.9 cm (type G)
arrangement	O/R
cuneiform text	photo: Fig. F.113; Sachs & Neugebauer (1956), p. 131
bibliography	Sachs & Neugebauer (1956) = SN
contents	unidentified system: procedures concerning lunar and solar motion

Obverse

P1 (about 1-4 lines missing)

1’ [xxxxx] ⌜xxx⌝ [xxxxxxxxxxxxxxxxxxxx]
2’ [xxxx] ⌜x⌝ **ninda$_2$**-*at* **an**-*u$_2$* 6.4.30 ⌜*u$_4$-me* x⌝ [xxxxxxxxxx]
3’ [xx] *u$_2$-maṭ-ṭu-u$_2$* 40 **ninda** *a-na* 6.4.30 **daḫ**-*ma* 6 ⌜**uš** 5.10⌝ [xxxxxxx]
4’ ⌜**giskim**⌝**.bi zu.še$_3$ ta** AN.BAR *šil-taḫ* **igi en igi.du$_8$.a**-*šu$_2$ ša$_2$* xx 1 **mu.**⌜**an.na**⌝ [xxxxx]
5’ *ip-pal-ka* 1 PAP.ḪAL ^{d}**utu** 13.22 PAP.ḪAL **u$_4$.sakar** *ina* 1.21 *u$_4$-mu ša$_2$ nis-ḫi* ⌜xx⌝ [xxx]
6’ **igi**.1.21.**gal$_2$.bi** 44.26.40 *ina* 1 **nim**-*ma* 59.15.33.20 *a-na* 6.40 ⌜**igi**⌝.[9.**gal$_2$.bi**]
7’ **il$_2$**-*ma* 6.35.3.42.13.20 : 6.35.3.42.13.20 *ina* 59.15.3⌜3.20 **nim**-*ma*⌝
8’ 59.8.58.16.17.46.40 *ni-is-ḫu* **sag.uš** *ša$_2$* ^{d}*ša$_2$-maš ša$_2$ u$_4$-mu*
9’ **giskim.bi zu.še$_3$** 59.8.58.16.17.46.40 *a-na* 6.4.30 **il$_2$**-*ma*
10’ 5.59.20 *ip-pal-ka* 59.8.58.16.17.46.40 **a.ra$_2$** 13.22 **du**-*ma*
11’ 13.10.37.54.53.49.37.46.40 ⌜*ni-is-ḫu* **sag.uš**⌝ *ša$_2$* **u$_4$.sakar**
12’ **g**[**iskim.b**]**i zu.še$_3$** 13.10.37.54.53.49.⌜37.46.40 **a.ra$_2$**⌝ 27.25 **du**-*ma*
13’ 6.43.54 **giskim.bi zu.še$_3$** *u$_4$*-⌜*mu* ^{d}*sin*⌝ *ana* **mul$_x$.mul$_x$te**-*u$_2$* *ina* 27 **u$_4$.meš** *ana* **mul$_2$.mul$_2$ dim$_4$**
14’ 59.8.58.16.17.46.4[0 **a.ra$_2$** 6].4.30 **du**-*ma* 1.20.03.5.20
15’ 12.22 *ša$_2$* ^{d}*sin ana* ^{d}**utu** *pa-nu*-⌜*u$_2$*⌝ [*a-na* 40 **ninda i**]**l$_2$**-*ma* 8.14.40 **u$_4$.da.zal.la$_2$ mu.an.na** *ša$_2$* ⌜*sin*⌝
16’ 8.14.40 *a-na* 1.20.03.5.2[0 **daḫ**-*ma* 1.20.11] ⌜$\frac{1}{3}$⌝ **uš** 1-*me* **danna** 1-*šu* **danna** 12 **u**[**š**]
17’ *qaq-qar ina* **mu**.[**an.na** xx 6 **a.r**]**a$_2$** 13 **du**-*ma* 1.18

Reverse

1 1-*me* **danna** ⌜56 **danna**⌝ [xxx *nis-ḫi*] ^{d}*sin ša$_2$* **mu.an.na** *kit-ti*
2 *ina* 1.20.12 ⌜**nim**⌝-*ma* 2.12 **tag$_4$** [xxx] ⌜**šu**$^{?}$⌝ **gar** 11 *u$_4$-mu re-e-ḫu*
3 *ina* 30 ⌜**mu.an.na**⌝**.meš** ⌜xxxxx⌝ [xxx 12].⌜2⌝2 **iti** *a-na* 30 **mu.an.na.meš il$_2$**-*ma*
4 ⌜**tag$_4$** 6$^{?}$.11 xx 30$^{?}$ x⌝ [xxx 30 *a-na* 6].⌜1⌝1 **daḫ** *šum-ma* 6.41
5 ⌜*ina* 30 **mu.an.na.meš**⌝ 4-*me* 1 ⌜**zi**$^{?}$⌝ [xx] *sin* **an**-*e* **bal** 6.41 *a-na*$^{!}$ 30 **šeš.meš**
6 **bar**-*ma* PAP.ḪAL ^{d}*sin tam-mar* **igi**.3⌜0.**gal$_2$.bi**⌝ 6.41 **a.ra$_2$** 2 **du**-*ma*
7 13.22 PAP.ḪAL ^{d}*sin ša$_2$ la* **u$_4$.da.zal.la$_2$**-*u*

Critical and philological notes

Obverse

1’ Only faint traces are visible.

2’ **ninda$_2$**-*at* **an**-*e* = *mindat šamê*, ‘the measure of the sky’ (?); cf. the discussion in SN. The traces after *u$_4$-me* might belong to MUL.

3’ *umaṭṭû* = 3 c. sg. pres. D *maṭû*, ‘to cause to be lacking; diminish; subtract’. The translation of SN, ‘it is less’, would require a G pres. *imaṭṭu*. Perhaps the subject is the quantity that was mentioned in the preceding break: ‘it causes to be lacking’. **daḫ**-*ma* = *tuṣṣabma*, ‘you append’ (2 m. sg. pres. G *waṣābu*: the use of this verb points to an early date. UŠ has the meaning 60 here. 5.10: only the lower parts are visible.

4’ **giskim.bi zu.še$_3$**: **giskim**(IGI+DUB) = 1. *giskimmu*, 2. *ittu* ‘sign’; **zu** = *idû*,‘to know’. **giškim zu** has the Akkadian equivalent *uddû*, ‘to mark; identify; know (exactly); interpret (signs)’, or the passive Dt, ‘to be known, recognisable, identifiable’ (*CAD* I p. 20 sub *idû*; *AHw* III, pp. 1455–1456, sub **wadû** D, Dt). Hence **giskim.bi zu.še$_3$** may be translated ‘in order to recognise/know/establish it’, or ‘in order for it to be recognised/known/established’. AN.BAR: replaces ^{mul} (SN). However, it is suspect that the determinative for stars (MUL, MUL$_2$) is lacking, and a star ‘Iron of the arrow(head)’, if *parzillu*, ‘iron’, is the correct reading of AN.BAR, is otherwise unknown. Before **mu.an.na** there are two crammed signs similar to 2 NA (partly overlapping), perhaps to be read *a-na*? (very uncertain).

5’ After *nis-ḫi* there are traces of a vertical wedge and, perhaps, a winkelhaken. PAP.ḪAL: the *CAD* M2, p. 306, lists *muttalliku*, ‘moving; roaming’ (ptc Gtn *alāku*) as an Akkadian reading, but the context requires a meaning ‘revolution’. The lexical list Idu II 360 (quoted *CAD* A1, p. 300 sub *alāku*) has an entry [**pa**]**-ap-ḫa-al** PAP.ḪAL = *i-tal-lu-ku*. It therefore appears that PAP.ḪAL can represent any form derived from the Gtn of *alāku*, perhaps *muttalliktu*, ‘going around’ > ‘revolution’, even though PAP.ḪAL is not attested elsewhere as a logogram for *muttalliktu*, and neither is the meaning ‘revolution’. Other possible readings of PAP.ḪAL (*pašāqu* = ‘to be narrow’ and its derivations; *purīdu* = ‘leg’) seem less likely to apply.

Obverse

P1 (Computations involving lunar and solar motion)

a *(about 1–4 lines missing)*
1'[...] ⸢...⸣ [...] **2'**[...] ⸢...⸣ the 'measure' of the sky, 6,4;30 ⸢days, ...⸣ [...] **3'**[...] diminishes it by [...]. You 'append' 40^{nin} to 6,4;30, it is 6 (times) ⸢60 (and) 5;10⸣ [...]

b (Computing the Sun's mean daily displacement)

4'In order to establish this: from the (first) appearance of the Iron? of the Arrow until its appearance of ... 1 year [...] **5'**it gives you. 1 revolution of the Sun is 13;22 revolutions of the crescent. In 1,21 days (in) which the displacement ⸢...⸣ [...] **6'**The reciprocal of 1,21 is 0;0,44,26,40, you deduct it from 1, it is 0;59,15,33,20. You 'raise' it to 0;0,6,40, the reciprocal of [9,0], **7'**it is 0;0,6,35,3,42,13,20. You ⸢deduct⸣ 0;0,6,35,3,42,13,20 from 0;59,15,33,20, it is **8'**0;59,8,58,16,17,46,40, the Sun's steady daily displacement. **9'**In order to establish this: you 'raise' 0;59,8,58,16,17,46,40 to 6,4;30, then **10'**it gives you 5,59;20.

c (Computing the Moon's mean daily displacement)

You multiply 0;59,8,58,16,17,46,40 by 13;22, it is **11'**13;10,37,54,53,49,37,46,40, the steady displacement of the crescent. **12'**In order to establish this: you multiply 13;10,37,54,53,49,⸢37,46,40 by⸣ 27;25, it is **13'**6,43;54.

d (Rule involving the Pleiades)

In order to establish this: the day when the Moon approaches the Pleiades, in 27 days it (again) approaches the Pleiades.

e (Computing the Moon's total displacement during 1^{yr})

14'You multiply 0;59,8,58,16,17,46,40 (error for 13;10,37,54,53,49,37,46,40) [by 6],4;30, it is 1,20,3;5,20. **15'**You 'raise' 12;22, by which the Moon is ahead of the Sun, [to 40^{nin}], it is 8;14,40, the *uddazallû*-coefficient of the year of the Moon. **16'**[You 'append'] 8;14,40 to 1,20,3;5,20, [it is 1,20,11] $\frac{1}{3}°$, (i.e.) 100^b (and) 60^b (and) 12° **17'**the position [...] in a year [...]. You multiply [6,0] ⸢by⸣ 13, it is 1,18,0, **R1**⸢you deduct⸣ 100^b (and) ⸢56^b⸣, [... the displacement] of the Moon for a true year, **2**from 1,20,12, there remains 2,12 [...] ⸢...⸣ you put down?, the remainder is 11 days.

f (Computations involving lunar motion during 30^{yr})

3In 30 ⸢years ...⸣ [...] you 'raise' [12;2]2 months to 30 years, and **4**⸢the remainder is 6,11? ... 30 ...⸣ [...] ... you append [30 to 6],11, ... it is 6,41. **5**⸢In 30 years⸣ the Moon transgresses the sky 401 [...]. You divide 6,41 into 30 parts, **6**and you see the revolutions for the Moon. The ⸢reciprocal of 30⸣: you multiply 6,41 by 0;2, it is **7**13;22, the revolutions for the Moon, which is without the *uddazallû*-coefficient.

6' The use of *našû* for multiplication points to an early date.
8' ^{d}*ša₂-maš*: this phonetic writing of *Šamaš* is very uncommon in astronomical texts.
12' **a.ra₂**, 'times': replaces *a-na* (SN).
13' **mul$_x$**, an unusual reading of AB₂, is here used as an alternative to **mul** or **mul₂**. **mul$_x$.mul$_x$**, **mul₂.mul₂** is translated here as Pleiades (*zappu*), but it may also denote the zodiacal sign Tau. **te** = *ṭeḫû* and **dim₄** = *sanāqu* are essentially synonymous verbs meaning 'to approach'.
14' 59.8.58.16.17.46.40: one expects 13.10.37.54.53.49.37.46.40 (cf. the commentary).
1.20.03.5.20: the 3 is preceded by a separation mark (transliterated as 0), apparently in order to prevent it from being combined with the 20. This practice is common in Uruk but rare in Babylon.
15' *a-na*: replaces **a.ra₂** (SN). *a-na* 40 **ninda** *tanašši*(**il₂**)*-ma*, 'you multiply it by 40^{nin}' ('raise it to 40^{nin}').
16' $100^b + 60^b + 12° = 160 \cdot 30° + 12° = 1{,}20{,}12°$.
18' $100^b + 56^b = 156 \cdot 40° = 1{,}18{,}0°$.

Reverse
1 **mu.an.na** *kit-ti* = *šatti kitti*, 'true year' (gen.), apparently a technical term for 12 synodic months.
2 Before **gar** there is a sign similar to ŠU. *re-e-ḫu* = *rēḫu*, 'remainder'.
3 After **meš** a vertical wedge and the upper parts of several signs are visible.
4 **tag₄**: omitted in SN. *šum-ma*, 'if', is unexpected; its syntactic function within the sentence is unclear. Perhaps read **tag**-*ma*, a learned writing for **tag₄**-*ma*, 'there remains'?
5 **zi**?: replaces ⸢UŠ⸣ 2[1] (SN), meaning uncertain.
6 **bar** = *zâzu*, 'to divide': replaces MAŠ (SN).

Commentary

A few cm of clay may be missing near the upper (lower) edge of the obverse (reverse). Near the lower (upper) edge a small trapezoid-shaped chunk appears to have been cut out. Obverse and reverse are inscribed with a single column of text. The reverse has only 7 lines of text, below which the tablet is uninscribed. There is no colophon (although there is ample space for it). Orthographic and notational features[693] point to an early Seleucid or late Achaemenid date. Other remarkable features of the tablet are the following. First, the purpose of the procedures is not connected with synodic tables. Second, they cannot be identified in terms of lunar systems A and B in any obvious manner. Some of the parameters seem to predate these systems. Third, the orthography and vocabulary are unusually learned. For instance, different synonyms, logographic or phonetic writings are used for the same word, e.g. *Šamaš*, 'Sun', is written as ^{d}**utu** (O5',15') and ^{d}*ša*$_2$-*maš* (O8'), the Moon as *uskaru*(**u$_4$.sakar**) = 'crescent' (O5',11') and ^{d}*sin* (O13',15'), and 'Pleiades' (or Tau?) as **mul$_x$.mul$_x$**(AB$_2$.AB$_2$) and **mul$_2$.mul$_2$** (O13'). By the same token, 'to go *x* times *y*' and 'to raise *x* to *y*' are both used for multiplication without there being any recognisable semantic differentiation (§2.2.4). The repeatedly occurring phrase 'In order to recognise/verify this', not attested elsewhere in the astronomical corpus, appears to introduce a computation or its verification. Some of the computations involve unusually large numbers. All this suggests that the text is the work of a skilled and ambitious scholar. The collection 80–6–17 to which the fragment belongs includes several other early or 'atypical' procedure texts.

P1 Computations involving lunar and solar motion

A few lines may be missing from the beginning of P1. This complex procedure has been divided into six parts. Part **a** becomes legible in O2', which mentions the term *mindat*? *šamê*, lit. perhaps 'measure of the sky'. The meaning of this term, which is not attested anywhere else, is unclear, but cf. the discussion in SN and Schnabel (1923), p. 260. As proposed by SN, the number 6,4;30^d = 365.5^d is an approximate value of the year, here said to produce a net subtractive displacement of the Sun of 0;40°. This is applied to the year, resulting in a corrected value year length of 6,5;10^d, which implies that the Sun is assumed to cover the 0;40° at a rate of 1° per day. As pointed out by Britton (2007b), p. 126, this may be one of the earliest explicit references to a year length in Babylonian astronomy.

Part **b** (O4'–10') contains a computation of the mean daily displacement of the Sun and a verification of the result. O4' mentions the '(first) appearance', i.e. helical rising, of a star called 'Iron(?) of the Arrow', perhaps Sirius, whose common name is 'The Arrow' (^{mul}**kak.si.sa$_2$**).[694] Perhaps the time between successive heliacal risings of Sirius is employed here as a definition of the year, or as a means of measuring its length. In O5' 1^{yr} is said to contain 13;22 revolutions of the Moon. The origin of this parameter is not explained here, but in O15' the related number 12;22, an approximation of the year expressed in synodic months, is called 'that by which the Moon is in front of the Sun'. This suggests that 13;22 = 12;22 + 1 results from the assumption that, each synodic month, the Moon covers 1 full revolution plus the synodic arc (σ) in order to catch up with the Sun, and that after 1^{yr} the synodic arcs add up to 1 full revolution.

As remarked by SN, 1,21 = 81 represents the whole number of days during which the Sun travels a whole number of degrees (namely 1,20 = 80°, which might therefore be restored in O5') corresponding to the assumption that the Sun travels 6,0° in 6,4;30^d (i.e. the 0;40° is ignored). The ratio 1,20/1,21 = 6,0/6,4;30 is the corresponding mean daily displacement, $\overline{v}_{\odot}$. In O6'–8' the exact value of $\overline{v}_{\odot}$, taking into account the 0;40°, is computed by correcting the approximate value in accordance with

$$\overline{v}_{\odot} = \left(1 - \frac{1}{1{,}21}\right) \cdot \left(1 - \frac{1}{9{,}0}\right) = \frac{1{,}20}{1{,}21} \cdot \left(1 - \frac{1}{9{,}0}\right) = \frac{1{,}20}{1{,}21} - \frac{0;40}{6{,}4;30} = \frac{6{,}0 - 0;40}{6{,}4;30}. \quad (5.8)$$

The result is correctly given as 0;59,8,58,16,17,46,40 and called the 'Sun's steady daily displacement' (O8'). A verification of the outcome is provided in O9'–10' in the form 0;59,8,58,16,17,46,40 · 6,4;30 = 5,59;20°. Due to rounding this differs from 6,0°.

Part **c** (O10'–12') concerns the Moon's mean daily displacement, which is said to be larger than that of the Sun by a factor 13;22, resulting in 13;22 · 0;59,8,58,16,17,46,40 = 13;10,37,54,53,49,37,46,40°/d. This is based on the mentioned assumption that 13;22 revolutions of the Moon correspond to 1 revolution of the Sun. O12' contains a multiplication by 27;25^d that might be interpreted as a verification of the result if 27;25^d is an approximation of the sidereal month (cf. §4.1.1). If so then the result should be 6,0 = 360°, but since 27;25^d is inaccurate, the outcome is 6,43;54° (O13'). This deviation is very conspicuous, which suggests that the computation may have to be interpreted differently.

Part **d** (O13') mentions that the Moon returns to the Pleiades (or: Tau?) in 27^d. The interpretation is not clear.[695] The instructions in part **e** (O14'–R2) concern the total distance traveled by the Moon during 1^{yr}. As pointed out by SN, the first factor in the multiplication in O14' must be corrected to 13;10,37,54,53,49,37,46,40°, the Moon's mean daily displacement, because the product with 6,4;30^d is 1,20,3;5,20°. In O15' 12;22 is multiplied by 0;40°, resulting in 8;14,40°, which is called the '*uddazallû*-coefficient of the lunar year'. Its astronomical significance is not clear. As remarked by SN,

[693] In particular the use of *waṣābu*, 'to append', for addition, *našû*, 'to raise' for multiplication, ^{d}**utu** for *Šamaš*, 'miles' (*bēru*) as a unit of time, and the older version of the 9.

[694] Alternatively it may be identical with the 'Tip of the Arrow', a Normal Star identified with θ Oph (Roughton, Steele & Walker 2004, p. 567) and located in the zodiacal sign Sco.

[695] The Pleiades are otherwise known from an intercalation rule in Mul.Apin (IIiiGapA8–11, IIii1–2), but there is no obvious connection.

one would rather expect a computation of the mean distance traveled by the Moon during the short interval 0;40°, i.e. 0;40 · 13;10,37,54,53,49,37,46,40° (= 8;47,5,16,35,53,5,11,6,40), since the outcome is added to the previous distance, resulting in 1,20,11;20°. This is converted into 'miles' and rounded (O16'), leading to 160^{b} + 12° (= 1,20,12°). In O17'–R1 the distance traveled by the Moon during 13 revolutions, 13 · 6,0 = 1,18,0° = 156^{b}, is subtracted from it, resulting in 1,20,12 − 1,18,0 = 2,12° for the mean distance traveled by the Moon in 0;22 sidereal revolutions = 1^{yr} − 13 revolutions. In R3 a connection appears to be established between this number and 1^{yr} − 12 synodic months, the well-known epact of the year (§3.3.2.1), whose value is stated to be 11^{d} (R2). If this is the correct interpretation then, as pointed out by SN, the nature of the connection is unclear, because 13 sidereal revolutions of the Moon are not equivalent with 12 synodic months.

The damaged part **f** (R3–7) deals with the interval of 30^{yr}. It appears that 12;22, the number of synodic months per year, is multiplied by 30, resulting in $6,11^{m}$. This number is not preserved, but in R4 the corresponding number of (sidereal) revolutions of the Moon is computed by adding 1 for each year, resulting in 6,41 = 401. This is verified by dividing 6,41 by 30, resulting in 13;22, known from O5' as the number of revolutions of the Moon per year. It is said to be 'without the *uddazallû*-coefficient' (R7), the meaning of which is not clear (from the rest of P1 it appears that the *uddazallû*-coefficient refers to the correction 0;40).

Appendix A
Other instructional texts from the first millennium BC

This table is a (certainly incomplete) list of Babylonian and Assyrian instructional texts from the first millennium BC concerning astronomy, astrology and mathematics. For tablets and fragments of Mul.Apin cf. Hunger & Pingree (1989), pp. 3–8; for tablets with instructions concerning temple rituals cf. Linssen (2004).

tablet	provenience	edition	content
NMAT procedure texts			
AO 6455 (*TU* 11)	Uruk	Brack-Bernsen & Hunger (2002)	GY: Lunar Six; month lengths; astrology
W 20030/16 (*BaM Beih* 2)	Uruk		GY: Moon
BM 36747+37018	Babylon		GY: Lunar Six
BM 36782	Babylon		GY: Lunar Six
BM 36855	Babylon		GY: Lunar Six
BM 37110	Babylon	Brack-Bernsen, Hunger & Walker (2008)	GY: Lunar Six
BM 37332	Babylon		GY: Lunar Six
BM 42282+42294	Babylon	Brack-Bernsen & Hunger (2008)	GY: Lunar Six
BM 43418	Babylon		GY: Lunar Six; eclipses
BM 57980 (*CT* 57, 292)	Babylon	Brack-Bernsen & Walker (2007)	GY: Lunar Six
BM 45990	Babylon		GY: Moon
BM 41004	Babylon	Brack-Bernsen & Hunger (2005/6)	GY: Moon, planets
BM 34560 (*LBAT* 1515)	Babylon		GY: planets
BM 45728	Babylon	Kugler *SSB* I, pp. 45–48; Britton (2003)	GY: planets
BM 47723	Babylon		GY?: planets
BM 36760	Babylon		GY?
BM 33564	Babylon		shadow lengths?
BM 33739	Babylon	Neugebauer & Sachs (1967), ad Text E	Moon's distance to ecliptic
BM 34639 (*LBAT* 1501)	Babylon	Neugebauer & Sachs (1967), ad Text E	lunar motion; *ziqpu* stars
BM 34790 (*LBAT* 1502)	Babylon	Neugebauer & Sachs (1967), ad Text E; Donbaz & Koch (1995), p. 76	lunar motion; *ziqpu* stars
BM 35408 (*LBAT* 1571a)	Babylon		eclipses; omens
BM 36766	Babylon		intercalations; unclear
BM 36854	Babylon		Moon, planets: periods
BM 37056+37074	Babylon		planets; astrology?
BM 76488	Babylon		planets; conjunctions
BM 76703	Babylon		Moon, Sun
MNB 1856 (*TBER* 82)	Babylon	Neugebauer & Sachs (1969), Text H	Moon, Mars
W 22554/6	Uruk	*SpTU* V, 269	instructions for observing planets, *ziqpu* stars
BM 34067+X (*LBAT* 1494)	Babylon		instructions for a gnonom-like instrument
BM 34719 (*LBAT* 1495)	Babylon		instructions for a gnonom-like instrument
Astrological procedure texts			
AO 6449 (*TU* 19)	Uruk	Hunger (1976b)	weather prediction
AO 6488 (*TU* 20)	Uruk	Hunger (1976b)	weather prediction
W 22246a	Uruk	*SpTU* I, 94	prediction of market prices
W 23293/14	Uruk	*SpTU* IV, 161	observation of eclipses, prediction of market prices
BM 32236+32336	Babylon		zodiacal signs, eclipses, watches
Mathematical and metrological problem texts			
AO 6484 (*TU* 33)	Uruk	*MKT* I, pp. 96–107	geometrical problems
AO 7848	Uruk	*MCT*, Text Y	geometrical, metrological problems
W 23291 (*SpTU* 4, 175)	Uruk	Friberg (1997)	metrology
W 23291-x	Uruk	Friberg, Hunger & al-Rawi (1990)	metrology
BM 34568	Babylon	*MKT* III, pp. 14–22	geometrical problems
BM 34724 (*LBAT* 1648)	Babylon	Friberg & Al-Rawi (to appear)	
BM 34800 (*LBAT* 1647)	Babylon	Friberg & Al-Rawi (to appear)	
BM 78822	Babylon	Jursa (1993/4)	metrology
MMA 86.11.404	Babylon	*CTMMA* II; *MCT*, Text X	unidentified computations

Appendix B
Mean values of the synodic time and the synodic arc for the Moon and the planets

The mean synodic time of a planet or the Moon, $\overline{\tau}$, is related to its sidereal period, $P_{\rm sid}$, and that of the Earth, $P_{\rm e} = 1^{\rm yr} = 365.25^{\rm d}$, according to

$$\overline{\tau} = \frac{P_{\rm sid}}{|P_{\rm sid}/P_{\rm e} - 1|}. \tag{B.1}$$

Values of $\overline{\tau}$ expressed in days ($^{\rm d}$) can be converted into mean tithis ($^{\overline{\tau}}$) with the help of Eq. (2.4) on p. 33. From $\overline{\tau}$ one immediately infers m, the whole number of years (revolutions of the Sun) contained in $\overline{\tau}$. The mean apparent angular velocity of an inner planet is the same as that of the Sun, which equals $360°/P_{\rm e}$; for an outer planet or the Moon it equals $360°/P_{\rm sid}$. Hence the mean total synodic arc, $\overline{\Sigma}$, equals

$$\overline{\Sigma} = \begin{cases} \dfrac{\overline{\tau}\cdot 360}{P_{\rm e}} = \dfrac{360}{|P_{\rm e}/P_{\rm sid} - 1|} & \text{(Mercury, Venus)} \\[2ex] \dfrac{\overline{\tau}\cdot 360}{P_{\rm sid}} = \dfrac{360}{|P_{\rm sid}/P_{\rm e} - 1|} & \text{(Moon, Mars, Jupiter, Saturn).} \end{cases} \tag{B.2}$$

From $\overline{\Sigma}$ one obtains the mean synodic arc as $\overline{\sigma} = \overline{\Sigma} - n \cdot 6{,}0°$, where n is the whole number of revolutions contained in $\overline{\Sigma}$. Results for the Moon and the 5 planets are compiled in Table B.1.

planet	$P_{\rm sid}$		$\overline{\tau}$			m	$\overline{\Sigma}$	n
Moon	$0.0748^{\rm yr}$	$27.32158^{\rm d}$	$0.0809^{\rm yr}$	$29.53059^{\rm d}$	$30^{\overline{\tau}}$	0	6,29;6,18,13°	1
Mercury	$0.2408^{\rm yr}$	$87.969^{\rm d}$	$0.317^{\rm yr}$	$115.877^{\rm d}$	$1{,}57;43{,}8^{\overline{\tau}}$	0	1,54;12,43°	0
Venus	$0.6152^{\rm yr}$	$224.701^{\rm d}$	$1.599^{\rm yr}$	$583.92^{\rm d}$	$9{,}53;12^{\overline{\tau}}$	1	9,35;32,20°	1
Mars	$1.8808^{\rm yr}$	$686.971^{\rm d}$	$2.135^{\rm yr}$	$779.96^{\rm d}$	$13{,}12;21^{\overline{\tau}}$	2	6,48;43°	1
Jupiter	$11.859^{\rm yr}$	$4335.57^{\rm d}$	$1.092^{\rm yr}$	$398.88^{\rm d}$	$6{,}45;13^{\overline{\tau}}$	1	33;9°	0
Saturn	$29.457^{\rm yr}$	$10759.22^{\rm d}$	$1.035^{\rm yr}$	$378.09^{\rm d}$	$6{,}24;6^{\overline{\tau}}$	1	12;39°	0

Table B.1: Periods, mean synodic time, mean synodic arc, and the number of whole revolutions of the Sun (m) and the Moon or the planet (n) per synodic cycle.

Appendix C
Multiple transitions of the zonal boundaries of a step function for the synodic arc

Most step functions for the synodic arc (§2.6.10) satisfy $\sigma_j < \alpha_j$, so that at most one zonal boundary is crossed between events $i-1$ and i, as visualised in Fig. 2.10 (p. 49). For Mercury systems A_1, A_2 and A_3 and Mars system A this condition is not satisfied in some zones, and two zonal boundary may be crossed. It is instructive to explore how this affects σ using a modern formalism.

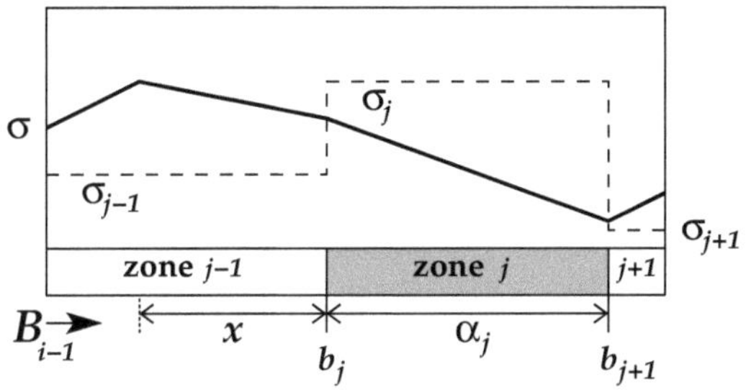

Fig. C.1: If $\sigma_j > \alpha_j$ the influence of zone $j+1$ becomes noticable in zone $j-1$, at a distance x left of b_j.

If $\sigma_j > \alpha_j$ multiple transition regions are formed, and σ deviates from the preliminary value σ_j throughout zone j. The boundaries of the additional transition regions and the associated interpolation coefficients are not mentioned in procedure texts (there is no need for that). They can be derived from the following considerations. If $\sigma_j > \alpha_j$ the influence of the next zone $j+1$ is felt throughout zone j, and there no longer is a region within zone j where $\sigma = \sigma_j$. The position where the influence of zone $j+1$ is first felt lies in zone $j-1$, at a distance x before b_j, where x satisfies $r_{j-1} \cdot (\sigma_{j-1} - x) = \alpha_j$, i.e.

$$x = \sigma_{j-1} - \frac{\alpha_j}{r_{j-1}}. \tag{C.1}$$

Hence two different interpolation coefficients apply in zone $j-1$: before $b_j - x$ it equals r_{j-1}, after it $r_{j-1} \cdot r_j$.[1] By the same token zone j will contain a transition at $b_{j+1} - \sigma_j + \alpha_{j+1}/r_j$ if $\sigma_{j+1} > \alpha_{j+1}$. Only if $\sigma_j < \alpha_j$ there will be a region within zone j where $\sigma = \sigma_j$, followed by a transition region beyond $b_{j+1} - \sigma_j$, where the interpolation coefficient equals r_j as in Fig. 2.10 (p. 49). If in the next zone $\sigma_{j+1} > \alpha_{j+1}$ then this transition region is followed by a second one within zone j to the right of $b_{j+1} - \sigma_j + \alpha_{j+1}/r_j$, where the interpolation coefficient equals $r_j \cdot r_{j+1}$. Hence a given zone j features two transitions if $\sigma_j < \alpha_j$ and $\sigma_{j+1} > \alpha_{j+1}$.

[1] It may happen that $r_{j-1} \cdot r_j = 1$, in which case σ is constant in that part of zone $j-1$, but the value will in general differ from σ_{j-1}.

Appendix D
Interpolation schemes for lunar system A

D.1 Interpolation scheme for computing W from Φ

The following table contains the interpolation scheme for computing W from Φ (§4.4.12). Φ_k and W_k are the control values, c_k is the interpolation coefficient, $|\Phi_{k+1} - \Phi_k|/\phi$ the length of interval k in units of ϕ, and $c_k \cdot |\Phi_{k+1} - \Phi_k|$ is the difference for W across interval k. The tabulated values of W_k have not been reduced to the range 0–6,0° by subtracting the appropriate multiple of 6,0°, i.e. the implied duration of 6^m is always $176^d + W_k$.

k	Φ_k [°]	W_k [°]	$\lvert\Phi_{k+1} - \Phi_k\rvert/\phi$	c_k	$c_k \cdot \lvert\Phi_{k+1} - \Phi_k\rvert$ [°]
1	2,13;20 ↓	6,15;5,11,6,40	1	28;42,30	8;30,22,13,20
2	2,13;2,13,20 ↓	6,6;34,48,53,20	1	26;42,30	7;54,48,53,20
3	2,12;44,26,40 ↓	5,58;40	1	24;40,50	7;18,45,55,33,20
4	2,12;26,40 ↓	5,51;21,14,4,26,40	1	22;42,30	6;43,42,13,20
5	2,12;8,53,20 ↓	5,44;37,31,51,6,40	1	20;42,30	6;8,8,53,20
6	2,11;51,6,40 ↓	5,38;29,22,57,46,40	1	18;42,30	5;32,35,33,20
7	2,11;33,20 ↓	5,32;56,47,24,26,40	1	16;42,30	4;57,2,13,20
8	2,11;15,33,20 ↓	5,27;59,45,11,6,40	1	14;42,30	4;21,28,53,20
9	2,10;57,46,40 ↓	5,23;38,16,17,46,40	1	12;42,30	3;45,55,33,20
10	2,10;40 ↓	5,19;52,20,44,26,40	1	10;42,30	3;10,22,13,20
11	2,10;22,13,20 ↓	5,16;41,58,31,6,40	1	8;42,30	2;34,48,53,20
12	2,10;4,26,40 ↓	5,14;7,9,37,46,40	1	6;30	1;55,33,20
13	2,9;46,40 ↓	5,12;11,36,17,46,40	1	4;42,30	1;23,42,13,20
14	2,9;28,53,20 ↓	5,10;47,54,4,26,40	1	2;42,30	0;48,8,53,20
15	2,9;11,6,40 ↓	5,9;59,45,11,6,40	1	0;42,30	0;12,35,33,20
16	2,8;53,20 ↓	5,9;47,9,37,46,40	1	1;17,30	0;22,57,46,40
17	2,8;35,33,20 ↓	5,10;10,7,24,26,40	1	3;17,30	0;58,31,6,40
18	2,8;17,46,40 ↓	5,11;8,38,31,6,40	1	5;17,30	1;34,4,26,40
19	2,8↓	5,12;42,42,57,46,40	1	7;10,50	2;7,39,15,33,20
20	2,7;42,13,20 ↓	5,14;50,22,13,20	1	9;17,30	2;45,11,6,40
21	2,7;24,26,40 ↓	5,17;35,33,20	1	11;23,20	3;22,28,8,53,20
22	2,7;6,40 ↓	5,20;58,1,28,53,20	1	13;17,30	3;56,17,46,40
23	2,6;48,53,20 ↓	5,24;54,19,15,33,20	1	15;17,30	4;31,51,6,40
24	2,6;31,6,40 ↓	5,29;26,10,22,13,20	1	17;17,30	5;7,24,26,40
25	2,6;13,20 ↓	5,34;33,34,48,53,20	1	19;17,30	5;42,57,46,40
26	2,5;55,33,20 ↓	5,40;16,32,35,33,20	1	21;17,30	6;18,31,6,40
27	2,5;37,46,40 ↓	5,46;35,3,42,13,20	1	23;17,30	6;54,4,26,40
28	2,5;20 ↓	5,53;29,8,8,53,20	1	25;4,10	7;25,40,44,26,40
29	2,5;2,13,20 ↓	6,0;54,48,53,20	1	27;17,30	8;5,11,6,40
30	2,4;44,26,40 ↓	6,9	1	29;17,30	8;40,44,26,40
31	2,4;26,40 ↓	6,17;40,44,26,40	1	31	9;11,6,40
32	2,4;8,53,20 ↓	6,26;51,51,6,40	1	32	9;28,53,20
33	2,3;51,6,40 ↓	6,36;20,44,26,40	1	33	9;46,40
34	2,3;33,20 ↓	6,46;7,24,26,40	1	34	10;4,26,40
35	2,3;15,33,20 ↓	6,56;11,51,6,40	1	35	10;22,13,20
36	2,2;57,46,40 ↓	7,6;34,4,26,40	1	36	10;40
37	2,2;40 ↓	7,17;14,4,26,40	1	37	10;57,46,40
38	2,2;22,13,20 ↓	7,28;11,51,6,40	1	38	11;15,33,20
39	2,2;4,26,40 ↓	7,39;27,24,26,40	1	39	11;33,20
40	2,1;46,40 ↓	7,51;0,44,26,40	1	40	11;51,6,40
41	2,1;28,53,20 ↓	8,2;51,51,6,40	1	41	12;8,53,20

k	Φ_k [°]	W_k [°]	$\lvert\Phi_{k+1}-\Phi_k\rvert/\phi$	c_k	$c_k \cdot \lvert\Phi_{k+1}-\Phi_k\rvert$ [°]
42	2,1;11,6,40 ↓	8,15;0,44,26,40	1	42	12;26,40
43	2,0;53,20 ↓	8,27;27,24,26,40	1	43	12;44,26,40
44	2,0;35,33,20 ↓	8,40;11,51,6,40	1	44	13;2,13,20
45	2,0;17,46,40 ↓	8,53;14,4,26,40	1	45	13;20
46	2,0↓	9,6;34,4,26,40	1	46	13;37,46,40
47	1,59;42,13,20 ↓	9,20;11,51,6,40	1	46;53,20	13;53,34,48,53,20
48	1,59;24,26,40 ↓	9,34;5,25,55,33,20	1	48	14;13,20
49	1,59;6,40 ↓	9,48;18,45,55,33,20	1	49	14;31,6,40
50a	1,58;48,53,20 ↓	10,2;49,52,35,33,20	3;25,37,30 } 5	50	50;46,17,46,40
50b	1,57;47,57,46,40 = m	10,53;36,10,22,13,20	1;34,22,30 } 5	50	23;18,8,53,20
51	1,58;15,55,33,20 ↑	11,16;54,19,15,33,20	1	49;40	14;42,57,46,40
52	1,58;33,42,13,20 ↑	11,31;37,17,2,13,20	1	48;4,10	14;14,34,4,26,40
53	1,58;51,28,53,20 ↑	11,45;51,51,6,40	1	47;2,30	13;56,17,46,40
54	1,59;9,15,33,20 ↑	11,59;48,8,53,20	1	45;2,30	13;20,44,26,40
55	1,59;27,2,13,20 ↑	12,13;8,53,20	1	43;2,30	12;45,11,6,40
56	1,59;44,48,53,20 ↑	12,25;54,4,26,40	1	40;55	12;7,24,26,40
57	2,0;2,35,33,20 ↑	12,38;1,28,53,20	1	38;51,15	11;30,44,26,40
58	2,0;20,22,13,20 ↑	12,49;32,13,20	1	36;51,15	10;55,11,6,40
59	2,0;38,8,53,20 ↑	13,0;27,24,26,40	1	34;51,15	10;19,37,46,40
60	2,0;55,55,33,20 ↑	13,10;47,2,13,20	1	32;37,55	9;40,7,24,26,40
61	2,1;13,42,13,20 ↑	13,20;27,9,37,46,40	1	30;17,30	8;58,31,6,40
62	2,1;31,28,53,20 ↑	13,29;25,40,44,26,40	1	29;2,30	8;36,17,46,40
63	2,1;49,15,33,20 ↑	13,38;1,58,31,6,40	1	27;2,30	8;0,44,26,40
64	2,2;7,2,13,20 ↑	13,46;2,42,57,46,40	1	25;2,30	7;25,11,6,40
65	2,2;24,48,53,20 ↑	13,53;27,54,4,26,40	1	23;1,40	6;49,22,57,46,40
66	2,2;42,35,33,20 ↑	14,0;17,17,2,13,20	1	20;51,15	6;10,44,26,40
67	2,3;0,22,13,20 ↑	14,6;28,1,28,53,20	1	18;51,15	5;35,11,6,40
68	2,3;18,8,53,20 ↑	14,12;3,12,35,33,20	1	16;51,15	4;59,37,46,40
69	2,3;35,55,33,20 ↑	14,17;2,50,22,13,20	1	14;44,35	4;22,5,55,33,20
70	2,3;53,42,13,20 ↑	14,21;24,56,17,46,40	1	12;17,30	3;38,31,6,40
71	2,4;11,28,53,20 ↑	14,25;3,27,24,26,40	1	11;2,30	3;16,17,46,40
72	2,4;29,15,33,20 ↑	14,28;19,45,11,6,40	1	9;2,30	2;40,44,26,40
73	2,4;47,2,13,20 ↑	14,31;0,29,37,46,40	1	7;2,30	2;5,11,6,40
74	2,5;4,48,53,20 ↑	14,33;5,40,44,26,40	1	5;8,20	1;31,21,28,53,20
75	2,5;22,35,33,20 ↑	14,34;37,2,13,20	1	2;51,15	0;50,44,26,40
76	2,5;40,22,13,20 ↑	14,35;27,46,40	1	0;51,15	0;15,11,6,40
77	2,5;58,8,53,20 ↑	14,35;42,57,46,40	1	1;8,45	0;20,22,13,20
78	2,6;15,55,33,20 ↑	14,35;22,35,33,20	1	3;8,45	0;55,55,33,20
79	2,6;33,42,13,20 ↑	14,34;26,40	1	5;20	1;34,48,53,20
80	2,6;51,28,53,20 ↑	14,32;51,51,6,40	1	7;42,30	2;17,2,13,20
81	2,7;9,15,33,20 ↑	14,30;34,48,53,20	1	8;57,30	2;39,15,33,20
82	2,7;27,2,13,20 ↑	14,27;55,33,20	1	10;57,30	3;14,48,53,20
83	2,7;44,48,53,20 ↑	14,24;40,44,26,40	1	12;57,30	3;50,22,13,20
84	2,8;2,35,33,20 ↑	14,20;50,22,13,20	1	15;5	4;28,8,53,20
85	2,8;20,22,13,20 ↑	14,16;22,13,20	1	17;8,45	5;4,48,53,20
86	2,8;38,8,53,20 ↑	14,11;17,24,26,40	1	19;8,45	5;40,22,13,20
87	2,8;55,55,33,20 ↑	14,5;37,2,13,20	1	21;8,45	6;15,55,33,20
88	2,9;13,42,13,20 ↑	13,59;21,6,40	1	23;22,5	6;55,25,55,33,20
89	2,9;31,28,53,20 ↑	13,52;25,40,44,26,40	1	25;42,30	7;37,2,13,20
90	2,9;49,15,33,20 ↑	13,44;48,38,31,6,40	1	26;57,30	7;59,15,33,20
91	2,10;7,2,13,20 ↑	13,36;49,22,57,46,40	1	28;57,30	8;34,48,53,20
92	2,10;24,48,53,20 ↑	13,28;14,34,4,26,40	1	30;57,30	9;10,22,13,20
93	2,10;42,35,33,20 ↑	13,19;4,11,51,6,40	1	32;58,20	9;46,10,22,13,20
94	2,11;0,22,13,20 ↑	13,9;18,1,28,53,20	1	35;8,45	10;24,48,53,20
95	2,11;18,8,53,20 ↑	12,58;53,12,35,33,20	1	37;8,45	11;0,22,13,20
96	2,11;35,55,33,20 ↑	12,47;52,50,22,13,20	1	39;8,45	11;35,55,33,20
97	2,11;53,42,13,20 ↑	12,36;16,54,48,53,20	1	41,15,25	12;13,27,24,26,40
98	2,12;11,28,53,20 ↑	12,24;3,27,24,26,40	1	43;42,30	12;57,2,13,20
99	2,12;29,15,33,20 ↑	12,11;6,25,11,6,40	1	44;57,30	13;19,15,33,20
100	2,12;47,2,13,20 ↑	11,57;47,9,37,46,40	1	46;57,30	13;54,48,53,20
101	2,13;4,48,53,20 ↑	11,43;52,20,44,26,40	1	48;41,15	14;25,33,20
102	2,13;22,35,33,20 ↑	11,29;26,47,24,26,40	1	49;42,55	14;43,49,37,46,40
103	2,13;40,22,13,20 ↑	11,14;42,57,46,40	4	50	59;15,33,20
104	2,14;51,28,53,20 ↑	10,15;27,24,26,40	1	49;51,15	14;46,17,46,40
105	2,15;9,15,33,20 ↑	10,0;41,6,40	1	48;51,15	14;28,31,6,40
106	2,15;27,2,13,20 ↑	9,46;12,35,33,20	1	47;51,15	14;10,44,26,40
107	2,15;44,48,53,20 ↑	9,32;1,51,6,40	1	46;41,40	13;50,7,24,26,40

k	Φ_k [°]	W_k [°]	$\lvert\Phi_{k+1}-\Phi_k\rvert/\phi$	c_k	$c_k\cdot\lvert\Phi_{k+1}-\Phi_k\rvert$ [°]
108	2,16;2,35,33,20 ↑	9,18;11,43,42,13,20	1	45;51,15	13;35,11,6,40
109	2,16;20,22,13,20 ↑	9,4;36,32,35,33,20	1	44;51,15	13;17,24,26,40
110	2,16;38,8,53,20 ↑	8,51;19,8,8,53,20	1	43;51,15	12;59,37,46,40
111a	2,16;55,55,33,20 ↑	8,38;19,30,22,13,20	0;30 } 1	42;51,15	6;20,55,33,20
111b	2,17;4,48,53,20 = M	8,31;58,34,48,53,20	0;30	42;51,15	6;20,55,33,20
112	2,16;55,55,33,20 ↓	8,25;37,39,15,33,20	1	41;51,15	12;24,4,26,40
113	2,16;38,8,53,20 ↓	8,13;13,34,48,53,20	1	40;51,15	12;6,17,46,40
114	2,16;20,22,13,20 ↓	8,1;7,17,2,13,20	1	39;51,15	11;48,31,6,40
115	2,16;2,35,33,20 ↓	7,49;18,45,55,33,20	1	38;51,15	11;30,44,26,40
116	2,15;44,48,53,20 ↓	7,37;48,1,28,53,20	1	37;48,20	11;12,5,55,33,20
117	2,15;27,2,13,20 ↓	7,26;35,55,33,20	1	36;51,15	10;55,11,6,40
118	2,15;9,15,33,20 ↓	7,15;40,44,26,40	1	35;51,15	10;37,24,26,40
119	2,14;51,28,53,20 ↓	7,5;3,20	1	34;51,15	10;19,37,46,40
120	2,14;33,42,13,20 ↓	6,54;43,42,13,20	1	33;51,15	10;1,51,6,40
121	2,14;15,55,33,20 ↓	6,44;41,51,6,40	1	32;51,15	9;44,4,26,40
122	2,13;58,8,53,20 ↓	6,34;57,46,40	1	31;51,15	9;26,17,46,40
123	2,13;40,22,13,20 ↓	6,25;31,28,53,20	1	30;51,15	9;8,31,6,40
124	2,13;22,35,33,20 ↓	6,16;22,57,46,40	0;8,45	30	1;17,46,40

D.2 Interpolation scheme for computing Λ from Φ

The following table contains the interpolation scheme for computing Λ from Φ (§4.4.13). Φ_k and Λ_k are the control values, c_k is the interpolation coefficient, $\lvert\Phi_{k+1}-\Phi_k\rvert/\phi$ the length of interval k in units of ϕ, and $c_k\cdot\lvert\Phi_{k+1}-\Phi_k\rvert$ the difference for Λ across interval k.

k	Φ_k [°]	Λ_k [°]	$\lvert\Phi_{k+1}-\Phi_k\rvert/\phi$	c_k	$c_k\cdot\lvert\Phi_{k+1}-\Phi_k\rvert$ [°]
1	2,0;38,8,53,20 ↓	−20;13,20	1	1	0;17,46,40
2	2,0;20,22,13,20 ↓	−19;55,33,20	1	3	0;53,20
3	2,0;2,35,33,20 ↓	−19;2,13,20	1	5	1;28,53,20
4	1,59;44,48,53,20 ↓	−17;33,20	1	7	2;4,26,40
5	1,59;27,2,13,20 ↓	−15;28,53,20	1	9	2;40
6	1,59;9,15,33,20 ↓	−12;48,53,20	1	11	3;15,33,20
7	1,58;51,28,53,20 ↓	−9;33,20	1	13	3;51,6,40
8	1,58;33,42,13,20 ↓	−5;42,13,20	1	14;8,45	4;11,28,53,20
9	1,58;15,55,33,20 ↓	−1;30,44,26,40	1	15;8,45	4;29,15,33,20
10a	1,57;58,8,53,20 ↓	2;58,31,6,40	0;34,22,30 } 1	16;8,45	2;44,26,53,53,20
10b	1,57;47,57,46,40 = m	5;42,58,0,33,20	0;25,37,30	16;8,45	2;2,35,19,26,40
11	1,57;55,33,20 ↑	7;45,33,20	1	17;8,45	5;4,48,53,20
12	1,58;13,20 ↑	12;50,22,13,20	32	18;8,45	2,52;2,57,46,40
13	2,7;42,13,20 ↑	3,4;53,20	1	18	5;20
14	2,8↑	3,10;13,20	1	17	5;2,13,20
15	2,8;17,46,40 ↑	3,15;15,33,20	1	16	4;44,26,40
16	2,8;35,33,20 ↑	3,20	1	15	4;26,40
17	2,8;53,20 ↑	3,24;26,40	1	14	4;8,53,20
18	2,9;11,6,40 ↑	3,28;35,33,20	1	13	3;51,6,40
19	2,9;28,53,20 ↑	3,32;26,40	1	12	3;33,20
20	2,9;46,40 ↑	3,36	1	11	3;15,33,20
21	2,10;4,26,40 ↑	3,39;15,33,20	1	10	2;57,46,40
22	2,10;22,13,20 ↑	3,42;13,20	1	9	2;40
23	2,10;40 ↑	3,44;53,20	1	8	2;22,13,20
24	2,10;57,46,40 ↑	3,47;15,33,20	1	7	2;4,26,40
25	2,11;15,33,20 ↑	3,49;20	1	6	1;46,40
26	2,11;33,20 ↑	3,51;6,40	1	5	1;28,53,20
27	2,11;51,6,40 ↑	3,52;35,33,20	1	4	1;11,6,40
28	2,12;8,53,20 ↑	3,53;46,40	1	3	0;53,20
29	2,12;26,40 ↑	3,54;40	1	2	0;35,33,20
30	2,12;44,26,40 ↑	3,55;15,33,20	1	1	0;17,46,40
31	2,13;2,13,20 ↑	3,55;33,20	8;8,45	0	0
32	2,15;27,2,13,20 ↑	3,55;33,20	1	1	0;17,46,40
33	2,15;44,48,53,20 ↑	3,55;15,33,20	1	2	0;35,33,20
34	2,16;2,35,33,20 ↑	3,54;40	1	3	0;53,20
35	2,16;20,22,13,20 ↑	3,53;46,40	1	4	1;11,6,40
36	2,16;38,8,53,20 ↑	3,52;35,33,20	1	5	1;28,53,20

k	Φ_k [°]	Λ_k [°]	$\lvert\Phi_{k+1}-\Phi_k\rvert/\phi$	c_k	$c_k\cdot\lvert\Phi_{k+1}-\Phi_k\rvert$ [°]
37a	2,16;55,55,33,20 ↑	3,51;6,40	0;30 } 1	6	0;53,20
37b	2,17;4,48,53,20 = M	3,49;57,46,40	0;30 } 1	6	0;53,20
38	2,16;55,55,33,20 ↓	3,49;20	1	7	2;4,26,40
39	2,16;38,8,53,20 ↓	3,47;15,33,20	1	8	2;22,13,20
40	2,16;20,22,13,20 ↓	3,44;53,20	1	9	2;40
41	2,16;2,35,33,20 ↓	3,42;13,20	1	10	2;57,46,40
42	2,15;44,48,53,20 ↓	3,39;15,33,20	1	11	3;15,33,20
43	2,15;27,2,13,20 ↓	3,36	1	12	3;33,20
44	2,15;9,15,33,20 ↓	3,32;26,40	1	13	3;51,6,40
45	2,14;51,28,53,20 ↓	3,28;35,33,20	1	14	4;8,53,20
46	2,14;33,42,13,20 ↓	3,24;26,40	1	15	4;26,40
47	2,14;15,55,33,20 ↓	3,20	1	16	4;44,26,40
48	2,13;58,8,53,20 ↓	3,15;15,33,20	1	17	5;2,13,20
49	2,13;40,22,13,20 ↓	3,10;13,20	1	18	5;20
50	2,13;22,35,33,20 ↓	3,4;53,20	32	18;8,45	2,52;2,57,46,40
51	2,3;53,42,13,20 ↓	12;50,22,13,20	1	17;8,45	5;4,48,53,20
52	2,3;35,55,33,20 ↓	7;45,33,20	1	16;8,45	4;47,2,13,20
53	2,3;18,8,53,20 ↓	2;58,31,6,40	1	15;8,45	4;29,15,33,20
54	2,3;0,22,13,20 ↓	−1;30,44,26,40	1	14;8,45	4;11,28,53,20
55	2,2;42,35,33,20 ↓	−5;42,13,20	1	13	3;51,6,40
56	2,2;24,48,53,20 ↓	−9;33,20	1	11	3;15,33,20
57	2,2;7,2,13,20 ↓	−12;48,53,20	1	9	2;40
58	2,1;49,15,33,20 ↓	−15;28,53,20	1	7	2;4,26,40
59	2,1;31,28,53,20 ↓	−17;33,20	1	5	1;28,53,20
60	2,1;13,42,13,20 ↓	−19;2,13,20	1	3	0;53,20
61	2,0;55,55,33,20 ↓	−19;55,33,20	1	1	0;17,46,40

Appendix E
Rising and setting times

E.1 Rising or setting time of a short arc near the ecliptic

Time differences between the rising or setting of the Sun and that of the Moon or a planet are a recurrent topic of Babylonian astronomy. In the Lunar Six modules of lunar system A (and B) the time between sunrise or sunset and moonrise or moonset near New Moon or Full Moon is computed by multiplying the Moon's elongation from the Sun (or anti-Sun) by a coefficient that is a function of the Sun's zodiacal position, and the same for the Moon's distance to the ecliptic (§4.5.18). The rising or setting time is then obtained by adding or subtracting the two contributions. A modern derivation of these coefficients proceeds by considering the sidereal time (θ) between the rising or setting of both ends of the arc. Consider an arc between two points, corotating with the sky, labeled 1 and 2. The sidereal time is $\theta = \tau + \alpha$, where τ is the hour angle, measured positively with the daily rotation from the southern meridian, and α is right ascension. Hence the rising or setting time of the arc is

$$\Delta\theta = \theta_2 - \theta_1 = \tau_2 - \tau_1 + \alpha_2 - \alpha_1, \tag{E.1}$$

The hour angle at rising or setting is defined by the condition that the object is at the horizon, so that its altitude is zero. This requires

$$\cos\tau = -\tan\phi\tan\delta, \tag{E.2}$$

where ϕ is the geographical latitude and δ is the declination. The negative solution, say $\tau_{\rm rise}$, represents the rising, and the positive solution, $\tau_{\rm set} = -\tau_{\rm rise}$, represents the setting.[2] If the corresponding ecliptical coordinates are (λ_1, β_1) and (λ_2, β_2) the right ascension is given by

$$\tan\alpha = \cos\varepsilon\,\tan\lambda - \sin\varepsilon\cdot\frac{\tan\beta}{\cos\lambda}. \tag{E.3}$$

By inserting α_1 and α_2 from this equation, and τ_1 and τ_2 from Eq. (E.2), into Eq. (E.1), we obtain $\Delta\theta$. It is now assumed that points 1 and 2 are close together and also close to the ecliptic, so that all angular differences, e.g. $\Delta\lambda = \lambda_2 - \lambda_1$, and β_1 and β_2, are small parameters. All trigonometric functions are expanded up to first order in β_1, β_2 and $\Delta\lambda$. Expanding the right-hand sides of Eqs. (E.2, E.3) yields

$$\cos\tau_2 \approx \cos\tau_1 - \frac{\tan\phi}{\cos^2\delta_1}\Delta\delta \tag{E.4}$$

$$\tan\alpha_2 \approx \tan\alpha_1 + \frac{\cos\varepsilon}{\cos^2\lambda_1}\Delta\lambda - \frac{\sin\varepsilon}{\cos\lambda_1}\Delta\beta. \tag{E.5}$$

By expanding the left-hand sides according to $\cos\tau_2 \approx \cos\tau_1 - \sin\tau\,\Delta\tau$ and $\tan\alpha_2 \approx \tan\alpha_1 + \Delta\alpha/\cos^2\alpha_1$ one obtains

$$\Delta\tau \approx \frac{\tan\phi}{\cos^2\delta_1\sin\tau_1}\Delta\delta \tag{E.6}$$

$$\Delta\alpha \approx \frac{\cos\varepsilon\cos^2\alpha_1}{\cos^2\lambda_1}\Delta\lambda - \frac{\sin\varepsilon\cos^2\alpha_1}{\cos\lambda_1}\Delta\beta. \tag{E.7}$$

The difference in declination is eliminated by expanding

[2] It follows immediately that $\Delta\theta_{\rm rise} + \Delta\theta_{\rm set} = 2(\alpha_2 - \alpha_1)$, i.e. the sum of the rising time and the setting time of the arc equals twice its rising time at *sphaera recta*, a historical term referring to the case $\phi = 0$.

$$\sin\delta_2 = \cos\varepsilon\sin\beta_2 + \sin\varepsilon\cos\beta_2\sin\lambda_2 \tag{E.8}$$

about the coordinates of point 1, leading to

$$\Delta\delta \approx \frac{\sin\varepsilon\cos\lambda_1}{\cos^2\delta_1}\Delta\lambda - \frac{\sin\varepsilon\cos^2\alpha_1}{\cos\lambda_1}\Delta\beta. \tag{E.9}$$

Upon substitution in Eq. (E.6), expressing $\Delta\theta$ in the form

$$\Delta\theta = q\Delta\lambda + r\Delta\beta, \tag{E.10}$$

and dropping the label 1, the coefficients q and r are obtained as

$$q = \frac{\tan\phi\sin\varepsilon\cos\lambda}{\cos^2\delta\sin\tau} + \frac{\cos\varepsilon\cos^2\alpha}{\cos\lambda} \tag{E.11}$$

$$r = \frac{\tan\phi\cos\varepsilon}{\cos^3\delta\sin\tau} - \frac{\sin\varepsilon\cos^2\alpha}{\cos\lambda}. \tag{E.12}$$

For given values of the geographical latitude (ϕ) and obliquity of the ecliptic (ε) q and r are functions of the ecliptical longitude, because $\cos\alpha$, $\cos\delta$ and $\sin\tau$ can be eliminated by inserting

$$\cos\alpha = \frac{1}{\sqrt{1+\cos^2\varepsilon\tan^2\lambda}} \tag{E.13}$$

$$\cos\delta = \sqrt{1-\sin^2\varepsilon\sin^2\lambda} \tag{E.14}$$

$$\sin\tau = \pm\sqrt{1-\tan^2\phi\tan^2\delta} = \pm\sqrt{1-\frac{\tan^2\phi\sin^2\varepsilon\sin^2\lambda}{1-\sin^2\varepsilon\sin^2\lambda}}. \tag{E.15}$$

In the last formula the plus sign signifies a setting, the minus sign a rising. The end result is

$$q(\lambda) = \frac{\cos\varepsilon}{1-\sin^2\varepsilon\sin^2\lambda}\left(1 \pm \frac{\tan\varepsilon\sin\phi\cos\lambda}{\sqrt{\cos^2\phi - \sin^2\varepsilon\sin^2\lambda}}\right) \tag{E.16}$$

$$r(\lambda) = \frac{\cos\varepsilon}{1-\sin^2\varepsilon\sin^2\lambda}\left(-\tan\varepsilon\cos\lambda \pm \frac{\sin\phi}{\sqrt{\cos^2\phi - \sin^2\varepsilon\sin^2\lambda}}\right). \tag{E.17}$$

Again the plus signs signify a setting, the minus signs a rising. Fig. E.1 shows q_{set} and r_{set} for Babylon ($\phi = 32.5°$) and Uruk ($\phi = 31.3°$) as functions of the ecliptical longitude (λ), assuming $\varepsilon = 23.45°$. As is readily verified, $q_{\text{rise}}(\lambda) = q_{\text{set}}(\lambda + 180°)$ and $r_{\text{rise}}(\lambda) = -r_{\text{set}}(\lambda + 180°)$. Also note that q_{rise}, q_{set}, r_{rise} and r_{set} are all positive definite. Now consider the case that point 1 is the Sun's centre, so that $\beta_1 = 0$ and $\Delta\beta = \beta_2 = \beta$. If we assume that $\Delta\lambda$ and β are positive, then $q_{\text{set}}\Delta\lambda$, $r_{\text{set}}\beta$ and $q_{\text{rise}}\Delta\lambda$ are positive, while $r_{\text{rise}}\beta$ is negative; if β is negative then $r_{\text{set}}\beta$ is negative and $r_{\text{rise}}\beta$ is positive. The algorithms underlying step 11 of the Lunar Six module (§4.5.18.10) are in full agreement with this.

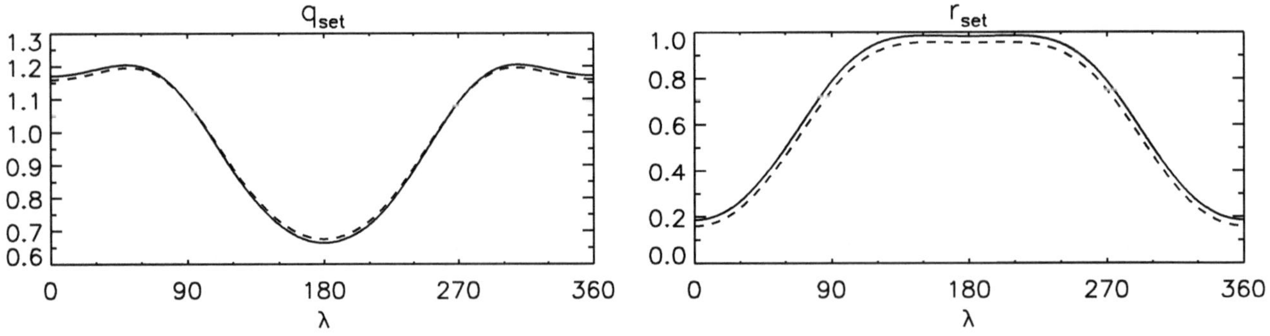

Fig. E.1: Coefficients q_{set} and r_{set} as a function of ecliptical longitude for Babylon (drawn) and Uruk (dashed).

E.2 Effects of refraction and use of disk rims on rising and setting times

Due to atmospheric refraction, all celestial objects appear to be closer to the zenith by an amount increasing from 0 (zenith) to a maximum at the horizon. If the observer is at the same elevation as the horizon, and assuming standard atmospheric circumstances, the apparent altitude of the object, h, satisfies (Meeus 1998)

$$h = h_0 + R(h). \tag{E.18}$$

Here h_0 is the true altitude and R the refraction angle, which decreases monotonically with increasing h from $R_0 \approx 0.57^\circ$ = 0;34° at the horizon ($h = 0^\circ$) to 0° at the zenith ($h = 90^\circ$). If the observer is elevated by a height H above the horizon there is a 'dip effect' resulting in a correction $\delta R = 0.0353^\circ\sqrt{H}$, where H is measured in meters.[3] Henceforth the dip effect is assumed to be absorbed in R_0. If the hour angle at rising or setting for the refracted case is τ_1, and that for the unrefracted case is τ_0, then the associated difference in sidereal time is

$$\Delta\theta = \tau_1 - \tau_0, \tag{E.19}$$

where

$$\cos\tau_1 = -\frac{\sin R_0 + \sin\phi\sin\delta}{\cos\phi\cos\delta}, \tag{E.20}$$

so that

$$\cos\tau_1 - \cos\tau_0 = -\frac{\sin R_0}{\cos\phi\cos\delta}. \tag{E.21}$$

By expanding $\cos\tau_1 \approx \cos\tau_0 - \sin\tau_0\,\Delta\tau$ and inserting Eq. (E.15) one obtains

$$\tau_1 - \tau_0 = \pm\frac{\sin R_0}{\sqrt{\cos^2\phi\cos^2\delta - \sin^2\phi\sin^2\delta}}, \tag{E.22}$$

where the plus sign is for a rising and the minus sign for a setting. If we restrict ourselves to the planets, and neglect their latitude, we can express δ in terms of the ecliptical longitude according to $\sin\delta = \sin\varepsilon\sin\lambda$. Furthermore, the angle R_0 is very small so that it is safe to approximate $\sin R_0 \approx R_0$. Hence we obtain

$$\Delta\theta \approx \pm s(\lambda)R_0, \tag{E.23}$$

where the plus is for a setting, the minus for a rising, and

$$s(\lambda) = \frac{1}{\sqrt{\cos^2\phi - \sin^2\varepsilon\sin^2\lambda}}. \tag{E.24}$$

Another application of this formula concerns the effect of using the rim of the lunar or solar disk instead of the disk centre in defining risings and settings. This is achieved by replacing R_0 in Eq. (E.23) by $\pm r$, where r is the disk radius measured in degrees, and the plus sign is for the lower rim, the minus sign for the upper rim with respect to the horizon. The combined effect of refraction and usage of the disk rim is the sum or the difference of the individual effects, depending on whether one is considering a setting or a rising, and whether the upper rim or the lower rim is used. The obvious reason why such different phenomena have the same effect on rising and setting times is that the Moon and the Sun are round, so that using disk rims instead of centres amounts to a shift in altitude irrespective of the angle between the ecliptic and the horizon. Formula (E.24) yields the factor by which altitude differences must be multiplied in order to find the resulting differences in rising and setting times, analogous to the coefficients q and r. In Fig. (E.2) its value for Babylon ($\phi = 32.5$) and for Uruk ($\phi = 31.3$) is plotted as a function of λ. Note that, as opposed to q and r, the amplitude of s is small compared to its mean value (1.26 = 1;16 for Babylon) and that it has two minima and maxima along the ecliptic.

E.3 Contributions to the disk correction in lunar system A

This appendix contains a preliminary assessment of several possible contributions to the 'disk correction' (Δ) in the Lunar Six modules of lunar system A (§4.4.19.10), assuming that it is a heuristic correction for systematic differences between

[3] This expression is the sum of the angle pertained by H as seen from the observer's horizon and a correction for the additional refraction resulting from the increased distance between observer and horizon (Seidelmann et al. 1992).

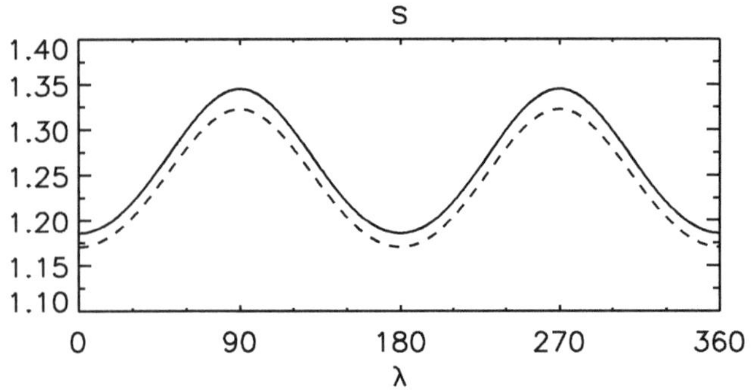

Fig. E.2: Coefficient s as a function of ecliptical longitude for Babylon (drawn) and Uruk (dashed).

computed and measured Lunar Six values. If this assumption is correct, it is not entirely certain that all the relevant contributions have been identified in the present analysis. For instance, the inaccuracies of the Lunar Six module itself will result in systematic differences between observed and predicted Lunar Sixes, but this is ignored for the moment. In other words, it is assumed that the lunar scheme correctly predicts the positions of the Moon and the Sun for disk centre.

The first contribution to be considered is the rim effect. In principle the Babylonian rim conventions can be inferred from recorded observations of Lunar Sixes, but attempts to do so have not yet been successful, due to remaining systematic errors. No combination of rim definitions and dip effect (cf. below) appears to succeed in reproducing simultaneously, in an average sense, all Lunar Six measurements to within 1° (Table E.1; Huber 2000; Steele 2000b).

	disk centre	uM uS
ŠU$_2$	+2.0	+1.3
NA	−1.1	−0.4
ME	−1.7	−1.1
GI$_6$	+1.7	+1.0
KUR	+2.1	+2.1
NA$_1$	+1.1	+1.1

Table E.1: Mean deviations [°] between modern computations (no dip effect) and recorded values of Lunar Six intervals defined using disk centre or upper (u) rims for both disks. In the latter case the deviations for ŠU$_2$, NA, ME and GI$_6$ are reduced but still significant. Adapted from Steele (2000b).

In the following analysis the rim convention is therefore considered to be a free parameter for ŠU$_2$, NA, ME and GI$_6$. The aim is to verify whether there is a choice of rims which, in combination with the dip effect, succeeds in reproducing the attested disk corrections for these Lunar Sixes. For NA$_1$ there is no choice but to use the lower (=front) rim of the Moon, since only that rim is visible. Most likely the upper (=back) rim of the Sun was used, because the first crescent may still be invisible when the Sun is partly above the horizon. This definition (fM bS) happens to also produce the best agreement for ŠU$_2$, NA, ME and GI$_6$. By the same token the upper rims of both Moon and Sun must have been used when measuring KUR. This may explain why there is no ‘disk correction’ in the Lunar Six module for KUR. As argued in Appendix E.2, the rim effect amounts to an altitude shift of the reference point for horizon crossings by

$$\Delta h = \pm r, \tag{E.25}$$

where r is the radius of the disk, i.e. $r_☽ \approx 0.27°$ or $r_\odot \approx 0.27°$. The plus sign applies to the upper rim, the minus sign to the lower rim. Refraction and dip effect (Appendix E.2) influence ŠU$_2$, NA, ME and GI$_6$, since for these intervals either the Moon or the Sun is rising while the other is setting, but they have no effect on NA$_1$ and KUR, when both are setting (NA$_1$) or rising (KUR) and therefore subject to identical effects. The resulting change in apparent altitude is

$$\Delta h = R_0 + \delta R \approx 0.57° + 0.0353°\sqrt{H}. \tag{E.26}$$

The third effect to be taken into account is the Moon’s parallax. When the Moon is at the horizon this results in an altitude shift (Meeus 1998)

$$\Delta h \approx -\pi_☽ \approx -0.95°. \tag{E.27}$$

An increase of the apparent altitude causes risings to occur sooner, settings later, and vice versa a decrease of the altitude. As shown in Appendix E.2, the change in the rising and setting times resulting from a vertical displacement Δh is $\Delta\tau = s \cdot \Delta h$, where $s \equiv 1/\sqrt{\cos^2\phi - \sin^2\varepsilon \sin^2\lambda}$, $\varepsilon = 23.5$ and $\phi = 32.5$ (Babylon). The yearly average of s for Babylon is 1.25. The sum of the contributions due to refraction, dip effect, rim effect and parallax is

$$\Delta \mathrm{L6} = s(\lambda)\,[2\sigma_1\,(R_0 + \delta R) + \sigma_2\,(\sigma_3 + \sigma_4)\,r - \sigma_5 \pi_☽], \tag{E.28}$$

where L6 stands for a Lunar Six interval, $r \approx 0;16°$ is the radius of the lunar and solar disks, and $\sigma_i = \pm 1$ represents the sign of correction i. In front of R_0 there is a plus sign ($\sigma_1 = +1$) if the rising precedes the setting (NA and ME), a minus sign ($\sigma_1 = -1$) for the opposite case (ŠU$_2$ and GI$_6$). The sign of the rim effect depends on two factors. If the Moon precedes the Sun (ŠU$_2$, ME) $\sigma_2 = +1$, if the reverse is true (NA, GI$_6$) $\sigma_2 = -1$. If the upper rim of the lunar disk is used then $\sigma_3 = -1$, if the lower rim is used $\sigma_3 = +1$. The same rules, applied to the solar disk, determine σ_4. The total factor for the rim effect is therefore as follows: $\sigma_2(\sigma_3 + \sigma_4) = +2$ if the *front* rim is used for the object crossing the horizon first and the *back* rim for the object crossing the horizon last, where back and front are defined with respect to the daily rotation. If front and back are exchanged then $\sigma_2(\sigma_3 + \sigma_4) = -2$. The rim effect vanishes if the same rims, defined in terms of the daily rotation, are used for both disks ($\sigma_3 + \sigma_4 = 0$). As is readily seen in Fig. 4.36, p. 176, the effect of parallax, through the Moon's lower apparent altitude, is to increase ŠU$_2$ and GI$_6$ and decrease NA and ME by the amount $s(\lambda) \cdot 0.95°$, i.e. $\sigma_5 = +1$ for ŠU$_2$ and GI$_6$, -1 for NA and ME. By inserting these rules, as far as they are determined at this point, one obtains the following total corrections:

$$\Delta \check{S}U_2 = s \cdot [-2(R_0 + \delta R) + (\sigma_3 + \sigma_4)\rho + 0.95] \tag{E.29}$$

$$\Delta NA = s \cdot [+2(R_0 + \delta R) - (\sigma_3 + \sigma_4)\rho - 0.95] \tag{E.30}$$

$$\Delta ME = s \cdot [+2(R_0 + \delta R) + (\sigma_3 + \sigma_4)\rho - 0.95] \tag{E.31}$$

$$\Delta GI_6 = s \cdot [-2(R_0 + \delta R) - (\sigma_3 + \sigma_4)\rho + 0.95]. \tag{E.32}$$

Here δR, σ_3 and σ_4 are considered as free parameters. The question is now whether any realisation of δR, σ_3 and σ_4 succeeds in reproducing the attested disk correction. The results, using $\delta R = 0$ (no dip effect) and $s \approx 1.25$ (Babylon), can be summarised as follows (Table E.2). For ŠU$_2$, NA, ME, GI$_6$ the best result is obtained if they are defined in terms of the front rim of the Moon and the back rim of the Sun, i.e. the Moon's first contact with the horizon and the Sun's last contact with the horizon. The total corrections for ME and GI$_6$ have the correct sign, but their magnitude is about a factor 2 too small. On the positive side, they are significantly larger than the corrections for ŠU$_2$ and NA ($\pm 0.44°$), as they should be. Hence the signs and relative magnitudes of the corrections agree with Δ, but the actual magnitudes do not. If the dip effect were included, the agreement with the attested corrections would become worse for ŠU$_2$ and NA, better for ME and GI$_6$.

Lunar Six	Δ	r(efraction)	p(arallax)	r+p	uM uS	ℓM uS	fM bS	bM fS	optimal rim choice	r+p+rim
ŠU$_2$	0	−1.44	+1.20	−0.24	−0.68	0	*+0.68*	−0.68	fM bS (= ℓM ℓS)	+0.44
NA	0	+1.44	−1.20	+0.24	+0.68	0	*−0.68*	+0.68	fM bS (= ℓM ℓS)	−0.44
ME	+2	+1.44	−1.20	+0.24	*+0.68*	0	*+0.68*	−0.68	fM bS (= uM uS)	+0.92
GI$_6$	−2	−1.44	+1.20	−0.24	*−0.68*	0	*−0.68*	+0.68	fM bS (= uM uS)	−0.92
KUR	0	0	−1.20	−1.20	0	−0.68	+0.68	−0.68	fM fS (= uM uS)	−1.20
NA$_1$	−2	0	−1.20	−1.20	0	−0.68	−0.68	+0.68	fM bS (= ℓM uS)	−1.88

Table E.2: Contributions to the Lunar Sixes due to refraction (no dip effect), parallax and rim effect [°]. M=Moon; S=Sun; u=upper rim; ℓ=lower rim; f=front rim; b=back rim. Rim definitions resulting in better agreement with the attested disk correction (Δ) are in italics. The final column contains the total correction for the optimal rim definition.

A satisfactory feature of having the rims defined in terms of the daily rotation instead of the horizon is that this results in approximately the same point on the lunar disk and also on the solar disk being used for measuring different Lunar Six intervals. It would be less satisfactory to always use, for instance, the upper rim of the Moon, because this amounts to approximately opposite points on the lunar disk at moonrise and moonset.

E.4 Temporal order of the Lunar Six intervals near Full Moon

If the Moon is sufficiently far away from the ecliptic during Full Moon the temporal sequence of the Lunar Six intervals ŠU$_2$, NA, ME and GI$_6$ (cf. §4.4.19) deviates from the most common one. Which sequence is realised depends on whether Full Moon occurs during the day or at night, and on how close Full Moon occurs to sunrise or sunset. Several constraints reduce the set of configurations to the ones shown in Figs. E.3 and E.4. First, ŠU$_2$ and NA always occur at consecutive sunrises, ME and GI$_6$ at consecutive sunsets. Second, the temporal sequence can deviate from the most common one only if at least one Lunar Six event occurs sufficiently close to Full Moon. Only then are the elongation (η) and the resulting contribution to the Lunar Six interval (Q) small, so that the contribution of the Moon's distance to the ecliptic (R) dominates. It therefore happens only rarely that two Lunar Six intervals are 'renamed' during the same Full Moon, and some combinations are prevented altogether. Third, the Moon's latitude changes by at most a few degrees in the course of 2–3 days. Sequences of Lunar Six intervals that would correspond to the Moon being on opposite sides of the ecliptic are therefore virtually excluded.

Ignoring pathological cases resulting from Full Moons very close to sunrise or sunset, approximate probabilities for the six configurations shown here, based on modern computations for the period 701 BC – 1 AD for Babylon and using disk centres, are, from top to bottom, 30.92%, 9.01%, 8.72%, 1.74%, 0.25%, 0.27% (Fig. E.3) and 31.31%, 7.85%, 8.11%, 1.06%, 0.38%, 0.37% (Fig. E.4).

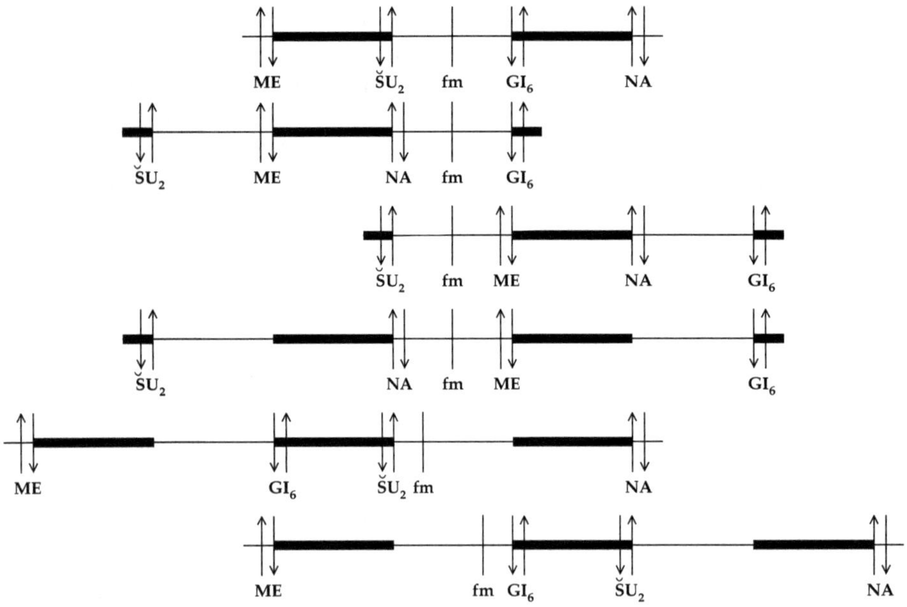

Fig. E.3: Possible sequences of Lunar Sixes for a Full Moon during day time. **Row 1**: most common case. **2**: $\check{S}U_2$ renamed. **3**: GI_6 renamed. **4**: $\check{S}U_2$ and GI_6 renamed. **5**: ME renamed (Full Moon near sunrise). **6**: NA renamed (Full Moon near sunset).

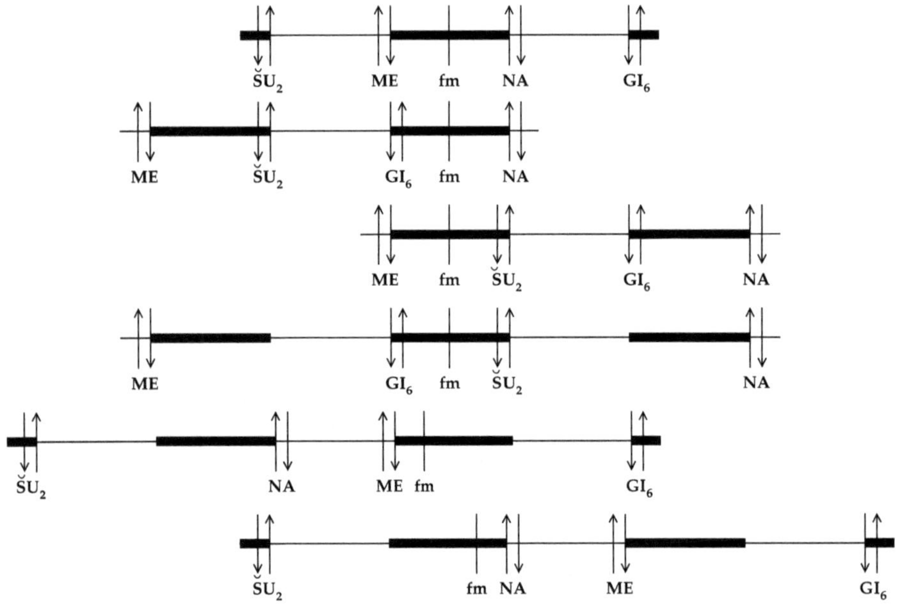

Fig. E.4: Possible sequences of Lunar Sixes for a Full Moon during the night. **Row 1**: most common case. **2**: ME renamed. **3**: NA renamed. **4**: ME and NA renamed. **5**: $\check{S}U_2$ renamed (Full Moon near sunset). **6**: GI_6 renamed (Full Moon near sunrise).

Appendix F

Photographs of the cuneiform tablets

Fig. F.1: No. **1** Reverse (procedures)

Fig. F.2: No. **3** Side X

Fig. F.3: No. **4** Side X

Fig. F.4: No. **2**. Left: side X; right: Y

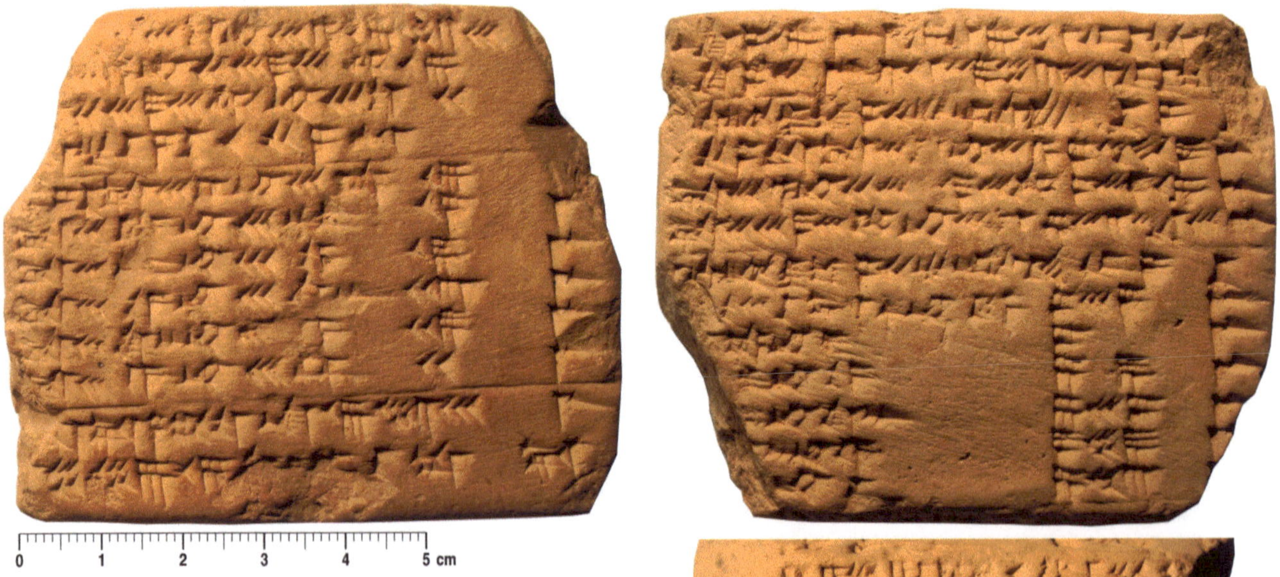

Fig. F.5: No. **5**. Left: obverse; right: reverse, lower edge

Fig. F.6: No. **6**. Left: obverse; right: reverse

Fig. F.7: No. **8** Reverse

Fig. F.8: No. **11** Side X

Fig. F.9: No. **10** Reverse (procedures)

Fig. F.10: No. 7 Reverse

Fig. F.11: No. **13**. Left: obverse; right: reverse

Fig. F.12: No. **9**. Left: obverse; right: reverse

Fig. F.13: No. **12**. Left: obverse; right: reverse

Fig. F.14: No. **14**. Left: obverse; right: reverse

Fig. F.15: No. **15**. Left: reverse (colophon; procedures); right: lower edge

Fig. F.16: No. **19**. Left: obverse; middle: reverse; right: lower edge

Fig. F.17: No. 18 Obverse

Fig. F.18: No. **18** Reverse

Fig. F.19: No. **16** Side X

Fig. F.20: No. **17** Reverse (procedures)

Fig. F.21: No. **20** Reverse

Fig. F.22: No. **21**. Left: obverse; right: reverse

Fig. F.23: No. **22**. Left: obverse; right: reverse

Fig. F.24: No. **23**. Left: obverse; right: reverse

Fig. F.25: No. **24** Reverse (procedures)

Fig. F.26: No. **34** Side X

Fig. F.27: No. **27** Side X

Fig. F.28: No. **28** Reverse (procedures)

Fig. F.29: No. **25**. Left: obverse; right: reverse

Fig. F.30: No. **26** Side X

Fig. F.31: No. **30** Side X

Fig. F.32: No. **29**. Left: obverse; right: reverse

Fig. F.33: No. **31** Side X (left: museum photo; right: excavation photo)

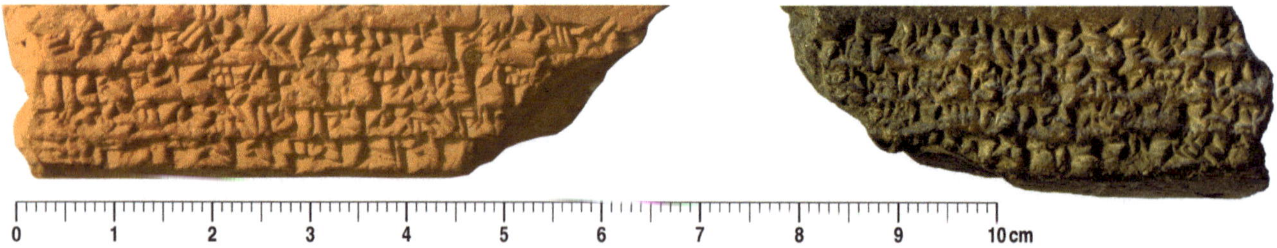

Fig. F.34: No. **35** Reverse (procedures, colophon)

Fig. F.35: No. **32**. Top: obverse; bottom: reverse

Fig. F.36: No. **36** Reverse (colophon, procedures)

Fig. F.37: No. **37**. Left: obverse; right: reverse

Fig. F.38: No. **33** Reverse (procedures, colophon)

Fig. F.39: No. 38. Left: side X; right: Y

Fig. F.40: No. 39. Left: obverse; right: reverse

Fig. F.41: No. **40** Side X

Fig. F.42: No. **51**. Left: side X; right: Y

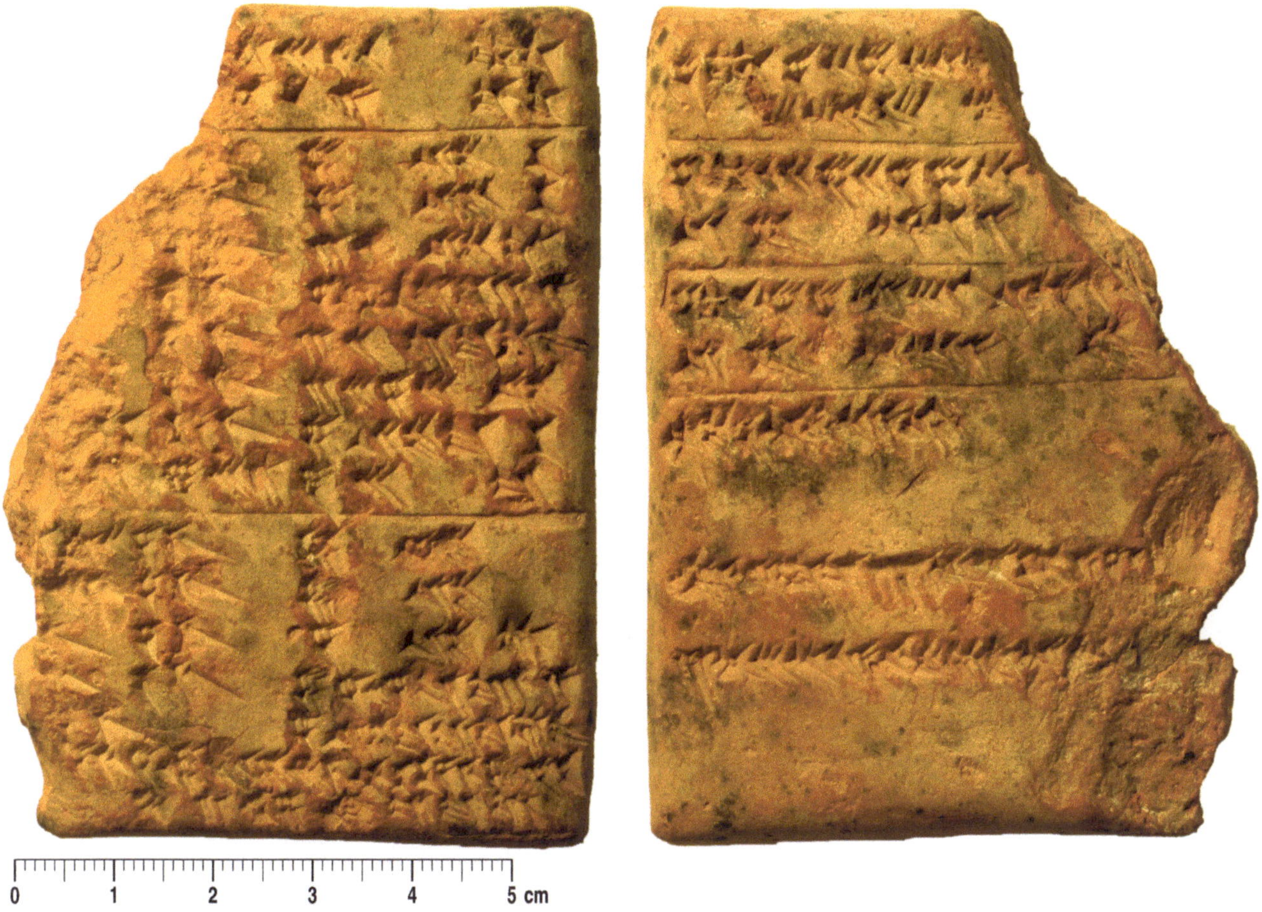

Fig. F.43: No. **41**. Left: obverse; right: reverse

Fig. F.44: No. **43**. Left: obverse; right: reverse

Fig. F.45: No. **46** Obverse

Fig. F.46: No. **46** Reverse

Fig. F.47: No. **44**. Left: obverse; right: reverse

Fig. F.48: No. **45** Reverse

Fig. F.49: No. **50** Side X

Fig. F.50: No. **47**. Left: obverse; right: reverse

Fig. F.51: No. **48**. Top: side X; bottom: Y

Fig. F.52: No. 52 Obverse

Fig. F.53: No. **52** Reverse

Fig. F.54: No. **53** Obverse

Fig. F.55: No. **53** Reverse

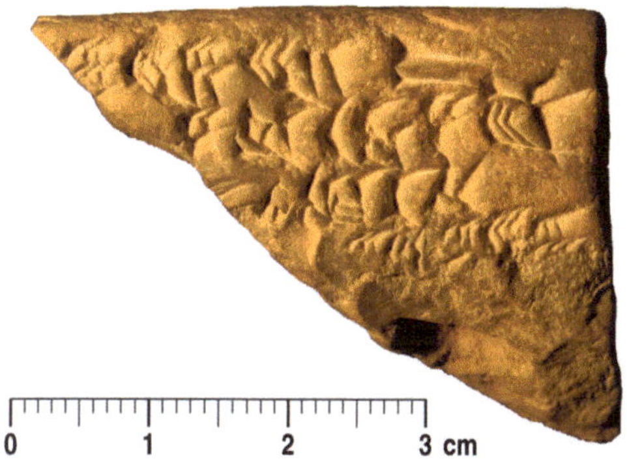

Fig. F.56: No. **54** Side X

Fig. F.57: No. **49**. Top: side X; bottom: Y

Fig. F.58: No. **55**. Left: obverse; right: reverse

Fig. F.59: No. **56**. Left: obverse; right: reverse

Fig. F.60: No. **57**. Left: obverse; right: reverse

Fig. F.61: No. **58**. Left: obverse; right: reverse; lower edge

Fig. F.62: No. **59**. Left: side X; middle: right edge of side X; right: side Y

Fig. F.63: No. **60** = No. **61**.C. Left: obverse; right: reverse

Fig. F.64: No. **61**.**B**. Left: obverse; right: reverse

Fig. F.65: No. **61.A** Obverse

Fig. F.66: No. **61**.**A** Reverse

Fig. F.67: No. **66**. Left: side X; right: Y

Fig. F.68: No. **61.D**. Left: obverse; right: reverse

Fig. F.69: No. **61.F** Obverse

Fig. F.70: No. **62** Obverse

Fig. F.71: No. **61**.E. Top: obverse; middle: lower edge; bottom: reverse

Fig. F.72: No. **63**. Top: obverse; bottom: reverse

Fig. F.73: No. **64**. Top: obverse?; bottom: reverse?

Fig. F.74: No. **65**. Left: obverse; right: reverse

Fig. F.75: No. **67**. Left: obverse; right: reverse

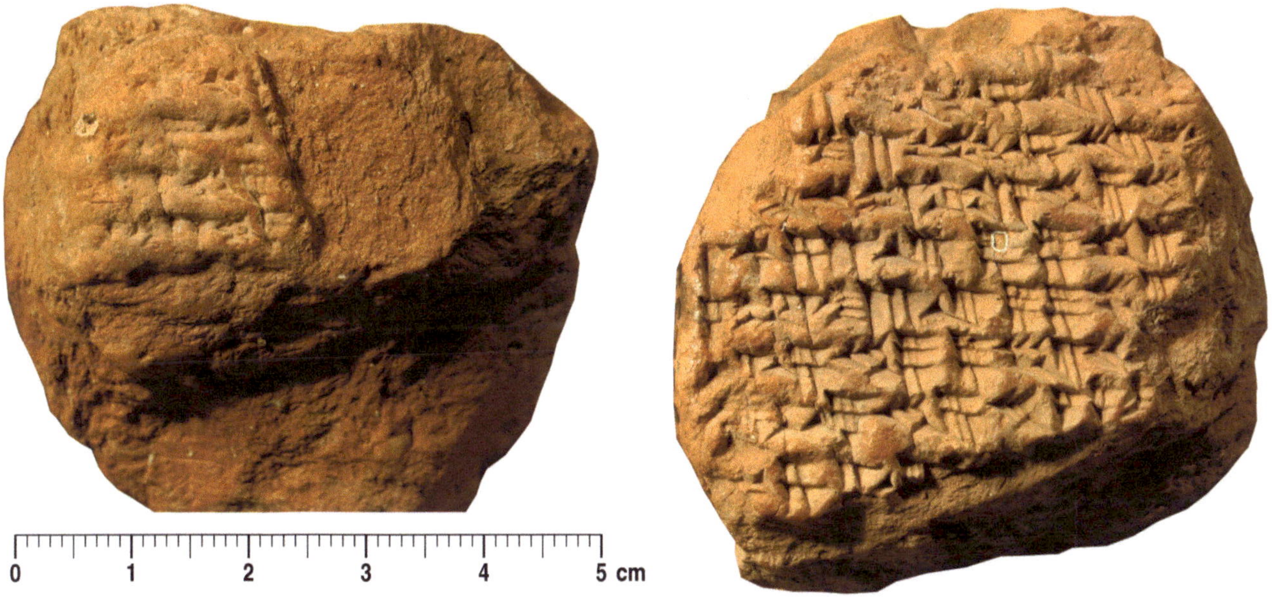

Fig. F.76: No. **68**. Left: obverse; right: reverse

Fig. F.77: No. **69**. Left: obverse; right: reverse

Fig. F.78: No. **70**. Left: obverse; right: reverse

Fig. F.79: No. **71** Side X

Fig. F.80: No. **75** Side X

Fig. F.81: No. **73**. Left: obverse; right: reverse

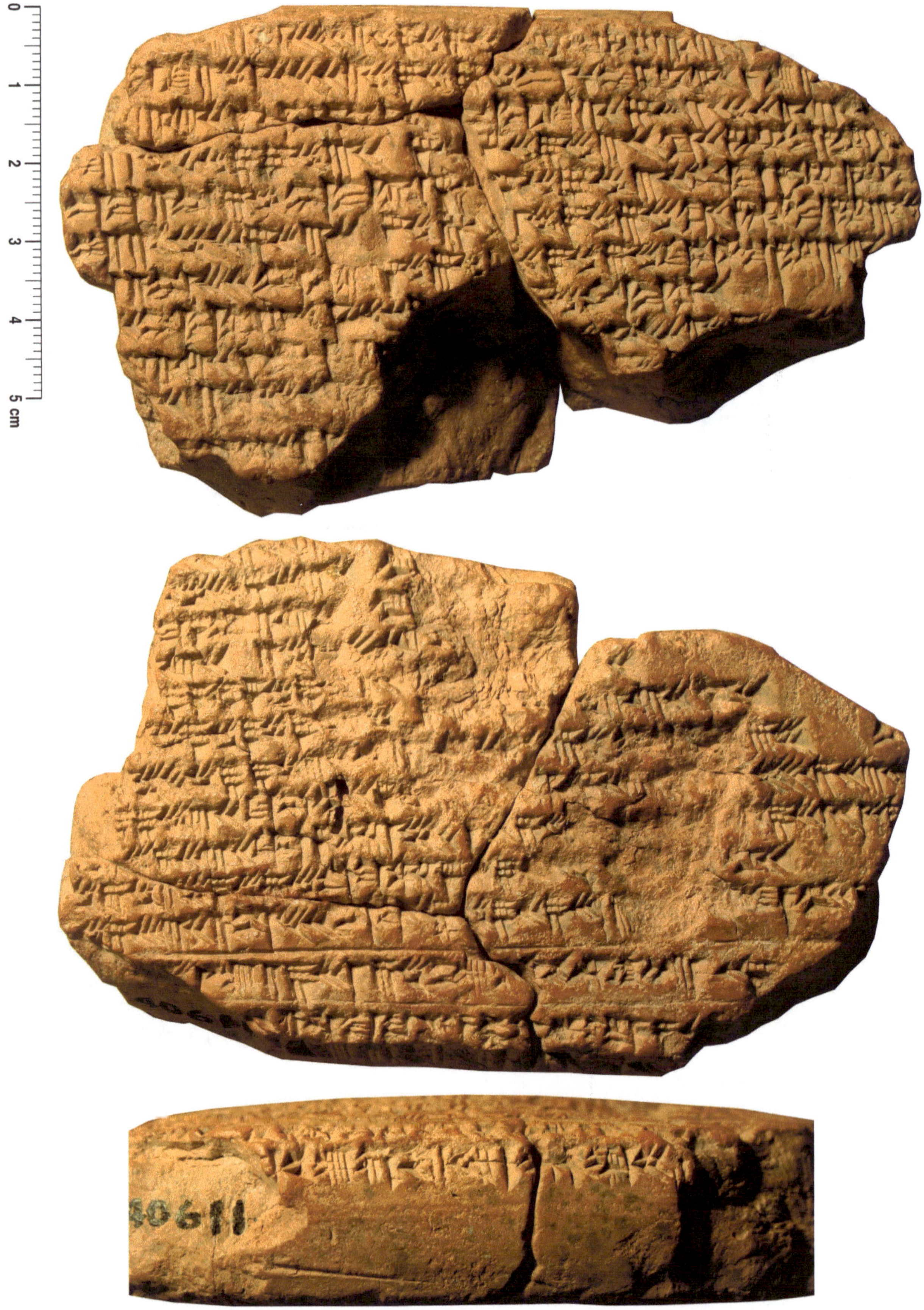

Fig. F.82: No. **72**. Left: obverse; right: reverse, lower edge

Fig. F.83: No. **74**. Top: obverse; bottom: reverse, right edge

Fig. F.84: No. **76**. Left: obverse; right: reverse

Fig. F.85: No. **77** Side X

Fig. F.86: No. **78** Obverse

Fig. F.87: No. **84** Side X

Fig. F.88: No. **80**. Left: obverse; right: reverse

Fig. F.89: No. **81**. Left: side X; right: Y

Fig. F.90: No. 79 Obverse

Fig. F.91: No. **79** Reverse (inset: line written perpendicularly between columns i and ii)

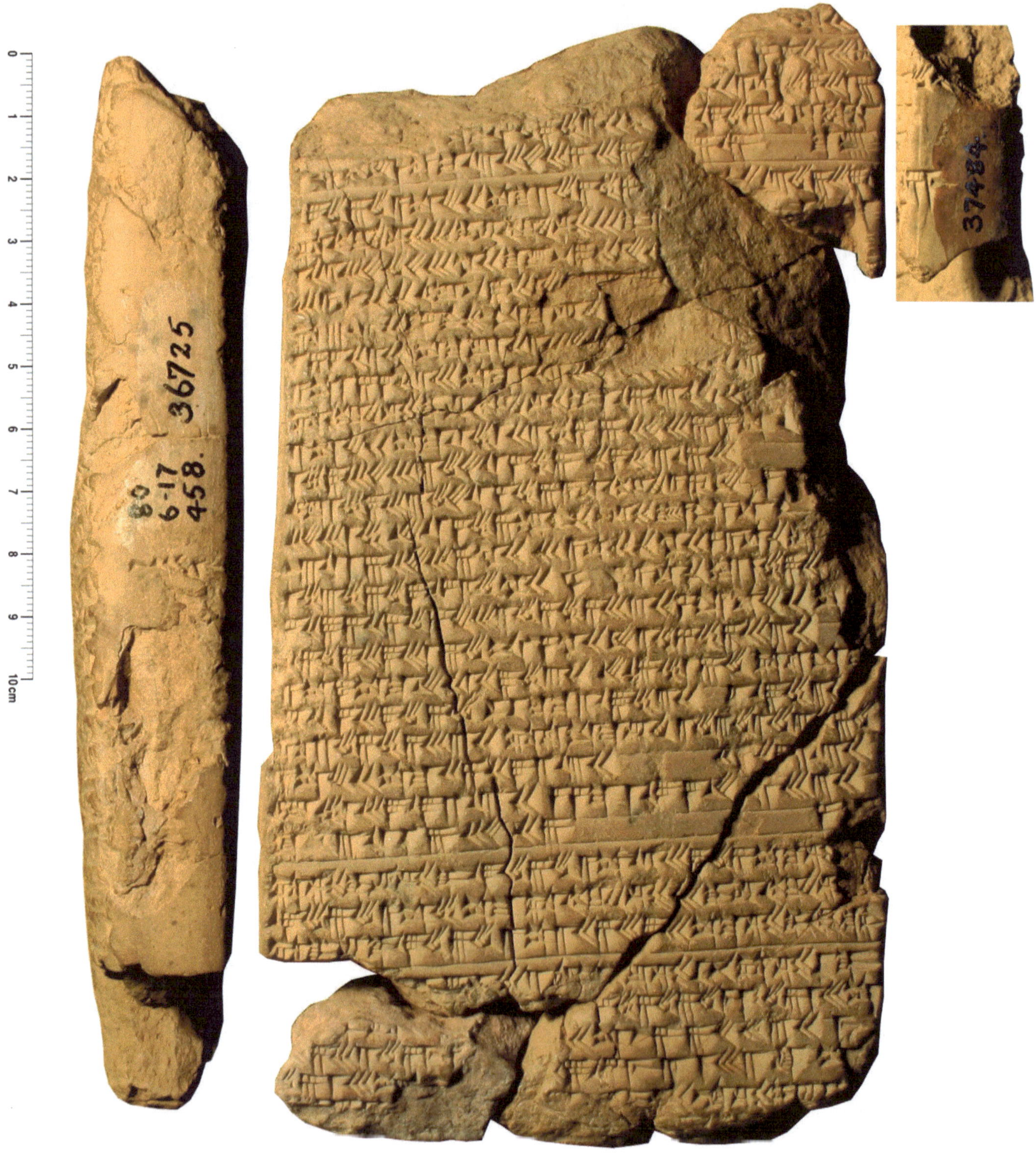

Fig. F.92: No. **82** Obverse

Fig. F.93: No. **82** Reverse

Fig. F.94: No. **83**. Top: obverse; bottom: reverse, right edge

Fig. F.95: No. **85**. Left: side X; right: Y

Fig. F.96: No. **86** Side X

Fig. F.97: No. 87 Side X

Fig. F.98: No. 88 Side X

Fig. F.99: No. 89 Side X

Fig. F.100: No. 90 Side X

Fig. F.101: No. **92**. Left: obverse; right: reverse

Fig. F.102: No. **93**. Left: obverse; right: reverse

Fig. F.103: No. **95**. Left: obverse; right: reverse

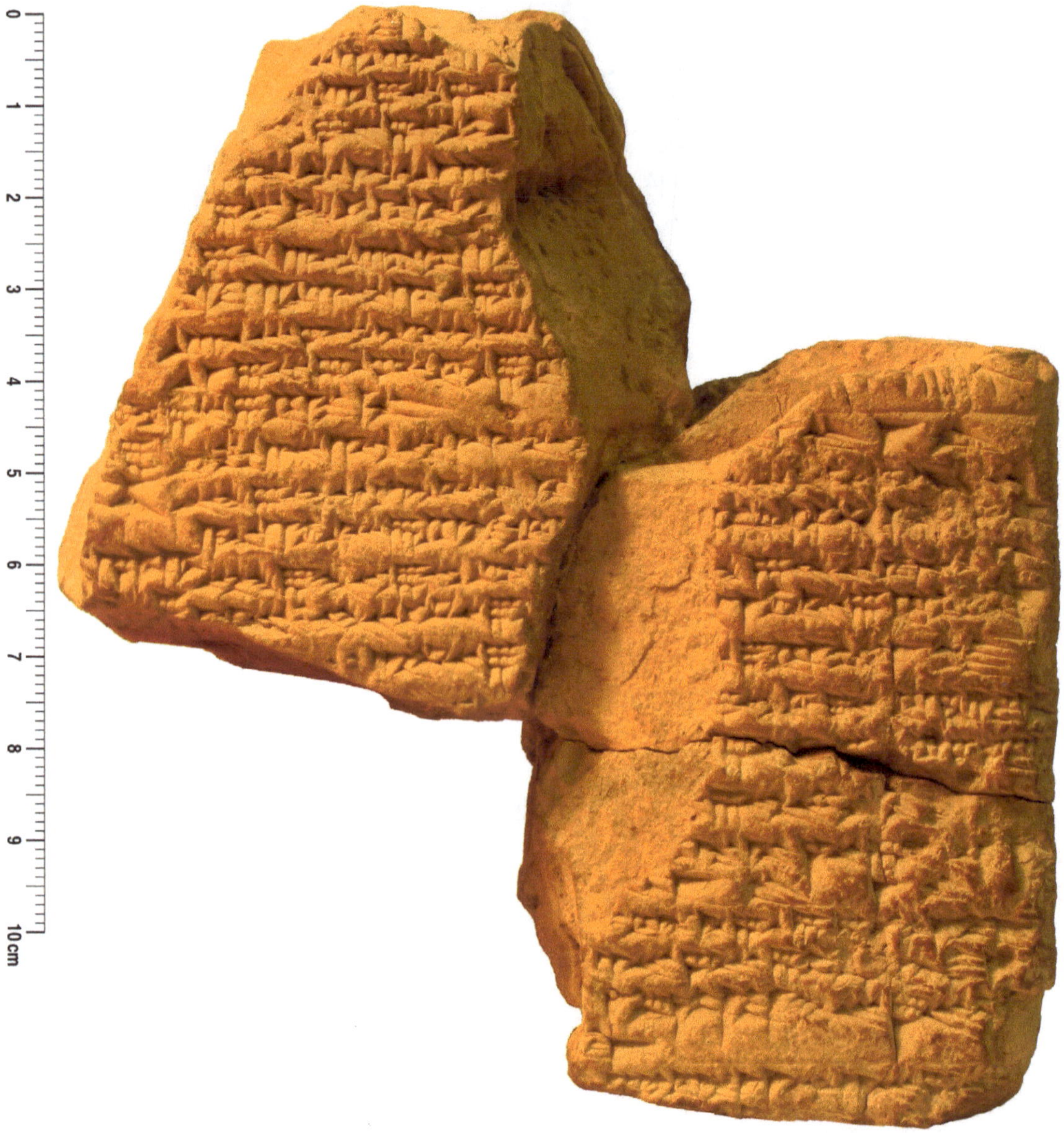

Fig. F.104: No. 97 Obverse

Fig. F.105: No. **97** Reverse

Fig. F.106: No. 91 Side X

Fig. F.107: No. 94 Reverse

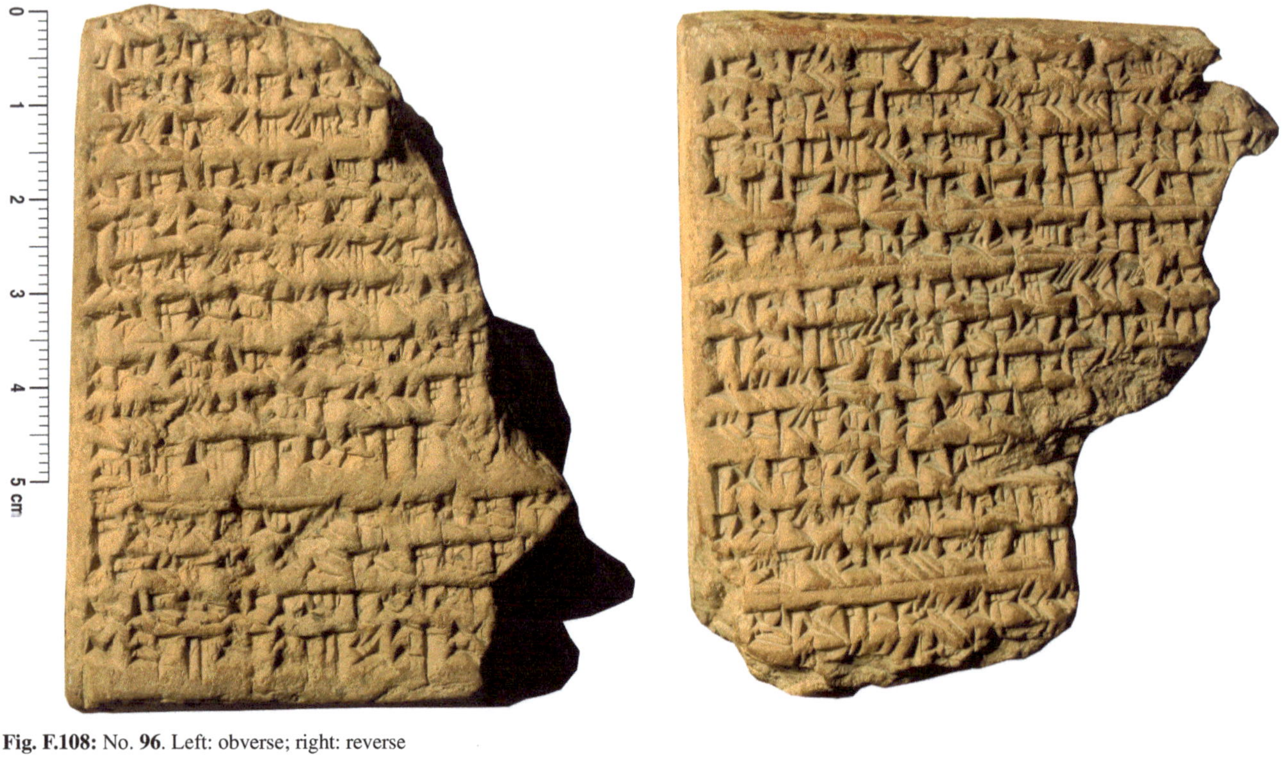

Fig. F.108: No. 96. Left: obverse; right: reverse

Fig. F.109: No. **98** Side X

Fig. F.110: No. **99** Side X

Fig. F.111: No. **100**. Left: obverse; right: reverse

Fig. F.112: No. **101** Obverse

Fig. F.113: No. **102**. Top: obverse; bottom: reverse

Glossary

Akkadian glossary

Mainly technical terms are included; for words not listed here cf. the *CAD*. The English term immediately following the Akkadian word usually conveys the basic, non-technical meaning. The list of attestations [in square brackets] is complete, unless introduced by 'e.g.'.

A

adi(**en**) 'until' [e.g. No. **82** O5',11',26'–28']

aḫāmiš 'together' [e.g. No. **18** Ri12',15'; No. **53** Oi31',ii16',18', Ri4,ii10; No. **61.A** O3,7,10,21,24,R5,8,17,19]

aḫāru cf. *uḫḫuru*

aḫāzu 'to take'; Š: cf. *šūḫuzu*

aḫu(**šeš**) 'half; part'; *ana n aḫḫē*(**šeš.meš**) *zâzu*(SE_3, **bar**) 'to divide into *n* parts' [No. **38** X11'; No. **52** Oi28,33,ii32, Ri12,19,26; No. **82** R11; No. **102** R5] This usage of **šeš**, not mentioned in the dictionaries (cf. *CAD* A I *aḫu* B 5), derives from the homophonous *aḫu*(**šeš**) A 'brother'.

alāku(**du**) 'to go'
  1 'to multiply': Q_1 **a.ra$_2$**|GAM|GAM_0 Q_2 *alāku* 'to multiply Q_1 by Q_2'; cf. §2.2.4.1 [e.g. No. **53** Oia2'–13']
  2 'to proceed': forward motion along the zodiac, with the planet or *qaqqaru*(**ki**), 'position', as subject [e.g No. **9** O2,4,5,9]

alla (*al*, *al-la*) 'beyond'
  1 *alla Q atāru*(**diri**) 'to exceed *Q*' [e.g. No. **37** O3,6–7]
  2 *alla Q iṣû*, *maṭû*(**la$_2$**) 'to be less than *Q*' [e.g. No. **53** Ri1–12]

allānuššu 'beyond it' [No. **91** X1'–6']

amāru(**igi**) 'to see'
  1 'to see' [*a-ma-ri*: No. **53** Rii11; *tam-mar*: e.g. No. **38** Y8',9'; No. **53** Oii17'; No. **102** R6]
  2 N: *nanmuru*(**igi**), 'to appear': first visible rising or setting of a planet (EF, MF, FA) [e.g. No. **18** Oi3',7',12',15']

amātu 'word; command' (in inv.) [e.g. No. **41** UE1]

ammatu(**kuš$_3$**) 'cubit' (unit of angle) [No. **18** Oi16',17'; No. **23** O9'; No. **25** O18'; No. **30** X2'; No. **40** X9',11'; No. **47** O1,2]; *ammatu rabītu*, 'large cubit': a cubit of 2;30°? [No. **9** O7]

ana tarṣi 'opposite' [e.g. No. **65** O10',11']

ana ūmi(**u$_4$**, **me**) 'to daylight'
  1 *ana ūmi*(**me**) *aṣû*$^?$(A)|*elû*(E), 'to rise to daylight': acronychal rising (AR) [e.g. No. **41** O6,14]
  2 *ana ūmi* is possibly the Akk. reading of DIŠ.UD in the Lunar Six module of lunar system A [e.g. No. **61.A** O10,12,24, 25,R8,10,20,21]

Antu spouse of the sky god (in inv., col.) [e.g. No. **41** UE1]

Anu(^{d}**diš**) the sky god (in inv., col.) [e.g. No. **33** R4]

apālu 'to answer': introduces the outcome of computations [No. **31** X7'–9'; No. **46** Oii3'–6', Rii6; No. **102** O5',10']

arḫu(**ab$_2$**, **iti**) 'month' [e.g. No. **95** Oii6'–8', Ri6'–7'; No. **97** Oii11'–13', 15', 20'–22']

arki(**egir**, *ar$_2$*) 'after' [e.g. No. **18** Oii9',13',Ri'17',23']

arkû(**egir**, *ar$_2$*) 'rear'
  1 'other; second' [e.g. No. **61.G** O5',6']
  2 'rear': in terms of the daily rotation of the sky [e.g. No. **47** R4',5']

aṣābu cf. *waṣābu*

aṣû(**e$_3$**, A?) 'to go out'
  1 G: 'to go out' (in col.) [No. **44** R7']
  2 Š: Q_1 *ana* Q_2 *šūṣû* 'to transform Q_1 into Q_2' [No. **53** Oii15'; No. **65** O15',R1,9]
  3 *aṣû* might be the Akk. reading of A in A **me** and *ana* **me** A, terms for AR

atāru(**diri**) 'to exceed' [e.g. No. **37** O3,6–7]

attalû(AN.MI) 'eclipse' [e.g. No. **53** Oib10'; No. **100** O5',6']

B

bābtu(**ka$_2$**-*tu$_2$*) 'ward' (*CAD* B sub *babtu* 1): zod. region, zone of a step function for the synodic arc, e.g. *bābtu ša$_2$* 30, 'ward of 30', is the zone where σ = 30° (Jupiter system A) [No. **18** Oi6',7', ii20'; No. **23** O5'; No. **25** O13; No. **26** X8']

banû 'to construct' [No. **83** R4]

barû(**igi.tab**) 'to verify' (in col.) [No. **53** Rii42]

bašû(**gal$_2$**, **tuk**, **ur**) 'to exist' [**gal$_2$**: e.g. No. **14** Oii7'; **tuk**: e.g. No. **53** Oii25',Rii14; **ur**: No. **96** O12']

Bēl(^{d}**en**) a name of Marduk (in inv., col.) [e.g. No. **22** UE0]

Bēltu(^{d}**gašan**) spouse of Bēl (in inv., col.) [*Bēltija*(^{d}**gašan**-*ia$_2$*): e.g. No. **52** E4]

bēru(**danna**) 'mile': unit of angle or time (cf. p. xxvi) [angle: e.g. No. **40** X10'; time: e.g. No. **53** Rii18,19]

biri 'between' [e.g. No. **82** R31,35]

birītu 'distance'
  1 distance (angular or temporal) [*bi-rit*: e.g. No. **13** R3'–6', 17',19',23']; *bi-ti-ka* is probably a phonetic abbrev. of *birītika* [No. **93** O2'–6']
  2 BE (=*bi$_2$*?) and *bi*, 'elongation', are probably phonetic abbrev. of *birītu* [BE: e.g. No. **61.A** O4,11,12,18,24–26; *bi*: cf. lunar system B, column *O*]

bītu(**e**$_2$) 'house'
  1 (zod.) position? [No. **46** Rii3,6]
  2 *bīt*(**e**$_2$) *attalê*(AN.MI), 'eclipse house': nodal region [No. **60** O1'–6',11'–14',16']

D

Dilbat 'Venus' [e.g. No. **46** Rii2]

E

elat 'beyond' (prep.) [No. **18** Oi3']

eli(*e*, **an.ta**) 'above' (prep.) [**an.ta**: e.g. No. **53** Oi30'; *e*: No. **53** Oii16']

elû(**e**$_{11}$, E, **nim**) 'to rise'
  1 'to come out'; said of computations) [**e**$_{11}$: e.g. No. **70** O2,3; **nim**: No. **61.G** O4'; No. **65** R5]
  2 E: 'to rise'; in E **me**, 'daylight rising'; *ana* **me** E, 'it rises to daylight' (= AR) [e.g. No. **18** Oi13']
  3 Š: *šūlû*(**e**$_{11}$, **nim**), 'to deduct'; in: Q_1 *ina*(**ta**) Q_2 *šūlû*, 'to deduct Q_1 from Q_2' [**e**$_{11}$: e.g. No. **67** O4,7,11,15,19, R2,6; **nim**: e.g. No. **58** R12']. This Akk. reading of **nim** is very probable though not proven beyond doubt; cf. §2.2.3.4.

elû perhaps the Akk. reading of **nim**, 'east'

emēdu cf. *nenmudu*

epēšu(**du**$_3$) 'to construct'
  1 'to compute' [e.g. No. **53** Oii15',20',27',28']
  2 'procedure' [e.g. No. **1** R1]; cf. also **du**$_3$.**du**$_3$

epūšu(**du**$_3$-*uš*) 'procedure' [No. **53** Oi14']

eqlu(**a.ša**$_3$) 'area' [No. **18** Oi21'; No. **38** Y6']

ereb Šamši(^{d}**utu.šu**$_2$.**a**) 'sunset; west' [No. **53** Rii23,24]

eṣēpu 'to double' (?) [No. **61.G** O22': meaning unclear]

eššu(**gibil**) 'new' [e.g. No. **5** R8; No. **46** Oii16']

etēqu(**dib**) 'to pass by'
  1 with respect to a position, star, planet, etc. [e.g. No. **42** O3,5,7]
  2 to not occur, be omitted (said of an expected phenomenon) [e.g. No. **42** O10,16,17]

G

gabba, *gabbi* 'all' [e.g. No. **46** Oii2',4']

giṭṭû(**im.gid**$_2$.**da**) '*giṭṭû*-tablet', one-columnar tablet with scholarly content [No. **9** R19; No. **44** R7']

Ḫ

ḫarrānu(**kaskal**[.**2**]) 'path' [No. **38** Y7'; No. **99** Xi'3,ii'5,6,iii'6]

ḫepû 'broken' [e.g. No. **53** Oii11'–13']

ḫīpu 'break' [e.g. No. **47** R2']

I

igigubbû(**igi.gub**) '*igigubbû*-coefficient'
  1 coefficient related to eclipse magnitude [No. **95** Ri'11'; No. **96** O5',R1]
  2 angular push [No. **46** Oii21',Ri31',ii1,6]
  3 a coefficient (meaning unclear) [No. **58** O1,2,6; No. **92** O3,12,R5'; No. **93** O8]

ilku meaning unclear [No. **14** Oii11']

imgiddû(**im.gid**$_2$.**da**) cf. *giṭṭû*

ina **1** 'in' [e.g. No. **32** O7,9,10,12]
  2 'from'; written *ina* or **ta** [*ina*: e.g. No. **65** O5',7',9'; No. **70** O2; **ta**: e.g. No. **74** R3,6,9,12,18,22]. In spatial expressions **ta**, 'from', represents *ultu*, perhaps also in some of the subtractions.

ina(**ta**) *libbi*(**ša**$_3$) cf. *libbu*

ištēn(1-*en*) 'one' [e.g. No. **53** Rii12',16']

itti(**ki**) *aḫāmiš* 'together' [e.g. No. **16** Xi'10'; No. **53** Oi31']

izuzzu(**gub**) 'to stand' [No. **53** Rii41]

J

jānu 'nothing; zero' [No. **52** Ri9]

K

kajjamānu(**sag.uš**) 'regular'
  1 *Kajjamānu*(^{d}**genna**, ^{d}**sag.uš**): 'The Steady One' = Saturn [e.g. No. **44** O9]
  2 'regular; steady', said in connection with mean motion [No. **102** O8',11']

kalu 'whole; all' [e.g. No. **53** Oii3',4']

kalû 'lamentation priest' (in col.) [e.g. No. **33** R4]

kapāpu 'to bend' [No. **46** Oii15']

kī **1** 'if' (conj.) [e.g. No. **52** Oi39,ii3–5,7–9; No. **93** O9',12',13']
  2 'like' (prep.) [e.g. No. **18** Oi5'; No. **25** O12']

kīma(**gin**$_7$) 'like, as' (prep.) [e.g. No. **18** Ri4']

kippatu(**gam**) 'circle; circumference': eclipse magnitude [No. **96** O5',R1,6,8; No. **95** Ri'11']

kiṣ(a)ru 'knot' [e.g. No. **43** Rii6'; No. **52** Rii23; No. **53** Oi20',33', ii16',18']

kullu 'to hold' [e.g. No. **46** Rii2]

kullumu 'to reveal' (in col.) [e.g. No. **28** R3]

kunnu(**gub**) 'confirmed' (verb. adj. D *kânu*): 'full (month)': said of the first lunar crescent if the preceding month is full [*kun-nu*: e.g. No. **53** Rii12,15,39; No. **93** O10'; **gub**: e.g. No. **53** Rii17]

L

lapān(*la*-**igi**) 'in front of' [*la*-**igi**: No. **16** Xi'7',10',12'; No. **53** Rii24; *la-pan*: No. **13** R11']

lē'u(^{giš}**da**) 'wooden board' (inlaid with wax; in col.) [e.g. No. **44** R6'; No. **53** Rii42]

libbu(**ša**$_3$) 'inner', in: *ina*(**ta**) *libbi*(**ša**$_3$), 'from' [e.g. No. **53** Oi2,22', Rii19; No. **66** X4']

libbû ša 'by means of which' [No. **53** Oi14',15',21',25'; No. **83** R4]. For this meaning, which is not mentioned in the dictionaries, cf. Ossendrijver (2010).

lumāšu 'constellation'
  1 'zodiacal sign' [e.g. No. **13** R32'; No. **53** Oii5'; No. **61.A** O6, 9,20,23; No. **97** Oii18',19']
  2 'zodiac' [e.g. No. **16** Xi'10'; No. **53** Oi29',Rii33,39]
  3 'zodiacal position' [No. **53** Oi26',Rii38]

M

maḫru 'front'
  1 *ana maḫri*(**igi**)-*šu*$_2$, 'forward' [e.g. No. **18** Oi2'; No. **97** R18']
  2 *ina maḫar*(**igi**), 'in front of' [e.g. No. **2** Xiii'5'; No. **41** O16]

maḫrû(**igi**-*u*$_2$) 'previous' [e.g. No. **61.G** O10',R2',16',17']

mālaku 'path' (of the Moon and the planets) [No. **29** O2'; No. **25** O17'; No. **30** X2'; No. **53** Oi20'; No. **80** O3]

malû(**diri**) 'to be full; fill up'; *malûššu* 'while it fills up' (inf. +loc. +poss. suff. 3 m. sg.) [*ma-lu-uš-šu*$_2$: No. **53** Ri4; No. **69** O2, No. **70** O3; **diri**-*šu*$_2$: No. **68** R8'; No. **76** R7'; *ma-lu-uš*: No. **64** O3]

manzāzu(**na**) 'stand point'
  1 **na** [*ša*$_2$ **u**$_4$.**1.kam**]: NA$_1$ (Lunar Six interval) [e.g. No. **52** Oi27, 31,36–38,40,41; No. **53** Rii15]
  2 **na**: NA (Lunar Six interval) [e.g. No. **61.A** O13,14,25,27]
  3 **na**: time interval between FA and LA of outer planet [No. **38** X8']. The reading *manzāzu*, suggested by Hunger (*ADRT* I, p. 21), is confirmed by BM 45821 (Al-Rawi & George 1991/2), a LB commentary to *EAE* T. 14, O10: 12 *man-za-za ša*$_2$ **u**$_4$ 1; O15:

12 **na** *ša*$_2$ ⌜**u**$_4$ 1⌝, '12°, NA for day 1'. Another proposed reading, *nanmurtu*, 'appearance', is incompatible with the phonetic complement *su* occasionally found in Astronomical Diaries and Goal-Year texts, e.g. BM 34034 (*LBAT* 1285) Rii5: **diri.še** 30 11 **na**-*su*, 'Month XII$_2$, (the first day of which coincides with the) 30th (of the previous month), its NA$_1$ (*manzāssu*) was 11'.

markasu(**dur**) 'band' [No. **20** R2; No. **40** X9'; No. **47** O1,2,5,6]

maṣṣartu(**en.nun**) 'watch'
 1 duration: in *maṣṣarti*(**en.nun**) *mūši*(**gi**$_6$), 'duration of the (entire) night' [No. **52** Oi42,ii10,14,Ri33]
 2 astronomical watch (1/3 of the night) [in Astronomical Diaries; not attested in mathematical astronomy]

maṣû 'to be sufficient' [No. **53** Rii11; No. **61.C** R12]

mašāḫu 'to measure' (in col.) [*im-šuḫ*: No. **36** R1]

mašû 'to forget' [No. **72** R13' (?)]

maṭû(**la**$_2$) 'to be lacking'
 1 'to decrease': [**la**$_2$-*u*$_2$: e.g. No. **53** Ri8; *ma-ṭu*: e.g. No. **53** Ri8]
 2 'to be less' [**la**$_2$-*u*$_2$: e.g. No. **53** Ri8; *ma-ṭu*: e.g. No. **53** Ri10; *i-maṭ-ṭu-u*$_2$: No. **67** O3,6,10,14,17,R1,5]
 3 'to be subtractive' [e.g. No. **53** Oii26',27' No. **81** Xii'9']
 4 D: *muṭṭû*(**la**$_2$) 'to diminish': Q_2 *ana* Q_1 *muṭṭû*, 'to diminish Q_1 by Q_2' (§2.2.3.3) [**la**$_2$: e.g. No. **61.A** O11,25; *u*$_2$-*maṭ-ṭu-u*$_2$: No. **102** O3'].

meḫertu(**gaba.ri**) 'counterpart' (lit.); 'value', said of time intervals [e.g. No. **18** Ri7',9',12',15'; No. **46** Oi9'; No. **53** Oii22'; No. **95** Oii'13',14']

middatu, *mindatu* 'measure': perhaps the Akk. reading of **ninda**$_2$ in **ninda**$_2$-*at* **an**-*e*, 'measure? of the sky' [No. **102** O2']

miḫiltu (=*miḫiṣtu*) 'stroke'
 1 elongation? [No. **53** Rii39]
 2 unclear [No. **43** Oiii'7'] *CAD* M *miḫiṣtu* 'stroke; cuneiform sign; writing'. In the LB metro-mathematical texts W 23291-x (Friberg et al. 1990, pp. 508, 524) and W 23291 (Friberg 1997, pp. 274–275) the meaning is 'side (of a rectangle)' or perhaps more generally '(written) number, quantity'.

mimma ša 'whatever' [e.g. No. **53** Ri3,9,20,23]

mīnû(**en.nam**) 'what'
 1 'what' (interrog.) [*mi-nu-u*$_2$: e.g. No. **13** R33',36'; **en.nam** (attested only in mathematical problems): No. **38** Y2' (?)]
 2 in: *mīnû kī*, 'whatever' [e.g. No. **52** Oii3,5,9,12,14,16; No. **53** Oi23']

minûtu(**šid**) 'number' [e.g. No. **82** O15'; No. **95** Oii6',8',15'–17']

mišlu(BAR) 'half' [e.g. No. **38** X8']

mīṭu(**la**$_2$) 'subtraction'; e.g. in *ṭīpu*(**tab**) *u mīṭu*(**la**$_2$) 'addition and subtraction', a term for difference [e.g. No. **82** O19',R13,16,29]

muḫḫu 'topside', in: *adi*(**en**) *muḫḫi*, 'until' [e.g. No. **9** O6]; *ina*/ *ultu*(**ta**) *muḫḫi*, 'from' [e.g. No. **9** O1,2]

mūšu(**gi**$_6$) 'night' [e.g. No. **53** Oii7',21',22'; No. **92** O5]

muttallik(t)u 'roaming' (lit.): perhaps the Akk. reading of PAP.ḪAL, 'revolution' (?); cf. p. 518 [No. **102** O5',R6,7]

muṭṭû(**la**$_2$) cf. *maṭû* **4**

N

nabalkutu(**bal**) 'to transgress' [*na-bal-kat*$_2$-*tu*$_4$: No. **5** R3,7; *nu-bal-kut*: No. **53** Rii15,16; **bal**: e.g. No. **93** O10',11']

naḫāsu(**la**$_2$) 'to retreat'
 1 'to be/become retrograde' [e.g. No. **18** Oi3',10',13',14']
 2 Q_1 *ina*(**ta**) Q_2 *naḫāsu*(**la**$_2$), 'to subtract Q_1 from Q_2' (§2.2.3.2) [e.g. No. **65** O9',16',21',R11]

nanmurtu(**igi**) 'appearance' (cognate noun of *amāru* N, 'to appear'): first appearance of a planet (EF, MF, FA) [e.g. No. **46** Oii21',22',25']

naparšudu 'to flee': to distance itself; said of the Moon after an eclipse [No. **99** Xii'7]

napḫaru(**pap**, PAP.PAP?) 'total' [**pap**: e.g. No. **13** R9'; No. **14** Oii6'; No. **27** X6'; No. **47** R2'; PAP.PAP: No. **9** O9; No. **13** R7',8';No. **19** LE5'; No. **25** O9'; No. **39** O4',6']

nappaltu '*nappaltu*-coefficient': meaning unclear [No. **53** Oii2',3', 7',8',10']. In NMAT attested as a term for the daily change of such functions as the shadow length, the length of daylight and the duration of visibility of the Moon [e.g. Mul.Apin II.ii.41–42, iii.15].

nasāḫu(**zi**) 'to tear out'
 1 'to move, displace itself' (along the zodiac) [e.g. No. **46** Oi17',18'; No. **61.A** O5]
 2 'to tear out': a subtractive operation (§2.2.3.1) [e.g. No. **52** Oi30,35,36,42,ii4,6,10,11; No. **53** Rii6,15]

nasāḫu(**zal**) 'to pass'; said of time [e.g. No. **18** Oi3'; No. **99** Xii'3]

naṣāru(**uru**$_3$) 'to watch'
 1 'to observe'; '(astronomical) watch' [in Astronomical Diaries etc.]
 2 'to protect' [*li-iṣ-ṣur*: No. **28** R2]

našû(GIŠ[.A], **il**$_2$) 'to raise'
 1 'to compute' [GIŠ: e.g. No. **61.A** O7,10,21,24; No. **92** O4,12; **il**$_2$: e.g. No. **53** Rii7,17,18]. This meaning of *našû* is not mentioned in the dictionaries. The object is either a fraction or the name of a quantity. In NB and LB administrative texts *našû*(GIŠ) is attested with the related meaning 'to withdraw (from an account), to draw (payments, compensation)'; cf. *CAD* N II, pp. 99–100, sub *našû* A3c2'b'. The entry **gu-ru** GIŠ-*tenû na-šu-u*, contained in several lexical lists, suggests that this usage of GIŠ derives from **guru**$_{17}$(GIŠ-*tenû*).
 2 'to multiply': Q_1 **a.ra**$_2$ Q_2 *našû*(**il**$_2$); cf. §2.2.4.2 [No. **102** O7', 9',15']

nēmettu(**uš**) 'station' (S1, S2, MS, ES) [e.g. No. **13** R6',10',19',22', 23']

nenmudu(**uš**) (N *emēdu*) 'to be/become stationary' [**uš**: e.g. No. **25** O6',15',16'; *nen-mud*: e.g. No. **9** O8; No. **18** Riii6']

nindanu(**ninda**) 'rod' = 0;1° (unit of arc). According to the *CAD*, A1 sub *akalu* (discussion at the end), the reading *nindanu* is not certain.

nipḫu(**kur**) 'flaring'
 1 'rising', in the sense of first appearance: in **kur** *ša*$_2$ **nim**, 'morning rising' (MF) [e.g. No. **5** O4,R4; No. **10** Riii15]
 2 'rising', daily: in **kur** *šamaš*$_2$, 'sunrise' [e.g. No. **52** Oii3,14, Ri32; No. **53** Oii21',Rii37]

nisḫu(**zi**) 'tearing out' (cognate noun of *nasāḫu*)
 1 'displacement'; usually along the zodiac [e.g. No. **41** O3,4; No. **97** Oii15']. The conventional translation 'velocity' (*CAD* N II) is incorrect and misleading, since there is no concept of instantaneous velocity in Babylonian astronomy.
 2 'that which is torn out' (lit.); subtraction (not mentioned in the dictionaries) [e.g. No. **52** Oi35,43, ii3,10,15]

P

palāḫu 'to honor' (in col.) [e.g. No. **28** R1]

palû(**bal**) 'turn': number period *Z* of a function [No. **18** Rii9'; No. **13** R33',36',37'; No. **45** Rii'9'; No. **83** R9; No. **96** R11; No. **95** Oii5',12']

pānu 'front', in: *ina pānāt*(**igi**-*at*$_2$), 'before' (temporal) [e.g. No. **32** O9,12, R3,8; No. **41** O9]

pû(**ka**) 'mouth; opinion': in *ša*$_2$ *pî*(**ka**) 2, 'according to a second opinion' [No. **18** Rii11',14',iii16'; No. **26** X5']; *ki-i pî*(**ka**), 'according to' [e.g. No. **18** Oi5']

pūt alpi(**sag.ki gu**$_4$) 'trapezoid' (< 'ox forehead') [No. **18** Oi20'; No. **38** Y4'] *pūtu*(**sag**[**.ki**]) 'side' (of trapezoid) [No. **18** Oi23'; No. **38** Y2',10'; No. **40** X2',3']

Q

qablītu(**murub$_4$**-*tu$_2$*) 'middle one' (f. sg. adj. used as a noun) [e.g. No. **32** O5,6,10, R5; No. **53** Oi20',21']

qablu(**murub$_4$**) 'middle' (noun) [e.g. No. **99** Xi'5,7, ii'5,6,8]

qablû(**murub$_4$**-*u$_2$*) 'middle' (m. sg. adj.); in the sense mean [No. **13** R13'–15',17'–18']

qabû(**e**, **du$_{11}$**[**.ga**]) 'to call; predict' [**e**: e.g. No. **61.A** O13,25–27; **du$_{11}$**[**.ga**]: No. **52** Oii8,9,Ri17,24,E38]

qaqqaru(**ki**) 'position': zod. position [e.g. No. **46** Oi5',7',15']

qatû 'final' [No. **53** Oii27',28',Ri1,2]

qātu(**šu**) 'hand'

1 *qātā*(**šu.2**) 'hands' (dualis), in: *ina qātē*(**šu$_2$.2**)-*ka* [e.g. No. **46** Rii2]

2 in: *šittā qātāti*, 'two thirds' [No. **38** X6',7']

qītu(**til**) 'end' (in col.) [e.g. No. **24** R1; No. **44** R6']

R

rabītu(**gal**[-*tu$_2$*]) 'large(st one)' (f. sg. adj.) [e.g. No. **9** O7; No. **31** X6', 8', 9']

rabû(**gal**) **1** 'to be/become large(r)' [e.g. No. **36** R3,4; No. **55** O3]

2 'large(st)' (m. sg. adj.) [e.g. No. **53** Oi19',ii16',17']

rabû(**šu$_2$**) 'to set'; 'setting'

1 daily setting, in **šu$_2$** *šamaš$_2$*, 'sunset' [e.g. No. **53** Oii21',28', Ri1,2]

2 'to set': last visible rising or setting of a planet (EL, ML, LA) [*i-rab-bi*: No. **9** O10; **šu$_2$**: e.g. No. **42** R4, 12]; in **šu$_2$** *ša$_2$* **šu$_2$**, 'western setting' (EL) [e.g. No. **46** Ri7,30,31, ii14]; **šu$_2$** *ša$_2$* **kur**|**nim**, 'eastern setting (ML) [e.g. No. **46** Ri29]

3 'west' [**šu$_2$**: e.g. No. **5** O4,5,R3,7; No. **9** O10; No. **42** O16,17]

râḫu(**tag$_4$**) 'to remain' [**tag$_4$**: e.g. No. **53** Oi2, ii4'; *re-ḫi*: e.g. No. **53** Rii37; No. **60** O2'; *i-re-ḫi*: No. **52** Oii10]. The stat. G in the subj. is usually realised as *re-ḫi*.

rēḫu 'remainder' [*re-e-ḫu*: No. **102** R2]

rêqu(**sud**) 'to be remote' [No. **13** R11'; No. **16** Xi17',10']

rīḫtu(**ib$_2$.tag$_4$**) 'remainder' [No. **14** Oii10'; No. **53** Rii37; No. **97** Oii14']

rupšu(**dagal**) 'width': said of the path of the Moon and the planets: *rupuš*(**dagal**) *ma-lak*, 'the width of the path' [e.g. No. **25** O17'; No. **53** Oi20']

S

sanāqu(**dim$_4$**) 'to approach' [No. **102** O13']

simānu, *simannu* 'duration; time interval' [e.g. No. **53** Oii23'–25', 27', 28', Ri1,2,ii17,22,27,28; No. **93** O12'–15']; *si-man ša$_2$ mūšu*(**gi$_6$**) *ittalak*(**du**), 'the time by which the night has progressed' (cf. *CAD* A1, pp. 311–312, sub *alāku* 3j2'a',b'; Neugebauer & Sachs 1967), often abbreviated to **gi$_6$ du**; cf. also §4.4.18. The meaning '(seasonal) hour', known from astrological texts (Rochberg 1989), is not attested in mathematical astronomy.

Ṣ

ṣabātu(**dib**) 'to hold'; *ina qātē*(**šu.2**) *taṣabbat*(**dib**), 'you hold ... in your hands' [e.g. No. **53** Rii17,20]; cf. also §2.3.1.

ṣalpu 'crossing': meaning unclear [No. **38** Y7',8'; No. **40** X4',5']

ṣeḫertu(**tur**-*tu$_2$*) 'small(est) one' (f. sg. adj.): minimum of a function [e.g. No. **41** O1]

ṣeḫēru(**tur**) 'to be/become small(er)' [e.g. No. **24** R2,3]

ṣeḫru(**tur**) 'small(est)' (m. sg. adj.) [e.g. No. **53** Oi19',ii17',19']

ṣiliptu(**bar.nun**) '*ṣiliptu*-coefficient'. In the Lunar Six procedures of lunar system A: **bar.nun** *ša$_2$ lu-maš*, '*ṣiliptu*-coefficient for the zodiac' (*q*); **bar.nun** *ša$_2$* **nim** *u* **sig**, '*ṣiliptu*-coefficient for height and depth' (*r*) [e.g. No. **53** Rii5,6,8,9,10,12,29,33; No. **93** O2'–6']. Since *q* and *r* are multiplicative factors, 'coefficient' is preferable as a pragmatic translation to 'correction' (*ACT*) since the latter has an additive connotation. The usual meaning 'diagonal', known from mathematical problem texts, is not attested in mathematical astronomy.

Š

šadû(**kur**) 'east' (abbrev. of **im.kur.ra**) [e.g. No. **42** O1,9,10]

šaḫāṭu 'to jump' [No. **98** X5]

šakānu(**gar**) 'put down'

1 *šakānu*(**gar**) +*ana*, 'put down as' [e.g. No. **60** O2',5',12',14']

2 *šakānu* +acc., 'put down (as)' [e.g. No. **53** Oi4,17',24', 25',30']

3 N: *naškunu*(**gar**), 'to occur', said of eclipses, etc. [e.g. No. **100** O5']

šalāmu(**gi**) 'to be/become intact'

1 G: said of a repeating synodic phenomenon [No. **9** R18]

2 G: said of the tablet (in inv.) [*liš-lim*: e.g. No. **41** UE1] I have adopted 'may it remain intact' as the translation of *lišlim* in all inv., but the alternative interpretation 'may it be successful' (*CAD* Š I, p. 217, *šalāmu* 4c) cannot be excluded.

3 D: *šullumu*(**gi**), 'to complete', said of time intervals in connection with the subdivision of the synodic cycle [No. **21** O2: **ta** 1 **me gi** 'After it has completed 1,0 days'; idem 6,8; No. **22** O6,9]. Cf. also *CAD* Š I *šalāmu* 11e, with numerous attestations from astrological omens.

šalmiš 'completely' [No. **83** R8]

šalšiš 'thirdly' [No. **18** Ri9'; No. **97** Oii12']

Šamaš(*šamaš$_2$*, **^{d}utu**) Sun; sun god [*šamaš$_2$*: e.g. No. **53** Oii4',7',9'; **^{d}utu**: e.g. No. **18** Ri13',15'; No. **53** Rii20]

šamû(**an**) 'heaven'

1 'sky' [e.g. No. **102** O2',R5]

2 'heaven' [e.g. No. **38** X5'–7']: the place of a planet when it is invisible at night (between LA and FA). Used with *uḫḫuru*(**zal**), 'to stay away (in heaven)'

šanîš 'secondly'; introduces alternative procedures [e.g. No. **18** Oii13',21', Ri6'; No. **65** O8']

šanû D: *šunnû*(**man**) 'to change' [e.g. No. **61.A** O26,R11,22]

šapālu(**sig**, **bur$_3$**) 'to be low; to descend' [e.g. No. **40** X9',10']

šapal 'below' (prep.) [probably the Akk. reading of **ki.ta** in No. **53** Oi30',ii16']

šaplānu(**ki.ta**) 'below' (adv.) [**ki.ta**: No. **60** O3'; **ki.ta**-*nu*: No. **60** O6'; *šap-la-nu*: No. **60** O13',15']

šaqû(**nim**, **la$_2$**) 'to be high; to ascend' [**nim**: e.g. No. **53** Rii6; **la$_2$**: e.g. No. **61.G** O23']

šarru(**lugal**) 'king' (in col.) [e.g. No. **12** Ri'11]

šaṭāru(**sar**) 'to write' [No. **53** Rii42; No. **83** R9 (meaning unclear)]

šiddu(**uš**) 'width' [e.g. No. **38** Y2']

Šiḫṭu(**^{d}gu$_4$.ud**) Mercury (lit. 'Attack; Jump') [e.g. No. **5** O11]

šiknu 'form' [No. **98** X4: meaning unclear]

šitqultu(**lal$_2$**) 'equinox' (cognate noun of *šaqālu* Gt, 'to be balanced') [e.g. No. **53** Oii20', 21', Rii20]

šūḫuzu 'to let proceed' (*aḫāzu* Š) [No. **40** X7'; No. **53** Rii23,25,27]

šullumu(**gi**) cf. *šalāmu* D

šūlû cf. *elû* Š

šumma(**be**) 'if' [**be**: e.g. No. **18** Oi7', ii7', 20'; No. **25** O13'; No. **26** X8'; *šum-ma*: No. **46** Rii6; No. **53** Oi22', 26'; No. **102** R4]

šunnû cf. *šanû* D

šuplu(**sig, bur$_3$**) 'depth'
 1 distance below the ecliptic [**sig**: e.g. No. **61.A** O9,10,23; **bur$_3$**: No. **61.D** R13]
 2 minimum value (of a function) [**sig**: e.g. No. **41** R5]

šūqu(**nim, la$_2$**) 'height'
 1 distance above the ecliptic [**nim**: e.g. No. **61.A** O8,9,10]
 2 maximum value [**nim**: e.g. No. **41** R3; **la$_2$**: e.g. No. **41** R5]

šurrû(**tab**) 'to begin' [No. **13** R29']

šūṣû cf. *aṣû* Š

šuššu(1-*šu*) '60' [e.g. No. **18** Oiii17',19']

šutābulu(ŠAR$_2$.ŠAR$_2$) 'to compute' (*abālu* Št) [No. **18** Oi5'; No. **25** O12'; No. **26** X5']; cf. *CAD* A I *abālu* A 10.

T

tāmartu(**igi.du$_8$.a**) 'appearance'
 1 probably a general term for synodic phenomenon, to be distinguished from *nanmurtu*(**igi**), '(first) appearance' [No. **13** R9'; No. **35** R3,4]
 2 '(first) appearance' [No. **102** O4']

tamṭītu 'deficit; subtraction'; cognate noun of *muṭṭû*, 'to diminish' [e.g. No. **52** Oi34,Ri27]

tardītu 'extension; addition'; cognate noun of *ruddû*, 'to extend' [No. **93** O7'; No. **100** O6']

tarṣa, *tarṣi*: cf. *ana tarṣi*.

târu(**gur**) 'to turn'
 1 'to recede': retrograde motion of the planets; often in hendiadys with *nasāḫu*, 'to move' [e.g. No. **18** Oii10',15',17']
 2 'to turn back' [e.g. No. **38**; No. **95** Oii5',8',9']
 3 'to become' (+*ana*) [e.g. No. **61.D** O14,28,R14,21]
 4 D: *turru*, 'to reject'; said of the first lunar crescent when the preceding month is hollow [e.g. No. **53** Rii16,17,40]

tašpiltu(*taš*) 'difference' [e.g. No. **82** O2'; No. **96** O13']

tērsītu 'computed table'; in col. of synodic tables [e.g. *ACT* 123,Zk; 123a,Zka]

Ṭ

ṭeḫû(**te**) 'to approach' [No. **56** O7?]

ṭepû(**tab**) 'to append, attach'
 1 'to add': asymmetric, identity-conserving addition (§2.2.2.1) [e.g. No. **67** O4,8,12,15,20, R3,7]
 2 'to increase, be increasing' [e.g. No. **53** Ri22–30]
 3 'to be additive' [e.g. No. **53** Oii27'; No. **79** Oii'1',2',20'–23']

ṭerītu 'squeezed; pinched' (f. sg. verb. adj. G *ṭerû*) [No. **18** Oi23'; No. **38** Y10']

ṭīpu(**tab**) 'appendage, attachment, addition'; *ṭīpu*(**tab**) *u mīṭu*(**la$_2$**), 'addition and subtraction': technical term for difference [e.g. No. **74** O7',13'; No. **82** O19', R13,16,29]

ṭuppu(**im**[**.dub**]) 'tablet' (in col.) [**im**: e.g. No. **45** Rii'6'; **im.dub**: e.g. No. **45** Rii'7']

ṭupšarru(^{lu_2}**umbisag**) 'scribe', in: *ṭupšar Enūma Anu Enlil*, 'scribe of *Enūma Anu Enlil*' = astronomer, astrologer [No. **7** R1]

U

ubānu(**šu.si**, *u*) finger (unit of arc) [**šu.si**: e.g. No. **53** Oii11'; No. **65** O20'; *u*: No. **47** O4]

uddazallû(**u$_4$.da.zal.la$_2$**) '*uddazallû*-coefficient' [e.g. No. **13** R15'; No. **14** Oii10'; No. **100** O6'; No. **102** O15']

uḫḫuru(**zal**) 'to stay away': said of planets during their invisibility: *uḫḫuru ina* **an**-*e* = 'to stay away in heaven' [No. **38** X5',6',7'; No. **29** O5'].

ultu(**ta**) 'from' (in spatial expressions) [e.g. No. **53** Oib2',6']. It is unclear whether apart from *ina* also *ultu* is a possible Akk. reading of **ta**, 'from', in subtractions.

ūmu(**me, u$_4$**) 'day'
 1 day [**me**: e.g. No. **53** Rii15,16; No. **61.A** O15,16,26,27; **u$_4$**: e.g. No. **95** Oii6–11]
 2 'day'; in the sense (mean) tithi [e.g. No. **41** O3–9,11–16]
 3 daylight; in: LA$_2$ *ša$_2$ ūmi*(**me**), 'duration of daylight' [e.g. No. **53** Rii18]; in: *ana* **me** E, 'rising to daylight' (AR) [e.g. No. **18** Oi13']
 4 *ūmū*(**me.meš, u$_4$.me, u$_4$.meš**) 'days; time' [e.g. No. **13** R4'; No. **28** R5]
 5 10th degree of a zod. sign [*u$_4$-me*: No. **61.A** R16; **me**: No. **61.B** O3'] This rare usage, known only from mathematical astronomy, is unexplained.

ūm bubbuli(**u$_4$.na$_2$.a, u$_4$.na$_2$.am$_3$**) 'day of disappearance'
 1 day when the lunar crescent is seen for the last time before New Moon [No. **53** Rii40 (?)]
 2 KUR (Lunar Six interval) [e.g. No. **52** Oii11; No. **53** Rii17]

uskaru(**u$_4$.sakar**) 'crescent' [No. **102** O5',11']

uṭṭatu(**še**) 'barleycorn': unit of angle (cf. p. xxvi)

W

(w)aṣābu(**daḫ**) 'to append': an additive operation (§2.2.2.2) [e.g. No. **102** O3',16']

Z

zappu(**mul$_2$**[**.mul$_2$**]) 'The Stars'
 1 Tau (zod. sign) [e.g. No. **18** Oii4',6',Ri22',iii8']
 2 Pleiades (constellation) [**mul$_x$.mul$_x$**: No. **102** O13' (?)]

zâzu(**bar**, SE$_3$) 'to divide': *ana n aḫḫē*(**šeš.meš**) *zâzu*, 'to divide into *n* parts' (§2.2.5) [**bar**: No. **82** R11; No. **102** R5; SE$_3$: No. **52** Oi28, 33,ii32,Ri12,19,27]

zibbātu(*zib*[**.me**]) 'tails': Psc (zod. sign) [e.g. No. **55** O7,9–11, R3']

Sumerograms and Akkadian abbreviations

a 1 Leo (zod. sign); **2** *māru*, 'son' (in col.)

A abbrev. of *aṣû*? (in A **me**, 'daylight rising?' = AR; *ana* **me** A, 'it rises? to daylight' = AR)

a.ra$_2$ **1** 'times' (§2.2.4.1); **2** *ana*, 'for' [No. **46** P20]

a.ša$_3$ *eqlu*

ab *Ṭebētu* (month X)

ab$_2$ *arḫu*

absin, absin$_0$(KI) Vir (zod. sign); Akk. *šer'u*, 'furrow'

alla Cnc (zod. sign); Akk. *alluttu*, 'crab'

an *šamû*

AN Mars (Akk. *Ṣalbatānu*)

AN.MI *attalû*

an.ta *eli*

apin *Araḫsamna* (month VIII)

ar$_2$ *arki, arkītu, arkû*

bal **1** *nabalkutu*; **2** *palû*

bar **1** *Nisannu* (month I); **2** (transliterated as 1/2) *mišlu*; **3** *zâzu*

BAR marker of eclipse possibility in column Ψ of lunar systems A, B; Akk. reading unclear

bar.nun *ṣiliptu*

BE **1** 'elongation'. Akk. reading probably *bi*$_2$, a phonetic abbreviation of *birītu*, 'distance'
2 marker of the impossibility of an eclipse in column Ψ of a lunar table (Akk. reading unknown)
3 *šumma*

bur$_3$(U) *šapālu*; *šuplu*

^{giš}**da** *lē'u*

dagal *rupšu*

daḫ *(w)aṣābu*

danna *bēru*

dib **1** *etēqu*; **2** *ṣabāṭu*

dim$_4$ *sanāqu*

diri **1** *atāru*; **2** abbrev. of **diri.še**; **3** *malû*

diri[**.še**] *Addaru arkû* (month XII$_2$)

^{d}**diš** *Anu*

DIŠ.UD function *R* (lunar system A); Akk. reading perhaps *ana ūmi*

du *alāku*

DU Morning Last (ML); Akk. reading unclear [No. **10** O1, R1,15]

du$_3$ *epēšu*, *epūšu*

du$_3$**.du**$_3$ 'procedure' (in mathematical procedures), Akk. *epēšu*, *epištu* or *nēpešu*? [No. **22** O14; No. **38** Y4']

du$_6$ *Tašrītu* (month VII)

du$_{11}$[**.ga**] *qabû*

dur *markasu*

e *eli*

e *qabû*

E abbrev. of *elû*(**e**$_{11}$); in E **me**, 'daylight rising' = AR (p. 58)

e$_2$ *bītu*

e$_3$ *aṣû* (G, Š)

e$_{11}$ *elû* (G, Š)

egir *arki*, *arkû*

en *adi*

^{d}**en** *Bēl*

en.nam *mīnû*

en.nun *maṣṣartu*

gaba.ri *meḫertu*

gal *rabû*, *rabītu*

gal-*tu*$_2$ *rabītu*

gal$_2$ *bašû*

$^{\text{lu}_2}$**gala** *kalû*

gam *kippatu*

GAM, GAM$_0$ 'times'; cf. **a.ra**$_2$

gan *Kislīmu* (month IX)

gar *šakānu*

GAR **ninda**

GAR.GAR *kamāru*

^{d}**gašan** *Bēltu*

genna *Kajjamānu* (Saturn)

gi *šullumu*

gi$_6$ **1** *mūšu*
2 GI$_6$ (Lunar Six interval), time between sunset and first moonrise after sunset (near Full Moon). The use of **gi**$_6$ may be explained by the fact that this interval occurs after sunset (**gi**$_6$ = *mūšu*, 'night'), as opposed to ME.

gi$_6$ **du** cf. *si-man ša*$_2$ **gi**$_6$ **du**

gibil *eššu*

gin$_7$ *kīma*

gir$_2$[**.tab**] Sco (zod. sign); Akk. *zuqaqīpu*, 'scorpion'

giskim.bi zu.še$_3$ 'in order to establish this'(?); cf. p. 518 [No. **102** O4',9',12',13']

GIŠ[.A] *našû* (the Sum. reading of GIŠ is probably **guru**$_x$).

GIŠ.GIŠ.ḪI 'you compute' (Akk. reading unknown); [No. **42** O8,14; No. **93** O12',13'; both from Uruk] In the duplicate No. **1** R3, *našû*(GIŠ), 'to compute', replaces GIŠ.GIŠ.ḪI, which must therefore have the same meaning, but if ḪI is a phonetic complement then *našû* would appear to be excluded.

gu **1** Aqr (zod. sign); Akk. *Gula*
2 meaning unclear [No. **18** Ri17']

gu$_4$ *Ajjaru* (month II)

[^{d}]**gu**$_4$[**.ud**] *Šiḫṭu*, 'Mercury'

gub **1** *kunnu*
2 'solstice' (Akk. reading unclear) [No. **52** Ri10; No. **53** Oii20',21', Rii17]
3 *izuzzu*

gur *târu* (G, D)

guru$_x$ *našû*(GIŠ)

ḪAB[-*rat*] 'disk; eclipse magnitude' (Akk. reading unclear) [ḪAB-*rat*: e.g. No. **53** Oib.10': ḪAB *attalê*(AN.MI), 'disk of the eclipse']

ḫun Ari (zod. sign)

ib$_2$**.tag**$_4$ *rīḫtu*

igi **1** *amāru* (G, N)
2 *nanmurtu*
3 *maḫru*, *maḫrû*
4 *pānu*: *ina pānāt*(**igi**-*at*)

igi.du$_8$**.a** *tāmartu*

igi.gub *igigubbû*

igi.tab *barû*

igi.*n***.gal**$_2$**.bi** reciprocal of *n* [No. **102** O6', R6]

il$_2$ *našû*

im[**.dub**] *ṭuppu*

im.gid$_2$**.da** *giṭṭû*, *imgiddû*

iti *arḫu*

ka *pû*

ka$_2$-*tu*$_2$ *bābtu*

kaskal[**.2**] *ḫarrānu*

ki **1** *itti*; **2** *qaqqaru*

KI IGI meaning and Akk. reading unclear [No. **38** Y3',6',11'?]

kimin 'ditto' [e.g. No. **18** Oiii3',8'; No. **46** Rii5,8]

ki.na$_2$**.a** meaning unclear [No. **61.C** R13]

ki.ta *šaplānu*, *šaplu*

ki.ta-*nu* *šaplānu*

kin *Ulūlu* (month VI)

kin.2 *Ulūlu arkû* (month VI$_2$)

kur 1 KUR (Lunar Six interval), time between last visible moonrise before sunrise and sunrise before the conjunction. The use of **kur** may reflect that this interval is delimited by the rising (**kur**) of the Moon and that of the Sun.
2 *napāḫu*, *nipḫu*
3 *šadû*, 'east'
4 *kašādu*

kuš$_3$ *ammatu*

la$_2$ **1** *naḫāsu*; **2** *maṭû*, *mīṭu*, *muṭṭû*; **3** *šaqû*, *šūqu*

LA$_2$ 'length; duration' (of time intervals); Akk. reading unknown [No. **61** P1,P2,P4 step 1; No. **79** Oi'9',10',11',Rii13; perhaps also LA$_2$-*ti* in No. **93** O3'–6']

lal$_2$ *šitqultu*

lugal *šarru*

man *šunnû*

maš[.maš] Gem (zod. sign); Akk. *māšu*, *tū'amū*

maš$_2$ Cap (zod. sign); Akk. *suḫurmāšu*, 'goat-fish'

me **1** *ūmu*
2 ME (Lunar Six interval), time interval between last moonrise before sunset and sunset (near Full Moon). The use of this logogram may be explained by the fact that ME occurs during day time (**me** = *ūmu*, 'day'), as opposed to GI$_6$.

mi *mīnû*

mu[.an.na] *šattu*

mu.du 'it proceeds' (?); 'progress' (?); Akk. reading unclear: a form of *alāku* or a cognate noun [No. **55** O6,11; No. **69** R2',3'?; No. **70** R6',7'; No. **94** Ri3]

mul$_2$**.babbar** *Peṣû*, 'Jupiter'

mul.mul, **mul**$_2$**[.mul**$_2$**]** Tau (zod. sign); cf. *zappu*

murub$_4$ *qablu*, *qablû*, *qablītu*

na NA, NA$_1$ (Lunar Six intervals), cf. *manzāzu*

ne *Abu* (month V)

nim **1** *šaqû*, *šūqu*
2 'east' (Akk. reading perhaps *elû*?)
3 *elû* (G, Š)

ninda(GAR) *nindanu* (?)

ninda$_2$ *middatu* (?)

nu tuk 'it does not exist'; Akk. reading probably *ul ibašši* on account of *ACT* 135 Ri24' **nu tuk**-*ši*

pa Sgr (zod. sign); Akk. *Pabilsag*

pap *napḫaru*

PAP.ḪAL 'period'; Akk. reading perhaps *muttalliktu*.

PAP.PAP 'total'; Akk. reading perhaps *napḫaru* [No. **9** O9; No. **13** R7',8'; No. **19** R5'; No. **25** O9'; No. **39** O4',8']

rin$_2$ Lib (zod. sign); Akk. *zibānītu*, 'scales'

sag[.ki] *pūtu*

sag.ki gu$_4$ *pūt alpi*

sag.me.gar *Sagmegar* = 'Jupiter'

sag.uš *kajjamānu*

sar *šaṭāru*

SE$_3$ 'to divide', Akk. reading perhaps *zâzu* (§2.2.5)

si *ubānu*

sig **1** *šapālu*, *šuplu*; **2** *Simānu* (month III)

sud *rêqu*

ša$_3$ *libbu*

šar$_2$**.šar**$_2$ *abālu* Št

še **1** *Addaru* (month XII); **2** *uṭṭatu*

šeš *aḫu*

šid *minûtu*

šu **1** *Dûzu* (month IV); **2** *qātu*

šu.si *ubānu*

šu$_2$ **1** *rabû*
2 ŠU$_2$ (Lunar Six interval), time between last moonset before sunrise and sunrise (near Full Moon). The use of **šu**$_2$ may be due to the fact that ŠU$_2$ is defined by the setting of the Moon (**šu**$_2$ = *rabû*, 'to set').

ta **1** *ina*; **2** *ultu*

tab **1** *ṭepû*, *ṭīpu*; **2** *šurrû*

tag$_4$ *râḫu*, *rēḫu*, *rīḫtu* (only in mathematics and mathematical astronomy; elsewhere **tag**$_4$ = *ezēbu*, 'to leave')

tar *tarṣa* (cf. *ana tarṣa*)

taš *tašpiltu*

til *qītu*

tuk *bašû*

tur *ṣeḫēru*; *ṣeḫru*.

tur-*tu*$_2$ *ṣeḫertu*.

u *ubānu*

U **bur**$_3$

u$_4$ *ūmu*

u$_4$**.da.zal.la**$_2$ *uddazallû*

u$_4$**.na**$_2$**.a**, **u**$_4$**.na**$_2$**.am**$_3$ **1** *ūm bubbuli*; **2** KUR (Lunar Six interval)

u$_4$**.sakar** *uskaru*

u$_4$**.1.kam** 'day 1'
1 first day of the month [e.g. No. **53** Rii15]
2 in *ša*$_2$ **u**$_4$**.1.kam**, 'of day 1': pertaining to New Moon (conjunction) [e.g. No. **65** R9,14,19]
3 NA$_1$ (Lunar Six interval) [e.g. No. **53** Rii11]

u$_4$**.14.kam** 'day 14'; in *ša*$_2$ **u**$_4$**.14.kam**, 'of day 14': pertaining to Full Moon (opposition) [e.g. No. **65** R9,13,14,19]

ud$_2$ cf. ZIZ$_2$

ugu *muḫḫu*

UL LA NI meaning unclear [No. **82** O28']

^{lu2}**umbisag** *ṭupšarru*

ur *bašû*

uru$_3$ *naṣāru*

uš **1** *nenmudu*; **2** *nēmettu*; **3** *šiddu*.

UŠ **1** (time) degree (°); Sum. and Akk. reading unknown [e.g. No. **97** Oii4']
2 UŠ.MEŠ: meaning unclear [No. **18** Ri13']

^{d}**utu** *Šamaš*

^{d}**utu.šu**$_2$**.a** *ereb Šamši*

zal **1** *uḫḫuru*; **2** *nasāḫu*

zi *nasāḫu*; *nisḫu*

ZIZ$_2$ *Šabāṭu* (month XI); the Sum. reading is **ud**$_2$

zu cf. **giskim.bi zu.še**$_3$

1-*šu* *šuššu*, 60

1-*en* *ištēn*, 1

2-*ta* *šittā*, 2

^{d}20 *Šamaš*, 'Sun, sun god'

^{d}30 *Sîn*, 'Moon, moon god'

Bibliography

Aaboe A. 1958, 'On Babylonian planetary theories', *Centaurus* 5, 209–277

—1964a, 'On period relations in Babylonian astronomy', *Centaurus* 10, 213–231

—1964b, 'A Seleucid table of daily solar (?) positions', *JCS* 18, 31–34

—1966, 'On a Babylonian schema for solar motion of the System A variety', *Centaurus* 11, 302–303

—1968, 'Some lunar auxiliary tables and related tablets from the Late Babylonian period', *KDVSMM* 36.12

—1969, 'A computed list of new moons for 319 B.C. to 316 B.C. from Babylon: BM 40094', *KDVSMM* 37.3

—1971, 'Lunar and solar velocities and the length of lunation intervals in Babylonian astronomy', *KDVSMM* 38.6

—1974, 'Scientific astronomy in antiquity', in: F.R. Hodson et al. (eds.), *The Place of Astronomy in the Ancient World, Philosophical Transactions, Royal Society of London* A 276, 21–42

—1977, 'A computed cuneiform text for Mercury from Babylon: BM 48147', in: Y. Maeyama, W.G. Saltzer (eds.), *ΠΡΙΣΜΑΤΑ. Natürwissenschaftliche Studien. Festschrift für Willy Hartner* (Steiner: Wiesbaden), 1–8

—1980, 'Observation and theory in Babylonian astronomy', *Centaurus* 24, 14–35

—1987, 'A Late-Babylonian Procedure Text for Mars, and Some Remarks on Retrograde Arcs', in: D.A. King, G. Saliba (eds.), *From Deferent to Equant: A Volume of Studies in the History of Science in the Ancient and Mediaeval Near East in Honor of E. S. Kennedy*, Annals of the New York Academy of Sciences 500 (New York), 1–14

—2001, *Episodes From the Early History of Astronomy* (Springer-Verlag: New York, Berlin, Heidelberg)

—2002, 'On Columns H and J in Babylonian Lunar Theory of System B', in: *UOS*, 1–4

Aaboe A., Britton J.P., Henderson J.A., Neugebauer O.N., Sachs A.J. 1991, 'Saros Cycle Dates and Related Babylonian Astronomical Texts', *TAPS* 81/6, 1–75

Aaboe A., Hamilton N.T. 1979, 'Contributions to the study of Babylonian lunar theory', *KDVSMM* 40.6

Aaboe A., Henderson J.A. 1975, 'The Babylonian theory of lunar latitude and eclipses according to System A', *AIHS* 25, 181–222

Aaboe A., Huber P. 1977, 'A text concerning subdivision of the synodic motion of Venus from Babylon: B.M. 37151', in: M. de Jong Ellis (ed.), *Essays on the Ancient Near East in Memory of J.J. Finkelstein*, Memoirs of the Connecticut Academy of Arts and Sciences, 19 (Archon Books: Hamden), 1–4

Aaboe A., Sachs A.J. 1966, 'Some dateless computed lists of longitudes of characteristic planetary phenomena from the Late-Babylonian period', *JCS* 20, 1–33

—1969, 'Two lunar texts of the Achaemenid period from Babylon', *Centaurus* 14, 1–22

Al-Rawi F.N.H., George A.R. 1991/2, 'Enūma Anu Enlil XIV and other early astronomical tables', *AfO* 38/39, 52–73

Balke Th.E. 2002, 'Die sumerischen Dimensionaladjektive nim und sig. Anmerkungen zur Polysemie und Grammatikalisation dimensionaler Ausdrücke im Sumerischen', in: O. Loretz, K.A. Metzler, H. Schaudig (eds.), *Ex Mesopotamia et Syria Lux. Festschrift für Manfried Dietrich zu seinem 65. Geburtstag*, *AOAT* 281, 31–53

Bassnett S. 2002, *Translation Studies* (Routledge: London, New York)

Beaulieu P.-A. 1993, 'The Impact of Month-lengths on the Neo-Babylonian Cultic Calendar', *ZA* 83, 66–87

—2000, 'The descendants of Sîn-lēqi-unninni', in: *Fs Oelsner*, 1–16

—2006, 'The Astronomers of the Esagil Temple in the Fourth Centrury BC', in: A.K. Guinan (ed.), *If a Man Builds a Joyful House. Assyriological studies in honour of Erle Verdun Leichty* (Brill: Leiden), 5–22

Bernsen L. 1969, 'On the construction of Column B in System A of the astronomical cuneiform texts', *Centaurus* 14, 23–28

Brack-Bernsen L. 1990, 'On the Babylonian lunar theory: A construction of Column Φ from horizontal observations', *Centaurus* 33, 39–56

—1997, *Zur Entstehung der babylonischen Mondtheorie* (Franz Steiner Verlag: Stuttgart)

—2002, 'Predictions of Lunar Phenomena in Babylonian Astronomy', in: *UOS*, 5–19

——2007, 'The 360-Day Year in Mesopotamia', in: J.M. Steele (ed.), *Calendars and Years. Astronomy and Time in the Ancient Near East* (Oxbow: Oxford), 83–100

Brack-Bernsen L., Hunger H. 1999, 'The Babylonian zodiac: speculations on its invention and significance', *Centaurus* 41, 280–292

——2002, 'TU 11. A collection of rules for the prediction of lunar phases and of month lengths', *SCIAMVS* 3, 3–90

——2005/6, 'On the "Atypical Astronomical Cuneiform Text E"', *AfO* 51, 96–107

——2008, 'BM 42282+42294 and the Goal-Year Method', *SCIAMVS* 9, 3–23

——2010, 'Methods for understanding and reconstructing Babylonian predicting rules', in: A. Imhausen, T. Pommerening (eds.), *Writings of Early Scholars in the Ancient Near East, Egypt, Rome and Greece*, 277–298

Brack-Bernsen L., Hunger H., Walker C. 2008, 'KUR – When the Old Moon Can Be Seen a Day Later', *Fs Slotsky*, 1–6

Brack-Bernsen L., Schmidt O. 1994, 'On the Foundations of the Babylonian Column Φ: Astronomical Significance of Partial Sums of the Lunar Four', *Centaurus* 37, 183–209

Brack-Bernsen L., Walker C. 2007, 'BM 57980: A Wrong Version of the Excellent Predicting Rule R', in: *Fs Hunger*, 35–41

Britton J.P. 1987, 'The Structure and Parameters of Column Φ', in: J.L. Berggren (ed.), *From ancient omens to statistical mechanics: essays on the exact sciences; presented to Asger Aaboe* (Copenhagen), 23–36

——1989, 'An early function for eclipse magnitudes in Babylonian astronomy', *Centaurus* 32, 1–52

——1990, 'A Tale of Two Cycles: Remarks on Column Φ', *Centaurus* 33, 57–69

——1999, 'Lunar anomaly in Babylonian astronomy', in: N.M. Swerdlow (ed.), *Ancient Astronomy and Celestial Divination*, 187–254

——2001, 'Remarks on a System A Text for Venus: ACT 1050', *AHES* 55, 525–554

——2002, 'Treatments of Annual Phenomena in Cuneiform Sources', in: *UOS*, 21–78

——2003, 'On Corrections for Solar Anomaly in Babylonian Lunar Theories', *Centaurus* 45, 46–58

——2007a, 'Studies in Babylonian Lunar Theory: Part I. Empirical Elements for Modeling Lunar and Solar Anomalies', *AHES* 61, 83–145

——2007b, 'Calendars, Intercalations and Year-Lengths in Mesopotamian Astronomy', in: J. Steele (ed.), 'Calendars and Years' (Oxbow: Oxford), 115–132

——2009, 'Studies in Babylonian Lunar Theory: Part II. Treatments of Lunar Anomaly', *AHES* 63, 357–431

——2010, 'Studies in Babylonian Lunar Theory: Part III. The Introduction of the Uniform Zodiac', *AHES* 64, 617–663

Britton J.P., Horowitz W., Steele J.M. 2007, 'BCM A1845-1982.2', *Fs. Hunger*, *WZKM* 97, 43–54

Britton J.P., Walker C.B.F. 1991, 'A 4th century Babylonian model for Venus: B.M. 33552', *Centaurus* 34, 97–118

Brown D. 2000, *Mesopotamian Planetary Astronomy-Astrology*, Cuneiform Monographs 18 (Styx: Groningen)

——2001, 'Misinformation on Mesopotamian Exact Sciences', in: T. Abusch et al., Proceedings of the 45th RAI (CDL Press: Bethesda), 79–89

Bruins E.M., Rutten M. 1961, *Textes Mathématiques de Suse*, Mémoires de la Mission Archéologique en Iran, Tome XXXIV (Paul Geuthner: Paris)

Clancier, Ph. 2009, *Les bibliothèques en Babylonie dans la deuxième moitié du Ier millénaire av. J.-C., AOAT* 363

Cohen M.E. 1993, 'The Cultic Calendars of the Ancient Near East' (CDL Press: Bethesda)

Damerow P. 2001, 'Kannten die Babylonier den Satz des Pythagoras? Epistemologische Bemerkungen zur Natur der babylonischen Mathematik', in: J. Høyrup, P. Damerow (eds.), *Changing Views on Ancient Near Eastern Mathematics*, *BBVO* 19, 219–310

Dietrich M. 1969, 'Die neubabylonischen Konjunktionen', in: M. Dietrich, W. Röllig (ed.), *Lišān mitḫurti: Festschrift Wolfgang Freiher von Soden zum 19. VI. 1968 gewidmet*, *AOAT* 1, 65–99

Donbaz V., Koch J. 1995, 'Ein Astrolab der dritten Generation: NV. 10', *JCS* 47, 63–84

Durand J.-M. 1981, *Textes babyloniens d'époque récente* (A.D.P.F.: Paris)

Edzard D. 1969, 'Eine altsumerische Rechentafel (OIP 14, 70)', in: M. Dietrich, W. Röllig (ed.), *Lišān mitḫurti: Festschrift Wolfgang Freiher von Soden zum 19. VI. 1968 gewidmet*, *AOAT* 1, 101–104

Epping J. 1889, *Astronomisches aus Babylon oder das Wissen der Chaldäer über den gestirnten Himmel, Stimmen aus Maria-Laach*, Ergänzungsheft 44 (Herder Verlag: Freiburg i.Br.)

——1890, 'Die babylonische Berechnung des Neumondes', *Stimmen aus Maria-Laach* 39, 225–240

Epping J., Strassmaier J.N. 1881, 'Zur Entzifferung der astronomischen Tafeln der Chaldäer', *Stimmen aus Maria-Laach* 21, 277–292

Falkenstein A. 1941, *Topographie von Uruk, I. Teil: Uruk zur Seleukidenzeit, Ausgrabungen der Deutschen Forschungsgemeinschaft in Uruk-Warka*, Band 3 (Leipzig)

Fatoohi L.J., Stephenson F.R. 1997/98, 'Angular measurements in Babylonian astronomy', *AfO* 44/45, 210–214

Fotheringham J.K. 1933, 'The Indebtedness of Greek to Chaldaean Astronomy', *QS* B 2, 28–44

Friberg J. 1990, 'Mathematik', *RlA* 7, 531–585

——1997, 'Seeds and reeds continued. Another metro-mathematical topic text from Late Babylonian Uruk', *BaM* 28, 251–365; Pls. 45–46

——2005, 'On the Alleged Counting with Sexagesimal Place Value Numbers in Mathematical Cuneiform Texts from the Third Millennium BC', *CDLJ* 2005:2

——2007, *A Remarkable Collection of Babylonian Mathematical Texts, Manuscripts in the Schøyen Collection: Cuneiform Texts* 1, Sources and Studies in the History of Mathematics and Physical Sciences (Springer)

Friberg J., Hunger H., al-Rawi F. 1990, 'Seeds and reeds: A metro-mathematical topic text from Late Babylonian Uruk', *BaM* 21, 483–557; Pls. 46–48

Goetsch H. 1968, 'Die Algebra der Babylonier', *AHES* 5, 79–160

Gray J.M.K. 2009, *A Study of Babylonian Goal-Year Planetary Astronomy*, PhD Thesis, Dept. of Physics, University of Durham

Hackl J. 2007, *Der subordinierte Satz in den spätbabylonischen Briefen*, *AOAT* 341

Hamilton N.T., Aaboe A. 1998, 'A Babylonian Venus Text Computed According to System A: ACT 1050', *AHES* 53, 215–221

Horowitz W. 2000, *Astronomical Cuneiform Texts in the Birmingham City Museum*, in: A.R. George, I.L. Finkel (eds.), *Wisdom, Gods and Literature. Studies in Assyriology in Honour of W.G. Lambert* (Winona Lake)

Høyrup J. 1993, 'On subtractive operations, subtractive numbers, and purportedly negative numbers in Old Babylonian mathematics', *ZA* 83, 42–60

——1996, 'Changing trends in the historiography of Mesopotamian mathematics: an insider's view', *History of Science* 34, 1–32

——2000, 'The finer structure of the Old Babylonian mathematical corpus', in: *Fs Oelsner*, 117–178

——2002, *Lengths, Widths, Surfaces. A Portrait of Old Babylonian Algebra and its Kin*, Sources and Studies in the History of Mathematics and Physical Sciences (Springer)

——2010, 'How to transfer the conceptual structure of Old Babylonian mathematics: Solutions and inherent problems. With an Italian parallel', in: A. Imhausen, T. Pommerening (eds.), *Writings of Early Scholars in the Ancient Near East, Egypt, Rome and Greece*, 385–417

Huber P.J. 1957, 'Zur täglichen Bewegung des Jupiter nach babylonischen Texten' *ZA* 52, 265–303

——1958, 'Über den Nullpunkt der babylonischen Ekliptik', *Centaurus* 5, 192–208

——1982, 'Astronomical dating of Babylon I and Ur 111', *Occasional Papers on the Near East* 1/4 (Malibu)

——2000, 'Babylonian Short-Time Measurements: Lunar Sixes', *Centaurus* 42, 223–234

——2007, 'A Lunar Six Text from 591 B.C.', *WZKM* 97, 213–217

Huber P.J., Steele J.M. 2007, 'Babylonian Lunar Six Tablets', *SCIAMVS* 8, 3–36

Hunger H. 1968, *Babylonische und assyrische Kolophone*, *AOAT* 2

——1976a, *Spätbabylonische Texte aus Uruk* I, *Ausgrabungen der Deutschen Forschungsgemeinschaft in Uruk-Warka* (Mann: Berlin)

——1976b, 'Astrologische Wettervorhersagen', *ZA* 66, 234–260

——1988, 'A 3456: eine Sammlung von Merkurbeobachtungen', in: E. Leichty et al., *A Scientific Humanist: Studies in Memory of Abraham Sachs* (Univ. Museum: Philadelphia), 201–223

——2003/4, 'A Procedure Text for Jupiter', *AfO* 50, 344–345

——2006, *Astronomical Diaries and Related Texts from Babylonia*, Volume VI. Goal-Year Texts (Verlag der Österreichischen Akademie der Wissenschaften: Wien)

Hunger H., Pingree D. 1989, *MUL.APIN: An Astronomical Compendium in Cuneiform*, *AfO Beih* 24 (F. Berger: Horn)

——1999, *Astral Sciences in Mesopotamia*, Handbuch der Orientalistik, Abt. 1 Band 44 (Brill: Leiden)

Hunger H., Sachs A. 2001, *Astronomical Diaries and Related Texts from Babylonia*, Volume V. Lunar and Planetary Texts (Verlag der Österreichischen Akademie der Wissenschaften: Wien)

Imhausen A. 2002, 'The Algorithmic Structure of the Egyptian Mathematical Problem Texts', in: *UOS*, 147–166

——2003, *Ägyptische Algorithmen. Eine Untersuchung zu den mittelägyptischen mathematischen Aufgabentexten*, Ägyptologische Abhandlungen 65 (Harrassowitz: Wiesbaden)

Jones A. 1983, 'The development and transmission of 248-day schemes for lunar motion in ancient astronomy', *AHES* 29, 1-36

——1999, *Astronomical Papyri from Oxyrhynchus*, Memoirs of the American Philosophical Society 233 (Philadelphia)

——2004, 'A study of Babylonian observations of planets near normal stars', *AHES* 58, 475–536

Jursa M. 1993/4, 'Zweierlei Maß', *AfO* 40/41, 71–73

——2005, *Neo-Babylonian Legal and Administrative Documents. Typology, Contents and Archives* (Ugarit-Verlag: Münster)

Kose A. 1998, *Uruk. Architektur IV*, *AUWE* 17

Kraus F. 1948, 'Die Istanbuler Tontafelsammlung', *JCS* 1, 93–119

Kessler K.-H. 1984, 'Eine Arsakidenzeitliche Urkunde aus Warka', *BaM* 15, 273–281

Kugler F.X. 1900, *Die babylonische Mondrechnung. Zwei Systeme der Chaldäer über den Lauf des Mondes und der Sonne* (Herder Verlag: Freiburg i.Br.)

——1907, *Sternkunde und Sterndienst in Babel*, Buch I: *Entwicklung der babylonischen Planetenkunde* (Aschendorffsche Verlagsbuchhandlung: Münster)

——1909, *Sternkunde und Sterndienst in Babel*, Buch II, Teil I *Natur, Mythus und Geschichte als Grundlagen babylonischer Zeitordnung nebst eingehenden Untersuchungen der älteren Sternkunde und Meteorologie* (Aschendorffsche Verlagsbuchhandlung: Münster)

—1910, *Im Bannkreis Babels. Panbabylonistische Konstruktionen und religionsgeschichtliche Tatsachen* (Aschendorffsche Verlagsbuchhandlung: Münster)

—1912, *Sternkunde und Sterndienst in Babel*, Buch II, Teil II, Heft 1 (Aschendorffsche Verlagsbuchhandlung: Münster)

—1913a, *Ergänzungen zum I. und II. Buch, Zur älteren babylonischen Topographie des Sternhimmels*, I (Aschendorffsche Verlagsbuchhandlung: Münster)

—1913b, *Ergänzungen zum I. und II. Buch, Astronomie und Chronologie der älteren Zeit*, II–VIII (Aschendorffsche Verlagsbuchhandlung: Münster)

—1924, *Sternkunde und Sterndienst in Babel*, Buch II, Teil II, Heft 2 (Aschendorffsche Verlagsbuchhandlung: Münster)

—1935: cf. Schaumberger (1935)

Kuhn T.S. 1970, *The Structure of Scientific Revolutions*, Second Edition (University of Chicago Press: Chicago)

Leichty E. 1986, *Catalogue of the Babylonian Tablets in the British Museum*, Vol. VI: *Tablets from Sippar* 1

Leichty E., Finkelstein J.J., Walker C.B.F. 1988, *Catalogue of the Babylonian Tablets in the British Museum*, Vol. VIII: *Tablets from Sippar* 3

Lenzen H.J. 1962, 'Bīt Rēš', in: *UVB* 18

Lindström G. 2003, *Uruk. Siegelabdrücke auf hellenistischen Tonbullen und Tontafeln*, *AUWE* 20

Linssen M.J.H. 2004, *The Cults of Uruk and Babylon. The Temple Ritual Texts as Evidence for Hellenistic Cult Practises* (Brill, Styx: Leiden)

Maeyama Y. 1978, 'On the Babylonian lunar theory', *AIHS* 28, 21–35

—1981, 'The basic problems of the Babylonian lunar theory', *AIHS* 31, 253–372

Marzahn J., Neumann H. (eds.) 2000, *Assyriologica et Semitica. Festschrift für Joachim Oelsner anläßlich seines 65. Geburtstages am 18. Februar 1997*, *AOAT* 252

McEwan G.J.P. 1981, *Priest and Temple in Hellenistic Babylonia*, *FAOS* 4 (Franz Steiner Verlag: Wiesbaden)

Mebert J. 2010, 'Die Venustafeln des Ammiṣaduqa und ihre Bedeutung für die astronomische Datierung der altbabylonischen Zeit', *AfO Beiheft* 31

Meeus J. 1998, *Astronomical Algorithms*, second edition (Willmann-Bell: Richmond)

Neugebauer O. 1935–1937, *Mathematische Keilschrifttexte*, I–III. Quellen und Studien zur Geschichte der Mathematik, Astronomie und Physik. Abteilung A: Quellen. 3. Band (Julius Springer: Berlin)

—1936, Review of J. Schaumberger, Drittes Ergänzungsheft zu *SSB*, *QS* B 3, 271–286

—1938a, 'Untersuchungen zur antiken Astronomie I', *QS* B 4, 29–33

—1938b, 'Untersuchungen zur antiken Astronomie II. Datierung und Rekonstruktion von Texten des Systems II der Mondtheorie', *QS* B 4 34–91

—1938c, 'Untersuchungen zur antiken Astronomie III. Die babylonische Theorie der Breitenbewegung des Mondes', *QS* B 4 193–346

—1938d, 'Untersuchungen zur antiken Astronomie V. Der Halleysche "Saros" und andere Ergänzungen zu UAA III', *QS* B 4 407–411

—1941, 'On a special use of the sign "zero" in cuneiform astronomical texts', *JAOS* 61, 213–215

—1945, 'Studies in ancient astronomy VII. Magnitudes of lunar eclipses in Babylonian mathematical astronomy', *Isis* 36/1, 103, 10–15 (reprinted in Neugebauer 1983)

—1947, 'Unusual writings in Seleucid astronomical texts', *JCS* 1, 217–218

—1948a, 'Solstices and Equinoxes in Babylonian Astronomy during the Seleucid Period', *JCS* 2, 209–222

—1948b, 'Arithmetical Methods for the Dating of Babylonian Astronomical Texts', Studies and essays presented to R. Courant on his 60th birthday, Jan. 8 1948 (Interscience Publishers: New York), 265–275

—1950, 'The alleged Babylonian discovery of the precession of the equinoxes', *JAOS* 70, 1–8 (reprinted in Neugebauer 1983)

—1951, 'The Babylonian Method for the Computation of the Last Visibilities of Mercury', *PAPS* 95, 110–116

—1952, 1957a, 1969, *The Exact Sciences in Antiquity* (Princeton University Press: Princeton); 1957a, 1969: second edition (Dover Publications: New York)

—1953, 'The Rising Times in Babylonian Astronomy', *JCS* 7, 100–102

—1954, 'Babylonian Planetary Theory', *PAPS* 98, 60–89

—1955, 1983, *Astronomical Cuneiform Texts*, Sources in the History of Mathematics and Physical Sciences 5 (Springer)

—1957b, 'Saros and Lunar Velocity in Babylonian Astronomy', *KDVSMM* 31.4

—1963, 'The Survival of Babylonian Methods in the Exact Sciences of Antiquity and Middle Ages', *PAPS* 107, 528–535.

—1975, *A History of Ancient Mathematical Astronomy* (Springer)

—1983, *Astronomy and History. Selected Essays* (Springer)

Neugebauer O., Sachs A. 1967, 'Some atypical astronomical cuneiform texts, I', *JCS* 21, 183–218

—1969, 'Some atypical astronomical cuneiform texts, II', *JCS* 22, 92–113

—1986, *Mathematical Cuneiform Texts*, American Oriental Series 29 (American Oriental Society: New Haven)

Oelsner J. 1982, 'Nochmals „Hipparenum (Plinius nat. hist. VI 123) = Nippur?"', *AoF* 9, 259–262

——1986, *Materialien zur babylonischen Gesellschaft und Kultur in hellenistischer Zeit*, Assyriologia VII (Budapest: Eötvös University)

——2000, 'Von Iqīšâ und einigen anderen spätgeborenen Babyloniern', in: S. Graziani (ed.), *Studi sul vicino oriente antico dedicati alla memoria di Luigi Cagni* (Naples), 797–814

——2001, 'HS 201 – Eine Reziprokentabelle der Ur III-Zeit', in: J. Høyrup, P. Damerow (eds.), *Changing Views on Ancient Near Eastern Mathematics*, *BBVO* 19, 53–58

Ossendrijver M. 2008, 'Astronomie und Astrologie in Babylonien', in: J. Marzahn (ed.), 'Babylon. Wahrheit', Catalogue to the Exhibition 'Babylon. Mythos und Wahrheit' in the Pergamon Museum, Berlin, 373–386.

——2010, 'Evidence for an Instrumental Meaning of *ina libbi*', *NABU* 44, 51

——2011a, 'Exzellente Netzwerke: die Astronomen von Uruk', in: G.J. Selz, K. Wagensonner (eds.), 'The Empirical Dimension of Ancient Near Eastern Studies', Wiener Offene Orientalistik Band 8, (LIT-Verlag: Vienna), 631–644

——2011b, 'Science in Action: Networks in Babylonian Astronomy', in: E. Cancik-Kirschbaum (ed.), *Proceedings of the Conference 'Babylon – Wissenskultur zwischen Orient und Okzident', Pergamon Museum, Berlin 26–28/6/2008*, 229–237

Pannekoek A. 1916, 'Calculation of dates in the Babylonian tables of planets', *Koninklijke Nederlandse Academie der Wetenschappen, Proceedings* 19, 684–703

——1941, 'Some remarks on the moon's diameter and the eclipse tables in Babylonian astronomy', *Eudemus* 1, 1–14

——1947, 'Planetary Theories', *Popular Astronomy* 55, 422–438

——1961, *A History of Astronomy* (Dover: New York)

Pedersén O. 1998, *Archives and Libraries in the Ancient Near East, 1500–300 BC* (CDL Press: Bethesda)

——2005, *Archive und Bibliotheken in Babylon*, *ADOG* 25

Pinches T.G., Sachs A. 1955, 'Late Babylonian astronomical and related texts' (Brown University: Providence)

Proust C. 2007, *Tablettes mathématiques de Nippur*, *Varia Anatolica* XVIII (IFEA - De Boccard: Istanbul)

——2009a, 'Numerical and Metrological Graphemes: From Cuneiform to Transliteration', *CDLJ* 2009:1

——2009b, 'Deux nouvelles tablettes mathématiques du Louvre: AO 9071 et AO 9072', *ZA* 99, 167–232

Rawlins D. 1991, 'Hipparchos's Ultimate Solar Orbit and the Babylonian Tropical Year', *DIO* 1.1, 49–66.

Ritter J. 1998, 'Jedem seine Wahrheit: die Mathematik in Ägypten und Mesopotamien', in: M. Serres (ed.), *Elemente einer Geschichte der Wissenschaften*, Suhrkamp Taschenbuch Wissenschaft 1355, 73–107

——2010, 'Translating rational-practice texts', in: A. Imhausen, T. Pommerening, *Writings of Early Scholars in the Ancient Near East, Egypt, Rome and Greece*, 349–383

Robson E. 1997, 'Three Old Babylonian Methods for Dealing with "Pythagorean" Triangles', *JCS* 49, 51–72

——2000, *Mesopotamian Mathematics, 2000–1600 BC. Technical Constants in Bureaucracy and Education*, *OECT* XIV (Clarendon Press: Oxford)

Rochberg F. 1988, *Aspects of Babylonian Celestial Divination: The Lunar Eclipse Tablets of Enūma Anu Enlil*, *AfO Beih* 22

——1989, 'Babylonian Seasonal Hours', *Centaurus* 32, 146–170

——1991, 'Between Observation and Theory in Babylonian Astronomical Texts', *JNES* 50, 107–120

——1998, *Babylonian Horoscopes*, *TAPS* 88, 1–164

——1999, 'Empiricism in Babylonian Omen Texts and the Classification of Mesopotamian Divination as Science', *JAOS* 119, 559–569

——2000, 'Scribes and Scholars: The *ṭupšar Enūma Anu Enlil*', in: *Fs Oelsner* 359–375.

——2004, *The Heavenly Writing. Divination, Horoscopy, and Astronomy in Mesopotamian Culture* (Cambridge University Press: Cambridge)

Ross M. (ed.) 2008, *From the Banks of the Euphrates. Studies in Honor of Alice Louise Slotsky* (Winona Lake)

Roughton N.A., Steele J.M., Walker C.B.F. 2004, 'A Late Babylonian Normal and *Ziqpu* Star Text', *AHES* 58, 537–572

Sachs A.J. 1948, 'A classification of the Babylonian astronomical tablets of the Seleucid period', *JCS* 2, 271–290

Sachs A.J., Hunger H. 1988–1996, *Astronomical Diaries and Related Texts from Babylonia*, I–III (Verlag der Österreichischen Akademie der Wissenschaften: Wien)

Sachs A.J., Neugebauer O.N. 1956, 'A procedure text concerning solar and lunar motion: BM 36712', *JCS* 10, 131–136

Schaumberger J. 1935, *Sternkunde und Sterndienst in Babel. Assyriologische, astronomische und astralmythologische Untersuchungen.* 3. Erganzungsheft (Verlag der Aschendorffschen Verlagsbuchhandlung: Münster)

Schmidt O. 1969, 'A mean value principle in Babylonian planetary theory', *Centaurus* 14, 267–286

Schnabel P. 1923, *Berossos und die babylonisch-hellenistische Literatur* (B.G. Teubner: Leipzig, Berlin)

——1924, 'Neue babylonische Planetentafeln', *ZA* 35, 99–112

——1927, 'Kidenas, Hipparch und die Entdeckung der Präzession', *ZA* 37, 1–60

Seidelmann P.K. (ed.) 1992, *Explanatory Supplement to the Astronomical Almanac* (University Science Books: Sausalito)

Serres M. 1998, *Elemente einer Geschichte der Wissenschaften*, Suhrkamp Taschenbuch Wissenschaft 1355

——1998, 'Gnomon: die Anfänge der Geometrie in Griechenland', in: M. Serres (ed.), *Elemente einer Geschichte der Wissenschaften*, Suhrkamp Taschenbuch Wissenschaft 1355, 109–175

Sidersky D. 1919, 'Le calcul chaldéen des néoménies', *RA* 16, 21–36

Spar I., Lambert W.G. 2004, *Cuneiform Texts in the Metropolitan Museum of Art II. Literary and Scholastic Texts of the First Millennium B.C.* (Brepols)

Steele J. 2000a, 'Eclipse Prediction in Mesopotamia', *AHES* 54, 421–454

——2000b, *Observations and Predictions of Eclipse Times by Early Astronomers* (Kluwer: Dordrecht)

——2000c, 'A 3405: An unusual astronomical text from Uruk', *AHES* 55, 103–135

——2001/2, 'The Meaning of BAR DIB in Late Babylonian Astronomical Texts, *AfO* 48/49, 107–112

——2002a, 'BM 36948: A Saturn ephemeris calculated using System A from Babylon', *JHA* 33, 261–264

——2002b, 'Some lunar ephemerides and related texts from Babylon', in: C. Wunsch (ed.), *Mining the Archives. Festschrift for Christopher Walker on the Occasion of his 60th Birthday* (ISLET: Dresden)

——2002c, 'A Simple Function for the Length of the Saros in Babylonian Astronomy', in: *UOS*, *AOAT* 297, 405–420

——2003, 'Planetary Latitudes in Babylonian Mathematical Astronomy', *JHA* 34, 269–289

——2005a, 'Four procedure texts concerning Jupiter's latitude and synodic motion from Babylon', *SCIAMVS* 6, 31–55

——2005b, 'A New Scheme from Uruk for the Retrograde Arc of Mars', *JCS* 57, 129–133

——2006a, 'Miscellaneous Lunar Tables from Babylon', *AHES* 60, 123–155

——2006b, 'Greek influence on Babylonian astronomy?', *Mediterranean Archaeology and Archaeometry*, Special Issue, Vol. 6.3

——2007a, 'Celestial Measurement in Babylonian Astronomy', *Annals of Science*, 64, 293–325

——2007b, 'The Length of the Month in Mesopotamian Calendars of the First Millennium BC', in: J.M. Steele (ed.), *Calendars and Years. Astronomy and Time in the Ancient Near East* (Oxbow: Oxford), 133–148

——2010, 'A New Scheme from Babylon for the Synodic Arc of Saturn', *JHA* 41, 261–268

Steele J., Imhausen A. (eds.) 2002, *Under One Sky. Astronomy and Mathematics in the Ancient Near East*, *AOAT* 297

Swerdlow N.M. 1998, *The Babylonian Theory of the Planets* (Princeton University Press: Princeton)

Tallquist K.L. 1905, *Neubabylonisches Namenbuch zu den Geschäftsurkunden aus der Zeit des Šamaššumukîn bis Xerxes* (Helsingfors)

Thureau-Dangin F. 1922, *Textes Cunéiformes du Louvre* VI: *Tablettes d'Uruk a l'usage des prêtres du Temple d'Anu au temps des Séleucides* (Geuthner: Paris)

——1938, *Textes Mathématiques Babyloniens*, Ex Oriente Lux, Deel 1 (Brill: Leiden)

van der Spek R. 1985, 'The Babylonian Temple during the Macedonian and Parthian Domination', *BiOr* 42, 541–562

van der Waerden B. 1941, 'Zur Babylonischen Planetenrechnung', *Eudemus* 1, 23–48

——1957, 'Babylonische Planetenrechnung', *Vierteljahrschrift der naturforschenden Gesellschaft in Zürich*, Jahrgang 102, Abh. 2

——1965, *Erwachende Wissenschaft II. Die Anfänge der Astronomie* (Noordhoff: Groningen); 1968 (2nd edition, Birkhäuser: Basel)

——1974, *Science Awakening II. The Birth of Astronomy*, revised edition of *Erwachende Wissenschaft II. Die Anfänge der Astronomie* (Noordhoff: Leiden)

van Dijk J. 1962, 'Die Inschriftenkunde', *UVB* 18, 39–62

von Soden W. 1969, *Grundriss der akkadischen Grammatik*, *AnOr* 33/47

von Soden W., Röllig W. 1991, *Das akkadische Syllabar*, *AnOr* 42

Walker C.B.F. 2004, 'Nos. 78–88. Astronomical Texts', in: *CTMMA* II, 315–345

Waschkies H.-J. 1989, *Anfänge der Arithmetik im alten Orient und bei den Griechen* (Verlag B.R. Grüner: Amsterdam)

Watson R., Horowitz W. 2011, *Writing Science Before the Greeks. A Naturalistic Analysis of the Babylonian Astronomical Treatise MUL.APIN* (Brill: Leiden, Boston)

Weidner E.F. 1925, 'Ein astrologischer Kommentar aus Uruk', *Studia Orientalia* 1, 347–358

Williams C.J. 2005, *Eclipse Theory in the Ancient World*, dissertation Brown University (UMI Dissertation Services: Ann Arbor)

Indices

Index of tablets and fragments

Concordance with *ACT*

Name index

Subject index